ENRICO FERMI

COLLECTED PAPERS

(NOTE E MEMORIE)

VOLUME I

ITALY 1921–1938

THE UNIVERSITY OF CHICAGO PRESS

ACCADEMIA NAZIONALE DEI LINCEI - ROMA

INTERNATIONAL STANDARD BOOK NUMBER: 0–226–24359–1
LIBRARY OF CONGRESS CATALOG CARD NUMBER: 60–12465

THE UNIVERSITY OF CHICAGO PRESS, CHICAGO AND LONDON
ACCADEMIA NAZIONALE DEI LINCEI, ROME, ITALY

 PUBLISHED 1962

THIRD IMPRESSION 1971
PRINTED IN THE UNITED STATES OF AMERICA

Enrico Fermi

For this volume Edoardo Amaldi, Enrico Persico, Franco Rasetti and Emilio Segrè (Editor-in-Chief) have been responsible for the editorial work as indicated in the introductions to the individual papers.

CONTENTS AND BIBLIOGRAPHY

(Papers preceded by an asterisk are not reprinted in these «Collected Papers»).

BOOKS

FOREWORD

On Enrico Fermi's death many of his friends and admirers wished to commemorate him in some permanent fashion. It was immediately apparent that the greatest and most durable monument to Fermi was his own work, a monument that will last as long as the pursuit of scientific knowledge. Hence this publication of his "Collected Papers" sponsored and financed by the Accademia Nazionale dei Lincei and the University of Chicago Press.

Fermi's life and work bear testimony to the international character of science in its noblest form. This is reflected in the number of journals of different nations and languages in which his writings appeared. Indeed, we can trace from the changes in his publication media some of his personal history and of the history of science from 1922 to 1954.

Since in these volumes we do not reprint every word that Fermi published, the manner of selection should be explained. Fermi occasionally wrote several versions of the same paper. We decided to publish only the most detailed one, regardless of language. Where he published different translations of the same paper, the choice was in the following order: English, Italian, German, French. Only for a few exceptionally important papers have we included an English translation not by Fermi himself. Any attempt at systematic translation by someone other than the author seemed inappropriate, especially in view of Fermi's well-known concern for the precise expression.

Original manuscripts and other important documents relative to Enrico Fermi are to be found at the "Domus Galilaeana" in Pisa and at the Library of the University of Chicago.

Volume I contains a complete list of all known writings by Fermi up to his departure from Italy in 1938, except his articles in the Enciclopedia Italiana. The papers marked with a star have not been reprinted, either because they are translations or abbreviations of papers which are reprinted, or because they are hastily composed accounts, often written by other persons, of Fermi's popular lectures. Fermi's books are not included in these "Collected Papers".

The Editorial Board was chosen by the sponsors from among Fermi's close friends and collaborators. The editors have checked the text, corrected some misprints or minor errors, and adapted the vector notation to American or German usage. They have also tried to provide each paper with a preface, a short historical introduction written by a person qualified to give an authoritative account of the origin of the paper. We hope that these introductions may be of interest to historians and students of the process of scientific invention. We sincerely thank everybody who collaborated in writing them.

The publishers of the journals in which these papers appeared have graciously given permission for their reproduction. We wish to thank them all, especially the Accademia Nazionale dei Lincei, Rome; the American Physical Society; the Consiglio Nazionale delle Ricerche, Rome; Nature, London; The Royal Society, London; the Società Italiana di Fisica; Zeitschrift für Physik.

The Editorial Board is sincerely grateful to Professor Giovanni Polvani, President of the Società Italiana di Fisica, for his warm support of this project.

Thanks are also due to the John Simon Guggenheim Memorial Foundation for a fellowship which allowed one of us (E. S.) to work on the compilation of this volume.

The Editorial Board

E. Amaldi, H. L. Anderson
E. Persico, F. Rasetti
C. S. Smith, A. Wattenberg
E. Segrè, *Chairman*

BIOGRAPHICAL INTRODUCTION

by EMILIO SEGRÈ.

Enrico Fermi was born September 29, 1901, in Rome at Via Gaeta N° 17, the third child of Alberto Fermi and Ida De Gattis. His family originally came from the North Italian town of Piacenza. He grew up in Rome, where he attended grammar school and later the Ginnasio Liceo Umberto I. He then studied at the University of Pisa as a Fellow of the Scuola Normale Superiore, and in 1922 he received his doctorate in physics. In 1923, with a Fellowship of the Italian government, he went to Göttingen to work under Max Born; and in 1924 he moved to Leyden to work under P. Ehrenfest. He returned to Rome the same year and in 1925 became "Libero Docente". He then held a lectureship at the University of Florence, and finally, in 1926, he was appointed Professor of Theoretical Physics at the University of Rome, where he remained until 1938.

In 1938 the Fascist racial laws, which affected his wife Laura Capon and her relatives and moreover deeply offended his sense of fairness, induced him to emigrate to the United States of America. Here, he was at first Professor of Physics at Columbia University in New York; then, during the Second World War he devoted all his activity to the development of nuclear energy on a large scale at Chicago, Illinois and at Los Alamos, New Mexico. At the end of the war, in 1946 he moved to the new Institute for Nuclear Studies at the University of Chicago, where he remained until his early and unexpected death on November 29, 1954.

These are the bare, external facts of Fermi's life. As is often the case with a man of science, they do not seem especially dramatic, although in his case the development of nuclear energy, more than emigration to America, certainly represented an event comparable in adventure to the discovery of a continent for an explorer.

The lamented circumstance of his early death has as a consequence that many of his contemporaries, collaborators and friends remember him well. In recounting their experiences with him they can try to reconstruct a living image of the man. This is highly desirable because many scientists have the natural desire to know personally, as it were, the major worthies of their craft. What physicist has not had the curiosity to learn what kind of a man Galileo, Newton or Maxwell was? How did he work? How did his ideas originate? These questions are not easily answered by the ordinary biographies, generally written by authors with little personal and scientific contact with their subject. In Fermi's case, however, the fact that he collaborated extensively with other scientists, who were often his own pupils, makes it easier to reconstruct his scientific portrait. For this reason, while the single papers here collected speak for themselves scientifically and give the finished product of his work, we have tried as often as possible to provide for each of them an introduction written by whoever is in a position to be well informed of the circumstances which led to that specific investigation. While these introductions can add but little to the scientific interest of the papers, we hope that they may contribute to a reconstruction of the picture of the scientist as we knew him.

If we examine Fermi's works as a whole, we find them still fresh and interesting, often written with rare pedagogical skill. In a certain sense, the papers make easy reading, so that an able student can profit greatly from their study, even as a beginner; at the same time, they often have such deep implications that they have inspired very difficult and recondite investigations. The typical style of Fermi's writings is a close reflection of his personal and intellectual history. We see it evolve as time goes on from the early papers written at Pisa, in almost complete isolation, on a variety of disconnected problems, to the late papers in Chicago when he was a mature physicist and a great leader in the field, investigating theoretically and experimentally, with numerous co-workers, a whole new chapter of physics.

In the following pages I shall try to render Fermi's image as it appears to me.

Fermi's interest in mathematics and physics manifested itself very early. He told me that when he was about ten years old he had seriously struggled to try to understand why a circle was represented by the equation $x^2 + y^2 = R^2$, and had finally succeeded only after great efforts. This episode gives us a measure of his development at that age.

A little later he must have made considerable progress because while still an adolescent he read, in Latin, a book entitled "Elementa Physicae Mathematicae" by "Andrea Caraffa e Societate Jesu, in Collegio Romano". This is an ample treatise of mathematical physics published by a Jesuit in 1840. Its two volumes comprise approximately 900 pages and they cover mechanics, optics and astronomy, using the resources of analysis. Fermi acquired the book in a used-book stall and must have studied it quite diligently because the book has numerous marginal notes and developments of the calculations in his handwriting.

We know that what little outside help Fermi received came mostly from Ingegnere A. Amidei who was among the first to notice the boy's extraordinary ability. In this connection we have the invaluable first-hand testimonial of Ing. Amidei as set forth in a letter by him to E. Segrè. He states that sentences quoted verbatim were noted by him in writing at the time of his conversations with Fermi. We quote here Ing. Amidei's letter:

"Antignano (Livorno).
"November 25th, 1958.

"Having had the opportunity of guiding and counseling Enrico Fermi in his studies during his youth from his 13th to the end of his 17th year, I deem it appropriate to hereby summarize his course of study during the above-mentioned period, which was from the autumn of 1914 to the autumn of 1918.

"In 1914 I was Principal Inspector of the Ministry of Railways, and my colleague was Chief Inspector Alberto Fermi. When we left the office we walked together part of the way home, almost always accompanied by the lad Enrico Fermi (my colleague's son), who was in the habit of meeting his father in front of the office. The lad, having learned that I was an avid student of mathematics and physics, took the opportunity of questioning me. He was 13 and I was 37.

"I remember very clearly that the first question he asked me was: 'Is it true that there is a branch of geometry in which important geometric properties are found without making use of the notion of measure?' I replied that this was very true, and that such geometry was known as 'Projective Geometry'. Then Enrico added, 'But how can such properties be used in practice, for example by surveyors or engineers?' I found this question thoroughly justified, and after having tried to explain some properties that had very useful applications, I told him that the next day I would bring him – as I did – a book on projective geometry by Professor Theodor Reye (1) that included an introduction which, in a masterful, artistic style succeeds, by itself, in explaining the usefulness of the results of this science.

"After a few days Enrico told me that besides the introduction he had already read the first three lessons and that as soon as he finished the book he would return it to me. After about two months he brought it back, and to my inquiry whether he had encountered any difficulties, replied 'none', and added that he had also demonstrated all the theorems and quickly solved all the problems at the end of the book (there are more than two hundred).

"I was very surprised and since I remembered that I had found certain problems quite difficult and given up trying to solve them since they would have taken too much time, I wanted to verify that Enrico had also solved those. He gave me the evidence.

"Thus it was certainly true that the boy, during the little free time that was left to him after he had fulfilled the requirements of his high school studies, had learned projective geometry perfectly and had quickly solved many advanced problems without encountering any difficulties.

"I became convinced that Enrico was truly a prodigy, at least with respect to geometry. I expressed this opinion to Enrico's father and his reply was, yes, at school his son was a good student, but none of his professors had realized that the boy was a prodigy.

"I then learned that Enrico studied mathematics and physics in second-hand books that he bought at Campo dei Fiori, hoping to find one treatise that would scientifically explain the motion of tops and gyroscopes, but he could never find an explanation, and so, mulling the problem over and over again in his mind, he succeeded in reaching an explanation of the various characteristics of the mysterious movements by himself. Then I suggested to him that to obtain a rigorous explanation, it was necessary to master a science known as 'Theoretical mechanics'; but in order to learn

(1) T. Reye, *Geometria di Posizione*. Tr. A. Faifofer, Venezia 1884.

it he would have to study trigonometry, algebra, analytical geometry and calculus, and I advised him not to try the problems of tops and gyroscopes, since he would be able to solve them easily once he had mastered the field that I had outlined. Enrico was convinced of the soundness of my advice and I supplied him with the books that I thought were most suitable for giving him clear ideas and a solid mathematical base.

" The books which I loaned him and the date of the loan are as follows:

" In 1914, for trigonometry, the treatise on plane and spherical trigonometry by Serret.

" In 1915, for algebra, the course on algebraic analysis by Ernesto Cesàro and, for analytic geometry, notes from lectures by L. Bianchi at the University of Pisa.

" In 1916, for calculus, the lectures by Ulisse Dini at the University of Pisa.

" In 1917, for theoretical mechanics, the 'Traité de mécanique' by S. D. Poisson.

" I also deemed it appropriate for him to study the 'Ausdehnungslehre' by H. Grassmann which has an introduction on the operations of deductive logic by Giuseppe Peano. These books were loaned to him in 1918.

" I thought it appropriate because it was my opinion that the Ausdehnungslehre (similar to vector analysis) is the most suitable tool for the study of different branches of geometry and theoretical mechanics. . . . .

" Enrico found vector analysis very interesting, useful and not difficult. From September, 1917, to July, 1918, he also studied certain aspects of engineering in books that I lent him.

" In July, 1918, Enrico received his diploma from the Liceo (skipping the third year) and thus the question arose whether he should enroll at the University in Rome. Enrico and I had some long discussions on this subject.

" First of all I asked him whether he preferred to dedicate himself to mathematics or to physics. I remember his reply and I transcribe it here literally: 'I studied mathematics with passion because I considered it necessary for the study of physics, *to which I want to dedicate myself exclusively*'. Then I asked him if his knowledge of physics was as vast and profound as his knowledege of mathematics and he replied: 'It is much wider and, I think, equally profound, because I've read all the best known books of physics' (*). I had already ascertained that when he read a book, even once, he knew it perfectly and didn't forget it. For instance, I remember that when he returned the calculus book by Dini, I told him that he could keep it for another year or so in case he needed to refer to it again. I received this surprising reply: 'Thank you, but that won't be necessary because I'm certain to remember it. As a matter of fact, after a few years I'll see the concepts in

(*) Persico says that one of them was the French translation of the large Treatise by Chwolson. During the Summer of 1918, Fermi went almost every day to the library of the " Istituto Centrale di Metereologia e Geodinamica " in order to study this book. Permission had been granted by the Director Prof. F. Eredia who had been Fermi's physics teacher at the Liceo Umberto I.

it even more clearly, and if I need a formula, I'll know how to derive it easily enough '.

" In fact, Enrico, together with a marvelous aptitude for the sciences, possessed an exceptional memory.

" I then considered that the proper moment had arrived to present a project that I had already considered in his behalf for a year, that is from the time when I advised him – and he immediately followed my suggestion – to study the German language, since I foresaw that it would be very useful for reading scientific publications printed in German without having to wait until they were translated into French or Italian. My plan was this: Enrico ought to enroll not at the University of Rome, but at the one in Pisa. First, however, he would have to win a competition to be admitted to the Scuola Normale at Pisa and attend (besides the courses in the School) the University of Pisa.

" Enrico recognized at once the soundness of my plan and decided to follow it, even though he knew that his parents would be opposed to it. Then I immediately went to Pisa to obtain the necessary information and the program for the competition to the Scuola Normale Superiore and immediately returned to Rome to examine it minutely with Enrico. I ascertained that he knew the subjects in the group of mathematics and physics perfectly, and I expressed my conviction that he would not only win the competition, but would also be the first among the applicants – as in fact he was.

" Enrico's parents did not approve of my plan, for understandable and human reasons. They said: ' We lost Giulio (Enrico's older brother, who died after a short illness in 1915) and now we are to allow Enrico to leave us to study at Pisa for four years while there is an excellent university here in Rome. Is this right? '

" I used the necessary tact to persuade them, a little at a time, that their sacrifice would open a brilliant career for their son and they finally agreed that my program should be carried out. Thus, as Enrico's wife wrote in her book ' Atoms in the Family ', ' at the end, the two allies – Fermi and Amidei – carried the day "

Thus far goes the Amidei letter.

Persico tells of experiments undertaken together with Fermi at about the same time as that covered by the Amidei letter [2]. They are noteworthy especially for the choice of problems, such as to determine with high precision the density of Rome's drinking water (Acqua Marcia). This is a rather different problem from the more common one of a boy wanting to build some sort of an electric motor or apparatus: today this would be most probably a radio; at the time we are considering, it would have been a wireless telegraph. This does not preclude the fact that, at the time, Fermi was an expert builder of electric motors, but obviously he was also interested in more sophisticated problems such as the investigation of the behaviour of gyroscopes.

(2) E. PERSICO, «Scientia», vol. 90, pag. 36 (1955).

The point reached by his scientific development at the age of 17, when he finished his secondary schools, is clearly demonstrated by the entrance examination for the "Scuola Normale Superiore" at Pisa. The theme was "Distinctive properties of sounds" and is dated November 14, 1918. Certainly the examiners expected an essay at a high school level as one reasonably would. Instead, after a few introductory sentences, we find in his examination sheet the partial differential equation of a vibrating reed and its solution with the help of a Fourier series. The examiner, Professor G. Pittarelli, a very kindhearted man and a good mathematician in his own right, must certainly have been stunned by the little essay. Fermi himself told me that Pittarelli, after having read the examination paper, called him to ascertain whether the candidate really understood what he had written. After questioning him, Pittarelli said that during his long teaching career he had never met anybody like Fermi and that certainly the boy had extraordinary talents. Fermi remembered these occurrences many years later with obvious pleasure and deep gratitude towards Pittarelli.

In his other studies it is clear that at the Ginnasio and Liceo he progressed brilliantly and without effort. He was the sort of model student who succeeds in everything. His professor of Italian was Giovanni Federzoni, and Fermi who had an exceptional memory, knew long excerpts from the "Divine Comedy" and other Italian poems by heart and for the rest of his life was able to quote them on appropriate occasions. "Orlando Furioso" by Ariosto apparently was one of his favorite readings even before he had to read it at school; already when he was about 12 years old he could recite entire cantos by heart. For him one reading was sufficient to commit a section to memory. He must have been bothered by Greek, which was compulsory in Italian schools. At Los Alamos I once complained that in a sort of nightmare I had dreamed of a Greek final examination at the Liceo: Fermi confessed that he had been subject to the same nightmare.

His literary tastes were very simple; his own writings, including popular lectures, are not noteworthy for elegant literary style, especially in his early years. He had little sensitivity to literary form; to him the content was the only thing that mattered. On the other hand he was extremely careful, almost pedantic, as far as scientific precision was concerned. This care for precision increased steadily with time and one notices a remarkable difference in style between the early papers and those of his more mature years.

With regard to his knowledge of languages, he learned German as a boy, as Ing. Amidei mentions. The complexities of German grammar fascinated him. I think in his youth he could sit down and write an article in German without any mistakes, and he spoke it fluently; he knew French as many Italians do, using it easily but not always correctly. More than any other language except Italian he used English. He learned everything that one can by application and study, but the muscles of his mouth never became accustomed to English sounds, and he always retained a strong Italian accent, which occasionally irritated him. In America, Fermi devoted more effort to his English pronunciation than most immigrants are wont to do, but the result always remained imperfect.

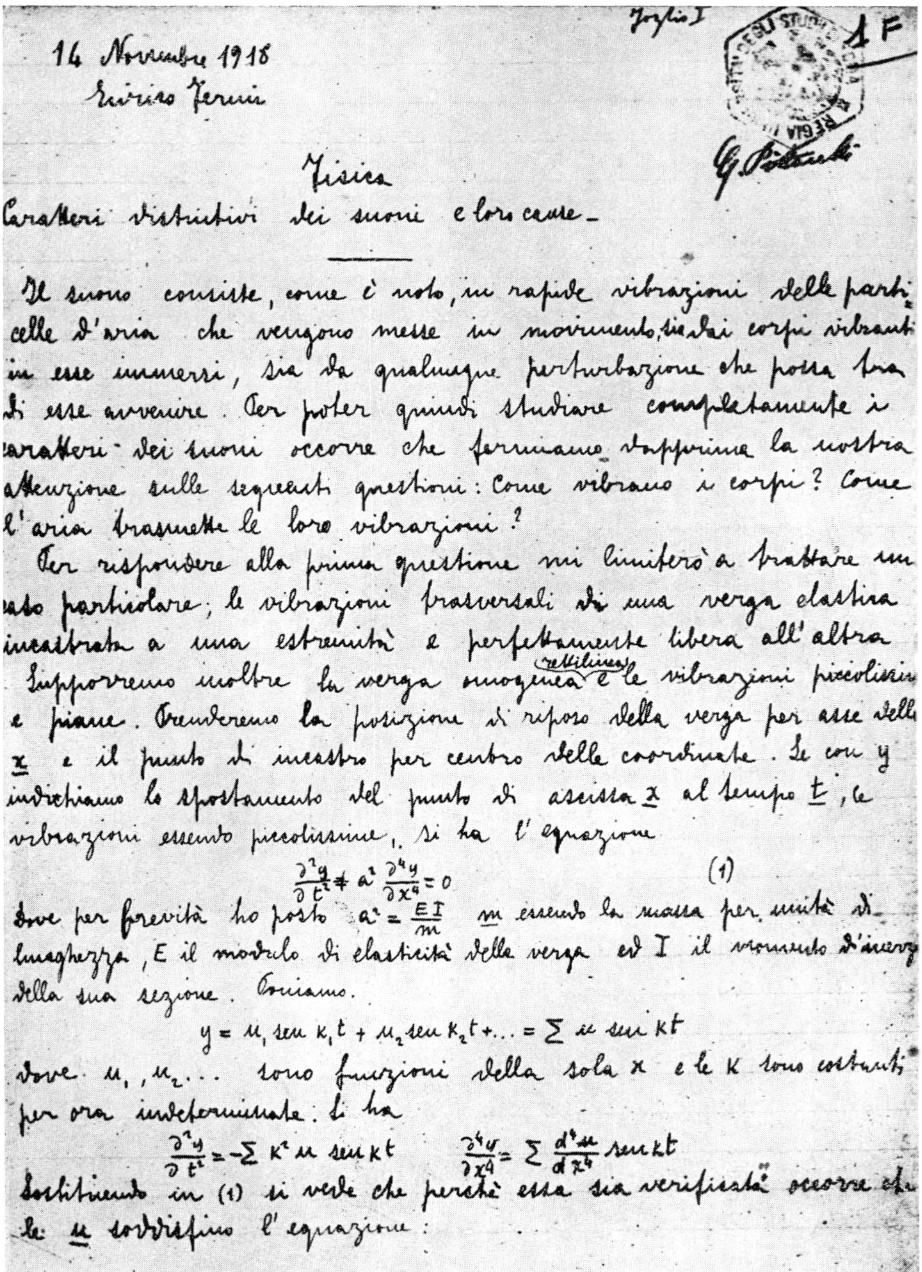

14 Novembre 1918
Enrico Fermi

Foglio I

1F

G. Pizzetti

Fisica

Caratteri distintivi dei suoni e loro cause_

Il suono consiste, come è noto, in rapide vibrazioni delle particelle d'aria che vengono messe in movimento, sia dai corpi vibranti in esse immersi, sia da qualunque perturbazione che possa fra di esse avvenire. Per poter quindi studiare completamente i caratteri dei suoni occorre che fermiamo dapprima la nostra attenzione sulle seguenti questioni: Come vibrano i corpi? Come l'aria trasmette le loro vibrazioni?

Per rispondere alla prima questione mi limiterò a trattare un caso particolare; le vibrazioni trasversali di una verga elastica incastrata a una estremità e perfettamente libera all'altra. Supporremo inoltre la verga omogenea (rettilinea) e le vibrazioni piccolissime e piane. Prenderemo la posizione di riposo della verga per asse delle x e il punto di incastro per centro delle coordinate. Se con y indichiamo lo spostamento del punto di ascissa x al tempo t, le vibrazioni essendo piccolissime, si ha l'equazione

$$\frac{\partial^2 y}{\partial t^2} + a^2 \frac{\partial^4 y}{\partial x^4} = 0 \qquad (1)$$

dove per brevità ho posto $a^2 = \frac{EI}{m}$ m essendo la massa per unità di lunghezza, E il modulo di elasticità della verga ed I il momento d'inerzia della sua sezione. Poniamo

$$y = u_1 \operatorname{sen} k_1 t + u_2 \operatorname{sen} k_2 t + \ldots = \sum u \operatorname{sen} kt$$

dove $u_1, u_2 \ldots$ sono funzioni della sola x e le k sono costanti per ora indeterminate. Si ha

$$\frac{\partial^2 y}{\partial t^2} = -\sum k^2 u \operatorname{sen} kt \qquad \frac{\partial^4 y}{\partial x^4} = \sum \frac{d^4 u}{dx^4} \operatorname{sen} kt$$

Sostituendo in (1) si vede che perchè essa sia verificata occorre che le u soddisfino l'equazione:

First page of Fermi's entrance examination for the admission to the "Scuola Normale Superiore" at Pisa.

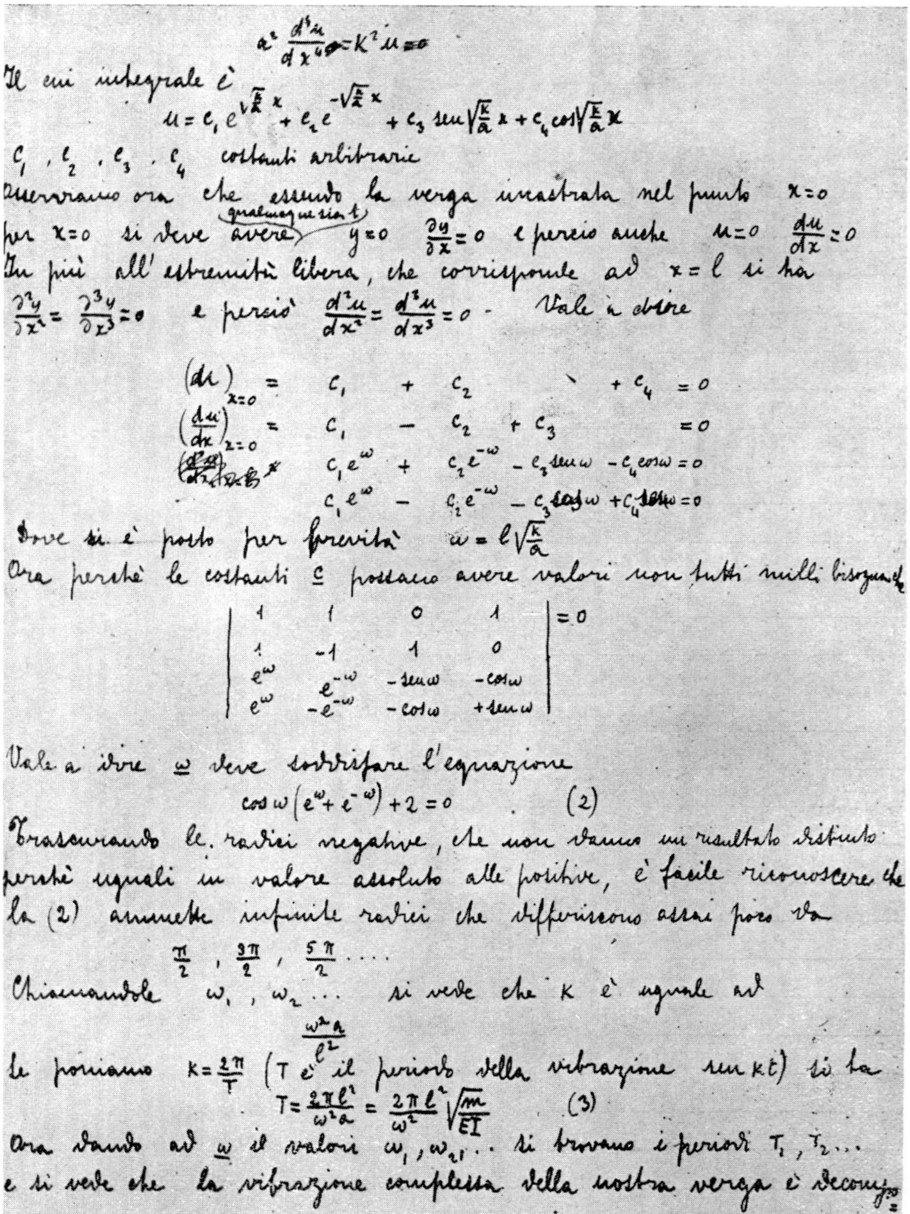

$$a^2 \frac{d^4u}{dx^4} - k^2 u = 0$$

Il cui integrale è

$$u = c_1 e^{\sqrt{\frac{k}{a}}\,x} + c_2 e^{-\sqrt{\frac{k}{a}}\,x} + c_3 \operatorname{sen}\sqrt{\frac{k}{a}}\,x + c_4 \cos\sqrt{\frac{k}{a}}\,x$$

c_1, c_2, c_3, c_4 costanti arbitrarie

Osserviamo ora che essendo la verga incastrata nel punto $x=0$ per $x=0$ si deve avere (qualunque sia t) $y=0$ $\frac{\partial y}{\partial x}=0$ e perciò anche $u=0$ $\frac{du}{dx}=0$

In più all'estremità libera, che corrisponde ad $x=l$ si ha $\frac{\partial^2 y}{\partial x^2} = \frac{\partial^3 y}{\partial x^3} = 0$ e perciò $\frac{d^2u}{dx^2} = \frac{d^3u}{dx^3} = 0$. Vale a dire

$$\begin{aligned}
(u)_{x=0} &= c_1 + c_2 + c_4 = 0 \\
\left(\frac{du}{dx}\right)_{x=0} &= c_1 - c_2 + c_3 = 0 \\
& c_1 e^{\omega} + c_2 e^{-\omega} - c_3 \operatorname{sen}\omega - c_4 \cos\omega = 0 \\
& c_1 e^{\omega} - c_2 e^{-\omega} - c_3 \cos\omega + c_4 \operatorname{sen}\omega = 0
\end{aligned}$$

Dove si è posto per brevità $\omega = l\sqrt{\frac{k}{a}}$

Ora perchè le costanti c possano avere valori non tutti nulli bisogna che

$$\begin{vmatrix} 1 & 1 & 0 & 1 \\ 1 & -1 & 1 & 0 \\ e^{\omega} & e^{-\omega} & -\operatorname{sen}\omega & -\cos\omega \\ e^{\omega} & -e^{-\omega} & -\cos\omega & +\operatorname{sen}\omega \end{vmatrix} = 0$$

Vale a dire ω deve soddisfare l'equazione

$$\cos\omega\,(e^{\omega} + e^{-\omega}) + 2 = 0 \qquad (2)$$

Trascurando le radici negative, che non danno un risultato distinto perchè uguali in valore assoluto alle positive, è facile riconoscere che la (2) ammette infinite radici che differiscono assai poco da

$$\frac{\pi}{2},\ \frac{3\pi}{2},\ \frac{5\pi}{2} \ldots$$

Chiamandole $\omega_1, \omega_2 \ldots$ si vede che k è uguale ad $\frac{\omega^2 a}{l^2}$

Se poniamo $k = \frac{2\pi}{T}$ (T è il periodo della vibrazione $\operatorname{sen} kt$) si ha

$$T = \frac{2\pi l^2}{\omega^2 a} = \frac{2\pi l^2}{\omega^2}\sqrt{\frac{m}{EI}} \qquad (3)$$

Ora dando ad ω il valori $\omega_1, \omega_2, \ldots$ si trovano i periodi $T_1, T_2 \ldots$ e si vede che la vibrazione complessa della nostra verga è decompo=

Second page of Fermi's entrance examination for the admission to the "Scuola Normale Superiore" at Pisa.

Fermi registered at the University of Pisa as a Fellow of the Scuola Normale Superiore in the fall of 1918. This school was founded by Napoleon in 1813 as a branch of the École Normale Supérieure of Paris. Its original purpose was the preparation of high-school teachers and the promotion of higher studies and research. Its pupils attend the University of Pisa, but in addition have special courses, mostly of the character of seminars. The pupils and professors live in the school as in a British college. Admission is by competition only and there are no fees of any kind. Several of the most distinguished literary and scientific figures of modern Italy studied there, and the roster of its alumni adds to the great prestige of the institution.

Fermi chose mathematics as his major subject but soon changed to physics. He conducted his studies at Pisa very independently. His grasp of physics was far above the local level of teaching and from his correspondence with Persico (3) we have some detailed information on how he spent his time. In February of 1919, as a freshman, he writes: "Since I have almost nothing to do for my class work and I have many books available, I try to enlarge my knowledge of mathematical physics and I shall try to do the same for pure mathematics". In the same correspondence he says that he has read Poincaré's "Théorie des tourbillons" and Appell's "Mécanique rationnelle" devoting himself especially to the methods of analytical mechanics. He also started studying Nernst's "Theoretische Chemie" and the "Lehrbuch der Allgemeinen Chemie" by Ostwald, chemical studies which he concluded "with some relief" a year later. We do not know where he learned the theory of relativity, although there is evidence that approximately at that time he had mastered Richardson's "Electron Theory of Matter", which expounds in fair detail whatever was known in 1916 concerning electromagnetic theory and atomic structure. A little later, probably in 1921, he read Weyl's "Raum, Zeit, Materie" and was impressed by the power and flexibility of variational methods which were used systematically in that book. He immediately assimilated the technique and the spirit of the method and proceeded to use it for new problems. Thus originated in his third University year his first published papers (N° 1, 2).

Among Fermi's papers at the University of Chicago is a very interesting document referring to his early university years. It is a small leather-bound note book which he filled between July and September, 1919, and which contains a sort of physics vademecum. It is divided in parts and, as was his habit, is written in pencil almost without erasure.

This booklet gives us a clear picture of Fermi's scientific preparation and intellectual progress in 1919 when he was between 18 and 19 years old. The first 28 pages contain a summary of analytical dynamics and are dated Caorso, July 12, 1919. In it he develops the theory of Hamilton and Jacobi, reaching very advanced topics with extreme concision, but equal clarity. There are no indications of his sources of information, but

(3) E. PERSICO, *loc. cit.*

very likely they are the works of Poincaré which he was studying around that time and also Appel.

Twenty-five pages on the electronic theory of matter follow; they are dated Rome, July 29, 1919, and they contain a résumé – as usual a very concise one – of the subject. In these pages he treats Lorentz's theory, special relativity, black-body theory, diamagnetism and paramagnetism. For this part there is also a bibliography listing several of the most important books on this subject, including Richardson's "Electron Theory of Matter" which we know he studied carefully.

Bohr's first papers on the hydrogen atom are also mentioned, although at that time they were little known and appreciated in Italy. The following section of 19 pages is dated Rome, August 10, 1919, and contains in greater detail the black-body theory according to Planck. This part is followed by an extensive bibliography on radioactive substances and their decay taken from Rutherford's "Radioactive Substances and Their Radiations". There are no comments, and it is dated Caorso, September, 1919. The following chapter, from pages 81 to 90, is devoted to Boltzmann's H–theorem and kinetic theory and is dated Caorso, September 14, 1919. We find the usual succinct but clear exposition of theory with some applications. The method used for establishing the H–theorem is the same as that used by Boltzmann which involves a detailed analysis of all collisions. The booklet, totaling 102 pages, concludes with two bibliographies taken from Townsend's book on gas discharges. They deal with electrical properties of gases and photoelectricity. The last notes are dated Rome, September 29, 1919, and are followed by a table of contents.

This booklet shows many of the author's characteristics in an embryonic stage. The choice of material is made with surprising discrimination, especially if one considers the author's age and the fact that he was essentially self-taught. Another characteristic is that Fermi, although never repulsed or frightened by any mathematical difficulty, does not seek elegant mathematics for its own sake. Whether a theory is easy or difficult does not seem to concern him; the important point is whether or not it illuminates the essential physical content of the situation. If the theory is easy, so much the better, but if difficult mathematics is necessary, he is quickly resigned to it. One also notices an appreciable difference between the sections in which the logical structure of the subject predominates over its experimental content and those of a more empirical character. In the first, one perceives the master's hand; in the second, the lack of experience and of a critical evaluation of the many papers quoted is apparent. All told, it is surprising that after one year of university work a student should be able to put together such a booklet, which would be very creditable even for a teacher with a long educational career behind him.

The physicists with whom Fermi came in contact at Pisa are: Professor L. Puccianti, then Professor of physics and Director of the physics laboratory, Dr. G. Polvani, Puccianti's first assistant and two other assistants: Drs. Pierucci and Ciccone. Puccianti was gifted with a keen mind but was a lazy person who, while still young, had done some interesting work in atomic spectroscopy. At the time Fermi went to Pisa, Puccianti had ceased doing original research, although he remained interested in a deep and critical understanding of classical physics. Fermi became a good friend of Polvani. Among Fermi's classmates we remember Rasetti, who was very close to Fermi, and Carrara, who also had a fellowship at the Scuola Normale Superiore.

Besides studying physics Fermi also acquired a vast and deep knowledge of mathematics. As a matter of fact, even though he occasionally showed scorn for certain parts of mathematics which he deemed too formal or too little imaginative or too critical, he certainly was not mathematically naive or ignorant of the most modern and subtle mathematical questions. If needed, Fermi could give the most rigorous proof of a theorem, and often, on an example, he would show a refined critical approach, abandoning it later for the sake of speed and simplicity. There is no doubt that from his studies at Pisa he derived a supply of mathematical facts, ideas and methods that he used for the rest of his life. Whenever he needed an ingenious and powerful mathematical method, Fermi always had it ready in some corner of his mind, even when it involved mathematical notions above and beyond the common knowledge of professional theoretical physicists.

The studies at the Scuola Normale Superiore proceeded brilliantly: he always obtained the highest possible grades, except in drawing and in some chemistry courses. This cost him no effort and left him, as we noted, ample leisure for his private studies.

He enjoyed enormous prestige among his fellow students and his superior ability was recognized by everybody; indeed, it was known that the Scuola Normale Superiore had an extraordinary man among its students.

The doctoral dissertation, by tradition, was on an experimental subject, and its content is essentially reproduced in papers N° 6 and 7. The day of the oral examination the classroom was full, and his colleagues expected a memorable performance. Instead, the examination, although excellent, was not exceptional. Anyway, he obtained his degree *cum laude* in July, 1922. Three days later Fermi passed also the examination required for the "Diploma" of the Scuola Normale with a thesis on probability (N° 38). On that occasion some of the mathematicians made numerous remarks and criticisms of his solution of a certain equation; however, again he obtained the maximum of grades, *cum laude*.

Immediately hereafter, Fermi returned to live with his parents in Rome and at that time became acquainted with Senator O. M. Corbino, Professor of experimental physics and Director of the Physics Institute at the University of Rome. Corbino, although already at that time considerably absorbed in politics and business, was the most open-minded among all Italian physicists in a position of authority. Gifted with a brilliant mind, a true love for science and a generous human approach, he exerted an influence on the development of physics in Italy far greater than that derived directly from his scientific contributions. As long as he lived, he remained the wise advisor, friend and protector of Fermi and of the whole group which gravi-tated around him. The generosity with which he gave this necessary help, and the obvious satisfaction and delight with which he followed first Fermi's successes and then those of the group which populated Rome's Institute, are the clear signs of his superior intellect and generous heart. To my knowledge there are no documents or direct evidence of the relations between Fermi and Corbino around 1922. The acquaintance, which had quickly grown into friendship, between Fermi and the three distinguished mathematicians Castelnuovo, Enriques and Levi-Civita dates back to the same time. These three scientists were all at that time on the Faculty of Sciences at the University of Rome and taught there. Although they were professional mathematicians, they followed with more interest than most Italian physicists the developments of theoretical physics, especially in the field of relativity. Fermi also met Professor Volterra, who was primarily interested in problems of classical mathematical physics, but did not become especially intimate with him. These cordial relations with the mathematicians are reflected in the lectures that Fermi held at the Mathematical Seminar of the University of Rome, several of which appeared in print in the "Periodico di Matematiche", directed by Enriques (N° 22, 29, 34, etc). At that time there was no physics seminar at Rome and research was much more active in pure mathematics than in physics.

During the winter of 1923, Fermi won a scholarship from the Ministero dell'Istruzione Pubblica to study abroad and chose to work with Max Born in Göttingen. The time was a critical period of incubation for the new quantum mechanics, and Göttingen was one of the major centers involved; in spite of this, Fermi did not profit very much from his stay there. It is not easy to understand why this happened: we may surmise that his love for concrete, well defined problems, and his distrust of questions that he considered too general and abstract, may have repelled him from the then current speculations. These were certainly somewhat nebulous, and even worse for him, commingled with philosophical overtones; nevertheless, later they were destined to bring about the new quantum mechanics. The preference shown by Fermi for concrete problems especially in his early years, is probably to be traced to his cultural formation, and to the fact that he was practically completely self-taught. Concrete problems probably gave him the immediate sensation of the importance and a check of the correctness of his work, whereas general and abstract questions were much more difficult to plumb, and moreover isolation was a great obstacle in assessing the value of the results obtained. In later years Fermi considerably changed his objects of research and was less inclined to confine himself to detailed ingenious investigations of concrete problems.

From a personal point of view, it is barely possible that during the Göttingen period the physicists in his age group, such as Heisenberg, Pauli, Jordan and others, all men of exceptional talents, who should have been his companions, may not have recognized his ability and quite unintentionally left him out of their intellectual community.

Anyway, Fermi remembered the following months, when he moved to Leyden with a Rockefeller Fellowship to work in P. Ehrenfest's Institute, as more fruitful. Ehrenfest was one of the greatest teachers of physics and a man of warm human interests. Very early he recognized Fermi's ability and encouraged the timid young Italian. Anyone who met Fermi in later years might be surprised perhaps that he ever needed encouragement, but Fermi himself often mentioned the fact. While he was aware of his superiority over the physicists he had met in Italy, he said he did not have any way to compare his capabilities with those of the great international names. Ehrenfest, who personally knew most of the international figures, gave him just such a standard of comparison. The interests prevailing among Dutch physicists at the time are reflected in some of Fermi's papers (N° 21). And on the other hand some eyewitnesses, such as Uhlenbeck, relate that Ehrenfest, who had delved deeply in statistical mechanics, was impressed by Fermi's ideas on the ergodic theorem (N° 11).

On the whole, the time spent abroad does not seem to have been of decisive importance for Fermi's development, as it might have been had the circumstances in Göttingen proved more favorable. Fermi was accustomed to intellectual isolation and to learning from books and journals more than from personal discussions (a habit which he reversed in later years). He returned to Italy without having absorbed either the "spirit of Copenhagen" or that of Göttingen. Heisenberg's great papers on matrix mechanics of 1925 did not appear sufficiently clear to Fermi, who reached a full understanding of quantum mechanics only later through Schrödinger's wave mechanics. I want to emphasize that this attitude of Fermi was certainly not due to the mathematical difficulty and novelty of matrix algebra; for him, such difficulties were minor obstacles; it was rather the physical ideas underlying these papers which were alien to him.

When he returned to Italy in 1924, he began teaching an introductory mathematical course for chemists at the University of Rome. In 1925 he obtained the "Libera Docenza" and moved to the University of Florence to teach mechanics and mathematical physics. The physics laboratory, directed by Prof. Garbasso, was located on the Arcetri hills near the villa where Galileo spent his last years. In Florence he found Rasetti, who was at the time Garbasso's assistant. In February, 1926, Fermi took part in the competition for a chair of Mathematical Physics at the University of Cagliari in Sardinia. The referees were Professors Levi-Civita, Volterra, Somigliana, Marcolongo and Guglielmo. Whereas the first two favored Fermi, the majority chose for the first place Professor G. Giorgi, and Fermi remained with his temporary appointment in Florence.

Fermi was disappointed by the result of the competition, which he considered unjust; moreover, he naturally wanted to improve his modest financial conditions and was impatient to see his ability recognized officially.

Fermi's early correspondence with some of his closest friends, such as Persico, shows evidence that he was worried about his career. I mention this fact because later Fermi seemed to consider such preoccupations on the part of young scientists as something unnatural or exaggerated or strange. Probably once he had overcome the difficulties connected with the beginning of his own career, he took a more detached and lofty attitude towards these human vicissitudes, but at an earlier time he had certainly been anxious.

The Florentine period saw Fermi's first major contribution to theoretical physics: the discovery of the statistical laws which govern particles subject to Pauli's exclusion principle, or fermions as they are called today (N° 30). Soon after the discovery of this statistics, Corbino succeeded in having a chair in theoretical physics, the first in Italy, established in Rome and the competition for it was judged in November, 1926, by Professors Maggi, Cantone, Garbasso, Majorana and Corbino. They unanimously agreed that Fermi was the outstanding candidate, saying that "they felt they could rest on him the best hopes for the affirmation and future development of theoretical physics in Italy". The other two successful candidates were his childhood friend, Prof. Persico, and Prof. Pontremoli. Fermi's actual appointment to Rome was opposed by an older member of the physics faculty, but Corbino readily overcame the obstacle. Persico replaced Fermi at Florence and in the fall of 1926 Fermi moved to Rome. In the Institute at Via Panisperna 89[a] he started one of the most fruitful periods of his scientific career. The decade from the discovery of his statistics to 1936 was probably the golden age of Fermi's life.

The old physics building in Via Panisperna, although built around 1880, was still perfectly adequate for scientific work at that time and compared favorably with other major European laboratories. The equipment was fair and mainly included instruments for optical spectroscopy with good modern Hilger spectographs and adequate subsidiary apparatus. The shop was old-fashioned with rather poor machines; the library, on the other hand, was excellent. The location of the Institute, surrounded as it was by a small park on a hill in a rather central part of Rome, was convenient and beautiful at the same time. The gardens landscaped with several palms and bamboo thickets, the silence prevailing, except at dusk when many sparrows populated the greenery, made it a most peaceful and attractive center of study. I believe that everybody who worked there until 1937, when the physics department moved to the new "Città Universitaria", has still in his heart an affectionate and poetic feeling for the old place. The upper floor of the building was occupied by Corbino's residence; the first floor contained the research laboratories and the offices of Corbino, Lo Surdo and Fermi, as well as the library; the ground floor contained the shop, the classrooms and some laboratories for the students; the basement contained the electric generators and other facilities. Fermi, with Rasetti, and later with their students occupied the whole south side of the first floor; Lo Surdo most of its north side. As time went on, however, the number of people associated with Fermi increased so much that they filled most of the first floor. Neighboring quarters were occupied by Prof. G. C. Trabacchi, who was the chief physicist at the Health Department (Sanità Pubblica). He had an excellent supply of instruments and materials which he generously loaned whenever we needed them. This earned him the nickname of "Divine Providence".

Corbino, who used to spend part of the morning in his study, often visited Fermi's quarters and stopped to talk about physics and other subjects with him and his students.

When Fermi arrived in Rome he must have felt a strong urge to find an adequate scientific environment. There was then in Rome, as mentioned above, a group of eminent mathematicians considerably older than Fermi, such as Volterra, Levi-Civita, Castelnuovo, Enriques and others. Fermi was especially friendly with the last three, but the excellent personal relations and the high opinion which they had of each other were not sufficient to create intellectual exchanges; these were hindered by the essentially different points of view, scientific interests, and the considerable age difference. Among the physicists, Corbino was the only one open-minded enough and with sufficient preparation to follow and appreciate the developments of modern physics, the rapid progress of theoretical physics in general, and especially of quantum theory. However, he was busy with his political and business activities and, while he was fully aware of the necessity of rejuvenating physics in Italy and of introducing new young workers to the field, his direct scientific activity was limited. On the other hand, his political and administrative help and the enthusiasm with which he fostered the development of a physics school in Rome were of paramount importance.

The first step following the arrival of Fermi in Rome was Rasetti's transfer from Florence to Rome. He became Corbino's first assistant (Aiuto) and immediately started to try to revive experimental work. At this point I might introduce some personal recollections in order to illustrate the formation of the physics school in Rome. In 1927 I met Rasetti through a friend and a schoolmate of mine, G. Enriques, the son of the mathematician. While we went mountain climbing together, I learned from Rasetti several physical theories; but even more important, I gained a clear impression that physics in Rome was awakening. As a matter of fact, in 1925 I had already heard some lectures by Fermi at the Mathematics Seminar of the University. They had made a deep impression on me because they showed me for the first time a young man who was deeply versed in subjects of which I had barely caught a glimpse (by reading the quantum theory book by Reiche) but had never met in my regular university studies. However I had not yet met Fermi personally and I did not have any way of comparing the young lecturer with the famous men I knew only through my readings.

During the summer of 1927, through Rasetti and G. Enriques, I came to know Fermi personally and I immediately had the sensation of having found an exceptional teacher. Fermi himself, in conversations, on hikes and at the seashore, asked me several simple questions in mathematics and classical physics, perhaps to discreetly test my knowledge and ability. In September, 1927, on my return from an expedition to the Alps where I had been with Rasetti, G. Enriques and others, I went to the International Physics Conference in Como and there had the unforgettable experience of seeing in person the great physicists whose names I had read in books: Lorentz, Rutherford, Planck, Bohr and many others, and then a group of extremely young men: Heisenberg, Pauli and Fermi. It was clear that, scientifically at least, Italy was represented by Fermi. During the short period of the Conference I also learned a considerable amount of physics because Rasetti and Fermi would point out to me the various celebrities and at the same time tell me their most important achievements. I returned to Rome in the fall of 1927, decided to follow my old desire of studying physics, and within a few months left the engineering studies which I had been pursuing until then.

On that occasion for the first time I saw Corbino's far-reaching influence. He overcame with ease the far from trivial administrative difficulties in order to save me the loss of credit for one year of engineering studies. Without anybody taking special notice of it, I had become Fermi's first pupil, at least in a formal sense. The school in Rome had been started.

In June, 1927, without my having heard anything about it, Corbino had made an appeal to one of the undergraduate physics classes for engineers to enlist some new physics students: Edoardo Amaldi, the son of the mathematician Prof. Ugo Amaldi, followed this invitation, and thus he also entered the small group of students who were to be personally tutored by Fermi with the hope of developing among them research physicists and possible collaborators. Perhaps later this would lead, eventually, to a renaissance of physics in Italy. However this goal was not explicitly formulated at the time.

After a few months of study I talked to my friend and schoolmate at the engineering school, Ettore Majorana, and he also joined our group. Majorana, in intellectual power, depth and extent of knowledge, was much above his new companions and in some respects – for instance as a pure mathematician – superior even to Fermi. His exceedingly original intellect and, unfortunately, also his natural pessimism and exaggerated self criticism, drove him to work alone and to lead a solitary life. On the whole he did not participate very much in our work, and he limited himself to helping us in difficulties or to stunning us with his original and novel ideas and methods, or with his ability as a lightning mental calculator. Later he isolated himself even more, and by 1935 he had disappeared from the University, seldom leaving his private residence.

Other students frequented the Institute and once in a while attended Fermi's private lectures.

These were completely improvised and informal. In the late afternoon we met in his office and often conversational topics gave rise to a lecture: for instance, we would ask what was known about capillarity and Fermi would improvise a beautiful lecture on the mathematical physics of capillarity. In this fashion we reviewed many subjects at an intermediate level, more or less corresponding to a beginning graduate course in an American university or to the famous "Introduction to Theoretical Physics" by Planck or, in English, to the books by Slater and Frank. Sometimes, however, the level became higher and Fermi would explain to us a paper which he had just read: in this way we became conversant with the famous papers by Schrödinger on wave mechanics. We never had a regular "course". If there was an entire field of which we knew nothing and we asked Fermi, he would limit himself to mentioning a good book to read. Thus when I asked for some instruction in thermodynamics, he told me to read the book by Planck. However, the readings he suggested were not always the happiest ones because he probably mentioned only the books which he had studied himself, and these were not necessarily the best suited from a pedagogical angle. After his lecture, we would write down notes on it, solve (or try hard to) the problems which he had given us, or others which we ourselves invented. The rest of our time we spent on experimental work.

The instruction was chiefly in theoretical physics and no distinction was made between future theoreticians or experimentalists. Fermi himself, who at the time worked mainly in theory, was also interested in experimental work. His knowledge and interests embraced all physics, and he diligently read several journals. He preferred concrete problems and distrusted theories which were too abstract or general, but any specific problem, in any field of physics: classical mechanics, spectroscopy, thermodynamics, solid state theory and several more, fascinated him and was a challenge to his ingenuity and physical sense. Often just talking to him one saw a beautiful explanation develop, simple and clear, that resolved some puzzling phenomenon.

At that time we repeatedly had occasion to witness the execution of a new and original piece of work. Naturally it is impossible to say how much preliminary work Fermi had done consciously or unconsciously. Certainly there were no written notes. What one saw was the development of a calculation at a moderate speed, but with exceedingly few errors, false starts or changes of direction. The work proceeded almost as in a lecture, although more slowly, and at the end the manuscript, or at least the equations, were ready to be copied for publication with little need of improvements. A curious characteristic of Fermi's working habit was the steady pace at which he proceeded. If there were easy passages he still proceeded quite slowly, and a simple-minded observer might have asked why he wasted so much time on such simple algebra; however, when difficulties arose which would have stopped a man of lesser ability for who knows how long, Fermi solved them without a change of speed. One had the impression that Fermi was a steamroller that moved slowly but knew no obstacles. The final result was always clear and often one was tempted to ask why it had not been found long ago since everything was so simple and natural. Once used, a method was stored in his memory and often adapted to problems which appeared to be very different from the one which first originated the physical idea or mathematical technique. For instance, it is interesting to note the evolution and successive applications of the "scattering length" in papers N° 95 and 107, or the recurrence of the statistical theme applied to both atoms and nuclei.

Already at that time (1928) Fermi made little use of books: Laska's collection of mathematical formulae and Landolt–Börnstein's tables of physical constants were almost the only reference books he had in his office. If he needed some complicated equation to be found in a book in the library, Fermi would often propose a wager, saying that he would derive the equation faster than we could find it in a book. Usually he won. The only treatise that I know he read after he came to Rome was Weyl's " Gruppentheorie und Quantenmechanik ".

The speed at which it was possible to form a young physicist at that school was incredible. Naturally, a good deal of the success was due to the immense enthusiasm aroused in the young people, never by exhortations or sermons, but by the eloquence of Fermi's personal example. After some time spent at the Institute in Via Panisperna, one became completely absorbed in physics, and in saying " completely " I am not exaggerating.

Fermi did not like to assign subjects for doctoral dissertations, or in general to suggest subjects of investigation. He expected the students to find one by themselves or to obtain one from some colleague who was more advanced in his studies. The reason for this, as he later told me, was that he did not easily find subjects simple enough for beginners: he generally thought of problems that interested him personally and were too difficult for students. Rasetti was very generous in teaching the experimental techniques that he knew or in loaning apparatus that he no longer used, but it was difficult to work with him because he had peculiar idiosyncrasies and irregular schedules of work. A strong and long-lasting personal friendship developed among all the participants in this adventure. Age differences were small: Fermi, the oldest, was 26 years old in 1927; Amaldi, the youngest, was 19. Corbino occasionally came to the evening lectures, but the event was rare. However, he took a strong interest in the welfare of the group, in the career problems and relations of the young men with the rest of the world. As was to be expected, word of what was happening in Rome spread quickly among the young Italian physicists, and soon we started to receive visits from G. Gentile Jr., B. Rossi, G. Bernardini, G. Racah, G. C. Wick, and later we had for longer periods L. Pincherle, R. Einaudi, E. Fubini, U. Fano and many others. By 1929, it became clear that whereas the theoretical situation was well in hand, it was necessary to strengthen our experimental activities. To import new experimental techniques to Rome, members of the group had to work in different laboratories to learn them on the spot. Thus Rasetti went to Pasadena, to Millikan's laboratory where he did important work on the Raman effect; I went to Amsterdam to Zeeman's laboratory to study forbidden spectral lines and Amaldi went to Debye's laboratory in Leipzig where he worked on X–ray diffraction of liquids. Later we supplemented these visits, which at first had the purpose of using facilities not available in Rome, for the completion of problems already started there, by trying entirely new fields. In this connection, Rasetti spent some time at Dahlem in Lise Meitner's laboratory, and I worked in Otto Stern's laboratory in Hamburg. The plan was successful, and without these periods of experimental training it would have been impossible later to perform the complex neutron work rapidly and efficiently. However, even from abroad, we kept in close contact with the group in Rome, and either by letter or during vacations we discussed theoretical problems with each other. There are many traces of this exchange in papers by Fermi, Majorana, Wick and others. During this period Fermi went abroad only for short visits.

He was by now accustomed to being rather isolated intellectually because only Majorana (and he was rather inaccessible as mentioned above) could speak with him about theory on an equal footing. On the other hand, about that time, the first victims of Nazi persecution, barred from their native countries and attracted by Fermi, began to arrive in Rome. Bethe, Placzek, Bloch, Peierls, Nordheim, London and others spent some time in Rome, often on their way to the United States. Among the Americans, we had a visit by Feenberg, whom Majorana especially liked. This mutual attraction manifested itself by their sitting in the library facing each other in silence because they knew no common language.

Life in Via Panisperna was very methodical. One worked from about 9 a.m. to 12 : 30 p.m. and from 3 p.m. to 7 p.m. Of course this schedule was self imposed and everybody was free to keep it as he wished. Work in the evening was practically unknown and on Sundays we often went for a hike in the vicinity of Rome or for a mountain trip. During the winter there was always some skiing expedition, and during the summer, a trip abroad or a vacation in the Alps. Already in 1930 Fermi had gone as far as Ann Arbor (Michigan), attracted there by its brilliant summer school of theoretical physics and by his Dutch friends Uhlenbeck and Goudsmit who had moved there.

The most significant personal events of this period were Fermi's marriage to Laura Capon, in July 1928, and his appointment by Mussolini to the Italian Academy, in 1929. This honor, though well deserved, was unexpected because Fermi's reputation at the time was limited to physicists and, according to tradition, at his age, academic honors were not yet due. The appointment came probably at the instigation of Corbino, who, although not a Fascist, had been a member of one of the early cabinets of Mussolini. This event changed Fermi's financial condition and had a beneficial effect on the subsequent development of physics in Italy, since its representative in the National Academy was undoubtedly the most qualified man available. However, even after his appointment to the Academy, Fermi wielded relatively little political influence because he was unwilling to sacrifice any time to occupations outside of physics and he did not like to participate in administrative matters or political affairs. He was able to resign his post at the "Enciclopedia Italiana Treccani" (see e.g. N° 83) and at the National Research Council, which he had taken on his arrival in Rome, mainly because of his financial position. He made some desultory attempts at increasing research facilities by the creation of new jobs and subsidies, but these attempts did not go farther than obtaining Mussolini's approval on a memorandum which was later pigeonholed. It took many more years before the younger generation could prevail in the important matter of university appointments.

In the meantime, the most important scientific event of those years, the formation of quantum mechanics, had taken place without important contributions from Italy, at least in the establishment of the principles, but Fermi had contributed to its applications in an important way. Fermi's statistics was born independently of quantum mechanics and certainly before Fermi himself had mastered it completely. In the original papers we can trace this effort to clarify and assimilate quantum mechanics (N° 37, 39, 42) between 1926 and 1931. Schrödinger's papers were the first to be understood and aroused great interest and enthusiasm. Fermi explained them quickly to his friends; later to Corbino, who remained skeptical for some time; and still later he spoke on the subject to the Mathematics Seminar, where the professional mathematicians, older in age and less familiar with the experimental background of physics, raised several ingenious objections to the interpretation of quantum mechanics commonly accepted.

Thus, for instance, paper N° 59 originated from a discussion in which Prof. Castelnuovo raised many questions. Fermi was inclined to be somewhat impatient with people who did not understand the new developments of quantum mechanics, but naturally he treated very differently silly objections, of which there were many, and genuine difficulties, like those Castelnuovo raised.

Once in a while he complained that even persons for whom he had the highest respect and admiration, such as Corbino, would occasionally show a skepticism towards quantum mechanics and its interpretation which he deemed based on lack of understanding. It must be said, however, that in his last years Fermi seemed less convinced that the current interpretation of quantum mechanics is the last word on the subject. This resistance against quantum theory was shown mainly by people older than Fermi and his group, because the younger physicists either understood or believed the new theories, and in any case learned to use them, even if they had not completely assimilated them.

The advent of quantum mechanics in Fermi's opinion, and in Corbino's also, signaled the completion of atomic physics. The fundamental questions were solved, and the future lay in the direction of the exploration of the nucleus, or of complicated structures leading ultimately to biology.

This represented a radical change in the research projects of the Institute. We must remember that our experimental tradition in Rome went back to the spectroscopic work of Puccianti in Pisa: all our successes obtained up to that time in experimental physics were on spectroscopic subjects; the available equipment was spectroscopic; and our knowledge was mainly in the field of atomic physics. Rutherford and the work of his school were rather alien to us.

Thus the change to nuclear subjects cost us considerable effort. It was not a whim or a desire to follow a fashion, but was the result of a deliberate plan which we debated with plenty of vigor. The first step towards its realization was Rasetti's trip to Dahlem to learn nuclear techniques. A nuclear physics conference at Rome in 1931, organized by the Italian Academy, helped to familiarize us with current problems, and we changed the subject of our readings to matters connected with the nucleus. Fermi, later in his career, at the end of the Second World War, experienced a similar change in his field of work when he left neutron physics, in which he was recognized as the greatest authority, to study high energy physics.

The first work on a nuclear subject is the review article N° 72, and the first great success was the beta ray theory (N° 76), where we find the coalescence of previous work on radiation theory with Pauli's idea of the neutrino. A little later, artificial radioactivity research was in full swing. As the history of that period is told in detail in the introductions to papers N° 78, 84 to 110 and 112 to 119, it is omitted here. The neutron work absorbed Fermi completely for the remainder of the time he was to spend in Italy.

Unfortunately, the political horizon clouded with ominous signs. Since 1936, approximately, at the Institute we had a feeling of impending catastrophe. The days of serene, undisturbed work were gone forever. From 1930 to 1936, during a period of relative exterior quiet, we have the peak of Fermi's activity and the prime of the school in Rome. The history of this period is recounted in the introductions to the papers of those years. There is little to add. The neutron work was carried out with the kind of extreme rapidity that was possible only for a small group working in complete harmony and without administrative encumbrances. In this connection it is worth mentioning the low cost of this research. The Consiglio Nazionale delle Ricerche, on recommendation of its secretary, G. Magrini, immediately appropriated 20.000 lire (1000 U.S.A. dollars at that time) which could be spent with complete freedom. This was our only subsidy, apart from our regular salaries. The neutron work also had an untoward consequence: it became absolutely impossible for Fermi, because of lack of time, to continue his extracurricular teaching of promising young men; neither was he able to spare time for foreign visitors; from 1934 to 1938, first they became rare, then practically disappeared.

The Ethiopian war marked the beginning of the decline of the work at the Institute, and the death of Prof. Corbino on January 23, 1937, brought further serious complications. Prof. Lo Surdo was appointed director and he did not understand or appreciate the merit of Fermi's work. The Institute itself was moved to the new university campus, to a larger and more modern building; however, this move cost invaluable time. All these events and the grave political situation materially hampered Fermi's work. Finally, the Fascist racial laws of 1938 directly affected his wife and possibly his two children. This fact, and probably his deep, although mute resentment against an injustice that offended his sense of fairness, were the final arguments that convinced him to leave Italy. He was not to return to his native country again until 1949.

Already in the past Fermi had had offers of permanent positions from the Eidgenössische Technische Hochschule in Zürich and from several American universities. Some of these offers included not only substantial economic advantages, but also the availability of research facilities far superior to those in Italy. There is no doubt that Fermi considered these offers very carefully and several times was seriously tempted to accept them, but for various reasons he ultimately decided against them. The events of 1938, however, gave him a final impetus, and the award of the Nobel prize in Stockholm gave him a pretext to leave Italy while avoiding possible political reprisals. From Stockholm he travelled directly to New York, and thus began a second period of his life.

"Lo primo tuo rifugio e 'l primo ostello" (Thy first refuge and first home – Dante, Divina Commedia, Paradiso XVII, 70) was Columbia University. Fermi might have described Columbia in these words, quoting Dante as he often did to show off, in a light fashion, among his Italian friends.

Indeed, one cannot properly speak of a refuge, because as soon as the rumor spread that Fermi was in the U.S.A. to stay, several universities made him excellent offers. At that time American universities had suddenly found a number of distinguished European scholars who had been dismissed from their positions at home and were looking for new positions. Hitler's persecutions and, on a smaller scale, Mussolini's, had forced many scientists, from famous ones to young men just beginning their careers, to leave their countries, and England and America with a liberal and generous sense of hospitality and an intelligent understanding of their own long-range interests, performed a good part of the rescue operations. This was an undertaking which, in a bleak period of human history, when the blackest and most hideous crimes recorded in history were committed, honors the scientific community and shows that even at a time of bestiality, mankind is open to noble deeds.

Fermi, however, was in many ways spared the difficulties of that time. He was sufficiently famous to find a suitable position without difficulty and he was young enough to adapt easily and rapidly to the new surroundings. This adjustment was rendered even easier by his previous acquaintance with the United States and by his fondness for the American way of life. At Columbia University he had personal acquaintances; he himself, Rasetti, Segrè and Amaldi had worked there. Besides the excellent local instruments, he also found copies of some of the instruments he had used in Rome; moreover there was a cyclotron. The study of neutrons had been one of the major subjects of investigation at Columbia and it would be easy to resume with larger means the work interrupted at Rome. Last but not least the head of the Physics Department, Professor G. B. Pegram, truly a Southern gentleman, had taken a personal interest in the Rome group and a warm friendship developed between him and his Roman guests. Thus there were good reasons for Fermi's choice of Columbia. As soon as he was settled, he tried to form a new group of co-workers. H. L. Anderson was among his first students there. However, neutron work in the style and tradition of Rome was nearing its end. Without the discovery of fission which suddenly changed the situation, and without the outbreak of the war, it is probable that Fermi would have left neutron physics and turned to another subject. Instead, the discovery of Hahn and Strassmann suddenly opened new horizons to nuclear physics and especially to neutronology. At the time, Fermi was without doubt the greatest expert on neutrons. His intuitive feeling for their behaviour was like that of a radio expert for the behaviour of circuits. He did not need to calculate in order to predict the results of an experiment, but true to his nature, after each experiment he performed numerous and detailed calculations which he saved and classified in an extremely orderly fashion. In this way he developed and increased a " thesaurus " of data, a mixture of theoretical and experimental results which became a most valuable and unique tool for his further work. He used to call the container of these sheets the " artificial memory ". The artificial memory started as a large envelope, later grew into a desk drawer, and finally filled an entire filing cabinet.

As soon as the announcement of the discovery of fission reached Fermi, he immediately saw the possibility of the emission of secondary neutrons and of a chain reaction, and at once he started experimenting with great alacrity in this direction. Nothing tells the story of the experiments which led to the pile better than the papers contained in the second volume of these Collected Papers. Most of the papers were unpublished and certainly not intended for publication in their present form; on the other hand, they represent a scientific chronicle of enormous interest, and taken together with their introductions, they give an exact account of the events of that period so important for both science and history. Anderson, who worked with Fermi from 1938 on and collaborated in a large portion of the experiments which led to the chain reaction, gives us the eyewitness report. I saw Fermi repeatedly during that period and we worked together again briefly during one of his visits to Berkeley, but the secrecy requirements of the time unfortunately prevented a free exchange of ideas. This, however, did not prevent us from discussing future projects. For instance, during a visit of mine to New York towards the end of 1939, while we were on a walk, along the Hudson River, near Leonia where he lived, we discussed plans for the research that was to lead to the discovery of Pu^{239} and to its use as a nuclear fuel.

During the last years of his stay in Italy, Fermi had developed a certain reticence. As far as I know, there were no special reasons for it at the time, but understandably it became extreme under the conditions of war work. Among the many scrupulous observers of security rules in atomic energy work, before and after the institution of the Manhattan Project, I doubt that there was anybody more discreet than Fermi. The legal status of "enemy alien" in which Fermi found himself also produced many anomalous situations, most of which were easily solved by common sense and the relative flexibility of the administrative and bureaucratic systems then prevailing. Some of these episodes were rather funny and did not leave any resentment on either side; mostly they were sources of amusing stories to be told among friends.

The chain reaction project was begun with a completely inadequate staff and equipment. Perhaps Fermi thought he might be able to repeat, on a somewhat larger scale, a work similar to the neutron research in Rome. He certainly did not realize, as very few of the scientists did, the colossal proportions in manpower and means that the project was to assume before its successful completion. Fermi was always reluctant to take administrative responsibilities so that in the Manhattan Project too, at least officially, he never became a prominent personality from the administrative or political point of view. Scientifically, he was undoubtedly the brightest star, and this was universally acknowledged. Because of his unparalleled knowledge of neutrons, he was consulted whenever serious difficulties arose. The sequence of researches directed towards the achievement of the chain reaction is a thing of beauty: the logical order of the experiments, the evaluation of the results and their use to direct the next step are to be followed in these Papers. Whoever has any scientific experience will be deeply impressed by a perusal of them. We find here a sequence of investigations connected with each other by an iron logic and a keen intuition; they stand equal to some of the major classical works of experimental science.

In the fall of 1942, during a visit to Chicago, Fermi locked me in a room with some of the reports (including numerical data and connected experimental results) on the exponential experiments and the theory of the pile.

I still remember how after a couple of hours of study I was dumbfounded by the success already reached and by the methods followed. Having read those reports, I saw clearly that a chain reaction with natural uranium was on the verge of being realized.

Indeed, on December 2, 1942, the pile under the Stadium of the University of Chicago became critical. This brought to an end another cycle of investigations. Four years had elapsed between this event and the discovery of fission, seven had passed since the discovery of slow neutrons, ten since the discovery of the neutron. Such was the unprecedented speed of this work.

Even in this period so overloaded with work, Fermi did not completely neglect his teaching activity. Instead of teaching classes or his private seminars, as in the past, he taught his co-workers during the work and many of them still remember with pride and pleasure the period of apprenticeship under Fermi's guidance. Once he had reached the main goal of the Chicago Project, Fermi typically enough, left to others the further development work in order to start something new and to lend his help where the difficulties were the greatest: at the time at Los Alamos.

The assignment of that laboratory, ably directed by J. R. Oppenheimer, was to make an atomic bomb with either U^{235} or Pu^{239}, as soon as they were delivered by other branches of the Project. The difficulties of this assignment were far from trivial; actually, at the beginning, nobody knew how to build an atomic bomb at all. Fermi did not have any specific scientific duty nor was he assigned any particular administrative responsibility; he was on the governing board of the laboratory with which the director consulted on all important matters, but otherwise he was a sort of oracle whose job was to solve problems above the ordinary capabilities of the staff, distinguished as it was. J. von Neumann, as a consultant, had a somewhat similar job.

Among the various activities at Los Alamos, Fermi took a direct interest only in the water boiler, a homogeneous reactor which had been built there; but in general he participated in all novel or unusual problems. I remember having listened in his office to discussions on hydrodynamics with von Neumann. They took the strange form of a competition in front of a blackboard to see who could first solve the problem; von Neumann, using his insuperable lightning analytical skill, usually won. Occasionally, such discussions could be interrupted by unexpected events. For instance, during one of these hydrodynamical sessions I once witnessed the arrival of a first class electronics expert who was confronted with a new and very difficult problem in circuitry. Within about twenty minutes Fermi concocted a circuit which would have solved the problem, but no one knew whether a tube with the necessary special characteristics even existed. Consultation of a tube manual disclosed that the required tube was really available and the apparatus was promptly built and worked satisfactorily.

The time in which the first atomic bomb would be ready for a test at Alamogordo was approaching. Several times we went to the New Mexico desert for preparations. Finally, shortly before daybreak on July 16, 1945, after a stormy night, we saw the light of the first atomic bomb. From a distance of about ten miles, through very dark glasses, we saw the blinding light of the explosion, an imposing and awful sight. The problem of measuring the energy release was immediately although crudely solved by Fermi. With the help of some confetti which he dropped, he measured the displacement produced when the front of the shock wave reached his observation point. Within a few hours, Fermi went with a tank to the crater of the explosion in order to measure the radioactivity of the sand.

That explosion signified in a way the end of the initial phase of the Los Alamos project. The major technical assignment had been accomplished, too late to influence the war in Europe in the decisive way that had been planned. Germany capitulated on the very day on which at Alamogordo we exploded a tremendous pile of ordinary explosives in preparation for some tests to be performed on the atomic bomb, which was expected to be ready in a couple of months. President Roosevelt had died suddenly in April and could not see the end of the enterprise he had so effectively fostered.

The success of the Alamogordo test was promptly communicated to President Truman, then at Potsdam in conference with Churchill and Stalin. The grave decision on the use of the bomb rested with the President. As usual, for serious policy decisions in the U.S., even before the success of the Alamogordo experiment, the President had appointed a restricted advisory committee, called the Interim Committee to counsel him on the employment of the bomb and on other matters pertaining to nuclear energy. Its recommendations had a purely advisory value because the final decision, according to the U.S. constitution, rested with the President as Commander in Chief. Earlier, several scientists who had been very active in the Manhattan Project, wrote eloquent pleas to the President recommending against the military use of the atomic bomb. Among them I might mention James Franck and Leo Szilard, who were particularly worried by the war use of atomic energy. Naturally, all this occurred in secrecy, without public knowledge. The Interim Committee appointed as scientific consultants A. H. Compton, Fermi, E. O. Lawrence and J. R. Oppenheimer. In a meeting held at Los Alamos about June 15, 1945, after long and painful deliberations, they recommended the military use of the bomb. This happened at Hiroshima and Nagasaki.

At the beginning of August, Japan surrendered, and thus ended the Second World War. The scientists at Los Alamos once again started to think about peacetime research, and there were many projects floating in the air. During the fall, plans for the future were taking shape. Fermi was still on a leave of absence from Columbia University, but just at that time the

University of Chicago planned to form a nuclear institute (the present Enrico Fermi Institute for Nuclear Studies) and repeatedly offered Fermi a position as its director. He absolutely refused a job that would have given him too much administrative work; fortunately Professor S. K. Allison, a distinguished scientist, a very able administrator, and a good friend of Fermi was appointed director. He supervised all the activities in the initial phases of the Institute and kept the job until after Fermi's death.

During the clear autumn days of Los Alamos, with the pressure of work relaxed, we resumed our old pastime of taking long walks, especially attractive in those beautiful and wild surroundings. Plans for the future were frequently our subject of conversation, sometimes in earnest, sometimes in a lighter vein. Among those that Fermi mentioned frequently was that in his mature years he would retire to a small college, with teaching duties only, and then write a book dealing exclusively with those theorems which are reputedly "well known" or "easily shown" and for that reason never really proved. He even began a list of such topics, but unfortunately the book was never written. For the immediate future, while the new, extraordinarily powerful neutron sources could have attracted Fermi to the exploitation of these new technical means in the neutron field where he was – as said before – the foremost world authority, he quoted instead, jokingly, a motto which he attributed to Mussolini, although probably it originated with D'Annunzio: "O rinnovarsi o perire" (Renew oneself or perish), and said that henceforth one had to turn to new chapters of physics in which the future lay. True to his word, when he returned to Chicago he worked on neutrons for a relatively short time, long enough to obtain brilliant results which themselves became starting points for a long series of investigations, but he immediately started to prepare himself for the new meson physics which was just beginning to develop.

In the last months of his stay at Los Alamos he also started to develop a keen interest in electronic calculating machines. In his Roman days, he had used a small mechanical hand calculator to compute numerically the ψ's (cf. paper N° 43) and in order to perform several other applications of the Thomas-Fermi method of calculating atomic properties. Numerical problems and numerical analysis were completely familiar to him and he recognized at once that the new powerful electronic computers, strongly advocated by von Neumann, were opening up new and unpredictable possibilities. The powerful computers at Los Alamos always had a strong attraction for him, and he experimented and worked with them for several summers. The problems connected with nuclear fusion had also interested him deeply. Already in 1946, he had reviewed them in a lecture course at Los Alamos in which he expounded many novel and original ideas.

On his return to Chicago, while waiting for the synchrocyclotron and the new Institute to be finished, he again worked with neutrons at the Argonne Laboratory, using the pile as a source. We see in this period the conclusion of the investigations on neutron diffraction and on the scattering length, the origin of which can again be traced back to the Roman period. At that time he also sought to re-establish a system of instruction somewhat similar to that used in Rome. He had a new group of young pupils, most of them returning from Los Alamos, where they had worked in the theoretical group of Bethe, in the experimental group of Segrè, or in other groups. We shall mention among them Agnew, Chamberlain, Chew, Goldberger, Rosenfeld, Woods, and, somewhat later, a new arrival from China, Yang. At Chicago Fermi took a very active part in all seminars and in many discussions; often, with a single remark, he sowed the seeds of further important developments. For instance, Maria Mayer in her classical paper on the shell model (" Phys. Rev. " 75, 1969 (1949)) generously acknowledges: " Thanks are due to Enrico Fermi for the remark ' Is there any indication of spin-orbit coupling? ' which is the origin of this paper ". In the meantime meson physics was developing and Fermi immediately recognized the importance of the Conversi-Pancini-Piccioni experiment which he was the first to interpret correctly. But by now the Chicago synchrocyclotron was about to work and in Berkeley it had already been shown that such a machine was a powerful artificial source of mesons. Thus we enter the last phase of Fermi's experimental work; this is a series of important investigations on the pion-nucleon interaction performed with Anderson and younger pupils, such as Rosenfeld, Orear, Yodh and others. It was at this time that they coined the new words, pion and muon, for π–meson and μ–meson.

In this last period, Fermi, who by now was no longer fond of long trips, spent his summer vacations at Los Alamos or at Brookhaven, a most welcome guest of these Laboratories, especially of the first, to which he was particulary attached. He also spent a summer in Berkeley and he attended all the High-Energy Conferences in Rochester. Wherever there were new physics problems and young physicists you were likely to see Fermi arrive, always ready with some new and fruitful idea, and also ready to receive information, challenge and inspiration from his younger colleagues. In this period we also note a change in Fermi's methods of keeping up with scientific developments. He read less and less and relied more and more on conversation and oral sources of information which were always plentiful. Many active physicists enjoyed and profited from discussing their problems with him and on his side he took notes of these conversations and inserted them in the artificial memory, which by now was very bulky and elaborately cross-indexed extending to all of physics. On the other hand, he barely glanced at the journals and completely stopped reading any physics books. He once said that Weyl's book on group theory and quantum mechanics was the last physics book he read. His old friends also noticed a contraction in the span of his interest. Even for him it was becoming impossible to keep up with all of physics. He began to limit himself to high-energy nuclear physics and to his direct research interests. He still kept in reserve his powerful resources for other branches of physics, if they became necessary, but he did not seek new ventures far afield.

Again he often spoke about what would happen with the passing years when old age would set in, but it is typical and meaningful that in 1946, at the end of the Los Alamos period, after a few minutes of reflection, in seriously estimating the work accomplished and the path of the future, he used the number 1/3, meaning that up to then he had accomplished about 1/3 of what he hoped to accomplish during his whole life. That the problem of aging was present to him, even if it did not worry him, is apparent from many conversations. In one of the last ones before he fell ill, I was telling him about some experiments on proton polarization that we had recently performed in Berkeley. He commented that they gave me the right to do nothing for about five years, without being labelled " senile ", because after a piece of work of some importance one had the " right " to stop producing new research for some time before one had to be classed as a " retired " or " finished " scientist.

In his last years I also noted an extreme desire to avoid any waste of time, almost as if he had forebodings that time would be his most precious commodity. He worked and behaved as if he now had an obsession to avoid squandering time or energy, or wasting any of his possibilities. He acted as if he had a task assigned, a goal to reach.

In 1954 he fell ill in an insidious form. In the summer, with an effort of his iron will, he still went to Europe, made some mountain trips, and gave a beautiful course on pion physics at the summer school in Varenna, but by now his health conditions were clearly alarming. He returned to Chicago in September and submitted to an exploratory operation to diagnose the nature of his illness, which had remained obscure despite repeated examinations. Unfortunately, the operation could only recognize a hopeless situation. Fermi realized this immediately and accepted it with Socratic serenity. I believe that neither the members of his family, nor the friends who visited him during his illness, will ever forget the deep impression they received from his conversation at that time. He died on November 29, 1954, shortly after his 53rd birthday. He gave to science all he had and with him disappeared the last universal physicist in the tradition of the great men of the 19th century, when it was still possible for a single person to reach the highest summits, both in theory and experiment, and to dominate all fields of physics.

Enrico Fermi at the age of 17.

The "Scuola Normale Superiore" at Pisa.

COLLECTED PAPERS

(NOTE E MEMORIE)

N° 1 and 2.

Paper N° 1 (related to N° 2) deals with the inert *mass* of a rigid system of electric charges, and derives it from the reaction on the moving charges of their own electromagnetic field. The result is that, in the most general case, the mass is expressed by a tensor. In the particular case of a spherically symmetric system, the tensor degenerates into a scalar equal to $(4/3)\, U/c^2$. This value, while in agreement with a well known calculation of the electromagnetic mass of a spherical homogeneous shell performed by Lorentz, contradicts Einstein's principle of equivalence (see papers 4 *a*, 4 *b*, 6).

Paper N° 2 determines, by the methods of general relativity, the effect of a uniform gravitational field on a system of electric charges. It turns out that the charges have a *weight* equal to that of a material mass U/c^2 (where U is the electrostatic energy of the system), in perfect agreement with Einstein's principle of equivalence between mass and energy.

E. PERSICO

1.

SULLA DINAMICA DI UN SISTEMA RIGIDO DI CARICHE ELETTRICHE IN MOTO TRASLATORIO

«Nuovo Cimento», *22*, 199–207 (1921).

§ 1. – Quando un sistema di cariche elettriche si muove di moto qualunque il campo elettrico da esso generato è diverso da quello a cui porterebbe la legge di Coulomb. Ora il campo elettrico prodotto da tutto il sistema, esercita sopra tutti gli elementi di carica del sistema delle forze. Il risultante di queste forze, vale a dire il risultante delle forze elettriche interne, sarebbe evidentemente nullo se valesse la legge di Coulomb, ma non lo è però più, almeno in generale, quando il sistema si muove, poiché in tale caso non vale più quella legge.

Tale risultante dà la reazione d'inerzia elettromagnetica e lo scopo del presente lavoro è appunto di calcolarlo nel caso di un sistema qualsiasi in moto traslatorio. Nel caso che il sistema sia una distribuzione sferica di elettricità superficiale, come si ammette nella maggior parte dei modelli elettronici, è noto che si trova [1] che tale risultante è dato, almeno in prima approssimazione, da

$$(1)\qquad -\frac{2\,e^2}{3\,Rc^3}\,\Gamma + \frac{2\,e^2}{3\,c^2}\,\dot{\Gamma}$$

dove e, R indicano la carica totale ed il raggio del sistema, c la velocità della luce, Γ e $\dot{\Gamma}$ l'accelerazione e la sua derivata rispetto al tempo. Per

(1) RICHARDSON, *Electron Theory of Matter*, Cap. XIII. La differenza tra le mie formule e quelle del Richardson proviene dal fatto che quest'ultimo usa le unità di Heaviside.

moti quasi-stazionarii il secondo termine di (1) diventa trascurabile con che (1) si riduce a

$$-m\Gamma \tag{2}$$

essendo m la massa elettromagnetica.

Nel § 2 si trova la generalizzazione della (1) al caso di sistemi qualsiasi, rispondenti per esempio ai modelli molecolari, sempre supponendo che la velocità sia trascurabile di fronte a quella della luce. Se F_i $(i = 1, 2, 3)$ sono le componenti del risultante in questione si trova

$$F_i = -\sum_k m_{ik}\,\Gamma_k + \sum_k \sigma_{ik}\,\dot{\Gamma}_k \tag{3}$$

essendo m_{ik}, σ_{ik} quantità dipendenti dalla costituzione del sistema. Non si può più quindi nel caso generale parlare di una massa elettromagnetica scalare, si introduce invece al suo posto la considerazione del tensore m_{ik}.

Il § 3 è dedicato allo studio dinamico della legge dei moti quasi-stazionarii:

$$K_i = \sum_k m_{ik}\,\Gamma_k \tag{4}$$

essendo K_i le componenti della forza esterna. Si mostra che con tale legge continuano a valere i principii fondamentali della forza viva e di Hamilton.

Nel § 4 infine la legge (4) dei moti quasi-stazionarii, valida soltanto per velocità piccole viene a mezzo del principio di relatività in senso stretto generalizzata al caso di velocità qualunque.

Resta con ciò completato lo studio della masse elettromagnetiche come masse d'inerzia. In un prossimo lavoro considererò, dal punto di vista della teoria generale della relatività, le masse elettromagnetiche come masse pesanti.

§ 2. – È noto [2] che la forza elettrica dovuta ad una carica puntiforme 1 in moto è la somma di due che, supponendo trascurabile il rapporto della velocità v della particella alla velocità c della luce sono: la prima $\mathbf{E}_1$, la forza data dalla legge di Coulomb; la seconda $\mathbf{E}_2$ ha l'espressione

$$\mathbf{E}_2 = \frac{\Gamma^* \cdot \boldsymbol{a}}{c^2 r}\,a - \frac{1}{c^2 r}\,\Gamma^*. \tag{5}$$

In questa formula r rappresenta la distanza tra la particella M ed il punto P in cui si calcola la forza, $\boldsymbol{a}$ un vettore di grandezza 1 e di orientazione MP. Γ^* infine è l'accelerazione della particella al tempo $t-(r/c)$. Se in M invece della carica 1 c'è la carica $\rho\,d\tau$ (ρ densità elettrica, $d\tau$ elemento di volume) la forza in P sarà $\rho\,d\tau\,(E_1 + E_2)$ e quindi la forza esercitata in P da tutte le cariche sarà $\int_\tau \rho\,(E_1 + E_2)\,d\tau$ ove l'integrazione si deve estendere a tutto lo spazio τ occupato da cariche. Se ora nel punto P si ha la carica $\rho'\,d\tau'$ su di esso agisce la forza $\rho'\,d\tau' \int \rho\,(E_1 + E_2)\,d\tau$.

(2) Vedi per esempio RICHARDSON, op. cit.

La forza agente su tutto il sistema è perciò

$$\mathbf{F} = \int\int \rho\rho' (\mathbf{E}_1 + \mathbf{E}_2) \, d\tau \, d\tau',$$

dove le due integrazioni debbono essere estese al medesimo campo. D'altra parte si ha evidentemente

$$\int\int \rho\rho' \, \mathbf{E}_1 \, d\tau \, d\tau' = 0$$

e quindi

$$\mathbf{F} = \int\int \rho\rho' \, \mathbf{E}_2 \, d\tau \, d\tau'.$$

Se ora indichiamo con Γ, $\dot{\Gamma}$ l'accelerazione e la sua derivata rispetto al tempo, al tempo t, potremo porre, se r è sufficientemente piccolo

$$\Gamma^* = \Gamma - \frac{r}{c} \dot{\Gamma}$$

si ottiene infine

$$\mathbf{F} = \int\int \left(\frac{\Gamma \cdot \boldsymbol{a}}{c^2 r} a - \frac{\Gamma}{c^2 r} \right) \rho\rho' \, d\tau \, d\tau' - \int\int \left(\frac{\dot{\Gamma} \cdot \boldsymbol{a}}{c^3} a - \frac{\dot{\Gamma}}{c^3} \right) \rho\rho' \, d\tau \, d\tau' . \tag{6}$$

Indichiamo con x_1, x_2, x_3, coordinate cartesiane ortogonali e siano (x_i) le coordinate di M, (x'_i) quelle di P. Le componenti di $\boldsymbol{a}$ sono $a_i = \frac{x'_i - x_i}{r}$. Scrivendo scalarmente la (6) si ottiene allora, osservando che, per l'ipotesi fatta del moto traslatorio le Γ_i e $\dot{\Gamma}_i$ sono costanti durante l'integrazione:

$$\mathrm{F}_i = - \sum_k m_{ik} \Gamma_k + \sum_k \sigma_{ik} \dot{\Gamma}_k . \tag{7}$$

Dove si è posto:

$$(8) \quad \begin{cases} m_{ii} = \dfrac{2\mathrm{U}}{c^2} - \displaystyle\int\int \frac{\rho\rho' (x'_i - x_i)^2}{c^2 r^3} \, d\tau \, d\tau' \\ m_{ik} = m_{ki} = - \displaystyle\int\int \frac{\rho\rho' (x'_i - x_i)(x'_k - x_k)}{c^2 r^3} \, d\tau \, d\tau' \qquad i \neq k \end{cases}$$

$$(9) \quad \begin{cases} \sigma_{ii} = \dfrac{e^2}{c^3} - \displaystyle\int\int \frac{\rho\rho' (x'_i - x_i)^2}{c^3 r^2} \, d\tau \, d\tau' \\ \sigma_{ik} = \sigma_{ki} = - \displaystyle\int\int \frac{\rho\rho' (x'_i - x_i)(x'_k - x_k)}{c^3 r^2} \, d\tau \, d\tau'. \qquad i \neq k. \end{cases}$$

In queste formule U rappresenta l'energia elettrostatica del sistema $= \frac{1}{2} \int\int \frac{\rho\rho'}{r} d\tau \, d\tau'$ ed e la carica elettrica totale $= \int \rho \, d\tau = \int \rho' \, d\tau'$.

Dalle espressioni (8) (9) risulta subito che se si cambiano gli assi (x_i) in altri (y_i) con la sostituzione ortogonale:

$$y_i = \sum_k \alpha_{ik} x_k$$

gli m_{ik} ed i σ_{ik} relativi ai nuovi assi sono:

$$m'_{ik} = \sum_{rs} \alpha_{ir} \alpha_{ks} m_{rs}$$

$$\sigma'_{ik} = \sum_{rs} \alpha_{ir} \alpha_{ks} \sigma_{rs} .$$

Tanto m_{ik} che σ_{ik} sono quindi tensori covarianti simmetrici. Per ciascuno di essi si avranno perciò tre direzioni principali ortogonali tali che prendendo gli assi ad esse paralleli si abbia $m_{ik} = 0$ oppure $\sigma_{ik} = 0$ quando $i \neq k$.

Gli assi principali dei tensori m, σ saranno però generalmente differenti. Nel caso che il sistema abbia simmetria sferica si possono effettuare le integrazioni in (8) (9) perché al posto di $\frac{(x'_i - x_i)(x'_k - x_k)}{r^2}$ si può porre il valor medio di questa espressione per tutte le possibili direzioni MP poiché ai due punti MP corrispondono in questo caso infinite coppie distinte soltanto per l'orientazione. Ora tale valor medio se $i = k$ è $\frac{2\pi}{4\pi}\int_0^\pi \cos^2\theta \sin\theta \, d\theta$; se invece $i \neq k$ è nullo.

Si ha perciò

$$m_{11} = m_{22} = m_{33} = \frac{4U}{3c^2} \quad ; \quad m_{23} = m_{31} = m_{12} = 0 ;$$

$$\sigma_{11} = \sigma_{22} = \sigma_{33} = \frac{2}{3}\frac{e^2}{c^3} \quad ; \quad \sigma_{23} = \sigma_{31} = \sigma_{12} = 0 .$$

Sostituendo questi valori in (7) si ottengono formule ben note nel caso che il sistema sia uno strato sferico omogeneo.

§ 3. – Riprendendo il caso generale osserviamo che per moti quasi-stazionarii alla (5) potrà sostituirsi:

$$F_i = -\sum_k m_{ik} \Gamma_k .$$

Se si pensa che sul sistema agisca una forza esterna (X_i) la forza totale sarà $(X_i + F_i)$. Se si suppone ora il sistema privo di massa materiale deve essere $X_i + F_i = 0$ e perciò

$$X_i = \sum_k m_{ik} \Gamma_k . \tag{10}$$

È agevole mostrare come con la legge del moto (10) si conservano il principio delle forze vive e quello di Hamilton. Infatti si indichi $V \equiv (V_1, V_2, V_3)$ la velocità e si moltiplichi (10) per V_i: sommando poi rispetto ad i si ottiene

$$\sum_i X_i V_i = \sum_{ik} m_{ik} V_i \frac{dV_k}{dt} .$$

Scambiando nelle seconda somma i con k ed osservando che $m_{ik} = m_{ki}$

$$\sum_i X_i V_i = \sum_{ik} m_{ik} V_k \frac{dV_i}{dt}$$

sommando

$$2\sum_{ik} \mathbf{X}_i \mathrm{V}_i = \Sigma\, m_{ik}\left(\mathrm{V}_i \frac{d\mathrm{V}_k}{dt} + \mathrm{V}_k \frac{d\mathrm{V}_i}{dt}\right) = \frac{d}{dt}\sum_{ik} m_{ik} \mathrm{V}_i \mathrm{V}_k .$$

Il primo membro è il doppio della potenza P delle forze esterne. Si ha perciò

$$(11) \qquad P = \frac{dT}{dt} \qquad \text{dove è} \quad T = \frac{1}{2}\Sigma\, m_{ik} \mathrm{V}_i \mathrm{V}_k .$$

Moltiplicando invece i due membri di (10) per δx_i e sommando si ottiene similmente

$$\sum_i \mathrm{X}_i\, \delta x_i = \frac{1}{2}\sum_{ik} m_{ik}\left(\frac{d^2 x_k}{dt^2}\,\delta x_i + \frac{d^2 x_i}{dt^2}\,\delta x_k\right)$$

$$= \frac{d}{dt}\left\{\frac{1}{2}\sum_{ik} m_{ik}\,(\dot{x}_k\,\delta x_i + \dot{x}_i\,\delta x_k)\right\} - \frac{1}{2}\sum_{ik} m_{ik}\,(\dot{x}_k\,\delta\dot{x}_i + \dot{x}_i\,\delta\dot{x}_k)$$

$$= \frac{d}{dt}\left\{\frac{1}{2}\sum_{ik} m_{ik}\,(\dot{x}_k\,\delta x_i + \dot{x}_i\,\delta x_k)\right\} - \delta \mathrm{T} .$$

Moltiplicando per dt e integrando tra due limiti t', t'' nei quali si suppongano nulle le variazioni δx_i si ottiene

$$(12) \qquad \int_{t'}^{t''}\left(\delta T + \sum_i \mathrm{X}_i\,\delta x_i\right) dt = 0$$

che esprime il principio di Hamilton.

Se invece che ad assi qualunque ci si riferisce agli assi principali del tensore m_{ik} le (10) prendono la forma semplice:

$$(13) \qquad \mathbf{X}_i = m_{ii}\,\Gamma_i .$$

§ 4. – Questa formula vale soltanto quando V/c è trascurabile. Per generalizzarla a velocità qualsiasi indichiamo con $\mathrm{S} \equiv (x_1, x_2, x_3, t)$ il sistema di riferimento indicato; con $\mathrm{S}^* \equiv (x, y, z, t)$ un sistema fisso rispetto a S con l'asse x orientato come la velocità del sistema ad un certo tempo $\bar{t}$ generico ma fissato; infine sia $\mathrm{S}' \equiv (x', y', z', t')$ un sistema con gli assi di spazio paralleli ad xyz che si muova relativamente ad S^* di moto uniforme con velocità eguale a quella del mobile al tempo $\bar{t}$, la cui grandezza sia v. Si avrà

$$(14) \qquad t' = \beta\left(t - \frac{v}{c^2}x\right) \;;\; x' = \beta\,(x - vt) \;;\; y' = y \;;\; z' = z \;;\; \beta = \frac{1}{\sqrt{1 - \frac{v^2}{c^2}}}$$

dove, fissato $\bar{t}$, v e quindi β sono costanti.

Supponiamo che le forze agenti sul nostro sistema derivino da un campo elettromagnetico esterno $(\mathbf{E}, \mathbf{H})$; siccome all'istante t il sistema ha velocità nulla rispetto ad S′ varranno per esso le (10) e, con evidente significato dei simboli, si avrà perciò:

$$e\mathrm{E}'_x = m_{xx}\,\Gamma'_x + m_{xy}\,\Gamma'_y + m_{xz}\,\Gamma'_z$$

$$e\mathrm{E}'_y = m_{yx}\,\Gamma'_x + m_{yy}\,\Gamma'_y + m_{yz}\,\Gamma'_z$$

$$e\mathrm{E}'_z = m_{zx}\,\Gamma'_x + m_{zy}\,\Gamma'_y + m_{zz}\,\Gamma'_z.$$

Si ha però

$$eE'_x = eE_x \quad ; \quad eE'_y = e\beta\left(E_y - \frac{v}{c}H_z\right) \quad , \quad eE'_z = e\beta\left(E_z + \frac{v}{c}H_y\right).$$

Se si pone perciò

$$\boldsymbol{k} = e\left(\mathbf{E} + \frac{1}{c}\mathbf{V}\times\mathbf{H}\right) \tag{15}$$

si trova

$$eE'_x = k_x \qquad eE'_y = e\beta\, k_y \qquad eE'_z = e\beta\, k_z\,.$$

D'altra parte:

$$\Gamma'_x = \frac{d^2x'\,dt' - d^2t'\,dx'}{dt'^3}$$

ma, al tempo $\bar{t}$, $\frac{dx'}{dt'} = 0$ quindi $\Gamma'_x = \frac{d^2x'}{dt'^2}$. Prendendo t per variabile indipendente ed osservando che $\frac{dx}{dt} = v$

$$\Gamma'_x = \beta^3\,\Gamma_x\,. \quad \text{Parimenti} \quad \Gamma'_y = \beta^2\,\Gamma_y \quad \Gamma'_z = \beta^2\,\Gamma_z\,.$$

Sostituendo

$$\left\{\begin{aligned} k_x &= m_{xx}\,\beta^3\,\ddot{x} + m_{xy}\,\beta^2\,\ddot{y} + m_{xz}\,\beta^2\,\ddot{z} \\ k_y &= m_{yx}\,\beta^2\,\ddot{x} + m_{yy}\,\beta\ddot{y} + m_{yz}\,\beta\ddot{z} \\ k_z &= m_{zx}\,\beta^2\,\ddot{x} + m_{zy}\,\beta\ddot{y} + m_{zz}\,\beta\ddot{z}. \end{aligned}\right. \tag{16}$$

Indicando con α_{xi} il coseno dell'angolo tra l'asse x e l'asse x_i si ha

$$k_i = \alpha_{xi}\,k_x + \alpha_{yi}\,k_y + \alpha_{zi}\,k_z\,.$$

D'altra parte essendo m_{io} covariante si ha per esempio

$$m_{xy} = \sum_r m_{rr}\,\alpha_{xr}\,\alpha_{yr}\,.$$

Parimenti

$$\ddot{x} = \sum_j \ddot{x}_j\,\alpha_{xj}\,.$$

Moltiplicando allora le (16) per α_{xi}, α_{yi}, α_{zi} e sommando si trova

$$k_i = \sum_{rj} m_{rr}\,\ddot{x}_j \left[\begin{aligned} &\beta^3\,\alpha_{xr}{}^2\,\alpha_{xj}\,\alpha_{xi} + \beta^2\,\alpha_{xr}\,\alpha_{yr}\,\alpha_{yj}\,\alpha_{xi} + \beta^2\,\alpha_{xr}\,\alpha_{zr}\,\alpha_{zj}\,\alpha_{xi} \\ &+ \beta^2\,\alpha_{yr}\,\alpha_{xr}\,\alpha_{xj}\,\alpha_{yi} + \beta\alpha_{yr}{}^2\,\alpha_{yj}\,\alpha_{yi} + \beta\alpha_{yr}\,\alpha_{zr}\,\alpha_{zr}\,\alpha_{zj}\,\alpha_{yi} \\ &+ \beta^2\,\alpha_{zr}\,\alpha_{xr}\,\alpha_{xj}\,\alpha_{zi} + \beta\alpha_{zr}\,\alpha_{yr}\,\alpha_{yj}\,\alpha_{zi} + \beta\alpha_{zr}{}^2\,\alpha_{zj}\,\alpha_{zi}\,. \end{aligned}\right]$$

Si ha però $\alpha_{xi} = \frac{\dot{x}_i}{v}$. Tenendo conto delle relazioni tra le α si trova da ultimo la cercata generalizzazione delle (13)

$$\begin{aligned} k_i = \beta \sum_{jr} \ddot{x}_j\, m_{rr} \Big\{ &(\beta - 1)^2 \frac{\dot{x}_i\,\dot{x}_j\,\dot{x}_r^2}{v^4} \\ &+ (\beta - 1)\left[(jr)\frac{\dot{x}_i\,\dot{x}_r}{v^2} + (ir)\frac{\dot{x}_j\,\dot{x}_r}{v^2}\right] + (ir)\,(jr)\Big\}. \end{aligned} \tag{17}$$

Dove

$$(jr) = 1 \quad \text{se } j = r \quad ; \quad (jr) = 0 \quad \text{se } j \neq r.$$

Nel caso della simmetria sferica posto $m_{11} = m_{22} = m_{33} = m$ si può effettuare la somma in (17) e si trova:

$$k_i = \beta m \ddot{x}_i + m\beta(\beta^2 - 1)\frac{\dot{x}_i}{v^2}\sum_j \dot{x}_j \ddot{x}_j$$

da cui ricordando che

$$\beta = \frac{1}{\sqrt{1 - \frac{v^2}{c^2}}}$$

si ritrova la notissima formula della dinamica elettronica

$$k_i = \frac{d}{dt}\frac{m\dot{x}_i}{\sqrt{1 - \frac{v^2}{c^2}}}.$$

Pisa, gennaio 1921.

N° 2.

For the introduction to this paper see N° 1.

2.

SULL'ELETTROSTATICA DI UN CAMPO GRAVITAZIONALE UNIFORME E SUL PESO DELLE MASSE ELETTROMAGNETICHE

« Nuovo Cimento », *22*, 176–188 (1921).

INTRODUZIONE.

Fine di questo scritto è la ricerca dell'alterazione prodotta da un campo gravitazionale uniforme sui fenomeni elettrostatici che hanno luogo in esso fatta sulle basi della teoria generale della relatività. Stabilita l'equazione differenziale che lega il potenziale elettrico alla densità delle cariche, e che corrisponde all'equazione di Poisson dell'elettrostatica classica, si riesce ad integrarla nel caso almeno che il campo di gravitazione sia sufficientemente poco intenso, ed il campo della gravitazione terrestre soddisfa largamente a tale condizione, trovando così le correzioni da apportarsi alla legge di Coulomb per la presenza del campo di gravità.

In una prima applicazione si studia la distribuzione dell'elettricità sopra una sfera conduttrice mostrando che per influenza del campo la sfera si polarizza.

La seconda applicazione è dedicata alla ricerca del peso di una massa elettromagnetica, vale a dire della forza che, per effetto di trovarsi in un campo di gravità, si esercita sopra un sistema di cariche elettriche rigido (per esempio sostenuto da un dielettrico rigido).

Si trova che tale peso è dato dal prodotto dell'accelerazione di gravità per u/c^2 dove u rappresenta l'energia elettrostatica delle cariche del sistema e c la velocità della luce. Si trova quindi che la massa pesante, cioè il rapporto tra il peso e l'accelerazione di gravità, per il nostro sistema non coincide almeno in generale, con la massa inerte, poiché, ad esempio per un sistema avente simmetria sferica, quest'ultima viene data da $(4/3)\ u/c^2$, con i simboli di prima.

È noto del resto come la relatività in senso stretto conduca a prendere $\Delta u/c^2$ come aumento della massa *inerte* di un sistema al quale viene comunicata l'energia Δu, e ciò può agevolmente mettersi in relazione con il risultato accennato.

Si mostra da ultimo come possa trovarsi un punto che, rispetto al peso del nostro sistema di cariche, gode delle stesse proprietà di cui gode il centro di gravità rispetto al peso di un ordinario sistema di masse materiali.

PARTE PRIMA

Elettrostatica di un campo di gravità.

§ 1. – Consideriamo una regione dello spazio sede di un campo uniforme di gravità e supponiamo che i fenomeni elettrostatici che pensiamo aver luogo in essa siano talmente poco intensi che si possa ritener trascurabile l'alterazione da essi prodotta sulla metrica della regione in discorso. L'elemento metrico della varietà spazio-tempo, relativo a tale regione potrà in tali ipotesi, porsi sotto la forma (1)

$$ds^2 = a\,dt^2 - dx^2 - dy^2 - dz^2 \tag{1}$$

essendo a funzione soltanto di z.

Le variabili t, x, y, z saranno anche indicate con x_0, x_1, x_2, x_3 ed i coefficienti della forma quadratica (1) con g_{ik}. Sia φ_i il potenziale vettore ed F_{ik} il campo elettromagnetico. Sarà

$$F_{ik} = \varphi_{ik} - \varphi_{ki} \tag{2}$$

riferendoci alla forma fondamentale (1).

Limitando le nostre considerazioni ai campi elettrostatici porremo $\varphi_1 = \varphi_2 = \varphi_3 = 0$ e, per brevità, $\varphi_0 = \varphi$. Ora si ha:

$$F_{ik} = \varphi_{ik} - \varphi_{ki} = \frac{\partial \varphi_i}{\partial x_k} - \frac{\partial \varphi_k}{\partial x_i}$$

cioè

$$\left\{\begin{array}{lll} F_{01} = \dfrac{\partial \varphi}{\partial x} & F_{02} = \dfrac{\partial \varphi}{\partial y} & F_{03} = \dfrac{\partial \varphi}{\partial z} \\ F_{23} = F_{31} = F_{12} = 0 & F_{ik} = -F_{ki} & F_{ii} = 0. \end{array}\right. \tag{3}$$

Si ha inoltre:

$$F^{(ij)} = \sum_{kh} g^{(ih)} g^{(jk)} F_{hk} = g^{(ii)} g^{(jj)} F_{ij}$$

da cui osservando che:

$$g^{(00)} = \frac{1}{a} \qquad g^{(11)} = g^{(22)} = g^{(33)} = -1$$

si ottiene:

$$\left\{\begin{array}{lll} F^{(01)} = -\dfrac{1}{a}\dfrac{\partial \varphi}{\partial x} & F^{(02)} = -\dfrac{1}{a}\dfrac{\partial \varphi}{\partial y} & F^{(03)} = -\dfrac{1}{a}\dfrac{\partial \varphi}{\partial z} \\ F^{(23)} = F^{(31)} = F^{(12)} = 0 & F^{(ik)} = -F^{(ki)} & F^{(ii)} = 0. \end{array}\right. \tag{4}$$

Nel caso che stiamo trattando può darsi all'azione la forma seguente,

$$W = \int_\omega \sum_{ik} F_{ik} F^{(ik)}\, d\omega + \int de \int \varphi\, dx_0 \tag{5}$$

(1) T. Levi-Civita, Nota II. *Sui ds^2 einsteiniani*, «Rend. Acc. Lincei», *27*, 1° sem. N° 7.

dove

$$d\omega = \sqrt{-\|g_{ik}\|}\, dx_0\, dx_1\, dx_2\, dx_3 = \sqrt{a}\, dx\, dy\, dz\, dt$$

è l'elemento di ipervolume della varietà e l'integrazione relativa a $d\omega$ va estesa ad una determinata regione della varietà mentre le integrazioni relative a de, dx_0 vanno estese rispettivamente a tutti gli elementi di carica elettrica le cui linee orarie passano per la regione considerata ed a quei segmenti delle dette linee che in essa si trovano.

§ 2. – Nella variazione di W, φ potrà variarsi arbitrariamente, con l'unica condizione che sul contorno della regione d'integrazione sia $\delta\varphi = 0$.

Invece le variazioni δx, δy, δz oltre alla condizione $\delta x = \delta y = \delta z = 0$ sul contorno potranno essere assoggettate anche ad altre da determinarsi nei varii casi particolari. Per esempio nell'interno di un corpo conduttore esse saranno completamente arbitrarie mentre in un dielettrico rigido dovranno rappresentare le componenti di uno spostamento virtuale rigido e così di seguito.

Portando in (5) i valori (3), (4) si trova:

$$W = -\frac{1}{2}\iiiint \frac{1}{\sqrt{a}}\left\{\left(\frac{\partial\varphi}{\partial x}\right)^2 + \left(\frac{\partial\varphi}{\partial y}\right)^2 + \left(\frac{\partial\varphi}{\partial z}\right)^2\right\} dx\, dy\, dz\, dt + \int de \int \varphi\, dt \tag{6}$$

da cui

$$\delta W = \iiiint \delta\varphi \left[\frac{1}{\sqrt{a}}\Delta_2\varphi + \frac{\partial\varphi}{\partial z}\frac{d\frac{1}{\sqrt{a}}}{dz} + \rho\right] dx\, dy\, dz\, dt \tag{7}$$

$$+ \iiiint \rho\left(\frac{\partial\varphi}{\partial x}\delta x + \frac{\partial\varphi}{\partial y}\delta y + \frac{\partial\varphi}{\partial z}\delta z\right) dx\, dy\, dz\, dt$$

come subito si ricava osservando che, per le ipotesi fatte, si ha, lungo una linea oraria, $dx = dy = dz = 0$ e che, essendo ρ la densità elettrica, $\rho\, dx\, dy\, dz = de$.

Perché δW sia identicamente nullo si trova intanto, siccome $\delta\varphi$ è arbitrario nell'interno del campo d'integrazione:

$$\Delta_2\varphi - \frac{d\log\sqrt{a}}{dz}\frac{\partial\varphi}{\partial z} = -\rho\sqrt{a}. \tag{8}$$

Dovrà oltre ciò aversi:

$$\iiiint \rho\left(\frac{\partial\varphi}{\partial x}\delta x + \frac{\partial\varphi}{\partial y}\delta y + \frac{\partial\varphi}{\partial z}\delta z\right) dx\, dy\, dz\, dt = 0, \tag{9}$$

per ogni sistema di valori di δx, δy, δz conforme ai vincoli supposti.

Nell'equazione (8) è contenuta la generalizzazione della legge di Poisson a cui la (8) si riduce se a è costante, cioè se manca il campo gravitazionale.

§ 3. – Se con G indichiamo l'accelerazione di gravità del campo considerato, cioè l'accelerazione con cui comincia a muoversi un punto materiale libero, si ha

$$G = -\frac{1}{2}\frac{da}{dz}.$$

Con ciò la (8) si scrive:

$$\Delta_2 \varphi + \frac{G}{a} \frac{\partial \varphi}{\partial z} = -\rho \sqrt{a}. \tag{11}$$

Per trovare la soluzione di (11), dato in ogni punto ρ, penseremo che le cariche elettriche siano contenute entro una piccola regione attorno all'origine delle coordinate. Porremo inoltre, nell'origine, $a = c^2$ (c = velocità della luce nell'intorno dell'origine) e supporremo la gravità così poco intensa da potersi trascurare quei termini che contengono il quadrato del rapporto lG/c^2 dove l rappresenta la massima lunghezza che interviene nel problema che si sta considerando. Sotto tali ipotesi può porsi:

$$\sqrt{a} = c + \frac{1}{2c} \frac{da}{dz} z = c\left(1 - \frac{G}{c^2} z\right).$$

La (11) può scriversi perciò:

$$\Delta_2 \varphi + \frac{G}{c^2} \frac{\partial \varphi}{\partial z} = -c\left(1 - \frac{G}{c^2} z\right)\rho. \tag{12}$$

L'integrale di questa equazione, come mostra senz'altro una verifica materiale è dato, con l'approssimazione indicata, da:

$$\varphi_P = \frac{c}{4\pi}\int\left(1 - \frac{G}{c^2}\right) z_M \, d\tau_M \left(\frac{1}{r} - \frac{G}{2c^2} \frac{z_P - z_M}{r}\right) \tag{13}$$

$$= \frac{c}{4\pi}\int \rho_M \, d\tau_M \left(\frac{1}{r} - \frac{G}{2c^2} \frac{z_P + z_M}{r}\right)$$

dove M è il punto generico del campo τ_M occupato da cariche elettriche, P il punto nel quale si calcola φ ed r la distanza MP.

Data la linearità dell'equazione (12) potrà naturalmente aggiungersi a (13) un qualsiasi integrale dell'equazione:

$$\Delta^2 \varphi + \frac{G}{c^2} \frac{\partial \varphi}{\partial z} = 0 \tag{12)*$$

ottenuta ponendo in (12) $\rho = 0$. Tale integrale rappresenterà il campo dovuto a cause esterne alle ρ_M. Per le applicazioni che abbiamo in vista conviene considerare una particolare soluzione di (12)* data da

$$\varphi = -cE_x^* x - cE_y^* y + \frac{c^2}{G} E_z^* e^{-\frac{G}{c^2} z} \tag{14}$$

essendo E_x^* E_y^* E_z^* delle costanti.

Nell'origine essendo E la forza elettrica si ha

$$E_x = -\frac{1}{c} F_{01} \qquad E_y = -\frac{1}{c} F_{02} \qquad E_z = -\frac{1}{c} F_{03}.$$

Risulta di qui che nell'origine la forza elettrica del campo esterno (14) ha per componenti:

$$E_x \; E_y^* \; E_z^*.$$

§ 4. – Calcoliamo ora il campo elettrico dovuto ad una carica e concentrata nell'origine delle coordinate. Da (13) si ha:

$$\varphi = \frac{ce}{4\pi}\left(\frac{1}{r} - \frac{G}{2c^2}\frac{z}{r}\right) \tag{15}$$

e questa formula dà come subito si riconosce ponendo G = 0 la generalizzazione della legge elementare di Coulomb. Ricordando le (3) si ottiene:

$$\left\{\begin{aligned} F_{01} &= \frac{ce}{4\pi}\left(\frac{x}{r^3} - \frac{G}{2c^2}\frac{zx}{r^3}\right) \\ F_{02} &= \frac{ce}{4\pi}\left(\frac{y}{r^3} - \frac{G}{2c^2}\frac{zy}{r^3}\right) \\ F_{03} &= \frac{ce}{4\pi}\left(\frac{z}{r^3} - \frac{G}{2c^2}\frac{z^2}{r^3} + \frac{G}{2c^2}\frac{1}{r}\right). \end{aligned}\right. \tag{16}$$

Possiamo riassumere le tre formule precedenti in un'unica vettoriale. Indicando infatti con $\mathbf{F}_0$ il vettore di componenti F_{01}, F_{02}, F_{03}, con $\boldsymbol{a}$ un vettore di grandezza 1 e di orientazione MP infine con $\mathbf{G}$ un vettore di grandezza G e di orientazione z le (16) possono scriversi:

$$F_0 = \frac{ce}{4\pi}\left\{\frac{\boldsymbol{a}}{r^2} - \frac{\mathbf{G}\cdot\boldsymbol{a}}{2c^2 r}\boldsymbol{a} + \frac{1}{2c^2 r}\mathbf{G}\right\}. \tag{17}$$

È interessante confrontare questa formula con quella che dà la forza elettrica esercitata da una carica elettrica e che, in assenza di campo gravitazionale, abbia accelerazione Γ, moto quasi stazionario e velocità trascurabile di fronte a quella della luce. Tale forza, con le stesse notazioni, è espressa da:

$$E = \frac{e}{4\pi}\left\{\frac{\boldsymbol{a}}{r^2} + \frac{\mathbf{G}\cdot\boldsymbol{a}}{c^2 r}\boldsymbol{a} - \frac{1}{c^2 r}\Gamma\right\}. \tag{18}$$

Di qui si vede che, ponendo in (18)

$$\Gamma = -\frac{\mathbf{G}}{2} \tag{19}$$

si ottiene

$$F_0 = cE.$$

In parole tale risultato può così enunciarsi, osservando che cE è la parte elettrica del campo elettromagnetico generato dalla carica in moto accelerato:

La parte elettrica (F_{01}, F_{02}, F_{03}) del campo elettromagnetico (F_{ik}) generato da una carica elettrica ferma in un campo uniforme di intensità G è uguale alla parte elettrica del campo elettromagnetico che, in assenza di campo gravitazionale, produrrebbe la stessa carica se si muovesse, con le condizioni indicate, con accelerazione G/2 in senso opposto al campo di gravitazione.

§ 5. – Studiamo ora come viene alterata dal campo gravitazionale la distribuzione dell'elettricità sopra un conduttore. Osserviamo perciò che essendo nell'interno del conduttore δx, δy, δz arbitrarii dovrà essere per (9) nell'interno $\varphi =$ costante e quindi (8) $\rho = 0$. L'elettricità è quindi tutta alla superficie. Supponiamo allora che il nostro conduttore sia una sfera col centro O nell'origine delle coordinate e raggio R.

Cerchiamo di soddisfare la condizione $\varphi =$ costante nell'interno assumendo per la densità elettrica superficiale in un punto generico M della superficie l'espressione:

$$\frac{e}{4\pi R^2} + \frac{e}{R} a \cos\theta \tag{20}$$

dove θ rappresenta l'angolo che il raggio vettore OM forma con l'asse z ed a una costante da determinarsi che supponiamo dell'ordine di grandezza di G/c^2. Il potenziale in un punto P interno sarà dato (13) da:

$$\varphi_P = \frac{c}{4\pi}\int_\sigma \left(\frac{e}{4\pi R^2} + \frac{e}{R} a\cos\theta\right)\left(\frac{1}{r} - \frac{G}{2c^2}\frac{z_P + z_M}{r}\right) d\sigma$$

dove l'integrazione deve estendersi a tutta la superficie σ della sfera. Trascurando termini di ordine superiore a G/c^2 si ottiene:

$$\varphi_P = \frac{ce}{16\pi^2 R^2}\int \frac{d\sigma}{r} + \frac{cea}{4\pi R}\int \frac{\cos\theta\, d\sigma}{r} - \frac{ce\, G z_P}{32\pi^2 R^2 c^2}\int \frac{d\sigma}{r} - \frac{ce\, G}{32\pi^2 R^2 c^2}\int \frac{z_M\, d\sigma}{r}. \tag{21}$$

Essendo P interno si ha però:

$$\int \frac{d\sigma}{r} = 4\pi R \qquad \int \frac{\cos\theta}{r} d\sigma = \frac{4}{3}\pi z_P \qquad \int \frac{z_M}{r} d\sigma = \frac{4}{3}\pi R\, z_P.$$

Si trova dunque:

$$\varphi_P = \frac{ce}{4\pi R} + \frac{c}{3}\left(\frac{e}{R} a - \frac{eG}{2\pi R c^2}\right) z_P. \tag{22}$$

Se dunque vogliamo φ_P costante dovremo porre

$$a = \frac{1}{2\pi}\frac{G}{c^2}.$$

Sostituendo questo valore in (20) si trova per la densità superficiale l'espressione:

$$\frac{e}{4\pi R^2}\left(1 + \frac{2G}{c^2} R\cos\theta\right). \tag{23}$$

Il fatto di trovarsi in un campo di gravitazione produce dunque nella sfera una polarizzazione di momento

$$\frac{2}{3}\frac{G}{c^2} e R^2.$$

PARTE SECONDA

Peso delle masse elettromagnetiche.

§ 6. – Supponiamo di avere un sistema di cariche sostenute da un sostegno rigido per modo che ai δx , δy , δz del § 2 bisogni attribuire la forma di componenti di uno spostamento rigido. Rimandando a dopo la considerazione degli spostamenti rotatorii consideriamo dapprima quelli traslatorii, vale a dire supponiamo δx , δy , δz siano arbitrarie funzioni del tempo ma non dipendano da x, y, z.

Cercheremo allora di soddisfare la (9) pensando che il potenziale φ_P in un punto generico P sia la somma del potenziale dato da (13) e di uno di tipo (14). Indicheremo questi due addendi con φ'_P e φ''_P e supporremo che il rapporto fra le derivate di φ''_P e φ'_P rispetto ad una direzione qualunque sia dell'ordine di grandezza lG/c^2 di cui abbiamo convenuto di trascurare i quadrati. Con ciò la (9) si scrive:

$$\int dt \left\{ \delta x \int_{\tau_P} \left(\frac{\partial \varphi'}{\partial x} + \frac{\partial \varphi''}{\partial x} \right) \rho_P \, d\tau_P + \delta y \int_{\tau_P} \left(\frac{\partial \varphi'}{\partial y} + \frac{\partial \varphi''}{\partial y} \right) \rho_P \, d\tau_P \right.$$

$$\left. + \delta z \int_{\tau_P} \left(\frac{\partial \varphi'}{\partial z} + \frac{\partial \varphi''}{\partial z} \right) \rho_P \, d\tau_P \right\} = 0 .$$

Questa equazione, dato che δx δy δz sono funzioni arbitrarie del tempo, tra di loro indipendenti, dà luogo alle tre equivalenti:

$$\int_{\tau_P} \left(\frac{\partial \varphi'}{\partial x} + \frac{\partial \varphi''}{\partial x} \right) \rho_P \, d\tau_P = \int_{\tau_P} \left(\frac{\partial \varphi'}{\partial y} + \frac{\partial \varphi''}{\partial y} \right) \rho_P \, d\tau_P \tag{24}$$

$$= \int_{\tau_P} \left(\frac{\partial \varphi'}{\partial z} + \frac{\partial \varphi''}{\partial z} \right) \rho_P \, d\tau_P = 0 .$$

Ora dall'espressione (13) di φ' si ottiene subito, osservando che

$$\frac{\partial r}{\partial x_P} = \frac{x_P - x_r}{r} :$$

$$\int_{\tau_P} \frac{\partial \varphi'}{\partial x} \rho_P \, d\tau_P = - \frac{c}{4\pi} \int_{\tau_P} \int_{\tau_M} \rho_P \, \rho_M \, d\tau_P \, d\tau_M \left\{ \frac{x_P - x_M}{r^3} - \frac{G}{2 c^2} \frac{(x_P - x_M)(z_P + z_M)}{r^3} \right\}$$

dove entrambi gli integrali vanno estesi al campo occupato da cariche. Scambiando nel secondo membro P con M, ciò che nulla altera, si ottiene:

$$\int_{\tau_P} \frac{\partial \varphi'}{\partial x} \rho_P \, d\tau_P = - \frac{c}{4\pi} \int_{\tau_M} \int_{\tau_P} \rho_M \, \rho_P \, d\tau_M \, d\tau_P \left\{ \frac{x_M - x_P}{r} - \frac{G}{2 c^2} \frac{(x_M - z_P)(z_M + z_P)}{r^2} \right\}$$

onde infine, prendendo la semisomma:

$$\int_{\tau_P} \frac{\partial \varphi'}{\partial x} \rho_P \, d\tau_P = 0 . \tag{25}$$

In modo in tutto analogo

$$\int_{\tau_P} \frac{\partial \varphi'}{\partial y} \rho_P \, d\tau_P = 0 . \tag{26}$$

D'altra parte similmente:

$$\int_{\tau_P} \frac{\partial \varphi'}{\partial z} \rho_P \, d\tau_P = - \int_{\tau_P}\int_{\tau_M} \rho_P \, \rho_M \, d\tau_P \, d\tau_M \left\{ \frac{z_P - z_M}{r^3} - \frac{G}{2c^2} \frac{(z_P - z_M)(z_P + z_M)}{r^3} + \frac{G}{2c^2} \frac{1}{r} \right\}$$

scambiando M con P

$$\int_{\tau_P} \frac{\partial \varphi'}{\partial z} \rho_P \, d\tau_P = - \frac{c}{4\pi} \int_{\tau_P}\int_{\tau_M} \rho_M \, \rho_P \, d\tau_M \, d\tau_P \left\{ \frac{z_M - z_P}{r^3} - \frac{G}{2c^2} \frac{(z_M - z_P)(z_M + z_P)}{r^3} + \frac{G}{2c^2} \frac{1}{r} \right\}$$

e prendendo la semisomma:

$$\int_{\tau_P} \frac{\partial \varphi'}{\partial z} \rho_P \, d\tau_P = - \frac{c}{4\pi} \frac{G}{2c^2} \int_{\tau_P}\int_{\tau_M} \frac{\rho_P \, \rho_M}{r} d\tau_P \, d\tau_M = - G \frac{u}{c^2} c \tag{27}$$

indicandosi con u l'energia elettrostatica del sistema (a prescindere dai termini di correzione gravitazionali). Per le ipotesi fatte sulle derivate di φ'' possiamo senz'altro scrivere, con la nostra approssimazione:

$$\left\{ \begin{aligned} \int_{\tau} \frac{\partial \varphi''}{\partial x} \rho \, d\tau &= - c E_x^* e ; \\ \int_{\tau} \frac{\partial \varphi''}{\partial y} \rho \, d\tau &= - c E_y^* e ; \\ \int_{\tau} \frac{\partial \varphi''}{\partial z} \rho \, d\tau &= - c E_z^* e ; \end{aligned} \right. \tag{28}$$

dove con $e = \int_{\tau} \rho \, d\tau$ si è indicata la carica totale del sistema. Sostituendo in (24) le espressioni ora ottenute si trova:

$$\left\{ \begin{aligned} e E_x^* &= 0 \\ e E_y^* &= 0 \\ e E_z^* &= - G \frac{u}{c^2} . \end{aligned} \right. \tag{29}$$

In queste formole è contenuto il nostro risultato. Esse ci dicono infatti che per mantenere in equilibrio il nostro sistema è necessario un campo esterno (E^*) il quale esercita sul sistema una forza data (in prima approssimazione) da eE^* che deve intendersi controbilanciare il peso del sistema, che vien dunque dato da $-eE^*$ ed ha perciò per componenti

$$0 , 0 , G \frac{u}{c^2} . \tag{30}$$

Giungiamo con questo alla conclusione che il *peso di una massa elettromagnetica ha sempre la direzione della verticale e grandezza eguale al peso di una massa materiale* u/c^2.

§ 7. – Nel paragrafo precedente abbiamo preso per δx, δy, δz le componenti di uno spostamento traslatorio. Se invece si prendono le componenti di uno spostamento rotatorio virtuale con l'asse passante per l'origine delle coordinate, si pone cioè

$$(31) \qquad \delta x = qz - ry \quad ; \quad \delta y = rx - pz \quad ; \quad \delta z = py - qx ,$$

l'integrale (9), a prescindere dalla parte dovuta al campo esterno φ'' diventa:

$$(32) \qquad \int d\tau \left\{ p \int_\tau \rho \left(y \frac{\partial \varphi}{\partial z} - z \frac{\partial \varphi}{\partial y} \right) d\tau + q \int_\tau \rho \left(z \frac{\partial \varphi}{\partial x} - x \frac{\partial \varphi}{\partial z} \right) d\tau + r \int_\tau \rho \left(x \frac{\partial \varphi}{\partial y} - y \frac{\partial \varphi}{\partial x} \right) d\tau \right\} .$$

Gli integrali tra parentesi si calcolano facilmente in base a (13) con artifici simili a quelli usati nel paragrafo precedente. Essi hanno i valori

$$(33) \qquad -\frac{G}{8\pi c} \int\int \frac{y_P}{r} \rho_P \rho_M \, d\tau_P \, d\tau_M \quad ; \quad + \frac{G}{8\pi c} \int\int \frac{x_P}{r} \rho_P \rho_M \, d\tau_P \, d\tau_M \quad ; \quad 0 .$$

E prendendo per origine il punto O′ definito dal punto O e dal vettore

$$O' - O = \frac{1}{2u} \int\int \frac{P - O}{r} \rho_P \rho_M \, d\tau_P \, d\tau_M ,$$

si riconosce subito che i tre integrali si annullano *qualunque* sia l'orientazione del sistema attorno ad O′. Segue di qui che rispetto alla nuova origine è nullo identicamente l'integrale (9), cioè è nullo il momento del peso rispetto a O′ qualunque sia l'orientazione del sistema: O′ gode dunque delle proprietà di centro di gravità.

Pisa, marzo 1921.

N° 3.

The origin of this paper is probably related to that of N° 2, although neither of the two papers contains an explicit reference to the other. It can be supposed, in fact, that Fermi, after the study of the relations between gravitational and electromagnetic fields, made in N° 2 under rather restrictive conditions, felt the opportunity of a more systematic treatment of this and other similar problems, by means of a system of space-time coordinates particularly fitted to follow the behaviour in time of phenomena happening in a small spatial region. He was so conducted to this paper, which, except in the last part, is essentially the demonstration of a theorem of absolute differential calculus. This theorem is of considerable interest for the applications, and is therefore reproduced in the most important treatises of absolute differential calculus (see e.g.: T. Levi-Civita, *Calcolo Differenziale Assoluto*, Roma 1925, p. 190; also, in English translation: *Absolute Differential Calculus*, Glasgow 1927, p. 167). Later, it has been extended by L. P. Eisenhart to a certain class of non-riemannian spaces (see L. P. Eisenhart, *Non-Riemannian Geometry*, New York 1927) and by P. Dienes to any linearly connected space (see « Rend. Acc. Linc. », *18*, (6), p. 369 (1933)). A possible extension to Weyl's space is also suggested in Fermi's paper. For a detailed derivation and discussion see also: L. O. Raifertaigh, *Fermi Coordinates*, Proc. Roy. Irish Acad., 59 A, p. 15 (1958). Extensive use of "Fermi coordinates" is made in J. L. Synge, *Relativity, the General Theory* (Amsterdam, 1960).

E. Persico

3.

SOPRA I FENOMENI CHE AVVENGONO IN VICINANZA DI UNA LINEA ORARIA

« Rend. Lincei », *31* (1), 21–23, 51–52, 101–103 (1922) (*).

NOTA I.

1. Per fare lo studio dei fenomeni che avvengono in vicinanza di una linea oraria, cioè, in linguaggio non relativistico, in una porzione di spazio, variabile eventualmente col tempo, ma sempre molto piccola in confronto alle divergenze dall'euclideità, della varietà spazio–tempo, converrà anzi tutto ricercare un opportuno riferimento tale che, in vicinanza della linea studiata, il ds^2 della varietà prenda una forma semplice. Per trovare questo riferimento, dobbiamo premettere qualche considerazione geometrica.

Sia data in una varietà riemanniana V_n, od anche in una varietà metricamente connessa nel senso di Weyl [1], una linea L. Associamo ad ogni punto P di L una direzione y, perpendicolare ad L, con la legge che la direzione $y + dy$, relativa al punto $P + dP$, si deduca da quella y relativa

(*) Presentate dal Corrispondente G. Armellini nella seduta del 22 gennaio 1922.

(1) Weyl, *Raum, Zeit, Materie*, p. 109. Berlin, Springer, 1921.

a P, nel seguente modo: Sia η la direzione tangente ad L in P; si trasportino parallelamente [2] y, η da P in P $+ d$P e siano $y + \delta y$, $\eta + \delta\eta$ le direzioni così ottenute, che per le proprietà fondamentali del trasporto parallelo saranno ancora ortogonali. Se L non è geodetica $\eta + \delta\eta$ non coinciderà con la direzione $\eta + d\eta$ della tangente ad L in P $+ d$P, e queste due direzioni individueranno in P $+ d$P una giacitura. Consideriamo in P $+ d$P l'elemento di S_{n-2} perpendicolare ad essa e ruotiamo rigidamente attorno a tale S_{n-2} tutta una particella circostante P $+ d$P, fino a che $\eta + \delta\eta$ non vada a sovrapporsi ad $\eta + d\eta$. Allora $y + \delta y$ andrà a finire in una posizione che prenderemo come direzione $y + dy$ relativa al punto P $+ d$P. Si intende bene come, fissata a piacere la direzione y in un punto di L, un processo di integrazione permetta di conoscerla per tutti i punti di L.

Cerchiamo ora le espressioni analitiche traducenti le operazioni indicate per una varietà riemanniana, che sono identiche a quelle valevoli per una varietà metrica di Weyl purché si abbia l'avvertenza di scegliere la « Eichung » in guisa che la misura di un segmento, che si muova rigidamente nelle vicinanze di L, sia costante. Sia

$$ds^2 = \sum_{ik} g_{ik}\, dx_i\, dx_k \qquad (i\,,k = 1\,,2\,,\cdots,n) \tag{1}$$

e siano $y_i\,, y^{(i)}\,; \eta_i\,, \eta^{(i)} = \frac{dx_i}{ds}$ i sistemi co– e controvarianti delle direzioni $y\,,\eta$. Avremo intanto

$$\frac{\delta\eta^{(i)}}{ds} = -\sum_{hl} \left\{ \begin{matrix} hl \\ i \end{matrix} \right\} \eta^{(h)} \frac{dx_l}{ds} = -\sum_{hl} \left\{ \begin{matrix} hl \\ i \end{matrix} \right\} \frac{dx_h}{ds} \frac{dx_l}{ds}\,,$$

è inoltre $\frac{d\eta^i}{ds} = \frac{d}{ds}\frac{dx_i}{ds} = \frac{d^2 x_i}{ds^2}$. Si trova dunque

$$\frac{\delta\eta^{(i)} - d\eta^{(i)}}{ds} = -\left(\frac{d^2 x_i}{ds^2} + \sum_{hl} \left\{ \begin{matrix} hl \\ i \end{matrix} \right\} \frac{dx_h}{ds} \frac{dx_l}{ds} \right) = -C^i.$$

Le C^i sono le componenti controvarianti del vettore **C**, curvatura geodetica, cioè di un vettore che ha l'orientazione della normale principale geodetica di L e grandezza eguale alla sua curvatura geodetica.

Si ha d'altra parte

$$\frac{\delta y^{(i)}}{ds} = -\sum_{hk} \left\{ \begin{matrix} hk \\ i \end{matrix} \right\} y^{(h)} \frac{dx_k}{ds}\,. \tag{2}$$

Ora, siccome y è perpendicolare ad L, lo spostamento, con cui da $y + \delta y$ si deduce $y + dy$, sarà parallelo alla tangente ad L e avrà grandezza eguale alla proiezione sopra y stesso di $\delta\eta - d\eta$; vale a dire, siccome y ha lunghezza 1, al prodotto scalare di $\delta\eta - d\eta$ per y, cioè

$$\sum_i (\delta\eta_i - d\eta_i)\, y^{(i)} = -ds \sum_i C_i\, y^{(i)}.$$

(2) T. LEVI-CIVITA, « Rend. Circ. Palermo », tomo XLII, p. 173 (1917).

Le sue componenti controvarianti si otterranno dunque moltiplicando la sua grandezza per le coordinate controvarianti della tangente ad L, cioè dx_i/ds. Esse son dunque, in ultima analisi, $-dx_i \sum_r C_r y^{(r)}$. Da (2) risulta ora immediatamente

$$\frac{dy^{(i)}}{ds} = -\sum_{hk} \left\{ \begin{matrix} h\,k \\ i \end{matrix} \right\} y^{(h)} \frac{dx_k}{ds} - \frac{dx_i}{ds} \sum_h C_h y^h. \tag{3}$$

La (3), scritta per $i = 1, 2, \cdots, n$, dà un sistema di n equazioni differenziali del primo ordine tra le n incognite $y^{(1)} y^{(2)} \ldots y^{(n)}$ che risultano così determinate, una volta che ne siano assegnati i valori iniziali. Sarebbe anche facile verificare formalmente dalle (3) che, se i valori iniziali delle $y^{(i)}$ soddisfano la condizione di perpendicolarità ad L, tale condizione resta verificata lungo tutta la linea.

2. In un punto P_0 di L assegnamo ora a piacere n direzioni $y_1, y_2, \cdots, y_n$ mutuamente ortogonali, con la condizione che y_n sia tangente ad L. Le direzioni $y_1, y_2, \cdots, y_{n-1}$ saranno perpendicolari ad L e potremo trasportarle lungo L con la legge assegnata al paragrafo precedente che, come è evidente dalla sua stessa definizione, conserva la loro ortogonalità. In tale modo veniamo ad associare ad ogni punto di L n direzioni mutuamente ortogonali, di cui l'ultima è quella della tangente ad L. Pensiamo ora la nostra V_n immersa in un S_N euclideo a un numero conveniente di dimensioni. Possiamo prendere come coordinate di un punto di V_n le coordinate cartesiane ortogonali della sua proiezione sopra l'S_n tangente a V_n in un punto generico P di L, aventi per origine P e per direzioni le direzioni $y_1, y_2, \cdots, y_n$ relative al punto P. Con tali coordinate l'elemento metrico di V_n in P prende la forma $ds^2 = dy_1^2 + dy_2^2 + \cdots + dy_n^2$; esse inoltre, come immediatamente si riconosce, sono geodetiche in P. Vale a dire, per le coordinate y si può nell'intorno di P porre, a meno di infinitesimi di ordine superiore al primo, $g_{ii} = 1$; $g_{ik} = 0$ $(i \neq k)$. È manifesto che di tali riferimenti ne avremo uno per ogni punto di L. Consideriamo ora un punto Q_0 di V_n che nel riferimento relativo al punto P_0 di L abbia coordinate $\bar{y}_1, \bar{y}_2, \cdots, \bar{y}_{n-1}, 0$. Per ogni altro punto P di L possiamo allora determinare un punto Q che, nel riferimento relativo a P, abbia le stesse coordinate che ha Q_0 nel riferimento relativo a P_0. Il punto Q percorrerà così una linea di decorso parallelo ad L. Vogliamo ora trovare la relazione che lega ds_Q a ds_P nell'ipotesi che Q sia infinitamente prossimo a P. Perciò osserviamo che lo spostamento che porta Q in $Q + dQ$ è composto degli spostamenti indicati al § 1 con δ e con $d - \delta$ e che il primo, essendo uno spostamento parallelo, fornisce, a meno di infinitesimi di ordine superiore, $\delta s_Q = ds_P$; il secondo è una rotazione che, come si è visto al § 1, dà $(d - \delta)\, s_Q = ds_P\, \mathbf{C} \cdot (Q - P)$, se con $\cdot$ si indica il simbolo del prodotto scalare e con $Q - P$ il vettore di origine P e termine Q. Inoltre ds_Q e $(d - \delta)\, s_Q$ hanno entrambi la direzione della tangente in L. Si ha dunque $ds_Q = \delta s_Q + (d - \delta)\, s_Q$; cioè

$$ds_Q = ds_P\,[1 + \mathbf{C} \cdot (Q - P)]. \tag{4}$$

Le traiettorie dei punti Q formano una $(n-1)^{\text{upla}}$ infinità di linee e quindi, almeno con opportune limitazioni, per ogni punto M di V_n passerà una di tali linee; così che potremo caratterizzare M mediante le coordinate del punto Q, $\bar{y}_1, \bar{y}_2, \cdots, \bar{y}_{n-1}$ corrispondenti alla linea passante per M, e l'arco s_P di linea L contato da un'origine arbitraria fino a quel punto P che corrisponde al Q coincidente con M.

Se M è infinitamente prossimo ad L, ds_Q sarà perpendicolare alla ipersuperficie $s_P =$ costante. Si avrà perciò

$$ds_M^2 = ds_Q^2 + d\bar{y}_1^2 + d\bar{y}_2^2 + \cdots + d\bar{y}_{n-1}^2 \,;$$

e, tenendo presente (4),

$$(5) \qquad ds_M^2 = [1 + \mathbf{C}\cdot(\mathrm{M}-\mathrm{P})]^2\, ds_P^2 + d\bar{y}_1^2 + d\bar{y}_2^2 + \cdots + d\bar{y}_{n-1}^2 \,.$$

Nelle vicinanze di L abbiamo con ciò trovata una espressione semplicissima del ds^2.

NOTA II.

3. Prima di passare all'applicazione fisica dei risultati ottenuti, vogliamo ancora fare qualche osservazione geometrica. È evidente intanto che le considerazioni precedenti, e quindi anche la formula (5) che ne è la conclusione, che per varietà qualunque sono valide solo vicino ad L, sono invece completamente rigorose per spazî euclidei. Associamo allora alla linea L della V_n una linea L* di uno spazio euclideo S_n, in cui indichiamo con x_i^* le coordinate cartesiane ortogonali. Se con degli asterischi indichiamo i simboli riferentisi alla linea L*, potremo scrivere per S_n la formula analoga a (5):

$$(5^*) \qquad ds_{M^*}^2 = [1 + \mathbf{C}^*\cdot(\mathrm{M}^*-\mathrm{P}^*)]\, ds_{P^*}^2 + d\bar{y}_1^{*2} + d\bar{y}_2^{*2} + \cdots + d\bar{y}_{n-1}^{*2} \,;$$

come nella (5) $\mathbf{C}$ è funzione di s_P, così nella $(5)^*$ $\mathbf{C}^*$ è funzione di s_{P^*}.

Siano $K^{(1)} K^{(2)}, \ldots, K^{(n-1)}$ le componenti controvarianti di $\mathbf{C}$ relative a $\bar{y}_1, \bar{y}_2, \cdots, \bar{y}_{n-1}$ e $K^{(1)*}, K^{(2)*}, \ldots, K^{(n-1)*}$ quelle di $\mathbf{C}^*$ relative alle $\bar{y}^*$. Cerchiamo se si possa determinare L* in modo che le funzioni $K^{(r)*}(s_P)$ diventino eguali alle $K^{(r)}(s_P)$. Cominceremo perciò a porre $s_P = s_{P^*}$, cioè a stabilire tra i punti di L e quelli di L* una corrispondenza biunivoca che conserva gli archi. Osserviamo poi che $K^{(r)*}$ è la proiezione di $\mathbf{C}^*$ sulla r^{esima} direzione y^*. È cioè

$$(6) \qquad K^{(r)*} = \sum_{i=1}^{i=n} y_{i|r}^* \frac{d^2 x_i^*}{ds_P^2} \qquad (r = 1, 2, \cdots, n-1).$$

Le $K^{(r)}$ sono poi funzioni note di s_P. La condizione $K^{(r)} = K^{(r)*}$ conduce dunque alle $(n-1)$ equazioni

$$(7) \qquad K^{(r)}(s_P) = \sum_{i=1}^{i=n} y_{i|r}^* \frac{d^2 x_i^*}{ds_P^2} \qquad (r = 1, 2, \cdots, n-1).$$

D'altra parte le (3), scritte per l'S_n, ci dànno $n(n-1)$ altre equazioni. Se a queste aggiungiamo l'altra

$$ds_P^2 = dx_1^{*2} + dx_2^{*2} + \cdots + dx_n^{*2}, \tag{8}$$

troviamo un sistema di $n - 1 + n(n - 1) + 1 = n^2$ equazioni tra le n^2 incognite x_i^*, $y_{i/r}^*$ che servono ad esprimerle in funzione di s_P. Possiamo così determinare le equazioni parametriche $x_i^* = x_i^*(s_P)$ della L^*. Con ciò la formula (5*) diventa identica alla (5), ossia abbiamo rappresentato per applicabilità i dintorni delia linea L^*, sui dintorni di L. Siccome poi L^* è in uno spazio euclideo, possiamo dire anche di aver disteso i dintorni di L in uno spazio euclideo, ossia di aver trovato coordinate che sono geodetiche contemporaneamente in tutti i punti di L.

NOTA III.

§ 4. Per mostrare l'applicazione dei risultati precedenti alla teoria della relatività, supporremo che V_n sia la V_4 spazio–tempo e che L sia una linea oraria, in vicinanza della quale ci proponiamo di studiare i fenomeni. Ponendo per brevità in (5) $ds_M = ds$, si trova in questo caso:

$$ds^2 = [1 + \mathbf{C}\cdot(M - P)]^2 ds_P^2 + d\bar{y}_1^2 + d\bar{y}_2^2 + d\bar{y}_3^2 .$$

Per evitare la comparsa di immaginarii e ristabilire l'omogeneità, conviene fare la seguente sostituzione di variabili:

$$s_P = vt \quad ; \quad \bar{y}_1 = ix \quad ; \quad \bar{y}_2 = iy \quad ; \quad \bar{y}_3 = iz ,$$

essendo v una costante con le dimensioni di una velocità, per modo che t abbia le dimensioni di un tempo. Si ottiene, così,

$$ds^2 = a\,dt^2 - dx^2 - dy^2 - dz^2 \tag{9}$$

dove

$$a = v^2 [1 + \mathbf{C}\cdot(M - P)]^2. \tag{10}$$

Da ora in avanti, con gli ordinarii simboli del calcolo vettoriale intenderemo riferirci allo spazio x, y, z. Ed è in questo senso che si può intendere il prodotto scalare che figura in (10), purché per **C** si intenda il vettore avente per componenti le componenti covarianti della curvatura geodetica della linea $x = y = z = 0$ e con $M - P$ il vettore di componenti x, y, z. Chiameremo x, y, z coordinate di spazio e t tempo. Per uniformità scriveremo talvolta x_0, x_1, x_2, x_3 al posto di t, x, y, z e chiameremo anche g_{ik} i coefficienti della forma quadratica (9).

§ 5. Sia (3) F_{ik} il campo elettromagnetico e $(\varphi_0, \varphi_1, \varphi_2, \varphi_3)$ il tensore di primo ordine «potenziale» di F_{ik}, in modo che sia $F_{ik} = \varphi_{ik} - \varphi_{ki}$.

(3) Per le notazioni e per la deduzione Hamiltoniana delle leggi della fisica, vedi WEYL., op. cit., pp. 186 e 208.

Poniamo $\varphi_0 = \varphi$ e chiamiamo $\boldsymbol{u}$ il vettore di componenti $\varphi_1, \varphi_2, \varphi_3$. Si avrà intanto:

$$\left.\begin{matrix} F_{01} \\ F_{02} \\ F_{03} \end{matrix}\right\} = \operatorname{grad} \varphi - \frac{\partial \boldsymbol{u}}{\partial t} \quad ; \quad \left.\begin{matrix} F_{23} \\ F_{31} \\ F_{12} \end{matrix}\right\} = -\operatorname{rot} \boldsymbol{u}\,,\ F_{ii} = 0\,,\ F_{ik} = -F_{ki}\,,$$

parimenti

$$\left.\begin{matrix} F^{01} \\ F^{02} \\ F^{03} \end{matrix}\right\} = \frac{1}{a}\left(-\operatorname{grad} \varphi + \frac{\partial \boldsymbol{u}}{\partial t}\right) \quad , \quad \left.\begin{matrix} F^{(23)} \\ F^{(31)} \\ F^{(12)} \end{matrix}\right\} = -\operatorname{rot} \boldsymbol{u}\,,\ F^{(ii)} = 0\,,\ F^{(ik)} = -F^{(ki)}$$

e quindi

$$\frac{1}{4} \sum_{ik} F_{ik} F^{(ik)} = \frac{1}{2} \left\{ \operatorname{rot}^2 \boldsymbol{u} - \frac{1}{a}\left(\operatorname{grad} \varphi - \frac{\partial \boldsymbol{u}}{\partial t}\right)^2 \right\}.$$

Sia $d\omega$ l'elemento di ipervolume di V_4. Avremo

$$d\omega = \sqrt{-\,\| g_{ik} \|}\, dx_0\, dx_1\, dx_2\, dx_3 = \sqrt{a}\, dt\, d\tau$$

dove $d\tau = dx\, dy\, dz$ è l'elemento di volume dello spazio.

Si ha anche:

$$\Sigma \varphi_i\, dx_i = \varphi\, dt + \boldsymbol{u} \cdot d\mathrm{M} \qquad d\mathrm{M} = (dx\,, dy\,, dz).$$

Prescindendo dall'azione del campo metrico, la cui variazione è nulla perché lo riguardiamo come dato *a priori* dalla (9), l'azione prenderà la seguente forma:

$$W = \frac{1}{4} \int_\omega \sum_{ik} F_{ik} F^{(ik)} d\omega + \int_e de \int \sum_i \varphi_i\, dx_i + \int_m dm \int ds$$

$$\begin{pmatrix} de = \text{elemento di carica elettrica} \\ dm = \text{elemento di massa} \end{pmatrix}.$$

Introducendo le notazioni indicate, si trova

$$W = \frac{1}{2} \iint \left\{ \operatorname{rot}^2 \boldsymbol{u} - \frac{1}{2}\left(\operatorname{grad} \varphi - \frac{\partial \boldsymbol{u}}{\partial t}\right)^2 \right\} \sqrt{a}\, dt\, d\tau + \iint (\varphi + \boldsymbol{u} \cdot \mathbf{V}_L)\, \rho\, d\tau\, dt + \iint \sqrt{a - V_M^2}\, k\, d\tau\, dt, \tag{11}$$

dove ρ, k sono rispettivamente le densità di elettricità e di materia, per modo che $de = \rho\, d\tau$, $dm = k\, d\tau$, $\mathbf{V}_L$ è la velocità delle cariche elettriche, $\mathbf{V}_M$ quella delle masse.

Gli integrali del secondo membro possono estendersi ad un campo arbitrario τ tra due tempi qualunque t_1, t_2. Si ha poi il vincolo che sul contorno del campo τ, e per i due tempi t_1, t_2, siano nulle tutte le variazioni.

All'infuori di queste condizioni, le variazioni di φ e di $\boldsymbol{u}$ sono completamente arbitrarie. Per contro, alle variazioni di x, y, z, considerate come coordinate di un elemento di carica o di massa, possono essere imposte ulteriori condizioni, traducenti i vincoli del particolare problema che si sta studiando. Scrivendo intanto che è nullo dW per una variazione qualunque $\delta\varphi$ di φ, si trova

$$o = -\iint \left(\operatorname{grad}\varphi - \frac{\partial \boldsymbol{u}}{\partial t}\right) \cdot \delta \operatorname{grad}\varphi \frac{dt\,d\tau}{\sqrt{a}} + \iint \delta\varphi\,\rho\,dt\,d\tau.$$

Trasformando il primo integrale con opportuna applicazione del teorema di Gauss, e tenendo presente che $\delta\varphi$ si annulla sul contorno, troviamo

$$o = \iint \delta\varphi \left\{ \rho + \operatorname{div}\left[\frac{1}{\sqrt{a}}\left(\operatorname{grad}\varphi - \frac{\partial \boldsymbol{u}}{\partial t}\right)\right]\right\} dt\,d\tau$$

e, siccome $\delta\varphi$ è arbitrario, abbiamo intanto l'equazione

$$\rho + \operatorname{div}\left\{\left[\frac{1}{\sqrt{a}}\left(\operatorname{grad}\varphi - \frac{\partial \boldsymbol{u}}{\partial t}\right)\right]\right\} = o. \tag{12}$$

In modo analogo, facendo variare $\boldsymbol{u}$, si trova

$$\rho V_L + \operatorname{rot}(\sqrt{a}\operatorname{rot}\boldsymbol{u}) - \frac{\partial}{\partial t}\left[\frac{1}{\sqrt{a}}\left(\operatorname{grad}\varphi - \frac{\partial \boldsymbol{u}}{\partial t}\right)\right] = o. \tag{13}$$

Queste due ultime equazioni permettono di determinare il campo elettro–magnetico, una volta assegnate le cariche ed il loro movimento.

Un altro gruppo di equazioni si può ottenere facendo variare in W le traiettorie delle cariche e delle masse. Siano δP_M la variazione della traiettoria delle masse, δP_L quella delle cariche. Indichiamo inoltre, essendo $\boldsymbol{u}$ un vettore funzione di punto e $\mathbf{V}$ un vettore, con $(\partial\boldsymbol{u}/\partial P)(\mathbf{V})$ il vettore di componenti $\frac{\partial u_x}{\partial x} V_x + \frac{\partial u_x}{\partial y} V_y + \frac{\partial u_x}{\partial z} V_z$ ed analoghe. Scrivendo che è nulla la variazione di W, si trova allora, coi soliti artifici:

$$\iint \left(\delta P_L \cdot \operatorname{grad}\varphi - \delta P_L + \left(\frac{\partial \boldsymbol{u}}{\partial t} + \frac{\partial \boldsymbol{u}}{\partial P}(\mathbf{V}_L)\right) + \mathbf{V}_L \cdot \frac{\partial \boldsymbol{u}}{\partial P}(\delta P_L)\right) \rho\,dt\,d\tau \tag{14}$$

$$+ \iint \delta P_M \cdot \left\{\frac{dt}{ds}\,\frac{\operatorname{grad} a}{2} + \frac{d}{dt}\left(\frac{dt}{ds}\mathbf{V}_M\right)\right\} k\,dt\,d\tau = o.$$

Se i δP ad un tempo non dipendono dai loro valori per altri tempi, dovrà essere nullo in (14) il coefficiente di dt. Si trova così:

$$\int \left\{\delta P_L \cdot \operatorname{grad}\varphi - \delta P_L \left[\frac{\partial \boldsymbol{u}}{\partial t} + \frac{\partial \boldsymbol{u}}{\partial P}(\mathbf{V}_L)\right] + \mathbf{V}_L \cdot \frac{\partial \boldsymbol{u}}{\partial \mathbf{P}}(\delta P_L)\right\} \rho\,d\tau \tag{15}$$

$$+ \int \delta P_M \cdot \left\{\frac{1}{2}\frac{dt}{ds}\operatorname{grad} a + \frac{d}{dt}\left(\frac{dt}{ds}\mathbf{V}_M\right)\right\} k\,d\tau$$

che deve essere verificata per tutti i sistemi di δP soddisfacenti ai vincoli.

N° 4 *a*, 4 *b*, 4 *c*, 10.

The results of papers N° 1 and 2 puzzled Fermi, because they contradict each other, if it is assumed, as required by general relativity, that the gravitational and inertial mass are the same. Moreover, the value (4/3) (U/c^2), obtained by Lorentz as well as by Fermi for the inert mass of a rigid, spherically symmetrical system of electric charges, was at variance with Einstein's principle of equivalence of mass and energy. It is now well known that the factor 4/3 can be interpreted as due to the part of the energetic tensor contributed by the internal non-electromagnetic stresses, whose existence must be assumed to assure the equilibrium of the charges. However, in the books known to Fermi, this discrepancy was not explained (he had evidently overlooked the explanation contained in M. von Laue, *Die Relativitätstheorie*, I, 3rd ed., 1919, p. 218), and so he found for it an explanation of his own, essentially equivalent to the former but obtained through Weyl's variational method.

Prof. Polvani remembers that the question was debated, one winter evening of 1922, in Pisa, while Fermi, Puccianti, Polvani and other friends walked through via San Frediano from the University to the Scuola Normale Superiore. Here the company parted without having reached any satisfactory conclusion. In the following two days, Fermi did not appear in the Institute of Physics. The third day he arrived with a paper, ready for publication, entitled «Correzione di una grave discrepanza . . . ». Puccianti, who had emphasized the need for a clarification, was enthusiastically happy.

This result, of which Fermi was particularly proud, was published by him, with minor alterations, in three different journals (N° 4 *a*, 4 *b*, 4 *c*). Subsequently, in collaboration with A. Pontremoli (a young physicist, then an assistant at the University of Rome, who later was to disappear tragically in the Nobile polar expedition of 1928) Fermi applied the same method to the calculation of the mass of the radiation contained in a cavity with reflecting walls, for which Abraham and others had found an expression containing the same factor 4/3 (see paper N° 10).

E. PERSICO.

4 *c*.

CORREZIONE DI UNA CONTRADDIZIONE TRA LA TEORIA ELETTRODINAMICA E QUELLA RELATIVISTICA DELLE MASSE ELETTROMAGNETICHE (*)

« Nuovo Cimento », *25*, 159–170 (1923).

§ 1. – La teoria delle masse elettromagnetiche fu studiata per la prima volta da M. Abraham [1] prima della scoperta della teoria della relatività. Abraham perciò, come era naturale, considerò nei suoi calcoli la massa di

(*) Sullo stesso argomento vedi due mie Note sui « Rend. Acc. Lincei » (5), 31, pp. 184, 306 (1922).

(1) ABRAHAM, *Theorie der Electricität*; RICHARDSON, *Electron Theory of Matter*. Cap. XI; LORENTZ, *The Theory of Electrons*, p. 37.

un sistema di cariche elettriche rigido nel senso della meccanica classica, e trovò che, nell'ipotesi che tale sistema avesse simmetria sferica, la sua massa era variabile con la velocità e precisamente eguale a (2) $\frac{4}{3}\frac{u}{c^2}$ (essendo u l'energia elettrostatica del sistema e c la velocità della luce) per velocità nulle o molto piccole, mentre per velocità v confrontabili con c intervenivano dei termini di correzione un po' complicati, dell'ordine di grandezza di v^2/c^2. Prima ancora della teoria della relatività, Fitz Gerald introdusse l'ipotesi che i corpi solidi subissero nella direzione del loro moto una contrazione nel rapporto

$$\sqrt{1 - \frac{v^2}{c^2}} : 1$$

e Lorentz rifece la teoria delle masse elettromagnetiche di Abraham, considerando invece che sistemi di cariche elettriche rigidi nel senso della meccanica classica, dei sistemi che subissero questa contrazione. Il risultato fu che la massa di quiete, ossia il limite della massa per velocità nulle, era sempre $\frac{4}{3}\frac{u}{c^2}$, mentre venivano alterati i termini correttivi dipendenti da $v^2 : c^2$. Le esperienze di Kaufmann, Bucherer e altri sulla massa delle particelle β dei corpi radioattivi, e delle particelle catodiche di grande velocità, si decisero nettamente a favore della teoria di Lorentz, così detta dell'elettrone contrattile, contro quella di Abraham, dell'elettrone rigido. E ciò fu in un primo tempo interpretato come prova della natura esclusivamente elettromagnetica della massa degli elettroni, perché si pensava che altrimenti la loro massa avrebbe dovuto essere costante. Scoperta in seguito la teoria della relatività, questa portò alla conseguenza che tutte le masse, fossero esse o no elettromagnetiche, dovevano variare con la velocità come quella dell'elettrone contrattile di Lorentz; per modo che le esperienze indicate vennero a lasciare indecisa la natura totalmente elettromagnetica o no della massa elettronica, venendo a costituire esclusivamente una conferma della teoria della relatività. D'altra parte la stessa teoria della relatività in senso stretto, e più ancora in seguito quella generale, condussero ad attribuire ad un sistema dotato dell'energia u la massa $u : c^2$, per modo che venne a sorgere una grave discrepanza tra la teoria elettrodinamica di Lorentz, che attribuisce ad una distribuzione sferica di elettricità la massa di quiete $\frac{4}{3}\frac{u}{c^2}$, e la teoria della relatività che le attribuisce invece la massa u/c^2. Ed una tale differenza (3) si presenta particolarmente grave, data la grande importanza della nozione di massa elettromagnetica come base della teoria elettronica della materia.

(2) Si dice ordinariamente che la massa elettromagnetica di uno strato elettrico sferico omogeneo di carica e, e di raggio r è $\frac{2}{3}\frac{e^2}{rc^2}$; se però si osserva che l'energia elettrostatica è $u = \frac{1}{2}\frac{e^2}{r}$, si trova appunto la massa $= \frac{4}{3}\frac{u}{c^2}$.

(3) Le esperienze di Kaufmann ecc. non possono naturalmente servire in questo caso a decidere quale dei due risultati è il giusto, perché esse permettono soltanto di misurare i termini correttivi che dipendono dalla velocità e che sono eguali secondo entrambi le teorie, mentre la differenza è invece tra le masse di quiete.

Tale discrepanza mi si presentò in particolar modo stridente in due recenti Note [4] in una delle quali, sulla base delle ordinarie teorie elettrodinamiche considerai le masse elettromagnetiche di sistemi a simmetria qualunque, trovando che in generale sono rappresentate da tensori invece che da scalari, che si riducono naturalmente a $\frac{4}{3}\frac{u}{c^2}$ nel caso della simmetria sferica; nell'altra invece, partendo dalla teoria generale della relatività, considerai il peso dei medesimi sistemi, che trovai in ogni caso eguale a $\frac{u}{c^2}$ G, essendo G l'accelerazione di gravità.

Nel presente lavoro dimostreremo precisamente: che la differenza tra i due valori della massa ottenuti nei due modi ha origine in un concetto di corpo rigido in contraddizione col principio di relatività che si applica nella teoria elettrodinamica (anche in quella dell'elettrone contrattile) e che conduce alla massa $\frac{4}{3}\frac{u}{c^2}$, mentre la nozione di corpo rigido più giustificata e conforme alla teoria della relatività conduce invece al valore u/c^2.

Notiamo ancora che la dinamica relativistica dell'elettrone fu svolta da M. Born [5] che però, partendo da un punto di vista non essenzialmente diverso dall'ordinario trovò naturalmente come massa di quiete $\frac{4}{3}\frac{u}{c^2}$.

Nelle nostre considerazioni prenderemo a base il principio di Hamilton come quello più adatto allo studio di un problema soggetto a vincoli alquanto complicati; invero il nostro sistema di cariche elettriche deve soddisfare ad un vincolo di natura diversa da quelli considerati nella meccanica ordinaria, dovendo esso, dipendentemente dalla propria velocità, presentare la contrazione di Lorentz, in conseguenza del principio di relatività. Notiamo però fin d'ora, ad evitar malintesi, che mentre la contrazione di Lorentz è dell'ordine di v^2/c^2, la sua influenza sopra la massa elettromagnetica verte sui termini principali di questa, cioè sulla massa di quiete ed ha perciò un'importanza assai maggiore, essendo apprezzabile anche per velocità piccolissime.

§ 2. – Consideriamo dunque un sistema di cariche elettriche, sostenute da un dielettrico rigido che, sotto l'azione di un campo elettromagnetico in parte dovuto al sistema stesso e in parte a cause esterne si muova di moto traslatorio descrivendo un tubo orario nello spazio tempo [6].

Vediamo con precisione che cosa debba intendersi per moto traslatorio rigido. Consideriamo perciò un qualunque sistema di riferimento di Lorentz–Einstein e supponiamo che per esso a un certo istante un punto del sistema di cariche abbia velocità nulla; diremo che il moto è traslatorio se con tali ipotesi nello stesso riferimento, per quell'istante, tutti i punti del sistema hanno velocità nulla. Ciò equivale a dire che le linee orarie dei punti del nostro sistema sono traiettorie ortogonali di una famiglia di spazi lineari;

(4) E. FERMI, «N. Cim.», VI, 22, pp. 176, 192 (1921).

(5) MAX BORN, «Ann. d. Phys.», 30, p. 1 (1909).

(6) In tutto il seguito si riguarda lo spazio tempo come euclideo, perché si intende che i campi elettromagnetici che in esso si considerano siano abbastanza poco intensi per non alterarne sensibilmente la struttura metrica.

ed infatti in un riferimento di Lorentz–Einstein in cui lo spazio sia uno degli spazii della famiglia e l'asse di tempo sia naturalmente perpendicolare ad esso tutto il sistema è in quiete al tempo zero, perché lo spazio taglia ortogonalmente le linee orarie di tutti i punti del sistema. Con questa definizione di moto traslatorio, che in sostanza è quella adottata da M. Born, la rigidità del sistema viene espressa dal fatto che la sua figura in questi spazi perpendicolari al tubo resta invariabile, ossia che tutte le sezioni rette del tubo sono tra di loro eguali.

Per poter applicare al caso nostro il principio di Hamilton ci occorre avere una variazione del movimento del nostro sistema conforme ai vincoli del problema, ossia alla rigidità, giustamente interpretata. Ora noi mostreremo che si giunge al valore $\frac{4}{3}\frac{u}{c^2}$ oppure a quello u/c^2 per la massa

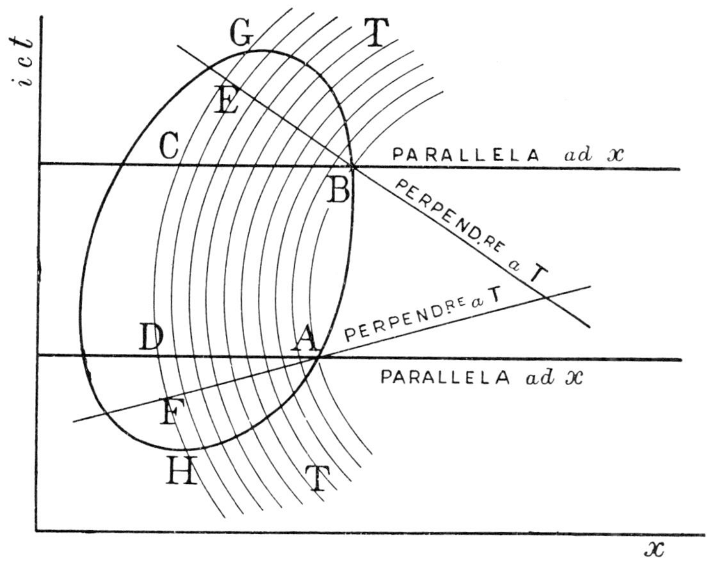

elettromagnetica, secondo che per tale variazione si prende una o l'altra delle due che andiamo ad illustrare e che distinguiamo con le lettere A e B. La variazione A però, come immediatamente si vedrà, è da scartarsi perchè in contraddizione col principio di relatività. Sia T il tubo orario descritto dal sistema. Nella figura lo spazio (x, y, z) è rappresentato su una sola dimensione dall'asse x, ed al tempo t è sostituito ict per aver una metrica definita.

Variazione A: si considera come variazione soddisfacente il vincolo della rigidità uno spostamento infinitesimo, rigido nell'ordinario senso cinematico, parallelo allo spazio (x, y, z), di ogni sezione del tubo parallela allo spazio medesimo. Nella figura otterremo dunque tale variazione spostando parallelamente all'asse x ogni sezione $t =$ cost del tubo di un segmento infinitesimo arbitrario. Se ci limitiamo a considerare spostamenti traslatorii avremo dunque δx, δy, δz funzioni arbitrarie del solo tempo, e $\delta t = 0$.

Variazione B: si considera come variazione soddisfacente al vincolo della rigidità uno spostamento infinitesimo perpendicolare al tubo di ogni sezione normale del tubo medesimo, rigido nell'ordinario senso cinematico. Nella figura otterremo tale variazione spostando parallelamente a sè, di un segmento arbitrario, ogni sezione normale del tubo.

Di tali due variazioni *quella* A *è in manifesta contraddizione col principio di relatività* e da scartarsi perchè, non essendo neppure invariante rispetto alle trasformazioni di Lorentz, essa è determinata invero dal particolare riferimento (t, x, y, z) che abbiamo scelto, non può esser l'espressione di nessuna nozione fisica, come quella della rigidità. La variazione B invece, oltre a soddisfare evidentemente la detta condizione di invarianza, poichè è costituita solo con elementi inerenti al tubo T e del tutto indipendenti dalla posizione degli assi di riferimento, è l'unica che si presenti spontanea, come quella che prende a base uno spostamento virtuale rigido nel sistema di riferimento rispetto al quale, all'istante che si considera, il sistema di cariche ha velocità nulla. Ora con un'osservazione superficiale potrebbe far tuttavia l'impressione che la differenza tra le conseguenze dei due sistemi di variazione A e B dovesse farsi sentire soltanto per velocità considerevoli, quando cioè il tubo T ha una notevole inclinazione sull'asse del tempo. Invece i calcoli che andiamo a sviluppare dimostreranno subito che la differenza si sente già per velocità nulle, e che appunto A dà come massa elettromagnetica $\frac{4}{3}\frac{u}{c^2}$ mentre B dà invece $\frac{u}{c^2}$.

§ 3. – Indichiamo, secondo la comodità, con (t, x, y, z) oppure con (x_0, x_1, x_2, x_3) le coordinate di tempo e di spazio e sia φ_i il quadripotenziale ed

$$F_{ik} = \frac{\partial \varphi_i}{\partial x_k} - \frac{\partial \varphi_k}{\partial x_i}$$

il campo elettromagnetico, **E** ed **H** le forze elettrica e magnetica, che si deducono da esso.

Il principio di Hamilton che riassume le leggi di Maxwell Lorentz e quelle della meccanica ci dice che [7]: l'azione complessiva, ossia la somma delle azioni del campo elettromagnetico e delle masse materiali ed elettriche, subisce variazione nulla per effetto di una variazione arbitraria delle φ_i e delle coordinate dei punti delle linee orarie delle cariche elettriche conforme ai vincoli e che si annulli sul contorno della regione di integrazione. Nel caso nostro non ci sono masse materiali, e gli unici elementi che si facciano variare sono le coordinate dei punti delle linee orarie delle cariche; basta perciò considerare solo l'azione delle cariche elettriche, cioè:

$$W = \sum_i \int de \int \varphi_i \, dx_i$$

dove de è l'elemento generico di carica elettrica ed il secondo integrale deve essere esteso a quell'arco di linea oraria descritta da de che è contenuto nel campo quadridimensionale G di integrazione. Per ogni sistema di variazioni δx_i conforme ai vincoli e *che si annulla sul contorno di* G deve dunque aversi

(7) WEYL., «Raum, Zeit, Materie», pp. 194–196; Berlin, Springer (1921).

$\delta W = 0$, cioè:

$$\sum_{ik} \int\int de \, F_{ik} \, \delta x_i \, dx_k = 0 \,. \tag{1}$$

Bisogna ora esaminare separatamente i risultati che si ottengono mettendo al posto dei δx_i i valori dati dal sistema di variazioni A o da quello B.

§ 4. - *Conseguenze del sistema di variazioni* A. — In questo caso il campo di integrazione si riduce semplicemente ad ABCD. Ed invero i campi BCG, ADH danno contributo nullo, poiché in essi tutti i δx_i si annullano dovendo esser nulli sul contorno di G, e quindi nei tratti BG, AH ed aver valore costante per t costante, ossia sulle parallele all'asse x. Se con t_1 e t_2 indichiamo i tempi di A e di B la (1) può scriversi, essendo $\delta t = 0$ ed i δx, δy, δz funzioni del solo tempo:

$$\sum_{ik} \int_{t_1}^{t_2} dt \, \delta x_i \int de \, F_{ik} \frac{dx_k}{dt} \qquad (i = 1, 2, 3) \quad (k = 0, 1, 2, 3) \,.$$

Siccome poi δx_i sono funzioni arbitrarie di t, ne ricaviamo le tre equazioni

$$\int de \sum_k F_{ik} \frac{dx_k}{dt} = 0$$

ossia:

$$\int de \left[E_x + \frac{dy}{dt} H_z - \frac{dz}{dt} H_y \right] = 0 \qquad \text{e le due analoghe.}$$

Se all'istante che si considera il nostro sistema ha velocità nulla nel riferimento (t, x, y, z) le tre equazioni si riassumono nell'unica vettoriale:

$$\int \mathbf{E} \, de = 0 \,. \tag{2}$$

A questa equazione saremmo giunti senza calcoli se, come si fa nelle trattazioni ordinarie e come in sostanza fa M. Born nel lavoro citato, avessimo supposto a priori nulla la forza totale agente sul sistema. Abbiamo appunto voluto dedurla col principio di Hamilton per mostrare il vizio della sua origine, poiché essa segue dal sistema di variazioni A che è in contraddizione col principio di relatività. Dalla (2) segue subito il valore $\frac{4}{3} \frac{u}{c^2}$ come massa elettromagnetica. Supponiamo infatti che $\mathbf{E}$ sia la somma di una parte $\mathbf{E}^{(i)}$ dovuta al sistema stesso, più un campo $\mathbf{E}^{(e)}$ uniforme dovuto a cause esterne. La (2) ci dà:

$$\int \mathbf{E}^{(i)} de + \mathbf{E}^{(e)} \int de = 0 \,.$$

Ora $\int de = e =$ carica; e quindi $\mathbf{E}^{(e)} \int de = \mathbf{F} =$ forza esterna. D'altra parte nel caso della simmetria sferica, sia il calcolo diretto, sia la ben nota considerazione del momento elettromagnetico (8) mostrano che:

$$\int \mathbf{E}^{(i)} de = -\frac{4}{3}\frac{u}{c^2}\,\mathbf{\Gamma},$$

essendo $\mathbf{\Gamma}$ l'accelerazione.

L'equazione precedente diventa dunque:

$$\mathbf{F} = \frac{4}{3}\frac{u}{c^2}\,\mathbf{\Gamma}$$

che confrontata con la legge fondamentale della dinamica del punto, $\mathbf{F} = m\mathbf{\Gamma}$, ci dà:

$$m = \frac{4}{3}\frac{u}{c^2}\,.$$

§ 5. – *Conseguenze del sistema di variazioni* B. — In questo caso le stesse considerazioni del § precedente dimostrano che il campo di integrazione si riduce ad ABEF, ossia alla regione compresa tra due sezioni normali del tubo T. Decomponiamola per mezzo di infinite sezioni normali in infiniti strati di spessore infinitesimo, e per calcolare il contributo di uno di questi all'integrale (1) riferiamoci al suo riferimento di quiete, prendendo lo spazio $(x\,,y\,,z)$ parallelo allo strato. Per esso sarà allora $\delta t = 0$, mentre $\delta x\,,\ \delta y\,,\ \delta z$ saranno costanti qualunque. Sarà inoltre $dx = dy = dz = 0$, perché la velocità di tutti i punti è nulla, $dt =$ altezza dello strato, che varierà da punto a punto, perché lo strato ha per basi due sezioni normali in generale non parallele. Se O è un punto generico ma fissato dello strato, per esempio l'origine delle coordinate, in cui dt ha il valore dt_0, $\mathbf{K}$ il vettore con l'orientazione della normale principale alla linea oraria passante per O e grandezza eguale alla sua curvatura, si ha manifestamente, essendo dt lo spessore nel punto generico P dello stato:

$$dt = dt_0\,[1 - \mathbf{K}\cdot(\mathrm{P}-\mathrm{O})]\,.$$

Siccome la velocità è nulla si ha semplicemente

$$\mathbf{K} = -\mathbf{\Gamma} : c^2\,,$$

e quindi:

$$dt = dt_0\left(1 + \frac{\mathbf{\Gamma}\cdot(\mathrm{P}-\mathrm{O})}{c^2}\right).$$

Sostituendo questi valori si trova che il contributo del nostro strato all'integrale (1) è:

$$-dt_0\left\{\delta x\int\left(1 + \frac{\mathbf{\Gamma}\cdot(\mathrm{P}-\mathrm{O})}{c^2}\right)\mathrm{E}_x\,de + \delta y\int\left(1 + \frac{\mathbf{\Gamma}\cdot(\mathrm{P}-\mathrm{O})}{c^2}\right)\mathrm{E}_y\,de \right.$$
$$\left. + \delta z\int\left(1 + \frac{\mathbf{\Gamma}\cdot(\mathrm{P}-\mathrm{O})}{c^2}\right)\mathrm{E}_z\,de\right\}.$$

Questa espressione deve annullarsi per tutti i valori di $\delta x\,,\ \delta y\,,\ \delta z$, otteniamo dunque da essa tre equazioni che si riassumono nell'unica vettoriale:

$$\int\left(1 + \frac{\mathbf{\Gamma}\,(\mathrm{P}-\mathrm{O})}{c^2}\right)\mathbf{E}\,de = 0\,. \tag{3}$$

(8) Richardson loc. cit.

Una corretta applicazione del principio di Hamilton ci ha dunque condotto alla (3) invece che alla (2). È ora facilissimo esaminare le conseguenze. Ponendo infatti:

$$\mathbf{E} = \mathbf{E}^{(i)} + \mathbf{E}^{(e)}$$

si trova:

$$\int \mathbf{E}^{(i)}\, de + \int \mathbf{E}^{(i)} \frac{\Gamma \cdot (\mathrm{P} - \mathrm{O})}{c^2}\, de + e\,\mathbf{E}^{(e)} + \mathbf{E}^{e} \int \frac{\Gamma \cdot (\mathrm{P} - \mathrm{O})}{c^2}\, de = 0\,.$$

Nel caso della simmetria sferica si ha come sopra

$$\int \mathbf{E}^{(i)}\, de = -\frac{4}{3}\frac{u}{c^2}\boldsymbol{\Gamma}\,;$$

sostituendo nella precedente si trova che $\mathbf{E}^{(e)}$ è confrontato solo con termini che contengono $\boldsymbol{\Gamma}$. Se trascuriamo dunque i termini [9] in Γ^2, possiamo trascurare l'ultimo integrale, ed otteniamo:

$$(4)\qquad -\frac{4}{3}\frac{u}{c^2}\Gamma + \int \mathbf{E}^{(i)} \frac{\boldsymbol{\Gamma} \cdot (\mathrm{P} - \mathrm{O})}{c^2}\, de + \mathbf{F} = 0\,.$$

Per calcolare l'integrale che ancora figura in (4) osserviamo che $\mathbf{E}^{(i)}$ è la somma della forza di Coulomb

$$= \int \frac{\mathrm{P} - \mathrm{P}'}{r^3}\, de'$$

(P' è il punto potenziante di carica de' ed $r = \overline{\mathrm{PP}'}$), e di un termine contenente Γ che può trascurarsi perché darebbe un contributo contenente Γ^2. Il nostro integrale diventa dunque:

$$\int\int \frac{\mathrm{P} - \mathrm{P}'}{r^3} \frac{\boldsymbol{\Gamma} \cdot (\mathrm{P} - \mathrm{O})}{c^2}\, de\, de'\,;$$

oppure scambiando P con P′, ciò che nulla altera, e prendendo la semisomma dei due valori così ottenuti:

$$\frac{1}{2}\int\int \frac{\mathrm{P} - \mathrm{P}'}{c r^3} [\boldsymbol{\Gamma} \cdot (\mathrm{P} - \mathrm{P}')]\, de\, de'\,.$$

Osserviamo che, nella nostra approssimazione Γ è costante per tutti i punti e può quindi portarsi fuori dagli integrali. Perciò la componente x dell'integrale precedente è:

$$\frac{1}{2c^2}\left\{ \Gamma_x \int\int \frac{(x - x')^2}{r^3}\, de\, de' + \Gamma_y \int\int \frac{(y - y')(x - x')}{r^3}\, de\, de' \right.$$
$$\left. + \Gamma_z \int\int \frac{(z - z')(x - x')}{r^3}\, de\, de' \right\}.$$

Ora, siccome il sistema ha simmetria sferica, ad ogni segmento P P′ ne corrispondono infiniti altri distinti solo per l'orientazione. Nei tre integrali potremo

(9) Propriamente il numero di cui si trascurano i quadrati è $\Gamma l/c^2$ essendo l la massima lunghezza che interviene nel problema. È evidente che tale approssimazione è nei casi comuni più che giustificata.

perciò sostituire

$$(x-x')^2, (x-x')(y-y'), (x-x')(z-z')$$

coi loro valori medii per tutte le possibili orientazioni di P P', che sono: $\frac{1}{3} r^2, 0, 0$.

Con ciò la componente x diventa:

$$\frac{\Gamma_x}{3c^2} \frac{1}{2} \int\int \frac{de\,de'}{r}.$$

Osserviamo ora che l'espressione

$$\frac{1}{2} \int\int \frac{de\,de'}{r}$$

non è altro che l'energia elettrostatica u; tornando allora alla scrittura vettoriale troviamo per l'integrale che figura in (4) l'espressione: $\frac{u}{3c^2} \mathbf{\Gamma}$. La (4) diventa così:

$$\frac{u}{c^2} \mathbf{\Gamma} = \mathbf{F} \tag{5}$$

che esprime appunto che la massa elettromagnetica è u/c^2.

N° 5.

This little article is remarkable in two respects. The date shows that already by 1923 Fermi was known well enough, at least in some circles, that his opinion was requested for an appendix to a translation of a book on relativity. I do not know how and why the publisher and the translator had the insight to request comment by a practically unknown young Ph. D. Also remarkable is the strange and almost prophetic premonition of things to come shown in this article. There were about a dozen more similar articles by older generation Italian physicists in the appendix to the same book and almost all of them were very sceptical and hostile to relativity. A perusal of these articles shows the state of mind of official physics in Italy at the time.

E. SEGRÈ.

5.

LE MASSE NELLA TEORIA DELLA RELATIVITÀ

Da A. Kopff, *I fondamenti della relatività Einsteiniana*, Ediz. italiana a cura di R. Contu e T. Bembo, Hoepli, Milano 1923, pp. 342–344.

La grandiosa importanza concettuale della teoria della relatività, come contributo ad una più profonda comprensione dei rapporti tra spazio e tempo, e le vivaci e spesso appassionate discussioni a cui essa ha in conseguenza dato luogo anche fuori degli ambienti strettamente scientifici, hanno forse un po' distolta l'attenzione da un altro suo risultato che, per esser meno clamoroso e, diciamolo pure, meno paradossale, ha tuttavia nella fisica conseguenze non meno degne di nota, ed il cui interesse è verosimilmente destinato a crescere nel prossimo svilupparsi della scienza.

Il risultato a cui accenniamo è la scoperta della relazione che lega la massa di un corpo alla sua energia. La massa di un corpo, dice la teoria della relatività, è uguale alla sua energia totale divisa per il quadrato della velocità della luce. Già un esame superficiale ci mostra come, almeno per la fisica quale la si osserva nei laboratori, l'importanza di questa relazione tra massa ed energia è tale da offuscare notevolmente quella delle altre conseguenze, quantitativamente lievissime, ma alle quali la mente si abitua con più sforzo. Valga un esempio: un corpo lungo un metro che si muovesse con la velocità, abbastanza rispettabile, di 30 km al minuto secondo (eguale press'a poco alla velocità del moto della terra attraverso gli spazii) apparirebbe sempre lungo un metro ad un osservatore trascinato dal suo moto, mentre ad un osservatore fermo apparirebbe lungo un metro meno cinque milionesimi di millimetro; come si vede il risultato, per strano e paradossale che possa parere, è tuttavia molto piccolo, ed è da ritenere che i due osservatori non si metteranno a litigare per così poco. La relazione tra massa ed energia invece ci porta senz'altro a delle cifre grandiose. Ad esempio se si riuscisse a mettere in libertà l'energia contenuta in un grammo di materia si otterrebbe un'energia maggiore di quella sviluppata in tre anni di lavoro ininterrotto da un motore di mille cavalli (inutili i commenti!). Si dirà con ragione che non appare possibile che, almeno in un prossimo avvenire, si trovi il modo di mettere in libertà queste spaventose quantità di energia, cosa del resto che non si può che augurarsi, perché l'esplosione di una così spaventosa quantità di energia avrebbe come primo effetto di ridurre in pezzi il fisico che avesse la disgrazia di trovar il modo di produrla.

Ma se anche una tale esplosione completa della materia non appare per ora possibile, sono però già in corso da qualche anno delle esperienze dirette ad ottenere la trasformazione degli elementi chimici uno nell'altro. Tale trasformazione, che si presenta naturalmente nei corpi radioattivi è stata recentemente ottenuta anche artificialmente da Rutherford che, bombardando con delle particelle α (corpuscoli lanciati con velocità grandissima dalle sostanze radioattive) degli atomi, è riuscito ad ottenerne la decomposizione. Ora a queste trasformazioni degli elementi uno nell'altro sono legati degli scambi energetici che la relazione tra massa ed energia ci permette di studiare in modo molto chiaro. Ad illustrarli valga ancora un esempio numerico. Si ha ragione di ritenere che il nucleo dell'atomo di elio sia costituito da quattro nuclei dell'atomo di idrogeno. Ora il peso atomico dell'elio è 4,002 mentre quello dell'idrogeno è 1,0077. La differenza tra il quadruplo della massa dell'idrogeno e la massa dell'elio, è dunque dovuta all'energia dei legami che uniscono i quattro nuclei di idrogeno per formare il nucleo dell'elio. Questa differenza è 0,029 corrispondente, secondo la relazione relativistica tra massa ed energia ad un'energia di circa sei miliardi di calorie per grammo-atomo di elio. Queste cifre ci dimostrano che l'energia dei legami nucleari è qualche milione di volte maggiore di quella dei più energici legami chimici e ci spiegano come contro il problema della trasformazione della materia, il sogno degli alchimisti, si siano per tanti secoli rotti gli sforzi degli ingegni più eletti, e come solo ora, adoprando i mezzi più energici a nostra disposizione, si sia riusciti ad ottenere questa trasformazione; in quantità del resto tanto minime da sfuggire alla più delicata analisi.

Bastino questi brevi accenni a dimostrare come la teoria della relatività, oltre a darci una interpretazione chiara delle relazioni tra spazio e tempo, sarà, forse in un prossimo avvenire, destinata ad esser la chiave di volta per la risoluzione del problema della struttura della materia, l'ultimo e più arduo problema della fisica.

N° 6.

I raggi Röntgen.

Questo lavoro e l'altro *Formazioni d'immagini coi raggi Röntgen* costituiscono essenzialmente la tesi di laurea di Fermi. Per quante ricerche siano state fatte all'Università di Pisa, non è stato possibile ritrovare la copia originale della Tesi di Laurea, né dai verbali di esame, né dai registri risulta quale fosse esattamente il titolo di essa. Però nell'*Annuario* dell'Università per l'anno 1922–23, a p. 412, nell'elenco dei laureati si legge: « E. FERMI: *Studi sopra i raggi Röntgen* », e questo è molto probabilmente il titolo della tesi.

G. POLVANI.

A review article essentially based on the introductory part of the dissertation discussed under N° 7.

F. RASETTI.

6.

I RAGGI RÖNTGEN

« Nuovo Cimento », *24*, 133–163 (1922).

NOTAZIONI.

e = carica elettronica (u. e. s.).
m = massa elettronica.
c = velocità della luce.
h = quanto di azione di Planck.
N = numero di Avogadro.
Z = numero atomico.
A = peso atomico.
λ = lunghezza d'onda.
ν = frequenza.
d = distanza tra i piani di un reticolato atomico.
n = ordine degli spettri.

§ 1. – LA SCOPERTA E LE PRIME PROPRIETÀ. — I raggi Röntgen o raggi X furono scoperti nel 1895 da W. C. Röntgen, che osservò che le parti del vetro di un tubo duro a raggi catodici che erano colpite da questi, emettevano a loro volta dei raggi che traversavano le pareti del tubo e molti corpi opachi alla luce visibile, impressionavano le lastre fotografiche e rendevano fluorescenti schermi di platinocianuro di bario e di altre sostanze.

Alla scoperta seguirono presto numerose ricerche atte a metterne in evidenza le prime proprietà; principalmente che i nuovi raggi si propagavano in linea retta senza subire né riflessioni né rifrazioni e che ionizzavano i gas attraversandoli. Questa seconda proprietà, importantissima di per sé, permise anche un modo di rivelazione sensibile e quantitativo dei nuovi raggi. Basta infatti far passare il fascio dei raggi in esame tra due piastre metalliche mantenute a potenziale differente, che la corrente di ionizzazione che tra di esse si stabilirà, e che si potrà misurare con un elettroscopio o con un elettrometro, ci darà la misura dell'energia del fascio, purché naturalmente esso sia totalmente assorbito dal gas tra le due piastre, ciò che in pratica si ottiene usando secondo i casi SO_2 oppure $CH_3 J$. Tale metodo di rivelazione, detto della camera di ionizzazione, può subire naturalmente numerose modificazioni, ed è l'unico che si presti a misure quantitative dell'intensità dei raggi X; esso è perciò in molti casi da preferirsi al metodo fotografico, nonostante la minore comodità.

Contemporaneamente a queste prime scoperte, si perfezionò anche la tecnica della produzione dei raggi. Attualmente i tubi a raggi X consistono essenzialmente in un'ampolla di vetro per lo più sferica, contenente un catodo concavo, che concentra i suoi raggi catodici sopra una piastra metallica ad alto punto di fusione, l'anticatodo, che emette i raggi X. Da anodo può servire lo stesso anticatodo oppure un elettrodo posto in un punto qualunque del tubo. Volendo una sorgente di raggi X molto intensa, è spesso necessario raffreddare l'anticatodo con una corrente d'acqua, per evitare che sia fuso dal bombardamento catodico. Quando poi si vogliano aver parecchio intense quelle specie di raggi che sono fortemente assorbite dal vetro dell'ampolla, conviene munire quest'ultima di una finestra di alluminio o di vetro al litio. Varie case costruttrici poi forniscono tubi con disposizioni particolari, atte a dar sorgenti di raggi di particolar costanza e intensità, sulle quali non ci fermiamo.

§ 2. – La natura dei raggi Röntgen. — Sulla natura dei raggi X si presentavano in principio spontanee tre ipotesi:

a) che essi fossero costituiti da una proiezione di corpuscoli elettrizzati analoga ai raggi catodici o ai raggi canale;

b) che essi fossero una proiezione di corpuscoli elettricamente neutri;

c) che essi fossero una propagazione di onde elettromagnetiche differenti dalla luce visibile soltanto per la lunghezza d'onda.

L'ipotesi *a*) fu subito scartata essendosi fin dai primi tempi dimostrato che i raggi X non subivano alcuna deviazione per effetto di campi elettrici o magnetici. Nel caso che fosse stata vera l'ipotesi *c*) i nuovi raggi avrebbero dovuto aver lunghezza d'onda estremamente piccola, perché solo così poteva spiegarsi l'assenza completa dei fenomeni di riflessione e di rifrazione. A verifica di questa ipotesi Walter e Pohl [1] tentarono un'esperienza di diffrazione attraverso ad una sottilissima fenditura. Tali esperienze non diedero alcun risultato chiaro. Se ne concluse tuttavia che, semmai, la lunghezza d'onda non poteva essere superiore a pochi Å (10^{-8} cm). Frattanto Barkla con una serie di ricerche che descriveremo tra poco, dimostrava che i raggi X emessi dagli ordinarii tubi sono costituiti da varie specie di raggi, dipendenti

(1) B. Walter e R. Pohl, «Ann. d. Phys.», *25*, 715 (1908).

dall'anticatodo, distinti per il loro coefficiente di assorbimento, di modo che si cominciò a parlare di una « criptocrosi » dei raggi Röntgen. La dimostrazione però della completa analogia tra i raggi X e la luce fu data solo nel 1912 da M. Laue. Già da una cinquantina d'anni aveva credito nella cristallografia l'ipotesi reticolare di Bravais, secondo la quale i cristalli sarebbero stati costituiti da reticolati tridimensionali di particelle, atomi o molecole, tali che le loro mutue distanze, valutate in base al numero di Avogadro, risultavano dell'ordine di grandezza di qualche Å. L'idea di Laue fu la seguente: Se i raggi Röntgen sono effettivamente, come risulterebbe dalle esperienze di Walter e Pohl, delle onde elettromagnetiche di piccolissima lunghezza d'onda, non è certamente possibile costruire artificialmente dei reticoli tanto fini da poterli diffrangere; se però è vera l'ipotesi reticolare di Bravais tali reticoli li troviamo naturalmente nei cristalli. Laue sviluppò allora matematicamente la teoria di questi reticoli a tre dimensioni, in modo non essenzialmente diverso dalla ordinaria teoria dei reticoli di diffrazione e concluse che se una sottile lamina cristallina si faceva attraversare da un sottile pennello di raggi Röntgen, sopra una lastra fotografica posta dalla parte opposta si sarebbe dovuto raccogliere oltre all'impressione dei raggi primarii anche una serie di macchie di diffrazione disposte attorno alla macchia primaria, con una simmetria subordinata a quella della lamina cristallina. L'esperienza, eseguita materialmente nel 1912 da Friedrich e Knipping [(2)] verificò nei più piccoli dettagli le previsioni teoriche di Laue, fornendo insieme la prova sperimentale della natura ondulatoria dei raggi X e della struttura reticolare dei cristalli e tracciando la via ad uno studio accurato degli uni e degli altri. Il dispositivo di Laue per l'osservazione dei nuovi fenomeni di diffrazione non era particolarmente adatto ad uno studio sistematico di questi, avendo il difetto di fornire tutti gli elementi in una sola volta in condizioni da poter solo con gravi difficoltà essere sceverati. W. H. e W. L. Bragg [(3)] perciò un anno dopo, partendo dalla stessa idea di Laue, idearono un dispositivo assai superiore a quello primitivo. Ne esponiamo rapidamente il principio. Consideriamo in un reticolato cristallino un piano reticolare, cioè un piano parallelo ad una faccia possibile del cristallo, e che contenga molti atomi. Siccome ognuno di tali atomi agisce come centro di diffrazione rispetto ad un treno d'onde incidenti, tutto il piano rifletterà una parte di tali onde. Ora il cristallo si può considerare costituito dalla sovrapposizione di tanti piani reticolari spaziati di intervalli uguali, d; così che un treno d'onde incidente verrà riflesso sopra ognuno di essi dando origine a tanti treni d'onde riflessi paralleli che interferiranno tra di loro. I Bragg dimostrarono che questi treni d'onda si trovavano in fase sommandosi solo quando fosse verificata la seguente relazione (relazione di Bragg):

$$n\lambda = 2\,d\cos\theta \tag{1}$$

(2) W. Friedrich, P. Knipping e M. Laue, « Ann. d. Phys. », *41*, 971 (1913).

(3) Per i lavori di W. H. e W. L. Bragg si consulti il loro ormai classico libro *X-Rays and Crystal Structure*. Bell and Sons, London 1918. Un altro importante trattato sopra questi argomenti è: Ledoux-Lebard, *La Physique des Rayons X*. Gauthier Villars. Paris, 1921.

essendo θ l'angolo di incidenza ed n un numero intero detto « ordine della riflessione ». Quando questa relazione non è verificata, i varii treni d'onda riflessi si distruggono per interferenza. Torneremo in seguito con più accuratezza sulla teoria di questa riflessione, ma per ora è sufficiente questo accenno. Supponiamo che un fascio monocromatico di raggi X caschi sopra una lamina cristallina montata su un cerchio graduato in modo da poter far variare l'angolo di incidenza θ. Essendo così fissati λ e d la (1) sarà soddisfatta solo per certi valori di θ, corrispondenti ai diversi valori dell'intero n; si avrà dunque una sensibile riflessione dei raggi incidenti per valori particolari di θ. Se osserviamo i raggi riflessi per mezzo di una camera di ionizzazione come fecero i Bragg, oppure fotograficamente come fecero De Broglie e Moseley, possiamo trovare i valori di θ e quindi, dalla (1) il rapporto $\lambda : d$. Otteniamo così dei veri spettri dei raggi X di varii ordini corrispondentemente ai varii valori che può prendere n. Si capisce che se la radiazione incidente non è monocromatica, si ottengono varie righe corrispondentemente ad ogni radiazione monocromatica in essa contenuta. Possiamo in questo modo ottenere degli spettri di una radiazione Röntgen del tutto analoghi a quelli della spettroscopia ordinaria. Torneremo del resto in seguito con più particolari sopra i dispositivi per ottenerli.

La (1) fornendoci così il rapporto $\lambda : d$ si presta a due modi di sfruttamento:

A) Tener costante d e far variare λ; studiare cioè le riflessioni di varie specie di raggi sopra una stessa faccia cristallina, in modo da poterle confrontare tra di loro.

B) Si tiene costante λ facendo variare d; ossia si studiano le riflessioni di una stessa radiazione monocromatica sopra le varie faccie di uno stesso cristallo, o sopra faccie di cristalli diversi. Questi studii furono con particolare successo seguiti dai Bragg. È chiaro come la conoscenza dei d, fornendoci elementi intimamente legati alla struttura del cristallo possa, almeno in casi semplici, condurre alla completa determinazione del reticolato cristallino, ciò che per parecchi cristalli fu effettivamente fatto per opera dei Bragg e di altri. Riservando al seguito i particolari di queste ricerche, ci limitiamo ad osservare che, una volta conosciuto il reticolato cristallino, la conoscenza del numero di Avogrado ci permette di avere d, e quindi anche λ in misura assoluta. Naturalmente in queste misure assolute si ha l'incertezza del numero di Avogadro, che non c'è invece in quelle relative. Vogliamo ora esaminare i risultati del metodo *A*). Si trovò intanto subito che lo spettro della radiazione Röntgen è costituito da un fondo continuo intenso, sul quale emergono delle righe molto nettamente definite, le cui lunghezze d'onda dipendono solo dal materiale di cui è fatto l'anticatodo e non dalla forma del tubo o dal modo di eccitazione.

Inoltre questi spettri Röntgen, o di alta frequenza costituiscono una proprietà atomica, nel senso che ogni elemento dà lo stesso spettro Röntgen, sia che si trovi nell'anticatodo allo stato elementare, o in qualunque combinazione chimica o che sia comunque soggetto alle più energiche azioni fisiche o chimiche.

Tale proprietà, praticamente assai importante, perché permette l'analisi spettrografica di elementi liquidi o gassosi, ha una vasta portata teorica in quanto dimostra che l'origine dei raggi X è nelle parti profonde dell'atomo che sono poco o nulla influenzate dalle circostanze esterne. E ciò costituisce una profonda differenza tra gli spettri d'alta frequenza e gli spettri ordinarii, i quali hanno origine nelle regioni corticali dell'atomo, e spiega come i primi abbiano una struttura assai più semplice e decifrabile che non i secondi.

Una notevole differenza di origine esisterebbe tra i raggi costituenti il fondo continuo e quelli costituenti le radiazioni caratteristiche, in quanto che questi ultimi si interpretano come dovuti alle perturbazioni causate nell'atomo dall'urto di un elettrone catodico, mentre i primi sarebbero invece onde di accelerazione, in questo caso negativa, emananti dall'elettrone catodico quando questo viene fermato dall'urto contro l'anticatodo. Cominciamo con l'esame delle radiazioni caratteristiche: Lo spettro Röntgen di un qualsiasi elemento è costituito da varie linee, raggruppabili in serie assai distanziate che, in ordine di lunghezza d'onda crescente o, come si dice, di durezza decrescente, si chiamano, con una denominazione di Barkla, che le aveva scoperte, nel suo studio sopra l'assorbimento, assai prima delle esperienze di Laue, serie K, L, M, ... La durezza di una stessa serie, per esempio della serie K cresce da elemento a elemento col numero atomico, così che per gli elementi leggeri è osservabile soltanto la serie K, siccome L ed M sono troppo molli per poter essere studiate, mentre per gli elementi pesanti la K diventa troppo dura e restano osservabili soltanto L ed M. Ogni serie è costituita da alcune righe di cui qualcuna doppia; le elencheremo parlando della loro teoria. Una relazione importantissima passa tra la lunghezza di onda delle righe Röntgen e il numero atomico dell'elemento emittente. Consideriamo per esempio la K_α che è la più intensa riga della serie K. Moseley [(4)], nel 1913 misurò le frequenze ν della K_α di 10 elementi presi in fila nel sistema periodico dal Ca al Cu e trovò verificata la seguente relazione:

$$\nu = \frac{3}{4} R (Z - 1)^2 \tag{2}$$

essendo R il numero di Rydberg. Tale relazione ci fornisce un criterio assai più sicuro che non il peso atomico, per determinare il numero atomico di un elemento, e basandosi su di essa si evitano le inversioni che si presentano nel sistema periodico tra Ni e Co, A e K etc.

La relazione di Moseley non è più rigorosamente soddisfatta per grossi numeri atomici. In genere anche per ogni altra riga si hanno relazioni approssimativamente lineari tra $\sqrt{\nu}$ e Z. Elementi isotopi, corrispondendo allo stesso numero atomico hanno anche le stesse righe. Parleremo in particolare di tali proprietà esponendo la teoria degli spettri, al quale proposito mostreremo anche la spiegazione delle divergenze tra le formule analoghe a quella di Moseley e l'esperienza.

Il fondo continuo, detto anche per la sua origine « Bremsstrahlung », è costituito da una radiazione di intensità piuttosto debole e di frequenza variabile con continuità da zero fino ad un certo massimo ν_m che si calcola

(4) H. G. MOSELEY, « Phil. Mag. », *26*, 1024 (1913).

con la formula di Planck–Einstein:

$$h\nu_m = eV$$

essendo V il potenziale di eccitazione. La formula è verificata molto bene, tanto che fornisce un metodo abbastanza preciso per la determinazione di h. In genere con un potenziale di eccitazione V non si possono produrre radiazioni di frequenza superiore a ν_m né nella Bremsstrahlung né nello spettro di righe; anzi per eccitare una determinata riga non basta il potenziale $h\nu/e$, perché facendo crescere il potenziale tutte le righe di una stessa serie compaiono in una volta sola quando il potenziale ha raggiunto il valore capace di eccitare il limite di questa serie, che ha frequenza maggiore di tutte le righe della serie. A questo proposito notiamo che mentre la serie K ha un solo limite, la serie L ne ha invece due, e le sue righe compaiono in due gruppi via via che questi due limiti vengono raggiunti.

§ 3. – L'ASSORBIMENTO. — I fenomeni or ora descritti trovano il loro riscontro nei fenomeni dell'assorbimento dei raggi X per parte dei materiali da essi traversati. Lo studio di questi fu in gran parte fatto da Barkla assai prima della scoperta di Laue. Cominciamo con l'osservare che l'assorbimento dei raggi X avviene per due ragioni essenzialmente diverse, in quanto che un corpo irradiato da raggi Röntgen emette a sua volta dei raggi Röntgen a spese dell'energia del fascio irradiante; ma questi raggi sono costituiti da una parte di lunghezza d'onda eguale a quella della radiazione primaria (raggi diffusi) e da una parte di lunghezza d'onda diversa (raggi di fluorescenza); studieremo perciò separatamente i due fenomeni.

Diffusione. – Si chiama coefficiente di diffusione (σ) il rapporto tra l'energia dei raggi diffusi dall'unità di volume e l'intensità della radiazione primaria. Se ρ è la densità del materiale σ/ρ sarà il coefficiente di diffusione relativo all'unità di massa, o coefficiente massico di diffusione. Barkla e Crowther (5) osservarono che, per una stessa radiazione primaria i coefficienti massici di diffusione sono per molti elementi leggeri press'a poco eguali tra di loro, salvo per l'idrogeno per il quale si trova un valore quasi doppio che per gli altri. Tutto ciò si accorda bene con la teoria della diffusione di Thomson (6). Secondo Thomson il meccanismo della diffusione è il seguente: sotto l'azione del campo elettrico oscillante prodotto dalla radiazione primaria, un elettrone del corpo diffondente si mette a vibrare e conseguentemente irradia energia della stessa frequenza. Le note formule di elettrodinamica dànno, con un calcolo facilissimo, che l'energia diffusa da un solo elettrone sotto l'azione dell'intensità primaria 1 è: $\frac{8\pi}{3}\frac{e^4}{m^2c^4}$. D'altra parte in un grammo di sostanza si trovano NZ/A elettroni; il coefficiente di diffusione massico è dunque: $\frac{8\pi N}{3}\frac{e^4}{m^2c^4}\frac{Z}{A}$. Esso risulta dunque indipendente dalla lunghezza d'onda e variando l'elemento disperdente varia come Z/A; ora è noto che Z/A è press'a poco = 1/2 per tutti gli elementi, salvo per

(5) C. G. BARKLA, « Phil. Mag. », 7, 543 (1904); J. A. CROWTHER, ibid., *14*, 653 (1907).
(6) J. J. THOMSON, *Conduction of Electricity through Gases*, Cambridge, 1903.

l'idrogeno, per il quale è = 1. Con ciò si spiegano dunque completamente i risultati di Barkla e Crowther; anche il valore teorico del coefficiente coincide abbastanza bene con quello sperimentale. Vediamo ora invece i punti nei quali si ha divergenza tra questa teoria e l'esperienza. In primo luogo la teoria di Thomson oltre a dar il modo di calcolare l'energia totale diffusa, permette anche di calcolarne la distribuzione e prevede precisamente che l'energia sia diffusa simmetricamente rispetto a un piano perpendicolare alla radiazione primaria e passante per il corpo diffondente. Sperimentalmente invece Barkla (7), Owen (8) ed altri trovarono risultati, invero discordi tra loro, che però concordavano nel trovar maggiore l'energia diffusa in senso concorde ai raggi primarii, che non quella in senso opposto. Wilson e Debye (9) cercarono di interpretar tale fatto con anomalie prodotte da fenomeni di diffrazione che avessero luogo o nei reticolati di microcristalli contenuti nel corpo disperdente, o addirittura nei reticolati atomici di cui son costituite le molecole.

Un'altra divergenza ben più grave tra la teoria di Thomson e l'esperienza consiste in ciò, che mentre la teoria prevede il coefficiente indipendente dall'elemento disperdente e dalla lunghezza d'onda, l'esperienza verifica questa proprietà solo per elementi leggeri e per lunghezze d'onda intermedie. In genere per lunghezze d'onda estremamente piccole il coefficente ha valore minore di quello che avrebbe secondo la teoria, e tende a zero insieme a λ. Mentre per λ grande σ/ρ cresce assai rapidamente. Teoricamente Debye (10) spiegò i grandi valori di σ/ρ per elementi pesanti e grandi lunghezze d'onda con l'osservazione che in tale caso molti elettroni vengono in certo modo a vibrare in blocco esercitando perciò un'azione proporzionale al quadrato del loro numero invece che al loro numero.

Più misteriosa si presenta invece la decrescita di σ/ρ per piccole lunghezze d'onda. Compton cercò di spiegarla ammettendo un elettrone anulare, invece che sferico, di raggio 185×10^{-10} cm. La stessa stranezza di tale ipotesi mostra quanto lontani siamo da una spiegazione. Probabilmente ci troviamo di fronte ad uno di quei casi in cui finisce la validità dell'elettrodinamica classica per cedere il posto alla teoria dei quanti.

Fluorescenza. – Quando un elemento è irradiato da una radiazione primaria Röntgen emette, oltre ad una radiazione diffusa, di frequenza eguale a quella primaria, anche delle radiazioni con le proprie frequenze caratteristiche per fluorescenza, quando almeno la radiazione primaria abbia durezza sufficiente. Si chiama coefficiente di fluorescenza (τ) il rapporto della energia emessa per fluorescenza dall'unità di volume di un elemento all'intensità della radiazione primaria. Come nel caso della diffusione τ/ρ si chiama coefficiente massico di fluorescenza. Il fenomeno della fluorescenza è in molte particolarità analogo alla emissione di raggi Röntgen sotto l'azione di un bombardamento catodico. Teniamo presente che per mezzo della costante h

(7) C. G. BARKLA, « Phil. Mag. », *15*, 228 (1927).
(8) E. A. OWEN, « Proc. Camb. Phil. Soc. », *16*, 161 (1911).
(9) P. DEBYE, « Ann. d. Phys. », *46*, 809 (1915).
(10) P. DEBYE, op. cit.

di Planck si ha tra quanto di energia ε, e frequenza ν, la relazione $\varepsilon = h\nu$. Ora abbiamo visto a suo luogo che per eccitare una riga Röntgen. per esempio la K_α non bastava che la particella catodica avesse energia $\varepsilon = eV = h\nu_{K_\alpha}$, ma occorreva un'energia maggiore, equivalente almeno alla frequenza del limite della serie K. In modo in tutto analogo per eccitare per fluorescenza l'emissione della K_α, non basta che la radiazione primaria abbia frequenza superiore alla riga da eccitare, ma deve aver frequenza almeno eguale al limite della serie K.

Lo stesso accade naturalmente anche per le altre serie, soltanto che per esse i limiti possono essere più d'uno. Vedremo a suo tempo come questi fatti si interpretino teoricamente. Esaminiamo intanto la variazione del coefficiente di fluorescenza in funzione della lunghezza d'onda. C'è intanto da aspettarsi che esso presenti delle discontinuità per le lunghezze d'onda corrispondenti ai limiti delle serie del corpo fluorescente. Nel tratto tra due limiti di assorbimento esso varia, secondo le misure di Glocker (11), come $\lambda^{2,6}$; cresce dunque rapidamente con λ. Esso cresce poi rapidissimamente col numero atomico dell'elemento fluorescente, essendo press'a poco proporzionale a Z^4. Se λ_A è la lunghezza d'onda del limite della serie K si ha, secondo Glocker:

$$(4) \qquad \begin{cases} \dfrac{\tau}{\rho} = 22{,}8 \cdot 10^{-6} Z^{4,28} \lambda^{2,8} & \text{per } \lambda > \lambda_A \\[2ex] \dfrac{\tau}{\rho} = 1120 \cdot 10^{-6} Z^{3,72} \lambda^{2,8} & \text{per } \lambda > \lambda_A . \end{cases}$$

I fenomeni ora descritti spiegano bene il seguente fatto osservato da Barkla: se si misurano i coefficienti di assorbimento della radiazione emessa da un anticatodo di numero atomico Z per parte di elementi di numero atomico via via crescente, si trova che tale coefficiente presenta una brusca diminuzione tra i numeri atomici dell'elemento assorbente $Z - 2$ e $Z - 1$. Ciò dipende dal fatto che la K_α che è la più intensa radiazione dell'elemento ha frequenza compresa tra i limiti della serie K degli elementi di numero atomico $Z - 2$ e $Z - 1$, ed è quindi capace di eccitare per fluorescenza soltanto la serie K di elementi di numero atomico $\leqq Z - 2$.

Di questi fenomeni ci si vale anche nei misuratori di durezza usati nella pratica medica. Quello di Wehnelt per esempio è costituito da un cuneo di alluminio posto accanto ad una lamina uniforme di argento. Si confrontano su uno schermo di platinocianuro di bario le intensità della radiazione da esaminare dopo che ha traversato l'apparecchio, e si cerca quale spessore di alluminio deve essere traversato per assorbire quanto la lamina d'argento. La durezza è maggiore quanto più grande è questo spessore.

Sulle leggi dell'assorbimento sono pure fondate le celebri esperienze di Barkla che lo condussero alla scoperta delle serie K ed L assai prima dei lavori di Laue. Consideriamo un fascio monocromatico di raggi X. Sia $\mu = \sigma + \tau$ il suo coefficiente complessivo di assorbimento in un certo mate-

(11) R. Glocker, «Phys. Zeits.», *19*, 66 (1918).

riale. Allora se W è la sua energia iniziale, dopo aver traversato lo spessore l del materiale assorbente, essa si ridurrà a:

$$W = W_o e^{-\mu l}. \tag{5}$$

Se in ascisse portiamo l e in ordinate W troviamo dunque una curva esponenziale. Tale conclusione cessa di essere valida se il fascio in esame non è monocromatico. Supponiamolo per esempio costituito da due radiazioni monocromatiche di energie iniziali $W_o^{(1)}$ e $W_o^{(2)}$ e di coefficienti di assorbimento rispettivamente μ_1 e μ_2. L'intensità dopo attraversato lo spessore l sarà:

$$W = W_o^{(1)} e^{-\mu_1 l} + W_o^{(2)} e^{-\mu_2 l}. \tag{6}$$

La nostra grafica (l, W) non sarà dunque più un'esponenziale e da essa potranno facilmente rilevarsi μ_1 e μ_2. Infatti da 4 suoi punti sostituendo in (6) le loro coordinate ricaviamo 4 equazioni dalle quali otteniamo le 4 incognite $W_o^{(1)}$, $W_o^{(2)}$, μ_1, μ_2. Si capisce che in pratica questa risoluzione può spesso farsi in modo assai più semplice.

Se per esempio μ_1, è abbastanza più grande di μ_2 avverrà che per l abbastanza grande il primo termine del secondo membro di (6) diventerà trascurabile di fronte al secondo e la grafica diventerà un'esponenziale dalla quale si ricaveranno subito $W_o^{(2)}$ e μ_2. Fisicamente l'essersi annullato il primo termine significa che sono stati completamente assorbiti i raggi di coefficinte μ_1, che cioè sono stati «filtrati» i raggi più penetranti. Tale metodo di analisi per filtrazione, che può evidentemente applicarsi anche a fasci comunque complessi, permise a Barkla di separare la serie K dalla L che è meno penetrante, ed è ancora da raccomandarsi per ottenere dei fasci approssimativamente omogenei.

Nelle ricerche sperimentali sopra l'assorbimento si può procedere in due modi. Quando non interessi distinguere tra coefficiente di diffusione e di fluorescenza si possono misurare, a mezzo di una camera di ionizzazione le intensità di un fascio di raggi prima e dopo che questo ha traversato uno spessore l della sostanza da esaminare. Se I_o ed I sono queste intensità si avrà $I/I_o = e^{-(\sigma+\tau) l}$, da cui si ricava $\sigma + \tau$. Se invece si desideri distinguere la diffusione dalla fluorescenza converrà esaminare in una qualunque direzione la radiazione uscente dal corpo colpito dai raggi primarii, mediante una camera di ionizzazione. Questa radiazione è costituita da una parte dovuta a fluorescenza, e da un'altra naturalmente più dura, dovuta alla diffusione. Queste due parti posson quindi separarsi per filtrazione attraverso opportune sostanze assorbenti; e così si ottengono separatamente σ e τ.

Prima di abbandonar l'argomento dell'assorbimento citiamo ancora un altro fenomeno che si presenta sempre ad accompagnare l'emissione per fluorescenza, ed è del tutto analogo all'effetto fotoelettrico.

Esso consiste nella emissione di elettroni per parte di un corpo colpito da raggi X. Anche questo fenomeno è regolato dalla relazione quantistica $\varepsilon = h\nu$, ε è l'energia cinetica dell'ettrone emesso, aumentata dell'energia necessaria a liberarlo dall'atomo in cui è contenuto. In questo fenomeno è da ricercarsi la causa della ionizzazione dei gas per parte dei raggi X che li attraversano.

§ 4. – La polarizzazione dei raggi Röntgen. — Come la luce ordinaria anche i raggi X possono esser polarizzati; anzi i raggi emessi dagli ordinari tubi lo sono per lo più parzialmente. Barkla (12) lo ha potuto dimostrare al modo seguente: un fascio di raggi X emesso da un tubo ordinario in direzione perpendicolare a quella dei raggi catodici, colpiva un corpo diffondente e si esaminava l'intensità della radiazione diffusa nelle varie direzioni perpendicolari ai raggi primarii; Barkla osservò che era maggiore l'intensità diffusa in direzione perpendicolare al piano del fascio primario e dei raggi catodici che non quella in direzione parallela al fascio catodico. Ciò, interpretato secondo la teoria di Thomson, significa che gli elettroni del corpo diffondente vibrano prevalentemente in senso parallelo ai raggi catodici, e quindi che il piano di polarizzazione dei raggi primarii è perpendicolare al fascio catodico. Questo si accorda bene con la teoria di Stokes e Wiechert secondo la quale la Bremsstrahlung è dovuta alle accelerazioni degli elettroni catodici al momento dell'urto; perché in questo caso appunto il campo magnetico dell'onda elettromagnetica emessa deve essere perpendicolare alle direzioni dei raggi catodici.

§ 5. – La teoria delle radiazioni caratteristiche (13). — La teoria delle righe d'alta frequenza, che ha permesso di rendersi conto con una precisione che difficilmente si sarebbe sperata, della struttura degli spettri Röntgen è basata sul modello atomico di Rutherford–Bohr che brevemente richiamiamo. Sono note le celebri esperienze di Rutherford sopra la diffusione delle particelle α che lo indussero ad ammettere contro il vecchio concetto di Thomson, che la parte elettricamente positiva dell'atomo occupasse un nucleo di dimensioni estremamente piccole rispetto a quelle di tutto l'atomo, e che avesse carica complessiva eguale al numero atomico dell'elemento, moltiplicato per la carica elettronica, e racchiudesse in sé praticamente tutta la massa dell'atomo. Secondo questa rappresentazione l'atomo sarebbe come una specie di sistema planetario in cui il Sole è rappresentato dal nucleo positivo e i pianeti dagli elettroni, che sono in numero eguale al numero atomico, o più, o meno secondo che l'atomo è neutro o ionizzato negativamente, o positivamente. Se a questo sistema planetario si applicano le leggi della elettrodinamica classica, si trovano dei risultati che non hanno nulla a che vedere coi risultati sperimentali. Fu nel 1913 che N. Bohr diede le leggi che regolano il moto dei sistemi atomici.

L'idea fondamentale di tali leggi è la seguente: calcoliamo secondo le solite leggi meccaniche ed elettrodinamiche le orbite meccanicamente possibili del nostro sistema, trascurando l'irradiazione. Secondo Bohr non tutte queste orbite possono effettivamente esser descritte dagli elettroni, ma solo alcune,

(12) C. G. Barkla, « Phil. Mag. », dal 1906 in poi.

(13) L'esposizione di queste teorie si può trovare, assai per disteso nel libro di Sommerfeld, *Atombau und Spektrallinien*, Vieweg, Braunschweig, 1921.

orbite statiche, formanti una successione discreta da scegliersi con una legge che diremo tra poco. Finché il sistema si trova sopra un'orbita statica, non irradia energia. Ogni tanto accade però che esso salta da un'orbita statica a un'altra, e allora si ha emissione di tanta energia quanta è la differenza tra le energie W_1 e W_2 delle due orbite. Secondo Bohr poi questa energia viene emessa in un solo quanto. Se quindi ν è la frequenza di emissione si ha:

$$h\nu = W_1 - W_2 . \tag{7}$$

Questa formula, una volta calcolate tutte le orbite statiche, e quindi le loro energie, permette di calcolare le frequenze di tutte le righe del nostro atomo.

Le leggi per la scelta delle orbite statiche da tutte quelle meccanicamente possibili sono alquanto complicate e furono formulate solo nel 1916 da Sommerfeld.

Ma per noi non occorre tale formulazione completa che serve solo per la considerazione delle orbite ellittiche. Limitandosi alle sole orbite circolari esse si riducono a richiedere che la quantità di moto areale di ogni elettrone sia un multiplo intero di $h/2\pi$. Descriviamo quali siano le orbite statiche nel caso dell'idrogeno perché, benché tale elemento non dia propriamente uno spettro Röntgen, esso costituisce tuttavia un modello analogamente al quale sono costruiti gli atomi più complessi. L'atomo di idrogeno è costituito dal nucleo e da un solo elettrone che meccanicamente potrà descrivere le ellissi kepleriane; limitiamoci dunque dapprima alle sole orbite circolari.

Un facile calcolo, in base alle leggi accennate, mostra che vi sono infinite orbite statiche circolari di raggi crescenti come i quadrati dei numeri interi, e di energie:

$$W_n = -\frac{2\pi^2 me^4}{h^2}\,\frac{1}{n^2},$$

dove n è il numero d'ordine dell'orbita. L'applicazione della formula (7) ci dà ora:

$$\nu = \frac{2\pi^2 me^4}{h^3}\left(\frac{1}{n^2} - \frac{1}{m^2}\right) \tag{8}$$

con n ed m numeri interi qualunque, come frequenze di tutte le righe dell'idrogeno. Facendo $n = 2 : m = 3, 4, 5 \cdots$ si ha la serie di Balmer; facendo $n = 1, m = 2, 3 \cdots$ si ha la serie ultravioletta di Lyman; facendo $n = 3$, $m = 4, 5 \cdots$ si ha la serie infrarossa di Paschen. Il coefficente $2\pi^2 me^4/h^3$ coincide, entro i limiti dell'errore sperimentale col numero R di Rydberg.

Chiamiamo i successivi anelli K, L, M ··· Se accanto alle orbite circolari si considerano anche quelle ellittiche, si trova che ad ogni orbita statica circolare ne corrispondono alcune ellittiche, di varie eccentricità, le quali hanno però eguale energia che le corrispondenti circolari, per modo che non danno luogo a nuove righe oltre quelle comprese nella formula (8). In generale l'energia W_n si ha nell'n^{esimo} cerchio e in $n - 1$ orbite ellittiche. Così che l'energia W_2 si ha nel solo cerchio K; l'energia W_2 nel cerchio L ed in una orbita ellittica e così via.

Le cose non vanno così semplicemente se nel calcolo delle orbite si tiene conto della variabilità della massa, prevista dalla teoria della relatività. Tenendo conto di questo si trova infatti che le energie di un'orbita circolare e delle corrispondenti ellittiche non sono rigorosamente eguali. Da ciò deriva che al posto di una semplice riga della formula (8) compaiono dei gruppi di righe vicinissime (Feinstruktur). La verifica quantitativa di queste previsioni teoriche costituisce insieme una brillante conferma della teoria di Bohr e di quella della relatività.

Sulla linea tracciata dal modello dell'atomo di idrogeno vediamo ora come si concepiscano i modelli almeno delle parti più interne degli atomi più complessi. Si ha intanto un nucleo positivo di carica Ze, attorno al quale girano Z elettroni disposti in anelli concentrici, denominati ordinatamente K, L, M.... Cominciamo a considerare gli elettroni dell'anello K. Essi risentono naturalmente poco l'azione degli elettroni degli anelli esterni e son soggetti principalmente all attrazione del nucleo e alla repulsione tra di loro. Il risultato di queste due azioni è che si può considerare soltanto l'azione del nucleo, purché invece che la carica eZ gli si attribuisca la carica $e\,(Z - z)$, dove z dipende dal numero di elettroni di cui si pensa costituito l'anello. L'energia dell'n^{esimo} anello che per l'idrogeno era $\frac{hR}{n^2}$ risulta ora $\frac{hR}{n^2}(Z - z)^2$. Già da questa rappresentazione piuttosto semplicistica possiamo ricavare dei risultati interessanti.

Invero la formula analoga alla (8) si scrive ora:

$$\nu = R\,(Z - z)^2\left(\frac{1}{n^2} - \frac{1}{m^2}\right). \tag{9}$$

Secondo che in questa facciamo $n = 1, 2, 3\cdots$ troviamo righe delle serie K, L, M···: la prima riga della serie K, la K_α si ha facendo $n = 1$, $m = 2$. Si ha dunque:

$$\nu_{K_\alpha} = (Z - z)^2\, R\, \frac{3}{4}$$

se l'anello K si pensa occupato da tre elettroni si ha press'a poco $z = 1$, e quindi:

$$\nu_{K_\alpha} = \frac{3}{4}\, R\,(Z - 1)^2$$

che è la relazione di Moseley; ed analogamente si ha la spiegazione delle relazioni approssimativamente lineari, che passano tra la radice quadrata della frequenza di una riga qualunque, ed il numero atomico. Attenendosi al modello ora esposto, si troverebbero però meno righe di quelle che effettivamente si osservano e si avrebbe una approssimazione assai grossolana della loro frequenza. Le differenze spariscono in buona parte se si tien conto della variabilità nella massa prevista dalla teoria della relatività.

Tale correzione è per gli atomi pesanti assai più importante che per l'idrogeno, perché le orbite degli atomi pesanti sono descritte con velocità assai maggiore. Così che mentre le « Feinstrukturen » delle righe dell'idrogeno sono soltanto osservabili con mezzi di alta risoluzione, quelle degli spettri di alta frequenza sono per lo più bene osservabili in condizioni ordinarie.

Per effetto relativistico l'anello K resta semplice, quello L si sdoppia in due L, L' ecc.... Possiamo ora dare un elenco delle righe dello spettro d'alta frequenza di qualunque elemento.

Serie K. – Costituita dalle righe che vengono emesse nel salto di un elettrone da un anello qualunque, sopra l'anello K. Ne diamo i nomi, secondo Sommerfeld, e le origini in ordine di durezza crescente: α' (L' — K), α (L — K), α_3 (? — K), α_4 (? — K), β (M — K), γ (N — K). Le righe più intense sono la doppia $\alpha\alpha'$; α_3 ed α_4 si osservano solo negli elementi più leggeri. La serie K si è potuta osservare a misurare solo per gli elementi tra il Sodio (Z = 11) e il Neodimio (Z = 60). Prima del Na è troppo poco penetrante; dopo il Neodimio è troppo dura per poter essere eccitata.

Serie L. – Essa è conosciuta per gli elementi dallo Z*n* (Z = 30) fino all'Uranio. I nomi e le origini delle sue righe più importanti sono: α' (M' — L), α (M — L), α_3 (?), β (M' — L'), γ (N — L), δ (N — L') ε (? — L), η (? — L). A proposito della serie L notiamo che, oltre ai due anelli L, L', si ha anche traccia di un terzo anello L, forse dovuto a perturbazioni tra gli anelli L e K.

Serie M. – È conosciuta per gli elementi dal Disprosio (Z = 66) all'Uranio. Le sue righe principali sono, secondo Stenström: α' (N' — M), α (N — M), β (N' — M'). In generale possiamo osservare che le serie L ed M sono più complicate e indecifrabili della K, fatto di cui ci si rende facilmente conto tenendo presente che si vengono via via considerando sistemi con maggior numero di elettroni ai quali le semplificazioni introdotte sono sempre meno strettamente applicabili.

Notiamo ancora che in qualche caso è stata osservata [14] una serie J, più dura della K, osservabile solo per elementi molto leggeri.

Il meccanismo dell'eccitazione dei raggi X per parte di un bombardamento catodico, o per fluorescenza, si pensa che sia il seguente. Un elettrone da un anello interno viene proiettato fuori dell'atomo a spese dell'energia della particella o dei raggi primarii; mentre è riassorbito, salta da un anello all'altro emettendo le righe. Con questa rappresentazione ci si spiega il comparire simultaneo di tutte le righe di una serie quando viene raggiunto il suo limite di eccitazione. Le frequenze di tali limiti infatti a meno del fattore — *h*, coincidono con le energie dei varii anelli, così per esempio solo quando i raggi catodici hanno energia sufficiente a liberare un elettrone dell'anello K posson cominciare a comparire le righe della serie K.

§ 6. – SPETTRI DEI RAGGI γ. — Questi raggi sono, come è noto, di natura analoga ai raggi X, dai quali si distinguono solo per la molto maggiore durezza. Ciò rende assai difficile la loro analisi spettrale che fu eseguita da Rutherford e Andrade [15] con un metodo particolare che non descriviamo. Il risultato fu che essi sono costituiti da una radiazione molto penetrante, che per estrapolazione si ritiene esser la serie K dell'elemento radio–attivo, da radiazioni più penetranti non analizzabili, e forse da un fondo continuo.

(14) C. G. BARKLA, *Bakerian Lecture*, 1914; R. LEDOUX-LEBARD, «C. R.», *163*, 608 (1919).

(15) E. RUTHERFORD e N. da C. ANDRADE, «Phil. Mag.», *27*, 855 (1919).

§ 7. – Spettrografi a raggi Röntgen. — Nello spettrografo più comune, a cristallo girante, il cristallo è montato sopra un asse, mosso da un movimento a orologeria e la registrazione è fatta sopra una pellicola fotografica avvolta a cilindro, con per asse l'asse di rotazione.

Da fenditura si fa spesso servire lo stesso anticatodo, che in tale caso, invece che a becco di flauto, come nei tubi ordinarii è tagliato quasi perpendicolarmente ai raggi catodici e viene osservato in direzione normale a questi, di modo che si presenta come una strettissima ellisse. È importante che l'anticatodo, oppure, se v'è, la fenditura distino dal cristallo quanto la pellicola fotografica, perché tale condizione, come osservano i Bragg, è necessaria per ottenere delle righe sottili anche usando un fascio aperto, il che è opportuno per utilizzare tutta la faccia cristallina e diminuire quindi il tempo di posa. Se invece che del metodo fotografico ci si vuol servire di quello della camera di ionizzazione, che è più raccomandabile specialmente nelle ricerche di analisi cristallina, per le quali occorre un confronto quantitativo delle intensità delle riflessioni dei varii ordini, sia il cristallo che la camera di ionizzazione si montano sopra un cerchio graduato e si fa una serie di letture girando sempre il cristallo di un certo angolo e la camera dell'angolo doppio. Questo è il metodo usato dai Bragg nelle loro classiche ricerche sulla struttura dei cristalli.

Quando si voglia ottenere rapidamente tutto lo spettro di un certo anticatodo senza bisogno di una particolare esattezza si può usare come cristallo riflettente una lamina di mica curvata a cilindro che viene tutta investita dal fascio da analizzare, per modo che l'angolo di incidenza varia da punto a punto. Si ha naturalmente la riflessione solo nei punti per i quali è verificata la relazione (1) di Bragg, ed i raggi riflessi vengono raccolti sopra una lastra fotografica. Questo metodo è dovuto a De Broglie [16]. La sig.na Brunetti ottenne con esso buoni risultati sostituendo alla mica una lamina di salgemma curvata a caldo.

I metodi fin qui descritti non si prestano spesso all'analisi cristallina, essendo per essi necessario che i cristalli abbiano facce un po' sviluppate. Debye e Scherrer [17] hanno perciò escogitato un importante metodo che permette l'analisi anche di polveri cristalline. Un sottile fascio di raggi X vien fatto cadere sopra un tubetto contenente la polvere cristallina. In essa si trovano naturalmente tutte le facce cristalline in tutte le possibili orientazioni, ve ne saranno perciò sempre di quelle che si troveranno sotto un angolo adatto a riflettere in raggi incidenti; ed i raggi riflessi segneranno, sopra una lastra posta perpendicolarmente ai raggi primarii dei cerchi concentrici. Questo metodo certamente prezioso per ricerche mineralogiche, non si presterebbe naturalmente a ricerche spettrometriche per le quali è necessario ottenere righe fini. Una comoda variante di questo metodo è descritta da Bohlin [18]. Per misure spettrometriche di grandi lunghezze d'onda, che sono assorbite da pochi mm. di aria è necessario operare nel vuoto interpo-

(16) M. De Broglie, « Journal de Phys. », *4*, 265 (1914).
(17) P. Debye e P. Scherrer, « Phys. Zeits. », *17*, 277 (1916).
(18) H. Bohlin, « Ann. d. Phys. », *61*, 421 (1920).

nendo sul passaggio dei raggi solo qualche sottile diaframma di alluminio. È pure importante osservare che dalla formula di Bragg: $n\lambda = 2d \sin\theta$ si ricava $\lambda < 2d$.

Per studiare le grandi lunghezze d'onda conviene dunque scegliere dei cristalli con d molto grande, quali il gesso ($2d = 15{,}2$ Å), la mica ($2d = 20{,}2$ Å). Il cristallo che presenta la magggior costante conosciuta è secondo Stenström il sale sodico del dietilglicolato di argento ($2d = 38$ Å). Misure spettrometriche sulle grandi lunghezze d'onda furono fatte per la prima volta da Moseley, e poi sistematicamente da Siegbahn [19] e dai suoi scolari. A questo proposito accenniamo anche, ad una importante serie di ricerche, in buona parte tutt'ora in corso, destinate ad esplorare la regione dello spettro tra i raggi Röntgen più molli e l'estremo ultravioletto.

La produzione di questi raggi si può fare con delle specie di tubi Röntgen con catodo rovente, che possono così lavorare sotto voltaggi bassissimi, anche poche decine di volt.

L'analisi si fa dall'estremo ultravioletto dei raggi così detti di Schumann ($\lambda = 1000$ Å) fino a 300 o 400 Å con dei reticoli artificiali; in tutto l'intervallo poi da 300 a 15 Å non si può disgraziatamente ricorrere che a metodi di filtrazione analoghi a quelli descritti usati da Barkla, che non permettono naturalmente la misura delle lunghezze d'onda.

Nelle misure su piccole lunghezze d'onda invece, anche usando cristalli di costante molto piccola quali il salgemma ($d = 2{,}814$) o la calcite ci si trova sempre di fronte alla difficoltà prodotta dalla scarsa deviazione dei raggi riflessi. È interessante a questo proposito il metodo escogitato da Rutherford e Andrade [20] nella loro analisi dei raggi γ.

Generalmente in uno spettro si nota una rapida diminuzione di intensità col crescere dell'ordine. Tale diminuzione è dovuta in parte all'imperfezione del cristallo e in parte all'irregolarità prodotta nel reticolato cristallino dall'agitazione termica. L'effetto di quest'ultima fu studiato per la prima volta da Debye [21], che dimostrò che col crescere della temperatura dovevano decrescere le intensità delle righe, specialmente per quelle di ordine elevato, senza che si avesse diminuzione del potere risolutivo.

Queste previsioni teoriche furono confermate dall'esperienza; per il salgemma per esempio già a 300° sparisce ogni fenomeno di diffrazione.

Facciamo un'ultima osservazione riguardo al potere risolutivo. A prima vista potrebbe sembrare che questo fosse enorme; è enorme infatti il numero degli elementi diffrangenti (qualche milione per mm. di spessore). Contro questo però vi sono molte ragioni che producono un allargamento delle righe.

In primo luogo l'impossibilità di usar fenditure molto strette; secondo il fatto di operare con onde sferiche invece che piane, che impedisce che

(19) Per queste e per le altre misure spettrometriche sistematiche dell'istituto fisico di Lund vedi i lavori di SIEGBAHN, FRIMAN e STENSTRÖM nel « Phil. Mag. » dal 1916 in poi.

(20) E. RUTHERFORD e N. da C. ANDRADE, « Phil. Mag. », *27*, 854 (1914).

(21) P. DEBYE, « Ann. d. Phys. », *43*, 49 (1914).

possano interferire tra loro atomi un po' lontani; terzo l'irregolarità naturale del cristallo e quarto infine lo spessore del cristallo. Il complesso di queste cause impedisce di raggiungere praticamente un potere risolutivo molto superiore all'1 per 1000.

§ 8. – L'ANALISI DEI CRISTALLI. — Passiamo da ultimo all'esame dell'applicazione dei fenomeni di diffrazione dei raggi X all'analisi cristallina. Ricordiamo perciò come il concetto moderno di molecola di un cristallo differisca notevolmente dal concetto di molecola in un liquido o in un gas. Invero in un fluido, nel quale le molecole, per effetto dell'agitazione termica si allontanano via via dalla posizione iniziale, esse possono definirsi come le più grosse particelle che restano unite nel movimento termico. In un corpo solido invece i moti termici consistono come è noto in vibrazioni attorno a posizioni medie di equilibrio; la nozione precedente viene perciò in questo caso a perdere ogni significato. Per i cristalli non è tuttavia difficile trovare una definizione soddisfacente. Teniamo presente perciò che in qualunque reticolato cristallino possono sempre trovarsi tre traslazioni non complanari tali che per effetto di ciascuna di esse il reticolato venga a ricoprirsi. Il complesso di tutti i punti che con una successione qualunque di tali traslazioni si deducono da solo punto assegnato è quel che si dice un reticolato semplice di Bravais. Esso non è altro dunque che il complesso delle intersezioni di tre famiglie di piani paralleli ed equidistanti, che non siano parallele a una stessa retta. Ma per la proprietà detta, che il cristallo si ricopra per ciascuna delle tre traslazioni fondamentali, esso verrà ad esser costituito da un certo numero di reticolati semplici. Chiameremo molecola il complesso degli atomi, i reticolati semplici dei quali costituiscono tutto il cristallo. Naturalmente in questo concetto resta ancora alquanta arbitrarietà, che può in parte togliersi, scegliendo la molecola con criterii di semplicità, ma in genere non si può eliminare completamente. Per esempio nel salgemma, che è fatto da un reticolato di cubi coi vertici occupati alternativamente da atomi di cloro e di sodio, per modo che ogni atomo di cloro è circondato da sei di sodio, e viceversa, si intenderà naturalmente la molecola costituita da un atomo di cloro e da uno dei sei atomi di sodio circostanti; manca però evidentemente ogni criterio per scegliere quali tra questi sei debba prendersi.

Da tutto questo risulta che il compito dell'analisi cristallina si divide naturalmente in due parti: la ricerca del reticolo semplice, ossia delle tre traslazioni fondamentali, e la ricerca della molecola, ossia di quel complesso di atomi che con le tre traslazioni fondamentali generano tutto il cristallo.

La risoluzione del primo problema è abbastanza facile; Bravais e Schönflies hanno infatti dimostrato che per ogni particolare simmetria cristallina esistono soltanto pochi tipi di reticolati semplici; precisamente indicando con a, b, c le lunghezze delle traslazioni fondamentali, e con $\widehat{bc}$, $\widehat{ca}$, $\widehat{ab}$ gli angoli tra di esse compresi le caratteristiche di questi tipi sono date dalla seguente tabella:

		Sistema cristallino
1ª	Reticolato più generale	triclino.
2ª	$a = b \quad \widehat{ac} = \widehat{bc}$	monoclino.
3ª	$\widehat{ab} = \widehat{ac} = \frac{\pi}{2}$	
4ª	$a = b \quad \widehat{ac} = \widehat{bc} = \frac{\pi}{2}$	trimetrico.
5ª	id. con un punto nel centro di ogni parallelepipedo elementare	
6ª	$\widehat{ab} = \widehat{bc} = \widehat{ca} = \frac{\pi}{2}$	
7ª	id. con un punto nel centro di ogni parallelepipedo elementare	
8ª	$a = b = c \quad \widehat{ab} = \widehat{bc} = \widehat{ca}$	trigonale.
9ª	$a = b \quad \widehat{ab} = \frac{\pi}{6} \quad \widehat{ac} = \widehat{bc} = \frac{\pi}{2}$	esagonale.
10ª	$a = b \quad \widehat{ab} = \widehat{bc} = \widehat{ca} = \frac{\pi}{2}$	dimetrico.
11ª	id. con un punto nel centro di ogni parallelepipedo elementare	
12ª	$a = b = c \quad \widehat{bc} = \widehat{cb} = \widehat{ca} = \frac{\pi}{2}$	monometrico.
13ª	id. con un punto nel centro di ogni cubo elementare	
14ª	come la 12ª con un punto nel centro di ogni faccia del cubo elementare	

Data questa tabella la scelta è già limitata tra pochi casi possibili, una volta che dalla cristallografia già conosciamo il sistema al quale il cristallo in esame appartiene. Per decidere completamente basta la misura dei *d* di due o tre delle sue faccie.

Per esempio per il sistema monometrico ci troviamo nel caso 12, 13, 14 secondo che il rapporto dei *d* dei piani (010), (110) (111) è $1 : \sqrt{2} : \sqrt{3}$, $1 : \frac{1}{\sqrt{2}} : \sqrt{3}$, $1 : \sqrt{2} : \frac{\sqrt{3}}{2}$; con tre misure spettrometriche si può dunque decidere la questione. Nel caso degli altri sistemi si procederebbe con criterii analoghi. Determinato così il reticolato fondamentale, che possiamo per esempio intender costituito coi centri di gravità delle molecole, resta ancora da risolvere il problema più difficile; precisamente quale sia la distribuzione degli atomi attorno al centro della loro molecola. Per risolverla mancano criteri generali e bisogna volta per volta procedere con artifici speciali, tanto che finora non si è potuto riuscirvi che in un limitato numero di casi semplici. Per lo più la ricerca si conduce così: se il cristallo fosse costituito da un solo reticolato semplice tutti i piani reticolari paralleli sarebbero tra di loro eguali. Ciò non accadrà in genere più se il reticolato non è semplice ed una serie di piani paralleli sarà costituita da una successione di piani, diversi per densità e distanza tra di loro, ripetentesi con un certo periodo. Il primo punto da risolvere è di determinare le distanze intercedenti tra i piani di uno di questi periodi, e le densità dei varii piani. Questo primo problema può spesso risolversi con un metodo generale che andiamo ad illustrare in un caso particolare facilmente generalizzabile. Supporremo precisamente che il periodo della successione di piani che vogliamo analizzare sia costituito da due soli piani di densità K e K′. Se θ è l'angolo di incidenza si vede subito che la differenza di fase tra l'onda riflessa sul primo piano e quella riflessa sopra un piano a profondità p è $2\pi \frac{2p\cos\theta}{\lambda}$. Ora per i piani di densità K si ha $p = md$, (m intero), per quelli di densità K′, $p = md + \delta$ (d e δ rappresentano le distanze da un qualunque piano K, risp. del successivo piano K e del piano K′ dello stesso periodo).

Per quanto si è visto a proposito della diffusione dei raggi questa è approssimativamente proporzionale alla densità del materiale diffondente. L'onda riflessa sul piano K avrà dunque ampiezza proporzionale alla parte reale di

$$\sqrt{K}\, e^{2\pi i\left(\nu t - \frac{2md\cos\theta}{\lambda}\right)}$$

quella riflessa nel piano K′ a:

$$\sqrt{K'}\, e^{2\pi i\left(\nu t - \frac{2(md+\delta)\cos\theta}{\lambda}\right)}.$$

Se N è il numero degli strati l'ampiezza risultante è:

$$\left(\sqrt{K} + \sqrt{K'}\, e^{-2\pi i \frac{2\delta\cos\theta}{\lambda}}\right) e^{2\pi i \nu t} \sum_{0}^{N-1} e^{-2\pi i \frac{2nd\cos\theta}{\lambda}}$$

o calcolando la somma, come progressione geometrica:

$$\left(\sqrt{K} + \sqrt{K'}\, e^{-2\pi i \frac{2\delta\cos\theta}{\lambda}}\right) e^{2\pi i \nu t} \frac{1 - e^{-2\pi i \frac{2Nd\cos\theta}{\lambda}}}{1 - e^{-2\pi i \frac{2d\cos\theta}{\lambda}}}$$

e passando dall'ampiezza all'intensità col liberarsi dagli immaginarii, si trova l'intensità proporzionale a:

$$\left\{K + K' + 2\sqrt{KK'}\cos\frac{4\pi\delta\cos\theta}{\lambda}\right\} \frac{\sin^2 \frac{2\pi N d\cos\theta}{\lambda}}{\sin^2 \frac{2\pi d\cos\theta}{\lambda}}. \tag{9}$$

Perché questa intensità sia considerevolmente grande occorre che il denominatore sia nullo, cioè che

$$n\lambda = 2d\cos\theta \qquad (n = \text{intero})$$

che è la relazione di Bragg. Soddisfatta questa, l'intensità varia da ordine a ordine come il primo fattore di (9) che può ora scriversi:

$$K + K' + 2\sqrt{KK'}\cos 2\pi n\frac{\delta}{d}\,. \tag{10}$$

Basta dunque confrontare tra di loro le intensità dei varii ordini, fare cioè in (10) $n = 1, 2, 3\cdots$ per ricavare dal loro confronto i rapporti $\frac{K'}{K}$ e $\frac{\delta}{d}$. Si intende che in questa teoria semplicistica abbiamo trascurato l'irregolarità del cristallo e i moti termici. Queste cause, come altrove dicemmo provocano un regolare decremento di intensità col crescere dell'ordine dello spettro, e nel confronto delle intensità bisognerà naturalmente, prima di applicare le (10), aver riguardo a questo decremento, ciò che in pratica si fa moltiplicando, prima di confrontarle, le intensità quali si ottengono dalla misura diretta, per 1 il primo ordine, per 5 il secondo, per 14 il terzo ecc. Supponiamo per esempio che sia $K = K'$; $d = 2\delta$. Allora (10) si annulla per $n = 1, 3, 5\cdots$ Spariranno cioè gli spettri di ordine dispari. Questo caso si presenta per esempio nel KCl, i cui piani (010) sono costituiti da una successione di piani equidistanti alternativamente di K e di Cl che, siccome K e Cl hanno peso atomico quasi eguale, hanno presso a poco la stessa densità. In queste riflessioni restano assai indeboliti gli spettri d'ordine dispari. Se invece pure avendosi $K = K'$, si avesse $d = 4\delta$ le intensità dei successivi ordini starebbero tra di loro come:

$$\frac{1}{2} : 1 : \frac{1}{2} : 0 : \frac{1}{2} \cdots$$

È evidente come il metodo esposto di analisi delle serie di piani sia, almeno teoricamente, applicabile anche quando i piani del periodo sono più di due. Si capisce però anche come in casi appena un po' complicati l'imprecisione stessa delle misure, e sopra tutto l'incertezza dovuta al naturale decremento di intensità di ordine in ordine, siano tali da renderne impossibile l'applicazione.

Una volta determinata al modo descritto la struttura del maggior numero di serie di piani, ciò che richiede naturalmente l'applicazione del metodo quantitativo della camera di ionizzazione, conviene procedere, puramente per tentativi, alla ricerca di un reticolato cristallino, col quale si spieghino le modalità di riflessione sopra le varie faccie. Questa ricerca, oltremodo penosa e difficile, non può naturalmente condursi a termine che in casi molto semplici, e fa meraviglia che si sia riusciti a determinare la struttura di cristalli di molecola alquanto complicata, come la calcite (Bragg).

È pur vero però che, nonostante numerosi sforzi restano ancora ignote le strutture di cristalli chimicamente assai semplici, quali il quarzo e lo zolfo.

Inoltre i metodi precedenti hanno l'inconveniente di non potersi applicare altro che a cristalli con facce di una certa estensione. Per l'analisi di minerali a struttura micro o criptocristallina si applica perciò il metodo di Debye e Scherrer.

Naturalmente non si può in questo caso pensare ad una analisi troppo accurata, perché i varii elementi vengono dati troppo mescolati tra di loro; tuttavia si possono sempre ottenere dei dati interessanti per lo studio di qualche minerale. Inoltre questo metodo dà anche qualche risultato interessante nella metallografia, mostrando chiaramente il cambiamento della struttura di un metallo nel passaggio da crudo a ricotto. Si osserva infatti in generale che il metallo crudo dà luogo a fenomeni di diffrazione assai più intensi e marcati rivelando l'esistenza di particelle cristalline molto più grosse e numerose che non il metallo ricotto.

N° 7.

Formazione di immagini coi raggi Röntgen.

Per questo lavoro vedi la notizia premessa all'altro: *I raggi Röntgen.* Qui aggiungiamo che i tubi speciali coi quali Fermi sperimentò la formazione delle immagini, furono costruiti nell'Istituto di Fisica della Università di Pisa da Fermi, aiutato validamente dal meccanico dell'Istituto stesso, sig. Orfeo Di Nasso.

G. POLVANI.

This paper contains the results of Fermi's first experimental work, which he presented as a dissertation at the University of Pisa in the spring of 1922. The following comment describes his activities in the Physics Department of that university.

For those unfamiliar with the organization of Italian universities, it must be explained that a physics student at the termination of the second year attains a sufficient amount of knowledge in the general field of physics to gain him acceptance in the department under a status similar to that of a graduate student in an American university. Thereafter he follows advanced courses and undertakes research work, frequently in the field where he will later prepare a dissertation.

In the fall of 1920, three students., Enrico Fermi, Nello Carrara and Franco Rasetti were thus admitted to the Physics Department at Pisa. Owing to the recently terminated First World War, there were no older students enjoying a similar status, hence the three, aged 19, represented the entire "graduate" group at that time. Professor Luigi Puccianti, director of the Physics Laboratory, allowed them freedom of initiative to a degree seldom granted to students in Italy or elsewhere. They were allowed to use the research laboratories at all times, received keys to the library and instruments cabinets, and were given permission to try any experiment they wished with the apparatus contained therein. Carrara and Rasetti, who in the previous year had come to recognize Fermi's immense superiority in the knowledge of mathematics and physics, henceforth regarded him as their natural leader, looking to him rather than to the professors for instruction and guidance.

A few weeks were happily spent opening the instrument cabinets, guessing at the possible use of the apparatus, and sometimes trying simple experiments. It turned out that the laboratory was best equipped for research in spectroscopy, a field to which Professor Puccianti had brought important contributions. Some x-ray equipment was also available, although of the type designed for lecture demonstration rather than research.

Fermi, after much reading of the pertinent literature, decided that x-rays were the field that offered the best chance for original research, and suggested that all three learn some of the technique. The tubes available were of the gas-filled type and were operated by a large induction coil, using a spark gap as rectifier. The first task that Fermi set for the group was to produce a Laue photograph, and after several attempts and mishaps, one was obtained. Although the poor collimation of the beam and the imperfect orientation of the crystal had contributed to produce an unprepossessing Laue pattern, the three students were thrilled by the result.

It soon appeared that sealed-in tubes were not fit for research and the experimenters decided to build their own tubes. The glass part was made to specification by a glass blower, while the physicists had to seal windows and electrodes. No diffusion pumps were available, hence the tubes were evacuated by means of rotary mercury pumps of the Gaede type. Considerable time was spent before these tubes could be satisfactorily operated, but eventually the K radiations of several elements were obtained and observed by Bragg reflection.

Early in 1922 Fermi decided to undertake some original work with the x-ray technique he had been learning, in order to prepare a dissertation. Since at that time he had already published or at least completed several important theoretical papers, it may be asked why he did not present a theoretical thesis. It must be explained that at the time in Italy theoretical physics was not recognized as a discipline to be taught in universities, and a dissertation in that field would have been shocking at least to the older members of the faculty. Physicists were essentially experimentalists, and only an experimental dissertation would have passed as physics. The nearest subject to theoretical physics, mechanics, was taught by mathematicians as a field of applied mathematics, with complete disregard for its physical implications. These circumstances explain why such topics as the quantum theory had gained no foothold in Italy: they represented a " no man' s land " between physics and mathematics. Fermi was the first in the country to fill the gap.

However, there is not the slightest doubt in the writer's mind that Fermi's experimental activities in his third and fourth university years were not prompted by the convenience of presenting an acceptable dissertation. He obviously enjoyed experimental work as much as theoretical abstraction, and especially the alternation of the two types of activities. He was from the first a complete physicist for whom theory and experiment possessed equal weights, even though for many years his fame was chiefly based on the theoretical contributions. But he never for a moment was one of those theoreticians who, to use a joking expression later much in use among the Rome group, " could not tell steel from aluminum ".

The present paper, as explained by Fermi in his introductory remarks, describes the experimental realization of a method suggested by Gouy for obtaining images with monochromatic x-rays by means of a curved crystal. Dardord (*) appears to have independently obtained similar results. Curved crystals had been used since the earliest times of x-ray spectroscopy as a substitute for rocking the crystal, but the experiments of Fermi and Dardord are the first where focusing was obtained. Curved-crystal focusing spectrometers acquired considerable importance in the following decades.

F. Rasetti.

7.

FORMAZIONE DI IMMAGINI COI RAGGI RÖNTGEN

«Nuovo Cimento», *25*, 63–68, (1923).

I Raggi Röntgen non subiscono né rifrazioni né riflessioni, almeno nel senso ordinario di questa parola, poiché la riflessione di diffrazione avviene solo sotto angoli di incidenza determinati. Ne segue che nell'ottica dei raggi X il problema di ottenere immagini non può, come nell'ottica ordinaria, risolversi per mezzo di lenti o di specchi sferici.

Gouy [1] ha suggerito teoricamente un metodo per ottenere delle immagini monocromatiche con raggi X, per mezzo di un cilindro di mica. Esso è, in poche parole, il seguente:

Consideriamo un cilindro circolare di mica e supponiamo che in un punto del suo asse sia una sorgente S di raggi Röntgen monocromatici. Essi si rifletteranno sulla mica in quei punti per i quali è verificata la relazione di Bragg, che si trovano evidentemente sopra cerchi sezione del cilindro. Ed i raggi riflessi sopra uno di questi cerchi si riuniranno evidentemente in un punto I

(*) M. R. Dardord, «Jour. Physique et Radium», *3*, 218 (1922).

(1) C. R. Gouy, «C. R.», *161*, 176 (1915).

dell'asse, simmetrico di S rispetto al piano del cerchio riflettente, dove si avrà un'immagine reale monocromatica di S.

Se S si trovasse nelle vicinanze dell'asse se ne formerebbe ancora un'immagine nelle vicinanze dell'asse [2]. Supponiamo ora di avere in vicinanza dell'asse una figura piana da cui punti escano raggi X monocromatici, e di porre una lastra fotografica nella posizione dove se ne forma l'immagine.

Sia r la distanza specchio oggetto, R il raggio del cilindro di mica, θ l'angolo di incidenza di Bragg, r' la distanza immagine specchio. Se proiettiamo il tutto sopra un piano perpendicolare all'asse del cilindro di mica le proiezioni di r ed r' saranno $r \cos \theta$, $r' \cos \theta$; e quindi per le solite formule degli specchi sferici sarà

$$\frac{1}{r \cos \theta} + \frac{1}{r' \cos \theta} = \frac{2}{R},$$

da cui

$$r' = \frac{R r}{2 r \cos \theta - R} .$$

Il coefficente lineare di ingrandimento dei segmenti perpendicolari ad r e all'asse del cilindro sarà:

$$\mu_1 = \frac{r'}{r} = \frac{R}{2 r \cos \theta - R} . \tag{1}$$

Se l'oggetto è in vicinanza dell'asse si ha dunque press'a poco $\mu_1 = 1$.

Per calcolare l'ingrandimento dei segmenti paralleli al piano dell'asse e di r indichiamo con φ e φ' gli angoli che le perpendicolari al piano dell'oggetto e della lastra fotografica formano rispettivamente con r, r'.

Si vede allora subito che l'ingrandimento cercato è:

$$\mu_2 = \frac{\cos \varphi'}{\cos \varphi} . \tag{2}$$

Supponiamo ora di fotografare una fenditura posta perpendicolarmente al piano di r e dell'asse con una lastra di mica piana di lunghezza l. Se h è la lunghezza della fenditura la lunghezza della sua immagine sarà $2l + k$. Se invece curviamo la mica in modo che l'immagine si formi in fuoco la lunghezza diventerà h. Le intensità delle due immagini staranno evidentemente press'a poco nel rapporto inverso delle loro lunghezze. Il loro rapporto è dunque

$$\frac{2l + h}{h} .$$

Se per esempio $h = 1$ cm. $l = 4$ cm. il rapporto è 9. L'intensità è dunque quasi decuplicata.

Descriverò ora il modo con cui mi è riuscito ottenere effettivamente queste immagini.

La sorgente dei raggi era costituita da un tubo della forma e dimensioni indicate press'a poco nella figura I.

Facevo il vuoto nel tubo con una pompa rotativa Cacciari, tipo Gaede. Il catodo K era concavo, di raggio di 6 o 7 cm. quando si desiderava concentrare il più possibile i raggi catodici sopra l'anticatodo; volendo invece

(2) Purché naturalmente, il cilindro sia limitato ad una regione sufficientemente piccola compresa tra due generatrici.

che tutta la superfice dell'anticatodo fosse colpita dai raggi, il catodo veniva fatto di raggio più piccolo. L'anticatodo era generalmente di ferro e qualche volta era tagliato quasi perpendicolarmente ai raggi catodici, per modo da poter fare a meno della fenditura. In altre esperienze invece era tagliato a becco di flauto, in modo da presentare una superfice estesa agli strumenti

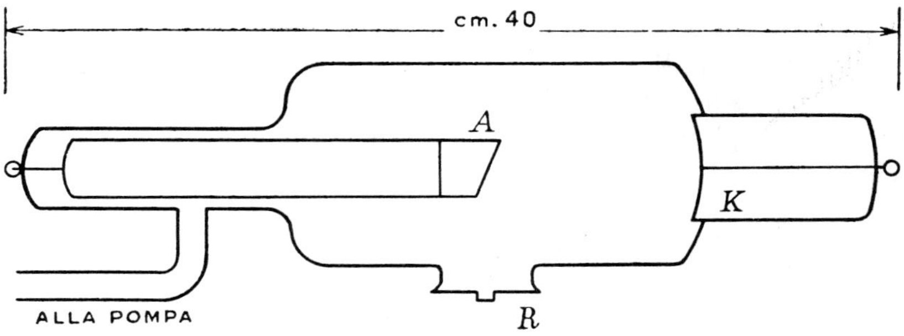

Fig. I.

di osservazione. Siccome le radiazioni caratteristiche del ferro sono considerevolmente assorbite dal vetro dell'ampolla trovai opportuno di fornire i tubi di una finestrella di alluminio R. Durante il funzionamento il tubo era mantenuto attaccato alla pompa, per modo che dopo un po' di tempo assumeva un regime abbastanza regolare. Il tubo era azionato da un grosso rocchetto di induzione con interruttore Wehnelt; nelle condizioni ordinarie la scintilla equivalente era di 10 o 12 cm.

Il tubo era racchiuso in una cassetta di legno foderata di piombo dello spessore di 6 mm dalla parte degli strumenti, e di 3 dalle altre parti.

Per ottenere delle immagini discretamente precise era necessario che la lastra di mica riflettente fosse il più possibile regolare. Essa veniva perciò accuratamente scelta tra molti campioni, e tuttavia non mi è mai riuscito di trovarne di quelle che, alla riflessione della luce, si mostrassero più regolari di un ordinario vetro da finestra. A questo fatto sono dovute le irregolarità e i baffi osservabili nelle immagini riportate. La mica veniva curvata legandola strettamente sopra un cilindro tornito di ottone. Sopra la parte convessa si stendeva poi uno strato di ceralacca spesso poco più di mezzo centimetro. Quando la ceralacca si era raffreddata si potevano togliere le legature e distaccare lo specchio dal cilindro. A questo modo mi è riuscito di ottenere degli specchi cilindrici relativamente precisi dato il limite imposto dalla naturale irregolarità delle lamine utilizzate. Essi avevano per lo più dimensioni di 4×6 cm. ma ordinariamente se ne limitava l'apertura riducendola a 4×2 per utilizzare le porzioni meno irregolari, che si giudicavano provando gli specchi alla riflessione della luce ordinaria. Lo specchio era montato su un cerchio graduato per poterlo mettere a posto. (L'angolo di cui si girava, per l'esame del 3° ordine della K_α del ferro era 16° 50′). La rivelazione dei raggi era fatta fotograficamente.

Eseguii dapprima alcune esperienze di orientamento con cristalli piani per accertare la natura dell'anticatodo e l'intensità delle riflessioni dei varii ordini. Risultò che veniva emessa la doppia K_α $K_{\alpha'}$ ($\lambda = 1{,}932$; $1{,}928$) poco risolubile nelle condizioni sperimentali in cui mi trovavo; la K_β ($\lambda = 1{,}748$). La K_γ era poco visibile per la scarsa intensità. Gli ordini più intensi erano il primo ed il terzo. Preferii lavorare sopra il terzo per non dover usare angoli di incidenza troppo vicini a 90°.

Esperimentai poi l'indicato metodo per ottenere immagini dapprima sopra l'anticatodo che funzionava anche da fenditura.

Le distanze anticatodo cristallo e cristallo immagine variavano tra 18 e 22 cm. La posa durava una diecina di minuti.

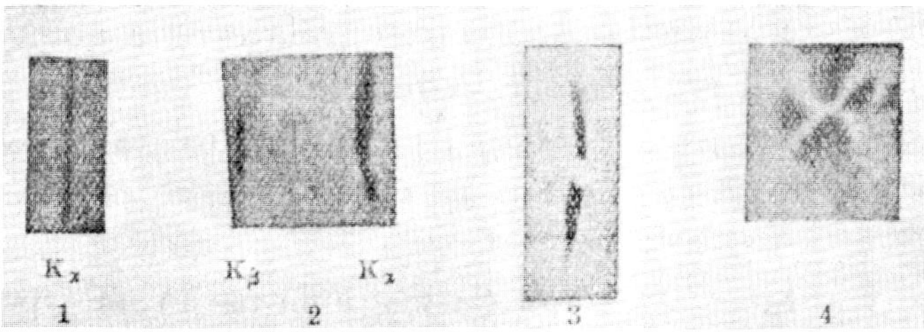

Fig. II, 1-4.

Potei subito constatare il fortissimo aumento di intensità che così si otteneva. Una grossolana idea di questo è data dalle figure II, 1 e II, 2 che rappresentano due fotografie del 3° ordine della K_α del ferro fatte press'à poco nelle stesse condizioni di posa e di funzionamento del tubo, la prima con mica piana e la seconda con mica curva. L'aumento di intensità anzi, specialmente usando specchi di apertura di 6 cm. era tale che abituando qualche minuto l'occhio al buio della stanza si riusciva a vedere distintamente le immagini sopra uno schermo di platino-cianuro di bario. Dalla fig. II, 2 è chiaramente visibile che l'intensità di emissione della parte centrale dell'anticatodo, dove venivano concentrati i raggi catodici è notevolmente maggiore di quella delle parti laterali. Ciò si vede perché il metodo delle immagini permette una osservazione della fenditura « alla Lockyer », permette cioè di osservare punto per punto ciò che accade nella fenditura. Per mettere ciò meglio in chiaro feci la seguente esperienza: Misi davanti alla finestra d'alluminio del tubo un filo di piombo del diametro di circa 1 mm. e spostai la lastra fotografica in modo da portarla nel punto dove si formava l'immagine della finestra di alluminio.

La fig. II, 3 dà il risultato di questa esperienza; in essa è chiaramente visibile l'intervallo prodotto nell'immagine dal filo di piombo.

La fig. II, 4 rappresenta infine un tentativo di ottener l'immagine di un oggetto a due dimensioni. L'anticatodo di ferro fu perciò tagliato a becco di flauto e vi furono fatti due solchi in croce nei quali furono incastrati due fili di rame in modo da formare una specie di X. Nella fig. II, 4 si vede l'immagine di questa X, naturalmente insieme a parecchie irregolarità dovute alla irregolarità del riflettore.

Questo lavoro fu eseguito nell'Istituto di Fisica della R. Università di Pisa nell'inverno 1922.

8.

RELAZIONE

letta dal Corrispondente G. Armellini (relatore), a nome anche del Socio T. Levi–Civita, nella seduta del 7 gennaio 1923, sulla Memoria del dott. Enrico Fermi, intitolata: *Sul peso dei corpi elastici.*

L'Autore immagina un corpo elastico immerso in uno spazio euclideo in condizioni di riposo, cioè in modo che le sue tensioni interne siano tutte nulle.

Suppone poi che il corpo venga tolto dallo spazio euclideo ed immerso in uno spazio statico einsteiniano, la cui metrica è determinata dalla forma data dal prof. Levi–Civita

$$ds^2 = g_{00}\, dt^2 - \Sigma\, g_{ik}\, dx_i\, dx_k\,,$$

dove le g sono funzioni delle sole x. L'Autore calcola quindi la variazione della energia specifica nell'intorno di un punto generico del corpo elastico quando esso passa dallo spazio euclideo allo spazio einsteiniano, e dimostra che ad essa corrisponde in generale una variazione del peso del corpo, variazione che non ha la direzione della gravità.

Poiché, però, i procedimenti adoperati sarebbero più adattati a dimostrare l'esistenza piuttosto che a calcolare l'ammontare di questa variazione di peso, l'Autore consacra il seguito della Memoria allo studio di due casi particolarmente semplici: quando cioè il corpo elastico sia una sfera od un cilindro circolare.

La condotta del lavoro rivela nell'Autore una completa padronanza dello strumento analitico. Non ostante che le formule generali – date nella prima parte della Memoria – fossero già state trovate dal Ricci, il quale però le ha soltanto esposte nelle sue lezioni all'Università di Padova senza mai pubblicarle, i relatori giudicano che il lavoro del dott. Fermi arreca un vero e sostanziale contributo all'argomento di cui tratta, e ne propongono quindi l'inserzione integrale negli Atti accademici.

SUL PESO DEI CORPI ELASTICI

« Mem. Lincei », Classe di Scienze Fisiche, Matematiche e Naturali, *14*, 114–124 (1923).

1. – Formule generali.

Intendendo in questo lavoro considerare soltanto fenomeni statici, supporremo che nello spazio–tempo valga la metrica che corrisponde a questo caso [1]; che sia cioè

$$ds^2 = g_{\infty}\, dt - \sum_{ik=1}^{3} \gamma_{ik}\, dx_i\, dx_k\,, \tag{1}$$

dove g_{∞} e le γ_{ik} sono funzioni solamente delle coordinate di spazio x_1, x_2, x_3. Consideriamo ora un corpo elastico che possa distendersi in uno spazio euclideo in posizione di riposo, in modo cioè che siano nulle tutte le tensioni elastiche tra le sue parti. Se lo spazio S di metrica

$$d\sigma^2 = \sum_{ik} \gamma_{ik}\, dx_i\, dx_k \qquad (i\,, k = 1\,, 2\,, 3) \tag{2}$$

fosse euclideo, potremmo in ∞^6 modi distendere in esso il nostro corpo in posizione di riposo. Siccome però ciò in generale non sarà, accadrà generalmente che nel nostro corpo, quando è immerso nello spazio S, avranno luogo delle tensioni elastiche; e l'energia di queste tensioni varierà generalmente spostando il corpo. Ne segue che il peso del nostro corpo, oltre che alla parte dovuta al peso della sua eventuale massa materiale ed al peso della sua energia elastica, consterà ancora di un'altra parte dovuta alle variazioni di energia elastica or ora accennate. Lo scopo di questo lavoro è appunto di calcolare quale sia questo peso addizionale. Faremo l'ipotesi che lo spazio S sia quasi euclideo, nella quale non è contenuto altro se non la limitazione che si fa sempre nella teoria dell'elasticità, che le deformazioni del corpo elastico siano molto piccole.

Pensiamo per un momento il nostro corpo elastico disteso in uno spazio euclideo ed indichiamo con u_1, u_2, u_3 delle coordinate cartesiane ortogonali in questo spazio. Quando il corpo si trova immerso nello spazio S, x_1, x_2, x_3 verranno ad essere funzioni di u_1, u_2, u_3 per modo che l'elemento metrico di S potrà anche esprimersi, per le u,

$$d\sigma^2 = \sum_{ik} a_{ik}\, du_i\, du_k\,, \tag{3}$$

essendo naturalmente

$$a_{ik} = \sum_{rs} \gamma_{rs} \frac{\partial x_r}{\partial u_i} \frac{\partial x_s}{\partial u_k}\,. \tag{4}$$

(1) T. Levi–Civita, « Rendiconti Accademia Lincei », serie 5ª, vol. XXVI (1917).

Indichiamo con α_1, α_2, α_3 le tre dilatazioni principali relative ad una particella del corpo. L'energia elastica specifica circostante alla particella sarà, come è noto,

$$W_0 = 4A(S\alpha_1)^2 - 4BS\alpha_2\alpha_3, \tag{5}$$

indicandosi con S la somma eseguita circolando sugli indici (1, 2, 3) ed essendo A e B costanti di elasticità che si esprimono per mezzo del modulo di elasticità E e del coefficiente σ di Poisson con le formule

$$A = \frac{E(1-\sigma)}{8(1+\sigma)(1-2\sigma)} \quad ; \quad B = \frac{E}{4(1+\sigma)}. \tag{6}$$

Ora le solite considerazioni di cinematica dei mezzi continui mostrano che α_1, α_2, α_3, sono le radici dell'equazione.

$$\begin{vmatrix} a_{11}-(1+x)^2 & a_{12} & a_{13} \\ a_{21} & a_{22}-(1+x)^2 & a_{23} \\ a_{31} & a_{32} & a_{33}-(1+x)^2 \end{vmatrix} = 0.$$

E quindi, per le note relazioni tra coefficienti e radici, si ha

$$\begin{cases} S(1+\alpha_1)^2 = Sa_{11} \\ S(1+\alpha_2)^2(1+\alpha_3)^2 = S\begin{vmatrix} a_{22} & a_{23} \\ a_{32} & a_{33}\end{vmatrix}. \end{cases} \tag{7}$$

Indicando al solito con ε_{ik}, 1 oppure 0, secondo che $i = k$ oppure $i \neq k$, si avrà, per le ipotesi fatte, $a_{ik} = \varepsilon_{ik} + \beta_{ik}$, essendo le β_{ik} piccolissime. Dalle (7) si ha allora.

$$2S\alpha_1 + S\alpha_1^2 = S\beta_{11}$$

$$S(1+\alpha_2)^2(1+\alpha_3)^2 = 3 + 2S\beta_{11} + S\begin{vmatrix} \beta_{22} & \beta_{23} \\ \beta_{32} & \beta_{33}\end{vmatrix}$$

da cui, essendo, sempre per le ipotesi fatte, le α molto piccole, si ricava approssimativamente

$$S\alpha_1 = \frac{1}{2}S\beta_{11} \quad ; \quad S\alpha_2\alpha_3 = \frac{1}{4}S\begin{vmatrix} \beta_{22} & \beta_{23} \\ \beta_{32} & \beta_{33}\end{vmatrix}, \tag{8}$$

e quindi (5)

$$W_0 = A(S\beta_{11})^2 - BS\begin{vmatrix} \beta_{22} & \beta_{23} \\ \beta_{32} & \beta_{33}\end{vmatrix}. \tag{9}$$

Siccome nei nostri calcoli ci serviremo del principio di Hamilton, occorre che troviamo la variazione di W_0, quando variano le x_i considerate come funzioni delle u_i; da (9) si ha:

$$\begin{aligned} \delta W_0 &= 2A(S\beta_{11})S\delta\beta_{11} - BS(\beta_{22}\delta\beta_{33} + \beta_{33}\delta\beta_{22} - 2\beta_{23}\delta\beta_{23}) \\ &= S[\delta\beta_{11}\{2A\beta_{11} + (2A-B)(\beta_{22}+\beta_{33})\} + 2B\beta_{23}\delta\beta_{23}]. \end{aligned} \tag{10}$$

Si ha poi da (4)

$$\delta\beta_{ik} = \delta a_{ik} = \sum_{rst} \frac{\partial\gamma_{rs}}{\partial x_t}\delta x_t \frac{\partial x_r}{\partial u_i}\frac{\partial x_s}{\partial u_k} + 2\sum_{rs}\gamma_{rs}\frac{\partial x_r}{\partial u_i}\frac{\partial \delta x_s}{\partial u_k}.$$

Per aver formule più semplici, possiamo prender le coordinate x coincidenti con le u, e la formula precedente diventa

$$(11) \qquad \delta\beta_{ik} = \sum_t \frac{\partial\gamma_{ik}}{\partial x_t}\,\delta x_t + 2\sum_s \gamma_{is}\,\frac{\partial\,\delta x_s}{\partial u_k}\,.$$

Come espressione dell'azione possiamo prendere

$$(12) \qquad W = \iint W_0\sqrt{g_{00}}\,d\tau\,dt + \iint k\sqrt{g_{00}}\,d\tau\,dt + \iint \sum_i F_i\,x_i\,dt\,d\tau\,,$$

nella quale il primo termine dà l'azione elastica; il secondo l'azione delle masse materiali di densità k; $d\tau$ è l'elemento di volume dello spazio; l'ultimo termine rappresenta infine l'azione di una forza esterna di componenti F_1, F_2, F_3 adatta a sostenere il corpo perchè non cada: forza che dunque altro non è se non l'eguale e contraria del peso del corpo. La variazione di W può ora immediatamente effettuarsi e siccome il principio di Hamilton richiede che sia $\delta W = 0$, troviamo l'equazione

$$\iint \sum_i F_i\,\delta x_i\,d\tau\,dt + \iint k\sum_i \frac{\partial\sqrt{g_{00}}}{\partial x_i}\,\delta x_i\,d\tau\,dt$$
$$+ \iint \left(W_0\sum_i \frac{\partial\sqrt{g_{00}}}{\partial x_i}\,\delta x_i + \delta W_0\sqrt{g_{00}}\right)dt\,d\tau = 0\,.$$

In δW_0, come risulta dalle (10) (11), sono contenute, oltre ai δx_i, anche le loro derivate, che però possono subito eliminarsi con una integrazione per parti. Siccome poi possiamo per esempio supporre le F_i costanti, annullando i coefficienti dei δx_i, troviamo delle equazioni adatte a determinarci la posizione del corpo, ossia, ciò che è equivalente, le β_{ik}. Fatto questo, restano ancora da determinare le F_i, cioè le componenti del peso del corpo. E questo si fa ora facilmente, supponendo le δx costanti. Si trova infatti, da (10) (11),

$$\delta\beta_{ik} = \sum_t \frac{\partial\beta_{ik}}{\partial u_t}\,\delta x_t\,;$$

e quindi

$$(13) \qquad \delta W_0 = \sum_t K_t\,\delta x_t\,,$$

avendo per brevità posto

$$(14) \qquad K_t = S\left[\frac{\partial\beta_{11}}{\partial u_t}\left\{2A\beta_{11} + (2A - B)(\beta_{22} + \beta_{33})\right\} + 2B\beta_{23}\,\frac{\partial\beta_{23}}{\partial u_t}\right].$$

Il principio di Hamilton ci dà dunque, annullando i coefficienti dei δx_i,

$$(15) \qquad \int F_t\,d\tau + \int (k + W_0)\,\frac{\partial\sqrt{g_{00}}}{\partial x_t}\,d\tau + \int K_t\sqrt{g_{00}}\,d\tau = 0\,.$$

Dalla quale risulta che il peso è:

$$\int (k + W_0)\,\frac{\partial\sqrt{g_{00}}}{\partial x_t}\,d\tau + \int K_t\sqrt{g_{00}}\,d\tau\,.$$

Esso consta dunque di due parti: la prima non è altro se non il peso delle masse materiali e dell'energia elastica, ed ha la direzione della gravità, la seconda, di componenti $\int K_t\sqrt{g_{00}}\,d\tau$, ha in generale direzione diversa ed è dovuta alla ragione accennata al principio di questa Nota. Essa può in particolare esser diversa da zero anche quando g_{∞} è costante, ossia quando non c'è forza di gravità nel senso ordinario.

2. – Caso di un cilindro con le generatrici parallele all'asse z.

Il metodo dato ora per calcolare questo peso addizionale è però, per lo più, inapplicabile perchè richiede la risoluzione di un complicato problema di teoria dell'elasticità. Mi propongo di mostrare, con due esempi particolarmente semplici, come in qualche caso si possa giungere a calcolare effettivamente questo peso addizionale dei corpi elastici. Supporremo, nel primo esempio, che il ds^2 dello spazio tempo sia

$$ds^2 = g_{\infty} dt^2 - dz^2 - \mathrm{E}\, dx^2 - 2\,\mathrm{F} dx\, dy - \mathrm{G} dy^2, \tag{16}$$

essendo g_{∞}, E, F, G funzioni soltanto di x e di y. Supporremo poi che il nostro corpo elastico sia costituito da un cilindro con le generatrici parallele all'asse z e che, disteso in posizione di riposo in uno spazio euclideo, abbia sezione circolare di raggio piccolo in confronto al raggio di curvatura della forma quadratica $\mathrm{E}dx^2 + 2\,\mathrm{F}dx\,dy + \mathrm{G}dy^2$. Quanto alle proprietà elastiche del nostro corpo, possiamo press'a poco indifferentemente supporre o che sia isotropo, ma di forma tanto allungata che sia trascurabile l'influenza degli estremi oppure che abbia lunghezza qualunque, ma sia costituito come da fibre rigide parallele all'asse. Siccome questa seconda ipotesi conduce a calcoli un po' più semplici della prima, ci atterremo ad essa.

Sia H la curvatura della forma quadratica $\mathrm{E}dx^2 + 2\,\mathrm{F}dx\,dy + \mathrm{G}\,dy^2$. Noi supporremo che H e g_{∞} non variino sensibilmente entro il nostro corpo, di modo che esso continuerà ad aver forma di cilindro circolare. Indichiamo con r la distanza di una particella generica del cilindro dall'asse, quando il cilindro è in posizione di riposo in uno spazio euclideo, e con $r+\varepsilon$ la sua distanza dall'asse quando il cilindro è immerso nel nostro spazio non euclideo. La lunghezza del cerchio col centro sull'asse passante per la nostra particella era $2\pi r$; ed ora, se lo spazio fosse euclideo, sarebbe $2\pi(r+\varepsilon)$. Siccome però ciò non è, la lunghezza di quel cerchio risulta, con la nostra approssimazione, $2\pi(r+\varepsilon)\left(1-\frac{\mathrm{H}r^2}{6}\right)$, di modo che gli allungamenti della nostra particella in direzione radiale e trasversale sono rispettivamente, trascurando nella nostra approssimazione il prodotto $\varepsilon\mathrm{H}$,

$$\frac{d\varepsilon}{dr} \quad ; \quad \frac{\varepsilon}{r} - \frac{\mathrm{H}r^2}{6}.$$

L'energia elastica specifica nell'intorno della nostra particella è dunque

$$\mathrm{W}_0 = 4\mathrm{A}\left\{\left(\frac{d\varepsilon}{dr}\right)^2 + \left(\frac{\varepsilon}{r} - \frac{\mathrm{H}r^2}{6}\right)^2\right\} + 4\mathrm{B}\frac{d\varepsilon}{dr}\left(\frac{\varepsilon}{r} - \frac{\mathrm{H}r^2}{6}\right), \tag{17}$$

essendo A e B costanti di elasticità legate al modulo di elasticità E ed al coefficiente di contrazione σ (naturalmente considerati nel piano perpendicolare alle fibre) dalle relazioni

$$\text{(18)} \qquad A = \frac{E}{8(1-\sigma^2)} \quad , \quad B = \frac{\sigma E}{4(1-\sigma^2)} \,.$$

Supponiamo che il nostro cilindro abbia altezza 1; allora l'energia corrispondente allo strato compreso tra i due cilindri coassiali di raggio r ed $r + dr$ è $2\pi r W_0 dr$. Sia R il raggio del cilindro: l'energia totale sarà allora

$$2\pi \int_0^R r W_0 \, dr \,,$$

cioè

$$8\pi \int_0^R r\,dr \left[A \left\{ \left(\frac{d\varepsilon}{dr}\right)^2 + \left(\frac{\varepsilon}{r} - \frac{Hr^2}{6}\right)^2 \right\} + B \frac{d\varepsilon}{dr} \left(\frac{\varepsilon}{r} - \frac{Hr^2}{6}\right) \right] .$$

Per l'equilibrio elastico dobbiamo determinare ε in funzione di r per modo che l'espressione precedente sia un minimo.

Abbiamo perciò, con le solite regole del calcolo delle variazioni, l'equazione differenziale

$$2A\left(\frac{\varepsilon}{r} - \frac{Hr^2}{6}\right) + B\frac{d\varepsilon}{dr} = \frac{d}{dr}\left\{ 2A \frac{d\varepsilon}{dr} r + B\left(\frac{\varepsilon}{r} - \frac{Hr^2}{6}\right) r \right\}$$

con le condizioni ai limiti

$$\text{per } r = 0 \qquad \varepsilon = 0$$

$$\text{per } r = R \qquad 2A\frac{d\varepsilon}{dr} + B\left(\frac{\varepsilon}{r} - \frac{Hr^2}{6}\right) = 0 .$$

Queste condizioni sono sufficienti a determinare ε in funzione di r, e ci forniscono precisamente

$$\text{(19)} \qquad \varepsilon = \frac{(2A - B)HR^2}{32A} r - \frac{(2A - 3B)H}{32A} r^3 .$$

Ci conviene ora sostituire questa espressione di ε nell'espressione dell'energia totale. Dopo qualche calcolo un po' laborioso, ma assolutamente privo di difficoltà si trova così per l'energia totale l'espressione:

$$\frac{\pi H^2 R^6}{3456 A^2} [224 A^3 - 108 A^2 B + 72 AB^2 + 27 B^3] ;$$

o, ponendo

$$\text{(20)} \qquad M = \frac{\pi}{3456 A^2} [224 A^3 - 108 A^2 B + 72 AB^2 + 27 B^3] \,,$$

otteniamo finalmente come energia totale

$$MH^2 R^6 .$$

Ed M è una costante che dipende esclusivamente dalle costanti elastiche del nostro corpo.

Indichiamo ora con m la massa totale del nostro cilindro e con F_x, F_y, F_z la tre componenti della equilibrante del peso. Per il principio di Hamilton, come prima, dovrà essere nulla la variazione dell'espressione

$$\sqrt{g_{\infty}}\,(m + MH_2 R^6) + F_x x + F_y y + F_z z .$$

Ora, tale variazione è

$$\left(\delta x \frac{\partial \sqrt{g_{00}}}{\partial x} + \delta y \frac{\partial \sqrt{g_{00}}}{\partial y}\right)(m + \mathrm{M}\mathrm{H}^2 \mathrm{R}^6) + 2\sqrt{g_{00}}\mathrm{M}\mathrm{H}\mathrm{R}^6\left(\frac{\partial \mathrm{H}}{\partial x}\delta x + \frac{\partial \mathrm{H}}{\partial y}\delta y\right)$$
$$- \mathrm{F}_x \delta x + \mathrm{F}_y \delta y + \mathrm{F}_z \delta z\,,$$

poichè g_{00} e H sono soltanto funzioni di x e di y.

Annullando i coefficienti di δx, δy, δz, troviamo ora

$$\text{(21)}\qquad \begin{cases} \mathrm{F}_x + \dfrac{\partial \sqrt{g_{00}}}{\partial x}(m + \mathrm{M}\mathrm{H}^2 \mathrm{R}^6) + 2\sqrt{g_{00}}\mathrm{M}\mathrm{H}\mathrm{R}^6 \dfrac{\partial \mathrm{H}}{\partial x} = 0 \\ \mathrm{F}_y + \dfrac{\partial \sqrt{g_{00}}}{\partial y}(m + \mathrm{M}\mathrm{H}^2 \mathrm{R}^6) + 2\sqrt{g_{00}}\mathrm{M}\mathrm{H}\mathrm{R}^6 \dfrac{\partial \mathrm{H}}{\partial y} = 0 \\ \mathrm{F}_z = 0\,. \end{cases}$$

Supponiamo, per comodità, che in vicinanza del nostro corpo la velocità della luce sia 1, cioè $g_{00} = 1$.

Allora $\dfrac{\partial \sqrt{g_{00}}}{\partial x}$, $\dfrac{\partial \sqrt{g_{00}}}{\partial y}$ non sono altro se non le componenti G_x e G_y dell'accelerazione di gravità. Risulta allora che le componenti del peso del nostro corpo sono

$$\mathrm{G}_x(m + \mathrm{M}\mathrm{H}^2 \mathrm{R}^6) + \mathrm{M}\mathrm{R}^6 \frac{\partial \mathrm{H}^2}{\partial x}$$
$$\mathrm{G}_y(m + \mathrm{M}\mathrm{H}^2 \mathrm{R}^6) + \mathrm{M}\mathrm{R}^6 \frac{\partial \mathrm{H}^2}{\partial y}$$
$$0$$

ossia, con scrittura vettoriale, che il suo peso è

$$\text{(22)}\qquad \mathbf{P} = \mathbf{G}(m + \mathrm{M}\mathrm{H}^2 \mathrm{R}^6) + \mathrm{M}\mathrm{R}^6 \operatorname{grad}(\mathrm{H}^2).$$

Esso consta dunque di due parti bene distinte. La prima non è altro se non il peso ordinario della massa e dell'energia elastica ed ha la direzione dell'accelerazione di gravità **G**. La seconda, dovuta invece alle deviazioni dell'euclideità del solo spazio, ha in generale direzione diversa dalla prima, determinata soltanto dalla curvatura H. Essa può dunque esistere anche quando la prima non esista: cioè un corpo elastico può pesare anche in un campo in cui non vi sia accelerazione di gravità propriamente detta.

3. – Caso di una sfera isotropa.

Il secondo esempio è il seguente: calcolare il peso di una sfera elastica isotropa di dimensioni molto piccole rispetto ai raggi di curvatura dello spazio. Anche ora conviene naturalmente cercare prima l'energia totale elastica della nostra sfera.

Consideriamo perciò, accanto allo spazio S di metrica (2), anche uno spazio euclideo S' e supponiamo che la nostra sfera successivamente sia distesa in posizione di riposo in S'; e poi posta in S.

Riferiamo lo spazio S′ a coordinate cartesiane ortogonali (ξ, η, ζ) con centro nel centro della sfera. Siccome poi tra i punti di S e di S′ vi è corrispondenza biunivoca (facendo corrispondere quelle coppie di punti che sono occupate dalla medesima particella della sfera elastica), potremo riferire anche lo spazio S alle coordinate ξ, η, ζ. Osserviamo poi che, per la approssimazione introdotta, alle tre direzioni principali ortogonali di curvatura dello spazio S, relative al centro della sfera, corrisponderanno in S′ tre direzioni, pure ortogonali, che potremo prendere per direzioni degli assi coordinati. L'elemento metrico di S sia ora riferito al nuovo sistema

$$(23)\qquad \begin{aligned} ds^2 = {} & (1 + \beta_{11})\, d\xi^2 + (1 + \beta_{22})\, d\eta^2 + (1 + \beta_{33})\, d\zeta^2 \\ & + 2\beta_{23}\, d\eta\, d\zeta + 2\beta_{31}\, d\zeta\, d\xi + 2\beta_{12}\, d\xi\, d\eta \end{aligned}$$

dove le β saranno molto piccole. Possiamo svilupparle in serie delle ξ, η, ζ, fermandoci ai termini quadratici. Osserviamo poi che, per le ipotesi fatte, mancheranno in questo sviluppo i termini lineari (se vi fossero, infatti, essi permetterebbero di distinguere un senso dall'opposto), e nei termini quadratici mancheranno quelli rettangolari (se vi fossero, infatti, la quadrica ottenuta eguagliando questo termine quadratico a una costante avrebbe il triedro principale non coincidente col triedro principale di curvatura, ciò che è contro la simmetria). Inoltre in $\beta_{23}, \beta_{31}, \beta_{12}$ mancherà anche il termine costante, perchè, per $\xi = \eta = \zeta = 0$, deve essere $\beta_{23} = \beta_{31} = \beta_{12} = 0$. Così che, in ultima analisi, il nostro sviluppo sarà

$$(24)\qquad \left\{ \begin{aligned} \beta_{11} &= \gamma_{11} + p_{11}\,\xi^2 + q_{11}\,\eta^2 + r_{11}\,\zeta^2 \\ \beta_{23} &= \phantom{\gamma_{11} + {}} p_{23}\,\xi^2 + q_{23}\,\eta^2 + r_{23}\,\xi^2 \end{aligned} \right.$$

e le analoghe circolando sugli indici. Le γ, p, q, r sono costanti.

Ora la (9) ci permette di calcolare l'energia specifica W_0, e si trova

$$W_0 = G + SL_1\,\xi^2 + SP_1\,\xi^4 + S\pi_1\,\eta^2\,\zeta^2,$$

essendo le G, L, P, π costanti formate con le γ, p, q, r. L'energia totale sarà $W = \int_\tau W_0\, d\tau$, esteso a tutto il volume della sfera.

Ora si ha

$$\int_\tau d\tau = \frac{4}{3}\pi R^3\ ;\ \int_\tau \xi^2\, d\tau = \frac{4}{15}\pi R^5\ ;\ \int_\tau \xi^4\, d\tau = \frac{4}{35}\pi R^7\ ;\ \int_\tau \eta^2\,\xi^2\, d\tau = \frac{4}{105}\pi R^7;$$

di modo che l'energia totale risulta

$$W = \frac{4}{3}\pi R^3 G + \frac{4}{15}\pi R^5 SL_1 + \frac{4}{35}\pi R^7 SP_1 + \frac{4}{105}\pi R^7 S\pi_1.$$

Dalla (9) e dalla (24) si trova poi, effettuando i calcoli, che

$$G = AS\,\gamma_{11}^2 + (2A - B)\, S\gamma_{22}\,\gamma_{33},$$

$$\begin{aligned} SL_1 = {} & 2\,AS\,\gamma_{11}\,(p_{11} + q_{11} + r_{11}) \\ & + (2A - B)\, S\, \{\gamma_{22}\,(p_{33} + q_{33} + r_{33}) + \gamma_{33}\,(p_{22} + q_{22} + r_{22})\}, \end{aligned}$$

$$\mathrm{SP}_1 = \mathrm{AS}\,(p_{11}^2 + q_{11}^2 + r_{11}^2) + (2\,\mathrm{A} - \mathrm{B})\,\mathrm{S}\,(p_{22}\,p_{33} + q_{22}\,q_{33} + r_{22}\,r_{33})$$
$$+ \mathrm{B}\,(p_{23}^2 + q_{23}^2 + r_{23}^2),$$

$$\mathrm{S}\pi_1 = 2\,\mathrm{AS}\,(q_{11}\,r_{11} + r_{11}\,p_{11} + p_{11}\,q_{11}) + (2\,\mathrm{AB})\,\mathrm{S}\,(q_{22}\,r_{33} + q_{33}\,r_{22} + r_{22}\,p_{33}$$
$$+ r_{33}\,p_{22} + p_{22}\,q_{33} + p_{33}\,q_{22}) + 2\,\mathrm{BS}\,(q_{23}\,r_{23} + r_{23}\,p_{23} + p_{23}\,q_{23}).$$

Sostituendo queste espressioni in (25) si trova W espresso per le γ, p, q, r. Ora per l'equilibrio bisogna determinare γ, p, q, r in modo che W sia minimo. Le γ, p, q, r son però soggette anche alla condizione che le curvature dello spazio abbiano il valore che debbono avere. Ora da (23) (24) risulta subito

$$(26)\quad (23\,,23) = -r_{22} - q_{33}\;;\quad (31\,,31) = -p_{33} - r_{11}\;;\quad (12\,,12) = -q_{11} - p_{22}.$$

La determinazione delle γ, p, q, r si può fare ora col metodo dei coefficienti indeterminati a_1, a_2, a_3, perchè, dato che le β_{ik} sono piccolissime, si può intendere che (23 , 23) (31 , 31) (12 , 12) siano le tre curvature principali. Si trova, con un calcolo un po' laborioso ma privo di ogni difficoltà

$$(27)\qquad \delta\left(\frac{\mathrm{W}}{4\,\pi\,\mathrm{R}^3} - \mathrm{S}a_1\,[(23\,,23) + r_{22} + q_{33}]\right)$$

$$= \mathrm{S}\,\delta\gamma_{11}\left[\frac{2\,\mathrm{A}}{3}\,\gamma_{11} + \frac{2\,\mathrm{A}-\mathrm{B}}{3}\,(\gamma_{22}+\gamma_{33}) + \frac{2\,\mathrm{AR}^2}{15}\,(p_{11}+q_{11}+r_{11})\right.$$
$$\left. + \frac{(2\,\mathrm{A}-\mathrm{B})\,\mathrm{R}^2}{15}\,(p_{33}+q_{33}+r_{33}+p_{22}+q_{22}+r_{22})\right]$$

$$+ \mathrm{S}\,\delta p_{11}\left[\frac{2\,\mathrm{AR}^2}{15}\,\gamma_{11} + \frac{(2\,\mathrm{A}-\mathrm{B})\,\mathrm{R}^2}{15}\,(\gamma_{22}+\gamma_{33}) + \frac{2\,\mathrm{AR}^4}{105}\,(3\,p_{11}+r_{11}+q_{11})\right.$$
$$\left. + \frac{(2\,\mathrm{A}-\mathrm{B})\,\mathrm{R}^4}{105}\,(3\,p_{33}+r_{33}+q_{33}+3\,p_{22}+r_{22}+q_{22})\right]$$

$$+ \mathrm{S}\,\delta q_{11}\left[\frac{2\,\mathrm{AR}^2}{15}\,\gamma_{11} + \frac{(2\,\mathrm{A}-\mathrm{B})\,\mathrm{R}^2}{15}\,(\gamma_{22}+\gamma_{33}) + \frac{2\,\mathrm{AR}^4}{105}\,(3\,q_{11}+r_{11}+p_{11})\right.$$
$$\left. + \frac{(2\,\mathrm{A}-\mathrm{B})\,\mathrm{R}^4}{105}\,(3\,q_{33}+r_{33}+p_{33}+3\,q_{22}+r_{22}+p_{22}) - a_3\right]$$

$$+ \mathrm{S}\,\delta r_{11}\left[\frac{2\,\mathrm{AR}^2}{15}\,\gamma_{11} + \frac{(2\,\mathrm{A}-\mathrm{B})\,\mathrm{R}^2}{15}\,(\gamma_{22}+\gamma_{33}) + \frac{2\,\mathrm{AR}^4}{105}\,(3\,r_{11}+p_{11}+q_{11})\right.$$
$$\left. + \frac{(2\,\mathrm{A}-\mathrm{B})\,\mathrm{R}^4}{105}\,(3\,r_{33}+p_{33}+q_{33}+3\,r_{22}+p_{22}+q_{22}) - a_2\right]$$

$$+ \frac{2\,\mathrm{BR}^4}{105}\,\mathrm{S}\,[\delta p_{23}\,(3\,p_{23}+q_{23}+r_{23}) + \delta q_{23}\,(3\,q_{23}+r_{23}+p_{23})$$
$$+ \delta r_{23}\,(3\,r_{23}+p_{23}+q_{23})] = 0.$$

Dobbiamo ora annullare tutti i coefficienti dei $\delta\gamma$, δp, δq, δr; e alle equazioni, così ottenute, aggiungere le (26). Otteniamo così 24 equazioni lineari ed omogenee tra le 24 incognite γ, p, q, r, a.

Dalle 3 che si ottengono annullando i coefficienti di δp_{23}, δq_{23}, δr_{23} si ricava intanto $p_{23} = q_{23} = r_{23} = 0$, da cui $\beta_{23} = 0$ ed analogamente $\beta_{31} = \beta_{12} = 0$.

Per determinar le rimanenti 15 incognite, poniamo

$$(28)\qquad \left\{\begin{array}{l} \dfrac{5}{R^2} S_1 = p_{11} + q_{11} + r_{11} \\ 2 A\gamma_{11} + (2A - B)(\gamma_{22} + \gamma_{33}) = C_1 \\ 3 p_{11} + r_{11} + q_{11} = P_1 \dfrac{7}{R^2} \\ \alpha_1 = a_1 \dfrac{15}{R^2} . \end{array}\right.$$

Le equazioni che si ottengono annullando le rimanenti variazioni sono

$$(29)\qquad \left\{\begin{array}{l} C_1 + 2AS_1 + (2A - B)(S_2 + S_3) = 0 \\ C_1 + 2AP_1 + (2A - B)(P_2 + P_3) = 0 \\ C_1 + 2AQ_1 + (2A - B)(Q_2 + Q_3) = \alpha_3 \\ C_1 + 2AR_1 + (2A - B)(R_2 + R_3) = \alpha_2 \end{array}\right.$$

e le analoghe. Per determinare da queste le S, P, Q, R, in funzione delle C, α, si trovano sempre sistemi del tipo

$$\begin{array}{l} 2Ax + (2A - B)(y + z) = X \\ 2Ay + (2A - B)(z + x) = Y \\ 2Az + (2A - B)(x + y) = Z \end{array}$$

che ha per soluzione

$$\begin{array}{l} x = MX + N(Y + Z) \\ y = MY + N(Z + X) \\ z = MZ + N(X + Y) \end{array}$$

avendo posto

$$(30)\qquad M = \frac{4A - B}{6AB - 2B^2} \quad ; \quad N = \frac{-2A + B}{6AB - 2B^2} .$$

Si ottiene così

$$(31)\qquad \left\{\begin{array}{l} -S_1 = MC_1 + N(C_2 + C_3) \\ -P_1 = MC_1 + N(C_2 + C_3 - \alpha_2 - \alpha_3) \\ -Q_1 = M(C_1 - \alpha_3) + N(C_2 + C_3 - \alpha_1) \\ -R_1 = M(C_1 - \alpha_2) + N(C_2 + C_3 - \alpha_1) \end{array}\right.$$

e analoghe.

D'altra parte, dalle posizioni (28) risulta subito $5 S_1 = 7(P_1 + Q_1 + R_1)$ e quindi (31)

$$(32)\qquad MC_1 + N(C_2 + C_3) = \frac{7}{8} N\alpha_1 + \frac{7}{16}(M + N)(\alpha_2 + \alpha_3)$$

e, sostituendo nelle (31) queste espressioni, si hanno le S, P, Q, R, espresse per le α. D'altra parte, dalle (28) si ha

$$r_{22} = \frac{1}{2R^2}(7R_2 - 5S_2)$$

e le 8 analoghe. Mettendo, poi al posto delle α_i, $\frac{15}{R^2} a_i$, troviamo infine

$$(33) \quad \begin{cases} p_{11} = \frac{105}{16 R^4} [-2 N a_1 + (7 N - M)(a_2 + a_3)] \\ q_{11} = \frac{105}{16 R^4} [6 N a_1 - (M + N) a_2 + (7 M - N) a_3] \\ r_{11} = \frac{105}{16 R^4} [6 N a_1 + (7 M - N) a_2 - (M + N) a_3] \end{cases}$$

e analoghe.

Sostituendo questi valori nelle (26). troviamo

$$(34) \quad \begin{cases} -\frac{16 R^4}{105} (23, 23) = 2(7 M - N) a_1 + (5 N - M)(a_2 + a_3) \\ -\frac{16 R^4}{105} (31, 31) = 2(7 M - N) a_2 + (5 N - M)(a_3 + a_1) \\ -\frac{16 R^4}{105} (12, 12) = 2(7 M - N) a_3 + (5 N - M)(a_1 + a_2). \end{cases}$$

Da queste tre equazioni si possono ricavare a_1, a_2, a_3, e, sostituendo i valori così ottenuti in (33) si otterrebbero le incognite cercate. Esse poi dovrebbero sostituirsi nell'espressione (25) di W, e così si otterrebbe l'energia elastica cercata. Possiamo abbreviare questo calcolo osservando che W è funzione omogenea di secondo grado delle γ, p, q, r, e che quindi, per il teorema di Eulero, si ha

$$2 W = \Sigma \gamma \frac{\partial W}{\partial \gamma} + \Sigma p \frac{\partial W}{\partial p} + \Sigma q \frac{\partial W}{\partial q} + \Sigma r \frac{\partial W}{\partial r}.$$

Ma dalla (27) risulta che tutte le derivate parziali di W sono zero, tranne quelle rapporto a $q_{11}, r_{11}, r_{22}, p_{22}, p_{33}, q_{33}$ che, a meno del fattore $4 \pi R^3$, sono eguali ad $a_3, a_2, a_1, a_3, a_2, a_1$; abbiamo dunque

$$W = 2 \pi R^3 (a_3 q_{11} + a_2 r_{11} + a_1 r_{22} + a_3 p_{22} + a_2 p_{33} + a_1 q_{33})$$

e, introducendo i valori (33)

$$W = \frac{105 \pi}{4 R} [(5 N - M) S a_2 a_3 + (7 M - N) S a_1^2].$$

Basta ora ricavare dalle (34) i valori delle a e sostituirli qui dentro, per trovare successivamente

$$a_1 = \frac{4 R^4}{105 (15 M - 7 N)(3 M + 2 N)} [-(13 M + 3 N)(23, 23) + (5 N - M)\{(31, 31) + (12, 12)\}],$$

$$S a_1^2 = \frac{16 R^8}{105^2 (15 M - 7 N)^2 (3 M + 2 N)^2} [(171 M^2 + 58 MN + 59 N^2) S (23, 23)^2 + 2(27 M^2 - 134 MN + 31 N^2) S (31, 31)(12, 12)],$$

$$Sa_2 a_3 = \frac{16\,R^8}{105^2\,(15\,M - 7\,N)^2\,(3\,M + 2\,N)^2}\,[(27\,M^2 - 134\,MN + 31\,N^2)\,S\,(23\,,\,23)^2$$

$$+ (198\,M^2 - 76\,MN + 54\,N^2)\,S\,(31\,,\,31)\,(12\,,\,12)],$$

$$W = \frac{4\,\pi R^7}{105\,(15\,M - 7\,N)^2\,(3\,M + 2\,N)^2}\,[(1170\,M^3 + 504\,M^2 N - 346\,MN^2 + 96\,N^3)$$

$$\cdot S\,(23\,,\,23)^2 + (180\,M^3 - 864\,M^2 N + 268\,MN^2 + 269\,N^3)\,S\,(31\,,\,31)\,(12\,,\,12)]\,,$$

Ci conviene trasformare ancora un momento questa espressione in modo da farvi figurare elementi invarianti.

Introduciamo perciò il tensore controvariante α^{ik} di Ricci, in luogo dei simboli di Riemann; osservando che nella (23) le β_{ik} sono piccolissime, si vede che praticamente si avrà

$$S\,(23\,,\,23)^2 = \sum_{ik} \alpha^{ik}\,\alpha_{ik}\,,$$

$$S\,(31\,,\,31)\,(12\,,\,12) = \frac{1}{2}\left[\left(\sum_i \alpha_i^i\right)^2 - S\,\alpha^{ik}\,\alpha_{ik}\right]\,.$$

Troviamo così, infine

$$W = R^7\left[\mu \sum_{ik} \alpha^{ik}\,\alpha_{ik} + \nu\left(\sum_i \alpha_i^i\right)^2\right] \tag{35}$$

dove si è posto

$$\left\{\begin{aligned} \mu &= \frac{2\,\pi\,(2160\,M^3 + 1872\,M^2 N - 960\,MN^2 - 77\,N^3)}{105\,(15\,M - 7\,N)^2\,(3\,M + 2\,N)^2} \\ \nu &= \frac{2\,\pi\,(180\,M^3 - 864\,M^2 N + 268\,MN^2 + 269\,N^3)}{105\,(15\,M - 7\,N)^2\,(3\,M + 2\,N)^2} \end{aligned}\right. . \tag{36}$$

Se, come in molti metalli, fosse $\sigma = 1/3$. si troverebbe

$$\mu = \pi E\,\frac{38669}{9274720} \qquad \nu = \pi E\,\frac{13171}{9274720}\,.$$

Per calcolare il peso, possiamo ora procedere in modo dal tutto analogo a quello seguito dall'esempio precedente. Con le notazioni di allora si trova che, vettorialmente, il peso è espresso da

$$\mathbf{G}\,(m + W) + R^7\left[\mu\,\mathrm{grad}\left(\sum_{ik} \alpha^{ik}\,\alpha_{ik}\right) + \nu\,\mathrm{grad}\left\{\left(\sum_i \alpha_i^i\right)^2\right\}\right]\,.$$

Il secondo termine rappresenta quella parte del peso che non ha la direzione della gravità.

N° 9.

It is not known how Fermi came to consider this problem. Many years later (around 1950) he gave this problem in an open book doctoral examination at the University of Chicago. Remarkably a student thought that this was probably a problem treated by Fermi in his early years and succeeded in finding it in the library. Rumor has it that he was the only one who gave the correct solution.

E. SEGRÈ

9.

SUL TRASCINAMENTO DEL PIANO DI POLARIZZAZIONE DA PARTE DI UN MEZZO ROTANTE

« Rend. Lincei », *32* (1), 115–118 (1923) (*).

1. Le esperienze di Fizeau, e le numerose teorie relativistiche e non relativistiche che ne son state fatte, hanno dimostrato che la luce che si propaga in un mezzo in moto è trascinata nel senso di questo non completamente, ma con un cofficiente di trascinamento eguale a $\frac{n^2-1}{n^2}$, essendo n l'indice di rifrazione del mezzo. Mi sono proposto lo studio di un problema analogo. Supponiamo che un mezzo trasparente ruoti attorno ad un asse e che della luce polarizzata rettilineamente si propaghi attraverso al mezzo parallelamente all'asse; se il piano di polarizzazione fosse trascinato completamente dalla rotazione del mezzo, esso dovrebbe evidentemente presentare una specie di potere rotatorio di grandezza $\omega n/c$, essendo ω la velocità angolare e c la velocità della luce, poichè a percorrere la lunghezza 1 la luce impiega il tempo n/c, e in questo tempo il mezzo ruota appunto dell'angolo $\omega n/c$.

Troveremo invece [1] che le cose vanno diversamente, e che anche in questo caso bisogna tener conto di un coefficiente di trascinamento, che si

(*) Presentata dal Corrispondente L. Puccianti nella Seduta del 4 febbraio 1923.

(1) J. J. Thomson ha studiato lo stesso problema (« Proc. Camb. Phil. Soc. », 1885), trovando un coefficiente di trascinamento eguale ad 1; ma la sua conclusione è errata, perchè in essa si considerano i campi elettrico e magnetico della luce come trasversali anche nel mezzo in moto, ciò che è in evidente contrasto con la formula (4) del suo lavoro. Un altra causa di errore nel lavoro di Thomson consiste nel porre $f=\frac{K}{4\pi}P$; essendo P, Q, R, conformemente alle (4) (5) (6), calcolate come somma della forza elettrica P′ e della forza elettromagnetica P″ agente sul corpo per effetto del suo moto; poichè quest'ultima agisce solo sopra lo spostamento elettrico del corpo e non su quello dell'etere. La formula corretta sarebbe $f=\frac{KP'}{4\pi}+\frac{K-1}{4\pi}P''$.

Il suo risultato è del resto evidentemente in contraddizione col fatto intuitivo che il coefficiente deve certo essere nullo, per corpi di indice di rifrazione = 1.

trova ancora eguale precisamente a quello di Fizeau, per modo che come potere rotatorio, invece che $\omega n/c$, si trova $\frac{n^2 - 1}{n^2} \frac{\omega n}{c}$.

2. Consideriamo un dielettrico isotropo di polarizzabilità elettrica k, per modo che, essendo n l'indice di rifrazione, sia

$$n^2 = 1 + 4\pi k . \tag{1}$$

Supponiamo poi che il dielettrico ruoti attorno all'asse x con velocità angolare ω. Essendo $\mathbf{E}$ ed $\mathbf{H}$ le forze elettrica e magnetica, le equazioni di Maxwell (2) si scriveranno per esso:

$$\text{rot}\,\mathbf{E} = -\frac{1}{c}\frac{\partial \mathbf{H}}{\partial t} \quad ; \quad \text{rot}\,\mathbf{H} = \frac{1}{c}\left(\frac{\partial \mathbf{E}}{\partial t} + 4\pi \boldsymbol{j}\right), \tag{2}$$

dove $\boldsymbol{j}$ rappresenta la densità di corrente (u.e.s.). Indicando con $\mathbf{S}$ la polarizzazione, si vede subito che $\boldsymbol{j}$ è la somma della corrente dovuta alla variazione della $\mathbf{S}$ col tempo $d\mathbf{S}/dt$, e della corrente dovuta alla convezione delle cariche elettriche libere $= -\mathbf{V}\,\text{div}\,\mathbf{S}$, se con $\mathbf{V}$ si indica la velocità del punto generico del dielettrico. Si ha dunque

$$\boldsymbol{j} = \frac{\partial \mathbf{S}}{\partial t} - \mathbf{V}\,\text{div}\,\mathbf{S} . \tag{3}$$

D'altra parte, $\mathbf{S}$ sarà dato (3) da

$$\mathbf{S} = k\left(\mathbf{E} + \frac{1}{c}\mathbf{V} \times \mathbf{H}\right). \tag{4}$$

3. Abbiamo così raccolto tutti gli elementi per lo svolgimento della nostra teoria; per semplificare un pò le cose limiteremo le nostre considerazioni ai punti posti assai vicino all'asse di rotazione x, e trascureremo i quadrati (4) di ω.

Cerchiamo se si possono soddisfare le (2) con le posizioni

$$E_x = E_x(x, y, z, t) \quad ; \quad E_y = E_y(x, t) \quad ; \quad E_z = E_z(x, t) ;$$
$$H_x = 0 \quad ; \quad H_y = H_y(x, t) \quad ; \quad H_z = H_z(x, t) ,$$

introducendo cioè, oltre alle ordinarie componenti trasversali dei vettori elettrico e magnetico, anche una componente longitudinale del vettore elettrico. Tenendo presente che le componenti di $\mathbf{V}$ sono $0, -\omega z, \omega y$, si ha dalla (4)

$$S_x = k\left\{E_x - \frac{\omega}{c}(y H_y + z H_z)\right\} \quad ; \quad S_y = k E_y \quad ; \quad S_z = k E_z$$

$$\text{div}\,\mathbf{S} = k\left\{\frac{\partial E_x}{\partial x} - \frac{\omega}{c}\left(y\frac{\partial H_y}{\partial x} + z\frac{\partial H_z}{\partial x}\right)\right\}$$

(2) Abbiamo scritto le equazioni di Maxwell senza metter esplicitamente in evidenza la costante dielettrica, perché di essa è tenuto conto nel termine $4\pi \boldsymbol{j}$.

(3) Ciò equivale manifestamente a far la teoria trascurando la dispersione, ponendo cioè l'indice di rifrazione per tutte le lunghezze d'onda eguale alla radice quadrata della costante dielettica.

(4) Propriamente il numero di cui si trascurano i quadrati è $\omega r/c$, essendo r la massima distanza, dall'asse, dei punti in considerazione.

e quindi per la (3), trascurando i termini in ω^2,

$$j_x = k \left\{ \frac{\partial E_x}{\partial t} - \frac{\omega}{c}\left(y \frac{\partial H_y}{\partial t} + z \frac{\partial H_z}{\partial t} \right) \right\} \quad ; \quad j_y = k \left\{ \frac{\partial E_y}{\partial t} + \omega z \frac{\partial E_x}{\partial x} \right\} ;$$

$$j_z = k \left\{ \frac{\partial E_z}{\partial t} - \omega y \frac{\partial E_z}{\partial x} \right\} .$$

Le (2), nella solita approssimazione, possono ora scriversi:

$$(5) \quad \left\{ \begin{aligned} \frac{\partial E_x}{\partial z} - \frac{\partial E_z}{\partial x} &= -\frac{1}{c}\frac{\partial H_y}{\partial t} \\ \frac{\partial E_y}{\partial x} - \frac{\partial E_x}{\partial y} &= -\frac{1}{c}\frac{\partial H_z}{\partial t} \end{aligned} \right. \qquad \left\{ \begin{aligned} n^2 \frac{\partial E_x}{\partial t} &= (n^2 - 1)\frac{\omega}{c}\left(y \frac{\partial H_y}{\partial t} + z \frac{\partial H_z}{\partial t} \right) \\ -\frac{\partial H_z}{\partial x} &= \frac{1}{c}\frac{\partial E_y}{\partial t} + \frac{4\pi k}{c}\left\{ \frac{\partial H_y}{\partial t} + \omega z \frac{\partial E_x}{\partial x} \right\} \\ \frac{\partial H_y}{\partial x} &= \frac{1}{c}\frac{\partial E_z}{\partial t} + \frac{4\pi k}{c}\left\{ \frac{\partial E_z}{\partial t} - \omega y \frac{\partial E_x}{\partial x} \right\} . \end{aligned} \right.$$

Dalla prima equazione del secondo gruppo si rileva che E_x contiene ω, e quindi le due ultime possono semplificarsi trascurando i termini in ω^2.

Tenendo presente (1), le (5) possono semplificarsi e scriversi:

$$(6) \quad \left\{ \begin{aligned} \frac{\partial E_x}{\partial z} - \frac{\partial E_z}{\partial x} &= -\frac{1}{c}\frac{\partial H_y}{\partial t} \\ \frac{\partial E_y}{\partial x} - \frac{\partial E_x}{\partial y} &= -\frac{1}{c}\frac{\partial H_z}{\partial t} \end{aligned} \right. \qquad \left\{ \begin{aligned} n^2 \frac{\partial E_x}{\partial t} &= (n^2 - 1)\frac{\omega}{c}\left(y \frac{\partial E_y}{\partial t} + z \frac{\partial H_z}{\partial t} \right) \\ -\frac{\partial H_z}{\partial x} &= \frac{n^2}{c}\frac{\partial E_y}{\partial t} \\ \frac{\partial H_y}{\partial x} &= \frac{n^2}{c}\frac{\partial E_z}{\partial t} . \end{aligned} \right.$$

Con facili operazioni di eliminazione si deduce da queste trascurando termini in ω^2:

$$(7) \qquad \frac{\partial^2 E_y}{\partial x^2} - \frac{n^2}{c^2}\frac{\partial^2 E_y}{\partial t^2} = \frac{(n^2 - 1)\,\omega}{c^2}\frac{\partial E_z}{\partial t} ;$$

$$\frac{\partial^2 E_z}{\partial x^2} - \frac{n^2}{c^2}\frac{\partial^2 E_z}{\partial t^2} = -\frac{(n^2 - 1)\,\omega}{c^2}\frac{\partial E_y}{\partial t} .$$

Proviamo a soddisfarle ponendo

$$E_y = \mathcal{E}_y\, e^{2\pi i \nu \left(t - \frac{n' x}{c} \right)} \quad ; \quad E_z = \mathcal{E}_z\, e^{2\pi i \nu \left(t - \frac{n' x}{c} \right)} ,$$

essendo $\mathcal{E}_y$, $\mathcal{E}_z$ costanti complesse: le (7) ci dànno

$$(7\ bis) \qquad 2\pi\nu\, \mathcal{E}_y (n^2 - n'^2) = (n^2 - 1)\,\omega i\, \mathcal{E}_z$$

$$2\pi\nu\, \mathcal{E}_z (n^2 - n'^2) = -(n^2 - 1)\,\omega i\, \mathcal{E}_y .$$

Per aver soluzioni non identicamente nulle, deve dunque essere:

$$4\pi^2 \nu^2 (n^2 - n'^2)^2 = (n^2 - 1)^2 \omega^2 ,$$

dalla quale si deduce

$$n'^2 = n^2 \pm \frac{(n^2 - 1)\,\omega}{2\pi\nu} ,$$

cioè, con la nostra solita approssimazione,

$$(8) \qquad n' = n \left(1 \pm \frac{(n^2 - 1)\,\omega}{4\pi\nu n^2} \right) .$$

I due valori n'_1 ed n'_2 che la (8) dà per n', sono i due indici di rifrazione delle onde polarizzate circolarmente nei due sensi, come subito si rileva dalle 7 (*bis*). Essi sono, come è noto, legati al potere rotatorio R dalla relazione $R = 2\pi(n'_1 - n'_2)\nu/c$. Tenendo dunque presenti i valori (8), si trova infine

$$R = \frac{n^2 - 1}{n^2}\,\frac{\omega n}{c}\,.$$

Il coefficiente di trascinamento del piano di polarizzazione è dunque, come si era annunziato, eguale a quello di Fizeau, e cioè

$$\frac{n^2 - 1}{n^2}\,.$$

N°. 10.

For the introduction to this paper see N°. 4.

10.

SULLA MASSA DELLA RADIAZIONE IN UNO SPAZIO VUOTO

E. FERMI e A. PONTREMOLI

« Rend. Lincei », *32* (1), 162–164 (1923) (*).

Recentemente uno di noi [1] ebbe a dimostrare, introducendo un più esatto concetto della rigidità, come l'elettrodinamica ordinaria consenta di giungere alla determinazione della massa di riposo di un elettrone non diversa da quella attribuitagli secondo la teoria della relatività, e che, come è noto, si ottiene semplicemente dividendo per il quadrato della velocità della luce l'energia del sistema.

Abbiamo osservato che una simile differenza, tra il valore determinato secondo l'elettrodinamica ordinaria o colla relatività, si presenta nel calcolo della massa della radiazione di uno spazio vuoto [2], e ci proponiamo di dimostrare che questo divario può eliminarsi mediante analoghe considerazioni.

Il procedimento sinora seguito per determinare colla elettrodinamica la massa della radiazione in una cavità consisteva anzitutto nel calcolare l'impulso elettromagnetico $\mathbf{G}_0$ per moti lenti e quasi stazionari, che risulta espresso [3], trascurando termini in v^2/c^2, da

$$\mathbf{G}_0 = \frac{4}{3} \frac{W_0}{c^2} \boldsymbol{v} ,$$

dove W_0 è l'energia della radiazione allo stato di quiete, $\boldsymbol{v}$ è la velocità attuale della cavità, c è la velocità della luce. Da questo si deduceva che la reazione di inerzia è

$$-\frac{d\mathbf{G}_0}{dt} = -\frac{4}{3} \frac{W_0}{c^2} \boldsymbol{\Gamma} .$$

dove $\boldsymbol{\Gamma}$ è l'accelerazione; donde una massa apparente della radiazione eguale a $\frac{4}{3}\frac{W_0}{c^2}$, mentre, secondo la teoria della relatività, essa dovrebbe essere semplicemente W_0/c^2.

(*) Presentata dal Socio O. M. Corbino nella Seduta del 18 febbraio 1923.

(1) E. FERMI, questi « Rendiconti », vol. XXXI, pp. 184 e 306 (1922), « Physikalische Zeit. », vol. XXIII (1922), p. 340.

(2) F. HASENÖHRL, « Ann. der Physik », vol. XV, p. 344 (1904), e vol. XVI, p. 589 (1905) K. VON MOSENGEIL, « Ann. der Physik », vol. XXII, p. 867 (1907): M. PLANCK, « Berlin. Sitzber. », p. 542 (1907); M. ABRAHAM, *Theorie der Elektrizität*, vol. II, p. 341 (1920).

(3) M. ABRAHAM, loc. cit., p. 345.

In questo procedimento è contenuta implicitamente l'asserzione che la forza esterna F sia eguale alla derivata dell'impulso elettromagnetico rispetto al tempo, cioè alla somma vettoriale delle forze elettromagnetiche $d\boldsymbol{\varphi}$ agenti sopra le singole parti del sistema; con ciò si viene a porre dunque:

$$\mathbf{F} = \int d\boldsymbol{\varphi}. \tag{1}$$

Ma ciò non è esatto, perchè, se si tiene conto della nozione di corpo rigido discussa da uno di noi nel lavoro citato, la forza esterna è data invece dalla relazione

$$\mathbf{F} = \int d\boldsymbol{\varphi}\left[1 + \frac{\mathbf{I}\cdot(\mathrm{P}-\mathrm{O})}{c^2}\right] \tag{2}$$

essendo (P — O) il vettore che ha termine nel punto P cui è applicata la forza $d\boldsymbol{\varphi}$ ed origine in un punto fisso O, che possiamo prendere per centro delle coordinate, interno al sistema.

Ora, $d\boldsymbol{\varphi}$ è la risultante della forza $d\boldsymbol{\varphi}_1$, esercitata dalla pressione di radiazione che sussisterebbe se la cavità fosse ferma, e di una forza $d\boldsymbol{\varphi}_2$, causata dalle perturbazioni di detta pressione, dovute al moto della cavità. Applicando la formula (1), si trova che, essendo evidentemente $\int d\boldsymbol{\varphi}_1 = 0$, perchè $d\boldsymbol{\varphi}_1$ è la forza esercitata da una pressione omogenea sopra una superficie chiusa, la forza esterna è

$$\mathbf{F} = \int d\boldsymbol{\varphi}_2. \tag{3}$$

Questa forza è precisamente quella calcolata come reazione d'inerzia dagli autori citati, onde

$$\int d\boldsymbol{\varphi}_2 = -\frac{4}{3}\frac{\mathrm{W}_0}{c^2}\boldsymbol{\Gamma}. \tag{4}$$

Invece per la formula (2), tenendo ancora presente che $\int d\boldsymbol{\varphi}_1 = 0$, si trova

$$\mathrm{F} = \int (d\boldsymbol{\varphi}_1 + d\boldsymbol{\varphi}_2)\left[1 + \frac{\boldsymbol{\Gamma}\cdot(\mathrm{P}-\mathrm{O})}{c^2}\right] = \int d\boldsymbol{\varphi}_1 \frac{\boldsymbol{\Gamma}\cdot(\mathrm{P}-\mathrm{O})}{c^2} + \int d\boldsymbol{\varphi}_2 + \int d\boldsymbol{\varphi}_2 \frac{\boldsymbol{\Gamma}\cdot(\mathrm{P}-\mathrm{O})}{c^2}.$$

Trascurando i termini in Γ^2 ed osservando che $d\boldsymbol{\varphi}_2$ è proporzionale a Γ, si può porre semplicemente

$$\mathbf{F} = \int d\boldsymbol{\varphi}_1 \frac{\boldsymbol{\Gamma}\cdot(\mathrm{P}-\mathrm{O})}{c^2} + \int d\boldsymbol{\varphi}_2. \tag{5}$$

La differenza tra (3) e (5) non è *a priori*, in questo caso, trascurabile, benchè contenga c^2 al denominatore, potendo $d\varphi_1/d\varphi_2$ divenire molto grande, come rapporto tra una forza e la sua perturbazione [4]. Infatti è $d\boldsymbol{\varphi}_1 = \boldsymbol{pn}\, d\sigma$, dove p è la pressione di radiazione eguale, come è noto, ad $\frac{1}{3}\frac{\mathrm{W}_0}{\mathrm{V}}$, V è il

(4) Nel caso delle masse elettromagnetiche si ha $d\boldsymbol{\varphi}$ eguale alla somma delle forze di Coulomb (che formano la parte preponderante) e delle forze dovute alla accelerazione. Per le prime vale evidentemente anche in questo caso la relazione $\int d\boldsymbol{\varphi}_1 = 0$; ancor esse quindi fanno sentire il loro effetto solo ove si applichi la formula (5) anziché la (3).

volume della cavità, $\boldsymbol{n}$ è un vettore unitario colla direzione della normale esterna relativa all'elemento $d\sigma$, di coordinate (x, y, z), della superficie della cavità.

La componente sull'asse x del primo integrale di (5) è dunque

$$\left[\int d\boldsymbol{\varphi}_1 \frac{\boldsymbol{\Gamma}\cdot(\mathrm{P}-\mathrm{O})}{c^2}\right]_x = \frac{\mathrm{W}_0}{3\,c^2\,\mathrm{V}}\int(\Gamma_x\, x + \Gamma_y\, y + \Gamma_z\, z)\cos\widehat{nx}\, d\sigma$$

$$= \frac{\mathrm{W}_0}{3\,c^2\,\mathrm{V}}\left[\Gamma_x\int x\cos\widehat{nx}\, d\sigma + \Gamma_y\int y\cos\widehat{nx}\, d\sigma + \Gamma_z\int z\cos\widehat{nx}\, d\sigma\right];$$

ma l'immediata applicazione del teorema di Gauss dimostra che

$$\int x\cos\widehat{nx}\, d\sigma = \mathrm{V} \quad , \quad \int y\cos\widehat{nx}\, d\sigma = \int z\cos\widehat{nx}\, d\sigma = 0 .$$

La nostra componente è perciò $\mathrm{W}_0\,\Gamma_x/3\,c^2$ e quindi

$$\int d\boldsymbol{\varphi}_1 \frac{\boldsymbol{\Gamma}\cdot(\mathrm{P}-\mathrm{O})}{c^2} = \frac{\mathrm{W}_0}{3\,c^2}\,\boldsymbol{\Gamma} .$$

Tenendo presente questa relazione e la (4), si vede che il rapporto tra gl'integrali del secondo membro della (5) è $-1/4$ e quindi effettivamente non trascurabile.

Sostituendo questi valori nella (5), si trova

$$\mathbf{F} = -\frac{\mathrm{W}_0}{c^2}\,\boldsymbol{\Gamma},$$

da cui la richiesta massa di quiete risulta, conformemente al principio di relatività, eguale a W_0/c^2.

N° 11 *a* and 11 *b*.

This paper is one of the group on analytical mechanics which Fermi completed during his stay in Göttingen. N° 11 *a* is a German translation of N° 11 *b*. The following information was kindly supplied by Dr. G. E. Uhlenbeck: Prof. P. Ehrenfest, who had delved deeply into the foundations of statistical mechanics, was impressed by the paper. He gave to Uhlenbeck, who was going to Rome for a while, a letter for Fermi with a number of questions and in this way Uhlenbeck met Fermi for the first time in the fall of 1924. They remained friends for life and the visits of Fermi to Ann Arbor were in part prompted by his friends Uhlenbeck and Goudsmit who were professors at the University of Michigan. Probably through this contact with Ehrenfest Fermi decided to go to Leyden for abouth three months.

The proof of the ergodic theorem given by Fermi is not considered rigorous from the mathematical point of view and it is difficult to make it rigorous. Some recent comments on the subject and bibliographical indications are given e.g. in C. L. SIEGEL, *Vorlesungen über Himmelsmechanik*, page 201, Springer, Berlin, 1956.

E. SEGRÈ.

11 *a*.

I. — BEWEIS, DASS EIN MECHANISCHES NORMALSYSTEM IM ALLGEMEINEN QUASI-ERGODISCH IST

« Phys. Zeits. », *24*, 261–265 (1923).

§ 1. Es sei ein mechanisches System gegeben; Γ sei sein Phasenraum, in dem die Koordinaten $q_1, q_2, \cdots, q_n$ und die zugehörigen Momente $p_1, p_2, \cdots, p_n$ durch einen Punkt dargestellt sind. Die von einem Punkt P_0 von Γ ausgehende Bahn des darstellenden Punktes wird völlig auf der durch P_0 gehende Energiefläche laufen. Wir wollen beweisen dass im allgemeinen diese Kurve unendlich nah an jeden Punkt der Energiefläche herankommt, d. h. dass unser System quasi–ergodisch [1] ist. Genauer gesagt, wollen wir folgendes beweisen.

Es seien auf einer Energiefläche E zwei beliebig kleine Gebiete gegeben, die, falls die Oberfläche aus mehreren Stücken besteht, auf demselben Stück liegen müssen, so dass es möglich ist, sie durch eine ganz auf E laufende Linie zu verbinden. Ich behaupte, dass es im allgemeinen Bahnen gibt, die durch beide Gebiete gehen.

(1) Bekanntlich würde das System ergodisch heissen, wenn die Bahn durch alle Punkte der Energiefläche gehen würde. Es sei hier erwähnt, dass Rosenthal [« Ann. d. Phys. », *42*, 179 (1913)] und Plancherel [« Ann. d. Phys. », *42*, 1061 (1913)] bewiesen haben, dass es keine ergodischen Systeme geben kann.

Um den Beweis durchzuführen, wollen wir zuerst einen Poincaréschen Satz verallgemeinern.

§ 2. Poincaré (2) hat nämlich bewiesen, dass ein kanonisches Normalsystem im allgemeinen, ausser dem Energieintegral F = konst. keine weiteren analytischen, eindeutigen, von der Zeit unabhängigen Integrale besitzen kann. Geometrisch bedeutet das, dass es im Phasenraum Γ, ausser den Energieflächen F = konst. keine andere Oberflächenfamilie gibt, mit der Eigenschaft, dass die Bahn des darstellenden Punktes immer auf derselben Oberfläche verläuft, auf der ihr Anfangspunkt liegt. Wir wollen diesen Satz verallgemeinern, indem wir beweisen werden, dass es ausser den F = konst. nicht nur keine weitere Oberflächenfamilie, sondern überhaupt keine weitere Oberfläche mit der Eigenschaft gibt, dass alle von einem ihrer Punkte ausgehende Bahnen immer auf der Oberfläche bleiben. Wir werden nämlich beweisen, dass, wenn eine analytische eindeutige Fläche diese Eigenschaft besitzt, sie im allgemeinen nur mit einer der F = konst. zusammenfallen kann, und zwar sobald $n > 2$ ist.

§ 3. Weil unser mechanisches System normal ist, können wir statt der $q_1, q_2, \cdots, q_n, p_1, p_2, \cdots, p_n$, neue Koordinaten $y_1, y_2, \cdots, y_n, x_1, x_2, \cdots, x_n$, einführen, so dass folgendes gelte:

1. Die Energie F ist von der Zeit unabhängig.

2. Ausser den Koordinaten x, y, enthält F auch einen Parameter μ, und ist nach Potenzen desselben für hinreichend kleine Werte von μ entwikkelbar, so dass:

$$F = F_0 + \mu F_1 + \mu^2 F_2 + \cdots \tag{1}$$

wo die F_i Funktionen der y und der x darstellen.

3. Die Lage des Systems, und daher auch F und die F_i, hängen nur periodisch (mit der Periode 2π) von den $x_1, x_2, \cdots, x_n$ ab.

4. F_0 ist von den x unabhängig.

Nehmen wir also an, dass die Oberfläche S, mit der Gleichung $\Phi(x, y, \mu) = 0$, die am Ende des § 2 besprochene Eigenschaft habe, und dass Φ analytisch, eindeutig, und in den x periodisch sei. Die Lage von S hängt von μ ab, was wir durch die Schreibweise S_μ bezeichnen werden. Wir können Φ nach Potenzen von μ entwickeln, und schreiben:

$$\Phi = \Phi_0 + \mu\Phi_1 + \mu^2 \Phi_2 + \cdots \tag{2}$$

wo die Φ_1 Funktionen der x und der y darstellen.

Wir bemerken aber, dass, wenn wir die Lage von S_μ für alle Werte von μ kennen, dadurch die Φ_i nicht völlig bestimmt sind. Und zwar, wenn wir S_μ für alle μ kennen, können wir z.B. y_1 folgendermassen darstellen:

$$y_1 = f_0 + \mu f_1 + \mu^2 f_2 + \cdots,$$

(2) H. POINCARÉ, *Méth. nouv. de la mécanique céleste*, Vol. 1, C. 5.

wo die f_i, bekannte Funktionen von $y_2, \cdots, y_n, x_1, x_2, \cdots, x_n$ sind. Setzen wir in Φ diesen Ausdruck statt y_1, so muss Φ identisch verschwinden. Wir können aber Φ nach Potenzen von μ entwickeln, und wir finden endlich:

$$\Phi_{y_1 = f_0 + \mu f_1 + \cdots} = [\Phi_0] + \mu \left| f_1 \frac{\partial \Phi_0}{\partial y_1} + \Phi_1 \right|$$

$$+ \mu^2 \left[\frac{f_1^2}{2} \frac{\partial^2 \Phi_0}{\partial y_1^2} + f_2 \frac{\partial \Phi_0}{\partial y_1} + f_1 \frac{\partial \Phi_1}{\partial y_1} + \Phi_2 \right] + \cdots = 0,$$

wo die Ausdrücke in eckiger Klammer für $y_1 = f_0$, d. h. auf der Oberfläche S_0 zu berechnen sind. Sämtliche Koeffizienten der Potenzen von μ müssen verschwinden; und so finden wir für die Φ_i folgende Bedingungen;

$$[\Phi_0] = 0,$$

$$[\Phi_1] = -f_1 \left[\frac{\partial \Phi_0}{\partial y_1} \right],$$

$$[\Phi_2] = -\frac{f_1^2}{2} \left[\frac{\partial^2 \Phi_0}{\partial y_1^2} \right] - f_2 \left[\frac{\partial \Phi_0}{\partial y_1} \right] - f_1 \left[\frac{\partial \Phi_1}{\partial y_1} \right].$$

. . . .

. . . .

Wegen der ersten dieser Bedingungen, muss Φ_0 auf S_0 verschwinden. Wählen wir Φ_0 ausserhalb S_0 ganz willkürlich, so wird dadurch Φ_1 auf S_0 durch die 2. Bedingung bestimmt, und bleibt ausserhalb S_0 ganz willkürlich. Nachdem wir Φ_1 usserhalb S_0 beliebig gewählt haben, wird Φ_2 auf S_0 durch die 3. Bedingung bestimmt usw.

Ich behaupte jetzt, dass man Φ_0 immer von den x unabhängig wählen kann. Dafür genügt es zu beweisen dass alle Punkte, die mit einem Punkte von S_0 die y–Koordinaten gemeinsam haben, ebenfalls zu S_0 gehören.

Ist nun P_0 ein Punkt von S_0 so muss S_0 die ganze Bahn des Systems für $\mu = 0$, die von P_0 ausgeht, enthalten. Weil aber F_0 die x nicht enthält, so haben alle Punkte dieser Bahn dieselben y–Werte wie P_0, während die x lineare Funktionen der Zeit sind. Im n dimensionalen Raum der x ist die Bahn also eine Gerade. Wegen der vorausgesetzten Periodizitätseigenschaften der x können wir aber sämtliche Punkte dieser Geraden in einen Würfel von der Seite 2π zurückführen. Dieser Würfel wird bekanntlich dicht mit Punkten erfüllt sein, wenn keine lineare ganzzahlige Beziehung zwischen

$$\frac{\partial F_0}{\partial y_1}, \frac{\partial F_0}{\partial y_2}, \cdots, \frac{\partial F_0}{\partial y_n}$$

besteht. Wir können nun immer annehmen, dass keine Beziehung wie:

$$m_1 \frac{\partial F_0}{\partial y_1} + m_2 \frac{\partial F_0}{\partial y_2} +, \cdots, + m_n \frac{\partial F_0}{\partial y_n} = 0 \tag{3}$$

identisch besteht, d. h. dass das ungestörte System nicht entartet ist. Würde nun Gl. (3) identisch auf S_0 gelten, so würde (3) eine Folge von $\Phi_0 = 0$ sein, und wir könnten daher, wegen der Willkürlichkeit von Φ_0, immer

$$\Phi_0 = \sum_i m_i \frac{\partial F_0}{\partial y_i}$$

setzen, woraus, wie behauptet, folgen würde, dass Φ_0 die x nicht enthält.

Ist keine Gleichung wie (3) auf S_0 identisch erfüllt, so ist im allgemeinen der Würfel dicht mit Punkten erfüllt, die zu S_0 gehören, und er muss daher in S_0 enthalten sein; also ist auch in diesem Falle der Beweis geliefert.

§ 4. Die Eigenschaft von S_μ lässt sich analytisch dadurch ausdrücken, dass die Gleichung (3)

$$[F, \Phi] = 0 \tag{4}$$

eine Folge der anderen

$$\Phi = 0 \tag{5}$$

ist. Wir können daher aus Gl. (5) z. B. y_1 als Funktion von $y_2, \cdots, y_n$, $x_1, x_2, \cdots, x_n, \mu$ finden, den gefundenen Wert in (4) setzen, und nach Potenzen von μ entwickeln. Dann müssen die Koeffizienten sämtlicher Potenzen von μ identisch verschwinden. Nun ist aber identisch $[F_0, \Phi_0] = 0$, weil F_0, so wie Φ_0, die x nicht enthält. Das erste Glied der Entwicklung von (4) ist daher $\mu([F_0, \Phi_1] + [F_1, \Phi_0])$, berechnet auf S_0. Auf S_0 muss also sein:

$$[F_0, \Phi_1] + [F_1, \Phi_0] = 0 . \tag{6}$$

Aus den Periodizitätseigenschaften von F_1 und Φ_1 geht hervor, dass wir

$$\Phi_1 = \sum_m A_m e^{\Theta_m} \quad ; \quad F_1 = \sum_m B_m e^{\Theta_m}$$

setzen können, wo A_m und B_m Funktionen der y darstellen, Θ_m eine Abkürzung für

$$i(m_1 x_1 + m_2 x_2 + \cdots + m_n x_n)$$

ist, und m statt der n Indizes $(m_1, m_2, \cdots, m_n)$ geschrieben ist. Gl. (6) wird jetzt, nach einer einfachen Rechnung:

$$\sum_m e^{\Theta_m}\left(A_m \sum_i m_i \frac{\partial F_0}{\partial y_i} - B_m \sum_i m_i \frac{\partial \Phi_0}{\partial y_i}\right) = 0 .$$

Diese Gleichung soll gelten, wenn wir z. B. statt y_1, den aus der Gleichung

$$\Phi_0(y_1, y_2, \cdots, y_n) = 0$$

gefundenen Wert setzen. Auf S_0 müssen also sämtliche Koeffizienten von e^{Θ_m} verschwinden, d. h.:

$$A_m \sum_i m_i \frac{\partial F_0}{\partial y_i} = B_m \sum_i m_i \frac{\partial \Phi_0}{\partial y_i} ; \tag{7}$$

hieraus folgt, dass, wenn in einem Punkte von S_0 $\Sigma m_i \frac{\partial F_0}{\partial y_i} = 0$ ist, auch $\Sigma m_i \frac{\partial \Phi_0}{\partial y_i} = 0$ sein muss, weil im allgemeinen (4) $B_m \neq 0$ ist.

(3) Für die Eigenschaften des hier benutzten Symbols $[F, \Phi]$ siehe Poincaré, l. c., $[F, \Phi]$ ist eine Abkürzung für den Ausdruck:

$$\sum_i \left(\frac{\partial F}{\partial y_i}\frac{\partial \Phi}{\partial x_i} - \frac{\partial F}{\partial x_i}\frac{\partial \Phi}{\partial y_i}\right).$$

(4) Es könnte wohl geschehen, dass auf S_0 eine der $B_m = 0$ wäre. Das würde aber, wie leicht ersichtlich, das Folgende nicht beeinflussen. Über diesen Punkt siehe auch Poincaré, l. c.

Jetzt müssen wir unterscheiden, ob die Anzahl der Freiheitsgrade = oder > 2 ist. Ist $n = 2$, so kann es wohl eintreten, dass S_0 die Gleichung $\frac{\omega_2}{\omega_1} - \alpha = 0$ habe, wo wir $\omega_i = \frac{\partial F_0}{\partial x_i}$ gesetzt haben und α eine beliebige irrationale Zahl darstellt. Dann ist es nicht möglich, zwei ganze Zahlen m_1 und m_2 zu finden, so dass $m_1 \omega_1 + m_2 \omega_2 = 0$, in einem Punkte von S_0 erfüllt sei; dann können wir auf S_0 setzen:

$$A_m = \frac{B_m}{m_1 \omega_1 + m_2 \omega_2} \sum_i m_i \frac{\partial \Phi_0}{\partial y_i},$$

weil der Nenner immer $\neq 0$ ist; wählen wir die A_m ausserhalb S_0 willkürlich, so ist Φ_1 bestimmt; und es lässt sich leicht einsehen, dass ein ähnliches Verfahren mindestens formal zur Bestimmung von $\Phi_2, \Phi_3, \cdots$ führen würde. Es bliebe noch die Frage der Konvergenz der so erhaltenen Reihen. Es ist aber nicht ausgeschlossen, dass für hinreichend schnell mit wachsenden m verschwindende B_m, und eventuell für besondere Werte von α die Reihen wirklich konvergieren. Hieraus sieht man, dass es, im Fall $n = 2$ im allgemeinen nicht ausgeschlossen ist, dass ausser den Energieflächen noch weitere, eventuell unendlich viele, Oberflächen gebe, mit der am Ende des § 2 besprochenen Eigenschaft.

§ 5. Anders steht es im Falle $n > 2$. Wir wollen z. B. $n = 3$ voraussetzen; unsere Schlussweise ist aber auch für $n = 4, 5, \cdots$ gültig.

Von den drei Verhältnissen $\frac{\omega_2}{\omega_1}, \frac{\omega_3}{\omega_1}, \frac{\omega_3}{\omega_2}$ kann im allgemeinen höchstens eines auf S_0 konstant sein, so dass wir immer annehmen dürfen, dass z. B. $\frac{\omega_2}{\omega_1}$ und $\frac{\omega_3}{\omega_1}$ auf S_0 nicht konstant sind. Dann ist S_0 dicht mit Punkten besetzt, für die z. B. $\frac{\omega_2}{\omega_1}$ rationale Werte hat; in irgendeinem solcher Punkte kann man zwei ganze Zahlen finden, für die $m_1 \omega_1 + m_2 \omega_2 = 0$. Dann ist in demselben Punkte auch $m_1 \frac{\partial \Phi_0}{\partial y_1} + m_2 \frac{\partial \Phi_0}{\partial y_2} = 0$; und daher:

$$\frac{\omega_2}{\omega_1} = \frac{\partial \Phi_0}{\partial y_2} \Big/ \frac{\partial \Phi_0}{\partial y_1}.$$

Da nun diese Gleichung in unendlich vielen, dicht auf S_0 verteilten Punkten gilt, so ist sie gewiss auf S_0 identisch befriedigt. In gleicher Weise beweist man, dass auch die Gleichung

$$\omega_3 : \omega_1 = \frac{\partial \Phi_0}{\partial y_3} : \frac{\partial \Phi_0}{\partial y_1}$$

auf S_0 identisch gilt. Also sind auf S_0 die partiellen Ableitungen von Φ_0 nach den y_i, denen von F_0 proportional. Das kann aber nur dann geschehen, wenn $\Phi_0 = 0$ eine der $F_0 =$ konst., sagen wir $F_0 = -c_0$, ist. Daher können wir, wegen der Willkürlichkeit von Φ_0 ausserhalb S_0, im ganzen Raum $\Phi_0 = F_0 + c_0$ setzen. Jetzt, da S_0 so bestimmt ist, kann im allgemeinen kein Ausdruck wie $\Sigma m_i \frac{\partial F_0}{\partial x_i}$ auf S_0 identisch verschwinden, wenn nur nicht $m_1 = m_2 = \cdots = m_n = 0$ ist. Auf S_0 können wir daher Gl. (7) durch $\Sigma m_i \frac{\partial F_0}{\partial y_i} = \Sigma m_i \frac{\partial \Phi_0}{\partial y_i}$

dividieren, und wir finden $A_m = B_m$, wenn nicht alle $m_i = 0$ sind. Hieraus folgt, dass auf S_0 die Differenz $\Phi_1 - F_1$ von den $x_1, x_2, \cdots, x_n$ nicht abhängen kann; und weil wir Φ_1 ausserhalb S_0 beliebig wählen können, so können wir im ganzen Raume setzen:

$$\Phi_1(x, y) = F_1(x, y) + f_1(y_1, \cdots, y_n) \quad , \quad \Phi_0(y) = F_0(y) + c_0 . \tag{8}$$

§ 6. Wir wollen nun annehmen, für ein gewisses r bewiesen zu haben, dass im ganzen Raum:

$$\left\{\begin{array}{lll} \Phi_r & = F_r & + f_r(y_1, \cdots, y_n), \\ \Phi_{r-1} & = F_{r-1} & + c_{r-1}, \\ \multicolumn{3}{c}{\cdots\cdots\cdots\cdots\cdots\cdots\cdots\cdots} \\ \Phi_1 & = F_1 & + c_1, \\ \Phi_0 & = F_0 & + c_0, \end{array}\right. \tag{9}$$

ich behaupte, dass dann ähnliche Gleichungen auch für $r + 1$ gelten. Da nun für $r = 1$ die (9) in die (8) übergeben, so wird es damit bewiesen sein, dass für irgendwelches r immer $\Phi_r - F_r =$ konst ist. Daher kann die S_μ für irgendwelches μ nur mit einer Energiefläche zusammenfallen.

Aus den (9) folgt, dass die ersten $r + 1$ Glieder der Entwicklung von (4) identisch verschwinden, und der Koeffizient von μ^{r+1} wird einfach:

$$[F_0, \Phi_{r+1}] + [F_1, f_r] + [F_{r+1}, F_0] = [F_0, \Phi_{r+1} - F_{r+1}] - [f_r, F_1] . \tag{10}$$

Dieser Ausdruck muss auf S_0 verschwinden. Behandeln wir nun diese Gleichung genau so wie Gl. (6), so finden wir, dass auf S_0 die partiellen Ableitungen von f_r, denen von F_0 proportional sind; da nun F_0 auf S_0 konstant ist, so muss auch f_r es sein; weil aber f_r ausserhalb S_0 beliebig ist, so können wir im ganzen Raum f_r einer Konstante c_r gleich setzen.

Jetzt wird (10) einfach $[F_0, \Phi_{r+1} - F_{r+1}] = 0$; setzen wir:

$$\Phi_{r+1} - F_{r+1} = \Sigma_m C_m e^{\Theta_m},$$

so wird es auf S_0:

$$\sum_m e^{\Theta_m} C_m \sum_i m_i \frac{\partial F_0}{\partial y_i} = 0,$$

d. h. es müssen sämtliche Koeffizienten von e^{Θ_m} verschwinden. $\sum_i m_i \frac{\partial F_0}{\partial y_i}$ kann im allgemeinen nur dann auf S_0 identisch verschwinden, wenn $m_1 = m_2 = \cdots = m_n = 0$; so dass nur $C_{0,0,\ldots,0} \neq 0$ sein kann. Setzen wir jetzt;

$$C_{0,0,\ldots,0} = f_{r+1}(y_1, \ldots, y_n),$$

so wird es auf S_0;

$$\Phi_{r+1} = F_{r+1} + f_{r+1}(y_1, \ldots, y_n) .$$

Diese Gleichung können wir aber, wegen der Willkürlichkeit von Φ_{r+1} ausserhalb S_0, auf den ganzen Raum ausdehnen. Und damit ist der Beweis vollständig erbracht.

§ 7. Wir sind jetzt imstande, durch eine einfache geometrische Überlegung zu beweisen, dass ein mechanisches Normalsystem im allgemeinen quasi-ergodisch ist. Ehe wir weiter gehen, wollen wir aber die Bedingungen (5) zusammenstellen, unter denen der verallgemeinerte Poincarésche Satz gültig ist:

1. Die Zahl der Freiheitsgrade muss $>$ 2 sein.
2. Das ungestörte System darf nicht entartet sein.
3. Aus keiner Gleichung wie z. B. $\frac{\omega_2}{\omega_1} =$ konst. darf identisch eine ähnliche, wie z.B. $\frac{\omega_4}{\omega_3} =$ konst. folgen.
4. Auf keiner Fläche $F_0 =$ konst. darf $\sum_i m_i \omega_i$ identisch verschwinden, wenn nicht

$$m_1 = m_2 = \cdots = m_n = 0 .$$

5. Auf keiner der $F_0 =$ konst. darf im allgemeinen aus $\sum_i m_i \omega_i = 0$ auch $B_m = 0$ folgen.

§ 8. Wir wollen jetzt wieder unser System durch die q und die p definieren. Sei σ ein beliebiges Gebiet auf einer Energiefläche E, und σ' das von den durch σ gehenden Bahnen besetzte Gebiet von E. Ich behaupte, dass σ' das ganze mit σ zusammenhängende Stück von E bedeckt. Wäre das nicht der Fall, so nennen wir σ'' das übrige Gebiet des mit σ zusammenhängenden Stückes von E, und S die Trennungsfläche von σ' und σ''. Nun kann keine Bahn des Systems, die einen Punkt P′ von σ' enthält, auch einen Punkt P″ von σ'' enthalten. Denn sonst würde, der repräsentative Punkt, der zur Zeit 0 in P′ ist, z. B. zur Zeit t in P″ sein; wegen der Kontinuität der Lösungen der mechanischen Gleichungen kann man dann aber ein hinreichend kleines, ganz in σ' enthaltenes Gebiet η' um P′ finden, so dass alle seine Punkte zur Zeit t sich in einem um P″ gelegenes, ganz in σ'' enthaltenes Gebiet η'' befinden. Durch η' gehen aber gewiss auch Bahnen, die durch σ gehen; die würden nun auch durch η'' gehen, und daher kann η'' nicht ganz in σ'' liegen.

Sei jetzt P ein auf S gelegener Punkt, P′ und P″ zwei in σ' bzw σ'' liegende Punkte. Nach Verlauf der Zeit t seien P , P′ P″ bzw. in P_1 , P'_1 , P''_2 angekommen; P'_1 , P''_1 werden bzw. in σ', σ'' liegen. Lassen wir nun P′ und P″ an P unendlich nah kommen, so werden auch P'_1 und P''_1, wegen der Kontinuität, P_1 zustreben, sie müssen aber immer in σ' bzw. σ'' bleiben, und das kann nur dann geschehen, wenn P_1 auf der Trennungsfläche σ liegt. Hieraus folgt, dass jede durch einen beliebigen Punkt P von S hindurchgehende Bahn ganz auf S liegt. Wir wissen aber, dass es ausser den Energieflächen im allge-

(5) Eigentlich sind diese Bedingungen für die Gültigkeit des Beweises hinreichend, aber nicht notwendig. Es wäre auch leicht, aus dem Beweise die für seine Gültigkeit wirklich nötigen Bedingungen zu finden.

meinen keine solche Flächen (6) gibt. Im allgemeinen muss daher σ' das ganze mit σ zusammenhängende Stück von E bedecken.

§ 9. Es seien nun σ und σ^* zwei auf demselben Stück von E gelegene, beliebig kleine Gebiete. Die durch σ gehenden Bahnen werden, wie bewiesen, das ganze Stück von E bedecken, und daher auch σ^*. Es gibt also gewiss Bahnen, die durch beide Gebiete hindurchgehen, d.h. das System ist quasi-ergodisch.

Göttingen, im April 1923.

(6) Anm. b. d. Korr.: Natürlich muss man für diese Flächen die im vorigen benutzten Regularitätseigenschaften annehmen.

11 *a*.

II. — ÜBER DIE EXISTENZ QUASI-ERGODISCHER SYSTEME

« Phys. Zeits. », *25*, 166–167 (1924).

Herr W. Urbanski hat in dieser "Zeitschr." *25*, 47 (1924) eine Kritik meines Beweises der Existenz quasi–ergodischer Systeme [1] veröffentlicht. Es werden nämlich zwei Punkte kritisiert:

1. Dass für die S–Fläche meines Beweises Regularitätseigenschaften angenommen werden, die nicht erfüllt sein können.

2. Dass aus meinem Beweise zwar die Existenz einer quasi–ergodischen Bahnkurve, nicht aber die von mehreren, hergeleitet werden kann.

Zum ersten Punkte möchte ich bemerken, dass die Kritik von Herrn Urbanski gewiss prinzipiell richtig ist. Die Schwierigkeit, dass die S–Fläche nicht regulär sein kann, ist von mir selbst in der Fussnote zu S. 255 meines Beweises bemerkt worden; ich möchte nur bemerken, dass die Schwierigkeit vielleicht nicht so gross ist, wie es Herrn Urbanski scheint. In der Tat ist es aus meinem Beweise leicht zu ersehen, dass die Funktion $\Phi\ (x\,,y\,,\mu)$ nur als Funktion von μ analytisch zu sein braucht; und daher werden die Eigenschaften der S–Fläche viel weniger eingeschränkt, als wenn Φ als Funktion sämtlicher Variabeln analytisch sein müsste.

Zu dem zweiten Punkte möchte ich bemerken, dass ich nur folgendes bewiesen zu haben gedenke: wenn σ und σ^* zwei beliebig kleine, auf einem zusammenhängenden Stück von E liegende Gebiete sind, gibt es immer mindestens eine Bahn des Systems, die durch beide Gebiete hindurchgeht. Dabei ist es notwendig, dass beide Gebiete endlich sind. Wenn in der Tat eines der beiden Gebiete, z. B. σ^* punktförmig wäre, so könnte es wohl geschehen, dass es auf einer periodischen Bahn läge. Und in diesem Falle würde nur diese geschlossene Bahn durch diesen Punkt gehen.

Als quasi–ergodisches System wird also bei meinem Beweise ein System gemeint, für das es Bahnen gibt, die beliebig nah an zwei beliebige Punkte einer zusammenhängenden Energiefläche herankommen, und nicht, wie es, wenn ich recht verstanden habe, Herr Urbanski meint, ein System, dessen sämtliche Bahnen die Energiefläche dicht erfüllen.

Ich vermute sogar, dass es überhaupt keine Systeme gibt, die quasi–ergodisch in diesem letzten Sinne sind. Solche Systeme würden z.B. keine periodischen Bahnen enthalten können, weil diese gewiss eine Ausnahme bilden würden.

Rom, 3. März 1924.

(1) FERMI, « Physik. Zeitschr. », *24*, 261 (1923).

12.

IL PRINCIPIO DELLE ADIABATICHE ED I SISTEMI CHE NON AMMETTONO COORDINATE ANGOLARI

«Nuovo Cimento», *25*, 171–175 (1923).

§ 1. – È nota l'importanza del principio delle adiabatiche di Ehrenfest [1] per la determinazione delle regole per la scelta delle orbite statiche di un sistema, nella teoria di BOHR. Tale principio, come si sa, può enunciarsi al modo seguente: Supponiamo che in un sistema meccanico, le forze oppure i vincoli vengano continuamente modificati in funzione del tempo ma lentissimamente in confronto ai periodi proprii del sistema, ossia, secondo l'espressione di Ehrenfest, adiabaticamente; il principio delle adiabatiche afferma che, se inizialmente il sistema si trovava in un'orbita quantisticamente privilegiata, esso vi si troverà anche alla fine della trasformazione.

Consideriamo per esempio un pendolo, e supponiamo di accorciare il filo assai lentamente in confronto al periodo del pendolo stesso. La frequenza ν del pendolo andrà allora piano piano aumentando, ma è facile riconoscere che anche l'energia u andrà aumentando, e precisamente in modo che il rapporto u/ν resta costante. Così che se inizialmente questo rapporto era un multiplo intero della costante h di Planck, esso resterà sempre tale, e quindi lo stato del sistema si manterrà durante tutta la trasformazione quantisticamente privilegiato. Per altri numerosi esempii rimandiamo alla memoria di Ehrenfest.

La base formale del principio delle adiabatiche è data da un teorema dovuto a Burgers [2]. Consideriamo un sistema che in certe coordinate generali $q_1, q_2, \cdots, q_f$ ammetta la separazione delle variabili [3]. Poniamo poi

$$I_K = \oint p_K \, dq_K \qquad (K = 1, 2, \cdots, f) \tag{1}$$

(1) EHRENFEST, «Ann. d. Phys.», *51*, 327 (1916).

(2) BURGERS, «Versl. Akad. van Wetensch.», Amsterdam 1916, 1917; «Ann. d. Phys.», *52*, 195 (1917).

(3) Per la validità delle considerazioni di Burgers basta più generalmente che il sistema ammetta coordinate angolari, ossia che si possano introdurre al posto delle q_K, p_K delle nuove variabili w_K, j_K tali che le q_K, espresse per le (w_K, j_K) siano periodiche a periodo 1 nelle w_K, e che l'energia, nelle nuove coordinate risulti funzione soltanto delle j. Allora, per le equazioni di Hamilton le j risultano costanti e le w funzioni lineari del tempo, e le q in funzione del tempo potranno svilupparsi in serie di Fourier ad f indici.

essendo le p_K i momenti coniugati canonicamente alle q_K, e l'integrale esteso, conformemente alle prescrizioni della teoria dei quanti, ad una oscillazione completa della coordinata q_K; di modo che le condizioni perché l'orbita in considerazione del sistema sia quantisticamente privilegiata sono:

$$\text{(2)} \qquad I_1 = n_1 h \quad ; \quad I_2 = n_2 h ; \cdots ; I_f = n_f h$$

essendo $n_1, n_2, \cdots, n_f$ dei numeri interi. Supponiamo ora di modificare adiabaticamente il nostro sistema, in modo però che ad ogni istante esso ammetta sempre la separazione delle variabili. Il teorema di Burgers ci dice allora che gli integrali $I_1, I_2, \cdots, I_f$ non variano durante la trasformazione, ossia che essi sono invarianti adiabatici. E dunque, se le condizioni (2) sono soddisfatte al principio della trasformazione, esse lo saranno ancora alla fine, e quindi il principio delle adiabatiche è soddisfatto.

In questa Nota mi propongo di far vedere, con un semplice esempio che se un sistema si trasforma adiabaticamente in un altro, e gli stati iniziale e finale ammettono entrambi la separazione delle variabili, ma gli stati intermedii no, gli I_K non sono più invarianti adiabatici, e quindi in questo caso il principio delle adiabatiche viene a perdere la sua base.

§ 2. – Consideriamo un punto materiale, mobile in un piano nell'interno di un rettangolo; supporremo che sopra il punto non agisca alcuna forza mentre si trova nell'interno del rettangolo, ma che quando urta le pareti rimbalzi elasticamente contro queste. Prendiamo i lati AB ed AC del rettangolo come assi coordinati x, y. È evidente allora che questo sistema ammette la separazione delle variabili in tali coordinate. Chiamando a, b le lunghezze dei lati AB, AC, la coordinata x oscilla infatti tra i valori o, a; quella y tra i valori o, b.

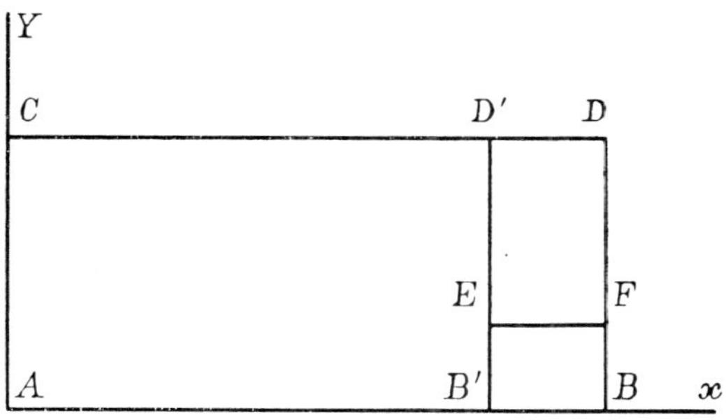

Inoltre se ad un certo istante le componenti della velocità sono u, v, esse ad un istante qualunque saranno $\pm u$, $\pm v$, dove bisogna prendere il segno + o quello — secondo che, all'istante considerato, la rispettiva coordinata cresce o decresce.

I momenti coniugati ad x e ad y saranno $\pm mu$, $\pm mv$, essendo m la massa del punto; si avrà dunque

$$\text{(3)} \qquad I_x = \oint (\pm mu)\, dx = \int_0^a mu\, dx + \int_a^0 (-mu)\, dx = 2\, mua$$

e parimenti

$$\text{(3')} \qquad I_y = 2\, mvb .$$

Vogliamo ora studiare come variano I_x ed I_y trasformando adiabaticamente il nostro sistema. Ci proponiamo precisamente di trasformare il rettangolo ABCD nell'altro AB′ CD′; ed osserviamo che tale trasformazione può farsi in tre modi:

1° si sposta il segmento BD parallelamente a se stesso, fino ad arrivare in B′ D′;

2° si sposta il segmento BB′ parallelamente a se stesso, fino ad arrivare in DD′, di modo che, in un istante intermedio, il nostro punto materiale può muoversi entro il poligono concavo AB′ EFDC;

3° si deforma in un modo qualsiasi la linea spezzata B′ BDD′ fino a portarla a coincidere col segmento di retta B′ D′. Escludendo dalle nostre considerazioni quest'ultimo caso che è evidentemente un po' complicato, ci limiteremo a discutere i primi due.

Quanto al primo osserviamo che in esso ad ogni istante il punto può sempre muoversi all'interno di un rettangolo, e quindi anche negli istanti intermedii è sempre possibile la separazione delle variabili; in ordine al teorema di Burgers dobbiamo dunque in questo caso aspettarci che I_x ed I_y restino invariati. Per I_y ciò è senz'altro evidente, perché né b, nè v variano durante la trasformazione e quindi, in ordine a (3′) neanche I_y. Quanto ad I_x, invece, a decresce durante la trasformazione, riducendosi da $a = \mathrm{AB}$, ad $a' = \mathrm{A\,B'}$; contemporaneamente però u cresce per effetto dei rimbalzi contro la parete mobile, ed un'immediata considerazione dimostra che le cose vanno appunto in modo che il prodotto au, e quindi anche I_x, resta costante [4], purchè naturalmente la trasformazione sia effettuata con sufficente lentezza.

Passando a considerare il caso 2° si riconosce invece facilmente che le cose vanno in modo diverso. Per quanto riguarda I_x infatti si vede subito che la componente x della velocità resta, a meno del segno, invariata, poichè essa potrebbe alterarsi in valore assoluto solo per effetto di un urto contro una parete mobile parallelamente all'asse x, mentre l'unica parete mobile, EF, si muove parallelamente ad y; invece a diminuisce da AB ad AB′. In complesso dunque I_x si riduce nel rapporto a'/a, e quindi non resta costante. Parimenti non resta costante neanche I_y; infatti b resta invariato mentre v cresce per effetto degli urti contro la parete mobile EF. Una immediata considerazione quantitativa dimostra che v, e quindi anche I_y, cresce nel rapporto a/a'.

Possiamo da queste considerazioni dunque concludere che gli integrali I_K sono invarianti adiabatici solo se negli stati intermedii il sistema permetta *sempre* la separazione delle variabili o almeno, in ordine ai teoremi di Burgers

(4) Infatti il numero degli urti contro la parete mobile BD nell' intervallo di tempo dt è evidentemente $\frac{u}{2a}dt$; d'altra parte, se V è la velocità della parete BD, l'aumento della velocità del punto ad ogni urto sarà 2 V, e quindi l'aumento di u nel tempo dt sarà:

$$du = 2\,\mathrm{V}\,\frac{u}{2a}\,dt = \frac{u}{a}\,\mathrm{V}\,dt = -\frac{u}{a}\,da$$

poichè evidentemente $-da = \mathrm{V}\,dt$. Integrando l'equazione precedente si trova appunto $ua =$ costante, come è affermato nel testo.

ammetta sempre un sistema di coordinate angolari, mentre ciò, almeno in generale, non è vero se negli stati intermedii il sistema non ha sempre un moto multiplamente periodico.

Anche dal punto di vista della teoria dei quanti, del resto, ci si rende abbastanza facilmente ragione di questo fatto. Si sa infatti che, secondo Bohr, la quantizzazione ben definita è possibile solo nel caso che il moto del sistema sia multiplamente periodico. Si capisce dunque che se negli stati intermedii della trasformazione il nostro sistema non si può quantizzare rigorosamente, questa imprecisione si trasmetta anche allo stato finale.

Göttingen, febbraio 1923.

13.

ALCUNI TEOREMI DI MECCANICA ANALITICA IMPORTANTI PER LA TEORIA DEI QUANTI

« Nuovo Cimento », *25*, 271–285 (1923).

§ 1.

Il principio delle adiabatiche di Ehrenfest [1] afferma, come è noto, che se un sistema meccanico si trova in un'orbita quantica, ed il suo meccanismo, forze o vincoli, viene alterato in modo infinitamente lento, il sistema si mantiene durante tutta la trasformazione in un'orbita quantisticamente privilegiata.

Perché questo principio abbia un senso determinato, è dunque evidentemente necessario che l'orbita finale del sistema, dipenda soltanto dal meccanismo finale, e non dall'aver seguito durante la trasformazione una od un'altra successione di meccanismi intermedii. Ora, che questo sia effettivamente il caso, è stato dimostrato da Burgers [2], almeno per quel tipo di sistemi che fin'ora è stato quasi unicamente preso in considerazione nella teoria dei quanti, cioè per quei sistemi che, o ammettono addirittura la separazione delle variabili, o per lo meno sono rappresentabili per mezzo di coordinate angolari [3], di modo che il loro movimento si può sempre considerare come risultante di moti periodici, in generale con tanti periodi quanti sono i gradi di libertà, o, in caso di degenerazione, con un numero inferiore. Ma proprio in questo momento, esaurito lo studio delle strutture atomiche relativamente più semplici, che appartengono a questo tipo [4], si vanno presentando con sempre maggior insistenza i problemi che non ammettono coordinate angolari, primo di tutti il problema dei tre corpi, che si presenta nello studio dell'atomo di elio, e la forma semplificata di problema dei quattro corpi che si presenta nello studio della molecola d'idrogeno. I tentativi fatti di ridurre con varii artifici lo studio di questi sistemi a quello dei sistemi con coordinate angolari, sono finora come è noto tutti falliti. Appare dunque desiderabile cercare se e fino a che punto si possa tentare un'estensione ai sistemi generali del principio delle adiabatiche, nella speranza che

(1) P. EHRENFEST, « Ann. d. Phys. », *51*, 237 (1916).

(2) BURGERS, « Versl. Akad. van Wetensch. », Amsterdam, 25 novembre 1916; « Ann. d. Phys. », *52*, 195 (1917); « Phil. Mag. », *33*, 514 (1917).

(3) Vedi per esempio SOMMERFELD, *Atombau und Spektrallinien*, III ed. Zusatz 7.

(4) Esse sono l'atomo di idrogeno e le sue varie perturbazioni (effetti Zeeman, Stark e Feinstruktur) ed il ione della molecola d'idrogeno H_2+, quando non vi siano rotazioni dei nuclei.

esso possa dar qualche informazione che possa servire alla ricerca delle regole per determinare le orbite privilegiate di questi sistemi più generali.

§ 2.

Anzitutto ci converrà fissare una classificazione dei sistemi che ci proponiamo di studiare. Ricorriamo perciò alla consueta rappresentazione geometrica dello stato del sistema mediante un punto di uno spazio Γ a $2f$ dimensioni, che ha per coordinate le f coordinate generali del sistema, $q_1, q_2, \cdots, q_f$, e gli f momenti ad esse coniugati $p_1, p_2, \cdots, p_f$. Da ogni punto di questo spazio passa una traiettoria determinata, corrispondente al moto del sistema che ha la posizione e le velocità iniziali determinate dal punto. Noi supporremo che forze e vincoli del sistema siano indipendenti dal tempo e che le forze abbiano un potenziale, di modo che esista un integrale della conservazione dell'energia. Chiameremo ipersuperfici E le ipersuperfici energia = costante; per ogni punto di Γ passa una delle E sulla quale, per l'integrale dell'energia, è situata la traiettoria che passa dal punto.

I sistemi meccanici così detti quasi-ergodici [5] hanno la proprietà che la traiettoria passa in generale infinitamente vicino ad ogni punto di E, di modo che la traiettoria riempie densamente una varietà a $2f - 1$ dimensioni.

Può darsi però che il nostro sistema, oltre all'integrale della conservazione dell'energia, ammetta qualche altro integrale primo uniforme e indipendente dal tempo, ed allora la varietà riempita dalla traiettoria avrà naturalmente un numero minore di dimensioni. Supponiamo dunque che il nostro sistema abbia in tutto m integrali primi uniformi e indipendenti dal tempo,

$$\Phi_1(p, q) = c_1 \quad ; \quad \Phi_2 = c_2 ; \cdots ; \Phi_m = c_m$$

essendo le c_i delle costanti arbitrarie. Allora per ogni punto di Γ passerà una varietà a $2f - m$ dimensioni G, intersezione delle m ipersuperfici $\Phi_i = c_i$; e la traiettoria passante da quel punto dovrà esser tutta contenuta in G.

Entro G non sarà più in generale possibile trovare una varietà subordinata che contenga tutta la traiettoria, anzi noi ammetteremo, in analogia ai sistemi quasi-ergodici, che per i sistemi che noi studiamo in generale tutta la G sia riempita densamente dalla traiettoria, ossia che questa passi infinitamente vicino a tutti i punti di G. Così che la traiettoria resterà caratterizzata, almeno nei suoi elementi statistici, dalla sola conoscenza dei valori di $\Phi_1, \Phi_2, \cdots, \Phi_m$ ad essa corrispondenti.

Chiameremo perciò questi valori *caratteristiche della traiettoria*.

Un sistema quasi-ergodico ha dunque una sola caratteristica, la sua energia.

Un sistema a energia indipendente dal tempo, che ammetta la separazione delle variabili ha in generale tante caratteristiche quanti gradi di libertà, corrispondenti alle f costanti a dell'integrale completo di Iacobi; un numero maggiore può esserci solo in casi di degenerazione, in cui esistano relazioni

(5) Chi scrive ha recentemente dimostrato che i sistemi meccanici normali sono in generale quasi-ergodici, così che questo è il caso più diffuso.

lineari a coefficienti interi tra le frequenze fondamentali. Consideriamo per esempio il moto di un punto mobile in un piano, attratto con forza proporzionale alla distanza da due rette ortogonali. Se i due coefficienti d'attrazione sono tra di loro incommensurabili il punto descrive nel piano una curva di Lissajous aperta.

E nello spazio Γ a quattro dimensioni il punto rappresentativo ricopre densamente una superficie G a due dimensioni. Il sistema ha dunque due caratteristiche; e come tali si possono prendere per esempio le energie delle proiezioni del movimento sopra le due rette ortogonali. Se invece i coefficienti d'attrazione sono commensurabili la curva di Lissajous degenera in una curva chiusa, e quindi la G diventa ad una sola dimensione, ciò che corrisponde a tre caratteristiche.

§ 3.

Noi supporremo ora di potere a nostro arbitrio variare le forze, oppure i vincoli del sistema, ossia ciò che, in complesso, con una felice denominazione dovuta a P. Hertz [(6)] chiameremo il *meccanismo* del sistema. Se variamo il meccanismo in modo infinitamente lento, abbiamo quel che si dice una trasformazione adiabatica; e nel § 5, troveremo facilmente un sistema di equazioni differenziali, che ci mostra come variano le caratteristiche del sistema, al variare adiabaticamente di un parametro μ, in funzione del quale è dato il meccanismo. Ma come già si è accennato, perché si possa parlar dell'applicazione del principio di Ehrenfest ad un certo sistema, è necessario che i valori che prendono le sue caratteristiche alla fine di una trasformazione adiabatica, non dipendano che dal meccanismo finale, e non dai meccanismi intermedii attraversati durante la trasformazione. Per studiare questa questione, noi supporremo in seguito, che il meccanismo, anziché da un solo parametro, dipenda da due, λ e μ. Invece che da un sistema di equazioni differenziali ordinarie, la dipendenza delle caratteristiche da λ e μ, sarà allora naturalmente espressa da un sistema di equazioni ai differenziali totali; e le condizioni perché i valori finali delle caratteristiche non dipendano dalla via seguita durante la trasformazione nel piano λ, μ, coincidono quindi con le condizioni di integrabilità illimitata di questo sistema. Noi dimostreremo che queste condizioni, per i sistemi quasi-ergodici sono effettivamente verificate, invece per i sistemi con più di una caratteristica in generale no, benché vi siano importanti classi di eccezioni.

§ 4.

Prima di passare allo studio delle trasformazioni adiabatiche ci conviene premettere alcune formule, che servono a calcolare la probabilità che, ad un istante qualunque, il punto rappresentativo si trovi in un certo ele-

(6) P. HERTZ, « Ann. d. Phys. », *33*, 225, 537 (1910). WEBER, GANS, « Repertorium der Physik », *1*, 2 (1916). A questi lavori rimandiamo per qualche schiarimento relativo alla parte statistica del testo.

mento di G. Indichiamo perciò, per uniformità, con $x_1, x_2, \cdots, x_{2f}$, invece che con $p_1, \cdots, p_f, q_1, \cdots, q_f$ le coordinate di Γ. Il nostro problema può allora così formularsi: cercare la probabilità che ad un certo istante $x_1, x_2, \cdots, x_{2f-m}$ abbiano valori compresi tra x_1 ed $x_1 + dx_1$, x_2 ed $x_2 + dx_2, \cdots, x_{2f-m}$ ed $x_{2f-m} + dx_{2f-m}$, mentre naturalmente le m rimanenti x prendono i valori necessarii perché il punto rappresentativo appartenga a G.

Ora si sa dalla meccanica statistica che, in conseguenza del teorema di Liouville, perché una distribuzione di punti nello spazio Γ sia stazionaria, è necessario che la loro densità in Γ abbia, sopra ciascuna delle G, valore costante.

Ora un elemento di volume di Γ può scriversi $dx_1, dx_2, \cdots, dx_{2f}$, ma anche, prendendo come nuove variabili $x_1, x_2, \cdots, x_{2f-m}\, \Phi_1, \Phi_2, \cdots, \Phi_m$ esso si può scrivere $\frac{1}{D} dx_1, \cdots, dx_{2f-m}\, d\Phi_1, \cdots, d\Phi_m$, avendo con D rappresentato il determinante funzionale $\frac{\partial(\Phi_1, \cdots, \Phi_m)}{\partial(x_{2f-m+1}, \cdots, x_{2f})}$. E siccome durante il movimento $d\Phi_1, d\Phi_2, \cdots, d\Phi_m$ restano naturalmente costanti, il detto elemento di volume risulta proporzionale a $\frac{1}{D} dx_1, \cdots, dx_{2f-m}$.

La probabilità cercata è dunque anch'essa proporzionale a quest'espressione; e siccome poi la probabilità totale è naturalmente $= 1$, si trova infine che la probabilità cercata è eguale a:

$$\frac{\frac{d\sigma}{D}}{\int \frac{d\sigma}{D}} \tag{1}$$

avendo per brevità posto $d\sigma = dx_1, dx_2, \cdots, dx_{2f-m}$, ed essendo l'integrale esteso a tutti quei valori di $x_1, x_2, \cdots, x_{2f-m}$ che corrispondono a punti di G.

Prima di abbandonare quest'argomento, vogliamo ancora dedurre una formula che ci servirà nel caso dei sistemi quasi-ergodici. In questo caso la G è un'ipersuperfice, e noi supporremo per semplicità che essa sia chiusa, e tale da essere incontrata in un sol punto dai raggi vettori uscenti da un polo interno ad essa, poiché la considerazione generale, pur non essendo essenzialmente diversa, porta a calcoli alquanto più complicati. Riferiamo lo spazio Γ a coordinate polari, caratterizzandone ogni punto per mezzo del suo raggio vettore, e della traccia di questo sopra la ipersfera di raggio 1 descritta attorno al polo. Indichiamo con H l'unica caratteristica, cioè la energia. In conformità a quanto si è detto, la probabilità che ad un certo istante il punto rappresentativo si trovi compreso entro un elemento di angolo solido $d\omega$ è proporzionale all'ipervolume compreso tra le due ipersuperfici $H(x_1, \cdots, x_{2f}) = H$ ed $H(x_1, \cdots, x_{2f}) = H + dH$, e l'angolo solido $d\omega$. Questo volume, a meno del fattore costante dH, è evidentemente $\frac{r^{2f-1}\, d\omega}{H_r}$, dove si è posto $H_r = \frac{\partial H}{\partial r}$; e dovendo al solito la probabilità totale essere $= 1$, si trova che la probabilità cercata è data da:

$$(2) \qquad \frac{r^{2f-1} \dfrac{d\omega}{H_r}}{\displaystyle\int r^{2f-1} \frac{d\omega}{H_r}}$$

essendo l'integrale esteso a tutta la sfera unitaria.

§ 5.

In questo paragrafo supporremo che il meccanismo del nostro sistema sia funzione di un parametro μ e ci proponiamo di studiare come variano le caratteristiche variando adiabaticamente questo parametro. Siccome il meccanismo dipende da μ, in generale anche le caratteristiche $\Phi_1, \Phi_2, \cdots, \Phi_m$, oltre che delle p e delle q, saranno funzioni anche di μ. Se dunque ad un certo istante il parametro μ varia di $d\mu$, la caratteristica Φ_i subirà corrispondentemente la variazione $\frac{\partial \Phi_i}{\partial \mu} d\mu$. E siccome poi la variazione è adiabatica, per aver la variazione effettiva di Φ_i bisognerà prendere il valor medio di questa espressione che, in conformità ai risultati del § precedente, risulta:

$$(3) \qquad d\mu \frac{\displaystyle\int \frac{\partial \Phi_i}{\partial \mu} \frac{d\sigma}{D}}{\displaystyle\int \frac{d\sigma}{D}}$$

che non è più funzione altro che di μ e delle $\Phi_1, \cdots, \Phi_m$. La dipendenza delle caratteristiche da μ in una trasformazione adiabatica viene perciò espressa dal sistema di equazioni differenziali ordinarie:

$$(4) \qquad \frac{d\Phi_1}{d\mu} = \frac{\displaystyle\int \frac{\partial \Phi_1}{\partial \mu} \frac{d\sigma}{D}}{\displaystyle\int \frac{d\sigma}{D}} \; ; \; \frac{d\Phi_2}{d\mu} = \frac{\displaystyle\int \frac{\partial \Phi_2}{\partial \mu} \frac{d\sigma}{D}}{\displaystyle\int \frac{d\sigma}{D}} \; ; \cdots ; \; \frac{d\Phi_m}{d\mu} = \frac{\displaystyle\int \frac{\partial \Phi_m}{\partial \mu} \frac{d\sigma}{D}}{\displaystyle\int \frac{d\sigma}{D}} .$$

Se si conoscono i valori delle Φ, per esempio per $\mu = 0$, l'integrazione di questo sistema ce ne fa conoscere i valori per μ qualunque. Nel caso particolare dei sistemi quasi-ergodici il sistema (4) si riduce all'unica equazione:

$$(5) \qquad \frac{dH}{d\mu} = \frac{\displaystyle\int \frac{H_\mu}{H_r} r^{2f-1} d\omega}{\displaystyle\int \frac{d\omega}{Hr} r^{2f-1}}$$

nella quale si è posto $H_\mu = \frac{\partial H}{\partial \mu}$.

§ 6.

Vogliamo ora studiare, in quali casi i valori finali delle caratteristiche siano indipendenti dalla via seguita nel passare adiabaticamente dal meccanismo iniziale a quello finale. Perciò rappresenteremo il meccanismo del sistema in funzione di due parametri λ e μ. Se si variano adiabaticamente questi due parametri rispettivamente di $d\lambda$ e di $d\mu$, la stessa considerazione del § precedente ci mostra che la corrispondente variazione delle caratteristiche è:

$$d\Phi_i = \frac{\int \frac{\partial \Phi_i}{\partial \lambda} \frac{d\sigma}{D}}{\int \frac{d\sigma}{D}} d\lambda + \frac{\int \frac{\partial \Phi_i}{\partial \mu} \frac{d\sigma}{D}}{\int \frac{d\sigma}{D}} d\mu \qquad (i = 1, 2, \cdots, m) \tag{6}$$

i coefficienti di $d\lambda$ e di $d\mu$ sono evidentemente funzioni soltanto di λ, μ, $\Phi_1, \cdots, \Phi_m$ e quindi le m equazioni (6) rappresentano un sistema di equazioni ai differenziali totali; se esso risulterà illimitatamente integrabile i valori finali delle Φ saranno effettivamente indipendenti dalla via seguita durante la trasformazione, altrimenti no.

Vogliamo dimostrare che, per i sistemi quasi-ergodici la condizione di integrabilità illimitata è soddisfatta. Per questi sistemi il sistema (6) si riduce infatti ad un'unica equazione ai differenziali totali, costruita analogamente alla (5)

$$dH = L d\lambda + M d\mu \tag{7}$$

dove si è posto

$$L = \frac{\int \frac{r^{2f-1} H_\lambda \, d\omega}{H_r}}{\int \frac{r^{2f-1} \, d\omega}{H_r}} \quad ; \quad M = \frac{\int \frac{r^{2f-1} H_\mu \, d\omega}{H_r}}{\int \frac{r^{2f-1} \, d\omega}{H_r}} \tag{8}$$

e quindi L ed M rappresentano delle funzioni di λ, μ ed H.

Per l'integrabilità illimitata della (7) è come si sa necessario e sufficiente che le derivate totali di L rispetto a μ, e di M rispetto a λ siano eguali, dunque deve essere

$$\frac{\partial L}{\partial \mu} + M \frac{\partial L}{\partial H} = \frac{\partial M}{\partial \lambda} + L \frac{\partial M}{\partial H} \, . \tag{9}$$

Per dimostrare che quest'eguaglianza è effettivamente soddisfatta, cominciamo col calcolarne il primo membro. Supponiamo perciò di dare ad H ed a μ delle variazioni indipendenti δH, $\delta\mu$, lasciando λ invariato; si avrà allora

$$\delta L = \frac{\partial L}{\partial H} \delta H + \frac{\partial L}{\partial \mu} \delta\mu . \tag{10}$$

D'altra parte, dalla prima delle (8) si rileva che:

$$\delta L = \frac{1}{\left(\int \frac{r^{2f-1} \, d\omega}{H_r}\right)^2} \left\{ \left(\int \frac{r^{2f-1} \, d\omega}{H_r}\right) \delta \int \frac{r^{2f-1} H_\lambda \, d\omega}{H_r} - \left(\int \frac{r^{2f-1} H_\lambda \, d\omega}{H_r}\right) \delta \int \frac{r^{2f-1} \, d\omega}{H_r} \right\} . \tag{11}$$

Nel calcolare le due variazioni di integrali, contenute in parentesi graffa, si possono senz'altro scambiare i segni δ ed $\int$, poiché dovendo l'integrale estendersi a tutta l'ipersfera unitaria, i suoi limiti non variano. Si ha perciò:

$$(12)\qquad \delta\int\frac{r^{2f-1}\,d\omega}{H_r}=(2f-1)\int\frac{r^{2f-2}\,\delta r\,d\omega}{H_r}-\int\frac{r^{2f-1}\,\delta H_r\,d\omega}{H_r^2}\,.$$

D'altra parte, restando invariata la traccia sopra la sfera unitaria, si ha:

$$\delta H=H_r\,\delta r+H_\mu\,\delta\mu$$

da cui

$$\delta r=\frac{\delta H}{H_r}-\frac{H_\mu}{H_r}\,\delta\mu$$

ed anche

$$\delta H_r=H_{rr}\,\delta r+H_{r\mu}\,\delta\mu=\frac{H_{rr}}{H_r}\,\delta H+\left(H_{r\mu}-\frac{H_{rr}\,H_\mu}{H_r}\right)\delta\mu\,.$$

Sostituendo in (12) queste espressioni di δr, δH_r si trova

$$\delta\int\frac{r^{2f-1}\,d\omega}{H_r}=\delta H\left\{(2f-1)\int\frac{r^{2f-2}\,d\omega}{H_r^2}-\int\frac{r^{2f-1}\,H_{rr}\,d\omega}{H_r^3}\right\}$$

$$-\,\delta\mu\left\{(2f-1)\int\frac{r^{2f-2}\,H_\mu\,d\omega}{H_r^2}+\int\frac{r^{2f-1}\,d\omega}{H_r^2}\left(H_{\mu r}-\frac{H_\mu\,H_{rr}}{H_r}\right)\right\}.$$

In modo del tutto analogo si trova:

$$\delta\int\frac{r^{2f-1}\,H_\lambda\,d\omega}{H_r}=\delta H\left\{(2f-1)\int\frac{r^{2f-2}\,H_\lambda\,d\omega}{H_r^2}+\int\frac{r^{2f-1}\,H_{\lambda r}\,d\omega}{H_r^2}\right.$$

$$\left.-\int\frac{r^{2f-1}\,H_\lambda\,H_{rr}}{H_r^3}\,d\omega\right\}+\delta\mu\left\{-(2f-1)\int\frac{r^{2f-2}\,H_\lambda\,H_\mu\,d\omega}{H_r^2}\right.$$

$$\left.+\int\frac{r^{2f-1}\,d\omega}{H_r}\left(H_{\lambda\mu}-\frac{H_{\lambda r}\,H_\mu}{H_r}\right)-\int\frac{r^{2f-1}\,H_\lambda\,d\omega}{H_r^2}\left(H_{\mu r}-\frac{H_\mu\,H_{rr}}{H_r}\right)\right\}.$$

Sostituendo in (11) queste due ultime espressioni, e confrontando poi con la (10) si trova infine:

$$\frac{\partial L}{\partial H}=\frac{1}{\left(\int\frac{r^{2f-1}\,d\omega}{H_r}\right)^2}\left[\left(\int\frac{r^{2f-1}\,d\omega}{H_r}\right)\left\{(2f-1)\int\frac{r^{2f-2}\,H_\lambda\,d\omega}{H_r^2}\right.\right.$$

$$\left.+\int\frac{r^{2f-1}\,H_{\lambda r}\,d\omega}{H_r^2}-\int\frac{r^{2f-1}\,H_\lambda\,H_{rr}}{H_r^3}\,d\omega\right\}$$

$$-\left(\int\frac{r^{2f-1}\,H_\lambda\,d\omega}{H_r}\right)\left\{(2f-1)\int\frac{r^{2f-2}\,d\omega}{H_r^2}\right.$$

$$\left.\left.-\int\frac{r^{2f-1}\,H_{rr}\,d\omega}{H_r^3}\right\}\right]$$

$$\frac{\partial L}{\partial \mu} = \frac{1}{\left(\int \frac{r^{2f-1}\, d\omega}{H_r}\right)^2} \left[\left(\int \frac{r^{2f-1}\, d\omega}{H_r}\right) \left\{ -(2f-1) \int \frac{r^{2f-2}\, H_\lambda\, H_\mu\, d\omega}{H_r^2} \right. \right.$$
$$+ \int \frac{r^{2f-1}\, d\omega}{H_r} \left(H_{\lambda\mu} - \frac{H_{\lambda r}\, H_\mu}{H_r}\right)$$
$$\left. - \int \frac{r^{2f-1}\, H_\lambda\, d\omega}{H_r^2} \left(H_{\mu r} - \frac{H_\mu\, H_{rr}}{H_r}\right) \right\}$$
$$- \left(\int \frac{r^{2f-1}\, H_\lambda\, d\omega}{H_r}\right) \left\{ (2f-1) \int \frac{r^{2f-2}\, H_\mu\, d\omega}{H_r^2} \right.$$
$$\left. \left. + \int \frac{r^{2f-1}\, d\omega}{H_r^2} \left(H_{\mu r} - \frac{H_\mu\, H_{rr}}{H_r}\right) \right\} \right].$$

Queste due ultime equazioni, insieme alla seconda delle (8) ci forniscono tutti gli elementi per calcolare il primo membro di (9). Formatolo effettivamente, si riconosce subito che in esso λ e μ figurano simmetricamente; esso sarà dunque certo eguale al secondo membro; e dunque la (9) è verificata.

Possiamo dunque concludere che per i sistemi quasi-ergodici, il valore che assume l'energia alla fine di una trasformazione adiabatica, non dipende affatto dai meccanismi intermedi della trasformazione.

§ 7.

Torniamo ora ad occuparci dei sistemi con più di una caratteristica. Perché anche per questi sistemi le caratteristiche finali fossero indipendenti dai meccanismi intermedii della trasformazione, dovrebbero essere verificate identicamente le condizioni di integrabilità illimitata del sistema ai differenziali totali (6). Se però, con un calcolo naturalmente alquanto più complicato, ma non essenzialmente dissimile da quello svolto nel § precedente, si formano effettivamente queste condizioni di integrabilità, si trova che in generale esse *non* sono soddisfatte. Piuttosto però che riportar qui questo lungo calcolo, preferiamo mostrare la cosa sopra l'esempio di un semplice sistema con due caratteristiche. L'esempio che scegliamo è molto analogo ad un altro, da me recentemente svolto in una nota sul principio delle adiabatiche.

Da un'origine O, tracciamo in un piano due assi ortogonali x, y. Presi poi nel primo quadrante due punti P, Q abbassiamo da essi le perpendicolari PA, PB, QC, QD sopra gli assi. Supporremo che P sia interno al rettangolo OC QD.

Pensiamo ora che nell'interno del poligono concavo A P B D Q C A si muova un punto materiale, sotto l'azione di nessuna forza; che però rimbalzi elasticamente urtando contro le pareti del poligono. I valori assoluti u, v delle componenti della velocità del punto sugli assi x, y restano evidentemente costanti durante il movimento, e quindi il sistema ha due caratteristiche. Supponiamo poi di mantenere fisso il punto Q (di coordinate a, b) e di poter invece muovere il punto P (di coordinate λ, μ). Avremo così effettivamente realizzato un sistema meccanico con due caratteristiche u, v e dipendente da due parametri λ, μ.

Con facili considerazioni, analoghe ad altre svolte nella nota sopra citata, si trova che, variando adiabaticamente la posizione del punto P, u e v variano con la seguente legge:

$$d \log u = \frac{2 \mu \, d\lambda}{ab - \lambda\mu} \quad ; \quad d \log v = \frac{2 \lambda \, d\mu}{ab - \lambda\mu} .$$

Evidentemente nessuna di queste due equazioni è illimitatamente integrabile; e perciò i valori che u e v assumono alla fine di una trasformazione dipendono anche dal cammino percorso dal punto P.

In generale dunque non è possibile applicare a sistemi con più caratteristiche il principio di Ehrenfest.

§ 8.

A questa regola esistono però alcune importanti classi di eccezioni, che ci proponiamo di studiare in questo paragrafo. La prima e più importante è quella dei sistemi con coordinate angolari, dei quali non solo si sa, in ordine ai teoremi di Burgers, che il principio di Ehrenfest è applicabile, nel senso che conduce in ogni caso a condizioni finali determinate, ma per i quali il detto principio resta anche *verificato dall'esperienza* come conseguenza logica delle condizioni di Sommerfeld, le quali hanno a loro sostegno tutta la teoria e l'esperienza fatte sopra l'atomo di idrogeno.

Un'altra notevole classe di eccezioni alle conclusioni del § 7, è la seguente: Supponiamo che delle m caratteristiche del nostro sistema una soltanto: l'energia, dipenda esplicitamente dai parametri λ, μ del meccanismo. Io dico che per questi sistemi, alla fine di ogni trasformazione adiabatica, l'energia assume un valore indipendente dai meccanismi intermedi, mentre le altre caratteristiche restano addirittura invariate.

Che tutte le caratteristiche, all'infuori dell'energia restino invariate risulta evidente dal fatto che, non contenendo esplicitamente i parametri, esse restano invariate in ogni processo elementare della trasformazione; od anche del resto dal sistema (6), poiché se Φ_i è una di queste caratteristiche: si ha per ipotesi $\frac{\partial \Phi_i}{\partial \lambda} = \frac{\partial \Phi_i}{\partial \mu} = 0$.

Per dimostrare che il valore finale dell'energia non dipende dalla via percorsa durante la trasformazione nel piano λ, μ, si potrebbe fare una considerazione del genere di quella svolta nel § 6. Ma è più semplice osservare che, per le ipotesi fatte, si può, con una trasformazione canonica *indipendente dai parametri*, fare in modo che le caratteristiche indipendenti dai parametri, vengano ad essere coordinate di Γ. Ciò fatto la considerazione del § 6, si può ripetere parola per parola, e le caratteristiche costanti figurano in essa semplicemente come dei parametri costanti.

I sistemi di questo tipo si presentano abbastanza spesso nelle applicazioni; per esempio appartengono ad esso tutti quei sistemi che posseggono, quali unici integrali primi uniformi e indipendenti dal tempo, oltre all'energia, qualche integrale della conservazione delle quantità di moto o delle quantità di moto areale, perché questi ultimi sono sempre indipendenti dai parametri del meccanismo.

§ 9.

Quanto ad una possibile applicazione di quanto si è detto, alla teoria dei quanti, osserviamo quanto segue:

Resta dalle nostre conclusioni esclusa la possibilità di un'estensione del principio di Ehrenfest ai sistemi non quasi-ergodici, salve le eccezioni indicate. Per i sistemi quasi-ergodici invece, oppure per le eccezioni studiate nel § 3, tale applicazione non è a priori esclusa, benché naturalmente non si possa per ora prevedere se l'esperienza ne debba confermare i risultati; comunque si potrebbe forse tentare se, per questa via, non si riuscisse ad avere qualche utile informazione sopra la forma delle regole per la determinazione delle orbite quantiche per i sistemi senza coordinate angolari. Naturalmente il solo principio di Ehrenfest, se anche l'esperienza dovesse confermarlo in questa sua applicazione più generalizzata, non è di per se stesso sufficiente alla determinazione di tali regole, permettendo solo, quando si conoscano le orbite privilegiate di un certo sistema, di dedurle anche per tutti quei sistemi che da esso possono dedursi con una trasformazione adiabatica. Esso potrebbe dunque forse servire, a parte la complicazione dei calcoli, a trovare le relazioni quantitative tra gli spettri di scintilla per esempio dei metalli alcalini e quelli d'arco dei gas nobili. Poiché i sistemi che emettono questi spettri differiscono solo per la carica del nucleo, e possono quindi con facilità trasformarsi adiabaticamente uno nell'altro.

Göttingen, Aprile 1923.

N° 14.

The problem of equilibrium between photons and electrons in a black cavity originated in Fermi's mind from reading Richardson's « Electron Theory of Matter ». For some time Fermi had in mind to make experiments, or to have somebody else in the Institute make experiments, on the discontinuity suggested by his theory. Then he dropped the matter.

G. POLVANI.

14.

SULLA TEORIA STATISTICA DI RICHARDSON DELL'EFFETTO FOTOELETTRICO

« Nuovo Cimento », *26*, 97-104 (1923).

Consideriamo una cavità in un corpo a temperatura uniforme.

In essa si troveranno degli elettroni, e la radiazione nera corrispondente a quella temperatura. Urtando contro le pareti della cavità gli elettroni potranno essere riassorbiti dal corpo, mentre nuovi elettroni, per l'effetto fotoelettrico della radiazione nera verranno dal corpo rimandati nella cavità. Tra questi due processi di assorbimento e di emissione degli elettroni dovrà sussistere equilibrio. Scrivendo appunto questo Richardson [1] è stato condotto ad un'equazione integrale avente per incognita la funzione $\varphi(\nu)$ che, a meno del potere emissivo del corpo, rappresenta il numero di elettroni che viene emesso per effetto fotoelettrico quando viene assorbita l'energia luminosa 1 di frequenza ν. Risolvendo l'equazione integrale si arriva all'interessante risultato che:

$$\begin{cases} \varphi(\nu) = 0 & \text{per } h\nu \leq w_0 \\ \varphi(\nu) = H\left(\dfrac{h}{\nu^2} - \dfrac{w_0}{\nu^3}\right) & \text{per } h\nu \geq w_0 . \end{cases}$$

Dove h è la costante di Planck, H una costante e w_0 è l'energia necessaria a liberare un elettrone dall'interno del corpo. Questo risultato è molto interessante, perché dà una chiara interpretazione del fatto che l'emissione fotoelettrica non avviene che quando la frequenza della luce supera un limite, w_0/h, ossia della legge di Einstein. Nella sua teoria tuttavia il Richardson,

(1) RICHARDSON, « Phil. Mag. », *23*, 594 (1912); oppure « Electron Theory of Matter », Cap. XVIII. – « Phil. Mag. », *27*, 176 (1914). Esperienze destinate alla verifica sperimentale della teoria, si trovano in lavori di Richardson, Compton e altri, nel « Phil. Mag. » (1914–17).

per semplificare un po' le cose, si serve, come formula per la radiazione del corpo nero, anziché della formula di Planck,

$$\frac{8\pi}{c^3}\frac{h\nu^3}{e^{h\nu/RT}-1}$$

di un'altra espressione, dovuta a Wien, precisamente

$$\frac{8\pi}{c^3}h\nu^3 e^{-h\nu/RT}.$$

Come si sa le formule di Planck e di Wien vengono a coincidere per frequenze elevate.

In questo lavoro ricercheremo le modificazioni che subisce la teoria di Richardson servendosi della formula di Planck; e mostreremo dapprima che, anche con la formula di Planck, vale rigorosamente il risultato che $\varphi(\nu)$ è zero per $h\nu < w_0$. In secondo luogo mostreremo che, mentre secondo Richardson $\varphi(\nu)$ per $h\nu > w_0$ si mantiene continua con la sua derivata prima, applicando la formula di Planck quest'ultima ha delle discontinuità, salvo eccezioni, per tutti i valori di ν per i quali $h\nu = nw_0$, essendo n un numero intero. Questo risultato, assai caratteristico ed inaspettato, può in certo modo interpretarsi forse dal punto di vista della teoria dei quanti; in quanto è facile comprendere che, via via che $h\nu$ raggiunge i varii multipli interi di w_0 comincino a farsi sentire nuove azioni che provocano le discontinuità di cui si è parlato. Dal punto di vista sperimentale, benché si possa ormai dire con sicurezza che né la presente teoria, né quella di Richardson danno un risultato concordante con l'esperienza, si potrebbe forse sperar di trovare la discontinuità qualitativa per

$$h\nu = 2\,w_0,$$

ma non pare che il materiale sperimentale finora raccolto sopra l'emissione fotoelettrica in funzione della lunghezza d'onda, sia sufficientemente esteso e sicuro da permettere un controllo del risultato teorico.

Da un altro punto di vista si può osservare che, come più volte e da più parti [2] è già stato messo in evidenza, tra la formola di Planck e quella di Wien esiste una spece di parallelismo, nel senso che la prima può considerarsi frutto dell'aver ammesso possibili i livelli di energia $nh\nu$; la seconda invece dell'aver ammesso solo il livello di energia $h\nu$. Da questo punto di vista meraviglia meno la differenza qualitativa che si trova applicando l'una o l'altra formula.

Ricordiamo per sommi capi il ragionamento di Richardson: sia T la temperatura assoluta, n il numero di elettroni esistenti per unità di volume nella cavità. Facili considerazioni termodinamiche conducono a scrivere:

$$n = AT^{3/2}\,e^{-w_0/RT} \tag{2}$$

essendo A una costante. La velocità degli elettroni è proporzionale a $T^{1/2}$, perciò il numero N di elettroni che penetrano dalla cavità nel corpo nell'unità

(2) Vedi per esempio le considerazioni recenti di De Broglie, *C. R.*, fasc. 19 (1922).

di tempo sarà proporzionale ad $n\,T^{1/2}$; ossia, indicando con B un coefficiente, che dipende anche dagli elementi geometrici della cavità si avrà:

$$N = BT^2 e^{-w_0/RT}. \tag{3}$$

Ora Richardson ammette che quando il corpo assorbe la energia luminosa dq di frequenza ν esso metta in libertà per effetto fotoelettrico $f(\nu)\,dq$ elettroni. Se con $L(\nu, T)$ indichiamo la legge di distribuzione dell'energia raggiante nello spettro del corpo nero, e con $\varepsilon(\nu)$ il potere assorbente del corpo la quantità totale di elettroni fotoelettrici liberati sarà proporzionale a:

$$\int_0^\infty L(\nu, T)\,\varepsilon(\nu) f(\nu)\,d\nu$$

ossia, ponendo

$$\varphi(\nu) = \varepsilon(\nu) f(\nu),$$

N risulterà proporzionale a:

$$\int_0^\infty L(\nu, T)\,\varphi(\nu)\,d\nu.$$

Indicando con K una costante si dovrà dunque avere

$$KT^2 e^{-w_0/RT} = \int_0^\infty L(\nu, T)\,\varphi(\nu)\,d\nu. \tag{4}$$

Ora Richardson ammette, come si è detto, che $L(\nu, T)$ abbia la forma data da Wien, cioè

$$L(\nu, T) = \frac{8\pi h}{c^3}\nu^3 e^{-h\nu/RT}.$$

La (4) si trasforma così nell'equazione integrale:

$$CT^2 e^{-w_0/RT} = \int_0^\infty \nu^3 e^{-h\nu/RT}\varphi(\nu)\,d\nu \qquad (C = \text{costante})$$

che ha appunto per soluzione le (1), dove si ponga

$$H = \frac{Ch}{R^2}.$$

Se invece della legge di Wien adottiamo quella di Planck, poniamo cioè:

$$L(\nu, T) = \frac{8\pi h}{h^3}\frac{\nu^3}{e^{h\nu/RT} - 1} \tag{5}$$

la (4) ci dà l'equazione integrale:

$$CT^2 e^{-w_0/RT} = \int_0^\infty \frac{\nu^3 \varphi(\nu)\,d\nu}{e^{h\nu/RT} - 1}. \tag{6}$$

Per trasformarla un poco cominciamo col porre

$$x = \frac{1}{RT};$$

allora la (6) diventa:

$$\frac{C}{R^2}\frac{e^{-w_0 x}}{x^2} = \int_0^\infty \frac{\nu^3 \varphi(\nu)\, d\nu}{e^{h\nu x} - 1} \,. \tag{6bis}$$

È facile dimostrare, con una verificazione diretta, che la soluzione della (6 *bis*) è la seguente

$$\varphi(\nu) = \sum_{j=1}^{\infty} \varphi_j(\nu) \tag{6''}$$

dove

$$\begin{cases} \varphi_j(\nu) = 0 \quad , & \text{per } \nu \leq j\dfrac{w_0}{h} \\ \varphi_j(\nu) = A_j\left(\dfrac{h}{\nu^2} - \dfrac{w_0}{\nu^3}\right) \quad , & \text{per } \nu \geq j\dfrac{w_0}{h} \,. \end{cases} \tag{6'''}$$

A_j ha sempre uno dei tre valori $\pm H/j^2$, o; e precisamente si ha $A_j = (-1)^n H/j^2$ se $j = p_1, p_2, \cdots, p_n$, dove i p_i sono numeri primi tutti diversi tra di loro; se invece j è divisibile per il quadrato di un numero primo si ha sempre $A_j = 0$. Si ha dunque

$$A_1 = H \quad ; \quad A_2 = -H/4 \quad ; \quad A_3 = -H/9 \quad ; \quad A_4 = 0 \, ;$$
$$A_5 = -H/25 \quad ; \quad A_6 = H/36 \, ; \cdots$$

Di qui si vede che la derivata di $\varphi(\nu)$ deve, come si è annunziato, essere discontinua per tutti i multipli di w_0/h, salvo quelli secondo 4, 8, 9, 12,··· Le considerazioni che seguono, benché sprovviste di un eccessivo rigore analitico, sono destinate a far vedere la ragione per la quale queste successive discontinuità si presentano. Moltiplichiamo ambo i membri della (6 *bis*) per $e^{w_0 x}\, dx$ ed integriamo poi tra $x = y$ ed $x = \infty$. Scambiando le quadrature nel secondo membro, ed effettuandole nel primo si trova:

$$\frac{C}{R^2 y} = \int_0^\infty \nu^3 \varphi(\nu)\, d\nu \int_y^\infty \frac{e^{w_0 x}\, dx}{e^{h\nu x} - 1} \,. \tag{7}$$

Ora il secondo integrale del secondo membro, come si vede subito è proprio soltanto quando $h\nu > w_0$. Dovrà dunque certamente $\varphi(\nu)$ essere $= 0$ tutte le volte che $h\nu \leq w_0$. E questo esprime appunto la legge di Einstein.

Volendo ulteriormente determinare $\varphi(\nu)$ anche per

$$\nu > \frac{w_0}{h} \,,$$

trovare cioè la formula corrispondente alla seconda delle (1), osserveremo che, per quanto or ora abbiamo visto basta intanto limitarsi ai valori di

$$\nu > \frac{w_0}{h} \text{ per i quali } \frac{h\nu}{RT}$$

è, almeno per temperature normali, molto grande. Trascurando dunque i quadrati di $e^{-h\nu/RT}$ scriveremo la (6 *bis*).

$$\frac{C}{R^2}\frac{e^{-w_0 x}}{x^2} = \int_{w_0/h}^\infty \nu^3 \varphi(\nu)\left(e^{-h\nu x} + e^{-2h\nu x}\right) d\nu \,. \tag{8}$$

Dove il limite inferiore dell'integrale è, in virtù della legge di Einstein, sostituito con w_0/h. Proviamo a risolverla ponendo:

$$\varphi(\nu) = H\left(\frac{h}{\nu^2} - \frac{w_0}{\nu^3}\right) + \psi(\nu) \tag{9}$$

dove supporremo che $\psi(\nu)/\varphi(\nu)$ abbia l'ordine di grandezza di $e^{-h\nu/RT}$.

Tenendo presente che $H = \frac{Ch}{R^2}$ si trova allora per $\psi(\nu)$ la equazione integrale (trascurando, per la nostra approssimazione il prodotto $\psi(\nu)\, e^{-h\nu/RT}$)

$$\int_{w_0/h}^{\infty} \nu^3 \psi(\nu)\, e^{-h\nu x}\, d\nu + \frac{ch}{R^2}\int_{w_0/h}^{\infty} (h\nu - w_0)\, e^{-2h\nu x}\, d\nu = 0$$

che, con ovvie trasformazioni, diventa:

$$\int_{w_0/h}^{\infty} \nu^3 \psi(\nu)\, e^{-h\nu x}\, d\nu + \frac{c}{4R^2}\,\frac{e^{-2w_0 x}}{x^2} = 0\,.$$

La sua soluzione è evidentemente:

$$\begin{cases} \psi(\nu) = 0 & \text{per } h\nu \leq 2w_0 \\ \psi(\nu) = -\dfrac{H}{4}\left(\dfrac{h}{\nu^2} - \dfrac{2w_0}{\nu^3}\right) & \text{per } h\nu > 2w_0\,. \end{cases} \tag{10}$$

Tenendo presente la (9) si trova dunque per $\varphi(\nu)$ la soluzione seguente:

$$\begin{cases} \varphi(\nu) = 0 & \text{per } h\nu < w_0 \\ \varphi(\nu) = H\left(\dfrac{h}{\nu^2} - \dfrac{w_0}{\nu^3}\right) & \text{per } w_0 < h\nu < 2w_0 \\ \varphi(\nu) = H\left(\dfrac{3}{4}\dfrac{h}{\nu^2} - \dfrac{w_0}{2\nu^3}\right) & \text{per } 2w_0 < h\nu\,. \end{cases} \tag{11}$$

Ed abbiamo così trovata la discontinuità anche nel punto

$$h\nu = 2w_0\,.$$

Possiamo però facilmente vedere che, spingendo convenientemente l'approssimazione si trova salvo eccezioni discontinuità in tutti i punti

$$h\nu = nw_0\,.$$

Infatti scriviamo le due equazioni integrali corrispondenti alla (8) che si ottengono trascurando rispettivamente la n-esima e la $n+1$-esima potenza di $e^{-h\nu/RT}$, cioè:

$$\begin{cases} \dfrac{C}{R^2}\,\dfrac{e^{-w_0 x}}{x^2} = \displaystyle\int_{w_0/h}^{\infty} \nu^3 \varphi(\nu)\left(e^{-h\nu x} + e^{-2h\nu x} + \cdots + e^{-(n-1)h\nu x}\right) d\nu \\ \dfrac{C}{R^2}\,\dfrac{e^{-w_0 x}}{x^2} = \displaystyle\int_{w_0/h}^{\infty} \nu^3 \varphi(\nu)\left(e^{-h\nu x} + e^{-2h\nu x} + \cdots + e^{-nh\nu x}\right) d\nu\,. \end{cases} \tag{12}$$

Indichiamo con $\varphi^*(\nu)$ la soluzione della prima, e con

$$\varphi^*(\nu) + \psi(\nu)$$

quella della seconda. Sostituendo nella seconda

$$\varphi^*(\nu) + \psi(\nu),$$

al posto di $\varphi(\nu)$, e tenendo presente che $\varphi^*(\nu)$ è soluzione della prima si trova:

$$\int_{w_0/h}^{\infty} \nu^3 \psi(\nu) \{ e^{-h\nu x} + e^{-2h\nu x} + \cdots + e^{-nh\nu x} \} d\nu + \int_{w_0/h}^{\infty} \nu^3 \varphi^*(\nu) e^{-nh\nu x} d\nu = 0 .$$

Moltiplichiamo questa equazione per $e^{nw_0 x}\, dx$ ed integriamo tra

$$x = y \text{ ed } x = \infty$$

scambiando le quadrature. Si ha:

$$\int_{w_0/h}^{\infty} \nu^3 \psi(\nu) d\nu \left\{ \int_y^{\infty} e^{-(h\nu - nw_0) x} dx + \cdots + \int_y^{\infty} e^{-(nh\nu - nw_0) x} dx \right\}$$

$$+ \int_{w_0/h}^{\infty} \nu^3 \varphi^*(\nu) d\nu \int_y^{\infty} e^{-n(h\nu - w_0) x} dx = 0 .$$

Ora l'integrale del secondo termine è certo proprio, perché

$$\nu > \frac{w_0}{h},$$

quelli del primo invece lo sono quando

$$h\nu > nw_0 .$$

Dovrà dunque $\psi(\nu)$ esser zero fino che

$$h\nu < nw_0 .$$

Da ciò ha origine la discontinuità nel punto

$$h\nu = nw_0 .$$

Roma, novembre 1922.

15.

GENERALIZZAZIONE DEL TEOREMA DI POINCARÉ SOPRA LA NON ESISTENZA DI INTEGRALI UNIFORMI DI UN SISTEMA DI EQUAZIONI CANONICHE NORMALI

«Nuovo Cimento», *26*, 105-115 (1923).

§ 1.

Consideriamo un sistema meccanico, con n gradi di libertà, caratterizzato da $2n$ variabili coniugate canonicamente,

$$x_1, x_2, \cdots, x_n \quad , \quad y_1, y_2, \cdots, y_n .$$

Ammettiamo poi le seguenti proprietà:

1ª L'energia F è indipendente dal tempo.

2ª Oltre alle variabili x, y, la F contiene anche un parametro μ, ed è una funzione analitica uniforme di queste variabili, che può, per valori sufficientemente piccoli di μ, svilupparsi in serie di potenze di questo parametro, così che:

$$F = F_0 + \mu F_1 + \mu^2 F_2 + \cdots \tag{1}$$

essendo le F_i funzioni soltanto delle x e delle y.

3ª La posizione del sistema, e per conseguenza anche F e le F_i, dipendono periodicamente con periodo 2π dalle $x_1, x_2, \cdots, x_n$.

4ª F_0 è funzione soltanto delle y.

Le equazioni canoniche di questo sistema si chiamano, come si sa, equazioni canoniche normali, e ad esse si riconducono tra l'altro tutti i problemi delle perturbazioni, studiati nella meccanica celeste. Con espressione presa dalla meccanica celeste chiameremo sistema imperturbato, quello a cui si riduce il nostro sistema per $\mu = 0$.

Poincaré [1] ha dimostrato che un sistema di equazioni canoniche normali, oltre all'integrale della conservazione dell'energia, non può in generale ammettere alcun altro integrale primo analitico uniforme e *indipendente dal tempo.*

(1) POINCARÉ, «Méthodes nouvelles de la mécanique céleste», Vol. 1, Cap. 5. A questa opera rimandiamo anche per le notazioni e gli schiarimenti relativi al testo.

Geometricamente ciò significa, che nello spazio Γ a $2n$ dimensioni, nel quale con un punto sono rappresentati posizione e momenti del sistema, all'infuori delle ipersuperfici $F =$ cost. non ci può in generale essere alcuna altra famiglia di ipersuperfici tali che la traiettoria del punto rappresentativo uscente da un punto di una di esse resti sempre sopra la superficie stessa.

Ci proponiamo in questo lavoro di generalizzare tale teorema dimostrando che se $n > 2$ all'infuori delle $F =$ cost. non esiste in generale neanche una sola ipersuperfice S_μ analitica uniforme, tale da contenere tutte le traiettorie del punto rappresentativo uscenti dai suoi punti. L'enunciato non geometrico può così formularsi: È in generale impossibile trovare una funzione $\Phi(x_1, \cdots, x_n y_1, \cdots, y_n, \mu)$ analitica uniforme (e quindi anche periodica nelle x) tale che, essendo inizialmente $\Phi = 0$, si mantenga durante tutto il movimento $\Phi = 0$, in virtù delle equazioni canoniche, a meno che da $\Phi = 0$ non segua identicamente $F =$ cost.

§ 2.

Supponiamo dunque che la superficie S_μ, di equazione $\Phi(x, y, \mu) = 0$ goda delle proprietà indicate. Noi dimostreremo allora che in generale S_μ deve coincidere con una $F =$ costante. Nella dimostrazione cercheremo il più possibile di attenerci all'analoga dimostrazione di Poincaré.

Sviluppiamo Φ in serie di potenze di μ, e sia:

$$\Phi = \Phi_0 + \mu\Phi_1 + \mu^2\Phi_2 + \cdots \tag{2}$$

dove le Φ_i, sono funzioni delle x e delle y. Osserviamo subito che, se anche è conosciuta la posizione di S_μ per tutti i valori di μ, le Φ_i non restano perciò completamente determinate. E precisamente Φ_0 è sottoposta alla sola condizione di doversi annullare sopra S_0, e del resto arbitraria, sceltala comunque, Φ_1 resta determinata sopra S_0 (dalla condizione che per μ infinitamente piccolo le due superfici S_μ, e, $\Phi_0 + \mu\Phi_1 = 0$ debbono differire per termini dell'ordine di μ^2) e del resto arbitraria; sceltala, all'infuori di S_0, a piacere, Φ_2 resta determinata sopra S_0 e del resto arbitraria, e così di seguito.

Io dico ora che Φ_0 può sempre scegliersi indipendente dalle x. Dopo quanto si è detto basta perciò dimostrare che tutti i punti che hanno le coordinate y a comune con un punto di S_0, appartengono pure ad S_0. Ora se P_0 è un punto di S_0, S_0 deve contenere tutta la traiettoria del punto rappresentativo, relativa al sistema imperturbato, che parte da P_0. Siccome F_0 non contiene le x, i punti di questa traiettoria avranno le y costanti, ed eguali a quelle di P_0, mentre le x saranno funzioni lineari del tempo, $x_i = \omega_i t + \alpha_i$, dove si è posto per brevità $\omega_i = \frac{\partial F_0}{\partial y_i}$. Nello spazio a n dimensioni delle x_i questa traiettoria è dunque una retta. Ora, per le ammesse proprietà di periodicità, possiamo riportare tutti i punti di questa retta nel cubo di lato 2π con lati paralleli agli assi e vertice nell'origine. Come si sa questo cubo resterà riempito densamente, purché non esista alcuna relazione lineare intera tra le ω_i.

Ammettiamo dapprima che sopra S non esista identicamente alcuna di tali relazioni. Allora il cubo resterà in generale riempito densamente, e quindi tutti i suoi punti apparterranno, come si era affermato, ad S_0. Se invece, essendo $m_1, m_2, \cdots, m_n$ degli interi valesse identicamente sopra S_0 la relazione $\Sigma m_i \omega_i = 0$, potremmo sempre ammettere che questa relazione non valga identicamente in tutto lo spazio, ed allora $\Sigma m_i \omega_i = 0$ è l'equazione di S_0, per l'arbitrarietà di Φ_0 possiamo dunque prendere $\Phi_0 = \Sigma m_i \omega_i$, e quindi, come si era detto, indipendente dalle x.

§ 3.

La proprietà di S_μ di contenere le traiettorie uscenti da tutti i suoi punti, può analiticamente tradursi dicendo che l'equazione [2]

$$[F, \Phi] = 0 \tag{3}$$

è una conseguenza dell'altra

$$\Phi = 0. \tag{4}$$

Se dunque dalla (4) ricaviamo per esempio y_1 come funzione di $y_2, \cdots, y_i, x_1, x_2, \cdots, x_n, \mu$, sostituiamo il valore trovato in (3) e sviluppiamo poi in serie di potenze di μ, i coefficenti di tutte le potenze debbono separatamente annullarsi.

Ora in questo sviluppo il primo termine $[F_0, \Phi_0]$ è identicamente nullo in tutto lo spazio, poichè F_0 e Φ_0 contengono soltanto le y; il coefficente di μ si riduce dunque a $[F_0, \Phi_1] + [F_1, \Phi_0]$, calcolato sopra S_0.

Dunque sopra S_0 deve essere:

$$[F_0, \Phi_1] + [F_1, \Phi_0] = 0. \tag{5}$$

Ora per la periodicità di F_1 e di Φ_1 possiamo scrivere: $\Phi_1 = \sum_m A_m e^{\theta m}$, $F_1 = \sum_m B_m e^{\theta m}$, dove A_m, B_m sono funzioni delle sole y, θ_m è un'abbreviazione dell'espressione $i(m_1 x_1 + m_2 x_2 + \cdots + m_n x_n)$, ed m rappresenta simbolicamente gli n indici $m_1, m_2, \cdots, m_n$ i quali nelle somme debbono andare da $-\infty$ a $+\infty$. Con un semplice calcolo la (5) si trasforma in:

$$\sum_m e^{\theta m} \left(A_m \sum_i m_i \frac{\partial F_0}{\partial y_i} - B_m \sum_i m_i \frac{\partial \Phi_0}{\partial y_i} \right) = 0. \tag{6}$$

Questa equazione deve diventare identica se al posto per esempio di y_1 si mette la sua espressione in funzione delle rimanenti y, ricavata dall'equazione $\Phi_0(y_1, y_2, \cdots, y_n) = 0$. E quindi sopra S_0 debbono singolarmente annullarsi tutti i coefficenti degli $e^{\theta m}$, cioè:

$$A_m \sum_i m_i \omega_i = B_m \sum_i m_i \frac{\partial \Phi_0}{\partial y_i}. \tag{7}$$

(2) Per il significato e le proprietà del simbolo $[F, \Phi]$ qui usato, vedi Poincaré, *loc. cit.* Cap. I.

Di qui segue che se in un punto di S_0 si ha $\sum_i m_i \omega_i = 0$, deve anche essere $\sum_i m_i \frac{\partial \Phi_0}{\partial y_i} = 0$, poiché in generale (3) sarà $B_m \neq 0$.

Dobbiamo a questo punto distinguere se il numero n dei gradi di libertà è uguale o maggiore di due. Se $n = 2$ può darsi benissimo che S_0 abbia l'equazione $\frac{\omega_2}{\omega_1} - \alpha = 0$, dove α è un numero irrazionale. Allora è impossibile trovare due interi m_1, m_2 tali che in un punto di S_0 sia

$$m_1 \omega_1 + m_2 \omega_2 = 0.$$

Da (7) si ricava dunque sopra S_0

$$A_m = \frac{B_m}{m_1 \omega_1 + m_2 \omega_2} \sum_i m_i \frac{\partial \Phi_0}{\partial y_i}$$

e si è sicuri che il denominatore non si annulla mai. Determinando poi a piacere le A_m all'infuori di S_0, resta determinato Φ_1, e potrebbe mostrarsi facilmente che con un procedimento analogo si possono determinare successivamente, almeno dal lato formale, anche $\Phi_2, \Phi_3, \dots$ Resta naturalmente ancora la questione della convergenza delle serie così ottenute. Ma non è naturalmente escluso che, se le B_m decrescono abbastanza rapidamente al crescere delle m, ed eventualmente per particolari valori di α, queste serie convergano effettivamente. Così che nel caso $n = 2$ non è escluso che, oltre le $F =$ costante, possano esistere anche altre, ed eventualmente infinite ipersuperfici con la proprietà discussa alla fine del § 1.

§ 4.

Altrimenti vanno le cose se $n > 2$; supporremo per esempio $n = 3$, ma il nostro ragionamento sarà evidentemente valido per qualunque $n > 2$. Dei tre rapporti $\frac{\omega_2}{\omega_1}$, $\frac{\omega_3}{\omega_1}$, $\frac{\omega_3}{\omega_2}$ può ben darsi che uno resti sopra S_0 costante, ma allora in generale non lo saranno gli altri due; così che, senza in alcun modo limitare la scelta di S_0, possiamo sempre supporre che su di essa né $\frac{\omega_2}{\omega_1}$, né $\frac{\omega_3}{\omega_1}$ siano costanti. Allora S_0 sarà coperta densamente di punti per i quali ω_2/ω_1 è razionale. In uno qualunque di questi punti ci saranno dunque due interi m_1, m_2 tali che $m_1 \omega_1 + m_2 \omega_2 = 0$; nel medesimo punto sarà quindi anche, in virtù di (7), $m_1 \frac{\partial \Phi_0}{\partial y_1} + m_2 \frac{\partial \Phi_0}{\partial y_2} = 0$. Quindi in quel punto sarà $\omega_1 : \omega_2 = \frac{\partial \Phi_0}{\partial y_1} : \frac{\partial \Phi_0}{\partial y_2}$. E siccome questa equazione deve valere in infiniti punti distribuiti densamente sopra S_0, essa deve essere certamente soddisfatta identicamente. Parimenti si dimostra che $\omega_3 : \omega_1 = \frac{\partial \Phi_0}{\partial y_3} : \frac{\partial \Phi_0}{\partial y_1}$. E quindi sopra S_0 si ha $\omega_1 : \omega_2 : \omega_3 = \frac{\partial \Phi_0}{\partial y_1} : \frac{\partial \Phi_0}{\partial y_2} : \frac{\partial \Phi_0}{\partial y_3}$, cioè le derivate parziali di Φ_0 e di F_0 rapporto alle y_i sono proporzionali. Ma ciò può accadere solo se $\Phi_0 = 0$

(3) Potrebbe invero darsi che S_0 fosse tale che su di essa una delle B_m si annullasse identicamente, ma allora le altre B_m sarebbero in generale differenti da zero, e ciò è sufficente per il seguito.

coincide con una delle $F_0 =$ costante, diciamo $F_0 = -c_0$; allora sopra S_0 si ha $\Phi_0 = F_0 + c_0$, e siccome l'estensione di Φ_0 all'infuori di S_0 è arbitraria, possiamo prendere in tutto lo spazio $\Phi_0 = F_0 + c_0$. Si ha ora quindi $\omega_i = \frac{\partial \Phi_0}{\partial y_i}$. Ora che S_0 è così determinato possiamo esser sicuri che in generale sopra S_0 non si annullerà identicamente $\Sigma m_i \omega_i$, a meno che tutte le m_i non siano $= 0$. Eccettuato questo caso possiamo dunque dividere i due membri di (7) per $\Sigma m_i \omega_i = \Sigma m_i \frac{\partial \Phi_0}{\partial y_i}$, e troviamo $A_m = B_m$; solo può essere $A_{00 \cdots 0} \neq B_{00 \cdots 0}$ e dunque la differenza fra Φ_1, ed F_1 può esser solo funzione delle y, diciamo $f_1(y_1, \cdots, y_n)$. Ciò sopra S_0; ma siccome all'infuori di S_0 Φ_1 è in nostro arbitrio, possiamo porre in tutto lo spazio $\Phi_1 = F_1 + f_1$.

In conclusione abbiamo dimostrato che:

$$(8) \qquad \begin{cases} \Phi_0 = F_0 + c_0 \\ \Phi_1 = F_1 + f_1(y_1, \cdots, y_n). \end{cases}$$

§ 5.

Vogliamo ora supporre di aver dimostrato che, per un certo r, si ha:

$$(9) \qquad \begin{cases} \Phi_0 = F_0 + c_0 \\ \Phi_1 = F_1 + c_1 \\ \cdots\cdots\cdots\cdots \\ \Phi_{r-1} = F_{r-1} + c_{r-1} \\ \Phi_r = F_r + f_r(y_1, \cdots, y_n) \end{cases}$$

dove le c_i sono costanti. Io dico allora che equazioni analoghe valgono anche per $r + 1$. Siccome poi per $r = 1$ le (9) coincidono con le (8) e sono dunque dimostrate, ne seguirà allora che per qualunque r è $\Phi_r - F_r =$ costante, e quindi la superfice S_μ coincide effettivamente con una delle $F =$ costante.

Dalla (9) segue subito, che i primi $r + 1$ termini dello sviluppo di (4) sono identicamente nulli in tutto lo spazio; e quindi il coefficente di μ^{r+1} si riduce semplicemente a

$$[F_0, \Phi_{r+1}] + [F_1, f_r] + [F_{r+1}, F_0] = [F_0, \Phi_{r+1} - F_{r+1}] + [F_1 f_r]$$

calcolato sopra S_0. Sopra S_0 si deve dunque avere:

$$(10) \qquad [F_0, \Phi_{r+1} - F_{r+1}] + [F_1, f_r] = 0.$$

Ragionando sopra questa equazione come sopra la (5), si trova che sopra S_0 le derivate parziali di F_0 e di f_r rapporto alle y_i debbono essere proporzionali; e siccome sopra S_0 F_0 è costante, lo dovrà essere anche f_r; siccome poi f_r all'infuori di S_0 è arbitraria, possiamo in tutto lo spazio porre f_r eguale a una costante c_r.

Ora la (10) si riduce $[F_0, \Phi_{r+1} - F_{r+1}] = 0$; poniamo $\Phi_{r+1} - F_{r+1} = \sum_m c_m e^{\theta m}$ allora la (10) diventa:

$$\Sigma e^{\theta m} C_m \sum_i m_i \omega_i = 0:$$

e quindi al solito sopra S_0 deve essere $C_m \sum_i m_i \omega_i = 0$ e perciò tutte le C ad eccezione di $C_{00,\ldots,0}$ debbono esser nulle sopra S_0; poniamo $C_{00,\ldots,0} = f_{r+1}(y_1, \cdots, y_n)$, allora, si ha sopra S_0

$$\Phi_{r+1} = F_{r+1} + f_{r+1}(y_1, \cdots, y_n)$$

e siccome Φ_{r+1} è, all'infuori di S_0, arbitrario, possiamo estendere questa equazione in tutto lo spazio. E così la nostra dimostrazione è completa.

§ 6.

Vogliamo infine raccogliere alcune condizioni sufficenti, ma non necessarie, per la validità della dimostrazione precedente:

1° Deve essere $n > 2$.

2° Nel sistema imperturbato non debbono esistere identicamente relazioni lineari intere tra le frequenze fondamentali ω_i.

3° Nessuna equazione del tipo $\omega_2/\omega_1 = \text{cost}$, deve essere conseguenza di un'altra dello stesso tipo, come per esempio $\omega_i/\omega_k = \text{cost}$.

4° Sopra nessuna delle $F_0 = \text{cost}$. può $\Sigma m_i \omega_i$ annullarsi identicamente, se non sia $m_1 = m_2 = \cdots = m_n = 0$.

5° Sopra nessuna delle $F_0 = \text{cost}$. deve in generale da $\Sigma m_i \omega_i = 0$ seguire anche $B_m = 0$.

Göttingen, aprile 1923.

N° 16.

This paper has historical interest showing for the first time the concern of Fermi for the entropy constant of the perfect monatomic gas. It is a first indication of the thoughts of Fermi on the subject. Indeed the division of phase space in cells of volume h^3 was a subject of deep meditations on his part and he arrived later at the conclusion that a further link was missing, without being able to perform the decisive step. (Cfr. N° 19).

E. SEGRÈ

16.

SOPRA LA TEORIA DI STERN DELLA COSTANTE ASSOLUTA DELL'ENTROPIA DI UN GAS PERFETTO MONOATOMICO

« Rend. Lincei », *32* (2), 395–398 (1923) (*).

1. La costante assoluta dell'entropia di un gas perfetto monoatomico fu determinata per la prima volta da Sackur [1] e da Tetrode [2] con un metodo statistico, che consisteva essenzialmente nel porre, con Boltzmann, l'entropia proporzionale al logaritmo della probabilità, calcolando il valore assoluto della probabilità con l'ammettere che il volume delle cellette, in cui è perciò necessario dividere lo spazio delle fasi, e che ha le dimensioni del cubo di una azione, fosse precisamente eguale alla terza potenza della costante h di Planck. Essi trovarono in questo modo che l'entropia di un gas monoatomico è data dalla formula:

$$S = Nk \left\{ \frac{5}{2} \log T - \log p + \log \frac{(2\pi m)^{3/2} k^{5/2} e^{5/2}}{h^3} \right\} \tag{1}$$

dove N è il numero degli atomi, k la costante assoluta di Boltzmann, T la temperatura assoluta, p la pressione ed m la massa degli atomi.

Che, nonostante la verifica sperimentale della (1), questo modo di dedurla non sia a molti apparso soddisfacente, è dimostrato dal gran numero di lavori teorici, che furono fatti in seguito, con lo scopo di trovarne una dimostrazione migliore. Di tutti questi tentativi, quello che senza dubbio ha meglio raggiunto il suo scopo è quello di Stern [3]; esso è basato sul seguente principio: consideriamo il nostro gas come vapore di un corpo solido; possiamo calcolarne la tensione massima in due modi: 1° con la teoria cinetica dei gas, ed in tale modo otteniamo un risultato completamente

(*) Presentata alla seduta del 2 dicembre 1923 dal Socio O. M. Corbino.

(1) O. SACKUR, « Ann. d. Phys. », *40*, 67 (1913).

(2) H. TETRODE, « Ann. d. Phys. » *38*, 434 (1912).

(3) O. STERN, « Phys. ZS. », *14*, 629 (1913); « ZS. f. Elecktrochem. », *25*, 66 (1919).

determinato, senza costanti arbitrarie; 2° con la termodinamica, ed in tale caso ci figura come costante arbitraria precisamente la costante assoluta dell'entropia del gas, poiché quella del corpo solido si può ottenere per mezzo del teorema di Nernst. Col confronto delle due espressioni, Stern giunge a determinare la costante assoluta; ed il suo metodo ha il vantaggio di non fare sopra il gas perfetto nessuna di quelle poco legittime ipotesi, che sono necessarie agli altri Autori, come per esempio quella di una quantizzazione del gas stesso, della quale non si vede chiaramente la ragione.

2. Un punto però, che anche nella teoria di Stern non resta molto chiaro, è il seguente: Stern trova come densità del vapore saturo per via cinetica la seguente espressione:

$$\rho = \frac{(2\pi m)^{3/2}\,\tilde{\nu}^3}{(kT)^{3/2}}\, e^{-\frac{w}{kT}} \tag{2}$$

dove $\tilde{\nu}$ è la media geometrica delle frequenze elastiche fondamentali del corpo solido, w l'energia necessaria a far evaporare un atomo dal corpo solido, allo zero assoluto. Col metodo termodinamico egli trova invece:

$$\rho = \frac{(2\pi m)^{3/2}\,\tilde{\nu}^3}{(kT)^{3/2}}\, e^{-\frac{w+\frac{3h\bar{\nu}}{2}}{kT}} \tag{3}$$

essendo $\bar{\nu}$ la media aritmetica delle dette frequenze.

La differenza negli esponenti di e viene da Stern spiegata con l'ipotesi così detta dell'energia allo zero assoluto (Nullpunktsenergie), secondo la quale si ammette che, anche allo zero assoluto, l'energia media di un oscillatore non sia eguale a zero, ma a mezzo quanto, cioè ad $h\nu/2$. Ammettendo questo, il calore di vaporizzazione di un atomo non sarebbe più w, bensì, siccome ogni molecola ha tre gradi di libertà, $w - (3\,h\nu/2)$ e così verrebbe compensata la differenza dei due esponenti.

In questo lavoro mi propongo di dimostrare che questa ipotesi innaturale non è affatto necessaria per spiegare la differenza suddetta, ma che basta perciò modificare leggermente la deduzione cinetica della (2), tenendo conto che le molecole del corpo solido possono muoversi solamente sopra orbite quantiche.

3. Per giungere cineticamente alla densità del vapore saturo occorre anzitutto calcolare la probabilità di uno stato in cui, fissato il volume V, N_g atomi appartengano al gas, ed N_s al corpo solido: essendo $N = N_g + N_s$ il numero totale degli atomi presenti. Osservando che gli atomi del solido, a differenza di quelli del gas, sono distinguibili tra di loro per la loro posizione, si riconosce facilmente, ricordando i principii della meccanica statistica, che tale probabilità è:

$$P = \binom{N}{N_g} N_s!\int e^{-\frac{W}{kT}}\,d\omega = \frac{N!}{N_g!}\int e^{-\frac{W}{kT}}\,d\omega \tag{4}$$

dove $d\omega$ è l'elemento di volume dello spazio delle fasi, che ha per coordinate le coordinate ed i momenti x_i, p_i $(i = 1, 2, \cdots, 3\,N_g)$ delle molecole del gas, e le coordinate ed i momenti x_j, p_j ($j = 1, 2, \cdots, 3N_s$), degli atomi della parte solida; W è l'energia corrispondente allo stato $d\omega$. Quanto alle molecole del gas non c'è nulla da dire; invece riguardo alle molecole del solido modifichiamo un pò le considerazioni di Stern, osservando che le molecole del corpo solido possono descrivere soltanto delle orbite statiche (4) le quali, in questo caso, sono caratterizzate dal fatto, che l'energia corrispondente ad ogni frequenza elastica deve essere $n_j h\nu_j$ essendo n_j, un intero, e ν_j la frequenza corrispondente. Conformemente alla solita statistica della teoria dei quanti, attribuiamo a ciascuno di questi moti elementari il peso h, ponendo cioè:

$$dx_j\, dp_j = h \tag{5}$$

di modo che

$$d\omega = h^{3N_s} \prod_1^{3N_g} dx_i\, dp_i .$$

L'energia W si calcola come somma dell'energia cinetica delle molecole del gas, eguale $\Sigma p_i^2/2m$, più l'energia del corpo solido $= \Sigma n_j h\nu_j$, più l'energia potenziale delle molecole del gas eguale a wN_g, essendo w appunto l'energia di vaporizzazione di una molecola allo zero assoluto. Si ha dunque:

$$W = \sum_1^{3N_g} \frac{p_i^2}{2m} + \sum_1^{3N_s} n_j h\nu_j + wN_g . \tag{7}$$

Osservando ancora che nella (4) i $6N_s$ integrali relativi a dx_j, dp_j; vanno naturalmente sostituiti con delle sommatorie, la (4) diventa:

$$P = \frac{N!}{N_g!} h^{3N_s} e^{-\frac{wN_g}{kT}} \prod_1^{3N_g} \left(\int dx_i\right) \left(\int_{-\infty}^{\infty} e^{-\frac{p^2}{2mkT}} dp\right)^{3N_g} \prod_{j=1}^{3N_s} \left(\sum_{n_j=0}^{\infty} e^{-\frac{n_j h\nu_j}{kT}}\right). \tag{8}$$

Osservando ora che se x_1, x_2, x_3 sono coordinate cartesiane della prima molecola del gas si ha:

$$\int dx_1 \int dx_2 \int dx_3 = V ,$$

essendo V il volume del gas, si trova:

$$\Pi \int dx_i = V^{N_g} .$$

Si ha inoltre:

$$\int_{-\infty}^{\infty} e^{-\frac{p^2}{2mkT}} dp = \sqrt{2\pi m\, kT} ;$$

$$\sum_{n_j=0}^{\infty} e^{-\frac{n_j h\nu_j}{kT}} = \frac{1}{1 - e^{-\frac{h\nu_j}{kT}}} = \frac{kT}{h\nu_j}\left(1 + \frac{h\nu_j}{2kT} + \cdots\right).$$

(4) Naturalmente in queste considerazioni, si tratta dell'atomo considerato come punto materiale, e i moti di cui si parla sono l'agitazione termica.

Applicando la formula di Stirling, si trova dunque:

$$\log P = \log N! - N_g \log \frac{N_g}{e} - \frac{N_g w}{kT} + N_g \log V + \frac{3}{2} N_g \log (2\pi m\, kT)$$

$$+ 3 N_s \log \frac{kT}{\bar{\nu}} + \sum_{1}^{3 N_s} \log \left(1 + \frac{h\nu_j}{2 kT} + \cdots \right)$$

ossia al limite per T molto grande:

$$\log P = \log N! - N_g \log \frac{N_g}{e} - \frac{N_g w}{kT} + N_g \log [V (2\pi m\, kT)^{3/2}]$$

$$+ 3 N_s \log \frac{kT}{\bar{\nu}} + \frac{3 N_s h\bar{\nu}}{2 kT} .$$

Nello stato di equilibrio P deve essere un massimo: e perciò deve annullarsi la sua derivata rispetto ad N_g, dove si deve osservare però che $N_s = N - N_g$ ed N è una costante. Si trova dunque:

$$0 = -\log N_g - \frac{w + 3 h\bar{\nu}/2}{kT} + \log [V (2\pi m\, kT)^{3/2}] - \log \left(\frac{kT}{\bar{\nu}} \right)^3 .$$

Di qui si ricava subito che la densità del gas è:

$$\frac{N_g}{V} = \frac{(2\pi m)^{3/2} (\bar{\nu}^3)}{(kT)^{3/2}} e^{-\frac{w + 3 h\bar{\nu}/2}{kT}}$$

in completa concordanza con la (3).

N° 17 *a* and 17 *b*.

Footnotes included in the otherwise identical German version, not reproduced here, indicate that Fermi was acquainted with an earlier attempt by Herzfeld (1) to solve the same problem by a somewhat cruder approach, and also that he was indebted to Max Born for a suggestion. Brillouin (2) has discussed and compared the different treatments that various authors have used to avoid the divergence of the partition function for the quantum levels of an atom.

F. RASETTI

17 *a*

SULLA PROBABILITÀ DEGLI STATI QUANTICI

«Rend. Lincei» *32* (2), 493–495 (1923) (*).

1. Dato un atomo, suscettibile di assumere varî stati quantici, il problema della determinazione della probabilità che esso si trovi in un assegnato stato quantico si risolve di solito ammettendo che le probabilità *a priori* di tutti i suoi stati possibili, siano eguali (3), e da questa ammissione segue, con gli ordinarî metodi della statistica, che la probabilità cercata è proporzionale a

$$(1) \qquad e^{-\frac{w_i}{kT}}$$

essendo w_i l'energia dello stato che si considera.

Vi sono però alcuni casi in cui queste considerazioni debbono evidentemente venire modificate. Se per esempio si considera un atomo, per fissare le idee un atomo di idrogeno, si ha che l'energia del suo i^{esimo} stato quantico è

$$(2) \qquad w_i = -\frac{A}{i^2}$$

e quindi, secondo le considerazioni precedenti, la probabilità dell'i^{esimo} stato dovrebbe essere

$$(3) \qquad p_i = -C\, e^{\frac{A}{kT\, i^2}}$$

essendo C una costante da determinare con la condizione che $\sum_1^\infty p_i = 1$. Siccome però

$$(4) \qquad \sum_1^\infty e^{\frac{A}{kT\, i^2}}$$

è evidentemente divergente, risulterebbe C = 0.

(1) K. F. HERZFELD, «Ann. d. Phys.», *51*, 261 (1916).

(2) L. BRILLOUIN, *Les statistiques quantiques et leurs applications*, Vol. 2. Les Presses Universitaires de France, Paris, 1930.

(*) Presentata nella seduta del 16 dicembre 1923 dal Socio O. M. Corbino.

(3) Vedi per esempio N. BOHR, «ZS. f. Phys.», *13*, 117 (1923).

Questa difficoltà viene di solito eliminata ammettendo che in un gas, ad una certa pressione, non siano possibili quegli stati quantici per i quali il raggio dell'atomo è maggiore della distanza media tra atomo ed atomo, cosicché la somma (4), invece che da 1 ad ∞, deve essere estesa solo ad un numero finito di termini.

In questo lavoro cercheremo di eliminare la difficoltà osservata in un modo un po' più preciso. Cercheremo perciò di fare il calcolo termodinamico dell'equilibrio tra i varî stati quantici, tenendo conto del volume delle molecole; e vedremo che questa considerazione ci condurrà ad un'espressione quantitativa del fatto, qualitativamente evidente, che gli atomi di maggiori dimensioni sono di formazione più difficile, perché maggiormente disturbati dagli urti contro gli altri atomi.

2. Consideriamo dunque l'equilibrio chimico tra varî stati quantici di uno stesso atomo, e supponiamo che le energie di essi siano

$$w_1, w_2, \cdots w_r$$

ed i rispettivi volumi atomici

$$v_1, v_2, \cdots v_r .$$

Siano poi $n_1, n_2, \cdots, n_r$ i numeri di atomi del $1, 2, \cdots, n^{\text{esimo}}$ stato; $n_1 + n_2 + n_3 + \cdots + n_r = n$ rappresenterà il numero totale degli atomi, che supponiamo racchiusi in un recipiente di volume unitario. Per calcolare il nostro equilibrio chimico dobbiamo prima di tutto calcolare l'energia libera del nostro miscuglio di stati quantici; e lo stato di equilibrio si troverà poi facilmente cercando quei valori di $n_1, n_2, \cdots, n_r$ per i quali l'energia libera è minima. Se F è l'energia libera, U l'energia totale, T la temperatura assoluta, S l'entropia, si ha

$$F = U - TS . \tag{5}$$

Ora nel nostro caso l'energia è la somma dell'energia cinetica degli atomi, e della loro energia di formazione; si ha dunque.

$$U = \frac{3}{2} nkT + \Sigma w_i n_i . \tag{6}$$

Il calcolo dell'entropia è alquanto più complesso, e lo faremo solo ammettendo, in analogia a quel che si fa a proposito dell'equazione di van der Waals, che il contributo all'entropia totale, di ciascuna specie di atomi, sia eguale all'entropia che avrebbero gli stessi atomi, se da soli occupassero un recipiente di volume eguale a quello del recipiente dato, diminuito del volume occupato dalle molecole presenti, che, come nell'equazione di van der Waals, possiamo per esempio prendere eguale al quadruplo del volume totale delle molecole presenti. Tale volume risulta cioè $1 - 4 \Sigma n_i v_i$; e quindi l'entropia risulta data da

$$S = \sum_1 n_i k \left\{ \frac{3}{2} \log T - \log \frac{n_i}{1 - 4 \Sigma n_i v_i} + A \right\} . \tag{7}$$

Se supponiamo che $\Sigma n_i v_i$ sia piccolo, possiamo un po' semplificare l'espressione (5) dell'energia libera, e scrivere

$$(8) \qquad F = \frac{3}{2} nkT + \Sigma w_i n_i - kT \sum_i n_i \left\{ \frac{3}{2} \log T - \log n_i - 4 \Sigma n_i v_i + A \right\}.$$

La condizione di equilibrio si otterrà, come si è già detto, cercando i valori delle n_i per i quali l'energia libera è un minimo, dove osserviamo però che le n_i debbono sempre verificare la condizione

$$n = n_1 + n_2 + \cdots + n_r .$$

Si trova dunque che

$$(9) \qquad n_i = C e^{-\frac{w_i}{kT}} e^{-4 n v_i}$$

essendo C una costante che deve essere determinata dalla condizione che $\Sigma n_i = n$.

Se si confronta quest'ultima formula con la (1) si vede subito che la differenza tra di esse consiste nel fattore e^{-4nv_i} che figura nella (9) e non, nella (1) e che può in un certo senso interpretarsi come una probabilità *a priori* dello stato quantico i^{esimo}, la quale naturalmente decresce col crescere della pressione, e del volume dell'atomo.

3. Consideriamo ora per esempio il caso dell'atomo di idrogeno; per esso abbiamo $-w_i = A/i^2$; il volume atomico lo vogliamo prendere eguale a quello di una sfera di raggio eguale a quello dell'atomo; siccome dunque il raggio è proporzionale ad i^2, il volume atomico risulterà proporzionale ad i^6. Sia $v = Bi^6$. Osserviamo anche che vi sono i atomi possibili (*) che corrispondono al numero quantico totale i (corrispondentemente ai varî valori che può prendere il quanto azimutale), e quindi si vede che la costante C delle (9) va determinata in modo che

$$(10) \qquad n = C \sum_1^{\infty} i \, e^{\frac{A}{kT i^2}} e^{-4 n B i^6} .$$

È evidente, e del resto facile da verificare, che la serie converge, e che resta quindi eliminata la difficoltà a cui accennavamo in principio. Per renderci in qualche modo un'idea dell'influenza esercitata dal fattore dipendente dal volume possiamo fare il seguente calcolo numerico. Per l'atomo di idrogeno B è uguale pressapoco a $5 \cdot 10^{-25}$. Se supponiamo per esempio di operare alla pressione di un'atmosfera si ha $n = 2{,}7 \cdot 10^{19}$ e quindi, come ordine di grandezza, $4 n B = 5 \cdot 10^{-5}$. Ne segue che, mentre per $i = 1$ si ha $e^{-4 n v_i} = e^{-0{,}00005}$, quindi vicinissimo ad 1, già per $i = 6$ si ha invece, $e^{-4 n v_i} = e^{-2{,}3} = \sim 0{,}1$: per $i = 10$ si ha $e^{-4 n v_i} = e^{-50} = \sim 10^{-22}$ cioè già un numero piccolissimo; e ciò ci dimostra che la causa di cui ci siamo occupati è effettivamente sufficiente a giustificare l'impossibilità dell'esistenza di numeri quantici elevati, finché la pressione non è estremamente piccola.

(*) Quando il lavoro fu scritto, non era chiaro l'effetto della quantizzazione spaziale sul numero degli stati quantici, e non si conosceva lo spin dell'elettrone. Il valore corretto della molteplicità del livello i dell'idrogeno è $2 i^2$ (N. d. R.).

N° 18.

A detailed theory of the change in the angular distribution of the optical resonance radiation that takes place with increasing density of the gas. The author accounts quantitatively for the experimental results of Wood, who observed in mercury vapor a transition from almost isotropic scattering at low gas densities, to regular reflection at very high densities.

F. RASETTI

18.

SOPRA LA RIFLESSIONE E LA DIFFUSIONE DI RISONANZA

« Rend. Lincei », *33*, (1), 90–93 (1924) (*).

1. Se si illumina un gas o vapore con luce di frequenza eguale a quella della sua riga di risonanza, si produce il fenomeno, scoperto da Wood (1), della così detta risonanza ottica, che consiste nel fatto che la luce primaria viene assai intensamente diffusa dal gas, per modo che bastano spesso degli strati estremamente sottili del gas per assorbire completamente la luce primaria. Ora si osserva questo fatto (2): finché la pressione del gas è sufficientemente piccola, quasi tutta la luce di risonanza viene diffusa irregolarmente in tutte le direzioni; quando invece la pressione diventa abbastanza grande, la maggior parte viene invece riflessa regolarmente, e solo una piccola parte viene sparpagliata in tutte le direzioni. Siccome non mi consta che di questo fenomeno sia ancora stata data una interpretazione ben chiara, cercherò in questo lavoro di svilupparne la teoria matematica, la quale, oltre a dare codesta interpretazione, permetterà anche di assegnare le condizioni sperimentali per le quali è più facile l'osservazione della riflessione di risonanza.

2. Supponiamo che il nostro gas sia contenuto in un recipiente prismatico a base quadrata di lato B coi lati paralleli agli assi y, z, illimitato dalla parte delle x positive, e nel senso opposto limitato dal piano $x = 0$. Supponiamo poi che la luce primaria, di frequenza v, abbia coseni di direzione l, m, n, ed investa la base del prisma, ed osserviamo la luce secondaria nella direzione di coseni α, β, γ, $(\alpha < 0)$. Sia k il coefficiente d'assorbimento; se il punto P di osservazione è abbastanza lontano dal gas ed r è la sua di-

(*) Presentata nella seduta del 13 gennaio 1924 dal Socio O. M. Corbino.

(1) WOOD, « Phil. Mag. », dal 1905 al 1912.

(2) WOOD, « Phil. Mag. », *23*, 689 (1912); DUNOYER, « C. R. », *156*, 1067 (1913).

stanza dal centro delle coordinate, si riconosce subito che, prescindendo dall'influenza dei bordi, l'ampiezza della luce secondaria in P è data da

$$\sum_h \frac{A}{r} e^{-\frac{k}{2}\left(\frac{1}{l}-\frac{1}{\alpha}\right)x_h} e^{2\pi i\nu\left(t-\frac{r}{c}\right)} e^{-\frac{2\pi i}{\lambda}\{(l-\alpha)x_h+(m-\beta)y_h+(n-\gamma)z_h\}}.$$

Dove l'indice h distingue i varî atomi del gas, e la somma si deve estendere a tutti questi atomi, A è un coefficiente di proporzionalità. Per passare dall'ampiezza all'intensità basta, come è noto, moltiplicare l'ampiezza precedente per il numero suo complesso coniugato. L'intensità è dunque

$$I = \frac{A^2}{r^2}\sum_{hj} e^{-\frac{k}{2}\left(\frac{1}{l}-\frac{1}{\alpha}\right)(x_h+x_j)} e^{\frac{2\pi i}{\lambda}\{(l-\alpha)(x_j-x_h)+(m-\beta)(y_j-y_h)+(n-\gamma)(z_j-z_h)\}}.$$

La somma deve naturalmente estendersi, per entrambi gli indici h, j, a tutti gli atomi presenti. Per effettuarla ci conviene spezzare I nella somma di due addendi I_1 ed I_2, dei quali il primo si ottenga come contributo dei termini della sommatoria per i quali $h = j$, ed il secondo sia il contributo dei termini per cui $h \neq j$. Avremo allora

$$I_1 = \frac{A^2}{r^2}\sum_h e^{-k\left(\frac{1}{l}-\frac{1}{\alpha}\right)x_h}$$

che potrà calcolarsi sostituendo la somma con l'integrale triplo

$$I_1 = \frac{NA^2}{r^2}\int_0^{\infty}\int_{-B/2}^{B/2}\int_{-B/2}^{B/2} e^{-k\left(\frac{1}{l}-\frac{1}{\alpha}\right)x}\,dx\,dy\,dz\,,$$

essendo N il numero di atomi per unità di volume. Calcolando l'integrale si trova

$$(1) \qquad I_1 = \frac{NA^2 B^2}{kr^2\left(\frac{1}{l}-\frac{1}{\alpha}\right)}.$$

Si ha poi

$$I_2 = \frac{A^2}{r^2}\sum_{h\neq j} e^{-\frac{k}{2}\left(\frac{1}{l}-\frac{1}{\alpha}\right)(x_h+x_j)} e^{\frac{2\pi i}{\lambda}\{(l-\alpha)(x_j-x_h)+\cdots\}}.$$

Anche quì possiamo ora sostituire la somma con un integrale sestuplo (3); si trova quindi

$$I_2 = \frac{A^2N^2}{r^2}\int\!\!\int\!\!\int\!\!\int\!\!\int\!\!\int e^{-\frac{k}{2}\left(\frac{1}{l}-\frac{1}{\alpha}\right)(x+\xi)} e^{\frac{2\pi i}{\lambda}\{(l-\alpha)(\xi-x)+\cdots\}}\,dx\,dy\,dz\,d\xi\,d\eta\,d\zeta$$

ed effettuando le quadrature

$$(2) \qquad I_2 = \frac{4A^2N^2\lambda^4}{\pi^4 r^2}\,\frac{\sin^2\frac{\pi B}{\lambda}(m-\beta)}{(m-\beta)^2}\,\frac{\sin^2\frac{\pi B}{\lambda}(n-\gamma)}{(n-\gamma)^2}\,\frac{1}{\frac{k^2}{4}\left(\frac{1}{l}-\frac{1}{\alpha}\right)^2+\frac{4\pi^2}{\lambda^2}(l-\alpha)^2}.$$

Si vede ora che mentre I_1 ha, come funzione di α, β, γ, un andamento regolare (poché $l > 0 > \alpha$), I_2 presenta invece un massimo assai pronunziato quando $m = \beta$, $n = \gamma$ e quindi $l = -\alpha$, purché, come vogliamo ammettere, sia $B \gg \lambda$. Questo massimo corrisponde ad una riflessione regolare;

(3) Questo non sarebbe più lecito se non si fossero già trattati a parte i termini per i quali $k = j$ che darebbero luogo a delle singolarità.

I_1 ci rappresenterà dunque la luce diffusa in tutte le direzioni, I_2 quella riflessa regolarmente. Siccome poi I_1 è proporzionale ad N, e I_2 ad N^2, si capisce anche come, al crescere della densità, I_2 vada poco a poco a prendere il sopravvento sopra I_1, come appunto sperimentalmente si osserva.

3. Per discutere però in un caso più concreto le circostanze di questo fenomeno, ci converrà modificare leggermente le premesse. Mentre fin'ora abbiamo preso come sorgente della luce primaria un punto all'infinito, prenderemo ora un piccolo disco circolare, pure all'infinito. Sia ρ_0 l'angolo sotto cui si vede il suo raggio; prendiamo poi le coordinate in modo che i coseni direttori del centro siano $-\cos\theta$, $-\sin\theta$, o (θ = angolo di incidenza) ed osserviamo nella direzione $\alpha = -\cos\theta$, $\beta = \sin\theta$, $\gamma = 0$ in cui si riflette il raggio proveniente dal centro. Per il punto della sorgente, di coordinate polari sferiche (4) ρ, φ avremo allora approssimativamente

$$l = \cos\theta + \rho\cos\varphi\sin\theta \quad ; \quad m = \sin\theta - \rho\cos\varphi\cos\theta \quad ; \quad n = -\rho\sin\varphi$$

e quindi applicando la (1) e la (2) si trova in questo caso

$$\frac{I_2}{I_1} = \frac{2N\lambda^4 k}{\pi^5 \rho_0^2 B^2 \left(k^2 + \frac{16\pi^2}{\lambda^2}\cos^4\theta\right)\cos\theta}$$

$$\times \int_0^{\rho_0}\int_0^{2\pi} \frac{\sin^2\left(\frac{\pi B}{\lambda}\rho\cos\varphi\cos\theta\right)\sin^2\left(\frac{\pi B}{\lambda}\rho\sin\varphi\right)}{\rho^3\cos^2\varphi\sin^2\varphi}\,d\rho\,d\varphi = \frac{2N\lambda^2 k}{\pi^3\rho_0^2\cos\theta\left(k^2 + \frac{16\pi^2}{\lambda^2}\cos^4\theta\right)}$$

$$\times \int_0^{\pi B\rho_0/\lambda}\int_0^{2\pi} \frac{\sin^2(x\cos\varphi\cos\theta)\sin^2(x\sin\varphi)}{x^3\cos^2\varphi\sin^2\varphi}\,dx\,d\varphi\,.$$

Distinguiamo ora secondo che $B\rho_0/\lambda$ è molto grande o molto piccolo. Nel primo caso potremo sostituire il limite superiore del primo integrale con ∞. Allora il valore dell'integrale doppio è $\pi^2\cos\theta$ e quindi

$$\frac{I_2}{I_1} = \frac{2N\lambda^2 k}{\pi\rho_0^2\left(k^2 + \frac{16\pi^2}{\lambda^2}\cos^4\theta\right)}\,. \tag{3}$$

Se invece $B\rho_0/\lambda \ll 1$, si potrà nell'integrale doppio sostituire gli archi ai seni; si trova quindi facilmente

$$\frac{I_2}{I_1} = \frac{2N B^2 k\cos\theta}{k^2 + \frac{16\pi^2}{\lambda^2}\cos^4\theta}\,. \tag{4}$$

4. Per esempio, supponiamo che sia $\rho_0 = 1/10$. Allora, escluso il caso che B sia straordinariamente piccolo, saremo in condizioni da applicare la (3), e troviamo che, perché I_2 sia dell'ordine di grandezza di I_1, condizione evidentemente necessaria per l'osservazione della riflessione di risonanza, e ponendo $k = 100$ deve essere come ordine di grandezza $N = 10^{20}$ ciò che corrisponde a una pressione di qualche atmosfera. Si spiega con ciò l'insuccesso di Dunoyer nell'osservazione della riflessione di risonanza per il sodio, il cui vapore non si può comodamente osservare che a pressioni abbastanza basse.

(4) S'intende nel piano all'infinito, con polo nel centro della sorgente.

N° 19.

This paper, written in January 1924, is chiefly important in showing that Fermi had been concerned with the Gibbs paradox and the absolute entropy constant (cfr. also N° 16) at least two years before he wrote his celebrated paper on the statistics of the ideal gas. He points out that in systems with identical particles, such as the helium atom, the complete period to which the Sommerfeld integral must be extended should correspond to the repetition of a state indistinguishable from the initial state, even though the two states differ classically in that the two particles have exchanged places. His remarks bear some analogy to the considerations on the exchange of the coordinates of identical particles in the wave function of a system, as were later formulated in quantum mechanics. We note again in this paper a trend of ideas (for instance at the end of section 1) which shows how he was groping for something missing (the exclusion principle). As soon as this was discovered by Pauli he was ready to develop the new statistics on a basis and from a point of view relatively independent of the contemporary development of quantum mechanics.

F. RASETTI

19.

CONSIDERAZIONI SULLA QUANTIZZAZIONE DEI SISTEMI CHE CONTENGONO DEGLI ELEMENTI IDENTICI

«Nuovo Cimento», *1*, 145–152 (1924).

Le regole date da Sommerfeld per la determinazione delle orbite quantiche dei sistemi che ammettono la separazione delle variabili, e che, come si sa, si riducono ad imporre che per tali orbite gli integrali delle fasi $\oint p\,dq$, siano tutti multipli interi della costante h di Planck, si sono mostrate in perfetto accordo con l'esperienza nella loro applicazione all'atomo di idrogeno, e rendono conto, entro i limiti dell'errore sperimentale, di tutti i fatti osservati nella spettroscopia di questo elemento; è noto d'altra parte che tutti i tentativi finora fatti, di estendere quelle regole anche al calcolo di elementi più complessi, non hanno portato altro che a risultati qualitativi, e che, malgrado numerosi sforzi, l'accordo quantitativo tra esperienza e teoria non si è potuto raggiungere nemmeno per l'elio che, dopo l'idrogeno, è il più semplice atomo.

Un tale insuccesso viene di solito attribuito al fatto che tali sistemi più complessi non ammettono la separazione delle variabili, e che il metodo delle perturbazioni, che è stato escogitato per poter generalizzare anche a loro le condizioni di Sommerfeld sia per una ragione o per l'altra insufficiente a calcolarli.

In questo lavoro mi propongo di mostrare come ci siano delle ragioni che inducono a credere, che l'insuccesso sia dovuto piuttosto all'insufficienza delle condizioni di Sommerfeld a calcolare le orbite statiche di quei sistemi che, indipendentemente dall'ammettere o no la separazione delle variabili, contengono alcune parti identiche (nel caso dell'atomo di elio per esempio i due elettroni sono tra di loro non distinguibili). Che delle modificazioni delle leggi quantistiche si possono rendere necessarie in conseguenza dell'essere il sistema da quantizzare costituito da elementi identici appare in qualche caso molto plausibile; consideriamo per esempio un anello occupato da tre elettroni disposti ai vertici di un triangolo equilatero. Siccome gli elettroni sono tra di loro indistinguibili è chiaro che, per riportare con una rotazione l'anello nella sua situazione iniziale, non sarà necessario farlo rotare di un angolo di 2π, ma basterà un angolo di $2\pi/3$. Vediamo così che, mentre se gli elettroni fossero tra di loro distinguibili il moto di rotazione dell'anello sarebbe periodico a periodo 2π, data invece la loro indistinguibilità esso diventa periodico a periodo $2\pi/3$. Se p è la quantità di moto areale dell'anello nella prima ipotesi $2\pi p$ dovrebbe essere un multiplo di h, nella seconda invece $2\pi p/3$ deve essere un multiplo intero di h, ossia il valore minimo di p è nel secondo caso tre volte più grande che nel primo. Considerazioni di questo genere applicate alla quantizzazione dei reticoli cristallini, nei quali, data l'identità tra tutti gli atomi, una traslazione eguale e parallela ad uno dei lati del parallelepipedo elementare, può considerarsi come un periodo, furono recentemente svolte da Compton [1] e altri che ne dedussero una teoria quantistica molto chiara della diffrazione dei raggi Röntgen.

1. Mi propongo ora di mostrare un esempio di sistema a variabili separate, per il quale appaia evidente l'insufficienza delle condizioni di Sommerfeld a calcolarne le orbite statiche.

Consideriamo perciò un gas perfetto, costituito da n molecole puntiformi, racchiuse in un volume v. Noi cercheremo di calcolare il valore assoluto dell'entropia di questo gas, facendo diverse ipotesi sopra il modo di quantizzarlo (per ottenere un valore finito dell'entropia di un gas perfetto è necessario quantizzarlo in un modo o nell'altro, poiché la trattazione classica ci condurrebbe sempre ad un valore infinito). In tutte queste ipotesi noi ridurremo sempre il nostro sistema a sistemi a variabili separate, che quantizzeremo con le regole di Sommerfeld, e riscontreremo tuttavia, che di tali ipotesi la sola che conduca ad un valore della costante dell'entropia d'accordo con il valore sperimentale è quella in cui il sistema fondamentale, a cui si ridurrà il nostro gas, contenga un solo atomo; mentre riducendolo a sistemi, pure a variabili separate, ma contenenti più di un atomo, si troveranno sempre delle maggiori o minori divergenze.

Il criterio che seguiremo nella trattazione del nostro problema sarà il seguente: divideremo anzitutto il volume v a disposizione del nostro gas in celle a forma di parallelepipedi rettangoli, e potremo a piacere per esempio

(1) A. H. COMPTON, « Proc. Nat. Acad. Sci. », 9, 359 (1923); G. BREIT, ibid., 9, 244 (1923).

dividere v in n parallelepipedi, mettendo in ciascuno di essi una sola molecola; oppure dividere v in $n/2$ parallelepipedi, e mettere in ciascuno 2 molecole, oppure infine dare a v la forma di un solo parallelepipedo, e lasciare in esso tutte le n molecole (2). Il sistema di alcune molecole di un gas perfetto racchiuse in un recipiente di forma di parallelepipedo è evidentemente a variabili separate, in coordinate cartesiane parallele agli spigoli, poiché, essendo il gas perfetto, sono da trascurare le azioni delle molecole tra di loro. Una volta quantizzato così il movimento del nostro gas, si può facilmente calcolarne l'entropia; si riscontrerà allora che il valore della costante dell'entropia d'accordo con l'esperienza si può ottenere soltanto dividendo il volume in n parallelepipedi, e mettendo in ciascuno di essi una sola molecola, e che invece quantizzando dei sistemi che contengano anche due sole molecole identiche, si troveranno sempre dei risultati discordanti.

2. Supponiamo che le celle parallelepipede in cui abbiamo diviso il nostro volume, abbiano lati a, b, c; lo stesso calcolo svolto nella citata memoria di Brody, ci conduce a scrivere che le energie quantisticamente possibili del moto di una molecola parallelo al lato a sono

$$w_n = \frac{n^2 h^2}{8 a^2 m} \tag{1}$$

essendo n un intero, m la massa delle molecole e h la costante di Planck. Il contributo all'entropia di questo grado di libertà si calcola con la nota formula

$$\sigma^* = \frac{\overline{w}}{T} + k \log\left(\Sigma e^{-\frac{w_i}{kT}}\right)$$

che si può facilmente dedurre con un calcolo diretto, e nella quale $\overline{w}$ rappresenta l'energia media del grado di libertà, eguale in questo caso a $kT/2$; sostituendo la sommatoria tra parentesi con un integrale, si trova subito nel nostro caso

$$\Sigma e^{-\frac{w_i}{kT}} = \int_0^\infty e^{-\frac{h^2 i^2}{8 a^2 mkT}} di = \frac{a}{h} \sqrt{2 \pi m} \sqrt{kT}$$

e quindi

$$\sigma^* = \frac{k}{2} \log T + k \log\left(\frac{a}{h} \sqrt{2 \pi mek}\right)$$

Computando i tre gradi di libertà della molecola (il gas è supposto monoatomico) si trova dunque che il contributo della molecola all'entropia totale è

$$\frac{3k}{2} \log T + k \log \frac{(2 \pi mke)^{3/2} v_c}{h^3}$$

avendo indicato con v_c il volume della cella.

(2) Un metodo simile a questo per il calcolo dell'entropia di un gas perfetto, fu svolto da Brody [« ZS. f. Phys. », *6*, 79 (1921)]. Per giungere alla formula esatta egli è però costretto a porre (pag. 82) $dE = 1$, essendo E una energia, con che la probabilità viene in sostanza a essere da lui definita come il numero degli stati quantici del gas (che viene quantizzato come contenuto in un recipiente parallelepipedo a pareti elastiche) che conducono ad una energia compresa tra E ed E + 1, di modo che la sua probabilità (form. 6) viene ad avere le dimensioni, anziché di un numero, dell'inverso di un'energia.

Come entropia complessiva di tutte le n molecole racchiuse in celle di volume v_c ci risulta alla fine, osservando che $pv = nkT$,

$$(2) \qquad S^* = kn \left\{ \frac{5}{2} \log T - \log p + \log \left[\frac{(2\pi m)^{3/2} k^{5/2} e^{3/2}}{h^3} \frac{nv_c}{v} \right] \right\}.$$

Per passare da questo valore all'entropia effettiva del gas, bisogna ancora notare che noi abbiamo supposto che in tutte le celle, che sono di eguale volume, fosse contenuto anche lo stesso numero di molecole.

Se indichiamo con P la probabilità che ciò avvenga, riconosciamo subito che per passare da (2) alla vera entropia del gas, bisogna, in conformità del principio di Boltzmann sottrarre ancora da (2) il logaritmo di P moltiplicato per k. Abbiamo dunque che l'entropia del gas è

$$(3) \qquad S = S^* - k \log P .$$

Supponiamo dapprima di aver diviso il volume v in n celle tutte eguali.

La probabilità P che in cella sia contenuta una sola molecola si calcola come rapporto del numero di modi in cui è possibile disporre le molecole in modo che ogni cella ne contenga una, che è evidentemente $n!$, al numero totale dei modi nei quali si possono distribuire le molecole in tutte le celle, che è n^n. Applicando la formula di Stirling si trova, con approssimazione sufficiente per il nostro caso

$$P = \frac{n!}{n^n} = \frac{(n/e)^n}{n^n} = e^{-n}$$

sostituendo dunque in (3) si deduce

$$(4) \qquad S = kn \left\{ \frac{5}{2} \log T - \log p + \log \frac{(2\pi m)^{3/2} k^{5/2} e^{5/2}}{h^3} \right\}$$

che è identica alla formula di Tetrode e Sackur, e quindi d'accordo con la esperienza. Concludiamo dunque che nel caso che si divida lo spazio in celle, per modo che ognuna di esse contenga un solo atomo si ottiene un risultato esatto per il valore della costante assoluta dell'entropia. Diversamente vanno le cose se in ogni cella è contenuto più di un atomo; supponiamo per esempio di dividere il volume v in $n/2$ celle di egual volume, e di porre conseguentemente in ciascuna delle celle 2 atomi. Per calcolare in questo caso la probabilità P, osserviamo che, il numero di modi di disporre n oggetti in $n/2$ celle, in modo che in ognuna vi siano solo 2 oggetti è evidentemente $n!/2^{n/2}$; il numero di tutte le disposizioni possibili è invece $(n/2)^n$. La probabilità P si può al solito valutare con la formula di Stirling, e si trova

$$P = \frac{n!}{2^{\frac{n}{2}} \left(\frac{n}{2}\right)^n} = \left(\frac{\sqrt{2}}{e}\right)^n.$$

Sostituendo dunque in (3) si troverebbe per l'entropia del gas l'espressione

$$(5) \qquad S = kn \left\{ \frac{5}{2} \log T - \log p + \log \frac{(2\pi m)^{3/2} k^{5/2} e^{5/2}}{h^3} + \log \sqrt{2} \right\}$$

che differisce dalla (4), e quindi dall'esperienza, di $\log \sqrt{2}$. Divergenze analoghe si trovano dividendo il volume in $n/3$, $n/4$, ... celle e mettendo in ognuna di esse rispettivamente 3, 4, ... molelecole. A titolo di esempio si può svolgere il calcolo per il caso che non si divida affatto il volume in celle, ma che si quantizzi il moto delle molecole considerandole tutte racchiuse in un recipiente a forma di parallelepipedo. In questo caso si ha evidentemente $P = 1$, e quindi $S = S^*$, e come si riconosce questa formula non ha nemmeno le dimensioni di quella di Tetrode e Sackur.

3. Per meglio mettere in evidenza ora, che la ragione delle divergenze osservate nei casi che in ogni cella sia contenuta più di una molecola, consiste proprio nel fatto, che si è quantizzato con le regole di Sommerfeld un sistema contenente degli elementi identici, facciamo la seguente considerazione:

Supponiamo di avere un miscuglio di due gas, di ciascuno dei quali, siano contenute $n/2$ molecole nel volume v; supporremo poi, per semplicità, che le due specie di molecole, pur essendo tra di loro distinte, abbiano però la stessa massa. Calcoliamo ora l'entropia di questa miscela dividendo il volume in $n/2$ celle eguali, e ponendo in ciascuna cella una molecola di una specie e una dell'altra. Se la nostra supposizione che la causa delle divergenze sia la presenza nella stessa cella di due molecole identiche, è giusta, dobbiamo in questo caso aspettarci un risultato esatto, poiché le due molecole presenti nella cella sono tra di loro diverse. Ed infatti per calcolare la nostra entropia potremo anche in questo caso applicare la (3), dove ora P rappresenta la probabilità che in ognuna delle $n/2$ celle sia contenuta una e una sola molecola di ciascuna delle due specie. Ora i casi possibili sono evidentemente $(n/2)^n$, e quelli favorevoli $[(n/2)\,!]^2$ e quindi con la formula di Stirling si trova

$$P = \frac{[(n/2)\,!]^2}{(n/2)^n} = e^{-n}.$$

Allora si trova come espressione dell'entropia della miscela

$$S = kn \left\{ \frac{5}{2} \log T - \log p + \log \frac{(2\pi m)^{3/2} e^{5/2} k^{5/2}}{h^3} + \log 2 \right\} \tag{6}$$

e ricordando che l'entropia della miscela di due gas perfetti è eguale alla somma delle entropie, che ciascuno di essi avrebbe se da solo occupasse tutto il volume disponibile, si riconosce immediatamente che il valore (6) coincide esattamente con quello voluto dalla termodinamica.

RIASSUNTO.

Viene discussa l'ipotesi, che l'insuccesso delle regole di Sommerfeld nel calcolo quantitativo degli atomi di numero atomico più grande di uno, più che una deficienza del metodo delle perturbazioni, oppure della meccanica, sia dovuto al fatto che questi atomi contengono almeno 2 elettroni non distinguibili tra di loro, e che le regole di Sommerfeld, anche nel caso della possibilità della separazione delle variabili, non siano applicabili nel caso che alcune parti del sistema siano tra di loro completamente identiche.

A sostegno di tale ipotesi si dimostra che la quantizzazione del moto di molecole eguali contenute in una cella a forma di parallelepipedo, conduce al risultato esatto per il valore assoluto dell'entropia del gas, solo nel caso che in ogni cella sia contenuta una sola molecola, mentre se il gas è una miscela di due specie di molecole e lo si quantizza racchiudendo le molecole in celle, per modo che in ciascuna cella siano contenute due molecole, però di specie diversa, si ottiene ancora il risultato esatto.

Roma, gennaio 1924.

N° 20.

A calculation of the effect of excited states of the atom on the ionization equilibrium, made possible by the use of Fermi's method (17 *a*, 17 *b*) for rendering the partition function of atomic states finite. Such refinements of the theory have become important in astrophysical applications.

F. RASETTI

20.

SULL'EQUILIBRIO TERMICO DI IONIZZAZIONE

« Nuovo Cimento », *1*, 153–158 (1924).

1. Nel 1921 Megh Nad Saha [1] ha proposto una teoria della ionizzazione dei gas per effetto della temperatura, nella quale il processo della trasformazione di un atomo neutro in un ione positivo e in un elettrone viene considerato in tutto e per tutto come una ordinaria reazione chimica di dissociazione. I consueti metodi termodinamici permettono allora di calcolare il grado di ionizzazione, come funzione della temperatura e della pressione, quando si conoscano le costanti chimiche dei componenti della reazione (atomi neutri, ioni positivi, elettroni), le quali si possono senz'altro ricavare dalla formula di Sackur e Tetrode. Indicando con c_a la concentrazione degli atomi neutri, e con c_e quella degli elettroni si trova così

$$\frac{c_a}{c_e^2} = \frac{h^3}{(2\pi\mu)^{3/2}} \frac{1}{(kT)^{3/2}} e^{\frac{w}{kT}} \tag{1}$$

essendo h la costante di Planck, μ la massa dell'elettrone, k la costante di Boltzmann, T la temperatura assoluta e w l'energia necessaria a ionizzare un atomo neutro. (Questa forma si deduce subito da quelle a pag. 41 della Memoria di Saha, esprimendo grado di ionizzazione x e la pressione P per mezzo delle concentrazioni; l'abbiamo scritta sotto questa forma perché più comoda per il seguito).

Il Saha ha fatto delle importanti applicazioni astrofisiche della (1). Egli ha cominciato dal dedurre dalla presenza delle righe d'arco o di scintilla nello spettro di una stella se in essa l'elemento corrispondente è allo stato neutro o ionizzato, ottenendo da ciò, per mezzo della (1), una valutazione, sia pure approssimativa della temperatura della stella, che si è mostrata in buon accordo con le temperature valutate per altra via.

Nella teoria del Saha si considera che per l'atomo allo stato neutro sia possibile soltanto uno stato. Ora è evidente, che se una tale ammissione può essere plausibile per temperature basse, per le quali gli stati

(1) M. N. SAHA, « Zeits. f. Phys. », *6*, 40 (1921).

quantici superiori non sono eccitati in modo sensibile, per temperature abbastanza elevate da produrre la dissociazione termica, oltre agli atomi ionizzati ve ne sarà anche un considerevole numero che avrà livelli di energia superiori al minimo; e di tali livelli di energia bisognerà naturalmente tener conto nel calcolo dell'equilibrio di ionizzazione. Questa difficoltà è del resto sentita anche dal Saha, che la discute a pagina 42 della sua Nota, concludendo che la termodinamica d'allora era insufficiente al calcolo dell'equilibrio chimico tra tutti questi livelli di energia, e che quindi il trascurarli era una necessaria ipotesi provvisoria. Se infatti in questo calcolo, conformemente alla ordinaria statistica della teoria dei quanti si prendessero i pesi statistici di tutti gli stati quantici eguali tra di loro, si troverebbe che la somma di tutti questi pesi è divergente, ciò che avrebbe di conseguenza il risultato assurdo che non vi è ionizzazione per quanto elevata sia la temperatura [2].

Recentemente ho avuto occasione di mostrare [3], che se l'ipotesi che tutti gli stati quantici abbiano lo stesso peso statistico è valida al limite per pressioni infinitamente piccole, essa non lo è più per pressioni finite, ed il peso statistico dei varii stati quantici è tanto più piccolo, quanto più grande è il volume dell'atomo quando esso si trova nello stato quantico in considerazione. Precisamente si trova che il peso statistico di uno stato quantico di volume v_i è

$$p_i = e^{-4nv_i} \tag{2}$$

essendo n il numero totale delle molecole (di qualsiasi specie) presenti nell'unità di volume.

Eliminata in questo modo la difficoltà della non convergenza della somma dei pesi statistici, è abbastanza facile fare il calcolo dell'equilibrio di ionizzazione tenendo conto di tutti i livelli di energia, e questo costituisce appunto lo scopo di questo lavoro.

2. La formula (1) dà la relazione tra la concentrazione degli atomi neutri e quella degli ioni, supponendo che la specie di atomi considerata abbia il peso statistico 1. Se invece essa avesse peso p varrebbe la formula ottenuta dalla (1) moltiplicando per p il secondo membro. Se si considera dunque uno stato quantico di energia w_i, e se con c_i si indica la concentrazione degli atomi in questo stato, si trova, tenendo presente la (2)

$$c_i = c_e^2 \frac{h^2}{(2\pi\mu)^{3/2}} \frac{1}{(kT)^{3/2}} e^{\frac{w_i}{kT} - 4nv_i} . \tag{3}$$

Se ora indichiamo con c_a la concentrazione complessiva di tutti gli atomi neutri in qualunque stato quantico essi si trovino, avremo

$$c_a = \Sigma c_i = \frac{h^3}{(2\pi\mu)^{3/2}} \frac{c_e^2}{(kT)^{3/2}} \Sigma e^{\frac{w_i}{kT} - 4nv_i} . \tag{4}$$

Questa formula deve esser sostituita alla (1).

(2) Vedi per esempio la discussione sopra la comunicazione di BECKER, « Phys. ZS. », *24*, 485 (1923).

(3) E. FERMI, « Rend. Lincei », *32*, 493 (1923).

Vogliamo ora passare a discutere quali siano le differenze tra la (1) e la (4), ma prima ancora osserviamo che se in questo calcolo si fosse preso come peso statistico 1 invece dell'espressione (2), la serie del secondo membro di (4) sarebbe $\Sigma e^{w_i/kT}$, e siccome le energie di ionizzazione sono tutte positive, tutti i termini di questa serie sono maggiori di 1, e quindi la serie divergente; il termine e^{-4nv_i} nell'esponente di e, ha per effetto di farla convergere, e di permetter quindi di tener conto, nel calcolo dell'equilibrio, di tutti gli stati quantici, senza, come viene sostanzialmente a fare il Saha, porre il peso statistico dello stato quantico di minima energia eguale ad uno, e tutti gli altri eguali a zero.

3. La differenza tra la (1) e la (4) consiste nel fatto che nella (4), al posto di

$$e^{\frac{w}{kT}} \tag{5}$$

figura l'espressione

$$\Sigma e^{\frac{w_i}{kT} - 4nv_i} . \tag{6}$$

Le differenze dipendono dunque dalle differenze tra queste due espressioni. Ora riconosciamo subito che esse saranno piccole per corpi con elevato potenziale di risonanza (gruppo zero), e maggiori per corpi a basso potenziale di risonanza (metalli alcalini). Se infatti formiamo il rapporto di (6) a (5) vediamo che esso è assai vicino ad 1 finché la temperatura non è tale che kT sia dell'ordine di grandezza del potenziale di risonanza. Quindi la differenza può essere di qualche entità solo per i metalli alcalini, e per temperature dell'ordine di grandezza di quelle che si hanno alla superficie delle stelle.

Per averne un'idea quantitativa ho eseguito il calcolo numerico della espressione (6) per il potassio e per la temperatura di 30.000 gradi. Vi è da notare che per i metalli alcalini i termini s sono semplici, mentre quelli $p, d, b, \cdots$ sono tutti doppi, così che al quanto totale i corrispondono $2i - 1$ livelli di energia (*). Quanto al volume v_i, che a rigor di termini sarebbe diverso per ciascuno di tali livelli, l'abbiamo posto eguale a quello del corrispondente stato quantico per l'atomo di idrogeno, approssimazione che si può giustificare con la considerazione che essa è certamente legittima per grandi numeri quantici, per i quali la orbita dell'elettrone di valenza del metallo è quasi eguale a quella dell'elettrone dell'idrogeno, mentre per piccoli numeri quantici, per i quali tale identificazione delle orbite del metallo e dell'idrogeno è certo assai inesatta, viene ad essere trascurabilissima l'influenza del volume, e quindi del tutto indifferente fare sopra il volume stesso una od un'altra ipotesi. Conformemente a quanto già feci nella Nota citata ho dunque preso il volume $v_i = 6{,}3 \cdot 10^{-25} \cdot i^6$; ciò premesso si può, con un pò di pazienza calcolare numericamente il valore della (6).

(*) In questo calcolo non è tenuto conto della molteplicità dei livelli dovuta alla quantizzazione spaziale (N.d.R.).

Si trova che per una densità eguale ad 1 : 1000 di quella dell'aria a 0° e un'atmosfera di pressione, e alla temperatura di 30 000 gradi si ha per il potassio.

$$\sum e^{\frac{w_i}{kT} - 4 n v_i} = 231 .$$

Nelle stesse condizioni il valore della (5) è invece.

$$e^{\frac{w}{kT}} = 5,34 .$$

La differenza è dunque in questo caso considerevole, perché, a parità di concentrazione degli ioni, la teoria di Saha prevede per gli atomi neutri una concentrazione circa 40 volte più piccola.

In modo forse più chiaro questo risultato si può esprimere, cercando quella temperatura che, secondo la formula di Saha, porta allo stesso grado di ionizzazione a cui porta la (4) a 30 000 gradi. Tale temperatura si ricava evidentemente dall'equazione

$$\frac{e^{\frac{w}{kT}}}{T^{3/2}} = \frac{231}{(30\,000)^{3/2}}$$

dove $w/k = 50\,200$. Da questa equazione si ricava presso a poco $T = 12\,100$.

Si vede dunque che per temperature elevatissime le differenze, per i metalli alcalini sono abbastanza rilevanti. Bisogna però tener conto che a queste temperature tali atomi sono quasi completamente ionizzati, così che il loro spettro d'arco non è più visibile, ciò che rende impossibile ogni confronto con i dati spettroscopici delle stelle più calde; ed infatti anche il Saha, nella determinazione delle temperature delle stelle più calde si è dovuto servire dei gas nobili, ad alto potenziale di ionizzazione.

Invece, a temperature più basse, dell'ordine di grandezza di quella del sole, le differenze tra la formula di Saha e la (4) sono già piuttosto piccole. Per esempio sempre per il potassio, alla concentrazione equivalente a un millesimo di atmosfera a temperatura normale, la (4) conduce per la temperatura di 6000 gradi a prevedere un numero di atomi superiore di circa il 25 °/₀ a quello previsto dalla formula di Saha, differenza questa, che se pur potrebbe considerarsi abbastanza grande per esperienze di laboratorio, è invece piccolissima data la necessaria incertezza di questi dati astrofisici.

Se si calcola la temperatura che, con la formula di Saha, conduce allo stesso grado di ionizzazione come la (4) per la temperatura di 6000 gradi, si trova circa 5870°. Basterebbe dunque un errore di 130° sulla misura della temperatura, per eguagliare i due risultati.

Possiamo dunque concludere questa discussione, dicendo che la correzione da apportarsi alla formula di Saha, per tener conto dell'equilibrio di tutti i livelli di energia dell'atomo, risulta generalmente assai piccola. Le differenze maggiori si hanno per i metalli alcalini, e per temperature il più possibile elevate.

Roma, febbraio 1924.

N° 21 *a* and 21 *b*

This paper is connected with Fermi's visit to Dutch universities. The problem of the intensities of spectral lines was investigated experimentally at that time in Utrecht, and there was also considerable interest from the theoretical point of view. A footnote in Fermi's paper indicates that he had discussed this topic with Professor P. Ehrenfest.

The Dutch version, not reproduced here, is a communication presented at a meeting of the "Nederlandsche Natuurkundige Vereeniging" on November 1, 1924. The more detailed Italian paper was published later.

F. RASETTI

21 *b*.

SOPRA L'INTENSITÀ DELLE RIGHE MULTIPLE

« Rend. Lincei », *1*, 120–124 (1925) (*).

1. I nuovi metodi elaborati nell'Istituto Fisico dell'Università di Utrecht per la misura dei rapporti di intensità delle componenti di una riga multipla (1), hanno permesso in questi ultimi tempi di raccogliere sopra questo argomento un materiale molto ricco e preciso, il quale ha condotto alla scoperta di regole empiriche di semplicità inaspettata (2). Tali regole si possono così riassumere: Dividiamo le righe di un multipletto in gruppi, ponendo in ciascun gruppo tutte quelle righe che hanno lo stesso livello di partenza (o di arrivo). Si trova che vale la seguente regola: la somma delle intensità di tutte le righe di un gruppo è proporzionale al quanto interno J, nella normalizzazione di Landé (3), del termine di partenza (o di arrivo). Per esempio per i doppietti *sp* degli alcalini si trova il rapporto 2/1, per i tripletti *sp* degli alcalino-terrosi i rapporti 5 : 3 : 1, ecc. Nel caso di multipletti complessi queste regole non sono sufficienti per la determinazione dei rapporti di intensità di tutte le componenti (4), ma anche per parecchi casi di questo genere si hanno delle buone misure di intensità.

Si presenta ora la questione dell'interpretazione teorica di questo materiale empirico. Allo stato attuale della teoria dell'atomo l'unica via per una tale interpretazione ci è data dal principio di corrispondenza. Ora è noto che Sommerfeld e Heisenberg (5) hanno stabilito per mezzo del principio

(*) Presentata nella seduta del 4 gennaio 1925 dal Socio O. M. Corbino.

(1) H. B. DORGELO, « ZS. f. Phys. », *13*, 206 (1923); *22*, 170 (1924).

(2) H. C. BURGER e H. B. DORGELO, « ZS. f. Phys. », *23*, 258 (1924).

(3) A. LANDÈ, « ZS. f. Phys. », *15*, 206 (1923).

(4) Dorgelo e Burger hanno anche tentato di generalizzare le loro regole a questi casi, ma come vedremo in seguito, § 3, noi dobbiamo ritenere inesatta la loro generalizzazione.

(5) A. SOMMERFELD e W. HEISENBERG, « ZS. f. Phys. », *11*, 130 (1922).

di corrispondenza una formula che dà il rapporto delle intensità delle componenti di una riga multipla. Questi Autori sono giunti al risultato che il rapporto delle intensità di tre componenti che corrispondono ai salti quantici $\Delta k = 1$, $\Delta j = +1, 0, -1$ è dato da

$$(1) \qquad (1 + \cos\theta)^2 : 2\sin^2\theta : (1 - \cos\theta)^2$$

dove θ rappresenta un valore medio tra gli angoli che il momento totale dell'atomo forma col momento dell'elettrone di valenza, nell'orbita iniziale e in quella finale. La formula (1) si presta soltanto a conclusioni di carattere qualitativo (Intensitätsregel di Sommerfeld).

Ci proponiamo in questo lavoro di dimostrare come una lieve modificazione delle considerazioni di Sommerfeld e Heisenberg possa invece condurre a stabilire delle formule, che salvo alcune eccezioni spiegabili, rendono conto quantitativamente dei risultati empirici. Ecco i punti sostanziali in cui le nostre considerazioni differiscono da quelle di Sommerfeld e Heisenberg.

I. Nella deduzione della (1) il salto quantico Δk, Δj che dà origine ad una componente del multipletto viene considerato come semplice, senza tener conto che realmente esso è il risultato di molti salti del quanto magnetico m, che in assenza del campo magnetico dànno tutti origine alla medesima componente; supponendo per esempio l'atomo introdotto in un campo magnetico H di intensità piccolissima, si riconosce che l'intensità della riga indecomposta si deve calcolare come somma delle intensità di tutte le sue componenti Zeeman [(6)].

II. Abbiamo determinato l'angolo θ e l'angolo φ (formato tra il momento d'impulso totale e H) al seguente modo. Ammettiamo con Landé che i quanti R, K, J, m rappresentino rispettivamente il momento del nocciolo, il momento dell'elettrone luminoso, il momento totale e la proiezione di quest'ultimo sopra H.

Si trova allora (J risultante di R e K)

$$\cos \widehat{KJ} = \frac{J^2 + K^2 - R^2}{2\,KJ} \quad , \quad \cos \widehat{HJ} = \frac{m}{J}\,.$$

Ora θ e φ rappresentano dei valori intermedii tra i valori che $\widehat{KJ}$ e $\widehat{HJ}$ assumono per l'orbita di partenza e per quella d'arrivo. Noi li abbiamo fissati al modo seguente:

$$\cos\theta = \frac{\bar{J}^2 + \bar{K}^2 - \bar{R}^2}{2\bar{J}\bar{K}} \quad ; \quad \cos\varphi = \frac{\bar{m}}{\bar{J}}$$

dove $\bar{R}$, $\bar{K}$, $\bar{J}$, $\bar{m}$ rappresentano le medie aritmetiche tra i due valori dei quanti corrispondenti, relativi al termine di partenza e a quello d'arrivo.

(6) Quantitativamente si ottiene a questo modo un risultato assai poco diverso da quello che si troverebbe considerando il salto come semplice, attribuendo però ad ogni salto il peso $\bar{J}$, eguale alla media dei due valori di J relativi al termine di partenza e a quello di arrivo; preferiamo però presentare il calcolo dell'intensità a questo modo, perché è teoricamente più chiaro, e rende inoltre conto della proibizione del salto quantico $J = 1/2 \rightarrow J = 1/2$. L'idea di procedere in questo modo mi venne da una conversazione col prof. P. Ehrenfest, al quale rinnovo qui i miei più vivi ringraziamenti.

Questo modo di calcolare θ e φ è certamente inesatto, ma esso rappresenta, come si vedrà, una approssimazione assai buona. Probabilmente il procedimento più corretto sarebbe quello di prendere come intensità la «media rettilinea» delle intensità tra le orbite estreme, poiché pare che tali medie rettilinee abbiano una particolare importanza nelle relazioni tra teoria classica e teoria dei quanti (7). Questo procedimento condurrebbe però a difficoltà che appaiono per ora insormontabili.

2. Sommerfeld e Heisenberg trovano per le intensità delle singole componenti Zeeman di un multipletto, a meno di un fattore che è lo stesso per tutte le righe dello stesso multipletto, i valori dati dalla seguente

TABELLA I.

ΔK	ΔJ	Δm	
1	± 1	1	$(1 \pm \cos\theta)^2\,(1 \pm \cos\varphi)^2$
1	± 1	0	$2\,(1 \pm \cos\theta)^2 \sin^2\varphi$
1	± 1	-1	$(1 \pm \cos\theta)^2\,(1 \mp \cos\varphi)^2$
1	0	± 1	$4 \sin^2\theta \sin^2\varphi$
1	0	0	$8 \sin^2\theta \cos^2\varphi$.

Bisogna introdurre ora per θ e per φ i valori (2) e sommare poi le intensità di tutte le componenti Zeeman che, per $H = 0$, dànno origine alla stessa riga indecomposta. Questo calcolo è un po' lungo, perché bisogna considerare separatamente i singoli casi, ma è del tutto privo di difficoltà.

Si trovano per le intensità le formule che sono date dalla seguente:

TABELLA II.

ΔK	ΔJ	*Intensità*
± 1	± 1	$\left\lvert \dfrac{2\,(K + J + 1)^2 - 2\,\overline{R}^2}{(2K + 1)\,(2J + 1)} \right\rvert^2 \dfrac{4J\,(8J^2 + 12J + 5)}{(2J + 1)^2}$
± 1	0	$\dfrac{4J^2\,(2K + 1)^2 - \{4J^2 + (2K + 1)^2 - 4\,\overline{R}^2\}}{J^2\,(2K + 1)^2}\,\dfrac{4J^2 - 1}{J}$
± 1	± 1	$\left[\dfrac{2\,\overline{R}^2 - 2\,(K - J + 1)^2}{(2K + 1)\,(2J - 1)} \right]^2 \dfrac{4\,(J - 1)\,(8J^2 - 4J + 1)}{(2J - 1)^2}$

NB. – *Per* K *e per* J *bisogna porre sempre i valori che corrispondono al termine con* K *più piccolo.*

3. Vogliamo ora confrontare queste formule con l'esperienza. Possiamo prevedere che per le combinazioni *sp*, per le quali la forma dell'orbita si altera molto durante il salto quantico, si avrà soltanto un accordo qualitativo, tra teoria ed esperienza, mentre l'accordo sarà assai migliore per le combinazioni *pd*, *db*.

Nel seguente confronto tra i valori teorici e i dati sperimentali, le intensità delle varie componenti del multipletto sono per comodità espresse in percentuali della riga più intensa.

(7) Vedi per esempio M. BORN, «ZS. f. Phys.», *26*, 379 (1924).

(sp)	Doppietti	teoria	100 : 26,5	(Metalli alcalini $1\,p - ms$)
		esp.	100 : 50	
	Tripletti	teoria	100 : 43 : 19	(Alcalino-terrosi $1\,p - ms$)
		esp.	100 : 60 : 20	

(pd)	Doppietti comp.	teoria	100 : 10 : 54	(Ca, $1\,p - 2\,d$)
		esp.	100 : 11 : 55	
	Tripletti comp.	teoria	100 : 17 : 1,5 : 55 : 15 : 23	(Ca, $1\,p - 2\,d$)
		esp.	100 : 18 : 1 : 54 : 19 : 25	

(db)	Tripletti comp.	teoria	100 : 6 : 0,15 : 68 : 8 : 46	(Sr, $1\,d - 3\,b$)
		esp.	100 : 9 : < 1 : 69 : 9 : 48	
	Tripletti con termine b	teoria	100 : 72 : 43	(Ca, $1\,d - 3\,b$)
	non risolto	esp.	100 : 72 : 45	

Come si vede, ove si escluda l'eccezione spiegabile delle combinazioni *sp* l'accordo si può dire ottimo.

Osserviamo ancora che la regola di Dorgelo e Burger citata al principio si può dedurre come legge limite delle nostre formule, ed è in pratica già eseguita con esattezza notevole per le combinazioni *pd*. Riteniamo dunque assai probabile che essa sia in realtà esatta. Riteniamo invece inesatta la generalizzazione che ne hanno tentato Dorgelo e Burger, e in particolare la tabella 5 del loro citato lavoro (pag. 265). Secondo tale tabella risulta infatti per esempio che il rapporto tra la componente più intensa e quella più debole di un tripletto composto *pd* è, per K grande, $12\,K^2 : 1$. Secondo il principio di corrispondenza si trova invece $\frac{256}{81} K^4 : 1$. Appare dunque che l'intensità della componente più debole decresca in realtà assai più rapidamente che secondo la tabella 5 di Dorgelo e Burger.

4. Aggiungiamo una osservazione relativa alle componenti Zeeman. Sommerfeld e Heisenberg, scegliendo l'angolo φ in modo diverso da noi, trovano per tali intensità dei risultati rappresentati graficamente alla fine del loro citato lavoro. Essi trovano che le intensità delle righe polarizzate parallelamente, anche osservando perpendicolarmente al campo, risulta minore in media del 20 % di quella delle componenti trasversali; e ciò li induce ad esprimere la supposizione, che tale differenza esista realmente, ciò che porterebbe alla strana conseguenza, che la luce emessa in un campo magnetico, se anche tanto debole da non produrre una separazione, dovrebbe essere polarizzata perpendicolarmente al campo. Calcolando col nostro metodo si trovano delle figure assai simili a quelle di Sommerfeld e Heisenberg, con la sola differenza che le componenti parallele risultano alquanto più intense, per modo che già per le combinazioni *pd* la differenza delle intensità complessive delle componenti parallele e perpendicolari non supera pochi per cento, e passando alle combinazioni superiori, per le quali il principio di corrispondenza merita maggiore fiducia, va sempre diminuendo. Noi riteniamo assai probabile che tale piccola differenza sia dovuta all'inesattezza della nostra applicazione del principio di corrispondenza, e che in realtà essa non esista. Ad ogni modo sarebbero desiderabili delle esperienze per chiarire questo importante punto.

N° 22, 29, 34.

The " Periodico di Matematiche " was a review directed by F. Enriques and devoted to the needs of the teachers of secondary schools. Enriques, although not a physicist, insisted strongly that the review should deal not only with mathematics but also, at an elementary level, with the most modern subjects of physics. It can be presumed, therefore, that these two articles were solicited from Fermi by Enriques, who had a very high opinion of the young physicist.

At that time Fermi gave several lectures before the Mathematics Seminar at the University of Rome and some of the articles correspond to the lectures. At that time there was no Physics Seminar at the University of Rome.

E. Segrè

22.

SUI PRINCIPI DELLA TEORIA DEI QUANTI

« Rend. Seminario matematico Università di Roma », *8*, 7–12 (1925).

La teoria dei quanti si basa, come si sa, su due principii fondamentali, che si possono enunciare come segue:

a) il moto di un sistema atomico è un moto meccanicamente possibile (cioè calcolabile mediante le solite leggi della meccanica e la legge di Coulomb); non però tutti i moti meccanicamente possibili possono essere effettivamente effettuati dal sistema, ma soltanto una successione discreta di essi, detti quantici o statici; cosicché in particolare anche l'energia w del sistema non potrà corrispondentemente assumere altro che una successione discreta di valori, detti livelli di energia, $w_1, w_2, \cdots, w_n$;

b) finché il sistema descrive un moto statico non si ha irradiazione di energia (contrariamente ai risultati della elettrodinamica classica). L'irradiazione dell'energia è invece sempre legata al processo, non meccanico, del salto dell'elettrone da un moto quantico ad un altro; se w e w' rappresentano le energie corrispondenti ai due moti tra i quali avviene il salto, l'energia irradiata sarà $w - w'$; si ammette che questa energia sia irradiata in un solo quanto. La frequenza con cui l'energia viene irradiata sarà dunque

$$\nu = \frac{w - w'}{h} . \tag{1}$$

Questi due principii possono considerarsi come ormai sperimentalmente accertati, specialmente dalle esperienze di Franck e della sua scuola sopra i potenziali di ionizzazione, con le quali esperienze si sono potuti misurare direttamente i livelli di energia, e verificare quindi la (1).

I principii *a*) e *b*) non sono però naturalmente sufficienti a risolvere tutti i problemi della fisica atomica; essi debbono perciò essere integrati con la soluzione di altri problemi, la cui soluzione è per ora soltanto incompleta od incerta.

Noi cercheremo in questa conferenza di esporre come si possa almeno tentare una soluzione sistematica di alcuni di questi problemi; e precisamente ci occuperemo dei seguenti:

1° quali sono le regole colle quali debbano scegliersi i moti quantici tra la successione continua dei moti meccanicamente possibili. La risoluzione completa di questo problema sarebbe evidentemente importantissima, perché permetterebbe poi, mediante la (1), di calcolare tutte le frequenze che l'atomo in considerazione può emettere, di risolvere cioè completamente il problema della determinazione teorica delle linee spettrali;

2° quale è la probabilità che, ad una data temperatura, un atomo si trovi in uno o nell'altro dei suoi moti meccanicamente possibili;

3° quale è la probabilità perché un atomo che si trova in un certo stato quantico, nell'intervallo di tempo dt passi ad un altro.

Questi tre problemi si possono, almeno parzialmente risolvere, ricorrendo a principii generali, e specialmente al cosiddetto principio delle adiabatiche di Ehrenfest, e al principio di corrispondenza di Bohr.

Chiamiamo *meccanismo* di un sistema meccanico il complesso delle forze e dei vincoli di esso; allora il moto delle parti del sistema dipenderà dal meccanismo e dalle condizioni iniziali del sistema. Possiamo ora pensare di alterare il meccanismo di un sistema in modo infinitamente lento: una tale trasformazione si dice una trasformazione adiabatica (per analogia alle trasformazioni omonime che si studiano nella teoria del calore). Le costanti che caratterizzano il movimento, per esempio l'energia, verranno in una trasformazione adiabatica a variare in funzione del meccanismo.

Consideriamo per esempio un oscillatore. Una variazione adiabatica del suo meccanismo si potrà ottenere facendo variare in modo infinitamente lento la forza che richiama la particella verso la sua posizione di equilibrio. Se ora chiamiamo x l'elongazione della particella, m la sua massa e ν la sua frequenza avremo

$$m\ddot{x} + 4\pi^2\nu^2 mx = 0 \tag{2}$$

l'energia sarà poi

$$w = \frac{1}{2} m\dot{x}^2 + 2\pi^2\nu^2 mx^2 = 4\pi^2\nu^2 m\overline{x^2} \tag{3}$$

indicandosi con $\overline{x^2}$ il valor medio di x^2. Se supponiamo dunque di dare a ν un incremento $d\nu$, l'energia varierà corrispondentemente di $4\pi^2\nu m x^2 d\nu$. Ma se tale variazione di ν avviene in un tempo molto lungo bisognerà sostituire x^2 con la sua media. Avremo perciò $dw = 4\pi^2\nu\, d\nu\, m\overline{x^2}$; e quindi

$$\frac{dw}{w} = \frac{d\nu}{\nu}$$

da cui risulta

$$\frac{w}{\nu} = \text{costante}.$$

Ne segue che se all'inizio della trasformazione si aveva $\frac{w}{\nu} = nh$, tale relazione si manterrà valida ancora durante tutta la trasformazione.

Ora si sa già dalla teoria della radiazione del corpo nero di Planck che i moti privilegiati di un oscillatore soddisfano appunto alla relazione $w = nh\nu$. Possiamo perciò concludere che se ad un certo istante un oscillatore si trova in un moto quantico, esso resterà in un moto quantico anche dopo una trasformazione adiabatica. Ora Ehrenfest ha generalizzato questa osservazione fatta per l'oscillatore, enunciando in generale il principio, che se un sistema si trova ad un istante iniziale in uno stato quantico, esso seguita a restare in un tale stato dopo una qualsiasi trasformazione adiabatica.

Per bene comprendere tutta la portata del principio di Ehrenfest. conviene ricordare un teorema dovuto a Burgers. Questo teorema si riferisce ai cosidetti sistemi a variabili separate, cioè a quei sistemi per i quali l'equazione alle derivate parziali di Jacobi, prendendo un opportuno sistema di coordinate generali $q_1, q_2, \cdots, q_f$ e di momenti $p_1, p_2, \ldots p_f$ ad esse coniugati, ammette una soluzione che si spezza nella somma di addendi dipendenti ciascuno da una sola delle coordinate, per modo che p_i è funzione della sola coordinata q_i. Ora Burgers dimostra che per tali sistemi gli integrali

$$(4) \qquad J_1 = \int p_1 dq_1 \; ; \; J_2 = \int p_2 dq_2 \; ; \cdots ; J_f = \int p_f dq_f$$

estesi ad una oscillazione completa della coordinata q_i, sono degli invarianti adiabatici, e cioè che il loro valore si mantiene costante se il sistema si assoggetta ad una trasformazione adiabatica. Tenendo presente il principio di Ehrenfest resta allora chiaro che le J saranno le quantità che debbono essere quantizzate. Ora per l'oscillatore si troverebbe facilmente $J = \frac{w}{h}$ e dunque per esso la condizione perché il moto sia quantico si può scrivere semplicemente $J = nh$. La naturale generalizzazione di questa condizione al caso generale dei sistemi a variabili separate ci conduce a scrivere le condizioni di Bohr–Sommerfeld per la determinazione delle orbite statiche di tali sistemi. Esse sono

$$(5) \qquad J_1 = n_1 h \; ; \; J_2 = n_2 h \; ; \cdots ; \; J_f = n_f h$$

dove le n rappresentano dei numeri interi. Sono note le brillanti applicazioni delle (5) alla determinazione teorica dello spettro dell'idrogeno, fin nelle sue più delicate particolarità, e si sa anche che da esse si è potuta ricavare l'interpretazione degli spettri di alta frequenza, e per lo meno qualitativamente, si sono potute spiegare gran parte delle particolarità degli spettri più complicati.

Siccome però tutti gli atomi, ad eccezione dell'idrogeno, possono rappresentarsi con sistemi a variabili separate solo in una prima, e piuttosto grossolana approssimazione, la determinazione quantitativa dei loro spettri non è ancora riuscita, e il metodo delle perturbazioni, così fecondo nell'astronomia, non è in questo caso riuscito a dare una soluzione soddisfacente nemmeno per l'atomo di elio, che è il più semplice dopo quello di idrogeno.

La soluzione del secondo dei problemi accennati in principio, quello cioè di determinare la probabilità che un atomo si trovi in uno dei suoi stati quantici, ci è anche essa indicata dal principio delle adiabatiche, Ehrenfest ha infatti dimostrato che se un sistema si trasforma in un altro con una trasformazione adiabatica, le probabilità a priori dei suoi stati quantici prima della trasformazione sono eguali a quelle degli stati quantici corrispondenti dopo la trasformazione. Ora, per esempio, per l'oscillatore si sa dalla teoria del corpo nero, che la probabilità a priori di tutti i suoi stati è eguale (dall'ammettere questo si deduce infatti la nota formula $\frac{h\nu}{e^{h\nu/k\mathrm{T}} - 1}$ per l'energia media dell'oscillatore). Possiamo perciò concludere intanto che per tutti i sistemi che si possono dedurre dall'oscillatore con una trasformazione adiabatica, le probabilità a priori di tutti gli stati quantici sono eguali tra di loro. È allora abbastanza naturale ammettere che una tale proprietà valga per tutti i sistemi atomici. Risolta così la questione preliminare della probabilità a priori, i soliti metodi della statistica ci conducono a porre la probabilità di uno stato quantico proporzionale a $e^{-w/k\mathrm{T}}$ essendo w la sua energia, e ci permettono perciò di risolvere completamente il problema secondo.

Per poter risolvere però il problema dell'intensità delle linee spettrali, non basta evidentemente sapere quale è il numero degli atomi che si trovano in un certo stato quantico, ma bisogna anche sapere quale è la probabilità che un atomo, che si trova in un certo stato quantico, salti in un altro stato determinato. Per la risoluzione di quest'ultimo problema è d'importanza fondamentale il principio di corrispondenza di Bohr. È questo un principio di natura assai diversa da tutti gli altri principii della fisica e della matematica; mentre infatti di solito tali principii contengono una affermazione di significato preciso, come per esempio $\mathrm{A} = \mathrm{B}$, oppure $\mathrm{A} > \mathrm{B}$, il principio di corrispondenza dice semplicemente A somiglia a B; e precisamente: l'intensità e lo stato di polarizzazione delle righe emesse da un atomo col meccanismo della teoria dei quanti, somigliano all'intensità e allo stato di polarizzazione delle righe che lo stesso atomo emetterebbe secondo il meccanismo classico. A base di questo principio sta l'osservazione fondamentale, fatta da Bohr, che al limite per numeri quantici infinitamente grandi si ha identità tra lo spettro emesso secondo la teoria dei quanti, e quello emesso secondo la teoria classica. È naturale dunque ammettere che a questa identità delle lunghezze d'onda faccia riscontro anche la identità dell'intensità e dello stato di polarizzazione. Si capisce poi come all'identità che si ha rigorosamente al limite, corrisponda, per numeri quantici finiti, la semplice somiglianza; queste sono le basi formali del principio di corrispondenza, il quale si è in un gran numero di casi dimostrato in ottimo accordo con l'esperienza, almeno fino a quel punto che l'imprecisione stessa della sua natura può permettere di attendersi. In qualche caso, del resto, il principio di corrispondenza permette di ottenere delle informazioni del tutto precise; e ciò si verifica, per esempio, nei casi nei quali qualche riga spettrale o qualche sua componente deve, classicamente, avere intensità nulla; in questo caso si può concludere che quella riga o quella sua componente mancano completamente anche dallo spettro quantistico. Tutte le conseguenze di questo genere sono state completamente verificate dall'esperienza.

Possiamo dunque concludere questa discussione dicendo, che se pur manca nella teoria dei quanti la risoluzione di alcuni problemi di importanza fondamentale, non manca tuttavia il tentativo di dedurre tutte le regole, logicamente, da alcuni principii generali.

N° 23 *a* and 23 *b*.

In this paper we find clearly the characteristics of the more mature style of Fermi: a fundamental idea, at the same time simple and clever, is applied to several interesting concrete problems of practical importance with the help of mathematical methods of sufficient approximation, but not better than warranted by the underlying physical hypotheses. The results agree with experiment, but only to an order of magnitude, and Fermi typically remarks that in one of the applications the agreement with experiments is better than should be expected.

Soon after the publication of this paper Bohr observed that Fermi's method when applied to the energy distribution of the electrons ejected disagrees with experiment and concluded: " Unter diesen Umständen kann es kaum als eine Stütze für die von Fermi benutzten Annahmen betrachtet werden, dass eine auf der Forderung der Energiebilanz gegründete Schätzung der Bremswirkung Resultate gibt, die angenähert mit den Versuchen übereinstimmen " («ZS. f. Phys.», *34*, 149 (1925)).

Later on, when quantum mechanics was developed, Fermi's method for the study of collisions with electrically charged particles found adequate theoretical justification through a theorem by Dirac (P. A. M. Dirac *Principles of Quantum Mechanics*, 1930, page 167) as remarked by E. J. Williams («Proc. Roy. Soc.», *139*, 163 (1933)). This author discussed carefully the validity limits of Fermi's method and, using the quantum mechanical values of the matrix elements, calculated the excitation and ionization probability in distant collisions.. A little later C. F. v. Weizsaecker used the same method for the calculation of Bremsstrahlung («ZS. f. Physik», *88*, 612 (1937)). Williams and Weizsaecker were working at the time at Copenhagen. The method under the name of Weizsaecker-Williams has found ever wider applications to atomic and nuclear problems, and Fermi himself used it on many occasions. Indeed its fundamental idea was a favorite one with him for his whole life.

Paper N° 23 *b* is a translation of paper N° 23 *a*, with a few minor additions.

E. Persico

23 *b*.

ÜBER DIE THEORIE DES STOSSES ZWISCHEN ATOMEN UND ELEKTRISCH GELADENEN TEILCHEN

«Z. Physik», *29*, 315–327 (1924).

Das elektrische Feld eines geladenen Teilchens, welches an einem Atom vorbeifliegt, wird harmonisch zerlegt, und mit dem elektrischen Feld von Licht mit einer passenden Frequenzverteilung verglichen. Es wird angenommen, dass die Wahrscheinlichkeit, dass das Atom vom vorbeifliegenden Teilchen angeregt oder ionisiert wird, gleich ist der Wahrscheinlichkeit für die Anregung oder Ionisation durch die äquivalente Strahlung. Diese Annahme wird angewendet auf die Anregung durch Elektronenstoss und auf die Ionisierung und Reichweite der α-Strahlen.

1. Wenn ein Atom, das sich im Normalzustand befindet, mit Licht von geeigneter Frequenz belichtet wird, kann es angeregt werden, d. h. kann es ein Quantum absorbieren, und dadurch zu einem höheren Energieniveau gehoben werden. Ist das Quantum des anregenden Lichtes grösser als die

Ionisierungsarbeit des Atoms, so kann es ionisiert werden, indem es ein Elektron der äusseren oder der inneren Atomschalen verliert.

Ähnliche Erscheinungen finden bei dem Stosse zwischen Atomen und Elektronen statt. Wenn man die Atome eines Gases mit Elektronen bombardiert, tritt für kleine Geschwindigkeiten der Elektronen die Anregung der Atome, und für grössere die Ionisation zuerst der äusseren, und dann der inneren Atomschalen ein.

Zweck der vorliegenden Arbeit ist, die Analogie zwischen diesen beiden Klassen von Erscheinungen zu präzisieren und die Erscheinungen bei dem Stosse quantitativ aus der Lichtabsorption abzuleiten.

Wenn ein elektrisch geladenes Teilchen in der Nähe eines Punktes vorüberfliegt, entsteht in diesem Punkte ein veränderliches elektrisches Feld. Wenn wir nun dieses Feld durch ein Fouriersches Integral in harmonische Komponenten zerlegen, so sehen wir, dass es gleich dem Felde ist, das in dem Punkte sein würde, wenn es mit Licht von einer passenden kontinuierlichen Frequenzenverteilung belichtet würde. Denken wir jetzt, dass ein Atom sich in diesem Punkte befinde: dann liegt es ziemlich nahe, anzunehmen, dass das elektrische Feld des geladenen Teilchens an dem Atom dieselben Anregungs- und Ionisierungserscheinungen verursacht wie das elektrische Feld des äquivalenten Lichtes. Wenn wir den Absorptionskoeffizienten des Atoms für das Licht als Funktion der Frequenz kennen, können wir die Wahrscheinlichkeit berechnen, dass ein elektrisch geladenes Teilchen, das mit gegebener Geschwindigkeit in einer gegebenen Entfernung von einem Atom vorbeifliegt, es ionisiere. Schon jetzt muss man aber eine Einschränkung zu dieser Korrespondenz zwischen dem elektrischen Felde des Lichtes und dem des Teilchens einführen. Man weiss ja, dass ein Körper mit der Geschwindigkeit v durch Stoss nur Erscheinungen verursachen kann, für die eine Energie kleiner als seine kinetische Energie nötig ist. Dagegen findet man dass in der harmonischen Zerlegung seines elektrischen Feldes alle Frequenzen, einschliesslich der höchsten, vorkommen. Man ist daher anzunehmen genötigt, dass alle Frequenzen, deren Quantum grösser als die kinetische Energie des Teilchens ist, keine Wirkung haben können, da die nötige Energie fehlt, um ein ganzes Quantum zu liefern.

Wir haben unsere Hypothese auf drei Erscheinungen angewendet, die eine experimentelle Bestätigung derselben ermöglichen, nämlich:

a) Die Anregung der Quecksilberlinie 2537 Å. Die optische Absorption dieser Linie ist quantitativ genügend bekannt; es gibt ausserdem Versuche, aus denen man die Grössenordnung der Wahrscheinlichkeit entnehmen kann, dass Quecksilberatome durch Stoss von langsamen Elektronen angeregt werden. Unsere Theorie gibt für diese Wahrscheinlichkeit einen etwas zu grossen Wert, doch von der richtigen Grössenordnung; natürlich schliesst aber die Unbestimmtheit der experimentellen Ergebnisse eine strenge Prüfung der Theorie aus.

b) Die Zahl der pro Zentimeterweg von den α-Teilchen des RaC erzeugten Ionenpaare. Wir haben diese Anwendung für das Helium durchgeführt, da man für dieses Gas mit genügender Annäherung den Absorptionskoeffizient als Funktion der Frequenz kennt, weil das Helium monoatomisch ist, und sein Atom nur aus dem K-Elektronenringe besteht. Die Übereinstimmung zwischen Theorie und Erfahrung ist befriedigend. Die früheren Theorien der Stossionisierung [1] können folgendermassen zusammengefasst werden: Wenn ein elektrisch geladenes Teilchen in der Nähe eines Atoms vorbeifliegt, werden seine Elektronen von ihm angezogen, so dass ihnen eine gewisse kinetische Energie übertragen wird, die mit der Annahme berechnet wird, dass die Elektronen im Innern des Atoms frei sind. Man nimmt weiter an, dass ein Elektron vom Atom abgerissen wird, sobald diese Energie seine Ionisierungsarbeit übertrifft.

c) Die Reichweite der α-Teilchen des RaC im Helium. Als Mechanismus der Bremsung der α-Teilchen haben wir den folgenden angenommen: Wenn das α-Teilchen durch das Helium fliegt, entsteht um es herum ein veränderliches elektrisches Feld, dessen Energie von den benachbarten Atomen absorbiert wird, als ob es das elektrische Feld einer Lichtwelle wäre. Die auf diese Weise absorbierte Energie wird natürlich auf Kosten der Energie des Teilchens geliefert, das dadurch gebremst wird. Auch für diesen Fall ist die Übereinstimmung zwischen Theorie und Erfahrung ganz befriedigend. Die Bremsung der α-Strahlen durch die Materie ist früher theoretisch von Thomson und Bohr [2] studiert worden. Nach Bohr sind die Elektronen im Atom von quasi-elastischen Kräften gebunden und werden von dem Teilchen angezogen und in Bewegung gesetzt; die Energie des Teilchens wird dabei natürlich um den Betrag der auf die Elektronen übertragenen Energie vermindert. Bohr findet eine gute Übereinstimmung mit der Erfahrung, indem er für die Frequenz der Bindung je nach dem Fall Werte der Grossenordnung der optischen oder der Röntgenfrequenzen annimmt.

Im zweiten Paragraphen dieser Arbeit werden einige Formeln für die Berechnung des dem elektrischen Felde des geladenen Teilchens äquivalenten Lichfeldes entwickelt, im dritten, vierten und fünften werden dann die Anwendungen *a*), *b*) und *c*) dargestellt.

2. Wir wollen zuerst die spektrale Energieverteilung berechnen, die dem von einem Teilchen mit der Ladung ε in einem Punkte P erzeugten Felde äquivalent ist. Sei v die Geschwindigkeit des Teilchens und b die kleinste Entfernung zwischen dem Teilchen und dem Punkte P. Die Komponenten der elektrischen Kraft, parallel und senkrecht zu der Geschwindigkeit des Teilchens, sind

$$(1)\qquad \left\{\begin{aligned} E_1 &= \frac{\varepsilon v \Theta}{(b^2 + v^2\Theta^2)^{3/2}}, \\ E_2 &= \frac{\varepsilon b}{(b^2 + v^2\Theta^2)^{3/2}}, \end{aligned}\right.$$

wo Θ die seit dem Augenblick der kleinsten Entfernung verflossene Zeit bedeutet. Ist nun T eine Zeit, die wir nachher ins Unendliche wachsen lassen

(1) J. J. THOMSON, « Phil. Mag. », *23*, 449, 1912; N. BOHR, ebenda *30*, 581 (1915); S. ROSSELAND, ebenda *45*, 65, (1923).

(2) N. BOHR, « Phil. Mag. », *25*, 10 (1913); *30*, 581 (1915).

werden, so können wir E_1 und E_2 in Fourierscher Reihe, für $-\frac{T}{2} < \Theta < \frac{T}{2}$, entwickeln. Wir finden

$$(2) \qquad \begin{cases} E_1 = \Sigma a_n \sin \dfrac{2\pi n}{T} \Theta , \\ E_2 = \Sigma b_n \cos \dfrac{2\pi n}{T} \Theta , \end{cases}$$

wo

$$(3) \qquad \begin{cases} a_n = \dfrac{2\varepsilon v}{T} \displaystyle\int_{-T/2}^{T/2} \dfrac{\Theta \sin \dfrac{2\pi n \Theta}{T} d\Theta}{(b^2 + v^2 \Theta^2)^{3/2}} , \\ b_n = \dfrac{2\varepsilon b}{T} \displaystyle\int_{-T/2}^{T/2} \dfrac{\cos \dfrac{2\pi n \Theta}{T} d\Theta}{(b^2 + v^2 \Theta^2)^{3/2}} . \end{cases}$$

Das elektrische Feld $a_n \sin (2\pi n \Theta/T)$ ist nun gleich dem elektrischen Felde einer Lichtwelle der Intensität $\frac{c}{4\pi} \frac{a_n^2}{2}$ und der Frequenz $\nu = n/T$. Sei nun $J(\nu) d\nu$ die mit der Wirkungsdauer T multiplizierte gesamte Lichtintensität des Frequenzenintervalls $d\nu$; wir finden

$$J(\nu) d\nu = \lim_{T=\infty} \frac{cT}{4\pi} \frac{a_n^2 + b_n^2}{2} T d\nu = \frac{c\varepsilon^2}{2\pi} \left\{ b^2 \left(\int_{-\infty}^{\infty} \frac{\cos 2\pi\nu\Theta \, d\Theta}{(b^2 + v^2 \Theta^2)^{3/2}} \right)^2 + v^2 \left(\int_{-\infty}^{\infty} \frac{\Theta \sin 2\pi\nu\Theta \, d\Theta}{(b^2 + v^2 \Theta^2)^{3/2}} \right)^2 \right\} d\nu .$$

Die in diesem Ausdruck vorkommenden Integrale können mit den modifizierten Besselschen (3) Funktionen zweiter Art, nullter und erster Ordnung ausgedrückt werden. Man hat in der Tat

$$\int_{-\infty}^{\infty} \frac{\cos \omega x \, dx}{(1 + x^2)^{3/2}} = 2 \omega K_1(\omega) ,$$

$$\int_{-\infty}^{\infty} \frac{x \sin \omega x \, dx}{(1 + x^2)^{3/2}} = - 2 \omega K_0(\omega) .$$

Durch diese Formeln finden wir sofort:

$$(4) \qquad J(\nu) = \frac{8\pi c \varepsilon^2 \nu^2}{v^4} \left\{ K_0^2 \left(\frac{2\pi\nu b}{v} \right) + K_1^2 \left(\frac{2\pi\nu b}{v} \right) \right\} ,$$

setzen wir zur Abkürzung

$$B(\omega) = K_0^2(\omega) + K_1^2(\omega) ,$$

so finden wir

$$(5) \qquad J(\nu) = \frac{8\pi c \varepsilon^2 \nu^2}{v^4} B \left(\frac{2\pi\nu b}{v} \right) .$$

(3) Siehe z. B. GRAY, MATHEWS and MACROBERT, *Bessel Functions*, London 1922 wo man auch numerische Tabellen dieser Funktionen finden kann. Über die Fouriersche Entwicklung dieser Kräfte siehe auch BOHR, l. c.

Da aber alle Frequenzen, deren Quantum grösser ist als die kinetische Energie des Teilchens, nach unserer Annahme keine Wirkung haben können, finden wir, dass das Vorüberfliegen des Teilchens folgender spektralen Verteilung äquivalent ist:

$$(6) \qquad J(\nu) = \begin{cases} \dfrac{8\pi c \varepsilon^2 \nu^2}{v^4} B\left(\dfrac{2\pi\nu b}{v}\right) & \text{für } h\nu \leq \dfrac{1}{2} mv^2, \\ 0 & \text{für } h\nu > \dfrac{1}{2} mv^2. \end{cases}$$

3. Wir wollen jetzt an einem einfachen Beispiel die Anwendung der allgemeinen Methode erläutern. Wir werden nämlich die Anregung des Resonanzlichtes durch Stoss studieren. Betrachten wir z. B. ein Quecksilberatom, und sei $\alpha(\nu)$ sein Absorptionskoeffizient für die Frequenz ν. Das bedeutet, dass das Atom, wenn es von der Lichtintensität $J(\nu)\, d\nu$ getroffen wird, im Mittel die Energie $J(\nu)\, \alpha(\nu)\, d\nu$ absorbiert, d. h. dass die Wahrscheinlichkeit der Absorption eines Quantums $h\nu$ ist:

$$(7) \qquad \frac{J(\nu)\, \alpha(\nu)\, d\nu}{h\nu}.$$

Da nun der Absorption eines Quantums der Absorptionslinie die Anregung des Atoms zum Energieniveau $2p$ entspricht, finden wir, dass (7) die Anregungswahrscheinlichkeit darstellen würde, falls nur die Frequenzen des Intervalls $d\nu$ wirken könnten. Sei nun $\Pi(b)$ die Anregungswahrscheinlichkeit des Zustandes $2p$, wenn das Teilchen in der Entfernung b vorüberfliegt; man erkennt sofort [(4)], dass

$$(8) \qquad \Pi(b) = 1 - e^{-\int \frac{J(\nu)\,\alpha(\nu)\,d\nu}{h\nu}}.$$

Da nun für den Fall der Resonanz $\alpha(\nu)$ nur in einem ganz schmalen Bereiche um die Resonanzfrequenz ν_0 von Null verschieden ist, können wir schreiben:

$$\int \frac{J(\nu)\,\alpha(\nu)\,d\nu}{h\nu} = \frac{J(\nu_0)\,\alpha}{h\nu_0}.$$

wo gesetzt worden ist

$$\alpha = \int \alpha(\nu)\, d\nu.$$

(4) In der Tat ist die Wahrscheinlichkeit, dass das Atom kein Quantum absorbiert.

$$1 - \Pi(b) = \Pi\left(1 - \frac{J(\nu)\,\alpha(\nu)\,d\nu}{h\nu}\right),$$

wo das Produkt über sämtliche unendlich kleine Frequenzintervalle $d\nu$ erstreckt werden muss. Wenn wir dann die Logarithmen der beiden Glieder nehmen, finden wir, im Limes:

$$\log[1 - \Pi(b)] = \Sigma \log\left(1 - \frac{J(\nu)\,\alpha(\nu)\,d\nu}{h\nu}\right) = -\int \frac{J(\nu)\,\alpha(\nu)\,d\nu}{h\nu},$$

woraus die Formel (8) folgt.

Formel (8) wird also in unserem Falle

$$(8^*) \qquad \Pi(b) = 1 - e^{-\frac{\alpha}{h\nu_0} J(\nu_0)} .$$

Hieraus kann man den äquivalenten Radius ρ des Atoms für die Anregung des Resonanzlichtes berechnen, d. h. den Radius, den das Atom unter der Annahme haben würde, dass jeder Stoss die Ionisation hervorriefe. Dieser Radius ist offenbar gegeben durch

$$\pi\rho^2 = 2\pi \int b\,db\, \Pi(b) .$$

Daher finden(5) wir, unter Berücksichtigung von (6), dass für $\frac{1}{2} mv^2 > h\nu_0$

$$(9) \quad \rho^2 = 2\int_0^\infty \left\{1 - e^{-\frac{8\pi\varepsilon^2 c\alpha\nu_0}{hv^4} B\left(\frac{2\pi\nu_0 b}{v}\right)}\right\} b\,db = \frac{v^2}{2\pi^2\nu_0^2}\int_0^\infty \left(1 - e^{-\frac{8\pi\varepsilon^2 c\alpha\nu_0}{hv^4} B(x)}\right) x\,dx .$$

Das Integral kann angenähert berechnet werden (6), und wir finden:

$$(9^*) \qquad \rho^2 = \frac{2\varepsilon^2 c\alpha}{\pi h\nu_0 v^2}\left(0{,}54 - \log\frac{8\pi\varepsilon^2 c\alpha\nu_0}{hv^4}\right) .$$

Experimentell wurde die Anregung der Resonanzlinie 2537 des Hg durch Elektronenstoss von Frl. Sponer (7) untersucht. Leider gibt aber diese Arbeit

(5) Tatsächlich wird ein Elektron, das mit kleiner Geschwindigkeit in der Nähe eines Atoms vorbeifliegt, vom elektrischen Felde des Atoms beträchtlich von seiner Bahn abgelenkt. Da wir aber nur die Grössenordnung der Erscheinung brauchen, wenden wir trotzdem Formel (6) an.

(6) Um einen angenäherten Ausdruck von

$$J(\alpha) = \int_0^\infty [1 - e^{-\alpha B(x)}]\, x\,dx$$

für $\alpha \ll 1$ zu bekommen, beachten wir, dass man für $x < 0{,}4$ mit genügender Annäherung hat: $B(x) = \frac{1}{x^2}$; für $\alpha \ll 1$ und $x > 0{,}4$ kann man weiter setzen:

$$1 - e^{-\alpha B(x)} = \alpha B(x) .$$

Daher können wir schreiben

$$J(\alpha) = \int_0^{0{,}4} \left(1 - e^{-\frac{\alpha}{x^2}}\right) x\,dx + \alpha\int_{0{,}4}^\infty B(x)\, x\,dx .$$

Für das zweite Integral finden wir durch numerische Berechnung den Wert $0{,}973\,\alpha$. Das erste kann mit Benutzung der asymptotischen Ausdrücke des Integrallogarithmus leicht berechnet werden, und man findet dafür den Wert $-\frac{\alpha}{2}\log\alpha - 0{,}705\,\alpha$. Daher erhält man endlich:

$$J(\alpha) = \left(0{,}268 - \frac{1}{2}\log\alpha\right)\alpha .$$

(7) HERTHA SPONER, «Zs. f. Phys.», 7, 185 (1921).

nur die Grössenordnung der Anregungswahrscheinlichkeit. Frl. Sponer findet nämlich, dass, wenn Elektronen mit einer Geschwindigkeit nicht viel grösser als 4,9 Volt gegen Hg-Atome stossen, nur einige Prozente Wahrscheinlichkeit vorhanden sind, dass die Atome angeregt werden. Um den Wert von ρ durch (9*) zu berechnen, nehmen wir für v den dem Potential von 8 Volt entsprechenden Wert. Den Wert von α können wir einer Arbeit von Füchtbauer [(8)] entnehmen. Füchtbauer findet, dass, wenn ein Quecksilberatom von Licht mit der Energieverteilung $J(\nu)$ getroffen wird, die Wahrscheinlichkeit der Absorption eines Quantums der Resonanzfrequenz $PJ(\nu_0)$ beträgt, wo $P = 8 \cdot 10^7$. Nun hat man offenbar $\alpha = P h\nu$ und daher $\alpha = 6 \cdot 10^{-4}$, (9*) gibt uns jetzt

$$\rho = 0{,}8 \cdot 10^{-8}.$$

Dieser Wert ist gewiss etwas grösser als der von Frl. Sponer gefundene. Nähme man auch an, dass die Ausbeute an unelastischen Stössen 9 Proz. wäre, würde man doch immer für ρ den Wert $0{,}4 \cdot 10^{-8}$ bekommen, d. h. ungefähr die Hälfte des von uns erhaltenen Wertes. Man kann sich aber leicht diese Abweichung erklären. Wenn Elektronen mit der Geschwindigkeit von einigen Volt in der Entfernung der Grössenordnung von 10^{8-} cm an einem Elektron vorbeifliegen, werden sie bereits sehr stark von ihm abgestossen. Da wir in der Berechnung von ρ diese Ablenkung vernachlässigt haben, konnte man vorhersehen, dass ρ zu gross herauskommen würde. Da aber die Unsicherheit der Messungen immerhin eine strenge Bestätigung der Theorie für diesen Fall ausschliesst, scheint uns eine angenähertere, und gewiss sehr komplizierte, Berechnung von ρ überflüssig zu sein, da wir bereits mit unseren vereinfachenden Annahmen die richtige Grössenordnung von ρ bekommen haben.

Die Erscheinungen der Stossionisation können in ganz ähnlicher Weise erklärt werden. Man weiss, dass bei allen Atomen, jenseits der Grenze der Hauptserie, ein kontinuierliches Spektrum sowohl in der Absorption wie in der Emission vorkommt, das dem Übergange des Leuchtelektrons vom Normalzustand zu einem Zustande entspricht, wo es vom Atom getrennt ist und dazu noch eine gewisse kinetische Energie besitzt. Dieses Spektrum besteht aus einer breiten Bande mit einer scharfen Kante nach Rot hin, an der Grenze der Hauptserie, und langsam nach den kurzen Wellenlängen abklingend. Ist nun die Geschwindigkeit des stossenden Elektrons so, dass die Frequenz $mv^2/2h$ jenseits dieser Absorptionsbande fällt, so können wir (9*) auch zur Berechnung des äquivalenten Radius des Atoms für die Ionisierung anwenden; fällt dagegen die Grenzfrequenz $mv^2/2h$ ins Innere der Absorptionsbande, so wird der Radius kleiner sein als nach (9*) und verschwinden, sobald $mv^2/2h$ gleich der Grenze der Hauptserie wird. Qualitativ ist ein solches Verhalten von ρ als Funktion von v durch die Erfahrung bestätigt [(9)]. Leider scheint es unmöglich, die Erscheinung quantitativ zu verfolgen, da man praktisch nichts über die Intensität des Absorptionsspektrums weiss.

(8) Füchtbauer, « Phys. ZS. », *21*, 322 (1920); Füchtbauer und Joos, ebenda S. 694.
(9) Nettleton, « Proc. Nat. Acad. Amer. », *10*, 140 (1924).

4. Günstiger liegen die Verhältnisse bei der Berechnung der von den α-Strahlen hervorgerufenen Ionisierung. In diesem Falle werden die sehr trägen α-Teilchen nur unbedeutend vom atomaren Felde abgelenkt. Die Ergebnisse über die Absorption in dem Gebiete der Röntgenstrahlen können weiter durch folgende Formel zusammengefasst werden [10]:

$$\alpha(\nu) = \frac{KZ^4}{\nu^3} + D\,, \tag{10}$$

wo $\alpha(\nu)$ den atomaren Absorptionskoeffizienten darstellt; Z bedeutet die Atomnummer, D den von der Zerstreuung herrührenden Betrag der Absorption und K einen Koeffizienten, der an den Grenzen der Röntgenserien Unstetigkeiten aufweist: für Werte von ν grösser als die Grenze ν_0 der K-Serie hat man $K = 0{,}6 \cdot 10^{30}$, und für $\nu < \nu_0$ $K = 0{,}1 \cdot 10^{30}$. Hieraus können wir entnehmen, dass der von der Ionisierung des K-Ringes herrührende Betrag der Absorption ist:

$$\alpha(\nu) = \begin{cases} 0 & \text{für} \quad \nu < \nu_0\,, \\ \dfrac{HZ^4}{\nu^3} & \text{für} \quad \nu > \nu_0\,. \end{cases} \tag{11}$$

Der Koeffizient H stellt natürlich die Unstetigkeit des Koeffizienten K beim Überschreiten der K-Grenze dar, man hat also

$$H = 0{,}5 \cdot 10^{30}.$$

Die Wahrscheinlichkeit $\Pi(b)$, dass das Vorüberfliegen eines α-Teilchens in der Entfernung b den K-Ring eines Atoms ionisiere, kann wieder durch (8) berechnet werden. Da nun die Masse des α-Teilchens sehr gross ist, können wir sie Grenzfrequenz $mv^2/2h$ durch ∞ ersetzen. Wir finden damit

$$\Pi(b) = 1 - e^{-\frac{8\pi c \varepsilon^2 HZ^4}{hv^4} \int\limits_{\nu_0}^{\infty} \frac{d\nu}{\nu^2} B\left(\frac{2\pi\nu b}{v}\right)}\,. \tag{12}$$

Den Radius können wir wieder nach der Formel berechnen:

$$\rho^2 = 2\int\limits_0^{\infty} b\,db\,\Pi(b)\,.$$

Wir finden also

$$\rho^2 = 2\int\limits_0^{\infty} b\,db \left\{1 - e^{-\frac{8\pi c \varepsilon^2 HZ^4}{hv^4} \int\limits_{\nu_0}^{\infty} \frac{d\nu}{\nu^2} B\left(\frac{2\pi\nu b}{v}\right)}\right\} = \frac{v^2}{2\pi^2\nu_0^2} \int\limits_0^{\infty} \xi\,d\xi \left\{1 - e^{-\alpha\xi \int\limits_{\xi}^{\infty} \frac{dx}{x^2} E(x)}\right\} \tag{13}$$

wo zur Abkürzung gesetzt worden ist:

$$\alpha = \frac{8\pi c \varepsilon^2 HZ^4}{h\nu_0 v^4}\,. \tag{14}$$

(10) Siehe z. B. H. A. KRAMERS, « Phil. Mag. », *46*, 836 (1923).

Die Integrale in (13) können für $\alpha \ll 1$ näherungsweise berechnet [11] werden und man findet

$$(13^*) \qquad \rho^2 = \frac{v^2}{2\pi^2 \nu_0^2}\,\alpha\left(0{,}45 - \frac{1}{6}\log\alpha\right).$$

Wir wollen diese Formel zur Berechnung der Ionisierung des Heliums durch die α-Strahlen des RaC anwenden. Da das Helium nur den K-Elektronenring besitzt und dieser abgeschlossen ist, versuchen wir, ob wir noch (13*) anwenden und $H = 0{,}5 \cdot 10^{30}$ annehmen können. Entsprechend der Ionisierungsspannung von 24,5 Volt nehmen wir weiter $\nu_0 = 6{,}0 \cdot 10^{15}$; für die α-Strahlen des RaC haben wir $v = 1{,}98 \cdot 10^9$. Wir finden also $\alpha = 0{,}0091$. Damit erhalten wir:

$$\rho^2 = 0{,}56 \cdot 10^{-16} \quad , \quad \rho = 0{,}75 \cdot 10^{-8}.$$

Die Zahl der pro Zentimeterweg von dem Teilchen erzeugten Ionenpaare ist natürlich $\pi\rho^2 n$, wo $n = 2{,}6 \cdot 10^{19}$ die Zahl der Atome pro Kubikzentimeter bei 15° darstellt.

Diese Zahl ist daher = 4800.

Experimentell findet man 4600.

Die Übereinstimmung zwischen diesen beiden Zahlen wäre ausserordentlich befriedigend. Man kann ja sagen, dass sie etwas genauer ist als man erwarten würde. Tatsächlich würde die grosse Unsicherheit, hauptsächlich im Koeffizienten H, eine Divergenz um 30 Proz. und vielleicht noch etwas mehr leicht erklären. [12]

(11) Man muss berechnen:

$$J = \int_0^\infty \xi\, d\xi \left\{ 1 - e^{-\alpha\xi \int_\xi^\infty \frac{dx}{x^2} B(x)} \right\}.$$

Wenn $\alpha < 0{,}1$, kann man J in ein Integral von 0 bis 0,4 und ein anderes von 0,4 bis ∞ zerlegen. Das erste kann berechnet werden, indem man für $\xi < 0{,}4$ mit genügender Annäherung

$$\int_\xi^\infty \frac{dx}{x^2} B(x) = \frac{1}{3\xi^3} - 0{,}64$$

ansetzt und dann die asymptotischen Ausdrücke für den Integrallogarithmus anwendet. Das zweite Integral kann näherungsweise geschrieben werden:

$$\alpha \int_{0,4}^\infty \xi^2\, d\xi \int_\xi^\infty \frac{dx}{x^2} B(x)\,.$$

Der Koeffizient von α, numerisch berechnet, ist 0,28. Wir finden dadurch:

$$J = \alpha\left(0{,}45 - \frac{1}{6}\log\alpha\right).$$

(12) Anm. b. d. Korr. Man muss auch bedenken, dass das theoretische und das experimentelle Ergebnis eigentlich nicht direkt miteinander vergleichbar sind, da bei dem ersten die sekundäre Ionisation nicht berücksichtigt wurde. Die letztere könnte auch viele Prozente der totalen Ionisation betragen. Jedenfalls konnte man aber bei der starken Extrapolation von H keine bessere Übereinstimmung erwarten.

5. Eine weitere experimentelle Bestätigung unserer Theorie kann man in der Theorie der Bremsung der α-Teilchen durch die Materie finden. Auch diese Anwendung machen wir für das Helium, da man für seine Absorption, wie gezeigt wurde, einen vernünftigen Ansatz machen kann.

Zuerst wollen wir den Energieverlust berechnen, den ein α-Teilchen beim Vorbeifliegen in der Entfernung b von einem Atom erfährt. Sei wieder $\Pi(b)$ die Wahrscheinlichkeit, dass der Vorbeigang des Teilchens in der Entfernung b die Ionisation verursacht, und $P(b, \nu)\,d\nu$ die Wahrscheinlichkeit der Ionisation, wenn nur die Frequenzen des Intervalls $d\nu$ anwesend wären. Die Wahrscheinlichkeit, dass die Ionisierung durch Absorption eines Quantums des Frequenzintervalls $d\nu$ geschehe, ist dann

$$\Pi(b)\frac{P(b,\nu)\,d\nu}{\int_0^\infty P(b,\nu)\,d\nu}.$$

Dieser Ionisierung entspricht der Energieverbrauch $h\nu$. Der mittlere Energieverlust des Teilchens ist daher

$$\Pi(b)\frac{\int_0^\infty P(b,\nu)\,h\nu\,d\nu}{\int_0^\infty P(b,\nu)\,d\nu}.$$

Wenn n die Zahl der Atome pro Volumeneinheit bedeutet fliegt das α-Teilchen während des Weges $d\nu$ in der Entfernung zwischen b und $b + db$ von $2\pi\, bdbn$ Atomen vorbei. Wenn T seine kinetische Energie bedeutet, finden wir also:

$$\frac{dT}{dx} = -2\pi n\int_0^\infty bdb\,\Pi(b)\frac{\int_0^\infty P(b,\nu)\,h\nu\,d\nu}{\int_0^\infty P(b,\nu)\,d\nu}. \tag{15}$$

Nun hat man

$$P(b,\nu) = \frac{\alpha(\nu)\,J(\nu)}{h\nu} = \begin{cases} \dfrac{8\pi c\varepsilon^2 HZ^4}{hv^4}\dfrac{1}{\nu^2}B\left(\dfrac{2\pi\nu b}{v}\right) & \text{für } \nu > \nu_0, \\ 0 & \text{für } \nu < \nu_0, \end{cases}$$

$$\Pi(b) = 1 - e^{-\int_0^\infty P(b,\nu)\,d\nu}$$

Formel (15) wird jetzt leicht, mit Hinzunahme von (14),

$$(16)\qquad \frac{dT}{dx} = -\frac{4\,c\,\varepsilon^2\,HZ^4\,n}{\nu_0^2\,v^2}\int_0^\infty \xi\,d\xi\left\{1 - e^{-\alpha\xi\int_\xi^\infty \frac{dx}{x^2}B(x)}\right\}\frac{\int_\xi^\infty \frac{dx}{x}B(x)}{\alpha\xi\int_\xi^\infty \frac{dx}{x^2}B(x)}.$$

Ist α sehr klein, so kann man diesen Ausdruck näherungsweise berechnen [13], und man findet

$$(17)\qquad \frac{dT}{dx} = -\frac{c\,\varepsilon^2\,HZ^4\,n}{\nu^2\,v^2}\left(0{,}96 + \log\frac{h\nu_0\,v^4}{8\pi\,c\,\varepsilon^2\,HZ^4}\right);$$

da nun $T = \frac{1}{2}mv^2$, finden wir endlich

$$(17^*)\qquad \frac{dv}{dx} = -\frac{c\,\varepsilon^2\,HZ^4\,n}{m\,\nu_0^2\,v^3}\left(0{,}96 + \log\frac{h\,\nu_0\,v^4}{8\pi\,c\,\varepsilon^2\,HZ^4}\right).$$

Hieraus können wir leicht eine Formel ableiten, die uns den von den α-Teilchen zurückgelegten Weg angibt, während ihre Geschwindigkeit vom Anfangswerte v_0 zum Endwerte v herabsinkt. Man findet nämlich:

$$(18)\qquad x = \frac{m\,\nu_0^2}{c\,\varepsilon^2\,HZ^4\,n}\int_v^{v_0}\frac{v^3\,dv}{0{,}96 + \log\frac{h\,\nu_0\,v^4}{8\pi\,c\,\varepsilon^2\,HZ^4}} = 2{,}4\,\frac{m\,\nu_0}{nh}\int_{\frac{0{,}104\,h\nu_0\,v^4}{c\,\varepsilon^2\,HZ^4}}^{\frac{0{,}104\,h\nu_0\,v_0^4}{c\,\varepsilon^2\,HZ^4}}\frac{du}{\log u}.$$

(13) Um dieses Integral zu berechnen, muss man für $\xi < 0{,}4$ die angenäherten Ausdrücke benutzen:

$$\int_\xi^\infty \frac{dx}{x^2}B(x) = \frac{1}{3\xi^3} - 0{,}64 \quad , \quad \int_\xi^\infty \frac{dx}{x}B(x) = \frac{1}{2\xi^2} - 0{,}64\,.$$

Weiter muss man das Integral zerlegen in einen Teil von 0 bis 0,4, der leicht mit den asymptotischen Ausdrücken des Integrallogarithmus ausgedrückt werden kann. Um den zweiten Teil von 0,4 bis ∞ zu berechnen, braucht man nur zu bemerken, dass man für $\xi > 0{,}4$ und sehr kleines α setzen kann

$$1 - e^{-\alpha\xi\int_\xi^\infty \frac{du}{u^2}B(u)} = \alpha\xi\int_\xi^\infty \frac{du}{u^2}B(u)\,.$$

Es genügt jetzt, numerisch zu berechnen:

$$\int_{0,4}^\infty \xi\,d\xi\int_\xi^\infty \frac{dx}{x}B(x) = 0{,}35\,;$$

endlich finden wir, dass der Wert des gesuchten Integrals ist: $0{,}24 - \frac{1}{4}\log\alpha$ woraus man durch Anwendung von (15) sofort (17) bekommt.

Natürlich ist diese Formel nur gültig, wenn beide Grenzen des Integrals gross gegen eins sind, da sonst auch Formel (17*), aus der (18) abgeleitet worden ist, nicht mehr anwendbar ist, und man würde (16) auch für grosse Werte von α berechnen müssen. Trotzdem können wir mit (18) den von den α-Strahlen des RaC durch Helium zurückgelegten Weg berechnen, während ihre Geschwindigkeit auf die Hälfte des Anfangswertes herabsinkt. Setzen wir in (18)

$$m = 6{,}6 \cdot 10^{-24} \quad , \quad \nu_0 = 6{,}0 \cdot 10^{15} \quad , \quad n = 2{,}6 \cdot 10^{19} ,$$

$$\varepsilon = 2 \cdot 4{,}77 \cdot 10^{-10} = 9{,}54 \cdot 10^{-10} \quad , \quad v_0 = 1{,}98 \cdot 10^{9} \quad , \quad v = 0{,}99 \cdot 10^{9},$$

so finden wir

$$x = 0{,}56 \int_{18}^{288} \frac{du}{\log u} = 32 \,.$$

Nun ist bekanntlich die Reichweite der α-Strahlen von der Geschwindigkeit $v_0/2$ gleich $1/8$ der Reichweite [14] der α-Strahlen von der Geschwindigkeit v_0. Daher ist die Reichweite im Helium

$$32 + 5 = 37 \text{ cm}.$$

Experimentell findet man, in guter Übereinstimmung mit dem theoretischen Werte, eine Reichweite von 33 cm.

(14) Anm. b. d. Korr. Auch die Berechnung der Bremsung nach Formel (16) für kleine Werte von v (α gross) ergibt, dass die Reichweite der α-Strahlen mit der Geschwindigkeit $\frac{v_0}{2} = 0{,}99 \cdot 10^9$ ungefähr 5 cm beträgt.

24.

SOPRA L'URTO TRA ATOMI E NUCLEI DI IDROGENO

«Rend. Lincei», *1*, 77–80 (1925) (*).

Recentemente J. Franck [1] ha discusso il processo della ionizzazione per urto tra atomi e ioni che abbiano velocità corrispondenti all'energia di pochi volt. Energeticamente un tale urto può condurre come risultato finale alla ionizzazione dell'atomo, quando l'energia del moto relativo dell'atomo e dello ione sia maggiore dell'energia che è necessaria per ionizzare l'atomo; conviene però tenere presente che le condizioni in cui si verifica questa ionizzazione sono profondamente diverse da quelle in cui avviene la ordinaria ionizzazione per urto elettronico. Siccome infatti la massa degli ioni e degli atomi è qualche migliaio di volte più grande di quella degli elettroni, si ha che, a parità di energia cinetica, la velocità dei primi sarà assai più piccola di quella dei secondi, così che mentre gli elettroni che hanno energia sufficiente per ionizzare un atomo debbono avere delle velocità dell'ordine di grandezza di quella degli elettroni di valenza dell'atomo, gli ioni, pur avendo energia sufficiente, avranno una velocità assai più piccola. Ora appare che nella letteratura il processo della ionizzazione per urto, anche nel caso degli ioni lenti, si sia sempre considerato come legato a qualche cosa di discontinuo; lo scopo di questo lavoro è invece di dimostrare, sopra l'esempio facilmente seguibile dell'urto tra atomi e nuclei di idrogeno, che si può avere un processo di ionizzazione per urto completamente continuo.

Noi considereremo precisamente questo caso: supponiamo che un atomo e un nucleo di idrogeno si vengano incontro con una velocità assai piccola in confronto a quella dell'elettrone dell'atomo, per modo che il moto dell'elettrone si possa considerare come alterato adiabaticamente dall'avvicinarsi del nucleo. Cominciamo con l'osservare che essendo il moto dell'elettrone regolato dal principio delle adiabatiche, esso dovrà ad ogni istante essere un moto quantistico per il sistema intermedio, che è costituito dall'elettrone attirato dai due nuclei che, con la nostra approssimazione si debbono, istante per istante, considerare come fermi. Tenendo poi presente che finché il nucleo è abbastanza lontano dall'atomo, quest'ultimo che si trova nel suo stato normale e quindi in un'orbita 1_1 si disporrà, come è noto dalla teoria dell'effetto Stark, con la sua orbita perpendicolare alla congiun-

(*) Presentata nella seduta del 4 gennaio 1925 dal Socio O. M. CORBINO.

(1) J. FRANCK, «Zs. f. Phys.», *25*, 312, (1924).

gente i due nuclei, si riconosce che per qualunque posizione dei due nuclei, l'elettrone descriverà un cerchio posto in un piano perpendicolare alla retta congiungente i due nuclei, e col centro sopra questa retta; inoltre il momento della quantità di moto dell'elettrone attorno a questa retta sarà costantemente $h/2\pi$. Supponiamo che ad un certo istante i due nuclei si trovino alla distanza $2a$. Sia poi x la distanza dal punto medio della congiungente i due nuclei, del piano del cerchio descritto dall'elettrone, ed r il raggio del cerchio medesimo. Se v è la velocità dell'elettrone ed m la sua massa, avremo

$$\frac{h}{2\pi} = mvr .$$

La forza centrifuga sarà dunque:

$$\frac{mv^2}{r} = \frac{h^2}{4\pi^2 m r^3} .$$

Questa forza deve essere equilibrata dall'attrazione dei due nuclei; le componenti di questa attrazione parallele alla congiungente e ad r saranno dunque risp. o ed $\frac{h^2}{4\pi^2 m r^3}$. Esprimendo questo si trovano fra x ed r le due equazioni:

$$(1) \qquad \begin{cases} \dfrac{e^2(a-x)}{\{(a-x)^2+r^2\}^{3/2}} - \dfrac{e^2(a+x)}{\{(a+x)^2+r^2\}^{3/2}} = 0 \\[2ex] \dfrac{e^2 r}{\{(a-x)^2+r^2\}^{3/2}} + \dfrac{e^2 r}{\{(a-x)^2+r^2\}^{3/2}} = \dfrac{h^2}{4\pi^2 m r^3} \end{cases}$$

dove e rappresenta la carica elettronica. Le radici di questo sistema di equazioni ci dànno le posizioni che può avere il cerchio descritto dall'elettrone per ogni posizione dei due nuclei. Riconosciamo subito che le (1) ammettono una soluzione per la quale è $x = 0$, mentre r è una funzione di a che per $a = 0$ ha valore $\rho/2$, essendo $\rho = \frac{h^2}{4\pi^2 e^2 m} = 0{,}53 \cdot 10^{-8}$ cm il raggio dell'atomo normale di idrogeno, e al crescere di a va crescendo fino all'infinito. Per tale soluzione i due nuclei occupano posizioni simmetriche rispetto al piano del cerchio descritto dall'elettrone. Quando $a < 0{,}65\,\rho$ si trova che questa è l'unica soluzione reale delle (1) (poiché r è essenzialmente positivo). Quando invece sia $a > 0{,}65\,\rho$, cioè i nuclei siano distanti più di 1,30 volte il raggio dell'atomo di idrogeno, si trova che oltre alla soluzione precedente ce ne sono due altre, per le quali x assume valori eguali ed opposti, e r valori eguali. Le chiameremo soluzioni asimmetriche. Per a molto grande esse sono date da $x = \pm a$, $r = \rho$, e corrispondono dunque al caso che l'atomo sia nel suo stato normale e molto lontano dall'altro nucleo. Al decrescere di a, le due soluzioni asimmetriche si vanno avvicinando, e per $a = 0{,}65\,\rho$ vengono a coincidere tra di loro e anche con la soluzione simmetrica, dando luogo ad una radice tripla $x = 0$, $r = 0{,}92\,\rho$. Se a decresce ancora si ha, come si è detto, soltanto la soluzione simmetrica.

Possiamo dunque concludere, che finché i due nuclei sono ancora molto lontani e si vanno avvicinando, il cerchio descritto dall'elettrone corrisponde alla soluzione asimmetrica. Si trova che in queste condizioni il nucleo viene respinto dall'atomo. Proseguendo i due nuclei ad avvicinarsi, sempre respingendosi, arriva un momento in cui a diventa $= 0{,}65\ \rho$, ed allora il cerchio è quello che corrisponde alla soluzione tripla; diminuendo ancora a il cerchio diviene quello che corrisponde alla soluzione simmetrica che è l'unica esistente per $a < 0{,}65\ \rho$. I due nuclei si respingono sempre e viene un momento in cui essi cominciano ad allontanarsi; a comincia a crescere ed il cerchio è sempre quello che corrisponde alla soluzione simmetrica fino ad $a = 0{,}65\ \rho$, dove si ha la soluzione tripla. Al crescere ulteriore di a possono avvenire due casi: o il cerchio corrisponde ad una delle soluzioni asimmetriche, ed allora il risultato finale dell'urto è evidentemente un nucleo e un atomo di idrogeno; in questo caso non si ha dunque ionizzazione. Ma può avvenire anche che al crescere di a oltre $0{,}65\ \rho$ l'elettrone continui a muoversi sopra l'orbita simmetrica. In questo caso i due nuclei continuano per un po' ad essere respinti (fino ad $a = 0{,}86\ \rho$) e per distanze maggiori vengono attratti.

Se l'energia totale del sistema, nel moto relativo al centro di gravità è negativa, e cioè se l'atomo e lo ione hanno prima dell'urto una energia cinetica relativa più piccola del lavoro di ionizzazione Rh dell'idrogeno (R numero di Rydberg) il prodotto finale dell'urto è un ione H^+ simmetrico metastabile, che poi si scinde ancora in un atomo e in un nucleo. Se invece l'energia dell'urto è maggiore di Rh i due nuclei si allontanano fino all'infinito, e l'elettrone resta nel piano mediano descrivendo un cerchio di raggio sempre crescente con velocità tendente a zero.

Il risultato finale dell'urto è dunque due nuclei e un elettrone tutti separati. Si vede dunque che quando l'energia relativa dell'urto è maggiore dell'energia di ionizzazione si può avere la ionizzazione per mezzo di un processo continuo.

N° 25.

The author remarks that for a plane molecule, one of the principal moments of inertia equals the sum of the other two. This relation must apply to all triatomic molecules, and should appear as a relation between the rotational constants of different band systems. A good verification is afforded by the infra-red absorption bands of water vapor. This paper was written in Florence.

F. RASETTI.

25.

UNA RELAZIONE TRA LE COSTANTI DELLE BANDE INFRAROSSE DELLE MOLECOLE TRIATOMICHE

« Rend. Lincei », *1*, 386–387 (1925) (*).

Le bande infrarosse, che come è noto sono dovute a salti delle molecole tra stati quantici che corrispondono alla stessa configurazione degli elettroni, e che differiscono tra di loro solamente per il diverso stato di rotazione e di oscillazione dei nuclei, ci permettono in modo assai semplice di misurare il momento d'inerzia della molecola nel suo stato normale. La teoria di queste bande [1] conduce infatti alla formula

$$\Delta\nu = \frac{h}{4\pi^2 I} \tag{1}$$

dove h è la costante di Planck, I il momento di inerzia della molecola e $\Delta\nu$ la differenza costante di frequenza tra le righe della banda infrarossa.

Se la molecola è biatomica, essa ha un solo momento di inerzia, e corrispondentemente un solo sistema di bande infrarosse; se invece essa è pluriatomica essa ha tre momenti di inerzia principali, e quindi anche tre sistemi di bande infrarosse, che corrispondono a tre diversi $\Delta\nu$. I tre valori di $\Delta\nu$ ci permettono di determinare i tre momenti di inerzia principali della molecola.

Se ora si ha un sistema di punti situati in un piano, si riconosce immediatamente che gli assi principali di inerzia sono uno perpendicolare al piano, e gli altri due contenuti in esso, e che il momento di inerzia relativo all'asse principale perpendicolare al piano è uguale alla somma degli altri due mo-

(*) Presentata dal Socio A. Garbasso nella seduta del 15 marzo 1925.

(1) SOMMERFELD, *Atombau und Spektrallinien*, ed. 3; Cap. 7.

menti principali di inerzia [2]. Nel caso di una molecola triatomica i tre atomi giacciono certamente tutti nello stesso piano, e quindi tra i tre momenti di inerzia principali deve passare la relazione $I_1 = I_2 + I_3$. Siccome poi per la (1) si ha che i tre momenti principali di inerzia sono inversamente proporzionali ai $\Delta\nu$ dei tre sistemi di bande infrarosse, troviamo che tra questi deve passare la relazione

$$\frac{1}{\Delta\nu_1} = \frac{1}{\Delta\nu_2} + \frac{1}{\Delta\nu_3}. \tag{2}$$

Questa relazione, sulla quale non mi risulta che sia mai stata richiamata l'attenzione, può servire sia come conferma della teoria delle bande infrarosse, sia ancora come prova dell'ordinamento attribuito a un sistema di bande infrarosse di una sostanza triatomica.

Disgraziatamente il materiale empirico per la verifica della (2) è assai scarso, essendo completamente noto un solo sistema di bande infrarosse di una sostanza triatomica, e precisamente quello del vapore di acqua.

Holjer Witt [3] è infatti recentemente riuscito a riconoscere nello spettro d'assorbimento infrarosso del vapore acqueo tre sistemi di bande, con le differenze di frequenza (in numeri d'onda) 16,8 ; 24,5 ; 56.

Ora si ha

$$\frac{1}{17,0} = \frac{1}{24,5} + \frac{1}{56}.$$

Si vede dunque che la (2) è bene verificata entro i limiti dell'errore sperimentale.

(2) Basta perciò prendere come assi delle x e delle y i due assi principali che sono situati nel piano, e come asse delle z il terzo asse principale. Allora i tre momenti di inerzia sono risp. $\Sigma m (x^2 + y^2)$, Σmx^2, Σmy^2. Da qui risulta subito la proprietà accennata.

(3) H. WITT, «Zs. f. Phys.», *28*, 249 (1924).

N° 26, 27, 28 (1) and 28 (2).

These papers are important for several reasons: they describe Fermi's second venture in the experimental field after several years of theoretical work; they show his ingenuity in dealing with unfamiliar techniques; and constitute the first instance of an investigation of atomic spectra by means of radiofrequency fields, a technique that was to receive numerous applications many years later.

Wood and Ellett, and Hanle, had announced their remarkable discovery of the effects of weak magnetic fields on the polarization of the resonance radiation of mercury. Rasetti had observed these effects in Florence. When Fermi came to that university a few months later, he was greatly interested in the phenomenon, whose only theory at the time was a classical one based on the concept of Larmor precession. Fermi pointed out that since the mercury resonance line showed an anomalous Zeeman effect with a Landé factor of 3/2, the mercury atom should more likely precess with a frequency 3/2 times higher than the Larmor frequency. The choice between the two alternatives might be decided by investigating the behavior of the polarization under magnetic fields of the intensity of about one gauss and frequency of a few megacycles per second, in approximate resonance with the precession frequency of the atom.

Rasetti had experience with the spectroscopic technique, but neither of the experimenters had any with radiofrequency circuits. However, Fermi calculated the characteristics of a simple oscillator circuit which should produce fields of the proper strengths and frequencies. Fortunately some triodes were discovered in an instrument cabinet and pronounced by Fermi apt to operate the projected circuit. The laboratory also possessed several hot wire ammeters to measure the current in the coils, in order that the magnetic field strength could be determined. Had these instruments not been available, the experiment could not have been performed, since the research budget of the laboratory was exceedingly meager and did not allow the purchase of costly equipment. Another consequence of this financial situation was the fact that the building was never heated, since it was easy to calculate that one month's heating would have absorbed the entire annual budget of the Physics Department. The temperature in the building from December to March ranged from 3° to 6° C. Unfortunately in the spring when the experiments were performed the room temperature had risen to 12° C, more comfortable for the inhabitants, but somewhat too high to ensure a sufficiently low density of the saturated mercury vapor.

Inductance coils and other simple parts were built by the experimenters, and when the circuit was assembled, it instantly worked as Fermi had predicted. The experiments were readily performed; unfortunately the accuracy was poor, due to the high temperature and the photographic method employed for measuring the polarization. Still the results clearly showed that the precession frequency of the atom agreed with the prediction based on the Landé factor.

A full account of the theory and experimental results is given in (28), while (27) is a condensed German version where the calculations have been omitted, and (26) is a brief preliminary announcement.

The interest of this experiment from the quantum-mechanical standpoint was emphasized by Jordan [1].

F. RASETTI

(1) P. JORDAN, *Anschauliche Quantentheorie*, Springer, Berlin 1936, pp. 45-46.

26.

EFFECT OF AN ALTERNATING MAGNETIC FIELD ON THE POLARISATION OF THE RESONANCE RADIATION OF MERCURY VAPOUR

E. FERMI and F. RASETTI

« Nature » (London), *115*, 764 (1925).

Recently, A. Ellett (« Nature », December 27, 1924, p. 931) and W. Hanle (« Zs. f. Phys. », *30*, 93, 1924) observed the depolarising effect of a weak magnetic field on resonance radiation. When the intensity of the field was sufficiently small they found, not only partial depolarisation, but also a rotation of the plane of polarisation. This is accounted for, on the classical point of view, by the superposed effect of the Larmor rotation and of the damped vibrations of the oscillator.

The same classical views suggest that the depolarising action of a high frequency alternating magnetic field of constant amplitude will vanish with increasing frequency. The effect should be well observable with fields of 2 or 3 gauss, and frequencies between 10^6 and 10^7.

We have performed the experiment, and have detected the presence of the expected phenomenon. A strong increase of the polarisation was actually observed in passing from a frequency of $1 \cdot 5 \times 10^6$ to one of 5×10^6, though the amplitude of the field remained constant.

We are carrying out further experiments in order to determine the quantitative features of the effect.

Istituto Fisico dell'Università, Firenze, Italy, April 3.

N° 28 (1) and 28 (2).

For the introduction to these papers see N° 26.

28 (1).

EFFETTO DI UN CAMPO MAGNETICO ALTERNATO SOPRA LA POLARIZZAZIONE DELLA LUCE DI RISONANZA(*)

E. FERMI e F. RASETTI

«Rend. Lincei», *1*, 716–722 (1925 (**).

Ci siamo proposti in questo lavoro di studiare l'effetto di un campo magnetico alternato ad alta frequenza sopra la polarizzazione della luce di risonanza; nella presente Nota tratteremo della teoria del fenomeno, in una successiva renderemo invece conto di alcune esperienze che ci hanno permesso di confermare i risultati previsti.

È noto che se si illumina un vapore, p. e. quello di mercurio, con la luce di una sua riga di assorbimento, si ha una intensissima diffusione selettiva; ed in questo consiste il fenomeno della risonanza ottica. Per renderci conto di questo fatto, possiamo pensare ad un modello classico, che in questo caso rende abbastanza bene conto delle circostanze del fenomeno. Supponiamo dunque che il nostro vapore contenga degli elettroni legati elasticamente, capaci di vibrare con frequenza eguale a quella della luce illuminante. Sotto l'azione del campo elettrico alternato della luce, gli elettroni si metteranno in risonanza e diffonderanno perciò molto fortemente la luce primaria. Se ora supponiamo che questa sia polarizzata, gli elettroni oscilleranno in direzione parallela a quella del suo vettore elettrico. Se osserviamo la luce diffusa in una direzione perpendicolare a quest'ultima, la troveremo dunque polarizzata, col suo vettore elettrico parallelo a quello della luce primaria. E ciò si osserva effettivamente nel caso del mercurio, illuminandolo con la riga 2536.7, almeno quando la densità del vapore sia sufficientemente piccola, poiché altrimenti si hanno delle perturbazioni di carattere secondario. È bene osservare a questo proposito che non per tutti gli elementi le cose tornano così bene come per il mercurio; precisamente le cose vanno secondo le previsioni della teoria classica solo nel caso che la riga di risonanza abbia un effetto Zeeman normale, o per lo meno si

(*) Vedi la nota preliminare in «Nature», 16 maggio 1925.

(**) Presentata dal Socio A. Garbasso nella seduta del 5 giugno 1925.

scinda in un tripletto, se anche di separazione differente da quella di Lorentz. Nel caso invece di effetti Zeeman anomali, come per esempio per le righe D del sodio, la polarizzazione presenta delle particolarità alquanto più complicate.

Esaminiamo ora quale effetto debba produrre sopra la polarizzazione della luce di risonanza, l'essere il vapore risonante immerso in un campo magnetico, abbastanza debole però perché la separazione Zeeman da esso prodotta non sia più grande della larghezza della riga con cui si illumina. Consideriamo per esempio il caso, che è quello che ci interesserà in seguito, che il campo magnetico sia perpendicolare al raggio incidente e al suo vettore elettrico, e che la radiazione di risonanza si osservi nella direzione del campo. Per fissare le idee, supporremo che il raggio primario e il campo magnetico siano orizzontali e ad angolo retto tra di loro, e che il vettore elettrico sia verticale. Il campo magnetico produrrà una scissione della riga di assorbimento in un tripletto Zeeman; ora osserviamo che se il vapore invece di assorbire, emettesse, nella direzione da cui viene la luce primaria si osserverebbero le componenti circolari come luce polarizzata rettilineamente col vettore elettrico verticale, mentre la componente rettilinea darebbe luogo a luce polarizzata col vettore elettrico orizzontale. Se dunque, come nel nostro caso, si illumina il vapore con luce con il vettore elettrico verticale, saranno assorbite solo le componenti circolari; le quali daranno luogo per diffusione, osservate nella direzione del campo, a due componenti circolari in senso opposto, le cui frequenze saranno praticamente indistinguibili, per modo che si osserverà una radiazione di risonanza non polarizzata. L'effetto del campo magnetico è dunque in questo caso di distruggere la polarizzazione della luce di risonanza; questo fenomeno fu in realtà scoperto da Wood e Ellett [1].

Recentemente W. Hanle [2] e A. Ellett [3] hanno fatto, l'uno sopra il mercurio e l'altro sopra il sodio, delle interessantissime osservazioni sopra l'effetto di un campo magnetico estremamente debole. Essi hanno osservato che in questo caso non soltanto la depolarizzazione prodotta dal campo non è completa, ma si ha anche una rotazione del piano di polarizzazione della luce diffusa, nel senso della rotazione di Larmor. Per il mercurio questi fenomeni si hanno per campi dell'ordine di grandezza di un Gauss.

Col modello precedentemente usato ci possiamo facilmente rendere conto della ragione di questo fatto. Consideriamo un oscillatore che abbia ricevuto un impulso in direzione verticale, e oscilli liberamente nel campo magnetico orizzontale. Se l'oscillatore non fosse smorzato, sotto l'azione del campo magnetico la sua linea d'oscillazione si metterebbe a rotare, restando costante l'ampiezza dell'oscillatore; per modo che un osservatore longitudinale al campo vedrebbe luce non polarizzata. Le cose vanno invece in modo diverso se l'oscillatore è smorzato, e la rotazione di Larmor è tanto lenta, che in una piccola frazione di giro l'ampiezza si smorzi già

(1) R. W. WOOD e A. ELLETT, « Proc. Roy. Soc. », *103*, 396 (1923).

(2) W. HANLE, « Zs. f. Phys. », *30*, 93 (1924).

(3) A. ELLETT, « Journ. Opt. Soc. Amer. », *10*, 427 (1925).

notevolmente; si capisce che in questo caso la depolarizzazione deve essere soltanto parziale e la linea d'oscillazione risultare in media rotata rispetto alla direzione del campo elettrico eccitante, ciò che dà luogo appunto alla rotazione del piano di polarizzazione della luce diffusa.

Passiamo ora a esaminare il caso, oggetto del nostro studio, che l'atomo risonante si trovi in un campo magnetico, non più costante ma alternato. Se l'oscillatore riceve un impulso dalla forza elettrica eccitante, la sua linea d'oscillazione si metterà dapprima a rotare, per la precessione di Larmor, nel senso corrispondente a quello del campo magnetico al dato istante; quando poi il campo si inverte, essa comincia a rotare in senso opposto, e così di seguito. Se la frequenza del campo alternato è molto grande in confronto alla frequenza di Larmor, accadrà che la direzione dell'oscillatore andrà avanti e indietro scostandosi sempre di poco da quella iniziale. Dobbiamo dunque attenderci che per frequenze grandissime, l'effetto depolarizzante del campo, pur restando costante l'intensità media di quest'ultimo, debba diventare piccolissimo.

Da quanto precede risulta evidente che le variazioni del grado di polarizzazione in funzione della frequenza saranno particolarmente notevoli per frequenze del campo alternato dell'ordine di grandezza di quella di Larmor. Occorre naturalmente anche, per osservare le variazioni della polarizzazione, che il campo sia abbastanza intenso per produrre, a bassa frequenza, una depolarizzazione considerevole. Per questo è necessario che il periodo della rotazione di Larmor sia al massimo dell'ordine di grandezza del tempo medio di smorzamento, il quale, per la riga 2536.7 del mercurio risulta da varie considerazioni [(4)] eguale a circa 10^{-7} secondi (il tempo di smorzamento dell'ampiezza sarà dunque doppio, cioè $2 . 10^{-7}$ secondi).

Dovremo dunque operare con campi di intensità di qualche Gauss (poiché la frequenza di Larmor che corrisponde al campo di un Gauss è $1{,}4 \cdot 10^6$), e di frequenza dell'ordine di cinque milioni.

Proponiamoci dunque di studiare il comportamento di un oscillatore risonante situato in un campo magnetico alternato. Supponiamo che il vettore elettrico E della luce eccitante abbia la direzione dell'asse delle x, e il campo magnetico quella dell'asse delle z. Il movimento dell'oscillatore avverrà allora nel piano xy.

Ora osserviamo che nel caso pratico della realizzazione delle esperienze di risonanza, la luce eccitante è sempre relativamente espansa, per modo che la sua larghezza $\Delta\nu$, misurata in frequenza, è di un ordine di grandezza alquanto superiore alla larghezza che avrebbe la riga se fosse espansa per l'unica ragione dello smorzamento dei treni d'onda. Avremo quindi $\Delta\nu$ molto maggiore di σ, indicandosi con σ il coefficiente di smorzamento dell'ampiezza del nostro oscillatore. Siccome nel caso nostro σ, la frequenza di Larmor n_L, e la frequenza n del campo magnetico sono dello stesso ordine di grandezza, avremo che $\Delta\nu$ sarà di ordine di grandezza maggiore di tutte queste quantità. Ora è noto che una riga spettrale di larghezza $\Delta\nu$

(4) L. A. TURNER, « Phys. Rev. », *23*, 464 (1924); R. C. TOLMAN, « Phys. Rev. », *23*, 693 (1924); W. HANLE, loc. cit.

è equivalente a un insieme di treni d'onda rigorosamente monocromatici, limitati, e contenenti ciascuno un numero di onde dell'ordine di grandezza di $\nu/\Delta\nu$. Avremo dunque che la nostra luce eccitante sarà equivalente a un insieme di treni d'onda incoerenti, ciascuno di durata $1/\Delta\nu$, assai breve in confronto al periodo di smorzamento dell'oscillatore $1/\sigma$, e, per le nostre ipotesi, anche a $1/n_L$ e a $1/n$. Siccome le equazioni del movimento dell'oscillatore sono lineari, avremo che ciascuno di questi treni d'onda produrrà il suo effetto indipendentemente dall'azione degli altri. Ora l'azione di un solo treno d'onda consiste in un impulso di durata assai breve, per modo che durante essa sarà trascurabile l'effetto dello smorzamento e della rotazione di Larmor. Ricevuto questo impulso nella direzione del campo elettrico eccitante, l'oscillatore continuerà a muoversi sotto l'azione soltanto della sua forza elastica e della forza esercitata dal campo magnetico esterno. La radiazione da esso diffusa sarà dunque quella emessa da un oscillatore al quale è stato dato un impulso parallelamente all'asse x, e che poi ha continuato a muoversi sotto l'azione delle forze indicate. Se non vi fosse né campo magnetico né smorzamento, l'oscillatore, ricevuto l'impulso iniziale, seguiterebbe a muoversi lungo l'asse delle x con ampiezza eguale a quella iniziale A_0; l'effetto dello smorzamento è di ridurre l'ampiezza dell'oscillazione, così che al tempo t essa sarà $A = A_0 e^{-\sigma t}$; l'effetto del campo magnetico $H = H_0 \cos(2\pi nt - \alpha)$ sarà di fare rotare la linea d'oscillazione con velocità angolare

$$\omega = \frac{eH}{2mc} = \frac{eH_0}{2mc} \cos(2\pi nt - \alpha)$$

e quindi al tempo t l'angolo di cui avrà rotato l'asse di oscillazione sarà

$$\Omega = \int_0^t \omega\, dt = \frac{eH_0}{4\pi mcn} \{\sin(2\pi nt - \alpha) + \sin\alpha\} = \frac{2n_0}{n} \sin \pi nt \cos(\pi nt - \alpha)$$

dove con $n_0 = \dfrac{eH_0}{4\pi mc}$, si è indicata la frequenza di Larmor corrispondente al valor massimo H_0 del campo magnetico. Avremo dunque che al tempo t le componenti dell'ampiezze parallele all'asse x e all'asse y saranno

$$(1) \qquad A_x = A_0 e^{-\sigma t} \cos\Omega \quad ; \quad A_y = A_0 e^{-\sigma t} \sin\Omega .$$

Le componenti X e Y dell'intensità diffusa al tempo t in direzione z, polarizzate col vettore elettrico parallelo risp. all'asse delle x e a quello delle y, saranno proporzionali ai quadrati delle ampiezze corrispondenti, si avrà cioè, a meno di un fattore costante

$$(2) \qquad X = e^{-2\sigma t} \cos^2\Omega \quad ; \quad Y = e^{-2\sigma t} \sin^2\Omega .$$

Siccome per quanto si è osservato, i vari impulsi che equivalgono all'azione della luce eccitante sono tra di loro incoerenti, l'intensità si otterrà sommando le intensità dovute all'azione dei singoli impulsi. Le intensità diffuse I_x e I_y sono dunque proporzionali alle espressioni che si ottengono integrando le (2) rispetto al tempo e a tutti i valori della fase α. Si avrà dunque a meno di un fattore

$$(3)\quad \begin{cases} I_x = \int_0^{2\pi}\int_0^{\infty} X\, dt d\alpha = \int_0^{2\pi}\int_0^{\infty} e^{-2\sigma t} \cos^2\left[\frac{2 n_0}{n} \sin \pi n t \cos(\pi n t - \alpha)\right] dt d\alpha \\ \qquad = \int_0^{2\pi}\int_0^{\infty} e^{-2\sigma t} \cos^2\left[\frac{2 n_0}{n} \sin \pi n t \cos \varphi\right] dt d\varphi \\ I_y = \int_0^{2\pi}\int_0^{\infty} Y\, dt d\alpha = \int_0^{2\pi}\int_0^{\infty} e^{-2\sigma t} \sin^2\left[\frac{2 n_0}{n} \sin \pi n t \cos \varphi\right] dt d\varphi\,. \end{cases}$$

Per valutare queste espressioni osserviamo che si ha

$$(4)\quad \begin{cases} I_x + I_y = \int_0^{2\pi}\int_0^{\infty} e^{-2\sigma t}\, dt d\varphi = \frac{\pi}{\sigma} \\ I_x - I_y = \int_0^{2\pi}\int_0^{\infty} e^{-2\sigma t} \cos\left[\frac{4 n_0}{n} \sin \pi n t \cos \varphi\right] dt d\varphi\,. \end{cases}$$

Indicando con $J_0(x)$ la funzione di Bessel di ordine zero, e tenendo presente la nota formula

$$2\pi J_0(a) = \int_0^{2\pi} \cos(a \cos \varphi)\, d\varphi$$

troviamo

$$I_x - I_y = 2\pi \int_0^{\infty} e^{-2\sigma t} J_0\left(\frac{4 n_0}{n} \sin \pi n t\right) dt\,.$$

Prendendo per nuova variabile $x = \pi n t$ abbiamo dunque

$$(5)\quad I_x - I_y = \frac{2}{n} \int_0^{\infty} e^{-\frac{2\sigma}{\pi n} x} J_0\left(\frac{4 n_0}{n} \sin x\right) dx\,.$$

Prima di tutto cerchiamo una espressione di questa differenza per n molto piccolo. Tenendo presente la nota formula

$$\frac{1}{\sqrt{a^2 + b^2}} = \int_0^{\infty} e^{-ax} J_0(bx)\, dx$$

si trova che per n molto piccolo si può scrivere

$$(6)\quad \lim_{n=0} (I_x - J_y) = \frac{\pi}{\sigma \sqrt{1 + \frac{4\pi^2 n_0^2}{\sigma^2}}}\,.$$

Questa formula ci permette di trovare il grado di polarizzazione per frequenze molto piccole. Esso risulta dato da

$$\lim \frac{I_x - I_y}{I_x + I_y} = \frac{1}{\sqrt{1 + \frac{4\pi^2 n_0^2}{\sigma^2}}}\,.$$

Come si vede da questa espressione, si trova che nel caso di frequenze molto piccole il grado di polarizzazione risulta eguale a zero quando n_0 e quindi anche H_0, è molto grande; in questo caso si ha dunque una depolarizzazione completa, mentre per n_0 eguale a zero, e quindi $H_0 = 0$, si trova il grado di polarizzazione eguale ad uno, cioè polarizzazione completa.

Tornando alla espressione generale (5) ci conviene trasformarla leggermente per renderla adatta al calcolo numerico. Osservando perciò che la funzione $J_0\left(\frac{4 n_0}{n} \sin x\right)$ è periodica a periodo π, si vede facilmente che la (5) si può trasformare in

$$I_x - I_y = \frac{2}{n} \frac{e^{2\sigma/n}}{e^{2\sigma/n} - 1} \int_0^{\pi} e^{-\frac{2\sigma}{\pi n} x} J_0\left(\frac{4 n_0}{n} \sin x\right) dx \,. \tag{7}$$

In base a quest'ultima espressione abbiamo calcolato numericamente il grado di polarizzazione, ponendo $\sigma = 5 \cdot 10^6$, per campi di 1,7; 2,5; 3,2 Gauss. Il risultato di questo calcolo è rappresentato dalle curve della figura, dove in ascisse sono riportati i valori della frequenza, in milioni al secondo, e in ordinate, il rapporto I_y/I_x.

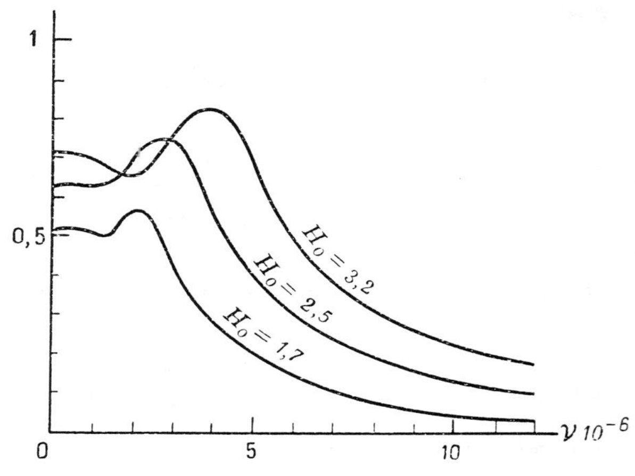

Come si rileva dalle curve, in tutti i casi si hanno per basse frequenze valori oscillanti della polarizzazione, per frequenze maggiori si ha poi un aumento della polarizzazione, che diventa completa per frequenze grandissime.

L'aumento della polarizzazione incomincia tanto più presto, quanto più è piccolo il campo.

28 (2).

ANCORA DELL'EFFETTO DI UN CAMPO MAGNETICO ALTERNATO SOPRA LA POLARIZZAZIONE DELLA LUCE DI RISONANZA (*)

E. FERMI e F. RASETTI

« Rend. Lincei », *2*, 117–120 (1925) (**).

In una Nota precedente [1] abbiamo calcolato come varia l'effetto di un campo magnetico alternato sulla polarizzazione della luce di risonanza, in funzione della frequenza. Abbiamo eseguite delle esperienze allo scopo di verificare le previsioni teoriche, sopra la risonanza del mercurio per la riga 2536.7.

Il dispositivo da noi usato per questo scopo constava di una parte ottica, per la produzione della risonanza, e di una parte elettrica destinata a generare il campo magnetico alternato.

La luce eccitante era ottenuta per mezzo di una lampada a mercurio, raffreddata mediante immersione in acqua. La riga 2536.7 veniva isolata mediante un monocromatore a quarzo; sul cammino dei raggi si trovava un grosso cristallo birifrangente di spato d'Islanda, che separava la luce in due fasci polarizzati ad angolo retto. Di questi, quello col vettore elettrico verticale veniva introdotto nella lampada di risonanza, che era costituita da un recipiente di vetro munito di due finestre, chiuse con lastre piane di quarzo, e situate ad angolo retto tra di loro. Di esse una serviva per l'entrata dei raggi eccitanti, e l'altra per l'osservazione della luce di risonanza; le lastre di quarzo erano tagliate parallelamente all'asse ottico e disposte in modo da non alterare la polarizzazione della luce. La lampada di risonanza conteneva una goccia di mercurio, ed era unita ad una pompa di Gaede, che funzionava continuamente durante le esperienze. La luce di risonanza veniva osservata in direzione perpendicolare a quella della luce eccitatrice; mediante un altro spato, interposto sul cammino di osservazione, si formavano due immagini del vapore risonante, polarizzate una orizzontalmente e l'altra verticalmente, che venivano fotografate con una camera a quarzo.

L'apparecchio per la produzione del campo oscillante constava essenzialmente di un circuito a triodi generatore di oscillazioni persistenti, e di

(*) Lavoro eseguito nell'Istituto Fisico della R. Università di Firenze.

(**) Presentata dal Socio A. Garbasso nell'adunanza del 5 giugno 1925.

(1) Questi « Rendiconti », vol. I (1925), p. 716, n. 12.

un altro circuito ad esso accoppiato del quale facevano parte due bobine, situate ai lati della lampada di risonanza, per la produzione del campo magnetico alternato.

Abbiamo usato dei triodi della « Western Electric Company », col filamento ricoperto di ossidi; la tensione di placca era di circa 300 volt ed era fornita da una dinamo, la corrente di placca per ogni triodo era di 40 o 50 milliampère. Abitualmente usavamo due di questi triodi in parallelo.

Il circuito generatore era costituito unendo il negativo del filamento col punto di mezzo di una bobina L, i cui estremi erano uniti attraverso ad un condensatore variabile C_1. Una delle armature del condensatore era congiunta direttamente alla griglia, e l'altra era unita alla placca attraverso alla dinamo. Il circuito destinato alla produzione del campo magnetico alternato conteneva in serie: una spira avvicinata alla bobina L, che funzionava da secondario di trasformatore, un condensatore variabile C_2, un amperometro a filo caldo, privo di shunt, due bobine di dieci spire ciascuna per la produzione del campo magnetico, e una bobina che si poteva inserire o chiudere in corto circuito, per mezzo della quale era possibile accordarsi in un campo di frequenze sufficientemente esteso.

Le capacità C_1 e C_2 potevano variare all'incirca da 50 a 500 millesimi di microfarad. La bobina L poteva secondo i casi avere una induttanza di 12 o 40 microhenry. Data la considerevole capacità tra la placca e la griglia degli audion, non era possibile spingersi comodamente a una frequenza superiore a $5 \cdot 10^6$. Nelle esperienze abbiamo usato delle frequenze variabili tra uno e cinque milioni.

L'ondametro che ci ha servito per la misura delle frequenze, ci è stato gentilmente tarato dal dott. Mazza dell'Istituto di Chimica inorganica dell'Università di Firenze.

Le considerazioni teoriche svolte nella Nota precedente, ci facevano ritenere opportuno di studiare l'effetto con campi magnetici di qualche Gauss. Per produrli, con le nostre bobine, erano necessarie correnti dell'ordine di grandezza di mezzo ampère, che si potevano produrre assai bene col nostro apparecchio.

Il campo magnetico terrestre era neutralizzato mediante una grossa bobina opportunamente disposta.

Alla temperatura dell'ambiente, circa 12°, la polarizzazione della luce di risonanza si dimostrò scarsa, anche in assenza di qualunque campo magnetico. Per questo ritenemmo utile di raffreddare la lampada di risonanza. Un sistema di raffreddamento a evaporazione di etere diede dei risultati discreti, in quanto riuscì a fare aumentare fortemente la polarizzazione in assenza di campo magnetico, senza però riuscire a mantenerla costante da una esperienza all'altra; fatto che ci rese impossibile di spingere la precisione delle misure, fino a poter studiare alcuni dei dettagli previsti dalla teoria.

Osserviamo ancora che siccome la riga 2536.7 del mercurio presenta un effetto Zeeman anomalo, in quanto essa si spezza in un tripletto con una separazione eguale a tre mezzi di quella normale, la precessione prodotta da un campo magnetico si deve calcolare eguale a 3/2 di quella di Larmor. Quindi le curve disegnate nella Nota precedente e indicate come corrispondenti a campi di 1,7; 2,5; 3,2 Gauss, si avranno in realtà, per il mercurio, per campi di intensità 2/3 di queste. Queste tre intensità, che corrispondevano a intensità efficaci di corrente nelle bobine di 0.32, 0.46, 0.6 amp., sono quelle con le quali abbiamo eseguite le esperienze. Su ciascuna lastra vennero fatte sei fotografie, con pose, eguali per ciascuna lastra, di due o tre minuti; ciascuna era costituita da due immagini della luce di risonanza polarizzate ad angolo retto. La prima fotografia era fatta senza campo magnetico, le altre cinque con campi di intensità massima costante e con frequenze di 1, 2, 3, 4, 5 milioni. Abbiamo fatto tre lastre con campi prodotti da correnti di 0.32 e di 0.46 amp., e due con campi corrispondenti a 0.6 amp.

Già l'osservazione ad occhio delle lastre mostrava nettamente, per correnti 0.32 e 0.46 amp., l'aumento della polarizzazione per grandi frequenze; ma per ottenere dati più precisi abbiamo fotometrato le nostre lastre, con un fotometro a pila termoelettrica. Riportiamo qui i risultati delle fotometrie.

I numeri rappresentano i rapporti degli annerimenti della componente polarizzata con il vettore elettrico orizzontale, e di quella col vettore elettrico verticale; le frequenze sono date in milioni.

Osserviamo a questo proposito che nelle nostre fotografie non si aveva proporzionalità tra annerimento e intensità, come sarebbe voluto dalla legge di Schwarzschild, e quindi i rapporti dànno solamente una indicazione qualitativa del grado di polarizzazione. Per la stessa ragione non sono confrontabili tra di loro altro che i dati della stessa lastra, perché tra lastra e lastra si hanno delle differenze di intensità di illuminazione e di posa.

Numero della lastra	Amp.	Frequenze				
		1,2	2	3	4	5
6	0,32	0,71	0,69	0,64	0,59	0,53
7	—	—	0,75	0,72	0,67	0,51
8	—	0,58	0,59	0,52	0,39	0,36
5	0,46	0,76	0,74	0,77	0,68	0,57
9	—	0,72	0,71	0,78	0,65	0,53
10	—	0,74	0,75	0,74	0,67	—
11	0,60	0,84	0,85	0,92	0,85	0,79
12	—	0,76	0,73	0,80	0,79	0,76

Come si può rilevare da questa tabella, la precisione non è sufficiente per la verificazione dettagliata delle particolarità delle curve disegnate nella Nota precedente. Quello che invece si può concludere con sicurezza, è che nelle lastre corrispondenti a 0.32 amp. la polarizzazione cresce con la frequenza, in modo più rapido per frequenze elevate; in quelle fatte con correnti di 0.46 amp., la polarizzazione si mantiene press'a poco costante fino a frequenze di circa tre milioni, e per frequenze superiori comincia a crescere, benché meno che nel caso precedente; infine, per le fotografie fatte con correnti di 0.6 amp., le oscillazioni della polarizzazione restano entro i limiti dell'errore di osservazione.

Confrontando questi risultati con le curve ottenute teoricamente nella Nota precedente, si vede che le previsioni sono confermate, almeno entro i limiti della nostra approssimazione.

N° 29.

For the introduction to this paper see N° 22.

29.

SOPRA LA TEORIA DEI CORPI SOLIDI

« Periodico di Matematiche », 5, 264–274 (1925).

Il carattere che distingue, dal punto di vista cinetico, un corpo solido da un fluido, consiste essenzialmente nel fatto, che mentre gli atomi di quest'ultimo nel loro moto termico possono allontanarsi assai considerevolmente dalla posizione da essi inizialmente occupata, quelli del primo oscillano invece continuamente attorno a una posizione di equilibrio. La ragione di questo diverso comportamento si deve naturalmente ricercare nel differente rapporto che esiste nei due casi tra l'energia cinetica del moto termico degli atomi, e l'energia delle forze che tendono a mantenerli nelle loro posizioni di equilibrio.

I problemi fondamentali della teoria cinetica dei corpi solidi si possono ridurre sostanzialmente a due. Il primo si propone la determinazione del moto termico degli atomi, ammettendo provvisoriamente che le forze che agiscono tra di essi siano quasi elastiche, mentre il secondo ha invece per oggetto la determinazione della natura effettiva di queste forze.

Noi ci occuperemo anzitutto del primo problema, e la questione che tratteremo più particolarmente sarà quella della determinazione dell'energia dei moti termici, e quindi anche dei calori specifici, in funzione della temperatura. Questo problema, che anche storicamente fu il primo ad essere trattato, è particolarmente adatto a mostrare i successivi stadi dello sviluppo della teoria di cui ci stiamo occupando, e a far vedere come essa dovette successivamente perfezionarsi per rendere conto dei nuovi fatti sperimentali che venivano man mano scoperti.

I calori specifici dei corpi solidi variano con la temperatura. Vi è però un intervallo di temperature, che comprende, per la maggior parte degli elementi solidi, anche la temperatura ordinaria, nel quale il calore specifico si mantiene abbastanza approssimativamente costante. Ora le prime misure furono naturalmente eseguite in questo intervallo e portarono alla scoperta della legge di Dulong e Petit, che afferma, come è noto, che il calore specifico degli elementi solidi è inversamente proporzionale al loro peso atomico, o, che è lo stesso, che se invece che all'unità di massa ci si riferisce al grammo atomo, il calore specifico risulta eguale per tutti gli elementi.

Questa legge che, quando si effettuino le misure a temperatura ordinaria, presenta solo pochissime eccezioni, trovò una facile interpretazione teorica.

Consideriamo infatti N atomi di un corpo che, in prima approssimazione, riguarderemo come punti materiali; essi avranno complessivamente 3 N gradi di libertà. Conformemente ai principî della meccanica statistica classica ad ogni grado di libertà corrisponde l'energia cinetica media $kT/2$, essendo T la temperatura assoluta e k la costante di Boltzmann; l'energia cinetica dei nostri N atomi sarà dunque $3\,NkT/2$. Siccome poi le forze che agiscono tra di essi sono in prima approssimazione delle forze elastiche, si ha che in media l'energia cinetica è uguale a quella potenziale.

L'energia termica totale del nostro corpo sarà dunque

$$W = 3\,NkT. \tag{1}$$

Dall'espressione dell'energia risulta immediatamente quella del calore specifico, che è dato da

$$c = \frac{dW}{dT} = 3\,Nk\,.$$

Secondo questa teoria esso risulta dunque eguale per tutti gli elementi e indipendente dalla temperatura; anche numericamente il suo valore coincide bene con la costante della legge di Dulong e Petit.

Esperienze successive dimostrarono però che la legge di Dulong e Petit perde completamente la sua validità per temperature molto basse, poiché a tali temperature il calore specifico cessa di essere costante, e diminuisce tendendo a zero quando la temperatura si avvicina allo zero assoluto.

Il primo a dare una interpretazione teorica di questo risultato fu Einstein. Egli osservò che essendo gli atomi di un corpo legati tra di loro da forze quasi elastiche, essi sono equiparabili a oscillatori, e, come prima approssimazione, li considerò equivalenti complessivamente a 3 N oscillatori, tutti della stessa frequenza ν. Secondo la teoria dei quanti la legge della equipartizione dell'energia perde la sua validità per oscillatori di frequenza abbastanza elevata, e l'espressione dell'energia media di un oscillatore invece di essere, come secondo la meccanica statistica classica, kT, ha invece il valore

$$\frac{h\nu}{e^{h\nu/kT} - 1}\,.$$

L'energia media dei 3 N oscillatori equivalenti al nostro corpo solido sarà dunque

$$W = \frac{3\,N\,h\nu}{e^{h\nu/kT} - 1}\,. \tag{2}$$

Questa formula, e quella che da essa può dedursi per il calore specifico, si mostrarono in abbastanza buon accordo con l'esperienza, specialmente per temperature non estremamente basse; i valori di ν calcolati dall'andamento dei calori specifici corrispondono anche abbastanza bene con le frequenze dei così detti Reststrahlen, che sono radiazioni luminose, situate nell'infrarosso estremo, le quali derivano appunto dalle vibrazioni degli atomi del corpo.

Per temperature molto vicine allo zero assoluto tuttavia anche la teoria di Einstein si dimostrò insufficiente; mentre infatti in realtà per temperature molto basse il calore specifico dei corpi solidi tende a zero come T^3, secondo la teoria di Einstein invece esso dovrebbe annullarsi assai più rapidamente.

Per correggere questa ultima divergenza occorreva modificare una ipotesi troppo semplicista della teoria di Einstein. In questa teoria si parte infatti dall'ipotesi che tutte le frequenze caratteristiche del corpo solido siano tra di loro eguali; il sistema elastico costituito da tutti gli atomi del corpo sarà invece in realtà capace di vibrare secondo frequenze assai numerose; nel calcolo dell'energia termica del corpo solido bisogna quindi tener conto separatamente di tutte queste frequenze caratteristiche.

Un primo metodo per il calcolo di tali frequenze fu escogitato da Debye, e benché esso debba far uso di una ipotesi a prima vista alquanto insoddisfacente, diede tuttavia risultati assai buoni; successivamente il metodo di Debye fu perfezionato e completato per opera di Born e Karman i quali ebbero anche modo di giustificare le ipotesi fatte da Debye. Il principio del metodo usato da Debye è il seguente: Si fa in esso anzitutto astrazione, in una prima approssimazione, dalla struttura discontinua del corpo solido, che viene considerato come un mezzo elastico omogeneo e continuo. Se ora si considera un pezzo di questo corpo, esso sarà capace di vibrare con certe frequenze caratteristiche dipendenti dalla sua forma, dalle sue dimensioni e dalle sue proprietà elastiche, allo stesso modo che una corda tesa può vibrare con frequenze che dipendono dalla sua lunghezza e dalla sua tensione. Naturalmente, per il problema che ci interessa, non è necessaria la determinazione dettagliata di tutte queste frequenze, ma basta conoscere la distribuzione, avere cioè una espressione asintotica per il numero delle frequenze caratteristiche che cadono in un assegnato intervallo di frequenze.

Ora si trova che il numero delle frequenze proprie comprese tra ν e $\nu + d\nu$ è dato, per un sistema di vibrazioni che si propagano con velocità v in un corpo di volume V, da $\frac{4\pi V}{v^3}\nu^2 d\nu$. Siccome poi nel caso dell'elasticità si hanno tre sistemi di propagazioni, due trasversali con velocità v_1 e uno longitudinale con velocità v_2, possiamo concludere che il numero delle frequenze proprie comprese tra ν e $\nu + d\nu$ è dato da $4\pi V\left(\frac{2}{v_1^3}+\frac{1}{v_2^3}\right)\nu^2 d\nu$. Attribuendo a ciascuna di queste frequenze l'energia media $h\nu : (e^{h\nu/kT} - 1)$, troviamo come espressione dell'energia termica di tutto il nostro corpo solido

$$W = 4\pi V\left(\frac{2}{v_1^3}+\frac{1}{v_2^3}\right)\int \frac{h\nu^3\, d\nu}{e^{h\nu/kT} - 1}. \tag{3}$$

Se la schematizzazione da noi fatta nel considerare il nostro corpo come un corpo elastico continuo, fosse in realtà esatta, l'integrale precedente dovrebbe evidentemente venir esteso tra zero e infinito, poiché sarebbe possibile la propagazione di vibrazioni elastiche di frequenza comunque grande.

È evidente invece che nel nostro caso ciò non può avvenire, poiché mentre il corpo continuo è capace di vibrare con infinite frequenze caratteristiche, il corpo reale discontinuo ha invece soltanto 3 N gradi di libertà. Di fronte a questa difficoltà Debye usò un artificio, che benché in linea di principio assai insoddisfacente, portò tuttavia a brillanti risultati, e fu del resto, almeno in parte, giustificato in seguito dalla teoria più perfezionata di Born e Karman. Debye ammise che in un corpo elastico reale non potessero propagarsi vibrazioni di frequenza superiore a un certo limite ν_{max}, e calcolò questo limite in modo che il numero totale delle frequenze caratteristiche inferiori ad esso, e cioè

$$4\pi V\left(\frac{2}{v_1^3}+\frac{1}{v_2^3}\right)\int_0^{\nu_{max}}\nu^2\,d\nu$$

risultasse eguale a 3 N, ossia al numero dei gradi di libertà del corpo.

Con questa ipotesi il limite superiore dell'integrale (3) diventa ν_{max} e otteniamo una espressione per l'energia termica, e quindi anche per il calore specifico, in ottimo accordo con l'esperienza. Osserviamo anche che l'accordo tra questa teoria e l'esperienza è tanto più significativo, in quanto la formula di Debye contiene quali unici parametri le costanti elastiche del corpo, le quali possono venir determinate con misure elastiche dirette, indipendenti completamente dalle misure del calore specifico.

Un perfezionamento ulteriore alla teoria dei calori specifici fu, come già abbiamo accennato, apportato da Born e Karman. Partendo dallo stesso concetto di Debye, e cioè di calcolare l'energia termica del corpo solido come somma delle energie di tanti oscillatori, di frequenze eguali alle frequenze caratteristiche del corpo, essi applicarono un metodo assai più corretto per il calcolo di queste frequenze, poiché considerarono il corpo quale esso è in realtà, e cioè costituito da un reticolato di atomi. Senza entrare nei dettagli del calcolo di un sistema elastico costituito da un reticolato di masse esercitanti tra di loro delle forze proporzionali agli spostamenti relativi, vogliamo far vedere sopra un esempio schematico, l'influenza che ha la struttura discontinua del corpo sopra la propagazione delle onde elastiche.

Studieremo per questo la propagazione di una perturbazione elastica in una fila di punti di massa eguale, spaziati tra di loro di un intervallo costante a. Supporremo che sopra ciascun atomo agiscano soltanto i due che gli sono più vicini, e che la forza che si esercita tra due atomi attigui sia proporzionale al loro spostamento relativo. Supponiamo dunque che lungo la nostra fila si propaghi una onda elastica sinusoidale, di frequenza ν e di velocità v. Lo spostamento di una particella si potrà scrivere

$$\delta = A \sin 2\pi\nu\left(t-\frac{x}{v}\right)$$

dove con x si rappresenta la ascissa della particella, che è data da na, essendo n un numero intero. Ci proponiamo di trovare la relazione che passa tra ν e v; faremo vedere precisamente che, a differenza di quello che accadrebbe in una sbarra elastica a struttura continua, la velocità di propagazione non è costante, ma dipende dalla frequenza.

Lo spostamento della particella per cui $n=1$, è dato da $A \sin 2\pi\nu\left(t-\frac{a}{v}\right)$ e quello della particella $n=0$, è dato invece da $A \sin 2\pi\nu t$. La forza agente tra queste due particelle, che per ipotesi è proporzionale al loro spostamento relativo, potrà dunque scriversi

$$kA\left\{\sin 2\pi\nu\left(t-\frac{a}{v}\right)-\sin 2\pi\nu t\right\}$$

essendo k il coefficiente di proporzionalità. Sulla particella $n = 0$, agirà anche un'altra forza, dovuta all'azione della particella per cui $n = -1$. Tale forza sarà espressa analogamente da $kA \left\{ \sin 2\pi n \left(t + \frac{a}{v} \right) - \sin 2\pi\nu t \right\}$.

Scrivendo che il prodotto della massa m della particella per cui $n = 0$, per la sua accelerazione $-4\pi^2\nu^2 \sin 2\pi\nu t$ è uguale alla somma delle forze che agiscono su di essa, troviamo

$$-4\pi^2\nu^2 mA \sin 2\pi\nu t = kA \left\{ \sin 2\pi\nu \left(t - \frac{a}{v} \right) + \sin 2\pi\nu \left(t + \frac{a}{v} \right) - 2 \sin 2\pi\nu t \right\}$$

$$= -2kA \sin 2\pi\nu t \left\{ 1 - \cos \frac{2\pi\nu a}{v} \right\} = -4kA \sin 2\pi\nu t \sin^2 \frac{\pi\nu a}{v}.$$

Da questa equazione si ricava, con immediate riduzioni

$$\nu = \frac{1}{\pi} \sqrt{\frac{k}{m}} \sin \frac{\pi\nu a}{v}. \tag{4}$$

Questa è appunto la relazione cercata tra la velocità di propagazione e la frequenza. Il caso della propagazione in un mezzo continuo, si può ottenere da quello studiato come caso limite per $a = 0$. In questo caso, sostituendo l'arco al seno si vede che ν viene a sparire dalla formula precedente, a quindi la velocità di propagazione v risulta indipendente dalla frequenza, come appunto deve essere in mezzi a struttura continua.

Osserviamo ancora che siccome il seno non può mai essere maggiore di 1 si ricava dalla (4)

$$\nu \leq \frac{1}{\pi} \sqrt{\frac{k}{m}}$$

ciò che ci dice che non può avvenire nella nostra fila di particelle una propagazione di onde elastiche di frequenza maggiore di $\frac{1}{\pi} \sqrt{\frac{k}{m}}$. Questo fatto giustifica in un certo senso l'introduzione, fatta arbitrariamente da Debye, della frequenza limite.

Passiamo ora ad accennare come possano essere interpretati, dal punto di vista della teoria di cui abbiamo indicate le basi, alcuni dei principali fenomeni della fisica dei corpi solidi.

Se le forze che agiscono tra gli atomi del corpo, fossero rigorosamente elastiche, la posizione media occupata da ciascun atomo non dovrebbe spostarsi per effetto del moto termico. In realtà però le forze interatomiche sono soltanto approssimativamente elastiche. La conseguenza di questo è che l'atomo in movimento si sposta anche in media dalla posizione che occuperebbe se fosse fermo, per modo che per effetto del moto termico viene a variare la distanza media tra gli atomi. Questo dà luogo alla dilatazione termica. Per vedere schematicamente come ciò accada, consideriamo un unico punto, attratto da una forza quasi elastica verso una posizione di equilibrio. Supporremo per semplicità il nostro punto mobile sopra una retta, e sia $x = 0$ l'ascissa della posizione di equilibrio. Prendiamo come espressione dell'energia potenziale della forza quasi elastica, $U = kx^2 + hx^3$; se mancasse il termine cubico le forza sarebbe esattamente elastica, e noi supporremo perciò che il termine cubico sia assai piccolo in confronto di quello quadratico. La forza agente sopra la nostra particella avrà per espressione $-\frac{\partial u}{\partial x} = -2kx - 3hx^2$; l'equazione del moto sarà

$$mx'' = -2hx - 3hx^2 .$$

Prendiamo i valori medi dei due termini della equazione precedente. Se si ammette che la velocità resti limitata, si avrà evidentemente $\overline{x''} = 0$. Abbiamo dunque

$$2h\overline{x} + 3k\overline{x^2} = 0$$

ossia

$$\overline{x} = -\frac{3}{2}\frac{h}{k}\overline{x^2} \tag{5}$$

si vede dunque che l'effetto della non perfetta elasticità del legame, consiste nel rendere differente da zero il valore medio dell'ascissa, ossia nel produrre uno spostamento della posizione media della particella. È chiaro come un ragionamento di questo genere sia suscettibile di essere esteso al caso di un reticolato atomico, e si capisce dunque come avviene che l'agitazione termica ha per effetto una dilatazione delle sostanze. Osserviamo ancora che dalla eguaglianza (5) risulta $\overline{x}$ proporzionale a $\overline{x^2}$, cioè in prima approssimazione alla energia del movimento; avremo dunque che la dilatazione termica sarà proporzionale alla energia termica del corpo. E questo si verifica in realtà, perché l'esperienza dimostra che il calore specifico, e il coefficiente di dilatazione termica hanno, in funzione della temperatura, un andamento del tutto simile.

La non completa elasticità del legame tra atomo e atomo, in un corpo solido presenta una notevole importanza per la spiegazione della conduzione termica degli isolanti, della piroelettricità, e di parecchi altri fenomeni. La ristrettezza del tempo ci vieta però di occuparci di questi argomenti.

Vogliamo invece accennare rapidamente agli studi che sono stati fatti sulla natura delle forze che agiscono tra atomo e atomo. Questo studio, dovuto principalmente all'opera di Born e della sua scuola, ha condotto fino ad ora a risultati definitivi solo per il caso delle sostanze a struttura salina. La ragione di questo, dipende dal fatto che nei corpi non salini, si hanno, secondo ogni probabilità, degli elettroni che migrano da un atomo all'altro descrivendo delle orbite intrecciate; ne segue in questo caso una straordinaria complicazione delle forze interatomiche, che fino ad ora sono restate indecifrabili.

I reticolati cristallini dei sali sono invece costituiti da ioni i quali, anche come costituenti del cristallo, conservano tuttavia sempre la loro individualità. Se per esempio consideriamo un pezzo di salgemma, il suo reticolato cristallino è costituito da ioni positivi disposti alternativamente con ioni negativi di cloro. Ora un ione di sodio è costituito da un nucleo positivo con carica undici volte maggiore di quella dell'elettrone, circondato da dieci elettroni negativi, così che il complesso ha una carica negativa di grandezza eguale a quella dell'elettrone. A grande distanza dallo ione la forza elettrica da esso prodotta sarà dunque con grande approssimazione eguale a quella di una carica puntiforme positiva, mentre molto vicino allo ione il suo potenziale elettrico dipenderà fortemente anche dalla sua struttura complessa. Lo ione negativo di cloro, è invece costituito da un nucleo positivo con carica 17 volte quella dell'elettrone, circondato da 18 elettroni. A grande distanza il suo potenziale elettrico sarà dunque quello di una carica puntiforme negativa, mentre molto vicino si fa sentire anche in questo caso la influenza della struttura complessa. Se ora supponiamo di porre un ione di cloro e uno di sodio in presenza uno dell'altro, si avrà che fino a che sono molto lontani essi si attireranno come due cariche puntiformi, una positiva e una negativa, con la legge di Coulomb; quando poi i due ioni si avvicinano molto la forza che agisce tra di essi dipenderà in modo essenziale dalla disposizione degli elettroni. Pur non potendosi in questo caso, fare un calcolo esatto della forza agente tra i due ioni, per la nostra ignoranza dei dettagli della loro struttura elettronica, si può tuttavia prevedere che quando gli elettroni che circondano i due ioni vengono ad avvicinarsi moltissimo, la forza repulsiva tra di essi finirà ad un certo punto per prevalere, di modo che nella immediata vicinanza i due ioni si respingeranno. La forza che agisce tra i due ioni può dunque venir rappresentata come la somma di una forza attrattiva inversamente proporzionale al quadrato della distanza, e di una forza repulsiva che acquista valori assai grandi quando i due ioni si trovano vicinissimi uno all'altro, e decresce rapidissimamente con la distanza.

Una brillante conferma sperimentale di queste vedute si ha nel calcolo dell'energia che tiene insieme un reticolato di ioni. Le esperienze di Laue, Bragg e altri sopra i fenomeni di diffrazione dei raggi Röntgen hanno infatti permesso in questi ultimi tempi di determinare esattamente la posizione degli atomi nei reticolati cristallini di numerose sostanze. Calcoli assai laboriosi permettono allora di calcolare l'energia potenziale delle forze interatomiche, la quale non è altro che l'energia che è necessaria per separare il cristallo negli ioni che lo costituiscono, e può essere anche misurata direttamente. La concordanza tra i valori misurati e calcolati di questa energia si è dimostrata praticamente perfetta. Questa conferma, come pure il successo ottenuto nel calcolo della compressibilità e delle altre costanti elastiche dei sali, hanno oramai messa fuori di dubbio la natura elettrica delle forze che tengono insieme gli atomi di queste sostanze.

N° 30 and 31.

It is unfortunate that little is known of the circumstances that led Fermi to some of his most significant contributions in the theoretical field, in contrast to the adequate information that we possess concerning his experimental work. The reasons are fairly obvious. Even when the experiments were not performed in collaboration with other workers, their progress could be followed almost day by day by his colleagues. As a theoretician, Fermi was entirely self-sufficient, some of his work was done at home in the early hours of the morning, and occasionally even his closest associates had little information on the problem that had occupied his mind until he presented to them, in an informal lecture, the finished product of his meditations.

These remarks apply to the present paper, probably his most famous theoretical contribution, where he formulated the theory of an ideal gas of particles obeying the Pauli exclusion principle, now designated in his honor as "fermions".

There is conclusive evidence to show that Fermi had been concerned with the problem of the absolute entropy constant at least since January 1924, when he wrote a paper (20) on the quantization of systems containing identical particles. He had also been discussing these problems with Rasetti several times in the following year. He told much later to Segrè that the division of phase space into finite cells had occupied him very much and that had not Pauli discovered the exclusion principle he might have arrived at it in a roundabout way from the entropy constant (cfr. N° 20). As soon as he read Pauli's article on the exclusion principle, he realized that he now possessed all the elements for a theory of the ideal gas which would satisfy the Nernst principle at the absolute zero, give the correct Sackur-Tetrode formula for the absolute entropy in the limit for low density and high temperature, and be free of the various arbitrary assumptions that it had been necessary to introduce in statistical mechanics in order to derive a correct entropy value. He does not seem to have been greatly influenced by Einstein's theory based on Bose's treatment of the black-body radiation as a photon gas, although he points out the analogy between the two forms of statistics. Apparently it took Fermi but a short time to develop the theory in the detailed and definitive form in which it was published in the German version.

It is noteworthy that Fermi did not enclose his ideal gas in a box, according to the current treatment, but placed the particles in a three-dimensional harmonic oscillator potential. In this way he obtained a spherically symmetric, monotonically decreasing gas density. For large radii, the density was always sufficiently low to ensure validity of the classical Boltzmann approximation and hence to define the temperature for the entire gas. Apart from this detail, the treatment is substantially identical with the one now found in most textbooks. He derived all the essential properties of the functions of the degeneracy parameter F (A) and G (A), gave their asymptotic expansions both for large and small values of A, and treated in detail the case of total degeneracy. The artifice of placing the molecules in a potential field led naturally later to the statistical theory of the atom as a degenerate electron gas; the main difference between the two problems being that in the latter case the potential is not given, being partly determined by the electrons themselves.

It is well known that Dirac [1] developed this type of statistics independently of Fermi. Dirac's paper was presented at the Royal Society on August 26, 1926, whereas Fermi's short note in «Rendiconti della Reale Accademia dei Lincei» was presented at the meeting of that Academy on February 7, and his longer paper was received by the editor of the «Zeitschrift für Physik» on March 26 of the same year. Thus both of Fermi's publications antedate Dirac's by an appreciable time. Dirac was the first to show that the two types of statistics,

(1) P.A.M. DIRAC, «Proc. Roy. Soc.», A-112, 661 (1926).

now usually designated as Bose-Einstein and Fermi-Dirac, are related to the two possibilities of eigenfunctions of a system being either symmetric or antisymmetric with respect to the exchange of the coordinates of two identical particles. These two alternative possibilities had been independently pointed out by Heisenberg, although he did not develop the theory of an ideal gas of particles described by these eigenfunctions.

The first published application of Fermi's statistics to the degenerate electron « gas » in a metal seems to be due to Pauli (2) who used it to explain the weak, temperature-independent paramagnetism of the alkalies. The usefulness of the degenerate Fermi gas in accounting for many other, hitherto unexplained properties of metals was first pointed out by Sommerfeld (3), who illustrated part of his work at the Como Conference of 1927. Fermi was fully aware of the importance of his statistics for the electrons in a metal. In particular the fact that the specific heat of the electron gas in a metal was vanishingly small at ordinary temperatures had been an object of theoretical and experimental study by Corbino in Rome.

That Fermi was concerned with the application of the new statistics to the electrons in metals is evidenced by the following excerpts from the discussions of the Como congress.

First Fermi answers an objection raised by H. A. Lorentz in the discussion of the paper presented by Sommerfeld (4):

LORENTZ: « Ich erlaube mir zu der Auseinandersetzung, die Prof. Sommerfeld von dieser schönen Theorie gegeben hat, eine kurze Bemerkung zu machen. Es wird angenommen, dass in einem gegebenen Metallstück nur *ein* Elektron in einem bestimmten Zustand, sagen wir in dem Nullzustand, sein kann. Gesetzt nun, wir hätten zwei gleiche Metallstücke, und in jedem derselben ein Elektron in jenem Zustande. Man würde erwarten, dass dann, wenn die beiden Körper durch einfache Zusammenstellung, ohne Aenderung ihrer Zustände, zu einem einzigen vereinigt werden, in diesem *zwei* Elektronen sich in dem Nullzustand befinden werden, und nicht wie angenommen wird, nur eines ».

FERMI: « Die von Prof. Lorentz erwähnte Schwierigkeit kann, wie mir scheint, durch die folgende Betrachtung beseitigt werden. Es sei ein Gefäss gegeben, das durch eine Wand in zwei gleichen Teilen A und B getrennt ist. Setzen wir jetzt ein Molekül in einem der Teilräume A oder B, so kann man seine hin und her Bewegung quantisieren und man findet für seine Geschwindigkeit gewisse Quantenwerte, dessen Dichte vom Volumen des Raumes abhängt, in welchem sich das Molekül hin und her bewegt. Fällt nun die Wand zwischen A und B so dass der Raum verdoppelt wird, so wird auch die Dichte der Quantenzustände doppel wie früher ».

« Wir wollen jetzt annehmen, dass in jedem Teil A und B des Gefässes N Moleküle enthalten sind. Ihre Geschwindigkeitsverteilung wird dann durch die Forderung beschränkt, dass jeder Quantenzustand höchstens einmal besetzt werden kann. Fällt nun die Wand, so wird die Anzahl der Moleküle verdoppelt, da aber auch die Anzahl der Plätze verdoppelt ist, so findet jedes Molekül sein Platz genau wie früher ».

During the same discussion, he briefly reports on his own work on metals (5):

FERMI: « Ich möchte zur Arbeit des Herrn Prof. Sommerfeld bemerken, dass ich selbst eine Arbeit angefangen habe über deren Ergebnis ich kurz berichten will. Es handelt sich nämlich um die Berechnung der Verdampfungsenergie eines Metalles. Als Modell für das Metall wird ein Jonengitter angenommen wo die Elektronen sich wie die Moleküle eines Gases bewegen. Dieses Gas wird natürlich mit der neuen Statistik anstatt der alten berechnet. Man findet dann, dass die Elektronen in der Nähe der positiven Jonen sich etwas zusammendrängen, sodass man die Theorie des Metalles ungefähr auf dieselbe Grundlage setzt wie die Debye'sche starker Elektrolyte. Als direktes Ergebnis der Theorie kann man die

(2) W. PAULI jr., « ZS. f. Physik », *41*, 81 (1927).

(3) A. SOMMERFELD, « Naturwissenschaften », *15*, 825 (1927); « ZS. f. Physik. », *47*, 1 and 43 (1928).

(4) Atti del Congresso Internazionale dei Fisici, 11–20 settembre 1927; vol. 2, pp. 470–471. Zanichelli, Bologna, 1928.

(5) Loc. cit., p. 471.

Energie berechnen, die nötig ist, um das Metall in Jonen und Elektronen zu trennen. Da man aber die Energie kennt welche frei wird, wenn man ein Elektron und ein Jon zu einem neutralen Atom vereinigt, kann man daraus auch die Energie entnehmen, welche nötig ist um ein neutrales Atom vom Metall zu entfernen, d. h. die Verdampfungsenergie. Die numerischen Rechnungen sind noch nicht vollständig fertig. Doch kann man jetzt schon sagen, dass aus dieser Theorie die Grössenordnung der Verdampfungsenergie richtig heraus kommt ».

Finally, in the discussion of the paper presented by Bohr on quantum mechanics and radiation theory, Fermi made the following comments (6):

FERMI: « La nuova teoria dei quanti, di cui il prof. Bohr ci ha illustrate le basi concettuali, getta nuova luce sopra il problema dell'irradiazione. È mio desiderio richiamare l'attenzione sopra alcuni aspetti di questo problema. La questione della reazione della radiazione nella meccanica quantistica può esser trattata con il seguente metodo che permette di ottenere facilmente dei risultati concreti, benché esso sia in buona parte ipotetico. Consideriamo per semplicità un sistema nel quale siano ad un certo istante eccitati i due stati quantici nel senso di Schrödinger; secondo l'interpretazione statistica ciò vorrà dire che se si facesse una esperienza per decidere in quale di questi due stati l'atomo si trova effettivamente, esisterebbe un rapporto definito per la probabilità di trovarlo nell'uno o nell'altro di essi.

« Si può allora facilmente calcolare il momento elettrico del sistema il quale viene ad essere una funzione del tempo e dà, secondo la teoria classica dell'irradiazione, origine a una irradiazione con frequenza determinata secondo la relazione di Bohr dalle energie dei due stati, e a una reazione di radiazione che nella teoria classica è rappresentata da una forza avente in prima approssimazione la grandezza

$$\frac{2}{3c^3} e \dddot{p} \cdot$$

« Questa forza può formalmente rappresentarsi come derivante da un potenziale variabile col tempo

$$V_1 = -\frac{2}{3c^3} e (\dddot{p}_x x + \dddot{p}_y y + \dddot{p}_z z) \qquad (1)$$

« Si può fare l'ipotesi che per tener conto della reazione della radiazione basti aggiungere al potenziale del sistema il potenziale (1). Effettivamente si trova che l'ipotesi conduce a risultati plausibili: si trova infatti che, come conseguenza di tale aggiunta al potenziale del sistema, esso passa dallo stato di maggiore energia a quello di energia minore in un tempo dell'ordine di grandezza delle vite medie atomiche. Si può anche facilmente seguire il modo in cui avviene tale passaggio tra i due stati quantici e calcolare quindi per es. la larghezza e la forma della riga spettrale che viene emessa nel salto quantico (7).

« Circa i nuovi metodi statistici nella meccanica dei quanti, vorrei fare qualche considerazione. È noto che la teoria dei quanti permette di determinare in modo del tutto naturale le dimensioni delle celle in cui è necessario dividere lo spazio delle fasi secondo la statistica di Boltzmann e Maxwell: se però in base a tale determinazione si vuole costruire la statistica di un gas perfetto si ci accorge che essa non è sufficiente allo scopo, poiché quando le dimensioni del recipiente in cui il gas è racchiuso diventano grandi gli stati quantici diventano più fitti per modo che alla fine viene a cessare ogni effetto della loro discontinuità.

« Per superare queste difficoltà sono stati recentemente fatti due tentativi, uno per parte di Einstein e uno da me: nel caso di Einstein si ammette tra le molecole del gas una dipendenza statistica del tipo di quella proposta da Bose per il caso dei quanti di luce; nel caso mio si è invece applicato a tutto il gas, considerato come un unico sistema costituito da tutte le molecole (che sono eguali tra di loro) il principio di esclusione di Pauli.

« I rapporti tra queste due statistiche sono stati chiariti per mezzo della nuova meccanica dai lavori di Heisenberg, Dirac e Winter: essi han dimostrato che se si ha un sistema che contiene delle particelle identiche tra di loro i suoi termini si dividono in gruppi per modo

(6) Loc. cit., pp. 594-596.

(7) E. FERMI, « Rend. Lincei », 5, 795 (1927).

che non è in alcun modo possibile ottenere dei passaggi tra due termini che appartengano a gruppi differenti. Uno di questi gruppi soddisfa alla statistica di Bose-Einstein e l'altro al principio di esclusione di Pauli e quindi alla statistica proposta dall'autore. L'esperienza ha fino ad ora dimostrato che gli elettroni di un atomo e anche i corpuscoli positivi soddisfano sempre al principio di esclusione.

« Applicando al gas di elettroni nell'interno di un metallo tale statistica, Pauli ha potuto spiegare il fatto che il paramagnetismo dei metalli alcalini solidi è considerevolmente minore di quanto corrisponderebbe al valore del momento magnetico dell'elettrone, e il Prof. Sommerfeld ci ha mostrato come si possa in base ad essa render conto anche di molte altre proprietà della conduzione metallica.

« Si può anche cercare in base alle stesse ipotesi di costruire una teoria dei metalli capace di render conto delle forze che tengono insieme la compagine del metallo. Basta perciò considerare gli ioni positivi del metallo disposti ai vertici del reticolato cristallino del metallo e calcolare poi la distribuzione degli elettroni di valenza sotto l'azione delle forze elettrostatiche con metodo simile a quello applicato da Debye e Hückel nella loro teoria degli elettroliti forti, ed applicando naturalmente la nuova statistica al posto di quella classica. I calcoli numerici necessari per questa teoria sono però assai lunghi e non sono ancora completi ».

F. Rasetti.

30.

SULLA QUANTIZZAZIONE DEL GAS PERFETTO MONOATOMICO

« Rend. Lincei », *3*, 145–149 (1926) (*).

1. Nella termodinamica classica si prende come calore specifico a volume costante di un gas perfetto monoatomico (riferendosi a una sola molecola) $c = 3\,k/2$. È chiaro però che se si vuole, anche per un gas ideale, ammettere la validità del principio di Nernst, bisogna ritenere che la precedente espressione di c sia soltanto una approssimazione per temperature elevate, e che in realtà c tenda a zero per $T = 0$, in modo che si possa estendere fino allo zero assoluto l'integrale esprimente il valore dell'entropia senza l'indeterminazione della costante. E per rendersi conto del come possa avvenire una tale variazione di c, è necessario ammettere che anche i moti del gas perfetto debbano essere quantizzati. Si capisce poi come una tale quantizzazione, oltre che sul contenuto di energia del gas, avrà anche una influenza sopra la sua equazione di stato, dando così origine ai così detti fenomeni di degenerazione del gas perfetto per basse temperature.

Lo scopo di questo lavoro è di esporre un metodo per effettuare la quantizzazione del gas perfetto che, a noi pare, sia il più possibile indipendente da ipotesi non giustificate sopra il comportamento statistico delle molecole del gas.

(*) Presentata dal Socio A. Garbasso nella seduta del 7 febbraio 1926.

Recentemente sono stati fatti numerosi tentativi di arrivare a stabilire l'equazione di stato del gas perfetto [8]. Le formule date dai vari autori e la nostra, differiscono tra di loro, e dalla equazione di stato classica, soltanto per temperature molto basse e per densità assai elevate; disgraziatamente sono queste le stesse circostanze nelle quali sono maggiormente importanti anche le deviazioni delle leggi dei gas reali da quelle dei gas perfetti; e siccome, in condizioni comodamente realizzabili sperimentalmente, le deviazioni dalla equazione di stato $pV = kT$ dovute alla degenerazione del gas, pur non essendo affatto trascurabili, sono sempre alquanto più piccole di quelle dovute all'essere il gas reale e non perfetto, le prime sono state fino ad ora mascherate da queste ultime; pur non essendo affatto escluso che, con una conoscenza più approfondita delle forze che agiscono tra le molecole di un gas reale, si possa, in un avvenire più o meno prossimo, separare tra di loro le due deviazioni, arrivando così a decidere sperimentalmente tra le diverse teorie della degenerazione dei gas perfetti.

2. Per poter effettuare la quantizzazione del moto delle molecole di un gas perfetto è necessario mettersi in condizioni tali da poter applicare al loro movimento le regole di Sommerfeld; e questo può naturalmente farsi in infiniti modi che, del resto, conducono tutti allo stesso risultato. Si può per es. supporre il gas racchiuso in un recipiente parallelepipedo a pareti elastiche, quantizzando il moto, triplamente periodico, della molecola che rimbalza tra le sei faccie del recipiente; o, più generalmente, si possono assoggettare le molecole a un qualsiasi sistema di forze tale che il loro moto venga ad essere periodico e possa quindi essere quantizzato. L'ipotesi che il gas sia perfetto ci autorizza in tutti questi casi a trascurare le forze agenti tra le molecole, per modo che il moto meccanico di ciascuna di esse viene a svolgersi come se le altre non esistessero. Si può tuttavia riconoscere che la semplice quantizzazione, con le regole di Sommerfeld, del moto delle molecole, considerate come completamente indipendenti le une dalle altre, non è sufficiente per ottenere dei risultati corretti; in quanto che, pur trovandosi così un calore specifico che tende a zero per $T = 0$, si ha però che il suo valore, oltre che dalla temperatura e dalla densità, viene anche a dipendere dalla quantità totale del gas, e tende, per qualunque temperatura, al limite $3\, k/2$ quando, pur restando costante la densità, la quantità totale del gas tende all'infinito. Appare dunque necessario ammettere che occorra qualche complemento alle regole di Sommerfeld, per il calcolo di sistemi che, come il nostro, contengono degli elementi non distinguibili tra di loro [9].

Per avere un suggerimento circa l'ipotesi più plausibile da farsi, conviene che esaminiamo come vanno le cose in altri sistemi che, al pari del nostro gas perfetto, contengono degli elementi indistinguibili; e precisamente vogliamo esaminare il comportamento degli atomi più pesanti dell'idrogeno,

(8) Vedi per esempio A. Einstein, «Sitzber. d. Pr. Akad. d. Wiss.», *22*, 261 (1924); *23*, 3, 18 (1925); M. PLANCK, «Sitzber. d. Pr. Akad. d. Wiss.», *23*, 49 (1925).

(9) E. FERMI, «N. C.», *1*, 145 (1924).

i quali tutti contengono più di un elettrone. Se consideriamo le parti profonde di un atomo pesante, siamo in condizioni tali che le forze agenti tra gli elettroni sono molto piccole in confronto di quelle esercitate dal nucleo. In queste circostanze la applicazione pura e semplice delle regole di Sommerfeld condurrebbe a prevedere che, nello stato normale dell'atomo, un numero considerevole di elettroni dovesse trovarsi in una orbita di quanto totale 1. In realtà si osserva invece che l'anello K è già saturato quando contiene due elettroni, e parimenti l'anello L si satura quando contiene 8 elettroni, etc. Questo fatto è stato interpretato da Stoner [10], e in modo ancora più preciso da Pauli [11], al modo seguente: caratterizziamo una orbita elettronica possibile in un atomo complesso mediante 4 numeri quantici; n, k, j, m, che hanno risp. i significati di quanto totale, quanto azimutale, quanto interno e quanto magnetico. Date le diseguaglianze alle quali debbono soddisfare questi 4 numeri, si trova che, per $n = 1$, esistono solo due terne di valori di k, j, m; per $n = 2$, ne esistono 8, etc. Per rendersi conto del fatto precedentemente osservato, basta dunque ammettere che nell'atomo non possano esistere due elettroni le cui orbite siano caratterizzate dagli stessi numeri quantici; bisogna in altre parole ammettere che una orbita elettronica sia già «occupata» quando contiene un solo elettrone.

3. Ci proponiamo ora di ricercare se una ipotesi simile non possa dare dei buoni risultati anche nel problema della quantizzazione del gas perfetto: ammetteremo dunque che nel nostro gas ci possa essere al massimo una molecola il cui movimento sia caratterizzato da certi numeri quantici, e faremo vedere che questa ipotesi conduce a una teoria perfettamente conseguente della quantizzazione del gas perfetto, e che in particolare rende ragione della prevista diminuzione del calore specifico per basse temperature, e conduce al valore esatto per la costante dell'entropia del gas perfetto.

Riservandoci di pubblicare, in una prossima occasione, i dettagli matematici della presente teoria, ci limitiamo in questa Nota ad esporre i principî del metodo seguito, e i risultati.

Dobbiamo anzitutto mettere il nostro gas in condizioni tali che il moto delle sue molecole sia quantizzabile. Come si è visto questo può farsi in infiniti modi; siccome però il risultato è indipendente dal modo particolare che si adotta, noi sceglieremo quello che è più comodo per il calcolo; e precisamente ammetteremo che sulle nostre molecole agisca una attrazione verso un punto fisso O, di intensità proporzionale alla distanza r della molecola da O; per modo che ogni molecola verrà a costituire un oscillatore armonico spaziale, di cui indichiamo con ν la frequenza. L'orbita della molecola sarà caratterizzata dai suoi tre numeri quantici s_1, s_2, s_3, che sono legati alla sua energia per mezzo della relazione.

$$w = h\nu (s_1 + s_2 + s_3) = h\nu\, s \,. \tag{1}$$

(10) E. C. STONER, «Phil. Mag.», *48*, 719. (1924).

(11) W. PAULI, «Zs. f. Phys.», *31*, 765 (1925).

L'energia di una molecola può dunque prendere tutti i valori multipli interi di $h\nu$, ed il valore $sh\nu$ può essere preso in $Q_s = \frac{1}{2}(s+1)(s+2)$ modi.

L'energia zero può dunque realizzarsi in un modo solo, l'energia $h\nu$ in 3 modi, l'energia $2h\nu$ in 6 modi, etc. Per rendersi conto della influenza della ipotesi da noi fatta, che a determinati numeri quantici non possa corrispondere più di una molecola, consideriamo il caso estremo di avere N molecole allo zero assoluto. A questa temperatura il gas deve trovarsi nello stato di energia minima. Se dunque non ci fosse nessuna limitazione al numero delle molecole che possono avere una certa energia, tutte le molecole si troverebbero nello stato di energia zero, e tutti e tre i numeri quantici di ciascuna di esse sarebbero nulli. Invece, per la nostra ipotesi, non ci può essere più di una molecola con tutti e tre i numeri quantici nulli; se è quindi $N = 1$, l'unica molecola occuperà il posto di energia zero; se è invece $N = 4$, una delle molecole occuperà il posto di energia zero, e le altre tre i posti di energia $h\nu$; se è $N = 10$, una delle molecole occuperà il posto di energia zero, altre tre i tre posti di energia $h\nu$, e le sei rimanenti i sei posti di energia $2h\nu$, etc.

Supponiamo ora di dover distribuire tra le nostre N molecole l'energia complessiva $W = Eh\nu$ (E = numero intero); e indichiamo con $N_s \leq Q_s$ il numero delle molecole di energia $sh\nu$. Si trova facilmente che i valori più probabili delle N_s sono

$$N_s = \alpha Q_s/(e^{\beta s} + \alpha) \tag{2}$$

dove α e β sono delle costanti dipendenti da W e da N. Per trovare la relazione tra queste costanti e la temperatura, osserviamo che, per effetto della attrazione verso O, la densità del nostro gas sarà una funzione di r, che deve tendere a zero per $r = \infty$. Per conseguenza, per $r = \infty$ debbono cessare i fenomeni di degenerazione, e in particolare la distribuzione delle velocità, facilmente deducibile da (2), deve trasformarsi nella legge di Maxwell. Si trova così che deve essere

$$\beta = h\nu/kT\,. \tag{3}$$

Siamo ora in grado di dedurre da (2) la funzione $n(L)\,dL$, che ci rappresenta, per un dato valore di r, la densità delle molecole di energia compresa tra L ed $L + dL$ (Analogo della legge di Maxwell), e da questa possiamo dedurre l'energia cinetica media $\overline{L}$ delle molecole a distanza r, la quale è funzione, oltre che della temperatura, anche della densità n. Si trova precisamente

$$\overline{L} = \frac{3h^2 n^{2/3}}{4\pi m}\, P\left(\frac{2\pi m kT}{h^2 n^{2/3}}\right). \tag{4}$$

Dove con $P(x)$, si è indicata una funzione, di definizione analitica un po' complicata, che, secondo che x è molto grande o molto piccolo, si può calcolare con le formule asintotiche

$$P(x) = x(1 + 2^{-5/2}x^{-3/2} + \cdots) \quad ; \quad P(x) = \frac{1}{5}\sqrt[3]{\frac{9\pi}{2}}\left\{1 + \frac{5}{9}\sqrt[3]{\frac{4\pi^4}{3}}\,x^2 + \cdots\right\}. \tag{5}$$

Per dedurre da (4) l'equazione di stato applichiamo la relazione del viriale. Si trova allora che la pressione è data da

$$p = \frac{2}{3} n \overline{L} = \frac{h^2 n^{5/3}}{2 \pi m} P \left(\frac{2 \pi m k T}{h^2 n^{2/3}} \right) . \tag{6}$$

Al limite per temperature elevate, cioè per piccola degenerazione, l'equazione di stato prende dunque la forma

$$p = n k T \left\{ 1 + \frac{1}{16} \frac{h^3 n}{(\pi m k T)^{3/2}} + \cdots \right\} . \tag{7}$$

La pressione risulta dunque maggiore di quella prevista dalla equazione di stato classica. Per un gas perfetto del peso atomico dell'elio, alla temperatura di 5° assoluti, e alla pressione di 10 atmosfere la differenza sarebbe del 15 %.

Da (4) e (5) si può anche dedurre l'espressione del calore specifico per basse temperature. Si trova

$$c_v = \sqrt[3]{\frac{16 \pi^8}{9}} \frac{m k^2}{h^2 n^{2/3}} T + \cdots \tag{8}$$

Parimenti possiamo trovare il valore assoluto dell'entropia. Effettuando i calcoli si trova, per alte temperature,

$$S = n \int_0^T \frac{1}{T} d\overline{L} = n \left\{ \frac{5}{2} \log T - \log p + \log \frac{(2 \pi m)^{3/2} k^{5/2} e^{5/2}}{h^3} \right\} \tag{9}$$

che coincide col valore dell'entropia dato da Tetrode e da Stern.

N° 31.

For the introduction to this paper see N° 30.

31.

ZUR QUANTELUNG DES IDEALEN EINATOMIGEN GASES(*)

« Z. Physik », *36*, 902–912 (1926).

Wenn der Nernstsche Wärmesatz auch für das ideale Gas seine Gültigkeit behalten soll, muss man annehmen, dass die Gesetze idealer Gase bei niedrigen Temperaturen von den klassischen abweichen. Die Ursache dieser Entartung ist in einer Quantelung der Molekularbewegungen zu suchen. Bei allen Theorien der Entartung werden immer mehr oder weniger willkürliche Annahmen über das statistische Verhalten der Moleküle, oder über ihre Quantelung gemacht. In der vorliegenden Arbeit wird nur die von Pauli zuerst ausgesprochene und auf zahlreiche spektroskopische Tatsachen begründete Annahme benutzt, dass in einem System nie zwei gleichwertige Elemente vorkommen können, deren Quantenzahlen vollständig übereinstimmen. Mit dieser Hypothese werden die Zustandsgleichung und die innere Energie des idealen Gases abgeleitet; der Entropiewert für grosse Temperaturen stimmt mit dem Stern–Tetrodeschen überein.

In der klassischen Thermodynamik wird die Molekularwärme (bei konstantem Volumen)

$$c = \frac{3}{2} k \tag{1}$$

gesetzt. Will man aber den Nernstschen Wärmesatz auch auf das ideale Gas anwenden können, so muss man (1) bloss als eine Näherung für grosse Temperaturen ansehen, da c im Limes für $T = 0$ verschwinden muss. Man ist deshalb genötigt, anzunehmen, dass die Bewegung der Moleküle idealer Gase gequantelt sei; diese Quantelung äussert sich bei niedrigen Temperaturen durch gewisse Entartungserscheinungen, so dass sowohl die spezifische Wärme als auch die Zustandsgleichung von ihren klassischen Ausdrücken abweichen werden.

Zweck der vorliegenden Arbeit ist, eine Methode für die Quantelung des idealen Gases darzustellen, welche nach unserem Erachten möglichst unabhängig von willkürlichen Annahmen über das statistische Verhalten der Gasmoleküle ist.

(*) Vgl. die vorläufige Mitteilung, « Lincei Rend. » (6), *3*, 145 (1926).

In neuerer Zeit wurden zahlreiche Versuche gemacht, die Zustandsgleichung idealer Gase festzustellen(1). Die Zustandsgleichungen der verschiedenen Verfasser und unsere unterscheiden sich voneinander und von der klassischen Zustandsgleichung $pV = NkT$ durch Glieder, welche nur bei sehr niedrigen Temperaturen und grossen Drucken beträchtlich werden; leider sind die Abweichungen der realen Gase von den idealen gerade unter diesen Umständen am grössten, so dass die an sich gar nicht unbedeutenden Entartungserscheinungen bisher nicht beobachtet werden konnten. Es ist jedenfalls nicht unmöglich, dass eine tiefere Kenntnis der Zustandsgleichungen der Gase gestattet wird, die Entartung von den übrigen Abweichungen von der Gleichung $pV = NkT$ zu trennen, so dass eine experimentelle Entscheidung zwischen den verschiedenen Entartungstheorien möglich wird.

Um die Quantenregeln auf die Bewegung der Moleküle unseres idealen Gases anwenden zu können, kann man verschiedenartig verfahren; das Endergebnis bleibt jedoch immer dasselbe. Z. B. können wir uns die Moleküle in ein parallelepipedisches Gefäss mit elastisch reflektierenden Wänden eingeschlossen denken; dadurch wird die Bewegung des zwischen den Wänden hin und her fliegenden Moleküles bedingt periodisch und kann deshalb quantisiert werden; allgemeiner kann man sich die Moleküle in ein derartiges äusseres Kraftfeld eingesetzt denken, dass ihre Bewegung bedingt periodisch werde; die Annahme, dass das Gas ideal ist, erlaubt uns, die mechanischen Wirkungen der Moleküle aufeinander zu vernachlässigen, so dass ihre mechanische Bewegung sich nur unter dem Einfluss der äusseren Kraft vollzieht. Es ist jedoch ersichtlich, dass die unter der Annahme der vollständigen Unabhängigkeit der Moleküle voneinander ausgeführte Quantelung der Molekularbewegung nicht hinreichend ist, uns Rechenschaft von der erwarteten Entartung zu geben. Das erkennt man am besten an dem Beispiel der in einem Gefäss eingeschlossenen Moleküle dadurch, dass, wenn die linearen Dimensionen des Gefässes wachsen, die Energiewerte der Quantenzustände jedes einzelnen Moleküls immer dichter werden, so dass für Gefässe makroskopischer Dimensionen bereits jeder Einfluss der Diskontinuität der Energiewerte praktisch verschwindet. Dieser Einfluss hängt ausserdem von dem Volumen des Gefässes ab, auch wenn die Zahl der im Gefäss enthaltenen Moleküle so gewählt wird, dass die Dichte konstant bleibt.

Durch eine quantitative Berechnung dieses Sachverhaltes (2) kann man sich überzeugen, dass man nur dann eine Entartung der erwarteten Grössenordnung erhält, wenn man das Gefäss so klein wählt, dass es im Mittel nur noch ein Molekül enthält.

Wir sprechen deshalb die Vermutung aus, dass zur Quantelung idealer Gase eine Zusatzregel zu den Sommerfeldschen Quantenbedingungen nötig sei.

(1) Vgl. z. B. A. EINSTEIN, Berl. Ber. 1924, S. 261; 1925, S. 318; M. PLANCK, ebenda 1925, S. 49. Unsere Methode ist der Einsteinschen insofern verwandt, als die Annahme der statistischen Unabhängigkeit der Moleküle bei beiden Methoden verlassen wird obgleich die Art der Abhängigkeit bei uns ganz anders ist wie bei Einstein, und das Endergebnis für die Abweichungen von der klassischen Zustandsgleichung sogar entgegengesetzt gefunden wird.

(2) E. FERMI, «Nuovo Cim.», *1*, 145 (1924).

Nun wurde kürzlich von W. Pauli [3], im Anschluss an eine Arbeit von E. C. Stoner [4], die Regel aufgestellt, dass, wenn in einem Atom ein Elektron vorhanden ist, dessen Quantenzahlen (die magnetischen Quantenzahlen eingeschlossen) bestimmte Werte haben, so kann im Atom kein weiteres Elektron existieren, dessen Bahn durch dieselben Zahlen charakterisiert ist. Mit anderen Worten ist eine Quantenbahn (in einem äusseren magnetischen Felde) bereits von einem einzigen Elektron vollständig besetzt.

Da diese Paulische Regel sich in der Deutung spektroskopischer Tatsachen als äusserst fruchtbar erwiesen hat [5], wollen wir versuchen, ob sie nicht etwa auch für das Problem der Quantelung idealer Gase brauchbar sei.

Wir werden zeigen, dass dies in der Tat der Fall ist, und dass die Anwendung der Paulischen Regel uns erlaubt, eine vollständig konsequente Theorie der Entartung idealer Gase darzustellen.

Wir werden also im folgenden annehmen, dass höchstens ein Molekül mit vorgegebenen Quantenzahlen in unserem Gase vorhanden sein kann: als Quantenzahlen kommen dabei nicht nur die Quantenzahlen in Betracht, welche die inneren Bewegungen des Moleküls, sondern auch die Zahlen welche seine Translationsbewegung bestimmen.

Zuerst müssen wir unsere Moleküle in ein passendes äusseres Kraftfeld einsetzen, so dass ihre Bewegung bedingt periodisch werde. Das kann auf unendlich viele Weisen geschehen; da aber das Resultat von der Wahl des Kraftfeldes nicht abhängt, wollen wir die Moleküle einer zentralen elastischen Anziehung nach einem festen Punkte O (Koordinatenursprung) unterwerfen, so dass jedes Molekül einen harmonischen Oszillator bilden wird. Diese Zentralkraft wird unsere Gasmenge in der Umgebung von O halten; die Gasdichte wird mit der Entfernung von O abnehmen und für unendliche Entfernung verschwinden. Sei ν die Eigenfrequenz der Oszillatoren, dann ist die auf die Moleküle wirkende Kraft durch

$$4\pi^2 \nu^2 mr$$

gegeben, wo m die Masse der Moleküle und r ihre Entfernung von O darstellt. Die potentielle Energie der Anziehungskraft ist dann

$$u = 2\pi^2 \nu^2 mr^2 . \tag{1}$$

Die Quantenzahlen des von einem Molekül gebildeten Oszillators seien s_1, s_2, s_3. Zur Charakterisierung des Moleküls sind eigentlich diese Quantenzahlen nicht hinreichend, denn dazu müsste man auch die Quantenzahlen der inneren Bewegungen angeben. Wir wollen uns aber auf einatomige Moleküle beschränken, und weiter wollen wir annehmen, dass alle in unserem Gase vorkommenden Moleküle sich im Grundzustand befinden, und dass dieser Zustand einfach (magnetisch unzerlegbar) ist. Dann brauchen wir uns um die inneren Bewegungen nicht zu kümmern, und wir können unsere

(3) W. PAULI jr., «ZS. f. Phys.», *31*, 765 (1925).
(4) E. C. STONER, «Phil. Mag.», *48*, 719 (1924).
(5) Vgl. z. B. F. HUND, «ZS. f. Phys.», *33*, 345 (1925).

Moleküle einfach als Massenpunkte ansehen. Die Paulische Regel lautet deshalb für unseren Fall: Es kann in der ganzen Gasmenge höchstens ein Molekül mit vorgegebenen Quantenzahlen s_1, s_2, s_3, vorhanden sein.

Die totale Energie dieses Moleküls wird durch

(2) $$w = h\nu (s_1 + s_2 + s_3) = h\nu s$$

gegeben. Die totale Energie kann deshalb ein beliebiges ganzzahliges Vielfaches von $h\nu$ sein; der Wert $sh\nu$ kann jedoch auf viele Weisen realisiert werden. Jede Realisierungsmöglichkeit entspricht einer Lösung der Gleichung

(3) $$s = s_1 + s_2 + s_3 ,$$

wo s_1, s_2, s_3 die Werte $0, 1, 2, 3, \cdots$ annehmen können. Bekanntlich hat Gleichung (3)

(4) $$Q_s = \frac{(s+1)(s+2)}{2} .$$

Lösungen. Die Energie Null kann deshalb nur auf eine einzige Art realisiert werden, die Energie $h\nu$ auf drei, die Energie $2h\nu$ auf sechs usw. Ein Molekül mit der Energie $sh\nu$ werden wir einfach ein «s»-Molekül nennen.

Nach unseren Annahmen können nun in der ganzen Gasmenge höchstens Q_s «s»-Moleküle vorkommen; also höchstens ein Molekül mit der Energie Null, höchstens drei Moleküle mit der Energie $h\nu$, höchstens sechs mit der Energie $2h\nu$ usw.

Um die Folgen dieses Tatbestandes klar übersehen zu können, wollen wir den extremen Fall betrachten, dass die absolute Temperatur unseres Gases Null sei. Sei N die Zahl der Moleküle. Beim absoluten Nullpunkt muss sich das Gas in dem Zustand kleinster Energie befinden. Wäre nun keine Einschränkung für die Zahl der Moleküle einer gegebenen Energie vorhanden, so würde sich jedes Molekül im Zustand der Energie Null ($s_1 = s_2 = s_3 = 0$) befinden. Nach dem Vorhergehenden kann aber höchstens ein Molekül mit der Energie Null vorkommen; wäre deshalb $N = 1$, so würde das einzige Molekül beim absoluten Nullpunkt den Zustand der Energie Null besetzen; wäre $N = 4$, so würde ein Molekül den Zustand der Energie Null, die drei übrigen die drei Plätze mit der Energie $h\nu$ besetzen; wäre $N = 10$, so würde sich ein Molekül am Platze mit der Energie Null befinden, drei andere an den drei Plätzen mit der Energie $h\nu$, und die sechs übrigen in den sechs Plätzen mit der Energie $2h\nu$ usw.

Beim absoluten Nullpunkt zeigen deshalb die Moleküle unseres Gases eine Art schalenförmigen Aufbau, der eine gewisse Analogie zur schalenartigen Anordnung der Elektronen in einem Atom mit mehreren Elektronen aufweist.

Wir wollen jetzt untersuchen, wie sich eine gewisse Energiemenge

(5) $$W = E h\nu .$$

(E = ganze Zahl) zwischen unseren N Molekülen verteilt.

Sei N_s die Zahl der Moleküle, die sich in einem Zustand mit der Energie $sh\nu$ befinden. Nach unseren Annahmen ist

(6) $$N_s \leq Q_s .$$

Man hat weiter die Gleichungen

$$\Sigma N_s = N\,, \tag{7}$$

$$\Sigma s N_s = E\,, \tag{8}$$

welche ausdrücken, dass die Gesamtzahl bzw. die Gesamtenergie der Moleküle gleich N bzw. $Eh\nu$ ist.

Jetzt wollen wir die Zahl P solcher Anordnungen unserer N Moleküle berechnen, dass sich N_0 auf Plätzen mit der Energie Null, N_1 auf Plätzen mit der Energie $h\nu$, N_2 auf Plätzen mit der Energie $2\,h\nu$ usw. befinden. Zwei Anordnungen sollen dabei als gleich angesehen werden, wenn die von den Molekülen besetzten Plätze dieselben sind: zwei Anordnungen, welche sich nur durch eine Permutation der Moleküle auf ihren Plätzen unterscheiden, sind deshalb als eine gleiche Anordnung anzusehen. Würde man zwei solche Anordnungen als verschieden ansehen, so würde man P mit der Konstante N! multiplizieren müssen: man könnte aber leicht einsehen, dass dies auf das folgende keinen Einfluss haben würde. Im oben erklärten Sinne ist die Zahl der Anordnungen von N_s Molekülen auf den Q_s Plätzen der Energie $sh\nu$ durch

$$\binom{Q_s}{N_s}$$

gegeben. Wir finden deshalb für P den Ausdruck

$$P = \binom{Q_0}{N_0}\binom{Q_1}{N_1}\binom{Q_2}{N_2}\cdots = \Pi\binom{Q_s}{N_s}. \tag{9}$$

Man bekommt die wahrscheinlichsten Werte der N_s, indem man das Maximum von P mit den Einschränkungen (7) und (8) sucht. Durch Anwendung des Stirlingschen Satzes kann man, mit für unseren Fall genügender Annäherung, schreiben:

$$\log P = \Sigma \log\binom{Q_s}{N_s} = -\,\Sigma\left(N_s \log\frac{N_s}{Q_s - N_s} + Q_s \log\frac{Q_s - N_s}{Q_s}\right). \tag{10}$$

Wir suchen also die Werte der N_s, welche (7) und (8) genügen, und für welche log P ein Maximum wird. Man findet

$$\alpha e^{-\beta s} = \frac{N_s}{Q_s - N_s}\,,$$

wo α und β Konstante darstellen. Die vorige Gleichung gibt uns:

$$N_s = Q_s \frac{\alpha e^{-\beta s}}{1 + \alpha e^{-\beta s}}. \tag{11}$$

Die Werte von α und β können durch die Gleichung (7) und (8) bestimmt werden, oder umgekehrt kann man α und β als gegeben ansehen; dann bestimmen (7) und (8) die Gesamtzahl und die Gesamtenergie unserer Moleküle. Wir finden nämlich

$$\left\{\begin{aligned} N &= \sum_0^\infty Q_s \frac{\alpha e^{-\beta s}}{1 + \alpha e^{-\beta s}}, \\ \frac{W}{h\nu} = E &= \sum_0^\infty s Q_s \frac{\alpha e^{-\beta s}}{1 + \alpha e^{-\beta s}}. \end{aligned}\right. \tag{12}$$

Die absolute Temperatur T des Gases ist eine Funktion von N und E oder auch von α und β. Diese Funktion kann nach zwei Methoden bestimmt werden, welche jedoch zum selben Resultat führen. Man könnte z. B. nach dem Boltzmannschen Prinzip die Entropie

$$S = k \log P$$

setzen und dann die Temperatur nach der Formel

$$T = \frac{dW}{dS}$$

berechnen. Diese Methode hat jedoch, wie alle auf dem Boltzmannschen Prinzip beruhenden Methoden, den Nachteil, dass man für ihre Anwendung einen mehr oder weniger willkürlichen Ansatz für die Zustandswahrscheinlichkeit braucht. Wir ziehen deshalb vor, folgendermassen zu verfahren: Beachten wir, dass die Dichte unseres Gases eine Funktion der Entfernung ist, welche für unendliche Entfernung verschwindet. Für unendlich grosses r werden deshalb auch die Entartungserscheinungen aufhören, und die Statistik unseres Gases in die klassische übergehen. Insbesondere muss für $r = \infty$ die mittlere kinetische Energie der Moleküle $3\,kT/2$ werden, und ihre Geschwindigkeitsverteilung in die Maxwellsche übergehen. Wir können also die Temperatur aus der Geschwindigkeitsverteilung in dem Gebiet unendlich kleiner Dichte bestimmen; und da die ganze Gasmenge auf konstanter Temperatur ist, werden wir zugleich die Temperatur auch für die Gebiete hoher Dichte kennen. Zu dieser Bestimmung werden wir uns sozusagen eines Gasthermometers mit einem enendlich verdünnten idealen Gase bedienen.

Zuerst müssen wir die Dichte der Moleküle mit einer kinetischen Energie zwischen L und L + dL in der Entfernung r berechnen. Die totale Energie dieser Moleküle wird nach (1) zwischen

$$L + 2\pi^2 \nu^2 m r^2 \quad \text{und} \quad L + 2\pi^2 \nu^2 m r^2 + dL$$

liegen. Nun ist die totale Energie eines Moleküls gleich $sh\nu$. Für unsere Moleküle muss s deshalb zwischen s und $s + ds$ liegen, wo

$$s = \frac{L}{h\nu} + \frac{2\pi^2 \nu m}{h} r^2 \quad , \quad ds = \frac{dL}{h\nu} \,. \tag{13}$$

Betrachten wir jetzt ein Molekül, dessen Bewegung durch die Quantenzahlen s_1, s_2, s_3 charakterisiert ist. Seine Koordinaten x, y, z sind durch

$$\left\{ \begin{array}{l} x = \sqrt{H s_1} \cos(2\pi\nu t - \alpha_1) \quad , \quad y = \sqrt{H s_2} \cos(2\pi\nu t - \alpha_2) \,, \\ z = \sqrt{H s_3} \cos(2\pi\nu t - \alpha_3) \end{array} \right. \tag{14}$$

als Funktionen der Zeit gegeben. Dabei ist

$$H = \frac{h}{2\pi^2 \nu m} \tag{15}$$

gesetzt worden: α_1, α_2 und α_3 bedeuten Phasenkonstanten, welche mit gleicher Wahrscheinlichkeit jedes beliebige Wertesystem annehmen können. Hieraus und aus den Gleichungen (14) folgt, dass $|x| \leq \sqrt{Hs_1}$, $|y| \leq \sqrt{Hs_2}$, $|z| \leq \sqrt{Hs_3}$, und dass die Wahrscheinlichkeit, dass x, y, z zwischen den

Grenzen x und $x + dx$, y und $y + dy$, $z + dz$ liegen, folgenden Ausdruck hat:

$$\frac{dx\,dy\,dz}{\pi^3 \sqrt{(\mathrm{H}\,s_1 - x^2)(\mathrm{H}\,s_2 - y^2)(\mathrm{H}\,s_3 - z^2)}}.$$

Wenn wir nicht die einzelnen Werte von s_1, s_2, s_3, sondern nur ihre Summe kennen, so ist unsere Wahrscheinlichkeit durch

$$\frac{1}{\mathrm{Q}_s}\,\frac{dx\,dy\,dz}{\pi^3}\,\Sigma\,\frac{1}{\sqrt{(\mathrm{H}\,s_1 - x^2)(\mathrm{H}\,s_2 - y^2)(\mathrm{H}\,s_3 - z^2)}} \tag{16}$$

ausgedrückt; die Summe ist auf alle ganzzahligen Lösungen der Gleichung (3) zu erstrecken, die den Ungleichungen

$$\mathrm{H}\,s_1 \geq x^2 \quad , \quad \mathrm{H}\,s_2 \geq y^2 \quad , \quad \mathrm{H}\,s_3 \geq z^2$$

genügen. Wenn wir die Wahrscheinlichkeit (16) mit der Anzahl N_s der «s»–Moleküle multiplizieren, so bekommen wir die Zahl der «s»–Moleküle, die im Volumenelement $dx\,dy\,dz$ enthalten sind. Unter Berücksichtigung von (11) finden wir also, dass die Dichte der «s»–Moleküle am Orte x, y, z durch

$$n_s = \frac{\alpha\,e^{-\beta s}}{1 + \alpha\,e^{-\beta s}}\,\frac{1}{\pi_3}\,\Sigma\,\frac{1}{\sqrt{(\mathrm{H}\,s_1 - x^2)(\mathrm{H}\,s_2 - y^2)(\mathrm{H}\,s_3 - z^2)}}$$

gegeben ist. Für hinreichend grosses s kann man die Summe durch ein zweifaches Integral ersetzen; nach Ausführung der Integrationen finden wir

$$n_s = \frac{2}{\pi^2\,\mathrm{H}^2}\,\frac{\alpha\,e^{-\beta s}}{1 + \alpha\,e^{-\beta s}}\,\sqrt{\mathrm{H}\,s - r^2}\,.$$

Mit Benutzung von (13) und (15) finden wir jetzt, dass die Dichte der Moleküle mit einer kinetischen Energie zwischen L und L + dL am Orte x, y, z folgenden Ausdruck hat:

$$n(\mathrm{L})\,d\mathrm{L} = n_s\,ds = \frac{2\pi\,(2m)^{3/2}}{h^3}\sqrt{\mathrm{L}}\,d\mathrm{L}\,\frac{\alpha\,e^{-\frac{2\pi^2 \nu m \beta r^2}{h}}\,e^{-\frac{\beta \mathrm{L}}{h\nu}}}{1 + \alpha\,e^{-\frac{2\pi^2 \nu m \beta r^2}{h}}\,e^{-\frac{\beta \mathrm{L}}{h\nu}}}\,. \tag{17}$$

Diese Formel muss mit dem klassischen Ausdruck des Maxwellschen Verteilungsgesetzes verglichen werden:

$$n^*(\mathrm{L})\,d\mathrm{L} = \mathrm{K}\sqrt{\mathrm{L}}\,d\mathrm{L}\,e^{-\mathrm{L}/k\mathrm{T}}\,. \tag{17'}$$

Man sieht dann, dass im Limes für $r = \infty$ (17) in (17') übergeht, wenn man nur

$$\beta = \frac{h\nu}{k\mathrm{T}} \tag{18}$$

setzt. Jetzt kann (17) folgendermassen geschrieben werden:

$$n(\mathrm{L})\,d\mathrm{L} = \frac{(2\pi)\,(2m)^{3/2}}{h^3}\sqrt{\mathrm{L}}\,d\mathrm{L}\,\frac{\mathrm{A}\,e^{-\mathrm{L}/k\mathrm{T}}}{1 + \mathrm{A}\,e^{-\mathrm{L}/k\mathrm{T}}}, \tag{19}$$

wo

$$\mathrm{A} = \alpha\,e^{-\frac{2\pi^2 \nu^2 m r^2}{k\mathrm{T}}}\,. \tag{20}$$

Die Gesamtdichte der Moleküle in der Entfernung r wird jetzt

$$(21) \qquad n = \int_0^\infty n(\mathrm{L})\, d\mathrm{L} = \frac{(2\pi m k\mathrm{T})^{3/2}}{h^3} \mathrm{F}(\mathrm{A}),$$

wo gesetzt worden ist:

$$(22) \qquad \mathrm{F}(\mathrm{A}) = \frac{2}{\sqrt{\pi}} \int_0^\infty \frac{\mathrm{A}\sqrt{x}\, e^{-x}\, dx}{1 + \mathrm{A}\, e^{-x}}.$$

Die mittlere kinetische Energie der Moleküle in der Entfernung r ist

$$(23) \qquad \overline{\mathrm{L}} = \frac{1}{n} \int_0^\infty \mathrm{L}\, n(\mathrm{L})\, d\mathrm{L} = \frac{3}{2} k\mathrm{T} \frac{\mathrm{G}(\mathrm{A})}{\mathrm{F}(\mathrm{A})},$$

wo

$$(24) \qquad \mathrm{G}(\mathrm{A}) = \frac{4}{3\sqrt{\pi}} \int_0^\infty \frac{\mathrm{A}\, x^{3/2}\, e^{-x}\, dx}{1 + \mathrm{A}\, e^{-x}}.$$

Mittels (21) kann man A als Funktion von Dichte und Temperatur bestimmen; wenn man den gefundenen Wert in (19) und (23) einsetzt, so bekommt man die Geschwindigkeitsverteilung und die mittlere kinetische Energie der Moleküle als Funktion von Dichte und Temperatur.

Zur Aufstellung der Zustandsgleichung wenden wir den Virialsatz an. Nach diesem ist der Druck durch

$$(25) \qquad p = \frac{2}{3} n \overline{\mathrm{L}} = nk\mathrm{T} \frac{\mathrm{G}(\mathrm{A})}{\mathrm{F}(\mathrm{A})}$$

gegeben: der Wert von A ist wieder aus (12) als Funktion von Dichte und Temperatur zu entnehmen.

Ehe wir weitergehen, wollen wir einige mathematische Eigenschaften der eingeführten Funktionen F (A) und G (A) darstellen.

Für $\mathrm{A} \leqq 1$ kann man beide Funktionen durch die konvergierenden Reihen

$$(26) \qquad \begin{cases} \mathrm{F}(\mathrm{A}) = \mathrm{A} - \dfrac{\mathrm{A}^2}{2^{3/2}} + \dfrac{\mathrm{A}^3}{3^{3/2}} - \cdots, \\[2ex] \mathrm{G}(\mathrm{A}) = \mathrm{A} - \dfrac{\mathrm{A}^2}{2^{5/2}} + \dfrac{\mathrm{A}^3}{3^{5/2}} - \cdots \end{cases}$$

darstellen. Für grosses A hat man die asymptotischen Ausdrücke

$$(27) \qquad \begin{cases} \mathrm{F}(\mathrm{A}) = \dfrac{4}{3\sqrt{\pi}} (\log \mathrm{A})^{3/2} \left[1 + \dfrac{\pi^2}{8(\log \mathrm{A})^2} + \cdots\right], \\[2ex] \mathrm{G}(\mathrm{A}) = \dfrac{8}{15\sqrt{\pi}} (\log \mathrm{A})^{5/2} \left[1 + \dfrac{5\pi^2}{8(\log \mathrm{A})^2} + \cdots\right]. \end{cases}$$

Es gilt weiter die Beziehung

$$(28) \qquad \frac{d\,\mathrm{G}(\mathrm{A})}{\mathrm{F}(\mathrm{A})} = d \log \mathrm{A}.$$

Wir müssen noch eine andere, durch die Beziehungen

$$\text{(29)} \qquad P(\Theta) = \Theta \frac{G(A)}{F(A)} \quad , \quad F(A) = \frac{1}{\Theta^{3/2}}$$

definierte Funktion $P(\Theta)$ einführen. Für sehr grosses bzw. sehr kleines Θ kann $P(\Theta)$ mit den Näherungsformeln

$$\text{(30)} \qquad \begin{cases} P(\Theta) = \Theta \left\{ 1 + \dfrac{1}{2^{5/2}\,\Theta^{3/2}} + \cdots \right\} & \text{bzw.} \\ P(\Theta) = \dfrac{3^{2/3}\,\pi^{1/3}}{5 \cdot 2^{1/3}} \left\{ 1 + \dfrac{5 \cdot 2^{2/3}\,\pi^{4/3}}{3^{7/3}}\,\Theta^2 + \cdots \right\} \end{cases}$$

berechnet werden. Mit Benutzung von (29), (28), (27) erkennt man weiter, dass

$$\text{(31)} \qquad \int_0^{\epsilon} \frac{d\,P(\Theta)}{\Theta} = \frac{5}{3}\,\frac{G(A)}{F(A)} - \frac{2}{3} \log A \,.$$

Wir sind jetzt imstande, aus der Zustandsgleichung (15) und der Gleichung (23) den Parameter A zu eliminieren, und wir finden den Druck und die mittlere kinetische Energie der Moleküle als explizite Funktionen von Dichte und Temperatur:

$$\text{(32)} \qquad p = \frac{h^2\,n^{5/3}}{2\,\pi m}\,P\left(\frac{2\,\pi m\,kT}{h^2\,n^{2/3}}\right),$$

$$\text{(33)} \qquad \overline{L} = \frac{3}{2}\,\frac{h^2\,n^{2/3}}{2\,\pi m}\,P\left(\frac{2\,\pi m\,kT}{h^2\,n^{2/3}}\right).$$

Im Grenzfall schwacher Entartung (T gross und n klein) nimmt die Zustandsgleichung folgende Form an:

$$\text{(34)} \qquad p = n\,kT\left\{1 + \frac{1}{16}\,\frac{h^3\,n}{(\pi m\,kT)^{3/2}} + \cdots\right\}.$$

Der Druck ist also grösser als nach der klassischen Zustandsgleichung ($p = nkT$). Für ein ideales Gas mit dem Atomgewicht von Helium, bei $T = 5°$ und einem Druck von 10 Atm. beträgt der Unterschied etwa 15 Proz.

Im Grenzfall grosser Entartung nehmen (32) und (33) die Form

$$\text{(35)} \qquad p = \frac{1}{20}\left(\frac{6}{\pi}\right)^{2/3}\frac{h^2\,n^{5/3}}{m} + \frac{2^{4/3}\,\pi^{8/3}}{3^{5/2}}\,\frac{m\,n^{1/3}\,k^2\,T^2}{h^2} + \cdots$$

$$\text{(36)} \qquad \overline{L} = \frac{3}{40}\left(\frac{6}{\pi}\right)^{2/3}\frac{h^2\,n^{2/3}}{m} + \frac{2^{1/3}\,\pi^{8/3}}{3^{2/3}}\,\frac{m\,k^2\,T^2}{h^2\,n^{2/3}} + \cdots$$

an. Man erkennt hieraus, dass die Entartung einen Nullpunktsdruck und eine Nullpunktsenergie zur Folge hat.

Aus (36) kann man auch die spezifische Wärme für tiefe Temperaturen berechnen. Man findet

$$\text{(37)} \qquad c_v = \frac{d\overline{L}}{dT} = \frac{2^{4/3}\,\pi^{8/3}}{3^{2/3}}\,\frac{m\,k^2\,T}{h^2\,n^{2/3}} + \cdots$$

Man erkennt, dass die spezifische Wärme beim absoluten Nullpunkt verschwindet, und zwar dass sie für tiefe Temperaturen der absoluten Temperatur proportional ist.

Zuletzt wollen wir zeigen, dass unsere Theorie zum Stern-Tetrodeschen Wert für die absolute Entropie des Gases führt. Durch Anwendung von (33) findet man in der Tat

$$S = n \int_0^T \frac{d\bar{L}}{T} = \frac{3}{2} nk \int_0^\Theta \frac{P'(\Theta)\, d\Theta}{\Theta} .$$

(31) gibt uns jetzt

(38) $$S = nk \left\{ \frac{5}{2} \frac{G(A)}{F(A)} - \log A \right\} ,$$

wo der Wert von A wieder aus (21) zu entnehmen ist. Für hohe Temperaturen finden wir deshalb mit Anwendung von (26)

$$A = \frac{nh^3}{(2\pi m kT)^{3/2}} \quad , \quad \frac{G(A)}{F(A)} = 1 .$$

(38) gibt uns dann

$$S = nk \left\{ \log \frac{(2\pi m kT)^{3/2}}{nh^3} + \frac{5}{2} \right\} = nk \left\{ \frac{3}{2} \log T - \log n + \log \frac{(2\pi m)^{3/2} k^{3/2} e^{5/2}}{h^3} \right\} ,$$

was mit dem Entropiewert von Stern und Tetrode übereinstimmt.

N° 32.

A calculation, based on the vector model and the correspondence principle, of the intensities of lines violating the selection rule for J, expected when the emitting atom is placed in a strong magnetic field. A quantitative application is made to the $^1S - ^3P$ combinations in the alkaline earths.

F. RASETTI.

32.

SOPRA L'INTENSITÀ DELLE RIGHE PROIBITE NEI CAMPI MAGNETICI INTENSI

« Rend. Lincei », *3*, 478–483 (1926) (*).

Nello spettro emesso da un atomo perturbato da un intenso campo elettrico o magnetico si osservano spesso delle righe o delle serie di righe che non apparivano nello spettro dell'atomo imperturbato. Questo fenomeno ci viene spiegato al modo seguente dal principio di corrispondenza di Bohr [1]: È noto che le frequenze di tutte le righe che un atomo può emettere, si possono rappresentare come differenze di termini spettroscopici; non però a tutte le possibili differenze di termini corrispondono delle righe che possono venire emesse dall'atomo. Il criterio dato da Bohr perché una combinazione di due termini sia possibile è che nello sviluppo in serie di Fourier del moto dell'elettrone luminoso figuri la componente armonica che corrisponde al salto tra i due termini; e in questo caso la intensità della riga è approssimativamente eguale alla intensità con cui la componente armonica corrispondente verrebbe emessa secondo la elettrodinamica classica, cioè proporzionale al quadrato dell'ampiezza della componente medesima. Si capisce ora come la perturbazione dell'orbita dell'elettrone luminoso per opera di un campo esterno possa generare nel suo moto delle componenti armoniche che mancavano nel moto imperturbato; e corrispondentemente a queste appariranno nello spettro delle nuove righe che non esistevano in assenza di campo perturbatore.

In questo lavoro mi propongo di studiare da questo punto di vista l'intensità delle nuove righe che appaiono in campi magnetici intensi [2].

(*) Presentata dal Socio A. Garbasso nella seduta del 21 marzo 1926.

(1) N. BOHR, « Zs. f. Phys. », *9*, 27 (1922).

(2) Per l'osservazione sperimentale di tali righe si veda: F. PASCHEN e E. BACK, « Physica », *1*, 261 (1921).

A questo problema è strettamente legato quello della distruzione della metastabilità di certi stati quantici nel campo magnetico. Vi sono certi stati da cui non è possibile altra uscita che attraverso a qualche salto quantico proibito. In assenza di perturbazioni dunque l'atomo che si trovi in uno di questi stati non ha alcun mezzo per uscirne, per modo che la sua vita media è infinita; se però l'atomo è soggetto a una opportuna perturbazione, per. es. a un campo magnetico, la vita media dello stato metastabile diventa finita.

Limiteremo le nostre considerazioni ai così detti termini di primo grado (3) (erster Stufe). A questo tipo appartengono, come si sa, gli spettri dei metalli alcalini, dei metalli terrosi, e tutti i termini normali degli alcalino-terrosi. Come modello meccanico dei termini di primo grado si può prendere la rappresentazione vettoriale di Landé (4). Secondo questa una orbita stazionaria dell'atomo è rappresentata, oltre che dal quanto totale n, da tre altri numeri quantici (5) K, R, J, ai quali vengono attribuiti risp. i significati di momento dell'elettrone luminoso, momento del nocciolo e risultante dei due precedenti, cioè momento totale dell'atomo. Se l'atomo si trova in un campo magnetico H bisogna introdurre un altro quanto, M, che è la proiezione di J sulla direzione di H. Il carattere cinematico dell'orbita può, in assenza di campo magnetico, descriversi al modo seguente (6): Nel piano perpendicolare a k l'elettrone descrive un moto a rosetta, cioè un moto periodico con una precessione uniforme nel suo piano. Il piano dell'orbita non resta fisso nello spazio, perché tutto l'atomo, e quindi anche i vettori k ed r, hanno un moto di precessione uniforme attorno a j, che, in assenza di forze esterne, conserva direzione invariabile nello spazio per il principio della conservazione delle quantità di moto areali; l'azimuth di questa precessione, φ_j, è la coordinata angolare coniugata a j. Quando l'atomo è posto in un campo magnetico esterno la direzione di j non resta più invariabile, poiché tanto k che r vengono ad assumere un moto di precessione attorno al campo magnetico. Ora per render conto dell'effetto Zeeman anomalo e dell'effetto Paschen Back bisogna ammettere che k sia sollecitato a precedere con la frequenza normale di Larmor

$$\Omega = \frac{e\mathrm{H}}{4\pi mc} = 1{,}4\cdot 10^{6}\cdot \mathrm{H} \tag{1}$$

mentre r sia sollecitato a precedere con frequenza doppia. Di questo fatto hanno recentemente tentato di dare una spiegazione Uhlenbeck e Goudsmit (7) cambiando il significato modellistico di r, che viene da essi interpretato come momento complessivo dei moti di rotazione degli elettroni; per quanto

(3) A. LANDÉ e W. HEISENBERG, «Zs. f. Phys.», *25*, 279 (1924).

(4) A. LANDÉ, ibid., *15*, 189, 1923. Nel caso di termini di primo grado k viene a coincidere con l, così che basta considerare le interazioni tra k ed r; W. HEISENBERG, «Zs. f. Phys.», *32*, 841 (1925); H. N. RUSSELL e F. A. SAUNDERS, «Astr. Journ.», *61*, 38 (1925).

(5) In tutto questo lavoro indicheremo con le lettere maiuscole R, K, J, M, i numeri quantici, e con le minuscole r, k, j, m, le quantità di moto corrispondenti.

(6) A. SOMMERFELD e W. HEISENBERG, «Zs. f. Phys.», *11*, 131 (1922).

(7) G. E. UHLENBECK e S. GOUDSMIT, «Naturw.», *13*, 953 (1925).

ardita possa sembrare, questa ipotesi rende certamente conto di un numero considerevole di fatti spettroscopici. L'effetto complessivo di questa diversa azione del campo sopra k ed r, finché H è piccolo, consiste semplicemente nell'imprimere a tutto l'atomo un moto di precessione uniforme attorno ad H, con una frequenza anomala espressa da $g\,\Omega$, dove g è il fattore di separazione; l'azimuth φ_m di questa precessione è la coordinata angolare coniugata a m. Quando invece i momenti esercitati da H su k ed r diventano dello stesso ordine di grandezza della interazione tra r e k, il carattere uniforme di questa precessione viene ad esser turbato, poiché r tende a precedere più rapidamente di k; contemporaneamente viene anche turbata la precessione attorno ad j. La disuniformità di queste precessioni ha per effetto che nello sviluppo in serie di Fourier del moto dell'elettrone luminoso vengono ad apparire delle nuove componenti armoniche, e da ciò ha origine l'apparizione delle nuove righe. Prima di passare al calcolo della loro intensità vogliamo esporre brevemente i risultati raggiunti.

In assenza di campi magnetici, o in campi magnetici deboli si trova che J ed M possono in un salto quantico variare soltanto di ± 1 oppure 0. Inoltre non può avvenire per il quanto J la combinazione $1/2 \to 1/2$, e per il quanto M non può avvenire la combinazione $0 \to 0$ quando sia contemporaneamente $\Delta J = 0$. In campi magnetici forti, si trova che non valgono più le regole di selezione per J, e si possono osservare sia la transizione $1/2 \to 1/2$ che anche quelle per cui $\Delta J = 2$, $\Delta J = 3$, ecc. La dipendenza della intensità delle nuove righe dal campo H, è diversa da riga a riga. Un primo tipo è costituito dalle righe che corrispondono al passaggio $1/2 \to 1/2$, oppure a $\Delta J = 2$, e per queste l'intensità è approssimativamente proporzionale ad H^2; il secondo tipo è quello delle righe per cui $\Delta J = 3$, e per esse l'intensità è proporzionale ad H^4, ecc. Per le righe del primo tipo l'intensità è data da

$$I = \alpha\, I_0\, \Omega^2/\omega^2 \tag{2}$$

dove I_0 è l'intensità del multipletto, ω la sua separazione [8], ed α un coefficiente numerico esprimibile in funzione dei numeri quantici.

Alla funzione Hamiltoniana H del nostro modello atomico, nel caso di multipletti che obbediscano alla regola degli intervalli si può dare la forma [9]

$$\frac{1}{h} H = \omega j^2 + \Omega \left\{ m \left(\frac{3}{2} + \frac{r^2 - k^2}{2 j^2} \right) + \frac{\cos \varphi_j}{2 j^2} \sqrt{(j^2 - m^2)\,[(r+k)^2 - j^2]\,[j^2 - (r-k)^2]} \right\}. \tag{3}$$

In assenza di campo j, m, φ_j, φ_m sono coordinate angolari. Quando il sistema è perturbato dal campo si passerà da queste alle nuove coordinate

(8) Il significato preciso di ω per multipletti che obbediscono alla regola degli intervalli (LANDÉ, loc. cit.) è tale che la differenza di due termini del multipletto aventi i quanti interni J e J′ sia $\Delta\nu = \omega\,(J^2 - J'^2)$.

(9) W. PAULI, «Zs. f. Phys.», *20*, 371 (1924); W. HEISENBERG, ibid., *26*, 291 (1924). Un calcolo simile a questo si trova in LANDÉ, ibid., *30*, 329 (1924). A questo rimandiamo per alcuni dettagli del seguente calcolo di perturbazioni.

angolari J, M, Φ_j, Φ_m mediante la trasformazione canonica definita dalla funzione trasformatrice

$$S = J\varphi_j + M\varphi_m + \Omega S_1(J, M, \varphi_j, \varphi_m).$$

Applicando i metodi della teoria delle perturbazioni [10] si trova

$$S_1 = -\frac{\sin\varphi_j}{4\omega J^3}\sqrt{(J^2 - M^2)[(K+R)^2 - J^2][J^2 - (K-R)^2]}$$

la trasformazione canonica risulta dunque data da

$$(4)\quad \begin{cases} j = J - \dfrac{\Omega}{\omega} A\cos\Phi_j \quad ; \quad m = M ; \\ \varphi_j = \Phi_j + \dfrac{\Omega}{\omega} B\sin\Phi_j \quad ; \quad \varphi_m = \Phi_m + \dfrac{\Omega}{\omega} C\sin\Phi_j \end{cases}$$

dove si è posto

$$(5)\quad \begin{cases} A = \dfrac{1}{4J^3}\sqrt{(J^2 - M^2)[(K+R)^2 - J^2][J^2 - (K-R)^2]} ; \\ B = \dfrac{\partial A}{\partial J} \quad ; \quad C = \dfrac{\partial A}{\partial M}. \end{cases}$$

Le formule (4) e (5) risolvono completamente la parte meccanica del nostro problema. Troviamo in particolare in esse la conferma del fatto che in un campo magnetico di intensità finita le precessioni attorno a j e ad H non sono uniformi. Dobbiamo ora fare lo sviluppo in serie di Fourier delle coordinate dell'elettrone luminoso. Per far questo possiamo servirci di alcune formule stabilite da Sommerfeld e Heisenberg [6] per l'analogo sviluppo in serie di Fourier in campi infinitamente deboli. È chiaro infatti che quelle formule conserveranno la loro validità anche nel nostro caso, purché per $o_j t$ e $o_m t$ si sostituisca risp. φ_j e φ_m. Naturalmente i coefficienti C_{srq} e D_{srq} delle formule (9) di Sommerfeld e Heisenberg, che nel loro caso sono costanti, nel caso nostro saranno invece funzioni del tempo; poiché essi sono funzioni di j che, in virtù della prima delle (4) è a sua volta funzione del tempo. Per esprimere esplicitamente le C e le D in funzione di j basta porre nelle formule (10) e (7) di Sommerfeld e Heisenberg per Θ e θ i valori che risultano dal loro significato geometrico, e cioè

$$(6)\qquad \cos\Theta = M/j \quad ; \quad \cos\theta = (j^2 + K^2 - R^2)/2Kj.$$

Dobbiamo ora sostituire nelle formule (9) di Sommerfeld e Heisenberg per j, φ_j, φ_m i loro valori (4). Troviamo allora, nella nostra approssimazione

$$(7)\quad \begin{cases} X + iY = \displaystyle\sum_{srq} \left\{ C_{srq} - \frac{\Omega A}{\omega}\frac{\partial C_{srq}}{\partial J}\cos\Phi_j \right\} \cdot \\ \qquad \cdot \left\{ 1 + \dfrac{i\Omega}{\omega}(qB + C)\sin\Phi_j \right\} e^{i(so_n t + ro_k t + q\Phi_j + \Phi_m)} \\ Z = \displaystyle\sum_{srq} \left\{ D_{srq} - \frac{\Omega A}{\omega}\frac{\partial D_{srq}}{\partial J}\cos\Phi_j \right\} \cdot \\ \qquad \cdot \left\{ 1 + \dfrac{i\Omega}{\omega} qB\sin\Phi_j \right\} e^{i(so_n t + ro_k t + q\Phi_j)}. \end{cases}$$

(10) M. BORN e W. PAULI, «Zs. f. Phys.», *10*, 137 (1922).

Sviluppando le precedenti espressioni in serie di Fourier rispetto alla coordinata angolare Φ_j si trova che esse contengono, limitandosi alla nostra approssimazione, soltanto termini in $e^{\pm 2i\Phi_j}$, $e^{\pm i\Phi_j}$, 1; questi termini corrispondono ai salti quantici $J \rightarrow J \pm 2$, $J \rightarrow J \pm 1$, $J \rightarrow J$. Le componenti armoniche corrispondenti ai salti $J \rightarrow J \pm 2$ e $1/2 \rightarrow 1/2$ hanno ampiezza proporzionale a Ω/ω; le corrispondenti probabilità di transizione saranno perciò proporzionali a Ω^2/ω^2, ciò che giustifica la (2). Se avessimo spinta la nostra approssimazione fino a comprendere anche i termini in Ω^2/ω^2, avremmo trovato, nello sviluppo in serie precedente anche dei termini corrispondenti al salto quantico $J \rightarrow J \pm 3$. La ampiezza delle righe corrispondenti risulta dunque proporzionale ad H^2 e quindi la loro intensità ad H^4.

A titolo di esempio consideriamo le transizioni proibite $J = 1/2 \rightarrow J = 1/2$. Per queste si ha una sola componente Zeeman, per cui $M = 0$, $\Delta M = 0$; le componenti armoniche corrispondenti si troveranno perciò solo nello sviluppo di Z; dalle (6) risulta inoltre in questo caso $\cos \Theta = 0$. Le formule (10) e (7) di Sommerfeld e Heisenberg ci dànno allora

$$D_{s,r,0} = 0 \quad ; \quad D_{s,1,1} + D_{s,1,-1} = \frac{1}{2} a_s \cos \theta \ ;$$

$$D_{-s,-1,1} + D_{-s,-1,-1} = \frac{1}{2} \bar{a}_s \cos \theta \, .$$

Le componenti armoniche cercate risultano dunque

$$-\frac{\Omega}{4\omega}\left(B \cos \theta + A \frac{\partial \cos \theta}{\partial J}\right)\left\{a_s e^{i(so_n + o_k)t} + \bar{a}_s e^{-i(so_n + o_k)t}\right\}.$$

La corrispondente probabilità di transizione è perciò proporzionale a

$$\frac{\Omega^2}{\omega^2}\left(B \cos \theta + A \frac{\partial \cos \theta}{\partial J}\right)^2 |a_s|^2 \tag{8}$$

parimenti la probabilità di transizione per il salto quantico non proibito tra i termini omologhi per cui $\Delta J = 1$ risulta proporzionale a

$$(1 + \cos \theta)^2 |a_s|^2. \tag{9}$$

Come esempio concreto calcoliamo la vita media dello stato metastabile $2p_3$ degli elementi del secondo gruppo del sistema periodico. Dallo stato $2p_3$ l'unica via di uscita è quella che porta allo stato normale 1 S attraverso il salto proibito 1 S — $2p_3$. Le espressioni (8) e (9) saranno dunque inversamente proporzionali alle vite medie τ_3 e τ_2 degli stati $2p_3$ e $2p_2$. Come valori di (8) e (9) prenderemo le medie aritmetiche tra i loro valori corrispondenti agli stati iniziale e finale. Si trova così che (8) e (9) sono risp. proporzionali a $0{,}006\, \Omega^2/\omega^2$ e 1,6; risulta dunque

$$\tau_3/\tau_2 = 270\, \omega^2/\Omega^2.$$

Per. es. per il magnesio si ha $\omega = 3 \cdot 10^{11}$. In un campo di 50.000 Gauss si ha poi $\Omega = 7 \cdot 10^{10}$. Risulta dunque $\tau_3 = 5000\, \tau_2$. Se ammettiamo che τ_2 sia dell'ordine di grandezza di 10^{-7} secondi troviamo infine $\tau_3 = 5 \cdot 10^{-4}$.

33.

ARGOMENTI PRO E CONTRO LA IPOTESI DEI QUANTI DI LUCE

« Nuovo Cimento », *3*, 47–54 (1926).

Vi fu un periodo di tempo, dopo la scoperta di Maxwell della teoria elettromagnetica della luce, in cui si credette definitivamente risolto il problema della struttura della luce, che si interpretava come un fenomeno ondulatorio consistente nella propagazione di vettori elettrici e magnetici, corrispondenti allo spostamento dell'etere della teoria di Fresnel. Negli ultimi venticinque anni però lo studio accurato degli scambi energetici tra radiazione e materia ha condotto alla scoperta di un gruppo di fenomeni, il cui numero e la cui importanza sono andati fino ad oggi sempre crescendo, la cui esistenza appare difficilmente conciliabile con la struttura continua della luce; tanto da indurre alcuni scienziati, e primo tra di essi Einstein, a contrapporre alla teoria ondulatoria una teoria di tipo corpuscolare, la così detta teoria dei quanti di luce, rinnovando così, con le modificazioni corrispondenti ai tempi mutati, l'antica questione sulla struttura della luce, già dibattuta tra Newton e Huyghens.

Allo stato attuale della scienza si può dire che manchi una teoria che renda conto in modo soddisfacente dei fenomeni ottici; in quanto che, mentre la teoria ondulatoria si adatta perfettamente alla spiegazione della interferenza, della diffrazione, della polarizzazione e di tutto quel complesso di fenomeni che costituiscono la così detta ottica classica, essa appare insufficiente a renderci conto di quello che accade tutte le volte che si esamina l'azione della luce sopra un singolo atomo; d'altra parte la teoria dei quanti di luce, costruita appositamente per spiegare questa ultima classe di fenomeni, non riesce a render conto dei fenomeni interferenziali, nemmeno nei casi più semplici, altro che con adattamenti del tutto innaturali, che sembra improbabile possano corrispondere alla essenza reale del fenomeno.

I risultati dello studio dell'assorbimento dell'energia luminosa da parte della materia, possono riassumersi nella seguente legge di Einstein.

Tutte le volte che un atomo o una molecola assorbe dell'energia luminosa, la quantità di energia che viene assorbita non può prendere valori qualsiasi, ma è legata alla frequenza della luce assorbita dalla relazione

$$W = h\nu \tag{1}$$

dove h è la costante di Planck ($= 6{,}54 \cdot 10^{-27}$ erg. sec.) e ν la frequenza. Questa legge trova applicazione in un numero assai grande di fenomeni; per ricordare soltanto i principali accenneremo all'effetto fotoelettrico che consiste nel fatto che una superficie metallica, illuminata con luce di frequenza sufficiente emette degli elettroni. Orbene, si trova sperimentalmente che l'energia cinetica massima degli elettroni emessi è legata alla frequenza dalla relazione

$$w = h\nu - w_o \tag{2}$$

dove w_o rappresenta l'energia necessaria per estrarre un elettrone dal metallo, così che l'energia totale comunicata all'elettrone dalla luce risulta appunto eguale, conformemente alla legge di Einstein, a $w + w_o = h\nu$. Naturalmente, perché possa avvenire l'emissione fotoelettrica, occorre che l'energia comunicata all'elettrone sia sufficiente ad estrarlo dal metallo, occorre cioè che sia $h\nu \geq w_o$; si trova così la interpretazione del fatto che per produrre l'effetto fotoelettrico è efficace solamente la luce di frequenza superiore ad un certo limite.

Un fenomeno del tutto analogo all'effetto fotoelettrico dei corpi solidi si osserva anche illuminando un gas con luce di frequenza conveniente. Si trova infatti che la luce può strappare dalla molecola del gas un elettrone, e che tra l'energia cinetica dell'elettrone emesso e la frequenza passa ancora la relazione (2), che ci dice che l'energia comunicata alla molecola è, anche in questo caso, $h\nu$.

Ora, dal punto di vista della teoria ondulatoria, apparirebbe naturale ammettere che l'energia trasmessa a un elettrone dalla luce dovesse essere tanto maggiore quanto più elevata la sua intensità, e dovesse essere in particolare estremamente piccola per luci molto deboli; sorprende dunque il risultato sperimentale che l'energia assorbita da ogni elettrone non dipende in alcun modo dalla intensità della luce, ma soltanto dalla sua frequenza. Ma un calcolo quantitativo di queste circostanze, convince addirittura della impossibilità di conciliare i fatti che risultano dallo studio dell'effetto fotoelettrico con l'idea della distribuzione continua della energia in tutto il fascio di luce. Se infatti si calcola in questa ipotesi l'energia che può essere assorbita da un atomo, anche ammettendo che esso assorba tutta la energia che cade su di esso, si trova, nel caso di illuminazioni con luce debole, che dovrebbero passare parecchi giorni prima che l'atomo possa riuscire ad accumulare tanta energia quanta occorre per strappargli un elettrone!.... Si riscontra invece sperimentalmente che, per quanto debole possa essere l'illuminazione l'effetto fotoelettrico comincia appena la luce cade sulla sostanza.

I fenomeni dell'effetto fotoelettrico sembrano invece suggerire spontaneamente l'ipotesi dei quanti di luce. Secondo questa ipotesi la luce è costituita di corpuscoli propagantisi, nel vuoto, con la velocità della luce, i quali hanno un contenuto di energia proporzionale alla frequenza, ed espresso precisamente da

$$\varepsilon = h\nu\ . \tag{3}$$

Ci si rende allora conto del come avviene che tutte le volte che viene assorbita della luce l'energia assorbita sia data da (1), e si capisce anche come questa energia debba essere indipendente dalla intensità. Infatti la differenza tra una luce intensa e una luce debole consiste solo nel fatto che i quanti della prima sono più fitti di quelli della seconda, ma quando un atomo viene colpito da un quanto esso assorbe in ogni caso tutta la sua energia. Come al quanto di luce si attribuisce una energia, data da (3), così gli si deve anche attribuire una quantità di moto. Per trovarne il valore ricordiamo il risultato, confermato dalla esistenza della pressione di radiazione, che ad una propagazione di energia luminosa W è legata la quantità di moto elettromagnetica W/c; ad un quanto di energia $h\nu$ dovremo dunque attribuire anche la quantità di moto

$$q = \frac{h\nu}{c} \cdot \tag{4}$$

Naturalmente, come contro la teoria ondulatoria si ha l'obiezione che essa non spiega l'effetto fotoelettrico, così contro la teoria dei quanti di luce, come contro tutte le teorie corpuscolari, si può obiettare che esse non spiegano i fenomeni interferenziali.

È caratteristico, a questo proposito, osservare come l'attribuire al quanto la quantità di moto (4) riesca tuttavia a rendere conto di certe particolarità che, a prima vista, apparirebbero potersi soltanto interpretare per mezzo del meccanismo ondulatorio. A titolo di esempio possiamo mostrare come la teoria dei quanti di luce possa rendere conto dell'effetto Doppler.

Ricordiamo perciò che, secondo la teoria dei quanti, l'emissione della luce da parte di un atomo è legata al salto dell'atomo tra due orbite di energia differente, e che la frequenza di emissione si calcola dividendo per h la differenza delle energie. Consideriamo allora un atomo A che abbia due orbite quantistiche le cui energie siano risp. 0 e w. La frequenza emessa dall'atomo fermo, corrispondente al salto tra queste due orbite, sarà

$$\nu_0 = \frac{w}{h} \cdot \tag{5}$$

Supponiamo ora (fig. 1) che l'atomo, prima di emettere, quando cioè si trova nello stato di energia interna w, si muova con velocità V. La sua energia sarà la somma dell'energia cinetica e di quella interna, cioè verrà data da

$$\frac{1}{2} m V^2 + w \, . \tag{6}$$

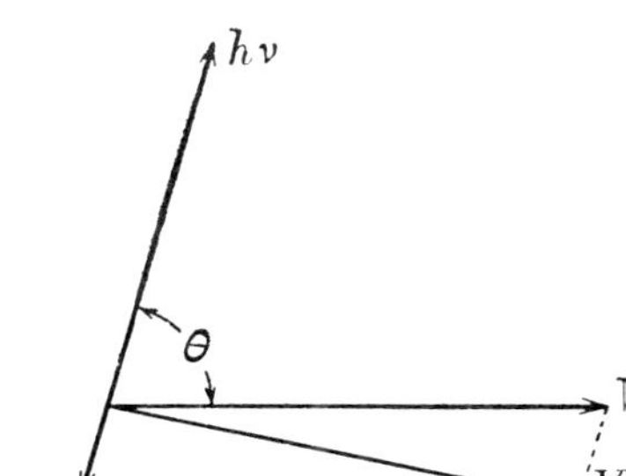

Fig. 1.

Supponiamo ora che, ad un certo istante, l'atomo salti dallo stato di energia interna w allo stato di energia interna 0, emettendo un quanto di frequenza ν in una direzione formante l'angolo θ con V. Per il principio della conservazione della quantità di moto, l'atomo dovrà subire il rinculo del quanto emesso, per modo che la sua velocità verrà ad essere variata, e diventerà V′. Indichiamo con v la differenza (vettoriale) tra V′ e V. Per la conservazione della quantità di moto si dovrà avere, tenendo presente (4)

$$mv = \frac{h\nu}{c}. \tag{7}$$

Dopo avvenuta l'emissione, l'energia dell'atomo si riduce alla sola energia cinetica, cioè, trascurando il quadrato di v e tenendo presente (7), a

$$\frac{1}{2} m \mathrm{V}'^2 = \frac{1}{2} m (\mathrm{V}^2 + v^2 - 2 \mathrm{V} v \cos\theta)$$

$$= \frac{1}{2} m \mathrm{V}^2 - mv \mathrm{V} \cos\theta = \frac{1}{2} m \mathrm{V}^2 - \frac{h\nu}{c} \mathrm{V} \cos\theta .$$

Per il principio della conservazione dell'energia, l'energia del quanto emesso deve essere eguale alla differenza delle energie dell'atomo prima e dopo l'emissione; si deve cioè avere

$$h\nu = \frac{1}{2} m \mathrm{V}^2 + w - \left(\frac{1}{2} m \mathrm{V}^2 - \frac{h\nu}{c} \mathrm{V} \cos\theta\right) = w + \frac{h\nu}{c} \mathrm{V} \cos\theta$$

di qui si ricava, con ovvie semplificazioni, tenendo presente (5)

$$\nu = \frac{\nu_0}{1 - \frac{\mathrm{V}}{c} \cos\theta}. \tag{8}$$

La formula precedente coincide con quella data dalla teoria ondulatoria dell'effetto Doppler per il cambiamento di frequenza della luce emessa da una sorgente in moto.

A questo proposito osserviamo ancora che, con considerazioni del genere di quelle esposte è riuscito a Compton, Duane e altri di rendere conto dei fenomeni di diffrazione di un reticolo, ciò che forse, può dare un indizio di una via per una spiegazione quantistica dei fenomeni interferenziali.

Altri tentativi, per la spiegazione quantistica dell'interferenza ben lontani dall'aver raggiunto un assetto definitivo, sono dovuti a De Broglie, secondo il quale il quanto verrebbe accompagnato da un sistema di onde virtuali aventi, per così dire, la funzione di guidarlo.

Parallelamente a questi tentativi di render conto dei fenomeni di diffrazione partendo dalla concezione corpuscolare dei quanti di luce, sono stati anche fatti numerosi tentativi di spiegare l'effetto fotoelettrico per mezzo della concezione ondulatoria. Di questi il più autorevole è quello fatto recentemente da Bohr, Kramers e Slater. Secondo questi autori un atomo, illuminato con luce di frequenza ν eguale ad una delle sue frequenze caratteristiche, avrebbe una probabilità, proporzionale alla intensità della luce di eccitarsi, aumentando la propria energia interna di $h\nu$. Parimenti un atomo che si trovi in uno stato eccitato emetterebbe un sistema di onde che trasmetterebbe agli atomi circostanti la probabilità di venire alla loro volta eccitati. Si capisce come, per restare d'accordo coi fatti osservabili, bisogna determinare tali probabilità in modo che resti statisticamente soddisfatto il principio della conservazione dell'energia; secondo le predette idee di Bohr, Kramers e Slater dunque il principio della conservazione dell'energia non sarebbe valido in ogni singolo processo atomico, ma lo sarebbe solo, *in media*, sopra un numero molto grande di tali processi.

Questo punto particolare della teoria è stato negli ultimi tempi assoggettato ad un controllo sperimentale. Il risultato è stato nettamente contrario alla concezione di Bohr, Kramers e Slater, ed ha portato alla dimostrazione che i principî della conservazione dell'energia e della quantità di moto conservano il loro valore anche in ogni singolo processo atomico.

Il processo atomico utilizzato in queste esperienze è la diffusione di raggi X da parte di elementi leggeri, e precisamente il così detto effetto Compton. Non mi dilungo in considerazioni sopra questo fenomeno, perché esso è stato oggetto di un articolo riassuntivo della sig.na Brunetti pubblicato recentemente su questo giornale. Basterà ricordare come il meccanismo della diffusione dei raggi X da parte degli elettroni di sostanze leggere, che possono praticamente considerarsi come liberi, sia molto diverso secondo la teoria ondulatoria e secondo quella dei quanti di luce. Secondo la prima teoria un elettrone viene scosso dalle vibrazioni della luce primaria e diventa perciò a sua volta centro di un sistema di onde secondarie (di frequenza eguale a quella delle primarie) che costituiscono appunto i raggi X diffusi. Secondo la teoria dei quanti di luce invece, l'elettrone, inizialmente fermo, viene urtato da un quanto che, in seguito a questo urto, viene deviato dando luogo alla radiazione diffusa. Per il principio della conservazione della quantità di moto, al mutamento di direzione del quanto deve corrispondere un impulso trasmesso all'elettrone il quale dunque dovrà mettersi in moto, cosicché, secondo questa teoria, ad ogni quanto diffuso deve corrispondere un elettrone di rimbalzo. Siccome inoltre il quanto perde un po' della sua energia, che viene ceduta sotto forma di energia cinetica all'elettrone di rimbalzo, si avrà che la frequenza del quanto diffuso dovrà essere un po' minore di quella del quanto incidente. È noto come l'esperienza confermi completamente questa ultima conclusione. Restava da vedere se effettivamente ad ogni quanto diffuso corrispondesse un elettrone di rimbalzo e se tra la direzione in cui viene diffuso il quanto e quella in cui è proiettato l'elettrone passasse effettivamente la relazione voluta dalla teoria. I risultati delle esperienze, eseguite da Geiger e Bothe in Germania, e, in modo ancora assai più espressivo, da Compton e Simon in America, e per i cui dettagli rimandiamo al precitato articolo della Sig.na Brunetti, sono stati a conferma completa della teoria quantistica estrema. Naturalmente data la enorme importanza concettuale di esperienze di questo genere, sarebbe desiderabile che esse venissero ripetute in diverse condizioni, in modo da mettere in evidenza il più possibile le varie particolarità del fenomeno.

Possiamo dire concludendo che, per quanto semplici ed espressivi possano apparire i risultati della teoria dei quanti di luce, essa lascia tuttavia inspiegato un vasto campo di fenomeni; e che altrettanto insufficiente si presenta la teoria ondulatoria. Se si considera tutto lo spettro, dai raggi X e gamma alle onde Hertziane, si può osservare come nel primo campo siano esclusivamente sviluppati i fenomeni che suggeriscono il meccanismo quantistico, poiché anche la diffrazione nei cristalli di Laue può agevolmente interpretarsi da questo punto di vista; nel secondo invece si riscontrano praticamente solo i fenomeni che si interpretano col meccanismo ondulatorio. Queste particolarità si possono naturalmente mettere in relazione col fatto che il quanto dei raggi X, la cui frequenza è enorme, è molto grande; il quanto delle onde Hertziane, di frequenza relativamente bassa, è invece praticamente infinitesimo. Si capisce dunque come tutti i fenomeni di tipo discontinuo debbano apparire in modo assai più marcato nel primo che nel secondo caso.

La zona, per così dire, di confine tra il meccanismo quantistico e quello ondulatorio, è costituita dalla regione visibile, dove sono bene sviluppati sia i fenomeni di tipo interferenziale che quelli di tipo fotoelettrico. Si capisce naturalmente come un tale dualismo tra teoria ondulatoria e teoria dei quanti di luce non possa essere altro che un adattamento provvisorio, in attesa di poter formare tutto il campo dell'ottica, dalle onde radiotelegrafiche ai raggi Roentgen in un unico edificio armonico.

N° 34.

For the introduction to this paper see N° 22.

34.

PROBLEMI DI CHIMICA, NELLA FISICA DELL'ATOMO (*)

« Periodico di Matematiche », *6*, 19–26 (1926).

Una delle proprietà più caratteristiche dell'atomo è la sua attitudine a formare combinazioni chimiche; e il problema del rendersi conto di queste proprietà fu perciò uno dei primi che si presentarono agli studiosi quando al principio di questo secolo, incominciò a svilupparsi la fisica dell'atomo.

Già da una cinquantina di anni, e precisamente fin dalle prime esperienze sopra il passaggio dell'elettricità attraverso ai gas, i fisici furono condotti alla scoperta di uno dei costituenti fondamentali della materia, l'elettrone, cioè l'elemento di elettricità negativa; e di questo elettrone essi riuscirono a determinare la carica elettrica e la massa giungendo al risultato fondamentale che tutti gli elettroni sono eguali, indipendentemente dalla sostanza da cui essi vengono estratti. Più difficile si dimostrò invece lo studio delle parti elettricamente positive dell'atomo. Una prima ipotesi sulla loro struttura fu quella di Thomson, il quale pensava che l'elettricità positiva fosse distribuita in modo uniforme in una sfera di diametro eguale a quello dell'atomo (cioè dell'ordine di grandezza di 10^{-8} cm) e che gli elettroni fossero immersi in questa sfera e occupassero delle posizioni di equilibrio sotto l'azione delle mutue repulsioni e della attrazione verso il centro della sfera. L'ipotesi di Thomson dovette in seguito essere abbandonata perché in contrasto con numerosi fatti sperimentali, e particolarmente con le esperienze di Rutherford sopra il passaggio delle particelle alfa attraverso alla materia; furono appunto queste esperienze che condussero Rutherford a proporre quella ipotesi sulla struttura dell'elettricità positiva, sulla quale ancora si basano le teorie atomiche moderne.

Secondo il modello atomico di Rutherford la parte positiva dell'atomo, detta nucleo, è concentrata in una regione di dimensioni assai più piccole di quelle dell'atomo, tanto da potersi con grande approssimazione considerare come un punto, e la massa del nucleo è praticamente eguale alla massa totale dell'atomo, poiché risulta che la massa dell'elettrone è eguale a circa 1/1800 della massa del nucleo più leggero, cioè di quello dell'idrogeno. La carica elettrica del nucleo risulta eguale a un multiplo intero di quella dell'elettrone,

(*) Conferenza tenuta al Congresso della « Mathesis », Milano 30 ottobre 1925.

e il coefficiente, detto numero atomico, coincide col numero d'ordine del posto che l'atomo in considerazione occupa nel sistema periodico di Mendeleieff. Siccome complessivamente l'atomo deve essere neutro, il nucleo sarà circondato da un numero di elettroni eguale al numero atomico dell'atomo. Cosicché l'atomo di idrogeno, di numero atomico 1, sarà costituito da un nucleo di carica eguale, salvo il segno, a quella dell'elettrone, e da un solo elettrone; l'atomo di elio, di numero atomico 2, da un nucleo di carica 2, e da due elettroni; infine per esempio l'atomo di uranio sarà costituito da un nucleo di carica 92 volte quella dell'elettrone e da 92 elettroni. Sotto molti punti di vista dunque il modello atomico di Rutherford ricorda la struttura del sistema solare; il nucleo ha in esso la funzione del Sole, e gli elettroni quella dei pianeti.

Se è merito di Rutherford l'aver tracciate le linee generali della struttura dell'atomo, dobbiamo a Bohr la conoscenza delle leggi fondamentali che ne regolano i movimenti. Per formarci una idea chiara di queste leggi riferiamoci all'atomo di idrogeno che, come si è accennato, è il più semplice, essendo costituito dal nucleo e da un solo elettrone.

Siccome l'elettrone viene attirato dal nucleo secondo la legge di Coulomb, esso dovrebbe descrivere attorno a questo, secondo le leggi della meccanica ordinaria, una ellissi di tipo *Kepleriano*, di cui il nucleo dovrebbe occupare uno dei fuochi. Ora secondo i risultati della elettrodinamica classica, una carica elettrica in moto non uniforme dovrebbe irradiare nello spazio della energia elettromagnetica, e in conseguenza di questa irradiazione l'orbita dell'elettrone dovrebbe andare continuamente restringendosi, per modo che l'elettrone finirebbe per cadere sopra il nucleo. Siccome però in realtà tutto questo non accade, Bohr fu costretto ad ammettere che nella fisica dell'atomo le leggi della elettrodinamica classica dovessero subire delle modificazioni; e l'effetto di codeste modificazioni, la cui forma precisa è tutt'ora ignota, e la cui determinazione costituisce per il momento il problema più importante della fisica atomica, egli cercò di riassumere in alcune leggi di cui ora vogliamo esporre le parti essenziali, limitandoci agli enunciati, e tralasciando le ragioni che ad esse hanno condotto, perché queste ci porterebbero lontani dal tema di questa conferenza. Secondo Bohr, di tutti i moti meccanicamente possibili dell'atomo, cioè, nel caso dell'idrogeno, di tutte le ellissi *Kepleriane*, soltanto una successione discreta può venire effettivamente descritta dall'atomo; e finché esso descrive uno di questi moti, detti moti stazionari, non ha luogo alcuna irradiazione di energia. Se con $w_1, w_2, \cdots$ indichiamo le energie degli stati stazionari, avremo dunque che normalmente il contenuto di energia dell'atomo potrà soltanto avere uno di questi valori.

L'irradiazione dell'energia luminosa da parte dell'atomo è legata invece ad un processo discontinuo, detto salto quantico, col quale l'atomo passa da un moto stazionario ad un altro. Supponiamo per esempio che il salto avvenga tra due stati stazionari di energie w_1 e w_2; verrà allora messa in libertà la differenza tra le energie dei due stati, $w_2 - w_1$, e questa energia verrà emessa in un solo quanto, per modo che la sua frequenza si potrà calcolare con la formula

$$h\nu = w_2 - w_1$$

essendo h la costante di Planck. Questa formula permette di calcolare completamente le frequenze ottiche che un certo atomo può emettere, cioè il suo spettro, una volta che siano noti i suoi moti stazionari, e quindi anche le loro energie. La determinazione teorica degli stati stazionari è però disgraziatamente assai spesso un problema irrisolubile, cosicché nella maggior parte dei casi bisogna accontentarsi di risultati qualitativi. L'unico caso in cui riesce la determinazione completa degli stati stazionari è quello dell'idrogeno; le orbite stazionarie dell'idrogeno si possono dividere in gruppi, caratterizzati dall'avere lo stesso asse maggiore, e quindi anche la stessa energia. L'asse maggiore delle ellissi dell'n^{esimo} gruppo è proporzionale ad n^2; e all'n^{esimo} gruppo appartengono n ellissi, tra cui anche un cerchio, distinte tra di loro dalla diversa eccentricità. Nel caso degli atomi più complicati si ha sempre un elettrone esterno che descrive orbite simili a quelle dell'unico elettrone dell'idrogeno; i suoi movimenti vengono però in questo caso perturbati dalla azione degli altri elettroni: e una conseguenza di questo fatto è che le orbite dello stesso gruppo vengono ad avere delle energie un po' differenti tra di loro. Siccome poi l'orbita dell'elettrone esterno può anche avere differenti inclinazioni rispetto a quelle degli altri elettroni, il numero dei possibili livelli di energia viene a crescere ulteriormente. Se per esempio ci riferiamo al caso tipico dei metalli alcalini, si trova che per essi al primo gruppo di orbite appartiene un solo livello di energia, al secondo tre, al terzo cinque, e così di seguito. Ma anche questi livelli di energia non sono semplici, vale a dire non possono essere realizzati in un solo modo; e ciò si riconosce perturbando l'atomo con una causa esterna, per esempio ponendolo in un campo magnetico; si riconosce allora che quei livelli che prima erano apparsi semplici si dividono ciascuno in un numero variabile di livelli, e questo fatto ha, come conseguenza spettroscopica, l'effetto Zeeman. Si riconosce così che, nel caso dei metalli alcalini ci sono 2 orbite semplici appartenenti al primo gruppo, 8 al secondo, 18 al terzo, 32 al quarto, e così di seguito.

È ora assai notevole il fatto che i numeri di queste orbite si trovano in una relazione assai stretta con la struttura del sistema periodico di Mendelejeff. Per renderci conto di questo osserviamo che per passare da un atomo a quello successivo nel sistema periodico occorre, a prescindere dall'aumento di una unità della carica del nucleo, aggiungergli un elettrone. Ora si trova che questo elettrone può occupare soltanto una delle orbite possibili, che sia stata ancora lasciata libera dagli elettroni precedenti. Siccome dunque nel primo gruppo di orbite ce ne sono soltanto due, avremo che alla loro occupazione corrisponderà la formazione dei due primi elementi del sistema periodico, l'idrogeno e l'elio; nel secondo gruppo ci sono 8 posti, che corrispondono alla formazione della seconda orizzontale del sistema periodico, e cioè degli elementi Li, Be, B, C, N, O, F, Ne; al completamento del secondo gruppo di orbite corrisponde dunque la formazione del gas nobile neon, che è perciò caratterizzato da una stabilità chimica particolare. Nel terzo gruppo di orbite ci sono 18 posti che corrispondono al formarsi degli elementi del terzo e quarto periodo, dal Na al Ni, e così di seguito. Ma oltre al renderci conto delle varie particolarità del sistema periodico, come per esempio della formazione del gruppo anomalo delle terre rare, queste idee sopra la distribuzione degli elettroni nell'atomo, ci permettono anche di renderci conto di gran parte delle sue proprietà chimiche caratteristiche, e delle sue proprietà spettroscopiche, sia nel campo della luce visibile che in quello dei raggi Roentgen; la verificazione di tutti questi fatti costituisce una importante conferma della esattezza della distribuzione degli elettroni tra le varie orbite.

Ma vogliamo ora passare all'esame di un'altra classe di fenomeni, che si presentano nello studio della fisica atomica, e che hanno una analogia spiccatissima con delle vere e proprie reazioni chimiche.

Abbiamo detto che un atomo può trovarsi in stati diversi, caratterizzati dai valori delle loro energie; dati allora due stati di energie w_1 e w_2, potremo domandarci quanti sono, ad una certa temperatura, gli atomi che si trovano nel primo, e quanti quelli che si trovano nel secondo di questi stati; è evidente la analogia di questo problema con quello della determinazione di un equilibrio chimico. La risposta alla domanda che ci siamo posta ci viene data dalle ordinarie leggi statistiche che ci dicono in questo caso che, purché i due livelli di energia siano semplici, indicando con n_1 e n_2 il numero degli atomi che si trovano rispettivamente nel primo e nel secondo livello si ha

$$\frac{n_1}{n_2} = e^{-\frac{w_2 - w_1}{kT}} \tag{1}$$

essendo k la costante di Boltzmann e T la temperatura assoluta.

Questa semplice relazione ci conduce a delle conseguenze importanti. Supponiamo infatti che w_1 sia, per il nostro atomo, il livello energetico di energia minima, e w_2 un altro livello qualunque. Nella maggioranza dei casi si trova allora che il rapporto $\frac{w_2 - w_1}{kT}$, anche per temperature di qualche migliaio di gradi, è sempre abbastanza grande, dell'ordine di grandezza di 20 o 30; così che il rapporto n_1/n_2 risulta dell'ordine di $e^{20} = 5 \cdot 10^8$. Si vede dunque che in queste condizioni n_1 è assai più grande di n_2, vale a dire la quasi totalità degli atomi si trova nel livello di energia minima. Vi è tuttavia un certo numero di elementi, per esempio il Tallio, in cui la differenza tra le energie dello stato di energia minima e di quello precedente, è assai più piccola, cosicché l'esponente $\frac{w_2 - w_1}{kT}$ diventa dell'ordine di uno già per temperature di un migliaio di gradi. Ed è appunto in questi casi che si è potuta fare una verifica sperimentale della (1), diventando, in queste condizioni, n_1 ed n_2 dello stesso ordine di grandezza.

Ma vogliamo ora occuparci di un altro fenomeno atomico, che presenta una somiglianza ancora più spiccata con una ordinaria reazione chimica. Consideriamo per questo un atomo A; se da esso si estrae un elettrone e esso si trasformerà nello ione positivo A^+; e inversamente lo ione positivo, riunendosi all'elettrone, ridarà l'atomo. Possiamo allora considerare la reazione di ionizzazione

$$A = A^+ + e \tag{2}$$

che è di tipo completamente analogo a una ordinaria reazione chimica di dissociazione. Megh Nad Saha ha applicato alla reazione (2) quegli stessi metodi termodinamici che si applicano al calcolo delle reazioni chimiche tra sostanze gassose, e le formule da lui date hanno avuto brillanti conferme sperimentali, e sono servite, anche nel campo della astrofisica, a interpretare parecchi fenomeni che erano prima restati incomprensibili. Una delle applicazioni più caratteristiche della teoria di Saha è la seguente, dovuta a Fowler, che ha permesso, sia pure in modo grossolano, di verificare le leggi dell'equilibrio di ionizzazione fino a temperature di qualche milione di gradi. È stato scoperto recentemente che quando una particella α (che come si sa non è altro che un nucleo di elio proiettato con velocità grandissima dalle sostanze radioattive) attraversa la materia, essa successivamente cattura e perde degli elettroni. Fowler ha intepretato questo fenomeno al modo seguente. Invece di considerare la particella α proiettata attraverso agli elettroni contenuti negli atomi attraversati dalla particella, possiamo pensare che la particella sia ferma, e che essa venga investita da uno sciame di elettroni, aventi velocità eguale, salvo il verso, a quella della particella. È chiaro allora che la particella si verrà a trovare press'a poco nelle stesse condizioni in cui sarebbe se fosse immersa in un gas di elettroni, a una temperatura tale che la loro velocità media fosse eguale a quella dello sciame, temperatura che, nel caso concreto, risulta appunto di qualche milione di gradi. Possiamo allora considerare a questa temperatura, la reazione, analoga alla (2)

$$\alpha + e = (\alpha e)$$

dove col simbolo α si è indicata la particella isolata, e con quello $(\alpha\, e)$ la particella accompagnata da un elettrone. Calcolando questo equilibrio chimico con le leggi del Saha, Fowler riuscì a ritrovare, in notevole corrispondenza coi risultati sperimentali, il rapporto tra il tempo in cui la particella viaggia isolata e quello in cui si accompagna con un elettrone.

Un altro problema, interessante assai da presso la chimica, che si presenta nello studio della fisica atomica, è quello della struttura delle molecole. Consideriamo per semplicità una molecola biatomica; può darsi che alcuni degli elettroni rotino alternativamente attorno all'uno e all'altro dei due nuclei, descrivendo delle orbite intrecciate, e allora il composto si dice omeopolare; se invece gli elettroni si separano in un gruppo che rota attorno all'uno, e in un gruppo che rota attorno all'altro nucleo la molecola si dice polare. A questo secondo tipo appartengono specialmente le molecole dei sali, e il suo studio è assai più avanzato che non quello dei composti omeopolari.

Consideriamo per esempio un atomo di cloro e uno di sodio. L'atomo di cloro ha molta tendenza ad assorbire un elettrone, perché, con esso, forma uno ione negativo, Cl^- assai stabile; invece il Na perde assai facilmente un elettrone, perché il ione che ne risulta, Na^+, è molto stabile. Si capisce dunque che quando i due atomi si troveranno in presenza uno dell'altro, il sodio cederà uno dei suoi elettroni al cloro, in modo che i due atomi si trasformeranno nei due ioni Cl^- e Na^+. Una volta formati, questi due ioni di segno contrario si attireranno per la legge di Coulomb, avvicinandosi finché la repulsione dei lori elettroni corticali non diventerà tale da controbilanciare l'attrazione. Avrà così luogo la formazione della molecola di cloruro di sodio.

Arcetri, 1925.

N° 35.

The authors, after listing the various successful applications of the electron spin hypothesis in explaining atomic spectra, point out an energetic difficulty. If one calculates the energy of the magnetic field, assuming that the magnetic dipole of the electron has a radius of the order of magnitude of the classical electron radius e^2/mc^2, one finds a value considerably greater than mc^2. Hence one must attribute a much larger radius to the magnetic structure than to the electric charge distribution. The ideas expressed in this paper were almost entirely contributed by Fermi.

F. RASETTI.

35.

SOPRA L'ELETTRONE ROTANTE

F. RASETTI ed E. FERMI

«Nuovo Cimento», *3*, 226–235 (1926).

Gli elementi misurabili dell'elettrone sono la sua carica elettrica e la sua massa; di quest'ultima si è tentato, come è noto, nella elettrodinamica classica, di dare una interpretazione che la riducesse alla prima. Ciò si è fatto considerando l'elettrone come una distribuzione di elettricità, superficiale o di volume, con simmetria sferica, la cui massa elettromagnetica si identificava generalmente con la massa totale dell'elettrone. Nonostante queste idee sulla sua struttura, nella teoria dell'atomo l'elettrone è stato fino ad ora quasi sempre considerato come un punto materiale. Solo negli ultimi tempi è stata fatta da Uhlenbeck e Goudsmit (1) l'ipotesi che la ragione di alcuni fenomeni spettroscopici, in particolare dell'effetto Zeeman anomalo, dovesse ricercarsi in un elemento della struttura dell'elettrone. Questi autori ammettono precisamente che l'elettrone sia animato da un moto di rotazione attorno a se stesso, in modo da possedere sia una quantità di moto areale, che un momento magnetico. Il presente lavoro è dedicato alla discussione di questa ipotesi, e si dimostrerà in particolare che essa, se verrà confermata dagli ulteriori fatti sperimentali, renderà molto probabile che all'elettrone si debbano attribuire delle dimensioni lineari notevolmente più grandi di quanto si era fino ad ora ritenuto.

Qualitativamente il fenomeno di Zeeman trova la sua interpretazione nel così detto teorema di Larmor, che dice: La perturbazione prodotta nel movimento di un sistema meccanico costituito da punti materiali che abbiano tutti la stessa massa m e la stessa carica elettrica e da un campo magnetico uniforme di intensità H, consiste in prima approssimazione in una precessione uniforme di tutto il sistema attorno alla direzione del campo, con frequenza $\nu_L = eH/4\pi mc$. Intimamente legato col teorema di Larmor è il ri-

(1) G. E. UHLENBECK e S. GOUDSMIT, «Naturw.», *13*, 953 (1925); «Nature», *117*, 264 (1926); BOHR, «Nature», *117*, 265 (1926).

sultato che per il sistema del tipo considerato si ha un rapporto tra momento magnetico e meccanico dipendente soltanto dalla carica e dalla massa dei punti, e dato precisamente da $e/2\,mc$.

Se gli elettroni di un atomo si considerano come punti materiali, il teorema di Larmor si può applicare anche all'atomo, poiché il nucleo si può considerare come fermo; cosicché, da questo punto di vista, ci si deve attendere che le frequenze delle righe emesse dall'atomo nel campo magnetico siano le combinazioni tra le frequenze proprie dell'atomo imperturbato e la frequenza di Larmor. Ad ogni riga di frequenza ν dell'atomo imperturbato dovrebbero dunque corrispondere tre righe con le frequenze $\nu - \nu_L$, ν, $\nu + \nu_L$. Notoriamente questa forma normale dell'effetto Zeeman si osserva soltanto in un numero molto limitato di righe. Nella maggioranza dei casi si ha invece una decomposizione in più di tre componenti, la cui separazione è generalmente differente da quella normale; ciò che costituisce il così detto effetto Zeeman anomalo. Si trova inoltre che, mentre per campi magnetici deboli lo spostamento di ogni componente dalla posizione della riga imperturbata è proporzionale ad H, col crescere del campo si manifesta una deformazione della configurazione Zeeman, per modo che al limite per campi molto forti essa tende a trasformarsi nel tripletto normale (Effetto Paschen–Back).

Per il principio di corrispondenza, anche nella teoria di Bohr le frequenze delle righe emesse dall'atomo perturbato da un campo magnetico, si debbono calcolare come frequenze di combinazione tra le frequenze proprie dell'atomo e la frequenza di precessione; è chiaro dunque che per rendersi conto dell'effetto Zeeman anomalo bisogna ammettere che la velocità di precessione dell'atomo sia diversa da quella normale, con la quale deve coincidere al limite per campi molto forti.

Un modello atomico che, sia pure in modo incompleto e in parecchi punti insoddisfacente, cerca di render conto di queste particolarità è stato sviluppato da Landé. Landé(2) distingue nell'atomo l'elettrone luminoso, che è quello al quale è dovuta l'emissione delle righe, e che si muove generalmente in un'orbita notevolmente lontana dal nucleo, e l'insieme di tutti gli altri elettroni più vicini al nucleo, cioè il nocciolo (Rumpf). Ciascuno di questi due elementi possiede una quantità di moto areale, che Landé chiama rispettivamente K ed R. La risultante J di K ed R è la quantità di moto areale di tutto l'atomo, la quale, in assenza di forze esterne, conserva direzione invariabile nello spazio. Tra elettrone e nocciolo agiscono delle forze dipendenti dalla orientazione di ques'ultimo rispetto al piano dell'orbita elettronica, cioè dall'angolo formato tra K ed R. Queste forze sono tali da dar luogo a una precessione di tutto l'atomo attorno all'asse J. Questa precessione ha un effetto caratteristico sulla struttura dello spettro emesso dall'atomo imperturbato; è chiaro infatti che la frequenza di ogni riga che l'atomo emetterebbe se non avesse la detta precessione, si combinerà con la frequenza della precessione, in modo da dar luogo ad una riga multipla.

Studiamo ora quale sarà l'effetto di porre questo atomo in un campo magnetico. Se si ammette che, sia l'elettrone luminoso, che il nocciolo siano

(2) A. LANDÉ, «Zs. f. Phys.», *15*, 189 (1923).

sollecitati a precedere attorno al campo con la frequenza normale di Larmor, l'effetto consisterà nell'imprimere a tutto l'atomo questa precessione, e come risultato finale si troverà l'effetto Zeeman normale. Per spiegare l'effetto Zeeman anomalo e l'effetto Paschen–Back, Landé ammette che l'elettrone luminoso sia sollecitato a precedere colla frequenza normale, e che invece il nocciolo sia sollecitato con frequenza doppia. Per questo è necessario anche ammettere che il rapporto tra il momento magnetico e il momento meccanico dell'elettrone luminoso sia quello normale, mentre per il nocciolo sia doppio. Finché le forze che si esercitano tra elettrone e nocciolo sono grandi in confronto alle forze esercitate dal campo H, nocciolo ed elettrone resteranno ancora legati nel loro moto, in modo che tutto l'atomo precederà con una frequenza intermedia tra le due; per campi deboli si avrà dunque una precessione, e quindi un effetto Zeeman anomalo. Quando invece l'azione del campo predomina su quella tra elettrone e nocciolo, ciascuno di questi elementi precederà indipendentemente dall'altro, con la sua frequenza di precessione; e siccome quella dell'elettrone è normale, ne risulterà per campi forti la separazione Zeeman normale, cioè l'effetto Paschen–Back. Questo schema rende conto, non solo qualitativamente, ma anche quantitativamente, delle principali caratteristiche degli effetti Zeeman anomali osservati (3). Nonostante questi successi, esso è stato sempre considerato come insufficiente e provvisorio, poiché, siccome il nocciolo è esso pure costituito di elettroni, non ci si rende conto della ragione per cui la sua frequenza di precessione debba essere doppia di quella normale. Un altro inconveniente della teoria di Landé è il seguente: Se R si interpreta come momento del nocciolo, bisogna ammettere che il suo valore coincida con il momento totale J' dello ione dell'atomo, poiché lo ione è appunto quello che resta dell'atomo quando gli si tolga l'elettrone luminoso; si riscontra invece che ad ogni determinato valore J' del momento dello ione, possono corrispondere due valori per R, che differiscono da J' per $\pm 1/2$.

È appunto per ovviare a questi inconvenienti che Uhlenbeck e Goudsmit hanno introdotta l'ipotesi dell'elettrone rotante. Osserviamo che nella teoria di Landé l'interpretazione di R come momento del nocciolo è alquanto arbitraria; probabilmente essa era stata scelta, nonostante tutti gli inconvenienti a cui dà luogo, per mancanza di altri elementi dell'atomo a cui attribuire R, una volta che si partiva dall'ipotesi di trascurare a priori la struttura dell'elettrone; è in essa invece che Uhlenbeck e Goudsmit ricercano di significato di R. Essi ammettono precisamente che l'elettrone roti su se stesso, e possegga pertanto sia un momento meccanico che un momento magnetico. Naturalmente per un sistema di questo tipo il rapporto tra momento magnetico e meccanico dipenderà dalla distribuzione della carica e della massa; per trovarsi in accordo coi fatti sperimentali si ammette che queste distribuzioni siano tali da dare un rapporto doppio di quello di Larmor. Si ammette anche che

(3) Lo schema di Landé vale soltanto per i così detti spettri di primo grado (erster Stufe); per gli altri se ne hanno delle generalizzazioni (H. N. RUSSELL e F. A. SAUNDERS, «Astr. Journ.», *61*, 38 (1925); W. HEISENBERG, «Zs. f. Phys.», *32*, 841 (1925). Per il nostro scopo basta però limitarsi al caso degli spettri di primo grado.

lo stato di rotazione di tutti gli elettroni sia lo stesso, e che essi differiscano solamente per la differente orientazione, che viene determinata dalle relazioni quantistiche col resto dell'atomo e con eventuali campi esterni. R si interpreta come somma vettoriale di tutti i momenti degli elettroni dell'atomo.

Con questa interpretazione si ottengono i seguenti vantaggi:

1) Sparisce la difficoltà fondamentale della anomalia magnetica di R; siccome infatti per il moto di rotazione di ogni elettrone il rapporto tra momento magnetico e momento meccanico è doppio del rapporto di Larmor, l'elettrone, in un campo magnetico esterno, sarà sollecitato a precedere con frequenza doppia di quella normale.

2) Si capisce anche come, quando si aggiunge un nuovo elettrone allo ione positivo, in modo da formare l'atomo neutro, secondo la orientazione di questo elettrone rispetto allo ione, si possano avere per l'atomo dei diversi valori di R, come infatti si riscontra.

3) Ci si rende chiaramente conto della circostanza osservata da Stoner (4) e Pauli (5), che per poter costruire uno schema unitario della formazione successiva degli elementi, con riguardo alle loro proprietà spettroscopiche, occorre attribuire all'elettrone un altro grado di libertà oltre a quelli di un punto materiale, e del quale era fino ad ora incomprensibile l'origine.

4) Nelle precedenti teorie lo spettro dell'idrogeno occupava una posizione singolare nel sistema periodico, essendo interpretato in modo completamente diverso dagli spettri degli atomi con un solo elettrone di valenza (Metalli alcalini). Col nuovo schema si stabilisce invece una completa analogia; in quanto che, pur restando inalterata la posizione dei termini data dalla teoria relativistica di Sommerfeld, viene modificata la loro nomenclatura (6); ne consegue una modificazione nelle regole di selezione e di calcolo delle intensità delle varie componenti della struttura fina, che si mostra in accordo sensibilmente migliore coi risultati delle misure.

5) Parimenti la nuova teoria rende conto delle strutture dei livelli Roentgen, e in particolare del fatto, fino ad ora inesplicabile, della validità generale della formula relativistica per il calcolo delle separazioni spettrali.

Da quanto abbiamo esposto si riconosce che l'ipotesi dell'elettrone rotante illumina parecchi dei punti che prima erano dei più oscuri nella teoria dell'atomo. Diverse sono però le obiezioni che si possono sollevare contro di essa; la prima di queste si è presentata agli stessi suoi autori. Essi partono infatti da un modello molto particolare di elettrone, che suppongono come una distribuzione sferica superficiale di elettricità, la cui massa è calcolata come dovuta solamente alla sua energia elettrostatica; il moto di rotazione di questo sistema viene calcolato con le ordinarie regole della quantizzazione di un rotatore. A questo modo si trova effettivamente un rapporto tra momento meccanico e momento magnetico doppio di quello normale di Larmor; si trova però che la velocità periferica all'equatore risulta notevolmente

(4) E. C. STONER, « Phil. Mag. », *48*, 719 (1024).

(5) W. PAULI, « Zs. f. Phys. », *31*, 765 (1925).

(6) G. E. UHLENBECK e S. GOUDSMIT, « Physica », *5*, 266 (1925); A. SOMMERFELD e A. UNSÖLD, « Zs. f. Phys. », *36*, 259 (1926).

maggiore della velocità della luce. Un altro inconveniente dell'ipotesi dell'elettrone rotante è stato notato da Kronig (7). È noto che generalmente i nuclei degli atomi contengono degli elettroni; salvo nel caso particolare che i momenti magnetici dei singoli elettroni si neutralizzino, il nucleo dovrebbe dunque possedere un momento magnetico risultante, il quale dovrebbe estrinsecarsi in un paramagnetismo dell'atomo dipendente dal nucleo, del quale non si osserva traccia.

Ma più di queste difficoltà, delle quali la prima dipende dalla scelta di un modello molto particolare per l'elettrone, e l'altra non è insormontabile, perché non ci sono difficoltà sostanziali per ammettere che i momenti magnetici degli elettroni nucleari si neutralizzino, ci pare grave quella che deriva dalle seguenti considerazioni, indipendenti in larga misura da ogni idea particolare sopra la struttura dell'elettrone.

È essenziale per le applicazioni spettroscopiche che l'elettrone rotante possegga un momento magnetico dell'ordine di grandezza del magnetone di Bohr. Intorno all'elettrone esisterà allora un campo magnetico il quale, a distanza molto grande in confronto alle dimensioni lineari della struttura dell'elettrone, coinciderà praticamente con quello dovuto a un ordinario dipolo puntiforme, mentre, quando la distanza è dell'ordine di grandezza delle dimensioni lineari delle predette strutture, potranno aversi deviazioni notevoli. Otterremo dunque un limite inferiore per l'energia magnetica dell'elettrone calcolando l'energia magnetica localizzata all'esterno di una sfera di raggio R dell'ordine di grandezza delle dimensioni lineari dell'elettrone, e identificando questo campo col campo di un dipolo puntiforme.

La intensità H del campo magnetico di un dipolo puntiforme è data da

$$H^2 = \frac{\mu^2}{r^6}(3\cos^2\theta + 1)$$

dove r rappresenta la distanza dal dipolo, θ la colatitudine e μ il momento magnetico. L'energia magnetica contenuta nell'elemento di volume $d\tau$ è dunque

$$\frac{H^2}{8\pi}d\tau = \frac{\mu^2}{8\pi r^6}(3\cos^2\theta + 1)\,d\tau\,. \tag{1}$$

L'energia W contenuta all'esterno di una sfera di raggio R è allora

$$W = \int_\tau \frac{H^2}{8\pi}d\tau = \frac{\mu^2}{8\pi}\int_0^\pi\int_R^\infty \frac{1}{r^6}(3\cos^2\theta + 1)\,2\pi r^2\sin\theta\,d\theta\,dr$$

ed eseguendo le integrazioni si trova

$$W = \frac{\mu^2}{3R^3}\,.$$

A questa energia deve corrispondere una massa che, con la teoria della relatività, si calcola dividendo W per c^2 (secondo la elettrodinamica classica si trova un valore dello stesso ordine di grandezza). L'energia totale dell'elettrone sarà certamente maggiore, perché all'espressione precedente va

(7) R. de L. KRONIG, «Nature», *117*, 550 (1926).

aggiunta l'energia magnetica contenuta entro alla sfera di raggio R, e inoltre l'energia delle altre strutture elettroniche.

Troviamo dunque come limite inferiore per la massa dell'elettrone, ponendo per μ un magnetone di Bohr, cioè $0{,}92 \cdot 10^{-20}$

$$m = \frac{\mu^2}{3\,c^2\,R^3}\,.$$

Di qui si deduce, tenendo presente che l'elettrone ha la massa $0{,}9 \cdot 10^{-27}$, come limite inferiore per il raggio R

$$R = 3{,}3 \cdot 10^{-12}. \tag{2}$$

Questo valore è circa 20 volte più grande di quanto ordinariamente si ritiene sia il raggio elettronico. In realtà mancano misure dirette di quest'ultimo; tuttavia l'inconveniente è grave, perché si sa che nel nucleo è contenuto un numero molto considerevole di elettroni. D'altra parte le dimensioni lineari della struttura del nucleo sono note con sufficiente precisione dalle misure sulla deviazione delle particelle alfa nel passaggio attraverso alla materia, e risultano, come è noto, dell'ordine di 10^{-12} cm. Come si vede i due fatti appaiono abbastanza inconciliabili, ove non si ammetta che l'elettrone, entrando a far parte della struttura nucleare, cambi considerevolmente di natura.

Ci pare che la deduzione precedente del limite inferiore (2) per il raggio dell'elettrone si presti difficilmente ad obbiezioni gravi; perché essa cadesse in difetto bisognerebbe ammettere l'illegittimità del calcolo dell'energia magnetica con la formula (1), il che, se pare plausibile nell'interno della struttura elettronica, dove forse la stessa nozione di campo magnetico perde ogni significato, appare invece più difficilmente ammissibile all'esterno di questa struttura. Un'altra via di uscita potrebbe consistere nel non ritenere valida la relazione relativistica tra massa ed energia; o infine nell'ammettere che le strutture magnetiche fossero notevolmente più grosse di quelle elettriche.

Ci pare, da questa discussione, di poter concludere che, nonostante le gravi difficoltà energetiche a cui abbiamo eccennato, l'ipotesi dell'elettrone rotante non debba per questo venir abbandonata. Naturalmente noi non riteniamo che essa debba venir presa troppo alla lettera, nel senso che ci si debba veramente figurare l'elettrone come un corpo macroscopico carico di elettricità e rotante attorno a se stesso, poiché quello che è essenziale per le applicazioni è che l'elettrone possegga un momento meccanico e un momento magnetico, indipendentemente da rappresentazioni modellistiche troppo particolari sopra l'origine di questi.

In ogni modo, la questione non potrà considerarsi chiusa finché non si avranno delle esperienze più dirette a conferma o contro l'ipotesi dell'elettrone rotante.

N° 36.

This paper is a very simple and perspicuous application of Born's wave theory of collisions, which had been published a few months before. Fermi's paper is reported upon by A. Sommerfeld in *Atombau und Spektrallinien* (« Ergänzungsband », p. 237, 1928) with the following comment: « Die besondere Eleganz dieses Fermischen Beispiels besteht darin, dass – ebenso wie in der Theorie des Beugungsgitters – keine besondere Annahme über die Beschaffenheit des Gitters – hier über die potentielle Energie V – nötig ist, sofern keine quantitative Angaben über die relative Intensität der verschiedenen Beugungsbilder – hier über die Ergiebigkeit der verschiedenen elastischen und inelastischen Stösse – verlangt werden ».

E. Persico.

36.

ZUR WELLENMECHANIK DES STOSSVORGANGES

« Z. Physik », *40*, 399–402 (1926).

In dieser Arbeit wird die Theorie des Stosses zwischen einem Massenpunkt und einem Rotator entwickelt. Diese Theorie lässt sich in einfacher Weise auf die optische Theorie der Diffraktion eines Gitters zurückführen.

In einer kürzlich erschienenen Abhandlung hat M. Born [1] die Grundlagen der Anwendung der Schrödingerschen Wellenmechanik [2] auf den Stossvorgang entwickelt. In der vorliegenden Arbeit möchte ich ein besonders einfaches Stossbeispiel darstellen, welches, wie mir scheint, etwas zur klaren Übersicht der Verhältnisse bei dem Stosse dienen kann.

Wir werden nämlich den Zusammenstoss zwischen einem Massenpunkt und einem Rotator mit raumfester Achse untersuchen.

Um uns einer anschaulichen dreidimensionalen Vorstellung bedienen zu können, werden wir annehmen, dass der stossende Punkt sich in einer Ebene bewegen kann. Die Verallgemeinerung unserer Betrachtung auf den Fall eines freien Punktes wird aber ohne weiteres einleuchten.

Seien also x und y die kartesischen Koordinaten des Punktes (wir wählen den Koordinatenursprung in der Nähe des Rotators) und φ das Azimut des Rotators. Wir fassen Punkt und Rotator zu einem einzigen System (mit drei Freiheitsgraden) zusammen. Als allgemeine Koordinaten für dieses System wählen wir

$$\xi = \sqrt{m}\,x \quad , \quad \eta = \sqrt{m}\,y \quad , \quad \zeta = \sqrt{J}\,\varphi ,$$

(1) M. Born, « ZS. f. Phys. », *37*, 863 (1926); *38*, 803 (1926).

(2) E. Schrödinger, « Ann. d. Phys. », *79*, 361, 489 (1926); *81*, 437; *80*, 109 (1926); vgl. ferner L. de Broglie, Thése, Paris 1925.

wo m die Masse des Punktes, J das Trägheitsmoment des Rotators darstellt. Dadurch hat man den Vorteil, dass man als Linienelement im Konfigurationsraum $ds^2 = d\xi^2 + d\eta^2 + d\zeta^2$ nehmen kann, so dass die Metrik im Konfigurationsraum die gewöhnliche kartesiche Metrik wird. Die Schrödingersche Wellengleichung lautet dann

$$\frac{\partial^2 \psi}{\partial \xi^2} + \frac{\partial^2 \psi}{\partial \eta^2} + \frac{\partial^2 \psi}{\partial \zeta^2} + \frac{8\pi^2}{h^2}(\mathrm{E} - \mathrm{V})\psi = 0 \,. \tag{1}$$

Die potentielle Energie V besteht bloss aus der Wechselwirkungsenergie von dem Punkt und dem Rotator und verschwindet, wenn der Punkt vom Rotator weit entfernt ist; d. h. V hat einen von Null verschiedenen Wert nur in der Umgebung der ζ-Achse; weiter hängt offenbar die Wechselwirkungsenergie periodisch mit der Periode 2π vom Winkel φ ab, d. h. V hängt periodisch mit Periode $2\pi\sqrt{\mathrm{J}}$ von ζ ab. Als Grenzbedingung für ψ in unserem Falle muss man ebenfalls, nach Schrödinger, annehmen, dass ψ periodisch mit der Periode 2π von φ, d. h. also periodisch mit der Periode $2\pi\sqrt{\mathrm{J}}$ von ζ abhängt.

Um die Lösung von (1) übersehen zu können, denken wir an die optische Analogie zur Wellengleichung (1). In den von der ζ-Achse entfernten Gebieten, wo V verschwindet, ist (1) die Wellengleichung in einem optisch-homogenen Medium; in der Umgebung der ζ-Achse hat das Medium eine Anomalie des Brechungsindex, welche periodisch von ζ abhängt. Optisch ist dies aber nichts weiter als ein lineares Gitter mit der Periode $2\pi\sqrt{\mathrm{J}}$. Fällt nun eine ebene Welle auf unser Gitter, so hat man Beugungsstrahlen von verschiedenen Ordnungen; jede Diffraktionsordnung entspricht, wie wir sehen werden, einem verschiedenen Typus von Stoss (unelastische Stösse, elastischer Stoss, Stösse zweiter Art).

Aus der elementaren Gittertheorie entnehmen wir folgendes Ergebnis:

Sei Θ der Winkel zwischen der Ausbreitungsrichtung der einfallenden Wellen und der ζ-Achse; die Richtungen der Beugungsstrahlen n-ter Ordnung bilden alle denselben Winkel Θ_n mit der ζ-Achse; und zwar hat man

$$2\pi\sqrt{\mathrm{J}}(\cos\Theta_n - \cos\Theta) = n\lambda \,. \tag{2}$$

Wir wollen zuerst annehmen, dass die Wechselwirkungsenergie verschwindet, d. h. wir beschränken uns auf die von der ζ-Achse entfernten Gebiete. Für diesen Fall setzen wir ψ als Produkt einer Funktion $\psi_1(\xi, \eta)$ und einer Funktion $\psi_2(\zeta)$ an. Gleichung (1) spaltet sich dann in die beiden

$$\frac{\partial^2 \psi_1}{\partial \xi^2} + \frac{\partial^2 \psi_2}{\partial \eta^2} + \frac{8\pi^2}{h^2}\mathrm{E}_1\psi_1 = 0 \,, \tag{3}$$

$$\frac{d^2 \psi_2}{d\zeta^2} + \frac{8\pi^2}{h^2}\mathrm{E}_2\psi_2 = 0 \,, \tag{4}$$

wo E_1 und E_2 Konstanten darstellen, deren Summe $\mathrm{E}_1 + \mathrm{E}_2 = \mathrm{E}$. Für diesen Fall, da keine Wechselwirkung besteht, kann man E_1 als Energie des Punktes, E_2 als Energie des Rotators ansetzen. Gleichung (4) besitzt die Lösungen

$$\psi_2 = e^{\pm \frac{2\pi i}{h}\sqrt{2\mathrm{E}_2}\,\zeta} \,. \tag{5}$$

Da ferner ψ_2 periodisch mit Periode $2\pi\sqrt{J}$ von ζ abhängen muss, hat man die Bedingung $\frac{1}{h}\sqrt{2E_2}\,2\pi\sqrt{J} = k =$ ganze Zahl, d. h.

$$E_2 = \frac{h^2 k^2}{8\pi^2 J} \quad \text{(Quantelung des Rotators).} \tag{6}$$

Damit (3) eine überall endliche Lösung besitzt, muss ferner $E_1 > 0$ sein (kinetische Energie des Punktes positiv!), und dann besitzt (3) folgende Lösung

$$\psi_1 = e^{\pm\frac{2\pi i}{h}\sqrt{2E_1}(\xi\cos\chi+\eta\sin\chi)} \quad (\chi = \text{konstanter Winkel}). \tag{7}$$

Man hat also folgende Lösung für Gleichung (1), ohne Wechselwirkung:

$$\begin{aligned}\psi = \psi_1(\xi,\eta)\,\psi_2(\zeta) &= e^{\pm\frac{2\pi i}{h}\sqrt{2E_1}(\xi\cos\chi+\eta\sin\chi)\pm\frac{2\pi i}{h}\sqrt{2E_2}\,\zeta} \\ &= e^{\pm\frac{2\pi i}{h}\sqrt{2(E_1+E_2)}\{\xi\cos\chi\sin\Theta+\eta\sin\chi\sin\Theta+\zeta\cos\Theta\}},\end{aligned} \tag{8}$$

wo

$$\cos\Theta = \frac{\sqrt{E_2}}{\sqrt{E_1+E_2}} \quad , \quad \sin\Theta = \frac{\sqrt{E_1}}{\sqrt{E_2+E_1}}. \tag{9}$$

Mechanisch entspricht Lösung (8), nach Born (l. c.), dem Falle, dass der Rotator sich im Quantenzustand mit der Energie E_2 befindet und der Massenpunkt sich geradlinig mit der kinetischen Energie E_1 in einer Richtung bewegt, welche den Winkel χ mit der x-Achse bildet. Optisch ist aber Lösung (8) die Amplitude einer ebenen Welle im Konfigurationsraum mit der Wellenlänge

$$\lambda = \frac{h}{\sqrt{2(E_1+E_2)}} \tag{10}$$

und den Richtungskosinus $\cos\chi\sin\Theta$, $\sin\chi\sin\Theta$, $\cos\Theta$.

Der Einfluss der Wechselwirkung V in Gleichung (1) besteht nun, wie oben bemerkt, darin, dass die ebene Welle gebeugt wird; und zwar bilden die gebeugten Strahlen mit der ζ-Achse einen der Winkel Θ_n. Die Amplitude eines gebeugten Strahles ist also proportional zu [man betrachte (2), (6), (9), (10)]:

$$e^{\frac{2\pi i}{\lambda}\{\xi\cos\chi^*\sin\Theta_n+\eta\sin\chi^*\sin\Theta_n+\zeta\cos\Theta_n\}} = e^{\frac{2\pi i}{h}\sqrt{2E_1^*}(\xi\cos\chi^*+\eta\sin\chi^*)}\,e^{\frac{2\pi i}{h}\sqrt{2E_2^*}\,\zeta},$$

wo

$$\left\{\begin{aligned} E_2^* &= \frac{1}{2}\frac{h^2}{\lambda^2}\cos^2\Theta_n = \frac{h^2(n+k)^2}{8\pi^2 J}, \\ E_1^* &= \frac{1}{2}\frac{h^2}{\lambda^2}\sin^2\Theta_n = E - E_2^* = E_1 - (E_2^* - E_2).\end{aligned}\right. \tag{11}$$

Wieder mechanisch gedeutet, entspricht dies aber folgendem:

Der Rotator ist vom k-ten Quantenzustand mit der Energie E_2 in den $(n+k)$-ten mit Energie $E_2^* = \frac{h^2(n+k)^2}{8\pi^2 J}$ übergegangen. Der Massenpunkt bewegt sich in einer Richtung, die den beliebigen Winkel χ^* mit der x-Achse bildet, und seine kinetische Energie ist E_1^*, d. h. sie ist die Energie E_1, ver mindert um den Betrag $E_2^* — E_2$. Dieser Prozess ist aber nichts weiter als ein Zusammenstoss, wo der stossende Punkt in eine beliebige Richtung "gestreut" worden ist. Weiter hat man folgende drei Fälle zu unterscheiden:

$n = 0$. Die kinetische Energie des Punktes und die Energie des Rotators sind unverändert geblieben (elastischer Zusammenstoss).

$n > 0$. Der Rotator ist nach einem Quantenzustand mit grösserer Energie auf Kosten der kinetischen Energie des Punktes gesprungen (unelastischer Stoss).

$n < 0$. Der Rotator ist nach einem Zustand mit kleinerer Energie gesprungen, und die Energiedifferenz findet man als kinetische Energie des Punktes wieder (Stoss zweiter Art).

Nach Born gibt die Intensität der zerstreuten Strahlen ein Mass für die Wahrscheinlichkeit der Stösse der verschiedenen Arten.

N° 37.

Coming back from Florence to Rome for the 1926 summer vacations, Fermi found his friends Pontremoli and Persico busy studying and discussing the first papers by Schrödinger on wave mechanics, which had just appeared in « Annalen der Physik ». He, who had already acquired a deep insight in their content, brought to these discussions his decisive elucidating contribution. A good many of the discussions aimed at finding in the ideas, so unusual, of wave mechanics, some link, no matter how vague, with the concepts that had controlled the work of atomic physicists until a few months before, e. g. the concepts of kinetic and potential energy of an electron, and that of adiabatic invariance of quantum states. From one of these discussions originated the present paper, whose chief ideas are due to Fermi.

The expressions for E_{pot} and E_{kin}, which are here justified only by analogy and without any specification of their physical meaning, are the ones that, a little later, when the probabilistic interpretation of wave mechanics was developed, acquired the exact meaning of « expectation value » of the potential and kinetic energies.

E. PERSICO.

37.

IL PRINCIPIO DELLE ADIABATICHE E LA NOZIONE DI FORZA VIVA NELLA NUOVA MECCANICA ONDULATORIA

E. FERMI ed E. PERSICO

« Rend. Lincei », *4* (II), 452–457 (1926) (*).

1. L'analogia, da lungo tempo riconosciuta, tra le leggi della meccanica del punto materiale e quelle dell'ottica geometrica, ha – secondo alcune recenti ipotesi che si sono mostrate feconde di conseguenze – una origine più profonda e una portata più vasta di quanto poteva finora ritenersi. Precisamente, E. Schrödinger [1] ha recentemente sviluppato una nuova meccanica guidato dall'idea che anche i fenomeni meccanici abbiano natura ondulatoria, e che, analogamente a quanto avviene per le leggi dell'ottica geometrica che cessano di essere valide nei sistemi ottici di piccole dimensioni, per i quali è necessario invece tener conto della natura ondulatoria della luce, così pure le leggi della ordinaria meccanica siano approssimazioni valide soltanto per sistemi di dimensioni grandi rispetto alla grandezza meccanica che corrisponde alla lunghezza d'onda, mentre per sistemi di dimensioni comparabili con questa lunghezza intervengono fenomeni più complicati analoghi a quelli della diffrazione ottica.

(*) Presentata dal Socio O. M. Corbino nella seduta del 7 novembre 1926.

(1) « Ann. d. Phys. », *79*, pp. 361, 489, 734 (1926); *80*, p. 437; *81*, p. 109 (1926).

Precisando questa idea, Schrödinger è riuscito a raggiungere in modo naturale e coerente quei risultati che la vecchia teoria dei quanti otteneva per mezzo di postulati slegati e talora contradditori, e a conseguire in modo preciso dei risultati nuovi che la vecchia teoria forniva solo vagamente per mezzo del principio di corrispondenza. A questa nuova meccanica è stato dato il nome di meccanica ondulatoria.

Scopo del presente lavoro è anzitutto l'estensione del principio delle adiabatiche alla meccanica ondulatoria. Nella teoria ordinaria dei quanti il principio delle adiabatiche afferma, come è noto, che se si altera adiabaticamente il meccanismo di un sistema atomico che si trovi inizialmente in uno stato quantico, il sistema resta durante tutta la trasformazione in uno stato quantico. Questo principio, nella quantistica ordinaria, ha la sua base matematica in un teorema di meccanica analitica dovuto a Burgers, il quale dimostra che gli $\oint p dq$ sono invarianti adiabatici. In questa Nota dimostreremo che un principio analogo può dedursi come conseguenza dei principii fondamentali della nuova meccanica. In una seconda parte del lavoro (§ 3), estenderemo poi alla meccanica ondulatoria le nozioni classiche di energia cinetica ed energia potenziale.

2. Consideriamo un sistema meccanico ad f gradi di libertà, soggetto a forze derivanti da un potenziale V. Supporremo che il sistema sia in un determinato stato quantico (il k–esimo) vale a dire che sia presente la sola oscillazione di frequenza $\nu_k = \frac{E_k}{h}$: allora lo scalare di campo ψ che caratterizza lo stato del sistema sarà dato da

$$\psi = u_k e^{\frac{2\pi i}{h} E_k t} \tag{1}$$

dove u_k è la k–esima autofunzione (Eigenfunktion) normalizzata della equazione differenziale (*)

$$\Delta u + \frac{8\pi^2}{h^2}(E - V) u = 0 \tag{2}$$

ed E_k è il relativo autovalore. Supponiamo esplicitamente tutte le E_k diverse tra di loro, escludendo così il caso dei sistemi degeneri, cui il principio delle adiabatiche non si applica. Immaginiamo ora che il sistema venga trasformato adiabaticamente, vale a dire che il potenziale si trasformi passando da V a V_1: la trasformazione sia fatta però con infinita lentezza rispetto alla frequenza ν_k con cui varia ψ. (Il caso in cui si trasformano adiabaticamente anche i vincoli del sistema si può ridurre notoriamente a questo). Vogliamo dimostrare che dopo la trasformazione il sistema possiede ancora una sola frequenza, e precisamente la k–esima (naturalmente diversa dalla precedente) cioè si trova ancora nello stato quantico k.

(*) In questa nota, gli operatori Δ e grad sono riferiti a uno spazio il cui elemento lineare è $ds^2 = 2\, E_{kin}\, dt^2$, come usava Schrödinger nelle sue prime memorie (N. d. Red.).

Per la dimostrazione potremo evidentemente decomporre la variazione continua di V in una successione di infinite variazioni infinitesime brusche δV. Potremo per esempio, prendere per δV la variazione che subisce il potenziale in un tempo finito T (la quale, per la supposta infinita lentezza della trasformazione, risulterà appunto infinitesima) e farla avvenire in un istante che dovremo scegliere *a caso* in questo intervallo di tempo, poiché, per la validità del principio delle adiabatiche, è necessario che non ci sia alcuna relazione fra le variazioni del potenziale e il moto proprio del sistema, per escludere casi di risonanza. Prenderemo poi T molto grande rispetto ai periodi propri del sistema.

Consideriamo ora singolarmente l'effetto di una generica di queste variazioni. Prendiamo per comodità l'origine dei tempi nel mezzo dell'intervallo T entro il quale deve avvenire la variazione, e sia τ l'istante in cui essa avviene: si dovrà poi fare una media facendo variare τ da $-\frac{T}{2}$ a $\frac{T}{2}$, e il nostro assunto si ridurrà a dimostrare che, se prima della discontinuità lo scalare di campo aveva la forma (1), anche dopo la discontinuità esso possederà ancora la sola frequenza k-esima. Contrassegnamo con un asterisco i valori che assumono le varie grandezze dopo il tempo τ, cosicché, per esempio, il potenziale sarà V per $t < \tau$ e $V^* = V + \delta V$ per $t > \tau$. Le autofunzioni e gli autovalori dell'equazione (2) subiranno essi pure al tempo τ una brusca variazione che si può calcolare dalla variazione del potenziale con le formule date da Schrödinger per la teoria delle perturbazioni. Naturalmente, essendo δV infinitesimo, basta limitarsi alla prima approssimazione. Si trova così:

$$u_k^* = u_k + \sum_{1}^{\infty}{}'_i \gamma_{ki} u_i \quad , \quad E_k^* = E_k + \varepsilon_k$$

dove

$$\gamma_{ki} = \frac{\int \rho u_k u_i \, \delta V \, dx}{E_k - E_i} \quad , \quad \varepsilon_k = \int \rho u_k^2 \, \delta V \, dx .$$

(Le γ e la ε sono manifestamente infinitesimi del primo ordine, e l'accento sulla sommatoria indica che in essa manca il termine per cui $i = k$). Le notazioni sono quelle di Schrödinger.

La soluzione ψ^* sarà necessariamente una combinazione lineare di soluzioni fondamentali, cioè sarà del tipo

$$\psi^* = \sum_{1}^{\infty}{}_k c_k u_k^* e^{\frac{2\pi i}{h} E_k^* t} .$$

I coefficienti di questa combinazione risultano determinati dalla condizione che per $t = \tau$ sia $\psi^* = \psi$ (continuità dello scalare di campo !). Si verifica facilmente che tale condizione è soddisfatta (a meno di infinitesimi di ordine superiore) prendendo

$$\psi^* = u_k^* e^{\frac{2\pi i}{h}(E_k^* t - \varepsilon_k \tau)} - \sum_{1}^{\infty}{}'_i \gamma_{ki} u_i^* e^{\frac{2\pi i}{h}(E_i^* t + [E_k^* - E_i^*]\tau)} .$$

Facciamo dunque la media di questa espressione rispetto a τ per $-\frac{T}{2} < \tau < \frac{T}{2}$. Gl' esponenziali che figurano nella sommatoria hanno ciascuno per media zero, essendosi supposto T assai grande rispetto ai periodi propri $\frac{h}{E_i}$ del sistema. Quanto al primo termine di ψ^* si noti che il fattore dipendente da τ, $e^{-\frac{2\pi i}{h}\varepsilon_k \tau}$, può scriversi, a meno di infinitesimi di ordine superiore, $1 - \frac{2\pi i}{h}\varepsilon_k \tau$, e la sua media tra $-\frac{T}{2}$ e $\frac{T}{2}$ è dunque 1. Risulta perciò in definitiva

$$\psi^* = u_k^* e^{\frac{2\pi i}{h} E_k^* t}$$

con che resta dimostrato il principio delle adiabatiche.

È poi manifesto che, se si fosse partiti, invece che dalla soluzione iniziale (1), dalla soluzione più generale

$$\psi = \sum_{1}^{\infty}{}_k \lambda_k u_k e^{\frac{2\pi i}{h} E_k t}, \tag{3}$$

si sarebbe giunti alla conclusione che i λ_k non vengono alterati da una trasformazione adiabatica del sistema.

Notiamo incidentalmente che, data l'omogeneità dell'equazione, resta in ψ una costante moltiplicativa arbitraria, che supporremo fissata in modo che risulti $\int \psi\bar{\psi}\rho\, dx = 1$, il che corrisponde, data la supposta normalizzazione delle u, alla condizione $\sum_{1}^{\infty}{}_k \lambda_k^2 = 1$.

3\. L'espressione già data per la variazione ε_k di E_k, nel caso della soluzione semplice (1), si può evidentemente mettere sotto la forma

$$\varepsilon_k = \int \delta V\, \psi\bar{\psi}\, \rho\, dx$$

indicando con $\bar{\psi}$ il coniugato di ψ. Ciò suggerisce naturalmente l'idea di definire come energia potenziale del sistema in un dato istante la quantità

$$E_{pot} = \int V\, \psi\bar{\psi}\, \rho\, dx\,, \tag{4}$$

il che appare tanto più opportuno se si pensa che, secondo Schrödinger, nel caso di un solo elettrone $\psi\bar{\psi}$ è proporzionale alla densità elettrica in ciascun punto, per modo che il prodotto $V\psi\bar{\psi}$ rappresenta l'espressione classica per la densità di energia potenziale.

Assumeremo la (4) come definizione dell'energia potenziale anche nel caso che lo scalare di campo abbia la forma generale (3).

Cerchiamo ora di definire nel modo più opportuno il concetto di energia cinetica. Lasciamoci guidare anche qui dalla considerazione di un caso particolare, e precisamente consideriamo la soluzione ψ corrispondente al caso classico del *punto* materiale non soggetto a forze ($V = 0$). Tale soluzione sarà del tipo (3), in cui saranno rappresentati soltanto i termini per i quali E_k è assai prossimo al valore $\frac{1}{2}mv^2$ dell'energia cinetica classica, con tali relazioni di fase da distruggersi per interferenza dovunque, eccetto in una regione ristretta, che corrisponde all'estensione della particella materiale. Adunque ψ soddisferà assai prossimamente all'equazione (2) in cui si faccia $V = 0$, $E = E_{\text{kin}} = \frac{1}{2}mv^2$ cioè a

$$\Delta\psi + \frac{8\pi^2}{h^2} E_{\text{kin}}\,\psi = 0 .$$

Moltiplicando per $\bar{\psi}\,\rho\,dx$ e integrando a tutto lo spazio si ottiene, con facili integrazioni per parti sul primo termine,

$$-\int \operatorname{grad}\psi\cdot\operatorname{grad}\bar{\psi}\rho\,dx + \frac{8\pi^2}{h^2}E_{\text{kin}}\int\psi\bar{\psi}\,\rho\,dx = 0$$

ed, essendosi supposta ψ normalizzata in modo che l'ultimo integrale sia uguale ad 1, si ottiene

$$E_{\text{kin}} = \int \frac{h^2}{8\pi^2}\operatorname{grad}\psi\cdot\operatorname{grad}\bar{\psi}\rho\,dx . \tag{5}$$

Questa espressione, che nel caso particolare ora trattato corrisponde a quella classica, verrà da noi assunta come espressione generale della forza viva.

A giustificazione delle definizioni introdotte mostreremo ora che la somma dell'energia cinetica e dell'energia potenziale così definite, per un sistema generico, si mantiene costante (conservazione dell'energia). Calcoliamo perciò la quantità $E_{\text{pot}} + E_{\text{kin}}$ in base alle (4) e (5), attribuendo a ψ l'espressione generale (3). Si trova subito

$$E_{\text{pot}} + E_{\text{kin}} = \sum_{1}^{\infty}{}_{rs}\,\lambda_r\lambda_s\, e^{\frac{2\pi i}{h}(E_r - E_s)t}\int\left\{ V u_r u_s + \frac{h^2}{8\pi^2}\operatorname{grad} u_r\cdot\operatorname{grad} u_s\right\}\rho\,dx .$$

L'integrale, trasformato per mezzo del teorema di Gauss, diventa

$$\int u_r\left\{ V u_s - \frac{h^2}{8\pi^2}\Delta u_s\right\}\rho\,dx$$

e tenendo presente che u_s soddisfa la (2) ove si faccia $E = E_s$, si trova

$$E_{\text{pot}} + E_{\text{kin}} = \sum_{1}^{\infty}{}_{rs}\,\lambda_r\lambda_s\, e^{\frac{2\pi i}{h}(E_r - E_s)t}\, E_s\int u_r u_s\,\rho\,dx .$$

Essendo l'ultimo integrale uguale a zero o a 1 secondo che $r \neq s$ o $r = s$, troviamo in definitiva

$$E_{\text{pot}} + E_{\text{kin}} = \sum_{1}^{\infty}{}_{s}\,\lambda_s^2\, E_s = \text{cost} .$$

Resta così dimostrato il principio della conservazione dell'energia, e si trova anche l'espressione dell'energia totale corrispondente allo scalare di campo (3), in funzione dei coefficienti λ.

N°. 38 *b*.

Come è detto nella *Nota biografica*, Fermi oltre che la Laurea in Fisica, conseguì lo stesso anno, 1922, il Diploma di abilitazione della Scuola Normale Superiore di Pisa, con una Tesi che trattava probabilisticamente della vita delle comete ed era divisa in due parti: nella prima, lemmatica, veniva istituito un teorema di calcolo delle probabilità, nella seconda si applicava il teorema al caso delle comete.

All'esame di Diploma, nella discussione della Tesi, alcune osservazioni furono mosse da parte di matematici a Fermi circa la risoluzione di una certa equazione. Può darsi che proprio queste critiche siano state la causa del perché, contrariamente alla consuetudine di pubblicare negli « Annali della Scuola Normale Superiore » di Pisa le Tesi di Abilitazione, quella di Fermi, pur meritandogli i pieni voti assoluti e la lode, non venne pubblicata.

La parte lemmatica venne da Fermi pubblicata poi (1926) con lievi varianti, nel « Nuovo Cimento », *3*, 313 (1926), mentre la seconda rimase inedita.

Dopo la morte di Fermi, il manoscritto della Tesi fu da me fatto ricercare per lunghi anni senza fortuna, nell'Archivio della Scuola Normale Superiore di Pisa, e finalmente, nel 1959, fu ritrovato dal Segretario della Scuola, dott. Mario Nencetti. La Tesi è ora pubblicata nella Collana di opuscoli inediti o rari della Domus Galilaeana di Pisa.

G. POLVANI.

38 *b*.

UN TEOREMA DI CALCOLO DELLE PROBABILITÀ ED ALCUNE SUE APPLICAZIONI

Tesi di Abilitazione della Scuola Normale Superiore di Pisa.
Presentata il 20 giugno 1922.

§ 1. Il teorema di cui vogliamo occuparci è relativo alle proprietà delle somme di molti addendi incoerenti di ciascuno dei quali sia nota la distribuzione statistica. Il teorema fondamentale sopra tali somme è dovuto a Laplace [1]. Lo enunciamo insieme all'accenno di una sua dimostrazione dalla quale dovremo prender le mosse per stabilire il nuovo teorema.

Sia n un numero grandissimo e $y_1 y_2 \cdots y_n$ rappresentino n quantità incognite, di ciascuna delle quali sia però conosciuta la distribuzione statistica; si sappia cioè che la probabilità che y_i abbia valore compreso tra y_i ed $y_i + dy_i$ è $\varphi_i(y_i)\,dy_i$, essendo φ_i una funzione conosciuta, che dovrà intanto aver certamente la proprietà:

$$\int_{-\infty}^{\infty} \varphi_i(y)\,dy = 1 . \tag{1}$$

che esprime che y_i avrà certamente un valore tra $-\infty$ e $+\infty$.

(1) Théorie analytique des probabilités, Oeuvres, VII, pag. 309.

Supporremo in più che la distribuzione statistica di y_i non sia alterata dai valori che possano eventualmente assumere le altre y, che cioè le y_i siano tra di loro completamente incoerenti. Ammettiamo poi che la media di y_i sia zero, che cioè:

$$\bar{y}_i = \int_{-\infty}^{\infty} y\, \varphi_i(y)\, dy = 0. \tag{2}$$

Poniamo infine la media dei quadrati di y_i

$$\overline{y_i^2} = \int_{-\infty}^{\infty} y^2\, \varphi_i(y)\, dy = k_i^2 \tag{3}$$

e supponiamo che, per i qualunque, k_i^2 sia trascurabile rispetto a $\sum_1^n k_i^2$.

Con queste ipotesi è valido il teorema di Laplace, che dice che:

La probabilità che valgano insieme le diseguaglianze:

$$x \leq \sum_1^n y_i \leq x + dx \tag{4}$$

è espressa da:

$$F(x)\, dx = \frac{1}{\sqrt{2\pi \sum_1^n k_i^2}}\, e^{-\frac{x^2}{2\sum_1^n k_i^2}}\, dx. \tag{5}$$

Per dimostrarlo indichiamo con r un numero $\leq n$ e sia $F(r, x)\, dx$ la probabilità che valgano le diseguaglianze:

$$x \leq \sum_1^r y_r \leq x + dx. \tag{6}$$

Sia ora p un valore qualunque. Cerchiamo la probabilità che valgano insieme le diseguaglianze:

$$\sum_1^{r-1} y_i < p < \sum_1^r y_i \tag{7}$$

che cioè per effetto dell'aggiunta di y_r a $\sum_1^{r-1} y_i$, questa somma oltrepassi il valore p. Tale probabilità è evidentemente:

$$\int_0^{\infty} d\xi\, F(r-1, p-\xi) \int_{\xi}^{\infty} \varphi_r(y)\, dy.$$

Analogamente la probabilità che valgano insieme le diseguaglianze:

$$\sum_1^{r-1} y_i > p > \sum_1^r y_i \tag{8}$$

è:

$$\int_0^\infty d\xi\, F(r-1, p+\xi) \int_\xi^\infty \varphi_r(y)\, dy.$$

La differenza di queste due probabilità è evidentemente eguale alla differenza tra la probabilità che sia $\sum_1^r y_i > p$, e quella che sia $\sum_1^{r-1} y_i > p$ cioè a:

$$\int_p^\infty F(r, x)\, dx - \int_p^\infty F(r-1, x)\, dx$$

abbiamo dunque:

$$\int_p^\infty F(r, x)\, dx - \int_p^\infty F(r-1, x)\, dx = \int_0^\infty d\xi\, F(r-1, p-\xi) \int_\xi^\infty \varphi_r(y)\, dy$$

$$- \int_0^\infty d\xi\, F(r-1, p+\xi) \int_{-\infty}^{-\xi} \varphi_r(y)\, dy.$$

Nel secondo membro si possono invertire le integrazioni con le formule:

$$\int_0^\infty d\xi \int_\xi^\infty dy = \int_0^\infty dy \int_0^y d\xi \quad ; \quad \int_0^\infty d\xi \int_{-\infty}^{-\xi} dy = \int_{-\infty}^0 dy \int_0^{-y} d\xi$$

con che esso diventa, cambiando ancora nel secondo termine ξ in $-\xi$

$$\int_{-\infty}^\infty \varphi_r(y)\, dy \int_0^y F(r-1, p-\xi)\, d\xi.$$

Poniamo, in via d'approssimazione:

$$F(r-1, p-\xi) = F(r-1, p) - \xi \frac{\partial F(r-1, p)}{\partial p}.$$

L'espressione precedente diventa così:

$$F(r-1, p) \int_{-\infty}^\infty \varphi_r(y)\, dy \int_0^y d\xi - \frac{\partial F(r-1, p)}{\partial p} \int_{-\infty}^\infty \varphi_r(y)\, dy \int_0^y \xi\, d\xi$$

$$= F(r-1, p) \int_{-\infty}^\infty y\, \varphi_r(y)\, dy - \frac{1}{2} \frac{\partial F(r-1, p)}{\partial p} \int_{-\infty}^\infty y^2 \varphi_r(y)\, dy$$

cioè, ricordando (2) e (3):

$$-\frac{k_r^2}{2} \frac{\partial F(r-1, p)}{\partial p}.$$

Otteniamo così l'eguaglianza:

$$\int_p^\infty F(r, x)\, dx - \int_p^\infty F(r-1, x)\, dx = -\frac{k_r^2}{2} \frac{\partial F(r-1, p)}{\partial p}. \tag{9}$$

Derivandola rispetto a p troviamo:

$$(10) \qquad -F(r,p)+F(r-1,p)=-\frac{k_r^2}{2}\frac{\partial^2 F(r-1,p)}{\partial p^2}.$$

Cambiamo in questa $r-1$ in r, p in x, e poniamo, nella nostra approssimazione:

$$F(r+1,x)-F(r,x)=\frac{\partial}{\partial r}F(r,x).$$

La (10) ci dà allora per $F(r,x)$ l'equazione differenziale:

$$(11) \qquad \frac{\partial}{\partial r}F(r,x)=\frac{k_{r+1}^2}{2}\frac{\partial^2}{\partial x^2}F(r,x).$$

Cambiando r nell'altra variabile:

$$(12) \qquad t=\int_0^{r+1}k_i^2\,di$$

la (11) diventa:

$$(13) \qquad \frac{\partial F}{\partial t}=\frac{1}{2}\frac{\partial^2 F}{\partial x^2}.$$

Si ha poi evidentemente la condizione che, per t qualunque sia:

$$(14) \qquad \int_{-\infty}^{\infty}F\,dx=1$$

e che, per $t=0$ F abbia valore diverso da zero solo quando $|x|$ è infinitesimo. È noto che queste condizioni sono più che sufficienti a determinare F. Esse si soddisfano ponendo:

$$F=\frac{1}{\sqrt{2\pi t}}e^{-\frac{x^2}{2t}}.$$

Sostituiamo a t il suo valore che, nel nostro grado di approssimazione è $\sum_1^r k_i^2$, e troviamo:

$$(15) \qquad F(r,x)=\frac{1}{\sqrt{2\pi\sum_1^r k_i^2}}e^{-\frac{x^2}{2\sum_1^r k_i^2}}.$$

Si ha poi evidentemente $F(x)=F(n,x)$, e quindi:

$$F(x)=\frac{1}{\sqrt{2\pi\sum_1^n k_i^2}}e^{-\frac{x^2}{2\sum_1^n k_i^2}} \qquad \text{C.V.D.}$$

§ 2. Conserviamo le notazioni ed ipotesi fatte al principio del paragrafo precedente e supponiamo in più che tutte le $\varphi_i(y)$ siano tra di loro eguali

(Le indicheremo sopprimendo l'indice). Indichiamo poi con a un valore positivo qualunque. Vale allora il:

TEOREMA. – *La probabilità che una almeno delle quantità:*

$$y_1 \;,\; y_1 + y_2 \;,\; y_1 + y_2 + y_3 , \cdots , \sum_1^n y_n$$

superi a è espressa da:

$$\frac{2}{\sqrt{\pi}} \int_{\frac{a}{\sqrt{2nk^2}}}^{\infty} e^{-x^2}\, dx$$

purché a sia sufficientemente grande rispetto a k.

In particolare se n tende all'infinito tale probabilità tende ad 1, cioè alla certezza.

Per dimostrarlo indichiamo con $F(r, x)\, dx$ $(x < a)$ la probabilità che siano verificate le diseguaglianze (6) e che inoltre tutte le r quantità:

$$y_1 , y_1 + y_2 , \cdots , \sum_1^r y_i \tag{16}$$

siano inferiori ad a. Le stesse considerazioni del paragrafo precedente ci mostrano intanto che $F(r, x)$ soddisferà ancora all'equazione differenziale (11) che, nel caso attuale può scriversi:

$$\frac{\partial F}{\partial r} = \frac{k^2}{2} \frac{\partial^2 F}{\partial x^2} . \tag{17}$$

Saranno cambiate invece le condizioni ai limiti.

Osserviamo infatti che

$$\int_{-\infty}^{a} F(r, x)\, dx$$

esprime la probabilità che nessuna delle quantità (16) superi a, e quindi:

$$-\int_{-\infty}^{a} F(r+1, x)\, dx + \int_{-\infty}^{a} F(r, x)\, dx$$

esprime la probabilità che, per effetto dell'aggiunta di y_{r+1} , $\sum_1^r y_i$ venga a superare a. Un calcolo analogo ad uno svolto nel paragrafo precedente ci mostra che questa probabilità è:

$$\int_0^{\infty} F(r, a - \xi)\, d\xi \int_{\xi}^{\infty} \varphi(y)\, dy$$

cioè con la nostra approssimazione trascurando ξ di fronte ad a :

$$F(r, a) \int_0^{\infty} d\xi \int_{\xi}^{\infty} \varphi(y)\, dy$$

ossia invertendo le quadrature:

$$F(r, a)\int_0^\infty \varphi(y)\,dy\int_0^y d\xi = F(r, a)\int_0^\infty y\varphi(y)\,dy.$$

Ponendo ora:

$$h = \int_0^\infty y\varphi(y)\,dy \tag{18}$$

troviamo:

$$\int_{-\infty}^{a}\{(F(r+1, x) - F(r, x)\}\,dx = -h\,F(r, a).$$

Con la nostra solita approssimazione possiamo però porre:

$$F(r+1, x) - F(r, x) = \frac{\partial F(r, x)}{\partial r}$$

e l'equazione precedente diventa:

$$\frac{\partial}{\partial r}\int_{-\infty}^{a} F(r, x)\,dx = -hF(r, a). \tag{19}$$

In definitiva la nostra funzione incognita F deve soddisfare all'equazione differenziale (17) nell'intervallo $-\infty, a$; all'equazione (19) nell'estremo a; deve poi annullarsi con le sue derivate, nell'estremo $-\infty$ e, per $r = 0$ aver valore diverso da zero soltanto per $|x|$ piccolissimo, per modo però che l'area compresa tra di essa e l'asse delle x sia $= 1$.

È facile dimostrare che, almeno quando, come nel nostro caso, h è positivo queste condizioni sono sufficienti ad individuare F.

Osserviamo perciò che, moltiplicando la (17) per dx, ed integrandola poi tra $-\infty$ ed a si trova:

$$\frac{k^2}{2}\left(\frac{\partial F}{\partial x}\right)_a = \frac{\partial}{\partial r}\int_{-\infty}^{a} F(r, x)\,dx$$

con ciò la (19) diventa:

$$\frac{k^2}{2h}\left(\frac{\partial F(r, x)}{\partial x}\right)_a + F(r, a) = 0. \tag{19'}$$

Per il nostro scopo basta dunque evidentemente provare che, se una funzione $\Phi(r, x)$ è $= 0$ per $r = 0$ e soddisfa alle equazioni:

$$\frac{\partial\Phi}{\partial r} = \frac{k^2}{2}\frac{\partial^2\Phi}{\partial x^2} \quad ; \quad \frac{k^2}{2h}\left(\frac{\partial\Phi}{\partial x}\right)_{x=a} + \Phi(r, a) = 0 \tag{20}$$

e, per $x = -\infty$ è sempre nulla essa è certo identicamente nulla. Infatti si ha:

$$\int_{-\infty}^{a}\left(\frac{\partial\Phi}{\partial x}\right)^2 dx = \int_{-\infty}^{a}\frac{\partial}{\partial x}\left(\Phi\,\frac{\partial\Phi}{\partial x}\right)dx - \int_{-\infty}^{a}\Phi\,\frac{\partial^2\Phi}{\partial x^2}\,dx$$

cioè, per le (20):

$$\int_{-\infty}^{a}\left(\frac{\partial\Phi}{\partial x}\right)^2 dx = \left(\Phi\,\frac{\partial\Phi}{\partial x}\right)_{-\infty}^{a} - \frac{2}{k^2}\int_{-\infty}^{a}\Phi\,\frac{\partial\Phi}{\partial r}\,dx$$

$$= \Phi(r\,,a)\left(\frac{\partial\Phi}{\partial x}\right)_{x=a} - \frac{1}{k^2}\frac{\partial}{\partial r}\int_{-\infty}^{a}\Phi^2\,dx = -\,\frac{2h}{k^2}\,\Phi^2(r\,,a) - \frac{1}{k^2}\frac{\partial}{\partial r}\int_{-\infty}^{a}\Phi^2\,dx$$

ossia:

$$\int_{-\infty}^{a}\left(\frac{\partial\Phi}{\partial x}\right)^2 dx + \frac{2h}{k^2}\,\Phi^2(r\,,a) + \frac{1}{k^2}\frac{\partial}{\partial r}\int_{-\infty}^{a}\Phi^2(r\,,x)\,dx = 0\,. \tag{21}$$

Supponiamo ora dunque che per qualche valore di r e di x, Φ potesse essere differente da zero; allora per qualche valore $\bar r$ di r $\int_{-\infty}^{a}\Phi^2\,dx$ sarebbe certamente positivo; siccome poi per $r=0$ è $\Phi=0$, e quindi $\int_{-\infty}^{a}\Phi^2(0\,,x)\,dx=0$, vi sarà certamente tra zero ed $\bar r$ qualche valore di r per il quale $\frac{d}{dr}\int_{-\infty}^{a}\Phi^2(r\,,x)\,dx$ è positivo. Ora in (21) i primi due termini non possono esser negativi, il primo è, almeno in qualche caso positivo, ciò che è assurdo. Sarà dunque certamente sempre $\Phi(r\,,x)=0$. C.V.D.

Premesso questo basterà che per F noi troviamo una soluzione qualsiasi che soddisfi alle condizioni imposte perché siamo sicuri che essa è la soluzione da noi cercata.

Proviamo se le nostre condizioni possano soddisfarsi ponendo:

$$F(r\,,x) = \frac{1}{k\sqrt{2\pi r}}\,e^{-\frac{x^2}{2rk^2}} - \frac{1}{k\sqrt{2\pi}}\int_0^r \frac{u(\rho)\,e^{-\frac{(a-x)^2}{2(r-\varrho)k^2}}}{\sqrt{r-\rho}}\,d\rho \tag{22}$$

essendo u (ρ) una funzione da determinarsi.

Con questa posizione resta senz'altro verificata l'equazione differenziale (17) e le condizioni limiti per $x=-\infty$ e per $r=0$. Resta dunque da determinare $u(\rho)$ in modo da soddisfare anche la (19).

Ora da (22) si ha:

$$F(r\,,a) = \frac{1}{k\sqrt{2\pi r}}\,e^{-\frac{a^2}{2rk^2}} - \frac{1}{k\sqrt{2\pi}}\int_0^r \frac{u(\rho)\,d\rho}{\sqrt{r-\rho}}$$

$$\int_{-\infty}^{a}F(r\,,x)\,dx = \frac{1}{k\sqrt{2\pi r}}\int_{-\infty}^{a}e^{-\frac{x^2}{2rk^2}}\,dx - \frac{1}{k\sqrt{2\pi}}\int_0^r\frac{u(\rho)\,d\rho}{\sqrt{r-\rho}}\int_{-\infty}^{a}e^{-\frac{(a-x)^2}{2(r-\varrho)k^2}}\,dx = \tag{23}$$

$$= \frac{1}{\sqrt{\pi}}\int_{-\infty}^{\frac{a}{k\sqrt{2r}}}e^{-x^2}\,dx - \frac{1}{2}\int_0^r u(\rho)\,d\rho$$

e quindi:

$$\frac{\partial}{\partial r}\int_{-\infty}^{a} F(r, x)\, dx = -\frac{a e^{-\frac{a^2}{2rk^2}}}{2k\sqrt{2\pi r^3}} - \frac{1}{2} u(r)$$

con ciò la (19) diventa:

$$\frac{e^{-\frac{a^2}{2rk^2}}}{k\sqrt{2\pi r}}\left(h - \frac{a}{2r}\right) = \frac{h}{k\sqrt{2\pi}}\int_0^r \frac{u(\rho)\, d\rho}{\sqrt{r-\rho}} + \frac{u(r)}{2} \tag{24}$$

cioè un'equazione integrale di seconda specie per la funzione incognita $u(\rho)$. Per quanti sforzi abbia fatto non mi è riuscito di risolverla altro che in via d'approssimazione. Ci occuperemo tra poco di questo. Vogliamo prima dimostrare senza approssimazioni che si ha:

$$\int_0^\infty u(r)\, dr = 1 .$$

Indicando perciò con θ una quantità positiva arbitraria, moltiplichiamo ambo i membri di (24) per $\sqrt{\theta}\, e^{-\theta r} dr$ ed integriamo poi tra $r = 0$ ed $r = \infty$. Si trova:

$$\frac{\sqrt{\theta}\, h}{k\sqrt{2\pi}}\int_0^\infty \frac{e^{-\theta r - \frac{a^2}{2rk^2}}}{\sqrt{r}}\, dr - \frac{a\sqrt{\theta}}{2k\sqrt{2\pi}}\int_0^\infty \frac{e^{-\theta r - \frac{a^2}{2rk^2}}}{r^{3/2}}\, dr$$

$$= \frac{h\sqrt{\theta}}{k\sqrt{2\pi}}\int_0^\infty e^{-\theta r} dr \int_0^r \frac{u(\rho)\, d\rho}{\sqrt{r-\rho}} + \frac{\sqrt{\theta}}{2}\int_0^\infty e^{-\theta r} u(r)\, dr$$

$$= \frac{h\sqrt{\theta}}{k\sqrt{2\pi}}\int_0^\infty u(\rho)\, d\rho \int_\rho^\infty \frac{e^{-\theta r} dr}{\sqrt{r-\rho}} + \frac{\sqrt{\theta}}{2}\int_0^\infty e^{-\theta r} u(r)\, dr$$

$$= \frac{h}{k\sqrt{2}}\int_0^\infty e^{-\theta \rho} u(\rho)\, d\rho + \frac{\sqrt{\theta}}{2}\int_0^\infty e^{-\theta r} u(r)\, dr .$$

Si ha inoltre:

$$\sqrt{\theta}\int_0^\infty \frac{e^{-\theta r - \frac{a^2}{2rk^2}}}{\sqrt{r}}\, dr = 2\int_0^\infty e^{-x^2 - \frac{a^2\theta}{2k^2x^2}}\, dx = \sqrt{\pi}\, e^{-\frac{a\sqrt{2\theta}}{k}} .$$

Passando al limite per $\theta = 0$ l'equazione precedente diventa dunque:

$$\frac{h}{k\sqrt{2}} = \frac{h}{k\sqrt{2}}\int_0^\infty u(\rho)\, d\rho .$$

Da cui:

$$\int_0^\infty u(\rho)\, d\rho = 1 \qquad \text{C.V.D.} \tag{25}$$

Già a questo punto possiamo ricavare un risultato interessante. Dalla (23) si ha infatti:

$$\lim_{r=\infty} \int_{-\infty}^{a} \mathrm{F}(r, x)\, dx = \lim_{r=\infty} \frac{1}{\sqrt{\pi}} \int_{-\infty}^{\frac{a}{k\sqrt{2r}}} e^{-x^2} dx - \frac{1}{2} \int_{0}^{\infty} u(r)\, dr = 0. \tag{26}$$

Se ricordiamo il significato di F (r, x) questo risultato può leggersi:

La probabilità che uno almeno dei valori (16) sia maggiore di a diventa certezza quando r tende ad infinito. E notiamo che questo risultato vale indipendentemente dalle approssimazioni che andiamo a fare per risolvere la (24).

Passiamo ora alla risoluzione approssimata della (24).

Osserviamo perciò che, come dimostra una verificazione immediata

$$w(r) = \frac{a e^{-\frac{a^2}{2rk^2}}}{k\sqrt{2\pi r^3}} \tag{27}$$

è soluzione dell'equazione integrale di seconda specie

$$\frac{e^{-\frac{a^2}{2rk^2}}}{k\sqrt{2\pi r}}\left(h + \frac{a}{2r}\right) = \frac{h}{k\sqrt{2\pi}} \int_{0}^{r} \frac{w(\rho)\, d\rho}{\sqrt{r-\rho}} + \frac{1}{2} w(r) \tag{28}$$

che in altro non differisce dalla (24) che per il segno entro la parentesi del primo membro. Ora per le ipotesi che abbiamo fatto tutte le volte che r è abbastanza grande perché $e^{-\frac{a^2}{2rk^2}}$ non sia piccolissimo $a/2r$ è trascurabile di fronte ad h, e quindi potremo assumere $w(r)$ come soluzione approssimata della (24), ponendo

$$u(r) = \frac{a e^{-\frac{a^2}{2rk^2}}}{k\sqrt{2\pi r^3}}. \tag{29}$$

Si verifica immediatamente dalla (29) che è $\int_{0}^{\infty} u(r)\, dr = 1$.

Dalla (23) si ricava ora:

$$\int_{-\infty}^{a} \mathrm{F}(r, x)\, dx = \frac{1}{\sqrt{\pi}} \int_{-\infty}^{\frac{a}{k\sqrt{2r}}} e^{-x^2} dx - \frac{1}{2} \int_{0}^{r} \frac{a e^{-\frac{a^2}{2\rho k^2}}}{k\sqrt{2\pi\rho^3}} d\rho$$

$$= \frac{1}{\sqrt{\pi}} \int_{-\infty}^{\frac{a}{k\sqrt{2r}}} e^{-x^2} dx - \frac{1}{\sqrt{\pi}} \int_{\frac{a}{k\sqrt{2r}}}^{\infty} e^{-x^2} dx = 1 - \frac{2}{\sqrt{\pi}} \int_{\frac{a}{k\sqrt{2r}}}^{\infty} e^{-x^2} dx.$$

E quindi:

$$1 - \int_{-\infty}^{a} F(r, x)\, dx = \frac{2}{\sqrt{\pi}} \int_{\frac{a}{k\sqrt{2r}}}^{\infty} e^{-x^2}\, dx. \tag{30}$$

Ricordando ora il significato di $F(r, x)$ si riconosce subito che

$$1 - \int_{-\infty}^{a} F(r, x)\, dx$$

rappresenta la probabilità che una almeno delle espressioni (16) sia maggiore di a. La formula (30) dimostra perciò completamente il teorema enunciato.

§ 3. Il teorema ora dimostrato è intanto suscettibile di una applicazione immediata ad un celebre problema di calcolo delle probabilità:

Pietro e Paolo fanno un gioco d'azzardo. In ogni partita ciascuno ha probabilità 1/2 di vincere; la posta è sempre di k lire. Ora Pietro è infinitamente ricco, Paolo invece possiede soltanto a lire. Se ad un certo momento Pietro riesce a vincere tutta la sostanza di Paolo, quest'ultimo è rovinato e deve smettere il gioco. Siamo dunque nel caso contemplato nel teorema precedente, e possiamo concludere che certamente, dopo un sufficiente numero di giocate Pietro manderà in rovina Paolo; inoltre, se a è assai più grande di k la probabilità che in n giocate si verifichi questo fatto è:

$$\frac{2}{\sqrt{\pi}} \int_{\frac{a}{k\sqrt{2n}}}^{\infty} e^{-x^2}\, dx.$$

§ 4. Vogliamo ora svolgere l'applicazione del teorema precedente ad una questione astronomica.

Consideriamo una cometa ellittica la cui orbita intersechi quella di Giove. L'orbita cometaria sarà naturalmente perturbata dall'azione di Giove, e ciò particolarmente quando Giove e la cometa si passano molto vicino. Ora può accadere che in queste continue trasformazioni l'orbita della cometa finisca per trasformarsi in parabolica od iperbolica; allora essa si allontanerà negli spazii sottraendosi per sempre all'attrazione di Giove e del Sole. Mi propongo di studiare quale sia la probabilità che ciò avvenga entro un certo tempo.

La teoria dell'influenza di Giove sopra le orbite cometarie non è ch'io sappia mai stata studiata da questo punto di vista; e si è soltanto considerata (2) per cercare in essa la spiegazione della cattura di comete con orbita parabolica, quando per caso esse passino vicino a Giove.

(2) TISSERAND, *Traité de mecanique céleste*, Tomo IV, pp. 198–216; CALLANDREAU, «Ann. de l'observatoire», t. 22; A. NEWTON, «Mem. of the Nat. Acad. of Sci.», t. 6.

Faremo le seguenti ipotesi semplificative, in sostanza quelle del problema ristretto dei tre corpi:

La cometa ha massa infinitesima, di modo che non perturba né Giove né il Sole.

La massa di Giove (m) è trascurabile di fronte a quella del Sole (M). Ci intendiamo con ciò autorizzati a supporre il Sole fisso ed a considerare l'orbita della cometa perturbata considerevolmente soltanto quando essa passa nell'immediata vicinanza di Giove.

L'orbita di Giove è circolare.

L'orbita della cometa è complanare a quella di Giove.

Indichiamo con u la velocità di Giove; con V la velocità della cometa quando attraversa l'orbita di Giove relativa ad una terna d'assi che si muovano lungo questa orbita con velocità u; con θ l'angolo che la direzione di V forma con l'orbita di Giove. Se v è la velocità assoluta della cometa, quando sta attraversando l'orbita di Giove si avrà:

$$v^2 = u^2 + V^2 + 2\, u V \cos \theta. \tag{31}$$

Supponiamo che una volta, mentre la cometa traversa l'orbita di Giove essa si trovi a passare assai vicino a questo pianeta. Essa avrà allora una forte perturbazione. Sia b la minima distanza a cui passerebbero i due astri se non si attraessero. Per le ipotesi che abbiamo fatte, affinché la perturbazione sia notevole b dovrà essere assai piccolo a confronto dei raggi di curvatura delle due orbite imperturbate, di modo che, durante questo «urto» la cometa descriverà sensibilmente nel suo moto relativo attorno a Giove, un'orbita iperbolica kepleriana.

§ 5. Consideriamo dunque questo moto relativo, riferendoci a coordinate polari (r, φ) aventi per polo Giove e l'asse polare parallelo alla direzione da cui viene la cometa.

Siccome il moto è kepleriano avremo intanto

$$\frac{1}{r} = A - B \cos (\varphi - \varphi_0) \tag{32}$$

essendo A, B, φ_0 delle costanti.

Inoltre per $\varphi = 0$ deve essere r infinito, cioè

$$A - B \cos \varphi_0 = 0\,. \tag{33}$$

Deve poi essere

$$b = \lim_{r=\infty} r \sin \varphi = \lim_{\varphi=0} \frac{\sin \varphi}{A - B \cos (\varphi - \varphi_0)} = - \frac{1}{B \sin \varphi_0}\,. \tag{34}$$

La costante delle aree è poi evidentemente Vb, e quindi per le note formule del moto kepleriano si ha

$$A = \frac{m}{V^2 b^2}\,. \tag{35}$$

Da (33) e (34) possiamo ora ricavare le altre due costanti. Si trova precisamente

$$\text{(36)} \qquad \operatorname{tg} \varphi_0 = -\frac{V^2 b}{m} \quad , \quad B = \frac{1}{b}\sqrt{1 + \frac{m^2}{b^2 V^4}} \, .$$

Sia ora ψ l'angolo tra la direzione in cui si avvicina la cometa e quella in cui essa si allontana. Si avrà evidentemente:

$$\psi = 2 \varphi_0 - \pi$$

e quindi

$$\text{(37)} \qquad \operatorname{tg} \frac{\psi}{2} = -\operatorname{cotg} \varphi_0 = \frac{m}{V^2 b} \, .$$

Possiamo concludere che la perturbazione consiste nel lasciare inalterato V e nell'alterare θ dell'angolo ψ dato dalla (37).

Ci converrà ora calcolare la media dei quadrati di ψ. Osserviamo perciò che si ha:

$$\psi = 2 \operatorname{arc\,tg} \frac{m}{V^2 b}$$

e quindi

$$\int_{-\infty}^{\infty} \psi^2 \, db = 4 \int_{-\infty}^{\infty} \left(\operatorname{arc\,tg} \frac{m}{V^2 b}\right)^2 db = \frac{4m}{V^2} \int_{-\infty}^{\infty} \left(\operatorname{arc\,tg} \frac{1}{x}\right)^2 dx = \frac{8m}{V^2} \int_{0}^{\infty} \left(\operatorname{arc\,tg} \frac{1}{x}\right)^2 dx$$

ponendo

$$h = \int_{0}^{\infty} \left(\operatorname{arc\,tg} \frac{1}{x}\right)^2 dx = \sim 2{,}5$$

si ha dunque

$$\text{(38)} \qquad \int_{-\infty}^{\infty} \psi^2 \, db = \frac{8\,mh}{V^2}$$

ora, essendo b assai piccolo la probabilità che il suo valore sia compreso tra b e $b + db$ è evidentemente

$$\frac{db}{2\pi R \sin\theta}$$

essendo R il raggio dell'orbita di Giove.

La media dei quadrati di ψ è perciò

$$\text{(39)} \qquad \overline{\psi^2} = \int_{-\infty}^{\infty} \psi^2 \frac{db}{2\pi R \sin\theta} = \frac{4\,mh}{\pi R V^2 \sin\theta} \, .$$

§ 6. Nel suo moto attorno al Sole la costante delle forze vive della nostra cometa è data da:

$$\frac{v^2}{2} - \frac{M}{R} = W \, .$$

Ora è ben noto che un'orbita kepleriana è ellittica, parabolica od iperbolica secondo che per essa la costante delle forze vive è negativa, nulla o positiva;

ora ricordando (31) si trova per la nostra cometa:

$$W = \frac{1}{2}\left(u^2 + V^2 + 2uV\cos\theta - 2\frac{M}{R}\right)$$

siccome però per Giove si ha la relazione:

$$\frac{u^2}{R} = \frac{M}{R^2}$$

possiamo scrivere

$$2W = V^2 + 2uV\cos\theta - \frac{M}{R}.$$

Siccome poi nelle successive perturbazioni V non viene alterato e cambia soltanto θ, perché ci sia la possibilità che la cometa possa diventare iperbolica è necessario che W, attualmente negativo, possa per opportuni valori di θ diventare positivo. Deve dunque intanto essere

$$V^2 + 2uV > \frac{M}{R}$$

osserviamo però che

$$u = \sqrt{\frac{M}{R}}$$

quindi la diseguaglianza precedente si può scrivere:

$$\left(V + \sqrt{\frac{M}{R}}\right)^2 > \frac{2M}{R}$$

dalla quale (*) e si riduce in ultima analisi a

$$V > (\sqrt{2} - 1)\sqrt{\frac{M}{R}} = (\sqrt{2} - 1)\,u. \tag{40}$$

Noi supporremo dunque senz'altro verificata questa diseguaglianza. Inoltre per qualche valore di θ, W deve esser certamente negativo, altrimenti l'orbita cometaria non potrebbe essere ellittica; sarà dunque:

$$V^2 - 2uV < \frac{M}{R}$$

dalla quale come sopra

$$V < (\sqrt{2} + 1)\sqrt{\frac{M}{R}} = (\sqrt{2} + 1)\,u. \tag{41}$$

Supponiamo dunque che V soddisfi (40) e (41) ed indichiamo con θ_0 quel particolare valore di θ per il quale l'orbita della cometa è iperbolica, ossia si ha $W = 0$, cioè

$$V^2 + 2uV\cos\theta_0 = \frac{M}{R}$$

e quindi

$$\cos\theta_0 = \frac{\frac{M}{R} - V^2}{2uV} = \frac{u^2 - V^2}{2uV}. \tag{42}$$

(*) A questo punto, nel manoscritto di Fermi vi è un rigo in bianco, che avrebbe dovuto contenere ovviamente lo sviluppo del quadrato dell'ultima formula (N. d. R.).

Quando θ è più grande di θ_0 si ha $W < 0$, e quindi la cometa descrive orbita ellittica; quando invece θ è minore di θ_0 l'orbita è iperbolica.

Ora noi supporremo che l'orbita inizialmente sia ellittica e molto allungata, di modo che θ sia assai prossimo a θ_0, e precisamente un po' maggiore. Chiamiamo θ^* questo valore iniziale.

Ogni volta che la cometa oltrepassa l'orbita di Giove θ si altera di ψ; la media dei quadrati di ψ dipende invero, come mostra la (39) da θ, siccome però abbiamo supposto che θ si mantenga sempre molto prossimo a θ_0 potremo porre:

$$\overline{\psi^2} = \frac{4\,mh}{\pi\,\mathrm{RV}^2 \sin\theta_0} \tag{43}$$

se dopo un po' di volte θ finisse per diventare $< \theta_0$ la cometa diventerebbe iperbolica e si allontanerebbe per sempre. Siamo dunque nelle condizioni di poter applicare il teorema del § 2. Dovremo perciò porre $a = \theta^* - \theta_0$; $k^2 = \dfrac{4\,mh}{\pi\,\mathrm{RV}^2 \sin\theta_0}$. Ed il teorema allora dimostrato ci dice che:

La probabilità che la cometa sia trasformata in iperbolica dopo aver attraversato n volte l'orbita di Giove è:

$$\frac{2}{\sqrt{\pi}} \int_{\frac{\theta^* - \theta_0}{\sqrt{\frac{8\,mhn}{\pi\,\mathrm{RV}^2 \sin\theta_0}}}}^{\infty} e^{-x^2}\,dx \tag{44}$$

e tende dunque ad 1 quando n tende ad infinito.

A tutto rigore si potrebbe obbiettare che i calcoli precedenti cadrebbero in difetto se il valore di V fosse tale che, quando l'orbita è parabolica, la cometa impiegasse lo stesso tempo di Giove ad andare da A in B, essendo A il punto in cui la cometa entra nell'orbita di Giove, e B il punto in cui essa ne esce.

Nella figura 1, S è il Sole, A J B l'orbita di Giove, A K B l'orbita della cometa. È facile però riconoscere che questo caso non può certo verificarsi se la cometa descrive la sua traiettoria di moto diretto. Infatti se v è la velocità assoluta in A della cometa sopra la sua orbita parabolica si ha

$$v^2 = u^2 + \mathrm{V}^2 + 2\,u\,\mathrm{V}\cos\theta_0$$

e quindi da (42)

$$v^2 = 2\,u^2$$

cioè:

$$v > u\,. \tag{45}$$

Ora la velocità della cometa non è costante, però in tutto il tratto AKB essa è sempre maggiore che negli estremi A e B, perciò la diseguaglianza (45) vale a più forte ragione per tutto il tratto A K B. D'altra parte, se il moto è diretto si ha l'arco A K B è più corto dell'arco A J B, e siccome esso è descritto anche con maggior velocità è certo che la cometa giungerà in B prima di Giove.

Se invece il moto della cometa fosse retrogrado, ed essa descrivesse per esempio l'orbita A K′ B′ nel senso indicato dalla freccia si avrebbe

$$\text{arco A K}' \text{B}' > \text{arco A J B}'$$

e quindi, benché valga ancora la (45) è evidente che, per un particolare valore del parametro dell'orbita della cometa può accadere che i due astri impieghino il medesimo tempo ad arrivare da A a B′; ciò naturalmente potrà avvenire soltanto per un valore particolare di V.

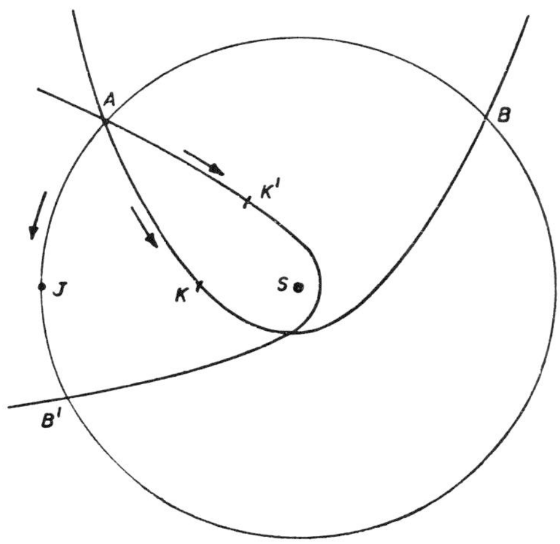

Fig. 1.

Ora se ciò avvenisse potrebbe accadere che la cometa, dapprima ellittica incontrasse Giove al suo passaggio da A e venisse trasformata in parabolica; allora però essa incontrerebbe di nuovo Giove al suo passaggio da B e potrebbe eventualmente avere una nuova perturbazione che la trasformasse ancora in ellittica.

Per questa considerazione noi intendiamo escluso dai nostri calcoli questo valore particolare di V.

§ 7. Vogliamo da ultimo considerare la possibilità che, prima di essere trasformata in iperbolica la cometa possa urtare Giove e quindi essere distrutta. Quale è la probabilità che ciò si verifichi?

Cerchiamo per questo dapprima la probabilità che la cometa, traversando una volta l'orbita di Giove si scontri con questo. Indichiamo per questo con ρ la somma dei raggi di Giove e della cometa. Perché avvenga lo scontro è evidentemente necessario che la distanza perielica di Giove dalla cometa, calcolata secondo le formule del moto kepleriano sia inferiore a ρ.

Sia δ questa distanza perielica; dalle formule del § 5 si rileva

$$\frac{1}{\delta} = A + B$$

e quindi da (35) e (36)

$$\frac{1}{\delta} = \frac{m}{V^2 b^2} + \frac{1}{b}\sqrt{1 + \frac{m^2}{V^4 b^2}}.$$

Se vogliamo che avvenga lo scontro dovrà essere $\delta < \rho$ e quindi

$$\frac{m}{V^2 b^2} + \frac{1}{b}\sqrt{1 + \frac{m^2}{V^4 b^2}} > \frac{1}{\rho}$$

moltiplicando questa diseguaglianza per la quantità, certamente positiva

$$\rho\left(\frac{1}{b}\sqrt{1 + \frac{m^2}{V^4 b^2}} - \frac{m}{V^2 b^2}\right)$$

troviamo

$$\frac{\rho}{b^2} > \frac{1}{b}\sqrt{1 + \frac{m^2}{V^2 b^2}} - \frac{m}{V^2 b^2}$$

e sommando le due ultime diseguaglianze:

$$\left(\frac{2m}{V^2} + \rho\right)\frac{1}{b^2} > \frac{1}{\rho}$$

da cui infine

(45) $$|b| < \sqrt{\rho^2 + \frac{2m\rho}{V^2}}$$

ricordiamo ora che la probabilità che il valore di b sia compreso tra b e $b + db$ è $\frac{db}{2\pi R \sin\theta_0}$, e quindi la probabilità p che avvenga lo scontro in un solo attraversamento dell'orbita di Giove è espressa da:

(46) $$p = \frac{1}{\pi R \sin\theta_0}\sqrt{\rho^2 + \frac{2m\rho}{V^2}}\,.$$

Noi supporremo p piccolissimo, ciò che equivale evidentemente a supporre trascurabile il raggio di Giove di fronte a quello della sua orbita.

Cerchiamo ora la probabilità che lo scontro avvenga l'n^{esima} volta che la cometa attraversa l'orbita di Giove. Perciò è evidentemente necessario che lo scontro non sia avvenuto prima; e la probabilità di questo è evidentemente $(1 - p)^{n-1}$, ossia nella nostra approssimazione

$$e^{-pn}.$$

Che la cometa non sia ancora stata trasformata in iperbolica; e, avendo supposto p estremamente piccolo, ricordando (44) e ponendo per brevità:

$$\frac{\theta^* - \theta_0}{\sqrt{\dfrac{8mh}{\pi R V^2 \sin\theta_0}}} = H$$

possiamo ritenere la probabilità che questo si verifichi, data da:

$$1 - \frac{2}{\sqrt{\pi}}\int_{\frac{H}{\sqrt{n}}}^{\infty} e^{-x^2}\,dx = \frac{2}{\sqrt{\pi}}\int_0^{\frac{H}{\sqrt{n}}} e^{-x^2}\,dx\,.$$

Ed infine che lo scontro avvenga effettivamente, per il che si ha la probabilita p.

The Physics Institute of the University of Rome, located in Via Panisperna 89 a.

In definitiva la probabilità che lo scontro avvenga la n^{esima} volta è

$$\frac{2e^{-pn}p}{\sqrt{\pi}}\int_0^{\frac{H}{\sqrt{n}}} e^{-x^2}\,dx$$

e quindi la probabilità che lo scontro avvenga non importa quale volta sarà la somma dell'espressione precedente da $n = 1$ fino ad $n = \infty$, ossia sostituendo la somma con un integrale

$$\frac{2p}{\sqrt{\pi}}\int_0^{\infty} e^{-pn}\,dn\int_0^{\frac{H}{\sqrt{n}}} e^{-x^2}\,dx\,.$$

Conviene in questa espressione invertire le integrazioni con la formula:

$$\int_0^{\infty} dn\int_0^{\frac{H}{\sqrt{n}}} dx = \int_0^{\infty} dx\int_0^{\frac{H}{x^2}} dn$$

si trova così per la probabilità cercata, l'espressione:

$$\frac{2p}{\sqrt{\pi}}\int_0^{\infty} e^{-x^2}\,dx\int_0^{\frac{H}{x^2}} e^{-pn}\,dn = \frac{2}{\sqrt{\pi}}\int_0^{\infty} e^{-x^2}\left(1 - e^{-\frac{pH}{x^2}}\right)dx$$

$$= 1 - \frac{2}{\sqrt{\pi}}\int_0^{\infty} e^{-x^2 - \frac{pH}{x^2}}\,dx = 1 - e^{-2\sqrt{pH}}\,.$$

La probabilità che lo scontro non avvenga mai è dunque:

$$e^{-2\sqrt{pH}}.$$

N° 39.

One of Fermi's early applications of wave mechanics. He shows that Schroedinger's expression for the average current density leads to the correct value for the orbital magnetic moment of an electron in a central potential field.

This appears to be the first work done by Fermi after being appointed Professor of Theoretical Physics at the University of Rome.

F. RASETTI

39.

QUANTUM MECHANICS AND THE MAGNETIC MOMENT OF ATOMS

« Nature » (London), *118*, 876 (1926).

In a recent paper E. Schrödinger (« Ann. d. Phys. », *81*, 109, 1926), put forward a hypothesis connecting the field scalar ψ of his undulatory mechanics with the electric density of current. His formulæ in the case of a system with only one electron (expressing j in electromagnetic units) reduce to:

$$\rho = e\psi\bar{\psi}\,, \tag{1}$$

$$j = \frac{eh}{4\pi imc}(\psi \operatorname{grad} \bar{\psi} - \bar{\psi} \operatorname{grad} \psi), \tag{2}$$

where p and j are respectively the electric density and density of current. The field scalar must be normalised in such a way that

$$\int \psi\bar{\psi}\, dr = 1, \tag{3}$$

dr being the element of volume.

Evidence in favour of the expression for ρ is given by the fact that the calculation of the intensity of emission of the atom, based on it, is in accordance both with experiment and with the results of Heisenberg's quantum mechanics. Expression (2) for j is assumed by Schrödinger as the simplest way of satisfying the equation of continuity of electricity.

It is desirable, therefore, to find some further evidence for this expression. I propose to show that it leads to the right expression for the magnetic moment of a hydrogen-like atom. We will consider only the part of the magnetic moment which, in the old quantum theory, was supposed to be due to the orbital motion of the electron, not taking into account the magnetic moment of the spinning electron, which can be considered separately.

The wave equation for a hydrogen-like atom in a magnetic field (without considering the spinning electron) was integrated by Fock (« Zeit. f. Phys. », *28*, 242, 1926). His expression for ψ can be written in the form

$$\psi = f(\theta\,, r)\, e^{in_1\varphi}\, e^{\frac{2\pi i E t}{h}} \tag{4}$$

where r, θ, φ are polar co-ordinates; $f(\theta, r)$ is a *real* function of r and θ; n_1 is an integer corresponding to the magnetic quantum number of the old theory.

Is is easily seen from (2) and (4) that j is a vector perpendicular at each point to the plane passing through the point and the polar axis, the magnitude of which is

$$(5) \qquad j = \frac{he}{4\pi imc} \frac{1}{R} \left(\psi \frac{\partial \bar{\psi}}{\partial \varphi} - \bar{\psi} \frac{\partial \psi}{\partial \varphi} \right) = - \frac{he}{2\pi mc R} n_1 f^2,$$

R being the distance of the point from the axis. The lines of current are circles situated in planes perpendicular to the axis and with their centres on the axis.

Let us now consider the current passing through an element $d\sigma$ of a meridian plane; its intensity is given by $jd\sigma$ and its circuit is a circle with radius R. The magnetic moment of this current is therefore $\pi R^2 \, jd\sigma$ (in the direction of the field).

The magnetic moment of the atom becomes, putting for j the expression (5),

$$(6) \qquad \mu = \pi \int R^2 jd\sigma = - \frac{hen_1}{2mc} \int Rf^2 \, d\sigma .$$

The last integral can be evaluated by substituting (4) in (3), and observing that $d\tau = 2\pi \, R d\sigma$; we find then $2\pi \int Rf^2 d\sigma = 1$. Putting this value into (6) we find at last

$$(7) \qquad \mu = - \frac{he}{4\pi mc} n_1 .$$

That is, the component of the magnetic moment in the direction of the field is the product of a Bohr magneton $\left(- \frac{he}{4\pi mc}\right)$ and the magnetic quantum number n_1, as it was expected.

It is noteworthy that the magnetic moment (7) arises in a certain way through the action of the field. In the absence of a magnetic field, the two states with n_1 equal in absolute value but of opposite signs, have the same energy; every linear combination of these two states is therefore a quantum state. Now it is easily seen from (4) that we can combine the two states in such a way that ψ becomes the product of $e^{2\pi i \frac{E}{h} t}$ and a real function of r, θ, φ containing $\cos n_1 \varphi$ or $\sin n_1 \varphi$, as a factor). In this case the density of current j would vanish everywhere, and the magnetic moment would reduce to zero.

Physics Institute of the University,
Rome, November 14.

N° 40 *a* and 40 *b*.

This paper describes a verification of the Boltzmann distribution of thallium atoms between the ground state and the first excited state.

In view of Rasetti's previous experience with the technique of anomalous dispersion in gases, Fermi suggested that the method be used to verify the Boltzmann distribution for atomic levels, since the previous observations based on absorption did not readily lend themselves to quantitative measurements.

Although the results did not attain great accuracy (the Boltzmann exponent agreeing with the theoretical value only to within 20 percent), the experiment can be regarded as the first quantitative verification of the Boltzmann formula for the thermal equilibrium of atoms among different energy levels.

This is the first experimental work performed by Fermi in the Institute of Rome.

F. RASETTI.

40 *b*.

UNA MISURA DEL RAPPORTO *h/k* PER MEZZO DELLA DISPERSIONE ANOMALA DEL TALLIO

E. FERMI e F. RASETTI

« Rend. Lincei », 5, 566–570 (1927) (*).

È noto che nella teoria dei quanti la legge statistica di Boltzmann può mettersi sotto la forma

$$n_i/n_l = p_i/p_l \cdot e^{-\frac{w_i - w_l}{kT}} = p_i/p_l \cdot e^{-\frac{h}{k}\frac{\Delta\nu}{T}} \tag{1}$$

dove n_i ed n_l indicano le densità degli atomi in due stati i ed l, aventi pesi statistici p_i e p_l ed energie w_i e w_l, in modo che $\Delta\nu = (w_i - w_l)/h$ rappresenta la differenza dei termini spettroscopici corrispondenti. In questo lavoro ci siamo proposti una verifica della legge espressa dalla (1); ciò ci ha naturalmente portati ad una misura del rapporto delle due costanti universali h e k.

Come giustificazione sperimentale indiretta della (1) si possono considerare l'andamento dei calori specifici dei corpi solidi e dell'idrogeno a basse temperature; la distribuzione di intensità negli spettri di bande; il calcolo spettroscopico del paramagnetismo di alcuni ioni.

Si hanno inoltre alcune osservazioni spettroscopiche qualitative sull'equilibrio termico tra i diversi stati quantici di un atomo, basate sul fatto che vapori ad alta temperatura assorbono delle righe che a bassa temperatura appaiono soltanto in emissione. Ciò si osserva assai frequentemente nell'arco.

(*) Presentata dal Socio O. M. Corbino nella seduta del 3 aprile 1927.

Osservazioni più precise furono fatte da W. Grotrian [1] sul vapore di tallio contenuto in tubi scaldati ad alta temperatura. Il termine fondamentale del tallio è doppio; i due livelli che lo costituiscono 6^2P_1 e 6^2P_2 hanno tra di loro una differenza di 7792 numeri d'onda, per modo che nel vapore a temperatura relativamente bassa la quasi totalità degli atomi si troverà nello stato 6^2P_1, e si osserveranno quindi in assorbimento soltanto le righe che partono da questo stato; in particolare la riga $6^2P_1 — 7^2S_1$, λ 3776. Ad alta temperatura invece una certa percentuale degli atomi verrà a trovarsi nello stato 6^2P_2 e quindi appariranno in assorbimento le righe che partono da questo stato; in particolare la riga verde $6^2P_2 — 7^2S_1$, λ 5350. Grotrian osservò effettivamente queste circostanze.

Essendo nostro scopo una verifica quantitativa della (1) abbiamo ritenuto più opportuno, invece dell'assorbimento, misurare la dispersione anomala in vicinanza delle predette righe del tallio, essendo il rapporto dei coefficienti di dispersione anomala delle due righe proporzionale al rapporto delle densità n_1 ed n_2 degli atomi negli stati 6^2 P_1 e 6^2 P_2. Abbiamo scelto come materiale per le nostre esperienze il tallio, perché esso riunisce in sè le seguenti circostanze.

1° Possedere, accanto al livello normale, un altro livello con differenza di energia dal precedente tale che, in un intervallo di temperature comodamente realizzabile, varî considerevolmente il rapporto del numero degli atomi nei due stati.

2° Avere per il detto intervallo di temperature una tensione di vapore sufficientemente grande.

3° Avere righe che partono dai due stati in una regione spettrale comodamente accessibile.

Per la misura della dispersione anomala ci siamo serviti dell'ordinario metodo interferenziale di Puccianti. I due raggi di un interferometro di Jamin venivano fatti passare entro due tubi di quarzo chiusi alle estremità da due lastre di vetro. Un avvolgimento di nichel–cromo attorno ad uno dei tubi permetteva di scaldarlo fino a 1250°. In questo tubo si metteva un pezzetto di tallio; lasciando nel tubo dell'aria alla pressione di qualche cm. di Hg, la velocità di distillazione del tallio verso le parti fredde del tubo veniva ridotta talmente che il pezzetto di tallio impiegava circa un'ora a distillare completamente.

Per valutare la temperatura veniva misurata l'energia elettrica dissipata nella spirale di riscaldamento; la temperatura come funzione di questa energia, nello stato stazionario, era stata determinata precedentemente per mezzo di una coppia termoelettrica platino–platino rodio, tarata dalla Physikalisch Technische Reichsanstalt di Berlino.

Le frangie di interferenza venivano proiettate sulla fenditura di uno spettrografo a prismi di Hilger di circa tre metri di distanza focale, avente per la riga verde una dispersione di 5,73 Å per mm. Una parte della luce, prima della fenditura, veniva deviata per riflessione sopra una lastra di vetro,

(1) W. GROTRIAN, «ZS. f. Phys.», *12*, 218 (1922).

in modo che un'altra immagine delle frangie si formava sulla fenditura di uno spettrografo a quarzo di Hilger, che dava per la riga λ 3776 una dispersione di 10,5 Å per mm. Davanti a ciascuna fenditura si trovava una lente cilindrica per concentrare maggiormente la luce.

Per fotografare la linea verde abbiamo usate delle lastre Hauff Flavin. Servendoci come sorgente della luce di un arco elettrico, potevamo fare delle fotografie contemporanee delle frangie in vicinanza delle due righe con pose di due o tre minuti primi.

Per mezzo di questo dispositivo abbiamo fatte sei serie di coppie di fotografie della dispersione anomala del vapore di tallio e abbiamo per ciascuna fotografia misurata la dispersione anomala. Per fare le misure nel modo più semplice, abbiamo disposto l'interferometro in modo che le frangie a grande distanza dalle righe di assorbimento fossero orizzontali. Per il caso della forte dispersione anomala della riga ultravioletta si misuravano le distanze dalla riga delle intersezioni di una decina di frangie successive con una retta orizzontale. Si riconosce facilmente che le inverse di queste successive distanze formano una progressione aritmetica la cui ragione è l'inversa del coefficiente di dispersione anomala d_1 il quale, secondo la teoria classica della dispersione, è legato al numero N_1 degli oscillatori elettronici contenuti nell'unità di volume e alla lunghezza l dello strato di vapore attraversato, dalla relazione

$$N_1 l = \frac{4\pi m c^2}{e^2} \frac{d_1}{\lambda_1} = 4{,}54 \cdot 10^{21} \frac{d_1 \,\text{Ang}}{\lambda_1^2 \,\text{Ang}} . \tag{2}$$

Il corrispondente coefficiente d_2 della dispersione anomala in vicinanza della riga verde si misurava semplicemente osservando la distanza dalla riga di assorbimento alla quale si ha il passaggio di una frangia.

La tabella seguente dà i risultati delle misure.

T_{ass}	d_1 Ang	d_2 Ang	d_1/d_2
1281	254	0.26*	978*
1343	552	0.72	768
1383	656	1.03	637
1433	960	2.09	459
1463	1680	3.85	436
1496	2100	5.01*	419*

I numeri indicati con un asterisco, relativi alle fotografie fatte alle temperature estreme, sono incerti, avendosi in un caso una dispersione anomala troppo piccola, nell'altro le frangie confuse da correnti convettive nell'interno dei tubi, inevitabili ad alte temperature.

Siccome il rapporto n_1/n_2 delle densità degli atomi nei due stati $6^2\,P_1$ e $6^2\,P_2$ è proporzionale al rapporto dei coefficienti di dispersione anomala d_1/d_2, dovremo avere, in virtù della (1),

$$d_1/d_2 = A e^{h\,\Delta\nu/kT}$$

dove A è una costante. Cioè, prendendo i logaritmi decimali

$$\log d_1/d_2 = \log A + 0{,}4343 \frac{h\,\Delta\nu}{k} \cdot \frac{1}{T} \cdot$$

Si vede dunque che log d_1/d_2 deve essere una funzione lineare di $1/T$. Nel diagramma della fig. 1 sono state riportate in ascisse le inverse delle temperature assolute e in ordinate log d_1/d_2.

Se si tien conto che le misure di temperatura possono contenere un errore fino a 10°, si riconosce che i punti trovati sperimentalmente possono considerarsi, entro i limiti dell'errore sperimentale, appartenenti a una retta. La retta I del grafico è stata ottenuta dai punti sperimentali col metodo dei minimi quadrati, attribuendo, per le ragioni spiegate sopra, un peso 1/3 alle

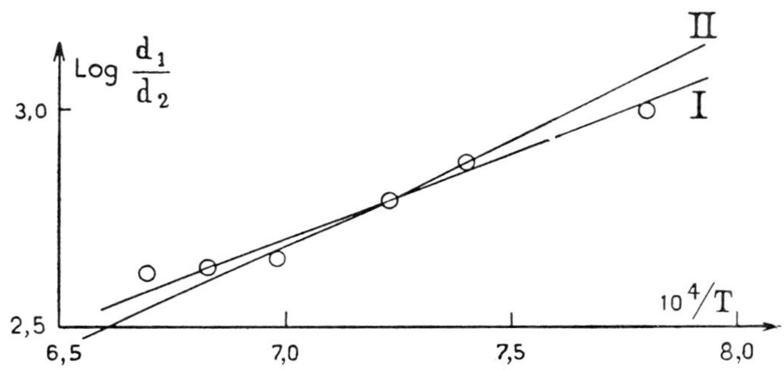

Fig. 1.

misure fatte alle temperature estreme. Il suo coefficiente angolare dovrebbe essere eguale a $0{,}4343\, h\,\Delta\nu/k$. Si trova così dalle nostre misure

$$h\,\Delta\nu/k = 9{,}1 \cdot 10^3$$

e siccome $\Delta\nu = 7792\, c$ risulta

$$h/k = 3{,}9 \cdot 10^{-11} ;$$

il suo valore effettivo è invece

$$h/k = 4{,}8 \cdot 10^{-11} .$$

Come si vede le nostre misure di h/k hanno dato un errore inferiore al 20 %; errore che, dato il genere di misure, può considerarsi soddisfacente.

La retta II del grafico ha una inclinazione corrispondente al valore effettivo del rapporto h/k; la sua equazione corrisponde a porre

$$d_1/d_2 = 0{,}22\; e^{11200/T}$$

dove si può valutare che l'errore del coefficiente 0,22 sia di ± 0,04.

Per mezzo della (2) e dell'equazione corrispondente che lega N_2 e d_2 risulta allora

$$N_1/N_2 = (0{,}44 \pm 0{,}08)\; e^{11200/T} .$$

Se si considera che il rapporto p_1/p_2 dei pesi statistici degli stati 6^2P_1 e 6^2P_2 è 1/2, si vede che i numeri degli elettroni di dispersione per atomo sono gli stessi in ambedue gli stati entro i limiti della nostra approssimazione.

Le nostre misure permettono anche di assegnare un limite inferiore al valore effettivo di questo numero, ammettendo che l'intero tratto di tubo compreso nella stufa fosse pieno di vapore saturo alla temperatura indicata. Si trova in questo modo

$$N_1/n_1 = N_2/n_2 > 1/10$$

il valore reale potrebbe essere benissimo il doppio di questo ed eventualmente anche di più.

Prendiamo questa occasione per rettificare alcuni dati di Mac Lennan (2) sopra il valore dell'indice di rifrazione del vapore di tallio, che risultano enormemente più grandi del vero. Infatti, secondo i risultati di Mac Lennan, il vapore di tallio, già alla temperatura di 540°, dovrebbe avere una rifrattività dell'ordine di grandezza di quella dell'aria a pressione atmosferica; ciò che, tenuto conto che il vapore di tallio a quella temperatura ha una pressione dell'ordine di 1/1000 di mm, corrisponderebbe ad una refrattività atomica circa 100.000 volte più grande di quella che si trova per gli altri vapori metallici. Dalle nostre misure risulta invece una rifrattività dell'ordine di grandezza normale. Inoltre Mac Lennan dà una curva di dispersione con caratteristiche particolari, delle quali non abbiamo trovato alcuna traccia nella forma delle nostre frangie di interferenza.

(2) MacLennan, «Proc. Roy. Soc.», *100*, 191 (1922).

N° 41.

In 1927, on the occasion of the first centenary of Alessandro Volta's death, « L'Energia Elettrica » (the journal of the « Associazione Elettrotecnica Italiana ») published a special issue dedicated to the most important scientific developments originated by the invention of the voltaic cell. O. M. Corbino, who edited the issue, asked Fermi for an article on the electro– and magneto–optical effects and their interpretations.

E. PERSICO.

41.

GLI EFFETTI ELETTRO E MAGNETOOTTICI E LE LORO INTERPRETAZIONI

Fascicolo speciale de « L'Energia Elettrica » nel 1° centenario della morte di A. Volta, Uniel, Roma (1927) 109–120.

INTRODUZIONE.

È accaduto molto spesso, nella storia delle scienze naturali, che i progressi di esse più importanti dal lato speculativo, fossero legati al fondersi insieme di due capitoli ritenuti dapprima sostanzialmente distinti. Così la teoria cinetica della materia, rappresentandoci il calore come una estrinsecazione dei moti meccanici delle molecole, permise di arrivare ad una comprensione profonda delle leggi della termodinamica, interpretata come meccanica di un numero di corpi tanto grande da obbedire a leggi statistiche. Così la scoperta della pila, mostrando per la prima volta una intima connessione tra i fenomeni elettrici e quelli chimici, fu il punto di partenza per lo studio della elettrochimica, tanto fecondo sia per la chimica pura che per la fisica della materia.

Un altro campo di confine tra due rami della fisica è quello della magneto–ottica e della elettroottica, cioè lo studio degli effetti sopra la emissione e sopra la propagazione della luce che si osservano quando la sostanza luminosa od il corpo attraversato dalla luce si trovano in un campo elettrico o magnetico. Che la presenza di un campo elettrico o magnetico debba alterare i fenomeni luminosi appare oggi, con la guida delle attuali idee sopra la natura della luce e la struttura della materia, senz'altro comprensibile. È infatti ormai bene accertato che le ultime particelle costituenti la materia sono corpuscoli elettrizzati alcuni positivamente (nuclei e protoni) ed altri negativamente (elettroni). Sono questi corpuscoli che, con i loro movimenti, dànno origine alla emissione delle onde elettromagnetiche costituenti la luce e determinano il modo di propagarsi della luce stessa attraverso ai corpi. Ora la presenza di un campo esterno, elettrico o magnetico, esercitando delle forze sopra tali corpuscoli, viene ad alterarne i movimenti e conseguentemente essa altererà anche i fenomeni luminosi. È interessante osservare però che il primo fenomeno magnetoottico fu scoperto da Faraday nel 1845, cioè trenta anni prima che si affermasse la teoria elettromagnetica della luce e che incominciassero a formarsi idee concrete sopra la struttura elettrica della materia. Così che questi fenomeni costituirono forse uno dei primi impulsi a ricercare una intima connessione tra luce ed elettricità da una parte e tra elettricità e materia dall'altra, e ci dànno anche oggi uno dei più validi mezzi per lo studio della struttura dell'atomo e della molecola. Prima di passare ad uno studio sistematico dei fenomeni elettro e magnetoottici vogliamo dare uno sguardo d'insieme al loro sviluppo.

Le prime idee concrete sopra una possibile influenza del magnetismo sopra la propagazione della luce pare risalgano a Sir J. F. W. Herschel, il quale fece anche qualche esperienza per mettere in evidenza effetti di questo tipo, senza tuttavia riuscirvi. M. Faraday studiò a lungo la stessa questione e, dopo numerosi tentativi riuscì, nel 1845, a scoprire l'effetto che da lui ha preso il nome. Se si manda della luce polarizzata rettilineamente attraverso ad un corpo isotropo trasparente che sia situato in un campo magnetico, in modo che le linee di forza siano parallele alla direzione di propagazione della luce si osserva che il piano di polarizzazione della luce che ha attraversato il corpo è rotato attorno alla direzione di propagazione della luce, rispetto al piano di polarizzazione della luce incidente.

Un fenomeno che presenta una certa analogia con l'effetto Faraday fu scoperto, circa 30 anni dopo da Kerr il quale mostrò che lo stato di polarizzazione della luce viene alterato dalla riflessione sopra una superficie speculare di ferro intensamente magnetizzato (Effetto magnetoottico di Kerr). Nel frattempo incominciava a svilupparsi lo studio della spettroscopia, e sorse il problema di ricercare una eventuale influenza di un campo magnetico sopra le frequenze delle righe spettrali di un gas luminoso contenuto in esso. Nel 1896 Zeeman fece vedere che le righe spettrali emesse da un vapore luminoso in un campo magnetico si scindevano in un numero variabile di componenti la cui distanza andava crescendo col crescere del campo magnetico. Quasi contemporaneamente riuscì a Lorentz di svolgere una teoria che permise di interpretare le caratteristiche principali del fenomeno, ed ebbe particolare importanza in quanto condusse alla dimostrazione che le particelle elettrizzate ai cui movimenti è dovuta l'emissione della luce, sono quegli stessi elettroni che costituiscono i raggi catodici. Dopo la scoperta di Zeeman lo studio sperimentale e teorico degli effetti di Zeeman è stato proseguito fino ad oggi con grande alacrità e costituisce attualmente, a prescindere dalla sua importanza in sé, uno dei più potenti metodi di indagine spettroscopica.

Due anni dopo Macaluso e Corbino scoprivano un fenomeno che può considerarsi come l'anello di congiunzione tra gli effetti Faraday e Zeeman, e consiste in una rotazione magnetica straordinariamente intensa che si osserva nei vapori in vicinanza delle loro righe d'assorbimento.

Un ultimo tipo di fenomeno magnetoottico, la birifrangenza magnetica, fu infine scoperto da Cotton che fece vedere che un corpo isotropo, posto in un campo magnetico, diventa birifrangente, e si comporta come un cristallo uniassico con l'asse parallelo alla direzione del campo. Parallelamente a questi effetti magnetoottici abbiamo gli effetti elettroottici. Conviene però osservare che, date le differenti proprietà di simmetria dei campi elettrico e magnetico (simmetria polare per il campo elettrico e assiale per quello magnetico) non potremo attenderci un analogo elettrico del potere rotatorio magnetico. I fenomeni elettroottici si riducono perciò sostanzialmente a due tipi: Effetto elettroottico di Kerr, che è l'analogo elettrico della birifrangenza magnetica ed effetto Stark che è l'analogo dell'effetto Zeeman e consiste in una separazione in componenti delle righe spettrali emesse da una sostanza che si trovi immersa in un intenso campo elettrico.

Fondamenti teorici della magnetoottica.

L'effetto di un campo magnetico sopra i fenomeni luminosi deve ricondursi in ultima analisi all'azione delle forze che il campo magnetico esercita sopra le particelle elettrizzate contenute nella materia. Un campo magnetico **H** esercita sopra un corpuscolo elettrizzato di carica e che si muove con velocità **V** una forza espressa, con notazione vettoriale, da

$$\frac{e}{c}\mathbf{V}\times\mathbf{H}. \tag{1}$$

Esiste un semplice teorema, dovuto a Larmor, il quale permette di valutare in modo assai espressivo l'influenza esercitata da una forza del tipo (1) sopra il moto di un sistema di punti aventi tutti la stessa carica e e la stessa massa m e soggetti, oltre che alle mutue azioni Coulombiane, soltanto ad una forza centrale (A questo tipo appartengono in realtà buona parte dei sistemi atomici). Consideriamo un tale sistema in assenza di campo magnetico e riferiamone il movimento, invece che ad assi fissi, ad una terna di assi coll'origine nel centro delle forze centrali, rotante con velocità angolare ω attorno ad un asse passante per il centro medesimo. Il moto del sistema di punti rispetto agli assi rotanti potrà descriversi, come è noto, come un movimento rispetto ad assi fissi, purché si considerino agenti sopra ogni punto oltre alle forze effettivamente ad esso applicate anche le forze centrifughe $m\omega^2 r$ (essendo r la distanza del punto dell'asse di rotazione), e le forze centrifughe composte (forze di Coriolis) che hanno per espressione

$$2\,m\mathbf{V}\times\omega. \tag{2}$$

Se supponiamo la velocità angolare $\boldsymbol{\omega}$ abbastanza piccola perché se ne possano trascurare i quadrati, riconosciamo che si possono trascurare le forze centrifughe pure e considerare agenti sui punti del sistema, oltre alle forze effettive, soltanto le forze centrifughe composte (2). Confrontiamo ora il moto del sistema di punti senza campo magnetico col moto nel campo magnetico, e riferiamo il primo movimento ad assi rotanti con velocità angolare ω, il secondo ad assi fissi. Nel primo caso dovremo considerare agenti sopra i punti del sistema le forze centrifughe composte (2), nel secondo le forze elettromagnetiche (1).

Si riconosce però immediatamente che, ove si ponga

$$\omega = \frac{e}{2\,mc}\,\mathrm{H} \tag{3}$$

le forze (1) e (2) vengono a coincidere, e vengono dunque a coincidere anche i due movimenti considerati. Abbiamo così il teorema di Larmor:

La perturbazione prodotta da un campo magnetico esterno sopra il moto di punti materiali, aventi tutti la stessa carica elettrica e e la stessa massa m e soggetti, oltre che alle mutue repulsioni, soltanto ad una forza centrale, consiste semplicemente in una rotazione uniforme di tutto il sistema attorno alla direzione del campo magnetico con la velocità angolare (3) (Rotazione di Larmor).

La frequenza corrispondente alla velocità angolare (3), cioè

$$\nu_{\mathrm{L}} = \frac{e\mathrm{H}}{4\,\pi\,mc} \tag{4}$$

si dice frequenza di Larmor.

Osserviamo ancora che un sistema contenente delle cariche elettriche in movimento verrà ad essere, per effetto di questo movimento, circondato da un campo magnetico, in modo che si comporterà come un piccolo magnete. Il suo momento μ, per il caso di sistemi ai quali è applicabile il teorema di Larmor, è legato da una relazione assai semplice alla quantità di moto areale p del sistema; si trova precisamente

$$\mu = \frac{ep}{2\,mc}\,. \tag{5}$$

Il meccanismo dell'emissione.

La emissione della luce da parte di un corpo incandescente, o reso, con qualsiasi metodo, luminoso, dipende dal movimento dei corpuscoli elettrizzati contenuti nel corpo. Le idee sopra il meccanismo preciso della emissione sono tuttavia state soggette a successivi cambiamenti; in quanto che al primo sorgere della teoria elettronica della materia si pensava che nei corpi fossero contenuti dei corpuscoli elettrici trattenuti da forze elastiche in posizioni di equilibrio. Allontanando uno di questi corpuscoli dalla posizione di equilibrio esso descrive attorno ad essa delle oscillazioni isocrone ed irradia perciò nello spazio energia elettromagnetica con la frequenza di tali oscillazioni. In seguito questa teoria fu abbandonata e sostituita dalla teoria dell'atomo di Bohr; secondo questa un atomo è costituito da un nucleo elettricamente positivo nel quale è concentrata praticamente tutta la massa atomica e che quindi sta fermo (purché naturalmente l'atomo non abbia un movimento d'insieme), e da un certo numero di elettroni negativi (tutti eguali tra loro) che girano attorno ad esso; in modo che l'atomo viene ad essere paragonabile ad un minuscolo sistema planetario, in cui il nucleo ha la funzione del sole e gli elettroni quella dei pianeti. Secondo Bohr gli elettroni descrivono delle orbite meccanicamente possibili; però, tra tutti i moti meccanicamente possibili essi possono descrivere soltanto una successione discreta di moti (detti stati stazionari o quantici) ed il processo dell'emissione è legato, in modo incomprensibile dal punto di vista della elettrodinamica classica, al passaggio dell'atomo da uno stato stazionario ad un altro. La differenza delle energie dei due stati stazionari viene emessa normalmente dall'atomo sotto forma di radiazione, la cui frequenza si calcola dividendo tale differenza di energia per la costante universale h (costante di Planck).

Nonostante la grande differenza tra il meccanismo di Bohr per la irradiazione ed il meccanismo classico, molte caratteristiche sono comuni ai due; vale anzi in generale il così detto principio di corrispondenza di Bohr il quale afferma che, in prima approssimazione, l'irradiazione di un atomo dedotta dai principi quantistici coincide con la radiazione che il sistema emetterebbe secondo la elettrodinamica classica. Questo principio giustifica in parte il fatto che molte conclusioni dedotte considerando la luce come emessa da un oscillatore classico restano valide in modo assai generale. Questo avviene in particolare nel caso dell'effetto Zeeman in cui larga parte dei risultati della prima teoria fattane da Lorentz considerando la luce come emessa da un oscillatore armonico si ritrovano nella teoria fatta sulla base dell'atomo di Bohr.

Effetto Zeeman.

Per questa ragione, nella descrizione dell'effetto Zeeman ci riferiremo dapprima al modello semplicistico della teoria di Lorentz, indicando quali risultati della teoria rispondano all'esperienza e quali no. Consideriamo perciò il movimento di un corpuscolo elettrizzato di carica e e massa m, che sia attratto da una forza proporzionale alla distanza verso un punto fisso O (origine delle coordinate) e sia inoltre soggetto all'azione di un campo magnetico H che supponiamo nella direzione dell'asse delle z. Indichiamo con ν_0 la frequenza propria dell'oscillatore in assenza del campo H. L'effetto del campo, per il teorema di Larmor, consiste in una rotazione uniforme di tutto il sistema attorno all'asse delle z, avente la frequenza (4). Ora il moto dell'oscillatore imperturbato, può sempre decomporsi in tre moti oscillatorii, uno lineare parallelo all'asse z, e gli altri due circolari paralleli al piano xy, uno in senso positivo e l'altro in senso negativo. Per effetto della rotazione di Larmor il primo di questi moti resta evidentemente inalterato, mentre le frequenze dei due moti rotatori vengono modificate, poiché la frequenza di Larmor si somma alla frequenza propria per il moto circolare che avviene in senso positivo, mentre si sottrae per il moto che avviene in senso negativo. Cosicché, per effetto del campo, il moto dell'oscillatore si decompone in tre moti.

a) Un moto lineare con frequenza ν_0, parallelo all'asse z;

b) un moto circolare con frequenza $\nu_0 + \nu_L$ parallelo al piano xy ed in senso positivo.

c) un moto circolare con frequenza $\nu_0 - \nu_L$ parallelo al piano xy ed in senso negativo;

Esaminiamo ora l'effetto ottico di questa decomposizione del movimento dell'oscillatore. Le cose vanno diversamente secondo la direzione dalla quale si osserva la luce emessa dall'oscillatore. I due casi più tipici si hanno per l'osservazione longitudinale, cioè parallela al campo magnetico e quindi all'asse delle z, e per l'osservazione trasversale cioè perpendicolare al campo magnetico.

Un osservatore situato sopra un punto molto lontano sopra l'asse z, non vedrà il movimento parallelo alla direzione di osservazione e percepirà soltanto i due movimenti circolari *b*) e *c*). La luce emessa longitudinalmente apparirà dunque, per effetto del campo magnetico, separata in due componenti polarizzate circolarmente nei versi opposti ed aventi le frequenze $\nu_0 + \nu_L$ e $\nu_0 - \nu_L$.

Invece un osservatore trasversale, situato per esempio sull'asse x, vedrà il moto *a*) in tutta grandezza mentre le proiezioni dei due moti oscillatori *b*) e *c*) gli si presenteranno come due moti oscillatori lineari, paralleli all'asse delle y. Cosicché la luce emessa in direzione trasversale apparirà separata in tre componenti polarizzate rettilineamente, una parallelamente al campo [1], avente la frequenza ν_0 della riga imperturbata, e le altre due polarizzate perpendicolarmente al campo con le frequenze $\nu_0 + \nu_L$ e $\nu_0 - \nu_L$. Osservando la luce in direzione obliqua la componente *a*) appare sempre polarizzata rettilinearmente e le componenti *b*) e *c*) appaiono invece polarizzate ellitticamente. Prima di passare ad un confronto di questa teoria con l'esperienza converrà che diamo una idea dell'ordine di grandezza del fenomeno. Ponendo nella (4) per e, m, c, i loro valori numerici per l'elettrone e ponendo H = 50.000 Gauss, cioè uno dei massimi campi magnetici raggiungibili, si trova $\nu_L = 7 \cdot 10^{10}$; per luci di 5000 A si trova dunque uno spostamento delle componenti estreme dalla riga imperturbata di circa 0,58 A.

Si vede dunque che, anche disponendo di campi magnetici assai intensi, l'effetto previsto è sempre molto piccolo e richiede per la sua osservazione mezzi spettrali di potere risolutivo alquanto elevato e righe spettrali che non siano naturalmente espanse.

Le prime esperienze di Zeeman furono fatte esaminando per mezzo di un reticolo la luce emessa da una fiamma di sodio situata tra le espansioni polari di un potente elettromagnete. L'osservazione longitudinale si faceva attraverso ad un foro praticato nelle espansioni polari, e quella trasversale direttamente attraverso l'interferro. Il potere risolutivo, nella disposizione originale di Zeeman, non era sufficiente per veder la riga spezzata in componenti. Zeeman poté soltanto osservare che la riga si allargava per effetto del campo, e che nel caso dell'osservazione longitudinale i suoi bordi erano polarizzati circolarmente nei due versi opposti, mentre, osservando trasversalmente il centro della riga appariva polarizzato parallelamente ed i bordi perpendicolarmente al campo magnetico. Già da questi primi risultati tuttavia Zeeman poté concludere che l'ordine di grandezza dell'allargamento e le caratteristiche generali qualitative del fenomeno erano in accordo con le previsioni della teoria di Lorentz, purché si ammettesse che i corpuscoli elettrici

(1) Contrariamente all'uso più comune e conformemente ad una opportunissima usanza che va affermandosi in molti lavori recenti, chiamiamo piano di polarizzazione della luce quello in cui vibra il vettore elettrico, e non quello del vettore magnetico.

oscillanti avessero carica negativa e massa qualche migliaio di volte più piccola delle masse degli atomi.

Effetti Zeeman normali ed anomali.

In seguito i mezzi sperimentali per lo studio dell'effetto Zeeman furono rapidamente perfezionati in modo da poter osservare una per una le componenti in cui si scinde una riga spettrale per effetto di un campo magnetico Si constatò allora che alcune righe presentavano un effetto Zeeman perfettamente conforme alla teoria di Lorentz. (Effetto Zeeman normale) altre invece presentavano una decomposizione in componenti assai più complicata che prese il nome generico di effetto Zeeman anomalo. In generale le righe semplici

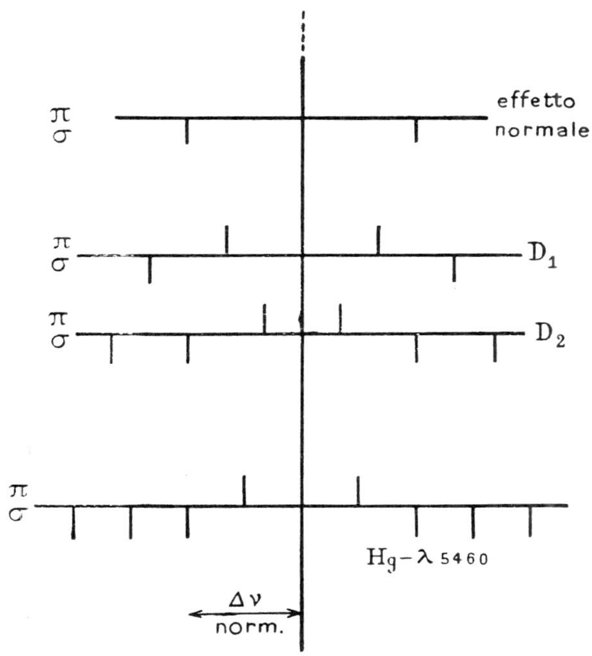

Fig. 1.

presentano l'effetto Zeeman normale mentre quelle che appartengono ad una molteplicità (doppietto, tripletto, ecc.) presentano l'effetto anomalo. Così per esempio le righe della serie di Balmer dell'idrogeno hanno l'effetto normale, mentre le righe D_1 e D_2 del sodio, che formano un doppietto, hanno un effetto anomalo; precisamente, osservando trasversalmente, la D_1 si spezza in 4 componenti, due parallele e due perpendicolari al campo, mentre la D_2 si spezza in 9 componenti due parallele e 4 perpendicolari al campo. Altre righe possono spezzarsi in un numero di componenti anche molto maggiore. Esempi di effetti Zeeman di vari tipi sono date dalla fig. 1.

Il modo di decomporsi di una riga spettrale nel campo magnetico, presenta un legame assai stretto con la serie spettrale a cui appartiene la riga. Questa circostanza, assai utile per la spettroscopia pratica, fu riconosciuta fin dai primi tempi della scoperta di Zeeman. Preston stabilì infatti la regola che righe appartenenti alla stessa serie spettrale presentano lo stesso tipo di effetto Zeeman; parimenti hanno lo stesso tipo Zeeman righe omologhe di elementi chimici differenti appartenenti alla stessa verticale del sistema periodico di Mendelejeff (Regola di Preston).

Si deve a Voigt una prima teoria degli effetti Zeeman anomali, basata sul calcolo della perturbazione magnetica di un sistema di due oscillatori. Tale teoria ha dato per gli spettri di dubletti, dei risultati concordanti con l'esperienza per il caso delle righe D del sodio; ciononostante sappiamo ormai che tale concordanza deve ritenersi di natura piuttosto casuale e che alla teoria di Voigt non si può attribuire altro che un valore fenomenologico.

Teoria quantistica dell'effetto Zeeman.

Per renderci conto delle idee che si hanno attualmente sopra l'interpretazione dell'effetto Zeeman normale ed anomalo, bisogna che ne cerchiamo la spiegazione nell'ambito delle teorie quantistiche e del modello atomico di Bohr–Rutherford.

La prima teoria quantistica dell'effetto Zeeman normale, dovuta a Debye, può brevemente descriversi come segue.

Consideriamo un atomo di Bohr in un campo magnetico H. Risulta dalle condizioni di Sommerfeld, che la quantità di moto areale degli elettroni attorno ad un asse parallelo al campo e passante per il nucleo, deve essere un multiplo intero di $h/2\pi$; essendo cioè s un numero intero (numero quantico magnetico), essa deve essere $sh/2\pi$.

D'altra parte abbiamo visto che tra quantità di moto areale e momento magnetico esiste la relazione di proporzionalità (5). Possiamo dunque concludere che la componente del momento magnetico dell'atomo parallela ad H è

$$\mu = s \frac{eh}{4\pi mc} \tag{6}$$

essa risulta cioè eguale ad un multiplo intero della costante universale

$$\frac{eh}{4\pi mc} = 0{,}91 \cdot 10^{-20} \tag{7}$$

che si dice magnetone di Bohr. L'energia magnetica dell'atomo nel campo H sarà espressa da

$$W_{\text{mag}} = \mu\text{H} = s\text{H} \frac{eh}{4\pi mc}. \tag{8}$$

Consideriamo ora un salto quantico dell'atomo che conduca all'emissione di una riga, la cui frequenza si calcola dividendo per h la differenza tra le energie dei due stati quantici iniziale e finale; se con s ed s^1 indichiamo i numeri quantici magnetici degli stati iniziale e finale, troveremo, nella differenza delle energie, un termine magnetico espresso da

$$\Delta W_{mag} = (s - s^1)\, H \frac{eh}{4\pi mc}$$

e quindi alla frequenza ν_0 della riga in assenza di campo magnetico dovrà aggiungersi, per effetto del campo, il termine $\Delta W_{mag}/h$, essa diventerà cioè

$$\nu_0 + (s - s^1) \frac{eH}{4\pi mc}. \tag{9}$$

Siccome poi, dal principio di selezione, si deduce che $(s - s^1)$ può prendere solo i valori 0, $+1$, -1 concludiamo che la nostra riga si spezzerà in tre componenti, aventi una la frequenza della riga imperturbata e le altre due la frequenza stessa cresciuta o diminuita della frequenza di Larmor (4). Questo risultato coincide con quello della teoria di Lorentz; anche lo stato di polarizzazione delle tre componenti, che può dedursi dal principio di corrispondenza, coincide con quello dato dalla teoria di Lorentz.

Il problema della interpretazione teorica dell'effetto Zeeman anomalo si presentò invece di risoluzione alquanto più difficile. Osserviamo infatti che, se si fa astrazione dalla struttura dell'elettrone, e lo si considera come un punto materiale, si può senz'altro applicare all'atomo di Bohr il teorema di Larmor e si è quindi condotti forzatamente all'effetto Zeeman normale. In un primo tempo furono fatte perciò, specialmente per opera di Landé ed Heisenberg, alcune ipotesi provvisorie che permisero, nonostante la loro ingiustificabilità, di raggiungere risultati importanti. Tali ipotesi consistono sostanzialmente nel distinguere l'atomo in due parti; l'elettrone luminoso, cioè l'elettrone più esterno dell'atomo al quale è dovuta la emissione della luce ed il nocciolo costituito da tutto il resto dell'atomo; e nell'ammettere che l'elettrone luminoso si comporti magneticamente in modo normale, mentre il nocciolo abbia una anomalia magnetica, inesplicabile nell'ambito della meccanica, consistente nel fatto che esso, per effetto di un campo magnetico, dovrebbe tendere a precedere con una velocità angolare doppia di quella voluta dal teorema di Larmor. Ammessa questa ipotesi si capisce come l'azione di un campo magnetico sopra tutto l'atomo consista, in prima approssimazione, nel farlo precedere con una frequenza intermedia tra quella normale di Larmor con cui tenderebbe a precedere l'elettrone luminoso e quella doppia della normale con cui tenderebbe a precedere il nocciolo. Così che l'atomo viene ad assumere una velocità di precessione diversa da quella normale, ciò che dà origine all'effetto Zeeman anomalo. Riuscì a Landé, in base a queste idee, di costruire una completa teoria degli effetti anomali; tale teoria permette di calcolare in modo completo l'effetto Zeeman di una riga, una volta che si sappia la posizione della riga nelle serie spettrali dell'atomo, e ciò anche nel caso delle molteplicità spettrali complicatissime che si trovano ad esempio negli spettri del ferro e del cromo. Nella teoria di Landé trova anche la sua espressione completa una regola, dedotta empiricamente da Runge, la quale afferma che la differenza tra le frequenze di una componente Zeeman e della riga imperturbata è sempre in rapporto razionale con la frequenza di Larmor.

Effetto Paschen–Back.

Inoltre può interpretarsi, per mezzo della teoria di Landé, un fenomeno, scoperto da Paschen e Back, consistente nel fatto che, mentre per campi magnetici deboli lo spostamento di ogni componente Zeeman dalla posizione della riga imperturbata è proporzionale al campo H, tale proporzionalità non si ha più con campi forti, e le varie componenti vanno spostandosi invece in modo tale che l'effetto Zeeman, per campi fortissimi finisce in ogni caso per diventare normale.

Per vedere come tale effetto possa spiegarsi con la teoria di Landé, osserviamo che, quando il campo magnetico diventa fortissimo, la differenza tra le azioni magnetiche sull'elettrone luminoso e sul nucleo diventa tale che ciascuno di questi due elementi si mette a precedere attorno al campo con la sua propria velocità di precessione (normale per l'elettrone e doppia della normale per il nocciolo). Siccome però l'emissione della luce è determinata dal moto dell'elettrone luminoso si capisce come l'effetto Zeeman debba venir determinato dal moto di precessione di questo e diventi quindi normale per campi magnetici assai intensi.

L'elettrone rotante.

Ciò che resta assai insoddisfacente nella teoria di Landé è la inesplicabilità della anomalia magnetica del nocciolo. Infatti, essendo anche il nocciolo costituito da elettroni soggetti alla forza centrale del nucleo, anche ad esso dovrebbe potersi applicare il teorema di Larmor, ed esso dovrebbe dunque comportarsi come magneticamente normale. La chiave per la risoluzione di questa contraddizione è da ricercarsi in una ipotesi recentemente proposta da Uhlenbeck e Goudsmit. Secondo questi autori si ammette che l'elettrone, oltre alla carica elettrica, abbia anche un momento magnetico della grandezza di un magnetone di Bohr (7). E che esso abbia inoltre, attorno al suo centro, un momento di quantità di moto di grandezza $h/4\pi$ con la direzione del momento magnetico. Siccome tutto questo potrebbe interpretarsi formalmente pensando l'elettrone come un corpuscolo carico di elettricità rotante attorno a sé stesso tale ipotesi è stata denominata ipotesi dell'elettrone rotante. Se pensiamo ora di porre un elettrone in un campo magnetico H, il campo eserciterà sul momento magnetico dell'elettrone una coppia meccanica. E siccome l'elettrone, avendo una quantità di moto areale, si comporta come un piccolo giroscopio, tale coppia avrà per effetto di imprimere all'asse dell'elettrone un moto di precessione attorno alla direzione del campo magnetico.

Fatti i calcoli, la velocità di precessione risulta doppia di quella (3) del teorema di Larmor per modo che la teoria degli effetti Zeeman può svolgersi ora in perfetta analogia alla teoria di Landé, eliminando però la inesplicabile anomalia magnetica del nocciolo ed attribuendo invece la anomalia dell'effetto Zeeman alla struttura interna dell'elettrone.

L'ipotesi dell'elettrone rotante, oltre che degli effetti Zeeman anomali, permette anche di trovare spiegazioni del tutto naturali di numerosi altri punti della fisica atomica e della spettroscopia, che apparivano incomprensibili prima di essa; perciò, benché ne manchino fino ad oggi conferme dirette, si ritiene generalmente che essa, almeno nelle sue linee generali, corrisponda alla realtà.

Esperienza di Stern e Gerlach.

Come si è visto la teoria dei quanti permette di spiegare in modo assai completo tutti i tipi di effetto Zeeman dai più semplici ai più complicati. Una conferma assai impressionante della realtà della spiegazione quantistica ci è fornita dalla esperienza di Stern e Gerlach, che ora vogliamo descrivere. Osserviamo perciò che, secondo la teoria quantistica, un atomo dotato di un momento magnetico, quando è posto in un campo magnetico si orienta rispetto a questo in modo che la componente del momento parallela al campo può prendere soltanto certi valori discreti i quali, nel caso degli effetti Zeeman normali, differiscono tra di loro di un magnetone di Bohr (cfr. formula (6)), in altri casi di quantità in rapporto razionale semplice col magnetone di Bohr.

Stern e Gerlach sono riusciti a verificare assai direttamente questa conseguenza della teoria quantistica, al modo seguente.

Consideriamo un atomo posto in un campo magnetico H ed indichiamo con μ la componente del momento magnetico dall'atomo parallela al campo. Se il campo H è uniforme, le forze magnetiche agenti sopra l'atomo avranno risultante nullo e non tenderanno quindi a modificarne il moto d'insieme. Ciò non è più vero quando il campo è inomogeneo; si vede infatti facilmente che in questo caso sopra l'atomo viene ad agire una forza che ha per espressione

$$\mu \frac{\delta H}{\delta z} \tag{10}$$

avendo indicata con la z la direzione del campo nel punto occupato dall'atomo.

L'esperienza di Stern e Gerlach consiste nel far passare un sottile fascetto di atomi attraverso ad un campo magnetico fortemente inomogeneo.

Il fascio di atomi si ottiene facendo evaporare lentamente una sostanza in un ambiente in cui sia stato fatto un vuoto tanto spinto che gli atomi evaporati possano attraversare tutto il recipiente senza subire urti; opportune limitazioni permettono di ottenere un fascio assai sottile. Il campo inomogeneo è fornito da un elettromagnete una delle cui espansioni polari presenta un solco, e l'altra uno spigolo che si affaccia contro il solco. Con questa disposizione il campo magnetico prende un valore assai grande in vicinanza dello spigolo, e va decrescendo rapidamente lontano da esso. Osserviamo ora che, secondo la teoria dei quanti, il momento magnetico μ della (10) può prendere solo certi valori discreti, e quindi anche le forze agenti sui diversi atomi del fascetto avranno valori differenti e discreti. Ciò ha per effetto di deviare in modo diverso i vari atomi del fascio, così che questo si dividerà in tanti fasci parziali corrispondenti ciascuno ad un valore di μ. Stern e Gerlach poterono osservare effettivamente una tale divisione del fascetto e poterono così non solo confermare qualitativamente la teoria quantistica dell'effetto Zeeman, ma anche trovare i valori di μ in ottimo accordo quantitativo con le previsioni teoriche per gli elementi per i quali fu possibile eseguire l'esperienza.

L'EFFETTO MAGNETICO SOPRA LA RISONANZA OTTICA.

L'effetto Zeeman, anche quando si abbiano a disposizione dei campi magnetici di intensità grandissima, è sempre assai piccolo e richiede per la sua osservazione mezzi spettroscopici di elevato potere risolutivo. In campi magnetici di pochi Gauss poi esso diventa talmente piccolo che nessuno dei mezzi di alta risoluzione ora noti sarebbe capace di metterlo in evidenza. Tuttavia è riuscito a R. W. Wood di osservare in modo indiretto effetti Zeeman dovuti a campi anche inferiori ad un Gauss, come per esempio il campo magnetico terrestre; e ciò osservando gli effetti esercitati da un debole campo magnetico sopra la risonanza ottica del mercurio. Le esperienze di Wood possono interpretarsi comodamente col modello classico secondo cui la luce viene emessa da un oscillatore armonico isotropo; noi ci atterremo perciò nella loro descrizione ad una tale rappresentazione.

Se si illumina un bulbo di quarzo contenente del vapore di mercurio con la luce di lunghezza d'onda 2536 A emessa da una lampada a mercurio si constata che tale luce viene diffusa in modo assai intenso dal vapore di mercurio (fenomeno della risonanza ottica). L'interpretazione classica di questo fenomeno è la seguente: L'atomo di mercurio contiene degli oscillatori aventi la frequenza della riga 2536; quando l'atomo si illumina con luce di questa stessa frequenza tali oscillatori si mettono in risonanza e diventano perciò a loro volta centri di emissione di luce della stessa frequenza, la quale costituisce appunto la luce diffusa.

Se supponiamo che la luce incidente sia polarizzata e che il suo vettore elettrico vibri parallelamente all'asse delle z avremo che anche gli oscillatori atomici saranno sollecitati a vibrare in questa stessa direzione e quindi la luce di risonanza emessa da essi avrà anche essa il piano di polarizzazione parallelo all'asse delle z. Ora Wood constatò che effettivamente in qualche caso i risultati erano conformi a questa previsione teorica mentre in altri casi, senza che apparentemente nulla di essenziale fosse cambiato nei suoi dispositivi, le cose andavano in modo diverso. Dopo alcune vane ricerche Wood riuscì finalmente a riconoscere che la causa di questi risultati contraddittori era da ricercarsi nella diversa orientazione dei suoi apparecchi rispetto al campo magnetico terrestre o ai debolissimi campi magnetici dovuti a pezzi in ferro del suo apparecchio; per renderci conto di questo esaminiamo come deve presentarsi il fenomeno della risonanza quando l'atomo risonante si trova in un campo magnetico. Come abbiamo visto a proposito della teoria di Lorentz dell'effetto Zeeman, il campo magnetico ha per effetto di scindere il moto dell'oscillatore in tre moti armonici di frequenze leggermente differenti, uno lineare parallelo al campo e due circolari in senso positivo e negativo attorno al campo. Se ora il vettore elettrico della luce eccitante vibra parallelamente al campo, di queste tre vibrazioni resterà eccitata soltanto quella lineare e tutto andrà quindi come in assenza di campo; se invece il vettore elettrico della luce vibra perpendicolarmente al campo, resteranno eccitate le due componenti circolari, per modo che da una linea di osservazione perpendicolare al campo la luce apparirà polarizzata rettilineamente, mentre, osservando nella direzione del campo si vedranno le due componenti polarizzate circolarmente, quando i mezzi di risoluzione usati siano sufficienti a separarle, se invece la differenza delle due componenti è più piccola del potere risolutivo, ciò che in pratica accade sempre con campi magnetici piuttosto deboli, le due componenti circolari si sovrapporranno in modo che la luce apparirà completamente depolarizzata.

L'esperienza conferma in modo completo questi risultati teorici sopra l'azione depolarizzante di un campo magnetico, dandoci così un mezzo di riconoscere la presenza di un effetto Zeeman anche quando la sua larghezza sia assai minore della larghezza delle righe spettrali.

Polarizzazione rotatoria magnetica.

Come abbiamo visto nella introduzione, il piano di polarizzazione di un raggio di luce che si propaga in un mezzo trasparente isotropo parallelamente alle linee di forza di un campo magnetico in cui il mezzo si trova immerso, subisce una rotazione attorno alla direzione di propagazione via via che la luce avanza nel mezzo (Effetto Faraday).

L'angolo α di cui rota il piano di polarizzazione può esprimersi per corpi non ferromagnetici con la formula

$$\alpha = C l H$$

dove l è la lunghezza del cammino percorso dalla luce nel campo magnetico, H l'intensità del campo e C una costante (detta costante di Verdet). Tale costante dipende, oltre che dalle varie sostanze, anche dalla lunghezza d'onda; si suole dare ad essa il segno + o —, secondo che la rotazione avviene nel verso della corrente che genera il campo magnetico o in quello opposto. Esprimendo la rotazione in minuti di arco, la costante di Verdet ha per l'acqua e per la luce D del sodio il valore 0,013; per il solfuro di carbonio 0,04.

Possiamo giungere ad una spiegazione teorica di questo fenomeno in base agli stessi principi che ci hanno spiegato l'effetto Zeeman, e cioè in base al teorema di Larmor. Astraendo da ogni considerazione quantitativa, possiamo infatti interpretare il fenomeno al modo seguente. Ricordiamo anzitutto che l'esistenza di un potere rotatorio, sia esso naturale o magnetico, equivale ad una differenza nella velocità di propagazione della luce polarizzata circolarmente in sensi opposti attorno all'asse che presenta il potere rotatorio. Se infatti abbiamo un raggio di luce polarizzata rettilineamente che si propaga parallelamente all'asse, questo può sempre decomporsi in due raggi di eguale ampiezza, polarizzati circolarmente nei due versi opposti. Se ora questi due raggi avessero lo stesso indice di rifrazione essi si troverebbero, in ogni punto dell'asse, con la stessa differenza di fase, per modo che, interferendo tra di loro, ricostituirebbero di nuovo una vibrazione rettilinea, sempre parallela alla direzione iniziale; invece, se i due raggi polarizzati circolarmente hanno indice di rifrazione diverso, la loro differenza di fase va mutando via via che ci si avanza lungo l'asse, cosicché i due raggi ricostruiscono sempre per interferenza una vibrazione rettilinea, la cui direzione varia però da punto a punto dell'asse; ed in ciò consiste appunto il potere rotatorio.

Per renderci conto del potere rotatorio magnetico, basterà dunque che mostriamo che il campo produce una differenza tra gli indici di rifrazione dei due raggi polarizzati circolarmente. Osserviamo per questo che la dispersione di un corpo dipende dal fatto che l'ampiezza dello spostamento dei suoi elettroni sotto l'azione del campo elettrico della luce, varia con la frequenza di questo; tale variazione è particolarmente rapida quando la frequenza della luce è vicina ad una riga o banda di assorbimento del corpo, poiché in questo caso si produce un fenomeno di risonanza (dispersione anomala). Ora ricordiamo che la perturbazione prodotta da un campo magnetico sul moto di un sistema di elettroni, consiste nell'imprimere a questo un moto rotatorio attorno alla direzione del campo che, almeno nei casi più semplici, può calcolarsi per mezzo del teorema di Larmor. Se quindi riferiamo il moto degli elettroni ad un sistema di assi che roti con la frequenza di Larmor, rispetto ad essi il moto elettronico avverrà come se gli assi fossero fissi e non ci fosse il campo magnetico. Se i nostri elettroni si trovano sotto l'azione di luce di frequenza ν polarizzata circolarmente, la frequenza con cui gira il vettore elettrico rispetto al sistema degli assi rotanti sarà $\nu - \nu_L$ oppure $\nu + \nu_L$, secondo che il senso della polarizzazione rotatoria coincide o è opposto a quello della rotazione di Larmor. Secondo che si verifica l'uno o l'altro di questi casi, la frequenza da cui vengono sollecitati gli elettroni è $\nu \mp \nu_L$, e quindi essi vengono a reagire sopra la luce in modo differente dando luogo alla differenza d'indice di rifrazione della luce polarizzata circolarmente nei due sensi, cioè al potere rotatorio magnetico.

Nel caso delle sostanze paramagnetiche può aversi anche un potere rotatorio magnetico derivante dall'orientamento magnetico delle molecole.

Questa teoria ci permette anche di interpretare un fenomeno scoperto da Macaluso e Corbino nel 1898, consistente in un potere rotatorio magnetico anormalmente intenso e rapidamente variabile con la frequenza che si osserva in vicinanza delle righe di assorbimento di un vapore. Si poté arrivare, in queste condizioni, ad osservare rotazioni magnetiche fino a 700° (Le prime osservazioni di Macaluso e Corbino furono fatte osservando con un reticolo, la luce passata attraverso ad una fiamma di sodio interposta tra due nicol). Questo fenomeno ci appare infatti come una conseguenza della dispersione anomala che si ha in vicinanza di una riga d'assorbimento; essa ha appunto per conseguenza una considerevole differenza di indice di rifrazione per le due frequenze $\nu - \nu_L$ e $\nu + \nu_L$ cioè un forte potere rotatorio.

La luce nei corpi ferromagnetici.

Un altro caso in cui si osservano poteri rotatori magnetici assai forti, benché dovuti a cause assai differenti, è quello in cui si fa passare la luce attraverso ad una laminetta di ferro o di altri corpi ferromagnetici magnetizzata e tanto sottile da essere trasparente. Nel ferro magnetizzato a saturazione si possono osservare in questo modo delle rotazioni del piano di polarizzazione che arrivano fino a 200.000 gradi per cm.

Un altro fenomeno che presenta un intimo legame con quello qui descritto è il fenomeno magneto–ottico di Kerr, che consiste in una alterazione dello stato di polarizzazione della luce per effetto di una riflessione sopra una superficie speculare di ferro intensamente magnetizzato.

Si capisce come questo fenomeno possa presentare una varietà di aspetti assai notevole secondo l'angolo di incidenza, la polarizzazione della luce e la orientazione della forza magnetica; in genere l'effetto consiste nel trasformare in luce ellittica la luce polarizzata rettilineamente. Il caso più semplice si ha studiando la riflessione normale di luce polarizzata rettilineamente sopra uno specchio di ferro magnetizzato perpendicolarmente alla sua superficie. In questo caso la luce riflessa, come hanno mostrato alcune ricerche del Righi, è polarizzata ellitticamente, ed inoltre l'asse maggiore dell'ellisse è rotato rispetto al piano di polarizzazione della luce incidente. Questo fatto è dovuto al diverso indice di rifrazione, ed al diverso assorbimento che ha nel ferro magnetizzato la luce polarizzata circolarmente nei due sensi opposti.

Birifrangenza elettrica e magnetica.

Quando un corpo isotropo viene inserito in un campo elettrico o magnetico esso diventa birifrangente, e si comporta otticamente come un cristallo uniassico, il cui asse ottico abbia la direzione nel campo.

La birifrangenza elettrica fu scoperta da Kerr nel 1875 e si chiama perciò spesso effetto Kerr. Le prime esperienze furono fatte col vetro: una lastra di vetro veniva interposta fra due nicol incrociati, e le venivano applicati due elettrodi in modo da poter produrre in essa un campo elettrico, le cui linee di forza formavano un angolo di 45° coi piani di polarizzazione di entrambi i nicol. In assenza di campo elettrico i due nicol producono l'estinzione della luce; il campo elettrico invece, rendendo birifrangente il vetro permette il passaggio di un po' di luce attraverso al sistema.

In un primo tempo si pensò che il fenomeno della birifrangenza elettrica non fosse sostanzialmente distinto da quello della birifrangenza dovuta ad una compressione, poiché la birifrangenza elettrica veniva interpretata come dovuta alla deformazione del vetro prodotta per elettrostrizione dal campo elettrico. In seguito però si riconobbe che tale interpretazione era errata, poiché fu osservata la birifrangenza elettrica anche nei liquidi e nei gas per i quali naturalmente non può parlarsi di una elettrostrizione anisotropa.

La birifrangenza elettrica può misurarsi per mezzo della differenza tra gli indici di rifrazione n_s ed n_o dei raggi straordinario e ordinario.

Tale differenza è proporzionale al quadrato del campo elettrico, per modo che si può porre

$$n_s - n_o = B\lambda\, E^2$$

dove E è il campo (in unità elettrostatiche), λ la lunghezza d'onda; B si dice costante dell'effetto Kerr, B dipende dai vari corpi ed anche dalla lunghezza d'onda. Per il solfuro di carbonio si ha $B = 3{,}7 \cdot 10^7$.

Anche la birifrangenza magnetica, scoperta da Cotton, soddisfa alle stesse leggi. La sua costante, per il nitrobenzene ha il valore $2{,}53 \cdot 10^{-12}$.

Secondo Langevin sia la birifrangenza elettrica che quella magnetica sono dovute, nella maggior parte dei casi, alla azione orientatrice del campo elettrico o di quello magnetico sopra le molecole del corpo. La orientazione magnetica delle molecole di sostanze paramagnetiche è determinata dal fatto che tali molecole hanno un momento magnetico il quale tende ad orientarsi parallelamente al campo; d'altra parte l'agitazione termica tende invece a dare alle molecole una distribuzione di orientazione uniforme, per modo che, ad ogni temperatura, viene a stabilirsi un equilibrio statistico tra queste due azioni che ha per conseguenza un addensamento degli assi magnetici nella direzione del campo.

L'azione orientatrice del campo elettrico può essere dovuta a due cause diverse: infatti le molecole di alcune sostanze (per esempio dell'acqua) hanno un momento elettrico, e per queste il meccanismo dell'orientamento elettrico è in tutto analogo a quello dell'orientamento magnetico delle molecole paramagnetiche; ma anche quando le molecole non abbiano un momento elettrico permanente, si può tuttavia in certi casi avere un orientamento elettrico.

Ammettiamo infatti che le molecole del corpo che si considera siano elettricamente anisotrope ed abbiano quindi tre assi principali di polarizzabilità elettrica differente. Se poniamo una tale molecola in un campo elettrico la sua energia verrà allora a dipendere dalla orientazione, e sarà precisamente minima o massima secondo che il campo elettrico è parallelo all'asse di massima o minima polarizzabilità della molecola.

Il campo elettrico tenderebbe perciò a disporre tutte le molecole coll'asse di massima polarizzabilità parallelo a sé; per effetto della agitazione termica si viene anche in questo caso a stabilire un equilibrio statistico, che ha per risultato un addensamento degli assi di massima polarizzabilità nella direzione del campo. Come conseguenza dell'orientamento, qualunque sia la sua causa, la costante dielettrica del corpo viene a dipendere dalla direzione; a questa anisotropia della costante dielettrica corrisponde naturalmente anche una anisotropia dell'indice di rifrazione, cioè una birifrangenza.

La teoria dell'orientamento si presta anche alla interpretazione della birifrangenza elettrica osservata da Block nelle nebbie, e della birifrangenza magnetica delle sospensioni di ferro colloidale, osservata da Maiorana.

Effetto Stark.

Dopo la scoperta di Zeeman che le righe spettrali vengono scisse in componenti da un campo magnetico, si presentava naturale la ricerca di una eventuale scissione delle righe per effetto di un campo elettrico. Ciononostante fu soltanto nel 1913 che riuscì a Stark di mettere in evidenza per la prima volta un tale fenomeno. Ciò è dovuto sostanzialmente alla difficoltà di produrre un intenso campo elettrico in seno ad un gas o vapore che emetta la luce, poiché un gas emittente è quasi sempre anche un conduttore elettrico. Tale difficoltà fu superata a breve distanza di tempo e con metodo differente da Stark e da Lo Surdo.

L'apparecchio di Stark è sostanzialmente un tubo a raggi canale, pieno di idrogeno; immediatamente dietro al catodo bucato si trova un elettrodo secondario. Tra il catodo e questo elettrodo viene stabilita una forte differenza di potenziale la quale, data la brevissima distanza tra i due elettrodi, non provoca la scarica disruptiva. I raggi canale, uscendo dai fori del catodo, vengono a trovarsi nel forte campo elettrico dovuto alla differenza di potenziale tra catodo ed elettrodo secondario, per modo che su di essi può osservarsi la scissione elettrica delle righe spettrali.

L'apparecchio di Lo Surdo consiste invece semplicemente in un tubo ad idrogeno del tipo dei tubi di Plücker, nel quale però la parte prossima al catodo è costituita da un sottile capillare. Ciò ha per effetto di produrre, vicino al catodo, una forte caduta di tensione, ed è appunto in questa che si osserva il fenomeno.

Le ricerche eseguite con questi metodi hanno dimostrato che le righe della serie di Balmer dell'idrogeno si scindono nel campo elettrico in numerose componenti, polarizzate alcune parallelamente altre perpendicolarmente al campo.

L'ordine di grandezza dell'effetto Stark nell'idrogeno è notevolmente più grande di quello dell'effetto Zeeman, infatti con campi elettrici intensi si può ottenere comodamente una scissione delle righe in cui la distanza tra le componenti esterne supera i 20 A.

Al pari dell'effetto Zeeman, anche l'effetto Stark ha avuta una notevole influenza sopra lo sviluppo delle nostre conoscenze sulla struttura dell'atomo; e ciò specialmente perché è stato possibile, per mezzo della teoria di Bohr, rendere conto in modo completo, sia qualitativo, che quantitativo dell'effetto Stark dell'idrogeno.

Il principio generale di tale interpretazione, dovuta a Epstein e Schwarzschild, è il seguente: Un atomo di idrogeno in un campo elettrico esterno è un sistema meccanico costituito da un elettrone mobile sotto l'azione della attrazione Coulombiana verso il nucleo (che può al solito con grande approssimazione considerarsi come fisso) e della forza costante derivante dall'azione del campo elettrico esterno. Il moto di questo sistema può facilmente integrarsi (riferendosi a coordinate paraboliche aventi l'asse nella direzione del campo esterno) e si possono poi applicare ad esso le condizioni di Sommerfeld per la determinazione degli stati stazionari. Si trova così che un livello energetico che, in assenza di campo, appariva come semplice, viene, sotto l'azione del campo elettrico a spezzarsi in parecchi livelli ravvicinati tra di loro, la cui distanza dal livello originario cresce, in prima approssimazione, in modo proporzionale al campo. Siccome poi le frequenze delle righe spettrali sono proporzionali (per il fattore $1/h$) alle differenze delle energie degli stati stazionari, si avrà che anche le righe verranno a spezzarsi in tante componenti. Il principio di corrispondenza permette poi di valutare l'intensità di tutte queste componenti.

Per renderci conto in modo più intuitivo dell'effetto del campo elettrico sopra lo spettro dell'idrogeno conviene che esaminiamo la perturbazione del moto Kepleriano dell'elettrone, dovuta al campo elettrico esterno. Se non ci fosse il campo esterno l'ellisse Kepleriana conserverebbe posizione invariabile nello spazio (a prescindere dalla lentissima precessione dovuta alla perturbazione relativistica). Il campo elettrico ha per effetto di imprimere all'ellisse un moto relativamente lento periodico; precisamente se consideriamo il secondo fuoco dell'ellisse (cioè quello non occupato dal nucleo) esso si muove descrivendo, in un piano perpendicolare al campo esterno una ellisse armonica con frequenza proporzionale al campo elettrico. Se decomponiamo armonicamente il moto elettronico troviamo allora che le frequenze delle componenti armoniche sono le frequenze di combinazione tra la frequenza fondamentale del moto Kepleriano e la frequenza della perturbazione secolare; ed è appunto a questa combinazione delle due frequenze che corrisponde lo spezzamento delle righe in componenti. (Conviene però osservare che, se non ci fosse la determinazione quantistica delle orbite nel campo esterno, l'effetto Stark dovrebbe presentarsi come un allargamento delle righe, invece che come una scissione in componenti). La fig. 2 dà lo schema delle compo-

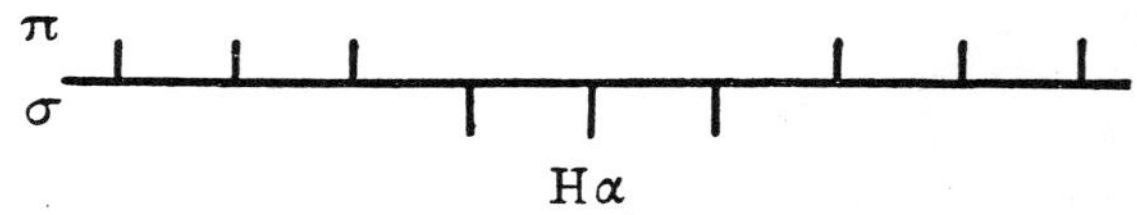

Fig. 2.

nenti in cui si spezza la riga H_α dell'idrogeno. Le righe successive della serie di Balmer si separano in un numero di componenti che va crescendo col numero d'ordine della riga.

Effetto Stark sopra gli spettri degli elementi.

L'azione del campo elettrico sopra lo spettro dell'atomo di idrogeno è notevolmente diversa da quella sopra gli spettri degli altri atomi. Nel caso degli altri atomi l'effetto Stark può consistere sia, come per l'idrogeno, in una decomposizione delle righe in componenti, sia in uno spostamento delle righe dalla posizione che esse occupano in assenza del campo sia infine nell'apparire di nuove righe. In tutti questi casi però la distanza delle componenti dalla riga imperturbata (o, nel caso di righe che appaiono sotto l'azione del campo, dalla loro posizione limite per campo nullo) è proporzionale al quadrato dell'intensità del campo e non direttamente proporzionale a questa, come avveniva per l'idrogeno.

La spiegazione di questo comportamento singolare dell'idrogeno rispetto agli altri elementi è da ricercarsi nella circostanza seguente. Se consideriamo l'orbita Kepleriana descritta dall'elettrone dell'idrogeno vediamo subito che essa, tutte le volte che la sua eccentricità è differente da zero, possiede un momento elettrico diverso da zero anche in media. Invece il moto degli elettroni di valenza degli altri atomi può in prima approssimazione descriversi come un moto periodico piano con sovrapposta una precessione di frequenza comparabile con la frequenza del moto periodico, per modo che la media del momento elettrico dell'atomo viene ad essere zero. Osserviamo ora che l'aumento dell'energia di un atomo in un campo elettrico esterno è il prodotto scalare dell'intensità del campo per il momento elettrico dell'atomo. In media tale energia sarà dunque nulla per tutti gli atomi fuorché per l'idrogeno. Ora è appunto a tale energia che è dovuto, in prima approssimazione, l'effetto Stark, e ciò ci spiega dunque il differente comportamento dell'idrogeno e degli altri elementi. A tutto rigore si può osservare che, per effetto della perturbazione relativistica, anche l'orbita dell'idrogeno è animata da un movimento di precessione, siccome però tale movimento è lentissimo si può facilmente riconoscere che il suo effetto diventa trascurabile già per campi elettrici assai piccoli.

Le considerazioni precedenti ci conducono anche a prevedere che l'effetto del campo elettrico debba essere tanto più forte, quanto minore è la velocità di precessione dell'orbita elettronica; ora la frequenza di tale precessione corrisponde alla differenza tra la frequenza del termine spettrale che si considera e la frequenza del termine balmeriano che ad esso corrisponde. Dobbiamo dunque attenderci che l'effetto del campo elettrico sarà tanto maggiore quanto più piccola è la differenza tra i termini dello spettro che si considera e i termini corrispondenti dello spettro dell'idrogeno.

Tale induzione teorica si trova pienamente confermata dall'esperienza. Si vede infatti che l'elio, i cui termini sono assai vicini a quelli dell'idrogeno, presenta un effetto Stark di entità non molto minore di quella dell'idrogeno mentre per esempio le righe del sodio presentano un effetto Stark tanto piccolo che la sua osservazione è riuscita solo recentemente a Ladenburg, con l'uso di mezzi delicatissimi.

Abbiamo detto che, in alcuni casi il campo elettrico può produrre l'apparizione di nuove righe spettrali. Tale fatto fu osservato per la prima volta nello spettro dell'elio dalla Sig.na Brunetti. Per rendercene conto teoricamente; osserviamo come, secondo la teoria di Bohr, le frequenze delle righe di un atomo si calcolino come differenze di termini spettrali; non però a tutte le differenze di termini corrisponde una riga che l'atomo può effettivamente emettere; perché la riga sia emessa effettivamente occorre che nello sviluppo in serie di Fourier del moto dell'elettrone luminoso sia rappresentata, con ampiezza diversa da zero, la componente armonica corrispondente ad essa.

Ora il campo elettrico, perturbando lievemente il moto dell'elettrone luminoso, ne altera lo sviluppo in serie di Fourier, per modo che talune componenti, che nel modo imperturbato avevano ampiezza nulla, vengono a comparire nel moto perturbato. Al comparire di queste componenti armoniche corrisponde appunto il comparire delle nuove righe.

Dobbiamo da ultimo accennare ad una caratteristica manifestazione dell'effetto Stark che consiste nel fatto che le righe emesse da un vapore vengono, allargate o spostate quando si cresce la pressione dell'ambiente in cui si trova il vapore emittente. Secondo Stark tale effetto deve interpretarsi col fatto che gli atomi del corpo emittente, trovandosi nelle vicinanze di altri atomi, vengono ad essere immersi nel campo elettrico generato da questi; e tale campo produce l'effetto Stark sopra le righe emesse. Naturalmente data la variabilità di questi campi elettrici molecolari l'effetto si manifesta come un allargamento delle righe invece che come separazione in componenti. Dobbiamo naturalmente attenderci che la presenza di ioni, che sono circondati da un campo elettrico assai intenso, produca un allargamento alle righe particolarmente notevole; ciò è confermato dal fatto che i maggiori allargamenti si osservano nelle scariche condensate. In particolari condizioni si possono così ottenere allargamenti delle righe della serie di Balmer così grandi che esse arrivano a sovrapporsi tra di loro, dando luogo ad una specie di spettro continuo.

Le righe in cui effetto Stark è più debole che nell'idrogeno presentano anche un effetto di pressione più piccolo.

N° 42.

In this short paper Fermi tries to introduce in the frame of wave mechanics the radiation reaction on an emitting atom. The procedure he employs is clever but, admittedly, rather arbitrary. Such endeavours as this where very soon outdated by Dirac's radiation theory, which appeared almost at the very time of the writing of this paper (« Proc. Roy. Soc. », *114*, 243 (1927)) and to which Fermi later brought some noteworthy contributions (see papers N° 50, 66, 67).

E. PERSICO.

42.

SUL MECCANISMO DELL'EMISSIONE NELLA MECCANICA ONDULATORIA

« Rend. Lincei », *5*, 795–800 (1927) (*).

Lo scopo di questo lavoro è lo studio del meccanismo dell'emissione della luce nella meccanica ondulatoria; tale studio ci porterà in particolare alla determinazione della larghezza naturale delle righe di emissione e della loro forma.

È noto come, nella teoria di Schroedinger [1] tutte le volte che in un atomo si trovano ad essere eccitati contemporaneamente due stati quantici i ed l il momento elettrico dell'atomo contiene una componente variabile col tempo con la frequenza $(E_i - E_l)/h$; secondo la teoria elettro-magnetica classica, questa variazione del momento elettrico ha per conseguenza una irradiazione di luce avente la stessa frequenza. Noi faremo vedere in questo lavoro come la reazione della radiazione emessa abbia per effetto di alterare le intensità di eccitazione dei due stati quantici, in senso tale che, trascorso un certo tempo, resta eccitato soltanto lo stato di energia più piccola; ed allora si arresta anche il processo dell'emissione. Cosicché il meccanismo di un salto quantico può descriversi al modo seguente. Supponiamo che, inizialmente, sia eccitato, nell'atomo che si considera, soltanto uno stato quantico di energia E_i. In queste condizioni il momento elettrico dell'atomo è costante e quindi l'atomo non può irradiare energia. Supponiamo ora che una piccola perturbazione esterna produca una piccola eccitazione di un altro stato di energia E_l minore di E_i. Comincerà allora subito una debole irradiazione con la frequenza $(E_i - E_l)/h$; e la reazione di questa irradiazione avrà per effetto di aumentare l'intensità di eccitazione dello stato l ai danni di quella dello stato i. Il processo proseguirà finché l'atomo non è passato completamente dallo stato i nello stato l.

(*) Presentata dal Socio O. M. Corbino nella seduta del 1° maggio 1927.

(1) E. SCHROEDINGER, « Ann. d. Phys. », *79*, 361, 489, 734 (1926); *80*, 437 (1926); *81*, 109 (1926).

Indichiamo con ψ lo scalare di campo della teoria di Schroedinger, e con $\bar{\psi}$ il suo complesso coniugato. Secondo Schroedinger la densità elettrica sarà allora espressa da

$$e\psi\bar{\psi} \tag{1}$$

essendo e la carica elettronica, ed avendo normalizzato ψ in modo che sia

$$\int_\tau \psi\bar{\psi}\, d\tau = 1 . \tag{2}$$

Il momento elettrico del sistema resterà allora espresso da

$$\boldsymbol{p} = e\int_\tau \boldsymbol{r}\, \psi\bar{\psi}\, d\tau \tag{3}$$

dove con $\boldsymbol{r}$ si è indicato il raggio vettore dell'elemento di volume $d\tau$.

Se il potenziale V_0 del sistema è indipendente dal tempo l'equazione di Schroedinger

$$\Delta\psi - \frac{8\pi^2 m}{h^2} V_0 \psi - \frac{4\pi i m}{h}\frac{\partial\psi}{\partial t} = 0 \tag{4}$$

ammetterà come soluzione generale una combinazione lineare a coefficienti costanti di soluzioni fondamentali del tipo

$$u_k \exp(2\pi i \nu_k t) \quad ; \quad \nu_k = E_k/h \tag{5}$$

dove le u_k non dipendono da t e sono soluzioni dell'equazione

$$\Delta u_k + \frac{8\pi^2 m}{h^2}(E_k - V_0)\, u_k = 0 \tag{6}$$

soddisfacenti alle condizioni di normalizzazione

$$\int u_k^2\, d\tau = 1 .$$

Le E_k sono gli autovalori della (6). In questo caso particolare ψ prenderà dunque la forma

$$\varphi = \Sigma \lambda_k u_k \exp(2\pi i \nu_k t) \quad ; \quad \Sigma \lambda_k \bar{\lambda}_k = 1 \tag{7}$$

essendo le λ dei parametri costanti. Il momento elettrico (3) prenderà con ciò l'espressione

$$\boldsymbol{p} = e\,\Sigma \bar{\lambda}_k \lambda_s \boldsymbol{r}_{ks} \exp 2\pi i(\nu_k - \nu_s)\, t \tag{8}$$

dove con $\boldsymbol{r}_{ks}$ si sono indicati i vettori

$$\boldsymbol{r}_{ks} = \int \boldsymbol{r} u_k u_s\, d\tau . \tag{9}$$

Di qui si riconosce in particolare che se tutte le λ, ad eccezione di una, si annullano, il momento elettrico diventa costante e quindi non si ha irradiazione. Nel caso generale esso risulta invece somma di funzioni sinusoidali del tempo aventi le frequenze $\nu_k - \nu_r$; e questa variazione di p dà luogo alla emissione di luce avente le stesse frequenze, la cui intensità è espressa, conformemente alla elettrodinamica classica, da

$$2\,\ddot{p}^2/3\,c^3. \tag{10}$$

Nelle considerazioni fin qui svolte abbiamo trascurata la reazione della radiazione emessa; per tener conto anche di questa, noi vogliamo ammettere che essa, conformemente alla elettrodinamica classica, sia rappresentata da una forza

$$\frac{2}{3}\frac{e}{c^3}\dddot{\boldsymbol{p}}. \tag{11}$$

Per poter introdurre nella equazione di Schroedinger la considerazione di questa forza dobbiamo naturalmente rappresentarla come derivante da un potenziale, il quale risulta evidentemente

$$\mathrm{V} = -\frac{2}{3}\frac{e}{c^3}\boldsymbol{r}\cdot\dddot{\boldsymbol{p}} = \frac{16\,\pi^3 i}{3}\frac{e^2}{c^3}\sum_{ks}(\boldsymbol{r}\cdot\boldsymbol{r}_{ks})(\nu_k - \nu_s)^3 \exp 2\,\pi i(\nu_k - \nu_s)\,t. \tag{12}$$

Per tener conto della reazione di radiazione dovremo dunque nella (4) aggiungere V al potenziale V_0. Questa aggiunta ha per effetto che, pur potendosi sempre scrivere ψ sotto la forma (7), le λ cessano di essere delle costanti e vengono a variare lentamente col tempo. Per determinare come avviene una tale variazione (2), sostituiamo nella (4) al posto di ψ l'espressione (7), e al posto di V_0, $\mathrm{V}_0 + \mathrm{V}$. Ricordando le (6) troviamo

$$\Sigma\,\dot{\lambda}_k u_k \exp 2\,\pi i\nu_k t = \frac{2\,\pi i}{h}\mathrm{V}\sum_s \lambda_s u_s \exp 2\,\pi i\nu_s t. \tag{13}$$

Moltiplicando i due membri di questa equazione per $u_k\,d\tau$ ed integrando a tutto lo spazio, si trova, tenendo presente che le u_k sono un sistema di funzioni ortogonali

$$\dot{\lambda}_k = \frac{2\,\pi i}{h}\sum_s \lambda_s \exp\left[2\,\pi i(\nu_s - \nu_k)\,t\right]\int \mathrm{V} u_k u_s\,d\tau. \tag{14}$$

Ponendo al posto di V la sua espressione (12) ricaviamo

$$\dot{\lambda}_k = -\frac{32\,\pi^4}{3}\frac{e^2}{c^3 h}\sum_{\alpha\beta s}\lambda_\alpha\bar{\lambda}_\beta\lambda_s(\nu_\alpha - \nu_\beta)^3 r_{\alpha\beta}\cdot r_{sk}\exp 2\,\pi i(\nu_\alpha - \nu_\beta + \nu_s - \nu_k)\,t. \tag{15}$$

Una perturbazione secolare dei valori delle λ è evidentemente data soltanto dai termini del secondo membro, per i quali $\alpha \neq \beta$ e per i quali il fattore esponenziale si riduce a una costante, cioè ai termini per cui $\nu_\alpha - \nu_\beta + \nu_s - \nu_k = 0$. Se, come vogliamo ammettere, non esistono relazioni

(2) Certamente questo procedimento per l'applicazione alla meccanica quantistica della nozione classica di reazione di radiazione è alquanto arbitrario; esso resta tuttavia in buona parte giustificato dai risultati a cui conduce.

razionali tra le ν_i, questa ultima equazione equivale a porre $\alpha = k$, $\beta = s$, e la (15) si scrive dunque, tenendo conto delle sole variazioni secolari

$$\dot{\lambda}_k = -\frac{32\pi^4}{3}\frac{e^2}{c^3 h}\sum_s \lambda_k \lambda_s \bar{\lambda}_s (\nu_k - \nu_s)^3 r_{ks}^2 . \tag{16}$$

Il caso per noi più interessante è quello che corrisponde all'emissione di una sola riga spettrale, cioè quello in cui tutte le λ, ad eccezione di due sole, per esempio λ_1 e λ_2, si annullano. In questo caso, indicando con A la costante reale

$$A = \frac{32\pi^4}{3}\frac{e^2}{c^3 h}(\nu_2 - \nu_1)^3 r_{12}^2 \tag{17}$$

le (17) e le equazioni che si deducono da esse cambiando i in $-i$ ci danno

$$\dot{\lambda}_1 = A\lambda_1 \lambda_2 \bar{\lambda}_2 \quad ; \quad \dot{\lambda}_2 = -A\lambda_2 \lambda_1 \bar{\lambda}_1 \tag{18}$$

$$\dot{\bar{\lambda}}_1 = A\bar{\lambda}_1 \lambda_2 \bar{\lambda}_2 \quad ; \quad \dot{\bar{\lambda}}_2 = -A\bar{\lambda}_2 \lambda_1 \bar{\lambda}_1 . \tag{19}$$

Da esse si deduce subito l'integrale primo $\lambda_1 \bar{\lambda}_1 + \lambda_2 \bar{\lambda}_2 = \text{cost}$; in virtù della seconda delle (7) la costante deve prendere il valore 1, in modo che si ha

$$\lambda_1 \bar{\lambda}_1 + \lambda_2 \bar{\lambda}_2 = 1 . \tag{20}$$

Dalle (18) (19) si deduce poi facilmente che gli argomenti dei complessi λ_1, λ_2 non variano col tempo e che, per i loro moduli, vale l'equazione

$$\frac{d}{dt}|\lambda_1|^2 = 2A|\lambda_1|^2(1 - |\lambda_2|^2)$$

da cui si deduce

$$2At = \int \frac{d|\lambda_1|^2}{|\lambda_1|^2(1 - |\lambda_1|^2)} = \log\frac{|\lambda_1|^2}{1 - |\lambda_1|^2} + \text{cost}. \tag{21}$$

Scegliendo l'origine dei tempi in modo che la costante di integrazione risulti zero, e ricordando la (21) si trova allora

$$|\lambda_1|^2 = \frac{e^{2At}}{e^{2At} + 1} \quad ; \quad |\lambda_2|^2 = \frac{1}{e^{2At} + 1} \tag{22}$$

da cui si deduce subito

$$|\lambda_1 \lambda_2| = 1/2 \cosh At . \tag{23}$$

Se nelle (22) si pone $t = -\infty$, si trova $\lambda_2 = 1$, $\lambda_1 = 0$; ponendo invece $t = +\infty$ si ha $\lambda_2 = 0$, $\lambda_1 = 1$. Si vede cioè che, come avevamo annunziato, l'effetto della reazione di radiazione è tale da far passare l'atomo dallo stato quantico 2 allo stato 1; queste considerazioni precisano dunque il meccanismo del salto quantico.

Per maggiore semplicità vogliamo ora ammettere che, all'istante iniziale, le λ siano reali; allora, per quanto si è detto, esse resteranno reali durante tutto il processo; l'espressione (8) per la componente del momento elettrico corrispondente alla emissione diventerà allora

$$p_{12} = er_{12} \frac{\cos 2\pi(\nu_2 - \nu_1)t}{\cosh At}. \tag{24}$$

Siamo ora in grado di calcolare la larghezza e la forma delle righe di emissione. Per far questo basta infatti che sviluppiamo l'espressione precedente del momento elettrico, per mezzo di un integrale di Fourier, nelle sue componenti armoniche. Si trova subito

$$\left\{\begin{aligned} p_{12} &= 2\, er_{12} \int_0^\infty d\nu \cos 2\pi\nu t \int_0^\infty \frac{\cos 2\pi(\nu_2 - \nu_1 - \nu)t + \cos 2\pi(\nu_2 - \nu_1 + \nu)t}{\cosh At}\, dt \\ &= \frac{2\, er_{12}}{A} \int_0^\infty d\nu \cos 2\pi\nu t \left\{ F\left[\frac{2\pi(\nu_2 - \nu_1 - \nu)}{A}\right] + F\left[\frac{2\pi(\nu_2 - \nu_1 + \nu)}{A}\right] \right\} \end{aligned}\right. \tag{25}$$

dove, con F (ω) si è indicata la funzione

$$F(\omega) = \int_0^\infty \frac{\cos \omega x}{\cosh x}\, dx. \tag{26}$$

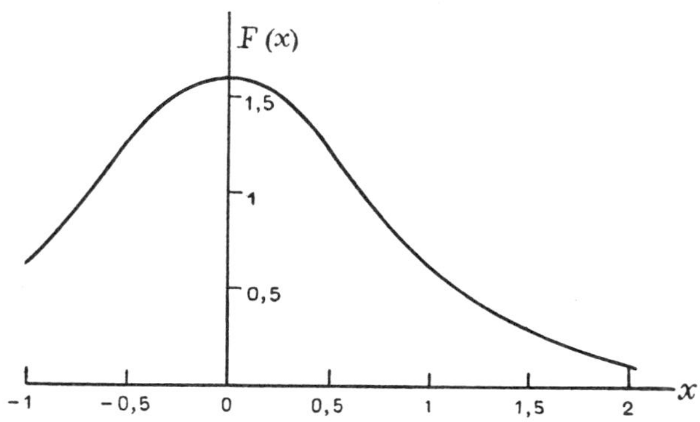

La F (ω) è rappresentata nel grafico annesso; abbiamo poi calcolato numericamente i seguenti valori:

$F(0) = 1.57 = \pi/2$; $F(0.2) = 1.50$; $F(0.5) = 1.18$; $F(1) = 0.62$;
$F(2) = 0.12$; $F(3) = 0.02$; $F(0.56) = \pi/2\sqrt{2}$.

Notiamo in particolare che la F (ω) prende valori assai piccoli per valori assoluti dell'argomento un po' grandi. In conseguenza di ciò la (25) può scriversi con grande approssimazione

$$p_{12} = \frac{2\, er_{12}}{A} \int_0^\infty F\left(\frac{2\pi\Delta\nu}{A}\right) \cos 2\pi\nu t\, d\nu \tag{27}$$

dove con $\Delta\nu = \nu_2 - \nu_1 - \nu$ si è indicata la distanza tra la frequenza che si considera ed il centro della riga di emissione. La intensità della riga di emissione a distanza $\Delta\nu$ dal suo centro risulta dunque proporzionale

$$F^2\left(\frac{2\pi\Delta\nu}{A}\right).$$

Questo risultato ci definisce la forma della riga. Se definiamo la larghezza della riga come il doppio di quella distanza dal suo centro per cui la intensità si riduce a metà si trova che tale larghezza è espressa da

$$\delta\nu = \frac{0.56}{\pi} A = 0.18\,A. \tag{28}$$

Naturalmente nelle effettive righe spettrali, oltre al loro allargamento naturale dovuto alla causa considerata in questo lavoro, deve tenersi conto dell'allargamento dovuto all'effetto Doppler e alla pressione. Tali effetti producono anzi in generale allargamenti delle righe assai maggiori della loro larghezza naturale, così che una misura diretta di questa per le righe di emissione si presenta praticamente, impossibile, ed il suo valore può soltanto dedursi indirettamente da una misura della costante A.

Dalla teoria della dispersione di Schroedinger si può dedurre

$$r_{12}^2 = \frac{h}{8\pi^2 m}\,\frac{D}{\nu_2 - \nu_1}$$

dove D è il numero degli elettroni di dispersione per ogni atomo, il quale può, come è noto, dedursi da una misura di dispersione anomala (3) e risulta in genere dell'ordine di grandezza dell'unità. Sostituendo questo valore in (17) si trova il valore di A che, sostituito a sua volta in (28) ci dà per la larghezza della riga il valore

$$\delta\nu = 0.18 \cdot \frac{4\pi^2}{3}\,\frac{e}{c^3 m}\,D\,(\nu_2 - \nu_1)^2 \quad ; \quad \delta\lambda = 6.7 \cdot 10^{-5}\,D\,\text{Å}ng.$$

Per esempio per la riga D_2 del Na si ha $D = 2/3$; si trova dunque $\delta\lambda = 4.5 \cdot 10^{-5}$ Å.

(3) Il valore di D può anche dedursi, per le righe D del sodio, dal lavoro di MINKOWSKI «Zs. f. Phys.», 36, 839 (1926), sulla larghezza delle righe di assorbimento.

N° 43–48.

After Fermi had developed the statistics of a gas of particles obeying the Pauli exclusion principle, he thought of applying this method to the completely degenerate state of the electrons in an atom, pointing out that the number of electrons being but moderately large, the results cannot be expected to attain high accuracy. Already the discussion at the Como conference quoted at N° 30, 31 foreshadows this type of thought, which matured in the fall of 1927. A differential equation for the average potential as a function of distance can then be readily established. The solution satisfying the appropriate boundary conditions cannot be expressed in closed form and Fermi solved it numerically with a small and primitive hand calculator. This numerical work took him probably a week. E. Majorana who was a lightning-fast calculator and a very skeptical man decided to check the numerical work. He did this by transforming the equation into a Riccati equation and solving the latter numerically. The result agreed exactly with the one obtained by Fermi. In the present paper Fermi gives a graph expressing the result of numerical calculations. He also shows that the total ionization potential of the atom can be found without integrating the differential equation, and results proportional to $Z^{7/3}$.

Fermi was unaware that essentially identical conclusions had previously been reached by Thomas (1). Thomas' paper was presented at the Cambridge Philosophical Society on November 6, 1926 while Fermi's note was presented at the meeting of December 4, 1927 of the Accademia dei Lincei. Thus Thomas' work antedates Fermi's by approximately a year. Fermi became acquainted with Thomas' paper only late in 1928, when it was pointed out to him by one (now unidentified) of the foreign theoreticians visiting Rome. This fact explains why no reference to Thomas' work is found even in the subsequent papers.

This statistical model, now generally designated as the Thomas–Fermi model, proved of great usefulness in evaluating various properties of atoms, in particular calculating approximate one-electron eigenfunctions, whenever great accuracy can be sacrificed in favor of simplicity and ease of calculation. Among other important applications, one may mention its use in the theory of the stopping power of matter for fast charged particles as developed by Bloch, Bethe and others.

Fermi made the first applications of this method and some more were made at the time by his associates. They are described in the following papers.

In N° 44 the author calculates the average number of electrons with given value of the angular momentum as a function of atomic number. The theoretical curves are compared with the experimental data. The theory predicts fairly accurately the values of Z at which orbits with angular momentum 0, 1, 2 and 3 begin to be occupied and hence the beginning of the 4 *f* rare earth series. Much later when Fermi was at Columbia University the same method was used by M. G. Mayer (2) to predict the beginning of the 5 *f* series.

In paper N° 45 the statistical potential is used to calculate the Rydberg correction for the S levels of all elements which can be deduced from an approximate solution of the Schroedinger equation for an S electron of zero energy. The experimental values fluctuate closely about the theoretical curve.

Paper N° 46 is a short note summarizing the results described at length in N° 44. Papers N° 47 and 48 are slightly variant German versions of papers N° 43, 44 and 45.

F. RASETTI.

(1) L. H. THOMAS, « Proc. Cambridge Phil. Soc. », *23*, 542 (1927).

(2) M. G. MAYER, « Phys. Rev. », *60*, 184 (1941).

43.

UN METODO STATISTICO PER LA DETERMINAZIONE DI ALCUNE PROPRIETÀ DELL'ATOMO

« Rend. Lincei », *6*, 602–607 (1927) (*).

Lo scopo del presente lavoro è quello di mostrare alcuni dei risultati sopra la distribuzione degli elettroni in un atomo pesante che si possono ottenere trattando tali elettroni, dato il loro considerevole numero, con un metodo statistico; ossia, in altre parole, considerandoli come un gas di elettroni che circonda il nucleo.

Naturalmente questo gas di elettroni viene a trovarsi in condizioni di degenerazione completa, per modo che ad esso non può applicarsi la statistica classica; si deve invece usare la forma di statistica proposta dall'autore [3] e basata sopra l'applicazione del principio di esclusione di Pauli alla teoria dei gas. Ciò ha per effetto che l'energia cinetica degli elettroni nelle condizioni in cui essi vengono a trovarsi nell'interno dell'atomo risulta effettivamente molto maggiore di quanto essa dovrebbe essere secondo il principio dell'equipartizione dell'energia e praticamente indipendente dalla temperatura, almeno finché questa non oltrepassa certi limiti.

In questa Nota noi mostreremo anzitutto come si possa calcolare statisticamente la distribuzione degli elettroni attorno al nucleo; e in base ad essa calcoleremo poi l'energia necessaria a ionizzare completamente l'atomo, cioè a strappargli tutti gli elettroni. Il calcolo della distribuzione degli elettroni attorno al nucleo permette inoltre di determinare l'andamento del potenziale alle varie distanze dal nucleo e quindi di conoscere il campo elettrico in cui vengono a trovarsi gli elettroni dell'atomo. Spero di poter in un prossimo lavoro mostrare l'applicazione di ciò al calcolo approssimato delle energie di legame dei singoli elettroni e ad alcune questioni relative alla struttura del sistema periodico degli elementi.

Per determinare la distribuzione degli elettroni dobbiamo anzitutto cercare la relazione tra la loro densità e il potenziale elettrico in ogni punto. Se V è il potenziale, l'energia di un elettrone sarà $-eV$ e quindi, secondo la statistica classica, la densità degli elettroni dovrebbe risultare proporzionale a $e^{eV/kT}$. Invece, secondo la nuova statistica, la relazione tra la densità e la temperatura è la seguente:

$$n = \frac{(2\pi mkT)^{3/2}}{h^3} F\left(\alpha e^{\frac{eV}{kT}}\right) \tag{1}$$

(*) Presentata nella seduta del 4 dicembre 1927 dal Socio O. M. Corbino.

(3) E. FERMI, « Zs. f. Phys. », *36*, 902 (1926).

dove α è costante per tutto il gas; la funzione F nel nostro caso (degenerazione completa) ha l'espressione asintotica

$$F(A) = \frac{4}{3\sqrt{\pi}} (\log A)^{3/2}. \tag{2}$$

Si trova dunque per il caso nostro

$$n = \frac{2^{7/2}\, \pi m^{3/2}\, e^{3/2}}{3\, h^3}\, v^{3/2} \tag{3}$$

dove

$$v = V + \frac{kT}{e} \log \alpha \tag{4}$$

rappresenta, a meno di una costante additiva, il potenziale. Osserviamo ora che siccome nel caso nostro si tratta di un gas di elettroni, bisogna tener conto del fatto [4], che il peso statistico degli elettroni è 2 (corrispondentemente alle due possibilità di orientazione dell'elettrone rotante); e quindi per la densità degli elettroni si deve, in realtà, prendere un valore eguale al doppio del valore (3); si ha cioè:

$$n = \frac{2^{9/2}\, \pi m^{3/2}\, e^{3/2}}{3\, h^3}\, v^{3/2}. \tag{5}$$

Se valesse nel nostro caso la statistica classica, si avrebbe l'energia cinetica media degli elettroni $= \frac{3}{2} kT$. Invece secondo la nuova statistica essa risulta data da

$$L = \frac{3}{2} kT G\left(\alpha e^{\frac{eV}{kT}}\right) / F\left(\alpha e^{\frac{eV}{kT}}\right)$$

dove G rappresenta una funzione che, nel caso della degenerazione completa, prende l'espressione asintotica

$$G(A) = \frac{8}{15\sqrt{\pi}} (\log A)^{5/2}.$$

Si trova dunque per il nostro caso

$$L = \frac{3}{5} ev. \tag{6}$$

Osserviamo ora che la densità elettrica in un punto risulta evidentemente data da $-ne$ e quindi il potenziale v soddisfa all'equazione

$$\Delta v = 4\pi ne = \frac{2^{13/2}\, \pi^2\, m^{3/2}\, e^{5/2}}{3\, h^3}\, v^{3/2}. \tag{7}$$

Nel caso nostro esso sarà poi evidentemente funzione soltanto della distanza r dal nucleo; la (7) può scriversi perciò

$$\frac{d^2 v}{dr^2} + \frac{2}{r}\frac{dv}{dr} = \frac{2^{13/2}\, \pi^2\, m^{3/2}\, e^{5/2}}{3\, h^3}\, v^{3/2}. \tag{8}$$

(4) W. PAULI, «Zs. f. Phys.», *41*, 81 (1927).

Se con Z indichiamo il numero atomico del nostro atomo si avrà poi evidentemente

$$\lim_{r=0} rv = Ze \tag{9}$$

$$\int n d\tau = 4\pi \int_0^\infty r^2\, n dr = Z \qquad (d\tau = \text{elemento di volume}).$$

Quest'ultima equazione, tenendo presente (5), può scriversi:

$$\frac{2^{13/2}\,\pi^2\, m^{3/2}\, e^{5/2}}{3\, h^3} \int_0^\infty v^{3/2}\, r^2\, dr = Ze. \tag{10}$$

Il potenziale v si otterrà dunque cercando una funzione che soddisfi all'equazione (8) con le due condizioni (9) e (10).

Per semplificare la ricerca di v cambiamo le variabili r, v, in altre due x, ψ ad esse proporzionali ponendo

$$r = \mu x \quad , \quad v = \gamma\psi \tag{11}$$

dove si ha

$$\mu = \frac{3^{2/3}\, h^2}{2^{13/3}\, \pi^{4/3}\, me^2\, Z^{1/3}} \quad , \quad \gamma = \frac{2^{13/3}\, \pi^{4/3}\, m Z^{4/3}\, e^3}{3^{2/3}\, h^2}\,. \tag{12}$$

Le equazioni (8), (9) e (10) diventano allora

$$\begin{cases} \psi'' + \dfrac{2}{x}\psi' = \psi^{3/2} \\ \lim\limits_{x=0} x\psi = 1 \\ \displaystyle\int_0^\infty \psi^{3/2}\, x^2\, dx = 1\,. \end{cases} \tag{13}$$

Queste equazioni si semplificano ulteriormente ponendo

$$\varphi = x\psi. \tag{14}$$

Esse diventano infatti

$$\begin{cases} \varphi'' = \varphi^{3/2}/\sqrt{x} \\ \varphi(0) = 1 \\ \displaystyle\int_0^\infty \varphi^{3/2}\sqrt{x}\, dx = 1\,. \end{cases} \tag{15}$$

È facile vedere che l'ultima condizione resta senz'altro verificata se φ si annulla per $x = \infty$. Resta dunque da cercare una soluzione della prima delle (15) con le condizioni ai limiti $\varphi(0) = 1$, $\varphi(\infty) = 0$.

Non essendomi riuscito di trovare l'integrale generale della prima delle (15), l'ho risolta numericamente. Il grafico della figura 1 rappresenta $\varphi(x)$; per x prossimo a zero si ha

$$\varphi(x) = 1 - 1.58\,x + \frac{4}{3}\,x^{3/2} + \cdots \tag{16}$$

Resta così risolto il problema della determinazione del potenziale elettrico dell'atomo a una data distanza dal nucleo. Esso risulta dato da

$$v = \gamma\,\frac{\varphi(x)}{x} = \frac{\gamma\mu}{r}\,\varphi(x) = \frac{Ze}{r}\,\varphi\left(\frac{r}{\mu}\right). \tag{17}$$

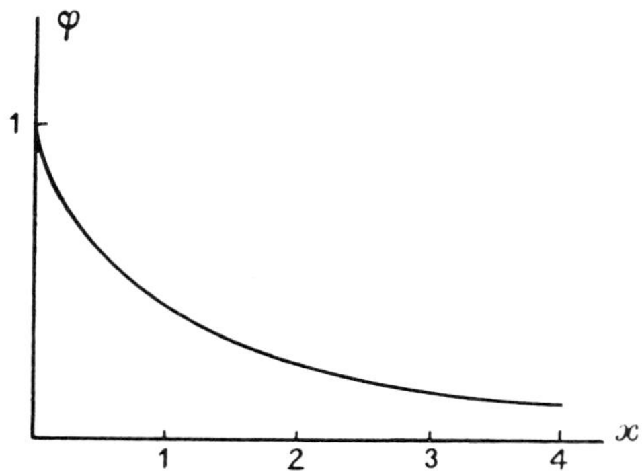

Fig. 1.

Si può dunque dire che il potenziale in ogni punto è uguale a quello prodotto da una carica efficace

$$Ze\,\varphi\left(\frac{r}{\mu}\right).$$

Passiamo ora a calcolare l'energia totale dell'atomo; questa dovrebbe calcolarsi come somma dell'energia cinetica di tutti gli elettroni e dell'energia potenziale del nucleo e degli elettroni. È più semplice però tener presente che in un atomo l'energia totale è eguale, salvo il segno, all'energia cinetica (ciò che del resto nel nostro caso può verificarsi con un facile calcolo). Si ha dunque

$$W = -\int L\,n\,d\tau$$

e tenendo presenti (5) (6) (11) (12) (14) si trova

$$W = -\frac{3}{5}\int_0^\infty r^2\,nv\,dr = -\frac{2^{13/3}\,3^{1/3}\,\pi^{4/3}\,me^4\,Z^{7/3}}{5\,h^2}\int_0^\infty \frac{\varphi^{5/2}}{\sqrt{x}}\,dx.$$

L'ultimo integrale può valutarsi tenendo presente che φ soddisfa alle (15) e a (16); si trova

$$\int_0^\infty \frac{\varphi^{5/2}}{\sqrt{x}}\,dx = -\frac{5}{7}\left(\frac{d\varphi}{dx}\right)_{x=0} = \frac{5}{7}\,1.58$$

e quindi si ha

$$W = -1.58 \frac{2^{13/3}\, 3^{1/3}\, \pi^{4/3}\, m e^4\, Z^{7/3}}{7\, h^2} = -1.58 \frac{2^{1/3}\, 3^{1/3}}{7\, \pi^{2/3}} Rh\, Z^{7/3}$$

cioè

$$W = -1.54\, Rh\, Z^{7/3} \tag{18}$$

dove con R si è indicato il numero di Rydberg, per modo che $-Rh$ è l'energia dello stato fondamentale dell'idrogeno.

La (18) ci dà l'energia necessaria per strappare da un atomo tutti i suoi elettroni. Naturalmente, dati i criteri statistici da cui essa è stata dedotta, essa comincia ad esser valida soltanto per valori considerevoli di Z; troviamo infatti che per l'idrogeno la (18) dà $W = -1.54\, Rh$, mentre si ha in realtà $W = -Rh$; la divergenza è dunque del 54 %. Per l'elio l'energia per produrre la ionizzazione completa è eguale evidentemente alla somma delle energie di ionizzazione dell'He e dell'He^+; si ha cioè

$$-W = (1.8 + 4)\, Rh = 5.8\, Rh$$

dalla teoria risulta invece $1.54 \cdot 2^{7/3}\, Rh = 7.8\, Rh$; la divergenza si riduce dunque per questo caso al 35 %. Per gli elementi immediatamente successivi all'elio (Li, Be, B, C), la quasi totalità dell'energia atomica è dovuta ai due soli elettroni K (nel carbonio circa l'86 %) e quindi il metodo statistico deve certamente dare ancora delle considerevoli divergenze. Per il C si trova infatti ancora una divergenza prossima al 34 %.

È da attendersi invece che per elementi di peso atomico considerevole le differenze tra la teoria statistica e i dati empirici vengano molto a ridursi; disgraziatamente mancano dati per un confronto preciso e si può basarsi soltanto sopra una grossolana valutazione dei numeri di schermo per le varie orbite; tale valutazione mostra tuttavia un accordo assai migliore.

N° 44.

For the introduction to this paper see N° 43.

44.

SULLA DEDUZIONE STATISTICA DI ALCUNE PROPRIETÀ DELL'ATOMO. APPLICAZIONE ALLA TEORIA DEL SISTEMA PERIODICO DEGLI ELEMENTI

« Rend. Lincei », 7, 342-346 (1928) (*).

In una Nota pubblicata recentemente in questi « Rendiconti » ebbi occasione di mostrare come si potessero ottenere dei risultati di media sopra la distribuzione degli elettroni attorno al nucleo di un atomo pesante considerando che gli elettroni formino attorno al nucleo una specie di atmosfera gassosa per il cui calcolo possono applicarsi delle considerazioni statistiche; data la densità considerevole di questo gas di elettroni, esso si trova alle temperature ordinarie in condizioni di degenerazione completa. Lo scopo di questa seconda Nota è di mostrare l'applicazione del metodo statistico alla teoria del sistema periodico degli elementi.

La teoria elettronica del sistema periodico si basa, come è noto, sopra la classificazione in gruppi degli elettroni atomici dovuta a Bohr [1] e Stoner [2] e sul principio di esclusione di Pauli. Un elettrone viene classificato mediante i suoi due numeri quantici totale e azimutale n e k; n può prendere tutti i valori interi da 1 a ∞, k, secondo la classificazione originaria di Bohr prende i valori interi da 1 ad n, secondo la nuova meccanica i valori da 0 ad $n - 1$. È noto d'altra parte che i risultati della nuova meccanica possono, in prima approssimazione, ottenersi per mezzo della teoria quantica ordinaria purché al numero k si attribuiscano invece che i valori interi $0, 1, 2, \cdots, n - 1$, i valori precedenti aumentati di 1/2.

Siccome nel presente lavoro noi ci serviremo dei metodi della vecchia teoria dei quanti, noi prenderemo dunque per k i valori 1/2, 3/2, 5/2, 7/2, $\cdots$, corrispondenti agli elettroni $s, p, d, f, \cdots$. Dal principio di esclusione, segue poi che in un atomo può aversi al massimo un elettrone per ogni stato

(*) Presentata dal Socio O. M. Corbino nella seduta dell'8 gennaio 1928.

(1) N. BOHR, « Zs. f. Phys. », 9, 1 (1922).

(2) E. C. STONER, « Phil. Mag. », *48*, 719 (1924).

quantico completamente definito. Ora si trova che fissati n e k esistono $4k$ stati quantici completamente definiti che hanno n e k come quanti totale e azimutale e quindi in un atomo si possono avere al massimo $4k$ elettroni n_k.

Nello stato fondamentale dell'atomo i suoi Z elettroni (Z = numero atomico) occuperanno gli Z posti liberi di minore energia. Ora, se il movimento di ogni elettrone avvenisse sotto l'azione di una unica carica elettrica concentrata nel nucleo, come nel caso dell'idrogeno, l'energia dipenderebbe soltanto, a meno delle piccole correzioni dovute alla relatività e all'elettrone rotante, dal numero quantico totale n. In realtà il moto di ogni elettrone ha luogo invece sotto l'azione della carica nucleare Ze ridotta dall'azione di schermo degli elettroni circostanti al nucleo. Per modo che l'elettrone può considerarsi soggetto all'azione di una carica efficace Z_{eff} che varia con la distanza r dell'elettrone dal nucleo e precisamente decresce da Ze ad e quando r cresce da 0 ad ∞. Ciò ha per effetto che, a parità di n, gli elettroni che hanno k più piccolo (le cui orbite hanno maggiore eccentricità e penetrano quindi nell'interno dell'atomo fino a minore distanza dal nucleo) vengono a trovarsi soggetti a una carica efficace più grande e quindi hanno energie di legame maggiori.

Avviene anzi spesso che gli elettroni aventi k grande sono legati meno fortemente di elettroni aventi n più grande ma k più piccolo. Questo fatto ha conseguenze importanti sopra la struttura del sistema periodico; esso infatti permette di rendere conto qualitativamente del fatto che nella successione degli elementi alcuni gruppi elettronici corrispondenti a valori elevati di k vengono occupati dopo altri gruppi aventi n più grande e k più piccolo. Così, per esempio, i 14 posti $4f$ ($n = 4$, $k = 7/2$) vengono occupati successivamente negli elementi con numeri atomici da 58 a 71 costituenti il gruppo delle terre rare; e in questi elementi si hanno già occupati i posti s aventi $n = 6$. Allo stesso modo i posti $3d$ ($n = 3$, $k = 5/2$) cominciano ad essere occupati nello Scandio (Z = 21) quando già sono occupati i posti $4s$. A queste circostanze sono legate, come è noto, le varie particolarità che si riscontrano nel sistema periodico.

Per la previsione precisa di queste particolarità sarebbe necessario conoscere il modo di variare della carica efficace alle varie distanze dal nucleo; ora, la teoria statistica svolta nella Nota precedente, ci ha permesso di risolvere, almeno in via di approssimazione, questo problema. Vogliamo ora vedere come si possano applicare quei risultati al calcolo della formazione dei diversi gruppi elettronici nell'atomo. La via più diretta per la risoluzione di questo problema sarebbe quella di calcolare l'energia delle varie orbite elettroniche nel campo di forza dovuto all'azione del nucleo e degli altri elettroni dell'atomo. Siccome però questo studio conduce a difficoltà considerevoli, ho preferito procedere per un'altra via, che, pur conducendo a risultati meno completi, permette tuttavia di raggiungere più rapidamente lo scopo. Nella Nota precedente è stata calcolata la distribuzione statistica degli elettroni alle varie distanze dal nucleo; è facile anche calcolare, in base alla teoria della degenerazione dei gas, la distribuzione delle velocità degli elettroni stessi e quindi si può dedurre quanti elettroni hanno una determinata quantità di moto areale, ossia un dato quanto azimutale k.

Dai risultati della Nota precedente, usando le stesse notazioni, risulta che la densità n degli elettroni a distanza r dal nucleo è

$$(1) \qquad n = \frac{2^{9/2}\,\pi m^{3/2}\,e^{3/2}}{3h^3}\,V^{3/2} = \frac{2^{9/2}\,\pi m^{3/2}\,Z^{3/2}\,e^3}{3h^3}\,\frac{1}{r^{3/2}}\,\varphi^{3/2}\left(\frac{r}{\mu}\right).$$

D'altra parte si deduce facilmente dalla teoria della degenerazione (3), tenendo presente che gli elettroni hanno peso statistico 2, che la densità degli elettroni di velocità compresa tra v e $v+dv$ è, nel caso della degenerazione completa

$$(2) \qquad n(v)\,dv = \begin{cases} \dfrac{8\pi m^3}{h^3}\,v^2\,dv & \text{per} \quad v < \left(\dfrac{3}{8\pi}\right)^{1/3} \dfrac{hn^{1/3}}{m} \\ 0 & \text{per} \quad v > \left(\dfrac{3}{8\pi}\right)^{1/3} \dfrac{hn^{1/3}}{m} \end{cases}$$

di qui risulta che il numero di elettroni aventi quantità di moto areale compresa entro i limiti p e $p+dp$ e contenuti nell'elemento di volume $d\tau$ è

$$(3) \qquad dN = 8\pi\,d\tau\,\frac{m^3}{h^3}\,\frac{p\,dp}{m^2 r^2}\sqrt{\left(\frac{3}{8\pi}\right)^{2/3}\frac{h^3 n^{2/3}}{m^2} - \frac{p^2}{m^2 r^2}}\,.$$

Tenendo poi presente che si ha $p = \frac{h}{2\pi}\,k$, e ponendo per n la sua espressione (1) e per $d\tau$ l'espressione $4\pi r^2\,dr$, si ha:

$$dN = 4k\,dk\,dr\sqrt{\frac{8mZe^2}{h^2}\,\frac{1}{r}\,\varphi\left(\frac{r}{\mu}\right) - \frac{k^2}{\pi^2 r^2}}\,.$$

Per ottenere il numero N_k degli elettroni aventi quanto azimutale nell'intervallo dk basta integrare l'espressione precedente rispetto ad r per tutti i valori per cui il radicale è reale. Ponendo $\xi = \mu r$ si trova subito, tenendo anche presente l'espressione di μ (Cfr. Nota I)

$$(4) \qquad N_k = \left(\frac{48}{\pi^2}\right)^{1/3} Z^{1/3}\,k\,dk\,\Phi\left(\frac{2^{4/3}\,k^2}{3^{2/3}\,\pi^{2/3}\,Z^{2/3}}\right)$$

dove si è posto

$$(5) \qquad \Phi(x) = \int \frac{d\xi}{\xi}\sqrt{\xi\varphi(\xi) - x}$$

essendo l'integrale esteso a tutti i valori positivi di ξ per cui il radicale è reale. La $\Phi(x)$ può calcolarsi numericamente in base ai valori di φ già calcolati nella Nota I; si trova che i suoi valori sono dati da

$$\Phi(0{,}49) = 0 \quad ; \quad \Phi(0{,}4) = 0.36 \quad ; \quad \Phi(0{,}3) = 0{,}87 \quad ; \quad \Phi(0{,}2) = 1{,}48 \quad ;$$
$$\Phi(0{,}1) = 2{,}20 \quad ; \quad \Phi(0) = 3{,}2$$

si ha poi $\Phi(x) = 0$ per $x > 0{,}49$.

(3) E. FERMI, «Zs. f. Phys.», *36*, 902 (1926).

Resta così risolto il problema della determinazione del numero degli elettroni che hanno un dato quanto azimutale k, in funzione del numero atomico Z dell'atomo; basta infatti perciò porre in (4) $dk = 1$ e si trova

$$N_k = \left(\frac{48}{\pi^2}\right)^{1/3} Z^{1/3} k\Phi\left(\frac{2^{4/3}\,k^2}{3^{2/3}\,\pi^{2/3}\,Z^{2/3}}\right). \tag{6}$$

La figura rappresenta i risultati della teoria e il loro confronto con l'esperienza; in essa si hanno quattro coppie di grafici; ciascuna coppia è costi-

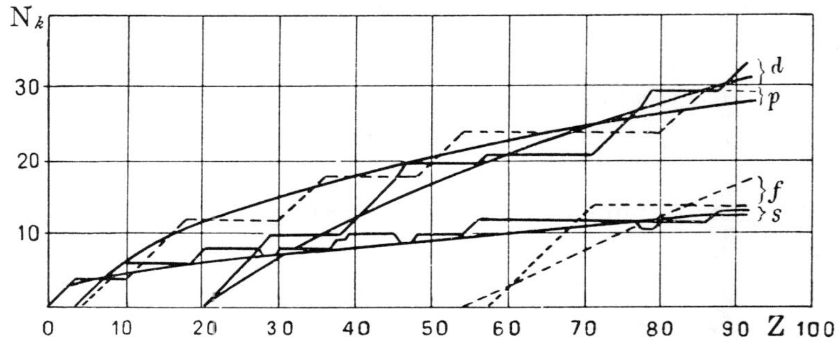

Fig. 1.

tuita da una curva continua e da una spezzata ed è relativa ad un valore del quanto azimutale (le quattro coppie corrispondono a

$$k = \frac{1}{2}\ ,\ \frac{3}{2}\ ,\ \frac{5}{2}\ ,\ \frac{7}{2},$$

cioè agli elettroni s, p, d, f). La curva continua rappresenta il numero degli elettroni dell'atomo che hanno il quanto azimutale indicato in funzione di Z (che è riportato in ascisse) quale risulta dalla teoria; invece le spezzate rappresentano lo stesso numero quale è in realtà. Si vede che l'accordo, almeno nelle linee generali è soddisfacente; naturalmente, per la stessa natura statistica della teoria svolta non si potrebbe attendere da essa di render conto particolarmente dei vari zig-zag delle curve empiriche, ma soltanto del loro andamento complessivo il quale viene in realtà rappresentato assai fedelmente dalla teoria. Notiamo in particolare che dalla teoria risulta per quale numero atomico debbono cominciare a comparire nell'atomo gli elettroni s, p, d, f. Tali numeri atomici sono quelli per cui le quattro curve teoriche del grafico hanno ascissa eguale ad 1, e cioè rispettivamente 1, 5, 21, 55. Ora gli elettroni s appaiono in realtà già nell'idrogeno (A = 1); gli elettroni p incominciano nel *Boro* (Z = 5); quelli d nello *Scandio* (Z = 21) e segnano l'inizio della anomalia del primo grande periodo; infine gli elettroni f cominciano ad apparire nel *Cerio* (Z = 58) e segnano l'inizio del gruppo anomalo delle terre rare. Si vede dunque che la teoria permette di prevedere con notevole esattezza i numeri atomici per i quali incominciano le varie anomalie del sistema periodico degli elementi.

N° 45.

For the introduction to this paper see N° 43.

45.

SULLA DEDUZIONE STATISTICA DI ALCUNE PROPRIETÀ DELL'ATOMO. CALCOLO DELLA CORREZIONE DI RYDBERG PER I TERMINI *S*

« Rend. Lincei », 7, 726–730 (1928) (*).

In due Note precedenti [1] ho mostrato che si può considerare l'insieme degli elettroni che circondano il nucleo di un atomo pesante come una specie di atmosfera di gas di elettroni in condizioni di completa degenerazione. Applicando allo studio di questa un metodo statistico si arriva a determinare la distribuzione degli elettroni attorno al nucleo e l'andamento del potenziale elettrico nell'interno dell'atomo come funzione della distanza dal nucleo.

Quest'ultimo risultato permette di determinare l'energia di qualsiasi orbita elettronica quando ne siano noti i numeri quantici totale e azimutale. Basta per questo procedere al modo seguente: Consideriamo un atomo di numero atomico Z e, tra i suoi Z elettroni, fissiamo la nostra attenzione sopra quello di cui vogliamo calcolare l'energia. Gli altri Z — 1 elettroni verranno invece considerati tutti insieme come una atmosfera da trattarsi col metodo statistico.

Dai risultati della Nota I si ricava, conservando le stesse notazioni, che il potenziale elettrico a distanza r dal nucleo è, per l'atomo di numero atomico Z — 1,

$$\frac{(Z-1)e}{r}\varphi\left(\frac{r}{\mu}\right).$$

Per passare dall'atomo di numero Z — 1 a quello di numero Z bisogna aggiungere una carica elettrica elementare al nucleo e un elettrone. Si riconosce allora che, in prima approssimazione, l'energia potenziale dell'elettrone aggiunto sarà:

$$V(r) = -\frac{e^2}{r}\left\{1 + (Z-1)\varphi\left(\frac{r}{\mu}\right)\right\} \tag{1}$$

dove per μ va messo naturalmente il valore che si ottiene dalla formula (12) della Nota I ponendo Z — 1 al posto di Z.

(*) Presentata dal Socio O. M. Corbino nella seduta del 6 maggio 1928.

(1) E. FERMI, « Rend. Lincei », 6, 602 (1927); 7, 342 (1928).

Determinata così l'energia potenziale dell'elettrone come funzione della sua distanza dal nucleo, si ha la possibilità di determinare i livelli energetici dei suoi stati quantici. Si può perciò ricorrere al metodo classico di Sommerfeld o, con maggiore esattezza, alla meccanica ondulatoria di Schroedinger.

In questo lavoro mi sono proposto di calcolare il valore dei termini *s* degli spettri di tutti gli elementi o, ciò che è lo stesso, la correzione di Rydberg delle serie *s*. In un altro lavoro Rasetti [2] ha calcolato i termini M_3 dello spettro dei raggi X.

L'equazione di Schroedinger per i termini *s* (quanto azimutale $k = 0$) si riduce a

$$\frac{d^2\psi}{dr^2} + \frac{2}{r}\frac{d\psi}{dr} + \frac{8\pi^2 m}{h^2}[E - V(r)]\psi = 0. \tag{2}$$

Questa equazione si trasforma subito nell'altra

$$z'' + z\left[-\eta + \frac{1}{x}\{1 + (Z - 1)\varphi(\gamma x)\}\right] = 0. \tag{3}$$

Dove si è posto

$$\begin{cases} r = \dfrac{h^2}{8\pi^2 me^2}x & \gamma = \dfrac{h^2}{8\pi^2 me^2 \mu} = \left\{\dfrac{16(Z-1)}{9\pi^2}\right\}^{1/3} \\ -\eta = \dfrac{h^2}{8\pi^2 me^4}E & z = x\psi. \end{cases} \tag{4}$$

Una autofunzione è determinata da una soluzione della (3) finita e continua nell'intervallo da o a ∞ della x. L'autovalore corrispondente η determina il livello energetico corrispondente e il numero degli zeri dell'autofunzione dà il numero quantico n dello stato. Per determinare la correzione di Rydberg basterà perciò riferirsi alla soluzione limite per $\eta = 0$ (cioè per $n = \infty$) e confrontare questa soluzione con la soluzione dell'equazione corrispondente per il caso dell'idrogeno.

Se nella (3) si pone $\eta = 0$ si trova l'equazione

$$z'' + \frac{z}{x}\{1 + (Z - 1)\varphi(\gamma x)\} = 0 \tag{5}$$

e l'equazione corrispondente per l'idrogeno è $(Z = 1)$

$$z'' + \frac{z}{x} = 0. \tag{6}$$

Siccome per x piuttosto grande l'energia potenziale dell'elettrone dell'idrogeno viene a coincidere con quella (1) del nostro elettrone, poiché φ tende a zero, le due autofunzioni, da un certo punto in poi, verranno ad avere andamento parallelo. Se supponiamo allora che per esempio l'n^{esimo} zero della soluzione di (5) coincida un l'n'^{esimo} zero della soluzione di (6), si potrà concludere che la correzione di Rydberg è

$$\alpha = n - n'. \tag{7}$$

(2) F. Rasetti, « Rend. Lincei », *7*, 915 (1928); « Zs. f. Phys. », *49*, 546 (1928).

Se, come avviene naturalmente nella maggior parte dei casi, gli zeri n^{esimo} ed n'^{esimo} non coincidono esattamente, è facile con un ovvio processo di interpolazione, determinare tuttavia la correzione di Rydberg.

La soluzione di (6) che si annulla per $x = 0$ è, come subito si riconosce,

$$z = \sqrt{x}\, J_1 (2 \sqrt{x}) \tag{8}$$

dove con J_1 si è indicata la funzione di Bessel di ordine uno, che è tabulata e di cui sono noti gli zeri. Il problema si riduce dunque a determinare una soluzione di (5) che si annulli per $x = 0$ o anzi, poiché ciò basta per il nostro scopo, a determinare i valori di x per cui questa soluzione si annulla. Per far questo conviene usare due metodi differenti per la ricerca della prima radice e delle successive; e ciò perché per $x = 0$ il secondo termine di (5) ha una singolarità.

Osserviamo ora che per $\gamma x < 0.3$ si può porre, commettendo un errore assai piccolo, $\varphi(\gamma x) = 1 - \gamma x$; con ciò (5) si riduce a

$$z'' + \frac{z}{x} \{Z - (Z - 1) \gamma x\}. \tag{9}$$

La soluzione di questa equazione è

$$z = e^{-\sqrt{\gamma (Z-1)}\, x} \left[x + \frac{2\sqrt{\gamma(Z-1)} - Z}{1!\,2!} x^2 + \frac{(2\sqrt{\gamma(Z-1)} - Z)(2 \cdot 2 \sqrt{\gamma(Z-1)} - Z)}{2!\,3!} x^3 + \cdots\right]. \tag{10}$$

Per mezzo di questa espressione, che converge assai rapidamente, si può agevolmente determinare il primo valore $x_1 > 0$ per cui si ha $z(x_1) = 0$.

Per determinare le successive radici dell'equazione $z(x) = 0$ ricorriamo invece alla seguente trasformazione.

$$z = e^{i \int y dx} \tag{11}$$

usata da Wentzel [(3)], che trasforma la (5) nell'equazione di Riccati

$$iy' = y^2 - f^2(x) \tag{12}$$

ove si è posto

$$f^2(x) = \frac{1}{x} \{1 + (Z - 1) \varphi(\gamma x)\}. \tag{13}$$

È facile risolvere la (12) per approssimazioni successive col metodo indicato da Wentzel; si trova così che l'n^{esima} radice x_n dell'equazione $z(x) = 0$ è data dalla seguente equazione

$$(n + 1)\pi - \frac{1}{4} \frac{f'(x_1)}{f^2(x_1)} = \int_{x_1}^{x_n} f(x)\, dx - \frac{1}{4} \frac{f'(x_n)}{f^2(x_n)} - \frac{1}{8} \int_{x_1}^{x_n} \frac{f'^2}{f^3} dx + \cdots \tag{14}$$

Siccome la funzione f è definita per mezzo della φ di cui non è nota l'espressione analitica, le quadrature che figurano in (14) debbono essere effettuate numericamente.

(3) G. Wentzel, «Zs. f. Phys.», *38*, 518 (1926).

A titolo di esempio accenniamo allo svolgimento del calcolo per il caso $Z = 55$ (Cesio). Dalla (10) risulta che, per questo valore di Z la prima radice di $z(x) = 0$ è $x_1 = 0.070$; dalla (14) si trova poi che la settima radice è $x_7 = 23.4$. Si ha dunque che $2\sqrt{x_7} = 9.68$ è prossima alla terza radice 10.17 dell'equazione $J_1(\xi) = 0$. Quindi la correzione di Rydberg è prossima a $7 - 3 = 4$. Con un procedimento di interpolazione si trova che essa è precisamente 4.16.

Il confronto tra la teoria e l'esperienza (*) è rappresentato nel grafico della figura. In ascisse sono riportati i numeri atomici e in ordinate le correzioni di Rydberg. La curva è ottenuta teoricamente e i punti rappresentano i valori empirici. Si vede che questi ultimi dipendono da Z in modo piuttosto irregolare, come era da attendersi dato che si tratta di spettri,

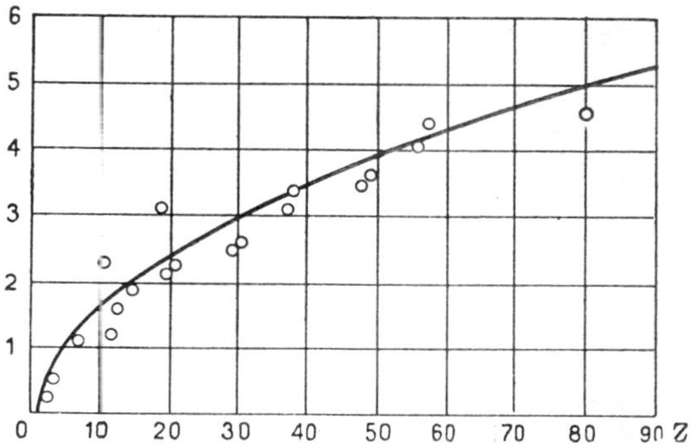

Fig. 1.

ottici. La teoria, data la sua stessa natura, non può render conto delle irregolarità che si hanno nel passaggio da un elemento all'altro; si vede tuttavia che essa rappresenta in modo soddisfacente l'andamento generale della correzione di Rydberg come funzione di Z.

L'accordo generale appare anche migliore se si osserva che i dati sperimentali si hanno quasi esclusivamente per elementi con carattere spiccatamente elettropositivo e si ha ragione di ritenere che gli elementi elettronegativi diano dei punti situati al di sopra della curva teorica, come ad esempio i punti corrispondenti ai due gas nobili Neon e Argon.

(*) I due punti sperimentali che si riferiscono ai gas nobili Ne ($Z = 10$) ed A ($Z = 18$) sono errati. I valori corretti, rispettivamente 1,36 e 2,20, giacciono leggermente al di sotto della curva teorica anziché molto al di sopra di essa (N. d. R.).

N° 49.

Fermi was invited to attend a restricted conference held in Leipzig under the chairmanship of P. Debye in 1928. This was considered an honor and presented an opportunity to report to a selected international audience the work performed in Rome. Fermi chose to present a review of the statistical model of the atom and its applications to various atomic problems by him and his collaborators. After a brief sketch of the theory, a numerical table of the function $\varphi(x)$ is given. Next, the appearance of electrons with given values of the orbital angular momentum in the periodic system is discussed.

There follows an account of calculations of X-ray levels, where Rasetti's results for the M-levels are reported. The results for the Rydberg correction af the S-levels, discussed in detail in N° 45, 48 are summarized. A brief account is given of calculations by Gentile and Majorana on the doublet separations for optical and X-ray levels, and on the relative intensities of the S-P lines in the alkali spectra. The behaviour of the $4f$ eigenfunctions is studied in the range of Z between 55 and 60; the charge distribution rapidly contracts with increasing Z, thus explaining the properties of the rare earths. The electron affinity of the halogens is the last problem discussed. This work had been the main activity of the Rome Institute in 1928.

F. RASETTI.

49.

ÜBER DIE ANWENDUNG DER STATISTISCHEN METHODE AUF DIE PROBLEME DES ATOMBAUES

Falkenhagen, Quantentheorie und Chemie. Leipziger Vortraege, 95–111 (1928).
Hirzel, Leipzig, 1928.

In einigen neulich erschienenen Noten habe ich eine Methode zur Berechnung von verschiedenen Eigenschaften der Atome dargestellt: die Atomelektronen werden als ein Elektronengas statistisch behandelt, das eine Art Atmosphäre um den Kern bildet. Ich möchte in dieser Mitteilung über die Ergebnisse berichten, welche bisher von mir und meinen Mitarbeitern erzielt worden sind, und über die möglichen Entwicklungen der dabei angewendeten Methode.

Wir werden also auf die genaue Lösung des Problems des Atombaues eines Atoms mit vielen Elektronen verzichten, indem wir die individuelle Behandlung aller Variablen durch eine Statistik der Gesamtheit der Elektronen ersetzen. Die Elektronen werden also als ein entartetes Gas betrachtet (die Dichte der Atomelektronen ist so gross, dass die Entartungsbedingung bei gewöhnlichen Temperaturen immer erfüllt ist), welches in der Nähe des Kernes von der Anziehung nach dem Kerne und der Abstossung der Elektronen voneinander in Gleichgewicht gehalten wird.

Als erstes Problem werden wir die Dichteverteilung dieses Gases um den Kern und den Verlauf des elektrischen Potentials im Inneren des Atoms als Funktion der Entfernung r vom Kerne bestimmen. Die Kenntnis dieses Potentials erlaubt die Berechnung beliebiger Termwerte, und zwar hat es sich gezeigt, dass nicht nur die Röntgenterme, sondern auch die optischen von der Theorie recht befriedigend dargestellt werden. In diesem Bericht werden wir auf die folgenden Punkte eingehen:

§ 1. Theorie der Bildung der Elektronengruppen im periodischen System. § 2. Berechnung der Röntgenterme. § 3. Berechnung optischer Termwerte. § 4. Berechnung der Intervalle der optischen und Röntgenmultipletts. § 5. Intensitätsverhältnisse der Serienlinien. § 6. Theorie der Gruppe der seltenen Erden. § 7. Theorie der Elektronenaffinität der Halogene. § 8. Mögliche Entwicklungen der statistischen Methode.

Seien also n und V die Dichte der Elektronen und das elektrische Potential in der Entfernung r vom Kerne. Zwischen diesen beiden Funktionen bestehen zwei Beziehungen; die erste liefert der Poissonsche Satz: $\Delta V = -4\pi\rho = 4\pi ne$. Da V nur von r abhängt, kann diese Gleichung in der Form geschrieben werden:

$$\Delta V = \frac{d^2 V}{dr^2} + \frac{2}{r}\frac{dV}{dr} = 4\pi ne \tag{1}$$

wobei e die Elementarladung darstellt.

Eine zweite Beziehung zwischen n und V liefert der statistische Zusammenhang zwischen Dichte und potentieller Energie; und zwar nach der klassischen Statistik würde die Boltzmannsche Beziehung

$$n = n_0 e^{\frac{eV}{kT}}$$

zwischen diesen beiden Grössen bestehen. Da aber das Elektronengas, wie gesagt, vollständig entartet ist, besteht zwischen n und V die folgende Beziehung

$$n = \frac{2^{9/2}\pi m^{3/2} e^{5/2}}{3h^3} V^{3/2} \tag{2}$$

m bedeutet die Masse des Elektrons.

Man beachte, dass die Temperatur in dieser Gleichung nicht mehr vorkommt; eine Folge der Elektronengasentartung ist also die Temperaturunabhängigkeit der Elektronenverteilung. Aus (1) und (2) können wir n eliminieren, und wir finden:

$$\frac{d^2 V}{dr^2} + \frac{2}{r}\frac{dV}{dr} = \frac{2^{13/2}\pi^2 m^{3/2} e^{5/2}}{3h^3} V^{3/2}. \tag{3}$$

Diese Gleichung bestimmt das Potential als Funktion von r. Man hat für V die zwei folgenden Grenzbedingungen

$$\left\{\begin{array}{ll} \lim\limits_{r=0} rV = Ze & \\ \int n\, d\tau = Z & (d\tau = \text{Volumenelement}). \end{array}\right. \tag{4}$$

Hierin ist das Integral über den ganzen Raum zu erstrecken. Z bedeutet die Ordnungszahl des Atoms. Die erste Gleichung besagt, dass das Potential in der Nähe des Kernes, wo die Elektronenabschirmung verschwindet, in die Form Ze/r übergeht. Die zweite Gleichung drückt aus, dass die Gesamtanzahl der Elektronen Z ist; mit Hilfe von (2) kann diese zweite Gleichung als Bedingung für V geschrieben werden.

Durch die Ansätze

$$x = r/\mu \quad , \quad \varphi = x\,V/\gamma \tag{5}$$

worin

$$\mu = \frac{3^{2/3}\,h^2}{2^{13/3}\,\pi^{4/3}\,me^2\,Z^{1/3}} \quad , \quad \gamma = \frac{2^{13/3}\,\pi^{4/3}\,me^3\,Z^{4/3}}{3^{2/3}\,h^2} \text{ ist,} \tag{6}$$

gehen (3) und (4) in die folgenden Gleichungen über

$$\frac{d^2\varphi}{dx^2} = \frac{\varphi^{3/2}}{\sqrt{x}} \tag{7}$$

$$\varphi(0) = 1 \quad , \quad \int_0^\infty \varphi^{3/2}\sqrt{x}\,dx = 1. \tag{8}$$

Man bestätigt ohne Mühe, dass die letzte Bedingung damit äquivalent ist, dass φ für $x = \infty$ verschwinden muss. Man hat also $\varphi(x)$ aus der Gleichung (7) mit den Grenzbedingungen $\varphi(0) = 1$, $\varphi(\infty) = 0$ zu bestimmen. Die folgende Tabelle gibt die numerisch bestimmten Werte von $\varphi(x)$ an.

x	$\varphi(x)$	x	$\varphi(x)$	x	$\varphi(x)$	x	$\varphi(x)$
0,00	1,000	0,8	0,485	8	0,037	38	0,0013
0,01	0,985	0,9	0,453	9	0,029	40	0,0011
0,02	0,972	1,0	0,425	10	0,024	45	0,00079
0,03	0,959	1,2	0,375	12	0,017	50	0,00061
0,04	0,947	1,4	0,333	14	0,012	55	0,00049
0,05	0,935	1,6	0,297	16	0,0093	60	0,00039
0,10	0,882	1,8	0,268	18	0,0072	65	0,00031
0,15	0,836	2,0	0,244	20	0,0056	70	0,00026
0,20	0,793	2,5	0,194	22	0,0045	75	0,00022
0,25	0,758	3,0	0,157	24	0,0037	80	0,00018
0,30	0,721	3,5	0,130	26	0,0031	85	0,00015
0,35	0,691	4,0	0,108	28	0,0026	90	0,00012
0,40	0,660	4,5	0,093	30	0,0022	95	0,00011
0,5	0,607	5,0	0,079	32	0,0019	100	0,00010
0,6	0,562	6	0,059	34	0,0017		
0,7	0,521	7	0,046	36	0,0015		

Durch die Kenntnis der Funktion $\varphi(x)$ wird das Potential im Inneren des Atoms bestimmt. Man bekommt nämlich aus (5) und (6)

$$V = \frac{Ze}{r}\,\varphi\left(\frac{r}{\mu}\right). \tag{9}$$

Aus (2) und (9) findet man auch die Dichte n der Elektronen

$$n = \frac{2^{9/2}\,\pi m^{3/2}\,Z^{3/2}\,e^3}{3\,h^3}\,\frac{1}{r^{3/2}}\,\varphi^{3/2}\left(\frac{r}{\mu}\right). \tag{10}$$

Wir wollen jetzt auf die Anwendungen der dargestellten Methode kurz eingehen.

§ 1. Theorie der Bildung der Elektronengruppen im periodischen System. – In einem Atom mit der Atomzahl Z, welches sich im Normalzustand befindet, kann man bekanntlich nach Bohr und Stoner angeben, wie viel Elektronen einer gegebenen Quantenbahn angehören. Wenn man zum Atom ein neues Elektron hinzufügt, um das nächstfolgende Element zu bilden, so nimmt dieses zwischen den im Sinne des Paulischen Verbots noch frei bleibenden Quantenbahnen diejenige an, welche die kleinste Energie hat. Nun sind bei den leichteren Atomen die Bahnen kleinster Energie auch diejenigen mit kleinster Hauptquantenzahl n, so dass die Elektronengruppen mit wachsender Hauptquantenzahl in den ersten Reihen des periodischen Systems ganz regelmässig nacheinander besetzt werden. Bei den schweren Atomen kommt es dagegen sehr oft vor, dass eine Bahn mit grosser Hauptquantenzahl stärker gebunden ist als eine mit kleinerem n. Das hat bekanntlich die Unregelmässigkeiten des periodischen Systems zur Folge; es kommt nämlich vor, dass ein Elektronenring mit einer gegebenen Hauptquantenzahl anfängt besetzt zu werden, wenn die unteren Elektronenringe noch nicht vollständig gebildet sind. Um diese Verhältnisse theoretisch vorhersagen zu können, ist es nötig, die Energie einer beliebigen Bahn berechnen zu können. Diese Energie wird aber von der Abschirmung durch die anderen Elektronen geliefert; die statistiche Theorie gibt nun die angenäherte Verteilung der Elektronen und erlaubt also eine angenäherte Auswertung ihrer Abschirmung. Man kann z. B. sehr leicht die Anzahl der Elektronen mit einer gegebenen azimutalen Quantenzahl k berechnen, welche in einem Atom mit der Atomzahl Z im Normalzustand vorhanden sind.

Aus Formel (10) kann man die Dichte des Elektronengases in einer beliebigen Entfernung vom Kerne entnehmen. Aus der Entartungstheorie folgt, dass, wenn man die Dichte des Gases in einem Punkte kennt, in diesem Punkte auch die Geschwindigkeitsverteilung der Elektronen bestimmt ist. Hieraus kann man ohne Mühe berechnen, wie viel Elektronen in einem Volumenelement vorhanden sind, deren Impulsmomente zwischen p und $p + dp$ liegen. Wenn wir die Impulsmomente mit der Einheit $h/2\pi$ messen, also $p = \frac{h}{2\pi}k$ setzen, finden wir, durch Integration über das ganze Atomvolumen, dass die Gesamtanzahl der Elektronen, deren Impulsmoment zwischen zwei gegebenen Grenzen k und $k + dk$ liegt

$$dN_k = \left(\frac{48}{\pi^2}\right)^{1/3} Z^{1/3}\, k\, dk\, \Phi\left(\frac{2^{4/3}\, k^2}{2^{2/3}\, \pi^{2/3}\, Z^{2/3}}\right) \tag{11}$$

ist, wo die Funktion Φ durch die folgende Formel definiert ist:

$$\Phi(A) = \int \sqrt{x\varphi(x) - A}\,\frac{dx}{x} \tag{12}$$

und das Integral über alle reellen Werte der Wurzel zu erstrecken ist. Die folgende Tabelle gibt die Werte von $\Phi(A)$ an. Für $A > 0{,}49$ ist immer $\Phi(A) = 0$, da in diesem Falle die Wurzel keine reellen Werte hat.

A	φ (A)	A	φ (A)
0,49	0,00	0,2	1,43
0,4	0,36	0,1	2,2
0,3	0,88	0,0	3,2

Nun bekommt man bekanntlich den besten Anschluss der alten Quantentheorie an die Wirklichkeit, wenn man für die azimutale Quantenzahl k halbzahlige Werte annimmt, und zwar $k = \frac{1}{2}, \frac{3}{2}, \frac{5}{2}, \frac{7}{2}$ für die *s*–, *p*–, *d*–, *f*–Elektronen setzt. Wir bekommen also die Anzahl der Elektronen im Atom mit einer gegebenen azimutalen Quantenzahl, indem wir in (11) den halbzahligen Wert von k einsetzen und für dk den Wert 1. Wir finden folglich:

$$N_k = \left(\frac{48}{\pi^2}\right)^{1/3} Z^{1/3}\, k\Phi\left(\frac{2^{4/3}\, k^2}{3^{2/3}\, \pi^{2/3}\, Z^{2/3}}\right). \tag{13}$$

In der Fig. 1 ist der Vergleich von Theorie und Erfahrung graphisch dargestellt. In der Figur sind vier Paar Kurven gezeichnet; jedes Paar bezieht sich auf einen Wert von k. (Die vier Paare beziehen sich auf die *s*–, *p*–, *d*–, *f*–Elektronen, also auf $k = \frac{1}{2}, \frac{3}{2}, \frac{5}{2}, \frac{7}{2}$) Jedes Paar besteht aus

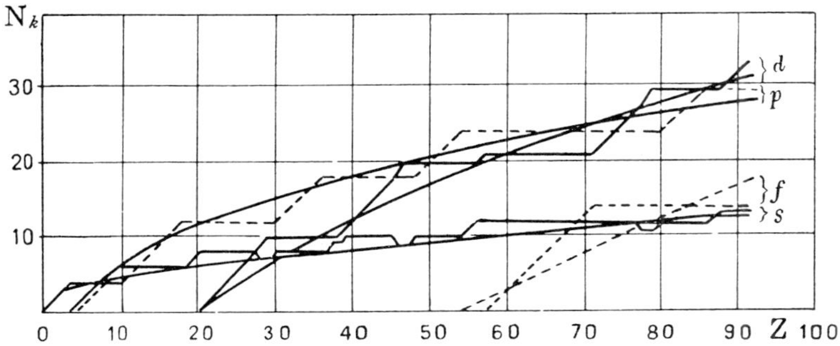

Fig. 1.

einer kontinuierlichen Kurve, welche die Anzahl N_k der k–Elektronen als Funktion der Ordnungszahl nach der Theorie (Formel (13)) darstellt, und aus einer Zickzacklinie, welche N_k erfahrungsgemäss nach der Stonerschen Tabelle angibt.

Die Theorie kann natürlich, wegen ihres statistischen Charakters, die feineren Besonderheiten der empirischen Kurven nicht wiedergeben; man sieht jedoch, dass der allgemeine Verlauf der entsprechenden theoretischen und empirischen Kurven ganz derselbe ist.

Aus der Theorie können wir die Werte der Ordnungszahlen entnehmen, für welche eine neue Atomgruppe des periodischen Systems beginnt. So z. B. sind die Ordnungszahlen, für welche die s–, p–, d–, f–Elektronen nach der Theorie erst im Atom erscheinen müssen, diejenigen, für welche die Ordinaten der vier theoretischen Kurven den Wert 1 haben; aus Fig. 1 liest man für diese Ordnungzahlen die Werte: 1, 5, 21, 55 ab. Nun ist in Wirklichkeit ein s–Elektron bereits im Wasserstoff ($Z = 1$) vorhanden; die p–Elektronen erscheinen erst im Bor ($Z = 5$); die d–Elektronen im Scandium ($Z = 21$), wo die Anomalie der ersten grossen Periode beginnt: die f–Elektronen erscheinen erst im Cerium ($Z = 58$), wo die Gruppe der seltenen Erden anfängt. Die Theorie gibt also angenähert die Ordnungszahlen an, bei welchen die verschiedenen Anomalien des periodischen Systems beginnen.

§ 2. Berechnung der Röntgenterme. – Die Ermittlung des elektrischen Potentials im Innern des Atoms erlaubt die Berechnung der Bindungsenergien der Elektronen in den verschiedenen Quantenzuständen. Dabei hat man allgemein zu bemerken, dass das Problem der Termberechnung für ein Atom mit Z Elektronen nach der Quantenmechanik ein 3 Z–dimensionales Problem darstellt (oder noch besser ein 4 Z–dimensionales, wenn man nach Dirac das Eigenmoment der Elektronen [1] mitberechnet). Dieses Problem kann jedoch in erster Näherung auf ein dreidimensionales Problem reduziert werden, indem man als Modell der ersten Näherung jedes Elektron separat behandelt unter der Wirkung des Potentials des Kernes und der übrigen $Z - 1$ Elektronen, für welche die statistische Raumverteilung angenommen wird. Nun ist nach (9) das elektrische Potential in der Entfernung r vom Kerne für ein Atom mit der Ordnungszahl $Z - 1$

$$\frac{(Z-1)\,e}{r}\,\varphi\left(\frac{r}{\mu}\right). \tag{14}$$

Um von diesem Atom zu unserem Atom mit der Ordnungszahl Z überzugehen, müssen wir die Kernladung um eine Elementarladung vermehren und ein Elektron zur Elektronenwolke hinzufügen. Die potentielle Energie dieses Elektrons wird in erster Näherung die Summe der Energie

$$-\frac{(Z-1)\,e^2}{r}\,\varphi\left(\frac{r}{\mu}\right)$$

(1) Vergl. den Vortrag von Herrn P. A. M. Dirac, S. 85 ff.

sein, welche vom Potential (14) herrührt, und der Energie

$$-\frac{e^2}{r},$$

welche von der Vermehrung der Kernladung herkommt. Man hat also die potentielle Energie

$$V(r) = -\frac{e^2}{r}\left\{1 + (Z-1)\,\varphi\left(\frac{r}{\mu}\right)\right\}. \tag{15}$$

Es werden mit diesem Verfahren zwei Annäherungen gemacht. Zuerst wird die Polarisierbarkeit des Atomrumpfes vernachlässigt; diese ist aber hauptsächlich für die Terme wichtig, deren Bahnen ausserhalb des Rumpfes verlaufen (vgl. hierzu § 8). Zweitens werden die Resonanzeffekte (2) im Sinne der Heisenbergschen Zwei–Körper–Theorie vernachlässigt. Die Resonanz rührt aber meistens von den Quantenbahnen her, welche fast dieselben Energiesprünge wie die des Leuchtelektrons haben; sie ist also nicht sehr gross für die optischen Bahnen von grosser Hauptquantenzahl, und auch nicht für die Röntgenterme, da die Störungspotentiale in keinem Fall in bezug auf das Potential (15) gross sind.

Wegen des statistischen Charakters der Theorie werden wir eine bessere Übereinstimmung mit der Erfahrung für die Elektronenbahnen erwarten, welche in den tiefen Gebieten des Atoms verlaufen, da die statistische Verteilung der Elektronen in diesen Gebieten eine bessere Darstellung der wirklichen Verteilung gibt. Wir werden also eine bessere Übereinstimmung für die Röntgenterme als für die optischen erwarten; übrigens kann die statistische Theorie nur eine genaue Darstellung von den Atomeigenschaften geben, welche regelmässig von der Atomzahl abhängen.

Bei den Röntgentermen hat das grösste Interesse die Berechnung derjenigen, welche eine grosse Abschirmung haben, weil man bei den anderen, z. B. bei den *K*–Termen, auch ohne eine genaue Kenntnis der Elektronenverteilung zu einer guten Abschätzung des Termwertes gelangen kann. Dagegen ist die Abschirmung z. B. bei den M–Termen sehr gross; bei vielen Atomen ist sie sogar grösser als die effektive Kernladung. Ihr Wert hängt also in hohem Masse von der Elektronenverteilung ab.

Nach der statistischen Theorie wurden die M_3–Terme von Rasetti berechnet. Die Rechnung wurde mit der folgenden Methode durchgeführt. Der Feldskalar ψ der Schrödingerschen Wellengleichung für die M_3–Terme des Wasserstoffs hat, als Funktion von r ein sehr steiles Maximum, so dass seine Werte praktisch in einer ziemlich dünnen sphärischen Schicht konzentriert sind. Qualitativ bleibt die Verteilung dieselbe, auch wenn man das Newtonsche Potential des Wasserstoffs durch das Potential (15) ersetzt. Die Wellenfunktion und der Termwert werden deshalb in erster Näherung von dem Verlauf des Potentials in einer dünnen sphärischen Schicht bestimmt. In einer solchen Schicht kann man immer (15) angenähert in der Form

$$V(r) = A + \frac{B}{r} + \frac{C}{r^2} \tag{16}$$

(2) Vgl. den Vortrag von Herrn LONDON, S. 59 ff.

schreiben, wo A, B, C Konstante darstellen, welche so bestimmt werden sollen, dass der Ausdruck (16) in der genannten Schicht das Potential (15) möglichst genau darstellt. Für die Form (16) des Potentials kann man aber die Wellengleichung genau lösen, so dass man eine erste Näherung für den Energiewert bekommt. Die zweite Näherung kann ohne Mühe störungstheoretisch gewonnen werden; die Korrektion des Termes ist aber immer sehr klein.

Rasetti hat seine Ergebnisse graphisch dargestellt. In der Fig. 2 sind als Abszissen die Ordnungszahlen der Atome eingetragen, als Ordinaten in üblicher Weise die Quadratwurzel der M_3–Terme. Die Sterne geben

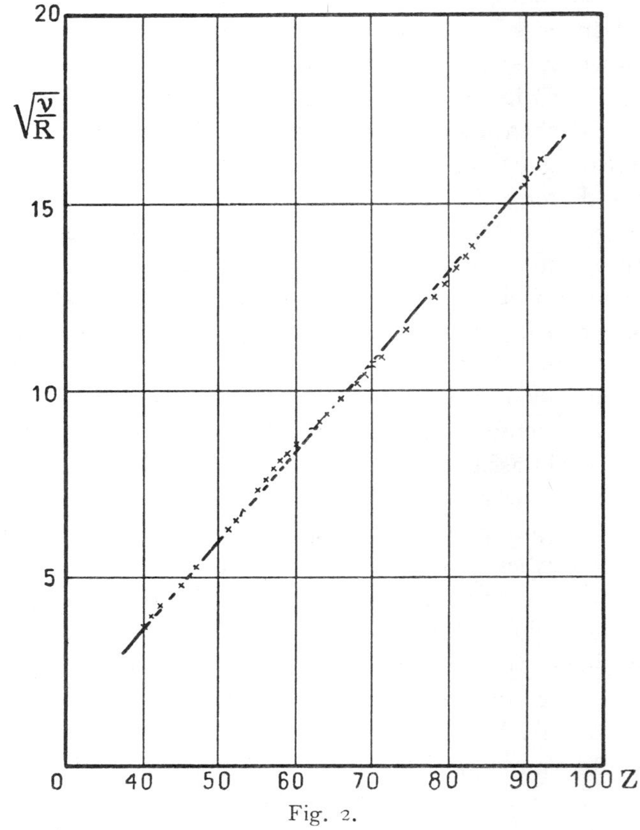

Fig. 2.

die experimentellen, die Kurve die theoretischen Werte an; es wurde eine kleine Korrektion für den in der Theorie vernachlässigten Einfluss der Relativität und des elektronischen Eigenmoments eingeführt. Man sieht, dass man kaum eine bessere Übereinstimmung von Theorie und Erfahrung hätte erwarten können.

§ 3. Berechnung optischer Termwerte. – Die Regelmässigkeit der optischen Energieniveaus als Funktion der Atomzahl ist bekanntlich viel weniger ausgeprägt als für die Röntgenterme. In der Fig. 3 sind als Abszissen die Atomnummer, als Ordinaten die Rydbergkorrektionen der *s*–Terme eingetragen. Die Punkte geben die empirischen Werte der Rydbergkorrektion an; für die Atome, welche mehr als eine *s*–Termreihe haben, wurde ein Mittelwert genommen. Man sieht, dass die empirischen Punkte ziemlich unregelmässig verteilt sind, jedoch liegen sie alle in der Nähe einer glatten Kurve (*). Die kontinuierliche Kurve in der graphischen Darstellung gibt die theoretisch mit dem statistischen Potential (15) gewonnenen Werte der Rydbergkorrektion der *s*–Terme an. Man sieht, dass die Theorie den allgemeinen Verlauf ganz richtig wiedergibt; dagegen werden auch in diesem Falle die feineren Unterschiede von Atom zu Atom von der statistischen Behandlung nicht gegeben. Das hat darin seinen Grund, dass die *s*–Bahnen, um uns in der Sprache der alten Quantentheorie auszudrücken, einen Teil haben, welcher ausserhalb des Atoms verläuft, und einen Teil, welcher nah an den Kern herankommt. Die Rydbergkorrektion hängt also teilweise von dem im Inneren des Atoms verlaufenden Teil der Bahn ab, teilweise aber auch von dem in der äusseren Schicht verlaufenden Teil. Die Unregelmässigkeiten werden also in dieser äusseren Schicht erzeugt, da die äussere Schicht sich unregelmässig von Atom zu Atom nach seinem chemischen Charakter ändert. Die statistische Theorie kann natürlich diese Unregelmässigkeiten nur ungenau darstellen.

Die theoretische Kurve der Fig. 3 wurde mit der folgenden Methode gewonnen: Es wurde die Lösung der Schrödingerschen Wellengleichung

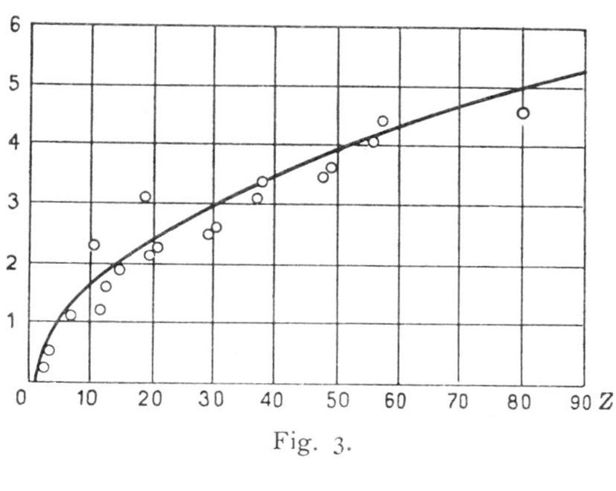

Fig. 3.

der *s*–Terme mit dem Ausdruck (15) des Potentials gesucht, welche dem Wert Null der Energie, also der Hauptquantenzahl ∞ entspricht. Diese Lösung kann ohne grosse Schwierigkeit gewonnen werden, indem man seinen ersten Teil (r klein) mit einem asymptotischen Ausdruck berechnet, und diesen weiter nach grossem r mit Hilfe der von Wentzel (3) benutzten

(*) The experimental points referring to the noble gases Ne (Z = 10) and A (Z = 18) are erroneous. The correct values, 1,36 and 2,20 respectively, lie slightly below the theoretical curve instead of much above it (Editor's note).

(3) G. WENTZEL, «Zs. f. Phys.», *38*, 518 (1926).

Transformation der Wellengleichung in eine Riccatische Gleichung fortsetzt. Die so erhaltene Lösung wird mit der entsprechenden Lösung für den Fall des Wasserstoffs verglichen. Da das Potential (15) für grosses r in das Newtonsche Potential übergeht, so ist für grosses r die Frequenz der Knotenstellen der beiden Lösungen dieselbe. Für kleines r sind dagegen die Knoten der ersten Lösung dichter als die der zweiten, weil das Potential im Inneren unseres Atoms rascher wächst als für den Fall des Wasserstoffs. Die Differenz der Anzahl der Nullstellen der beiden Lösungen gibt ohne weiteres die Differenz ihrer (unendlich grossen) Quantenzahlen, also die Rydbergkorrektion.

§ 4. Berechnung der Intervalle der optischen und Röntgenmultipletts. – Die statistische Theorie der Multiplettintervalle wurde von G. Gentile und E. Majorana durchgeführt. Diese Verfasser haben die Intervalle der M_3–Röntgendubletts und die der $6\,{}^2P$ und $7\,{}^2P$–Dubletts des Cäsiums berechnet. Die Methode war die folgende: Man muss zuerst die Wellengleichung mit dem statistischen Potential (15) numerisch lösen. Dieser Teil der Arbeit ist etwas lang, weil man den Eigenwert nicht kennt, so dass man ihn versuchsweise in der Nähe des empirisch bekannten Energiewertes suchen muss. Sobald man die Eigenfunktion berechnet hat, kann man die vom elektronischen Eigenmoment herrührende Spaltung mit der wohlbekannten Formel

$$\Delta\nu = \frac{e^2 h\left(k + \frac{1}{2}\right)}{8\pi^2 m^2 c^2} \, \overline{\frac{1}{r^3}}$$

berechnen, wo

$$\overline{\frac{1}{r^3}} = \int \frac{1}{r^3} \psi^2 \, d\tau$$

den Mittelwert von $1/r^3$ darstellt, und ohne weiteres mit einer numerischen Integration erhalten werden kann.

Die empirischen Werte der M_3–Dubletts wurden von der Theorie mit grosser Genauigkeit gegeben, da der Fehler nie grösser als 10 % war. Schlechter ist die Übereinstimmung für den Fall der optischen Dubletts des Cäsiums. Es hat sich nämlich gezeigt, dass das Intervall dieser Dubletts sehr stark von dem Verlauf des Potentials in der äusseren Schicht des Atoms abhängt. Wenn man die Spaltung mit (15) berechnet, findet man ungefähr das Doppelte des experimentellen Wertes, wenn man aber das Potential in den äusseren Teilen des Atoms ein wenig ändert, um davon Rechnung zu tragen, dass der Rumpf der Alkalimetalle edelgasähnlich ist, also mehr zusammengezogen als die Rümpfe der anderen Atome (vgl. auch § 8), wird der Fehler viel kleiner.

§ 5. Intensitätsverhältnisse der Serienlinien. – Das Intensitätsverhältnis der beiden ersten Linien der Hauptserie des Cäsiums wurde ebenfalls von Gentile und Majorana berechnet. Dieses Problem hat deshalb Interesse, weil das Intensitätsverhältnis der beiden ersten Linien der Hauptserien der Alkalimetalle experimentell von der Grössenordnung 100 ist, während es nach den gewöhnlichen Theorien viel kleiner sein sollte.

Dieses Verhältnis wurde von Gentile und Majorana mit der folgenden Methode bestimmt: Es wurden die Eigenfunktionen $\psi_1\, \psi_2\, \psi_3$ der drei Terme 6 S, 6 P, 7 P mit dem statistischen Potential (15) numerisch berechnet. Das Intensitätsverhältnis wird dann bekanntlich durch die beiden Matrizenelemente

$$\int \psi_1\, \psi_2\, x d\tau \quad \text{und} \quad \int \psi_1\, \psi_3\, x d\tau$$

welche die Schwingungsamplituden bestimmen, berechnet.

Das Intensitäsverhältnis der zwei Linien 6 S — 6P und 6 S — 7 P ergab sich mit dieser Methode ungefähr gleich 125. Dieser Wert stimmt gut mit dem experimentellen überein.

§ 6. Theorie der Gruppe der seltenen Erden. – Die Gruppe der seltenen Erden entsteht bekanntlich durch die Besetzung der 14 $4f$–Elektronenbahnen; sie ist durch die äusserst grosse chemische Ähnlichkeit seiner 14 Elemente charakterisiert. Diese Ähnlichkeit wird bekanntlich dadurch erklärt, dass der Unterschied von zwei seltenen Erden in einer verschiedenen Anzahl von Elektronen in der N–Elektronenschale besteht, also in einer inneren Schale des Atoms, welche seine äusseren (chemischen) Eigenschaften nur wenig beeinflusst. Diese Erklärung bedarf aber einer Ergänzung; es ist nämlich nicht klar, in welchem Sinne die $4f$–Bahnen als innere Bahnen bezeichnet werden sollen. Wenn man z. B. die Elemente betrachtet, welche im periodischen System vor den seltenen Erden kommen, also die Elemente 55 Cs, 56 Ba, 57 La, so findet man, dass die $4f$–Bahnen für diese Elemente optische Bahnen sind, deren Bindungsenergie klein, und deren Radius grösser ist als der Atomradius. Man sollte also meinen, dass mit dem Wachsen der Atomzahl von 57 (Lanthan) zu 58 (Cerium) der Radius der $4f$–Bahn sich ein wenig zusammenzieht, und dass ihre Energie ein wenig zunehmen sollte, so dass das hinzugefügte Elektron die $4f$–Bahn besetzt. Man würde also erwarten, dass die $4f$–Elektronen der seltenen Erden, wenn sie auch eine sehr kleine Hauptquantenzahl haben (N–Schale!), eine Bahn beschreiben sollten, welche schwach gebunden ist und in der äusseren Schicht des Atoms kreist; dann würde aber natürlich die Erklärung für die chemische Ähnlichkeit der seltenen Erden nicht mehr bestehen.

Ich habe also das Verhalten der mit dem statistischen Potential (15) numerisch berechneten $4f$–Bahnen für Atomzahlen zwischen 55 und 60 untersucht; diese Untersuchung liefert eine natürliche Erklärung für die besprochene Schwierigkeit.

Aus dem Vergleich der $4f$–Eigenfunktionen für die Werte 55 und 60 von Z sieht man, dass die $4f$–Eigenfunktion sich sehr rasch ändert, während die Atomzahl von 55 bis 60 wächst. Und zwar für $Z = 55$ hat die Eigenfunktion ihre von Null beträchtlich verschiedenen Werte in einem

Gebiet ausserhalb des Atoms, während die Eigenfunktion für Z = 60 die von Null verschiedenen Werte im Inneren des Atoms hat, und zwar sehr nahe an dem Kern (etwa 0,3 Å vom Kerne entfernt.) Genauer ist der Sachverhalt der folgende: Beide Eigenfunktionen, als Funktionen von r betrachtet, haben zwei Maxima; ein erstes Maximum in der Nähe des Kernes, und ein zweites ausserhalb des Atoms. Für die Eigenfunktion mit Z = 55 ist aber das erste Maximum sehr niedrig und das zweite gross; für die andere Eigenfunktion (Z = 60) ist die Sache umgekehrt, also das erste Maximum gross und das zweite klein. Die Bindungsenergie der $4f$–Elektronen nimmt mit wachsendem Z zu; nicht aber sehr schnell. Ich kann ihren Wert nicht genau angeben, weil es für den Fall Z = 60 (seltene Erden) sehr stark von der Relativität beeinflusst wird.

Zusammenfassend können wir also sagen, dass die $4f$–Bahnen der seltenen Erden keine sehr stark gebundene Bahnen sind; sie sind aber von den chemischen Verhältnissen deshalb nur sehr wenig beeinflusst, weil sie sehr tief im Inneren des Atoms verlaufen.

Wir können vielleicht die Sache etwas klarer darstellen, indem wir an die Bohrschen Bahnen denken; nach Bohr sind die $4f$–Bahnen Kreisbahnen. Für das Atom Z = 55 ist die $4f$–Bahn ein Kreis, welcher ausserhalb des Atoms in der Nähe des äussersten Randes des Rumpfes läuft. Wenn Z wächst, zieht sich der Radius der Bahn zusammen, so dass der Kreis im Inneren des Rumpfes zu liegen kommt; hier aber wächst die effektive Kernladung, weil die Elektronenabschirmung kleiner wird; das hat eine weitere Zusammenziehung der Bahn zur Folge und also ein weiteres Wachsen der effektiven Kernladung usw. Man sieht auf diese Weise, wie das plötzliche Zusammenziehen der $4f$–Bahn zustande kommt, welches der Bildung der seltenen Erden entspricht.

§ 7. THEORIE DER ELEKTRONENAFFINITÄT DER HALOGENE. – Wenn man zu einem neutralen Atom ein neues Elektron hinzufügt, um ein negatives Ion zu bilden, ist die darauf wirkende Kraft sehr von der Kraft verschieden, welche auf das Leuchtelektron eines neutralen Atoms wirkt. In dem ersten Falle (neutrales Atom + Elektron) ist nämlich der Rumpf des Systems ungeladen; in dem zweiten (Ion + Elektron = neutrales Atom) hat der Rumpf die Ladung 1. Wenn das Elektron sich in grosser Entfernung vom Atom befindet, ist also in dem ersten Falle die darauf wirkende Kraft nur diejenige, welche von der Polarisation des Atoms herrührt; in dem zweiten ist dagegen der Rumpf elektrisch geladen, also seine Wirkung viel grösser. Wenn das Elektron sich dem Kerne nähert, und in das Innere des Rumpfes eindringt, wächst die effektive Kernladung, welche in beiden Fällen für $r = 0$ den Wert Ze erreicht.

In dem Falle der Elektronenaffinität der Halogene wird ein Elektron zu einem neutralen Atom zugefügt, um ein negatives Halogenion zu bilden. Dieses Elektron wird also teilweise von der Kraft angezogen, die von der Polarisation des Atoms herrührt, teilweise auch, wenn es sich im Inneren des Atoms befindet, von der Anziehung nach dem Kerne, die im Inneren des Atoms nicht vollständig von der Wirkung der übrigen Elektronen abgeschirmt wird. Diese letzte Wirkung ist sehr leicht zu berechnen: in der Tat ist das Potential dieser Anziehung, wenn man die, durch das hinzugefügte Elektron verursachte Deformation des Atoms vernachlässigt, nach (9) gegeben; die potentielle Energie des Elektrons ist also

$$- \frac{Ze^2}{r} \varphi\left(\frac{r}{\mu}\right). \tag{17}$$

Die Wirkung der Polarisation ist viel schwieriger zu berechnen; ich habe diese Wirkung vorläufig vernachlässigt (vgl. § 8), da man auch ohne sie die Möglichkeit der Bildung negativer Halogenionen verstehen kann. Nur der Wert der Elektronenaffinität wird nach der Theorie durch ihre Vernachlässigung etwas kleiner als der experimentelle Wert.

Ich habe die Berechnungen für den Fall des Jods ($Z = 53$) durchgeführt. Beim Jod hat man eine unvollständige edelgasähnliche Elektronenschale, wo ein Elektron fehlt. Sie besteht nämlich aus zwei $5s$– und fünf $5p$–Elektronen; es ist also noch ein $5p$–Platz frei. Wenn man also ein neues Elektron hinzufügt, geht dieses auf den freien $5p$–Platz.

Ich habe die Eigenfunktion dieses Platzes mit dem Potential (17) numerisch berechnet; man findet, wie es sein muss, dass seine Energie negativ ist, dass also das Elektron an das Atom gebunden wird. Es ist bemerkenswert dabei, dass man, wenn man die $6s$–, $6p$–oder $5d$–Bahnen zu berechnen sucht, findet, dass keine von diesen eine negative Energie besitzt, also dass die $5p$–Bahn die einzige ist, in welcher das Elektron gebunden werden kann. Der Wert der Bindungsenergie, welcher die Elektronenaffinität misst, kommt aus der Theorie etwas klein heraus; theoretisch findet man 2,2 Volt, während die experimentelle Elektronenaffinität des Jodatoms etwa 3,6 Volt ist. Die Ursache für den kleinen theoretischen Wert ist in der Vernachlässigung der Polarisation zu suchen.

§ 8. MÖGLICHE ENTWICKLUNGEN DER STATISTISCHEN METHODE. – In den vorigen Paragraphen habe ich einige Anwendungen der statistischen Betrachtung der Elektronengasatmosphäre darzustellen gesucht, welche den Kern eines Atoms umgibt.

Diese Betrachtungen können in folgender Hinsicht verfeinert werden. Ich habe oft bemerkt, dass die statistische Elektronenverteilung (10) die Verhältnisse der äusseren Schicht des Atoms nur ungenau darstellt. Das hat darin seinen Grund, dass, wenn man mit V das mittlere Potential an einem Orte darstellt, die potentielle Energie eines Elektrons an jenem Orte nicht genau $= -eV$ ist, sondern etwas grösser, weil man in der Berechnung der Energie eines Elektrons, nur die Wirkung des Kernes und der übrigen Elektronen, und nicht die des betrachteten Elektrons mitrechnen muss. So z. B. wenn das Elekton am äussersten Rande des Atoms ist, steht es unter der Wirkung der effektiven Kernladung Eins, und nicht Null, die aus dem Ausdruck $-eV$ der Energie resultieren würde. Das hat zur Folge, dass die Elektronengashülle etwas mehr zusammengezogen ist als nach dem Verteilungsgesetz (10). So würde z. B. nach (10) die Elektronenwolke mit rasch abnehmender Dichte sich bis ins Unendliche ausdehnen. Wenn man aber die Theorie in dem hier erklärten Sinne verfeinert, findet man, dass die Atmosphäre begrenzt bleibt. Ich hoffe, dass es mit einer Weiterführung dieser Rechnungen möglich sein wird, die Wirkung der Polarisation des Atoms durch das Feld des Leuchtelektrons zu berechnen, wodurch eine Verfeinerung in der Theorie der Termaufspaltungen und der Elektronenaffinität sich ergeben wird.

Bisher haben wir die statistische Methode nur auf die Berechnung der Atome angewendet; es ist aber im Prinzip möglich, sie auch auf komplexe Systeme zu übertragen; nämlich auf die Moleküle und auf die festen Körper. Natürlich sind die mathematischen Schwierigkeiten für dieses Problem viel grösser, hauptsächlich weil das System die sphärische Symmetrie verliert. Z. B. in dem Fall einer zweiatomigen Molekel hat man die Gleichung (1) für den Fall zu lösen, dass das Potential zwei Quellpunkte in den beiden Kernen hat. Das Problem kann sehr viel durch die Bemerkung vereinfacht werden, dass man das Potential in der Nähe der beiden Kerne (wo es mit dem Potential eines einzigen Atoms übereinstimmt) und in grosser Entfernung von den Kernen (wo es mit dem Potential eines Atoms mit der Kernladung, die gleich der Summe der Ladungen der beiden Kerne ist, identisch wird) schon kennt. Wenn man die statistische Verteilung der Elektronen in einem Molekül berechnen kann, wird man imstande sein, die elektronischen Terme seines Bandenspektrums auszuwerten.

Auch für den Fall der festen Körper geht die sphärische Symmetrie natürlich verloren. Auch für diesen Fall, mindestens für die einfachsten Gittertypen, wird es aber wahrscheinlich möglich sein, die Verteilungsfunktion der Elektronen zu gewinnen. Ich hoffe, dass es dann möglich sein wird, viele Konstanten der festen Körper theoretisch zu berechnen.

E. Fermi, « Rend. Lincei », *6*, 602 (1927); *7*, 342 (1928); *7*, 726 (1928). « Zeitschr. f. Phys. », *48*, 73; *49*, 550 (1928) – F. Rasetti, « Rend. Lincei », *7*, 915 (1928); « Zeitschr. f. Phys. », *49*, 546 (1928). – G. Gentile und E. Majorana, « Rend. Lincei », Im Erscheinen.

N° 50, 52, 64, 65, 66, 67, 70.

During the winter of 1928-29 Fermi started studying the quantum theory of radiation. His first step was to look at the papers by Dirac (Proc. Roy. Soc. A », *114*, 243, 710, 1927) on this subject and to understand the results obtained by him. The method used by Dirac did not appeal to Fermi, who preferred, as he did very often, to recast the theory in a form mathematically more familiar to him. He did this by the method indicated in this paper and, having obtained the Hamiltonian and the general formulas, he made innumerable examples and applications, the most significant of which, together with his formulation, formed the subject of a course given in April 1929 at the Institut Poincaré in Paris and in a more complete form at the Summer School of Theoretical Physics at Ann Arbor, Michigan (N° 60 *a*) during the summer of 1930. Fermi did not study, at least at the time, the papers on quantum electrodynamics by Pauli and Heisenberg nor was he very familiar at the time with the formalism of second quantization.

He tried very hard to find some solution to the known difficulties of the divergencies of quantum electrodynamics and he filled several notebooks with attempts to find some form of Hamiltonian which would be satisfactory in this respect. The attempts did not bear fruit, but some interesting results are to be found in papers N° 64, 65, 67 and 70.

While doing this work Fermi taught his results to several of his pupils and friends including Amaldi, Majorana, Racah, Rasetti and Segrè. Every day when work was over he gathered the various people mentioned above around his table and started to elaborate before them, first the basic formulation of quantum electrodynamics and then, one after the other, a long series of applications of the general principles to particular physical problems. A striking feature of Fermi's method of working on a theoretical problem in public (so to speak) and of teaching at the same time, was the way in which he could say out loud what he was thinking, proceeding at a steady unhesitating pace; never going extremely fast, but never failing to make progress. The notebooks containing the original manuscript of this work on quantum electrodynamics are kept at the Domus Galilaeana in Pisa.

E. AMALDI.

50.

SOPRA L'ELETTRODINAMICA QUANTISTICA (*)

« Rend. Lincei », *5*, 881-997 (1929).

In questa Nota vengono scritte le equazioni della elettrodinamica in forma quantistica anche per il caso che nel campo vi siano dei corpuscoli elettrici. Viene data per il momento soltanto la forma che corrisponde a corpuscoli di piccola velocità. In alcune Note seguenti verranno mostrate le applicazioni principali.

1. L'elettrodinamica quantistica, per il caso dell'assenza di cariche elettriche, è stata trattata in una forma presso ché definitiva in un recente lavoro di Pauli e Jordan [1]. Questi autori decompongono il campo di radia-

(*) Presentata dal Socio O. M. Corbino nella seduta del 5 maggio 1929.

(1) W. PAULI e P. JORDAN, « ZS. f. Phys. », *47*, p. 151 (1928).

zione nelle sue componenti armoniche ed applicano, per ciascuna componente, delle condizioni di quantizzazione simili a quelle valide per un oscillatore; essi riescono anche a dedurne delle leggi sulla non commutatività del prodotto delle componenti del campo elettrico e di quello magnetico che, anche nel loro aspetto esterno, soddisfano alla condizione di invarianza relativistica.

Più complicato è il problema della costruzione di una elettrodinamica quantistica per il caso, che praticamente è il più importante, in cui nel campo vi sono anche delle cariche elettriche. Una soluzione parziale di questo problema è stata data da Dirac [2] nel suo fondamentale lavoro sopra la teoria dell'irraggiamento. La elettrodinamica di Dirac è però incompleta per il fatto che essa considera soltanto il campo elettromagnetico di radiazione, e cioè un campo che si possa rappresentare come somma di onde elettromagnetiche piane. In realtà il campo elettromagnetico che si produce in vicinanza di un corpuscolo elettrico in movimento è alquanto più complicato, per modo che la teoria di Dirac corrisponde alla realtà soltanto per quanto riguarda le regioni dello spazio abbastanza lontane dai corpuscoli elettrici che generano il campo. P. e. non si potrebbe trattare, per mezzo di essa, il problema dell'interazione tra due elettroni dello stesso atomo o di atomi vicini tenendo conto che il campo si propaga da uno all'altro per potenziali ritardati; similmente non si può trattare, con la teoria di Dirac, il problema della teoria quantistica della massa elettromagnetica, ecc.

In questa Nota mostrerò una via che può condurre alla costruzione di una elettrodinamica quantistica completa; per il momento scriveremo le equazioni soltanto per il caso che la velocità dei corpuscoli elettrici che generano il campo sia piccola a confronto della velocità della luce, e, in alcune Note seguenti, mostreremo le applicazioni di esse alla teoria quantistica delle masse elettromagnetiche e ad alcuni altri fenomeni.

Nel caso che le particelle elettriche abbiano grandi velocità si può applicare la teoria relativistica di Dirac dell'elettrone rotante; si riesce così a scrivere le equazioni dell'elettrodinamica in forma che soddisfa alla condizione di invarianza relativistica, benché non mi riesca ancora di vedere completamente le conseguenze delle equazioni così ottenute.

2. Per scrivere le equazioni dell'elettrodinamica quantistica, dobbiamo cercare dapprima di porre le equazione dell'elettrodinamica classica in una forma che si presti ad essere tradotta in forma quantistica. Per fare questo, invece di prendere come variabili che descrivono il campo elettromagnetico i valori dei potenziali scalare e vettore in ogni punto dello spazio, noi prenderemo i coefficienti dello sviluppo di essi in integrale di Fourier. Questi coefficienti, che sono naturalmente delle funzioni del tempo soltanto, ci descrivono il potenziale scalare e quello vettoriale per ogni istante in tutto il campo che si considera.

(2) P. A. M. DIRAC, «Proc. Roy. Soc.», *A 114*, 243 (1927).

Per semplificare la considerazione conviene supporre che il campo elettromagnetico che si considera sia racchiuso entro una cavità di volume finito Ω. Facendo poi allontanare le sue pareti a distanza infinita ci si riconduce al caso della radiazione libera in tutto lo spazio.

Siano V, ed **U** il potenziale scalare e il potenziale vettore che soddisfano alle equazioni:

$$\Delta V - \frac{1}{c^2}\frac{\partial^2 V}{\partial t^2} = -4\pi\rho \tag{1}$$

$$\Delta \mathbf{U} - \frac{1}{c^2}\frac{\partial^2 \mathbf{U}}{\partial t^2} = -4\pi\rho\frac{\boldsymbol{v}}{c} \tag{2}$$

in cui ρ e $\boldsymbol{v}$ rappresentano rispettivamente la densità delle cariche elettriche e la loro velocità.

Potremo supporre che **U** e V, entro lo spazio Ω, possano rappresentarsi nel modo seguente, come somma di componenti sinusoidali nelle coordinate x, y, z:

$$V = \sqrt{\frac{8\pi}{\Omega}}\, c \sum_s Q_s \cos\left(\frac{2\pi\alpha_s\cdot\mathbf{X}}{\lambda_s} + \beta_s\right) \tag{3}$$

$$\mathbf{U} = \sqrt{\frac{8\pi}{\Omega}}\, c \sum_s \boldsymbol{q}_s \sin\left(\frac{2\pi\alpha_s\cdot\mathbf{X}}{\lambda_s} + \beta_s\right) \tag{4}$$

in cui Q_s e $\boldsymbol{q}_s$ sono rispettivamente uno scalare e un vettore che dipendono unicamente dal tempo, α_s è un vettore che dà la direzione dell'onda s^{esima}, $\mathbf{X}$ è un vettore che ha per componenti le tre coordinate x, y, z del punto generico dello spazio, β_s è una costante di fase; infine il fattore di normalizzazione $\sqrt{\frac{2\pi}{\Omega}}\, c$ è stato scelto in modo da semplificare il più possibile le formule che incontreremo nel seguito.

È noto che, al limite per Ω molto grande, il numero delle componenti armoniche del tipo precedente che ha lunghezza d'onda compresa tra λ e $\lambda + d\lambda$, è dato da

$$\frac{4\pi\Omega}{\lambda^4}\, d\lambda \tag{5}$$

o, introducendo al posto della lunghezza d'onda la frequenza corrispondente,

$$\nu = \frac{c}{\lambda} \tag{6}$$

si può scrivere la (5) nella forma equivalente

$$\frac{4\pi\Omega}{c^3}\, \nu^2\, d\nu\,. \tag{7}$$

Le (3) e (4) possono risolversi, nel modo consueto, rispetto ai coefficenti Q_s e $\boldsymbol{q}_s$ dello sviluppo; e si trova

$$c\sqrt{2\pi\Omega}\, Q_s = \int V \cos\Gamma_s\, d\tau \tag{8}$$

$$c\sqrt{2\pi\Omega}\, \boldsymbol{q}_s = \int \mathbf{U} \sin\Gamma_s\, d\tau \tag{9}$$

dove si è posto per brevità

$$\Gamma_s = \frac{2\pi\alpha_s \cdot \mathbf{X}}{\lambda_s} + \beta_s \tag{10}$$

e

$$d\tau = dx\, dy\, dz\,. \tag{11}$$

Vogliamo ora stabilire una equazione differenziale a cui soddisfano le Q_s e le $\boldsymbol{q}_s$ e che equivale alle equazioni di Maxwell (1) e (2).

Per far questo moltiplichiamo i due membri di (1) per $\cos \Gamma_s\, d\tau$ ed integriamo a tutto il volume Ω. Con ovvie trasformazioni supponendo che sulla superficie che limita Ω il potenziale si annulli, e che il campo elettromagnetico si annulli in modo tale che si possano omettere certi integrali estesi a tale superficie, si trova

$$-4\pi\int \rho \cos \Gamma_s\, d\tau = \int \Delta V \cdot \cos \Gamma_s\, d\tau - \frac{1}{c^2}\int \frac{\partial^2 V}{\partial t^2} \cos \Gamma_s\, d\tau =$$

$$= +\int V \cdot \Delta \cos \Gamma_s \cdot d\tau - \frac{1}{c^2}\frac{\partial^2}{dt^2}\int V \cos \Gamma_s\, d\tau$$

e siccome da (10) risulta

$$\Delta \cos \Gamma_s = -\frac{4\pi^2}{\lambda^2} \cos \Gamma_s$$

si ricava, tenendo presente la (8) e la (6),

$$\ddot{Q}_s + 4\pi^2 \nu_s^2 Q_s = \sqrt{\frac{8\pi}{\Omega}}\, c \int \rho \cos \Gamma_s\, d\tau\,. \tag{12}$$

Se supponiamo che le cariche elettriche siano concentrate in punti di cariche $e_1, e_2, e_3, \cdots$ aventi coordinate $\mathbf{X}_1, \mathbf{X}_2, \mathbf{X}_3, \cdots$, l'integrale del secondo membro di (12) può senz'altro effettuarsi, e l'equazione (12) diventa

$$\ddot{Q}_s + 4\pi^2\nu_s^2 Q_s = \sqrt{\frac{8\pi}{\Omega}}\, c \sum_i e_i \cos\left(\frac{2\pi\alpha_s\, \mathbf{X}_i}{\lambda_s} + \beta_s\right) = \sqrt{\frac{8\pi}{\Omega}}\, c \sum_i e_i \cos \Gamma_{si} \tag{13}$$

dove la somma deve essere estesa a tutte le cariche elettriche del campo.

Operando in modo simile sopra le componenti $\boldsymbol{q}_s$ dello sviluppo del potenziale vettore si trova per esse l'equazione.

$$\ddot{\boldsymbol{q}}_s + 4\pi^2 \nu_s^2 \boldsymbol{q}_s = \sqrt{\frac{8\pi}{\Omega}} \sum_i e_i\, \mathbf{X}_i \sin \Gamma_{si}\,. \tag{14}$$

Indicando con χ_s, w_{s1}, w_{s2} le componenti del vettore $\boldsymbol{q}_s$ rispettivamente nella direzione $\boldsymbol{\alpha}_s$ e in due altre direzioni $\mathbf{A}_{s1}$ ed $\mathbf{A}_{s2}$ perpendicolari tra di loro e perpendicolari alla direzione $\boldsymbol{\alpha}_s$, possiamo scrivere

$$\boldsymbol{q}_s = \boldsymbol{\alpha}_s\, \chi_s + \mathbf{A}_{s1}\, w_{s1} + \mathbf{A}_{s2}\, w_{s2} \tag{15}$$

dove con $\mathbf{A}_{s1}$ ed $\mathbf{A}_{s2}$ si sono indicati i versori delle due direzioni corrispondenti. L'equazione (14) si può ora sostituire con le tre equazioni scalari equivalenti:

$$\ddot{\chi}_s + 4\pi^2 \nu_s^2 \chi_s = \sqrt{\frac{8\pi}{\Omega}} \sum_i e_i\, (\boldsymbol{\alpha}_s \cdot \mathbf{X}_i) \sin \Gamma_{si} \tag{16}$$

$$(17)\quad \begin{cases} \ddot{w}_{s1} + 4\pi^2 \nu_s^2 w_{s1} = \sqrt{\dfrac{8\pi}{\Omega}} \sum\limits_i e_i (\mathbf{A}_{s1} \cdot \mathbf{X}_i) \sin \Gamma_{si} \\ \ddot{w}_{s2} + 4\pi^2 \nu_s^2 w_{s2} = \sqrt{\dfrac{8\pi}{\Omega}} \sum\limits_i e_i (\mathbf{A}_{s2} \cdot \mathbf{X}_i) \sin \Gamma_{si} . \end{cases}$$

Dalle equazioni (13) e (16) si deduce subito una relazione tra Q_s e χ_s. Basta per questo derivare la (13) rispetto al tempo e sommarla poi membro a membro con la (16) moltiplicata per $2\pi\nu_s$. Si trova così

$$(18)\quad \left(\frac{d^2}{dt^2} + 4\pi^2 \nu_s^2\right)(2\pi\nu_s \chi_s + \dot{Q}_s) = 0 .$$

Questa equazione è evidentemente soddisfatta se si pone

$$(19)\quad 2\pi\nu_s \chi_s + \dot{Q}_s = 0 .$$

Questa equazione, come immediatamente si verifica, è identica alla nota relazione

$$(20)\quad \operatorname{div} \mathbf{U} + \frac{1}{c}\frac{\partial V}{\partial t} = 0$$

che lega tra di loro il potenziale scalare e il potenziale vettore; in virtù di (18) tale relazione, pur non essendo una conseguenza delle equazioni differenziali, resta tuttavia verificata per tutti i valori del tempo a condizione che, all'istante iniziale, essa sia verificata insieme alla sua derivata prima.

3. Siamo ora in grado di scrivere le equazioni del movimento delle cariche elettriche e le equazioni del campo elettromagnetico in una forma che consente la loro immediata traduzione nelle equazioni della meccanica quantistica. In questa Nota ci limiteremo, come si è detto, a svolgere la teoria non relativisticamente; vale a dire, supporremo che le velocità delle cariche elettriche siano sempre piccole a confronto della velocità della luce.

Come coordinate per determinare le posizioni delle cariche prenderemo

$$\mathbf{X}_1, \mathbf{X}_2, \cdots, \mathbf{X}_i, \cdots.$$

E come coordinate che determinano il campo elettromagnetico

$$Q_s, \chi_s, w_{s1}, w_{s2} .$$

Queste ultime, per mezzo delle equazioni (3), (4), (15) determinano in tutti i punti del campo i potenziali scalare e vettore. Tra le coordinate Q_s e χ_s passa la relazione (19). Possiamo tuttavia considerare queste due variabili come indipendenti, poiché abbiamo visto che l'equazione (19) è una conseguenza delle equazioni differenziali, quando almeno si ammetta, come vogliamo fare, che essa sia verificata, insieme alla sua derivata, all'istante iniziale. La (19) si riduce dunque soltanto a una condizione da imporre alle costanti di integrazione.

Accanto alle coordinate.

$$\mathbf{X}_i, Q_s, \chi_s, w_{s1}, w_{s2}$$

considereremo anche altre variabili

$$p_i, P_s, \tilde{\omega}_s, \omega_{s1}, \omega_{s2}$$

che rappresenteranno i momenti coniugati ad esse.

Possiamo ora verificare che, nella nostra approssimazione non relativistica, le equazioni del movimento e dell'elettrodinamica si possono derivare dalla seguente funzione Hamiltoniana

$$(21)\qquad H = \sum_i \frac{p_i^2}{2m_i} + \sum_s \left[\frac{1}{2}\{\omega_{s1}^2 + \omega_{s2}^2 + \tilde{\omega}_s^2 - P_s^2\} + 2\pi^2 \nu_s^2 \{w_{s1}^2 + w_{s2}^2 + \chi_s^2 - Q_s^2\}\right] + c\sqrt{\frac{8\pi}{\Omega}} \sum_{is} e_i Q_s \cos \Gamma_{si} - \sqrt{\frac{8\pi}{\Omega}} \sum_{is} \frac{e_i}{m_i} [(\alpha_s \chi_s + \mathbf{A}_{s1} w_{s1} + \mathbf{A}_{s2} w_{s2}) \cdot \boldsymbol{p}_i] \sin \Gamma_{si} .$$

Ciò si verifica immediatamente per quanto riguarda le equazioni del movimento; infatti la parte che contiene le coordinate $\mathbf{X}_i$ delle cariche elettriche e i loro momenti $\boldsymbol{p}_i$, in virtù delle (3) e delle (4) si riduce a

$$\sum_i \frac{p_i^2}{2m_i} + \Sigma e_i V_i - \Sigma \frac{e_i}{c} \frac{\boldsymbol{p}_i}{m_i} \cdot \mathbf{U}_i + (\text{parte indipendente da } \boldsymbol{p}_i \text{ e } \mathbf{X}_i)$$

che non è altro che la consueta Hamiltoniana di un sistema di punti materiali carichi elettricamente e posti in un campo elettromagnetico.

Per verificare che dalla Hamiltoniana (21) seguono anche le equazioni dell'elettromagnetismo, formiamoci ad esempio le equazioni canoniche per la coppia di variabili coniugate Q_s, P_s. Tali equazioni risultano

$$\frac{dQ_s}{dt} = \frac{\partial H}{\partial P_s} = -P_s \quad ; \quad \frac{dP_s}{dt} = -\frac{\partial H}{\partial Q_s} = +4\pi^2 \nu_s^2 Q_s - c\sqrt{\frac{8\pi}{\Omega}} \sum_i e_i \cos \Gamma_{si} .$$

Eliminando da queste due equazioni il momento P_s ritroviamo la (13); in modo simile, tenendo conto che, in prima approssimazione, i momenti $\boldsymbol{p}_i$ coniugati alle coordinate $\mathbf{X}_i$ delle cariche, sono le quantità di moto $m\dot{\mathbf{X}}_i$, si verificherebbe che, dalla nostra Hamiltoniana, seguono anche le (16) e (17). Siccome le equazioni (13), (16) e (17) sono equivalenti alle equazioni dell'elettrodinamica (1) e (2), resta dimostrato che anche tali equazioni possono dedursi dalla Hamiltoniana (21).

5. Avendo così scritto le equazioni dell'elettrodinamica e del movimento delle cariche sotto forma canonica, resta immediata la loro traduzione in equazioni di meccanica quantistica. Basta per questo considerare l'espressione H come un operatore, in cui i momenti, secondo l'ordinaria regola, siano equivalenti all'operazione di derivazione rispetto alla coordinata corrispondente e di moltiplicazione per $-\frac{h}{2\pi i}$. L'equazione di Schroedinger sarà allora la seguente

$$(22)\qquad H\psi = \frac{h}{2\pi i} \frac{\partial \psi}{\partial t} .$$

Mostreremo nelle Note seguenti l'applicazione di questa equazione.

N° 51.

This paper records Fermi's contribution to a discussion by letter which he carried on with Levi-Civita about the effect of the increase in mass of a planet caused by the fall of meteorites. The average velocity of the meteorites, which had been neglected in Levi-Civita's previous work, is taken into account by Fermi under various simple hypotheses concerning their velocity distribution in the frame of reference of the fixed stars.

E. SEGRÈ.

51.

SUL MOTO DI UN CORPO DI MASSA VARIABILE (*)

« Rend. Lincei », *9*, 984–986 (1929).

Viene precisato in questa Nota l'effetto della caduta di meteoriti, sopra le equazioni del movimento di un pianeta.

... I risultati dell'effetto della caduta dei meteoriti sopra le equazioni del movimento di un pianeta, mi pare si possano precisare al modo seguente.

Sia $\mathbf{V}$ la velocità (vettoriale) del pianeta; m la sua massa (funzione del tempo t) e v la velocità di uno dei meteoriti; $\overline{\boldsymbol{v}}$ il valore medio (vettoriale) della velocità dei meteoriti che cadono sul pianeta nel tempo dt. Nelle sue Note (1) è sostanzialmente contenuta l'equazione seguente

$$\frac{d\,(m\mathbf{V})}{dt} = \mathbf{F} + \frac{dm}{dt}\,\overline{\boldsymbol{v}}\,. \tag{1}$$

Il problema si riduce dunque al calcolo di $\overline{\boldsymbol{v}}$, che Lei ha supposto senz'altro eguale a zero. La determinazione di $\overline{\boldsymbol{v}}$ dipende naturalmente dalle ipotesi che si fanno sopra la legge di distribuzione delle velocità dei meteoriti. L'ipotesi più naturale è forse ammettere che essi abbiano velocità media nulla rispetto allo spazio delle stelle fisse; e che la legge di distribuzione delle velocità sia Maxwelliana.

Più generalmente ammettiamo, per ora, che si abbia una legge di distribuzione delle velocità dei meteoriti che ammetta simmetria sferica; per modo che la probabilità che un meteorita abbia velocità appartenente a un elemento dw dello spazio rappresentativo delle velocità sia $f(v)\,dw$ essendo f una funzione, per ora arbitraria, della sola grandezza della velocità de meteorita.

(*) Presentata dal Socio T. Levi-Civita nella seduta del 1° giugno 1929.

(1) Questi « Rendiconti », vol. VIII, pp. 329–333, 621–622 9(128).

La probabilità di un urto del pianeta con un meteorita di velocità v è evidentemente, proporzionale a $f(v)\,dw$ e, inoltre, alla velocità relativa

$$g = |\mathbf{v} - \mathbf{V}|$$

del pianeta e del meteorita.

Il valor medio della velocità dei meteoriti che cadono sul pianeta è dunque

$$\bar{\mathbf{v}} = \frac{\int \mathbf{v}\,|\mathbf{v} - \mathbf{V}|\,f\,dw}{\int |\mathbf{v} - \mathbf{V}|\,f\,dw}.$$

È evidente, per simmetria, che $\bar{\mathbf{v}}$ deve essere parallelo a $\mathbf{V}$. Riducendoci dunque a calcolare solo la componente nella direzione di $\mathbf{V}$, che è l'unica diversa da zero, ed usando coordinate polari in modo che si possa porre $dw = 2\pi \sin\theta\, d\theta\, v^2\, dv$, troviamo

$$\bar{v} = \frac{\int_0^\infty \int_0^\pi v\cos\theta \sqrt{V^2 + v^2 - 2Vv\cos\theta}\, f(v)\, 2\pi \sin\theta\, d\theta\, v^2\, dv}{\int_0^\infty \int_0^\pi \sqrt{V^2 + v^2 - 2Vv\cos\theta}\, f(v)\, 2\pi\sin\theta\, d\theta\, v^2\, dv}.$$

L'integrazione rispetto a θ si effettua subito e dà come risultato

$$-\bar{v} = \frac{\int_0^V v^3 f(v)\, dv \left[\frac{2}{3} v - \frac{2}{15}\frac{v^3}{V^2}\right] + \int_V^\infty v^3 f(v)\, dv \left[\frac{2}{3} V - \frac{2}{15}\frac{V^3}{v^2}\right]}{\int_0^V v^2 f(v)\, dv \left[2V + \frac{2}{3}\frac{v^2}{V}\right] + \int_V^\infty v^2 f(v)\, dv \left[2v + \frac{2}{3}\frac{V^3}{v^2}\right]}.$$

Se per $f(v)$ si pone l'espressione Maxwelliana

$$f(v) = e^{-h^2 v^2}$$

il calcolo degli integrali non si può fare coi mezzi ordinari, e bisognerebbe ricorrere alle tabelle dei valori della funzione

$$\Theta(x) = \int_x^\infty e^{-x^2}\, dx.$$

Il risultato è invece estremamente semplice se si ammette, con ipotesi certamente non rispondente alla realtà, che la grandezza della velocità di tutti i meteoriti sia la stessa. In questa ipotesi si trova infatti

$$\bar{v} = \begin{cases} -\dfrac{v^2}{5V}\,\dfrac{5V^2 - v^2}{3V^2 + v^2} & \text{per } v \leq V \\[2ex] -\dfrac{V}{5}\,\dfrac{5v^2 - V^2}{3v^2 + V^2} & \text{per } v \geq V. \end{cases}$$

Se, in particolare, ammettiamo che la velocità dei meteoriti sia molto maggiore di quella del pianeta, risulta, dalle formule precedenti, l'espressione semplicissima

$$\bar{\boldsymbol{v}} = -\frac{\mathbf{V}}{3}\,.$$

È interessante vedere la forma che prende l'equazione del movimento (1) in questa ipotesi. Ponendo in essa per $\bar{\boldsymbol{v}}$ il valore precedente, si trova

$$\frac{d\,(m\mathbf{V})}{dt} = \mathbf{F} - \frac{1}{3}\,\mathbf{V}\,\frac{dm}{dt}$$

che si può anche scrivere

$$\frac{1}{m^{1/3}}\,\frac{d}{dt}\,(m^{4/3}\,\mathbf{V}) = \mathbf{F}\,.$$

Questa forma, come si vede, non è molto dissimile da quella ordinaria, alla quale naturalmente si riduce se la massa non dipende dal tempo.

N° 52.

For the introduction to this paper see N° 50.

52.

SULLA TEORIA QUANTISTICA DELLE FRANGE DI INTERFERENZA (*)

«Rend. Lincei», *10*, 72–77 (1929); «Nuovo Cimento», *7*, 153–158 (1930).

In questo lavoro viene svolta, in base alla teoria di Dirac dell'irraggiamento, la teoria di un fenomeno di interferenza e, precisamente, delle frange di Lippmann. Si vede dal procedimento che lo stesso metodo potrebbe applicarsi in generale alla trattazione di qualsiasi fenomeno di interferenza. Il risultato coincide con quello della teoria elettromagnetica classica.

Il punto di vista fondamentale della teoria di Dirac (1) dell'irraggiamento, consiste nel trattare un atomo che emette oppure assorbe luce, e la radiazione nello spazio come un unico sistema, che ha come coordinate, oltre alle coordinate dell'atomo, anche certe altre variabili che servono a specificare i valori dei potenziali del campo elettromagnetico in ogni punto dello spazio. Se, per semplificare la considerazione, pensiamo la radiazione racchiusa entro una cavità, sappiamo che la radiazione potrà vibrare soltanto secondo certe frequenze caratteristiche, dipendenti dalla forma e dalle dimensioni della cavità. Si possono allora prendere come coordinate del campo di radiazione, per ciascuna vibrazione fondamentale, il valore del campo elettrico istantaneo in uno dei ventri, oppure una grandezza ad esso proporzionale, per modo che per ogni frequenza fondamentale si ha una coordinata. Se si fanno allontanare all'infinito le pareti della cavità le vibrazioni fondamentali della radiazione finiscono per tendere a certe distribuzioni asintotiche, in modo che, al limite, cessa ogni influenza delle pareti della cavità sopra il fenomeno e ci si riconduce al caso dello spazio libero.

È noto che la teoria di Dirac rende conto in modo soddisfaciente di tutto il gruppo di fenomeni connessi con gli scambi energetici tra la luce e la materia (emissione, assorbimento, vita media degli atomi, fenomeni di dispersione, ecc. . . .). Essa invece deve essere in qualche modo integrata per rendere conto anche dell'azione che esercitano tra di loro due atomi vicini o due elettroni dello stesso atomo (propagazione dell'azione per potenziali ritardati, ecc. . . .). Un tentativo di perfezionare la teoria di Dirac in questo

(*) Presentata dal Socio O. M. Corbino nella seduta del 1° giugno 1929.

(1) P. A. M. DIRAC «Proc. Roy. Soc.», *A 114*, 243, 710 (1927).

senso è stato da me fatto in una Nota recente pubblicata in questi « Rendiconti ». Senza occuparmi di questo raffinamento della teoria vorrei far vedere in questa Nota, sopra un esempio concreto, che la teoria di Dirac rende conto anche dei fenomeni di interferenza; per modo che si può ben dire che essa comprende in sé tutte le proprietà dell'irradiazione.

Il criterio in base al quale si può costruire una teoria dell'interferenza è il seguente. Gli elementi del problema sono: un atomo emittente, A, che funziona da sorgente di luce; un interferometro, che possiamo, schematicamente, pensare costituito da un insieme di schermi opachi, di superficie riflettenti, oppure che delimitano mezzi di differente indice di rifrazione; infine un altro atomo B che, assorbendo più o meno intensamente la luce emessa da A, ci indica in quali regioni dello spazio la radiazione emessa da A arriva con intensità maggiore (frange chiare) e in quali punti essa arriva con intensità minore (frange oscure). Naturalmente, secondo il criterio fondamentale della teoria di Dirac, i due atomi A e B e il campo di radiazione dovranno venir considerati come un unico sistema, che deve venir trattato con i metodi della meccanica quantistica.

Consideriamo dapprima il caso che non ci sia l'interferometro. Supponiamo che, all'istante iniziale, l'atomo A sia eccitato e l'atomo B sia invece nello stato fondamentale; supporremo infine che, all'istante iniziale, non vi sia radiazione nello spazio circostante ai due atomi. Si trova allora che l'effetto dell'interazione tra i due atomi e il campo di radiazione è il seguente. Supponiamo che sia trascorso, dall'istante iniziale un certo tempo t grande in confronto alla vita media dell'atomo A, in modo che sia praticamente nulla la probabilità che esso sia ancora eccitato. Si trova allora che se $t < \frac{r}{c}$ (r essendo la distanza tra i due atomi A e B) l'energia perduta dall'atomo A diseccitandosi si ritrova certamente in una delle componenti armoniche della radiazione. Se invece $t > \frac{r}{c}$ vi è sempre una certa probabilità di trovare l'energia perduta da A in una delle componenti della radiazione; vi è però anche una certa probabilità (inversamente proporzionale ad r^2) di non trovare energia di radiazione e di trovare invece eccitato l'atomo B. Come si vede in questi risultati che, per ristrettezza di spazio, non possiamo qui dimostrare, è contenuta la teoria della propagazione della luce con velocità eguale a c.

Dobbiamo ora analizzare quale influenza è esercitata, sopra il sistema costituito dai due atomi A e B e dalla radiazione, dalla presenza dell'interferometro. Osserviamo per questo che le vibrazioni fondamentali della radiazione racchiusa entro una cavità vengono alterate dalla presenza in essa di superficie riflettenti o rifrangenti, schermi, ecc. . . . ; in modo che, mentre nel caso che non vi sia l'interferometro, si possono prendere come vibrazioni fondamentali delle onde stazionarie piane, ciò non è più lecito quando vi sia l'interferometro, poiché verranno introdotte delle particolari condizioni ai limiti sulle superficie dei pezzi ottici che lo costituiscono.

Per svolgere la teoria sopra un esempio il più possibile semplice, vogliamo riferirci al caso delle frange di Lippmann che si formano per interferenza tra le onde luminose che arrivano sopra uno specchio piano e quelle che vengono riflesse all'indietro dallo specchio. Dovremo dunque considerare il sistema costituito da un atomo emittente, un atomo assorbente e la radiazione contenuta entro una cavità una delle cui pareti è costituita dallo specchio la cui presenza dà origine alle frange di Lippmann; dovremo poi passare al limite allontanando a distanza infinita tutte le pareti della cavità, ad eccezione, naturalmente, di quella costituita dallo specchio, che resta invece fissa.

Prendiamo un sistema di assi x, y, z in modo che gli assi y e z si trovino nel piano dello specchio. Per semplicità supporremo che i due atomi A e B si trovino tutti e due sull'asse delle x; siano x e x' le loro ascisse. Ammetteremo anche che le direzioni di vibrazione dei due atomi, almeno per quanto riguarda gli stati quantici che ci interessano, coincidano con la direzione y.

Indichiamo con n ed n' i numeri che caratterizzano gli stati quantici dei due atomi e con $n_1, n_2, n_3, \cdots, n_s, \cdots$ i numeri che caratterizzano gli stati quantici delle diverse componenti armoniche della radiazione. La funzione di Schroedinger sarà in generale una combinazione lineare

$$\Psi = \Sigma\, a_{nn'\, n_1\, n_2 \cdots n_s \cdots}\, \psi_{nn'\, n_1\, n_2 \cdots n_s \cdots}$$

delle funzioni corrispondenti agli stati quantici $n, n', n_1, n_2, \cdots, n_s, \cdots$ I quadrati dei moduli delle a,

$$\left| a_{nn'\, n_1\, n_1 \cdots n_s \cdots} \right|^2$$

rappresentano la probabilità che i due atomi si trovino negli stati n ed n' e le diverse componenti di radiazione negli stati $n_1, n_2, \cdots, n_s, \cdots$

Se non vi fosse interazione tra i due atomi e il campo di radiazione, le a sarebbero costanti. L'interazione ha invece per effetto che le a cessano di essere costanti ma vengono a dipendere dal tempo, soddisfacendo alla equazione differenziale

$$\dot{a}_n = \frac{2\pi i}{h} \sum_m \mathrm{H}_{nm}\, a_m\,. \tag{1}$$

(Per semplicità tutti gli indici delle a sono stati riassunti con una sola lettera, n oppure m); H_{nm} sono gli elementi della matrice di perturbazione che rappresenta l'interazione tra i due atomi e la radiazione. Le equazioni differenziali (1) permettono di determinare i valori delle a conoscendo i loro valori iniziali.

Supporremo che inizialmente l'atomo A sia eccitato nello stato quantico 2 ($n = 2$), invece l'atomo B si trovi nello stato fondamentale $n' = 1$; supporremo inoltre che inizialmente non vi sia radiazione nel campo, cioè $n_1 = n_2 = n_3 = \cdots n_s = \cdots = 0$. Ciò vuol dire che all'istante iniziale $t = 0$ si può porre:

$$a_{2,1,0,0,\cdots,0,\cdots} = 1 \tag{2}$$

e tutte le altre a eguali a zero; poiché è certo che, all'istante iniziale, i due atomi e la radiazione si trovano rispettivamente negli stati $2, 1, 0, 0, \cdots, 0 \cdots$

Se non vi fosse la perturbazione dovuta all'interazione, questa situazione non muterebbe col tempo, L'effetto della perturbazione rende invece, come si è detto, le a dipendenti dal tempo, per modo che alcune delle a che hanno valore iniziale nullo, possono, dopo un certo tempo, avere valore diverso da zero.

A noi interessa particolarmente di conoscere, in funzione del tempo il valore della grandezza

$$a_{1,2,0,0,\cdots,0,\cdots}$$

poiché il quadrato del suo modulo rappresenta la probabilità che il primo atomo si sia diseccitato, passando dallo stato $n = 2$ allo stato $n = 1$, e che la energia da lui emessa sia stata assorbita dal secondo atomo il quale sia passato così dallo stato $n' = 1$ allo stato $n' = 2$.

Ci manca lo spazio per sviluppare qui il calcolo di $a_{1,2,0,0,\cdots,0,\cdots}$, in base alle equazioni differenziali (1) e alle condizioni iniziali (2). Dobbiamo qui limitarci a indicare il risultato, riservandoci di tornare in altro luogo sopra questo argomento. Ammettendo per semplicità che il primo atomo A abbia una vita media brevissima, e che invece la vita media del secondo atomo B sia lunghissima, si trova

$$(3)\qquad a_{1,2,0,\cdots,0,\cdots} = \begin{cases} 0 & \text{per } t < t_1 \\[2ex] -\dfrac{1}{x-x'}\,\dfrac{16\pi^3 e^2}{c^2 h}\,\dfrac{\nu Q\,\nu' Q'}{2\pi i(\nu'-\nu)+\dfrac{1}{2\tau}}\,e^{\frac{2\pi i\nu'(x-x')}{c}} & \text{per } t_1 < t < t_2 \\[2ex] -\dfrac{16\pi^3 e^2}{c^2 h}\,\dfrac{\nu Q\,\nu' Q'}{2\pi i(\nu'-\nu)+\dfrac{1}{2\tau}}\left[\dfrac{1}{x-x'}\,e^{\frac{2\pi i\nu'(x-x')}{c}} - \dfrac{1}{x-x'}\,e^{\frac{2\pi i\nu'(x+x')}{c}}\right] & \text{per } t > t_2\,. \end{cases}$$

In questa equazione ν e ν' sono le frequenze delle righe emesse dai due atomi nel salto quantico tra gli stati 2 ed 1. Q e Q' sono due grandezze il cui quadrato è proporzionale all'intensità con cui i due atomi emettono le dette righe (propriamente Q e Q' sono gli elementi 2, 1 delle matrici che rappresentano la componente y del momento elettrico dei due atomi), τ è la vita media dell'atomo A. τ per ipotesi è molto piccolo, e quindi la riga emessa dal primo atomo è molto larga; invece, essendo la vita media del secondo atomo assai lunga, la riga che esso è capace di assorbire è assai stretta. Abbiamo posto infine nelle (3)

$$(4)\qquad t_1 = \frac{x-x'}{c} \quad ; \quad t_2 = \frac{x+x'}{c}$$

per modo che t_1 rappresenta l'istante in cui un'azione, partita da A all'istante 0 con velocità c, arriva in B; e t_2 rappresenta l'istante in cui essa vi ritorna dopo essersi riflessa sopra lo specchio.

Rileviamo dalla formula (3) che, fino all'istante t_1, l'atomo B si eccita. Tra gli istanti t_1 e t_2 esso ha una certa probabilità di essere eccitato (questo effetto, classicamente, si interpreta come effetto dell'onda proveniente direttamente dall'atomo emittente A). Dopo l'istante t_2 la probabilità che l'atomo B sia eccitato viene di nuovo bruscamente a variare e può, secondo i casi, aumentare o diminuire (effetto dell'onda, partita da A, che ritorna in B dopo la riflessione sopra lo specchio).

Se consideriamo, ciò che praticamente interessa, un istante successivo a t_2, per il quale è dunque applicabile la terza delle formule (3), riconosciamo subito che la probabilità di eccitazione dell'atomo B, dipende notevolmente dalla sua distanza x' dal piano dello specchio. Si vede immediatamente che le massime probabilità di eccitazione si hanno quando i due fattori esponenziali $e^{\frac{2\pi i\nu'(x\pm x')}{c}}$ hanno fase opposta, cioè quando

$$2\pi n + \pi = \frac{2\pi\nu'(x+x')}{c} - \frac{2\pi\nu'(x-x')}{c} = \frac{4\pi\nu'}{c}x' \quad (n = \text{numero intero})$$

ossia quando

$$x' = \frac{\lambda'}{2}\left(n + \frac{1}{2}\right).$$

Queste posizioni corrispondono, anche classicamente, ai ventri delle onde stazionarie. Similmente si trova che i nodi, in cui la probabilità di eccitazione dell'atomo B è minima, corrispondono invece alle posizioni

$$x' = \frac{\lambda'}{2}n\,.$$

Si ritrovano così in modo completo i risultati che dà in questo caso la teoria classica dell'interferenza.

N° 53.

A calculation, by wave-mechanical methods, of the strength of the coupling between the orbital angular momentum of the electrons and the molecular axis for the He_2 molecule. The levels investigated are those designated as the 4 *d* complex, where $n = 4$, $l = 2$ and the purpose is to determine the energy differences of the states where the axial component of the electronic angular momentum is 0, 1 or 2. The results are in good agreement with the experimental value. This work was discussed at length with E. Majorana who had recently joined the Rome Institute.

F. Rasetti.

53.

SUL COMPLESSO 4 *d* DELLA MOLECOLA DI ELIO (*)

« Rend. Lincei », *10*, 515–517 (1929); « Nuovo Cimento », 7, 159–161 (1930).

La teoria delle perturbazioni permette di calcolare, per la molecola di elio, l'energia di accoppiamento tra la retta congiungente i due nuclei della molecola e la quantità di moto areale degli elettroni. Se ne deducono le differenze tra i termini $4\,d\sigma$, $4\,d\pi$, $4\,d\delta$ in accordo coi valori sperimentali.

Nelle molecole degli atomi di piccolo peso atomico, particolarmente nelle molecole H_2 ed He_2, si osservano numerosi gruppi di termini spettroscopici, caratterizzati dallo stesso valore del numero quantico totale n e azimutale l. Si ha così, per esempio, nell'elio il gruppo dei termini $4\,d$, aventi numero quantico totale 4 e numero quantico azimutale 2. Il comportamento di questi termini, studiato particolarmente da W. Weizel (1) e G. H. Dieke (2), presenta le seguenti particolarità. Finché il quanto di rotazione R dei nuclei è piccolo, e cioè per rotazione lenta, il quanto azimutale l si orienta rispetto alla retta congiungente i nuclei, in modo che la sua proiezione λ sopra questa retta assume un valore intero. Nel caso dei termini $4\,d$, aventi $l = 2$, la proiezione può prendere i valori 0, 1, 2, e si hanno in corrispondenza i termini indicati con i simboli

$$4\,d\sigma\,,\ 4\,d\pi\,,\ 4\,d\delta\,.$$

Quando invece la rotazione diventa rapida (R grande), viene a rompersi l'accoppiamento tra l e la retta congiungente i nuclei. Ciò ha per effetto uno spezzamento in due dei termini $4\,d\pi$ e $4\,d\delta$ (σ–type doubling); infine, al limite per R molto grande, l finisce per accoppiarsi completamente con R.

(*) Presentata dal Socio O. M. Corbino nella seduta del 3 novembre 1929.

(1) W. Weizel, « Zs. f. Phys. », *56*, 727 (1929).

(2) G. H. Dieke, « Zs. f. Phys. », *57*, 71 (1929).

In questa Nota mi sono proposto di calcolare le differenze tra i termini $4\,d\sigma$, $4\,d\pi$, $4\,d\delta$ del complesso. Ciò può farsi abbastanza facilmente per mezzo della teoria delle perturbazioni della meccanica quantistica, approfittando del fatto che, avendo i termini in questione dei numeri quantici totale e azimutale abbastanza elevati, le autofunzioni dell'elettrone luminoso differiscono per essi abbastanza poco dalle autofunzioni dell'atomo di idrogeno; per modo che si ha in tali autofunzioni una prima approssimazione molto comoda. La legittimità di questo processo è confermata dal fatto che i termini in questione possono rappresentarsi abbastanza bene per mezzo di una formula di Rydberg e la correzione di Rydberg risulta molto piccola. Noi prenderemo dunque come autofunzioni imperturbate per i tre termini $4\,d\sigma$, $4\,d\pi$, $4\,d\delta$ della molecola di elio le corrispondenti autofunzioni dell'atomo di idrogeno; e cioè:

$$(1)\qquad \begin{cases} \psi_0 = (3\cos^2\theta - 1)\, r^2\left(1 - \dfrac{r}{12 r_0}\right) e^{-\frac{r}{4 r_0}} \\[2ex] \psi_1 = \sin\theta\cos\theta\, e^{\pm i\varphi}\, r^2\left(1 - \dfrac{r}{12 r_0}\right) e^{-\frac{r}{4 r_0}} \\[2ex] \psi_2 = \sin^2\theta\, e^{\pm 2 i\varphi}\, r^2\left(1 - \dfrac{r}{12 r_0}\right) e^{-\frac{r}{4 r_0}} \end{cases}$$

in cui $r_0 = 0{,}53 \cdot 10^{-8}$ cm. rappresenta il raggio della prima orbita dell'idrogeno.

Il potenziale perturbatore V deriva principalmente dal fatto che i due nuclei, invece di essere riuniti nel centro di gravità della molecola, hanno uno dall'altro una certa distanza a. Siccome ciascuno dei due nuclei è accompagnato da un elettrone, la sua carica effettiva sarà e invece di $2\,e$. Vi è inoltre un terzo elettrone, oltre all'elettrone di valenza; siccome però il campo dovuto ad esso verosimilmente non differisce molto da un campo centrale, noi non ne considereremo qui l'azione. Come potenziale perturbatore avremo dunque da considerare il potenziale dovuto a due cariche positive e concentrate in due punti distanti di a uno dall'altro, ed aventi nel loro punto di mezzo una carica negativa $-2\,e$. Questo potenziale può svilupparsi in potenze inverse di r; il primo termine dello sviluppo è:

$$(2)\qquad \mathrm{V} = \frac{e a^2}{4 r^3}(3\cos^2\theta - 1)\,.$$

Questa espressione rappresenta bene il potenziale perturbatore per valori abbastanza grandi di r; siccome però le autofunzioni (1) hanno valori molto piccoli per r piccolo, potremo senz'altro prendere (2) come potenziale perturbatore.

È noto dalla teoria delle perturbazioni che le perturbazioni delle energie dei nostri tre termini, dovute al potenziale perturbatore V, sono date da:

$$\varepsilon_0 = -e\,\frac{\int \mathrm{V}\,|\psi_0|^2\,d\tau}{\int |\psi_0|^2\,d\tau} \quad ; \quad \varepsilon_1 = -e\,\frac{\int \mathrm{V}\,|\psi_1|^2\,d\tau}{\int |\psi_1|^2\,d\tau} \quad ; \quad \varepsilon_2 = -e\,\frac{\int \mathrm{V}\,|\psi_2|^2\,d\tau}{\int |\psi_2|^2\,d\tau}\,.$$

Sostituendo negli integrali le espressioni (1) e (2) delle ψ e di V, ed effettuando le quadrature si trova:

$$(4) \qquad \varepsilon_0 = -2\frac{e^2 a^2}{13440 r_0^3} \quad ; \quad \varepsilon_1 = -\frac{e^2 a^2}{13440 r_0^3} \quad ; \quad \varepsilon_2 = 2\frac{e^2 a^2}{13440 r_0^3} \,.$$

Siccome nella molecola di elio si ha, con grande approssimazione, $a = 2 r_0$, abbiamo

$$\frac{e^2 a^2}{13440 r_0^3} = \frac{e^2}{2 r_0} \frac{1}{1680} \,.$$

Esprimendo le energie in numeri di onde si ha

$$\frac{e^2}{2 r_0} = 109737 \text{ numeri di onde}$$

poiché $e^2/2 r_0$ rappresenta il numero di Rydberg. Abbiamo dunque infine

$$\varepsilon_0 = -130 \quad ; \quad \varepsilon_1 = -65 \quad ; \quad \varepsilon_2 = 130 \text{ numeri di onde.}$$

Empiricamente sono conosciute le differenze tra i termini $4 d\delta$ e $4 d\sigma$, e tra i termini $4 d\pi$ e $4 d\sigma$. Tali differenze risultano dalla nostra teoria espresse da

$$\varepsilon_2 - \varepsilon_0 = 260 \quad ; \quad \varepsilon_1 - \varepsilon_0 = 65 \,.$$

I corrispondenti valori sperimentali sono invece

$$300 \qquad\qquad 75 \,.$$

Come si vede l'accordo tra teoria ed esperienza è abbastanza buono, siccome la divergenza è di circa il 15 %. Questa differenza residua è verosimilmente dovuta alle ipotesi semplificative della teoria e probabilmente in buona parte al fatto che si è trascurata l'influenza perturbatrice di uno degli elettroni.

N° 54 *a* and 54 *b*.

The purpose of the paper is to explain the observations reported by Rasetti and by Füchtbauer and collaborators, showing that the intensity ratios of the two components of the S-P doublets in the alkalies sometimes differ appreciably from the value 2 predicted by the elementary theory. Large discrepancies from the value 2 are observed for the second and higher doublets of the series in the heavier alkalies.

This had been the subject of Rasetti's thesis and was in the tradition of Puccianti of Pisa who had used the anomalous dispersion of vapors to measure line intensities. The anomaly reported above had remained a puzzle for many years.

One day studying Pauli's theory of the spin Fermi found the qualitative reason for the effect. He explained it in detail to Segré at 2 P.M. challenging him to have the calculation completed by 4 P.M., after which time Fermi would do the work (as he did).

The simplicity of the assumptions, the directness of approach and the exceptionally lucid treatment make this paper a classic among the early applications of perturbation theory to atomic spectroscopy. Its theme and methods made a deep impression on Fermi and the line of thought of perturbations in spectra recurs repeatedly in his work.

Paper N° 54 *b* is an Italian version of N° 54 *a*.

F. RASETTI.

54 *a*.

ÜBER DAS INTENSITÄTSVERHÄLTNIS DER DUBLETTKOMPONENTEN DER ALKALIEN

« Z. Physik », 59, 680–686 (1930).

Nach den bekannten Intensitätsregeln sollte das Intensitätsverhältnis der Komponenten für alle Dublette der Hauptserie der Alkalien 2: 1 sein. Die experimentell gefundenen Verhältnisse weichen, hauptsächlich bei Cäsium, sehr weit von dem theoretischen Werte ab. In vorliegender Arbeit wird diese Unstimmigkeit zwischen Theorie und Experiment erklärt.

Nach den zuerst von Dorgelo und Burger aufgestellten und später durch die Quantenmechanik begründeten Intensitätsregeln sollte das Intensitätsverhältnis der beiden Komponenten für alle Dublette der Hauptserie der Alkalien 2 : 1 sein. Dagegen weichen die experimentell bestimmten Verhältnisse in einigen Fällen sehr weit von diesem theoretischen Werte ab, und zwar hat das Intensitätsverhältnis auch für die Linien desselben Elements einen verschiedenen Wert. Die Abweichungen des beobachteten Verhältnisses von 2 wachsen sehr stark mit der Atomnummer des Elements und sind also am grössten für Cs. Übrigens ist das Verhalten des ersten Dubletts der Hauptserie von dem der anderen Dublette sehr verschieden. Für das erste Dublett findet man Verhältnisse sehr nahe an 2, vielleicht etwas kleiner. Für das zweite Dublett ist dagegen das Verhältnis grösser als 2, und die Abweichungen sind in einigen Fällen sehr erheblich. So z. B. findet Rasetti [1] für das zweite Dublett des Cs aus sehr genauen Messungen nach der Methode der anomalen Dispersion das Intensitätsverhältnis 3,85. Wesentlich übereinstimmende Ergebnisse wurden auch von anderen Forschern nach verschiedenen Methoden erhalten. So finden z. B. Füchtbauer und Wolff [2] in einer kürzlich erschienenen Arbeit die Verhältnisse 3,3 und 4,7 für das zweite und dritte Dublett des Cs.

In der vorliegenden Arbeit möchte ich zeigen, wie man diese Resultate theoretisch deuten kann. Wir werden nämlich beweisen, dass man das Intensitätsverhältnis 2 findet, wenn man in die Integrale, welche quantenmechanisch die Intensitäten bestimmen, die ungestörten Eigenfunktionen einsetzt. Setzt man dagegen in dieselben Integrale die gestörten Eigenfunktionen ein, so findet man Intensitätsverhältnisse in Übereinstimmung mit der Erfahrung. Die grosse Störung des Intensitätsverhältnisses des zweiten Dubletts des Cs (3,85 statt 2) hat darin ihren Grund. Bei allen Alkalidämpfen ist die Intensität des ersten Dubletts der Hauptserie etwa 100 mal grösser als die Intensität des zweiten Dubletts. Sei nun ψ_i die ungestörte Eigenfunktion irgendeines Terms i, dann kann die gestörte Eigenfunktion bekanntlich in der Form

$$\psi_i + \sum_k{}' \alpha_{ik} \psi_k$$

geschrieben werden, wo die α_{ik} kleine Koeffizienten darstellen und die Summe auf alle Zustände des Atoms, mit Ausnahme des i–ten Zustandes, erstreckt werden muss. In den gestörten Eigenfunktionen des zweiten P–Terms hat man deshalb auch ein Glied, welches aus dem Produkt der Eigenfunktion des ersten P–Terms mit einem kleinen Koeffizienten besteht. Wenn man diese gestörten Eigenfunktionen mit denen des S–Terms kombiniert, um die Intensitäten zu berechnen, hat dieses Glied eine grosse Wirkung, obgleich es einen kleinen Koeffizienten hat, weil die Kombinationen des S–Terms mit dem ersten P–Term sehr intensiv sind. Das ist der Grund der grossen Intensitätsanomalie.

In der Ausführung der Theorie werden wir die Paulische Methode für die Behandlung des rotierenden Elektrons an Stelle der Diracschen benutzen. Für unseren Fall ist nämlich die Annäherung der einfacheren Paulischen Methode bereits genügend. Als ungestörtes Problem werden wir, wie üblich, ein Elektron in einem zentralsymmetrischen Felde betrachten, mit Vernachlässigung der Wirkung des Eigenmoments des Elektrons. Durch diese kommt die Störungsenergie

(1) $$v = \frac{\mu_0}{2mc} \frac{1}{r} \frac{dV}{dr} \mathbf{M} \cdot \boldsymbol{\sigma}$$

hinzu. Hier ist r = Radiusvektor, μ_0 = Bohrsches Magneton, V = elektrisches Potential, M = mechanisches Moment der Elektronenbahn. $\boldsymbol{\sigma}$ ist ein q–Vektor, dessen Komponenten durch die Matrizen

(2) $$\sigma_x = \begin{vmatrix} 0 & 1 \\ 1 & 0 \end{vmatrix} \quad , \quad \sigma_y = \begin{vmatrix} 0 & -i \\ i & 0 \end{vmatrix} \quad , \quad \sigma_z = \begin{vmatrix} 1 & 0 \\ 0 & -1 \end{vmatrix}$$

(1) F. Rasetti, «Cim.», *1*, 115 (1924).
(2) C. Füchtbauer und H. W. Wolff, «Ann. d. Phys.», *3*, 359 (1929).

gegeben sind. Wenn man das Eigenmoment des Elektrons mit berücksichtigt, spaltet sich die Eigenfunktion für jeden Quantenzustand in zwei Funktionen

$$\psi = \begin{vmatrix} \psi_\alpha(x, y, z) \\ \psi_\beta(x, y, z) \end{vmatrix}$$

der Ortskoordinaten x, y, z. So z. B. findet man für den zweifach entarteten $^2S_{1/2}$-Term die beiden Eigenfunktionen

$$\begin{vmatrix} F(r) \\ 0 \end{vmatrix} , \quad \begin{vmatrix} 0 \\ F(r) \end{vmatrix} . \tag{3}$$

Als ungestörte Eigenfunktionen für den zweifach entarteten $n\,^2P_{1/2}$-Term erhält man

$$\begin{vmatrix} \frac{1}{\sqrt{3}} f_n(r)\, P_0 \\ \sqrt{\frac{2}{3}} f_n(r)\, P_1 \end{vmatrix} , \quad \begin{vmatrix} -\sqrt{\frac{2}{3}} f_n(r)\, P_{-1} \\ \frac{1}{\sqrt{3}} f_n(r)\, P_0 \end{vmatrix} , \tag{4}$$

wo

$$P_1 = \sqrt{\frac{3}{8\pi}} \sin\Theta\, e^{i\varphi} \quad , \quad P_0 = \sqrt{\frac{3}{4\pi}} \cos\Theta \quad , \quad P_{-1} = \sqrt{\frac{3}{8\pi}} \sin\Theta\, e^{-i\varphi} \tag{5}$$

die drei Kugelfunktionen erster Ordnung darstellen. Für den vierfach entarteten $n\,^2P_{3/2}$-Term findet man die vier ungestörten Eigenfunktionen:

$$\begin{vmatrix} f_n(r)\, P_1 \\ 0 \end{vmatrix} , \quad \begin{vmatrix} -\sqrt{\frac{2}{3}} f_n(r)\, P_0 \\ \frac{1}{\sqrt{3}} f_n(r)\, P_1 \end{vmatrix} , \quad \begin{vmatrix} \frac{1}{\sqrt{3}} f_n(r)\, P_{-1} \\ \sqrt{\frac{2}{3}} f_n(r)\, P_0 \end{vmatrix} , \quad \begin{vmatrix} 0 \\ f_n(r)\, P_{-1} \end{vmatrix} . \tag{6}$$

Die Stärke des Überganges von einem Zustand mit der Eigenfunktion $\begin{vmatrix} \psi_\alpha \\ \psi_\beta \end{vmatrix}$ zu einem Zustand mit der Eigenfunktion $\begin{vmatrix} \psi'_\alpha \\ \psi'_\beta \end{vmatrix}$ wird bekanntlich durch

$$a = \left(\int z \overline{\begin{vmatrix} \psi_\alpha \\ \psi_\beta \end{vmatrix}} \begin{vmatrix} \psi'_\alpha \\ \varphi'_\beta \end{vmatrix} d\tau \right)^2 = \left[\int z \left(\bar\psi_\alpha \psi'_\alpha + \bar\psi_\beta \psi'_\beta \right) d\tau \right]^2 \tag{7}$$

bestimmt.

Wendet man diese Formel auf die Übergänge $n\,^2P_{1/2} \rightarrow {}^2S_{1/2}$ und $n\,^2P_{3/2} \rightarrow {}^2S_{1/2}$ an, indem man für die Eigenfunktionen ihre ungestörten Ausdrücke (3), (4), (6) einsetzt, und beachtet man, dass $z = r \cos\Theta$, so findet man für die Stärke der beiden Übergänge:

$$\left\{ \begin{aligned} a_{n,1} &= \frac{8\pi}{9} \left[\int_0^\infty r^3 F(r)\, f_n(r)\, dr \right]^2 \text{ für } n\,^2P_{1/2} \rightarrow {}^2S_{1/2} , \\ a_{n,2} &= \frac{16\pi}{9} \left[\int_0^\infty r^3 F(r)\, f_n(r)\, dr \right]^2 \text{ für } n\,^2P_{3/2} \rightarrow {}^2S_{1/2} . \end{aligned} \right. \tag{8}$$

Die beiden a stehen also in diesem Falle in dem Verhältnis 2 : 1.

Wir wollen jetzt die gestörten Eigenfunktionen für die beiden Zustände $n\,{}^2\mathrm{P}_{1/2}$ und $n\,{}^2\mathrm{P}_{3/2}$ berechnen, um mit Hilfe von diesen das Intensitätsverhältnis wieder zu bestimmen. Wir wenden deshalb die wohlbekannte Störungsformel:

$$\psi_{\text{gestört}} = \psi_i - \sum_k{}' \frac{\int \bar{\psi}_k v \psi_i \, d\tau}{\mathrm{E}_i - \mathrm{E}_k}$$

an, wo für v der Ausdruck (1) als Operator eingesetzt und die Summe auf alle Werte von k, mit Ausnahme von $k = i$, erstreckt werden muss. Durch eine Rechnung, welche keine Schwierigkeiten bietet, findet man als gestörte Eigenfunktionen

für $n\,{}^2\mathrm{P}_{1/2}$:

$$\begin{vmatrix} \frac{1}{\sqrt{3}} f_n'(r)\,\mathrm{P}_0 \\ \sqrt{\frac{2}{3}} f_n'(r)\,\mathrm{P}_1 \end{vmatrix}, \quad \begin{vmatrix} -\sqrt{\frac{2}{3}} f_n'(r)\,\mathrm{P}_{-1} \\ \frac{1}{\sqrt{3}} f_n'(r)\,\mathrm{P}_0 \end{vmatrix},$$

für $n\,{}^2\mathrm{P}_{3/2}$:

$$\begin{vmatrix} f_n''(r)\,\mathrm{P}_0 \\ 0 \end{vmatrix}, \quad \begin{vmatrix} -\sqrt{\frac{2}{3}} f_n''(r)\,\mathrm{P}_0 \\ \frac{1}{\sqrt{3}} f_n''(r)\,\mathrm{P}_1 \end{vmatrix}, \quad \begin{vmatrix} \frac{1}{\sqrt{3}} f_n''(r)\,\mathrm{P}_{-1} \\ \sqrt{\frac{2}{3}} f_n''(r)\,\mathrm{P}_0 \end{vmatrix}, \quad \begin{vmatrix} 0 \\ f_n''(r)\,\mathrm{P}_{-1} \end{vmatrix},$$

wo

$$(9) \qquad \left\{ \begin{aligned} f_n'(r) &= f_n(r) - \frac{h\mu_0}{2\pi mc} \sum_m{}' \frac{\int_0^\infty r \frac{d\mathrm{V}}{dr} f_n f_m \, dr}{\mathrm{E}_n - \mathrm{E}_m} f_m(r), \\ f_n''(r) &= f_n(r) + \frac{h\mu_0}{4\pi mc} \sum_m{}' \frac{\int_0^\infty r \frac{d\mathrm{V}}{dr} f_n f_m \, dr}{\mathrm{E}_n - \mathrm{E}_m} f_m(r). \end{aligned} \right.$$

Die Summen sind auf alle m, mit Ausnahme von $m = n$, erstreckt.

Die in (9) auftretenden Integrale können, auch ohne eine genaue Kenntnis der Eigenfunktionen, nach der folgenden Methode berechnet werden. Bekanntlich ist die Dublettaufspaltung des $n\,{}^2\mathrm{P}$–Terms durch

$$(10) \qquad \Delta_n = -\frac{3\mu_0 h}{4\pi mc} \int_0^\infty r \frac{d\mathrm{V}}{dr} f_n^2(r) \, dr$$

ausgedrückt. Die Funktionen $f_n(r)$ sind alle für kleines r einander proportional und unterscheiden sich voneinander nur für grosses r. Da aber das für die Auswertung der in (9) und (10) auftretenden Integrale wichtige Gebiet nur aus den sehr kleinen Werten von r besteht, können wir für dies Gebiet $f_m(r) = k_{mn} f_n(r)$ setzen, wo k_{mn} einen konstanten Proportionalitätsfaktor darstellt. Aus (10) finden wir

$$k_{mn} = \sqrt{\frac{\Delta_m}{\Delta_n}}.$$

Wir haben also:

$$-\int_0^\infty r\frac{dV}{dr}f_n f_m\, dr = -k_{mn}\int_0^\infty r\frac{dV}{dr}f_n^2\, dr = k_{mn}\frac{4\pi mc}{3\mu_0 h}\Delta_n = \frac{4\pi mc}{3\mu_0 h}\sqrt{\Delta_n\Delta_m}\,.$$

Setzen wir diese Ausdrücke in (9) ein, so finden wir:

$$(11)\qquad \begin{cases} f_n'(r) = f_n(r) - \dfrac{2}{3}\sum_m{}' \dfrac{\sqrt{\Delta_n\Delta_m}}{E_n - E_m} f_m(r)\,, \\[2ex] f_n''(r) = f_n(r) + \dfrac{1}{3}\sum_m{}' \dfrac{\sqrt{\Delta_n\Delta_m}}{E_n - E_m} f_m(r)\,. \end{cases}$$

Wir können jetzt die Intensitäten mit den Formeln (8) berechnen, wobei wir aber für die Übergänge $n\,{}^2P_{1/2} \rightarrow {}^2S_{1/2}$, $f_n'(r)$ statt $f_n(r)$, und für die Übergänge $n\,{}^2P_{3/2} \rightarrow {}^2S_{1/2}$, $f_n''(r)$ statt $f_n(r)$ einsetzen müssen. Wir bekommen also:

$$a_{n,1} = \frac{8\pi}{9}\left(B_n - \frac{2}{3}\sum_m{}' \frac{\sqrt{\Delta_n\Delta_m}}{E_n - E_m} B_m\right)^2,$$

$$a_{n,2} = \frac{16\pi}{9}\left(B_n + \frac{1}{3}\sum_m{}' \frac{\sqrt{\Delta_n\Delta_m}}{E_n - E_m} B_m\right)^2,$$

wo gesetzt wurde:

$$(12)\qquad B_n = \int_0^\infty r^3 F(r)\, f_n(r)\, dr\,.$$

Als Intensitätsverhältnis finden wir also statt 2:

$$\frac{a_{n,2}}{a_{n,1}} = 2\left(\frac{1 + \dfrac{1}{3}\sum_m{}' \dfrac{\sqrt{\Delta_n\Delta_m}}{E_n - E_m}\dfrac{B_m}{B_n}}{1 - \dfrac{2}{3}\sum_m{}' \dfrac{\sqrt{\Delta_n\Delta_m}}{E_n - E_m}\dfrac{B_m}{B_n}}\right)^2.$$

Es bleiben nur noch die Verhältnisse der B zu bestimmen. Aus der Definition (12) folgt, dass B_m/B_n gleich der Wurzel aus dem Intensitätsverhältnis J_m/J_n der beiden Übergänge $m\,{}^2P \rightarrow {}^2S$ und $n\,{}^2P \rightarrow {}^2S$ (ohne Berücksichtigung der Dublettaufspaltung) ist. Also

$$\frac{B_m}{B_n} = \sqrt{\frac{J_m}{J_n}}\,.$$

Als Endformel finden wir somit das Intensitätsverhältnis:

$$(13)\qquad \frac{a_{n,2}}{a_{n,1}} = 2\left(\frac{1 + \dfrac{1}{3}\sum_m{}' \dfrac{\sqrt{\Delta_n\Delta_m}}{E_n - E_m}\sqrt{\dfrac{J_m}{J_n}}}{1 - \dfrac{2}{3}\sum_m{}' \dfrac{\sqrt{\Delta_n\Delta_m}}{E_n - E_m}\sqrt{\dfrac{J_m}{J_n}}}\right)^2,$$

wo alle im zweiten Glied vorkommenden Grössen direkt der Beobachtung zugänglich sind.

Als Beispiel wollen wir das Intensitätsverhältnis der beiden Komponenten für das zweite Dublett des Cs, $7\,^2P \to 6\,^2S$, berechnen. Wir haben:

$$\begin{array}{llll} \Delta_6 = 554 & & E_6 = -19951 \\ \Delta_7 = 181 & & E_7 = -9551 \\ \Delta_8 = 80 & & E_8 = -5659 \\ \Delta_9 = 43 & & E_9 = -3747 \end{array}$$

Wir haben also

$$\sum_m{}' \frac{\sqrt{\Delta_m \Delta_7}}{E_7 - E_m} \sqrt{\frac{J_m}{J_7}} = 0{,}032 \sqrt{\frac{J_6}{J_7}} - 0{,}031 \sqrt{\frac{J_8}{J_7}} - 0{,}010 \sqrt{\frac{J_9}{J_7}} - \cdots$$

Praktisch wichtig ist nur das von der Störung der 6 P–Terme herrührende Glied. J_6/J_7 ist meines Wissens nie gemessen worden. Das entsprechende Verhältnis ist für Kalium (3) ungefähr 200, und für Natrium (4) 120. Theoretisch wurde J_6/J_7 von Gentile und Majorana (5) für Cs berechnet; sie finden $I_6/I_7 = 125$. Hier haben wir diesen Wert benutzt. Wir finden dann:

$$\sum_m{}' \frac{\sqrt{\Delta_m \Delta_7}}{E_7 - E_m} \sqrt{\frac{J_m}{J_7}} = 0{,}022 \sqrt{125} = 0{,}36 \,.$$

Aus (13) bekommen wir also:

$$\frac{a_{7,2}}{a_{7,1}} = 2 \left(\frac{1{,}12}{0{,}76}\right)^2 = 4{,}3 \,.$$

Dieses Verhältnis ist etwas grösser als das beobachtete Verhältnis 3,85. Die Differenz ist aber sehr gut erklärlich, wenn man bedenkt, dass wir nur die erste Näherung der Störungstheorie ausgeführt haben. Für das zweite Dublett des Kaliums findet man in ähnlicher Weise das Verhältnis 2,16. Auch für diesen Fall ist der theoretische Wert etwas grösser als der von Rasetti bestimmte Wert 2,10. Für das erste Dublett der Hauptserie findet man theoretische Verhältnisse, die etwas kleiner als 2 sind. Der Unterschied ist jedoch viel zu klein, um beobachtet werden zu können.

Zusammenfassend dürfen wir sagen dass die grossen Abweichungen des Intensitätsverhältnisses von 2 durch die hier dargestellte Theorie in der richtigen Grössenordnung wiedergegeben werden.

(3) W. PROKOFJEW und G. GAMOW, «ZS. f. Phys.», *44*, 887 (1927), geben 111 als Verhältnis der Anzahl der Dispersionselektronen: hieraus berechnet sich der angegebene Wert.

(4) A. FILIPPOV und W. PROKOFJEW, «ZS. f. Phys.», *56*, 458 (1929).

(5) G. GENTILE und E. MAJORANA, «Lincei Rend.», *8*, 229 (1928).

N° 55, 57 *a*, 75.

There had always been a strong spectroscopic tradition in Pisa, Florence and Rome and most experimental work had been in that field. For some time, however, it had been felt by Fermi that physicists would be ready in the near future to attack the problems of nuclear structure and that efforts should be directed to that end.

The study of hyperfine structure was a very natural bridge between the two domains. Pauli's interpretation of the effect as a consequence of a nuclear spin and magnetic moment had attracted Fermi's attention. However, the phenomenon at the time looked rather puzzling and although Pauli's explanation seemed plausible, there were legitimate doubts whether it was complete.

Fermi had mastered Dirac's theory of the electron and read Hargreaves' paper (« Proc. Roy. Soc. », *124*, 568 (1929)), but noted that the latter was incomplete on an essential point because it neglected the spin-spin interaction. He repeated the calculation correctly using Dirac's theory. The main result is contained in Eq. 2 of paper N° 57 which is very important for the analysis of hyperfine structure. This equation may be derived by methods much simpler than those used originally by Fermi in this paper.

During the following years the experimental material concerning the hyperfine structure of spectral lines increased considerably. With this increase there were some apparent cases of discrepancies in the value of the magnetic moment of nuclei when measured by using different spectral terms. Fermi and his friends were very conscious of the importance of perturbations in spectroscopic phenomena such as the appearance of forbidden lines, intensity anomalies and others. It was natural to try to explain also the hyperfine structure anomalies as an effect of perturbations. This required an improvement of the original Fermi formula Eq. 2 of paper N° 57 and a careful analysis of the perturbations induced by neighbouring terms.

This additional work was performed by Fermi and Segrè in 1933 and is reported in paper N° 75 *a, b*. The version N° 75 *b* is more complete than N° 75 *a*, which omits some of the calculations. Paper N° 55 is a preliminary letter to the Editor of « Nature ».

E. SEGRÈ.

55.

MAGNETIC MOMENTS OF ATOMIC NUCLEI

« Nature » (London), *125*, 16 (1930).

The hyperfine structures of atomic spectra are considered as due to the interaction of the nuclear spin with the electronic orbital and spin moments. A theoretical calculation of this interaction enables one to obtain information on the magnitude of the magnetic moments of the nuclei from the separation of the hyperfine structures. This can be done for the case of the alkali atoms. Hargreaves (« Proc. Roy. Soc. », *124*, 568; 1929) has calculated the separation due to a nuclear moment $h/4\pi$ for the case of atoms with only one electron. In his calculations, however, the interaction between the electronic and nuclear spins, which is of the same order of magnitude as the other terms, has been neglected. I have therefore carried out the calculations with the following improvements. For the s-terms, which is the most important case, since they give the largest contribution to the separation, I have used Dirac's theory of the electron, since the simpler Pauli's theory gives a wrong result. For the p-terms I have used Pauli's method, taking into account the interaction of the nuclear and electronic spins. For the p-terms it is not necessary to evaluate numerically the eigenfunctions, since the constants involved in the formulae can be derived from empirical data on the separation of the electronic spin doublets. For the s-terms this is of course impossible, and I have calculated the eigenfunctions by the statistical method.

If the mechanical and the magnetic moments of the nucleus are respectively $kh/2\pi$ and μ_0, one finds that the s-terms split into two components with a separation

$$\Delta = \frac{2k+1}{k} \frac{8\pi}{3} \mu\mu_0 \psi^2(0) , \tag{1}$$

where μ is Bohr's magneton and $\psi(0)$ is the value of the normalised eigenfunction at the origin. The separation of the p-terms is much smaller. Each line of the principal series of the alkalis splits, therefore, into two components with the separation Δ, each of them having a finer structure, which is not resolved, and gives rise to small differences in the observed separation for the different lines. The ratio of the intensities of the two components is $(k+1)/k$, the weaker component being shifted towards the violet. The ratio of the intensities is also $3, 2, 5/3 \cdots 1$ for $k = 1/2, 1, 3/2, \cdots, \infty$.

For cæsium one finds, in wave numbers,

$$\Delta = 146 \frac{\mu_0}{\mu} \frac{2k+1}{k} ,$$

and for sodium

$$\Delta = 13 \cdot 4 \frac{\mu_0}{\mu} \frac{2k+1}{k} .$$

The observed values are, for cæsium (D. A. Jackson « Proc. Roy. Soc. », *121*, 432; 1928) $\Delta = 0 \cdot 3$, and for sodium (H. Schüler, « Naturwiss., » *16*, 512; 1928) $\Delta = 0 \cdot 06$ wave numbers. From this, on the assumption of the values $1/2, 1, 3/2, \cdots, \infty$ for k, we obtain the following values for the ratio μ/μ_0 of the Bohr magneton to the magnetic moment of the nucleus:

$k =$	1/2	1	3/2	∞
Cæsium	1950	1460	1300	980
Sodium	890	670	600	450

From the observed ratio of the intensities in sodium one should expect a probable value for k of 1 or 1/2.

The uncertainty of the given values arises from the lack of precision of the empirical data and from the application of the statistical method to the evaluation of the eigenfunctions. This latter source of error might be evaluated to within 20 or 30 per cent.

Further details will be published later.

Physical Institute of the University, Rome, Dec. 4 (1929).

N° 56, 58, 60, 61, 62.

Fermi avendo padroneggiato finalmente in modo completo la meccanica quantistica desiderava che essa venisse apprezzata dai fisici italiani, specialmente da quelli della giovane generazione e quindi decise, caratteristicamente, di insegnarla di persona. Di questa sua opera di maestro in senso lato testimoniano il suo libro (*Introduzione alla fisica atomica*, Zanichelli, Bologna 1928) che peraltro è scritto ai primi albori della meccanica ondulatoria e, successivamente, i numerosi articoli di divulgazione N° 56, 60, 61, 62 scritti intorno a quel tempo per varie occasioni. A quell'epoca egli era già famoso anche tra il pubblico profano e aveva molte richieste di conferenze. Quasi ogni anno parlò alla Società Italiana per il Progresso delle Scienze che al suo convegno annuale tenuto in città diverse riuniva un considerevole numero di studiosi di campi svariati.

Di questa opera quasi propagandistica di Fermi, ci sono tracce in sue lettere private a Segrè in cui esprime il dispiacere che menti anche illuminate e aperte della generazione precedente non riescano ad apprezzare prontamente l'importanza degli sviluppi teorici della meccanica quantistica. A un livello più alto il lavoro N° 59 fa parte dello stesso sforzo.

E. SEGRÈ.

56.

I FONDAMENTI SPERIMENTALI DELLE NUOVE TEORIE FISICHE

«Atti Soc. It. Progr. Sci.», 18ª Riunione, vol. 1, 365–371 (1929).

Nella conferenza precedente sono state mostrate le profonde innovazioni concettuali che stanno a fondamento dei principii della nuova fisica. Esse, a chi vi si trovi di fronte per le prime volte, saranno sembrate certamente talmente vaste e sconvolgenti da far sorgere spontanea la curiosità di sapere quali siano i risultati che le nuove teorie hanno permesso di raggiungere; poiché una simile rivoluzione nei concetti più fondamentali della fisica può giustificarsi soltanto con molti, vari ed importanti risultati. Io cercherò di esporre in modo sintetico questi risultati. Per la ristrettezza del tempo non potrò spiegare la catena logica che conduce dalle premesse della teoria alle sue applicazioni; dovrò invece limitarmi a metterne in evidenza il significato e la portata.

Nella vecchia teoria dei quanti si erano potuti interpretare, e, in parte, anche prevedere un grande numero di fenomeni, tanto da non lasciar dubbio che si fosse sulla buona strada. Non si poteva però illudersi che bastassero perfezionamenti e ritocchi per correggerla delle sue manchevolezze; poiché la teoria conteneva in sé delle contraddizioni logiche che, poco osservate nei primi tempi, nella foga di scoprire nuovi risultati, diventavano col passare del tempo, sempre più sgradevoli.

Citerò l'esempio della notissima discussione tra la teoria elettromagnetica ondulatoria e la teoria dei quanti di luce. Anche la stessa teoria dell'atomo di Bohr era del resto piena di contraddizioni; in quanto che, mentre da una parte si applicava la meccanica ordinaria al calcolo del movimento degli elettroni, se ne negava d'altra parte la validità in moltissimi casi e non si avevano chiare idee sopra le leggi che avrebbero dovuto sostituirsi ad essa.

Nonostante queste manchevolezze, certamente gravissime, si era riusciti a interpretare qualitativamente, tra l'altro, quasi tutti i fenomeni della spettroscopia; in molti casi anzi si arrivava perfino a calcoli quantitativi, come per esempio, per l'atomo di idrogeno. Per gli altri atomi invece si ottenevano soltanto risultati approssimati; e ciò non solo per difetto dei metodi matematici di approssimazione, ma anche per un'evidente insufficienza delle premesse fisiche per il calcolo completo dei fenomeni.

Prima di parlare dei nuovi risultati che si sono ottenuti con la meccanica quantistica dirò che essa ha permesso di confermare tutti i risultati della vecchia teoria dei quanti che erano d'accordo con l'esperienza, per mezzo di una teoria completamente priva di interne contraddizioni. Nello stesso ordine di fenomeni, dove la vecchia teoria dava e poteva dare soltanto risultati qualitativi, la nuova ha portato la precisione quantitativa. Valga ad esempio il calcolo quantitativo dell'energia di ionizzazione dell'atomo di elio, problema che si è tentato a lungo e vanamente di risolvere con la vecchia teoria; il calcolo fatto col metodo di Schrödinger e spinto fino alla quarta approssimazione ha invece già permesso di ottenere teoricamente un valore discosto da quello sperimentale di poco più dell'1 ‰. Come questo, si potrebbero citare altri numerosi esempi in cui la meccanica quantistica porta alla precisione quantitativa, dove la vecchia teoria dava soltanto risultati qualitativi. Così nella teoria degli spettri di bande, nella teoria degli atomi con più elettroni, e perfino nella stessa teoria dell'atomo di idrogeno, che era sempre stata considerata, per così dire, la roccaforte della vecchia teoria dei quanti. Notiamo incidentalmente che il fatto che la vecchia teoria dei quanti, pur non arrivando a dare risultati quantitativamente esatti, avesse tuttavia permessso di raggiungere conclusioni assai importanti, non sembra oggi casuale; si può infatti dimostrare che essa rappresenta una prima approssimazione dei risultati della meccanica quantistica, prima approssimazione che in parecchi casi si avvicina già abbastanza alla realtà.

Ma non vogliamo qui soffermarci sopra quelle conclusioni della meccanica quantistica, che possono in qualche modo considerarsi come un perfezionamento di quelle della vecchia teoria, per passare invece all'esame di quanto vi è in essa di essenzialmente nuovo. Tra questi risultati, uno dei più caratteristici ed importanti è la spiegazione della possibilità delle combinazioni chimiche omeopolari. Le molecole, come tutti sanno, si possono suddividere in polari, rappresentate tipicamente dalle molecole saline e in omeopolari, che comprendono gli altri tipi di legame chimico. Le molecole polari si possono considerare costituite schematicamente dall'insieme di due o più ioni, tenuti insieme dall'attrazione elettrostatica che si esercita tra di essi. È impossibile però spiegare mediante una attrazione elettrostatica molecole, quali per esempio quella di idrogeno, che sono costituite dalla riunione di due atomi neutri. E anche se si tiene conto delle forze mutue che si esercitano fra i due nuclei e i due elettroni dei due atomi di idrogeno si trova che tali forze sono assolutamente insufficienti a spiegare la stabilità del legame della molecola H_2. A questo punto interviene la meccanica quantistica. Secondo questa non esiste alcun criterio per distinguere tra di loro i due elettroni dei due atomi. Non posso qui entrare in dettagli per la brevità del tempo disponibile. Accennerò soltanto al risultato, che può interpretarsi come se avvenisse un continuo scambio tra gli elettroni dei due atomi. Si trova che, per effetto di questo ha origine tra i due atomi una forza attrattiva, sufficientemente grande per spiegare come la molecola stia insieme. Sopra queste stesse basi non si è solamente riusciti a spiegare la possibilità della molecola d'idrogeno, ma anche di un grande numero di altre molecole e si è riusciti inoltre a chiarire il significato della valenza chimica dal punto di vista elettronico, anche per quegli atomi, come per esempio, il cloro, che hanno diverse valenze, alcune elettropositive e altre elettronegative.

Naturalmente esistono alcuni composti chimici che non possono propriamente considerarsi né polari né omeopolari, ma rappresentano dei casi intermedi tra questi due.

Un'altra applicazione molto recente, che non trova, essa pure, corrispondente nella teoria classica è quella fatta da Gamow alla disintegrazione dei nuclei radioattivi. Consideriamo un punto che si trovi in una posizione di equilibrio stabile; per esempio, un punto pesante mobile lungo una linea avente in A un minimo e dalle due parti di A due massimi B e C di eguale altezza. Supponiamo che inizialmente il punto si trovi nella posizione di equilibrio stabile A ed imprimiamogli una certa velocità, non sufficiente però per fargli scavalcare le due punte B e C, il punto, secondo la meccanica classica, si metterà a compiere delle oscillazioni intorno ad A, che hanno ampiezza tanto maggiore quanto maggiore è l'energia che è stata impressa al punto. In ogni modo esso non potrà mai uscire, per quanto si aspetti, dal tratto B C, scavalcando una delle due punte B oppure C.

Le cose vanno in modo differente secondo la meccanica quantistica. Per rendercene conto basta che pensiamo alla sua interpretazione ondulatoria, datane da Schrödinger. Secondo tale interpretazione, una differenza di potenziale corrisponde a una differenza di indice di rifrazione, in modo che, grossolanamente, possiamo pensare che la linea su cui si muove il punto rappresenti il grafico dell'indice di rifrazione. Ora si sa che quando la luce incontra una brusca variazione dell'indice di rifrazione una parte di essa viene riflessa; la frazione riflessa è tanto più grande quanto maggiore è la variazione dell'indice di rifrazione. Per quanto sia grande la variazione, però, una parte della luce viene sempre trasmessa.

Supponiamo ora di avere della luce nel tratto compreso tra B e C. Essa si rifletterà sopra le due variazioni di potenziale rappresentate da B e C. Se le riflessioni potessero considerarsi perfette la luce verrebbe riflessa alternativamente tra B e C senza uscir mai da questo intervallo. Questo caso corrisponderebbe alla meccanica classica, secondo cui il punto materiale non può uscire dal tratto B C. Abbiamo detto però che le riflessioni non sono perfette e che una parte, se pure molto piccola vien sempre trasmessa, per modo che la luce seguita per un po' a riflettersi tra B e C, ma ad ogni riflessione la quantità che si riflette diventa minore, perché una piccola parte della luce viene trasmessa all'esterno di B C. In questo modo, aspettando un tempo sufficientemente lungo, tutta la luce che li trovava tra B e C verrà infine a passare all'esterno. Ciò si interpreta nella meccanica quantistica di un punto materiale dicendo che la probabilità che il punto si trovi tra B e C, la quale corrisponde, nel paragone ottico, alla energia luminosa contenuta sul tratto B C, va diminuendo col tempo e finisce col ridursi a 0. Il caso delle particelle α è, secondo Gamow, simile a quello qui descritto. La particella è legata al nucleo in modo tale che, secondo la meccanica classica, non ne potrebbe mai uscire anche aspettando un tempo infinitamente lungo. Ma siccome abbiamo visto che, nella meccanica quantistica un punto materiale può traversare in determinate circostanze, una soglia di potenziale più alta della sua energia cinetica, troviamo che esiste una certa probabilità che la particella α abbandoni il nucleo e che l'atomo si disintegri.

La teoria, svolta su queste basi da Gamow, ha permesso così di spiegare, almeno nelle sue linee essenziali, il processo della disintegrazione radioattiva, ritrovando tra l'altro la legge di Geiger, che collega tra di loro la vita media con la velocità di emissione della particella α.

Un'altra teoria caratteristica della meccanica quantistica, è quella dell'effetto Ramsauer.

Se si proiettano degli elettroni molto veloci contro degli atomi gli elettroni li traversano senza venire molto deviati dalla loro traiettoria, per effetto della loro grande velocità.

Se la velocità degli elettroni diminuisce, la deviazione prodotta nel passaggio attraverso l'atomo diventa maggiore. Sembrerebbe anzi che, via via che diminuisce la velocità, la deviazione dovesse sempre aumentare. Ramsauer ha invece trovato che, per il caso di molti atomi, particolarmente quelli dei gas nobili, le cose vanno in modo affatto diverso, e precisamente la deviazione che subiscono gli elettroni traversando gli atomi col decrescere della velocità degli elettroni, va in un primo tempo aumentando, come si sarebbe indotti a prevedere, ma, quando la velocità è molto piccola, torna di nuovo a diminuire, tanto che gli elettroni molto lenti possono traversare gli atomi senza venire praticamente deflessi. Questo fenomeno, incomprensibile dal punto di vista della meccanica classica, trova una naturalissima interpretazione nella meccanica ondulatoria.

Risulta infatti che a un elettrone in moto corrisponde, nella meccanica ondulatoria, un sistema di onde di lunghezza d'onda tanto minore quanto più grande è la velocità del punto. Se l'elettrone è molto lento, la lunghezza d'onda corrispondente è assai grande e, per elettroni lentissimi la lunghezza d'onda diventa notevolmente più grande delle dimensioni di un atomo. Ora è chiaro, pensando al paragone ottico, che un sistema di onde viene poco disperso da un ostacolo, come sarebbe nel caso nostro l'atomo, piccolo in confronto alla lunghezza d'onda. Ciò corrisponde appunto al fatto che gli elettroni lenti vengono poco deviati dagli atomi.

Un'altra teoria che presenta notevole analogia con quella dell'effetto Ramsauer, ma che ha già e soprattutto avrà forse in futuro importanza molto maggiore è relativa al calcolo del cammino libero medio degli elettroni nei metalli. È noto che la teoria della conduzione dei metalli ha compiuto negli ultimi anni un grande progresso sopratutto in seguito ai lavori di Sommerfeld, che ha ripreso valendosi delle moderne statistiche, l'antica concezione del gas elettronico di Drude, ottenendo così la spiegazione di svariati fenomeni della conduzione metallica che sembravano, colle vecchie teorie, condurre a contraddizioni insolubili.

Uno dei problemi più essenziali per la teoria dei metalli era quello del calcolo della loro conducibilità elettrica. Per risolvere questo problema è necessario conoscere il valore del cammino libero medio degli elettroni interni del metallo. È facile convincersi che se gli atomi del metallo fossero disposti in un reticolo perfettamente regolare, la diffusione degli elettroni interni al metallo per effetto degli urti contro gli atomi dovrebbe ridursi a o. Infatti, considerando al solito le proprietà di un elettrone analoghe a quelle di un'onda che si muova in un mezzo di indice di rifrazione variabile, abbiamo una diffusione nulla se il reticolato è perfetto. Alla diffusione nulla corrisponde cammino libero medio infinito e quindi conducibilità elettrica infinita. Naturalmente, anche se consideriamo un cristallo costruito in modo perfetto, i suoi atomi non saranno situati esattamente ai vertici del reticolo altro che allo zero assoluto; per tutte le altre temperature essi si scosteranno di un po' dalle posizioni di equilibrio dando luogo a un reticolato tanto più irregolare quanto più elevata è la temperatura. La diffusione delle onde che rappresentano gli elettroni dei metalli è dovuta appunto a questa irregolarità ed è perciò più intensa per temperature elevate che per temperature basse.

Sopra queste basi è stato possibile svolgere la teoria della conducibilità elettrica, spiegandone la dipendenza dalla temperatura. Si noti anche che questa teoria rende conto in modo naturale del fatto che la conducibilità elettrica delle leghe è in generale minore di quella dei metalli puri. Infatti nel mescolare i metalli si viene ad avere un reticolato atomico alquanto irregolare, poiché sono mescolati irregolarmente gli atomi dei due metalli che costituiscono la lega.

A questa irregolarità del reticolo atomico corrisponde una forte diffusione degli elettroni e quindi una bassa conducibilità. Sempre a proposito della diffusione degli elettroni da parte di un reticolato cristallino ricorderò anche la celebre esperienza di Davisson e Germer, di cui ha già parlato il prof. Persico, e che porta forse la prova più diretta e impressionante della teoria ondulatoria della materia.

Già prima che Heisenberg e Schrödinger fondassero la nuova meccanica, era stato enunciato da Pauli, in base all'osservazione della struttura del sistema periodico degli elementi il così detto principio di esclusione. L'importanza di questo principio, per la comprensione dei fenomeni atomici, può considerarsi fondamentale. Esso dà infatti la ragione per cui gli elettroni di un atomo non si precipitano tutti nell'orbita più stabile che è quella più prossima al nucleo ma si dispongono intorno ad esso in varie orbite dando luogo a tutte le diverse particolarità del sistema periodico degli elementi. Anche la teoria degli spettri degli atomi e delle molecole con più elettroni di valenza si appoggia in gran parte sopra il principio di Pauli, che ha infine permesso di costruire la teoria del gas elettronico sul quale si fonda la moderna teoria dei metalli. Il principio di Pauli afferma che ogni orbita elettronica quantisticamente possibile può contenere al massimo un solo elettrone. Anche questo principio non è comprensibile altro che con l'aiuto della meccanica quantistica. Per mezzo di questa teoria si può infatti dimostrare che se un gruppo di elettroni soddisfa inizialmente al principio di Pauli, il principio resta automaticamente soddisfatto anche in seguito.

L'interpretazione delle proprietà spettroscopiche ha anche reso necessario attribuire agli elettroni un momento magnetico (ipotesi dell'elettrone rotante di Uhlenbeck e Goudsmit); se non si ammettesse questa proprietà dell'elettrone resterebbe inspiegata tutta la struttura del sistema periodico, poiché i periodi, invece che di otto atomi, verrebbero a essere costituiti da quattro soltanto. Ora nella vecchia teoria dei quanti il momento magnetico dell'elettrone doveva essere postulato come una proprietà indipendente non collegata in alcun modo alle altre proprietà dell'elettrone. Un grande progresso a questo proposito è stato fatto dalla teoria dell'elettrone rotante di Dirac, basata sui concetti della meccanica quantistica.

Non potendo qui entrare in dettagli sopra questa teoria, mi limiterò ad accennare che il momento magnetico dell'elettrone non si presenta in essa come una ipotesi supplementare, ma come una conseguenza naturale della fusione della meccanica quantistica e della teoria della relatività.

Accenniamo da ultimo al fatto che anche i due gruppi di fenomeni dei quanti di luce e della teoria ondulatoria trovano oggi, nella meccanica quantistica, una interpretazione unitaria e armonica grazie al principio, introdotto da Dirac, di considerare come un unico sistema il campo di radiazione e gli atomi emittenti o assorbenti.

Ho cercato di dare con questo una idea sommaria delle più importanti applicazioni della meccanica quantistica. Tali risultati, torno a ripeterlo, sono ottenuti con un processo rigorosamente logico, senza i continui compromessi e le continue ipotesi supplementari che hanno caratterizzato lo sviluppo della vecchia teoria dei quanti e si ha perciò oggi molto più che in passato, la sensazione di una costruzione teorica stabile.

57 *a*.

SUI MOMENTI MAGNETICI DEI NUCLEI ATOMICI (*)

«Mem. Accad. d'Italia», *1*, (Fis.) 139–148 (1930).

Le strutture iperfine degli spettri derivano da una interazione tra il moto degli elettroni atomici e il momento magnetico intrinseco del nucleo. In questo lavoro vengono dedotti, dai valori sperimentali delle strutture iperfine i momenti magnetici dei nuclei di Sodio, Rubidio e Cesio.

1. Le strutture iperfine delle righe spettrali derivano nella maggior perte dei casi, come è noto, da una interazione tra il momento magnetico del nucleo e il moto degli elettroni atomici. Questa interazione può grossolanamente descriversi nel modo seguente: gli elettroni che girano intorno al nucleo, dànno origine a un campo magnetico nel quale il nucleo viene a trovarsi immerso; se il nucleo porta un momento magnetico, si avranno delle differenti energie secondo l'orientazione dell'asse del nucleo rispetto al campo magnetico prodotto dagli elettroni. In conseguenza ogni livello energetico verrà a spezzarsi in due o più livelli, molto vicini tra di loro, e si avrà perciò anche uno spezzamento delle righe in alcune componenti. La presenza delle strutture iperfine nello spettro di un atomo viene così ad essere un indice del fatto che il nucleo di quell'atomo porta un momento magnetico. È evidente come, da un confronto delle strutture delle righe di un atomo con le strutture calcolate teoricamente, sia possibile dedurre il valore del momento magnetico del nucleo. Per tutti gli atomi con più di un elettrone di valenza il calcolo presenta tuttavia delle difficoltà molto notevoli; relativamente più semplice è invece il caso degli atomi alcalini.

Le strutture iperfine degli spettri dei metalli alcalini sono state finora osservate per il Sodio, il Rubidio e il Cesio. Per il Potassio non è stata trovata una struttura delle righe, ciò che farebbe pensare che il nucleo di K non abbia momento magnetico; le righe del Litio presentano una struttura dovuta probabilmente a cause diverse.

Jackson [1] ha esaminate le strutture dei primi tre doppietti della serie principale del Cesio; egli trova che tutte queste righe sono costituite da due componenti, aventi tra di loro all'incirca la stessa separazione, e cioè 0,3 numeri d'onde. Anche le righe D del Sodio, studiate da Schüler [2], consistono ciascuna di due componenti; la distanza tra di esse è circa 0,06 numeri d'onde. Risultati simili sono stati trovati da Filippow e Groos [3]

(*) Presentata nell'Adunanza del 30 gennaio 1930.

(1) D. A. JACKSON, «Proc. Roy. Soc.», *121*, 432 (1928).

(2) H. SCHÜLER, «Naturwiss.», *16*, 512 (1928).

(3) FILIPPOW e GROOS, «Naturwiss.», *17*, 121 (1929).

per il Rubidio; le righe della serie principale di questo elemento consistono ciascuna di due componenti con una separazione di 0,114 numeri d'onde.

La circostanza che per esempio, tutte le righe della serie principale del Cesio hanno all'incirca la stessa separazione, si interpreta facilmente ammettendo che la maggior parte di tale separazione derivi da una struttura iperfina del termine fondamentale $6\,^2S_{1/2}$, che è comune a tutte le righe della serie principale; le piccole differenze osservate da Jackson tra le separazioni delle diverse righe, derivano invece probabilmente da strutture non risolte dei termini p.

Indichiamo con k il momento meccanico intrinseco del nucleo, misurato con l'unità $h/2\pi$; secondo le ordinarie regole per la composizione dei vettori quantistici dovremo allora attenderci che il termine fondamentale dei metalli alcalini (avente momento elettronico totale $1/2$) si spezzi in due termini con momento totale (momento nucleare + momento elettronico)

$$f_1 = k + \frac{1}{2} \qquad \text{ed} \qquad f_2 = k - \frac{1}{2} \cdot$$

Ciò è in accordo col fatto sperimentale che la struttura fina di tutte le righe delle serie principali degli alcalini consiste in una separazione in due componenti, quando si ammetta che una eventuale struttura più complicata dei termini p sia troppo minuta per poter venire risolta. Dalla semplice osservazione della separazione, non è perciò possibile dedurre il valore di k, poiché si trovano sempre dei doppietti, qualunque sia il valore di k. La determinazione di k può invece farsi mediante una misura del rapporto di intensità delle due componenti della struttura fina; si trova infatti dalle ordinarie regole di intensità, che possono facilmente estendersi, per mezzo della meccanica quantistica, anche al caso delle strutture iperfine, che il rapporto di intensità delle due componenti è dato da:

$$\frac{2f_1 + 1}{2f_2 + 1} = \frac{k+1}{k} \tag{1}$$

ed ha dunque i valori $3, 2, 5/3, \cdots, 1$, secondo che $k = 1/2, 1, 3/2, \cdots, \infty$. Così per esempio per il caso del sodio, Schüler valuta il rapporto di intensità a 2 o 2,5; se ne può dunque dedurre che il valore più probabile di k, per il nucleo di Na è 1. Per il caso del Cs Filippow e Groos trovano che le due componenti hanno approssimativamente la stessa intensità, dal che si deduce che il valore di k per il nucleo del Cs deve essere abbastanza grande e che non può essere, come ritiene Jackson, $k = 1/2$.

2. Il nostro problema consiste ora nel calcolo teorico delle strutture fine di un atomo alcalino; da un confronto della separazione teorica con quella sperimentale si potrà poi determinare il valore del momento magnetico del nucleo. In un lavoro pubblicato recentemente Hargreaves [4] ha trattato la teoria della struttura iperfina per atomi con un solo elettrone. Egli ammette che il momento del nucleo sia $k = 1/2$ e applica la teoria dell'elettrone di Pauli–Darwin. Nella teoria di Hargreaves vien trascurata

(4) J. HARGREAVES, « Proc. Roy. Soc. », *124*, 568 (1929).

l'energia magnetica di interazione tra il momento nucleare e il momento intrinseco dell'elettrone. I termini derivanti da questa interazione sono però come si può facilmente verificare, dello stesso ordine di grandezza di quelli che derivano dall'interazione tra il momento orbitale e il momento del nucleo, e se ne deve perciò tener conto.

Per i termini *s* ho usato la teoria dell'elettrone di Dirac; ciò è necessario, poiché l'approssimazione della teoria di Pauli–Darwin non è in questo caso sufficiente. Per i termini *p* si ottiene lo stesso risultato sia applicando la teoria di Pauli–Darwin, tenendo però conto anche dell'interazione tra i momenti magnetici del nucleo e dell'elettrone, sia applicando la teoria di Dirac. Sia per i termini *p* che per quelli *s* ho stabilito le formule per un valore qualsiasi del momento nucleare k.

Per i termini *p* non è necessario un calcolo numerico dell'autofunzione dell'elettrone, poiché le costanti che intervengono nelle formule possono venir calcolate con sufficiente approssimazione, conoscendo la separazioni dei doppietti della serie principale. Un simile procedimento è evidentemente inapplicabile per i termini *s* e occorre perciò per essi una valutazione numerica dell'autofunzione. Ho fatto questa valutazione per i termini fondamentali del Na e del Cs, per mezzo del metodo statistico per calcolare il potenziale nell'interno dell'atomo. Per il Rb ho usato invece dei dati che mi sono stati gentilmente comunicati dal prof. Hartree, e che sono stati trovati col suo metodo del « self consistent field ».

Prima di passare alla dimostrazione delle formule, daremo qui i risultati e ne discuteremo le principali conseguenze.

Prima di passare alla dimostrazione delle formule, daremo qui i risultati e ne discuteremo le principali conseguenze.

L'azione del momento proprio del nucleo sui termini *s* consiste in una separazione in due termini con differenza di energia:

$$\Delta(\mathrm{S}) = \frac{8\pi}{3} \frac{2k+1}{k} \mu\mu_0 \psi^2(0) \tag{2}$$

in cui: μ_0 = magnetone di Bohr, μ = momento magnetico del nucleo, $\psi(0)$ = = valore dell'autofunzione normalizzata dell'elettrone nell'origine.

Il comportamento dei due termini P è differente. Per i termini $^2\mathrm{P}_{1/2}$ si ha una separazione in due componenti aventi la differenza di energia:

$$\Delta(^2\mathrm{P}_{1/2}) = \frac{2k+1}{k} \frac{8}{3} \mu\mu_0 \overline{\frac{1}{r^3}} \tag{3}$$

in cui

$$\overline{\frac{1}{r^3}} = \int \frac{1}{r^3} \psi^2 \, d\tau \tag{4}$$

rappresenta il valore medio di $1/r^3$. I termini $^2\mathrm{P}_{3/2}$ si dividono invece in 2, 3 oppure 4 componenti, secondo che $k = 1/2$, $k = 1$ oppure $k > 1$. In quest'ultimo caso, le differenze dei quattro termini dal termine imperturbato sono:

$$\frac{8}{5} \mu\mu_0 \overline{\frac{1}{r^3}} \; ; \quad \left(\frac{1}{3} - \frac{1}{k}\right) \frac{8}{5} \mu\mu_0 \overline{\frac{1}{r^3}} \; ; \quad -\left(\frac{1}{3} + \frac{4}{3k}\right) \frac{8}{5} \mu\mu_0 \overline{\frac{1}{r_3}} \; ; \quad -\left(1 + \frac{1}{k}\right) \frac{8}{5} \mu\mu_0 \overline{\frac{1}{r^3}} \, . \tag{5}$$

I pesi statistici di questi quattro termini sono rispettivamente $2k+4$, $2k+2$, $2k$, $2k-2$. Per $k=1$ manca l'ultimo dei quattro termini (5); per $k=1/2$ mancano gli ultimi due termini (5).

Per la valutazione della separazione (2) dei termini *s* è necessario, come è stato già detto, calcolare numericamente l'autofunzione. Ho fatto ciò per i termini fondamentali 6 S del Cs, 3 S del Na e 5 S del Rb, servendomi per quest'ultimo dei dati comunicatimi da Hartree; si trova:

$$\begin{array}{ll} \text{per il Cs} & \psi^2(0) = 2{,}7 \cdot 10^{25} \\ \text{per il Na} & \psi^2(0) = 2{,}45 \cdot 10^{24} \\ \text{per il Rb} & \psi^2(0) = 8{,}8 \cdot 10^{24} \end{array}$$

Dalla (2) otteniamo, ponendo per il magnetone di Bohr il suo valore, e esprimendo la separazione in numeri d'onde:

$$\text{(6)} \qquad \left\{ \begin{array}{ll} \text{per il Cs} & \Delta(6\,S) = 146\,\frac{\mu}{\mu_0}\,\frac{2k+1}{k} \\ \text{per il Na} & \Delta(3\,S) = 13{,}4\,\frac{\mu}{\mu_0}\,\frac{2k+1}{k} \\ \text{per il Rb} & \Delta(5\,S) = 48\,\frac{\mu}{\mu_0}\,\frac{2k+1}{k}\,. \end{array} \right.$$

Per il calcolo delle separazioni dei termini *p* si noti che la differenza di energia tra i due termini $^2P_{3/2}$ e $^2P_{1/2}$ è notoriamente:

$$\delta = -\frac{3\mu_0 h}{4\pi mc}\,\overline{\frac{1}{r}\frac{dV}{dr}} \tag{7}$$

dove V rappresenta il potenziale elettrico. Troviamo dunque:

$$\overline{\frac{1}{r}\frac{dV}{dr}} = -\frac{4\pi mc}{3\mu_0 h}\,\delta\,. \tag{8}$$

Nella vicinanza del nucleo si ha però $V = Ze/r$; dunque:

$$\frac{1}{r}\frac{dV}{dr} = -Ze\,\frac{1}{r^3}\,.$$

Poiché ora i termini importanti per la costruzione del valore medio (8) corrispondono a valori molto piccoli di r, troviamo, con sufficiente approssimazione:

$$\overline{\frac{1}{r^3}} = -\frac{1}{Ze}\,\overline{\frac{1}{r}\frac{dV}{dr}} = \frac{4\pi mc}{3\mu_0 Zeh}\,\delta$$

cioè:

$$\overline{\frac{1}{r}} = \frac{\delta}{3Z\mu_0^2}\,. \tag{9}$$

L'errore che si commette applicando questa formula è in genere piuttosto piccolo; per esempio per il Cs esso può valutarsi attorno al 4 %.

Ponendo il valore (9) nelle formule (3) e (5), si trova che la struttura iperfina del termine $^2P_{1/2}$ ha separazione

$$\Delta(^2P_{1/2}) = \frac{2k+1}{k}\,\frac{8\mu}{9Z\mu_0}\,\delta \tag{10}$$

e che le distanze dal termine imperturbato dei quattro termini in cui si scinde $^2P_{3/2}$ sono

$$(11)\qquad \frac{8\,\mu\delta}{15\,Z\mu_0}\ ;\quad \left(\frac{1}{3}-\frac{1}{k}\right)\frac{8\,\mu\delta}{15\,Z\mu_0}\ ;\quad -\left(\frac{1}{3}+\frac{4}{3\,k}\right)\frac{8\,\mu\delta}{15\,Z\mu_0}\ ;$$

$$-\left(1+\frac{1}{k}\right)\frac{8\,\mu\delta}{15\,Z\mu_0}\,.$$

Per esempio nel caso $k = 1/2$, mancano gli ultimi due dei termini (11) e la differenza dei primi due si riduce a:

$$\frac{64}{45\,Z}\,\frac{\mu}{\mu_0}\,\delta\,.$$

È interessante confrontare tra di loro quantitativamente le differenze che risultano dalla teoria per la separazione dei termini *s* e dei termini *p*. Così, per esempio, nel caso del Cesio la differenza dei termini 6 $^2P_{3/2}$ e 6 $^2P_{1/2}$ è 554 numeri d'onde, e si trova quindi dalla (10) che la struttura iperfina del termine 6 $^2P_{1/2}$ è $9\,\frac{\mu}{\mu_0}\,\frac{2k+1}{k}$. Confrontando questo valore con la separazione del termine fondamentale 6 S, data dalla (6), si trova che la separazione del termine 6 S è più di dieci volte maggiore della separazione del termine 6 $^2P_{1/2}$. Questo fatto giustifica completamente il risultato sperimentale che tutte le righe della serie principale dei metalli alcalini hanno una struttura iperfina consistente in due sole componenti, poiché la risoluzione non è sufficiente a risolvere la molteplicità maggiore derivante dalla struttura dei termini *p*.

Dai valori delle separazioni della struttura iperfina, che sono noti sperimentalmente (0,30 per il Cs, 0,06 per il Na e 0,114 per il Rb) si può, per mezzo delle (6), calcolare il valore del rapporto μ_0/μ tra il magnetone di Bohr ed il momento magnetico del nucleo, quando si conosca il valore di *k*. La tabella seguente ci mostra i valori che si trovano per questo rapporto, attribuendo a *k* i valori $1/2\,,\ 1\,,\ 3/2\,,\cdots,\infty$.

$k =$	1/2	1	3/2	∞
Cs	1.950	1.460	1.300	980
Na	890	670	600	450
Rb	1.700	1.250	1.100	840

Come si è già detto, il valore di *k* può determinarsi conoscendo il rapporto di intensità delle due componenti della struttura iperfina; per il sodio il valore più probabile è $k = 1$. Troviamo dunque dalla tabella come valore più probabile del momento magnetico del nucleo di sodio $\mu_0/670$.

I possibili errori nei dati della tabella derivano in parte dall'incertezza delle misure delle separazioni, in parte dalla valutazione statistica delle autofunzioni. A quest'ultima causa si potrebbe attribuire un errore del 20 o del 30 %; verosimilmente la precisione delle misure è invece molto maggiore.

3. Presenta un certo interesse confrontare la più semplice ipotesi che si può fare sopra il magnetismo dei nuclei con i risultati che abbiamo ottenuti. Per analogia col comportamento degli elettroni si potrebbe per esempio ammettere, come fa Jackson per il caso $k = 1/2$, che il momento magnetico di un nucleo sia dato dall'espressione

$$2k \frac{Zeh}{4\pi ac}$$

in cui a rappresenta la massa del nucleo. Il rapporto del magnetone di Bohr al momento magnetico risulterebbe allora

$$\gamma = 1.840 \frac{M}{2kZ} = 920 \frac{M}{kZ}$$

in cui M rappresenta il peso atomico. Per il Cs si ha $M/Z = 2{,}41$, e risulterebbe dunque $\mu_0/\mu = 2.220/k$. Dal confronto con i risultati della tabella si otterrebbe per k un valore prossimo a 2, il che non sarebbe inammissibile. L'ipotesi sembra però insostenibile per il caso del sodio. Per questo elemento si ha infatti $M/Z = 2.09$. Si otterrebbe dunque $\mu_0/\mu = 1.920/k$. Dal confronto con la tabella si dedurrebbe per k un valore da 3 a 4. Un valore così elevato di k è però in contrasto con i dati di Schüler sopra il rapporto di intensità delle due componenti della struttura iperfina.

4. In questa ultima parte del lavoro vogliamo mostrare brevemente la via seguita per il calcolo teorico delle strutture iperfine.

Incominciamo dal caso della separazione dei termini S. Scriviamo perciò l'equazione di Dirac, che dà la Hamiltoniana relativistica per l'elettrone, nella forma:

$$\text{(12)} \qquad W = eV - c\alpha \cdot \left(\boldsymbol{p} - \frac{e}{c}\mathbf{U}\right) - (\beta + 1)mc^2$$

in cui $\boldsymbol{p}$ rappresenta la quantità di moto dell'elettrone, cioè l'operatore $\frac{h}{2\pi i}$ grad; β e α sono rispettivamente un q-scalare e un q-vettore i quali, come operatori, sono rappresentati dalle matrici:

$$\text{(13)} \qquad \left|\begin{matrix} 1 & 0 & 0 & 0 \\ 0 & 1 & 0 & 0 \\ 0 & 0 & -1 & 0 \\ 0 & 0 & 0 & -1 \end{matrix}\right| = \beta \quad ; \quad \left|\begin{matrix} 0 & 0 & 0 & 1 \\ 0 & 0 & 1 & 0 \\ 0 & 1 & 0 & 0 \\ 1 & 0 & 0 & 0 \end{matrix}\right| = \alpha_x ;$$

$$\left|\begin{matrix} 0 & 0 & 0 & -i \\ 0 & 0 & i & 0 \\ 0 & -i & 0 & 0 \\ i & 0 & 0 & 0 \end{matrix}\right| = \alpha_y \quad ; \quad \left|\begin{matrix} 0 & 0 & 1 & 0 \\ 0 & 0 & 0 & -1 \\ 1 & 0 & 0 & 0 \\ 0 & -1 & 0 & 0 \end{matrix}\right| = \alpha_z .$$

Il potenziale elettrico V, nel caso di un atomo alcalino, dipende soltanto dal raggio vettore r. Il potenziale vettore $\mathbf{U}$ deriva dal momento magnetico del nucleo. Precisamente se si indicano con μ_x, μ_y, μ_z, le componenti di questo momento magnetico, il potenziale vettore prodotto da esso sarà $\mu \times \boldsymbol{r}/r^3$, ed avrà dunque per componenti:

$$(14)\qquad U_x = \frac{1}{r^3}(z\,\mu_y - y\,\mu_z) \quad ; \quad U_y = \frac{1}{r^3}(x\,\mu_z - z\,\mu_x)\,;$$
$$U_z = \frac{1}{r^3}(y\,\mu_x - x\,\mu_y)\,.$$

Naturalmente anche μ_x, μ_y, μ_z debbono considerarsi come operatori che soddisfano alle seguenti relazioni

$$(15)\qquad \mu_y\,\mu_z - \mu_z\,\mu_y = \frac{i\mu}{k}\,\mu_x \quad ; \quad \mu_z\,\mu_x - \mu_x\,\mu_z = \frac{i\mu}{k}\,\mu_y\,;$$
$$\mu_x\,\mu_y - \mu_y\,\mu_x = \frac{i\mu}{k}\,\mu_z\,.$$

e che hanno per autovalori $\mu, \frac{\mu}{k}(k-1), \cdots, -\frac{\mu}{k}(k-1), -\mu$.

Ponendo in [12] le espressioni [14] di $\mathbf{U}$ si trova:

$$(16)\qquad W = eV - c\alpha\cdot\boldsymbol{p} - (\beta+1)\,mc^2$$
$$+\frac{e}{r^3}\left[\alpha_x(z\,\mu_y - y\,\mu_z) + \alpha_y(x\,\mu_z - z\,\mu_x) + \alpha_z(y\,\mu_x - y\,\mu_y)\right].$$

L'ultimo termine rappresenta la perturbazione dovuta all'azione del momento magnetico del nucleo. Come energia di perturbazione troviamo dunque:

$$(17)\qquad w = \frac{e}{r^3}\left[\alpha_x(z\,\mu_y - y\,\mu_z) + \alpha_y(x\,\mu_z - z\,\mu_x) + \alpha_z(y\,\mu_x - x\,\mu_y)\right].$$

Per il calcolo delle separazioni prodotte da questa perturbazione, applicheremo, adattandolo al caso nostro, l'ordinario procedimento della teoria delle perturbazioni. Osserviamo per questo che, prescindendo dal momento proprio del nucleo, ogni autofunzione consiste di quattro funzioni delle sole coordinate di posizione x, y, z (poiché gli operatori (13) sono matrici di quarto ordine). Le prime due di queste quattro funzioni, ψ_1 e ψ_2, sono molto piccole in confronto alle altre due ψ_3 e ψ_4; in prima approssimazione si ha infatti [(5)]:

$$(18)\qquad \psi_1 = i\frac{h}{4\pi mc}\left\{\left(\frac{\partial}{\partial x} - i\frac{\partial}{\partial y}\right)\psi_4 + \frac{\partial}{\partial z}\psi_3\right\};$$
$$\psi_2 = i\frac{h}{4\pi mc}\left\{\left(\frac{\partial}{\partial x} + i\frac{\partial}{\partial y}\right)\psi_3 - \frac{\partial}{\partial z}\psi_4\right\}.$$

ψ_3 e ψ_4 soddisfano, in prima approssimazione all'ordinaria equazione di Schrödinger non relativistica. Prescindendo dall'azione del momento nucleare i termini s sono doppiamente degeneri; ogni termine s avrà perciò due autofunzioni indipendenti. Come tali possiamo prendere per esempio:

$$(19)\qquad \begin{cases} \Psi_1 = \left[\frac{hi}{4\pi mc}\cos\theta\psi'(r), \frac{hi}{4\pi mc}\sin\theta\, e^{i\varphi}\psi'(r), \psi(r), 0\right] \\ \Psi_2 = \left[\frac{hi}{4\pi mc}\sin\theta e^{-i\varphi}\psi'(r), -\frac{hi}{4\pi mc}\cos\theta\psi'(r), 0, \psi(r)\right]. \end{cases}$$

(5) Cfr., per esempio, C. G. DARWIN, «Proc. Roy. Soc.», *118*, 654 (1928).

r, θ e φ sono coordinate polari; ψ è una funzione di r soltanto, che, in prima approssimazione, coincide con l'ordinaria funzione di Schrödinger per il termine s considerato.

Quando si considera anche la degenerazione derivante dalle $2k+1$ possibili orientazioni del nucleo; il termine s diventa $2(2k+1)$ volte degenere. La autofunzione verrà allora a dipendere anche da un'altra coordinata, per esempio, μ_z che rappresenta il momento nucleare. Siccome questa ha $2k+1$ autovalori, troviamo che ogni autofunzione viene ora a essere costituita da $4(2k+1)$ funzioni di x, y e z. Così, per esempio, per $k=1$ avremo 12 funzioni, che possono opportunamente raggrupparsi nel quadro:

$$\begin{vmatrix} \psi_{1,1} & \psi_{2,1} & \psi_{3,1} & \psi_{4,1} \\ \psi_{1,0} & \psi_{2,0} & \psi_{3,0} & \psi_{4,0} \\ \psi_{1,-1} & \psi_{2,-1} & \psi_{3,-1} & \psi_{4,-1} \end{vmatrix} \tag{20}$$

α_x, α_y, α_z e β operano sopra le colonne di questo schema: μ_x, μ_y, μ_z operano invece sopra le righe. Nel caso $k=1$, gli autovalori di μ_x, μ_y e μ_z sono μ, 0, $-\mu$. Per soddisfare alle (15), si possono prendere per μ_x, μ_y, μ_z gli operatori rappresentanti dalle matrici seguenti:

$$\mu_x = \mu \begin{vmatrix} 0 & \frac{1}{\sqrt{2}} & 0 \\ \frac{1}{\sqrt{2}} & 0 & \frac{1}{\sqrt{2}} \\ 0 & \frac{1}{\sqrt{2}} & 0 \end{vmatrix} ; \quad \mu_y = \mu \begin{vmatrix} 0 & -\frac{i}{\sqrt{2}} & 0 \\ \frac{i}{\sqrt{2}} & 0 & -\frac{i}{\sqrt{2}} \\ 0 & \frac{i}{\sqrt{2}} & 0 \end{vmatrix} ; \tag{21}$$

$$\mu_z = \mu \begin{vmatrix} 1 & 0 & 0 \\ 0 & 0 & 0 \\ 0 & 0 & -1 \end{vmatrix} .$$

Per ogni termine s troviamo $2\times 3 = 6$ autofunzioni imperturbate indipendenti; possiamo prendere per esempio le sei autofunzioni seguenti:

$$\begin{vmatrix} \Psi_1 \\ 0 \\ 0 \end{vmatrix} , \begin{vmatrix} 0 \\ \Psi_1 \\ 0 \end{vmatrix} , \begin{vmatrix} 0 \\ 0 \\ \Psi_1 \end{vmatrix} , \begin{vmatrix} \Psi_2 \\ 0 \\ 0 \end{vmatrix} , \begin{vmatrix} 0 \\ \Psi_2 \\ 0 \end{vmatrix} , \begin{vmatrix} 0 \\ 0 \\ \Psi_2 \end{vmatrix} .$$

Per brevità abbiamo indicato con un solo simbolo, Ψ_1, Ψ_2, 0, 0, le quattro funzioni di una riga; il significato di Ψ_1 e Ψ_2 è dato dalle (19) mentre il simbolo 0 sta per la riga 0, 0, 0, 0.

Per il calcolo della separazione del termine dobbiamo ora, secondo l'ordinario procedimento del calcolo delle perturbazioni per i sistemi degeneri, formarci la matrice di perturbazione; gli autovalori di questa matrice ci daranno direttamente gli spostamenti dei vari termini in cui si spezza il termine imperturbato. Per mezzo di un calcolo alquanto lungo, ma che non presenta ormai alcuna difficoltà, si trova la matrice di perturbazione seguente:

$$w = \frac{8\pi}{3}\mu\mu_0\psi^2(0)\begin{vmatrix} 1 & 0 & 0 & 0 & 0 & 0 \\ 0 & 0 & 0 & -\sqrt{2} & 0 & 0 \\ 0 & 0 & -1 & 0 & \sqrt{2} & 0 \\ 0 & -\sqrt{2} & 0 & -1 & 0 & 0 \\ 0 & 0 & \sqrt{2} & 0 & 0 & 0 \\ 0 & 0 & 0 & 0 & 0 & 1 \end{vmatrix}. \tag{22}$$

Risolvendo l'equazione secolare corrispondente a questa matrice, si trova che essa ha per autovalori

$$w_1 = \frac{8\pi}{3}\mu\mu_0\psi^2(0) \qquad \text{(quadruplo)}$$

e

$$w_2 = -\frac{16\pi}{3}\mu\mu_0\psi^2(0) \qquad \text{(doppio)}$$

il termine si spezza dunque in un termine superiore, quattro volte degenere, che corrisponde dunque a un momento totale $f_1 = 3/2$ e in un termine inferiore doppio e che ha quindi come momento totale $f_2 = 1/2$.

La separazione è:

$$w_1 - w_2 = 8\pi\mu\mu_0\psi^2(0) \tag{23}$$

che coincide con la (2) per il caso particolare $k = 1$.

In modo del tutto simile si potrebbe dimostrare le formule che dànno le separazioni per un valore qualsiasi di k, sia per i termini s che per i termini p.

In Paper N° 57 *b*, which is a German, less detailed version of N° 57 *a*, sent to « Zeitschr. für Physik » on December 18 th, 1929, there is an additional section proving Eq. (3), (5), (7) of paper N° 57 *a*. We report it here for completeness, only having changed the number of some Eqs. in order to facilitate the reference to paper N° 57 *a*. The German vector notation is used.

. . . .

5. Es bleibt nur noch die Theorie der Aufspaltungen der p–Terme [Formeln (3) und (5)] (*) anzudeuten. Wie gesagt, habe ich für die p–Terme die Paulische Theorie an Stelle der Diracschen benutzt [1]. Ich habe die Rechnungen eingehend für den Fall $k = 1/2$ durchgeführt. Als Hamiltonfunktion erster Näherung nehmen wir die folgende:

$$W_0 = \frac{1}{2m}p^2 + eV - \frac{1}{2mc^2}(W_0 - eV)^2 + \frac{\mu_0}{2mc}\frac{1}{r}\frac{dV}{dr}(\boldsymbol{\sigma}, \mathbf{M}). \tag{24}$$

(*) The equations are numbered consistently with Paper N° 57 *a* (*Editors Note*).

(1) Für die $^2P_{1/2}$–Terme habe ich die Rechnungen auch nach der Diracschen Methode ausgeführt. Das Ergebnis ist mit dem der Paulischen Methode identisch.

Die beiden ersten Glieder sind die gewöhnliche nichtrelativistische Hamiltonfunktion ohne Berücksichtigung des elektronischen Eigenmoments. Das dritte Glied stellt die Störung durch die Relativität dar; das vierte Glied gibt die Wirkung der Kopplung zwischen Bahnimpulsmoment **M** und Elektronspin $\boldsymbol{\sigma}$ an. Da $\boldsymbol{\sigma}$ bekanntlich durch die zweireihigen Matrizen

$$\sigma_x = \begin{vmatrix} 0 & 1 \\ 1 & 0 \end{vmatrix} \quad , \quad \sigma_y = \begin{vmatrix} 0 & -1 \\ 1 & 0 \end{vmatrix} \quad , \quad \sigma_z = \begin{vmatrix} 1 & 0 \\ 0 & -1 \end{vmatrix}$$

dargestellt werden kann, spaltet sich jede Eigenfunktion, wenn man die Wirkung des Kernspins vernachlässigt, in zwei Funktionen von x , y , z allein. Als ungestörte Eigenfunktionen finden wir, wie in der gewöhnlichen Paulischen Theorie, für den zweifach entarteten $^2P_{1/2}$–Term

$$\left[\frac{1}{\sqrt{3}} f(r) P_0 , \sqrt{\frac{2}{3}} f(r) P_1\right] \quad , \quad \left[-\sqrt{\frac{2}{3}} f(r) P_{-1} , \frac{1}{\sqrt{3}} f(r) P_0\right] \tag{25}$$

wo

$$P_1 = \sqrt{\frac{3}{8\pi}} \sin\Theta e^{i\varphi} \quad , \quad P_0 = \sqrt{\frac{3}{4\pi}} \cos\Theta \quad , \quad P_{-1} = \sqrt{\frac{3}{8\pi}} \sin\Theta e^{-i\varphi}$$

die drei Kugelfunktionen erster Ordnung darstellen. Für den vierfach entarteten $^2P_{3/2}$–Term finden wir die vier Eigenfunktionen

$$[f(r) P_1 , 0] \quad ; \quad \left[-\sqrt{\frac{2}{3}} f(r) P_0 , \frac{1}{\sqrt{3}} f(r) P_1\right] \quad ; \tag{26}$$

$$\left[\frac{1}{\sqrt{3}} f(r) P_{-1} , \sqrt{\frac{2}{3}} f(r) P_0\right] \quad ; \quad [0 , f(r) P_{-1}] .$$

Die durch die Wirkung des magnetischen Moments des Kernes hinzukommende Störung besteht in der Wechselwirkung des Kernmoments mit dem Bahnimpuls **M** und mit dem Elektronspin. Die erste Wechselwirkungsenergie ist:

$$\frac{e}{mcr^3} (\boldsymbol{\mu} , \mathbf{M})$$

und die zweite, welche in der Arbeit von Hargreaves vernachlässigt wurde,

$$\frac{\mu_0}{r^3} (\boldsymbol{\sigma} , \boldsymbol{\mu}) + \frac{3\mu_0}{r^5} (\boldsymbol{r} , \boldsymbol{\sigma}) (\boldsymbol{r} , \boldsymbol{\mu}) .$$

Als gesamte Störung finden wir also

$$w = \frac{e}{mcr^3} (\boldsymbol{\mu} , \mathbf{M}) - \frac{\mu_0}{r^3} (\boldsymbol{\sigma} , \boldsymbol{\mu}) + \frac{3\mu_0}{r^5} (\boldsymbol{r} , \boldsymbol{\sigma}) (\boldsymbol{r} , \boldsymbol{\mu}) . \tag{27}$$

Wie im vorigen Falle, ist $\boldsymbol{\mu}$ ein q–Vektor, dessen Komponenten den Bedingungen (15) genügen müssen. In dem von uns betrachteten Falle $k = 1/2$ spaltet sich jede Eigenfunktion in vier Funktionen von x, y , z allein. Diese vier Funktionen schreiben wir in folgender Anordnung:

$$\begin{vmatrix} \psi_{1,1} & \psi_{2,1} \\ \psi_{1,2} & \psi_{2,2} \end{vmatrix} .$$

Die den Elektronspin darstellenden Operatoren operieren dann auf die Spalten dieses Schemas. Die das Kernmoment darstellenden Operatoren

$$(28) \qquad \mu_x = \mu \begin{vmatrix} 0 & 1 \\ 1 & 0 \end{vmatrix} \quad , \quad \mu_y = \mu \begin{vmatrix} 0 & -i \\ i & 0 \end{vmatrix} \quad , \quad \mu_z = \mu \begin{vmatrix} 1 & 0 \\ 0 & -1 \end{vmatrix}$$

operieren dagegen auf die Reihen. Es ist z.B.

$$\mu_x \begin{vmatrix} \psi_{11} & \psi_{21} \\ \psi_{12} & \psi_{22} \end{vmatrix} = \mu \begin{vmatrix} \psi_{12} & \psi_{22} \\ \psi_{11} & \psi_{21} \end{vmatrix} \quad ; \quad \sigma_x \begin{vmatrix} \psi_{11} & \psi_{21} \\ \psi_{12} & \psi_{22} \end{vmatrix} = \begin{vmatrix} \psi_{21} & \psi_{11} \\ \psi_{22} & \psi_{12} \end{vmatrix} .$$

Der ursprünglich zweifach entartete $^2P_{1/2}$–Term wird jetzt durch Betrachtung der beiden Orientierungsmöglichkeiten des Kernspins vierfach entartet. Als ungestörte Eigenfunktionen können wir z.B. die folgenden vier nehmen:

$$\begin{vmatrix} \frac{1}{\sqrt{3}} f P_0 , \sqrt{\frac{2}{3}} f P_1 \\ 0 \quad , \quad 0 \end{vmatrix} , \quad \begin{vmatrix} 0 \quad , \quad 0 \\ \frac{1}{\sqrt{3}} f P_0 , \sqrt{\frac{2}{3}} f P_1 \end{vmatrix} , \quad \begin{vmatrix} -\sqrt{\frac{2}{3}} f P_{-1} , \frac{1}{\sqrt{3}} f P_0 \\ 0 \quad , \quad 0 \end{vmatrix} ,$$

$$\begin{vmatrix} 0 \quad , \quad 0 \\ -\sqrt{\frac{2}{3}} f P_{-1} , \frac{1}{\sqrt{3}} f P_0 \end{vmatrix} .$$

Um die von der Störungsfunktion (27) herrührende Spaltung des $^2P_{1/2}$–Terms zu berechnen, müssen wir die die Störungsenergie darstellende vierreihige Matrix bilden. Dies geschieht jetzt ohne Schwierigkeit. Man beachte die Formeln

$$M_x\ P_1 = -\frac{h\sqrt{2}}{4\pi} P_0 \quad ; \quad M_y\ P_1 = -i\,\frac{h\sqrt{2}}{4\pi} P_0 \quad ; \quad M_z\ P_1 = \frac{h}{2\pi} P_1 \ ,$$

$$M_x\ P_0 = \frac{h\sqrt{2}}{4\pi}(P_{-1} - P_1) \ ; \quad M_y\ P_0 = i\,\frac{h\sqrt{2}}{4\pi}(P_1 + P_{-1}) \ ; \quad M_z\ P_0 = 0 ,$$

$$M_x\ P_{-1} = \frac{h\sqrt{2}}{4\pi} P_0 \quad ; \quad M_y\ P_{-1} = -i\,\frac{h\sqrt{2}}{4\pi} P_0 \quad ; \quad M_z\ P_{-1} = -\frac{h}{2\pi} P_{-1} .$$

Man findet dann für den $^2P_{1/2}$–Term die folgende Störungsmatrix

$$\frac{8}{3}\,\mu\mu_0\,\overline{\frac{1}{r^2}} \begin{vmatrix} 1 & 0 & 0 & 0 \\ 0 & -1 & -2 & 0 \\ 0 & -2 & -1 & 0 \\ 0 & 0 & 0 & 1 \end{vmatrix}$$

mit den Eigenwerten:

$$\frac{8}{3}\,\mu\mu_0\,\overline{\frac{1}{r^3}}\ \text{(dreifach)} \quad \text{und} \quad -8\,\mu\mu_0\,\overline{\frac{1}{r^3}}\ \text{(einfach)}.$$

Für den $^2P_{3/2}$–Term findet man in ähnlicher Weise die Störungsmatrix:

$$\frac{8}{5}\mu\mu_0\overline{\frac{1}{r^3}}\left|\begin{matrix} 1 & 0 & 0 & 0 & 0 & 0 & 0 & 0 \\ 0 & 1/3 & 0 & 0 & 2/\sqrt{3} & 0 & 0 & 0 \\ 0 & 0 & -1/3 & 0 & 0 & -4/3 & 0 & 0 \\ 0 & 0 & 0 & -1 & 0 & 0 & 2/\sqrt{3} & 0 \\ 0 & 2/\sqrt{3} & 0 & 0 & -1 & 0 & 0 & 0 \\ 0 & 0 & -4/3 & 0 & 0 & -1/3 & 0 & 0 \\ 0 & 0 & 0 & 2/\sqrt{3} & 0 & 0 & 1/3 & 0 \\ 0 & 0 & 0 & 0 & 0 & 0 & 0 & 1 \end{matrix}\right|$$

mit den Eigenwerten:

$$\frac{8}{5}\mu\mu_0\overline{\frac{1}{r^3}}\,\text{(fünffach)} \quad \text{und} \quad -\frac{8}{3}\mu\mu_0\overline{\frac{1}{r^3}}\,\text{(dreifach)}.$$

Es ergibt sich also, dass der $^2P_{1/2}$–Term sich in einen oberen, dreifach entarteten und in einen unteren einfachen Term spaltet; die Aufspaltung ist durch Formel (3) gegeben. Der $^2P_{3/2}$–Term spaltet sich in einen oberen, fünffach entarteten und in einen unteren, dreifach entarteten Term; die Aufspaltung ist (5).

Es bleibt nur Formel (7) zu beweisen. Für diesen Zweck bemerken wir, dass der Energieunterschied zwischen den beiden Termen $^2P_{3/2}$ und $^2P_{1/2}$ durch die Wirkung des elektronischen Eigenmoments, bekanntlich

$$\delta = -\frac{3\mu_0 h}{4\pi mc}\overline{\frac{1}{r}\frac{dV}{dr}}$$

ist. Wir finden also

$$\overline{\frac{1}{r}\frac{dV}{dr}} = -\frac{4\pi mc}{3\mu_0 h}\delta. \tag{29}$$

In der Nähe des Kernes hat man aber $V = Ze/r$, also

$$\frac{1}{r}\frac{dV}{dr} = -Ze\frac{1}{r^3}.$$

Da nun die für die Bildung des Mittelwerts (29) wichtigen Glieder von der nächsten Nähe des Kerns herrühren, finden wir mit genügender Annäherung

$$\overline{\frac{1}{r^3}} = -\frac{1}{Ze}\overline{\frac{1}{r}\frac{dV}{dr}} = \frac{4\pi mc}{3\mu_0 Zeh}\delta,$$

d. h.

$$\overline{\frac{1}{r^3}} = \frac{\delta}{3Z\mu_0^2};$$

diese Formel ist mit (7) identisch. Für den Fall des ersten *p*–Terms des Cs ist der aus (7) sich ergebende Wert nur etwa 4 % kleiner als der genaue Wert.

Zusatz bei der Korrektur.

1. Inzwischen habe ich auch für die $^2P_{3/2}$–Terme Formel (5) mit der Diracschen Theorie des Elektrons bestätigen können.

Auch für die *p*–Terme kann ich jetzt die Hyperfeinstrukturaufspaltungen für einen beliebigen Wert des Eigenmoments *k* des Kernes angeben.

Für die $^2P_{1/2}$–Terme findet man eine Aufspaltung in einem oberen $(2k+2)$–fach und in einem unteren $2k$–fach entarteten Term; die Aufspaltung ist:

$$\Delta\,(^2P_{1/2}) = \frac{2k+1}{k}\,\frac{8}{3}\,\mu\mu_0\,\overline{\frac{1}{r^3}}\,,$$

Für die $^2P_{1/2}$–Terme bekommt man eine Aufspaltung in zwei, drei oder vier Terme, je nachdem $k = 1/2$, $k=1$ oder $k > 1$. In diesem letzteren Falle sind die Abstände der vier Terme vom ungestörten Term die folgenden:

$$\frac{8}{5}\,\mu\mu_0\,\overline{\frac{1}{r^3}}\quad,\quad \left(\frac{1}{3}-\frac{1}{k}\right)\frac{8}{5}\,\mu\mu_0\,\overline{\frac{1}{r^3}}\,,$$

$$-\left(\frac{1}{3}+\frac{4}{3k}\right)\frac{8}{5}\,\mu\mu_0\,\overline{\frac{1}{r^3}}\quad,\quad -\left(1+\frac{1}{k}\right)\frac{8}{5}\,\mu\mu_0\,\overline{\frac{1}{r^3}}\,.$$

Die statistischen Gewichte dieser vier Terme sind

$$2k+4\quad,\quad 2k+2\quad,\quad 2k\quad,\quad 2k-2.$$

Für $k=1$ fehlt der letztere und für $k=1/2$ fehlen die beiden letzteren dieser vier Terme. Alle diese Ergebnisse sind mit den gewöhnlichen Regeln für die Kopplung der Quantenvektoren in Einklang.

N° 59.

During the years 1926-29 quantum mechanics had reached Rome only through printed works, however, it had been completely mastered by Fermi. Matrix mechanics (Heisenberg) initially seemed too different from ordinary mathematical physics and at first could not be deeply understood and appreciated, but the full importance of Schrödinger's work and of wave mechanics were immediately grasped. Also Born's interpretation of the field scalar was easily accepted (see, however, N° 42 where the original Schrödinger interpretation is still used). Dirac's papers were read as soon as they appeared, but occasionally it took considerable time to digest them completely. Assimilation of a paper, for Fermi, often meant obtaining the same results by different means and making some new important application. It was indeed a process of rediscovery.

Fermi very early explained wave mechanics to his friends and pupils in informal talks; later he gave some lectures in the Mathematical Colloquium at the University of Rome. These lectures were followed by lively discussions in which, among others, Professors Castelnuovo, Enriques and Levi-Civita participated. The main difficulties were connected with the correct interpretation of the uncertainty principle.

The following paper develops one of the points made by Fermi and shows an interesting aspect of the way in which problems of quantum mechanics should be formulated and of the type of answers obtainable.

E. SEGRÈ.

59.

L'INTERPRETAZIONE DEL PRINCIPIO DI CAUSALITÀ NELLA MECCANICA QUANTISTICA(*)

« Rend. Lincei », *11*, 980–985 (1930) e « Nuovo Cimento », 7, 361–366 (1930).

Lo scopo di questa Nota è di precisare fino a che punto si può, secondo la meccanica quantistica, parlare di causalità, e in che senso deve intendersi la affermazione che la meccanica quantistica non conduce a una determinazione degli avvenimenti futuri. Da questo stesso programma risulta che in questo lavoro non saranno contenuti che in minima misura risultati nuovi; ma verranno solo espresse, in una forma nuova, relazioni già conosciute. Per maggiore chiarezza di esposizione mi limiterò a considerare il caso della meccanica di un punto mobile lungo una retta; le considerazioni qui svolte potrebbero però facilmente estendersi al caso di sistemi più complicati.

Lo stato del sistema meccanico costituito da un punto mobile lungo una retta si determina, nella meccanica classica, indicando a un certo istante, per esempio $t = 0$, la posizione e la velocità o, che è lo stesso, la quantità di moto. Basta cioè, per individuare lo stato, dare, per $t = 0$, l'ascissa x del

(*) Presentata dal Socio O. M. Corbino nella seduta del 30 maggio 1930.

punto e la sua quantità di moto $p = m\dot{x}$. Se, come vogliamo supporre, è conosciuta l'energia potenziale U (x) della forza che agisce sopra il punto, la conoscenza di x e p all'istante $t = 0$ determina univocamente i valori di queste due grandezze per tutti gli istanti passati e futuri; in questo senso si intende la affermazione che nella meccanica classica vale il principio di causalità; vale cioè la proprietà che la determinazione dello stato del sistema a un certo istante è sufficiente per determinare lo stato del sistema stesso in qualsiasi istante del passato o dell'avvenire; ben inteso, supponendo che il sistema considerato non venga perturbato dall'azione di altri sistemi non completamente conosciuti. In tutto quello che segue parleremo sempre di principio di causalità classico in questo senso preciso.

Naturalmente, sempre restando nel caso classico, non è necessario, per la determinazione dello stato del sistema, misurare proprio i valori di x e di p all'istante zero; si può anche per esempio misurare i valori di due funzioni indipendenti $f(x, p)$ e $g(x, p)$ della x e della p; dal valore di queste si possono poi, volendo, ricavare i valori di x e di p.

Dobbiamo ora vedere come queste considerazioni vanno modificate quando le si trasportano nella meccanica quantistica. Dobbiamo perciò esaminare che cosa occorre conoscere di un sistema, secondo la meccanica quantistica, per poter dire di averne determinato lo stato. Heisenberg ha fatto vedere che non è possibile una misura contemporanea esatta di x di p; che anzi, se si effettua una misura esatta della x, la quantità di moto p resta necessariamente del tutto indeterminata, e che viceversa resta completamente indeterminata la x se si misura con esattezza completa la p. In altre parole per un sistema avente, come il nostro, un solo grado di libertà, si può pensare di effettuare con precisione assoluta la misura di una sola grandezza fisica, per esempio x, oppure p, oppure una qualsiasi funzione $g(x, p)$. Dobbiamo dunque intendere che effettuare una simile misura determini lo stato del sistema a un certo istante, poiché ci rappresenta il massimo che noi possiamo sapere sopra la situazione del sistema a quell'istante. Ciò porta però con sé la conseguenza che la determinazione dello stato di un sistema può venir fatta in modi *essenzialmente differenti*, dipendentemente dalla particolare scelta della funzione $g(x, p)$ che viene misurata; mentre nel caso classico, in cui si ammette possibile la misura di due funzioni indipendenti f e g di x e p, vengono sempre ad essere determinate, in ultima analisi, queste due variabili.

Tornando al caso della meccanica quantistica, se, misurando una grandezza fisica $g(x, p)$ troviamo che essa ha il valore g', possiamo facilmente dedurre quale è lo scalare di campo corrispondente. Esso si ottiene infatti, come è noto [1], come soluzione dell'equazione

$$g(x, p)\,\psi(x) = g'\,\psi(x) \tag{1}$$

dove nel primo membro $g(x, p)$ deve intendersi come l'operatore che si ottiene sostituendo in $g(x, p)$ alla variabile p l'operazione $\frac{h}{2\pi i}\frac{d}{dx}$. Questa

(1) P. A. M. DIRAC, «Proc. Roy. Soc.», *113*, 621 (1927).

equazione per esempio, nel caso che la grandezza $g(x, p)$ sia l'energia del sistema, si riduce alla ordinaria equazione di Schroedinger.

Se invece di $g(x, p)$ si fosse misurata un'altra grandezza $G(x, p)$ si sarebbe trovato uno scalare di campo differente; in modo che lo scalare di campo, che rappresenta lo stato del sistema viene a dipendere, oltre che dal sistema, anche dal modo particolare che noi seguiamo per determinarlo. Secondo la meccanica quantistica lo stato del sistema a un dato istante può dunque venire definito in due modi differenti; e cioè: *a*) misurando il valore di una grandezza fisica $g(x, p)$; *b*) dando, per l'istante considerato, una funzione (generalmente complessa) della x,

$$\psi(x) = \rho(x)\, e^{i\theta(x)}$$

che rappresenta lo scalare di campo a quell'istante.

È facile convincersi che questi due metodi sono completamente equivalenti tra di loro. Abbiamo già visto infatti (1) che, conoscendo il valore g' che assume la grandezza $g(x, p)$ si può, a meno di una costante moltiplicativa inessenziale, determinare a quell'istante lo scalare di campo $\psi(x)$. Ci proponiamo ora di mostrare inversamente, che, data una qualsiasi funzione complessa $\psi(x) = \rho e^{i\theta}$ che rappresenti a un certo istante lo scalare di campo, esiste sempre un operatore reale $g(x, p)$ tale che

$$g(x, p)\, \psi(x) = 0 \tag{2}$$

per modo che si può dire che, se lo scalare di campo è $\psi(x)$ la grandezza $g(x, p)$ ha il valore 0. Così che lo stato del sistema che corrisponde allo scalare di campo $\psi(x)$ può anche definirsi come quello per cui $g(x, p) = 0$. Naturalmente, perché questa considerazione abbia un significato fisico, è necessario che la grandezza $g(x, p)$ sia reale. Per dimostrare l'esistenza della grandezza $g(p, x)$, facciamo vedere che si può porre

$$g(x, p) = \left(p - \frac{h}{2\pi}\frac{d\theta}{dx}\right)^2 + \frac{h^2}{4\pi^2}\frac{1}{\rho}\frac{d^2\rho}{dx^2}. \tag{3}$$

Questa grandezza è infatti evidentemente reale, poiché ρ e θ sono due funzioni reali della x. Si verifica poi immediatamente che

$$g\left(x, \frac{h}{2\pi i}\frac{d}{dx}\right)\rho\, e^{i\theta} = 0 \tag{4}$$

cioè che la grandezza g soddisfa alla (2). Naturalmente la grandezza (3) non è l'unica che soddisfa alla (2). Per esempio soddisfano alla (2) anche le grandezze reali $g^2, g^3, \cdots$. Per noi basta però aver dimostrato che esiste almeno una grandezza $g(x, p)$ soddisfacente alla (2).

Dobbiamo ora passare alla trattazione del problema specifico al quale è dedicata questa Nota; e cioè al problema di determinare quali elementi dello stato del sistema ad un istante futuro è possibile determinare per mezzo di esperienze fatte all'istante attuale.

Ci proponiamo di far vedere che, data una qualsiasi grandezza fisica $G(x, p)$, *è sempre possibile conoscere il valore che essa avrà al tempo qualsiasi* t_0 *per mezzo di una esperienza, opportunamente scelta, da eseguirsi al tempo* $t = 0$.

Per vedere questo supponiamo che, se si misura la grandezza G all'istante t_0, si trovi per essa il valore G'. Da questo fatto si può dedurre, per mezzo della (1), quale è lo scalare di campo del sistema al tempo t_0. Esso viene infatti determinato dall'equazione

$$G\left(x, \frac{h}{2\pi i}\frac{d}{dx}\right)\psi(x, t_0) = G'\,\psi(x, t_0)\,. \tag{5}$$

Determinata così la ψ al tempo t_0 possiamo conoscere il valore per un istante qualsiasi. Se infatti con $H(x, p)$ si indica l'energia del sistema, la funzione $\psi(x, t)$ soddisfa all'equazione di Schroedinger

$$H\left(x, \frac{h}{2\pi i}\frac{\partial}{\partial x}\right)\psi(x, t) = -\frac{h}{2\pi i}\frac{\partial\psi}{\partial t}\,. \tag{6}$$

E siccome questa è di primo ordine rispetto al tempo, essa permette di determinare completamente la funzione ψ conoscendo i valori che essa assume all'istante $t = t_0$. Si può dunque per mezzo nella (6) conoscere quale deve essere lo scalare di campo $\psi(x, 0)$ all'istante $t = 0$. Nella funzione $\psi(x, 0)$ così determinata, figurerà naturalmente come parametro il valore G' che la grandezza $G(x, p)$ ha per ipotesi al tempo t_0; metteremo ciò in evidenza scrivendo $\psi(x, 0, G')$ in luogo di $\psi(x, 0)$.

Abbiamo già fatto vedere (3) che è sempre possibile determinare una grandezza fisica reale $g(x, p)$ tale che

$$g\left(x, \frac{h}{2\pi i}\frac{d}{dx}\right)\psi(x, 0, G') = 0\,. \tag{7}$$

Ciò significa che se, misurando la grandezza $g(x, p)$ si trova che essa ha il valore zero, l'autofunzione è $\psi(x, 0, G')$. Se dunque, mediante una misura effettuata all'istante $t = 0$ si trova $g(x, p) = 0$, possiamo concludere che, misurando al tempo t_0 la grandezza $G(x, p)$ si troverà per essa il valore G'.

Siccome nella $\psi(x, 0, G')$ interviene il parametro G', lo stesso parametro sarà contenuto anche in $g(x, p)$, ciò che possiamo esprimere scrivendo $g(x, p, G')$. La condizione necessaria e sufficiente perché al tempo t_0 si trovi $G(x, p) = G'$ è dunque che al tempo $t = 0$ si trovi

$$g(x, p, G') = 0\,.$$

Risolvendo questa equazione rispetto a G', troviamo

$$A(x, p) = G'\,.$$

Possiamo dunque concludere che il valore G' che avrà la grandezza $G(x, p)$ al tempo t_0 si ottiene misurando all'istante $t = 0$ la grandezza $A(x, p)$. Resta con questo risolto il problema che ci eravamo proposti.

Possiamo illustrare l'applicazione di questo metodo, sopra un'esempio concreto estremamente semplice. Consideriamo un punto mobile sopra una retta, sul quale non agisca alcuna forza, e cerchiamo quale è la grandezza che si deve misurare al tempo $t = 0$ per determinare che valore avrà l'ascissa del punto al tempo t_0. Supponiamo per questo che al tempo t_0 il punto abbia ascissa x_0. La funzione di Schroedinger al tempo t_0 sarà allora evidentemente

$$\psi(x, t_0) = \delta(x - x_0) \tag{8}$$

essendo δ il simbolo della funzione impropria di Dirac. Nel caso che sul nostro punto non agiscano forze, la sua Hamiltoniana si riduce a $H = p^2/2m$; l'equazione di Schroedinger (6) diventa dunque

$$\frac{\partial^2 \psi}{\partial x^2} = -\frac{4\pi m i}{h}\frac{\partial \psi}{\partial t}. \tag{9}$$

Dobbiamo cercarne una soluzione che, per $t = t_0$ si riduca alla (8). È facile verificare che tale soluzione può porsi a meno di una costante moltiplicativa nella forma [2]

$$\psi(x, t) = \lim_{\alpha=0} \frac{1}{\sqrt{t_0 - t}} e^{-\frac{(x-x_0)^2}{t_0-t}\left[\alpha + \frac{\pi i m}{h}\right]}$$

e si ha quindi, a meno di una costante moltiplicativa,

$$\psi(x, 0) = e^{-\frac{\pi i m}{h t_0}(x-x_0)^2}.$$

Per determinare $g(x, p, x_0)$ dobbiamo dunque porre nella (3)

$$\theta = \frac{\pi m}{h t_0}(x - x_0)^2 \quad ; \quad \rho = 1.$$

Si ha dunque

$$g(x, p, x_0) = \left\{p + \frac{h}{2\pi}\frac{2\pi m}{h t_0}(x - x_0)\right\}^2 = \left\{p + \frac{m}{t_0}(x - x_0)\right\}^2.$$

Se, misurando g al tempo zero si trova che esso ha valore zero, se cioè, per $t = 0$, si trova

$$\left\{p + \frac{m}{t_0}(x - x_0)\right\}^2 = 0 \tag{10}$$

il valore della ascissa al tempo t_0 sarà certamente x_0. Risolvendo la (10) rispetto ad x_0 si trova

$$x_0 = x + \frac{t_0}{m} p \tag{11}$$

e cioè si arriva al risultato che, per conoscere l'ascissa al tempo t_0 basta misurare al tempo 0 la grandezza $x + \frac{t_0}{m} p$. Questo risultato era classicamente evidente, poiché p/m non è altro che la velocità del punto e questa è costante durante il tempo t_0 poiché, in assenza di forze, il moto è uniforme. In questo caso dunque siamo arrivati a un risultato banale. In casi più complessi invece il risultato della meccanica quantistica sarebbe essenzialmente diverso da quello della meccanica classica.

(2) E. H. KENNARD, «Zs. f. Phys.», *44*, 326 (1927).

Da quanto precede possiamo dedurre le conclusioni seguenti: Tutto quello che è possibile conoscere dello stato di un sistema a un dato istante, mediante esperienze fatte a quell'istante, è possibile conoscerlo anche mediante opportune esperienze fatte ad un qualsiasi istante antecedente o seguente a quello che si considera. In questo senso dunque l'indeterminazione del sistema non viene per così dire a crescere col passare del tempo. Sarebbe però errato concludere da questo che i rapporti di causalità validi secondo la meccanica quantistica siano identici a quelli che valgono nelle teorie classiche. In queste ultime si può infatti, con opportune misure fatte sul sistema al tempo zero, prevedere il valore di qualsiasi grandezza fisica (funzione di x e p) a qualsiasi tempo. Invece, secondo la meccanica quantistica si può, al tempo zero fare una misura che permetta di conoscere il valore che avrà una determinata grandezza a un determinato tempo. Però se si volesse conoscere il valore di un'altra grandezza fisica o eventualmente anche della stessa grandezza a un istante diverso occorrerebbe effettuare all'istante zero una misura differente, incompatibile, almeno in generale, con la precedente.

60.

ATOMI E STELLE

«Atti Soc. It. Progr. Sci.», 19ª Riunione, Bolzano-Trento 1930, vol. 1, 228–235.

In questa conferenza cercherò di passare in rassegna rapidamente le vedute attuali sopra la struttura dei corpi; sia di quelli che si nascondono alla osservazione diretta per la loro estrema piccolezza, e cioè gli atomi, le molecole, e i loro elementi costitutivi, elettroni e nuclei, sia di quelli la cui osservazione è resa incerta dalla enorme grandezza e lontananza.

Naturalmente, data la grande vastità di questo programma, la mia esposizione dovrà per necessità limitarsi a pochi fatti essenziali; ed è questo che mi dà l'ardire di dir qualche parola anche sulla struttura delle stelle, per quanto questo argomento esca dal campo usuale dei miei studi. La nostra esposizione riuscirà più chiara incominciando dal parlare dei corpi piccoli per passare poco alla volta alla considerazione di quelli più grandi. Verremo così però ad allontanarci dall'ordine storico; poiché, date le nostre possibilità di osservazione, vennero per prime comprese le proprietà dei corpi aventi dimensioni usuali e da queste si potè solo in seguito, per mezzo di laboriose indagini arrivare a riconoscere prima l'esistenza e poi le proprietà di particelle costituenti assai più piccole.

Anche non volendo ricordare le speculazioni dei Greci, Democrito e i suoi seguaci, che giunsero all'ipotesi che la materia fosse costituita da atomi, e cioè da tante particelle staccate una dall'altra, poiché non arrivavano a comprendere come una materia continua potesse essere compressibile, la nozione dell'esistenza degli atomi e delle molecole venne introdotta nella scienza moderna per due vie differenti. E cioè per via chimica dove l'ipotesi dell'esistenza di una particolare specie di atomo per ogni elemento chimico, e della possibilità che alcuni atomi si raggruppino a formare le molecole delle diverse sostanze spiega in modo immediato le leggi delle proporzioni definite e multiple; se poi l'ipotesi dell'esistenza degli atomi si completa con l'ipotesi di Avogadro, essa spiega anche la legge che i rapporti dei volumi di due o più gas che entrano in combinazione tra di loro sono sempre dei numeri semplici. D'altra parte anche i fisici arrivavano ad ammettere l'esistenza degli atomi e delle molecole per via totalmente diversa e cioè principalmente per spiegare le proprietà dei gas. La teoria cinetica ammette infatti che un gas sia costituito da un numero enorme di molecole, in moto disordinato in tutte le direzioni, che si urtano continuamente e rimbalzano elasticamente le une sulle altre. La proprietà dei gas di esercitare una pressione sopra le pareti del recipiente che li contiene si spiega come risultato dei numerosissimi urti delle molecole del gas contro le pareti del recipiente. Questo bombardamento delle molecole contro le pareti è tanto fitto che è impossibile distinguere uno dall'altro i singoli urti e si ha l'effetto come di una pressione continua. I moti delle molecole sono intimamente legati alla temperatura del corpo; si trovò infatti che la temperatura assoluta è proporzionale all'energia cinetica media dei movimenti delle molecole.

Ma lo spirito dello scienziato moderno è essenzialmente quantitativo; e così i fisici non si accontentarono di convincersi che l'ipotesi molecolare permette di spiegare questa e quella legge, ma si posero immediatamente le seguenti domande: Quante sono le molecole contenute in una data quantità di materia? Quale è la massa di una singola molecola e quali le sue dimensioni? Il trovare la risposta a queste domande non era facile. Ciò dipende dal fatto che per spiegare la maggior parte dei fenomeni, e certamente i più appariscenti, basta ammettere che le molecole esistano e siano in numero tanto grande da sfuggire individualmente all'osservazione e da obbedire a leggi statistiche. Il risultato allora, nella maggior parte dei casi, è indipendente dal numero delle molecole e quindi non ci può fornire un mezzo per determinare questo numero. Vi sono però alcuni casi in cui il numero delle molecole ha una effettiva influenza sopra i fenomeni; e si può arrivare anche, in particolari circostanze, a rilevare l'azione esercitata da una singola molecola o da un gruppo di poche molecole. Lo studio di questi fenomeni permette di arrivare alla determinazione del numero di molecole contenute in una data quantità di gas. Si chiama numero di Avogadro il numero di molecole contenuto in un grammo molecola di qualsiasi sostanza. Secondo le misure più attendibili esso è $N = 6{,}06 \cdot 10^{23}$; e la più bella conferma della teoria molecolare è che il valore di questo numero, misurato con sette o otto metodi completamente indipendenti, risulta sempre lo stesso, entro i limiti dell'errore sperimentale. Le dimensioni delle molecole sono naturalmente diverse da caso a caso; esse hanno però sempre l'ordine di grandezza di 10^{-8} cm.

Riconosciuta così in modo definitivo l'esistenza delle molecole, venne il problema della loro struttura. Si sa dalla chimica che le molecole sono in genere costituite da alcuni atomi; ma quale è la struttura di questi? Sarebbe interessante analizzare attraverso a quali esperienze e considerazioni si arrivò dapprima a sospettare che gli atomi non fossero gli « indivisibili » degli antichi ma contenessero delle particelle più piccole e infine a determinare il numero, la disposizione e le proprietà di queste. Ma ragioni di tempo mi consentono soltanto una esposizione di fatti che oggi si ritengono accertati o almeno molto probabili.

Le particelle costituitive della materia sono di due tipi: particelle cariche di elettricità negativa, o elettroni; particelle cariche positivamente, o nuclei.

Gli elettroni sono corpuscoli tutti eguali tra di loro; leggerissimi, avendo massa 1800 volte più piccola di quella dell'atomo più leggero, l'idrogeno. Essi hanno tutti la stessa carica negativa $-e = -4{,}77 \cdot 10^{-10}$ u. e. s.

A differenza degli elettroni, esistono invece molte diverse specie di nuclei. Intanto per ciascuno dei 92 elementi chimici si ha almeno un particolare tipo di nucleo. E per molti elementi chimici si hanno anzi alcuni tipi diversi di nuclei, ciascuno dei quali dà in realtà origine a un diverso tipo di elemento. Solo che gli elementi che ne risultano hanno proprietà così enormemente simili tra di loro che è praticamente impossibile separarli uno dall'altro. Per caratterizzare un nucleo occorre darne la carica elettrica e la massa. La carica elettrica, sempre positiva, è un multiplo intero di quella carica $e = 4{,}77 \cdot 10^{-10}$ u. e. s. che abbiamo già incontrato nell'elettrone. Essa sarà dunque $Z\,e$ dove Z è un numero intero che prende il nome di numero atomico del nucleo. I numeri atomici variano da 1 a 92.

Passiamo ora a descrivere la struttura di un atomo e incominciamo dall'atomo più semplice, l'idrogeno. Esso è costituito da un nucleo di numero atomico $Z = 1$ e da un elettrone che è costretto a restare nelle vicinanze del nucleo dalla attrazione elettrostatica dovuta al fatto che nucleo ed elettrone hanno cariche elettriche di segno opposto. Invece un atomo qualunque contiene in generale un nucleo solo ed un numero di elettroni tale da neutralizzarne la carica positiva. Per esempio l'atomo di ferro è costituito da un nucleo di numero atomico 26 circondato da 26 elettroni. Questo modello atomico, proposto da Rutherford ricorda molto nella sua struttura il sistema solare. Il nucleo positivo al centro dell'atomo corrispondente al sole e gli elettroni che lo circondano corrispondono ai pianeti.

Ora, se si calcola questo minuscolo sistema planetario atomico, con i procedimenti della ordinaria meccanica, di quella stessa meccanica che, applicata al vero sistema planetario, dà risultati di così mirabile precisione, si trovano invece risultati in completo disaccordo, qualitativo e quantitativo, con l'esperienza. Questo fatto non può meravigliarci. Basta riflettere che le leggi della meccanica e dell'elettrodinamica sono state dedotte e controllate in base ad esperienze fatte sempre sopra corpi di dimensioni usuali o grandi; l'applicare queste stesse leggi ad un sistema atomico di dimensioni cento milioni di volte più piccole di un centimetro costituisce quindi una estrapolazione arditissima e a priori assai incerta.

Riconosciuta l'inapplicabilità delle leggi ordinarie ai sistemi atomici sorse il problema di trovare quali fossero le leggi a cui questi sistemi obbediscono. La ricerca di queste leggi incominciò circa 20 anni fa e forse soltanto oggi si avvicina alla sua conclusione. Nei primi tempi si cercò di arrivare alle nuove leggi partendo dalle antiche, valide per i sistemi ordinarii, modificandole qua e là in modo da adattarle ai sistemi atomici (Teoria di Bohr). Poi poco alla volta si riconobbe la necessità di un mutamento più essenziale nelle stesse premesse della meccanica e si arrivò così alla costruzione della nuova meccanica quantistica che va oggi applicandosi con successo a un numero sempre crescente di fenomeni. Mi manca qui però il tempo per accennare ad essa con qualche particolare.

Prima di passare a dir qualche parola sulla struttura di corpi più complessi dell'atomo vogliamo ancora dire qualche cosa sopra le particelle costitutive dell'atomo; gli elettroni e i nuclei. Hanno essi stessi una struttura complessa oppure no? Per l'elettrone e per il nucleo più leggero, cioè quello di idrogeno, detto anche protone, non si hanno fino ad oggi indizi seri di una struttura complessa. Ci sono invece buone ragioni per ritenere che tutti gli altri nuclei siano costituiti da aggregati di elettroni e protoni. Sorge così per i fisici il nuovo problema della struttura del nucleo che si presenta oggi come uno dei più affascinanti campi della fisica dell'avvenire.

Ma lasciando da parte questi problemi, ancora estremamente lontani dalla loro soluzione, dobbiamo passare ad occuparci delle forze che agiscono tra gli atomi e li riuniscono insieme per formare sia aggregati di pochi atomi, quali sono le molecole; sia aggregati assai più numerosi, come i corpi solidi. Le forze che agiscono tra gli atomi si possono ridurre sostanzialmente a due tipi, tra i quali non mancano tipi di transizione. Il primo ci è dato dalle forze elettrostatiche; il secondo dalle così dette forze di risonanza o omeopolari.

Le forze elettrostatiche possono anche esercitarsi tra atomi neutri, purché essi siano polarizzati; ma in questo caso si tratta di forze assai piccole che non sono normalmente sufficienti a tenere insieme gli atomi per formare molecole o corpi solidi. Invece le forze elettrostatiche agiscono intensamente tra gli ioni, cioè tra quegli atomi che, avendo perduto oppure acquistato uno o più elettroni, sono restati carichi positivamente oppure negativamente. Sia dall'esperienza che dalla teoria risulta che alcuni atomi (quelli dei metalli) hanno tendenza a perdere elettroni, formando ioni positivi; altri atomi invece (quelli dei metalloidi) hanno tendenza ad aggregarsi elettroni formando ioni negativi. Se allora poniamo un atomo di un metallo, sodio per esempio, in presenza di un atomo di un metalloide, per esempio cloro, il Cl avido di elettroni si approprierà uno degli elettroni del sodio. Il cloro diventerà così uno ione negativo e il sodio uno ione positivo; i due ioni, avendo carica di nome opposto si attrarranno fino a venire in contatto formando la molecola NaCl. Questo è, in poche parole, il meccanismo delle forze polari; il meccanismo delle forze omeopolari, che sono quelle che tengono insieme molecole del tipo H_2, N_2, O_2, non formate da ioni, ma da atomi neutri è assai più recondito e non può venir spiegato senza considerazioni alquanto complicate.

Ci limiteremo pertanto a dire che le forze omeopolari sono una caratteristica conseguenza della nuova meccanica atomica e che non hanno un corrispondente nella meccanica ordinaria. Esse sono state studiate principalmente negli ultimi anni e, insieme alle forze polari permettono di interpretare la formazione dei principali tipi di composti chimici.

Sostanzialmente simili alle forze che tengono insieme gli atomi di una molecola, sono le forze che li raggruppano per formare i corpi solidi. Tra questi i più caratteristici sono i cristalli (cristalli propriamente detti e corpi a struttura microcristallina); poiché i corpi solidi amorfi non differiscono qualitativamente da liquidi estremamente viscosi. I cristalli invece sono caratterizzati dalla disposizione mirabilmente regolare delle loro molecole, in reticolati formati da serie di piani equidistanti (le loro distanze sono dell'ordine di un centomilionesimo di centimetro). Anche tra i cristalli, come abbiamo visto per le molecole, dobbiamo distinguere i cristalli polari da quelli omeopolari; i primi sono costituiti da ioni positivi e negativi, tenuti insieme dalle attrazioni elettrostatiche; i secondi sono invece tenuti insieme, almeno in parte, dalle forze di risonanza a cui abbiamo già accennato. Un cristallo polare molto tipico è il salgemma costituito da atomi, o meglio ioni di cloro e sodio, i primi con carica negativa e i secondi positiva, disposti alternativamente nei vertici di un reticolo cubico. In generale i sali formano cristalli polari. Cristalli non polari sono invece per esempio il diamante e molti cristalli metallici.

Ci manca qui il tempo di parlare con qualche dettaglio dei corpi liquidi e gassosi e ci limitiamo perciò ad accennare che ciò che, dal punto di vista molecolare, maggiormente li distingue dai solidi è il fatto che le loro molecole, in seguito all'agitazione termica vagano per tutta la massa del corpo allontanandosi molto dalla posizione iniziale; mentre l'agitazione termica degli atomi in un corpo solido consiste in una vibrazione intorno ad una posizione di equilibrio. Nei liquidi le distanze tra molecola e molecola sono dello stesso ordine di grandezza delle dimensioni molecolari; così che le varie molecole sono quasi continuamente in contatto. Invece nei gas esse sono assai più distanziate tra di loro; ed in un gas sufficientemente rarefatto si può anzi considerare ciascuna molecola per conto suo, praticamente pochissimo perturbata dall'azione delle altre.

Abbiamo così passata in rapida rassegna la struttura dei corpi quali essi si presentano in condizioni ordinarie alla nostra osservazione; e per condizioni ordinarie intendo qui quelle che si possono realizzare nei nostri laboratori. Ma al di fuori dei nostri laboratori, ed in luoghi inaccessibili all'esperienza diretta dell'uomo, la materia può trovarsi in condizioni così profondamente diverse da quelle ordinarie che la sua struttura viene ad esserne assai notevolmente alterata. Condizioni di questo genere hanno luogo verosimilmente, per l'enorme temperatura e l'enorme pressione, nell'interno delle stelle. Ma prima di parlare dell'interno diciamo due parole della superficie.

Lo stato fisico della superficie delle stelle ci viene in buona parte rivelato dall'esame dello spettro della luce che esse ci inviano. Questo spettro, opportunamente interpretato, può dirci infatti molte cose; esso ci rivela intanto quali sono gli elementi chimici presenti alla superficie della stella, poiché riconosciamo nello spettro le righe che sappiamo esserne caratteristiche. Ma lo studio dello spettro ci informa anche sopra la temperatura della stella; così per esempio si trova che la presenza con grande intensità delle righe dell'idrogeno e dell'elio è indice di temperatura elevata; mentre la presenza delle righe emesse dai metalli neutri e ancor più dalle molecole indica invece una temperatura relativamente bassa. Un altro metodo che consente la determinazione della temperatura esterna della stella è il seguente. Se riscaldiamo un corpo fino a renderlo appena incandescente esso ci appare rosso; aumentando la temperatura la sua colorazione si sposta verso il bianco e se potessimo portarne la temperatura fino oltre una decina di migliaia di gradi la sua luce finirebbe con l'apparire bluastra. Ciò dipende dal fatto che nella luce emessa da un corpo incandescente la percentuale della luce di piccola lunghezza d'onda (violetta) va crescendo con la temperatura, in modo che la colorazione è tanto più violacea quanto più alta è la temperatura. E siccome si conosce la legge secondo cui la colorazione dipende dalla temperatura si può, inversamente, dal colore dedurre la temperatura.

Da questi studi è risultato che la temperatura della superficie delle stelle può arrivare fino a venti o trentamila gradi. Sembra anzi che le stelle subiscano una specie di processo evolutivo incominciando la loro esistenza come corpi di grandi dimensioni (stelle giganti) e di temperatura relativamente bassa; le dimensioni vanno poi decrescendo mentre la temperatura cresce fino a un massimo, oltrepassato il quale la stella incomincia a raffreddarsi mentre le sue dimensioni continuano a decrescere (stelle nane).

Come si intravede anche da questi pochi cenni, lo stato fisico della superficie delle stelle, almeno di molte stelle, è relativamente ben conosciuto. Se invece vogliamo dir qualche cosa dell'interno di esse siamo necessariamente costretti a ricorrere ad ipotesi molto arrischiate. Tutte le teorie sopra l'interno delle stelle sono pertanto assai incerte; esse però concordano nel fatto che nell'interno delle stelle la temperatura deve essere elevatissima; essa è stata valutata dell'ordine di grandezza di decine di milioni di gradi.

A una temperatura così straordinariamente elevata le proprietà della materia diventano assai differenti da quelle che noi conosciamo. Basta riflettere che l'energia cinetica dei moti di agitazione termica, essendo proporzionale alla temperatura, risulta centomila volte maggiore che non lo sia a temperatura ordinaria. Si capisce allora come non possano esistere molecole; poiché, se per caso a un certo momento due atomi si riunissero per formare una molecola, essi verrebbero immediatamente strappati uno dall'altro dagli urti violentissimi contro gli altri corpuscoli. Ma nemmeno la compagine dell'atomo è abbastanza resistente da non infrangersi sotto urti così impetuosi. Consideriamo per esempio un atomo di ferro; in condizioni normali esso è costituito da un nucleo accompagnato da 26 elettroni. Ma sotto l'azione del bombardamento quasi tutti questi elettroni gli vengono strappati e verosimilmente soltanto un paio di essi restano col nucleo mentre gli altri 24 si disperdono nell'ambiente. In conclusione possiamo dunque raffigurarci la materia nell'interno di una stella costituita da una miscela disordinatissima contenente alcuni nuclei accompagnati da due o tre elettroni, e molti elettroni liberi moventisi in tutte le direzioni con velocità dell'ordine di quelle dei raggi catodici. Il tutto in un ambiente compenetrato da una radiazione di enorme frequenza e la cui intensità è tanto grande da produrre una pressione che, secondo Eddington, può in alcuni casi arrivare a spezzare la compagine stessa della stella.

Ma tutto quanto ho detto è forse puro risultato di immaginazione: nessuno ne ha mai avuto e probabilmente nessuno ne potrà mai avere esperienza diretta. D'altra parte la mente umana ha bisogno di figurarsi anche le cose molto lontane e molto riposte; e forse, senza questa tendenza, quel piccolo numero di fenomeni che la scienza degli uomini è riuscita a comprendere sarebbe anche più ristretto.

61.

I FONDAMENTI SPERIMENTALI DELLA NUOVA MECCANICA ATOMICA

« Periodico di Matematiche », *10*, 71–84 (1930).

Quando, nei primi anni del nostro secolo, i fisici incominciarono a interessari seriamente ai problemi atomici, due questioni principali si presentarono al loro studio. Occorreva infatti in primo luogo ricercare quale fossero gli elementi costitutivi dell'atomo; in secondo luogo scoprire le leggi e il comportamento di tali elementi. Il primo problema, già in parte affrontato dai fisici del secolo scorso, trovò la sua risoluzione definitiva grazie principalmente ai lavori di Rutherford. Secondo i suoi risultati, che sono poi sempre stati confermati dalle indagini ulteriori, dobbiamo distinguere nell'atomo un nucleo centrale, cioè un corpuscolo carico di elettricità positiva, ed un certo numero di elettroni carichi negativamente. Ciò che distingue gli uni dagli altri gli atomi dei diversi elementi chimici, è il numero di elettroni che essi contengono, e che prende il nome di *numero atomico*. Così nell'atomo di idrogeno, che ha numero atomico 1, abbiamo il nucleo e un solo elettrone; siccome l'atomo, nel suo complesso, deve essere elettricamente neutro, la carica positiva del nucleo di idrogeno dovrà essere eguale alla carica negativa del suo unico elettrone. L'atomo di elio ha numero atomico 2, e contiene quindi due elettroni e il nucleo; siccome l'atomo deve essere neutro, il nucleo di elio dovrà avere carica doppia, salvo il segno, di quella dell'elettrone. Se si considera infine l'atomo più complesso, l'uranio, si trova che esso contiene 92 elettroni, in modo che la carica elettrica del nucleo di uranio dovrà equivalere a 92 volte la carica elettrica dell'elettrone. In tutti gli atomi la massa del nucleo è sempre molto grande (qualche migliaio di volte) in confronto alla massa di tutti gli elettroni dell'atomo presi insieme; per modo che il centro di gravità dell'atomo viene praticamente a coincidere col nucleo. In conseguenza, se tutto l'atomo non è animato da un movimento di insieme, il nucleo starà fermo, mentre gli elettroni gireranno intorno ad esso in orbite più o meno complicate.

Trovati così gli elementi di cui sono costituiti gli atomi, restava il più arduo problema di determinare le leggi a cui essi soddisfano. Questo secondo problema è ancora in evoluzione, e soltanto oggi si incomincia ad avere la sensazione di non essere lontani dalla sua soluzione.

L'ipotesi più naturale che si è indotti a fare quando ci si trova di fronte a qualche cosa di sconosciuto, è che il suo comportamento sia analogo a quello di cose simili meglio note. Così la prima ipotesi sul comportamento dell'atomo che abbiamo descritto, sarà che il suo nucleo e i suoi elettroni si muovono, sotto l'azione delle forze attrattive e repulsive che si esercitano tra di essi, secondo la legge di Coulomb, obbedendo ai principî della meccanica classica. Secondo questa ipotesi l'atomo ci si presenterà dunque come un minuscolo sistema planetario (avente il diametro dell'ordine di grandezza di un centomilionesimo di cm.) in cui il nucleo, che sta fermo nel centro dell'atomo, corrisponde al Sole, mentre gli elettroni che girano intorno ad esso corrispondono ai pianeti. Siccome poi gli elettroni non sono altro che delle cariche elettriche in movimento, potremo pensare che essi producano degli effetti magnetici ed elettromagnetici, calcolabili per mezzo delle consuete leggi dell'elettrologia.

Non sfuggirà a nessuno che in questo complesso di ipotesi è contenuta una ardita e pericolosa estrapolazione. Le leggi della meccanica e dell'elettrologia sono state infatti dedotte in base ad esperienze fatte su corpi aventi dimensioni enormi a confronto di quelle di un atomo; e generalizzare tali leggi alla fisica dell'atomo potrà essere lecito soltanto dopo che esse, anche nel campo dell'atomo, saranno state assoggettate a un accurato controllo sperimentale.

Ora è senz'altro evidente che le leggi dell'elettrologia e della meccanica non possono essere valide senza modificazioni nel mondo atomico. Consideriamo per esempio il caso di un atomo di idrogeno; il suo unico elettrone, attirato dal nucleo con forza inversamente proporzionale al quadrato della distanza, descriverà attorno ad esso, secondo le leggi della meccanica, un'orbita ellittica, di cui il nucleo occupa uno dei fuochi. Si sa d'altra parte dalle leggi dell'elettrologia che una carica elettrica, che non si muova di moto rettilineo ed uniforme, irradia sempre nello spazio una certa quantità di energia elettromagnetica; concludiamo dunque che anche l'elettrone che gira intorno al nucleo dell'atomo di idrogeno dovrebbe perdere continuamente energia irradiandola nello spazio sotto forma di onde elettromagnetiche. In conseguenza di questa perdita di energia, l'orbita dell'elettrone dovrebbe andare via via restringendosi, fino a che l'elettrone andrebbe a cadere nel nucleo. Siccome questo fatto, che porterebbe a un completo cambiamento di natura dell'atomo, certamente non avviene, si deve dedurre, che, o le leggi della meccanica, o le leggi dell'elettrologia, oppure le une e le altre, non sono applicabili senza modificazioni nel mondo atomico.

Nel 1913 Niels Bohr di Copenaghen tentò di risolvere queste difficoltà proponendo le sue celebri ipotesi per il calcolo dei fenomeni atomici. Secondo queste ipotesi il calcolo delle orbite degli elettroni di un atomo viene ancora fatto in base alle leggi della meccanica ordinaria: si ammette però che, non tutte le orbite prevedibili per mezzo della meccanica possano venire effettivamente descritte dagli elettroni, ma soltanto una successione discreta, che costituisce la successione degli stati quantici dell'atomo. Ad ogni stato quantico corrisponde anche naturalmente una determinata energia dell'atomo, la cui energia è pertanto suscettibile di assumere soltanto certi valori; discreti, detti livelli energetici.

Bohr ammette che quando l'atomo si trova in uno stato quantico, esso non irradi energia; ciò è in contraddizione coi principî ordinari dell'elettrodinamica, poiché, secondo questi, al movimento degli elettroni dell'atomo dovrebbe necessariamente corrispondere una emissione di onde elettromagnetiche. Invece secondo Bohr l'emissione di radiazioni da parte dell'atomo avviene con un processo discontinuo che prende il nome di salto quantico; egli ammette infatti che, quando l'atomo non si trova nello stato quantico di minima energia, esso possa passare bruscamente dallo stato in cui si trova a uno stato di energia inferiore emettendo, sotto forma di radiazione, la differenza tra le energie dei due stati. Secondo la ordinaria elettrodinamica, la frequenza della radiazione emessa dall'atomo, dovrebbe essere eguale a quella dei movimenti degli elettroni interni all'atomo; invece secondo Bohr la frequenza emessa si calcola semplicemente dividendo la quantità di energia emessa nel salto quantico (cioè la differenza tra le energie dei due stati tra cui avviene il salto) per la costante universale h di Planck. La frequenza così calcolata è differente da quella dei moti descritti dagli elettroni, e si può dimostrare che le due frequenze vengono a coincidere solo nel caso limite di frequenze molto piccole.

Il complesso delle ipotesi di Bohr, che abbiamo qui brevemente elencate, si presenta assai insoddisfacente dal lato logico, poiché si ha in esso una mescolanza dei concetti classici e dei nuovi concetti quantistici, senza che sia chiaro il criterio che limita l'applicabilità degli uni e degli altri. Ciò nonostante le ipotesi di Bohr hanno costituito il cardine dello sviluppo della fisica atomica fino agli ultimissimi anni, e si sono mostrate di una fecondità enorme, conducendo qualitativamente all'interpretazione degli spettri di tutti gli atomi e di alcune loro proprietà chimiche; e da esse si è tratto continuamente suggerimento per escogitare nuove esperienze che hanno valso a portare sempre maggior luce nelle nostre conoscenze sul mondo atomico.

Restava tuttavia sempre un notevole disagio concettuale che si andava sempre più accentuando man mano che la teoria veniva applicata a fenomeni sempre più complessi; poiché la soluzione dei diversi problemi si otteneva sempre a prezzo di qualche compromesso tra la teoria classica e quella quantistica, oppure di ipotesi supplementari logicamente in disaccordo tra di loro. Citerò il notissimo esempio della controversia tra la teoria elettromagnetica ondulatoria e la teoria dei quanti di luce, secondo cui è necessario attribuire alla luce una specie di natura corpuscolare. Le due teorie avevano ciascuna il proprio campo di applicazione; precisamente la teoria ondulatoria serviva per l'interpretazione dei fenomeni di interferenza ed affini, mentre la teoria dei quanti di luce veniva applicata quando si dovevano studiare gli scambi energetici tra luce e materia. E non sembrava possibile arrivare alla costruzione di una teoria unica che spiegasse insieme le due classi di fenomeni.

È stato principalmente da questo disagio concettuale che si è maturata poco per volta nei fisici la convinzione che fosse necessario riprendere fin dalle basi lo studio della fisica dell'atomo, costruendo una nuova meccanica atomica, invece di cercare di adattare, con continue ipotesi supplementari la ordinaria meccanica ai fenomeni dell'atomo. Dopo svariati tentativi infruttuosi, che ben presto si mostrarono in disaccordo con l'esperienza, si arrivò infine alla costruzione di una meccanica atomica, che sembra rendere conto in modo assai soddisfacente di tutti i fenomeni dell'atomo che oggi conosciamo.

Che i tempi fossero maturi per la costruzione della meccanica atomica, è, del resto, avvalorato in qualche modo dal fatto che ad essa si è giunti quasi contemporaneamente seguendo vie affatto diverse in due luoghi differenti. Heisenberg a Copenaghen vi arrivava partendo dal concetto che nella nuova meccanica dovessero trovar posto soltanto grandezze accessibili, almeno in linea teorica, a una misura effettiva (grandezze direttamente osservabili); mentre Schroedinger a Zurigo, perfezionando e sviluppando alcune idee espresse per primo da Louis de Broglie di Parigi, costruiva la nuova meccanica basandosi sopra una analogia formale tra la meccanica classica e l'ottica geometrica. Come si vede da questo accenno, le due vie per la costruzione della nuova meccanica non hanno apparentemente nulla a che vedere tra di loro; e anche gli strumenti matematici di cui esse si servono sono totalmente differenti, poiché Heisenberg usa la teoria delle matrici infinite, mentre Schroedinger si serve solo delle ordinarie equazioni differenziali. Ciò nonostante si è potuto dimostrare che le due teorie sono matematicamente equivalenti tra di loro, e cioè conducono, in tutti i casi possibili, agli identici risultati. Così che ormai non occorre più distinguere tra le due teorie, ma si applica, caso per caso, quella di esse che è matematicamente più comoda. Siccome in quasi tutti i casi il metodo più facile a seguire e più accessibile all'intuito è quello di Schroedinger, accennerò qui ad esso in poche parole. Si può stabilire tra l'ottica geometrica e la meccanica una analogia molto stretta, facendo corrispondere alla traiettoria di un punto materiale il percorso di un raggio di luce. Se il punto materiale si muove senza che su di esso agiscano forze, cioè se esso attraversa una regione dello spazio in cui il potenziale è costante, la sua traiettoria è una linea retta: similmente un raggio di luce che attraversi una regione di indice di rifrazione costante, segue un cammino rettilineo. Dunque a una regione in cui il potenziale è costante corrisponde, nel paragone ottico, una regione in cui è costante l'indice di rifrazione. In una regione dello spazio in cui il potenziale è variabile, un punto materiale descrive una traiettoria curva; e similmente in una regione dello spazio in cui l'indice di rifrazione varia da punto a punto, i raggi di luce seguono dei cammini curvi (si pensi per esempio al fenomeno del miraggio). Data una qualsiasi distribuzione di potenziale, si può sempre immaginare che, in una regione dello spazio, l'indice di rifrazione vari da punto a punto in modo tale che il cammino percorso dalla luce risulti identico a quello percorso da un punto materiale sotto l'azione del dato potenziale. Resta così sviluppata in modo completo l'analogia tra la meccanica classica e l'ottica geometrica. È noto però che tutte le volte che la luce incontra degli ostacoli aventi dimensioni dell'ordine di grandezza della lunghezza d'onda, le leggi dell'ottica geometrica perdono la loro validità e avvengono i fenomeni di diffrazione, che possono spiegarsi soltanto ammettendo una natura ondulatoria della luce. Ora Schroedinger osserva che anche la meccanica classica si mostra in difetto quando si cerca di applicarla a sistemi di dimensioni molto piccole, come sono gli atomi; in questo caso infatti avvengono i fenomeni peculiari della teoria dei quanti. Egli cerca perciò di spingere oltre l'analogia che abbiamo osservata tra la meccanica e l'ottica ed arriva a costruire la così detta meccanica ondulatoria, analoga all'ottica ondulatoria invece che all'ottica geometrica. Naturalmente come l'ottica ondulatoria, quando viene applicata a problemi di dimensioni grandi, finisce per dare gli stessi risultati dell'ottica geometrica, così anche la meccanica ondulatoria di Schroedinger, applicata allo studio della meccanica di corpi di dimensioni apprezzabili, dà risultati che coincidono con quelli della meccanica classica. Le differenze si manifestano invece quando si studiano, con la meccanica ondulatoria, problemi di dimensioni estremamente piccole come per esempio i problemi della fisica dell'atomo. E in questo campo la nuova meccanica portò ben presto a risultati brillantissimi, poiché i diversi fenomeni atomici, e, in primo luogo, il fatto che l'atomo è suscettibile soltanto di certi stati energetici discreti, risultarono spontaneamente come conseguenze matematiche della teoria, senza bisogno di alcuna ipotesi supplementare; come si è detto, si poté poi dimostrare che la meccanica ondulatoria di Schroedinger è identica all'altra forma della nuova meccanica ideata da Heisenberg.

Da quanto si è detto potrà forse sembrare molto strana l'idea di considerare i fatti meccanici come una manifestazione di un fenomeno ondulatorio. La stessa idea però può venir suggerita anche da un altro tipo di fenomeni. Abbiamo già accennato infatti alla teoria dei quanti di luce che, per la spiegazione di un certo gruppo di fenomeni, era indotta a postulare una specie di natura corpuscolare della luce, e cioè, in qualche modo, a ricondurre i fenomeni dell'ottica a quelli di uno sciame di corpuscoli; anche questo fatto servì certamente a suggerire che l'analogia tra i fenomeni meccanici e quelli ondulatori è più profonda di quanto non sembri a prima vista. Poco dopo la pubblicazione della teoria di Schroedinger, si ebbe del resto, per opera di Davisson e Germer, una brillante conferma sperimentale di questa profonda analogia. Questi fisici poterono infatti osservare che, lanciando un fascetto di elettroni contro un cristallo di nichel, gli elettroni venivano rflessi all'indietro solo in certe direzioni ben determinate. Questo non è altro che un fenomeno di diffrazione degli elettroni, completamente analogo, anche dal lato quantitativo, ai fenomeni di diffrazione presentati dai raggi X che colpiscono un cristallo. In seguito queste esperienze di diffrazione dei fascetti di elettroni nei cristalli furono ripetute anche da altri in svariate condizioni, e risultarono sempre fenomeni analoghi a quelli che si osservano adoprando, invece di un fascetto di elettroni, un raggio di luce di opportuna lunghezza d'onda.

Ma certamente la migliore conferma del fatto che la nuova meccanica ci abbia veramente avvicinati alla comprensione dei fenomeni atomici, sta nell'esame dell'imponente numero di fatti sperimentali svariati che essa ha permesso di interpretare. Senza entrare in particolari circa la spiegazione dei diversi fenomeni, mi propongo qui di passare in rassegna alcuni dei principali di essi.

Dirò anzi tutto che la nuova meccanica ha permesso di confermare tutti i numerosissimi risultati qualitativi che erano stati già interpretati con la vecchia teoria dei quanti; ciò è stato possibile senza alcun bisogno di quelle ipotesi supplementari continue che erano necessarie nella vecchia teoria, ma semplicemente con una applicazione conseguente dei postulati fondamentali della teoria. Nei casi in cui si è potuto sormontare le difficoltà matematiche, arrivando ad un calcolo quantitativo si è sempre trovato l'accordo quantitativo con i risultati sperimentali; valga l'esempio del calcolo dell'energia dell'atomo di elio, che è stato spinto fino all'approssimazione di oltre l'1 ‰ al valore sperimentale. Come questo si potrebbero citare altri numerosi esempi in cui la nuova meccanica ha portato la precisione quantitativa dove la vecchia teoria dava solo dei risultati qualitativi. Così nella teoria degli spettri delle molecole, nella teoria degli atomi con più elettroni, e perfino nella stessa teoria dell'atomo di idrogeno che era sempre stata considerata per così dire la rocca forte della vecchia teoria dei quanti. Notiamo incidentalmente che il fatto che la vecchia teoria, pur non arrivando quasi mai a dare risultati quantitativi, avesse tuttavia permesso di raggiungere conclusioni assai importanti, non sembra oggi casuale; si può infatti dimostrare che la vecchia teoria dei quanti rappresenta una prima approssimazione della nuova meccanica, prima approssimazione che in parecchi casi già si avvicina considerevolmente alla realtà.

Ma non vorrei qui soffermarmi sopra quelle conclusioni della meccanica quantistica che possono in qualche modo considerarsi soltanto come un perfezionamento dei risultati della vecchia teoria; preferisco invece passare all'esame di quanto vi è in essa di essenzialmente nuovo anche dal lato qualitativo.

Nel 1921 Ramsauer scoprì un fenomeno che sembrò al suo tempo inesplicabile. Se si fanno passare degli elettroni attraverso a un gas, le deviazioni che essi subiscono per effetto degli urti contro le molecole del gas dipendono, come è naturale, dalla velocità degli elettroni; elettroni molto veloci non vengono quasi deflessi dall'effetto degli urti contro le molecole, mentre gli elettroni un po' più lenti subiscono una deviazione maggiore. Si penserebbe naturalmente che la deviazione dovesse crescere sempre col diminuire della velocità. Ramsauer scoprì invece che accade l'opposto, e che cioè, quando la velocità degli elettroni scende al di sotto di un certo limite, essi vengono tanto meno deviati quanto più piccola è la loro velocità; in modo che si può dire che elettroni di piccolissima velocità possono attraversare le molecole del gas senza venir in alcun modo deviati dalla loro traiettoria. L'interpretazione di questo fenomeno, incomprensibile nella meccanica classica, si presenta assai semplice e spontanea nella meccanica ondulatoria. Abbiamo detto infatti che, secondo questa teoria, l'elettrone deve farsi corrispondere a un treno di onde; la lunghezza d'onda dipende dalla velocità dell'elettrone e risulta tanto maggiore quanto minore è la velocità. Ad elettroni molto lenti corrisponderanno quindi onde di grande lunghezza Ora la deviazione che l'elettrone subisce nell'urtare contro una molecola corrisponde alla diffrazione che subisce il treno d'onde quando investe la molecola. È ben noto che la diffrazione subìta da un treno di onde che urta un ostacolo di dimensioni molto piccole è considerevole finché la lunghezza d'onda è dell'ordine di grandezza delle dimensioni dell'ostacolo, mentre diventa trascurabile per lunghezze d'onda molto grandi rispetto a tali dimensioni. Segue quindi che, quando la velocità dell'elettrone diventa piccola, e quindi diventa grande la lunghezza delle onde che ad esso corrispondono, l'effetto degli urti dovrà farsi sempre più piccolo, appunto come risulta dalle esperienze di Ramsauer.

Un'altra applicazione della meccanica ondulatoria, che presenta notevole analogia con quella che abbiamo qui esposto, ma che ha ben maggiore importanza, consiste nel calcolo del cammino libero medio degli elettroni liberi nell'interno dei metalli e nella conseguente determinazione teorica della conducibilità elettrica. La teoria dei metalli ha compiuto negli ultimi anni dei progressi assai importanti, specialmente in seguito ai lavori di Sommerfeld, che ha ripreso, valendosi delle moderne teorie statistiche, la concezione di Drude del gas di elettroni. Sommerfeld ha così potuto interpretare molti fenomeni che sembravano irrisolubili alla luce delle vecchie teorie.

Uno dei problemi più essenziali per la teoria dei metalli era naturalmente quello del calcolo della loro conducibilità elettrica; e per risolverlo era necessario conoscere il valore del cammino libero medio degli elettroni liberi nell'interno del metallo, poiché la resistenza elettrica risulta inversamente proporzionale al cammino libero medio. Ora è facile convincersi che, se gli atomi del metallo fossero disposti in un reticolato perfettamente regolare, la diffusione degli elettroni per effetto degli urti contro di essi dovrebbe annullarsi. Considerando infatti, secondo la concezione di Schroedinger, le proprietà di un elettrone analoghe a quelle di un'onda che si muova in un mezzo di indice di rifrazione variabile, troviamo che la diffusione prodotta da un reticolato perfetto è nulla, così come è nulla la diffusione della luce prodotta da un cristallo perfetto. Alla diffusione nulla corrisponde cammino libero medio degli elettroni infinito, e quindi conducibilità elettrica infinita. Se anche consideriamo però un cristallo costruito in modo perfetto, i suoi atomi non saranno situati perfettamente nei vertici del reticolo, altro che alla temperatura dello zero assoluto. Per tutte le altre temperature essi si discosteranno più o meno da tali posizioni, per effetto dell'agitazione termica, e daranno perciò luogo a un reticolato tanto meno regolare quanto più elevata è la temperatura. L'intensità della diffusione degli elettroni crescerà perciò col crescere della temperatura e si avrà quindi in corrispondenza un aumento della resistenza elettrica. Sopra queste basi è stato possibile costruire una teoria della conducibilità elettrica che si mostra in soddisfacente accordo con l'esperienza. Si osservi anche che, in base ai principî esposti, si spiega il fatto che la conducibilità elettrica delle leghe metalliche è in generale minore di quella dei metalli puri. Infatti, nel mescolare i metalli, si viene a costituire un reticolato atomico alquanto irregolare. A questa irregolarità corrisponde, come abbiamo spiegato, una forte diffusione degli elettroni e quindi una forte resistenza elettrica.

La nuova meccanica ha permesso di raggiungere dei risultati di grande interesse nello studio dei sistemi che contengono due o più corpuscoli identici tra di loro. Tali sono per esempio tutti gli atomi, ad eccezione dell'idrogeno; essi contengono infatti almeno due elettroni e tutti gli elettroni sono eguali. Similmente, in una molecola costituita da due atomi eguali, come per esempio

H_2, N_2, O_2, abbiamo, oltre a tutti gli elettroni, che sono eguali tra di loro anche i due nuclei.

Consideriamo per esempio un atomo che contenga due elettroni. Abbiamo osservato che ogni elettrone può trovarsi soltanto in certi determinati stati quantici. A priori si penserebbe possibile che tutti e due gli elettroni del nostro sistema potessero, in determinate condizioni, trovarsi nello stesso stato quantico. Ora Pauli, dall'osservazione di svariati fatti spettroscopici, era arrivato, già prima che Heisenberg e Schroedinger fondassero la nuova meccanica, ad enunciare il suo principio di esclusione che afferma che non può mai accadere che due elettroni si trovino nello stesso stato quantico. Questo principio è importantissimo e certamente tutte le proprietà della materia resterebbero sostanzialmente modificate, se esso non fosse verificato. Esso ci dà in particolare la ragione per cui tutti gli elettroni di un atomo non si precipitano nell'orbita più stabile, che è quella più vicina al nucleo, ma si dispongono intorno ad esso in orbite svariate, determinando così le particolarità chimiche, che differenziano gli elementi uno dall'altro. Nella vecchia teoria dei quanti il principio di Pauli non poteva in alcun modo venire spiegato; era necessario ammetterlo come una ipotesi supplementare, non sempre conciliabile con le altre ipotesi della teoria. La nuova meccanica ha permesso invece di dimostrare che se si ammette che, ad un certo istante iniziale, ogni stato quantico sia occupato al massimo da un solo elettrone, non può mai accadere, negli istanti successivi, che due elettroni vengano a trovarsi nel medesimo stato quantico; in altre parole, se si ammette che il principio di Pauli sia verificato allo stato iniziale, si è sicuri che esso sarà sempre verificato in seguito.

Dalla nuova meccanica risulta anche spontaneamente un'altra importante proprietà dell'elettrone e cioè quella di possedere un momento magnetico. Questa proprietà invece doveva venire postulata nella vecchia teoria. Anche l'esistenza del momento magnetico ha, sulle proprietà della materia, una importanza assai maggiore di quanto non possa sembrare a prima vista. Se non se ne ammettesse l'esistenza infatti si troverebbe per esempio che l'idrogeno invece di essere, come è, un gas chimicamente molto attivo, si comporterebbe come un gas inerte, press'a poco analogo all'elio. In modo simile verrebbero totalmente alterate le proprietà chimiche di tutte le sostanze; il sistema periodico degli elementi avrebbe i suoi periodi di quattro invece che di otto, così che per esempio il boro, che è una sostanza solida di grande durezza, sarebbe invece un gas nobile. Similmente sarebbe un gas nobile il fluoro, che è invece una delle sostanze chimicamente più attive che si conoscono.

Notevolissimi sono anche gli studi, basati sopra la nuova meccanica, relativi alle proprietà dei composti chimici omeopolari. È noto che i composti chimici si possono classificare in due grandi categorie: composti polari e composti omeopolari. I più caratteristici esempi di composti polari ci sono dati dalle molecole dei sali, degli acidi e delle basi. Per esempio si trova che la molecola di cloruro di sodio, NaCl, è costituita da uno ione positivo di sodio, cioè da un atomo di sodio che, avendo perduto un elettrone, è restato carico di elettricità positiva, e da uno ione negativo di cloro, cioè da un atomo di cloro che si è appropriato l'elettrone perduto dall'atomo di sodio, restando perciò carico negativamente. I due ioni, carichi di elettricità di segno opposto, si attraggono, per la legge di Coulomb; è appunto questa attrazione elettrostatica che serve a tenere insieme la molecola. Non è invece possibile spiegare per mezzo di un semplice gioco di attrazioni elettrostatiche l'esistenza delle molecole, come per esempio la molecola di idrogeno H_2, o di ossigeno O_2, le quali non sono costituite da due ioni, ma da due atomi, entrambi allo stato neutro; a questa seconda classe di composti chimici si dà il nome di composti omeopolari. Non mi è qui possibile spiegare in poche parole attraverso a quale serie di considerazioni è stato possibile, per mezzo della meccanica quantistica, dimostrare che, in determinate condizioni, anche due atomi, entrambi elettricamente neutri, si possono attirare riunendosi per formare una molecola omeopolare; mi limito ad accennare che tale possibilità è determinata dal fatto che gli elettroni dei due atomi sono assolutamente identici tra di loro, per modo che, senza in alcun modo alterare il sistema, si possono scambiare tra di loro due elettroni uno appartenente al primo atomo e l'altro al secondo. È stato così possibile abbozzare una teoria quantistica delle valenze chimiche omeopolari, che rende conto in modo soddisfacente dei principali fatti osservati. Così per esempio risulta dalla teoria che tra due atomi di idrogeno si esercita una forza attrattiva che determina la formazione della molecola di idrogeno; mentre non esiste una simile forza attrattiva tra due atomi di elio, per modo che, almeno in condizioni normali, non si ha la formazione di una molecola di elio.

Accennerò per concludere a un certo numero di risultati che la nuova meccanica ha permesso di raggiungere nel campo della fisica nucleare, e cioè nello studio di quei corpuscoli, carichi di elettricità positiva che si trovano nel centro dell'atomo ed hanno dimensioni alcune migliaia di volte minori di esso. Abbiamo detto più sopra che gli elettroni hanno tutti un loro momento magnetico proprio. Ora è risultato da studi recenti che anche i nuclei atomici, o meglio alcuni di essi, hanno pure un momento magnetico, che è circa mille volte più piccolo del momento magnetico degli elettroni. L'esistenza di questo momento si manifesta in svariati fenomeni. Così per esempio si osserva che le righe spettrali di molti elementi, specialmente di quelli aventi peso atomico elevato, sono costituite da alcune componenti, assai vicine tra di loro, e distinguibili una dall'altra solo con mezzi spettroscopici di elevatissima risoluzione. Queste diverse righe spettrali corrispondono alle diverse orientazioni che il momento magnetico del nucleo può assumere rispetto al resto dell'atomo. Un'altra manifestazione, anche più appariscente, si ha negli spettri delle molecole costituite da due atomi eguali; si trova infatti che essi sono costituiti in molti casi da successioni di righe aventi alternativamente intensità forte e debole. La nuova meccanica ha permesso di spiegare queste alternanze di intensità osservate già da molto tempo. Risulta infatti che i momenti magnetici dei due nuclei degli atomi che costituiscono la molecola si possono disporre uno rispetto all'altro parallelamente oppure antiparallelamente; le righe forti o deboli corrispondono all'una o all'altra di queste due possibili orientazioni dei due nuclei. In altre parole la molecola può trovarsi in due stati che hanno proprietà notevolmente diverse tra di loro; e cioè lo stato in cui i due momenti nucleari sono paralleli e quello in cui essi sono antiparalleli. Così per esempio l'ordinario idrogeno è un miscuglio di molecole coi nuclei paralleli (ortoidrogeno) e molecole coi nuclei antiparalleli (paraidrogeno). In condizioni normali i numeri delle molecole di questi due tipi stanno nel rapporto 3:1. Solo a bassissime temperature si ha una prevalenza delle molecole del paraidrogeno, che hanno una energia un po' minore, sopra quelle dell'ortoidrogeno. Così se si mantiene per alcuni giorni dell'idrogeno allo stato liquido, le sue molecole poco alla volta si vanno tutte trasformando in molecole di paraidrogeno. Quando si fa rievaporare questo idrogeno liquido, si ottiene per qualche tempo del paraidrogeno gassoso quasi puro le cui proprietà fisiche (per esempio il calore specifico) sono diverse dalle proprietà dell'ordinario idrogeno, miscuglio di para e orto–idrogeno.

Oltre a questi fenomeni che descrivono, per così dire, delle proprietà esterne dei nuclei, ne sono noti, già da molto tempo, altri che ci mettono in presenza di fatti che avvengono nell'interno del nucleo. Intendo parlare dei fenomeni radioattivi, che consistono in una disintegrazione del nucleo atomico che proiettando all'esterno una particella α oppure un elettrone, si trasforma in un nucleo differente.

La nuova meccanica ha permesso anche di costruire una teoria, se pure necessariamente molto schematica, dei fenomeni della disintegrazione radioattiva, spiegando come possa accadere che una particella α resti in alcuni casi legata nell'interno di un nucleo per anni o per secoli e poi venga proiettata all'esterno senza l'intervento di alcuna ragione apparente.

Con i nuovi metodi dunque, dopo aver studiato con successo i problemi della fisica dell'atomo, si incomincia ora a trattare la fisica dei nuclei. E come le leggi classiche dovettero venir modificate e approfondite per poter essere applicate allo studio dell'atomo e della molecola, così c'è da attendersi che sia necessario modificare le leggi che valgono per l'atomo prima di poter ottenere una teoria soddisfacente dei fenomeni del nucleo. E così, estendendo il proprio studio a una classe sempre più larga di fenomeni, la Fisica farà forse nell'avvenire un nuovo passo verso la comprensione sempre più profonda delle leggi della natura.

62.

LA FISICA MODERNA

«Nuova Antologia», *65*, 137–145 (1930).

Gli ultimi trenta anni hanno rappresentato per la fisica un periodo di intenso sviluppo, determinato principalmente dalle ricerche sopra la struttura della materia. Il pensiero scientifico dallo studio dei fenomeni macroscopici, si è portato principalmente a quello dei fenomeni che avvengono nelle ultime ed infime particelle che costituiscono i diversi corpi. Le leggi classiche della meccanica e dell'elettrologia, che regolano le proprietà di tutti i corpi di dimensioni apprezzabili, che ci spiegano il movimento degli astri, e ci permettono di comprendere il funzionamento di una macchina, non furono più trovate sufficienti, nel passare allo studio del mondo atomico. Esse dovettero essere integrate e sostituite da nuove leggi più adatte ad eliminare le difficoltà che via via si andavano presentando nella esplorazione di nuovi fenomeni sconosciuti.

Tutta questa evoluzione avvenne attraverso un tumultuoso succedersi e avvicendarsi di ipotesi e di teorie, spesso in contrasto tra di loro, le quali, producendo discussioni e scambi di idee tra le diverse scuole, aprivano la via a nuove indagini e a nuove scoperte.

Ed è stato un periodo bello per chi lo ha vissuto; bello appunto per la sua grande varietà, per le continue sorprese che esso ha riservato al ricercatore e per le conseguenze insperate che, in un tempo relativamente assai breve, ci hanno portato ad una concezione abbastanza chiara del mondo misterioso dell'atomo e della molecola.

Un esame superficiale della situazione della nostra scienza può forse destare l'impressione che tra gli indirizzi della fisica nel nostro secolo e nel secolo scorso esista una discontinuità; e ciò sia in quanto si consideri la scienza di per sé, sia in quanto la si consideri dal punto di vista delle sue applicazioni. Alla fine dell'ottocento le teorie fisiche avevano infatti raggiunto un grado di perfezione formale assai elevato, in singolare contrasto con lo sviluppo rapido, ma confuso e tumultuoso, che ha caratterizzato l'evoluzione della fisica atomica. D'altra parte, salve non molte eccezioni, questa non ha ancora dato luogo a quelle grandiose applicazioni tecniche e pratiche che sono state invece il frutto più appariscente degli studi del secolo scorso. Io credo però che questa discontinuità sia soltanto apparente e che la fisica moderna rappresenti la continuazione logica dell'opera scientifica del diciannovesimo secolo.

Cerchiamo di formarci un quadro della situazione della fisica alla fine del settecento. La meccanica poteva ormai considerarsi praticamente completa, grazie all'opera degli insigni scienziati Galileo e Newton e dei loro continuatori. Invece gli altri rami della fisica erano in parte ancora assai incompleti, come l'ottica, e in parte avevano di poco sorpassati i primi rudimenti, come l'elettricità. Ma forse più importante dell'eredità di nozioni trasmessa all'ottocento dai secoli precedenti, fu l'eredità del metodo scientifico. I grandi maestri del seicento e del settecento, mentre da un lato avevano insegnato che lo studio della Natura può farsi solo attraverso all'esperienza, avevano d'altro lato costruito in gran parte i mirabili strumenti matematici che consentono di interpretarne e di farne fruttare i risultati. Gli studiosi del diciannovesimo secolo seppero mettere a gran profitto questi insegnamenti costruendo uno dei più mirabili edifici che siano ricordati nella storia delle scienze.

La personalità che domina nel campo della fisica i primi anni dell'ottocento è Alessandro Volta, al cui genio è dovuto se lo studio dei fenomeni elettrici poté diventare concreto e conclusivo e raggiungere in pochi decenni la conoscenza delle leggi fondamentali dell'elettrologia. Fino alla scoperta del grande comasco l'elettricità veniva prodotta per mezzo delle macchine elettrostatiche, le quali potevano bensì generare una corrente elettrica, ma tanto poco intensa che le sue proprietà più interessanti sfuggivano necessariamente all'osservazione. La pila di Volta diede invece ai fisici uno strumento capace di produrre delle correnti elettriche di intensità relativamente grande. Fu così possibile accorgersi che queste creano attorno a sé un campo magnetico; e che, d'altra parte, con opportuni dispositivi, si può produrre corrente elettrica per mezzo di un campo magnetico. Su questi principî un oscuro studioso italiano, il Pacinotti, costruì il primo motore elettrico, che da lui prese il nome di anello di Pacinotti. Esso può funzionare come motore, cioè può produrre energia meccanica per mezzo della corrente elettrica; e viceversa, messo in moto da una sorgente esterna di energia, può generare una corrente.

Sono queste le basi delle grandi applicazioni tecniche dell'elettricità nel secolo scorso. La dinamo permette infatti di ricavare dalla energia delle acque cadenti o da quella dei combustibili, le enormi quantità di corrente elettrica che illuminano le nostre città, azionano i motori delle nostre industrie e ci rendono in mille modi più facile e più comoda la vita.

Un'altra conquista della fisica nei primi decenni del secolo diciannovesimo fu la scoperta del principio della conservazione dell'energia. Questo principio è ormai talmente radicato nel pensiero delle persone, anche più lontane dalla scienza, da sembrar strano che esso sia stato chiaramente compreso solo poco più di un secolo fa. Si deve al medico Mayer l'aver affermato che il calore, ritenuto fino allora un misterioso fluido, è invece una forma di energia. In essa si trasforma per esempio l'energia meccanica di una ruota in movimento, che viene frenata dagli attriti; non si ha dunque sparizione di energia, ma semplicemente trasformazione di essa. Questo principio rappresenta uno dei fondamenti della tecnica delle macchine termiche, che trasformano il calore in energia meccanica; accanto ad esso dobbiamo ricordare il secondo principio della termodinamica, scoperto nello stesso periodo da Carnot. Esso, ponendo una limitazione alla possibilità di trasformare calore in lavoro meccanico, ci insegna contemporaneamente come deve essere costituita una macchina termica per dare un rendimento il più possibile elevato.

Con le affermazioni di Carnot e di Mayer nasce un nuovo ramo della fisica, fecondo di applicazioni pratiche: la termodinamica. Essa è uno strumento di grande potenza per prevedere e calcolare le trasformazioni dell'energia, ed i suoi risultati, indipendenti da ipotesi particolari sopra la struttura dei corpi, presentano un grado di attendibilità assai prossimo alla certezza.

Tuttavia la termodinamica non ci spiega l'intimo meccanismo dei fenomeni, ma ci dà, in certo modo, soltanto delle regole per calcolarli. A colmare questa lacuna doveva affermarsi la teoria cinetica del calore. Essa ci insegna che il calore è una forma di energia meccanica: l'energia dei moti invisibili delle molecole e degli atomi che costituiscono i corpi. Si ha così una interpretazione immediata della trasformabilità del calore in energia meccanica. Con le teorie cinetiche, vediamo incominciare a manifestarsi nella fisica la tendenza a passare dallo studio dei fenomeni macroscopici verso quello del mondo molecolare ed atomico.

Dato il numero straordinariamente grande delle molecole che costituiscono un corpo, e la loro enorme piccolezza, perde importanza il comportamento delle singole molecole, ed i fenomeni vengono ad essere determinati sostanzialmente dalle leggi statistiche a cui esse obbediscono. Il concetto di probabilità e le leggi dei grandi numeri vengono perciò a dominare questo ramo della fisica. Il secondo principio della termodinamica trova nelle leggi statistiche la sua naturale spiegazione: esso risulta dipendente dalla circostanza che un sistema tende sempre a trasformarsi verso stati di probabilità via via crescente.

Lo sviluppo della termodinamica e delle teorie cinetiche nella fisica, e i contemporanei progressi della chimica fecero sorgere e prendere rapidamente grande importanza ad un altro ramo della scienza: la chimica fisica. Essa ci dà uno degli esempi più brillanti di quanto sia fecondo lo studio della zona di confine fra due scienze differenti, e di quanto proficuo sia il far convergere gli sforzi di uomini di formazione culturale diversa. Nella chimica fisica il fenomeno atomico viene a prendere sempre maggiore rilievo e ben si può dire che questa scienza è stata una delle cause determinanti lo sviluppo della fisica atomica: in particolare il concetto di ione, cioè di atomo carico di elettricità, introdotto dalla chimica fisica per spiegare le proprietà degli elettroliti, suggerisce spontaneamente l'idea che l'atomo sia costituito da corpuscoli elettrizzati; questa ipotesi è il fondamento delle teorie moderne sopra la struttura dell'atomo.

Che i corpuscoli elettrici fossero uno degli elementi fondamentali della struttura atomica, veniva anche indicato dai fenomeni che si osservano in un gas rarefatto attraversato dalla scarica elettrica. Lo studio di questa, iniziato nella seconda metà del secolo scorso, portò tra l'altro alla scoperta dei raggi catodici: essi sono proiezioni velocissime di corpuscoli carichi di elettricità negativa, detti elettroni. Sono questi che, insieme ad altri corpuscoli carichi positivamente, i nuclei, costituiscono la compagine dell'atomo.

Tra gli studi che precedettero e determinarono l'evoluzione della fisica moderna, ha un posto di grande importanza la teoria delle onde elettromagnetiche, enunciata in questo stesso periodo da Maxwell, e verificata dalle esperienze di Hertz. Questa teoria e queste esperienze, che nel campo tecnico vennero applicate da Marconi nell'invenzione mirabile della radiotelegrafia, ebbero una duplice importanza per la fisica pura. Da una parte infatti, col mostrare l'analogia profonda tra le proprietà delle onde elettromagnetiche e di quelle luminose, esse suggerirono l'ipotesi, di cui si trovarono in seguito continue conferme, che la luce fosse costituita da onde elettromagnetiche di piccola lunghezza d'onda. D'altra parte la teoria di Maxwell, richiamando l'attenzione sopra l'energia irradiata da corpuscoli elettrici in movimento, costituì uno dei punti di partenza per lo studio delle proprietà delle particelle che costituiscono l'atomo.

La storia delle scienze fisiche nel diciannovesimo secolo si conclude con la scoperta di due fenomeni di eccezionale portata. I raggi X, scoperti da Röntgen, ci facevano conoscere un nuovo tipo di radiazioni di frequenza circa mille volte maggiore di quella della luce ordinaria; queste, per la loro proprietà di traversare corpi opachi, e per l'azione da esse direttamente esercitata sopra certi tessuti, trovarono immediatamente un fecondissimo campo di applicazione nella medicina.

La radioattività, d'altra parte, ci metteva in presenza di nuove imprevedibili proprietà della materia le quali, come si scoperse poco alla volta nel seguito, hanno la loro origine negli stessi nuclei degli atomi, cioè nella particella centrale e più profonda dell'edificio atomico.

Vediamo così che quasi tutti i rami della fisica, i quali si svilupparono maggiormente verso la fine del secolo scorso, venivano a convergere verso lo stesso problema: la struttura dell'atomo.

Nella fisica dell'atomo la collaborazione e l'integrazione tra i risultati della teoria e quelli dell'esperienza sono state continue e fecondissime, portando, in trenta anni di studi, da poche idee vaghe, imprecise ed incerte sopra il numero, le dimensioni e perfino la stessa esistenza reale degli atomi, ad una conoscenza relativamente sicura della struttura interna di ciascuno di essi, e dei movimenti delle particelle che lo costituiscono.

Abbiamo già detto come, fin dalla fine del secolo scorso, vari indizi suggerissero l'ipotesi che, nell'interno dell'atomo, fossero contenuti dei corpuscoli elettrici più piccoli. Non è questo il luogo per descrivere le indagini lunghe, difficili e pazienti che permisero di analizzare l'intima natura di questi corpuscoli, portandoci così alla conoscenza dell'elettrone negativo e del nucleo positivo, cioè dei due elementi costitutivi dell'atomo. Il nucleo ne occupa il centro, ed è circondato da alcuni elettroni che girano intorno ad esso. La struttura dell'atomo viene così a ricordare, in dimensioni enormemente piccole, quella del sistema dei pianeti che girano intorno al Sole.

Gli elettroni, che in questo paragone corrispondono ai pianeti, sono tutti eguali tra di loro, qualunque sia la loro origine. Così sono eguali quelli che noi possiamo estrarre da un atomo di ferro o da uno di ossigeno; e questi a loro volta sono identici agli elettroni che vengono proiettati dal Sole e, arrivando sull'atmosfera terrestre, producono il fenomeno dell'aurora boreale. Esistono invece molti nuclei differenti: ogni elemento chimico ha il suo particolare nucleo; anzi, per parecchi elementi si conoscono più specie di nuclei differenti.

Se la scoperta degli elementi costitutivi dell'atomo e della loro distribuzione fu difficile e laboriosa, ancora più ardua fu la ricerca delle leggi a cui obbediscono, nei loro movimenti, questi corpuscoli. Ci se ne rende facilmente conto se si pensa che l'atomo è oltre mille volte più piccolo del più piccolo corpo di cui sia ancora possibile apprezzare la forma al microscopio; e le dimensioni del nucleo sono verosimilmente ancora diecimila volte più piccole. Si comprende quindi la necessità di ricorrere sempre in questi studi a metodi indiretti: fra questi il più potente è l'analisi spettrale della luce emessa dall'atomo. La spettroscopia fu definita la musica degli atomi: è una musica di non sempre facile interpretazione, il cui studio serve a dirci qualche cosa sopra gli strumenti che l'hanno emessa.

Nei primi tentativi si cercò di comprendere l'atomo in base a quelle stesse leggi che sappiamo valide nella fisica dei fenomeni macroscopici. Si riconobbe ben presto però che esse non sono applicabili a corpi di dimensioni così estremamente piccole; sorse allora la teoria di Bohr, che cercava, modificando le leggi ordinarie, di adattarle ai nuovi problemi, e che permise di raggiungere un grande numero di risultati, specialmente qualitativi. Quasi sempre però, quando si cercava di raggiungere la precisione quantitativa, la teoria di Bohr si mostrava insufficiente; così che i fisici dovettero convincersi che non bastava modificare le vecchie leggi, ma era necessario in gran parte sostituirle con delle nuove.

Desidero qui dire due parole sopra il criterio che è servito di guida nella costruzione di queste nuove leggi, e che è stato uno dei fattori più importanti e caratteristici della fisica del nostro secolo. Questo criterio, pur essendo in sé quasi evidente, ha bisogno, in alcuni casi, di un certo coraggio per essere applicato. Nella fisica di oggi si considera che abbiano un significato preciso soltanto le grandezze suscettibili di una determinazione sperimentale; ben inteso anche se questa possa farsi solo per mezzo di esperienze di esecuzione difficilissima, o anche praticamente impossibile, purché l'impossibilità sia tecnica e non teorica. Ora esistono alcune nozioni che sfuggono ad una precisazione rigorosa e che perciò, dal punto di vista della fisica moderna, debbono venire modificate, anche quando esse ci derivano dall'intuito comune. Così per esempio nella costruzione della nuova meccanica dell'atomo gli stessi concetti di posizione e di velocità hanno dovuto subire delle modificazioni essenziali. Le conseguenze di esse, come è naturale, diventano trascurabili per la meccanica dei corpi comuni, e conducono invece a risultati importanti quando si considerano sistemi di dimensioni piccolissime.

Per illustrare l'applicazione di questo criterio, non mi riferirò all'esempio della fisica atomica, ma a quello della teoria della relatività. Fu appunto in questo interessante capitolo della fisica moderna che esso ha avuta la sua prima affermazione. La teoria della relatività è sorta, come è noto, per spiegare l'insuccesso di alcune esperienze che cercavano di mettere in evidenza un effetto, sopra i fenomeni terrestri, del movimento di traslazione della terra attraverso allo spazio. Le apparenti contraddizioni a cui portavano queste esperienze trovarono la loro risoluzione in una critica e in un precisamento di una nozione che sembra tra le più primitive e radicate nel nostro intuito: voglio dire il concetto di contemporaneità di due avvenimenti. Tutti noi siamo portati a prima vista a ritenere che l'affermazione che due fenomeni sono contemporanei abbia sempre un significato preciso, del tutto indipendente dalle difficoltà e dai metodi da seguire per provarla. Nei casi più comuni noi consideriamo come contemporanei due fenomeni, se li percepiamo nel medesimo istante. Diciamo così per esempio che due persone parlano contemporaneamente, se udiamo insieme i suoni delle loro voci. È evidente però che questa verificazione della contemporaneità è assai grossolana, poiché viene trascurato in essa il tempo che il suono impiega per giungere a noi. La verificazione può farsi in modo assai più preciso se, invece che del suono, ci serviamo della luce, considerando contemporanei due fenomeni che vediamo nel medesimo istante. Ma anche questo metodo ha bisogno di una correzione, poiché anche la luce impiega un certo tempo se pure assai piccolo, per giungere a noi. Si potrebbe creder possibile di tener conto di questo, correggendo il piccolo errore nei risultati; ma anche per questa via sorge una difficoltà: è facile convincersi infatti che un osservatore fermo dovrebbe portare una correzione differente da quella di un osservatore in movimento.

La conclusione di questa critica è che la contemporaneità di due avvenimenti non ha un senso assoluto, ma solo relativo all'osservatore che la constata; due fenomeni possono risultare contemporanei per un osservatore e non contemporanei per un altro. Einstein ha sviluppato matematicamente i risultati di questo principio e su di esso ha basata la sua teoria della relatività.

Le conseguenze di questa teoria sono sempre estremamente piccole dal lato quantitativo, e diventerebbero considerevoli soltanto per corpi animati da velocità prossime a quella della luce. Ciò non diminuisce a mio parere l'enorme valore speculativo della teoria della relatività; questa ci mostra infatti per la prima volta, nella storia della scienza, la revisione di un concetto che si presenta alla nostra mente con quasi altrettanta evidenza dei principî fondamentali della logica.

Mentre, come abbiamo detto, i risultati sperimentali dedotti dalla teoria della relatività sono quantitativamente assai piccoli, le teorie sopra la struttura dell'atomo hanno invece indirizzato gli sperimentatori verso un campo di indagine assai fecondo, ciò che ha permesso di accumulare un complesso di risultati veramente imponente. Non solo fenomeni che già erano conosciuti hanno trovata la loro interpretazione e il loro posto, ma è stato possibile verificare con l'esperienza nuovi fatti previsti dalla teoria. Citerò tra questi una recente esperienza di due fisici americani, Davisson e Germer; essa ha provato che una proiezione di corpuscoli ha, sotto molti punti di vista, le stesse proprietà che si credevano caratteristiche delle radiazioni costituite da onde analoghe a quelle luminose.

Le nuove teorie hanno esercitato una notevole influenza anche al di fuori della fisica: soprattutto nella chimica. Poiché esse spiegano il meccanismo delle forze che, riunendo tra di loro alcuni atomi determinano la formazione delle molecole delle diverse sostanze, oppure tengono insieme la compagine di un cristallo. Si tratta per ora di studi che sono soltanto ai loro inizi, ma che promettono di svilupparsi assai rapidamente; e non è facile prevedere fin dove essi potranno condurre.

Ed ora che lo studio dell'atomo e della molecola è bene avviato verso la sua soluzione, l'attenzione dei fisici incomincia a rivolgersi, con sempre maggiore insistenza, verso un nuovo problema: quello della struttura del nucleo. Il nucleo, l'abbiamo già detto, è circa diecimila volte più piccolo dell'atomo; inoltre i suoi elementi sono serrati uno all'altro con forze così grandi, che per spezzarlo occorre concentrare su di esso delle quantità di energia relativamente enormi; così che si conoscono fino ad ora pochi casi isolati in cui si sia riusciti a produrre artificialmente una disintegrazione del nucleo. Vero è che la natura stessa, con i fenomeni radioattivi, ci ha dato un esempio di disintegrazione spontanea di alcuni nuclei. Si tratta però di un fenomeno che noi solo possiamo per ora osservare, senza che si sia riusciti in alcun modo a produrne una accelerazione o un rallentamento, o, comunque ad influenzarlo. Difficilissimo dunque il nuovo problema, e scarsi fino ad oggi i mezzi per il suo studio; ma gli studiosi non debbono lasciarsi sgomentare dalle difficoltà: altrimenti la scienza non potrebbe progredire.

E qui sorgerà spontanea per molti una domanda: quali conseguenze pratiche sono derivate o potranno derivare da un così grande aumento delle nostre conoscenze sulla struttura intima della materia? Il rivoluzionamento della tecnica, determinatosi sotto l'impulso delle scoperte scientifiche del secolo scorso, ha prodotto, nel pensiero di molti uomini, l'impressione che la principale misura dei risultati scientifici sia data dalle immediate applicazioni pratiche che essi permettono. Se, con questo criterio, vogliamo giudicare la scienza di oggi, dobbiamo confessare che le sue applicazioni pratiche sono state fino ad ora alquanto scarse.

Occorre però riflettere che i progressi tecnici che derivano dallo svilupparsi di un capitolo della scienza pura, seguono sempre con un considerevole ritardo l'evoluzione di questa. Gli uomini che si occupano di scienza pura, per una necessaria divisione del lavoro, raramente sono gli stessi che studiano i problemi della tecnica. E prima che le idee degli scienziati si chiariscano, si consolidino e si diffondano in modo da poter venire utilizzate, debbono necessariamente passare alcuni anni e, in qualche caso, alcune decine di anni. Quando Volta scoperse la pila, né lui né i suoi contemporanei potevano intravvedere quali sarebbero state le conseguenze della sua opera; e dovettero trascorrere decine e decine di anni prima che l'elettricità, dall'essere oggetto di studio di pochi scienziati, il cui lavoro poteva sembrare lontano dai problemi e dai bisogni della vita pratica, diventasse il fattore dominante delle meravigliose invenzioni che hanno caratterizzato la fine del secolo scorso.

Non deve dunque sorgere nel pubblico l'impressione che l'opera degli scienziati vada ormai allontanandosi dalla vita, per confondersi nell'inseguimento di idee astruse e puramente astratte. È ben vero che molte volte la loro opera non è direttamente determinata dalla volontà di portare un contributo a una questione tecnica, ma piuttosto da un intimo bisogno di spingere sempre più oltre i confini che limitano la nostra conoscenza. Sorgeranno tuttavia prima o poi degli altri uomini di ingegno più pratico, i quali troveranno il modo di utilizzare le nuove conoscenze per un ulteriore sviluppo della civiltà.

E i giovani che oggi, negli studi, si affacciano alla soglia della vita, ricordino che il nostro paese ha bisogno sia di coloro che, attraverso ad un lavoro paziente, gettino la fertile semenza delle idee scientifiche, sia di coloro che da questa sappiano far germogliare una messe di risultati fecondi per la vita degli uomini.

N° 63.

Paper N° 63 is an extension of the statistical method developed in papers N° 43, 44, 45 and 48 to the case of ionized atoms. Particular attention is paid to the calculation of the Rydberg corrections for the *s* terms of atoms ionized three times.

The agreement between the results of the calculation and the experimental results is better than that found for neutral atoms, because the different properties of the ions depend on the atomic number of the ion in a more regular manner than the corresponding properties of neutral atoms.

E. SEGRÈ.

63.

SUL CALCOLO DEGLI SPETTRI DEGLI IONI (*)

«Mem. Accad. d'Italia», *1* (Fis.) 149–156 (1930); «N. Cimento», *8*, 7–14 (1931).

Il metodo statistico proposto dall'autore per il calcolo delle proprietà degli atomi neutri, viene esteso in questo lavoro al caso degli atomi ionizzati, e viene applicato in particolare alla determinazione delle correzioni di Rydberg per i termini *s* degli spettri degli atomi tre volte ionizzati.

In alcuni recenti lavori [1] ho mostrato che numerose proprietà degli atomi, qualunque sia il loro numero atomico, possono essere dedotte con discreta approssimazione per mezzo di una considerazione statistica, in cui gli elettroni che circondano il nucleo vengono trattati come una nuvola di gas elettronico, in condizioni di degenerazione completa, la cui distribuzione di densità in funzione della distanza dal nucleo viene determinata statisticamente. Lo scopo del presente lavoro è di estendere quei metodi al calcolo degli atomi ionizzati.

In una prima parte verranno esposti i metodi generali per il calcolo della distribuzione statistica degli elettroni in uno ione; e quindi per la determinazione del potenziale elettrico in funzione della distanza dal nucleo. Il

(*) Presentata nell'Adunanza del 21 marzo 1930–VIII.

(1) E. FERMI, «Rend. Lincei», *6*, 602 (1927) (citata nel seguito come nota I); *7*, 726 (1928); «Zs. f. Phys.», *48*, 73 (1928); «Zs. f. Phys.», *49*, 550 (1928). Cfr. anche F. RASETTI, «Rend. Lincei», *7*, 915 (1928); G. GENTILE e E. MAJORANA, «Rend. Lincei», *8*, 229 (1928).

Questi lavori sono riassunti nella monografia di E. FERMI, in *Falkenhagen, Quantentheorie und Chemie*, Lipsia 1928. Le applicazioni del metodo statistico ai problemi nucleari sono illustrate in una Memoria di E. FERMI, *Sopra i momenti magnetici dei nuclei atomici*, «Memorie della R. Acc. d'Italia», *1*, 139 (1930).

risultato verrà poi applicato al calcolo degli spettri degli ioni, e precisamente alla determinazione della correzione di Rydberg per i termini *s* degli atomi tre volte ionizzati.

§ 1. – Indichiamo con V il potenziale elettrico, che sarà funzione della distanza r del punto che si considera dal nucleo. La densità degli elettroni alla distanza r dal nucleo dipende dal valore del potenziale; precisamente si è dimostrato nella Nota I che essa è legata in generale al potenziale dalla relazione

$$n = \frac{2^{9/2} \pi m^{3/2} e^{3/2}}{3 h^3} (V - \alpha)^{3/2} \tag{1}$$

in cui α rappresenta una costante. La formola (1) è valida soltanto per $V > \alpha$, nel qual caso essa dà un risultato reale; per $V < \alpha$ si ha invece $n = 0$. Se poniamo

$$V - \alpha = v \tag{2}$$

v ci rappresenterà ancora, a meno di una costante additiva, il potenziale elettrico; la (1) diventerà allora

$$n = \frac{2^{9/2} \pi m^{3/2} e^{3/2}}{3 h^3} v^{3/2} . \tag{3}$$

Questa formula è valida per $v > 0$; per $v < 0$ la densità degli elettroni è nulla. Ora si trova che, nel caso degli ioni, v risulta positivo fino ad una certa distanza r_0 dal nucleo; mentre per distanze maggiori esso diventa negativo. Ciò significa che la nuvola degli elettroni che circonda il nucleo si estende fino alla distanza r_0 da esso. Siccome v deve soddisfare all'equazione di Poisson, troviamo per esso, come nella Nota I, l'equazione differenziale

$$\Delta v = 4 \pi n e = \frac{2^{13/2} \pi^2 m^{3/2} e^{5/2}}{3 h^3} v^{3/2} \tag{4}$$

valida per $r < r_0$; mentre per $r > r_0$ varrà l'altra equazione

$$\Delta v = 0 . \tag{5}$$

Per r molto piccolo il potenziale deve tendere all'espressione limite Ze/r, in cui Z rappresenta il numero atomico; si avrà dunque, come prima condizione limite per la determinazione di v, la seguente:

$$\lim_{r=0} r v = Ze . \tag{6}$$

Se supponiamo poi che il nostro atomo sia ionizzato z volte, esso conterrà $Z - z$ elettroni. Si avrà dunque

$$\int n d\tau = Z - z .$$

E se teniamo conto che la nuvola di elettroni si estende fino alla distanza r_0, e che la sua densità è data dalla (3), possiamo scrivere la relazione precedente nella forma:

$$\frac{2^{13/2}\,\pi^2\,m^{3/2}\,e^{5/2}}{3\,h^3}\int_0^{r_0} v^{3/2}\,r^2\,dr = (Z - z)\,e\,. \tag{7}$$

Come nella Nota I possiamo cambiare le variabili r, v in altre due ad esse proporzionali, ponendo

$$r = \mu x \quad ; \quad v = \gamma\psi \tag{8}$$

dove

$$\mu = \frac{3^{2/3}\,h^2}{2^{13/3}\,\pi^{4/3}\,me^2\,Z^{1/3}} \quad ; \quad \gamma = \frac{2^{13/3}\,\pi^{4/3}\,me^3\,Z^{4/3}}{3^{2/3}\,h^2}\,. \tag{9}$$

Tenendo conto che v dipende soltanto da r, le equazioni (4), (5), (6), (7), si potranno scrivere allora nella forma:

$$\begin{cases} \psi'' + \frac{2}{x}\psi' = \psi^{3/2} & \text{per } x < x_0 \\ \psi'' + \frac{2}{x}\psi' = 0 & \text{per } x > x_0 \\ \lim\limits_{x=0} x\psi = 1 & \\ \int_0^{x_0} \psi^{3/2}\,x^2\,dx = 1 - \frac{z}{Z} & \end{cases} \tag{10}$$

$x = x_0$ rappresenta il valore della x per cui la funzione ψ si annulla.

Cambiando nuovamente variabile, col porre

$$\varphi = x\psi \tag{11}$$

le (10) diventano

$$\begin{cases} \varphi'' = \varphi^{3/2}/\sqrt{x} & \text{per } x < x_0 \\ \varphi'' = 0 & \text{per } x > x_0 \\ \varphi(0) = 1 & \\ \int_0^{x_0} \varphi^{3/2}\sqrt{x}\,dx = 1 - \frac{z}{Z} & \end{cases} \tag{12}$$

L'ultima condizione può facilmente trasformarsi tenendo conto della prima. Si trova infatti

$$\int_0^{x_0} \varphi^{3/2}\sqrt{x}\,dx = \int_0^{x_0} \varphi''\,x\,dx = \left| x\varphi' - \varphi \right|_0^{x_0}$$

e siccome $\varphi(0) = 1$, $\varphi(x_0) = 0$,

$$\int_0^{x_0} \varphi^{3/2}\sqrt{x}\,dx = x_0\,\varphi'(x_0) + 1\,.$$

L'ultima delle (12) diventa dunque

(13) $$- x_0 \varphi'(x_0) = \frac{z}{Z} \cdot$$

Per arrivare alla costruzione numerica della funzione φ definita dalle condizioni (12) (13) partiremo dalla soluzione, già determinata nella Nota I di una funzione $\varphi_0(x)$ (nella Nota I questa funzione è indicata semplicemente con φ) che soddisfa all'equazione differenziale

(14) $$\varphi_0'' = \frac{\varphi_0^{3/2}}{\sqrt{x}}$$

con le condizioni al limite.

(15) $$\varphi(0) = 1 \quad , \quad \varphi(\infty) = 0 .$$

Di questa funzione sono state date delle tabelle numeriche abbastanza estese nei lavori citati precedentemente. Le due funzioni φ_0 e φ soddisfano nel tratto da 0 ad x_0 alla stessa equazione differenziale; inoltre esse hanno entrambi il valore 1 per $x = 0$. Per valori di x abbastanza minori di x_0 la loro differenza relativa si mantiene abbastanza piccola, e potrà perciò venir determinata come soluzione dell'equazione alle variazioni dell'equazione differenziale (14). In altre parole potremo porre

(16) $$\varphi = \varphi_0 - k\eta$$

dove k rappresenta una costante, ed η è una soluzione dell'equazione alle variazioni

$$\eta'' = \frac{3}{2}\sqrt{\frac{\varphi_0}{x}}\,\eta$$

e che si annulla per $x = 0$. Il valore della derivata di η per $x = 0$ può fissarsi arbitrariamente, poiché, cambiandolo, la η viene ad alterarsi semplicemente per un fattore di proporzionalità; possiamo prendere per esempio $\eta'(0) = 1$. È facile verificare che l'equazione alle variazioni ha per integrale generale

$$\eta = A\left(\varphi_0 + \frac{x\varphi_0'}{3}\right)\left\{\int \frac{dx}{\left(\varphi_0 + \frac{x\varphi_0'}{3}\right)^2} + B\right\}$$

in cui A e B sono delle costanti arbitrarie. La soluzione particolare che soddisfa alle nostre condizioni limite è la seguente

(17) $$\eta = \left(\varphi_0 + \frac{x\varphi_0'}{3}\right)\int_0^x \frac{dx}{\left(\varphi_0 + \frac{x\varphi_0'}{3}\right)^2} .$$

Nonostante si sia così trovata una espressione della funzione cercata per mezzo di quadrature, conviene tuttavia, per la determinazione numerica, non servirsi della (17) ma determinare direttamente la η per mezzo dell'equazione differenziale a cui essa soddisfa. Nella tabella seguente sono raccolti i valori numerici della funzione determinati in questo modo:

x	$\eta(x)$	x	$\eta(x)$	x	$\eta(x)$	x	$\eta(x)$
0,0	0,00	1,4	2,23	8	103	20	2080
0,1	0,10	1,6	2,76	9	146	21	2480
0,2	0,21	1,8	3,36	10	201	22	2940
0,3	0,32	2,0	4,04	11	271	23	3460
0,4	0,44	2,5	6,1	12	359	24	4050
0,5	0,56	3,0	8,8	13	467	25	4710
0,6	0,70	3,5	12,3	14	598	26	5450
0,7	0,85	4,0	16,7	15	755	27	6280
0,8	1,01	4,5	22,1	16	942	28	7210
0,9	1,18	5	28,6	17	1160	29	8250
1,0	1,36	6	45,9	18	1420	30	9410
1,2	1,76	7	70,0	19	1730	31	10690

La determinazione della costante k viene fatta per mezzo della condizione (13) nel modo seguente. Presi alcuni valori arbitrari per k (in pratica il calcolo è stato fatto per $k = 10^{-3}, 10^{-4}, 10^{-5}, 10^{-6}, 10^{-7}$), si calcola per mezzo della (16) la funzione φ fino a quei valori di x per cui la differenza relativa fra φ e φ_0 si mantiene abbastanza piccola; per esempio non superiore a $1/10$. Da questi valori in poi la φ si prolunga, integrando numericamente l'equazione differenziale a cui essa soddisfa, fino a determinare il valore x_0 per cui $\varphi(x_0)$ si annulla; questa parte del calcolo è molto rapida e, con un po' di pratica, la si può fare in pochi minuti. Nella tabella seguente sono riportati in corrispondenza dei diversi valori di k, i valori dell'espressione $-x_0\,\varphi'(x_0)$ determinati nel modo indicato.

$$k = 10^{-3} \quad ; \quad 10^{-4} \quad ; \quad 10^{-5} \quad ; \quad 10^{-6} \quad ; \quad 10^{-7}$$

$$-x_0\,\varphi'(x_0) = 0{,}22 \quad ; \quad 0{,}10 \quad ; \quad 0{,}05 \quad ; \quad 0{,}02 \quad ; \quad 0{,}01.$$

Da questa tabella si ricava immediatamente per interpolazione il valore di k, una volta che sia dato il rapporto z/Z tra il grado di ionizzazione e il numero atomico; questo rapporto infatti, secondo la (13) deve appunto essere eguale a $-x_0\,\varphi'(x_0)$. Con questi procedimenti si può determinare molto rapidamente e con precisione sufficiente la funzione φ fino ad x_0. Dalla seconda della (12) risulta che per $x > x_0$ la φ diventa una funzione lineare di x; e siccome essa deve essere continua insieme alla sua derivata, la sua espressione sarà:

$$\varphi'(x_0)(x - x_0)$$

che, tenendo conto della (13), può anche scriversi

$$-\frac{z}{Zx_0}(x - x_0)\,. \tag{18}$$

Determinata la φ si ottiene immediatamente il potenziale elettrico alla distanza r dal nucleo. Dalle equazioni (8), (9) e (11) risulta infatti che esso è dato da

$$v = \frac{Ze}{r}\,\varphi\left(\frac{r}{\mu}\right). \tag{19}$$

Conviene tener presente però che se, conformemente alle ordinarie convenzioni, si vuole che il potenziale elettrico si annulli per distanza infinita, occorre alterare per una costante additiva l'espressione (19). Ponendo infatti in essa per φ l'espressione (18) valida per distanza grande, si trova

$$v = \frac{ze}{r} - \frac{ze}{\mu x_0}\,.$$

Dalla quale si riconosce che, per ottenere una espressione del potenziale che si annulli a distanza infinita, occorre aggiungere a v la costante

$$\frac{ze}{\mu x_0}$$

prendendo come espressione del potenziale

$$V = v + \frac{ze}{\mu x_0} = \frac{Ze}{r}\,\varphi\left(\frac{r}{\mu}\right) + \frac{ze}{\mu x_0}\,. \tag{20}$$

§ 2. I metodi che abbiamo qui esposto per la determinazione teorica del potenziale elettrico nell'interno di uno ione, possono venire applicati per il calcolo numerico dei suoi livelli energetici. Per mostrare questa applicazione sopra un esempio concreto, ho calcolato i termini s (o, ciò che è lo stesso, le loro correzioni di Rydberg) per gli spettri degli atomi tre volte ionizzati. Per poter calcolare i livelli energetici di un elettrone, occorre conoscere l'energia potenziale della forza che lo attrae verso il centro dell'atomo. Essa può calcolarsi facilmente per mezzo dell'espressione (20) del valore medio del potenziale elettrico alla distanza r dal nucleo; bisogna naturalmente tener conto del fatto che il potenziale elettrico che agisce sopra un elettrone quando esso si trova alla distanza r dal nucleo non coincide esattamente col potenziale (20), poiché ad esso non contribuisce l'azione dell'elettrone che si sta considerando; con un ragionamento simile a quello applicato per il caso degli atomi neutri, si trova, che, con sufficiente approssimazione, l'energia potenziale di un elettrone dell'atomo si può supporre espressa da

$$U = -\frac{e^2}{r}\left\{1 + (Z - 1)\,\varphi\left(\frac{r}{\mu}\right)\right\} - \frac{ze^2}{\mu x_0}\,. \tag{21}$$

Determinata così l'energia potenziale, si può scrivere senz'altro l'equazione di Schroedinger per i termini s. Essa risulta:

$$\frac{d^2\psi}{dr^2} + \frac{2}{r}\frac{d\psi}{dr} + \frac{8\pi^2 m}{h^2}\left[E - U(r)\right]\psi = 0\,. \tag{22}$$

Il calcolo della correzione di Rydberg può farsi allora nel modo seguente. Si effettua l'integrazione numerica dell'equazione precedente, cercando la soluzione che corrisponde al numero quantico infinito, e cioè dall'autovalore E = 0. Questa soluzione si confronta con la corrispondente soluzione dell'equazione di Schroedinger per un atomo idrogenoide avente carica nucleare eguale a $(z + 1)\, e$, eguale cioè alla carica efficace che agisce sopra l'elettrone luminoso del nostro ione, quando esso si trova all'esterno della nuvola degli altri elettroni circostanti al nucleo. Siccome, a partire da un certo valore r_0 di r in poi, l'energia potenziale dell'elettrone dell'atomo idrogenoide viene a coincidere con l'energia potenziale dell'elettrone luminoso del nostro ione, ne deriva che, per r abbastanza grande, le due autofunzioni avranno un andamento simile e in particolare i loro nodi si succederanno ad intervalli eguali. Se supponiamo per esempio che l'n^{esimo} nodo dell'autofunzione per lo ione coincida con l'n'^{esimo} nodo dell'autofunzione dell'atomo idrogenoide, la correzione di Rydberg sarà espressa semplicemente dalla differenza

$$\alpha = n - n' . \tag{23}$$

Se invece i nodi di una delle due autofunzioni occupano posizioni intermedie tra i nodi dell'altra autofunzione, si può tuttavia determinare la correzione di Rydberg con un evidente procedimento di interpolazione. Per i dettagli della descrizione di questo calcolo delle correzioni di Rydberg, rimandiamo al calcolo analogo svolto per il caso degli atomi neutri.

Nella figura è rappresentato il confronto fra i risultati teorici e quelli sperimentali. In ascisse sono riportati i numeri atomici; in ordinate i valori

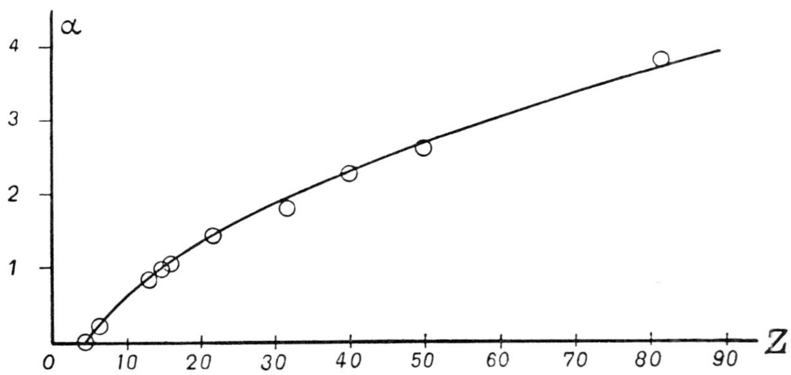

delle correzioni di Rydberg per i termini s degli atomi tre volte ionizzati. I circoletti rappresentano i valori sperimentali, che sono stati dedotti dai dati dell'opera di Grotrian, *Graphische Darstellung der Spektren*, Berlino 1928. La curva rappresenta i risultati della teoria; essa è stata disegnata per punti, effettuando il calcolo per i numeri atomici 23, 50, 82 e tenendo conto inoltre del dato evidente che la correzione di Rydberg si annulla per Z = 4.

Come si vede, l'accordo fra la teoria e l'esperienza è molto buono; esso risulta anche molto migliore di quello che si era trovato per il caso degli atomi neutri, ciò che può facilmente comprendersi per il fatto che le diverse proprietà degli ioni dipendono dal numero atomico in modo più regolare che non le corrispondenti proprietà degli atomi neutri, e vengono perciò rese più fedelmente da una teoria che, per il suo stesso carattere statistico, non è adatta per il calcolo delle differenze irregolari tra atomo e atomo.

N° 64.

For the introduction to this paper see N° 50.

64.

SOPRA L'ELETTRODINAMICA QUANTISTICA (*)

« Rend. Lincei », *12*, 431-435 (1930).

In una Nota pubblicata recentemente in questi « Rendiconti » [1] ho scritto in forma quantistica le equazioni dell'elettrodinamica; cioè le equazioni del sistema costituito dal campo elettromagnetico e da un numero qualunque di cariche elettriche puntiformi. Le equazioni scritte allora si riferivano al caso non relativistico; presupponevano cioè che le velocità delle cariche non fossero molto elevate. Esse possono tuttavia senza alcuna difficoltà essere scritte in forma relativistica, basandosi sopra la teoria di Dirac dell'elettrone rotante. È noto che recentemente anche W. Heisenberg e W. Pauli [2] hanno trattato il problema dell'elettrodinamica quantistica. Siccome però i metodi seguiti da questi Autori sono essenzialmente diversi dai miei, credo non inutile pubblicare anche i miei risultati.

La forma definitiva in cui verranno espressi i risultati di questo lavoro è particolarmente semplice. Troveremo infatti che la Hamiltoniana che, nel senso del principio di corrispondenza, rappresenta la naturale traduzione quantistica dell'elettrodinamica classica, si ottiene semplicemente aggiungendo alla Hamiltoniana della teoria dell'irradiazione di Dirac un termine che rappresenta l'energia elettrostatica del sistema di cariche elettriche; per modo che, nella presente forma, l'elettrodinamica quantistica viene a non essere in alcun modo più complicata della teoria di Dirac dell'irradiazione. Questa semplificazione si può raggiungere come vedremo mediante una opportuna espressione della condizione

$$\frac{1}{c}\frac{\partial V}{\partial t} + \operatorname{div} \mathbf{U} = 0 \tag{1}$$

che lega tra di loro i potenziali scalare e vettore e che, anche nella teoria di Heisenberg e Pauli costituisce uno degli elementi più caratteristici dell'elettrodinamica quantistica.

Nella Nota I abbiamo trovata l'espressione (21) che rappresenta l'Hamiltoniana del nostro sistema. Se, invece della meccanica classica, vogliamo rappresentare il moto dei punti per mezzo della Hamiltoniana di Dirac,

(*) Presentata dal Socio O. M. Corbino, è pervenuta all'Accademia il 29 settembre 1930.

(1) E. FERMI, « Rend. Lincei », *5*, 881 (1929). Citata nel seguito con I.

(2) W. HEISENBERG und W. PAULI, « Zs. f. Phys. », *56*, 1 (1929); *59*, 160 (1930)

possiamo verificare facilmente che, al posto della I dobbiamo usare l'Hamiltoniana seguente:

$$\begin{aligned}(2)\qquad H = &-c\sum_i \gamma_i\cdot p_i - \sum_i \delta_i m_i c^2 + \sum_i e_i c\sqrt{\frac{8\pi}{\Omega}}\sum_i Q_s \cos\Gamma_{si}\\ &+\sum_s e_i c\sqrt{\frac{8\pi}{\Omega}}\sum_s \gamma_i\cdot(\alpha_s\chi_s + A_{s1}w_{s1} + A_{s2}w_{s2})\sin\Gamma_{si}\\ &+\sum_s\left[\frac{1}{2}(\omega_{s1}^2+\omega_{s2}^2+\tilde\omega_s^2-P_s^2)+2\pi^2\nu_s^2(w_{s1}^2+w_{s2}^2+\chi_s^2-Q_s^2)\right].\end{aligned}$$

Le notazioni sono quelle della Nota I; δ_i e γ_i rappresentano un q–scalare e un q–vettore tali che $\delta_i, \gamma_{ix}, \gamma_{iy}, \gamma_{iz}$ sono i quattro operatori, rappresentabili con matrici del quarto ordine, che intervengono nella Hamiltoniana relativa all'i–esimo punto materiale; naturalmente le γ e la δ relative a uno dei punti sono permutabili con le γ e la δ relative a un altro dei punti.

Osserviamo in particolare che dall'Hamiltoniana (2) risulta

$$\left\{\begin{aligned}\dot Q_s &= \frac{\partial H}{\partial P_s} = -P_s \quad ; \quad \dot P_s = -\frac{\partial H}{\partial Q_s} = 4\pi^2\nu_s^2 Q_s - c\sqrt{\frac{8\pi}{\Omega}}\sum_i e_i\cos\Gamma_{si}\\ \dot\chi_s &= \frac{\partial H}{\partial\tilde\omega_s} = \tilde\omega_s \quad ; \quad \dot{\tilde\omega}_s = -\frac{\partial H}{\partial\chi_s} = -4\pi^2\nu_s^2\chi_s - c\sqrt{\frac{8\pi}{\Omega}}\sum_i e_i\gamma_i\cdot\alpha_s \sin\Gamma_{si}\end{aligned}\right.$$

e

$$\dot{\mathbf{X}}_i = \frac{\partial H}{\partial p_i} = -c\gamma_i$$

da cui

$$\frac{d}{dt}\cos\Gamma_{si} = 2\pi\nu_s\gamma_i\cdot\alpha_s\sin\Gamma_{si}\,.$$

Da questa equazione e dalle (3) risulta subito che anche dalla nuova Hamiltoniana (2) deriva l'equazione (18) I; e quindi che se l'espressione (19) I si annulla insieme alla sua derivata prima all'istante zero, essa resta sempre nulla in virtù delle equazioni differenziali. E resta quindi verificata la condizione (1), equivalente alla (19) I. Per mezzo delle (3), la (19) I si può scrivere

$$(4)\qquad 2\pi\nu_s\chi_s - P_s = 0$$

e la sua derivata, a meno di un fattore costante

$$(5)\qquad \tilde\omega_s - 2\pi\nu_s Q_s + \frac{c}{2\pi\nu_s}\sqrt{\frac{8\pi}{\Omega}}\sum_i e_i\cos\Gamma_{si} = 0\,.$$

In una interpretazione classica potremmo dunque dire che l'elettrodinamica ordinaria si ottiene integrando le equazioni canoniche dedotte dalla (2) e imponendo (4) e (5) come condizioni iniziali; ciò basta, poiché si è detto che se (4) e (5) sono verificate all'istante zero, esse lo sono anche automaticamente a un istante qualsiasi.

Per tradurre tutto questo nel linguaggio della meccanica quantistica, osserviamo, che, affinchè le due grandezze (4) e (5) possano avere simultaneamente il valore zero, è necessario che esse siano commutabili, poiché altrimenti il fatto che una delle due grandezze ha un valore determinato renderebbe di necessità indeterminato il valore dell'altra. Ora si verifica facilmente, in base alle ordinarie regole di commutazione, che i primi membri di (4) e (5) sono effettivamente commutabili; si può quindi anche quantisticamente attribuire ad essi allo stesso istante il valore determinato zero.

Al procedimento classico di integrazione delle equazioni canoniche con valori arbitrari delle costanti di integrazione, corrisponde, nella meccanica ondulatoria l'integrazione dell'equazione di Schroedinger corrispondente alla Hamiltoniana (2), scegliendo arbitrariamente la funzione che rappresenta lo scalare di campo

$$\psi = \psi(t, \boldsymbol{x}_i, \boldsymbol{\sigma}_i, w_{s1}, w_{s2}, \chi_s, Q_s) \tag{6}$$

($\boldsymbol{\sigma}_i$ rappresenta simbolicamente la coordinata interna « spincoordinate » dell'i.esimo corpuscolo) per il valore $t = 0$ del tempo. Se vogliamo invece soddisfare le condizioni (4) e (5) non possiamo più lasciare arbitraria questa funzione; resta invece determinato il modo secondo cui essa dipende dalle variabili χ_s e Q_s. Siccome infatti $\tilde{\omega}_s$, coniugata di χ_s, deve avere, secondo la (5) il valore

$$\tilde{\omega}_s = 2\,\pi\nu_s Q_s - \frac{c}{2\,\pi\nu_s}\sqrt{\frac{8\,\pi}{\Omega}}\sum_i e_i \cos\Gamma_{si}$$

risulta che ψ deve dipendere da χ_s nel fattore

$$e^{\frac{2\pi i}{h}\chi_s\left(2\,\pi\nu_s Q_s - \frac{c}{2\,\pi\nu_s}\sqrt{\frac{8\pi}{\Omega}}\sum_i e_i \cos\Gamma_{si}\right)}. \tag{7}$$

Dalla (4) segue in modo simile che Q_s deve intervenire soltanto nel fattore

$$e^{\frac{2\pi i}{h}2\,\pi\nu_s Q_s \chi_s} \tag{8}$$

il quale è del resto già contenuto nel fattore (7). In conclusione la soluzione corrispondente alle condizioni (4) e (5) deve avere la forma:

$$\psi = \left[\prod_s e^{\frac{2\pi i}{h}\chi_s\left(2\,\pi\nu_s Q_s - \frac{c}{2\,\pi\nu_s}\sqrt{\frac{8\pi}{\Omega}}\sum_i e_i \cos\Gamma_{si}\right)}\right]\varphi(t, \boldsymbol{x}_i, \boldsymbol{\sigma}_i, w_{s1}, w_{s2}). \tag{9}$$

Dobbiamo ora dimostrare che effettivamente si può soddisfare l'equazione di Schroedinger per mezzo della posizione (9). L'equazione di Schroedinger dedotta dall'Hamiltoniana (2) è

$$-\frac{h}{2\,\pi i}\frac{\partial\psi}{\partial t} = H\psi \tag{10}$$

dove H è naturalmente interpretato come operatore. Sostituendo nella (10) al posto di ψ l'espressione (9) si trova, con calcoli non difficili la seguente equazione a cui deve soddisfare φ:

$$-\frac{h}{2\,\pi i}\frac{\partial\varphi}{\partial t} = R\varphi \tag{11}$$

dove R rappresenta il seguente operatore.

$$(12)\quad R = -c\sum_i \boldsymbol{\gamma}_i \cdot \boldsymbol{p}_i - \sum_i \delta_i m_i c^2 + \sum_i e_i c \sqrt{\frac{8\pi}{\Omega}} \sum_s \boldsymbol{\gamma}_i \cdot (\mathbf{A}_{s1} w_{s1} + \mathbf{A}_{s2} w_{s2}) \sin \Gamma_{si}$$
$$+ \sum_s \left[\frac{1}{2}(\omega_{s1}^2 + \omega_{s2}^2) + 2\pi^2 \nu_s^2 (w_{s1}^2 + w_{s2}^2)\right] + \frac{c^2}{\pi\Omega} \sum_s \frac{1}{\nu_s^2} \left(\sum_i e_i \cos \Gamma_{si}\right)^2 .$$

A prescindere dall'ultimo termine, R coincide con l'Hamiltoniana della teoria dell'irradiazione di Dirac, in cui si trascura il potenziale scalare e la componente longitudinale del potenziale vettore, considerando solo il campo determinato delle componenti trasversali del potenziale vettore, e cioè il solo campo di radiazione. Dobbiamo discutere il significato dell'ultimo termine. Per questo lo trasformiamo nel modo seguente:

$$\frac{c^2}{\pi\Omega} \sum_s \frac{1}{\nu_s^2} \left(\sum_i e_i \cos \Gamma_{si}\right)^2 = \frac{e^2}{\pi\Omega} \sum_{sij} e_i e_j \frac{\cos \Gamma_{si} \cos \Gamma_{sj}}{\nu_s^2}$$
$$= \frac{c^2}{\pi\Omega} \sum_{ij} e_i e_j \sum_s \frac{\cos \Gamma_{si} \cos \Gamma_{sj}}{\nu_s^2} .$$

La somma rispetto ad s si può trasformare in un integrale e si trova, con calcoli privi di difficoltà

$$\sum_s \frac{\cos \Gamma_{si} \cos \Gamma_{sj}}{\nu_s^2} = \frac{\pi\Omega}{2c^2} \frac{1}{r_{ij}}$$

dove r_{ij} rappresenta la distanza tra i due punti i e j. Sostituendo troviamo

$$(13)\quad \frac{c^2}{\pi\Omega} \sum_s \frac{1}{\nu_s^2} \left(\sum_s e_i \cos \Gamma_{si}\right)^2 = \frac{1}{2} \sum_{ij} \frac{e_i e_j}{r_{ij}} .$$

La (13) ci dà dunque semplicemente la ordinaria espressione dell'energia elettrostatica; come nell'elettrostatica classica, l'espressione (13) diventa infinita nel caso di cariche elettriche puntiformi. Questo inconveniente, più che dalla elettrodinamica, deriva dalla imperfetta conoscenza della struttura dell'elettrone e potrebbe per esempio venir eliminato considerando elettroni di raggio finito. Noi lo elimineremo formalmente, come si suol fare anche nell'elettrostatica classica, escludendo dalla somma (13) i termini per cui $i = j$ che rappresentano in certo modo una costante addittiva infinitamente grande. Indicheremo ciò con un'apice al segno $\sum$. Per mezzo della (13), la (12) diventa:

$$(14)\quad R = -c\sum_i \boldsymbol{\gamma}_i \cdot \boldsymbol{p}_i - \sum_i \delta_i m_i c^2 + \sum_i e_i c \sqrt{\frac{8\pi}{\Omega}} \sum_s \boldsymbol{\gamma}_i \cdot (\mathbf{A}_{s1} w_{s1} + \mathbf{A}_{s2} w_{s2}) \sin \Gamma_{si}$$
$$+ \sum_s \left[\frac{1}{2}(\omega_{s1}^2 + \omega_{s2}^2) + 2\pi^2 \nu_s^2 (w_{s1}^2 + w_{s2}^2)\right] + \frac{1}{2} \sum_{ij}{}' \frac{e_i e_j}{r_{ij}} .$$

Osserviamo infine che la funzione φ che abbiamo sostituito allo scalare di campo ψ per mezzo della (9) può in tutte le considerazioni sostituirsi ad esso, da cui differisce per un fattore complesso di modulo 1. La φ soddisfa all'equazione (11) che è del tipo di una equazione di Schroedinger in cui però si deve prendere come Hamiltoniana R invece di H. Questa nuova Hamiltoniana R, come si legge immediatamente dalla (14) è costituita semplicemente aggiungendo l'ordinaria espressione dell'energia elettrostatica alla ordinaria Hamiltoniana dei termini di pura radiazione. Possiamo dunque concludere che, in questa forma, il problema di elettrodinamica quantistica non è in alcun modo più complicato di un ordinario problema di teoria della radiazione.

Naturalmente, come già abbiamo accennato, anche questa teoria conserva in sè due difetti fondamentali che però più che di orgine elettrodinamica, possono considerarsi derivanti dalla non completa conoscenza della struttura elettronica. Essi sono la possibilità che ha l'elettrone di Dirac di passare a livelli energetici con energia negativa ed il fatto che l'energia intrinseca ha valore infinito se si ammette l'elettrone esattamente puntiforme.

N° 65.

For the introduction to this paper see N° 50.

65.

LE MASSE ELETTROMAGNETICHE NELLA ELETTRODINAMICA QUANTISTICA

« Nuovo Cimento », *8*, 121–132 (1931).

SUNTO. — In questo lavoro viene discusso il problema delle masse elettromagnetiche nella elettrodinamica quantistica per mezzo di ipotesi che equivalgono sostanzialmente ad attribuire all'elettrone un raggio finito. Vengono in particolare illustrate alcune caratteristiche differenze tra il caso classico e quello quantistico.

§ 1. – L'energia del campo elettrostatico prodotto da una carica elettrica puntiforme è, come è noto, infinita; per conseguenza nella teoria elettrodinamica ordinaria la massa elettromagnetica di una carica elettrica esattamente puntiforme risulta essa pure infinita. Si è per questo assai spesso pensato che i corpuscoli elettrici (elettroni o protoni) non si possano a tutto rigore considerare come cariche puntiformi, ma che essi abbiamo una struttura di estensione finita. Una difficoltà in tutto simile a quella della elettrodinamica classica si incontra anche nella elettrodinamica quantistica, poiché, anche secondo questa teoria, una carica elettrica puntiforme viene necessariamente ad essere associata ad una energia infinita; né è possibile eliminare questa difficoltà alterando il valore dell'energia per una costante additiva infinita, poiché, come hanno mostrato Oppenheimer [1] e Waller [2], il termine infinito non è costante, ma varia al variare degli stati quantici; ed anche le sue variazioni sono infinite. Ciò che ha del resto il suo riscontro classico nel fatto che se consideriamo una carica elettrica puntiforme in moto,non solo è infinita l'energia elettrica, ma anche l'energia magnetica, che dipende dalla velocità e corrisponde, in prima approssimazione, all'energia cinetica elettromagnetica.

L'artificio che si presenta per primo per cercare di evitare questi inconvenienti, consiste nel tentar di introdurre anche nell'elettrodinamica quantistica qualche cosa di equivalente ad un raggio finito dell'elettrone. Se si cerca di far questo però, ci si trova di fronte a difficoltà assai maggiori di quelle che si incontrano nel corrispondente problema classico; così che non sembra possibile, almeno con lo schema attuale dell'elettrodinamica quan-

(1) J. OPPENHEIMER, « Phys. Rev. », *35*, 461 (1930).
(2) I. WALLER, « ZS. f. Phys. », *62*, 673 (1930).

tistica, fare la teoria di un elettrone di raggio finito, conservando la invarianza relativistica. Ha tuttavia un certo interesse studiare fino a che punto può costruirsi una teoria delle masse elettromagnetiche, poiché in essa si presentano alcune caratteristiche differenze dalla teoria classica, allo studio delle quali è principalmente dedicato il presente lavoro [(3)].

Il metodo che seguiremo in questo studio è quello da me esposto in una recente Nota [(4)], alla quale rimandiamo anche per le notazioni: esso consiste sostanzialmente nel considerare come variabili dinamiche che caratterizzano il campo elettromagnetico i coefficienti dello sviluppo in componenti armoniche dei potenziali scalare e vettore a un certo istante.

§ 2. – Per introdurre in un modo analiticamente semplice l'ipotesi che l'elettrone abbia estensione finita, osserviamo che, se così è, le varie parti dell'elettrone verranno a trovarsi in punti di egual fase, rispetto ad onde di lunghezza d'onda grande in confronto alle dimensioni elettroniche; mentre onde di lunghezza d'onda dell'ordine delle dimensioni elettroniche, o più piccola, avranno fase differente nei diversi punti dell'elettrone. Si comprende allora molto facilmente, e calcoli immediati lo confermano, che l'elettrone verrà ad interagire tanto meno intensamente con le diverse onde, quanto più piccola è la loro lunghezza d'onda, poiché, per onde assai brevi, le diverse parti dell'elettrone agiscono in disaccordo.

L'effetto di ciò può rappresentarsi dicendo che l'elettrone interagisce con le componenti armoniche di alta frequenza del campo come se esso avesse una carica elettrica efficace più piccola di quella che determina l'interazione con le componenti armoniche di bassa frequenza, e che naturalmente coincide con l'ordinario valore della carica elettronica quale è dato da misure statiche o quasi statiche. Ben inteso che queste variazioni della carica efficace non sono sensibili ancora per nessuna delle radiazioni che si possono produrre in laboratorio; esse acquisterebbero importanza soltanto per lunghezze d'onda, che fossero piccole frazioni di unità X.

Noi prenderemo dunque come Hamiltoniana per il sistema costituito dal campo elettromagnetico e dalla radiazione la stessa espressione usata nella nota A (formula (2)) in cui però al posto di e_i (carica della i–esima particella) porremo $e_i(\nu_s)$ o anche brevemente e_{is}; ν_s rappresenta la frequenza della componente armonica del campo con cui la particella interagisce. Ci limiteremo inoltre per semplicità al caso che si abbia nel campo un solo elettrone, e scriveremo perciò, come Hamiltoniana complessiva del campo e dell'elettrone

$$\text{(1)}\qquad \begin{aligned} H = &-c\boldsymbol{\gamma}\cdot\boldsymbol{p} - \delta mc^2 + c\sqrt{\frac{8\pi}{\Omega}}\sum_s e_s Q_s \cos\Gamma_s \\ &+ c\sqrt{\frac{8\pi}{\Omega}}\sum_s e_s \boldsymbol{\gamma}\cdot(\boldsymbol{\alpha}_s \chi_s + \mathbf{A}_{s1} w_{s1} + \mathbf{A}_{s2} w_{s2}) \sin\Gamma_s \\ &+ \sum_s \left\{ \frac{1}{2}(\omega_{s1}^2 + \omega_{s2}^2 + \tilde{\omega}_s^2 - P_s^2) + 2\pi^2\nu_s^2 (w_{s1}^2 + w_{s2}^2 + \chi_s^2 - Q_s^2) \right\}. \end{aligned}$$

(3) Cfr. anche W. HEISENBERG, « ZS. f. Phys. », *65*, 4 (1930).

(4) E. FERMI, « Rend. Lincei », *12*, 431 (1930). Citato nel seguito con A.

È facile verificare che, anche nell'ipotesi qui fatta, che l'elettrone interagisca con una diversa carica elettrica con le differenti componenti armoniche, a seconda della loro frequenza, è possibile ridurre l'autofunzione ad una forma in tutto analoga alla ((9) *A*). Ciò si fa in modo identico a quello seguito nella Nota *A* tenendo conto delle due relazioni

$$2\pi\nu_s\chi_s - P_s = 0\,, \tag{2}$$

e

$$\tilde{\omega}_s - 2\pi\nu_s Q_s + \frac{c}{2\pi\nu_s}\sqrt{\frac{8\pi}{\Omega}}\, e_s \cos\Gamma_s = 0\,, \tag{3}$$

che esprimono che è valida in ogni istante la relazione

$$\frac{1}{c}\frac{\partial V}{\partial t} + \operatorname{div}\mathbf{U} = 0\,, \tag{4}$$

che collega tra di loro i potenziali scalare e vettore. Per mezzo delle (2) e (3) resta fissato il modo con cui la ampiezza di probabilità $\psi(t, \boldsymbol{x}, \boldsymbol{\sigma}, w_{s1}, w_{s2}, \chi_s, Q_s)$ dipende dalle variabili χ_s e Q_s; si trova precisamente, come nella Nota *A*,

$$\psi = e^{\frac{2\pi i}{h}\sum_s \chi_s\left(2\pi\nu_s Q_s - \frac{c}{2\pi\nu_s}\sqrt{\frac{8\pi}{\Omega}}\, e_s\cos\Gamma_s\right)}\varphi(t, \boldsymbol{x}, \boldsymbol{\sigma}, w_{s1}, w_{s2})\,. \tag{5}$$

La funzione φ, che può, in qualche modo, considerarsi ancora come una ampiezza di probabilità, soddisfa ad una equazione del tipo

$$-\frac{h}{2\pi i}\frac{\partial\varphi}{\partial t} = R\varphi\,, \tag{6}$$

dove R, come nella A, rappresenta l'operatore

$$\begin{aligned} R = {} & -c\gamma\cdot\boldsymbol{p} - \delta mc^2 + c\sqrt{\frac{8\pi}{\Omega}}\sum_r e_r(\gamma\cdot\mathbf{A}_r)\,w_r\sin\Gamma_r \\ & + \sum_r\left(\frac{1}{2}\omega_r^2 + 2\pi^2\nu_r^2 w_r^2\right) + \frac{c^2}{\pi\Omega}\sum_s\frac{1}{\nu_s^2}\,e_s^2\cos^2\Gamma_s\,. \end{aligned} \tag{7}$$

In questa espressione di R è stata introdotta una piccola semplificazione rispetto alle notazioni della Nota *A*. Invece di considerare singolarmente le due coordinate w_{s1} e w_{s2} le abbiamo indicate con un solo indice, w_r. Naturalmente il nuovo indice r assume un numero di valori doppio di quelli del corrispondente indice s, date le due possibilità di polarizzazione per le componenti trasversali dell'onda.

§ 3. – Dobbiamo prima di tutto ricercare il significato fisico dell'ultimo termine di (7) per far vedere che esso corrisponde all'energia elettrostatica intrinseca dell'elettrone; tale energia non risulta più necessariamente infinita con le nostre ipotesi che equivalgono sostanzialmente ad attribuire all'elettrone un raggio finito.

La somma che figura nell'ultimo termine di (7) può trasformarsi in un integrale; basta per questo sostituire $\cos^2\Gamma_s$ col suo valor medio 1/2; esprimere

e_s come funzione della frequenza ν_s, col porre $e_s = e(\nu_s)$; sostituire infine $\sum_s$ con $\int_0^\infty \frac{4\pi\Omega}{c^3} \nu_s^2 \, d\nu_s$ dove $\frac{4\pi\Omega}{c^3} \nu_s^2 \, d\nu_s$ rappresenta il numero delle frequenze comprese tra ν_s e $\nu_s + d\nu_s$, per le onde longitudinali (indice s e non r). Si trova così:

$$\frac{c^2}{\pi\Omega} \sum_s \frac{1}{\nu_s^2} e_s^2 \cos^2 \Gamma_s = \frac{2}{c} \int_0^\infty e^2(\nu_s) \, d\nu \, . \tag{8}$$

Si dimostra senza difficoltà che per esempio ad una distribuzione sferica superficiale di carica elettrica corrisponde

$$e(\nu) = \frac{ce}{2\pi\nu R} \sin \frac{2\pi\nu R}{c} \tag{9}$$

Sostituendo questa espressione nella (8) si trova

$$\frac{2}{c} \int_0^\infty e^2(\nu) \, d\nu = \frac{e^2}{2R} \, , \tag{10}$$

che rappresenta appunto l'energia elettrostatica della distribuzione di elettricità considerata. Del resto l'identità tra l'espressione (8) e l'energia elettrostatica si dimostra senza difficoltà anche per il caso di una distribuzione di elettricità più complicata. Naturalmente il termine (8) diventa infinito per una carica elettrica puntiforme, per cui $e(\nu)$ non dipende dalla frequenza.

La presenza del termine (8) nella Hamiltoniana R, ed il suo significato fisico che abbiamo ora discusso, non debbono naturalmente farci credere che l'energia elettrostatica, divisa per c^2, rappresenti la massa elettromagnetica anche nell'elettrodinamica quantistica, come si trova in quella classica; ci resta infatti ancora da considerare l'effetto di un altro termine di R che dipende dalla presenza della carica elettrica; precisamente

$$c \sqrt{\frac{8\pi}{\Omega}} \sum_r e_r (\boldsymbol{\gamma} \cdot \mathbf{A}_r) \, w_r \sin \Gamma_r \, . \tag{11}$$

Prima di poter studiare l'azione di questo termine dobbiamo cercare alcune semplificazioni del nostro problema dipendenti dall'esistenza dell'integrale della quantità di moto.

§ 4. – Dalla forma (5) dell'autofunzione appare evidente che la quantità di moto $\boldsymbol{p}$ del corpuscolo non può mai assumere un valore determinato; sarebbe perciò necessario che ψ dipendesse dalle coordinate di posizione nella forma $e^{\frac{2\pi i}{h} \boldsymbol{p} \cdot \boldsymbol{x}}$, ciò che è escluso dall'espressione (5) della ψ. Che del resto $\boldsymbol{p}$ non possa assumere un valore determinato si riconosce anche dal fatto che $\boldsymbol{p}$ non commuta col primo membro della (3), che ha il valore determinato zero. Si può anche riconoscere facilmente la causa fisica di questa indeterminazione di $\boldsymbol{p}$; infatti l'elettrone è circondato da un campo elettromagnetico che lo accompagna nel suo movimento e che viene in qualche modo a costituire una parte integrante dell'elettrone stesso. Come quantità di moto dell'elettrone bisogna perciò necessariamente considerare anche il contributo di questo campo. Per ottenere la quantità di moto totale del sistema bisogna aggiungere anche la quantità di moto elettromagnetica del campo di radiazione. Tale quantità di moto totale dovrà naturalmente essere costante, e quindi commutare con H; siccome essa deve poi essere suscettibile di valori determinati, essa dovrà anche commutare con i primi membri di (2) e (3).

Si verifica senza difficoltà che la quantità di moto totale è rappresentata dal vettore seguente

$$\Pi = \boldsymbol{p} - \sqrt{\frac{8\pi}{\Omega}} \sum_s e_s \chi_s \boldsymbol{\alpha}_s \sin \Gamma_s + \frac{2\pi}{c} \sum_j \nu_j \boldsymbol{\alpha}_j (w_j w_{j^*} - w_j w_{j^*}) , \tag{12}$$

il quale soddisfa alle condizioni precedentemente indicate. L'ultimo termine di (12) rappresenta la quantità di moto della radiazione. In esso sono state usate le notazioni seguenti. L'indice j, rispetto al quale si effettua la somma, rappresenta soltanto metà delle componenti trasversali r della radiazione; l'altra metà è rappresentata dall'indice che abbiamo indicato con j^*. Per ogni componente j esiste una corrispondente j^* correlata ad essa dalle relazioni seguenti:

$$\nu_j = \nu_{j^*} \quad ; \quad \boldsymbol{\alpha}_j = \boldsymbol{\alpha}_{j^*} \quad ; \quad \mathbf{A}_j = \mathbf{A}_{j^*} \quad ; \quad \cos \Gamma_j = \sin \Gamma_{j^*} . \tag{13}$$

Per modo che le due componenti j e j^* hanno eguali le frequenze, le direzioni di propagazione e di polarizzazione e sono spostate di fase una rispetto all'altra di un quarto di periodo.

La verifica che il vettore Π commuta con i primi membri di (2) e (3) è immediata. Per far vedere anche che esso è costante, e quindi commuta con H, conviene scrivere H nella forma

$$\begin{aligned} H = & - c\boldsymbol{\gamma}\cdot\boldsymbol{p} - \delta mc^2 + c\sqrt{\frac{8\pi}{\Omega}} \sum_s e_s (Q_s \cos \Gamma_s + \boldsymbol{\gamma}\cdot\boldsymbol{\alpha}_s \chi_s \sin \Gamma_s) \\ & + c\sqrt{\frac{8\pi}{\Omega}} \sum_j e_j \boldsymbol{\gamma}\cdot\mathbf{A}_j (w_j \sin \Gamma_j + w_{j^*} \cos \Gamma_j) \\ & + \sum_j \left[\frac{w_j^2 + w_{j^*}^2}{2} + 2\pi^2 \nu_j^2 (w_j^2 + w_{j^*}^2)\right] + \sum_s \left[\frac{\tilde{\omega}_s^2 - P_s^2}{2} + 2\pi^2 \nu_s^2 (\chi_s^2 - Q_s^2)\right]. \end{aligned} \tag{14}$$

Si ottiene allora immediatamente la relazione

$$\Pi H - H \Pi = 0 , \tag{15}$$

tenendo conto di (2) e di (3) ed osservando che evidentemente

$$\frac{8\pi}{\Omega} \sum_s \frac{ce_s^2}{2\pi\nu_s} \boldsymbol{\alpha}_s \sin \Gamma_s \cos \Gamma_s = 0 ,$$

data l'isotropia della distribuzione delle componenti armoniche della radiazione.

§ 5. – Nel caso che Π abbia valore numerico determinato, si possono facilmente eliminare dall'ampiezza di probabilità le coordinate di posizione, con un procedimento simile a quello già usato per eliminare le coordinate Q_s e χ_s, per mezzo delle relazioni (2) e (3). Questa eliminazione è stata anche fatta da Heisenberg, per il suo caso, nel citato lavoro.

Osserviamo che se si attribuisce al primo membro della (12) un valore determinato (vettoriale) l'ampiezza di probabilità deve soddisfare l'equazione differenziale che si ottiene eguagliando $\Pi\psi$ all'operatore che figura nel secondo membro di (12) applicato a ψ. Tenendo conto della (5) questa equazione si trasforma nella seguente equazione per la φ:

$$\frac{2\pi i}{h}\Pi\varphi = \operatorname{grad}\varphi - \sum_j \frac{2\pi\nu_j}{c}\boldsymbol{\alpha}_j\left(w_j\frac{\partial\varphi}{\partial w_{j^*}} - w_{j^*}\frac{\partial\varphi}{\partial w_j}\right). \tag{15}$$

L'integrale generale di questa equazione è:

$$\varphi = e^{\frac{2\pi i}{h}\Pi\cdot\boldsymbol{x}}\, F\left(t\,,\boldsymbol{\sigma}\,,\sqrt{w_j^2+w_{j^*}^2}\,,\Gamma_j - \operatorname{arctg}\frac{w_j}{w_{j^*}}\right) = e^{\frac{2\pi i}{h}\Pi\cdot\boldsymbol{x}}\, F(t\,,\boldsymbol{\sigma}\,,k_j\,\vartheta_j) \tag{16}$$

dove si è posto

$$w_j = k_j\sin(\Gamma_j - \vartheta_j) \quad ; \quad w_{j^*} = k_j\cos(\Gamma_j - \vartheta_j)\,. \tag{17}$$

L'equazione differenziale a cui soddisfa la F si trova sostituendo in (6) l'espressione (16), ed osservando che l'operatore R, dato da (7) deve scriversi, tenendo conto del significato dell'indice j,

$$\begin{aligned} R = {} & c\boldsymbol{\gamma}\cdot\boldsymbol{p} - \delta mc^2 + c\sqrt{\frac{8\pi}{\Omega}}\sum_j e_j(\boldsymbol{\gamma}\cdot\mathbf{A}_j)(w_j\sin\Gamma_j + w_{j^*}\cos\Gamma_j) \\ & + \frac{1}{2}\sum_j\left[\frac{\omega_j^2+\omega_{j^*}^2}{2} + 2\pi^2\nu_j^2(w_j^2+w_{j^*}^2)\right] + \frac{2}{c}\int_0^\infty e^2(\nu)\,d\nu\,. \end{aligned} \tag{18}$$

Tenendo conto delle (17) si trova allora con calcoli privi di difficoltà:

$$-\frac{h}{2\pi i}\frac{\partial F}{\partial t} = GF\,, \tag{19}$$

dove G rappresenta l'operatore seguente

$$\begin{aligned} G = {} & -c\Pi\cdot\boldsymbol{\gamma} - \delta mc^2 + \sum_j ih\nu_j(\boldsymbol{\gamma}\cdot\boldsymbol{\alpha}_j)\frac{\partial}{\partial\vartheta_j} \\ & + \sum_j\left\{2\pi^2\nu_j^2k_j^2 - \frac{h^2}{8\pi^2}\left(\frac{\partial^2}{\partial k_j^2} + \frac{1}{k_j}\frac{\partial}{\partial k_j} + \frac{1}{k_j^2}\frac{\partial^2}{\partial\vartheta_j^2}\right)\right\} \\ & + c\sqrt{\frac{8\pi}{\Omega}}\sum_j e_j(\boldsymbol{\gamma}\cdot\mathbf{A}_j)\,k_j\cos\vartheta_j + \frac{2}{c}\int_0^\infty e^2(\nu)\,d\nu\,. \end{aligned} \tag{20}$$

Se in luogo di k_j e ϑ_j si introducono come variabili

$$\xi_j = k_j\cos\vartheta_j \quad ; \quad \eta_j = k_j\sin\vartheta_j\,, \tag{21}$$

l'operatore G prende la forma

$$G = -c\Pi\cdot\gamma - mc^2\,\delta + \sum_j ih\nu_j\,(\gamma\cdot\alpha_j)\left(\xi_j\frac{\partial}{\partial\eta_j} - \eta_j\frac{\partial}{\partial\xi_j}\right) \tag{22}$$

$$+ \sum_j\left\{2\pi^2\nu_j^2(\xi_j^2+\eta_j^2) - \frac{h^2}{8\pi^2}\left(\frac{\partial^2}{\partial\xi_j^2}+\frac{\partial^2}{\partial\eta_j^2}\right)\right\}$$

$$+ c\sqrt{\frac{8\pi}{\Omega}}\sum_j e_j\,(\gamma\cdot\mathbf{A}_j)\,\xi_j + \frac{2}{c}\int_0^\infty e^2(\nu)\,d\nu\,,$$

in cui è messa in evidenza la quantità di moto della radiazione (terzo termine) e la energia della radiazione (quarto termine).

§ 6. – Una ulteriore semplificazione dell'equazione (19) si può ottenere se facciamo l'ipotesi che il corpuscolo non abbia massa materiale, e che tutta la sua massa sia di origine elettromagnetica. Dovremo in questo caso porre nella (20) e nella (22) $m = 0$. Si verifica allora immediatamente che la grandezza

$$i\gamma_x\,\gamma_y\,\gamma_z\,, \tag{23}$$

commuta con G, e può considerarsi come un integrale suscettibile di valori numerici costanti. Elevandola al quadrato, e tenendo conto delle proprietà delle γ, si trova $(i\gamma_x\,\gamma_y\,\gamma_z)^2 = 1$; gli autovalori di $i\gamma_x\,\gamma_y\,\gamma_z$ sono dunque $+1$ e -1. Questo fatto ci permette di semplificare il nostro problema riducendoci da una teoria a quattro componenti, in cui per γ_x, γ_y e γ_z, si debbono prendere le matrici di Dirac, ad una teoria a due componenti in cui si debbono usare le matrici di Pauli. Più precisamente si trova senza difficoltà che se

$$i\gamma_x\,\gamma_y\,\gamma_z = 1\,,$$

deve prendersi

$$\gamma_x = \begin{vmatrix} 0 & 1 \\ 1 & 0 \end{vmatrix}\,, \quad \gamma_y = \begin{vmatrix} 0 & -i \\ i & 0 \end{vmatrix}\,, \quad \gamma_z = \begin{vmatrix} 1 & 0 \\ 0 & -1 \end{vmatrix}. \tag{24}$$

Se invece

$$i\gamma_x\,\gamma_y\,\gamma_z = -1\,,$$

deve prendersi

$$\gamma_x = \begin{vmatrix} 0 & -1 \\ -1 & 0 \end{vmatrix}\,, \quad \gamma_y = \begin{vmatrix} 0 & -i \\ i & 0 \end{vmatrix}\,, \quad \gamma_z = \begin{vmatrix} 1 & 0 \\ 0 & -1 \end{vmatrix}. \tag{25}$$

Si riconosce subito che questi due casi non differiscono sostanzialmente tra di loro, e corrispondono semplicemente alle due possibili orientazioni dello spin. Noi supporremo perciò che le γ abbiano l'espressione (24).

L'operatore G si scrive ora, facendo in esso $m = 0$ e prendendo l'asse delle z nella direzione della quantità di moto totale

$$G = -c\,\Pi\gamma_z + \sum_j ih\nu_j\,(\alpha_j\cdot\gamma)\,\frac{\partial}{\partial\vartheta_j} + \sum_j\left(2\pi^2\nu_j^2 k_j^2 - \frac{h^2}{8\pi^2}\Delta_j\right) \tag{26}$$

$$+ c\sqrt{\frac{8\pi}{\Omega}}\sum_j e_j\,(\mathbf{A}_j\cdot\gamma)\,\xi_j + \frac{2}{c}\int_0^\infty e^2(\nu)\,d\nu\,,$$

ove le γ rappresentano gli operatori (24).

§ 7. - Il problema della determinazione della massa elettromagnetica del corpuscolo resta così ridotto matematicamente alla ricerca degli autovalori dell'operatore (26). Questi dipendono naturalmente dal valore della quantità di moto Π, che figura in (26) come un parametro; per modo che 'energia viene ad essere una funzione della quantità di moto. Confrontando questa relazione, con l'espressione relativistiva dell'energia cinetica W in funzione della quantità di moto

$$W = \sqrt{c^2 \Pi^2 + M^2 c^4}, \tag{27}$$

si dovrebbe trovare la massa elettromagnetica M. Naturalmente si deve in ciò tener presente che la quantità di moto Π si può in genere interpretare come somma della quantità di moto dei quanti di radiazione eventualmente presenti nel campo elettromagnetico. Perché Π rappresenti la quantità di moto del corpuscolo soltanto, occorre cercare quelle soluzioni che corrispondono all'assenza di radiazione.

La ricerca esatta degli autovalori di G sembra essere connessa con considerevoli difficoltà matematiche, che per il momento non mi è riuscito di superare. L'unica via per cercare di formarsi una idea sulla loro struttura consiste perciò nella applicazione del metodo delle perturbazioni. Precisamente si spezza la Hamiltoniana G nella somma

$$G = G_0 + g, \tag{28}$$

di una Hamiltoniana imperturbata

$$G_0 = -c\,\Pi\gamma_z + \sum_j ih\nu_j\,(\alpha_j \cdot \gamma)\,\frac{\partial}{\partial \vartheta_j} + \sum_j \left(2\pi^2 \nu_j^2 k_j^2 - \frac{h^2}{8\pi^2}\Delta_j\right) + \frac{2}{c}\int_0^\infty e^2(\nu)\,d\nu, \tag{29}$$

e di un termine di perturbazione

$$g = c\sqrt{\frac{8\pi}{\Omega}} \sum_j e_j\,(\mathbf{A}_j \cdot \gamma)\,\xi_j. \tag{30}$$

Perché il metodo delle perturbazioni dia dei risultati attendibili, sarebbe naturalmente necessario che la perturbazione fosse piccola in confronto della separazione degli autovalori del problema imperturbato. Ora si rinonosce facilmente che questo non è il caso per il nostro problema, per modo che il metodo non può dare nel nostro caso nulla di più che qualche indicazione sopra il tipo di soluzione, da intendersi però con le debite riserve.

Ciò nonostante riporterò qui il risultato che si ottiene col metodo delle pertubazioni, poiché esso sembra indicare alcune particolarità della trattazione quantistica del problema delle masse elettromagnetiche essenzialmente diverse da quelle che si osservano nel caso classico.

Come soluzione di approssimazione zero dobbiamo prendere quella in cui tutte le componenti della radiazione hanno il numero quantico zero, poiché dobbiamo cercare una soluzione in cui non ci sia energia nel campo di radiazione (all'infuori, ben inteso, della energia di zero). Per questa soluzione particolare si ha evidentemente

$$\frac{\partial}{\partial \vartheta_j} = 0,$$

ciò che significa che, in approssimazione zero, è nulla la quantità di moto elettromagnetica. Naturalmente la soluzione è costituita da due funzioni, corrispondenti alle due componenti delle matrici di Pauli.

In approssimazione zero si può prendere eguale a zero la seconda di queste componenti: si ottiene così $\gamma_z = -1$. Per questo modo che, in approssimazione zero, l'energia risulta data, a meno della energia di zero (Nullpunkts-energie) e dell'energia elettrostatica $\frac{2}{c}\int_0^\infty e^2(\nu)\,d\nu$, da $c\Pi$, che coincide col valore dell'energia cinetica W che si ottiene dalla (27) quando, in approssimazione zero, si ponga in essa $M = 0$.

È facile vedere che la perturbazione di prima approssimazione dell'energia è nulla; essa è data infatti dal valor medio del termine di perturbazione g sopra l'autofunzione di approssimazione zero. Ora le autofunzioni dell'oscillatore di radiazione, avendo numero quantico zero, sono funzioni pari, per modo che per esse si ha $\xi_j = 0$, da cui si deduce $\bar{g} = 0$. Occorre quindi, per ottenere il primo termine di correzione dell'energia non nullo, procedere alla seconda approssimazione. Il termine correttivo di seconda approssimazione dell'energia è dato, secondo le formule generali della teoria delle perturbazioni, tenendo conto che la perturbazione di prima approssimazione dell'energia è, per il nostro caso, nulla, da

$$W_2 = \sum_k \frac{|g_{0k}|^2}{E_0 - E_k}.$$

Il suo calcolo è un po' laborioso, ma non presenta difficoltà sostanziali dato che le autofunzioni e gli autovalori di approssimazione zero sono completamente conosciuti. Senza riportare il calcolo per disteso ne indichiamo il risultato:

$$W_2 = \frac{2h}{c^2\Pi}\int_0^\infty e^2(\nu)\,\nu d\nu - \frac{2}{c}\int_0^\infty e^2(\nu)\,d\nu. \tag{31}$$

Aggiungendo l'energia (31) all'energia di approssimazione zero, data da

$$c\Pi + \frac{2}{c}\int_0^\infty e^2(\nu)\,d\nu + \text{energia di zero}, \tag{32}$$

si trova l'espressione dell'energia come funzione della quantità di moto dell'elettrone, in seconda approssimazione. Essa risulta, omettendo la costante additiva inessenziale rappresentata dalla energia di zero della radiazione

$$W = c\Pi + \frac{2h}{c^2\Pi}\int_0^\infty e^2(\nu)\,\nu d\nu. \tag{33}$$

Questa espressione deve essere confrontata con quella che si ottiene sviluppando in serie la (27) rispetto alle potenze di M, e fermandosi ai termini in M^2. Cioè

$$c\Pi + \frac{M^2 c^3}{2\Pi}. \tag{34}$$

Dal confronto risulta

$$\text{(35)} \qquad M^2 = \frac{4h}{c^5} \int_0^\infty e^2(\nu)\, \nu d\nu .$$

Ossia

$$\text{(36)} \qquad M = \sqrt{\frac{4h}{c^5} \int_0^\infty e^2(\nu)\, \nu d\nu} ,$$

come espressione della massa elettromagnetica quantistica nella nostra approssimazione.

Da questi risultati, che, per le ragioni che abbiamo già esposto, debbono tuttavia venir considerati con certe riserve, sembrano dedursi le conclusioni seguenti.

a) La massa elettromagnetica nell'elettrodinamica quantistica non è direttamente dipendente, come nella elettrodinamica classica, dall'energia elettrostatica della distribuzione di carica elettrica che costituisce il corpuscolo. Tale energia elettrostatica anzi viene, in ultima analisi, a sparire dalle formule; essa è contenuta infatti nella energia di approssimazione zero data da (32) (secondo termine). Il secondo termine della correzione dell'energia (31) è però eguale e contrario all'energia elettrostatica; così nella somma (33) questa viene a sparire.

b) L'espressione (36) che abbiamo trovata per la massa elettromagnetica differisce sostanzialmente da quella classica per la presenza del fattore h. Essa anzi si annullerebbe nel caso limite $h = 0$.

Naturalmente queste conclusioni portano in sé tutta l'incertezza del metodo di approssimazione per mezzo del quale esse sono state dedotte; il quale del resto costituisce per il momento l'unico modo per cercar di formarsi una idea della soluzione. Conclusioni assai più sicure si potrebbero raggiungere quando riuscisse di risolvere esattamente il problema della ricerca degli autovalori della Hamiltoniana (26).

N° 67.

For the introduction to this paper see N° 50.

67.

QUANTUM THEORY OF RADIATION (*)

« Rev. Mod. Phys. », *4*, 87–132 (1932).

INTRODUCTION.

Until a few years ago it had been impossible to construct a theory of radiation which could account satisfactorily both for interference phenomena and the phenomena of emission and absorption of light by matter. The first set of phenomena was interpreted by the wave theory, and the second set by the theory of light quanta. It was not until in 1927 that Dirac succeeded in constructing a quantum theory of radiation which could explain in an unified way both types of phenomena. In this article we shall develop the general formulas of Dirac's theory, and show its applications to several characteristic examples (Part I). In the second part of this work Dirac's relativistic wave equation of the electron will be discussed in relation to the theory of radiation. The third part will be devoted to the problems of the general quantum electrodynamics, and to the difficulties connected with it.

PART I. — DIRAC'S THEORY OF RADIATION

§ 1. – FUNDAMENTAL CONCEPT.

Dirac's theory of radiation is based on a very simple idea; instead of considering an atom and the radiation field with which it interacts as two distinct systems, he treats them as a single system whose energy is the sum of three terms: one representing the energy of the atom, a second representing the electromagnetic energy of the radiation field, and a small term representing the coupling energy of the atom and the radiation field.

If we neglect this last term, the atom and the field could not affect each other in any way; that is, no radiation energy could be either emitted or absorbed by the atom. A very simple example will explain these relations. Let us consider a pendulum which corresponds to the atom, and an oscillating string in the neighborhood of the pendulum which represents the radiation field. If there is no connection between the pendulum and the string, the two

(*) Lectures delivered at the Symposium for Theoretical Physics during the Summer Session of 1930 at the University of Michigan.

systems vibrate quite independently of each other; the energy is in this case simply the sum of the energy of the pendulum and the energy of the string with no interaction term. To obtain a mechanical representation of this term, let us tie the mass M of the pendulum to a point A of the string by means of a very thin and elastic thread *a*. The effect of this thread is to perturb slightly the motion of the string and of the pendulum. Let us suppose for instance that at the time $t = 0$, the string is in vibration and the pendulum is at rest. Through the elastic thread *a* the oscillating string transmits to the pendulum very slight forces having the same periods as the vibrations of the string. If these periods are different from the period of the pendulum, the amplitude of its vibrations remains always exceedingly small; but if a period of the string is equal to the period of the pendulum, there is resonance and the amplitude of vibration of the pendulum becomes considerable after a certain time. This process corresponds to the absorption of radiation by the atom.

If we suppose, on the contrary, that at the time $t = 0$ the pendulum is oscillating and the string is at rest, the inverse phenomenon occurs. The forces transmitted through the elastic thread from the pendulum to the string put the string in vibration; but only the harmonics of the string, whose frequencies are very near the frequency of the pendulum reach a considerable amplitude. This process corresponds to the emission of radiation by the atom.

§ 2. – Analytic representation.

Returning to the case of the atom and the radiation field, the first problem which we have to solve is the finding of a convenient set of coordinates to represent the system. The position of the atom may be described by means of any system of general coordinates; if we assume that the atom contains only one electron, we may choose, for instance, its Cartesian coordinates (and eventually also the spin coordinate). The state of the radiation field could be determined by the values of the components of the electric and the magnetic vectors at any point of the space. We could also represent the field by means of a scalar and a vector potential. In this case we must give at any point of space the values of the scalar potential V and of the three components U_x, U_y, U_z of the vector potential. In this representation the field is described by a continuous infinity of variables, which is very difficult to handle; furthermore such representation is inconvenient because the energy of the field expressed in terms of the variables, contains them in a very mixed form, even if we neglect, in a first approximation, the action of the atom on the field.

For these reasons it is often more convenient to represent the field in the following way. Instead of considering the radiation in infinite space, let us consider the radiation enclosed in a cavity of finite volume Ω with perfectly reflecting walls. If afterwards we let the cavity become infinite in every direction, we shall get as a limit the properties of radiation in free space.

The electromagnetic vibrations in a cavity of finite volume, just as the vibrations of an elastic body of finite volume, may be represented by the superposition of a discreet infinity of fundamental vibrations each one corresponding to a system of standing waves. The number of standing vibrations whose frequency lies between ν and $\nu + d\nu$ is given, for a very large volume Ω, by:

$$dN = \frac{8\pi}{c^3}\Omega\nu^2\, d\nu \tag{1}$$

c being the velocity of light.

It is to be noticed that a radiation field only and not a general electromagnetic field may be represented through a superposition of standing vibrations. The general quantum electrodynamics deals with the quantum theoretical representation of a general electromagnetic field; we shall discuss this theory in Part III of this article. At present we shall limit ourselves to the simple radiation theory, that is, we shall consider quantum theoretically only that part of the electromagnetic field which is responsible for the phenomena of radiation. The radiation field may be then represented as a superposition of ordinary plane electromagnetic waves; whereas for instance the Coulomb forces need a more general representation.

The electromagnetic field of a plane standing wave has a vector potential of the form:

$$\boldsymbol{U} = \boldsymbol{A}\, u(t) \sin\left(\frac{2\pi\nu}{c}(\alpha\cdot\boldsymbol{X}) + \beta\right). \tag{2}$$

The sine factor gives us the amplitude dependence on position; $\boldsymbol{X}$ is a vector with components x, y, z; $\boldsymbol{\alpha}$ is a unit vector giving the direction of the standing wave; $\boldsymbol{A}$ is a unit vector giving the direction of vibration of the electric force; since the wave is transversal, $\boldsymbol{A}$ and $\boldsymbol{\alpha}$ are at right angles. The factor $u(t)$ which gives the dependence on the time, is generally a sine function of t. However, this is not always the case; if there is an atom which either emits or absorbs radiation, the amplitude of the standing vibration may increase or decrease.

Now we represent the radiation field as the superposition of standing waves of the type (2), with frequencies $\nu_1, \nu_2, \cdots, \nu_s$. The number of frequencies lying between ν and $\nu + d\nu$ is given by (1). The directions $\boldsymbol{\alpha}_s$ and $\boldsymbol{A}_s$, of the standing waves and of polarization are distributed at random. We have also:

$$\boldsymbol{U} = \sum_s \boldsymbol{A}_s\, u_s(t) \sin\Gamma_s = \sum_s \boldsymbol{U}_s \tag{3}$$

where:

$$\Gamma_s = \frac{2\pi\nu}{c}(\boldsymbol{\alpha}_s\cdot\boldsymbol{X}) + \beta_s. \tag{4}$$

If there is neither emission nor absorption of radiation, the $u_s(t)$ are sine functions of the time; but in the general case, they may depend on t in any way. Now it is evident that if we know, at a given time t the values of all the u_s the vector potential throughout the space Ω is determined for that instant since this is given by (3). We may therefore take the u_s as coordinates representing the radiation field at any moment.

§ 3. – Electromagnetic energy of radiation field.

We have now to express the electromagnetic energy of the radiation field in terms of the coordinates u_s. The electric and magnetic forces, derived from the potential $\mathbf{U}$, are:

$$\mathbf{E} = -\frac{1}{c}\frac{\partial \mathbf{U}}{\partial t} \quad ; \quad \mathbf{H} = \operatorname{rot} \mathbf{U} .$$

From these equations we get, using (3) and (4),

$$(5) \qquad \begin{cases} \mathbf{E} = -\sum\limits_s \frac{1}{c} \mathbf{A}_s \dot{u}_s \sin \Gamma_s \\ \mathbf{H} = \sum\limits_s \frac{2 \pi \nu_s}{c} (\boldsymbol{\alpha}_s \times \mathbf{A}_s) u_s \cos \Gamma_s . \end{cases}$$

The electromagnetic energy contained in the space Ω is:

$$W_e = \Omega \frac{\overline{\mathbf{E}^2} + \overline{\mathbf{H}^2}}{8 \pi}$$

where the barred expressions represent mean values. We must calculate the space average of $\mathbf{E}^2$ and $\mathbf{H}^2$. It is evident that the mixed terms in the squares have the average zero. Remembering that $\overline{\sin^2 \Gamma_s} = \overline{\cos^2 \Gamma_s} = \frac{1}{2}$ and $(\boldsymbol{\alpha}_s \times \mathbf{A}_s)^2 = 1$, since $\mathbf{A}_s$ and $\boldsymbol{\alpha}_s$ are perpendicular unit vectors, we find

$$\overline{\mathbf{E}^2} = \frac{1}{2 c^2} \sum_s \dot{u}_s^2 \quad ; \quad \overline{\mathbf{H}_s^2} = \sum_s \frac{2 \pi^2 \nu_s^2}{c^2} u_s^2 .$$

Thus the electromagnetic energy is

$$(6) \qquad W_e = \frac{\Omega}{8 \pi c^2} \sum \left(\frac{1}{2} \dot{u}_s^2 + 2 \pi^2 \nu_s^2 u_s^2 \right) .$$

From this expression of the radiation energy we may easily obtain the equations which give the dependence of u_s on the time in Hamiltonian form. We introduce for this purpose a new variable v_s canonically conjugated to u_s, by means of the usual rules

$$v_s = \frac{\partial W_e}{\partial \dot{u}_s} = \frac{\Omega}{8 \pi c^2} \dot{u} .$$

The energy (6) becomes in the Hamiltonian form

$$(7) \qquad W_e = \sum_s \left(\frac{8 \pi c^2}{\Omega} \frac{v_s^2}{2} + \frac{\Omega}{8 \pi c^2} 2 \pi^2 \nu_s^2 u_s^2 \right) .$$

From this Hamiltonian we get the canonical equations

$$(8) \qquad \dot{u}_s = \frac{\partial W_e}{\partial v_s} = \frac{8 \pi c^2}{\Omega} v_s \quad ; \quad \dot{v}_s = -\frac{\partial W_e}{\partial u_s} = -\frac{\Omega}{8 \pi c^2} 4 \pi^2 \nu_s^2 u_s .$$

If we eliminate v_s from these equations, we get

$$(9) \qquad \ddot{u}_s + 4 \pi^2 \nu_s^2 u_s = 0 .$$

We find, as was to be expected, that u_s is a periodic function of the time with the frequency ν_s. We may also say that the canonical Eqs. (8) are equivalent to the Maxwell equations for the vacuum. It is convenient to avoid complicated factors in the Hamilton function (7) changing by constant factors the canonical variables u_s and v_s into two other conjugate variables q_s and p_s given by

$$u_s = \left(\frac{8\pi c^2}{\Omega}\right)^{1/2} q_s \quad ; \quad v_s = \left(\frac{\Omega}{8\pi c^2}\right)^{1/2} p_s . \tag{10}$$

The energy (7) takes now the form:

$$W_e = \sum_s \left(\frac{1}{2} p_s^2 + 2\pi^2 \nu_s^2 q_s^2\right) \tag{11}$$

which is the same as the Hamiltonian of a system of many independent oscillators with mass 1 and frequencies $\nu_1, \nu_2, \cdots, \nu_s$. The vector potential (3) in terms of the new variables q_s, takes the form:

$$\mathbf{U} = c\left(\frac{8\pi}{\Omega}\right)^{1/2} \sum_s \mathbf{A}_s q_s \sin \Gamma_s . \tag{12}$$

§ 4. – Hamiltonian of the atom and the radiation field.

We now must write the Hamiltonian for the atom which, added to (11), shall give us the Hamiltonian of the complex system of the atom and the radiation field.

The Hamiltonian function for an electron may be obtained to a first approximation from the ordinary relativistic Hamilton function for a point charge, that is:

$$0 = -\frac{1}{2m}\left\{\left(mc + \frac{W_a - eV}{c}\right)^2 - \left(\boldsymbol{p} - \frac{e\mathbf{U}}{c}\right)^2\right\} + \frac{mc^2}{2} \tag{13}$$

by neglecting the terms in $1/c^2$. We shall see later (Part II) how it is possible to use also Dirac's relativistic Hamiltonian of the spinning electron in the radiation theory.

If we neglect the terms in $1/c^2$ we get from (13)

$$W_a = \frac{p^2}{2m} + eV - \frac{e}{mc}(\mathbf{U}\cdot\boldsymbol{p}) . \tag{14}$$

The Hamiltonian of the complex system of the atom and the radiation field is obtained adding (11) to (14) and putting in (14) the expression (12) instead of $\mathbf{U}$. We obtain in this way:

$$\begin{aligned} H = {} & \frac{1}{2m} p^2 + eV + \sum_s \left(\frac{1}{2} p_s^2 + 2\pi^2 \nu_s^2 q_s^2\right) \\ & - \frac{e}{m}\left(\frac{8\pi}{\Omega}\right)^{1/2} \sum_s (\mathbf{A}_s \cdot \boldsymbol{p})\, q_s \sin \Gamma_s . \end{aligned} \tag{15}$$

The first and second term of (15) give us the Hamiltonian which describes the motion of the electron if we neglect the effect of the radiation on it. The third term is the Hamiltonian (11) of the radiation field. The last term

$$\mathcal{H} = -\frac{e}{m}\left(\frac{8\pi}{\Omega}\right)^{1/2} \sum_s (\mathbf{A}_s \cdot \boldsymbol{p})\, q_s \sin \Gamma_s \tag{16}$$

is the coupling term, since it contains both the coordinates of the radiation (q_s) and of the atom ($\boldsymbol{p}$ and $\mathbf{X}$, contained in Γ_s).

In some cases, particularly for the theory of dispersion and of the Compton effect, it is necessary to write the Hamiltonian with a little closer approximation. In developing the term $(1/2\,m)\,(\boldsymbol{p} - e\mathbf{U}/c)^2$ of (13), we have neglected

$$e^2\,\mathbf{U}^2/2\,mc^2\,. \tag{17}$$

If we keep this term, and introduce for $\mathbf{U}$ the expression (12), we must add to the Hamiltonian (15) the term

$$\mathrm{H}^{(2)} = \frac{4\,\pi e^2}{\Omega m} \sum_{s\sigma} (\mathbf{A}_s \cdot \mathbf{A}_\sigma)\, q_s\, q_\sigma \sin \Gamma_s \sin \Gamma_\sigma\,. \tag{18}$$

We shall see later the very peculiar connection of this term with the jumps from positive to negative mass which characterise Dirac's theory of the spinning electron.

§ 5. - Classical treatment.

It is important to notice, that the results of the classical theory of emission of electromagnetic radiation and particularly Larmor's formula can be derived in a classical way from the Hamiltonian (15). This may be seen if we derive from (15) the canonical equations; for instance, if we consider the pair of variables q_s, p_s we get the equations:

$$\dot{q}_s = \frac{\partial \mathrm{H}}{\partial p_s} = p_s \quad ; \quad \dot{p}_s = -\frac{\partial \mathrm{H}}{\partial q_s} = -4\,\pi^2\,\nu_s^2\, q_s + \frac{e}{m}\left(\frac{8\pi}{\Omega}\right)^{1/2} \sum_s (\mathbf{A}_s \cdot \boldsymbol{p}) \sin \Gamma_s\,.$$

If we eliminate p_s, we find for q_s:

$$\ddot{q}_s + 4\,\pi^2\,\nu_s^2\, q_s = \frac{e}{m}\left(\frac{8\pi}{\Omega}\right)^{1/2} \sum_s (\mathbf{A}_s \cdot \boldsymbol{p}) \sin \Gamma_s\,. \tag{19}$$

This is the equation for the forced vibrations of an oscillator of frequency ν_s. If we suppose for instance that at the time $t = 0$ there is no radiation in the field, i.e., that $q_s = p_s = 0$; but that there is an elextron moving with non-uniform motion, so that its momentum $\boldsymbol{p}$ varies, we see from (19) that after a cetain time q_s shall be different from zero; this means that there is a certain amount of energy in the s component of the radiation which has been emitted by the moving charge. The effect is of course bigger, if the motion of the charge is periodic with a period near to ν_s. It might actually be shown by this method that the amount of energy emitted per unit time by the moving charge is given, to a first approximation, by:

$$\frac{2}{3}\,\frac{e^2}{c^3}\,\mathrm{A}^2 \tag{20}$$

where A is the acceleration of the particle, in accordance with Larmor's result.

This shows us, that the classical treatment of (15) gives us the same results as the ordinary theory of radiation; we must now apply to (15) the quantum mechanical methods.

§ 6. – PERTURBATION THEORY.

For this we write down some general formulas of the perturbation theory of wave mechanics which we shall use later. Let

$$H = H_o + \mathcal{H} \tag{21}$$

be the Hamiltonian of a system with coordinates q and moments p. The Schrödinger equation is:

$$-\frac{h}{2\pi i}\frac{\partial\psi}{\partial t} = H\psi = (H_o + \mathcal{H})\,\psi \tag{22}$$

where H is an operator obtained from H with the substitution of $(h/2\pi i)$ $(\partial/\partial q)$ in place of p.

We consider now the unperturbed problem corresponding to the Hamiltonian H_o. The Schrödinger equation corresponding to it is:

$$-\frac{h}{2\pi i}\frac{\partial\Psi}{\partial t} = H_o\,\Psi\,. \tag{23}$$

Let

$$\varphi_1, \varphi_2, \cdots, \varphi_n \cdots$$

be the normalized eigenfunctions of the unperturbed problem and

$$E_1, E_2, \cdots, E_n \cdots$$

be the corresponding eigenvalues. The most general solution of (23) is then:

$$\Psi = \sum_n a_n\,\varphi_n(q)\,e^{-2\pi i E_n t/h} \tag{24}$$

where the a_n are constants. The physical meaning of the a_n is contained in the statement that

$$|\,a_n\,|^2 \tag{25}$$

is proportional to the probability of finding the system in the nth quantum state. We may also normalize the a_n in such a way that

$$\sum_n |\,a_n\,|^2 = 1\,.$$

Then $|\,a_n\,|^2$ gives us directly the probability that the system is in the nth quantum state; a_n is called the amplitude of probability for the nth quantum state. The solution ψ of the perturbed problem (22) can be developed in a series of eigenfunctions of the unperturbed problem; it can therefore be written in the form (24); only the a's are no longer constants, but are functions of the time t. Substituting (24) in (22), we find for the a's the differential equations:

$$\dot{a}_n = -\frac{2\pi i}{h} \sum_m \mathcal{H}_{nm} a_m e^{2\pi i (E_n - E_m) t/h} \tag{26}$$

where:

$$\mathcal{H}_{nm} = \int \bar{\varphi}_n \mathcal{H} \varphi_m dq \tag{27}$$

is the element n, m of the perturbation matrix, representing the pertubation energy $\mathcal{H}$; $\bar{\varphi}_n$ is the conjugate complex to φ_n. From (25) we see that the a's vary with time, so that also the probability of the different quantum states, which is given by (25) changes with time. This means that the effect of the perturbation is to induce transition probabilities among the quantum states of the unperturbed system.

§ 7. - Quantum mechanical treatment.

We must now apply these methods to the Hamiltonian (15) of an atom and the radiation field. As the unperturbed Hamiltonian, we take:

$$H_0 = \frac{1}{2m} p^2 + eV + \sum_s \left(\frac{1}{2} p_s^2 + 2\pi^2 \nu_s^2 q_s^2 \right). \tag{28}$$

The interaction energy (16) is considered as the perturbation energy.

The Hamiltonian H_0 of the unperturbed system is the sum of the terms

$$\frac{1}{2m} p^2 + eV \tag{29}$$

representing the energy of the atom, and terms like:

$$\frac{1}{2} p_s^2 + 2\pi^2 \nu_s^2 q_s^2 \tag{30}$$

representing the energy of the sth component of the radiation which is identical with the energy of an oscillator of the same frequency ν_s.

Let

$$u_n, u_{n_1}, u_{n_2}, \dots, u_{n_s}, \dots$$

be the Schrödinger functions for the atom (with Hamiltonian (29)) and for each component of radiation (with Hamiltonian (30)). For simplicity of writing we distinguish all these functions only by the index. The Schrödinger function corresponding to the unperturbed Hamiltonian H_0 is then given by the product

$$\varphi_{n, n_1, n_2, \dots, n_s \dots} = u_n u_{n_1} u_{n_2} \cdots u_{n_s} \cdots \tag{31}$$

The corresponding eigenvalue is the sum:

$$E_{n, n_1, n_2, \dots, n_s \dots} = E_n + E_{n_1} + \dots + E_{n_s} + \dots \tag{32}$$

Since the Hamiltonians (30) are of the oscillator type we have, as for the oscillator

$$E_{n_s} = h\nu_s\left(n_s + \frac{1}{2}\right).$$

We may also neglect the constant energy $h\nu_s/2$, which does not affect the phenomena, since the frequencies ν_s are constant, and only differences of energy are considered, and write simply

$$E_{n_s} = h\nu_s\, n_s\,. \tag{33}$$

Formula (32) now takes the form

$$E_{n,n_1,\cdots,n_s\cdots} = E_n + h\nu_1 n_1 + h\nu_2 n_2 + \cdots \tag{34}$$

The general form of the field scalar, corresponding to (24) is then:

$$\psi\,(\mathbf{X}\,,\, q_1\,,\, q_2\,,\cdots,\, q_s\cdots) \tag{35}$$

$$= \Sigma\, a_{nn_1\cdots n_s\cdots} u_n u_{n_1}\cdots u_{n_s}\cdots e^{-2\pi i\,(E_n + h\nu_1 n_1\cdots + h\nu_s n_s + \cdots)\,t/h}\,.$$

The physical meaning of the a's, according to (25), is the following:

$$\left| a_{n\,n_1\,n_2\cdots n_s\cdots} \right|^2$$

is the probability that the atom is in the quantum state n; the first component of radiation in the state n_1; the second in the state n_2. If we have for instance $a_{300\cdots0\cdots} = 1$ and all the other a's are $= 0$, we may say that it is certain that the atom is in the third quantum state, and no component of radiation is excited.

If we neglect the effect of the perturbation term $\mathcal{H}$, the a's are constant; the effect of the interaction term $\mathcal{H}$ is, according to the general formula (26), that the a's vary with the time. If we have, for instance, at the time $t = 0$, $a_{300\cdots0\cdots} = 1$ and all the other a's equal to zero, after a certain time t, some of the a's which were $= 0$, say $a_{2,1,0\cdots0\cdots}$ may have a value different from zero. There is therefore a finite probability of finding that, at the time t the atom is in the state 2, having jumped down from state 3 to state 2 and that the first component of the radiation is excited. This is the quantum theoretical mechanism of the radiation of energy.

We have now to write for our case the equations corresponding to (26) to find how the a's vary with time. For this we must find the expression of the matrix element of the perturbation $\mathcal{H}$, corresponding to a transition of the whole system from a quantum state $n\,,\, n_1\,,\, n_2\,\cdots$ to another $m\,,\, m_1\,,\, m_2\,\cdots$. This is given, according to (27) (31), since the u's are real, by:

$$\mathcal{H}_{nn_1\cdots n_s\cdots\,;\; mm_1\cdots m_s\cdots} \tag{37}$$

$$= \int\!\!\int\cdots\int\cdots u_n u_{n_1}\cdots u_{n_s}\cdots \mathcal{H}\, u_m u_{m_1}\cdots u_{m_s}\cdots d\mathbf{X}\, dq_1\cdots dq_s\cdots$$

where we must put for $\mathcal{H}$ the operator (16).

The integral (37) may be very easily calculated taking into account the following relations

$$\int u_{n_a} u_{m_a}\, dq_a = \delta_{n_a m_a} \tag{38}$$

which expresses the orthogonality of the u's;

$$(39)\qquad \int q_s u_{n_s} u_{m_s} dq_s = \begin{cases} 0 & \text{if } m_s \neq n_s \pm 1 \\ \left[\dfrac{h(n_s+1)}{8\pi^2 \nu_s}\right]^{1/2} & \text{if } m_s = n_s + 1 \\ \left[\dfrac{hn_s}{8\pi^2 \nu_s}\right]^{1/2} & \text{if } m_s = n_s - 1\,. \end{cases}$$

These equations may be easily verified, since the u_s are the well known eigenfunctions of a harmonic oscillator with mass 1 and frequency ν_s; (39) are the elements of the matrix representing the coordinate q_s for this oscillator. We must remember further that the operator $\boldsymbol{p}$, with components p_x, p_y, p_z, means $(h/2\pi i)$ grad; we put then

$$(40)\qquad \begin{cases} \mathbf{P}_{snm} = \int u_n \sin \Gamma_s \boldsymbol{p} u_m \, dx\, dy\, dz \\ \qquad = (h/2\pi i) \int \sin \Gamma_s u_n \operatorname{grad} u_m \, dx\, dy\, dz\,. \end{cases}$$

We find now easily that the matrix element (37) is always $= 0$ if more than one of the indices $m_1 m_2, \cdots, m_s \cdots$ of the radiation components is different from the corresponding $n_1 n_2, \cdots, n_s \cdots$. If only one of the $m_1 m_2, \cdots, m_s, \cdots$, say m_s, differs from the corresponding n_s, the result, according to (39) is different from zero only if $m_s = n_s \pm 1$. In this case we have:

$$(41)\qquad \mathcal{H}_{nn_1 \cdots n_s \cdots\,;\, mn_1 \cdots n_s \pm 1 \cdots} = -\frac{e}{m}\left(\frac{h}{\pi \Omega \nu_s}\right)^{1/2} (\mathbf{A}_s \cdot \mathbf{P}_{snm}) \begin{bmatrix} (n_s+1)^{1/2} \\ n_s^{1/2} \end{bmatrix}$$

where we must use the upper expression $(n_s + 1)^{1/2}$ if $m_s = n_s + 1$ and the lower $n_s^{1/2}$ if $m_s = n_s - 1$. Very important is the particular case that the dimensions of the atom are very small compared with the wave-length, so that the Γ_s, i.e., the phases of the radiation components may be considered as constants over all the space where the eigenfunctions of the electron are practically different from zero. In this case we may take $\sin \Gamma_s$ out of the integral in (40) and we get:

$$\mathbf{P}_{snm} = \frac{h}{2\pi i} \sin \Gamma_s \int u_n \operatorname{grad} u_m \, dx\, dy\, dz\,.$$

Remembering that:

$$\frac{h}{2\pi i} \int u_n \operatorname{grad} u_m \, d\tau$$

represents the momentum of the electron, it can be immediately proved that:

$$(42)\qquad \frac{h}{2\pi i} \int u_n \operatorname{grad} u_m \, d\tau = -2\pi i\, m \nu_{mn} \mathbf{X}_{nm}$$

where:

$$(43)\qquad \nu_{mn} = (\mathrm{E}_m - \mathrm{E}_n)/h$$

is the frequency corresponding to the jump from state m to state n;

$$(44)\qquad \mathbf{X}_{mn} = \mathbf{X}_{nm} = \int \mathbf{X} u_n u_m \, d\tau$$

is the element mn of the matrix representing the radius vector $\mathbf{X}$ (observe that the letter m is used in (42) both as index and as the mass of the electron; but since no confusion is possible we prefer not to introduce a new symbol). We obtain now:

$$\mathbf{P}_{snm} = -2\pi i\, m\nu_{mn}\, \mathbf{X}_{nm} \sin \Gamma_s\,; \tag{45}$$

(41) becomes then:

$$\mathcal{H}_{nn_1 \cdots n_s \cdots;\, mn_1 \cdots n_s \pm 1 \cdots} = 2\pi i e \left(\frac{h}{\pi\Omega}\right)^{1/2} \frac{\nu_{mn}}{\nu_s^{1/2}} (\mathbf{A}_s \cdot \mathbf{X}_{nm}) \begin{bmatrix} (n_s+1)^{1/2} \\ n_s^{1/2} \end{bmatrix} \sin \Gamma_s\,. \tag{46}$$

Now we may write at once the equations analogous to (26) for the variations of the a's as functions of the time. We get:

$$\dot{a}_{nn_1 \cdots n_s \cdots} = -\frac{2\pi i}{h} \sum_{mm_1 \cdots m_s \cdots} a_{mm_1 \cdots m} \cdots \mathcal{H}_{nn_1 \cdots;\, mm_1 \cdots}\, e^{-2\pi i (E_{mm_1} \cdots - E_{nn_1} \cdots)\, t/h}. \tag{47}$$

Remembering (46) (43) (34) we get with a few reductions

$$\dot{a}_{nn_1 \cdots n_s \cdots} = \frac{4\pi^{3/2} e}{(h\Omega)^{1/2}} \sum_{m,s} \frac{\nu_{mn}}{\nu_s^{1/2}} (\mathbf{A}_s \cdot \mathbf{X}_{nm}) \sin \Gamma_s \left[a_{mn_1 \cdots n_s+1 \cdots} (n_s+1)^{1/2} e^{-2\pi i (\nu_{mn}+\nu_s)t} + a_{mn_1 \cdots n_s-1 \cdots}\, n_s^{1/2} e^{-2\pi i (\nu_{mn}-\nu_s)t}\right]. \tag{48}$$

This is the fundamental equation of the radiation theory. In the applications we shall encounter equations which differ from (48) either because of the use of a higher degree of approximation, or because systems containing more than one electron are considered. When these cases come up we shall show the necessary modifications of the Eq. (48).

We shall now discuss some applications of the general theory that we have developed, with the chief purpose of showing that this theory may actually be considered as a satisfactory theory of radiation phenomena. For this we shall work out the following examples:

1. Emission from an excited atom and mean life.
2. Propagation of light in vacuum.
3. A case of interference: the Lippmann fringes.
4. The Doppler effect.
5. The Compton effect.

For applications to other problems see the bibliography.

§ 8. - Emission from an excited atom.

Let us consider an atom which at the time $t = 0$ is in an excited state; let us suppose that there is no radiant energy in the space surrounding it. We may consider, for the sake of simplicity only two states of the atom, numbered 1 and 2, and suppose that the atom at $t = 0$ is in the state 2. All this may be expressed by saying that for $t = 0$

$$a_{200 \cdots 0 \cdots} = 1 \tag{49}$$

and all the other a's are $= 0$. We know from experience that after a certain time the atom must go over to the state 1 of less energy and the energy difference must be found in the radiation field.

We will now show how it is possible to study this process with the fundamental Eq. (48).

We put

$$\nu_{21} = -\nu_{12} = \nu \quad ; \quad \mathbf{X}_{12} = \mathbf{X}_{21} = \mathbf{X}.$$

Eqs. (48) give then:

$$(50) \qquad \dot{a}_{1,0\cdots 1_s\cdots} = \frac{4\,\pi^{3/2}\,e}{(\Omega h)^{1/2}}\,\frac{\nu}{\nu_s^{1/2}}\,(\mathbf{A}_s\cdot\mathbf{X})\sin\Gamma_s\,a_{200\cdots 0\cdots}\,e^{-2\pi i(\nu-\nu_s)t}$$

$$(51) \qquad \dot{a}_{2,0\cdots 0\cdots} = \frac{4\,\pi^{3/2}\,e}{(\Omega h)^{1/2}}\sum_s\frac{\nu}{\nu_s^{1/2}}\,(\mathbf{A}_s\cdot\mathbf{X})\sin\Gamma_s\,a_{1,00\cdots 1_s\cdots}\,e^{-2\pi i(\nu_s-\nu)t}.$$

We try to solve these equations by

$$(52) \qquad a_{20\cdots 0\cdots} = e^{-\gamma t}$$

where γ is a constant which must be determined.

We substitute (52) in (50) and then integrate with respect to t, determining the integration constant by the initial conditions $a_{100\cdots 1_s\cdots} = 0$; we find:

$$(53) \qquad a_{1,00\cdots 1_s\cdots} = \frac{4\,\pi^{3/2}\,e}{(\Omega h)^{1/2}}\,\frac{\nu}{\nu_s^{1/2}}\,(\mathbf{A}_s\cdot\mathbf{X})\sin\Gamma_s\,\frac{e^{\{-2\pi i(\nu-\nu_s)-\gamma\}t}-1}{-2\pi i(\nu-\nu_s)-\gamma}.$$

We substitute (52) and (53) in (51) and multiply by $-e^{\gamma t}$ and find:

$$\gamma = \frac{16\,\pi^3\,e^2}{\Omega h}\sum_s\frac{\nu^2}{\nu_s}\,(\mathbf{A}_s\cdot\mathbf{X})^2\sin^2\Gamma_s\,\frac{1-e^{\{\gamma-2\pi i(\nu_s-\nu)\}t}}{-2\pi i(\nu-\nu_s)-\gamma}.$$

The sum may be calculated by the following method: since the phase, direction and polarization of the different radiation components are distributed at random, we may substitute for $(\mathbf{A}_s\,\mathbf{X})^2\sin^2\Gamma_s$ its mean value, taken over all phases, directions and polarizations; we replace then the sum by an integral over ν_s, multiplying by the factor:

$$(54) \qquad \frac{8\,\pi}{c^3}\,\Omega\nu_s^2\,d\nu_s$$

which gives, according to (1) the number of radiation components with frequency between ν_s and $\nu_s + d\nu_s$. We get then, observing

$$\overline{(\mathbf{A}_s\cdot\mathbf{X})^2} = \frac{1}{3}\,\mathrm{X}^2 \quad ; \quad \overline{\sin^2\Gamma_s} = \frac{1}{2}$$

that

$$\gamma = \frac{64\,\pi^4\,e^2}{3\,hc^3}\,\nu^2\,\mathrm{X}^2\int_0^\infty\frac{1-e^{\{\gamma-2\pi i(\nu_s-\nu)\}t}}{-2\pi i(\nu-\nu_s)-\gamma}\,\nu_s\,d\nu_s.$$

It may be proved that for small γ this integral has the value $\nu/2$; we obtain thus:

$$(55) \qquad \gamma = \frac{32\,\pi^2\,e^2}{3\,hc^3}\,\nu^3\,\mathrm{X}^2$$

which determines the constant γ. The relation of this constant with the mean life of the state 2 is easily found: the probability of finding the atom in the state 2 as shown by (52) is:

$$|a_{2,0\cdots 0\cdots}|^2 = e^{-2\gamma t}.$$

By definition of the mean life τ, this probability must be $e^{-t/\tau}$; we get then by comparison

$$\tau = \frac{1}{2\gamma} = \frac{3\,hc^3}{64\,\pi^4\,e^2\,\nu^3\,X^2}\,. \tag{56}$$

We may also deduce from this theory the form and the width of the emitted spectral line. For this purpose we observe that, after the emission has taken place, i.e., after a time t long with respect to the mean life, the exponential $e^{\{-2\pi i(\nu-\nu_s)-\gamma\}t}$ becomes negligible; then we get from (53):

$$a_{100\cdots 1_s\cdots} = \frac{4\,\pi^{3/2}\,e}{(\Omega h)^{1/2}}\,\frac{\nu}{\nu_s^{1/2}}\,(\mathbf{A}_s\cdot\mathbf{X})\sin\Gamma_s\,\frac{1}{2\,\pi i\,(\nu_s-\nu)+\gamma}\,. \tag{57}$$

The probability that the emitted quantum belongs to the s component is therefore:

$$|a_{10\cdots 1_s\cdots}|^2 = \frac{16\,\pi^3\,e^2}{\Omega h}\,\frac{\nu^2}{\nu_s}\,(\mathbf{A}_s\cdot\mathbf{X})^2\sin^2\Gamma_s\,\frac{1}{\gamma^2+4\,\pi^2\,(\nu_s-\nu)^2}\,. \tag{58}$$

The last factor

$$\frac{1}{\gamma^2+4\,\pi^2\,(\nu_s-\nu)^2} \tag{59}$$

represents the form of the emitted line; it is identical with the form which may be deduced in the classical theory for exponentially damped oscillators.

§ 9. - Propagation of light in vacuum.

This section and the next one will be devoted to the proof that the results of ordinary wave theory can be applied to the computation of the intensity of light both for the propagation in vacuum and for cases of interference.

This has been proved for a general case by Racah and more recently by a very general and direct method by Heisenberg, who does not use the Fourier analysis but calculates the amplitude of the field vectors directly.

We prefer however to show here by the use of two examples how the phase relations between the different components are effective in determining the propagation with finite velocity and the interference phenomena.

Let A and B be two atoms; let us suppose that at the time $t = 0$, A is in an excited and B in the normal state. After a certain time A emits its energy which may in turn be absorbed by the atom B which then becomes excited. Since the light needs a finite time to go from A to B, the excitation of B can take place only after the time r/c, r being the distance between the two atoms. We will show that all this may be deduced from the quantum theory of radiation. We simplify the problem by the assumption that the mean life of the first atom A is very short, in order that the light be emitted from A at a very definite time; we suppose further that the mean life of B is very long. The result is that the line emitted from the atom A is very broad and might be considered as a portion of a continuous spectrum; on the contrary the atom B absorbs a very sharp line. We must first modify slightly the fundamental Eqs. (48), for the case of two atoms in the radiation field. We use indices and magnitudes without dash for the first atom A, and dashed letters for the atom B.

$$\left| a_{nn'\, n_1 \cdots n_s \cdots} \right|^2$$

is the probability that A is in the state n, B in the state n', and the radiation components in the states $n_1 \cdots n_s \cdots$; the equations analogous to (48) for the case of two atoms, may be obtained by the same considerations as Eqs. (48). The right hand side will now consist of two terms, each one analogous to the right hand member of (48) and each one referring to one of the atoms. We get precisely:

$$\text{(60)} \quad \dot{a}_{nn'\, n_1 \cdots n_s \cdots} = \frac{4\pi^{3/2} e}{(\Omega h)^{1/2}} \sum_{ms} \frac{\nu_{mn}}{\nu_s^{1/2}} (\mathbf{A}_s \cdot \mathbf{X}_{nm}) \sin \Gamma_s \{ a_{mn'\, n_1 \cdots n_s + 1 \cdots} (n_s + 1)^{1/2} e^{-2\pi i (\nu_{mn} +}$$

$$+ a_{mn'\, n_1 \cdots n_s - 1 \cdots} n_s^{1/2} e^{-2\pi i (\nu_{mn} - \nu_s) t} \}$$

$$+ \frac{4\pi^{3/2} e}{(\Omega h)^{1/2}} \sum_{m's} \frac{\nu'_{m'n'}}{\nu_s^{1/2}} (\mathbf{A}_s \cdot \mathbf{X}'_{m'n'}) \sin \Gamma'_s \{ a_{nm'\, n_1 \cdots n_s + 1 \cdots} (n_s + 1)^{1/2} e^{-2\pi i (\nu'_{m'n'} + \nu_s) t}$$

$$+ a_{nm'\, n_1 \cdots n_s - 1 \cdots} n_s^{1/2} e^{-2\pi i (\nu'_{m'n'} - \nu_s) t} \}$$

Γ_s and Γ_s are the phases of the sth standing vibrations at the places of the first and second atom. We may suppose that the atom A is at the origin of the coordinates, and the atom B is on the x-axis at a distance r from the origin. We have then from (4):

$$\text{(61)} \qquad \Gamma'_s = \Gamma_s + \frac{2\pi\nu_s}{c} r \cos \theta_s$$

θ_s being the angle between the x-axis and the direction of the sth radiation component.

As before we consider for each atom only two states, 1 and 2, and we put

$$\nu_{21} = -\nu_{12} = \nu \quad ; \quad \nu'_{21} = -\nu'_{12} = \nu'$$

$$\mathbf{X}_{21} = \mathbf{X}_{12} = \mathbf{X} \quad ; \quad \mathbf{X}'_{21} = \mathbf{X}'_{12} = \mathbf{X}'.$$

For simplicity we admit further that both vectors $\mathbf{X}$ and $\mathbf{X}'$ reduce to the only y-component. At the time $t = 0$, the first atom is excited in the state 2, and the second atom is in the normal state 1; further there is no radiation in the field; we have thus:

$$a_{210 \cdots 0 \cdots} = 1$$

while all the other a's are 0 for $t = 0$. We must find what is the probability that, at the time t, the first atom has lost its energy and that this energy has been absorbed by the second atom; this probability is given by:

$$\left| a_{120 \cdots 0 \cdots} \right|^2.$$

We have shown in the preceding section that after a time long with respect to the mean life, the energy of the excited atom A is transferred to the radiation field according to (57). This formula can be applied also in our case, if we neglect the very small perturbation due to the presence of the atom B. We may write:

$$a_{1100\cdots 1_s\cdots} = \frac{4\pi^{3/2} e}{(\Omega h)^{1/2}} \frac{\nu}{\nu_s^{1/2}} (\mathbf{A}_s \cdot \mathbf{X}) \sin \Gamma_s \frac{1}{-2\pi i (\nu_s - \nu) + \gamma} . \tag{62}$$

We put now in (60): $n = 1$, $n' = 2$, $n_1 = n_2 = \cdots = n_s = \cdots = 0$ and we get:

$$\dot{a}_{120\cdots 0\cdots} = -\frac{4\pi^{3/2} e}{(\Omega h)^{1/2}} \sum_s \frac{\nu'}{\nu_s^{1/2}} (\mathbf{A}_s \cdot \mathbf{X}') \sin \Gamma'_s \, a_{110\cdots 1_s\cdots} \, e^{-2\pi i(-\nu'+\nu_s) t}$$

since the other terms are zero. Substituting (62) in this equation we obtain.

$$\dot{a}_{120\cdots 0\cdots} = -\frac{16\pi^3 e^2}{\Omega h} \nu\nu' \sum_s \frac{(\mathbf{A}_s\cdot\mathbf{X})(\mathbf{A}_s\cdot\mathbf{X}') \sin\Gamma_s \sin\Gamma'_s}{\nu_s [2\pi i(\nu - \nu_s) + \gamma]} e^{-2\pi i(\nu_s - \nu') t} .$$

We integrate with respect to t and remember that for $t = 0$, $a_{120\cdots 0\cdots} = 0$: We obtain thus:

$$a_{120\cdots 0\cdots} = \frac{16\pi^3 e^2}{\Omega h} \nu\nu' \sum_s \frac{(\mathbf{A}_s\cdot\mathbf{X})(\mathbf{A}_s\cdot\mathbf{X}') \sin\Gamma_s \sin\Gamma'_s}{\nu_s [2\pi i(\nu - \nu_s) + \gamma]} \frac{1 - e^{-2\pi i(\nu_s - \nu') t}}{2\pi i(\nu_s - \nu')} . \tag{63}$$

To valuate the sum over s we must transform it to an integral. For this we first substitute in the usual way the mean value for the expression $(\mathbf{A}_s\cdot\mathbf{X})(\mathbf{A}_s\cdot\mathbf{X}')\sin\Gamma_s\sin\Gamma'_s$. Remembering (61), the fact that $\mathbf{A}_s$ and α_s are perpendicular unit vectors, and that $\mathbf{X}$ and $\mathbf{X}'$ reduce to the only y–component, we find with some calculation:

$$\overline{(\mathbf{A}_s\cdot\mathbf{X})(\mathbf{A}_s\cdot\mathbf{X}')\sin\Gamma_s\sin\Gamma'_s} \tag{64}$$

$$= \frac{XX'}{4}\left[\frac{c}{2\pi\nu_s r}\sin\frac{2\pi\nu_s r}{c} + \left(\frac{c}{2\pi\nu_s r}\right)^2 \cos\frac{2\pi\nu_s r}{c} - \left(\frac{c}{2\pi\nu_s r}\right)^3 \sin^2\frac{2\pi\nu_s r}{c}\right].$$

The average is taken over all values of the phase, the direction and polarization. We suppose now that the distance r of the two atoms is very large compared with the wave-length; we may then neglect the square and the cube of the very small expression $c/2\pi\nu_s r$, and we write:

$$\overline{(\mathbf{A}_s\cdot\mathbf{X})(\mathbf{A}_s\cdot\mathbf{X}')\sin\Gamma_s\sin\Gamma'_s} = \frac{cXX'}{8\pi\nu_s r}\sin\frac{2\pi\nu_s r}{c} . \tag{65}$$

To calculate the sum (63) we must now substitute this average value, multiply by (54) and change the sum to an integral over ν_s. We get thus:

$$a_{120\cdots 0\cdots} = -\frac{1}{r}\frac{16\pi^3 e^2}{c^2 h}\nu X\nu' X' \int_0^\infty \frac{\sin(2\pi\nu_s r/c)\left(1 + e^{-2\pi i(\nu_s - \nu') t}\right)}{[2\pi i(\nu - \nu_s) + \gamma]\, 2\pi i(\nu_s - \nu')} d\nu_s \tag{66}$$

The integration can be effected by observing that because of the factor $(\nu_s - \nu')$ in the denominator the values of the integrand are concentrated in the neighbourhood of the value ν' for the variable ν_s. Since we have supposed that the mean life of the first atom is very short, the factor

$[2\pi i(\nu_s - \nu) + \gamma]^{-1}$ varies very regularly (that is the line excited by the first atom is so wide that it can be considered as a piece of continuous spectrum). We may therefore take this factor out of the integral putting into it $\nu_s = \nu'$. We may also extend the integration from $-\infty$ to $+\infty$, since, for the same reason, the terms that we add are negligible. At last we take $\xi = \nu_s - \nu'$ as a new variable. We find then:

$$(67) \qquad a_{120\cdots 0\cdots} = -\frac{1}{r}\,\frac{16\pi^3 e^2}{c^2 h}\,\frac{\nu X \nu' X'}{2\pi i(\nu-\nu')+\gamma}\int_{-\infty}^{+\infty}\frac{\sin(2\pi r/c)(\nu'+\xi)(1-e^{-2\pi i t\xi})}{2\pi i\xi}\,d\xi\,.$$

The integral may be written:

$$\int_{-\infty}^{+\infty}\frac{(\sin(2\pi r\nu'/c)\cos(2\pi r\xi/c)+\cos(2\pi r\nu'/c)\sin(2\pi r\xi/c))(1-\cos 2\pi t\xi + i\sin 2\pi t\xi)}{2\pi i\xi}\,d\xi$$

$$= \frac{1}{2\pi i}\sin\frac{2\pi r\nu'}{c}\int_{-\infty}^{\infty}\frac{d\xi}{\xi}\cos\frac{2\pi r}{c}\xi\,(1-\cos 2\pi t\xi)$$

$$+\frac{1}{2\pi}\cos\frac{2\pi r\nu'}{c}\int_{-\infty}^{\infty}\frac{d\xi}{\xi}\sin\frac{2\pi r}{c}\xi\sin 2\pi t\xi$$

$$+\frac{1}{2\pi i}\cos\frac{2\pi r\nu'}{c}\int_{-\infty}^{\infty}\sin\frac{2\pi r}{c}\xi\,(1-\cos 2\pi t\xi)\frac{d\xi}{\xi}$$

$$+\frac{1}{2\pi}\sin\frac{2\pi r\nu'}{c}\int_{-\infty}^{\infty}\cos\frac{2\pi r}{c}\xi\sin 2\pi t\xi\frac{d\xi}{\xi}\,.$$

The two first integrals are evidently zero, since the integrated functions are odd. The other integrals can be calculated at once by the integral formulas:

$$\int_{-\infty}^{\infty}\frac{\sin qx}{x}\,dx = \pi \quad ; \quad \int_{-\infty}^{\infty}\frac{\sin qx\cos px}{x}\,dx = \begin{cases}\pi & \text{for } q > p\\ 0 & \text{for } q < p.\end{cases}$$

We find that the integral in (67) is given by:

$$\begin{cases} 0 & \text{for } r/c > t\\ (1/2i)\,e^{2\pi i r\nu'/c} & \text{for } t > r/c\,.\end{cases}$$

Substituting in (67) we find:

$$(68) \qquad a_{120\cdots 0\cdots} = \begin{cases} 0 & \text{for } t < r/c\\ \dfrac{1}{r}\,\dfrac{8\pi^3 ie^2}{c^2 h}\,\dfrac{\nu X\,\nu' X'}{2\pi i(\nu-\nu')+\gamma}\,e^{2\pi i r\nu'/c} & \text{for } t > r/c\,.\end{cases}$$

The square modulus of $a_{1200\cdots}$ measures the probability of finding the second atom excited. This probability is therefore zero if $t < r/c$, i.e., until the time necessary for the light emitted by the first atom to reach the second.

After that time the probability that the second atom is excited, is:

$$| a_{120\cdots0\cdots} |^2 = \frac{1}{r^2} \left[\frac{16\,\pi^3 e^2}{c^2 h} \nu X \nu' X' \right]^2 \frac{1}{4\,\pi^2 (\nu' - \nu) + \gamma^2} .$$

We notice that this probability is inversely proportional to the square of the distance r; we conclude therefore that the theory gives correctly the velocity of propagation of light and the decrease of intensity with the distance from the source of light.

§ 10. - THEORY OF THE LIPPMANN FRINGES.

The Lippmann fringes are produced as light is reflected from a mirror perpendicular to the direction of propagation. They consist in the system of standing waves formed by the incident and the reflected waves. We will show now, how this phenomenon may be explained by the quantum theory of radiation.

We must consider a plane mirror S and two atoms, a first atom A which emits the light and a second atom B which absorbs it. We suppose that the atom A (light source) is very far from the mirror, so that the waves reaching the mirror are very nearly plane waves. On the contrary we suppose the atom B not very far from the mirror, and we will show that the probability of excitation of B depends periodically on its distance from the mirror exactly as is to be expected in the classical theory from the position of the maxima and minima of the standing waves.

We simplify the calculations by supposing that the straight line AB (which we take as x–axis) is perpendicular to the mirror. As origin of coordinates we take the intersection of this line with the mirror; the coordinates of the atom A are $x, 0, 0$, and those of B are $x', 0, 0$. Then $x >> x'$. We suppose further as in the preceding chapter, that the mean life of A is very short and that of B is very long; and that the vectors $\mathbf{X}$ and $\mathbf{X}'$ which determine the transition probability from the state 2 to state 1 for both atoms reduce to the only y–component.

We have always considered the radiation contained in a volume Ω; in our case it is convenient to take the mirror S as one of the walls limiting the space Ω. As Ω becomes infinite the wall S remains fixed and all the other walls are taken to infinite distance.

Every standing vibration constituting the radiation field must have S as a nodal plane. Its y–component must have therefore the form:

$$B_s = \frac{Y_s}{2^{1/2}} \left\{ \sin \left[\frac{2\,\pi\nu_s}{c_s} (\alpha_{sx}\, x + \alpha_{sy}\, y + \alpha_{sz}\, z) + \beta_s \right] \right.$$
$$\left. - \sin \left[\frac{2\,\pi\nu_s}{c_s} (- \alpha_{sx}\, x + \alpha_{sy}\, y + \alpha_{sz}\, z) + \beta_s \right] \right\} \tag{70}$$

where $1/2^{1/2}$ is a normalization factor; Y_s is the y–component of the unit vector $\mathbf{A}_s$.

By exactly the same method that we used for the deduction of (63) we find now a very similar equation, obtained by substituting in (63) for $\mathbf{A}_s \sin \Gamma_s$ and $\mathbf{A}_s \sin \Gamma'_s$ the values B_s and B'_s which are obtained by putting into (70) the coordinates $x, 0, 0$ and $x', 0, 0$, of the atoms A and B. This formula is:

$$a_{120\cdots0\cdots} = \frac{16\pi^3 e^2}{\Omega h} \nu\nu' \sum_s \frac{B_s X B'_s X'}{\nu_s} \frac{1}{-2\pi i(\nu_s - \nu) + \gamma} \frac{1 - e^{-2\pi i(\nu_s - \nu')t}}{-2\pi i(\nu_s - \nu')}.$$

The sum over s may be effected, as in the preceding chapter by taking first the mean value of $B_s B'_s$ over all the phases and orientations of the radiation components. We find thus the following formula corresponding to (65):

$$\text{(72)} \qquad \overline{B_s B'_s} = \frac{c}{8\pi\nu_s(x - x')} \sin \frac{2\pi\nu_s(x - x')}{c} - \frac{c}{8\pi\nu_s(x + x')} \sin \frac{2\pi\nu_s(x + x')}{c}.$$

We substitute this expression in (71), multiply by

$$(8\pi/c^3)\,\Omega\nu_s^2\,d\nu_s$$

and integrate over ν_s. We find thus:

$$\text{(73)} \qquad a_{120\cdots0\cdots} = R^- - R^+$$

where R^- and R^+ represent two terms equal to the right hand side of (68) where in place of r respectively $x - x'$ and $x + x'$ has been substituted; we get:

$$\text{(74)} \qquad R^- = \begin{cases} 0 & \text{for } t < \dfrac{x - x'}{c} \\ \dfrac{1}{x - x'} \dfrac{8 i\pi^3 e^2}{c^2 h} \dfrac{\nu X \nu' X'}{-2\pi i(\nu' - \nu) + \gamma} e^{2\pi i\nu'(x - x')/c} & \text{for } t > \dfrac{x - x'}{c} \end{cases}$$

$$R^+ = \begin{cases} 0 & \text{for } t < \dfrac{x + x'}{c} \\ \dfrac{1}{x + x'} \dfrac{8 i\pi^3 e^2}{c^2 h} \dfrac{\nu X \nu' X'}{-2\pi i(\nu' - \nu) + \gamma} e^{2\pi i\nu'(x + x')/c} & \text{for } t > \dfrac{x + x'}{c}. \end{cases}$$

The two terms R^- and R^+ clearly represent the effect of the incident and the reflected wave. Let us now consider a time $t > (x + x')/c$. Then both for R^- and R^+ the second expressions are valid and we get from (73)

$$\text{(75)} \quad a_{120\cdots0\cdots} = \frac{8 i\pi^3 e^2}{c^2 h} \frac{\nu X \nu' X'}{-2\pi i(\nu' - \nu) + \gamma} \left\{ \frac{1}{x - x'} e^{2\pi i\nu'(x - x')/c} - \frac{1}{x + x'} e^{2\pi i\nu'(x + x')/c} \right\}$$

It is evident that this expression has a large value if the two exponential factors have the same phase. The condition of equal phase is:

$$\frac{2\pi\nu'}{c}(x - x') = \frac{2\pi\nu'}{c}(x + x') - 2\pi n$$

where n is an integer. We get from this:

$$x' = \frac{n}{2}\frac{c}{\nu'} = n\frac{\lambda'}{2}$$

where $\lambda' = c/\nu'$ represents the wave-length corresponding to the frequency ν'. We see therefore that the places where the probability of excitation is small (dark fringes) are planes, parallel to the mirror and spaced by a half wave-length from each other; similarly we find that the places where the probability of excitation is strong (bright fringes) are the planes situated in the middle between two dark fringes. We may conclude that the results of the quantum theory of radiation describe this phenomenon in exactly the same way as the classical theory of interference.

§ 11. - THEORY OF THE DOPPLER EFFECT.

The change of frequency of the light emitted from a moving source is very simply explained by the wave theory of light. But it finds also a simple, through apparently very different, explanation in the light-quantum theory; it can be shown that the Doppler effect may be deduced from the conservation of energy and momentum in the emission process.

Let us consider an atom A with two energy levels w_1 and w_2; the frequency emitted by the atom when it is at rest is then

$$\nu = (w_2 - w_1)/h\ .$$

Let us now suppose that the atom is excited and that it moves with velocity V; its total energy is then:

$$w_2 + \frac{1}{2} m\mathrm{V}^2.$$

At a given instant the atom emits, on jumping down to the lower state, a quantum of frequency ν'; the recoil of the emitted quantum produces a slight change of the velocity, which after the emission becomes V'; the energy of the atom is then $w_1 + \frac{1}{2} m\mathrm{V}'^2$. We get therefore from the conservation of energy

$$h\nu' = \left(w_2 + \frac{1}{2} m\mathrm{V}^2\right) - \left(w_1 + \frac{1}{2} m\mathrm{V}'^2\right) = h\nu + \frac{1}{2} m\left(\mathrm{V}^2 - \mathrm{V}'^2\right). \tag{76}$$

The conservation of momentum gives:

$$m\mathbf{V}' = m\mathbf{V} - \frac{h\nu'}{c}$$

where the bold face letters mean vectors. Taking the square we get:

$$m^2\,\mathrm{V}'^2 = m^2\,\mathrm{V}^2 + \frac{h^2\,\nu'^2}{c^2} - 2\,m\mathrm{V}\,\frac{h\nu}{c}\cos\theta$$

θ being the angle between the velocity and the direction of emission. From this equation and (76) we get, neglecting terms in $1/c^2$:

$$\nu' = \nu\left(1 + \frac{\mathrm{V}}{c}\cos\theta\right) \tag{77}$$

which is the classic formula for the Doppler effect to a nonrelativistic approximation.

We will now work out the theory of the Doppler effect with Dirac's theory of radiation. We shall see that the interpretation of the Doppler effect in this theory is very similar to its interpretation in terms of light quanta; and it is due essentially to the changes in momentum due to the recoil of the emitted light. In all the examples which we have worked out till now, we have used the approximation (45), which is obtained by supposing that the portion of space where the electron moves is so small, that the phases Γ_s of the standing vibrations can be considered as constants in it. Now we shall see that this simplification can no longer be made if we wish our theory to represent also the impulse properties of light quanta. So that for the theory of the Doppler effect and the Compton effect it is necessary to consider that the phase Γ_s are actually variables.

We simplify the problem by considering the emitting atom as costituted by a proton (charge e, mass m_1, coordinates $x_1, y_1, x_1 \equiv \mathbf{X}_1$) and an electron (charge $-e$, mass m_2, coordinates $x_2, y_2, z_2 \equiv \mathbf{X}_2$). The Hamilton function for the system consisting of this atom and the radiation field is the obvious generalization of (15); it is:

$$H = \frac{p_1^2}{2m_1} + \frac{p_2^2}{2m_2} + eV + \sum_s \left(\frac{1}{2} p_s^2 + 2\pi^2 \nu_s^2 q_s^2\right) - \frac{e}{m_1}\left(\frac{8\pi}{\Omega}\right)^{1/2} \sum_s (\mathbf{A}_s \cdot \mathbf{p}_1) \sin \Gamma_{s1}$$
$$+ \frac{e}{m_2}\left(\frac{8\pi}{\Omega}\right)^{1/2} \sum_s (\mathbf{A}_s \cdot \mathbf{p}_2) \sin \Gamma_{s2} .$$

We take now as new coordinates: the coordinates

$$\xi = \frac{m_1 \mathbf{X}_1 + m_2 \mathbf{X}_2}{m_1 + m_2}$$

of the center of gravity, and the relative coordinates of the two particles:

$$\mathbf{X} = \mathbf{X}_1 - \mathbf{X}_2 .$$

The momenta conjugated to ξ and $\mathbf{X}$ are:

$$\eta = M\dot{\xi} \quad \text{and} \quad \mathbf{p} = m\left(\frac{\mathbf{p}_1}{m_1} - \frac{\mathbf{p}_2}{m_2}\right)$$

where $M = m_1 + m_2$ is the total mass of the atom and $m = m_1 m_2 / (m_1 + m_2)$ is the relative mass.

We make the assumption that the dimensions of the atom are very small compared with the wave-length. We may then substitute for Γ_{s1} and Γ_{s2} the value:

$$\Gamma_s = \frac{2\pi\nu_s}{c} (\alpha_s \cdot \xi) + \beta_s \tag{78}$$

of the phase in the center of gravity. The Hamiltonian of the system becomes then:

$$H = \frac{\eta^2}{2M} + \frac{p^2}{2m} + eV(\mathbf{X}) + \sum_s \left(\frac{1}{2} p_s^2 + 2\pi^2 \nu_s^2 q_s^2\right) \tag{79}$$
$$- \frac{e}{m}\left(\frac{8\pi}{\Omega}\right)^{1/2} \sum_s (\mathbf{A}_s \cdot \mathbf{p}) \sin \Gamma_s .$$

We consider now the last term

$$\mathcal{H} = -\frac{e}{m}\left(\frac{8\pi}{\Omega}\right)^{1/2}\sum_s (\mathbf{A}_s \cdot \boldsymbol{p}) \sin \Gamma_s \tag{80}$$

as the perturbation. The unperturbed Hamiltonian is:

$$\mathrm{H}_0 = \frac{\eta^2}{2\,\mathrm{M}} + \frac{p^2}{2\,m} + e\mathrm{V}(\mathbf{X}) + \sum_s \left(\frac{1}{2} p_s^2 + 2\,\pi^2 \nu_s^2 q_s^2\right). \tag{81}$$

The first term of (81) is the Hamiltonian of the translatory motion of the center of gravity; the corresponding eigenfunctions are:

$$\Omega^{-1/2}\, e^{2\pi i (\boldsymbol{\eta}_n \cdot \boldsymbol{\xi})/h} \tag{82}$$

where $\Omega^{-1/2}$ is a normalization factor; $\boldsymbol{\eta}_n$ represents the momentum of the atom, which can be supposed to assume discreet values, since the atom can move in a finite volume Ω.

The second and third term of H_0 represent the Hamiltonian of the internal coordinates of the atom; the corresponding eigenfunctions are $u_{n'}$. The last term of H_0 is the Hamiltonian of the radiation components whose eigenfunctions are:

$$u_{n_1}\, u_{n_2} \cdots u_{n_s} \cdots$$

The eigenfunctions of the unperturbed system are given by the product:

$$\varphi_{nn'n_1\cdots n_s\cdots} = \Omega^{-1/2}\, e^{2\pi i (\boldsymbol{\eta}_n \cdot \boldsymbol{\xi})/h}\, u_{n'}\, u_{n_1} \cdots u_{n_s} \tag{83}$$

and the corresponding eigenvalue, as in (34) is the sum

$$\mathrm{E}_{nn'n_1\cdots n_s\cdots} = \eta_n^2/2\,\mathrm{M} + \mathrm{E}_{n'} + h\nu_1 n_1 + \cdots + h\nu_s n_s + \cdots \tag{84}$$

The probability of the state $nn'\,n_1 \cdots n_s \cdots$ is given by the square of the modulus of the quantity:

$$a_{nn'n_1\cdots n_s\cdots}$$

The a's, according to the general formula (26) satisfy differential equations, analogous to (47)

$$\dot{a}_{nn'n_1\cdots n_s\cdots} = \frac{2\pi i}{h} \sum_{mm'm_1\cdots m_s\cdots} a_{mm'm_1\cdots m_s\cdots}\, \mathcal{H}_{nn'\cdots;\, mm'\cdots}\, e^{2\pi i (\mathrm{E}_{mm'\cdots} - \mathrm{E}_{nn'\cdots})t/h} \tag{85}$$

where $\mathcal{H}_{nn'\cdots;\, mm'\cdots}$ represents, according to (27), the matrix element of the perturbation energy (80) corresponding to the transition from the state with indices m, m', m_1, $\cdots$ to the state n, n', n_1, $\cdots$; we have:

$$\mathcal{H}_{nn'\cdots;\, mm'\cdots} = \int \cdots \int \cdots \bar{\varphi}_{nn'\cdots}\, \mathcal{H}\, \varphi_{mm'\cdots}\, d\boldsymbol{\xi}\, d\mathbf{X}\, dq_1 \cdots \tag{86}$$

Substituting in (86) the expressions (80) and (83) we see that the integral (86) splits in the product of integrals. We are interested in the factor containing the coordinates $\boldsymbol{\xi}$ of the center of gravity; this factor is, remembering (78):

$$\frac{1}{\Omega}\int e^{-2\pi i (\boldsymbol{\eta}_n \cdot \boldsymbol{\xi})/h} \sin\left\{\frac{2\pi\nu_s}{c}(\boldsymbol{\alpha}_s \cdot \boldsymbol{\xi}) + \beta_s\right\} e^{2\pi i (\boldsymbol{\eta}_m \cdot \boldsymbol{\xi})/h}\, d\boldsymbol{\xi} \tag{87}$$

where $d\xi$ represents the element of volume. Expressing the sine in terms of exponentials, (87) becomes:

$$\frac{e^{i\beta_s}}{2\,i\Omega}\int e^{(2\pi i/h)\,([\boldsymbol{\eta}_m-\boldsymbol{\eta}_n+(h\nu_s/c)\,\boldsymbol{\alpha}_s]\cdot\boldsymbol{\xi})}\,d\xi-\frac{e^{-i\beta_s}}{2\,i\Omega}\int e^{(2\pi i/h)\,([\boldsymbol{\eta}_m-\boldsymbol{\eta}_n-(h\nu_s/c)\,\boldsymbol{\alpha}_s]\cdot\boldsymbol{\xi})}\,d\xi\,.$$

The integrals have generally a value very near to zero; they are very different from zero only if

$$\boldsymbol{\eta}_m-\boldsymbol{\eta}_n\pm(h\nu_s/c)\,\boldsymbol{\alpha}_s=0 \tag{88}$$

since in this case one of the exponentials is equal to unity. Eq. (88) is the condition of conservation of momenta, since $\boldsymbol{\eta}_m$ and $\boldsymbol{\eta}_n$ are the momenta of the motion of the center of gravity of the atom and $(h\nu_s/c)\,\boldsymbol{\alpha}_s$ is the moment of the emitted quantum; the double sign arises from the fact that the s–component of the radiation field is a standing vibration which may be considered as resulting from two progressing waves moving in the opposite directions defined by the unit vectors $+\boldsymbol{\alpha}_s$ and $-\boldsymbol{\alpha}_s$.

We see therefore that the conservation of momentum in the emission process follows from the radiation theory. That also the energy principle must hold results from Eq. (85), since only the terms with

$$E_{mm'\ldots}=E_{nn'\ldots}$$

give an important contribution.

Now we have shown that the formulas for the Doppler change of frequency can be derived if we assume the conservation of energy and momentum. The Doppler formulas can therefore be deduced from the radiation theory. We could also deduce from this theory the formulas for the intensity of radiation in different directions; the results are identical with the results of the classical theory.

§ 12. – Scattering of radiation from free electrons.

The scattering of radiation from free electrons can be considered in two different approximations. The first approximation leads to the classical Thomson formula for the intensity of the scattered radiation with no change of frequency. The second approximation gives the phenomena of the Compton effect where the momenta of light quanta are considered. We shall see also that the theory of scattering carried out by help of Dirac's relativistic wave function for the electron is essentially different from the present nonrelativistic theory. In this section we shall always discuss the nonrelativistic theory.

It can be shown that the simple interaction term (16) between the radiation and the electron is not responsible for the scattering from free electrons, not even in first approximation. This is connected with the well-known fact that the free electron has no probability of spontaneous transition between two states n and m with different velocity. The interaction term (16) involves the spontaneous transitions; therefore it may be neglected for the case of the free electron.

The interaction term which is responsible for the scattering is the term (17) which we have hitherto neglected, since it is smaller than (16) and only becomes important if the effect of (16) is zero. Let m and n be two (translational) states of the free electrons. The corresponding eigenfunctions can be written:

$$\psi_n = \Omega^{-1/2}\, e^{2\pi i (\boldsymbol{p}_n \cdot \mathbf{X})/h} \quad ; \quad \psi_m = \Omega^{-1/2}\, e^{2\pi i (\boldsymbol{p}_m \cdot \mathbf{X})/h} \tag{89}$$

where $\boldsymbol{p}_n$ and $\boldsymbol{p}_m$ represent the momentum of the electron in the states n and m; $\Omega^{-1/2}$ is the normalization factor.

To get a scattering different from zero, we need consider the interaction term (17). By help of (12) this term can be written in the form (18). We must first calculate the matrix elements corresponding to (18); then we shall substitute them in a formula analogous to (26) in order to find out the variations of the a's with the time. This calculation may be carried out to two different approximations. In the first one the supposition is made, that the phase Γ_s of the waves can be regarded as a constant; this approximation yields simply Thomson's formula for the intensity of the scattered radiation and is equivalent to neglecting the momentum properties of light quanta. In order to get the theory of the Compton shift of wave-length it is necessary to consider the dependence of the phases Γ_s on the coordinates. We shall at present restrict ourselves to the first approximation and consider the Γ_s as constants.

By means of (38) and (39) it may be very easily proved that a matrix element of $H^{(2)}$,

$$H^{(2)}_{n, n_1, n_2, \ldots\, ;\, m, m_1, m_2, \ldots}$$

is only different from zero if the following conditions exist:

(a) $n = m$

(b) the numbers $n_1\, n_2 \cdots$ with the exception of only two, say n_s and n_σ, are equal to the corresponding indices $m_1\, m_2 \cdots$

(c) $m_s = n_s \pm 1$ and $m_\sigma = n_\sigma \pm 1$ (where the $\pm$ signs are incoherent).

If these conditions are satisfied we get:

$$H^{(2)}_{n_s, n_\sigma\, ;\, n_s \pm 1, n_\sigma \pm 1} = \frac{e^2 h\, (\mathbf{A}_s \cdot \mathbf{A}_\sigma)}{\pi m\, \Omega} \frac{\sin \Gamma_s \sin \Gamma_\sigma}{(\nu_s\, \nu_\sigma)^{1/2}} \cdot \left| \begin{matrix} (n_s + 1)^{1/2} \\ n_s^{1/2} \end{matrix} \right| \cdot \left| \begin{matrix} (n_\sigma + 1)^{1/2} \\ n_s^{1/2} \end{matrix} \right| \tag{90}$$

where in the last two terms the upper or the lower expression must be taken according to the + or — sign in $n_s \pm 1$ or in $n_\sigma \pm 1$. Putting these matrix elements in the general formula (26) we get the following differential equations for the a's:

$$\dot{a}_{n, n_1, n_s \cdots n_\sigma \cdots} = -\frac{2\, i e^2}{m \Omega} \sum_{s\sigma} \frac{(\mathbf{A}_s \cdot \mathbf{A}_\sigma)}{(\nu_s\, \nu_\sigma)^{1/2}} \sin \Gamma_s \sin \Gamma_\sigma \tag{91}$$

$$\times \Big[a_{n n_1 \cdots n_s + 1 \cdots n_\sigma + 1 \cdots} \left[(n_s + 1)\,(n_\sigma + 1)\right]^{1/2} e^{-2\pi i (\nu_s + \nu_\sigma)\, t}$$

$$+ a_{n n_1 \cdots n_s - 1 \cdots n_\sigma + 1 \cdots} \left[n_s\,(n_\sigma + 1)\right]^{1/2} e^{+2\pi i (\nu_s - \nu_\sigma)\, t}$$

$$+ a_{n n_1 \cdots n_s + 1 \cdots n_\sigma - 1 \cdots} \left[(n_s + 1)\, n_\sigma\right]^{1/2} e^{-2\pi i (\nu_s - \nu_\sigma)\, t}$$

$$+ a_{n n_1 \cdots n_s - 1 \cdots n_\sigma - 1 \cdots} (n_s\, n_\sigma)^{1/2} e^{+2\pi i (\nu_s + \nu_\sigma)\, t} \Big] .$$

To get the intensity of the scattered light, we make the assumption that at the time $t = 0$

$$a_{n,0,0\ldots n_s\ldots 0_\sigma\ldots} = 1 \tag{92}$$

while all the other a's are zero. This means physically that there is a certain amount of radiation in the field; this is the primary radiation of frequency ν_s and energy density

$$w_s = n_s h\nu_s/\Omega. \tag{93}$$

We must find the intensity of the radiation scattered in the component σ. The probability that a quantum is scattered in the radiation component σ is given by the square modulus of

$$a_{n,0\ldots n_s - 1,\ldots 1_\sigma\ldots}$$

If we limit ourselves to a very short time t, we might still assume (92) to be valid in first approximation; and we obtain therefore from (91):

$$\dot{a}_{n,0\ldots n_s - 1,\ldots 1_\sigma\ldots} = -\frac{2ie^2}{m\Omega}(\mathbf{A}_s\cdot\mathbf{A}_\sigma)\frac{\sin\Gamma_s\sin\Gamma_\sigma}{(\nu_s\nu_\sigma)^{1/2}} n_s^{1/2} e^{-2\pi i(\nu_s - \nu_\sigma)t} \tag{94}$$

since only the third term in the square brackets gives a result different from zero. Integration with respect to time, with the initial condition $a_{n,0,\ldots n_s - 1,\ldots 1_\sigma\ldots} = 0$ for $t = 0$ yields:

$$a_{n\ldots n_s - 1\ldots 1_\sigma\ldots} = \frac{e^2}{\pi m\Omega}\frac{(\mathbf{A}_s\cdot\mathbf{A}_\sigma)\sin\Gamma_s\sin\Gamma_\sigma}{(\nu_s\nu_\sigma)^{1/2}} n_s^{1/2}\frac{e^{-2\pi i(\nu_s - \nu_\sigma)t} - 1}{\nu_s - \nu_\sigma}.$$

The probability for the scattering of a quantum in the component σ, is the square modulus of this expression, i.e.,

$$\frac{4e^4}{\pi^2 m^2\Omega^2}\frac{(\mathbf{A}_s\cdot\mathbf{A}_\sigma)^2\sin^2\Gamma_s\sin^2\Gamma_\sigma}{\nu_s\nu_\sigma} n_s\frac{\sin^2\pi(\nu_s - \nu_\sigma)t}{(\nu_s - \nu_\sigma)^2}.$$

Summing over the index σ we obtain the probability for the scattering of a quantum in any component of radiation. This sum may be transformed into an integral by a method similar to that described in section 8. We first substitute for $(\mathbf{A}_s\cdot\mathbf{A}_\sigma)^2$, $\sin^2\Gamma_s$, $\sin^2\Gamma_\sigma$ their mean values 1/3, 1/2, 1/2. Then we multiply by the number of radiation components $(8\pi/c^3)\,\Omega\,\nu_\sigma^2\,d\nu_\sigma$ with frequency between ν_σ and $\nu_\sigma + d\nu_\sigma$ and integrate over ν_σ. We obtain:

$$\frac{8}{3\pi}\frac{e^4}{c^3 m^2}\frac{n_s}{\Omega\nu_s}\int_0^\infty \nu_s\frac{\sin^2\pi(\nu_s - \nu_\sigma)t}{(\nu_s - \nu_\sigma)^2}d\nu_\sigma.$$

The integral can be evaluated by observing that its values are all in the immediate neighborhood of $\nu_\sigma = \nu_s$. (The scattered radiation has the same frequency as the primary light). We can therefore extend the limits of integration from $-\infty$ to $+\infty$ and substitute ν_s for the first ν_σ. By the integral formula

$$\int_{-\infty}^{+\infty}\frac{\sin^2 kx}{x^2}dx = \pi k \tag{95}$$

we obtain:

$$N = \frac{8\pi}{3} \frac{e^4}{c^3 m^2} \frac{n_s}{\Omega} t \tag{96}$$

as the number of quanta scattered during the time t. The scattered energy is:

$$h\nu_s N = \frac{8\pi}{3} \frac{e^4}{c^3 m^2} w_s t . \tag{97}$$

This expression coincides with the well-known formula derived by Thomson from the classical theory of radiation.

The theory carried out to this approximation gives no account of the Compton shift of wave-length, which is due to the momentum of light quanta. A theory of the Compton effect may be obtained if we calculate the matrix element:

$$H^{(2)}_{n,n_1 \cdots ;\, m,m_1 \cdots} \tag{98}$$

without the assumption, which we have hitherto made, that the phase Γ_s of the radiation components can be considered as a constant over the space occupied by the electron. From (18) and the expressions (89) of the eigenfunctions of the electron we see that the factor in the matrix element (98) depending on x, y, z is:

$$(1/\Omega) \int e^{-2\pi i (\mathbf{p}_m \cdot \mathbf{X})/h} \sin \Gamma_s \sin \Gamma_\sigma \, e^{+2\pi i (\mathbf{p}_n \cdot \mathbf{X})/h} \, dx\, dy\, dz . \tag{99}$$

We write the sine functions in terms of exponentials and remember that

$$\Gamma_s = \frac{2\pi\nu_s}{c} (\alpha_s \cdot \mathbf{X}) + \beta_s \quad ; \quad \Gamma_\sigma = \frac{2\pi\nu_\sigma}{c} (\alpha_\sigma \cdot \mathbf{X}) + \beta_\sigma .$$

We see therefore that (99) splits in a sum of terms like:

$$\pm \frac{e^{\pm i\beta_s \pm i\beta_\sigma}}{4\Omega} \int e^{2\pi i ([\mathbf{p}_m - \mathbf{p}_n \pm (h\nu_s/c)\, \alpha_s \pm (h\nu_\sigma/c)\, \alpha_\sigma] \cdot \mathbf{X})} \, dx\, dy\, dz .$$

If the coefficient of $\mathbf{X}$ in the exponent is considerably different from zero, the integral over the space Ω vanishes, since the exponential is a rapidly varying function with mean value zero. We get a term different from zero only if the coefficient is practically zero. That is:

$$\mathbf{p}_m - \mathbf{p}_n \pm \frac{h\nu_s}{c} \alpha_s \pm \frac{h\nu_\sigma}{c} \alpha_\sigma = 0 . \tag{100}$$

This is simply the condition of conservation of momentum. The double signs arise from the fact that a stationary wave is the superposition of two progressive waves of opposite directions.

From the conservation of momentum, the Compton wave-length change could be deduced with a method very analogous to the one used in the ordinary theory of Compton effect. We will not enter into the details of this theory, from which even intensity formulas in nonrelativistic approximation can be derived.

PART II. — THEORY OF RADIATION AND DIRAC'S WAVE EQUATION

In the second part of this work we shall first show how the general formulas of the previous section can be derived if we take as a basis Dirac's relativistic wave equation for the electron, instead of Schrödinger's equation. After this we shall study the very peculiar role which is played in the theory of light scattered from free electrons by the states of negative energy, characteristic of Dirac's theory of the electron. We shall also discuss the possibility of radiative transitions from states of positive to states of negative energy of a Dirac electron. These transitions in reality certainly do not take place; nevertheless it has some interest to see how they are derived from the present theory, since a correct theory should find some way of preventing them.

§ 13. - DIRAC'S WAVE FUNCTION OF THE ELECTRON.

We shall in this paragraph collect some formulas on Dirac's wave function which will be of use later.

It is well known that in Dirac's relativistic theory the electron at a given time is specified by four coordinates. Three of them, x, y, z, are the ordinary positional coordinates; the fourth coordinate σ will represent some internal degree of freedom; we shall call σ the spin coordinate. While the coordinates x, y, z have a continuous range of variability from $-\infty$ to $+\infty$, the spin coordinate σ can assume four values only; it is no limitation of generality to call these four values 1, 2, 3, 4. The wave function ψ will depend on x, y, z, σ:

$$\psi = \psi(x, y, z, \sigma) . \tag{101}$$

Since the variable σ takes only a finite set of values, it is often convenient to write it as an index:

$$\psi = \psi_\sigma(x, y, z) . \tag{102}$$

The wave function is thus represented by a set of four functions $\psi_1(x, y, z)$, $\psi_2(x, y, z)$, $\psi_3(x, y, z)$, $\psi_4(x, y, z)$ of the space coordinates only.

In Dirac's theory of the electron two types of operators are to be considered. The operators of the first kind act on the dependence of ψ on the space coordinates; for instance:

$$x^2 , xy , p_x = \frac{h}{2\pi i} \frac{\partial}{\partial x} , \ldots$$

A second kind of operators acts on the dependence of ψ on σ. The most general type of linear operator of this kind is a linear substitution on the four $\psi_1, \psi_2, \psi_3, \psi_4$. Therefore these operators are represented by matrices of the linear substitution; they have four rows and four columns. We shall consider chiefly four operators of this kind, *viz.*,

$$(103)\quad \left\{\begin{aligned} \gamma_x &= \begin{vmatrix} 0 & 0 & 0 & 1 \\ 0 & 0 & 1 & 0 \\ 0 & 1 & 0 & 0 \\ 1 & 0 & 0 & 0 \end{vmatrix} ; \quad \gamma_y = \begin{vmatrix} 0 & 0 & 0 & -i \\ 0 & 0 & i & 0 \\ 0 & -i & 0 & 0 \\ i & 0 & 0 & 0 \end{vmatrix} ; \\ \gamma_z &= \begin{vmatrix} 0 & 0 & 1 & 0 \\ 0 & 0 & 0 & -1 \\ 1 & 0 & 0 & 0 \\ 0 & -1 & 0 & 0 \end{vmatrix} ; \quad \delta = \begin{vmatrix} 1 & 0 & 0 & 0 \\ 0 & 1 & 0 & 0 \\ 0 & 0 & -1 & 0 \\ 0 & 0 & 0 & -1 \end{vmatrix} . \end{aligned}\right.$$

For instance the effect of the operator γ_x applied to the eigenfunction $\psi = (\psi_1 , \psi_2 , \psi_3 , \psi_4)$ is to change it in $\gamma_x \psi = (\psi_4 , \psi_3 , \psi_2 , \psi_1)$. Similarly

$$(104)\quad \gamma_y \psi = (-i\psi_4 , i\psi_3 , -i\psi_2 , i\psi_1) \quad ; \quad \gamma_z \psi = (\psi_3 , -\psi_4 , \psi_1 , -\psi_2) ;$$
$$\delta\psi = (\psi_1 , \psi_2 , -\psi_3 , -\psi_4) .$$

The γ's and δ satisfy the well-known relations:

$$\gamma_x^2 = \gamma_y^2 = \gamma_z^2 = \delta^2 = 1 \quad ; \quad \gamma_y \gamma_z + \gamma_z \gamma_y = 0 \text{ and similar}$$
$$\gamma_x \delta + \delta\gamma_x = 0 \text{ and similar.}$$

We will very often summarize $\gamma_x , \gamma_y , \gamma_z$ by a q–vector (that is a vector whose components are q–numbers) $\boldsymbol{\gamma}$. It is well known that $\gamma_x , \gamma_y , \gamma_z , c\delta$ can be considered to transform as the four components of a four vector; our vector $\boldsymbol{\gamma}$ is the space component of this four vector.

Now we write down the well-known Dirac's relativistic Hamilton function for the electron in the form:

$$(105)\quad W = eV - c\left(\boldsymbol{\gamma} \cdot \left(\boldsymbol{p} - \frac{e\mathbf{U}}{c}\right)\right) - mc^2 \delta .$$

The product containing γ is to be considered as an ordinary scalar product; V and **U** are the scalar and the vector potential.

The Schrödinger equation corresponding to the Hamiltonian (105) can be written putting in evidence the four $\psi_1 \psi_2 \psi_3 \psi_4$. We obtain, remembering the meaning of the operators contained in (105) the four equations:

$$(106\text{ a})\quad (mc^2 + W - eV)\psi_1 = -\frac{ch}{2\pi i}\left[\frac{\partial\psi_4}{\partial x} - i\frac{\partial\psi_4}{\partial y} + \frac{\partial\psi_3}{\partial z}\right] + e\left[(U_x - iU_y)\psi_4 + U_z\psi_3\right]$$

$$(106\text{ b})\quad (mc^2 + W - eV)\psi_2 = -\frac{ch}{2\pi i}\left[\frac{\partial\psi_3}{\partial x} + i\frac{\partial\psi_3}{\partial y} - \frac{\partial\psi_4}{\partial z}\right] + e\left[(U_x + iU_y)\psi_3 - U_z\psi_4\right]$$

$$(106\text{ c})\quad (-mc^2 + W - eV)\psi_3 = -\frac{ch}{2\pi i}\left[\frac{\partial\psi_2}{\partial x} - i\frac{\partial\psi_2}{\partial y} + \frac{\partial\psi_1}{\partial z}\right] + e\left[(U_x - iU_y)\psi_2 + U_z\psi_1\right]$$

$$(106\text{ d})\quad (-mc^2 + W - eV)\psi_4 = -\frac{ch}{2\pi i}\left[\frac{\partial\psi_1}{\partial x} + i\frac{\partial\psi_1}{\partial y} - \frac{\partial\psi_2}{\partial z}\right] + e\left[(U_x + iU_y)\psi_1 - U_z\psi_2\right].$$

It can be proved that the dissymmetry in these equations is only apparent; it arises in some way from the fact that the spin coordinate has been referred to the z-axis, which has therefore a different treatment.

The energy W contains the intrinsic energy mc^2 of the electron; its values are therefore in the neighborhood of mc^2. As well known, the Hamiltonian (105) has also besides these « normal » eigenvalues, « anomalous » ones, which lie near the value $-mc^2$. These negative eigenvalues which have certainly for the electron no physical meaning, would correspond in some way to states of an electron with negative mass. They are supposed to be due to some fault either in the theory or in its interpretation but the tentative assumptions which have been made to get a correct theory cannot at present be claimed successful. In the following sections we shall see the importance of the negative states for the interpretation of actual phenomena, e.g., the scattering of light. Any theory which would try to get rid of the negative states by simply striking them away, should be very careful not to remove the scattering properties of the electron at the same time.

We will now consider for a moment some properties of the normal states with positive energy. Since the energy W lies near mc^2, it can be conveniently written:

$$W = mc^2 + E \tag{107}$$

where E represents the ordinary energy without the term representing the intrinsic energy. For the sake of simplicity we neglect in (106) the terms depending on the vector potential **U**. We see now that in Eqs. (106a) and (106b) the coefficient of ψ_1 and ψ_2 is very large: $(2\,mc^2 + E - eV)$, while ψ_3 and ψ_4 in the first side of (106 c) and (106 d) have a much smaller coefficient: $(E - eV)$. From this we infer that ψ_1 and ψ_2 are much smaller than ψ_3 and ψ_4. From (106 a) and (106 b), neglecting in a first nonrelativistic approximation $E - eV$ with respect to mc^2, we obtain:

$$\left\{\begin{aligned} \psi_1 &= \frac{ih}{4\pi mc}\left(\frac{\partial\psi_4}{\partial x} - i\,\frac{\partial\psi_4}{\partial y} + \frac{\partial\psi_3}{\partial z}\right) \\ \psi_2 &= \frac{ih}{4\pi mc}\left(\frac{\partial\psi_3}{\partial x} + i\,\frac{\partial\psi_3}{\partial y} - \frac{\partial\psi_4}{\partial z}\right) \end{aligned}\right. \tag{108}$$

(from these equations we see that ψ_1 and ψ_2 are smaller than ψ_3 and ψ_4 by a factor of the order of magnitude v/c). We substitute (108) in (106 c) and (106 d) always putting the U's equal to zero, and obtain both for ψ_3 and ψ_4 the ordinary Schrödinger equation

$$\left\{\begin{aligned} (E - eV)\,\psi_3 + \frac{h^2}{8\pi^2 m}\,\Delta\psi_3 &= 0 \\ (E - eV)\,\psi_4 + \frac{h^2}{8\pi^2 m}\,\Delta\psi_4 &= 0\,. \end{aligned}\right. \tag{109}$$

We see therefore that in the nonrelativistic approximation ψ_3 and ψ_4 are eigenfunctions of the ordinary Schrödinger problem, corresponding to the same eigenvalue. Therefore if there is no degeneration in the Schrödinger problem ψ_3 and ψ_4 can differ only by a constant factor, from the normalized eigenfunction corresponding to the eigenvalue E in Schrödinger's equation

(110) $$(E - eV)\, w + \frac{h^2}{8\pi^2 m} \Delta w = 0 .$$

We can take for instance either:

$$\psi_3 = w \quad , \quad \psi_4 = 0$$

or:

$$\psi_3 = 0 \quad , \quad \psi_4 = w .$$

These solutions correspond to the two possible orientations of the spin with respect to the z-axis. From these expressions for ψ_3 and ψ_4, and (108) we obtain the complete expression of the four components of ψ in the form:

(111) $$\begin{cases} \psi_\alpha = \left[\dfrac{ih}{4\pi mc} \dfrac{\partial w}{\partial z} , \dfrac{ih}{4\pi mc} \left(\dfrac{\partial w}{\partial x} + i \dfrac{\partial w}{\partial y} \right), w , 0 \right] \\ \psi_\beta = \left[\dfrac{ih}{4\pi mc} \left(\dfrac{\partial w}{\partial x} - i \dfrac{\partial w}{\partial y} \right), - \dfrac{ih}{4\pi mc} \dfrac{\partial w}{\partial z} , 0 , w \right] \end{cases}$$

ψ_α and ψ_β are the eigenfunctions corresponding to the two orientations o the spin.

We will now write also the expressions of the exact Dirac wave function for the case where there are no forces acting on the electron ($V = 0$, $\mathbf{U} = 0$) and the components of the momentum p_x , p_y , p_z , are therefore constants. Corresponding to these values of the momentum components, ψ must contain the space coordinates x , y , z in the exponential factor $e^{2\pi i (p_x x + p_y y + p_z z)/h}$. The four components of ψ will therefore be products of four constants B_1 , B_2 , B_3 , B_4 by this factor:

(112) $$\psi = [B_1 , B_2 , B_3 , B_4]\, e^{2\pi i (p_x x + p_y y + p_z z)/h} .$$

Putting in (105) this expression for ψ and taking $V = 0$, $\mathbf{U} = 0$, we get for the B's the following equations:

(113) $$\begin{cases} (mc^2 + W)\, B_1 + c\,(p_x - ip_y)\, B_4 + cp_z\, B_3 = 0 \\ (mc^2 + W)\, B_2 + c\,(p_x + ip_y)\, B_3 - cp_z\, B_4 = 0 \\ (-mc^2 + W)\, B_3 + c\,(p_x - ip_y)\, B_2 + cp_z\, B_1 = 0 \\ (-mc^2 + W)\, B_4 + c\,(p_x + ip_y)\, B_1 - cp_z\, B_2 = 0 . \end{cases}$$

It can be readily proved that these linear equations have not identically vanishing solutions only if:

$$W^2 = m^2 c^4 + c^2 p^2$$

that is:

(114) $$W = \pm\, (m^2 c^4 + c^2 p^2)^{1/2}$$

This is the ordinary relativistic relation between energy and momentum. The + sign corresponds to the ordinary positive values of the energy; the — sign to the anomalous negative energy values. For each of the two energy values (114) there are two linear independent solutions of (113) which correspond to the two possible orientations of the spin. They can be written in the form:

$$(115\text{ a})\quad \begin{cases} [B_1, B_2, B_3, B_4] = \left(1+\dfrac{c^2p^2}{(mc^2+W)^2}\right)^{-1/2}\left[-\dfrac{cp_z}{mc^2+W}, -\dfrac{c(p_x+ip_y)}{mc^2+W}, 1, 0\right] \\ [B_1, B_2, B_3, B_4] = \left(1+\dfrac{c^2p^2}{(mc^2+W)^2}\right)^{-1/2}\left[-\dfrac{c(p_x-ip_y)}{mc^2+W}, \dfrac{cp_z}{mc^2+W}, 0, 1\right] \end{cases}$$

for the positive energy values; and in the form:

$$(115\text{ b})\quad \begin{cases} [B_1, B_2, B_3, B_4] = \left(1+\dfrac{c^2p^2}{(mc^2-W)^2}\right)^{-1/2}\left[1, 0, \dfrac{cp_z}{mc^2-W}, \dfrac{c(p_x+ip_y)}{mc^2-W}\right] \\ [B_1, B_2, B_3, B_4] = \left(1+\dfrac{c^2p^2}{(mc^2-W)^2}\right)^{-1/2}\left[0, 1, \dfrac{c(p_x-ip_y)}{mc^2-W}, -\dfrac{cp_z}{mc^2-W}\right] \end{cases}$$

for the negative energy values.

The normalization factors $(1+c^2p^2/(mc^2\pm W)^2)^{-1/2}$ have been so chosen that the sum of the square modulus of the B's is unity. For vanishing momentum $\boldsymbol{p}$, the B's take the very simple expressions:

$$(115\text{ c})\quad \begin{cases} \left.\begin{array}{l}[B_1\, B_2\, B_3\, B_4] = [0\,0\,1\,0] \\ [B_1\, B_2\, B_3\, B_4] = [0\,0\,0\,1]\end{array}\right\} W = +mc^2 \\ \left.\begin{array}{l}[B_1\, B_2\, B_3\, B_4] = [1\,0\,0\,0] \\ [B_1\, B_2\, B_3\, B_4] = [0\,1\,0\,0]\end{array}\right\} W = -mc^2. \end{cases}$$

§ 14. - Radiation theory in nonrelativistic approximation.

We shall restrict ourselves to the case for which the electron of our atom can be considered in an electrostatic field of force, and the only nonelectrostatic forces are those due to the radiation field. In Eq. (105) we may therefore suppose that V is independent of the time, and represents the electrostatic potential of the atom, while **U** vanishes if we neglect the interaction of the atom and the radiation field; if we do not neglect this interaction we put for **U** the expression (12) of the vector pontential of radiation. The Hamilton function (105) of the electron becomes then:

$$(116)\qquad eV - c(\gamma\cdot\boldsymbol{p}) - mc^2\delta + ec\left(\frac{8\pi}{\Omega}\right)^{1/2}\sum_s(\gamma\cdot\mathbf{A}_s)\,q_s\sin\Gamma_s$$

where the last term represents the effect of the radiation field.

We get the Hamilton function of the complex system of the atom and the radiation field, adding to (116) the Hamiltonian (11) of the radiation. We get thus:

$$(117)\qquad H = eV - e(\gamma\cdot\boldsymbol{p}) - mc^2\delta + \sum_s\left(\frac{p_s^2}{2}+2\pi^2\nu_s^2q_s^2\right) + ec\left(\frac{8\pi}{\Omega}\right)^{1/2}\sum_s(\gamma\cdot\mathbf{A}_s)\,q_s\sin\Gamma_s.$$

The Hamiltonian function (117) can be split up into the sum of an unperturbed Hamiltonian

$$(118)\qquad H_0 = eV - c(\gamma\cdot\boldsymbol{p}) - mc^2\delta + \sum_s\left(\frac{p_s^2}{2}+2\pi^2\nu_s^2q_s^2\right)$$

and a perturbation term:

$$\mathcal{H} = ec\left(\frac{8\pi}{\Omega}\right)^{1/2} \sum_s (\boldsymbol{\gamma}\cdot\mathbf{A}_s)\, q_s \sin\Gamma_s \tag{119}$$

representing the interaction energy.

The eigenfunctions of the unperturbed problem represented by H_0 can be very easily written, since H_0 is the sum of a term containing the coordinates of the electron only, and of terms each containing the only variables, q_s, p_s of the sth radiation component. The unperturbed eigenfunctions are therefore products of eigenfunctions of the atom and the radiation oscillators as in formula (31)

$$\varphi_{nn_1 n_2 \cdots n_s \cdots} = u_n\, u_{n_1}\, u_{n_2} \cdots u_{n_s} \cdots \tag{120}$$

where the symbols are the same as in (31).

We must now calculate the matrix elements of the perturbation (119)

$$\mathcal{H}_{nn_1 n_2 \cdots n_s \cdots\,;\, mm_1 m_2 \cdots m_s \cdots}$$

Its calculation is practically identical with the calculation for the derivation of (41). It is found that the only matrix elements which don't vanish identically are:

$$\mathcal{H}_{nn_1 n_2 \cdots n_s \cdots\,;\, mn_1 n_2 \cdots n_s \pm 1 \cdots} = ec\left(\frac{h}{\pi\Omega}\right)^{1/2} \frac{1}{\nu_s^{1/2}} (\mathbf{D}_{snm}\cdot\mathbf{A}_s) \left| \begin{matrix} (n_s+1)^{1/2} \\ n_s^{1/2} \end{matrix} \right. \tag{121}$$

where the upper or the lower expression of the last factor must be taken according to the two possibilities $m_s = n_s + 1$ or $m_s = n_s - 1$. The vector $\mathbf{D}_{snm}$ has the following meaning:

$$\mathbf{D}_{snm} = \int \bar{u}_n\, \boldsymbol{\gamma} \sin\Gamma_s\, u_m\, d\omega \tag{122}$$

where integration must be extended over all the configuration space for the electron (that is: integrate over the space coordinates x, y, z from $-\infty$ to $+\infty$, and sum over the four values of the spin variable).

The expression of $\mathbf{D}_{snm}$ can be very much simplified if we make the assumption that the dimensions of the atom are much smaller than the wavelength. In this case we may consider $\sin\Gamma_s$ as a constant all over the space occupied by the atom and take it out of the integral (122). We get thus:

$$\mathbf{D}_{snm} = \sin\Gamma_s \int \bar{u}_n\, \boldsymbol{\gamma}\, u_m\, d\omega = \sin\Gamma_s\, \boldsymbol{\gamma}_{nm} \tag{123}$$

where $\boldsymbol{\gamma}_{nm}$ is the matrix element of the operator $\boldsymbol{\gamma}$.

We calculate the last factor by the nonrelativistic approximation (111) for the eigenfunctions. We take first both for u_n and u_m eigenfunctions of the type ψ_α (spin in the positive z-direction). We have:

$$u_m = \left[\frac{ih}{4\pi mc}\frac{\partial w_m}{\partial z}, \frac{ih}{4\pi mc}\left(\frac{\partial w_m}{\partial x} + i\frac{\partial w_m}{\partial y}\right), w_m, 0\right].$$

Let us calculate the x-component of the vector (123); remembering the meaning of γ_x we obtain:

$$\gamma_x u_m = \left[0, w_m, \frac{ih}{4\pi mc}\left(\frac{\partial w_m}{\partial x} + i\frac{\partial w_m}{\partial y}\right), \frac{ih}{4\pi mc}\frac{\partial w_m}{\partial z}\right].$$

We have also:

$$\bar{u}_n = \left[-\frac{ih}{4\pi mc}\frac{\partial w_n}{\partial z}, -\frac{ih}{4\pi mc}\left(\frac{\partial w_n}{\partial x} - i\frac{\partial w_n}{\partial y}\right), w_n, 0\right].$$

We get $\int \bar{u}_n \gamma_x u_m d\omega$, summing up the products of the corresponding four components of the two last expressions, and integrating over all space. We get:

$$\int \bar{u}_n \gamma_x u_m d\omega = -\frac{ih}{4\pi mc}\int\left(w_m\frac{\partial w_n}{\partial x} - w_n\frac{\partial w_m}{\partial x}\right)d\tau$$

$$-\frac{h}{4\pi mc}\int\left(w_m\frac{\partial w_n}{\partial y} + w_n\frac{\partial w_m}{\partial y}\right)d\tau$$

$$= \frac{ih}{2\pi mc}\int w_n\frac{\partial w_m}{\partial x}d\tau - \frac{ih}{4\pi mc}\int\frac{\partial}{\partial x}(w_n w_m)\,d\tau - \frac{h}{4\pi mc}\int\frac{\partial}{\partial y}(w_n w_m)\,d\tau.$$

The last two integrals can be transformed by Green's theorem into surface integrals extended over a surface at infinite distance; since the eigenfunctions w_n and w_m decrease very rapidly they vanish. We obtain therefore:

$$\int \bar{u}_n \gamma_x u_m d\omega = \frac{ih}{2\pi mc}\int w_n\frac{\partial w_m}{\partial x}d\tau.$$

Calculating with a similar method the components y and x we obtain vectorially:

$$\int \bar{u}_n \boldsymbol{\gamma} u_m d\omega = \frac{ih}{2\pi mc}\int w_n \operatorname{grad} w_m d\tau \tag{124}$$

which shows the affinity between the operator $\boldsymbol{\gamma}$ of Dirac's theory and the expression $-\boldsymbol{p}/mc = -\boldsymbol{v}/c$ of Schrödinger's theory (remember that $\boldsymbol{p} \sim [h/2\pi i]$ grad).

From (42) (observe that the u's in (42) are the Schrödinger eigenfunctions, so they correspond to our present w's) we obtain now:

$$\int \bar{u}_n \boldsymbol{\gamma} u_m d\omega = (2\pi i/c)\,\nu_{mn}\,\mathbf{X}_{nm} \tag{125}$$

where $\mathbf{X}_{nm}$ is the matrix element representing the radius vector in Schrödinger's approximation. From (123) and (121) we obtain at last:

$$\mathcal{H}_{mn_1\cdots n_s\cdots\,;\,mn_1\cdots n_s\pm 1\cdots} = 2\pi ie\left(\frac{h}{\pi\Omega}\right)^{1/2}\frac{\nu_{mn}}{\nu_s^{1/2}}(\mathbf{A}_s\cdot\mathbf{X}_{nm})\left[\frac{(n_s+1)^{1/2}}{n_s^{1/2}}\right]\sin\Gamma_s \tag{126}$$

which is identical to (46) derived previously.

This shows that in the present nonrelativistic approximation the results obtained for the radiation theory with Dirac's wave function are completely identical to those derived in Part I with Schrödinger's function.

We notice further that (125) has been derived on the assumption that the states n and m have their spins pointing in the same direction. If their spins point in opposite direction the result is zero. This fact can at first sight seem contradictory, since it means that there are no transitions between states with opposite spin directions. But we must remember that we have made the calculations on the hypothesis that the eigenfunctions of the Schrödinger problem are not degenerate. The effect of this is that there is no coupling between spin and orbital movement, as in s-terms; where changes of spin direction are observed, this is only due to the coupling of spin and orbit and if this coupling is loosened, as in Paschen-Back effect, no changes of spin direction actually occur.

§ 15. - DIRAC'S THEORY AND SCATTERING FROM FREE ELECTRONS.

The theory of scattering of light from free electrons has some interest, as we have said, because it shows in a very striking way the actual importance of the states with negative energy even for very real phenomena where these mysterious states do not explicitly appear. This theory can of course be carried out either to the approximation of the Compton effect or to the approximation giving simply Thomson's coefficient of scattering and no change of frequency. Since the essential features of the theory are conserved also if we neglect the change of frequency, we shall carry out the theory to this approximation (i.e., we shall not consider the momentum properties of light quanta). The exact theory leads to the intensity formula of Klein-Nishina.

The approximation introduced by neglecting the momentum properties of light quanta is equivalent, as we have often said, to considering the phase Γ_s of light constant over the place occupied by the electron. We may also suppose that the velocity of the electron during the process of scattering is always negligible, since we neglect the recoil of the scattered quanta. We can take therefore the eigenfunctions of the electron in the very simple form (115 c) corresponding to velocity zero. We shall indicate the four states (115 c) by the indices 1, 2, 3, 4. States 1, 2 are states of positive energy $+ mc^2$, with spin pointing in the direction $+ z$ and $- z$; states 3, 4 have negative energy $- mc^2$ with the spin pointing in the directions $\pm z$.

We will suppose that at the beginning ($t = 0$) the electron is in the state 1 and there are n_s quanta in the s-component of radiation (primary radiation). We can put therefore:

$$a_{1,0\cdots n_s,0\cdots} = 1 \tag{127}$$

while the other a's are zero. To find out the amount of radiation scattered into the σ-component, we must find the value of $a_{1,0\cdots n_s - 1,\cdots 1_\sigma\cdots}$ at the time t. Now the matrix element of the perturbation corresponding to this transition is zero, since two radiation components (s and σ) change their quantum number; the transition can therefore occur only through an intermediate state, which can combine both with the initial and the final state. It is easily seen that there are only four such states:

$$(3, n_s - 1, 0), (4, n_s - 1, 0), (3, n_s, 1), (4, n_s, 1). \tag{128}$$

For brevity only the quantum numbers of the electron and of the radiation components, s, σ have been indicated. The states $(2, n_s - 1, 0)$ and $(2, n_s, 1)$ have not been considered since it is immediately shown by the definition that $\gamma_{12} = 0$ so that these states do not combine.

The intermediate states (128) are states for which the electron has a negative energy; without these states of negative energy no scattering process would be possible. We will now show, that the scattering calculated with the intermediate states (128) actually gives Thomson's intensity formula.

For this we must first write the matrix elements corresponding to the transition from the initial state $(1, n_s, 0)$ to the intermediate states (128), and from these states to the final state $(1, n_s - 1, 1)$. From (115 c) we immediately find:

$$(129) \quad \begin{cases} \gamma_{x13} = 0 \quad ; \quad \gamma_{y13} = 0 \quad ; \quad \gamma_{z13} = 1 \\ \gamma_{x14} = 1 \quad ; \quad \gamma_{y14} = i \quad ; \quad \gamma_{z14} = 1 . \end{cases}$$

Interchanging of the indices changes the corresponding γ into its complex conjugate value: (e.g, $\gamma_{y41} = -i\gamma_{y14}$). From (121) (123) 129) we obtain the required matrix elements: they are:

$$(130) \quad \begin{cases} \mathcal{H}_{3,n_s-1,0\,;\,1,n_s,0} = ec\left(\frac{h}{\pi\Omega}\right)^{1/2} \frac{\sin\Gamma_s}{\nu_s^{1/2}} A_{sz}\, n_s^{1/2} \\ \mathcal{H}_{4,n_s-1,0\,;\,1,n_s,0} = ec\left(\frac{h}{\pi\Omega}\right)^{1/2} \frac{\sin\Gamma_s}{\nu_s^{1/2}} (A_{sx} - iA_{sy})\, n_s^{1/2} \\ \mathcal{H}_{3,n_s,1\,;\,1,n_s,0} = ec\left(\frac{h}{\pi\Omega}\right)^{1/2} \frac{\sin\Gamma_\sigma}{\nu_\sigma^{1/2}} A_{\sigma z} \\ \mathcal{H}_{4,n_s,1\,;\,1,n_s,0} = ec\left(\frac{h}{\pi\Omega}\right)^{1/2} \frac{\sin\Gamma_\sigma}{\nu_\sigma^{1/2}} (A_{\sigma x} - iA_{\sigma y}) \end{cases}$$

for the transitions from the initial states $(1, n_s, 0)$ to the intermediate states (128); and

$$(131) \quad \begin{cases} \mathcal{H}_{1,n_s-1,1\,;\,3,n_s-1,0} = ec\left(\frac{h}{\pi\Omega}\right)^{1/2} \cdot \frac{\sin\Gamma_\sigma}{\nu_\sigma^{1/2}} A_{\sigma z} \\ \mathcal{H}_{1,n_s-1,1\,;\,4,n_s-1,0} = ec\left(\frac{h}{\pi\Omega}\right)^{1/2} \frac{\sin\Gamma_\sigma}{\nu_\sigma^{1/2}} (A_{\sigma x} + iA_{\sigma y}) \\ \mathcal{H}_{1,n_s-1,1\,;\,3,n_s,1} = ec\left(\frac{h}{\pi\Omega}\right)^{1/2} \frac{\sin\Gamma_s}{\nu_s^{1/2}} A_{sz}\, n_s^{1/2} \\ \mathcal{H}_{1,n_s-1,1\,;\,4,n_s,1} = ec\left(\frac{h}{\pi\Omega}\right)^{1/2} \frac{\sin\Gamma_s}{\nu_s^{1/2}} (A_{sx} + iA_{sy})\, n_s^{1/2} \end{cases}$$

for the transitions from the intermediate states to the final state $(1, n_s - 1, 1)$.

From these matrix elements and the general formula (26) we easily calculate the amplitudes of probability for the intermediate states. If we limit ourselves to a very short interval of time we may still suppose (127) to be valid in first approximation, and we obtain from (26):

$$\dot{a}_{3,n_s-1,0} = -\frac{2\pi i}{h} \mathcal{H}_{3,n_s-1,0\,;\,1,n_s,0}\, a_{1,n_s,0}\, e^{2\pi i(-2mc^2 - h\nu_s)t/h}$$

(notice the very big change in the energy due to the transition of the electron from state 1 with positive energy $+mc^2$, to state 3 with negative energy $-mc^2$). Putting $a_{1,n_s,0} = 1$ and remembering (130) we obtain:

$$\dot{a}_{3,n_s-1,0} = -\frac{2\pi i}{h} ec\left(\frac{h}{\pi\Omega}\right)^{1/2} \frac{\sin\Gamma_s}{\nu_s^{1/2}} n_s^{1/2} A_{sz}\, e^{-2\pi i(2mc^2 + h\nu_s)t/h} .$$

Integrating and neglecting $h\nu_s$, in comparison with $2mc^2$ in the denominator we obtain:

$$a_{3,n_s-1,0} = \frac{e}{2mc}\left(\frac{h}{\pi\Omega}\right)^{1/2} \frac{\sin\Gamma_s}{\nu_s^{1/2}} n_s^{1/2} A_{sz}\, e^{-2\pi i(2mc^2+h\nu_s)t/h}. \tag{132}$$

Notice that the integration constant should have been determined with the condition $a_{3,n_s-1,0}=0$ for $t=0$; instead we have chosen the constant so that the mean value of $a_{3,n_s-1,0}$, is zero. This corresponds exactly to what is done also in the classical theory of dispersion of light from a harmonic oscillator; one considers in that case the motion of the oscillator to be represented simply by the forced vibrations and one neglects the vibrations with the characteristic frequency of the oscillator which are superposed on them. The justification of this classical proceeding lies in the well known fact that the vibrations of characteristic frequency are very rapidly damped by the reaction of radiation, so that, in the permanent state, only the forced vibrations remain. The justification in our case is quite similar; it could be shown that the effect of an integration constant added to (132) would be very rapidly damped from the reaction of states like $(1, n_s-1, 1)$ which has been neglected in our calculations.

The amplitudes of probability for the other intermediate states are deduced with exactly the same method as (132). They are:

$$\left\{\begin{aligned} a_{4,n_s-1,0} &= \frac{e}{2mc}\left(\frac{h}{\pi\Omega}\right)^{1/2} \frac{\sin\Gamma_s}{\nu_s^{1/2}} n_s^{1/2} (A_{sx}-iA_{sy})\, e^{-2\pi i(2mc^2+h\nu_s)t/h} \\ a_{3,n_s,1} &= \frac{e}{2mc}\left(\frac{h}{\pi\Omega}\right)^{1/2} \frac{\sin\Gamma_\sigma}{\nu_\sigma^{1/2}} A_{\sigma z}\, e^{-2\pi i(2mc^2-h\nu_\sigma)t/h} \\ a_{4,n_s,1} &= \frac{e}{2mc}\left(\frac{h}{\pi\Omega}\right)^{1/2} \frac{\sin\Gamma_\sigma}{\nu_\sigma^{1/2}} (A_{\sigma x}-iA_{\sigma y})\, e^{-2\pi i(2mc^2-h\nu_\sigma)t/h}. \end{aligned}\right. \tag{133}$$

We apply now once more the general formula (26) to the calculation of the amplitude of probability for the final state. We get:

$$\dot a_{1,n_s-1,1} = -\frac{2\pi i}{h}\left[\mathcal{H}_{1,n_s-1,1;\,3,n_s-1,0}\, a_{3,n_s-1,0}\, e^{-2\pi i(2mc^2+h\nu_\sigma)t/h} + \dots\right]$$

where similar terms for the other three intermediate states have been omitted. With (132) (133) and (131) we find now:

$$\dot a_{1,n_s-1,1} = -\frac{2i}{\Omega}\frac{e^2}{m}\left(\frac{n_s}{\nu_s\nu_\sigma}\right)^{1/2} \sin\Gamma_s \sin\Gamma_\sigma\,(A_s A_\sigma)\, e^{2\pi i(\nu_\sigma-\nu_s)t/h}. \tag{134}$$

This equation coincides exactly with equation (94) obtained in the theory of scattering made without using Dirac's wave equation. By exactly the same method used for (94) we deduce from (134) Thomson's formula for the intensity of scattered radiation.

The very profound difference between these two theories of scattering should be emphasized; in the first theory, deduced from the Schrödinger wave equation for the electron, the scattering effect was due to the presence in the Hamiltonian of the term (17). This term is quadratic (and not linear) in the vector potential, and therefore enables transitions for which a quantum jumps in a single act from one radiation component to another. In Dirac's wave equation only terms linear in the potentials are contained; this has the effect that no direct transitions between two states can occur, if more than one radiation component changes its quantum number. Therefore it would seem probable at first sight that Dirac's relativistic free electron has no scattering properties. We have shown however that this conclusion is wrong; the scattering properties come out if one properly takes into account also the negative states. Scattering appears as a sort of resonance (very far from the resonance line) of the quantum jump (of energy $2\,mc^2$) between the positive and the negative states.

§ 16. – Radiative transitions from positive to negative states.

We have seen in the preceding sections that a very great number of phenomena find their natural explanation in Dirac's theory of radiation. We will now briefly discuss some serious difficulties of this theory. They are mainly connected with difficulties in the theory of the electron.

It is well known that the most serious difficulty in Dirac's relativistic wave equation lies in the fact that it yields besides the normal positive states also negative ones, which have no physical significance. This would do no harm if no transition between positive and negative states were possible (as are, e. g., transitions between states with symmetrical and antisymmetrical wave function). But this is unfortunately not the case: Klein has shown by a very simple example that electrons impinging against a very high potential barrier have a finite probability of going over into a negative state.

Dirac has tried with a very keen hypothesis to overcome these difficulties. He postulates that there are in every portion of space an infinite number of electrons which fill nearly completely in the sense of Pauli's principle, all the states of negative energy; a transition from a positive to a negative state therefore occurs very seldom since only a few negative states are unoccupied. Dirac goes further with the hypothesis, as he postulates that the unoccupied, negative places, the « holes », are to be interpreted as protons; in fact it is easily seen that a hole behaves like a positive charge with positive mass. The quantum transition in which an electron jumps from a positive state into a hole would therefore correspond to a hypothetic process of annihilation of an electron and a proton, with radiation of the energy corresponding to their masses.

Oppenheimer, Dirac and Tamm have calculated the probability of transition from a positive to a negative state with radiation of the energy difference. From the standpoint that the negative states are equivalent to protons, their result gives the rate of annihilation of electrons and protons.

Without carrying out these calculations in any detail, we limit ourselves to some qualitative remarks. Let us discuss the probability of radiative transition between a state (1) in which the momentum of the electron is $\boldsymbol{p}_1 = 0$ and its energy $W_1 = + mc^2$ to a state (2) where its energy is negative $W_2 = - mc^2$ and its momentum is $\boldsymbol{p}_2 = 0$.

It is evident that the energy difference $W_1 - W_2 = 2\, mc^2$ can not be radiated in a single quantum, since the momentum condition $p_1 - p_2 = 2\, mc$ is not verified. It is however possible to obtain a finite probability of transition between the two states 1, 2 with emission of two quanta having both energy mc^2/h and opposite direction of motion. This process is of course consistent both with energy and momentum conservation.

The process will therefore happen in two steps. First step: a quantum of energy mc^2 is emitted and the electron receives the recoil going over to a state with momentum mc. The energy of the electron in this state is (114): $W = \pm\, 2^{1/2}\, mc^2$. This intermediate state does not satisfy also conservation of (unperturbed) energy; as we have often seen in preceding instances, the amplitude of probability for this state can not continuously increase with time, but it can nevertheless be different from zero though having very small oscillating values. From this intermediate state a direct transition to the final state, with emission of a quantum of energy mc^2 and momentum opposite to the momentum of the first quantum is possible; since this last state satisfies energy and momentum conservation, it is actually possible to show that its amplitude of probability steadily increases giving a finite rate of transition from initial to final state.

Carrying out the calculation, the required probability of transition per unit time results:

$$\pi e^4/m^2\, c^3\,. \tag{135}$$

If the electron in the negative state has not momentum zero, but has the energy $W' = - mc^2\, \alpha$, $(\alpha \geqq 1)$ the probability of transition becomes:

$$\frac{\pi e^4}{m^2\, c^3}\, \frac{1}{\alpha\,(\alpha + 1)} \left[\frac{\alpha^2 + 4\,\alpha + 1}{(\alpha^2 - 1)^{1/2}} \log\, [\alpha + (\alpha^2 - 1)^{1/2}] - \alpha - 3 \right]. \tag{136}$$

If we assume that all the negative states are empty, formula (136) summed over all negative states would give an infinite probability for the transition from a positive to a negative state: electrons could not remain in a positive state, not even for a very short time.

If we assume on the other hand the "hole" theory of protons, the theory of the transitions becomes very uncertain, since the electron is in that case surrounded by an infinite number of other (negative) electrons. The interaction effects of these electrons are neglected in the theory, though it is evident that they might have enormous effects. Dirac suggests that this interaction might be responsible for the difference in mass of the electron and the proton. If we tentatively try to apply (136) to the process of annihilation of an electron and a proton, putting for m some mean value between the masses of the electron and the proton, the rate of annihilation comes out much too rapid; matter would be destroyed in a very short time.

PART III. — QUANTUM ELECTRODYNAMICS

§ 17. - The electromagnetic field, whose interaction with matter we have hitherto considered, is not an electromagnetic field of the most general type, since a field of general type cannot be constructed by simply superposing plane electromagnetic waves. It can be immediately seen that in a plane electromagnetic wave $\operatorname{div} \mathbf{E} = 0$ and this equation holds also for **any** superposition of waves. Instead in a general electromagnetic field we have $\operatorname{div} \mathbf{E} = + 4\pi\rho$, ρ being the density of electricity; this shows that no field, where charges are present, can be represented as a superposition of electromagnetic waves.

An electromagnetic field of the most general type is represented by help of a scalar potential V and a vector potential **U** by the well-known relations:

$$\mathbf{E} = -\operatorname{grad} V - \frac{1}{c}\frac{\partial \mathbf{U}}{\partial t} \quad ; \quad \mathbf{H} = \operatorname{rot} \mathbf{U}. \tag{137}$$

V and **U** are classically connected to the density of charge and the velocity $\dot{\mathbf{X}}$ by:

$$\Delta V - \frac{1}{c^2}\frac{\partial^2 V}{\partial t^2} = -4\pi\rho \quad ; \quad \Delta \mathbf{U} - \frac{1}{c^2}\frac{\partial^2 \mathbf{U}}{\partial t^2} = -\frac{4\pi}{c}\rho\dot{\mathbf{X}}. \tag{138}$$

Further **U** and V are not completely independent of each other; they satisfy the relation:

$$\operatorname{div} \mathbf{U} + \frac{1}{c}\frac{\partial V}{\partial t} = 0 \tag{139}$$

which is closely connected to the equation of continuity for the electricity.

A general quantum theory of the electromagnetic field was constructed by Heisenberg and Pauli by a method in which the values of the electromagnetic potentials in all the points of space are considered as variables.

Independently the writer proposed another method of quantization of the electromagnetic field starting from a Fourier analysis of the potentials. Though Heisenberg and Pauli's method puts in evidence much more clearly the properties of relativistic invariance and is in many respects more general, we prefer to use in this article the method of the writer, which is more simple and more analogous to the methods used in the theory of radiation.

We will consider only a region of space of finite volume and we suppose that both scalar and vector potential at a given time can be represented by Fourier series of the type:

$$\left\{ \begin{aligned} V &= c\left(\frac{8\pi}{\Omega}\right)^{1/2} \sum_s Q_s \cos \Gamma_s \\ \mathbf{U} &= c\left(\frac{8\pi}{\Omega}\right)^{1/2} \sum_s \boldsymbol{u}_s \sin \Gamma_s \end{aligned} \right. \tag{140}$$

where Q_s and $\boldsymbol{u}_s$ are a scalar and a vector function respectively of the time only. The factor $c(8\pi/\Omega)^{1/2}$ has been put for convenience of normalization as in (12). Γ_s is given by (4). It is convenient to develop V and **U** in series of $\cos \Gamma_s$ and $\sin \Gamma_s$ respectively, since in this case Eq. (139) takes a much simpler form. It should finally be noticed that the number of characteristic frequencies between ν_s and $\nu_s + d\nu_s$ is to be taken equal to:

$$(4\pi/c^3)\,\Omega\nu_s^2\,d\nu_s \tag{141}$$

i.e., to the half of (1), since in our case the two possibilities of polarization for the transverse waves are taken in account by the fact that $\boldsymbol{u}_s$ is a vector.

As variables representing the field at a given time we take Q_s and the three components of the vector $\boldsymbol{u}_s$; it is convenient however to take these components in directions related to the form of the phase factor $\sin \Gamma_s$; we consider three mutually perpendicular unit vectors: $\boldsymbol{\alpha}_s$, which points in the direction of the wave, $\mathbf{A}_{s1}$ and $\mathbf{A}_{s2}$ perpendicular to that direction. Let χ_s, q_{s1}, q_{s2} be the components of $\boldsymbol{u}_s$ in the directions $\boldsymbol{\alpha}_s$, $\mathbf{A}_{s1}$, $\mathbf{A}_{s2}$; we have then:

$$\boldsymbol{u}_s = \boldsymbol{\alpha}_s\chi_s + \mathbf{A}_{s1}q_{s1} + \mathbf{A}_{s2}q_{s2}\,. \tag{142}$$

As variables describing the field we can take:

$$Q_s\,,\ \chi_s\,,\ q_{s1}\,,\ q_{s2}\,. \tag{143}$$

They depend only on the time.

It is very easy to deduce from (138) the differential equations that determine the dependence of the variables (143) on the time. Multiply both members of the first Eq. (138) by $\cos \Gamma_s\,d\tau$ and integrate over all the space. We suppose that the potential V vanishes on the very distant surface limiting our space Ω, so that certain integrals over that surface can be omitted; we obtain then by obvious transformations:

$$\begin{aligned}-4\pi\int\rho\cos\Gamma_s\,d\tau &= \int\Delta V\cdot\cos\Gamma_s\,d\tau - \frac{1}{c^2}\int\frac{\partial^2 V}{\partial t^2}\cos\Gamma_s\,d\tau\\ &= \int V\cdot\Delta\cos\Gamma_s\,d\tau - \frac{1}{c^2}\int\frac{\partial^2 V}{\partial t^2}\cos\Gamma_s\,d\tau\,.\end{aligned}$$

From (4) we obtain:

$$\Delta\cos\Gamma_s = -\frac{4\pi^2\nu_s^2}{c^2}\cos\Gamma_s\,.$$

We have therefore:

$$+4\pi c^2\int\rho\cos\Gamma_s\,d\tau = \left(\frac{d^2}{dt^2} + 4\pi^2\nu_s^2\right)\int V\cos\Gamma_s\,d\tau\,. \tag{144}$$

Putting for V its expression (140) and remembering that the functions $\cos \Gamma_s$ are orthogonal and satisfy the relation

$$\int\cos\Gamma_s\cos\Gamma_\sigma\,d\tau = \frac{1}{2}\Omega\delta_{s\sigma}$$

we obtain:

$$\int V\cos\Gamma_s\,d\tau = c\,(2\pi\Omega)^{1/2}\,Q_s\,.$$

From (144) we obtain therefore:

$$\ddot{Q}_s + 2\pi^2 \nu_s^2 Q_s = c\left(\frac{8\pi}{\Omega}\right)^{1/2} \int \rho \cos \Gamma_s \, d\tau . \tag{145}$$

This equation takes a much simpler form if we suppose that there are only point charges $e_1, e_2, e_3, \cdots$, at the points $\mathbf{X}_1, \mathbf{X}_2, \mathbf{X}_3, \cdots$ The integral in (145) becomes then a sum over the point charges and we obtain:

$$\ddot{Q}_s + 2\pi^2 \nu_s^2 Q_s = c\left(\frac{8\pi}{\Omega}\right)^{1/2} \sum_i e_i \cos \Gamma_{si} \tag{146}$$

where the sum has to be extended over all the charges; Γ_{si} is the value of the phase Γ_s at the place $\mathbf{X}_i$ of the ith charge

$$\Gamma_{si} = \frac{2\pi\nu_s}{c}(\alpha_s \cdot \mathbf{X}_i) + \beta_s . \tag{147}$$

By the same method we find a similar equation for the vector $\boldsymbol{u}_s$:

$$\ddot{\boldsymbol{u}}_s + 2\pi^2 \nu_s^2 \boldsymbol{u}_s = \left(\frac{8\pi}{\Omega}\right)^{1/2} \sum_i e_i \dot{\mathbf{X}}_i \sin \Gamma_{si} . \tag{148}$$

Remembering (142), we find that the three components of this vector equation in the three direction α_s, $\mathbf{A}_{s1}$, $\mathbf{A}_{s2}$ are:

$$\ddot{\chi}_s + 2\pi^2 \nu_s^2 \chi_s = \left(\frac{8\pi}{\Omega}\right)^{1/2} \sum_i e_i (\alpha_s \cdot \dot{\mathbf{X}}_i) \sin \Gamma_{si} \tag{149}$$

$$\left\{\begin{aligned} \ddot{q}_{s1} + 2\pi^2 \nu_s^2 q_{s1} &= \left(\frac{8\pi}{\Omega}\right)^{1/2} \sum_i e_i (\mathbf{A}_{s1} \cdot \dot{\mathbf{X}}_i) \sin \Gamma_{si} \\ \ddot{q}_{s2} + 2\pi^2 \nu_s^2 q_{s2} &= \left(\frac{8\pi}{\Omega}\right)^{1/2} \sum_i e_i (\mathbf{A}_{s2} \cdot \dot{\mathbf{X}}_i) \sin \Gamma_{si} . \end{aligned}\right. \tag{150}$$

The Eqs. (146), (149), (150) are equivalent to (138).

Take the derivative of (146) with respect to t and add it to (149) multiplied by $2\pi\nu_s$. Then we find:

$$\left(\frac{d^2}{dt^2} + 4\pi^2 \nu_s^2\right)(2\pi\nu_s \chi_s + \dot{Q}_s) = 0 . \tag{151}$$

This equation is evidently satisfied if:

$$2\pi\nu_s \chi_s + \dot{Q}_s = 0 . \tag{152}$$

It is immediately seen that this last equation is equivalent to (139). Eq. (152) does not follow directly from the differential equations (146), (149); though it results from (151) that, if at a given time (e.g., $t = 0$), (152) and its derivative with respect to time

$$2\pi\nu_s \dot{\chi}_s + \ddot{Q}_s = 0 \tag{153}$$

are both satisfied, then they are satisfied for all time.

We must now write in Hamiltonian form the equations that describe the motion of the particles and the variation of the electromagnetic field. For this we simply write the Hamilton function and then verify that the canonical equations that can be derived by it actually represent the motion of the particles and the Maxwell equations. The Hamilton function is the following:

$$(154)\quad H = -c\sum_i(\gamma_i\cdot p_i) - \sum_i m_i c^2\delta_i + \sum_i e_i c\left(\frac{8\pi}{\Omega}\right)^{1/2}\sum_s Q_s\cos\Gamma_{si}$$
$$+\sum_i e_i c\left(\frac{8\pi}{\Omega}\right)^{1/2}\sum_s(\gamma_i\cdot(\alpha_s\chi_s + \mathbf{A}_{s1}q_{s1} + \mathbf{A}_{s2}q_{s2}))\sin\Gamma_{si}$$
$$+\sum_s\left[\frac{1}{2}(p_{s1}^2 + p_{s2}^2 + \omega_s^2 - P_s^2) + 2\pi^2\nu_s^2(q_{s1}^2 + q_{s2}^2 + \chi_s^2 - Q_s^2)\right].$$

In this Hamilton function the variables are $\mathbf{X}_i$, and the spin coordinates, describing the motion of the particles; p_i are the momenta (vectors) conjugated to the coordinates $\mathbf{X}_i$; Q_s, χ_s, q_{s1}, q_{s2} are the coordinates describing the field and P_s, ω_s, p_{s1}, p_{s2} are their conjugated momenta. γ_i and δ_i represent operators analogous to Dirac's operators γ and δ of Eq. (105) operating on the spin coordinate of the ith particle. The structure of the Hamiltonian (154) is very simple. Remembering (140), its first four terms can be written:

$$(155)\quad \sum_i\left\{e_i V_i - c\left(\gamma_i\cdot\left(p_i - \frac{e_i}{c}\mathbf{U}_i\right)\right) - m_i c^2\delta_i\right\}$$

which is simply the repetition of Dirac's Hamilton function (105) for all the particles. The last term of (154) represents the Hamiltonian function of the electromagnetic field without interaction with the charges, and is analogous to (11).

From this we see clearly that the Hamiltonian (154) correctly represents the motion of the particles, since their coordinates are contained in (155) which is equivalent to Dirac's Hamilton function. We must show that also the Maxwell equations, or the equivalent equations (146), (149), (150) can be deduced from the Hamiltonian (154). For this we write the canonical equations derived from (154); we obtain:

$$(156)\quad \dot{Q}_s = \frac{\partial H}{\partial P_s} = -P_s\ ;\quad \dot{P}_s = -\frac{\partial H}{\partial Q_s} = 4\pi^2\nu_s^2 Q_s - \sum_i e_i c\left(\frac{8\pi}{\Omega}\right)^{1/2}\cos\Gamma_{si}.$$

If we eliminate P_s from these equations, we obtain:

$$\ddot{Q}_s + 4\pi^2\nu_s^2 Q_s = \sum_i e_i c\left(\frac{8\pi}{\Omega}\right)^{1/2}\cos\Gamma_{si}$$

which is identical with (146). Similarly the canonical equations for the pair of conjugate variablex χ_s, ω_s are:

$$(157)\quad \dot{\chi}_s = \frac{\partial H}{\partial\omega_s} = \omega_s\ ;\quad \dot{\omega}_s = -\frac{\partial H}{\partial\chi_s} = -4\pi^2\nu_s^2\chi_s^2 - c\left(\frac{8\pi}{\Omega}\right)^{1/2}\sum_i e_i(\gamma_i\cdot\alpha_s)\sin\Gamma_{si}.$$

Elimination of ω_s yields:

$$(158)\quad \ddot{\chi}_s + 4\pi^2\nu_s^2\chi_s = -c\left(\frac{8\pi}{\Omega}\right)^{1/2}\sum_i e_i(\gamma_i\cdot\alpha_s)\sin\Gamma_{si}.$$

Now we observe that the velocity of the ith particle in Dirac's relativistic theory is

$$(159)\quad \dot{\mathbf{X}}_i = -c\gamma_i.$$

(This results also from the Hamiltonian (154), since $\dot{X}_{ix} = \partial H/\partial p_{ix}$). Eq. (158) coincides therefore with (149). By the same method it can be proved that also the Eq. (150) for the transverse components of the vector potential can be derived from (154).

Eq. (152) which is equivalent to (139) can be written, remembering (156):

$$2\pi\nu_s \chi_s - P_s = 0 \tag{160}$$

and its derivative with respect to time is $2\pi\nu_s \dot{\chi}_s - \dot{P}_s$, which can be written by (156), (157):

$$\omega_s - 2\pi\nu_s Q_s + \frac{c}{2\pi\nu_s}\left(\frac{8\pi}{\Omega}\right)^{1/2} \sum_i e_i \cos \Gamma_{si} = 0\,. \tag{161}$$

We have proved by (151) that if (160), (161) are satisfied for $t = 0$, they are satisfied also automatically for any value of time.

§ 18. - In a classical interpretation we could therefore say that electrodynamics and motion of the points can be deduced by integration of the canonical equations corresponding to the Hamilton function (154); the initial values of the variables must satisfy the supplementary conditions (160), (161).

As we go over to the quantum mechanical interpretation, we must first observe that it is in general impossible that two functions of the variables of the system have simultaneously a well determined value, with the exception of the case that the two functions commute; so at first sight it would seem impossible to satisfy simultaneously (160) and (161). This is, however, possible in this special case, since the first members of (160) and (161) commute with each other as an immediate verification shows (remember that ω_s and χ_s are conjugate and therefore $\omega_s \chi_s - \omega_s \chi_s = h/2\pi i$; and similarly $P_s Q_s - Q_s P_s = h/2\pi i$, while all the other variables in (160) and (161) commute).

To the classical integration of a system of canonical equations corresponds in wave mechanics the integration of the Schrödinger equation:

$$H\psi = -\frac{h}{2\pi i}\frac{\partial \psi}{\partial t} \tag{162}$$

where H, given by (154) must be considered in the ordinary way as an operator acting on the function ψ of the coordinates only:

$$\psi = \psi(t, \mathbf{X}_i, \sigma_i, q_{s1}, q_{s2}, \chi_s, Q_s) \tag{163}$$

σ_i represent the spin coordinates.

If there were no condition limiting the initial values of the variables, then ψ for $t = 0$ could be chosen arbitrarily. But we have the conditions (160) (161). We will show that these conditions determine the form of the dependence of ψ on χ_s and Q_s. Indeed, ω_s, conjugate to χ_s, must have according to (161) the value:

$$\omega_s = 2\pi\nu_s Q_s - \frac{c}{2\pi\nu_s}\left(\frac{8\pi}{\Omega}\right)^{1/2} \sum_i e_i \cos \Gamma_{si}\,.$$

It results from this that χ_s can be contained in ψ only in a factor:

$$e^{2\pi i \omega_s \chi_s/h} = \exp\left[\frac{2\pi i}{h}\chi_s\left(2\pi\nu_s Q_s - \frac{c}{2\pi\nu_s}\left(\frac{8\pi}{\Omega}\right)^{1/2}\sum_i e_i \cos \Gamma_{si}\right)\right]. \tag{164}$$

By the same method we deduce from (160) that χ_s must be contained only in a factor:

$$e^{4\pi^2 i\nu\ \chi_s Q_s/h}$$

which is already contained in (164).

We see therefore that the form of ψ must be:

$$\psi = \exp\left[\frac{2\pi i}{h}\sum_s \chi_s\left(2\pi\nu_s Q_s - \frac{c}{2\pi\nu_s}\left(\frac{8\pi}{\Omega}\right)^{1/2}\sum_i e_i \cos\Gamma_{si}\right)\right] \times \varphi(t, \mathbf{X}_i, \sigma_i, q_{s1}, q_{s2}) . \tag{165}$$

If we substitute this expression for ψ in the Schrödinger Eq. (162) we obtain a new differential equation for φ. With some calculations it is found that this equation can be put in the form:

$$R\varphi = -\frac{h}{2\pi i}\frac{\partial\varphi}{\partial t} \tag{166}$$

which strongly resembles the form of a Schrödinger equation. The operator R is the following:

$$R = -c\sum_i(\gamma_i\cdot p_i) - \sum_i m_i c^2 \delta_i + \sum_i e_i c\left(\frac{8\pi}{\Omega}\right)^{1/2}(\gamma_i\cdot(\mathbf{A}_{s1}q_{s1} + \mathbf{A}_{s2}q_{s2}))\sin\Gamma_{si} + \sum_s\left\{\frac{1}{2}(p_{s1}^2 + p_{s2}^2) + 2\pi^2\nu_s^2(q_{s1} + q_{s2}^2)\right\} + \frac{c^2}{\pi\Omega}\sum_s\frac{1}{\nu_s^2}\left(\sum_i e_i\cos\Gamma_{si}\right)^2 . \tag{166}$$

This operator R can be considered as a sort of Hamilton function acting on φ. By this method therefore the coordinates Q_s and χ_s representing the scalar pontential and one component of the vector potential are completely eliminated both from the new amplitude of probability φ and from the new Hamilton function R. Not considering the last term in R, which we shall discuss later, the operator R is identical with the Hamilton function of Dirac's theory of radiation (117). (There are only some formal differences: in (117) only one electron, instead of many particles is considered; in (166) the two polarized components are considered separately with the indices $s1$ and $s2$, whereas in (117) there is only one index s).

We must find out the physical meaning of the last term in (166). This term is:

$$\frac{c^2}{\pi\Omega}\sum_s\frac{1}{\nu_s^2}\left(\sum_i e_i\cos\Gamma_{si}\right)^2 = \frac{c^2}{\pi\Omega}\sum_{ij} e_i e_j \sum_s \frac{\cos\Gamma_{si}\cos\Gamma_{sj}}{\nu_s^2} .$$

The sum over s can be transformed into an integral. (Take the mean value of $\cos\Gamma_{si}\cos\Gamma_{sj}$ over all directions of propagation and phases for the s–component; and then remember (141)). We find at last:

$$\sum_s\frac{\cos\Gamma_{si}\cos\Gamma_{sj}}{\nu_s^2} = \frac{\pi\Omega}{2c^2}\frac{1}{r_{ij}}$$

r_{ij} being the distance between the two points i and j. The last term of (166) takes therefore the very simple form:

$$\frac{1}{2}\sum_{ij}\frac{e_i e_j}{r_{ij}} \tag{167}$$

which is the ordinary classical expression for the electrostatic energy of our system of charges.

At this point we meet a very serious difficulty, since the electrostatic energy of point charges is infinite; every charge has an infinite electrostatic self-energy. We could try of course to avoid this difficulty, as it is very often done in classical electrostatics by simply neglecting in the sum (167) all the terms with $i = j$ which represent the self-energy of the charges. We shall see however that even this very crude proceeding is not sufficient to avoid infinite terms of non-electrostatic origin in the self-energy.

The problem which we meet now in quantum electrodynamics is identical with that of radiation theory since our new Hamilton function R is the Hamilton function of radiation theory plus the electrostatic energy. We have hitherto considered in the radiation theory as unperturbed system, the system obtained by neglecting the interaction between atom and radiation field. The interaction term had then simply the effect of determining transitions between different states of the unperturbed system which have the same or nearly the same unperturbed energy.

But we can ask whether there are quantum states for the complete problem. This problem is mathematically very difficult and can only be discussed by the method of successive approximations. However the second approximation still yields an infinite perturbation term in the energy levels and it seems therefore probable, that for point electrons there are no quantum states of the unperturbed problem. It could be noticed however that the application of the perturbation method is for this problem extremely uncertain, since the differences between the quantum states of the unperturbed problem are very small in comparison with the perturbation.

To all these difficulties no satisfactory answer has yet been given. One would be tempted to give the electron a finite radius; this would actually avoid infinite terms, as in the classical theory of electromagnetic masses. But this method is connected with serious difficulties for the relativistic invariance.

In conclusion we may therefore say that practically all the problems in radiation theory which do not involve the structure of the electron have their satisfactory explanation; while the problems connected with the internal properties of the electron are still very far from their solution.

BIBLIOGRAPHY.

We collect here some of the papers on quantum theory of radiation and quantum electrodynamics:

E. AMALDI, « Rend. Lincei », *9*, 876 (1929) (Theory of Raman effect).

F. BLOCH, « Phys. Zeits. », *29*, 58 (1928) (Reaction of radiation).

G. BREIT, « Phys. Rev. », *34*, 553 (1929) (Interaction of two electrons).

P. A. M. DIRAC, « Proc. Roy. Soc. », A *114* 243 (1927). (General theory); « Proc. Roy. Soc. », A *114*, 710 (1927) (Dispersion); « Proc. Camb. Phil. Soc. », *26*, 361 (1903) (Transition of the electron from positive to negative states).

E. FERMI, « Rend. Lincei », *10*, 72 (1929) (Lippmann fringes); « Rend. Lincei », *9*, 881 (1929); *12*, 431 (1930) (Quantum electrodynamics); « Annales Inst. Poincaré », 1929 (General theory of radiation); «Nuovo Cim. », *8*, 121 (1931) (Quantum theory of electromagnetic mass).
J. A. GAUNT and W. H. MCCREA, « Proc. Camb. Phil. Soc. », *23*, 930 (1927) (Quadrupole radiation).
M. GÖPPERT-MAYER, « Ann. d. Physik », *9*, 273 (1931) (Double transitions).
W. HEISENBERG, « Ann. d. Physik », *9*, 338 (1931) (Correspondence between classical and quantum theoretical properties of light); « ZS. f. Physik », *65*, 4 (1930) (Self-energy of the electron).
W. HEISENBERG and W. PAULI, « ZS. f. Physik », *56*, 1 (1929); *59*, 168 (1930) (Quantum electrodynamics).
P. JORDAN and W. PAULI, « ZS. f. Physik », *47*, 151 (1928) (Quantum electrodynamics of spaces without electric charges).
S. KIKUCHI, « ZS. f. Physik », *68*, 803 (1931) (Compton Effect).
L. LANDAU and R. PEIERLS, «ZS. f. Physik», *62*, 188 (1930). (Quantum electrodynamics).
J. R. OPPENHEIMER, « Phys. Rev. », *35*, 461 (1930) (Interaction of field and matter); « Phys. Rev. », *35*, 939 (1930) (Annihilation of electrons and protons).
G. RACAH, « Rend. Lincei », *11*, 837, 1100 (1930) (Interference).
L. ROSENFELD, « Ann. d. Physik », *5*, 113 (1930) (Quantization of wave fields).
E. SEGRÈ, « Rend. Lincei », *9*, 887 (1929) (Fluorescence).
I. TAMM, « ZS. f. Physik », *60*, 345 (1930) (Scattering of light in solid bodies); « ZS. f. Physik » *62*, 7 (1930) (Compton effect, annihilation of electrons and protons).
I. WALLER, « ZS. f. Physik », *58*, 75 (1929); *61*, 837 (1930) (Scattering); *62*, 673 (1930) (Self-energy of the electron).
V. WEISSKOPF, « Ann. d. Physik », *9*, 23 (1931) (Fluorescence).
V. WEISSKOPF and E. WIGNER, « ZS. f. Physik », *63*, 54 (1930); *65*, 18 (1930) (Width of spectral lines).

Nº 68.

This paper was suggested to Fermi by Rasetti's observation that the totally symmetric vibration of the CO_2 molecule manifests itself in the Raman spectrum with a group of 4 closely spaced lines, instead of one line as expected. When Rasetti returned to Rome from Pasadena where he had performed the experiment, this puzzling result was the object of much discussion with Fermi. Several unsatisfactory explanations were discarded. Finally, Fermi hit upon the correct interpretation and immediately performed the calculations by means of the perturbation theory. The effect was shown to be due to an accidental degeneracy of the first excited state of the above-mentioned vibration with the second excited state of the transverse vibration, whose frequency happens to be almost exactly half of the former. This became the first known case of a phenomenon that was later observed in several organic molecules, among them CCl_4, as discussed in another of Fermi's papers (Nº 71).

This explanation of the apparently anomalous Raman effect in CO_2 aroused considerable interest at the Leipzig conference on molecular structure, where it was reported by Rasetti [1]. The Raman effect was one of the main topics discussed at that conference.

F. RASETTI.

68.

ÜBER DEN RAMANEFFEKT DES KOHLENDIOXYDS

« Z. Physik », *71*, 250–259 (1931).

Das Ramanspektrum des CO_2 wird theoretisch erklärt; es wird gezeigt, dass das Auftreten von zwei Ramanlinien, wo man nach dem Modell des CO_2–Moleküls nur eine erwarten würde, auf eine zufällige quasi Entartung von zwei Schwingungsniveaus zurückgeführt werden kann.

Das Ramanspektrum des gasförmigen CO_2 besteht [2] aus zwei ziemlich starken Ramanlinien mit Frequenzen von 1285 und 1388 Wellenzahlen; in der Nähe von diesen beiden Linien werden noch zwei andere sehr schwache Trabanten beobachtet. Zweck dieser Arbeit ist, dieses Spektrum aus dem Modell des CO_2–Moleküls abzuleiten.

1. Beim CO_2 liegen bekanntlich die drei Atome in der Ruhelage auf einer Geraden, und zwar das C–Atom im Mittelpunkte zwischen den beiden O–Atomen (Fig. 1ᵃ). Fasst man die drei Atome als elastisch gebundene Massenpunkte auf, so hat dieses System mechanisch drei Grundfrequenzen ν_1, ν_2,

(1) F. RASETTI, in Debye, Leipziger Vortraege 1931, pp. 59–69. Hirzel, Leipzig, 1931.

(2) F. RASETTI, « Nature », *123*, 205 (1929); R. G. DICKINSON, R. T. DILLON u. F. RASETTI, « Phys. Rev. », *34*, 582 (1929).

ν_3, welche den Bewegungsformen [3] b, c, d in Fig. 1 entsprechen. Die Frequenz ν_2 ist zweifach entartet; die anderen sind einfach. Nun ist die Frage zu verstehen, wie man aus diesen drei Grundfrequenzen das Ramanspektrum erklären kann, weil es von vornherein unwahrscheinlich ist, dass die beobachteten Ramanlinien irgendwie aus den Rotationen des Moleküls oder aus elektronischen Niveaus herrühren.

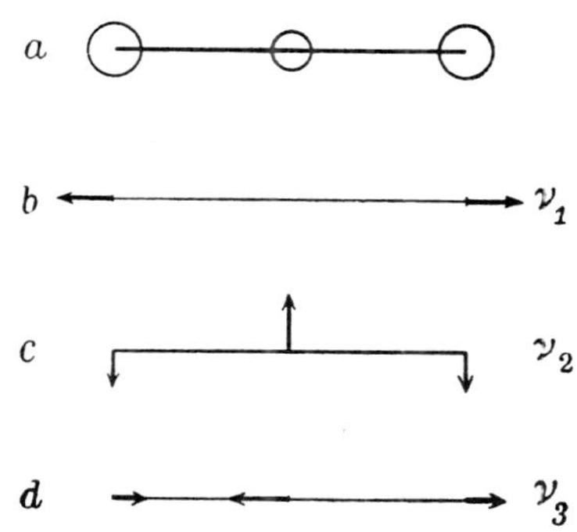

Fig. 1.

Von den drei Grundfrequenzen ν_1, ν_2, ν_3 sind offensichtlich nur die zwei letzten optisch aktiv, weil die erste wegen der Symmetrie der Bewegung, mit keinem schwingenden Dipol verbunden ist. In der ultraroten Absorption werden also nur ν_2 und ν_3, abgesehen von Kombinationen, beobachtet. In der Tat werden im Ultraroten zwei besonders intensive Banden bei 14,87 und 4,27 μ beobachtet [4]), entsprechend zu 673 bzw. 2350 Wellenzahlen; diese Frequenzen sind wahrscheinlich mit den Grundfrequenzen ν_2 und ν_3 zu identifizieren. Und zwar

$$\nu_2 = 673 \quad ; \quad \nu_3 = 2350,$$

weil es aus den Bewegungstypen (Fig. 1) plausibel erscheint, dass $\nu_3 > \nu_2$.

Die Frequenz ν_1 ist optisch inaktiv und kann deswegen aus dem ultraroten Spektrum nicht direkt ermittelt werden. Ihr Wert kann trotzdem mit Hilfe der folgenden Betrachtungen abgeschätzt werden. Die beiden Grundfrequenzen ν_1 und ν_3 sind in erster Näherung von der Stärke der elastischen Bindung zwischen dem C–Atom und den beiden O–Atomen, bestimmt. Falls man die Bindung zwischen den beiden O–Atomen vernachlässigt, werden also die beiden Grundfrequenzen ν_1 und ν_3 von einem einzigen elastischen Koeffizienten bestimmt. Sei kx die elastische Kraft zwischen einem O–Atom und dem C–Atom, wenn sie um x relativ zueinander verschoben sind. Seien weiter M und m die Massen von C und O. Dann hat man mit unseren vereinfachenden Hypothesen

$$(1) \qquad \nu_1 \cong \frac{1}{2\pi}\sqrt{\frac{k}{m}} \quad ; \quad \nu_3 \cong \frac{1}{2\pi}\sqrt{\frac{k}{m}}\sqrt{\frac{2m+\mathrm{M}}{\mathrm{M}}}.$$

Die Frequenzen ν_1 und ν_2 stehen also angenähert im Verhältnis

$$(2) \qquad \frac{\nu_3}{\nu_1} \cong \sqrt{\frac{2m+\mathrm{M}}{\mathrm{M}}} = \sqrt{\frac{2\cdot 16+12}{12}} = 1{,}91\,.$$

Aus dem angegebenen Wert von ν_3 findet man also

$$\nu_1 \cong 1230.$$

(3) D. M. Dennison, « Rev. Mod. Phys. », *3*, 280 (1931).

(4) E. F. Barker, « Astrophys. Journ. », *55*, 391 (1922); C. Schaefer u. B. Philipps, « ZS. f. Phys. », *36*, 641 (1926); A. Eucken, « ZS. f. Phys. », *37*, 14 (1926).

Dieser letzte Wert ist natürlich nur grob angenähert; wir werden sehen, dass ν_1 wahrscheinlich ungefähr 100 Wellenzahlen höher liegt.

2. Wir haben gesagt, dass nur die Frequenzen ν_2 und ν_3 optisch aktiv sind. Im Ramaneffekt muss man dagegen erwarten, dass nur die Frequenz ν_1 intensiv ist. Dies hat darin seinen Grund (5), dass die Bewegungstypen ν_2 und ν_3 (vgl. Fig. 1) symmetrisch verlaufen (d.h. nach einer halben Periode befindet sich das Molekül in einer Lage, welche symmetrisch zur Anfangslage ist). Dagegen ist der erste Bewegungstyp nicht symmetrisch und kann also im Ramaneffekt erscheinen. Und zwar tritt bekanntlich intensiv im Ramanspektrum nur die Grundfrequenz und nicht ihre Obertöne auf (6).

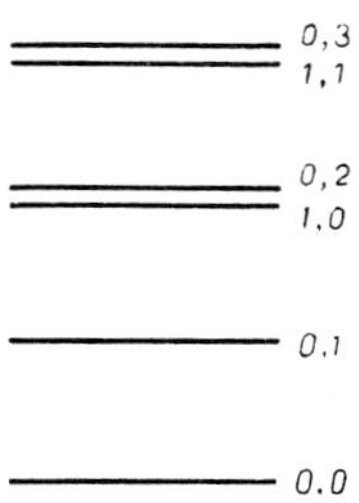

Fig. 2.

Man würde also ein Ramanspektrum erwarten, bestehend aus einer einzigen intensiven Linie in der Umgebung von 1230 Wellenzahlen. Nun liegt zwar die Frequenz der Ramanlinien nicht weit von diesem Wert entfernt; das Spektrum hat jedoch eine beträchtlich kompliziertere Struktur: zwei starke und zwei schwache Linien. Wir werden zuerst die starken Linien diskutieren; die schwachen Komponenten rühren von einem thermisch angeregten Schwingungszustand mit einem Quantum der Frequenz 673 her; wir werden diesen Punkt erst später untersuchen.

Für das Auftreten von zwei starken Frequenzen an Stelle von einer schlagen wir folgende Erklärung vor.

In Fig 2 ist ein theoretisches Schema der tieferen Schwingungsniveaus unseres Moleküls dargestellt. Das Schema wurde dadurch vereinfacht, dass die Entartung der Frequenz ν_2 vorläufig vernachlässigt worden ist; die zwei Oszillatoren mit den Frequenzen ν_1 und ν_2 wurden vorerst als exakt harmonisch und ungekoppelt angenommen; die Frequenz ν_3 kommt für die tieferen Niveaus nicht in Betracht, weil sie viel höher liegt. Die neben den Niveaus angeschriebenen zwei Zahlen bedeuten die Quantenzahlen der beiden Oszillatoren ν_1 und ν_2. Man würde eine Ramanlinie erwarten, entsprechend dem Übergang (1 , 0) — (0 , 0).

Die Möglichkeit, zwei starke Ramanlinien zu erhalten, entsteht durch die Tatsache, dass die beiden Niveaus (1 , 0) und (0 , 2) beinahe zusammenfallen. In der Tat ist $2\nu_2 = 1346$. ν_1 ist nicht exakt bekannt, liegt aber gewiss in der Nähe von diesem Wert. Hierdurch stören sich die beiden Zustände (1 , 0) und (0 , 2) sehr stark, so dass die Quantenzustände, welche, infolge der Kopplung zwischen den beiden Oszillatoren ν_1 und ν_2 aus ihnen entstehen, Eigenfunktionen entsprechen, welche lineare Kombinationen der ungestörten Eigenfunktionen (1 , 0) und (0 , 2) sind, und zwar sind die Koeffizienten von (1 , 0) für beide gestörte Eigenfunktionen von der Grössenordnung 1. Beide Zustände können also im Ramaneffekt intensiv mit dem Grundniveau kombinieren, weil beide einen grossen Prozentsatz der Eigen-

(5) C. MANNEBACK, « ZS. f. Phys. », *62*, 224 (1930).
(6) J. H. VAN VLECK, « Proc. Nat. Acad. Amer. », *15*, 754 (1929).

funktion (1, 0) enthalten, die mit (0, 0) stark kombiniert. Wir werden auch zeigen, dass der beobachtete Frequenzunterschied (103 Wellenzahlen) zwischen den beiden starken Ramanlinien mit dieser Hypothese recht gut verträglich ist.

3. Der Sachverhalt ist in Wirklichkeit wegen der Entartung der Frequenz ν_2 etwas komplizierter. Diese Entartung entsteht durch die Möglichkeit (3) für das C–Atom, in beliebigen Richtungen der Ebene senkrecht durch den Mittelpunkt der Molekülachse zu schwingen. Die Eigenfunktionen müssen also mit drei Zahlen numeriert werden, auch wenn die Quantenzahl der grossen Frequenz ν_3 als 0 festgesetzt wird, wie wir hier immer annehmen wollen.

Als Koordinaten für die Bestimmung der Bewegung nehmen wir die folgenden: x bedeutet die Projektion der Verschiebung der O–Atome auf der Molekülachse. ξ und η sind die Projektionen der Verbindungslinie des C–mit einem der O–Atome auf zwei senkrecht zueinander und zur Molekülachse stehende feste Richtungen. (Vgl. Fig. 3, wo AB die Gleichgewichtslage des Moleküls, C, O, O die augenblicklichen Lagen der drei Atome darstellen. Die Figur ist auf eine Ebene parallel zur Molekülachse und zu ξ projiziert.) x ist die Koordinate der Bewegung ν_1; ξ, η die Koordinaten der beiden entarteten Bewegungen mit der Frequenz ν_2; von einer vierten Koordinate für die Bewegung ν_3 wurde abgesehen, weil die Frequenz ν_3 nicht angeregt ist.

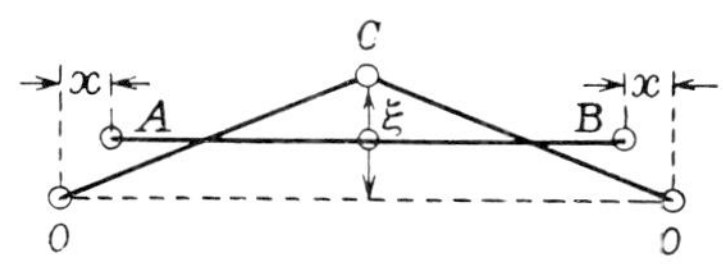

Fig. 3.

Die schwingende Masse für die Koordinate x ist

$$m_1 = 2\,m. \tag{3}$$

Für ξ, η ist die schwingende Masse m_2 gleich der reduzierten Masse zwischen $2\,m$ und M, d. h.

$$m_2 = \frac{2\,m\mathrm{M}}{2\,m + \mathrm{M}}\,. \tag{4}$$

Ein Quantenzustand des ungestörten Systems wird mit drei Quantenzahlen bezeichnet, etwa (n_1, n_2', n_2''). n_1, n_2', n_2'' sind die Quantenzahlen der Variablen x, ξ, η. Die ungestörte Energie ist (wenn man der Übersichtlichkeit wegen die Nullniveaus der Energie ein halbes Quantum höher nimmt als üblich, so dass die Nullpunktsenergien verschwinden)

$$\mathrm{E}_{n_1 n_2' n_2''} = h\nu_1\, n_1 + h\nu_2\,(n_2' + n_2''). \tag{5}$$

Die ungestörten Eigenwerte (0, 2, 0), (0, 1, 1), (0, 0, 2) sind also exakt entartet; ihr Wert ist (wegen der Tatsache, dass $\nu_1 \cong 2\,\nu_2$) sehr nahe an dem ungestörten Eigenwert (1, 0, 0).

Im ungestörten Problem hat die potentielle Energie die Form

$$kx^2 + \alpha\,(\xi^2 + \eta^2), \tag{6}$$

wo k und α Konstanten darstellen; der Wert von k kann angenähert durch (1) berechnet werden. α kann aus der Frequenz ν_2 mit der Gleichung

$$\nu_2 = \frac{1}{2\pi}\sqrt{\frac{\alpha(2m+M)}{2mM}} \tag{7}$$

gewonnen werden.

4. Wir wollen jetzt die Störungsenergie berechnen, und zwar nur bis auf Glieder dritter Ordnung in den Koordinaten. Aus Symmetriegründen findet man sofort, dass nur Glieder von den Typen

$$x^3 \quad \text{und} \quad x(\xi^2 + \eta^2)$$

vorkommen können. Das erste Glied bewirkt keine Kopplung zwischen den Oszillatoren, da es nur von x abhängt. Wir brauchen also dieses Glied nicht zu berücksichtigen, und wir könnten es eventuell in der ungestörten Energie mitberücksichtigen. Als Kopplungsglied nehmen wir also nur

$$U = \beta x(\xi^2 + \eta^2), \tag{8}$$

wo β eine Konstante darstellt.

Man kann auch mit plausiblen Annahmen die Grössenordnung von β abschätzen. Nehmen wir dafür an, dass die potentielle Energie des Moleküls für die Bewegungstypen ν_1 und ν_2 sich als Summe einer Funktion der Entfernung CO (Fig. 3), welche die Bewegung ν_1 bestimmt, und einer Funktion des Winkels zwischen CO und der Molekülachse, welche die Schwingung ν_2 bestimmt, darstellen lässt. In dieser Annahme, die eigentlich sehr willkürlich ist, findet man mit leichten Rechnungen, indem man (6) berücksichtigt:

$$\beta = \frac{k - 2\alpha}{a}, \tag{9}$$

wo a die Entfernung CO für die Gleichgewichtslage darstellt, deren Wert sich aus dem Trägheitsmoment (7) $70 \cdot 10^{-40}$ berechnen lässt. Man findet

$$a = 1{,}15 \cdot 10^{-8}.$$

5. Wir betrachten nunmehr die vier quasientarteten Eigenfunktionen $(1,0,0), (0,2,0), (0,0,2), (0,1,1)$ und berechnen die zugehörigen Matrixelemente der Kopplungsenergie (8). Dies geschieht ohne weiteres mit Hilfe der bekannten Formeln für die Matrix der Koordinate q eines Oszillators mit Masse m:

$$q_{n,n+1} = \sqrt{\frac{h}{8\pi^2 \nu m}}\sqrt{n+1}. \tag{10}$$

Alle anderen Matrixelemente verschwinden. Hierdurch finden wir

$$x_{01} = x_{10} = \sqrt{\frac{h}{8\pi^2 \nu_1 m_1}} \quad ; \quad (\xi^2)_{02} = \xi_{01}\,\xi_{12} = \frac{h}{8\pi^2 \nu_2 m_2}\sqrt{2} \tag{11}$$

(7) W. Houston, « Phys. Rev. », *37*, 227 (1931).

Mit Anwendung dieser Formeln finden wir die gewünschten Matrixelemente:

$$U(1\,0\,0\,,\,0\,2\,0) = U(1\,0\,0\,,\,0\,0\,2) = P. \tag{12}$$

$$U(1\,0\,0\,,\,0\,1\,1) = 0, \tag{13}$$

wo gesetzt wurde

$$P = \frac{\beta h^{3/2}}{8\sqrt{2}\,\pi^3\sqrt{\nu_1 m_1 \nu_2 m_2}}. \tag{14}$$

Die Eigenfunktion (0, 1, 1) wird also vom Störungsglied mit den anderen Eigenfunktionen nicht gekoppelt; da diese Eigenfunktion im Ramaneffekt inaktiv ist, brauchen wir sie also nicht zu berücksichtigen.

Die Energiematrix für die anderen Eigenfunktionen wird also

$$\begin{vmatrix} A & P & P \\ P & B & O \\ P & O & B \end{vmatrix}, \tag{15}$$

wo A den ungestörten Eigenwert (1, 0, 0) und B den ungestörten Eigenwert (0, 0, 2) darstellt.

Die Energiematrix (15) hat die drei Eigenwerte

$$B\,, \frac{A+B}{2} \pm \sqrt{2\,P^2 + \left(\frac{A-B}{2}\right)^2}, \tag{16}$$

welche linearen Kombinationen der drei Eigenfunktionen (1, 0, 0), (0, 2, 0), (0, 0, 2) entsprechen. Der Eigenwert B gehört zur linearen Kombination

$$(0\,,\,0\,,\,2) - (0\,,\,2\,,\,0), \tag{17}$$

welche die Eigenfunktion (1, 0, 0) nicht enthält. Da der Ramanübergang gerade durch (1, 0, 0) ermöglicht wird, wird also der Term (17) im Ramaneffekt nicht auftreten. Es bleiben also nur die zwei anderen Eigenwerte

$$\frac{A+B}{2} \pm \sqrt{2\,P^2 + \left(\frac{A-B}{2}\right)^2}. \tag{18}$$

Diese entsprechen den linearen Kombinationen

$$\frac{1}{P}\left[\frac{A-B}{2} \pm \sqrt{2\,P^2 + \left(\frac{A-B}{2}\right)^2}\right](1\,,\,0\,,\,0) + (0\,,\,2\,,\,0) + (0\,,\,0\,,\,2), \tag{19}$$

die tatsächlich die Eigenfunktion (1, 0, 0) enthalten, welche die Ramanübergänge nach dem Grundniveau ermöglicht.

Es scheint nun, dass das beobachtete Ramanspektrum von CO_2 sich recht gut durch die Annahme erklären lässt, dass die beiden ungestörten Niveaus A und B fast genau zusammenfallen. Setzen wir in (18) A = B, so finden wir die beiden Energien

$$A \pm \sqrt{2}\,P. \tag{20}$$

Die Kombination mit dem Grundniveau (0 , 0 , 0) gibt also zwei Ramanlinien mit den Frequenzen (20); also mit dem Frequenzunterschied

$$\Delta\nu = 2\sqrt{2}\ P. \tag{21}$$

Mit Hilfe von (14), (9), (7), (4), (3), (1) unter Zugrundelegung der angegebenen Werte von ν_2, ν_3, a kann P numerisch abgeschätzt werden; man findet

$$P = 40 \text{ Wellenzahlen.}$$

Aus (21) erhalten wir also

$$\Delta\nu = 113.$$

Dieser Wert stimmt sehr gut mit der experimentell gefundenen Aufspaltung von 103 Wellenzahlen; die Übereinstimmung ist sogar grösstenteils zufällig, da wir die Konstante β von Formel (8) nur grob abgeschätzt haben. Wichtig für uns war nur festzustellen, dass man für die Aufspaltung die richtige Grössenordnung erhält.

Für A = B wird (19):

$$\pm \sqrt{2}\ (1, 0, 0) + (0, 2, 0) + (0, 0, 2).$$

Man sieht also, dass mit dieser Hypothese der Prozentsatz der Eigenfunktion (1 , 0 , 0) in den beiden gestörten Eigenfunktionen derselbe ist (50 %). Man würde deshalb die gleiche Intensität für die beiden starken Ramanlinien erwarten. Nun ist aber die Ramanlinie mit grösserer Frequenz etwas stärker, was man gut durch die Annahme verstehen kann, dass A ein wenig grösser als B ist; es genügt ein Unterschied von ungefähr 10 Wellenzahlen, um den beobachteten Intensitätsunterschied zu erklären.

Durch diese Betrachtungen sind also die beiden lichtstärkeren Ramanlinien vollständig gedeutet.

6. Es bleibt noch, den Ursprung der beiden schwachen Ramanlinien zu diskutieren. Diese beiden Linien können, wie wir bereits angedeutet haben, auf die Wirkung der Moleküle zurückgeführt werden, die durch thermische Anregung ein Quantum von der Frequenz 673 enthalten [Zustand (0 , 1) des vereinfachten Niveauschemas von Fig. 2]. Der Boltzmannfaktor hat für diesen Zustand bei Zimmertemperatur den Wert 1/28; da aber das statistische Gewicht des angeregten Zustandes 2 ist, findet man im thermischen Gleichgewicht 1/14 Moleküle im Zustand (0 , 1).

Der Zustand (0, 1) kombiniert im Ramaneffekt mit dem Zustand (1 , 1). Da aber dieses Niveau sehr nahe an (0 , 3) liegt (Fig. 2), erhalten wir wieder durch denselben Mechanismus wie früher zwei Ramanlinien.

Zur quantitativen Berechnung muss man natürlich die Entartung der Frequenz ν_2 berücksichtigen. Der Zustand (1 , 1) ist zweifach: (1 , 1 , 0) und (1 , 0 , 1); der Zustand (0 , 3) ist vierfach: (0 ,3 , 0), (0 , 2 , 1), (0 , 1 , 2) und (0 , 0 , 3). Die Energiematrix, die diesen sechs Eigenfunktionen entspricht, kann wie früher ohne weiteres berechnet werden, indem man als Störungsenergie wieder (8) nimmt. Man findet die Energiematrix:

$$(22)\qquad \begin{vmatrix} A' & 0 & \sqrt{3}\,P & 0 & P & 0 \\ 0 & A' & 0 & P & 0 & \sqrt{3}\,P \\ \sqrt{3}\,P & 0 & B' & 0 & 0 & 0 \\ 0 & P & 0 & B' & 0 & 0 \\ P & 0 & 0 & 0 & B' & 0 \\ 0 & \sqrt{3}\,P & 0 & 0 & 0 & B' \end{vmatrix},$$

wo A′ und B′ die ungestörten Eigenwerte darstellen. P ist wieder durch (14) gegeben.

Die Matrix (22) spaltet sich in zwei gleiche Matrizen dritter Ordnung

$$(23)\qquad \begin{vmatrix} A' & \sqrt{3}\,P & P \\ \sqrt{3}\,P & B' & 0 \\ P & 0 & B' \end{vmatrix},$$

welche die Eigenfunktionen (1, 1, 0), (0, 3, 0), (0, 1, 2) und (1, 0, 1), (0, 0, 3), (0, 2, 1) koppeln; beide Gruppen geben natürlich dieselben Linien, so dass es genügt, nur die erste Gruppe zu berücksichtigen.

Die Matrix (23) hat die drei Eigenwerte

$$(24)\qquad B' \; ; \; \frac{A'+B'}{2} \pm \sqrt{4\,P^2 + \left(\frac{A'-B'}{2}\right)^2}\,.$$

Der Eigenwert B′ wird jedoch im Ramaneffekt nicht beobachtet, weil die entsprechende Eigenfunktion

$$(0\,,3\,,0) - \sqrt{3}\,(0\,,1\,,2)$$

(1, 1, 0) nicht enthält. Die beiden anderen Eigenwerte entsprechen den Eigenfunktionen

$$\frac{1}{P}\left\{\frac{A'-B'}{2} \pm \sqrt{4\,P^2 + \left(\frac{A'-B'}{2}\right)^2}\right\}(1\,,1\,,0) + \sqrt{3}\,(0\,,3\,,0) + (0\,,1\,,2)\,,$$

enthalten also die ungestörte Eigenfunktion (1, 1, 0), welche die Ramanübergänge ermöglicht.

Machen wir auch in diesem Falle die Hypothese $A' = B'$, so erhalten wir die beiden Eigenwerte $A' \pm 2\,P$.

Die Aufspaltung der zwei schwachen Ramanlinien ist also

$$(25)\qquad \Delta'\nu = 4\,P.$$

Ein Vergleich mit (21) ergibt das Verhältnis zwischen den Aufspaltungen der schwachen und der starken Ramanlinien. Man findet

$$(26)\qquad \frac{\Delta'\nu}{\Delta\nu} = \sqrt{2}$$

in vorzüglicher Übereinstimmung mit dem experimentellen Verhältnis 1,4.

Aus den Intensitäten der beiden schwachen Ramanlinien kann man auch für diesen Fall schliessen, dass A′ um einige Wellenzahlen höher als B′ liegt, was jedoch nur einen Effekt zweiter Ordnung auf die Werte der Aufspaltungen $\Delta\nu$ und $\Delta'\nu$ ausübt, so dass das Verhältnis (26) noch mit grosser Annäherung gültig bleibt.

Zusatz bei der Korrektur. Durch den in dieser Arbeit für das Beispiel des CO_2 erläuterten Mechanismus dürften wahrscheinlich auch bei anderen Molekeln, hauptsächlich bei solchen mit vielen Eigenfrequenzen, mehr Ramanlinien erscheinen, als man nach den üblichen Theorien erwarten würde.

Dies scheint z. B. bei CS_2 der Fall zu sein; beobachtet werden zwei Ramanfrequenzen von 655 und 800 Wellenzahlen. Da CS_2 dieselbe gestreckte Struktur wie CO_2 hat, würde man erwarten, dass nur eine von den drei Eigenfrequenzen in Ramaneffekt aktiv ist. Die Anwesenheit von zwei Ramanlinien rührt wahrscheinlich auch in diesem Falle von derselben quasi-Entartung ($\nu_1 \cong 2\nu_2$) wie bei CO_2 her. Die Intensitäten der beiden Ramanlinien 655 und 800 werden als 10 bzw. 2 geschätzt. Dieser ziemlich grosse Intensitätsunterschied zeigt, dass ν_1 wesentlich kleiner als $2\nu_2$ ist. Aus einem Intensitätsverhältnis 5 erhält man $2\nu_2 - \nu_1 \cong 100$ Wellenzahlen. Man würde also als ungestörte Werte der Eigenfrequenzen $\nu_1 = 678$ und $\nu_2 = 388$ erhalten; diese Werte sind natürlich ziemlich ungenau, weil das Intensitätsverhältnis nur geschätzt ist.

Es ist wahrscheinlich, dass dieselbe Erscheinung auch bei CCl_4 vorliegt. In diesem Falle sind theoretisch vier verschiedene Eigenfrequenzen zu erwarten, welche alle im Ramaneffekt aktiv sein müssen. Experimentell findet man, dass die grösste von diesen Frequenzen in zwei Komponenten von 758 und 789 cm^{-1} zerfällt. Dieses Verhalten kann dadurch erklärt werden, dass die Summe von zwei anderen beobachteten Grundfrequenzen $314 + 459 = = 773$ ist; diese Summe liegt also fast genau in der Mitte zwischen den beiden beobachteten hohen Ramanfrequenzen. Auch in diesem Falle liegt also eine zufällige quasi-Entartung vor, wodurch die Aufspaltung gedeutet werden kann; die Symmetrieverhältnisse des Moleküls lassen tatsächlich ein Kopplungsglied zu, das diese Aufspaltung ermöglichen kann.

N° 69.

As explained in a footnote, the experimental part of this work was performed by Rasetti, using his technique of exciting Raman spectra with the resonance line of mercury and subsequently absorbing the exciting line with mercury vapor, thus permitting the unhindered observation of weak Raman lines of low frequency shifts. By this technique, the weak, second-order Raman spectrum of rocksalt, which owing to the symmetry of the lattice does not possess a first-order spectrum, was photographed. Fermi was entirely responsible for the theory. He showed that the second-order spectrum should be a continuum, in agreement with experiment, extending up to twice the characteristic infrared absorption frequency of the crystal.

The theory of the effect was later developed in much greater detail by Born and Bradburn (1) who found even closer agreement with the observations on rocksalt. The objections raised by Raman and coworkers against the existence of a continuum and its interpretation seem totally unfounded.

F. RASETTI

69.

ÜBER DEN RAMANEFFEKT DES STEINSALZES (*)

E. FERMI und F. RASETTI

« Z. Physik », *71*, 689–695 (1931)

Es ist gelungen, mit einer besonders lichtstarken Methode, den Ramaneffekt des Steinsalzes zu beobachten. Das Ramanspektrum weicht von den gewöhnlich bei den Kristallen beobachteten Ramanspektren nicht nur wegen seiner Schwäche, sondern auch wegen seiner kontinuierlichen Struktur ab. Es wird eine qualitative Theorie für dieses Spektrum vorgeschlagen, in welcher es als ein Ramaneffekt zweiter Näherung gedeutet wird.

Die Symmetrie des Steinsalzgitters lässt bekanntlich erwarten (2), dass es keine der ultraroten Eigenschwingung bei 52 μ entsprechende Ramanlinie haben soll. Ein Ramaneffekt kann nur als Erscheinung höherer Ordnung erwartet werden.

In der Tat hatte man bisher mit den gewöhnlichen Methoden keinen Ramaneffekt beim Steinsalz oder bei den anderen ähnlich gebauten Alkalihalogenidkristallen gefunden. Es ist nun gelungen, durch sehr starke Anregung im Ultraviolett und ziemlich langer Expositionsdauer ein gut ausgeprägtes Ramanspektrum auch bei Steinsalz zu beobachten.

Die experimentelle Anordnung ist vielfach von einem von uns beschrieben worden (3). Als primäres Licht dient die Resonanzlinie λ 2537 des Quecksilbers,

(1) M. BORN and MARY BRADBURN, « Proc. Roy. Soc. », *A 188*, 161 (1947).

(*) In dieser Arbeit rührt der experimentelle Teil von F. RASETTI, der theoretische Teil von E. FERMI her.

(2) C. SCHAEFER, « ZS. f. Phys. », *54*, 153, (1929).

(3) F. RASETTI, ebenda, *66*, 646 (1930).

die von einer sehr intensiven wassergekühlten Quecksilberlampe emittiert wird. Das zerstreute Licht unveränderter Frequenz wird durch Quecksilberdampf wieder absorbiert; dieser Kunstgriff ist für die Beobachtung schwacher Ramanspektren in der Nähe der Anregungsfrequenz sehr wesentlich.

Als Kristall wurde ein Steinsalzprisma eines Spektrometers für das Ultrarot benutzt; seine Höhe betrug ungefähr 5 cm. Die Lampe war neben einer Seitenfläche des Prismas parallel zur Achse eingestellt, so dass man durch mehrfache Reflexionen an den inneren Flächen des Prismas eine sehr starke

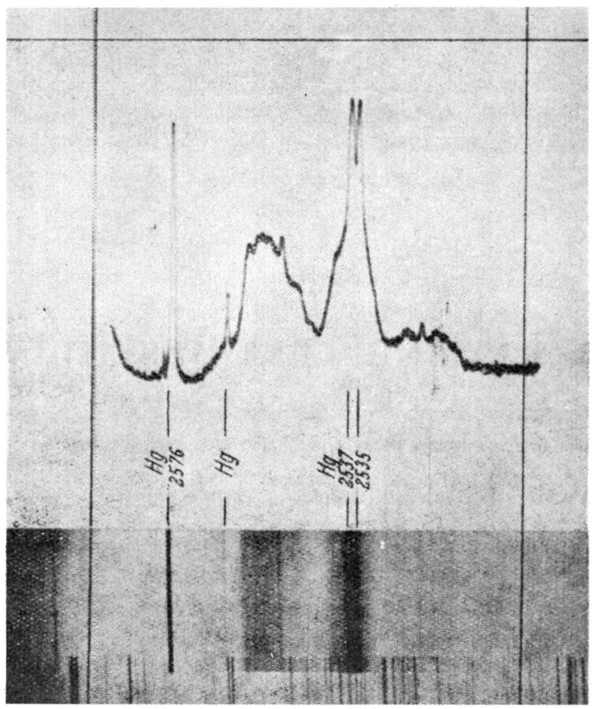

Fig. 1.

Beleuchtung erreichte; die Beobachtung geschah parallel zur Achse des Prismas, um eine möglichst grosse Dicke zu benutzen. Als Spektrograph wurde ein Hilgerscher Spektrograph E 315 verwendet; die Expositionsdauer betrug acht Stunden.

Das erhaltene Ramanspektrum ist in Fig. 1 wiedergegeben; die Intensitätsverteilung ist aus der Photometrie zu ersehen. Das Spektrum ist sehr schwach; unter denselben Bedingungen würde man etwa bei Kalkspat oder Quarz ein ziemlich starkes Ramanspektrum in wenigen Minuten erhalten. Im Gegensatz zu den gewöhnlichen Kristallen, deren Ramanspektren aus ziemlich scharfen Linien bestehen, hat Steinsalz ein kontinuierliches Ramanspektrum mit unregelmässig verteilten Maxima und Minima, das wesentlich aus einem Teil in der Nähe der anregenden Linie (von 0 bis 50 cm^{-1}) und einem Teil von 160 bis 365 cm^{-1} besteht; auch in der Lücke zwischen diesen beiden Spektralgebieten ist die Intensität nicht verschwindend, sondern nur etwas kleiner.

Wir wollen jetzt einige Betrachtungen entwickeln, welche die beobachteten Erscheinungen mindestens qualitativ erklären.

Bekanntlich kann jedes Kristallgitter mit einen Kontinuum von Frequenzen schwingen. Wesentlich verschieden ist aber das Schwingungsspektrum, je nachdem das Gitter einfach oder komplex ist. Und zwar hat im ersteren Falle das Gitter für jede (vektorielle) Wellenzahl drei Grundfrequenzen, welche drei Schwingungsrichtungen der Partikel entsprechen; diese Frequenzen sind Funktionen der drei Komponenten der Wellenzahl und verschwinden für verschwindende Wellenzahl. Im Falle eines komplexen Gitters, welches aus s einfachen Gittern zusammengesetzt ist, hat man für jede Wellenzahl 3 s im allgemeinen verschiedene Frequenzen; für verschwindende Wellenzahl schliessen sich drei von diesen Frequenzen an die Frequenz Null (akustische Schwingungen) an; die anderen schliessen sich dagegen an die ultraroten Frequenzen an, welche den Schwingungen der einzelnen einfachen Gitter gegeneinander entsprechen.

So hat man z. B. im Falle des Steinsalzes ein zusammengesetztes Gitter, welches aus den zwei gleichen flächenzentrierten Gittern der Cl– und der Na–Ionen besteht. Man hat also für jede Wellenzahl sechs Frequenzen, deren drei sich an die akustischen Schwingungen und drei an die ultrarote Reststrahlenfrequenz bei 52 μ anschliessen.

Man kann nun beweisen [(4)], dass nur solche Frequenzen in erster Näherung im Ramanspektrum eines Kristalls erscheinen, für welche die elastische Wellenzahl von der Grössenordnung der Wellenzahl des anregenden Lichtes ist; da diese letzte sehr klein in Bezug auf den reziproken Abstand aufeinanderfolgender Atome ist, kann man mit sehr guter Annäherung annehmen, dass die Wellenzahl der im Ramaneffekt wirksamen Eigenfrequenzen Null ist.

Um dies am besten zu sehen, wollen wir ein kleines Stück des Kristalls betrachten, dessen lineare Dimensionen klein in Bezug auf die optischen Wellenlängen, aber gross in Bezug auf die interatomaren Abstände angenommen werden, so dass man die Phasenunterschiede der Lichtwelle an verschiedenen Orten des Kristalls vernachlässigen kann.

Die Streuung des Lichtes wird bekanntlich in diesem Falle durch das vom Strahlungsfeld im Kristall induzierte elektrische Moment M bestimmt; dies hängt aber in erster Näherung von der elektrischen Polarisierbarkeit k des Kristalls nach der Formel

$$M = kE \tag{1}$$

ab, wo E die elektrische Feldstärke darstellt; M wurde auf die Volumeneinheit bezogen.

(4) L. MANDELSTAM, G. LANDSBERG, u. M. LEONTOWITSCH, «ZS. f. Phys.», *60*, 334 (1930); IG. TAMM, ebenda S. 345.

Sind einige elastische Schwingungen des Kristalls angeregt, so ist eigentlich die Polarisierbarkeit nicht an allen Orten des Kristalls dieselbe, weil sie etwas von den örtlich verschiedenen Verzerrungen des Kristalls beeinflusst wird; man muss also in (1) für k den Mittelwert über den ganzen Kristall nehmen. In erster Näherung kann die Änderung der Polarisierbarkeit an einem Orte proportional der Verzerrung des Kristallgitters an demselben Orte angesetzt werden; der Einfluss einer elastischen Welle kleiner Wellenlänge an Orten entgegengesetzter Phase wird also entgegengesetzt sein und sich durch die Mittelung über den ganzen Kristall fortheben. Falls aber der ganze Kristall elastisch mit derselben Phase schwingt (Wellenzahl Null oder praktisch Null), hat man eine zeitlich periodische Schwankung der mittleren Polarisierbarkeit, mit derselben Frequenz ω der elastischen Schwingung. Das induzierte elektrische Moment (1) des ganzen Kristalls, in harmonischen Komponenten analysiert, hat eine grosse Komponente, die mit der Frequenz ν des anregenden Lichtes schwingt; diese bedingt die Rayleighstreuung un veränderter Frequenz; plus schwächere Komponenten, die mit den Kombinationsfrequenzen $\nu \pm \omega$ schwingen, welche die Ramanstreuung bedingen. Diese klassische Behandlung des Problems kann bekanntlich quantentheoretisch gedeutet werden.

Es ergibt sich also, dass in erster Näherung nur Frequenzen mit der elastischen Wellenzahl Null vorkommen können. Bei einem einfachen Gitter, wo man nur Schwingungen vom akustischen Typus hat, gibt es also in erster Näherung keine Ramanlinie, sondern nur eine sehr kleine richtungsabhängige Verbreiterung der Rayleighlinie, welche davon herrührt, dass in Wirklichkeit, wie oben angedeutet, nicht nur die Wellenzahl Null, sondern auch Wellenzahlen von der Grössenordnung der optischen Wellenzahl wirksam sind. Bei zusammengesetzten Gittern hat man aber für die Wellenzahl Null ausser der Frequenz Null noch ultrarote Frequenzen, welche den Schwingungen der verschiedenen einfachen Gitter gegeneinander entsprechen. Die Anzahl dieser Frequenzen ist $3s - 3$; einige von diesen Frequenzen können koinzidieren oder im Ramaneffekt inaktiv sein.

Dieser letzte Fall kommt im besonderen immer dann vor, wenn die betreffende Eigenschwingung symmetrisch ist, in dem Sinne, dass die Polarisierbarkeit eine gerade Funktion der Verrückung ist, was immer mit leicht erkennbaren Symmetrieverhältnissen des Kristallgitters zusammenhängt.

So wird z. B. beim Flussspat nur eine Ramanfrequenz bei 321 cm beobachtet [(5)]. Das Flussspatgitter besteht aus drei kubischen flächenzentrierten Gittern, wovon eines mit Ca– und zwei mit F–Atomen besetzt sind. Wegen der Symmetrie dieses Gitters hat man zwei dreifach entartete ultrarote Frequenzen mit der Wellenzahl Null. Eine entspricht der Schwingung des Ca gegen die beiden F–Gitter. Diese ist mit einem schwingenden elektrischen Dipolmoment verbunden, ist aber symmetrisch; entsprechend wird sie in ultraroter Absorption, nicht aber als Ramaneffekt beobachtet. Die andere Eigenfrequenz entspricht einer Schwingung der beiden F–Gitter gegeneinan-

(5) F. RASETTI, «Nature», *127*, 626 (1931).

der, wobei das Ca–Gitter stehen bleibt. Diese Frequenz ist optisch inaktiv, ist aber unsymmetrisch und kommt also im Ramaneffekt vor.

Im Falle des Steinsalzes hat man drei zusammenfallende Frequenzen mit der Wellenzahl Null; die entsprechenden Schwingungen sind aber symmetrisch, und werden deswegen im Ramaneffekt nicht beobachtet. Damit ist natürlich nicht gesagt, dass man bei Steinsalz keinen Ramaneffekt beobachten kann, sondern nur, dass ein etwaiger Ramaneffekt viel lichtschwächer als bei anderen Kristallen sein muss; es handelt sich nämlich um einen Ramaneffekt zweiter Näherung.

Wir wollen also den Ramaneffekt zweiter Näherung kurz besprechen: Die Tatsache, dass der Ramaneffekt bei Kristallen gewöhnlich aus scharfen Linien und nicht aus einem Kontinuum besteht, ist dadurch bedingt, dass man in erster Näherung nur Frequenzen mit der Wellenzahl Null beobachtet. Die Polarisierbarkeit des Kristalls ist eine Funktion der elastischen Verrükkungen, welche in erster Näherung linear von diesen Verrückungen abhängt. Die linearen Glieder geben natürlich einen für den ganzen Kristall nicht interferierenden Anteil nur für die Wellenzahl Null. Falls aber, wie beim Steinsalz, die linearen Glieder verschwinden, muss man zur nächsten Näherung übergehen, also auch die quadratischen Glieder der Abhängigkeit der Polarisierbarkeit von den elastischen Verrückungen berücksichtigen. Das hat zur Folge, dass jetzt auch Schwingungen mit von Null verschiedener Wellenzahl wirksam sein können. Dies ist der Grund dafür, dass man im Ramaneffekt zweiter Näherung nicht scharfe Linien, sondern ein Kontinuum beobachtet.

Um diese Verhältnisse für den Fall des NaCl etwas genauer darzustellen, nehmen wir ein Koordinatensystem mit dem Zentrum in einem der Cl–Atome und den Achsen x, y, z parallel zu den Kristallachsen. Sei a die Entfernung zwischen einem Cl und dem nächsten Na–Atom. Dann liegen die Ruhelagen der Atome an den Stellen $x = n_1 a$, $y = n_2 a$, $z = n_3 a$, wo die n ganze Zahlen sind; und zwar ist das betreffende Atom ein Cl– oder ein Na–Atom, je nachdem $n_1 + n_2 + n_3$ gerade oder ungerade ist.

Für jede vektorielle Wellenzahl $f \equiv (f_1, f_2, f_3)$ hat das Steinsalzgitter sechs Eigenschwingungen mit sechs Frequenzen

$$\omega_i(f) \qquad (i = 1, 2, 3, 4, 5, 6).$$

Zu jeder Eigenschwingung gehört eine Normalkoordinate q_{if}. Die Verrükkung eines Atoms des Gitters drückt sich durch die Normalkoordinaten folgendermassen aus:

$$\xi_n = \sum_f \sum_{i=1}^{6} q_{if} \left| \begin{matrix} A_{if} \\ B_{if} \end{matrix} \right| \cos \left[2\pi a (n_1 f_1 + n_2 f_2 + n_3 f_3) + \begin{matrix} \varphi_{if} \\ \psi_{if} \end{matrix} \right], \tag{2}$$

wo man die oberen (A, φ)– oder die unteren (B, ψ)– Ausdrücke nehmen muss, je nachdem $n_1 + n_2 + n_3$ gerade oder ungerade ist, d.h. je nachdem es sich um ein Cl– oder um ein Na–Atom handelt. Zu summieren ist über alle Wellenzahlen des Debyespektrums.

Die Polarisierbarkeit $\varkappa$ eines Atoms $n \equiv (n_1, n_2, n_3)$ des Gitters, hängt offenbar von den relativen Verschiebungen der umliegenden Atome ab. Sie ist also eine Funktion der Grössen $\xi_{n+l} - \xi_n$, wo n und l eine Abkürzung für n_1, n_2, n_3 und l_1, l_2, l_3 darstellen. Wir wollen diese Funktion nach Potenzen dieser Variablen entwickeln, und zwar bis auf Glieder zweiter Ordnung; man findet:

$$\varkappa = \varkappa_0 + \sum_l \alpha_l (\xi_{n+l} - \xi_l) + \sum_{jl} \beta_{jl} (\xi_{n+j} - \xi_n)(\xi_{n+l} - \xi_n). \tag{3}$$

Nun ist offenbar, wegen der Symmetrie des Gitters, $\alpha_l = -\alpha_{-l}$; (3) wird also:

$$\varkappa = \varkappa_0 + \frac{1}{2} \sum_l \varkappa_l (\xi_{n+l} - \xi_{n-l}) + \sum_{jl} \beta_{jl} (\xi_{n+j} - \xi_n)(\xi_{n+l} - \xi_n). \tag{4}$$

Um die Polarisierbarkeit des ganzen Kristalls zu erhalten, müssen wir den Mittelwert von (4) über das ganze Gitter nehmen; wir ersetzen in (4) die Ausdrücke (2) für die Verschiebungen und beachten, dass die Atome $n + l$ und $n - l$ gleichartig sind. Dann finden wir für die mittlere Polarisierbarkeit $\bar{k}$ einen Ausdruck der Form

$$\bar{k} = k_0 + \sum_f \sum_{r,s=1}^{6} A_{rsf}\, q_{rf}\, q_{sf}, \tag{5}$$

weil sich alle anderen Glieder durch die Mittelung fortheben.

Setzen wir diesen Ausdruck für die mittlere Polarisierbarkeit in (1) ein und beachten, dass E mit der Frequenz ν des anregenden Lichtes schwingt und die Normalkoordinaten q_{if} mit den entsprechenden Eigenfrequenzen $\omega_i(f)$, so finden wir, dass das elektrische Moment M harmonische Komponenten mit den Kombinationsfrequenzen

$$\nu \pm 2\,\omega_i(f) \quad ; \quad \nu \pm [\omega_i(f) + \omega_k(f)] \quad ; \quad \nu \pm [\omega_i(f) - \omega_k(f)] \tag{6}$$

hat, wobei nur Eigenfrequenzen derselben (vektoriellen) Wellenzahl zu kombinieren sind. Die Frequenzen (6) entsprechen den Stokesschen und anti–Stokesschen Ramanlinien; sie sind hier klassich abgeleitet worden, man kann aber ohne Schwierigkeit zeigen, dass man auch mit einer quantentheoretischen Berechnung genau dieselben Frequenzen findet.

Das Ramanspektrum zweiter Näherung erstreckt sich also kontinuierlich von der Frequenz Null bis zum Doppelten der maximalen Eigenfrequenz, welche die Reststrahlenfrequenz, d. h. die Frequenz der Schwingung des Cl– und Na–Gitters gegeneinander ist. Aus der Grenze des Ramanspektrums bei 365 Wellenzahlen ergibt sich diese maximale Frequenz zu 183 Wellenzahlen, d. h. ungefähr 55 μ, in ziemlich guter Übereinstimmung mit der aus den Reststrahlen erhaltenen Wellenlänge von 52 μ.

Um die Intensitätsverteilung des Ramanspektrums zu berechnen, wären besondere Annahmen über die Beeinflussung der Polarisierbarkeit der Kristallatome durch die elastischen Verzerrungen notwendig. Es scheint deswegen vorläufig aussichtslos, die feineren Einzelheiten des Spektrums erklären zu wollen. Es sei nur bemerkt, dass die Frequenzen, welche zu derselben elastischen Wellenzahl gehören, zu sehr vielen verschiedenen Kombinationsfrequenzen Anlass geben. Für einige Frequenzen hat man übrigens Häufungen von Eigenschwingungen, welche Intensitätsmaxima im Ramanspektrum zur Folge haben, so dass man qualitativ recht gut das unregelmässige Aussehen des beobachteten Ramanspektrums verstehen kann.

N° 70.

For the introduction to this paper see N° 50.

70.

ÜBER DIE WECHSELWIRKUNG VON ZWEI ELEKTRONEN

H. BETHE und E. FERMI

« Z. Physik », *77*, 296–306 (1932)

Die Zusammenhänge zwischen den Wechselwirkungsformeln von Breit [1] und Møller [2] und der Quantenelektrodynamik werden diskutiert. Es wird gezeigt, dass die Breitsche Formel sich aus der Møllerschen (§ 1) und aus der Quantenelektrodynamik (§ 2) ableiten lässt, und dass aus der letzteren auch die Møllersche Formel folgt (§ 3).

Einleitung [*]. Für die Wechselwirkung zweier Elektronen existieren die beiden Ansätze von Breit [1] und Møller [2], die scheinbar von gänzlich verschiedenen Punkten ausgehen. Es erschien wünschenswert, die Beziehungen der beiden Ansätze zueinander und zur Quantenelektrodynamik zu untersuchen. Die Breitsche Differentialgleichung ist von Breit direkt aus der Quantenelektrodynamik abgeleitet worden, und Rosenfeld [3] hat gezeigt, dass die für die Breitsche Ableitung notwendige Formel der Quantenelektrodynamik sich aus dem von Møller angewandten Heisenbergschen [4] Korrespondenzverfahren ergibt. Wir werden die Breitsche Formel *direkt* aus der Møllerschen gewinnen, und für die Ableitung aus der Quantenelektrodynamik eine Form wählen, die ganz analog ist zu unserer Ableitung der Møllerschen Theorie, so dass die verschiedenen Näherungsverfahren besser vergleichbar werden.

Bekanntlich kann man aus der Quantenelektrodynamik genau die Coulombsche Wechselwirkungsenergie zwischen zwei Elektronen ableiten, wenn man nur die Kopplung der *longitudinalen* Wellen des elektromagnetischen Feldes mit den Elektronen berücksichtigt [5]. Jede Abweichung vom Coulomb-

(*) In dieser Arbeit wurde die deutsche Vektorschreibweise des Originales beibehalten (E. A.).

(1) G. BREIT, « Phys. Rev. », *34*, 553 (1929); *39*, 616 (1932).

(2) C. MØLLER, « ZS. f. Phys. », *70*, 786 (1931).

(3) L. ROSENFELD, ebenda *73*, 253, (1931).

(4) W. HEISENBERG, « Ann. d. Phys. », *9*, 338, (1931).

(5) E. FERMI, « Rev. of Mod. Phys. », *4*, 87 (1932), im folgenden als l. c. zitiert.

schen Gesetz muss daher durch die Wechselwirkung mit den *transversalen* Wellen zustandekommen, die wir im folgenden kurz als *Strahlungsfeld* bezeichnen. Die Kopplung zwischen Materie und Strahlungsfeld verursacht nun in *erster* Näherung nur solche Übergänge, bei denen sich der Quantenzustand der Elektronen ändert und ein Lichtquant absorbiert oder emittiert wird. Es interessieren uns aber die Matrixelemente der Wechselwirkungsenergie der Elektronen, welche einer Änderung des Quantenzustandes der Elektronen allein, ohne Änderung des Zustandes des Strahlungsfeldes, entsprechen, wobei wir insbesondere annehmen können, dass sowohl im Anfangs– wie im Endzustand keine Strahlung vorhanden sein soll. Solche Übergänge der Elektronen allein bekommt man offenbar erst durch einen Doppelprozess, wobei zuerst ein Quant emittiert und dann dasselbe Quant wieder absorbiert wird.

Die Ableitungen der Breitschen und der Møllerschen Formel aus der Quantenelektrodynamik unterscheiden sich nun wesentlich in folgendem: Im ersten Falle hat man die Coulombsche Wechselwirkungsenergie der Elektronen bereits in die « ungestörte Hamiltonfunktion » hineinzunehmen, die Störung ist dann lediglich die Wechselwirkung der Elektronen mit den transversalen Wellen des elektromagnetischen Feldes. Die Rechnung wird nur in nichtrelativistischer Näherung, d. h. nur bis zu Gliedern der Grössenordnung v^2/c^2, durchgeführt. Es werden ausserdem alle Übergänge des materiellen Teils des Systems vernachlässigt, mit denen eine permanente Emission von Strahlung verbunden ist. Die letzteren kann man natürlich mit der gewöhnlichen Strahlungstheorie behandeln.

Bei der Møllerschen Theorie ist dagegen die Coulombenergie ein Teil der *Störung* und tritt gleichberechtigt neben die Wechselwirkung mit dem Strahlungsfeld, die beiden Wechselwirkungen sind nämlich im extrem relativistischen Fall von gleicher Grössenordnung. In nullter Näherung bewegen sich die Elektronen unabhängig voneinander. Die Møllersche Theorie wird relativistisch exakt durchgeführt, dafür wird nach Potenzen von e entwickelt und Glieder von höherer Ordnung als e^2 vernachlässigt. Wenn man also die Breitsche Wechselwirkung aus der Møllerschen Theorie ableitet, so lässt sich nur rechtfertigen, dass man die Coulomb–Wechselwirkung in erster Näherung behandelt, geht man aber von der Quantenelektrodynamik aus, so ist die Coulombkraft im ungestörten System berücksichtigt, so dass ihre Wirkung exakt behandelt werden kann. Es sei bemerkt, dass in der Møllerschen (zweiten) Näherung keine Strahlung auftritt, und dass eine solche erst bei Berücksichtigung der Strahlungswechselwirkung in *dritter* Näherung herauskommen würde, wenn man ausschliesst, dass die Elektronen im Endzustand negative Energie haben.

§ 1. Ableitung der Breitschen Wechselwirkungsenergie aus der Møllerschen Theorie. – Um die Wechselwirkung zweier Elektronen zu berechnen, hat man nach Møller die retardierten Potentiale zu bilden, die

(6) Wir berücksichtigen bei dieser Ableitung die Coulombenergie *nicht* im ungestörten Problem, da sie in der Møllerschen Wechselwirkung mit enthalten ist. Wir können also von Quantenzuständen der einzelnen Elektronen sprechen.

von der Ladungsverteilung des ersten Elektrons erzeugt werden, und hat diese als Störung zu betrachten, welche auf das zweite Elektron wirkt. Wir wollen das Matrixelement der Wechselwirkungsenergie berechnen, welches einem Übergang des ersten Elektrons vom Zustand (6) n_1 zum Zustand n'_1 und des zweiten Elektrons von n_2 nach n'_2 entspricht, wobei die Gesamtenergie im Anfangs– und Endzustand gleich sein soll:

$$E_1 + E_2 = E'_1 + E'_2 \,. \tag{1}$$

Dem Übergang $n_1 \rightarrow n'_1$ entspricht die Ladungsverteilung

$$\rho_{n_1 n'_1}(\mathfrak{r}_1, t) = e_1 u'^*_1(\mathfrak{r}_1)\, u_1(\mathfrak{r}_1)\, e^{\frac{2\pi i}{h}(E'_1 - E_1)t}, \tag{2}$$

wo u_1 und u'_1 die Diracschen Eigenfunktionen des ersten Elektrons im Anfangs– und Endzustand sind und e_1 seine Ladung. Die Ladungsverteilung (2) erzeugt unter Berücksichtigung der Retardierung am Orte $\mathfrak{r}_2$ zur Zeit t das skalare Potential:

$$\varphi_{n_1 n'_1}(\mathfrak{r}_2, t) = \int \frac{\rho_{n_1 n'_1}\left(\mathfrak{r}_1, t - \frac{|\mathfrak{r}_2 - \mathfrak{r}_1|}{c}\right)}{|\mathfrak{r}_2 - \mathfrak{r}_1|}\, d\tau_1 \tag{3}$$

$$= e_1 e^{\frac{2\pi i}{h}(E'_1 - E_1)t} \int \frac{(u'^*_1(\mathfrak{r}_1)\, u_1(\mathfrak{r}_1))}{|\mathfrak{r}_2 - \mathfrak{r}_1|}\, e^{\frac{2\pi i}{hc}(E_1 - E'_1)|\mathfrak{r}_2 - \mathfrak{r}_1|}\, d\tau_1 \,.$$

Entsprechend erhält man für das Vektorpotential:

$$\mathfrak{A}_{n_1 n'_1}(\mathfrak{r}_2, t) = -e_1 e^{\frac{2\pi i}{h}(E'_1 - E_1)t} \int \frac{(u'^*_1(\mathfrak{r}_1)\, \gamma_1\, u_1(\mathfrak{r}))}{|\mathfrak{r}_2 - \mathfrak{r}_1|}\, e^{\frac{2\pi i}{hc}(E_1 - E'_1)|\mathfrak{r}_2 - \mathfrak{r}_1|}\, d\tau_1 \,. \tag{4}$$

$\gamma_{x_1}, \gamma_{y_1}, \gamma_{z_1}, \delta_1$ sind die Diracschen Operatoren des ersten Elektrons. γ_1 ist der Vektor mit den Komponenten $\gamma_{x_1}, \gamma_{y_1}, \gamma_{z_1}$.

Nun ist die Breitsche Formel für die Wechselwirkungsenergie, welche wir abzuleiten wünschen, nur exakt bis zu Gliedern der Ordnung $1/c^2$ einschliesslich. Es liegt daher nahe, die Exponentialfunktionen in (3), (4), welche die Retardierung des Potentials darstellen, nach $1/c$ zu entwickeln:

$$\varphi_{n_1 n'_1}(\mathfrak{r}_2, t) = e_1 e^{\frac{2\pi i}{h}(E'_1 - E_1)t} \int (u'^*_1(\mathfrak{r}_1)\, u_1(\mathfrak{r}_1)) \left[\frac{1}{|\mathfrak{r}_2 - \mathfrak{r}_1|} + \frac{2\pi i}{hc}(E_1 - E'_1) - \frac{2\pi^2}{h^2 c^2}(E_1 - E'_1)^2 |\mathfrak{r}_2 - \mathfrak{r}_1|\right] d\tau_1 \tag{5}$$

und $\mathfrak{A}$ entsprechend. Das zweite Glied in der eckigen Klammer in (5) verschwindet: denn entweder sind die Zustände $n_1\, n'_1$ identisch, dann ist $E_1 - E'_1 = 0$, oder sie sind voneinander verschieden, dann verschwindet das Volumenintegral wegen der Orthogonalität der Eigenfunktionen. Die Störungsenergie, die auf das zweite Elektron wirkt, ist nun:

$$V^{n'_1}_{n_1}(\mathfrak{r}_2, t) = e_2\, \varphi_{n_1 n'_1}(\mathfrak{r}_2, t) + e_2\, (\gamma_2 \cdot \mathfrak{A}_{n_1 n'_1}(\mathfrak{r}_2, t))$$

Wir bilden ihr Matrixelement, das dem Übergang $n_2 \rightarrow n_2'$ des zweiten Elektrons entspricht:

$$V_{n_1\, n_2}^{n_1'\, n_2'} = e_1 e_2 e^{\frac{2\pi i}{h}(E_1' + E_2' - E_1 - E_2)t} \int u_2'^* (\mathfrak{r}_2)\, u_1'^*(\mathfrak{r}_1) \left[\frac{1}{|\mathfrak{r}_2 - \mathfrak{r}_1|} - \frac{2\pi^2}{h^2 c^2}(E_1 - E_1')^2\, |\mathfrak{r}_2 - \mathfrak{r}_1| - \frac{(\gamma_1 \cdot \gamma_2)}{|\mathfrak{r}_2 - \mathfrak{r}_1|}\right] u_2(\mathfrak{r}_2)\, u_1(\mathfrak{r}_1)\, d\tau_1\, d\tau_2 . \tag{6}$$

Das erste Glied in der Klammer ist hier das gewöhnliche Coulombpotential, das zweite kommt von der Retardierung des skalaren Potentials, das dritte ist der Einfluss des (unretardierten) Vektorpotentials: Da die γ–Operatoren selbst die Grössenordnung v/c haben, brauchen wir die Retardierung des Vektorpotentials in unserer Näherung nicht zu berücksichtigen. Das erste und dritte Glied sind symmetrisch in den beiden Elektronen, das zweite ist es nicht, was an der Unsymmetrie der Møllerschen Methode liegt. Da aber die Gesamtenergie im Anfangs– und Endzustand gleich sein soll (1), können wir das Glied symmetrisieren, indem wir statt $(E_1 - E_1')^2$ schreiben: $-(E_1 - E_1')(E_2 - E_2')$. Damit wird (6) das Matrixelement von

$$V = e_1 e_2 \cdot \frac{1 - (\gamma_1 \cdot \gamma_2)}{|\mathfrak{r}_2 - \mathfrak{r}_1|} + \frac{2\pi^2 e_1 e_2}{h^2 c^2}\big(H_2 (H_1 |\mathfrak{r}_2 - \mathfrak{r}_1| - |\mathfrak{r}_2 - \mathfrak{r}_1| H_1) - (H_1 |\mathfrak{r}_2 - \mathfrak{r}_1| - |\mathfrak{r}_2 - \mathfrak{r}_1| H_1) H_2\big) , \tag{7}$$

wobei H_1 und H_2 die ungestörten Hamiltonfunktionen der beiden Elektronen ohne Wechselwirkung sind, also

$$H_1 = - e_1 \varphi_0 (\mathfrak{r}_1) - \frac{hc}{2\pi i}(\gamma_1 \cdot \mathrm{grad}_1) - e_1 (\gamma_1 \cdot \mathfrak{A}_0 (\mathfrak{r}_1)) - m_1 c^2 \delta_1 . \tag{8}$$

φ_0 und $\mathfrak{A}_0$ sind hier skalares und Vektorpotential irgendeines äusseren statischen Feldes, in dem die beiden Elektronen sich bewegen. Alle Bestandteile von H_1 sind mit $|\mathfrak{r}_2 - \mathfrak{r}_1|$ vertauschbar ausser dem Gradienten, daher ist

$$H_1 |\mathfrak{r}_2 - \mathfrak{r}_1| - |\mathfrak{r}_2 - \mathfrak{r}_1| H_1 = - \frac{hc}{2\pi i}\left(\gamma_1 \cdot \frac{\mathfrak{r}_1 - \mathfrak{r}_2}{|\mathfrak{r}_1 - \mathfrak{r}_2|}\right) = F$$

und

$$H_2 F - F H_2 = + \frac{h^2 c^2}{4\pi^2}\left(\frac{(\gamma_1 \cdot \gamma_2)}{|\mathfrak{r}_1 - \mathfrak{r}_2|} - \frac{(\gamma_1 \cdot \mathfrak{r}_1 - \mathfrak{r}_2)(\gamma_2 \cdot \mathfrak{r}_1 - \mathfrak{r}_2)}{|\mathfrak{r}_1 - \mathfrak{r}_2|^3}\right). \tag{9}$$

Setzt man dies in (7) ein, so erhält man genau die Breitsche Formel für die Wechselwirkungsenergie:

$$V = \frac{e_1 e_2}{|\mathfrak{r}_1 - \mathfrak{r}_2|}\left(1 - \frac{1}{2}(\gamma_1 \cdot \gamma_2) - \frac{1}{2}\frac{(\gamma_1 \cdot \mathfrak{r}_1 - \mathfrak{r}_2)(\gamma_2 \cdot \mathfrak{r}_1 - \mathfrak{r}_2)}{|\mathfrak{r}_1 - \mathfrak{r}_2|^2}\right). \tag{10}$$

§ 2. ABLEITUNG DES BREITSCHEN WECHSELWIRKUNGSGESETZES AUS DER QUANTENELEKTRODYNAMIK. – Als Ausgangspunkt für diese Ableitung nehmen wir dei Hamiltonfunktion des Systems Elektronen + elektromagnetisches Feld in der Form [l. c. (166)], die man erhält, wenn man mit Hilfe der Kontinuitätsgleichung die Koordinaten des skalaren Potentials und der longitudinalen Komponente des Vektorpotentials eliminiert. In dieser Form enthält die Hamiltonfunktion bereits die Coulombwechselwirkung der Partikeln. Indem man [l. c. (167)] beachtet, kann die Hamiltonfunktion in der Form geschrieben werden:

$$(11) \qquad R = \sum_s \left(\frac{1}{2} p_s^2 + 2\pi^2 \nu_s^2 q_s^2\right) + \sum_{i<j} \frac{e_i e_j}{r_{ij}} - \sum_i c(\gamma_i \cdot p_i) - \sum_i m_i c^2 \delta_i$$

$$+ c\sqrt{\frac{8\pi}{\Omega}} \sum_{is} e_i (\gamma_i \cdot A_s) q_s \sin \Gamma_{si}.$$

In dieser Formel sind die Bezeichnungen dieselben wie in l. c.; die Strahlungskomponenten q_s sind jedoch mit einem einzigen Index s und nicht mit zwei s_1 und s_2 numeriert; die Coulombwechselwirkung der Elektronen ist explizit geschrieben, indem von den unendlich grossen konstanten elektrostatischen Selbstenergien abgesehen ist.

Zur Ableitung der Breitschen Formel müssen wir die Coulombkraft im ungestörten System mit berücksichtigen. Wir betrachten also als ungestörte Hamiltonfunktion die Summe der Energie der Strahlung

$$(12) \qquad H_s = \sum_s \left(\frac{1}{2} p_s^2 + 2\pi^2 \nu_s^2 q_s^2\right)$$

und der Energie der materiellen Teilchen

$$(13) \qquad H_M = \sum_i - c(\gamma_i \cdot p_i) - m_i c^2 \delta_i + \sum_{i<j} \frac{e_i e_j}{r_{ij}}.$$

Als Störung fassen wir die Kopplungsenergie von Partikeln und Feld auf.

$$(14) \qquad H = c\sqrt{\frac{8\pi}{\Omega}} \sum_{is} e_i (\gamma_i \cdot A_s) q_s \sin \Gamma_{si}.$$

Die Zustände des ungestörten Systems sind charakterisiert durch eine Quantenzahl n, welche den Zustand des materiellen Teiles des Systems mit Hamiltonfunktion H_M bestimmt, und durch die Quantenzahlen $n_1, n_2 \cdots n_s \cdots$ der Strahlungsoszillatoren. Die entsprechenden Wahrscheinlichkeitsamplituden seien

$$(15) \qquad a_{n n_1 n_2 \cdots n_s \cdots}$$

Durch die Wirkung der Kopplungsenergie (14) variieren die a mit der Zeit entsprechend den bekannten Gleichungen:

$$(16) \qquad a_{n' n_1' n_2' \cdots n_s'} = -\frac{2\pi i}{h} \sum_{n n_1 n_2 \cdots} H_{n' n_1' n_2' \cdots}^{n n_1 n_2 \cdots} a_{n n_1 n_2 \cdots} e^{\frac{2\pi i}{h}(E_{n' n_1' \cdots} - E_{n n_1 \cdots}) t}.$$

Bekanntlich sind die einzigen von Null verschiedenen Matrixelemente von H die folgenden:

$$(17) \qquad H_{n' n_1 n_2 \cdots n_s \pm 1 \cdots}^{n n_1 n_2 \cdots n_s \cdots} = c\sqrt{\frac{8\pi}{\Omega}} Q_{n'n}(s) \sqrt{\frac{h}{8\pi^2 \nu_s}} \cdot \begin{cases} \sqrt{n_s + 1} \\ \sqrt{n_s}, \end{cases}$$

wo $Q_{n'n}(s)$ das Matrixelement $n'n$ der Grösse

$$(18) \qquad Q(s) = \sum_i e_i (\gamma_i \cdot A_s) \sin \Gamma_{si}$$

darstellt.

Wir wollen jetzt annehmen, dass am Anfang das materielle System sich in dem Zustand n befindet und kein Strahlungsoszillator angeregt ist. Es ist dann

$$|a_{n\,00\cdots0\cdots}| = 1\,.$$

Wir wollen sehen, wie durch die Wirkung der Störung eine Wahrscheinlichkeitsamplitude für den Zustand $n'\,00\cdots0\cdots$ entsteht. Nun hat die Störungsmatrix (17) kein Element, welches die beiden Zustände $n\,000\cdots$ und $n'\,000\cdots$ direkt koppelt. Ein solcher Übergang kann nur indirekt auf dem Umweg über ein Zwischenglied vorkommen; als solche kommen nach (17) die Zustände $m\,00\cdots1_s\cdots$ in Frage, welche mit Anfangs- und Endzustand kombinieren. Man hat zunächst durch Anwendung von (16) auf den Übergang $n\,000\cdots\to m\,00\cdots1_s\cdots$:

$$(19)\qquad \left\{\begin{aligned} &\dot{a}_{m\,0\cdots1_s\cdots} = -\frac{2\pi i}{h}\,\mathrm{H}^{n\,00\cdots0\cdots}_{m\,00\cdots1_s\cdots}\,a_{n\,00\cdots0\cdots}\,e^{2\pi i(\nu_{mn}+\nu_s)t}\,,\\ &\text{wo}\\ &\nu_{mn} = \frac{\mathrm{E}_m - \mathrm{E}_n}{h}\,. \end{aligned}\right.$$

Da $a_{n\,00\cdots0\cdots}$ praktisch konstant ist, können wir integrieren und wir finden:

$$(20)\qquad a_{m\,0\cdots1_s\cdots} = -\frac{\mathrm{H}^{n\cdots0\cdots}_{m\cdots1_s\cdots}}{h\,(\nu_{mn}+\nu_s)}\,a_{n\,00\cdots}\,e^{2\pi i(\nu_{mn}+\nu_s)t}\,.$$

Wir müssen jetzt wieder Gleichung (16) auf die Übergänge $m\,0\cdots1_s\cdots \to n'\,0\cdots0\cdots$ anwenden. Wir finden:

$$(21)\qquad \dot{a}_{n'\,0\cdots0\cdots} = -\frac{2\pi i}{h}\sum_{ms}\mathrm{H}^{m\,0\cdots1_s\cdots}_{n'\,0\cdots0\cdots}\,a_{m\,0\cdots1_s\cdots}\,e^{2\pi i(\nu_{n'm}-\nu_s)t}$$

$$= -\frac{2\pi i}{h}\sum_{ms}\left(-\frac{1}{h}\,\frac{\mathrm{H}^{n\cdots0\cdots}_{m\cdots1\cdots}\,\mathrm{H}^{m\cdots1_s\cdots}_{n'\cdots0\cdots}}{\nu_{mn}+\nu_s}\right)\cdot a_{n\cdots0\cdots}\,e^{2\pi i\nu_{n'n}t}\,.$$

Wir können die Grösse

$$(22)\qquad \mathrm{K}_{n'n} = -\frac{1}{h}\sum_{ms}\frac{\mathrm{H}^{n\cdots0\cdots}_{m\cdots1_s\cdots}\,\mathrm{H}^{m\cdots1_s\cdots}_{n'\cdots0\cdots}}{\nu_{mn}+\nu_s}$$

als ein Matrixelement auffassen, das die Übergänge zwischen den Zuständen $n0\cdots0\cdots \to n'\,0\cdots0\cdots$ direkt vermittelt. Die entsprechende Grösse K stellt die Korrektion dar, die man an der Coulombwechselwirkung anzubringen hat, um der Retardierung der Potentialen Rechnung zu tragen.

Mit Hilfe von (17) finden wir aus (22):

$$(23)\qquad \mathrm{K}_{n'n} = -\frac{c^2}{\pi\Omega}\sum_{ms}\frac{\mathrm{Q}_{n'm}(s)\,\mathrm{Q}_{mn}(s)}{\nu_s(\nu_s+\nu_{mn})}\,.$$

Wir müssen jetzt die Grösse K als Funktion der Koordinaten und Momente der Partikel ausdrücken. Dies geschieht leicht, wenn wir in erster Näherung im Nenner von (23) ν_{mn} in Vergleich mit ν_s vernachlässigen. (Man beachte, dass das Verhältnis ν_{mn}/ν_s von der Grössenordnung v/c ist). Wir bekommen dann:

$$
(24)\quad \begin{cases} K_{n'n} = -\frac{c^2}{\pi\Omega}\sum_{sm}\frac{1}{\nu_s^2}\sum_m Q_{n'm}(s)\,Q_{mn}(s)\,, \\ \qquad = -\frac{c^2}{\pi\Omega}\sum_s\frac{1}{\nu_s^2}\,[Q^2(s)]_{n'n}\,. \end{cases}
$$

Wir haben also

$$
(25)\qquad K = -\frac{c^2}{\pi\Omega}\sum_s\frac{1}{\nu_s^2}\,Q^2(s)\,.
$$

Es bleibt nur noch die Summe über s auszuführen, um zu beweisen, dass die gefundene Grösse K mit der Breitschen Weschselwirkungsfunktion identisch ist.

In (25) setzen wir für $Q(s)$ den Ausdruck (18) ein, und finden

$$
(26)\qquad K = -\sum_{ij} e_i e_j \frac{c^2}{\pi\Omega}\sum_s\frac{1}{\nu_s^2}(\gamma_i\cdot A_s)(\gamma_j\cdot A_s)\sin\Gamma_{si}\sin\Gamma_{sj}\,.
$$

Die Summe über s wird in bekannter Weise in ein Integral transformiert, indem man $\sum_s$ durch $\int_0^\infty \frac{8\pi}{c^3}\Omega\nu_s^2\,d\nu_s$ ersetzt und den Ausdruck

$$
(\gamma_i\cdot A_s)(\gamma_j\cdot A_s)\sin\Gamma_{si}\sin\Gamma_{sj}
$$

über alle Phasen, Ausbreitungs- und Polarisationsrichtungen mittelt. Wir finden also

$$
(27)\qquad K = -\sum_{ij} e_i e_j\cdot\frac{8}{c}\int_0^\infty \overline{(\gamma_i\cdot A_s)(\gamma_j\cdot A_s)\sin\Gamma_{si}\sin\Gamma_{sj}}\,d\nu_s\,.
$$

Die Mittelung kann jetzt ohne weiteres ausgeführt werden. Durch eine Rechnung, die keine prinzipiellen Schwierigkeiten bietet, finden wir, indem wir (147) l. c. beachten,

$$
(28)\qquad \overline{(\gamma_i\cdot A_s)(\gamma_j\cdot A_s)\sin\Gamma_{si}\sin\Gamma_{sj}} = \frac{(\gamma_i\cdot\gamma_j)}{4}\left(\frac{\sin\vartheta}{\vartheta}+\frac{\cos\vartheta}{\vartheta^2}-\frac{\sin\vartheta}{\vartheta^3}\right)
$$
$$
= \frac{(\gamma_i\cdot r_{ij})(\gamma_j\cdot r_{ij})}{4r_{ij}^2}\left(\frac{\sin\vartheta}{\vartheta}+3\frac{\cos\vartheta}{\vartheta^3}-3\frac{\sin\vartheta}{\vartheta^3}\right).
$$

wo

$$
(29)\qquad \vartheta = \frac{2\pi r_{ij}}{c}\nu_s\,.
$$

Aus (27) finden wir also, indem wir als Integrationsvariable ϑ statt ν_s einführen:

$$
(30)\qquad K = -\sum_{ij}\frac{e_i e_j}{\pi r_{ij}}\left\{(\gamma_i\cdot\gamma_j)\int_0^\infty\left(\frac{\sin\vartheta}{\vartheta}+\frac{\cos\vartheta}{\vartheta^2}-\frac{\sin\vartheta}{\vartheta^2}\right)d\vartheta\right.
$$
$$
\left.-\frac{(\gamma_i\cdot r_{ij})(\gamma_j\cdot r_{ij})}{r_{ij}^2}\cdot\int_0^\infty\left(\frac{\sin\vartheta}{\vartheta}+\frac{3\cos\vartheta}{\vartheta^2}-3\frac{\sin\vartheta}{\vartheta^3}\right)d\vartheta\right\}.
$$

Die beiden Integrale haben die Werte $+\pi/4$ und $-\pi/4$. Wir finden also

$$K = -\sum_{ij} \frac{e_i e_j}{4 r_{ij}} \left\{ (\gamma_i \cdot \gamma_j) + \frac{(\gamma_i \cdot r_{ij})(\gamma_j \cdot r_{ij})}{r_{ij}^2} \right\}. \tag{31}$$

In dieser Formel ist eine unendlich grosse konstante Selbstenergie der Elektronen enthalten (7). Sehen wir von dieser Konstanten ab, so finden wir als Ausdruck für die Wechselwirkung von zwei Elektronen:

$$-\frac{e_i e_j}{2 r_{ij}} \left\{ (\gamma_i \cdot \gamma_j) + \frac{(\gamma_i \cdot r_{ij})(\gamma_j \cdot r_{ij})}{r_{ij}^2} \right\}, \tag{32}$$

welche mit der Breitschen Wechselwirkung identisch ist.

§ 3. Ableitung der Møllerschen Formel aus der Quantenelektrodynamik. – Als Ausgangspunkt nehmen wir wieder die Hamiltonfunktion (11), wobei wir uns auf den Fall von zwei Partikeln 1 und 2 beschränken. Wir fassen aber diesmal die Coulombenergie $e_1 e_2/r_{12}$ als Störung auf, so dass die ungestörte Hamiltonfunktion

$$H_0 = \sum_s \left(\frac{1}{2} p_s^2 + 2\pi^2 \nu_s^2 q_s^2 \right) + \sum_{i=1}^{2} (-c (\gamma_i \cdot p_i) - mc^2 \delta_i) \tag{33}$$

ist. Die Quantenzustände des ungestörten Systems sind durch die Quantenzahlen $n_1 n_2$ der Elektronen und die Quantenzahlen der Strahlungsoszillatoren definiert. Wir interessieren uns für den Übergang:

$$n_1 n_2 0 \cdots 0 \rightarrow n'_1 n'_2 0 \cdots 0 \cdots \tag{34}$$

Dieser Übergang kann einmal auf direktem Wege durch die Coulombsche Wechselwirkung hervorgerufen werden oder aber durch den Mechanismus, den wir im vorigen Paragraphen diskutiert haben, auf dem Umweg über einen Zwischenzustand

$$m_1 m_2 0 \cdots 1_s \cdots \tag{35}$$

Dem ersten Prozess entsprechen die Matrixelemente

$$\left(\frac{e_1 e_2}{r_{12}} \right)_{n_1 n_2}^{n'_1 n'_2}, \tag{36}$$

dem zweiten Prozess nach (23)

$$K_{n_1 n_2}^{n'_1 n'_2} = -\frac{c^2}{\pi\Omega} \sum_{m_1 m_2 s} \frac{Q_{m_1 m_2}^{n'_1 n'_2}(s)\, Q_{n_1 n_2}^{m_1 m_2}(s)}{\nu_s (\nu_s + \nu_{m_1 n_1} + \nu_{m_2 n_2})}. \tag{37}$$

Es ist konsequent, im ersten Falle sich auf die erste Näherung zu beschränken und im zweiten Falle bis zur zweiten zu gehen, weil (36) und (37) beide proportional zu $e_1 e_2$ sind.

(7) Diese hebt die elektrostatische Selbstenergie genau auf. Man muss aber nicht denken, dass dadurch die Schwierigkeiten der unendlichen Selbstenergie beseitigt wären. Man findet wieder eine unendlich grosse Selbstenergie eines freien Elektrons, wenn man die Diagonalelemente von (23) berechnet und dabei ν_{mn} nicht vernachlässigt und als Zwischenzustand m auch Zustände negativer Energie betrachtet.

Wir wollen jetzt die Matrixelemente (37) berechnen. Wir bemerken zunächst, dass Q (s) [vgl. (18)] aus zwei Summanden Q (1 s) und Q (2 s) besteht, von denen jeder nur von den Koordinaten *eines* Elektrons abhängt. Da die ungestörten Eigenfunktionen Produkte von Eigenfunktionen der einzelnen Elektronen sind, können die Matrixelemente von Q (s) nur solchen Übergängen entsprechen, bei denen ein einziges Elektron springt. Dadurch reduziert sich die Summe über $m_1 m_2$ auf zwei Glieder: $m_1 m_2 = n_1 n'_1$ und $m_1 m_2 = n'_1 n_2$. (37) wird dann

$$(38) \qquad K^{n'_1 n'_2}_{n_1 n_2} = -\frac{c^2}{\pi\Omega} \sum_s \frac{Q_{n'_1 n_1}(1\,s)\, Q_{n'_2 n_2}(2\,s)}{\nu_s} \left(\frac{1}{\nu_s + \nu_{n'_1 n_1}} + \frac{1}{\nu_s + \nu_{n'_2 n_2}} \right).$$

Q (1 s) hat die Form [vgl. (18) l. c. (147)]:

$$(39) \qquad Q(1\,s) = \frac{e_1}{2i} (\gamma_1 \cdot A_s) \left[e^{\frac{2\pi i \nu_s}{c} (\alpha_s \cdot \mathfrak{r}_1) + i\beta_s} - e^{-\frac{2\pi i \nu_s}{c} (\alpha_s \cdot \mathfrak{r}_2) - i\beta_s} \right].$$

Die Eigenfunktionen der Zustände n_1 und n'_1 schreiben wir in die Form

$$(40) \qquad u_1 = \frac{a_1}{\sqrt{\Omega}} e^{\frac{2\pi i}{h} (\mathfrak{p}_1 \cdot \mathfrak{r}_1)} \quad , \quad u'_1 = \frac{a'_1}{\sqrt{\Omega}} e^{\frac{2\pi i}{h} (\mathfrak{p}'_1 \cdot \mathfrak{r}_1)} ,$$

wo die a vierkomponentige Konstanten darstellen, die auf Eins normiert sind.

Die Matrixelemente von Q (1 s) sind also nur dann von Null verschieden, wenn

$$(41) \qquad \frac{h\nu_s}{c} \alpha_s = \pm (\mathfrak{p}_1 - \mathfrak{p}'_1) .$$

Ihre Werte sind in diesem Falle

$$(42) \qquad \mp \frac{e_1}{2i} (\tilde{a}'_1 \gamma_1 a_1 \cdot A_s)\, e^{\mp i\beta_s} .$$

Für Q (2 s) bekommen wir einen entsprechenden Ausdruck. Da man nun eine Mittelung über die Phasen β_s vorzunehmen hat, liefern von Null verschiedene Beiträge nur die Glieder, für welche das Vorzeichen in Q (1 s) und Q (2 s) entgegengesetzt ist. Es ist also

$$(43) \qquad \mathfrak{p}_1 - \mathfrak{p}'_1 = -(\mathfrak{p}_2 - \mathfrak{p}'_2) = \pm \frac{h\nu_s}{c} \vec{\alpha}_s \quad , \quad \nu_s = \frac{c}{h} | \mathfrak{p}_1 - \mathfrak{p}'_1 | ,$$

$$(44) \qquad K^{n'_1 n'_2}_{n_1 n_2} = -\frac{e_1 e_2 h^2}{4\pi\Omega} \sum_s \frac{(\tilde{a}'_1 \gamma_1 a_1 \cdot A_s)(\tilde{a}'_2 \gamma_2 a_2 \cdot A_s)}{| \mathfrak{p}_1 - \mathfrak{p}'_1 |}$$

$$\cdot \left(\frac{1}{| \mathfrak{p}_1 - \mathfrak{p}'_1 | + \frac{(E'_1 - E_1)}{c}} + \frac{1}{| \mathfrak{p}_1 - \mathfrak{p}'_1 | + \frac{(E'_2 - E_2)}{c}} \right).$$

Wegen (43) besteht die Summe über s nur noch aus vier Gliedern, entsprechend dem doppelten Vorzeichen in (43) und zwei zueinander senkrechten Polarisationsrichtungen. Statt über s zu summieren, können wir also über alle Richtungen des Einheitsvektors A_s senkrecht zu $\mathfrak{p} - \mathfrak{p}'$ mitteln und mit 4 multiplizieren. Führen wir das aus und beachten wir noch, dass wegen des Energiesatzes

$$E_1 - E_1' = E_2' - E_2 ,$$

so finden wir:

$$(45)\quad K_{n_1 n_2}^{n_1' n_2'} = - \frac{e_1 e_2 h^2}{\pi\Omega}\left[(\tilde{a}_1' \gamma_1 a_1 \cdot \tilde{a}_2' \gamma_2 a_2) - \frac{(\tilde{a}_1' \gamma_1 a_1 \cdot (\mathfrak{p}_1 - \mathfrak{p}_1'))(\tilde{a}_2' \gamma_2 a_2 \cdot (\mathfrak{p}_1 - \mathfrak{p}_1'))}{(\mathfrak{p}_1 - \mathfrak{p}_1')^2}\right] \cdot \frac{1}{(\mathfrak{p}_1 - \mathfrak{p}_1')^2 - \left(\frac{E_1 - E_1'}{c}\right)^2} .$$

Wir müssen noch zu (45) das Matrixelement (36) addieren; dieses hat nach Møller den Wert:

$$(46)\quad \left(\frac{e_1 e_2}{r_{12}}\right)_{n_1 n_2}^{n_1' n_2'} = \frac{e_1 e_2 h^2}{\pi\Omega} \frac{(\tilde{a}_1' a_1)(\tilde{a}_2' a_2)}{(\mathfrak{p}_1 - \mathfrak{p}_1')^2} .$$

Die Summe ist:

$$(47)\quad \begin{cases} H_{n_1 n_2}^{n_1' n_2'} = K_{n_1 n_2}^{n_1' n_2'} + \left(\frac{e_1 e_2}{r_{12}}\right)_{n_1 n_2}^{n_1' n_2'} \\ = \frac{e_1 e_2 h^2}{\pi\Omega} \cdot \frac{(\tilde{a}_1' a_1)(\tilde{a}_2' a_2) - (\tilde{a}_1' \gamma_1 a_1\ \tilde{a}_2' \gamma_2 a_2)}{(\mathfrak{p}_1 - \mathfrak{p}_1')^2 - \left(\frac{E_1 - E_1'}{c}\right)} + \frac{e_1 e_2 h^2}{\pi\Omega\,|\,\mathfrak{p}_1 - \mathfrak{p}_1'\,|^2} \\ \cdot \frac{-\left(\frac{E_1 - E_1'}{c}\right)^2 (\tilde{a}_1' a_1)(\tilde{a}_2' a_2) + (\tilde{a}_1' \gamma_1 a_1 \cdot (\mathfrak{p}_1 - \mathfrak{p}_1'))(\tilde{a}_2' \gamma_2 a_2 \cdot (\mathfrak{p}_1 - \mathfrak{p}_1'))}{(\mathfrak{p}_1 - \mathfrak{p}_1') - \left(\frac{E_1 - E_1'}{c}\right)^2} \end{cases}$$

Der erste Summand der rechten Seite ist genau die Møllersche Wechselwirkungsenergie. Wir müssen also beweisen, dass der zweite Summand verschwindet. Wir schreiben zunächst diesen Summand in symmetrischer Form mit Benutzung des Energie– und Impulssatzes:

$$\text{Faktor} \cdot \left[+ \frac{E_1 - E_1'}{c}(\tilde{a}_1 a_1)\frac{E_2 - E_2'}{c}(\tilde{a}_2' a_2) - (\tilde{a}_1' \gamma_1 a_1 \cdot (\mathfrak{p}_1 - \mathfrak{p}_1'))(\tilde{a}_2' \gamma_2 a_2 \cdot (\mathfrak{p}_2 - \mathfrak{p}_2'))\right] \cdot$$

Um zu beweisen, dass dies verschwindet, genügt es zu zeigen, dass

$$(49)\quad (E_1 - E_1')\,\tilde{a}_1' a_1 = -c\,(\tilde{a}_1' \gamma_1 a_1 \cdot (\mathfrak{p}_1 - \mathfrak{p}_1'))$$

und eine entsprechende Gleichung für die zweite Partikel gilt. (49) folgt aber sofort aus der Diracgleichung

$$(50)\quad E_1 a_1 = -c\,(\mathfrak{p}_1 \gamma_1 a_1) - m_1 c^2 \delta_1 a_1$$

und der entsprechenden Gleichung für $\tilde{a}_1'$.

Man überzeugt sich übrigens leicht, dass die Møllersche Formel sich in genau der gleichen Weise ableiten lässt, wenn eine der Partikeln gebunden ist.

Es ist befriedigend, dass in unserer Näherung keine Strahlung emittiert wird. Man könnte an sich denken, dass Prozesse möglich sind, bei denen die beiden Elektronen ihre Zustände ändern und zwei Quanten emittiert werden.

Bei solchen Prozessen müsste man aber zweimal den Impulssatz befriedigen (der Impuls des einen Quants muss gleich der Impulsänderung des einen Elektrons sein, und dasselbe gilt für das andere Quant und das andere Elektron) und einmal den Energiesatz. Das ist aber unmöglich, sofern im Endzustand keines der Elektronen eine negative Energie hat.

Der eine von uns (H. Bethe) möchte der Rockefeller Foundation für die Gewährung eines Stipendiums danken, das ihm den Aufenthalt in Rom ermöglicht hat.

Rom, Institut für theoretische Physik.

N° 71.

In this article Fermi summarizes some of his earlier results on the theory of the Raman effect and extends it to cases not previously considered. After a brief introduction on the fundamental frequencies of molecules and their activity in the Raman effect, he expounds the theory of the CO_2 molecule essentially as given in the previous paper on the same subject (N° 68). He then remarks that an analogous effect is responsible for splitting one of the Raman lines of CCl_4 into a doublet. There follows a brief theory of the first and second-order Raman effect in crystals. The theory of the second-order spectrum of rocksalt is given, following the treatment previously published in German (N° 69). Finally, he considers the Raman spectra of calcite and sodium nitrate. The observations show one more Raman line than expected as a first-order spectrum from the knowledge of the structure of these crystals. He suggests that the line of highest frequency be attributed to the second-order spectrum, which for special reasons turns out to be concentrated into a fairly sharp line instead of the widespread continuum observed in the alkali halides. Fermi's interest in the theory of the Raman effect had been aroused mainly by Rasetti's experimental work in Pasadena and Rome.

F. RASETTI.

71.

L'EFFETTO RAMAN NELLE MOLECOLE E NEI CRISTALLI (*)

«Memorie Accad. d'Italia», *3* (1), (Fis.) 239–256 (1932)

RIASSUNTO. — Dopo un breve riassunto dei criteri generali per la interpretazione dell'effetto Raman nelle molecole e nei cristalli, vengono studiati in questa Memoria alcuni notevoli casi di eccezione, che permettono l'interpretazione degli spettri Raman dell'anidride carbonica e del tetracloruro di carbonio; spiegano l'apparire di un debolissimo spettro continuo nel salgemma, per il quale la teoria ordinaria farebbe prevedere l'assenza di effetto Raman; e consentono di ordinare completamente gli spettri Raman della calcite e del nitrato di sodio.

1. Nello spettro Raman di una molecola, sia appartenente ad un gas che a un liquido, si osservano normalmente alcune righe corrispondenti a tutte o a parte delle frequenze di oscillazione della molecola nello stato fondamentale elettronico; assai più raramente si osservano invece righe corrispondenti a una variazione dello stato di rotazione. Queste infatti, salve pochissime eccezioni, appaiono con notevole intensità solo nell'intorno della riga eccitatrice e vengono spesso mascherate dall'alone che la circonda; inoltre le si osservano bene soltanto nei gas, poiché, essendo assai ravvicinate una dall'altra, esse normalmente si confondono tra di loro nei liquidi. È noto

(*) Presentata nell'Adunanza dell'11 marzo 1932.

infine un solo caso in cui sia stata osservata con sicurezza in effetto Raman una transizione tra due livelli elettronici. In questo lavoro ci limiteremo perciò alla discussione dell'effetto Raman di oscillazione, che rappresenta di gran lunga il caso più frequente. E particolarmente ci occuperemo qui di alcuni casi di eccezione alle normali regole di selezione per l'effetto Raman, che conviene tenere presenti tutte le volte che si debba interpretare lo spettro Raman di una molecola oppure di un cristallo. Alcuni di questi casi sono già stati pubblicati in note preliminari [1].

Alla discussione di questi casi di eccezione, premettiamo un breve riassunto dei criteri che si seguono per stabilire le ordinarie regole di selezione per l'effetto Raman.

Per descrivere le oscillazioni di una molecola, conviene riferirne le deformazioni a un sistema di coordinate normali, $q_1, q_2, \cdots, q_f$ il cui numero è eguale al numero delle vibrazioni fondamentali della molecola.

Per mezzo di queste l'energia potenziale della deformazione elastica della molecola prende in prima approssimazione la forma

$$\sum_i 2\pi^2 \nu_i^2 q_i^2 \tag{1}$$

mentre l'energia cinetica ha pure la forma ortogonale

$$\frac{1}{2} \Sigma m_i \dot{q}_i^2 . \tag{2}$$

In seconda approssimazione compaiono nell'energia potenziale dei termini di terzo grado e di grado superiore nelle q; questi termini hanno per conseguenza una certa anarmonicità delle oscillazioni della molecola.

Per la produzione delle righe Raman sono essenziali come è noto le pulsazioni della polarizzabilità elettrica della molecola. Per rappresentarle conviene osservare che la polarizzabilità elettrica, e quindi tutte le componenti $\alpha_{xx}, \alpha_{xy}, \cdots$ del tensore che la rappresenta, sono funzioni delle coordinate q; sviluppando tali funzioni in serie di potenze si trova

$$\alpha_{xy} = \alpha_{xy0} + \sum_i \alpha_{xyi} q_i + \sum_{ij} \alpha_{xyij} q_i q_j + \cdots \tag{3}$$

con espressioni analoghe per le altre componenti del tensore.

Per lo studio di prima approssimazione, e cioè delle sole righe più intense, basta conservare nello sviluppo (3) i soli termini lineari e prescindere dalla anarmonicità delle oscillazioni. Siccome in un oscillatore armonico le coordinate sono rappresentate da matrici in cui sono diversi da zero solo i termini che corrispondono a transizioni di ± 1 del numero quantico di oscillazione, segue subito che, nella nostra approssimazione, potranno apparire in effetto Raman soltanto le frequenze fondamentali. Condizione necessaria poi perché una frequenza fondamentale si osservi effettivamente è che il coefficiente della corrispondente coordinata nello sviluppo (3) sia diverso da zero almeno per una delle componenti del tensore.

(1) E. FERMI, «Zs. f. Phys.», *71*, 250 (1931); E. FERMI e F. RASETTI, «Zs. f. Phys.», *71*, 689 (1931).

Ora accade in parecchi casi che per la simmetria particolare della molecola il coefficiente, per qualcuna delle coordinate, si annulli esattamente; la frequenza corrispondente è in questi casi proibita in effetto Raman. Senza entrare in una discussione della dipendenza di queste regole di selezione dagli elementi di simmetria della molecola (2), illustriamo la cosa sopra l'esempio semplice delle molecole del tipo $A\ B_2$ in cui, nella posizione di equilibrio, l'atomo A si trovi nel centro del segmento che congiunge i due atomi B. In una molecola di questo tipo abbiamo tre frequenze fondamentali, ν_1, ν_2, ν_3, di cui una è doppia. Le forme di vibrazione corrispondenti sono illustrate nella fig. 1. La frequenza ν_2 è doppia, perché la corrispondente vibrazione può aver luogo sia nel piano della figura, sia in un piano perpendicolare ad esso. Quando la molecola vibra secondo una delle due forme di vibrazione ν_2 o ν_3, tutte le componenti della sua polarizzabilità sono evidentemente, per simmetria, delle funzioni pari della elongazione (infatti in posizioni di fase opposta la polarizzabilità è evidentemente la stessa). Nello sviluppo della polarizzabilità in serie di potenze delle coordinate normali, mancheranno dunque tutti i coefficienti delle potenze dispari delle coordinate normali corrispondenti a ν_2 e ν_3; in particolare mancheranno dunque i termini lineari in tali coordinate, e le frequenze corrispondenti saranno perciò inattive in effetto Raman. Invece la frequenza ν_1 è attiva in effetto Raman. Con considerazioni di questo stesso tipo si vedrebbe anche subito che, in infrarosso, le frequenze ν_2 e ν_3 sono attive, mentre ν_1 è inattiva.

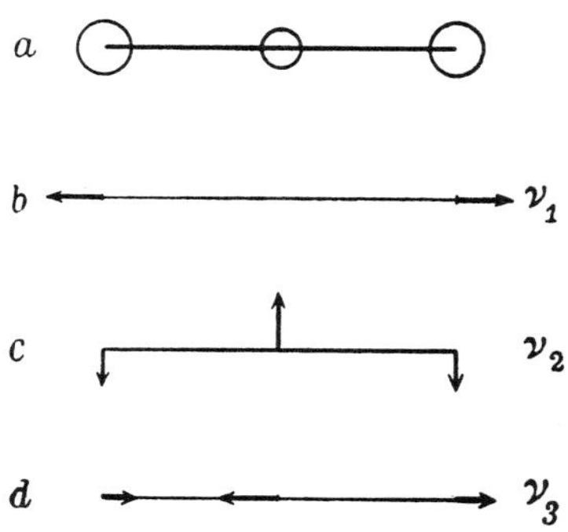

Fig. 1.

Con considerazioni analoghe a quelle che abbiamo svolte per questo esempio semplice, è effettivamente possibile in molti casi rendersi conto dell'origine delle righe più intense che si osservano in effetto Raman, possibilmente integrando le osservazioni con lo studio dello spettro di assorbimento infrarosso e della polarizzazione delle righe Raman e infrarosse.

Oltre alle righe corrispondenti alle frequenze fondamentali di oscillazione, si osservano talvolta negli spettri Raman delle altre righe, in genere assai poco intense che corrispondono a frequenze armoniche e di combinazione. Il loro apparire è dovuto a due cause, cioè:

a) la presenza nello sviluppo (3) di termini quadratici e di ordine superiore, che rendono possibile la transizione di più di una unità nei numeri quantici di oscillazione;

b) la anarmonicità delle oscillazioni.

Considerazioni simili a quelle ora svolte per il caso delle molecole, possono anche servir di guida per l'interpretazione degli spettri Raman dei cristalli. Ritorneremo su questo problema in seguito, parlando dell'effetto Raman del salgemma.

(2) G. PLACZEK, in « Leipziger Vorträge », 1931, edito a cura di P. Debye, Lipsia 1931.

Le regole di selezione a cui abbiamo accennato, e quelle che studieremo in seguito per il caso dei cristalli, trovano in alcuni casi delle eccezioni che conviene tener presenti tutte le volte che si deve interpretare uno spettro Raman. Discuteremo queste eccezioni sopra alcuni esempi caratteristici.

2. SPETTRO RAMAN DELL'ANIDRIDE CARBONICA. – Un notevole caso di eccezione alle ordinarie regole di selezione per l'effetto Raman delle molecole poliatomiche si può presentare quando casualmente una delle frequenze fondamentali venga quasi a coincidere con una frequenza di combinazione. Illustreremo questo fenomeno sopra l'esempio dell'anidride carbonica.

Lo spettro Raman dell'anidride carbonica gassosa (3), consiste di quattro righe. Due notevolmente intense, con frequenze di 1285 e 1388 cm^{-1}; e due assai più deboli aventi frequenze di 1265 e 1408 cm^{-1}. Siccome la molecola dell'anidride carbonica è del tipo rappresentato nella fig. 1, le ordinarie regole di selezione, che abbiamo già discusse per questo caso, farebbero prevedere l'esistenza di una sola riga Raman intensa, corrispondente all'unica frequenza fondamentale attiva ν_1. Dall'assorbimento infrarosso sono note le frequenze fondamentali ν_2 e ν_3, che hanno approssimativamente i valori $\nu_2 = 673$ e $\nu_3 = 2350$. Da questa ultima frequenza può anche grossolanamente valutarsi la frequenza ν_1; se infatti si ammette trascurabile l'azione esercitata dai due atomi di ossigeno uno sull'altro si dovrebbe avere, come risulta da un facile calcolo, tenendo conto che i pesi atomici del carbonio e dell'ossigeno sono 12 e 16,

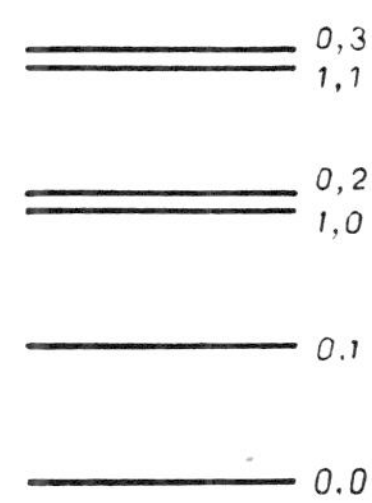

Fig. 2.

$$\nu_1 \cong \nu_3 \sqrt{\frac{12}{2 \cdot 16 + 12}} \cong 1230 \, cm^{-1}.$$

Effettivamente le frequenze delle quattro righe che si osservano non sono molto lontane da questo valore; occorre però cercare una spiegazione del perché si osservano quattro righe in luogo di una soltanto.

Occupiamoci dapprima delle due righe più intense; vedremo poi come si possano interpretare anche le due componenti deboli.

L'apparire di due righe intense in luogo di una sola è determinato dal fatto che casualmente la frequenza ν_1, viene quasi esattamente a coincidere col doppio della frequenza ν_2. Nella fig. 2, è rappresentato uno schema teorico dei livelli di oscillazione più profondi del CO_2. Lo schema è stato semplificato trascurando provvisoriamente la degenerazione della frequenza ν_2 e considerando le due oscillazioni ν_1 e ν_2 come esattamente armoniche e non accoppiate in alcun modo tra di loro. La frequenza ν_3, assai più elevata dalle altre due, non interviene nei livelli energetici più profondi. I numeri scritti accanto ai livelli nella figura sono i numeri quantici di oscillazione per le frequenze ν_1 e ν_2. Secondo le ordinarie regole di selezione ci si attenderebbe solo una riga Raman, corrispondente alla transizione (1,0)-(0,0).

(3) F. RASETTI, «Nature», *123*, 205 (1929).

Il fatto che si osservano invece due righe intense, dipende dalla circostanza che i due livelli energetici (1 , 0) e (0 , 2) sono casualmente assai prossimi uno all'altro e, in conseguenza, si perturbano molto intensamente. Per modo che le autofunzioni ad essi corrispondenti vengono in realtà ad essere date da combinazioni lineari delle autofunzioni imperturbate dei due stati con coefficienti che, per entrambi le autofunzioni, sono dell'ordine di grandezza dell'unità. Segue da ciò che tutti e due gli stati si possono combinare intensamente in effetto Raman collo stato fondamentale (0 , 0), poiché entrambi contengono una elevata percentuale dell'autofunzione imperturbata (1 , 0) che si combina intensamente con lo stato fondamentale.

In realtà la situazione è un po' più complicata per il fatto che la frequenza ν_2, come già si è osservato, è degenere. Occorre perciò numerare le autofunzioni con tre numeri quantici (anche quando si prescinda dalla considerazione della frequenza ν_3, ciò che naturalmente può farsi nel nostro caso). Per la determinazione delle oscillazioni della molecola scegliamo le seguenti coordinate:

x = proiezione dello spostamento degli atomi di ossigeno sulla direzione dell'asse molecolare.

ξ ed η, proiezioni del segmento che congiunge l'atomo C con uno dei due atomi O, su due direzioni perpendicolari tra di loro e all'asse della molecola. Nella fig. 3, AB rappresenta la posizione di equilibrio della molecola; C, O, O sono le posizioni istantanee del carbonio e dei due ossigeni; la figura è proiettata su un piano parallelo all'asse della molecola e a ξ.

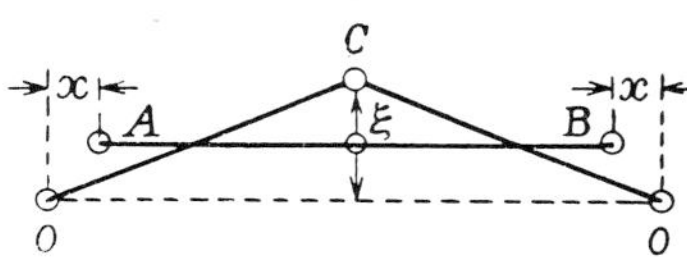

Fig. 3.

La coordinata x corrisponde alla frequenza ν_1; la massa vibrante ad essa relativa è

$$m_1 = 2\,m$$

dove m rappresenta la massa dell'atomo di ossigeno.

Le altre due coordinate ξ ed η corrispondono alla frequenza degenere ν_2; esse hanno una massa vibrante

$$m_2 = \frac{2\,m\mathrm{M}}{2\,m + \mathrm{M}}$$

dove M è la massa dell'atomo di carbonio.

Per caratterizzare uno stato quantico del sistema imperturbato occorre dare tre numeri quantici $(n_1\,,\,n_2'\,n_2'')$ relativi alle tre coordinate $x\,,\,\xi\,,\,\eta$. E l'energia imperturbata, a meno dell'energia di zero, è data da

$$\mathrm{E}_{n_1\,n_2'\,n_2''} = h\nu_1\,n_1 + h\nu_2\,(n_2' + n_2'')\,.$$

Siccome si ha $\nu_1 \cong 2\,\nu_2$, vengono a coincidere quasi esattamente tra di loro i quattro autovalori imperturbati (1 , 0 , 0) (0 , 2 , 0) (0 , 1 , 1) (0 , 0, 2).

Nel sistema imperturbato l'energia potenziale ha la forma

$$kx^2 + \alpha\,(\xi^2 + \eta^2)$$

in cui k ed α rappresentano delle costanti i cui valori possono determinarsi immediatamente conoscendo le frequenze ν_1 e ν_2.

Come perturbazione considereremo le deviazioni dell'energia potenziale dalla forma quadratica; e in primo luogo considereremo l'effetto dei termini di terzo grado nelle coordinate. Si riconosce subito che, per ragione di simmetria, i termini di terzo grado si riducono ai due soli tipi

$$x^3 \text{ e } x(\xi^2 + \eta^2).$$

Il primo termine, dipendendo dalla sola coordinata x, non rappresenta un accoppiamento tra diverse oscillazioni, ma semplicemente una anarmonicità della oscillazione ν_1; non ci occorre quindi tenerne conto, e potremmo eventualmente conglobarlo nell'energia potenziale imperturbata. Come energia potenziale perturbatrice abbiamo dunque solo da considerare un termine della forma

$$U = \beta x(\xi^2 + \eta^2)$$

in cui β rappresenta una costante.

Dobbiamo calcolarci gli elementi della matrice di perturbazione relativi all'incrocio dei quattro termini (1, 0, 0) (0, 2, 0) (0, 1, 1) (0, 0, 2). Ciò si fa immediatamente con le note formule, per rappresentare gli elementi di matrice della coordinata q di un oscillatore di massa m.

$$q_{n+1,n} = q_{n,n+1} = \sqrt{\frac{h}{8\pi^2 \nu m}} \sqrt{n+1}\,.$$

Da questa formula si deduce

$$x_{10} = x_{01} = \sqrt{\frac{h}{8\pi^2 \nu_1 m_1}} \quad , \quad (\xi^2)_{02} = \xi_{01}\,\xi_{12} = \frac{h}{8\pi^2 \nu_2 m_2}\sqrt{2}\,.$$

Da cui si ottengono immediatamente gli elementi di matrice desiderati

$$U(100, 020) = U(100, 002) = P, U(100, 011) = 0$$

dove si è posto

$$P = \frac{\beta h^{3/2}}{16\pi^3 \sqrt{\nu_1 m_1}\, \nu_2 m_2}\,.$$

Si riconosce dunque che l'autofunzione (0, 1, 1) non è accoppiata alle altre dalla perturbazione; e siccome essa, nel problema imperturbato, è inattiva per l'effetto Raman, possiamo non considerarla. Il pezzo della matrice rappresentante l'energia, che corrisponde all'incrocio delle altre tre autofunzioni è dunque

$$\begin{vmatrix} A & P & P \\ P & B & O \\ P & O & B \end{vmatrix}$$

in cui A rappresenta l'autovalore imperturbato (1, 0, 0) e B l'autovalore imperturbato delle due autofunzioni degeneri (0, 2, 0) (0, 0, 2).

La matrice precedente ha gli autovalori

$$B\,, \quad \frac{A+B}{2} \pm \sqrt{2P^2 + \left(\frac{A-B}{2}\right)^2}$$

i quali corrispondono ad autofunzioni che sono combinazioni lineari delle autofunzioni imperturbate (1, 0, 0) (0, 2, 0) (0, 0, 2). L'autovalore B corrisponde alla combinazione lineare

$$(002) - (020)$$

che non contiene la autofunzione « attiva » (1, 0, 0); la riga corrispondente alla transizione da questo livello al livello fondamentale non viene perciò osservata in effetto Raman. Restano così gli altri due autovalori

$$\frac{A+B}{2} \pm \sqrt{2 P^2 + \left(\frac{A-B}{2}\right)^2} \tag{4}$$

le cui autofunzioni sono le combinazioni lineari

$$\frac{1}{P}\left[\frac{A-B}{2} \pm \sqrt{2 P^2 + \left(\frac{A-B}{2}\right)^2}\right](100) + (020) + (002) \tag{5}$$

che, contenendo l'autofunzione attiva (1, 0, 0), sono tutte e due attive in effetto Raman.

Resta con ciò spiegato l'apparire di due righe Raman intense. L'effetto Raman che si osserva può spiegarsi ammettendo che i due livelli energetici imperturbati A e B coincidano quasi esattamente. Ponendo nella (4) $A = B$ si trovano i due livelli energetici

$$A \pm \sqrt{2}\, P \tag{6}$$

che combinandosi con lo stato fondamentale (0, 0, 0) dànno luogo alle due righe Raman intense; la loro differenza di frequenza risulta dunque

$$\Delta\nu = 2\sqrt{2}\, P\,. \tag{7}$$

Dalla separazione sperimentale $\Delta\nu = 103\ \text{cm}^{-1}$ si dedurrebbe dunque $P = 36\ \text{cm}^{-1}$; questo valore di P è plausibilissimo e conduce ad un valore per la costante B di un giusto ordine di grandezza. Nell'ipotesi $A = B$ le combinazioni lineari (5) diventano

$$\pm \sqrt{2}\,(100) + (020) + 002)\,.$$

La percentuale dell'autofunzione attiva (1, 0, 0) in queste due autofunzioni è del 50 %, ciò condurrebbe ad attendersi eguale intensità per le due righe Raman. Siccome sperimentalmente si trova che la riga 1388 è un pò più intensa della 1285, si conclude che B deve essere leggermente minore di A.

Resta ancora da discutere l'origine delle due righe deboli. Esse debbono ricondursi all'azione delle molecole nelle quali è eccitato per agitazione termica un quanto della frequenza $\nu_2 = 673$, che si trovano cioè nello stato (0, 1) dello schema semplificato della fig. 2. A temperatura ordinaria il fattore di Boltzmann ha per questo stato il valore 1/28; siccome il peso statistico è 2, vi è circa 1/14 delle molecole che si trovano in questo stato. Lo stato (0, 1) si combina in effetto Raman con lo stato (1, 1); e siccome questo viene a giacere assai prossimo al livello (0, 3) (cfr. fig. 2) troviamo nuovamente, con lo stesso meccanismo di prima, due righe Raman in luogo di una sola.

Per il calcolo quantitativo si deve naturalmente tener conto della degenerazione della frequenza ν_2. Si deve precisamente considerare la matrice dell'energia corrispondente all'intersezione dei sei stati quantici (1, 1, 0) (1, 0, 1) (0, 3, 0) (0, 2, 1) (0, 1, 2) (0, 0, 3). Con le stesse notazioni di prima si trova indicando con A′ e B′ i livelli energetici imperturbati, la seguente matrice

$$\begin{vmatrix} A' & 0 & \sqrt{3}\,P & 0 & P & 0 \\ 0 & A' & 0 & P & 0 & \sqrt{3}\,P \\ \sqrt{3}\,P & 0 & B' & 0 & 0 & 0 \\ 0 & P & 0 & B' & 0 & 0 \\ P & 0 & 0 & 0 & B' & 0 \\ 0 & \sqrt{3}\,P & 0 & 0 & 0 & B' \end{vmatrix}$$

La matrice precedente si spezza in due matrici eguali del terzo ordine che accoppiano le autofunzioni (1, 1, 0) (0, 3, 0) (0, 1, 2) oppure le altre (1, 0, 1) (0, 0, 3) (0, 2, 1). Tali matrici sono

$$\begin{vmatrix} A' & \sqrt{3}\,P & P \\ \sqrt{3}\,P & B' & 0 \\ P & 0 & B' \end{vmatrix}$$

ed hanno gli autovalori

$$B' \,, \quad \frac{A'+B'}{2} \pm \sqrt{4\,P^2 + \left(\frac{A'-B'}{2}\right)^2} \,. \tag{8}$$

L'autovalore B′ non viene osservato in effetto Raman, perché la corrispondente autofunzione

$$(030) - \sqrt{3}\,(012)$$

non contiene l'autofunzione attiva (1, 1, 0). Gli altri due autovalori corrispondono alle due autofunzioni

$$\frac{1}{P}\left\{\frac{A'-B'}{2} \pm \sqrt{4\,P^2 + \left(\frac{A'-B'}{2}\right)^2}\right\}(110) + \sqrt{3}\,(030) + (012)$$

che contengono l'autofunzione (1, 1, 0), che rende possibili le transizioni Raman. Se facciamo anche in questo caso l'ipotesi A′ = B′, troviamo i due autovalori A′ ± 2 P. La separazione delle due righe Raman deboli risulta dunque

$$\Delta\nu' = 4\,P. \tag{9}$$

Dal confronto con la (6) si ricava dunque

$$\frac{\Delta\nu'}{\Delta\nu} = \sqrt{2}$$

in ottimo accordo con il rapporto 1,4 che si osserva sperimentalmente tra le separazioni delle righe deboli e di quelle intense.

Fenomeni analoghi a quello che abbiamo qui discusso per il caso dell'anidride carbonica si presentano anche in altre molecole. Così se ne ha un nuovo esempio per il tetracloruro di carbonio. In questa molecola si hanno quattro frequenze fondamentali, tutte attive in effetto Raman.

Sperimentalmente si trova che la più alta di queste frequenze è scissa in due componenti con frequenza di 758 e 789 cm^{-1}. Questo fatto si spiega osservando che due altre frequenze fondamentali che si osservano in effetto Raman a 314 e 459 cm^{-1} hanno per somma 773. Questa somma giace quasi esattamente in mezzo alle due frequenze 758 e 789. Ci troviamo dunque di nuovo in presenza di una quasi degenerazione accidentale che determina la scissione in due della frequenza Raman indicata. Il tetracloruro di carbonio presenta anche un interessante esempio di una armonica che compare in effetto Raman. Si osserva infatti una debole riga Raman a 1540 cm^{-1}; essa corrisponde al doppio della frequenza spezzata in due dalla degenerazione accidentale.

3. EFFETTO RAMAN NEI CRISTALLI–SALGEMMA. – Le regole generali per determinare quali righe debbano apparire nello spettro Raman di un cristallo si possono dedurre in modo simile a quanto si fa per il caso delle molecole.

Naturalmente in un cristallo, dato che esso è sempre costituito da un numero enormemente grande di atomi, si ha sempre un numero enormemente grande di frequenze caratteristiche, disposte in modo da formare uno spettro continuo. Tuttavia in effetto Raman si osservano normalmente degli spettri costituiti da righe discrete; ciò dipende dal fatto che, come ora diremo, si osservano nello spettro Raman di un cristallo, almeno in prima approssimazione, solo quelle frequenze elastiche che corrispondono a vibrazioni elastiche del cristallo con lunghezza d'onda infinita.

In un cristallo costituito da un insieme di n reticoli atomici semplici le vibrazioni elastiche possono, come è noto, caratterizzarsi per mezzo della lunghezza d'onda e della direzione di propagazione di onde elastiche; per ogni lunghezza d'onda e direzione di propagazione, si hanno in genere $3n$ frequenze elastiche, di cui tre, per lunghezza d'onda infinita, tendono a zero; mentre le altre $3n - 3$ tendono a valori di frequenze infrarosse.

Per studiare nel modo più semplice come queste diverse frequenze elastiche agiscono in effetto Raman, considereremo un pezzetto di cristallo, che abbia dimensioni piccole in confronto alla lunghezze d'onda della luce e grande in confronto alle dimensioni interatomiche. Se nel cristallo è eccitata una vibrazione elastica, essa determinerà delle pulsazioni della polarizzabilità elettrica del cristallo, che determinano l'effetto Raman. Ora può ammettersi che in prima approssimazione la variazione della polarizzabilità elettrica in una certa posizione del cristallo sia proporzionale alla deformazione elastica del cristallo in quella particolare regione. Se quindi la lunghezza d'onda della vibrazione elastica che determina le variazioni della polarizzabilità è piccola in confronto alle dimensioni del pezzetto di cristallo, avremo che in punti di fase opposta rispetto alla vibrazione elastica, la variazione della polarizzabilità elettrica sarà opposta, così che in media l'effetto sopra tutto il cristallo verrà ad essere nullo. Un effetto non nullo si potrà avere in questa approssimazione solo dalle vibrazioni elastiche che hanno lunghezza d'onda comparabile con le dimensioni del cristallo, e cioè praticamente infinita. Una analisi più accurata dimostra che sono attive in effetto Raman vibrazioni che hanno lunghezza d'onda elastica dell'ordine di grandezza della lunghezza d'onda della luce, e quindi enormemente grande in confronto delle distanze interatomiche.

Ciò spiega come si osservino normalmente nell'effetto Raman dei cristalli delle frequenze discrete invece di uno spettro continuo; le frequenze che si osservano corrispondono infatti alle vibrazioni in cui i singoli reticoli semplici che costituiscono il reticolo cristallino si mantengono rigidi e vibrano uno contro l'altro. Naturalmente anche per queste vibrazioni si hanno delle ulteriori regole di selezione, in tutto analoghe a quelle che valgono per il caso delle molecole, determinate da particolari simmetrie dell'aggregato atomico del cristallo. In particolare la simmetria del cristallo può esser tale che tutte le frequenze siano proibite in effetto Raman, per modo che il cristallo non ha in prima approssimazione, alcun effetto Raman.

Un caso di questo genere si presenta nel salgemma. Il reticolo di questo cristallo è costituito come è ben noto da due reticoli cubici a facce centrate, uno occupato dagli atomi di cloro e l'altro dagli atomi di sodio. Le vibrazioni di lunghezza d'onda elastica infinita si ottengono facendo vibrare, uno contro l'altro, questi due reticoli. Si ha così una sola frequenza triplamente degenere. È facile convincersi che essa è inattiva in effetto Raman, poiché le variazioni della polarizzabilità del cristallo determinate da uno spostamento relativo dei due reticoli di cloro e di sodio sono evidentemente di ordine superiore al primo.

Segue che nel salgemma non si deve avere alcun effetto Raman, quando almeno ci si limiti alla osservazione del fenomeno intenso del primo ordine: e questa conclusione è completamente confermata dalla esperienza. È riuscito tuttavia a Rasetti, servendosi di un dispositivo assai luminoso e con esposizioni molto lunghe, di mettere in evidenza uno spettro Raman anche nel salgemma.

Lo spettro che così si osserva ha tuttavia un aspetto molto differente dagli spettri che comunemente si osservano nei cristalli. Mentre questi infatti sono per solito costituiti da righe discrete, lo spettro Raman del salgemma consiste in uno spettro continuo che, partendo dalla riga eccitatrice (nelle osservazioni di Rasetti la 2537 del mercurio) si estende con varie oscillazioni di intensità, fino a una distanza da essa di circa 365 cm^{-1}.

L'interpretazione teorica di questo effetto Raman si ottiene facilmente ammettendo che esso sia da interpretarsi come un effetto Raman del secondo ordine. Per riconoscere in modo facile che uno spettro Raman del secondo ordine deve essere continuo, invece che costituito da righe staccate, si osservi che le variazioni locali della polarizzabilità elettrica in un cristallo dipendono dalla deformazione locale del cristallo. Se in prima approssimazione si considera la variazione della polarizzabilità proporzionale alla deformazione, l'effetto si annulla in media nell'intero cristallo, salvo che per le vibrazioni elastiche di lunghezza d'onda infinita, e si hanno dunque, come già si è discusso, delle righe Raman discrete, corrispondenti appunto alle frequenze di oscillazione elastica del cristallo aventi lunghezza d'onda infinita; quando però, come nel salgemma, sono nulli i termini lineari nella variazione della polarizzabilità elettrica, si deve passare alla considerazione dei termini quadratici; e questi non sono più nulli nemmeno per onde elastiche di piccola lunghezza d'onda. Poiché, anche in punti di fase opposta, hanno lo stesso segno e quindi non si eliminano nella media su tutto il cristallo. Concludiamo dunque che mentre nell'effetto Raman di primo ordine sono attive soltanto le frequenze di lunghezza d'onda elastica infinita, nell'effetto del secondo ordine sono attive frequenze di tutte le lunghezze d'onda, per modo che lo spettro del secondo ordine viene ad essere uno spettro continuo, come effettivamente si osserva.

Per studiare la cosa un po' più precisamente per il caso del salgemma, introduciamo un sistema di coordinate x, y, z con origine in uno degli atomi di Cl. Se con a si indica la distanza tra atomi contigui di Cl e di Na, le coordinate delle posizioni di equilibrio degli atomi saranno $x = n_1 a$, $y = n_2 a$, $z = n_3 a$, dove n_1, n_2, n_3 sono numeri interi. Si ha precisamente un atomo di Cl o di Na a seconda che la somma $n_1 + n_2 + n_3$ è pari o dispari.

Per ogni numero d'onde elastico vettoriale $f \equiv (f_1, f_2, f_3)$ il reticolo cristallino ha sei vibrazioni caratteristiche con sei frequenze proprie

$$\omega_i(f) \qquad (i = 1, 2, 3, 4, 5, 6)$$

e a ciascuna di queste vibrazioni corrisponde una coordinata normale q_{if}. Lo spostamento di un atomo del reticolo, espresso in funzione delle coordinate normali, è

$$\xi_n = \sum_f \sum_{i=1}^{6} q_{if} \begin{vmatrix} A_{if} \\ B_{if} \end{vmatrix} \cos \left[2\pi a (n_1 f_1 + n_2 f_2 + n_3 f_3) + \begin{matrix} \varphi_{if} \\ \psi_{if} \end{matrix} \right] \tag{10}$$

dove si debbono prendere le espressioni superiori (A, φ) o quelle inferiori (B, ψ) a seconda che $n_1 + n_2 + n_3$ è pari o dispari, e cioè a seconda che si tratta di un atomo di cloro o di sodio. Si deve sommare sopra tutti i numeri d'onde dello spettro di Debye.

La polarizzabilità k di un atomo (n) dipende evidentemente dal suo spostamento relativo agli atomi circostanti; essa è dunque una funzione delle grandezze $\xi_{n+l} - \xi_n$, in cui n ed l sono abbreviazioni per le terne di numeri $n_1 n_2 n_3$ ed $l_1 l_2 l_3$. Sviluppando questa funzione in serie di potenze fino ai termini di secondo ordine si trova

$$k = k_0 + \sum_l \alpha_l (\xi_{n+l} - \xi_n) + \sum_{jl} \beta_{jl} (\xi_{n+j} - \xi_n)(\xi_{n+l} - \xi_n) . \tag{11}$$

Per la simmetria particolare del nostro reticolo (in cui ogni atomo si trova in un centro di simmetria) si avrà poi evidentemente $\alpha_l = -\alpha_{-l}$. L'espressione precedente diventa dunque

$$k = k_0 + \frac{1}{2} \sum_l \alpha_l (\xi_{n+l} - \xi_{n-l}) + \sum_{jl} \beta_{jl} (\xi_{n+j} - \xi_n)(\xi_{n+l} - \xi_n) . \tag{12}$$

La polarizzabilità media in tutto il cristallo si otterrà sostituendo per le ξ le loro espressioni (10) e mediando su tutto il reticolo, osservando che gli atomi $n+l$ ed $n-l$ sono dello stesso tipo. Si trova così senza difficoltà una espressione della forma

$$(13) \qquad \bar{k} = k_0 + \sum_f \sum_{r,s=1}^{6} A_{rsf}\, q_{rf}\, q_{sf}$$

poiché tutti gli altri termini si eliminano nella media.

A questa espressione delle variazioni della polarizzazione elettrica del cristallo corrispondono evidentemente in effetto Raman le frequenze

$$(14) \qquad \omega_i(f) + \omega_k(f) \quad , \quad \omega_i(f) - \omega_k(f) \quad , \quad 2\,\omega_i(f)$$

e cioè le frequenze di combinazione che si ottengono sommando o sottraendo due frequenze elastiche dello stesso numero d'onde f (vettoriale).

Lo spettro Raman di secondo ordine si estenderà dunque attraverso ad oscillazioni di intensità, che sarebbero assai difficili a calcolare quantitativamente, dalla frequenza zero fino a una frequenza eguale al doppio della frequenza massima. Il valore di questa, noto dalle osservazioni infrarosse, si aggira intorno ai 60 μ; mentre dallo spettro Raman si calcolerebbe una frequenza, abbastanza prossima ad essa, di circa 55 μ.

4. Lo spettro Raman della calcite. - Nello spettro Raman della calcite [4] si osservano sei frequenze: 155, 282, 714, 1088, 1438, 1741 cm^{-1}. L'interpretazione delle cinque frequenze più basse non presenta nessuna particolarità interessante e ci dedicheremo in questo paragrafo allo studio della frequenza 1741, che, come vedremo, presenta invece un certo interesse.

Nella calcite, come del resto in tutti i cristalli contenenti degli ioni complessi, CO_3, NO_3, SO_4, NH_4 ecc. le frequenze possono classificarsi in frequenze interne e frequenze esterne. Le prime, che sono in genere le frequenze più elevate, corrispondono a oscillazioni in cui le varie parti dello ione complesso vibrano le une contro le altre. Le altre invece corrispondono a vibrazioni degli ioni in cui questi vibrano gli uni contro gli altri senza subire deformazioni interne considerevoli. Le frequenze interne si riconoscono facilmente dalle esterne, sia perché sono più grandi, sia anche perché esse sono poco modificate dall'influenza del reticolo cristallino, e quindi si ritrovano, più o meno spostate, in tutti i cristalli che contengono uno stesso ione complesso, o anche in soluzioni contenenti lo stesso ione. Nella calcite dovremo dunque attenderci alcune frequenze interne corrispondenti alle vibrazioni dello ione CO_3, e alcune frequenze esterne.

Lo ione CO_3 ha una struttura piana, in cui l'atomo di carbonio si trova nel centro di un triangolo equilatero i cui vertici sono occupati dai tre atomi di ossigeno. Una semplice analisi delle frequenze elastiche di un siffatto sistema, dimostra che esse sono in numero di quattro, e cioè:

una frequenza ν_1 in cui il carbonio resta fermo mentre il triangolo dei tre atomi di ossigeno alternativamente si contrae e si allarga conservando la propria simmetria;

(4) Questi dati sono citati da K. W. F. Kohlrausch, *Der Smekal-Raman-Effekt*, Berlino 1931. Altre righe, indicate da Kohlrausch come incerte, debbono ritenersi dovute a impurità.

due frequenze ν_2 e ν_3 doppiamente degeneri consistenti in vibrazioni dei quattro atomi nel piano in cui essi sono contenuti;

una frequenza ν_4 consistente in una vibrazione dell'atomo di carbonio perpendicolarmente al piano dei tre atomi di ossigeno, i quali, conservando distanze inalterate tra di loro, eseguono una vibrazione complementare a quella dell'atomo di ossigeno, in modo che il centro di gravità del sistema resti fisso.

Si riconosce facilmente che se si considera lo ione CO_3 libero, debbono essere attive in effetto Raman le frequenze ν_1, ν_2, ν_3, mentre in infrarosso sono attive le frequenze ν_2, ν_3, ν_4.

I valori di queste quattro frequenze subiscono dei lievi cambiamenti quando lo ione CO_3 entra a far parte del legame cristallino. Così nel caso della calcite le tre frequenze attive in effetto Raman dello ione CO_3 si possono identificare al modo seguente $\nu_1 = 1088$, $\nu_2 = 714$, $\nu_3 = 1438$. La frequenza inattiva ν_4, che non si osserva in effetto Raman, è nota tuttavia dalle misure dell'assorbimento infrarosso ed è prossima a 870 cm^{-1}. La stessa frequenza si osserva del resto anche come riga Raman debolissima in altri cristalli contenenti lo ione CO_3, per esempio nell'aragonite, in cui la posizione dello ione nel reticolo cristallino ne determina delle lievi deformazioni, che hanno per effetto di rendere attiva in effetto Raman la frequenza ν_4 che, come si è detto, è inattiva nello ione libero, come lo è anche quando lo ione si trova incorporato nel reticolo cristallino della calcite. Le due frequenze 155 e 282 dello spettro Raman della calcite si interpretano senza difficoltà come dovute a vibrazioni esterne del reticolo; resta dunque solo da interpretare la frequenza più elevata 1741.

Il fatto che questa frequenza coincida abbastanza approssimativamente col doppio della frequenza inattiva ν_4 dello ione CO_3, rende plausibile ammettere che essa sia in qualche modo connessa con la armonica di ν_4. Vi è tuttavia una caratteristica difficoltà per una tale interpretazione; abbiamo visto infatti che in un cristallo ci si deve attendere nel secondo ordine uno spettro Raman continuo e non uno spettro di righe. Le considerazioni che ora faremo hanno lo scopo di dimostrare che può accadere che, in determinate circostanze, si osservino anche nell'effetto Raman dei cristalli delle righe corrispondenti ad armoniche superiori.

Cerchiamo di renderci conto del come dovrà presentarsi l'effetto Raman del secondo ordine di una frequenza interna del gruppo CO_3 della calcite.

Osserviamo prima di tutto che nel reticolo della calcite ogni cella cristallina contiene due molecole $CaCO_3$. Abbiamo quindi due reticoli semplici di ioni CO_3. I gruppi CO_3 appartenenti a questi due reticoli semplici sono situati in piani paralleli tra di loro (perpendicolari all'asse ternario), e differiscono perché rotati uno rispetto all'altro di un angolo di 60° nel loro piano. Quando nel reticolo cristallino si ha un'onda elastica che deformi gli ioni CO_3, le forze elastiche di richiamo che tendono a riportare gli ioni alla loro configurazione normale sono in prima approssimazione forze che si esercitano tra gli atomi appartenenti allo stesso ione e solo in ulteriore approssimazione debbono considerarsi anche delle forze dovute all'effetto della deformazione degli ioni circostanti, essi pure deformati.

In conseguenza la frequenza della vibrazione viene a dipendere poco dalla lunghezza d'onda della deformazione elastica; le differenze di frequenza che si hanno per una stessa vibrazione interna, corrispondentemente a differenti lunghezze d'onda elastiche sono dello stesso ordine di grandezza delle differenze di frequenza della stessa vibrazione interna, quando lo ione appartiene a reticoli cristallini differenti, e cioè di alcune decine di numeri d'onde.

Già questo fatto ha per conseguenza che gli spettri continui che, nell'effetto Raman di seconda approssimazione, corrispondono alle armoniche superiori di una vibrazione interna, debbono in realtà ridursi a strisce di abbastanza piccola larghezza. In determinate circostanze può poi accadere che questa larghezza si riduca ancora di molto, fino a che lo spettro continuo finisca per diventare una vera e propria riga per il motivo seguente.

Siccome nella cella del reticolo della calcite sono contenuti due ioni CO_3, ad ogni frequenza propria dello ione CO_3 corrisponderanno, per ogni numero d'onde elastico (vettoriale), due frequenze, non molto lontane dalla frequenza della corrispondente vibrazione nello ione libero. In determinate circostanze può avvenire, come mostreremo ora sopra un esempio, che mentre queste due frequenze dipendono, se pure non molto, dalla lunghezza d'onda elastica, la loro somma ne sia invece indipendente. In queste circostanze lo spettro continuo che normalmente corrisponde nell'effetto Raman del secondo ordine alla somma delle due frequenze indicate, viene evidentemente a ridursi ad una riga.

Un esempio molto semplice delle circostanze in cui può accadere che la somma di due frequenze risulti indipendente dalla lunghezza d'onda elastica è il seguente: consideriamo una fila di atomi (reticolo unidimensionale) equidistanti. Supponiamo che ogni atomo sia soggetto ad una forza elastica abbastanza intensa, che tende a mantenerlo nella sua posizione di equilibrio, forza che, in questo modello, corrisponde alle forze di richiamo *interne* dello ione; supporremo poi che sopra l'atomo si esercitino anche delle forze proporzionali alle variazioni della sua distanza dai due atomi attigui; queste siano però piccole in confronto alle precedenti, in modo da produrre solo delle lievi variazioni della frequenza di oscillazione degli atomi. Per rappresentare in qualche modo nel nostro modello unidimensionale la circostanza, che il reticolo della calcite contiene in ogni cella due ioni CO_3, ammetteremo che nella fila si debba considerare ogni cella come composta dall'insieme di due atomi, per il che gli atomi della fila debbono essere alternativamente distinguibili, benché essi abbiano la stessa massa (si osservi che, nel reticolo della calcite, la differenza tra due ioni CO_3 appartenenti a reticoli semplici diversi, consiste nella diversa orientazione del triangolo formato dai tre ossigeni).

Indichiamo con a la distanza tra due atomi successivi della fila, per modo che la periodicità del reticolo lineare sia $2a$. Siano poi: k il coefficiente della forza elastica che richiama ogni atomo verso la sua posizione di equilibrio e r il coefficiente della forza che lo collega ai due atomi circonvicini; per le nostre ipotesi si avrà $k \gg r$. Le equazioni del movimento dei nostri atomi si scrivono senza difficoltà: indicando con x_n lo spostamento dell'n^{esimo} atomo della fila si avrà

$$m\ddot{x}_n = -kx_n + r(x_{n+1} + x_{n-1} - 2x_n). \tag{15}$$

Questo sistema di equazioni si integra subito ponendo

$$x_n = A e^{2\pi i(\omega t - gn)} \tag{16}$$

in cui A rappresenta una ampiezza costante. Sostituendo nella (15) si trova

$$-4\pi^2 m\omega^2 = -k + 2r(\cos 2\pi g - 1)\,. \tag{17}$$

e cioè

$$\omega^2 = \frac{k}{4\pi^2 m} + \frac{r}{2\pi^2 m}(1 - \cos 2\pi g)\,. \tag{18}$$

Ricordando che per ipotesi $k \gg r$, possiamo estrarre la radice quadrata dalla (18) sviluppando rispetto a r e trascurando i termini in r^2. Si trova così l'espressione della frequenza ω di oscillazione in funzione di g:

$$\omega = \frac{1}{2\pi}\sqrt{\frac{k}{m}} + \frac{r}{2\pi\sqrt{km}}(1 - \cos 2\pi g)\,. \tag{19}$$

Osserviamo ora che, siccome il periodo del nostro reticolo è $2a$, onde elastiche per le quali i due g differiscano di $1/2$ hanno la stessa lunghezza d'onda. Consideriamo dunque due onde elastiche, i cui g abbiano i valori

$$g_1 = \frac{f}{2} \quad ; \quad g_2 = \frac{f+1}{2} \tag{20}$$

e che cioè corrispondano tutte e due allo stesso numero d'onde elastico $f/2a$. Le frequenze corrispondenti a queste due onde sono, secondo la (19)

$$\left\{\begin{aligned} \omega_1 &= \frac{1}{2\pi}\sqrt{\frac{k}{m}} + \frac{r}{2\pi\sqrt{km}}(1 - \cos \pi f) \\ \omega_2 &= \frac{1}{2\pi}\sqrt{\frac{k}{m}} + \frac{r}{2\pi\sqrt{km}}(1 + \cos \pi f)\,. \end{aligned}\right. \tag{20}$$

Nell'effetto Raman di secondo ordine si può osservare la somma di due frequenze che hanno lo stesso numero d'onde elastico. Tale somma nel caso nostro è

$$\omega_1 + \omega_2 = \frac{1}{\pi}\sqrt{\frac{k}{m}} + \frac{r}{\pi\sqrt{mk}} \tag{21}$$

e non dipende dunque dal numero d'onde elastico. Per modo che in queste circostanze si osserverà, in secondo ordine, una riga Raman ben definita invece che, come accade normalmente, uno spettro continuo.

Queste considerazioni rendono assai probabile che la riga 1741 della calcite debba interpretarsi come armonica superiore della frequenza inattiva dello ione CO_3. Del resto le considerazioni che nell'esempio precedente abbiamo svolto per il caso di un reticolo ad una sola dimensione possono facilmente generalizzarsi anche al caso di un reticolo di tre dimensioni. Senza entrare in dettaglio in questo calcolo accenno al suo risultato che è il seguente. Per ogni numero d'onde elastico (vettoriale) esistono due frequenze che derivano da ogni frequenza elastica interna; esse corrispondono alle due frequenze (20) dell'esempio precedentemente trattato e differiscono di poco dalla corrispondente frequenza che si avrebbe nello ione libero. La somma di queste due frequenze resta poi indipendente dal numero d'onde elastico quando possa ritenersi che sopra gli ioni di un dato reticolo semplice si esercitino forze elastiche dovute alla deformazione dei soli ioni dell'altro reticolo semplice. Ora nel caso della calcite gli ioni CO_3 più vicini a uno ione CO_3 determinato, appartengono all'altro reticolo semplice; a distanza poco maggiore ci sono tuttavia già degli ioni dello stesso reticolo semplice. Bisogna perciò ammettere che l'azione di questi ultimi, per qualche ragione che non conosciamo, venga ad essere piuttosto piccola.

Uno spettro sostanzialmente simile a quello della calcite si osserva anche nel nitrato di sodio, che ha una struttura cristallina del tutto analoga a quella della calcite. Anche in questo cristallo si osserva una riga Raman della frequenza di 1670 cm^{-1} che deve verosimilmente interpretarsi come una armonica della frequenza inattiva dello ione NO_3.

N° 72 *b.*

There were three important international conferences on nucler physics between 1931 and 1933. The first "Convegno di fisica nucleare", in Rome was held in October 11–18, 1931 and attended by about forty-five physicists. This conference was held under the auspices of the Accademia d'Italia at the instigation of Fermi and Corbino. Its proceedings are published under the title, "Fondazione Alessandro Volta, Convegno di fisica nucleare" (Reale Accademia d'Italia, Roma, 1932).

They contain only the reports, but no minutes of the discussions. No mention of the neutron or the neutrino is to be found in them. We know, however, that on that occasion there were conversations between Pauli and Fermi on the neutrino. (See the introduction to papers N° 76, 80).

The second conference was held in Paris July, 1932 in the larger frame of the Fifth International Conference on Electricity.

Fermi was invited to give a report and he chose as a subject the nucleus, reading paper N° 72 on July 7, 1932.

The neutron had barely been observed and while there was still uncertainty on the interpretations of the experiments of Bothe and Curie and Joliot, Ettore Majorana in the Institute in Rome had immediately understood that there was what he called "a neutral proton". Majorana proceeded then to develop a model of a nucleus built of protons and "neutral protons" only, and proceeded considerably far in the description of the forces between these particles. He told Fermi and several of his friends of this work. Fermi immediately grasped its importance and urged Majorana to publish it. Majorana demurred because he thought that the results acquired thus far by him were too incomplete. He said that he did not want to write a paper on the subject yet. Fermi asked for permission to report Majorana's results at the Paris Conference, naturally giving him credit for the new ideas. Majorana answered that he would give permission only if the ideas were attributed to an old professor of Electrical Engineering who, incidentally, was supposed to be present at the Conference. This was obviously impossible and for these reasons Majorana's ideas were made known only much later, when they had been discovered independently by other physicists.

In the Fermi report we see the uncertainty still prevailing on the subject of the neutron.

In the discussion following Fermi's communication, according to the "Comptes Rendus de la première section du Congrès International d'Electricité" (Gauthier-Villars, Paris, 1932), there was the following exchange with the polish physicist, Wertenstein:

"M. Wertenstein demande des explications au sujet d'une émission possible de rayonnement accompagnant les rayons β naturels des corps radioactifs et rétablissant la conservation de l'énergie; il ne conçoit pas que ce puissent être des neutrons pour des raisons de masse.

"M. Fermi répond que de tels neutrons ne sont pas ceux que l'on a decouverts mais auraient une masse très inférieure".

This exchange shows that the neutrino hypothesis by then had become known to physicists interested in beta decay.

Madame P. Curie presented also a report in which she described new experiments on the neutron and in the discussion following her report we find remarks by Fermi.

Finally, in 1933, there was a Solvay Conference at Bruxelles, from October 22 to the 29th. It was attended by about forty scientists. By that time nuclear physics had made gigantic strides, the neutron was well established, and the accelerating machines were beginning to give important results. Fermi participated in this conference and intervened repeatedly in the discussions. The name "neutrino" appears in remarks by Pauli.

The proceedings of this conference are published as "Institut International de Physique Solvay, Structure et propriétés des noyaux atomiques" (Gauthier-Villars, Paris, 1934).

E. SEGRÈ.

72 *b*.

LO STATO ATTUALE DELLA FISICA DEL NUCLEO ATOMICO

« Ric. Scientifica », *3* (2) 101–113 (1932)

Lo stato attuale della fisica del nucleo atomico può paragonarsi, sotto diversi aspetti, allo stato della fisica dell'atomo, trent'anni fa. A quell'epoca, infatti, nello studio delle proprietà dell'atomo ci si trovava dinanzi a una serie di fenomeni che non sembravano interpretabili colle teorie di allora, teorie basate essenzialmente sulla dinamica e sull'elettrodinamica classica, che hanno trovato il loro inquadramento naturale nella teoria dei quanti, prima sotto forma quasi esclusivamente qualitativa e oggi anche quantitativa, almeno nella maggior parte dei casi. Naturalmente le leggi quantiche non sono solamente valevoli per i fenomeni della scala atomica, ma anche per i fenomeni del mondo macroscopico; solamente, per questi ultimi, la loro importanza finisce col diminuire e le leggi classiche dànno una approssimazione perfettamente sufficiente. Ciò che determina la necessità di sostituire le leggi quantiche alle leggi classiche, è il cambiamento di dimensioni degli oggetti studiati quando si passa dalle dimensioni ordinarie alle dimensioni atomiche.

Passando dallo studio della fisica atomica a quello della fisica nucleare, noi ci troviamo dinanzi a un nuovo cambiamento della scala dei fenomeni: si discende infatti dalle dimensioni atomiche, dell'ordine di grandezze di 10^{-8} cm alle dimensioni nucleari dell'ordine di 10^{-13} cm e cioè si passa allo studio di oggetti che hanno all'incirca dimensioni 100.000 volte più piccole di quelle degli atomi. La natura dei fenomeni atomici ci fa supporre che leggi, che regolano il comportamento dei corpuscoli costituenti l'atomo, non siano più applicabili senza modificazioni profonde, allo studio del comportamento dei corpuscoli costituenti il nucleo atomico. Questa ipotesi, specialmente per ciò che riguarda il comportamento degli elettroni facenti parte dei nuclei atomici, sembra in realtà confermata dall'insieme delle conoscenze attuali sui fenomeni dei nuclei atomici. È per questo motivo che io cercherò, in questo rapporto, di fissare l'attenzione specialmente sui fenomeni che si interpretano difficilmente colla meccanica corpuscolare attuale.

1. – Pesi atomici e difetto di massa.

La scoperta degli isotopi e il risultato fondamentale che il peso atomico di tutti gli atomi è, con una grande approssimazione, un numero intero, hanno permesso di avanzare l'ipotesi che tutti i nuclei atomici siano costituiti da due tipi fondamentali di corpuscoli, gli elettroni e i nuclei atomici dell'idrogeno o protoni. Ammessa questa ipotesi, il numero dei protoni e degli elettroni costituenti un nucleo dato si ottiene immediatamente, conoscendo il peso atomico M e la carica elettrica del nucleo, che è uguale al numero atomico Z dell'elemento e prendendo per unità il valore assoluto della carica elettronica. Chiamando N_p e N_e il numero dei protoni e degli elettroni contenuti in un nucleo atomico, si ha:

$$N_p = M \qquad N_e = M - Z. \tag{1}$$

Per M si deve prendere naturalmente il peso atomico dell'atomo, arrotondandolo al numero intero più vicino. Ci si può chiedere se, oltre agli argomenti esposti, ce ne sono degli altri a fondamento dell'ipotesi che tutti i nuclei atomici siano costituiti da protoni e da elettroni. Ora che i nuclei atomici, o almeno qualche nucleo atomico, contengano effettivamente degli elettroni, è dimostrato dal fatto che noi osserviamo questi elettroni di origine nucleare nella radiazione β delle sostanze radioattive, e anche, al di fuori delle sostanze radioattive propriamente dette, nelle deboli radiazioni β emesse dal potassio e dal rubidio. La presenza di elettroni negli altri nuclei, benché sia molto probabile per ragioni di analogia, non è stata finora confermata direttamente. Una prova diretta dell'esistenza dei protoni nei nuclei atomici è data dalla disintegrazione artificiale: bombardando infatti un nucleo atomico di azoto, o di un altro elemento leggero, con particelle α dotate di grande velocità, si osservano dei protoni proiettati con una grande velocità, che debbono essere considerati come un prodotto di disintegrazione espulso dai nuclei di azoto sotto l'azione dell'urto con una particella α.

Determinata così l'esistenza dei protoni e degli elettroni nel nucleo atomico, o almeno in qualche nucleo, ci si può chiedere se questi sono veramente i soli corpuscoli elementari che intervengono nella struttura nucleare. È naturalmente impossibile di dare oggi a questa domanda una risposta sicura; e d'altra parte non ci si può dissimulare che l'interpretazione del nucleo atomico come un aggregato di protoni e di elettroni soli, incontra delle serie difficoltà, che indicheremo più innanzi.

Sovente si formano all'interno dei nuclei degli aggregati stabili di 4 protoni e di due elettroni, le particelle α o nuclei di elio, che conservano una certa individualità. Per quanto riguarda l'esistenza delle particelle α nell'interno dei nuclei atomici, noi ne abbiamo una prova diretta nelle radiazioni α delle sostanze radioattive; e ne troviamo una conferma indiretta nello studio dei pesi atomici dei diversi isotopi. Per lo studio della proprietà di un nucleo atomico, è importante di conoscere la sua massa con grande esattezza. Si chiama « difetto di massa » di un nucleo, la differenza fra la massa degli elettroni e dei protoni che lo costituiscono (calcolata prendendo come base il numero dei protoni e degli elettroni dato dalla formula (1)) e la massa effettiva del nucleo atomico. Se si ammette che il nucleo atomico è costituito da protoni e da elettroni, questa differenza di massa deve essere considerata come la massa equivalente, secondo la teoria della relatività, all'energia liberata, nella sintesi del nucleo atomico, dai protoni e dagli elettroni che lo costituiscono. Se si portano, su un grafico, in ascisse i pesi atomici e in ordinate i difetti di massa, si trova che il difetto di massa dipende regolarmente dal peso atomico e, in prima approssimazione, si può dire che gli sia proporzionale.

Questo fatto si può facilmente interpretare colla seguente ipotesi: ammettiamo che gli elettroni e i protoni, contenuti in un nucleo atomico, siano associati in gruppi di quattro protoni e due elettroni, riuniti nella configurazione di una particella α e formanti quindi un insieme stabile; il nucleo atomico nel suo insieme potrà quindi essere considerato come costituito da un insieme di particelle α, oltre a qualche protone ed elettrone in più, in numero insufficiente per poter formare altre particelle α. L'energia di formazione del nucleo atomico per opera di protoni e di elettroni consta quindi di due parti: una prima parte, la più grande, dovuta all'energia di formazione del numero massimo di particelle α che si possono formare coi protoni e cogli elettroni che costituiscono il nucleo; e una seconda parte, più piccola, che rappresenta l'energia di formazione del nucleo atomico dall'unione di particelle α e di protoni ed elettroni in più. Da questo punto di vista è interessante tracciare la curva dei difetti di massa, non rispetto alla massa totale degli elettroni e dei protoni che costituiscono il nucleo, ma rispetto alla massa delle particelle α e dei protoni ed elettroni eventualmente in più: si trovano in questo modo dei difetti di massa molto più piccoli, che si possono interpretare come energie di formazione del nucleo atomico pensato formato dall'unione delle particelle α e dei protoni ed elettroni in più. Fra tutti i tentativi d'interpretazione teorica della curva dei difetti di massa, si deve ricordare in primo luogo quello di Gamow: si tratta tuttavia di tentativi necessariamente basati su ipotesi alquanto primitive.

La relazione tra l'energia di formazione dei differenti nuclei atomici e le loro masse si potrebbe verificare, almeno in principio, per i nuclei di ogni famiglia radioattiva: infatti si conosce, con sufficiente approssimazione, l'energia liberata nei successivi processi di disintegrazione e si ha dunque un mezzo di costruire teoricamente la curva dei difetti di massa per gli elementi di ogni sostanza radioattiva; tuttavia la precisione colla quale si conosce il peso atomico non è sufficiente per poter fare una verifica di questo genere. Una tale precisione è ugualmente insufficiente per poter constatare le variazioni di peso atomico che si verificano nella formazione dell'isotopo dell'ossigeno O^{17} per disintegrazione artificiale dell'azoto.

II. – Momenti meccanici dei nuclei atomici e statistiche dei nuclei.

Una proprietà notevole di qualche nucleo atomico è quella di possedere un momento meccanico intrinseco. Quando si esprime questo momento, prendendo come unità $h/2\pi$, questo momento ha, come per le masse atomiche, un valore uguale a un numero intero o a un numero intero diviso per due.

L'esistenza del momento nucleare si manifesta in diversi fenomeni che, almeno in certi casi, permettono di misurarlo. I fenomeni principali nei quali si manifesta l'esistenza del momento nucleare, sono:

a) le intensità alternate negli spettri di bande;

b) la struttura iperfina delle righe spettrali degli atomi.

Le intensità alternate si osservano negli spettri delle molecole biatomiche, costituite dall'unione di due atomi uguali, per esempio nelle bande dell'N_2. Le successive righe di rotazione di queste bande appaiono con una intensità alternativamente debole e forte. Questa alternanza dell'intensità è stata spiegata da Heisenberg come dovuta al fenomeno quantico dello «scambio» tra i due nuclei atomici della molecola. Risulta dalla teoria che se i nuclei atomici di una molecola biatomica omonucleare non hanno un momento intrinseco, metà delle righe deve mancare nello spettro della molecola, il che si constata effettivamente, per esempio, nelle bande dell'elio e dell'ossigeno.

Se al contrario, il nucleo atomico ha un momento differente dallo zero, le righe che altrimenti mancherebbero, compaiono invece con intensità ridotta, dando così luogo all'alternanza caratteristica dell'intensità. Si può anche trovare facilmente una relazione fra il rapporto d'intensità delle linee forti e deboli successive e il valore J del momento nucleare (misurato di solito in unità $h/2\pi$). Il rapporto di intensità è infatti

$$\frac{J+1}{J}. \tag{2}$$

Da questa formula risulta, in particolare, che l'alternanza di intensità è tanto più forte quanto più piccolo è il momento del nucleo; per esempio per $J = 1/2$, il rapporto di intensità tra le linee successive è uguale 3, se al contrario $J = 1$, il rapporto di intensità è uguale a 2, ecc. Si comprende dunque facilmente come, misurando il rapporto tra le intensità delle righe alternate di una banda spettrale, si possa, per mezzo della formula (2), trovare il valore del momento del nucleo. Misure di questo genere sono state eseguite per parecchi nuclei atomici; esse sono praticamente possibili soltanto per nuclei atomici, non troppo pesanti, altrimenti le bande spettrali diventano troppo ridotte e non si può più misurare il rapporto delle intensità.

Con questo metodo si è potuto, per esempio, riconoscere che un protone è dotato di un momento meccanico uguale a $1/2$, eguale di conseguenza al momento meccanico intrinseco dell'elettrone: il che si può interpretare se si ammette che il protone sia, come l'elettrone, un corpuscolo elementare che obbedisce a una meccanica regolata da una equazione d'onde analoga a quella stabilita da Dirac nella sua teoria relativistica dell'elettrone. In modo analogo si può determinare, per esempio, che la particella α e il nucleo atomico dell'ossigeno hanno un momento meccanico nullo e che il nucleo atomico dell'azoto ha un momento meccanico uguale a 1: noi ritorneremo più innanzi sulle difficoltà caratteristiche dell'interpretazione teorica di quest'ultimo risultato.

Prima di passare alla discussione della determinazione dei momenti meccanici nucleari, che si può ottenere dalle osservazioni sulla struttura iperfina, indichiamo un'altra proprietà importante del nucleo che si può determinare attraverso l'osservazione dell'intensità alternata degli spettri. Le linee dello spettro trovano la loro origine nei passaggi da alcuni termini della molecola ad altri termini della molecola; le alternanze dell'intensità provengono, dal fatto che i termini della molecola, hanno delle funzioni d'onda alternativamente simmetriche e antisimmetriche rispetto allo scambio delle coordinate dei centri di gravità dei due nuclei atomici. Nel caso nel quale il nucleo atomico non ha un momento intrinseco, il suo stato è caratterizzato dalla semplice indicazione delle coordinate del centro di gravità e di conseguenza i termini simmetrici o antisimmetrici mancano alternativamente a seconda che per il tipo di nucleo considerato vale il principio di Pauli o la statistica di Bose-Einstein. La presenza del momento nucleare ha per effetto di far apparire i termini che altrimenti mancherebbero totalmente: questi hanno però una probabilità di esistenza più piccola degli altri, in modo che le righe corrispondenti sono meno intense e dànno origine così al fenomeno delle intensità alternate. Si può dedurre il tipo di statistica alla quale obbediscono i nuclei atomici, quando si conosce il carattere di simmetria dei termini che dànno luogo alle righe intense: ma per conoscerlo, è necessaria una analisi dei termini elettronici della molecola. Per esempio si è potuto dimostrare che il protone si comporta come l'elettrone e obbedisce al principio di Pauli, mentre invece i nuclei atomici dell'elio, dell'ossigeno e dell'azoto soddisfano alla statistica di Bose-Einstein.

Un secondo metodo molto importante per l'osservazione dei momenti meccanici dei nuclei atomici si ottiene dallo studio della struttura iperfina. Le righe spettrali emesse da qualche atomo, osservate con un alto potere risolutivo, appaiono divise in parecchie componenti molto vicine le une alle altre. Si è potuto trovare che questa divisione in più componenti è determinata dall'azione del momento meccanico del nucleo sulle orbite degli elettroni dell'atomo.

Questo fenomeno si può rappresentare nel modo seguente: gli elettroni dell'atomo nella loro rotazione attorno al nucleo atomico dànno origine a un campo magnetico H, nel quale il nucleo atomico si trova immerso. Ora è logico ammettere che sia nel nucleo che negli elettroni, il momento meccanico intrinseco è associato a un momento magnetico: nei nuclei il momento magnetico sarebbe molto più piccolo di un magnetone di Bohr. Secondo le diverse orientazioni dell'asse nucleare rispetto al campo H, si avranno diverse energie di accoppiamento tra il nucleo atomico e il sistema di elettroni e quindi una scissione dei termini, e quindi delle righe, in componenti molto vicine tra loro. Il numero delle componenti dipende oltreché dal termine dell'atomo che si considera, dal valore del momento meccanico del nucleo e può quindi, almeno in certi casi, servire per la sua determinazione. In particolare questa determinazione può farsi con certezza, quando è possibile l'osservazione dell'effetto Zeeman delle righe. Questo metodo di determinazione dei momenti meccanici dei nuclei è applicabile specialmente nel caso degli elementi molto pesanti, giacché in questi la struttura iperfina è generalmente più larga e, di conseguenza, più facilmente osservabile: ed inoltre le righe di questi elementi sono più fini e definite ed è quindi minore il loro allargamento per effetto Doppler. Con questo metodo si sono potuti misurare, per esempio, i momenti meccanici dei nuclei di bismuto, di tallio, di manganese e di molti altri elementi: e si è potuto così constatare che negli elementi con differenti isotopi, molto spesso questi vari isotopi hanno dei momenti meccanici nucleari differenti fra loro. È ciò che si verifica per il cadmio, per il quale qualche isotopo ha un momento uguale a uno, mentre altri hanno un momento uguale a 1/2. La discussione quantitativa della struttura iperfina permette anche di ottenere delle indicazioni sul momento magnetico che è associato al momento meccanico del nucleo. Se si potesse fare l'ipotesi che la sola causa di accoppiamento tra l'orientazione dell'asse nucleare e il sistema degli elettroni fosse l'energia del momento magnetico nucleare rispetto al campo magnetico generato dalla rotazione degli elettroni intorno al nucleo, si potrebbe anche, senza difficoltà di calcolo, dedurre direttamente il valore del momento magnetico del nucleo dalla larghezza di separazione della struttura iperfina. In realtà non è possibile di far dei calcoli di questo genere con grande precisione, a causa della conoscenza imperfetta delle autofunzioni delle diverse orbite elettroniche, che di solito si valutano con dei metodi numerici non molto precisi: inoltre si ha ragione di credere che il momento magnetico del nucleo atomico non è la sola causa dell'accoppiamento che determina la struttura iperfina: per conseguenza calcoli di questo genere non possono sinora che dare delle indicazioni sull'ordine di grandezza del momento magnetico del nucleo atomico. Questo ordine di grandezza è di circa 1 millesimo del magnetone di Bohr: esso è dunque paragonabile al momento magnetico che si dovrebbe attribuire a un protone se si ammettesse che esso è dato dalla stessa formula che dà il magnetone di Bohr, ove si sostituisca la massa elettronica colla massa protonica.

Prima di terminare questo argomento, noi dobbiamo ancora ricordare le difficoltà che si presentano nell'interpretazione teorica dei momenti e delle statistiche nucleari, quando si ammette che tutti i nuclei atomici siano costituiti dall'insieme di protoni e di elettroni. Siccome i protoni obbediscono, come gli elettroni, al principio di Pauli e hanno un momento meccanico intrinseco uguale a 1/2, si può dedurre, con considerazioni indipendenti da ipotesi particolari sulla struttura del sistema, che ogni insieme di protoni e e di elettroni deve:

a) seguire la statistica di Bose–Einstein o il principio di Pauli, a seconda che il numero di corpuscoli che lo costituiscono è pari o dispari;

b) avere un momento intrinseco intero o uguale a un intero dispari diviso per due, a seconda che il numero dei corpuscoli che lo costituiscono è pari o dispari.

Attualmente queste regole sono confermate effettivamente per qualche nucleo atomico; per esempio, la particella α che contiene quattro protoni e due elettroni e cioè, in totale, sei corpuscoli, ha un momento nullo e obbedisce alla statistica di Bose–Einstein.

Viceversa per altri nuclei atomici, queste regole sono in difetto: per esempio per il nucleo atomico dell'azoto. In questo, secondo la formula (1), si debbono avere quattordici protoni e sette elettroni e cioè, in totale, un numero dispari di corpuscoli. Secondo le regole esposte innanzi, per il nucleo atomico dell'azoto, dovrebbe valere il principio di Pauli, mentre le osservazioni di Rasetti sull'effetto Raman della molecola di azoto hanno messo fuor di dubbio che la statistica valida è quella di Bose–Einstein; inoltre si trova un momento intrinseco uguale a 1, mentre dal numero dei corpuscoli costituenti ci si dovrebbe attendere un momento uguale alla metà di un numero dispari. Si è concluso che questa anomalia proviene dal fatto che il nucleo atomico dell'azoto contiene un numero dispari di elettroni.

I nuclei atomici, per i quali si conoscono la statistica e il momento nucleare, sono troppo pochi, perché si possa affermare che si tratta di una regola generale. Le difficoltà dell'interpretazione della struttura nucleare sono così gravi da far dubitare dell'esattezza dell'ipotesi che i soli corpuscoli costituenti i nuclei siano i protoni e gli elettroni. D'altra parte non si deve giammai dimenticare che le idee fondamentali stesse della Meccanica quantica non sono probabilmente più applicabili allo studio dell'interno del nucleo atomico, specialmente per ciò che riguarda il modo di comportarsi degli elettroni.

Per quanto riguarda il momento magnetico nucleare, ci si può a prima vista meravigliare che un nucleo atomico, che contiene degli elettroni, che hanno un momento magnetico intrinseco di un magnetone di Bohr, abbia tuttavia un momento magnetico risultante dell'ordine di grandezza di un millesimo di magnetone di Bohr. Risulta tuttavia, anche dalla teoria ordinaria dell'elettrone, che il suo momento è uguale a un magnetone di Bohr solamente quando l'elettrone è libero, mentre per un elettrone fortemente legato, il momento è tanto più piccolo quanto meno estesa è l'orbita dell'elettrone stesso. Per questo stesso fatto l'energia dell'elettrone può crescere in modo che la sua massa può finire per non essere più trascurabile e per dare un contributo sensibile nel peso atomico dell'elemento: questo ci conferma ancora una volta che le concezioni della meccanica quantica ordinaria non sono applicabili allo studio della dinamica degli elettroni nel nucleo atomico. A proposito dei momenti e delle statistiche nucleari, ci resta ancora da indicare che per il caso del protone e della particella α è stato possibile di controllare, per mezzo di metodi indipendenti, i risultati ottenuti dall'osservazione delle bande spettrali. Per i protoni ciò è stato fatto per mezzo dell'osservazione dei calori specifici a basse temperature e in generale attraverso lo studio delle proprietà dell'orto e del para–idrogeno; per la particella α per mezzo delle esperienze di Chadwick e di Blackett sulla diffusione delle particelle α nell'elio, interpretata colla teoria di Mott sopra l'urto fra corpuscoli che obbediscono alla statistica di Bose–Einstein. Esperienze dello stesso genere sono state ripetute da Gerthsen coi raggi canali dell'idrogeno.

III. - Raggi α.

Le manifestazioni più apparenti dei fenomeni nucleari consistono essenzialmente nelle radiazioni e nelle trasformazioni radioattive. Noi ci limiteremo soltanto a discutere i risultati più recenti e più significativi per la interpretazione teorica dei fenomeni nucleari.

Cominciamo dalla discussione della radiazione α.

La teoria della emissione delle particelle α ha compiuto in questi ultimi anni un grande progresso, grazie ai lavori di Gamow. Ridotta alle sue caratteristiche essenziali, la teoria di Gamow si può descrivere nel modo seguente: si ammette innanzitutto che la forza esercitata da un nucleo atomico pesante su una particella α è regolata dalla legge repulsiva di Coulomb soltanto a partire da una certa distanza dal nucleo (dell'ordine di grandezza di 10^{-12} cm), e che in vicinanza del nucleo la forza diventa attrattiva, in modo da poter spiegare il fatto che le particelle α non abbandonano istantaneamente il nucleo atomico al quale esse appartengono. La curva rappresentativa dell'energia potenziale, in funzione della distanza r della particella dal centro del nucleo, ha allora l'andamento della curva indicata nella figura 1.

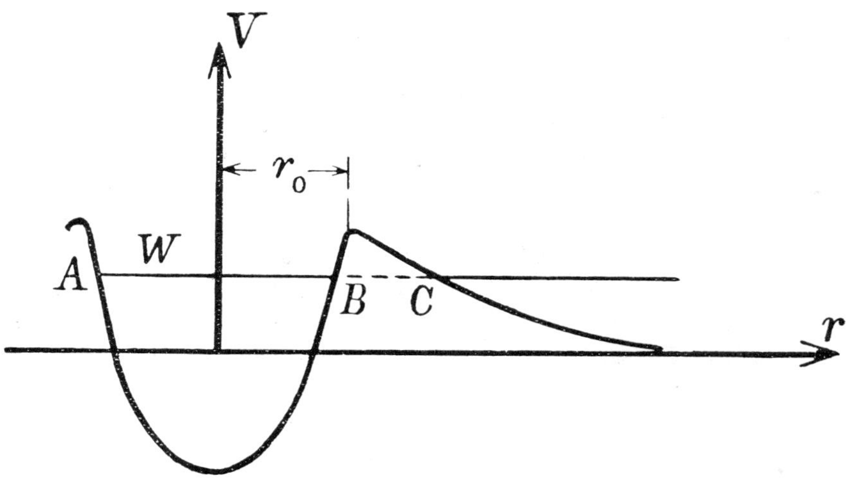

Fig. 1.

Supponiamo, con Gamow, che l'energia della particella α, contenuta all'interno di un nucleo atomico radioattivo corrisponda al livello W della figura. Secondo la meccanica classica, il movimento di una particella avente una energia W e posta all'interno del nucleo consisterebbe in una serie di oscillazioni tra i limiti A e B.

La situazione è invece differente dal punto di vista della meccanica ondulatoria: infatti, secondo la meccanica ondulatoria, tutte le volte che la particella arriva in B, essa ha una probabilità piccola, ma finita, di oltrepassare la zona BC, ove, secondo la meccanica classica, la sua energia cinetica sarebbe negativa. Essa può così raggiungere la zona al di là di C, ove la forza esercitata dal nucleo diventa repulsiva, in modo che la particella può abbandonare definitivamente il nucleo atomico. La probabilità che la particella α abbandoni il nucleo è dunque proporzionale al numero di volte che essa arriva in B nell'unità di tempo e alla probabilità che, in un urto contro B, essa sorpassi la « barriera di potenziale » tra B e C. Quest'ultima probabilità decresce assai rapidamente coll'altezza e lo spessore della barriera di potenziale. Su queste basi Gamow ha potuto calcolare le vite medie degli elementi radioattivi; facendo l'ipotesi che, per tutti gli elementi della stessa famiglia radioattiva, il raggio r_0, al quale si arresta la validità della legge di Coulomb, sia presso a poco costante, la teoria di Gamow dà una semplicissima interpretazione della relazione scoperta empiricamente da Geiger e Nuttall, tra la vita media di un elemento radioattivo e la velocità della particella che esso emette disintegrandosi.

Un fatto assai importante per la comprensione delle proprietà del nucleo è quello scoperto recentemente, da Rosenblum, e cioè che i raggi α emessi dai prodotti C delle tre famiglie radioattive sono costituiti da gruppi aventi delle velocità di emissione differenti. L'osservazione di questo fenomeno è stata resa possibile dall'uso del grande elettromagnete di Bellevue, che permette di deviare le traiettorie delle particelle α fino a far loro descrivere un semicerchio e per conseguenza di mettere a fuoco lo spettro delle velocità delle particelle α. I risultati di Rosenblum sono stati confermati da Lord Rutherford, per mezzo di un metodo elegante di misura della curva di distribuzione dei percorsi delle particelle α. Dei tre prodotti C, quello che è stato meglio analizzato è il Th C, nel quale sono state osservati con esattezza almeno quattro gruppi omogenei di particelle α.

Per l'interpretazione teorica di questi risultati, Gamow ha proposto la seguente spiegazione: ammettiamo che, quando la particella α lascia l'atomo C, il prodotto di disintegrazione, ossia l'atomo C″ sia in uno stato eccitato. L'energia colla quale la particella α abbandona il corpo C sarà evidentemente tanto più grande, quanto più profondo sarà il livello energetico del nucleo atomico residuo del corpo C″; e le differenze tra le energie di emissione dei differenti gruppi di particelle α del corpo C daranno direttamente le differenze tra i livelli energetici del nucleo del corpo C″. Sulla base di queste considerazioni Gamow ha costruito uno schema teorico dei livelli energetici del nucleo atomico del Th C″. Ora dovrebbe essere possibile di ritrovare le differenze tra questi livelli energetici fra le frequenze dello spettro γ del Th C″. Ma non sembra tuttavia che una tale corrispondenza tra i raggi γ del Th C″ e i raggi α del Th C sia stata sinora stabilita sperimentalmente in una maniera indiscussa. Vi è una analogia notevole tra le strutture fini dei raggi α dei corpi C e i gruppi di particelle α, di energia straordinariamente grande, che si osservano nei corpi C′, particolarmente nel Th C′ e nel Ra C′. Per quest'ultimo corpo (Ra C′), Lord Rutherford ha osservato nove gruppi di particelle α, aventi una energia di emissione notevolmente superiore alla normale; l'intensità di questi gruppi è estremamente piccola, corrispondente come ordine di grandezza, a circa una particella per ogni 10^5 particelle aventi una energia normale. Per la spiegazione di questi gruppi di particelle α, con una energia superiore alla normale, Gamow ha supposto che il nucleo atomico del Ra C′, formantesi dalla disintegrazione del Ra C per emissione di una particella β, possa restare in uno stato eccitato, dal quale può ritornare allo stato normale, sia per emissione di un quanto di radiazione γ, sia per disintegrazione con emissione di una particella α con una velocità superiore alla normale. Naturalmente, secondo, questa interpretazione, dovrebbe esistere una corrispondenza tra le frequenze dello spettro γ del Ra C′ e la differenza di energia tra i differenti gruppi di particelle α di questo elemento. Non è stato ancora possibile di stabilire una tale relazione.

IV. - Raggi β.

I raggi β emessi dalle sostanze radioattive possono essere classificati in raggi β primari, costituiti da elettroni direttamente emessi da un nucleo atomico che si disintegra, e in raggi β secondari, costituiti da elettroni che non provengono dal nucleo atomico, ma dalle orbite dell'atomo e che sono espulsi da questo atomo per effetto fotoelettrico, attraverso un processo detto di fotoelettricità interna, sotto l'azione di un quanto di radiazione γ emesso dal nucleo atomico.

Mentre le particelle α, che abbandonano il nucleo atomico che si disintegra, hanno delle velocità molto ben definite, i raggi β primari lasciano il nucleo atomico con velocità variabili entro limiti assai vari. Per esempio si è trovato che le particelle β del Ra E escono dal nucleo atomico con energie variabili in maniera continua da un minimo di 200.000 a un massimo di un milione di volt-elettrone. Si è a lungo supposto che questo « spettro continuo dei raggi β » non fosse che una apparenza, dovuta a perdite di energia subite dalla particella β dopo aver abbandonato il nucleo atomico. Al contrario le misure calorimetriche di Ellis e Wooster, e quelle più recenti di Meitner e Orthmann, nelle quali si misura direttamente l'energia totale delle particelle β attraverso il calore che esse sviluppano nell'interno di un recipiente che le assorbe integralmente, hanno messo fuor di dubbio che l'energia colla quale un nucleo emette la particella β varia effettivamente entro limiti piuttosto estesi. Questo fatto ha una grande importanza teorica, in quanto esso sembra in contraddizione con tutte le teorie del nucleo atomico, che mantengono il principio della conservazione dell'energia; infatti si potrebbe pensare che le differenze di energia delle particelle β fossero compensate da uno spettro continuo di radiazione γ: ricerche fatte per rivelare la esistenza di questa radiazione γ hanno dato risultati negativi.

Sembra dunque che per gli elettroni nucleari l'energia non ha più un valore ben definito, a meno che non si voglia pensare a una nuova forma di energia che sfugge ai nostri mezzi attuali di osservazione.

Si potrebbe pensare per esempio, secondo un suggerimento di Pauli, che nel nucleo atomico si trovino dei neutroni che sarebbero emessi contemporaneamente alle particelle β. Questi potrebbero passare attraverso grandi spessori di materia non perdendo che una piccola parte della loro energia e, quindi, sfuggire praticamente all'osservazione. L'esistenza del neutrone potrebbe senza dubbio dare una spiegazione molto semplice di certi fenomeni attualmente poco comprensibili, come la statistica e i momenti nucleari anormali di certi nuclei, e forse anche, la natura della radiazione penetrante.

I raggi β secondari, come si è detto, sono formati da elettroni che non provengono dall'interno del nucleo atomico. Questi vengono prodotti da un vero e proprio effetto fotoelettrico determinato dall'assorbimento di un quanto di radiazione γ da parte dell'atomo della sostanza radioattiva stessa nella quale si sono prodotti i raggi γ o anche da parte di un altro atomo di specie differente esposto all'azione dei raggi γ. Se si indica con K , L ,· · · i livelli energetici dell'atomo nel quale ha luogo l'effetto fotoelettrico, l'energia dei fotoelettroni emessi sarà $h\nu - K$, $h\nu - L$, ... Ora questo processo di emissione di raggi β secondari, nel quale si ha una vera e propria emissione di un quanto γ, seguita da un riassorbimento, con effetto fotoelettrico, da parte degli anelli elettronici dell'atomo, sembra si debba oggi spiegare con un meccanismo un po' differente: gli elettroni dell'atomo e particolarmente gli elettroni K, le orbite dei quali sono le più vicine al nucleo atomico, si trovano in interazione diretta col nucleo, in modo che a rigore il nucleo e gli elettroni che lo circondano debbono essere considerati come facenti parte di un unico sistema. Nel sistema complesso unico possono avvenire naturalmente delle transizioni tra stati colla stessa energia, determinati direttamente dall'accoppiamento delle due parti (nucleo atomico e sistema di elettroni), costituenti il sistema complesso. Si può così avere una transizione da uno stato nel quale il nucleo si trova in un livello energetico eccitato e un elettrone si trova nel livello K, a uno stato nel quale il nucleo si trova nel livello fondamentale e l'elettrone si trova in un livello eccitato appartenente allo spettro continuo.

Quando il nucleo atomico si trova in un livello energetico eccitato, vi sono dunque due possibilità di ritornare allo stato fondamentale:

a) emissione vera e propria di un quanto γ, che può essere in seguito riassorbito da un altro atomo;

b) passaggio diretto dell'energia dal nucleo a un elettrone degli anelli elettronici.

Si è trovato sperimentalmente che il rapporto tra le probabilità di questi due processi può variare entro limiti piuttosto estesi. Si è tentato anche di valutare teoricamente la probabilità del processo *b*) e di confrontarla con quella che è stata trovata sperimentalmente. Secondo i calcoli di Fowler, Casimir, ecc. sembra tuttavia che la probabilità teorica sia più piccola di quella sperimentale, ciò che proverebbe che l'accoppiamento fra il nucleo e gli elettroni atomici è in realtà più forte di quello che lascerebbero prevedere le teorie odierne.

V. - Raggi γ e livelli energetici nucleari.

La misura dell'energia cinetica dei raggi β secondari, che sono emessi per effetto fotoelettrico, o meglio, per passaggio diretto dell'energia dall'interno del nucleo all'interno del corpo elettronico che circonda il nucleo radioattivo, rappresenta il metodo di misura più conveniente dell'energia dei quanti di radiazione γ e quindi della loro frequenza. Problema importantissimo è la ricerca fino a qual punto le frequenze γ così trovate si possono disporre nello schema dei termini nucleari. Noi abbiamo già alluso alle informazioni che si possono avere sui livelli energetici dei nuclei atomici, esaminando la struttura delle particelle α e il comportamento delle particelle dotate di energia enormemente grande, almeno per quanto riguarda i corpi C′ e C″.

Indipendentemente da questo, diversi tentativi sono stati fatti per stabilire degli schemi semplici di livelli energetici, rappresentanti le frequenze degli spettri γ delle diverse sostanze: tuttavia non sembra che la precisione, colla quale sono conosciute attualmente le frequenze dei raggi γ, sia sufficiente per stabilire in modo univoco uno schema di termini: fatto dovuto essenzialmente alla mancanza di schemi teorici che possano servire di guida in tali ricerche. Va ricordato a questo proposito il tentativo di Rutherford e di Ellis che cercano di associare le diverse frequenze γ a transizioni simultanee di più di una particella α.

Particolarità importanti sull'assorbimento dei raggi γ molto duri sono state messe in evidenza in questi ultimi anni da Chao, Meitner e Hupfeld, ecc. Facendo delle esperienze coi raggi γ del Th C'', di 4,7 UX, essi hanno trovato che il coefficiente d'assorbimento delle varie sostanze, riferito a un numero fisso di elettroni, non è costante, ma cresce col peso atomico della sostanza assorbente. Per gli atomi leggeri l'assorbimento coincide con quello calcolato in base alle formule di Klein–Nishina, mentre per gli atomi pesanti esso è superiore. Sembra che questo fenomeno si debba attribuire a un assorbimento o a una diffusione dovuta al nucleo atomico, che si sovrappone alla diffusione degli elettroni atomici e che cresce in intensità coll'aumento del peso atomico del nucleo assorbente.

VI. – Disintegrazione ed eccitazione artificiale.

Mentre i fenomeni radioattivi ci mettono in presenza di trasformazioni del complesso nucleare, che si producono spontaneamente nei nuclei degli elementi molto pesanti e che non si è riusciti ad accelerare o a rallentare con alcun mezzo oggi conosciuto, si conosce da lungo tempo la possibilità di produrre artificialmente delle trasformazioni nei nuclei atomici degli elementi leggeri, bombardandoli con particelle α dotate di grande velocità.

Dal punto di vista teorico si comprende facilmente che questo fenomeno si presenta solo per i nuclei leggeri. Infatti quando una particella α urta un nucleo atomico pesante, la possibilità che essa entri all'interno, oltrepassando la barriera di potenziale che lo circonda, è praticamente trascurabile. Al contrario una particella che urta un nucleo atomico leggero ha una probabilità considerevole di penetrare nel suo interno, essendo qui la barriera di potenziale notevolmente più bassa.

Una particella α, urtando un nucleo atomico leggero, può produrre l'emissione di un protone o anche l'emissione di raggi γ. Siccome nei due processi la particella urtante può, o no, essere catturata dal nucleo atomico, noi dobbiamo distinguere i 4 seguenti casi:

a) emissione di un protone colla cattura della particella α;

b) emissione di un protone senza cattura;

c) emissione di un quanto γ con cattura;

d) emissione di un quanto γ senza cattura.

Il primo di questi casi è noto da lungo tempo: recenti ricerche hanno permesso di studiarlo con dettagli ancora maggiori e di dimostrare, in particolare, che in qualche caso i protoni emessi sono costituiti da gruppi di differenti velocità.

Per esempio nel caso del boro, Bothe ha potuto riconoscere l'esistenza di tre gruppi di protoni, caratterizzati da un potere penetrante differente. Nel caso nel quale la particella α resta catturata dal nucleo atomico urtato, i principi della conservazione dell'energia e della quantità di moto permettono di vedere come deve variare la velocità dei protoni emessi nelle direzioni formanti angoli differenti colla direzione di propagazione della particella α che provoca gli urti. Questa velocità deve infatti essere indipendente dalla direzione quando essa si riferisce al baricentro del sistema formato dal nucleo atomico urtato e dalla particella α urtante. Dalla osservazione del rapporto esistente tra la velocità del protone e la direzione di emissione, si può risalire alla velocità del baricentro e di conseguenza, conoscendo la massa del nucleo atomico urtato, determinare la velocità della particella α al momento dell'urto. Misure di questo genere, eseguite da Bothe nel caso dei due gruppi di protoni che si ottengono dalla disintegrazione artificiale del boro, hanno dato come risultato per la velocità della particella α, che produce l'urto, dei valori che non coincidono esattamente con quello che ci si potrebbe attendere; tuttavia queste differenze sono forse dovute alla inesattezza delle misure.

Risultati analoghi sono stati ottenuti da Pose per l'alluminio; secondo Pose occorrerebbero delle velocità ben definite delle particelle α per produrre l'emissione di ogni gruppo di protoni; ma questi risultati non sono stati ancora confermati.

Non si conosce con certezza alcun caso nel quale la disintegrazione artificiale si produca secondo lo schema *b*), nel quale il nucleo atomico urtato perde un protone senza catturare la particella α. Sembra tuttavia che un caso di questo genere esista nella disintegrazione artificiale del boro, almeno per uno dei tre gruppi di protoni che sono emessi dal suo nucleo.

Ci resta infine da parlare dei processi *c*) e *d*), nei quali il nucleo urtato emette un quanto di radiazione γ. Questi processi sono stati osservati, solo di recente, da Bothe e Becker, bombardando diversi elementi leggeri coi raggi α del polonio. Risulta che diversi elementi, sotto l'azione del bombardamento, emettono una radiazione γ, l'intensità della quale è assai variabile da elemento a elemento e ha un massimo pronunciato per il berillio. Una tale radiazione si presenta sia negli elementi che, sotto l'azione del bombardamento, emettono dei protoni (per esempio nell'alluminio e nel boro), sia negli elementi che non dànno luogo a emissione di protoni, come il litio e il berillio.

Questi fenomeni possono interpretarsi ammettendo che la particella α venga captata dal nucleo atomico, cedendo eventualmente la sua energia a qualche protone nucleare; e che la radiazione sia emessa durante la costituzione del nuovo nucleo così formato.

Un fatto notevolissimo che è stato messo in evidenza dalle misure di assorbimento di Bothe, di I. Curie e di Joliot, è che le radiazioni γ artificiali del berillio e del boro hanno un potere penetrante molto superiore a quello di tutte le radiazioni γ conosciute delle sostanze radioattive. Secondo le valutazioni, veramente un po' ipotetiche di I. Curie, la radiazione del berillio corrisponderebbe a una lunghezza d'onde di 0,7 UX, e cioè a una energia di circa 20 milioni di volt–elettrone.

Secondo recenti esperienze di I. Curie e Joliot, sembra che le radiazioni del berillio abbiano la proprietà di liberare dei protoni, di grande velocità, da tutte le sostanze contenenti idrogeno: fenomeno che, secondo questi autori sarebbe da interpretare come una specie di effetto Compton esercitato protoni anziché sugli elettroni liberi.

***

Dopo la redazione di questo manoscritto, sono state fatte importanti esperienze di fisica nucleare. Limitandomi ai lavori più caratteristici: citerò qui i seguenti:

Proseguendo le esperienze di Bothe, di I. Curie e di Joliot, Chadwick ha potuto dimostrare che le radiazioni del berillio possono mettere in movimento anche dei nuclei più pesanti dei protoni; in seguito a queste esperienze egli ha fatto l'ipotesi che le radiazioni del berillio non siano delle radiazioni γ, ma proiezioni di neutroni aventi massa uguale a quella del protone.

Un'altra esperienza interessante è quella di Cockcroft e di Walton, che hanno provocato la disintegrazione artificiale del litio, per urto con raggi canale di idrogeno di grande energia.

Infine, in un dominio parallelo a quello del quale ci siamo qui occupati, Steinke e Schindler hanno dimostrato che le radiazioni cosmiche possono staccare dal piombo delle particelle con un alto potere ionizzante, animate di una energia dell'ordine di cento milioni di volt.

N° 73.

In this paper Fermi discusses the influence of the ammonia molecule's rotation on the doubling of its levels. This doubling originates—according to Dennison and Hardy—in the oscillation by which the nitrogen atom crosses the plane determined by the three hydrogens. One observation, made by Barker a few years earlier, remained unexplained: the doubling of the levels mentioned above decreases with the increase in the rotational quantum number. Fermi, in this very simple paper, shows that the effect observed by Barker is due to the centrifugal force which acts in the sense of reducing the probability of the nitrogen crossing the plane of the hydrogen atoms, as the rotational velocity increases.

This paper, together with the articles N° 61, 62 and 64 on the accidental degeneracy of the carbon dioxide molecule's frequencies of oscillation and on the Raman effect in crystals, constitutes a series of investigations from the period 1931-33 in which Fermi attempted to explain various molecular phenomena. His interest in these studies is reflected in the book "Molecole e Cristalli", published by Casa Zanichelli in 1934.

Fermi's interest in this type of problem was always dictated by his desire to give a quantitative explanation to an experimental observation. The problems to which these researches refer were at that time at the center of the Institute's experimental activity. Papers N° 61, 62 and 64 refer especially to Rasetti's work.

The paper on the ammonia molecule is connected, instead, even if in an indirect way, to experiments conducted in Rome during that same period by G. Placzek and me, and to theoretical research by G. Placzek and E. Teller completed during that same period in Rome.

The interest in the study of complex molecules with the help of the Raman effect was stimulated by Placzek's presence at the Institute. He was attracted to Rome by the work of Fermi and Rasetti in this field and contributed a profound critical analysis of these phenomena.

It was in Rome at this time that Placzek began writing his article, which he completed shortly afterwards in Copenhagen, on "Rayleigh-Streuung und Raman-Effekt". The paper appeared in the "Handbuch der Radiologie" (Band VI-2) in 1934.

During the winter of 1931-32, Placzek and I had carried on experimental studies of the rotational spectrum of ammonia in the Raman effect, [1] and verified the selection rules for a "symmetrical-top rotator" which had been established for the general case by Placzek and Teller. [2] Although Fermi's work at this time has no direct connection with these investigations, it is nevertheless an essential part of the overall picture of the activity at the Institute during this period.

E. AMALDI.

73.

SULLE BANDE DI OSCILLAZIONE E ROTAZIONE DELL'AMMONIACA (*)

« Rend. Lincei », *16*, 179–185 (1932); « Nuovo Cimento », *9*, 277–283 (1932).

La molecola di ammoniaca, NH_3, ha, come è noto, una configurazione a piramide triangolare regolare, di cui il vertice principale è occupato dall'atomo di azoto, mentre i tre vertici del triangolo base sono occupati dai tre atomi di idrogeno. Conformemente a questa struttura l'ammoniaca ha

(1) E. AMALDI and G. PLACZEK, *Ramaneffekt des gasförmigen Ammoniaks*, « Naturwissenschaften », *20*, 521, (1932); « ZS. f. Phys. », *81*, 259, (1932).

(2) G. PLACZEK and E. TELLER, « ZS. f. Phys. », *81*, 209 (1932).

(*) Pervenuta all'Accademia il 29 luglio 1932. [Fermi era stato eletto Socio Corrispondente dell'Accademia dei Lincei il 27 luglio 1932 (Socio Nazionale il 6 maggio 1935) Questa nota e le posteriori sono quindi presentate da lui. N.d.R.].

quattro frequenze fondamentali, di cui due sono doppiamente degeneri, mentre le altre due sono semplici. Queste due ultime corrispondono a forme di vibrazione in cui la molecola conserva sempre la simmetria di piramide regolare, mentre si hanno variazioni dell'altezza e del lato di base della piramide. Le due frequenze semplici hanno approssimativamente i valori di 950 e 3330 cm^{-1}; e la prima di queste, di cui particolarmente ci interesseremo in questo lavoro, corrisponde a un movimento per cui, con approssimazione sufficiente, può ritenersi che vari solo l'altezza della piramide, mentre le dimensioni del triangolo di base restano invariate. Come coordinata per descrivere questa oscillazione potremo dunque prendere la distanza x dell'atomo di azoto dal triangolo dei tre idrogeni; ora è evidente che l'atomo di azoto ha due posizioni di equilibrio rispetto al triangolo degli idrogeni, situate simmetricamente rispetto al piano di questo; in conseguenza l'energia potenziale, U (x) considerata come funzione di x, sarà una funzione pari

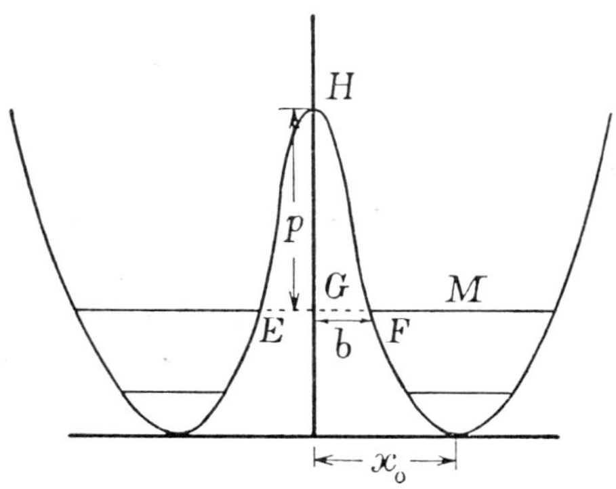

Fig. 1.

di x con due minimi corrispondenti alle due posizioni di equilibrio, per $x = \pm x_0$ (fig. 1). Ciò ha per conseguenza (3) che i livelli di oscillazione della molecola vengono ad essere doppi; lo sdoppiamento è dovuto alla possibilità che ha l'atomo di azoto di attraversare il piano occupato dagli idrogeni passando dall'intorno di una delle due posizioni di equilibrio all'intorno dell'altra. Per questo l'atomo deve oltrepassare una barriera di potenziale il cui culmine si trova in corrispondenza di $x = 0$; la separazione di due livelli è poi data, come è noto, dalla frequenza con cui avviene il passaggio dell'atomo attraverso alla barriera di potenziale. Essa andrà perciò aumentando nei successivi livelli di oscillazione, poiché la loro energia andrà sempre più approssimandosi alla sommità della barriera di potenziale.

In conseguenza di questa struttura dei livelli di oscillazione, tutte le righe di oscillazione e di rotazione vengono ad essere doppie; come si osserva effettivamente per esempio nella banda situata a circa 950 cm^{-1}, in cui le righe di rotazione presentano uno sdoppiamento di circa 30 cm^{-1}.

Lo scopo di questo lavoro è di studiare quale effetto sia esercitato dalla rotazione della molecola sopra lo sdoppiamento; sperimentalmente si osserva

(3) D. M. DENNISON e J. D. HARDY, «Phys. Rev.», *39*, p. 938 (1932).

infatti che esso non è costante nei diversi termini di rotazione, ma va decrescendo col crescere del numero quantico di rotazione riducendosi dal valore iniziale di circa 33 cm^{-1} fino al valore di circa 23 cm^{-1} corrispondente all'undicesimo livello di rotazione (4).

L'effetto della rotazione sopra lo sdoppiamento può qualitativamente interpretarsi osservando che, quando la molecola ruota, la forza centrifuga tende ad allontanare l'atomo di azoto del triangolo dei tre idrogeni; per modo che essa viene ad ostacolare il passaggio dell'atomo di azoto attraverso al triangolo dei tre idrogeni, agendo come una elevazione supplementare della barriera di potenziale che separa le due posizioni di equilibrio simmetriche. Ne risulta un abbassamento della probabilità di scambio tra le due posizioni di equilibrio, e quindi una diminuzione dello sdoppiamento, tanto maggiore quanto più rapida è la rotazione.

Questo modello schematico del fenomeno è dunque qualitativamente in accordo con quanto si osserva sperimentalmente. Ci proponiamo ora di far vedere che anche l'ordine di grandezza previsto dalla teoria per l'entità dell'effetto corrisponde bene a quanto si osserva.

Per determinare completamente la forma e le dimensioni della piramide triangolare che rappresenta la molecola di ammoniaca, occorrerebbe conoscerne i momenti principali di inerzia. Questi, per simmetria, si riducono a due soli: A e B, dove con B si è indicato il momento relativo all'asse di figura e con A il momento relativo a una retta passante per il centro di gravità e perpendicolare all'asse di simmetria. Di questi due momenti uno solo è noto dallo spettro di bande: da esso risulta $A = 2{,}80 \cdot 10^{-40}$. L'altro momento B può soltanto valutarsi con considerazioni indirette e, secondo Dennison, si deve prendere all'incirca $B = 4{,}76 \cdot 10^{-40}$. Nei calcoli seguenti, nei quali del resto ci interessano soltanto gli ordini di grandezza, ci appoggeremo sopra questi dati. Da essi risulta il lato del triangolo base della piramide

$$a = 1{,}69 \cdot 10^{-8}$$

e l'altezza della piramide

$$x_0 = 0{,}32 \cdot 10^{-8}.$$

Se supponiamo che la molecola si deformi, conservando la base invariata mentre varia l'altezza x, il momento di inerzia B non varierà; mentre il momento A sarà una funzione di x. Con considerazioni elementarissime si trova

$$A = 2{,}38 \cdot 10^{-40} + 4{,}10 \cdot 10^{-24} x^2 = A_0 + m x^2 \tag{1}$$

dove $m = 4{,}10 \cdot 10^{-24}$ è la massa ridotta dall'azoto rispetto al triangolo dei tre idrogeni.

Siccome la molecola ha ellisse d'inerzia di rotazione, la sua energia di rotazione sarà data, come è noto, da

$$W_R = \frac{h^2}{8\pi^2} \left\{ \frac{J(J+1)}{A} + \left(\frac{1}{B} - \frac{1}{A}\right) \Omega^2 \right\} \tag{2}$$

(4) E. F. Barker, «Phys. Rev.», *33*, p. 684 (1929); i risultati sperimentali di Barker sono stati riordinati da P. M. Badger, «Phys. Rev.», *35*, p. 1038 (1930).

dove J ed $\Omega \leq$ J sono numeri interi che rappresentano il momento totale della molecola e il momento di rotazione rispetto all'asse di figura, misurati in unità $h/2\pi$.

Per valori fissati di J e Ω, l'energia di rotazione viene dunque ad essere, secondo (1) e (2), una funzione della x. Ora si trova facilmente che, in prima approssimazione, la forza che si esercita tra il piano dei tre idrogeni e l'atomo di azoto, mentre la molecola è in rotazione, può considerarsi derivata da una energia potenziale:

$$V(x) = U(x) + W_R(x) + \text{costante} \tag{3}$$

ottenuta aggiungendo all'energia potenziale U (x) che si ha nella molecola ferma, l'energia (2) della rotazione; ciò equivale sostanzialmente a tener conto della forza centrifuga. Come massa vibrante dovremo poi prendere la massa ridotta m tra la massa dell'azoto e quella dei tre idrogeni.

Consideriamo dapprima l'oscillazione della molecola quando essa non ruota. Se come valor medio della frequenza di oscillazione prendiamo 950 cm^{-1}, il primo livello eccitato di oscillazione avrà una energia di $\frac{3}{2}\,950 = 1425\ \text{cm}^{-1}$, pari a $2{,}8 \cdot 10^{-13}$ erg, superiore all'energia che avrebbe la molecola se i suoi nuclei fossero fermi nella posizione di equilibrio. D'altra parte l'energia, potenziale della forza di richiamo, per oscillazioni nell'intorno della posizione di equilibrio x_0, può mettersi nella forma

$$U = 2\pi^2 m\nu^2 (x - x_0)^2 = 65700\,(x - x_0)^2 . \tag{4}$$

Segue dunque che l'ampiezza M della vibrazione che si avrebbe secondo la teoria classica per il primo livello di oscillazione è

$$\sqrt{\frac{2{,}8 \cdot 10^{-13}}{65700}} = 0{,}206 \cdot 10^{-8} .$$

Secondo la teoria classica si potrebbero avere due vibrazioni perfettamente equivalenti attorno alle due posizioni di equilibrio simmetriche rispetto al piano dei tre idrogeni; secondo la meccanica ondulatoria esiste invece una probabilità finita per il passaggio tra queste due forme di oscillazione attraversando la barriera di potenziate EHF che le separa. La semilarghezza GF $= b$ della barriera di potenziale è data evidentemente dalla semi distanza x_0 delle due posizioni di equilibrio, diminuita della ampiezza dell'oscillazione classica. Essa è cioè nel caso nostro

$$b = 0{,}32 \cdot 10^{-8} - 0{,}206 \cdot 10^{-8} = 0{,}114 \cdot 10^{-8} . \tag{5}$$

Per valutare anche l'altezza $p =$ GH della barriera di potenziale, possiamo utilizzare il dato che la separazione dovuta alla possibilità dell'azoto di attraversare la barriera di potenziale, quando J $= 0$, è di circa 33 cm^{-1}; ciò vuol dire che i passaggi dell'azoto da uno all'altro degli intorni delle due posizioni di equilibrio simmetriche avvengono con una frequenza di 33 cm^{-1}. Naturalmente, per un calcolo teorico di questa frequenza, sarebbe necessario conoscere esattamente la forma della barriera di potenziale; per una semplice valutazione dell'ordine di grandezza, quale è quella che qui ci interessa, ci basterà invece supporre arbitrariamente che la barriera di potenziale abbia forma di parabola dato che essa certamente non si discosterà di molto da tale configurazione. Prenderemo dunque come forma della barriera la seguente:

$$g(x) = p - \frac{p}{b^2}x^2 \quad , \quad |x| \leq b. \tag{6}$$

Per il nostro calcolo grossolano, possiamo calcolare la frequenza $\delta\nu$ con cui avvengono i passaggi attraverso alla barriera di potenziale con la formula

$$\delta\nu = \nu e^{-\sqrt{\frac{8\pi^2 m}{h^2}}\int_{-b}^{b}\sqrt{g(x)}\,dx} \qquad (h = \text{costante di Planck}); \tag{7}$$

né d'altra parte avrebbe molto scopo fare il calcolo con maggiore esattezza data l'incertezza sopra la forma effettiva della barriera di potenziale. Tenendo presente la (6) ed effettuando l'integrale si trova

$$\frac{\delta\nu}{\nu} = e^{-\frac{\pi^2 b}{h}\sqrt{2mp}} \tag{8}$$

ponendo in questa formula $\delta\nu = 33\ \text{cm}^{-1}$, $\nu = 950\ \text{cm}^{-1}$ e ponendo per b, m ed h i loro valori numerici si ricava l'altezza p della barriera di potenziale. Essa risulta

$$p = 4{,}6 \cdot 10^{-13}\ \text{erg} = 2350\ \text{cm}^{-1}.$$

Per calcolare l'effetto della rotazione della molecola, sostituiamo in (2) l'espressione (1) del momento di inerzia A, e sviluppiamo rispetto alle potenze di x, arrestandoci ai termini in x^2. Troviamo allora

$$W_R = \frac{h^2}{8\pi^2}\left\{\frac{J(J+1)}{A_o} + \left(\frac{1}{B} - \frac{1}{A_o}\right)\Omega^2\right\} - \frac{h^2 m\,[J(J+1) - \Omega^2]}{8\pi^2 A_o^2}x^2. \tag{9}$$

Sostituiamo questa espressione nella (3), e disponiamo della costante arbitraria in modo che nelle due posizioni di equilibrio U e V siano eguali tra di loro, in modo che, in prima approssimazione, non vari il livello energetico. Poniamo dunque:

$$V(x) = U(x) + \frac{h^2 m\,[J(J+1) - \Omega^2]}{8\pi^2 A_o^2}\left(x_o^2 - x^2\right); \tag{10}$$

l'ultimo termine di questa formula rappresenta l'elevazione della barriera di potenziale dovuta all'effetto della rotazione. Per calcolare la separazione $\delta\nu_R$ dei termini dovuta all'attraversamento della barriera di potenziale, potremo applicare ancora una formula analoga alla (7) dove però sotto il radicale si dovrà porre la somma di $g(x)$ e dell'ultimo termine di (10). Troviamo cioè

$$\delta\nu_R = \nu e^{-\sqrt{\frac{8\pi^2 m}{h^2}}\int\sqrt{g(x) + \frac{h^2 m\,[J(J+1)-\Omega^2]}{8\pi^2 A_o^2}(x_o^2 - x^2)}\,dx} \tag{11}$$

e sviluppando il radicale col trascurare i quadrati del termine aggiunto a $g(x)$ troviamo infine, tenendo presente la (7) e la (6)

$$\delta\nu_R = \delta\nu\, e^{-\frac{hm^{3/2}}{4\pi\sqrt{2}\,A_0^2}\frac{b}{\sqrt{p}}\int_{-b}^{b}\frac{x_0^2-x^2}{\sqrt{b^2-x^2}}dx\cdot[J(J+1)-\Omega^2]} \tag{12}$$

da cui, effettuando la quadratura.

$$\delta\nu_R = \delta\nu\, e^{-\frac{hm^{3/2}}{4\sqrt{2}\,A_0^2}\frac{b}{\sqrt{p}}\left(x_0^2-\frac{b^2}{2}\right)[J(J+1)-\Omega^2]} \tag{13}$$

Possiamo infine, nella nostra approssimazione, sviluppare l'esponenziale scrivendo

$$\delta\nu_R = \delta\nu - \delta\nu\,\frac{hm^{3/2}}{4\sqrt{2}\,A_0^2}\,\frac{b}{\sqrt{p}}\left(x_0^2-\frac{b^2}{2}\right)[J(J+1)-\Omega^2]. \tag{14}$$

Ponendo per le varie grandezze i loro valori numerici ed esprimendo le frequenze in cm^{-1}, si trova

$$\delta\nu_R = 33 - 0{,}083\,[J(J+1)-\Omega^2] \tag{15}$$

che ci esprime la separazione del termine con i numeri quantici di rotazione J ed Ω.

Per quanto si riferisce al confronto con l'esperienza, si osservi che, finché si prescinda dalla separazione qui discussa, dovuta alla possibilità di oltrepassare la barriera di potenziale, termini che differiscono soltanto per il valore di Ω vengono a dar luogo a righe coincidenti, poiché vale la regola di selezione $\Delta\Omega = 0$. Risulta invece da (15) che le righe corrispondenti a termini che differiscono soltanto per il valore di Ω sono un po' spostate le une rispetto alle altre. In pratica il potere risolutivo non è però sufficiente per osservare queste separazioni, e si osservano solamente gli addensamenti di queste righe che, come facilmente si vede, hanno luogo in corrispondenza di $\Omega = 0$. Ciò che in pratica si può vedere è dunque soltanto la separazione che si ottiene ponendo nella (15) $\Omega = 0$. Cioè

$$\delta\nu_J = 33 - 0{,}083\,J(J+1). \tag{16}$$

La separazione calcolata con questa formula per i successivi termini di rotazione coincide molto bene con quella che si osserva sperimentalmente. Per esempio per $J = 11$, che è il massimo numero quantico per cui si hanno osservazioni, risulta dalla (16) $\delta\nu_{11} = 22$ mentre sperimentalmente si trova una separazione di 23 cm^{-1}.

Naturalmente fenomeni del tipo di quello qui studiato per il primo termine eccitato della frequenza 950, si hanno anche negli altri termini, ma essi sono meno bene studiati, parte per la loro piccolezza e parte per la debole intensità delle bande ad essi corrispondenti. In particolare anche il termine fondamentale deve presentare una separazione analoga. Anzi la determinazione che abbiamo fatta della altezza della barriera di potenziale che separa le due posizioni di equilibrio, permette di fare una valutazione dell'ordine di grandezza di questa separazione: esso risulta all'incirca di 1 cm^{-1}.

N° 74.

This is my recollection of the circumstances that led to the paper "L'effetto del campo magnetico terrestre sulla radiazione penetrante".

During one of my frequent visits to Rome, Enrico Fermi and I had a discussion concerning the influence of the earth's magnetic field on the intensity distribution of cosmic rays. I was very much interested in this effect whose study offered the possibility of determining the sign of the charge of primary cosmic ray particles. Sometime before, starting from a first integral of the equations of motion established by Störmer, I had predicted the existence of an east-west asymmetry in the distribution of the cosmic-ray intensity. Experiments performed at Florence had failed to reveal the expected asymmetry. I was aware of the fact that this negative result might be ascribed to atmospheric absorption, if the energy losses of cosmic-ray particles in air were much greater than those due to ionization alone. However, before reaching any definite conclusion, it was necessary to improve the theory of geomagnetic effects. In my previous work I had only investigated the question as to whether cosmic ray particles of a given momentum could reach a given point of the earth's surface in a given direction. In order to make qualitative predictions it was also necessary to evaluate the relative intensities of the radiation in the various allowed directions. When I presented this problem to Fermi, he pointed out that Liouville's theorem was applicable to the case in question, and provided a very simple solution showing, in fact, that the intensity was the same in all allowed directions.

On the basis of this result, we re-examined the data obtained in the Florence experiment, as well as the results of the experiments of Clay and others on the latitude effect, and came to the conclusion that an abnormally large atmospheric absorption provided indeed the most likely interpretation of all available data. This interpretation was later to be proved correct when it was found that primary cosmic rays are for the most part protons which lose most of their energy by nuclear collisions. Assuming the numerical value of the atmospheric losses indicated by our analysis, we then concluded that in the vicinity of the equator the east-west effect should have been clearly observable. On the strength of this prediction I decided to organize a cosmic-ray expedition to Africa, which was carried out shortly thereafter and provided valuable data on the east-west effect whose existence had been in the meantime announced by Johnson and by Alvarez and Compton.

B. ROSSI.

74.

AZIONE DEL CAMPO MAGNETICO TERRESTRE SULLA RADIAZIONE PENETRANTE (*)

E. FERMI e B. ROSSI, «Rend. Lincei», *17*, 346–350 (1933).

Risultati sperimentali acquisiti in questi ultimi anni hanno dimostrato che la radiazione penetrante consiste di corpuscoli carichi elettricamente. Ancora discussa è invece la questione se questi corpuscoli rappresentino il fenomeno primario oppure si generino essi stessi nell'alta atmosfera per effetto di una radiazione di altra natura; l'argomento principale che può indurre a sospettare una tale natura secondaria della radiazione corpuscolare consiste nel fatto che la radiazione penetrante ha valore presso che indi-

(*) Presentata nella seduta del 5 febbraio 1933.

pendente dalla latitudine, tanto che una leggera diminuzione dell'intensità del fenomeno a basse latitudini è stata messa in evidenza solo negli ultimi tempi con esperienze di notevole accuratezza.

Ci proponiamo di discutere in questa Nota l'effetto del campo magnetico terrestre alle varie latitudini e alle varie altezze, nell'ipotesi che la radiazione sia dovuta primariamente all'effetto di corpuscoli elettrici (elettroni o protoni) provenienti dagli spazi cosmici, e distribuiti a grande distanza dai corpi celesti isotropicamente e uniformemente; e di confrontare i risultati teorici con quelli sperimentali.

Il movimento di un corpuscolo elettrizzato nel campo magnetico terreste è stato studiato da C. Stoermer in connessione coi problemi delle aurore boreali. La sua teoria, applicata ai problemi della radiazione penetrante, nell'ipotesi che i corpuscoli che arrivano sulla Terra siano elettroni [1] porta ai risultati seguenti:

Quando la loro energia supera $61,5 \cdot 10^9$ volt–e essi possono giungere su qualsiasi punto della Terra e da qualsiasi direzione; se la loro energia è più piccola, in un punto della Terra di latitudine magnetica λ, essi possono giungere soltanto da quelle direzioni il cui angolo col piano del meridiano magnetico (positivo verso oriente) supera il valore ϑ definito da

$$\sin \vartheta = \frac{K^2}{R^2} \cos \lambda - \frac{2 K}{R \cos \lambda} \tag{1}$$

$\Big(K^2 = \frac{300 M}{V}$; M = momento magnetico della terra $= 8,33 \cdot 10^{25}$; C. G. S.; R = raggio della Terra $= 6,38 \cdot 10^8$ cm; V = energia in volt–e$\Big)$ [2].

Dal calcolo risulta che questa condizione è necessaria; ammetteremo che essa sia anche, in generale, sufficiente (come lo è certamente in alcuni casi particolari), ossia che gli elettroni possano effettivamente giungere da tutte le direzioni, il cui angolo col meridiano magnetico è maggiore di ϑ [3].

La dipendenza dell'angolo limite ϑ da V per diverse latitudini magnetiche è la seguente

$\lambda = 0°$	$V = 10,3 \cdot 10^9$	$\vartheta = 90°$
	12,7	25°
	15,4	0°
	25	— 43°
	39,1	— 69° 30′
	61,5	— 90°

(1) Risultati sostanzialmente eguali (salvo il segno, quando v'è luogo di considerarlo) si otterrebbero nell'ipotesi che i corpuscoli della radiazione penetrante fossero protoni; ciò almeno finché le energie considerate sono superiori a qualche miliardo di volt–e, poiché in questo caso, per effetto delle correzioni relativistiche, la differenza tra le masse di quiete del protone e dell'elettrone determina solo una lieve differenza nelle equazioni del loro movimento.

(2) B. ROSSI, «Nuovo Cim.», *8*, 8 (1931).

(3) Da un calcolo pubblicato da LEMAITRE e VALLARTA, «Phys. Rev.», *43*, 87 (1933), mentre la presente Nota era già in corso di stampa risulterebbe che in realtà questo non è sempre vero; che esistono cioè «direzioni proibite» il cui angolo col piano del meridiano magnetico è maggiore di ϑ. Questo però non modifica sostanzialmente i risultati della nostra analisi.

$\lambda = 15^\circ$	$V = 9{,}45 \cdot 10^9$	$\vartheta = 90^\circ$
	11,1	28°
	13,4	0°
	18,9	$-36^\circ 30'$
	25	$-60^\circ 30'$
	31	-90°
$\lambda = 30^\circ$	$V = 6{,}6 \cdot 10^9$	$\vartheta = 90^\circ$
	7,5	30°
	8,6	0°
	10,3	-30°
	12,5	-60°
	13,6	-90°
$\lambda = 45^\circ$	$V = 3{,}27 \cdot 10^9$	$\vartheta = 90^\circ$
	3,53	30°
	3,85	0°
	4,24	-30°
	4,6	-60°
	4,75	-90°

D'altra parte si dimostra facilmente che se consideriamo uno sciame di elettroni (o protoni) aventi tutti la stessa energia, che si muovono in un campo magnetico qualsiasi, e seguiamo nel loro movimento quelli che all'istante iniziale si trovano in un dato elemento di volume e hanno velocità appartenenti a un dato elemento di angolo solido, il prodotto dell'elemento di volume nel quale essi sono contenuti, per l'elemento di angolo solido a cui appartengono le loro velocità resta costante durante il movimento dello sciame di corpuscoli. Da questa proprietà, analoga al teorema di Liouville, e dall'ipotesi che la distribuzione degli elettroni a distanza infinita dalla Terra sia uniforme ed isotropa segue immediatamente che l'intensità specifica della radiazione penetrante (numero di corpuscoli, per intervallo unitario di energia e per angolo solido unitario, che attraversano normalmente una superficie unitaria nell'unità di tempo) è per ogni punto (prescindendo dall'assorbimento atmosferico) la stessa in tutte le direzioni, da cui i corpuscoli possono effettivamente giungere.

Se non vi fosse l'atmosfera, l'intensità, sulla superficie della Terra di ciascuna componente monoenergetica della radiazione penetrante (di energia inferiore a $6{,}15 \cdot 10^{10}$ volt–e) dovrebbe, procedendo dal polo magnetico verso l'equatore, rimanere costante fino ad una determinata distanza dal polo (tanto più grande quanto maggiore è la sua energia); poi diminuire gradualmente (pel fatto che un cono di apertura via via crescente rimane «in ombra») fino, eventualmente, ad annullarsi. L'intensità *totale* della radiazione penetrante, ammesso che lo spettro d'energia di questa radiazione si estenda con continuità da zero fino ad un massimo, dovrebbe quindi presentare una diminuzione continua dal polo verso l'equatore.

Questo risultato si modifica quando si tenga conto dell'assorbimento atmosferico.

In prima approssimazione (valida nel caso in cui lo spessore della atmosfera sia da considerarsi infinitamente piccolo rispetto al raggio di curvatura delle traiettorie) possiamo trattare separatamente l'azione del campo magnetico terrestre e dell'assorbimento atmosferico. La distribuzione di intensità della radiazione penetrante, quale si produce per l'azione combinata di questi due fattori, resta così determinata nel seguente modo: in un dato punto ed in una data direzione possono giungere soltanto quei corpuscoli, la cui energia è ad un tempo maggiore di quella definita dall'equazione (1) e maggiore di quella necessaria per attraversare, nella direzione data, lo strato di atmosfera sovrastante; e tali corpuscoli giungono colla medesima intensità con cui giungerebbero in assenza sia dell'atmosfera che del campo magnetico terrestre.

Si riconosce immediatamente quale dev'essere la dipendenza della distribuzione così definita dall'altezza sul livello del mare, dalla latitudine magnetica e, per un dato luogo di osservazione, dalla direzione di provenienza [4]:

a) procedendo dal polo magnetico verso l'equatore, l'intensità totale della radiazione penetrante osservata ad una determinata altitudine deve rimanere costante fino a quella distanza dall'equatore, dove il campo magnetico terrestre comincia ad impedire l'arrivo di alcuni dei corpuscoli, che potrebbero attraversare lo strato di atmosfera sovrastante. A partire da questo punto, l'intensità deve diminuire gradatamente verso l'equatore;

b) ad una data latitudine magnetica, l'intensità della radiazione penetrante, misurata, *in una determinata direzione*, a partire dal limite della atmosfera verso il basso, deve mantenersi costante fino a quella distanza dal limite dell'atmosfera, dove cominciano a rimanere assorbiti i più lenti fra i corpuscoli ai quali il campo magnetico terrestre permette di raggiungere nella direzione data, il limite dell'atmosfera. Da questo punto deve cominciare una graduale diminuzione di intensità; e ad una distanza dal limite dell'atmosfera sufficientemente grande (tale cioè che possano raggiungerla soltanto corpuscoli così rapidi da non essere ostacolati dal campo magnetico terrestre) questa intensità deve coincidere con quella che si osserverebbe in assenza del campo magnetico.

Se, come generalmente avviene, si misura l'intensità *complessiva* in tutte le direzioni (e non l'intensità in una direzione data) la curva dell'intensità in funzione della distanza dal limite dell'atmosfera deve inizialmente presentare un decremento assai lento (anziché un tratto esattamente orizzontale) e raccordarsi attraverso ad un flesso con la curva normale di assorbimento (quella cioè che si osserverebbe in assenza del campo magnetico terrestre);

c) in tutti i punti nei quali l'intensità complessiva della radiazione penetrante è minore di quella che si osserverebbe in assenza del campo magnetico terrestre, il numero dei corpuscoli, che giungono da occidente del meridiano magnetico, deve essere minore del numero di quelli che giungono da oriente del meridiano stesso giacché, per quanto abbiamo veduto, le direzioni occidentali « entrano in ombra » prima di quelli orientali.

(4) Cfr. anche la recente Nota di E. REGENER, « Naturwiss. », *21*, 43 (1933) (nota aggiunta alla correzione delle bozze).

In realtà, una dipendenza dell'intensità dalla latitudine magnetica, del tipo di quella descritta in *a*) è stata effettivamente osservata da Clay [5], da Clay e Berlage [6] e da Compton [7]; e una curva di intensità in funzione dell'altezza, del tipo di quella descritta in *b*) è stata effettivamente osservata da Regener [8].

L'accordo dei risultati sperimentali colle previsioni teoriche è anche quantitativo se ammettiamo che le perdite di energia dei corpuscoli della radiazione penetrante nella materia siano di circa $13 \cdot 10^6$ volt–e per gr/cm². Con questa ipotesi si spiega anche perché nessuna dissimetria del tipo di quella descritta in *c*) sia stata osservata a Firenze [9]; alle nostre latitudini infatti ed in vicinanza al livello del mare l'atmosfera assorbe tutti i raggi sui quali si potrebbe far sentire l'effetto del campo magnetico terrestre.

Il valore, che siamo stati condotti ad attribuire alla perdita di energia dei corpuscoli della radiazione penetrante nella materia, è circa triplo di quello calcolato da Heisenberg [10] per elettroni di corrispondente energia. Data però l'incertezza delle valutazioni teoriche in questo campo, riteniamo di dover forse attribuire maggior peso all'accordo qualitativo che al disaccordo quantitativo e di essere quindi autorizzati a concludere che i risultati delle esperienze finora eseguite sono conformi, anche per quel che riguarda l'influenza del campo magnetico terrestre, all'ipotesi che la radiazione penetrante primaria sia una radiazione corpuscolare.

Per l'ulteriore controllo di questa ipotesi sono in programma le seguenti esperienze:

1° determinazione della dipendenza dell'intensità dall'altezza a diverse latitudini magnetiche. Il flesso dovrebbe presentarsi ad una quota tanto più bassa quanto più bassa è la latitudine magnetica;

2° misura della dipendenza dell'intensità dall'azimut eseguita ad una altezza sul livello del mare sufficientemente grande e ad una latitudine magnetica sufficientemente bassa perché sia sensibile l'azione del campo magnetico terrestre. Si dovrebbe trovare una dissimmetria dell'intensità rispetto al piano del meridiano magnetico, dal cui senso si potrebbe dedurre il segno della carica dei corpuscoli della radiazione penetrante.

(5) J. CLAY, «Amsterd. Proc.», *33*, 711 (1930).
(6) J. CLAY e H. P. BERLAGE, «Naturwiss.», *20*, 687 (1932).
(7) A. H. COMPTON, «Phys. Rev.», *41*, 111 (1932).
(8) E. REGENER, «Naturwiss.», *30*, 69 (1932).
(9) B. ROSSI, loc. cit.
(10) W. HEISENBERG, «Ann. d. Phys.», *13*, 430 (1932)

N° 75 *b*.

For the introduction to this paper see N° 57.

75 *b*.

SULLA TEORIA DELLE STRUTTURE IPERFINI (*)

E. FERMI ed E. SEGRÈ, «Mem. Accad. d'Italia», *4* (Fis.), 131–158 (1933).

RIASSUNTO. — In questo lavoro si studiano le strutture iperfini degli elementi: Li, Na, Cu, Ga, Rb, In, Cs, Ba, Au, Hg, Tl, Pb, Bi che sono ben conosciute sperimentalmente e relativamente accessibili alla teoria. Si discute in particolare fino a che punto le strutture iperfini osservate si possano interpretare teoricamente colla sola ipotesi di un momento magnetico nucleare. Si riconosce che, ove si tenga conto delle numerose perturbazioni anche tra configurazioni elettroniche diverse, l'ipotesi di un accoppiamento magnetico non è in alcun caso in contrasto coi risultati sperimentali, e che anzi essa basta a render conto, almeno nelle linee generali, di tutto il materiale osservato: pur non potendosi escludere che altre cause intervengano a portare delle piccole perturbazioni. Infine sono esaminate le conclusioni che si possono trarre dai dati sperimentali riguardo ai momenti magnetici nucleari.

§ 1. – INTRODUZIONE.

Le strutture iperfini (sif) delle righe spettrali, sono state già da tempo attribuite all'interazione tra il nucleo e il sistema degli elettroni [1]. L'ipotesi più spontanea per rappresentare questa interazione è di attribuire al nucleo un momento magnetico intrinseco associato al suo momento meccanico. Gli elettroni muovendosi intorno al nucleo determinano su questo, un campo magnetico rispetto al quale il nucleo può orientarsi in diversi modi acquistando diverse energie magnetiche. Ciò dà luogo ad uno spezzamento dei livelli energetici. In tal guisa si ottiene da ogni termine elettronico un multipletto di sif del tutto analogo ad un multipletto ordinario in perfetto accoppiamento di Russell–Saunders, purché tra i vettori che rappresentano i momenti della quantità di moto delle diverse parti dell'atomo si stabilisca la corrispondenza seguente:

Multipletto ordinario		Multipletto di sif
L	Corrisponde a	J
S	» »	I
J	» »	F

(*) Presentata nell'Adunanza del 10 marzo 1933.

Secondo le convenzioni ordinarie con I e con F indichiamo rispettivamente il momento meccanico intrinseco nucleare e il momento totale dell'atomo (risultante di J e di I).

In particolare per il multipletto di sif valgono la regola degli intervalli e le regole ordinarie pel calcolo degli effetti Zeeman come è stato dimostrato sperimentalmente per la prima volta da Goudsmit e Back [2]. Ciò permette di caratterizzare tutto un multipletto di sif con una sola costante A. Le posizioni dei singoli livelli di sif rispetto al centro di gravità del multipletto si ottengono dalla formula

$$A(J, I) = \frac{1}{2} A [F(F+1) - J(J+1) - I(I+1)] \tag{1}$$

con $I + J \geq F \geq |I - J|$.

Se si ammette che la sif sia dovuta esclusivamente a un momento magnetico nucleare, la costante A prende, classicamente, il valore

$$A = \frac{\mu H}{IJ} \tag{2}$$

μ essendo il momento magnetico nucleare ed H il campo magnetico medio determinato dai moti elettronici nel nucleo. Quantisticamente possiamo usare ancora la stessa formula (2) purché il significato di μ e di H sia precisato come segue: μ è il valore della componente z del momento magnetico nucleare quando sia $I_z = I$ ciò che classicamente corrisponde ad essere il momento nucleare parallelo all'asse z; H è similmente il valor medio della componente H_z del campo magnetico determinato dagli elettroni sul nucleo per lo stato in cui $M = J$.

Il campo H può esser calcolato qualora siano note le autofunzioni elettroniche. Si può risalire quindi per mezzo della (2), dai dati sperimentali su A ai valori del momento magnetico nucleare. La teoria potrà considerarsi soddisfacente se calcolando μ da diversi termini di atomi neutri o jonizzati contenenti lo stesso nucleo, si otterrà lo stesso risultato.

Vari Autori [3] hanno eseguito calcoli di tal genere per un elettrone in campo centrale in approssimazione non relativistica. Successivamente sono state apportate le dovute correzioni relativistiche [4] le quali specialmente per elementi pesanti producono un effetto cospicuo, infine Goudsmit ed altri [5] hanno discusso anche il caso di più elettroni. Il confronto quantitativo dei risultati di questi calcoli con l'esperienza, ha dato luogo a serie difficoltà [6] in quanto che calcolando il momento magnetico nucleare dalle sif di vari termini sembrava che risultassero valori di μ considerevolmente differenti tra di loro. Si era così venuta formando l'opinione che accanto ad un eventuale momento magnetico nucleare dovessero sussistere altre cause atte a produrre effetti dello stesso ordine di grandezza sulla struttura dei termini.

In questo lavoro ci siamo proposti di riprendere in esame integralmente la questione cercando di eliminare le varie cause di incertezza dovute ad approssimazioni insufficienti. Abbiamo così potuto constatare che le pretese discrepanze possono trovare una naturale giustificazione in perturbazioni trascurate nei calcoli finora compiuti. Tali perturbazioni non sono di solito valutabili quantitativamente con esattezza; tuttavia esse hanno ordini di grandezza perfettamente confrontabili con quelli delle divergenze tra la teoria elementare e i dati sperimentali, e provengono generalmente da configurazioni elettroniche diverse da quella a cui appartiene il termine in questione. I loro effetti sono notevoli o perché i livelli perturbatori sono molto prossimi o perché essi hanno costanti di accoppiamento A assai grandi rispetto a quelle dei termini perturbati.

Nel § 2 studieremo i problemi di calcolo numerico delle autofunzioni individuali degli elettroni per ciò che interessa le sif. Pei termini *s* importa solo il valore di $\psi(0)$ e mostreremo che questo è dato con approssimazione sufficiente da una formula semi-empirica di Landé-Goudsmit perfezionata con alcune correzioni in buona parte relativistiche, in parte anche di altra natura, che verranno indicate in particolare. Con ciò si eliminano molte delle incertezze dovute a metodi numerici non sufficientemente esatti, che finora esistevano nella valutazione delle sif derivanti da elettroni *s*.

Per elettroni *p* e *d* l'accoppiamento col nucleo è determinato dal valor medio di $1/r^3$; per calcolare questa grandezza non è necessaria una conoscenza esatta dell'autofunzione poiché essa è deducibile dalle separazioni del multipletto ordinario con l'applicazione di fattori di correzione, per tener conto del fatto che la separazione del multipletto ordinario, a differenza di quella del multipletto di sif, è determinata da $\overline{\frac{1}{r}\frac{\partial V}{\partial r}}$ e non da $\overline{1/r^3}$.

Queste correzioni in qualche caso sono considerevoli (30 %) ma la loro valutazione numerica non presenta notevole incertezza poiché esse sono precipuamente determinate dalla parte interna dell'autofunzione in cui il metodo statistico dà una sufficiente approssimazione.

Nei §§ 3 e 4 discuteremo le possibili perturbazioni e mostreremo in alcuni esempi tipici i loro effetti, sia nel caso che i termini che si perturbano appartengano alla stessa configurazione che a configurazioni diverse. Infine diamo dei criteri per riconoscere quando è lecito ammettere che un termine sia poco perturbato per modo che la sua costante di sif possa essere utilizzata per dedurne un valore attendibile del momento magnetico nucleare.

Il § 5 è dedicato all'esame sommario delle sif di tutti quegli elementi su cui si hanno dati sperimentali attendibili e che sono ancora dal punto di vista teorico abbastanza semplici per esser trattati con sufficiente approssimazione.

§ 2. - Metodi numerici per il calcolo delle costanti di accoppiamento dei singoli elettroni.

a) *Caso di un solo elettrone s*. - La costante di accoppiamento di un solo elettrone *s* col nucleo in campo centrale è data in approssimazione non relativistica da:

$$a = \frac{16\pi}{3}\frac{1}{J}\mu\mu_0\psi^2(0) \tag{3}$$

dove μ_0 è il magnetone di Bohr.

Essa del resto non può evidentemente applicarsi nel caso relativistico. Secondo la teoria di Dirac dell'elettrone infatti, l'ampiezza di probabilità per l'elettrone nell'origine tende all'infinito. Si trova in questo caso che $\psi^2(0)$ va sostituito in (3) coll'espressione

$$\frac{8\pi mc}{h}\int_0^\infty \mathrm{FG}\,dr = \psi^2_{\mathrm{eff}}(0) \tag{4}$$

in cui F e G sono le autofunzioni di Darwin [7] collegate alla $4\,\psi$ di Dirac dalle relazioni

$$\left\{\begin{array}{ll} \psi_1 = i\cos\theta \mathrm{F}(r) & \psi_1 = i\sin\theta e^{-i\varphi}\mathrm{F}(r) \\ \psi_2 = i\sin\theta e^{i\varphi}\mathrm{F}(r) & \psi_2 = -i\cos\theta\mathrm{F}(r) \\ \psi_3 = \mathrm{G}(r) & \psi_3 = 0 \\ \psi_4 = 0 & \psi_4 = \mathrm{G}(r) \end{array}\right. \tag{5}$$

dove si deve prendere l'una o l'altra quaterna a seconda dell'orientazione dello spin.

Pel calcolo dell'integrale (4) non è sufficiente conoscere le autofunzioni nelle sole vicinanze dell'origine, perché anche le parti esterne delle autofunzioni hanno importanza decisiva pel tramite del fattore di normalizzazione. Ora ciò che conta maggiormente per la valutazione di detto fattore, è proprio l'ultima arcata dell'autofunzione, quella cioè in cui più particolarmente si fanno sentire le incertezze della valutazione numerica. Ne consegue una notevole imprecisione nei computi fatti applicando direttamente il metodo statistico.

Vari Autori [8] hanno usato una formula semiempirica:

$$\psi^2(0) = \frac{1}{\pi a^3}\,\frac{\mathrm{Z}(1+z)^2}{n^3_{\mathrm{eff}}} \tag{6}$$

(a = raggio di Bohr; z = grado di jonizzazione; n_{eff} = numero quantico efficace) perfezionata eventualmente con un fattore di correzione relativistica.

Ci siamo proposti di fare un'indagine sistematica sulla applicabilità della [6] che, oltre alla sua notevole semplicità, presenta altresì il vantaggio di essere indipendente dagli errori che possono derivare dal calcolo statistico della parte esterna dell'autofunzione in luogo della quale si usa n_{eff} ricavabile direttamente dai dati spettroscopici empirici.

Abbiamo potuto dimostrare che la formula (6) dà risultati corretti per elementi leggeri e per termini ben rappresentabili con una formula di Rydberg.

Più in generale, tenendo conto anche delle correzioni relativistiche $\psi^2(0)$ si può esprimere colla formula empirica

$$\psi^2_{\mathrm{eff}}(0) = \frac{1}{\pi a^3(1-\beta^2\mathrm{Z}^2)^2}\,\frac{\mathrm{Z}}{2\mathrm{R}h}\,\frac{d\mathrm{E}}{dn} \tag{7}$$

in cui $\beta=1/139$ (*) ed E è l'energia espressa in funzione del numero quantico n. La giustificazione di questa formula si trova nell'appendice I. La funzione E (n) è definita naturalmente solo per valori interi dell'argomento; essa tuttavia può essere interpolata con una funzione derivabile usando per esempio della formula di Rydberg od altre analoghe. La correzione relativistica, rappresentata dal fattore $\frac{1}{(1-\beta^2 Z^2)^2}$ ha carattere di interpolazione e il suo valore numerico è quasi eguale a quello dato da Racah e Breit [4]. Salvo questo fattore la (7) coincide colla (6) se i termini sono rappresentabili con una formula di Rydberg $E=-\frac{Rh}{(n-s)^2}$. Se la correzione di Rydberg s non è costante la (7) assume la forma

$$\psi_{\text{eff}}^2(0)=\frac{2{,}16\cdot 10^{24}}{(1-\beta^2 Z^2)^2}\,\frac{Z(1+z)^2}{n_{\text{eff}}^3}\left(1-\frac{ds}{dn}\right). \tag{8}$$

Oltre al fattore relativistico vi è questa formula l'ulteriore fattore correttivo $\left(1-\frac{ds}{dn}\right)$ che può in vari casi determinare correzioni del 10 %. Introducendo il valore di $\psi_{\text{eff}}^2(0)$ dato dalla (8) nella (3) ed esprimendo la costante di sif in numeri d'onde si ottiene la seguente formula

$$a=15{,}5\,\frac{\mu}{\mu_0 I}\,\frac{1}{(1-\beta^2 Z^2)^2}\,\frac{Z(1+z)^2}{n_{\text{eff}}^3}\left(1-\frac{ds}{dn}\right) \tag{9}$$

che serve praticamente per i calcoli, nei quali, dove si hanno dati sufficienti, abbiamo posto ds/dn eguale alla differenza delle correzioni di Rydberg per il termine considerato e quello successivo.

b) *Caso di un elettrone p. o d.* – Per un elettrone p o d la costante di sif è data da (9)

$$a=\frac{2l(l+1)}{Ij(j+1)}\,\mu\mu_0\,\overline{\frac{1}{r^3}} \tag{10}$$

e dipende quindi sostanzialmente da $\overline{1/r^3}$. Per valutare questa grandezza conviene giovarsi dei dati sulla struttura grossa, la cui separazione totale è data da:

$$\delta=\frac{\mu_0^2}{hce}(2l+1)\,\overline{\frac{1}{r}\frac{dV}{dr}} \tag{11}$$

in cm^{-1}.

Conoscendo δ si ricava immediatamente il valor medio di $\frac{1}{r}\frac{dV}{dr}$. Nell'integrale che determina questo valor medio preponderano i termini provenienti da regioni prossimi al nucleo (r piccolo) in cui V è computabile assai esattamente col metodo statistico. Al solito però le parti esterne dell'autofunzione sono essenziali per la loro influenza sul fattore di normalizzazione, e quindi fanno pesare la loro imprecisione su tutto il calcolo. Eseguendo il rapporto tra $\overline{\frac{1}{r}\frac{dV}{dr}}$ e $\overline{1/r^3}$ si può, invece, far sparire dai calcoli tale fattore di normalizzazione con evidente notevole vantaggio.

(*) La costante β è $2\pi e^2/hc$ il cui valore numerico è assai prossimo a $1/137$ (Nota d. Red.).

A tal uopo in una prima grossolana approssimazione si può porre, trascurando per la parte centrale dell'atomo l'azione di schermo degli elettroni,

$$V = \frac{Ze}{r}, \tag{12}$$

risulta allora

$$\eta = -\frac{\overline{\frac{1}{r}\frac{dV}{dr}}}{\overline{1/r^3}} = Ze. \tag{13}$$

Questa approssimazione dà in qualche caso risultati buoni; in altri è necessario usare per V l'espressione statistica più esatta

$$V = \frac{e}{r}\left[(Z-1)\,\varphi\,(r/\mu) + 1\right] \tag{14}$$

da cui si ricava

$$-\frac{1}{r}\frac{dV}{dr} = \frac{e}{r^3}\left[(Z-1)\,(\varphi - x\varphi') + 1\right] \tag{15}$$

in cui x è eguale ad r diviso per l'unità di lunghezza propria del metodo statistico μ.

Il rapporto η è dato da

$$\eta = e\left[(Z-1)\,\frac{\int(\varphi - x\varphi')\,\frac{\psi^2}{r^3}\,d\tau}{\int\frac{\psi^2}{r^3}\,d\tau} + 1\right]. \tag{16}$$

La valutazione numerica degli integrali che compaiono nella (16) può farsi con notevole precisione col metodo statistico, poiché a causa del fattore $1/r^3$ i termini preponderanti negli integrali provengono dalle parti interne dell'atomo. Si ottiene così rispetto alla (13) una correzione che è in generale piccola per gli elettroni p, maggiore per quelli d, dato che le loro orbite sono meno penetranti.

Eseguendo i calcoli si ha per esempio per un elettrone $3\,d$ del rame

$$\eta = 19{,}6\,e = Z^*\,e \tag{17}$$

anziché $\eta = 29\,e$ come risulterebbe dalla formula approssimata (13); la correzione è quindi in questo caso del 30 % rispetto alla formula approssimata.

A questa correzione bisogna, secondo Breit [9], aggiungerne un'altra, sensibile specialmente per elettroni p di atomi pesanti, per tener conto che l'energia delle due orbite $p_{1/2}$ e $p_{3/2}$ è differente e quindi anche il loro fattore di normalizzazione non è lo stesso.

Da (10), (11), e (17) si ricava, per il calcolo pratico,

$$A = \frac{2l(l+1)}{Ij(j+1)(2l+1)}\,\frac{\mu}{\mu_0}\,\frac{\delta}{Z^*}. \tag{10 bis}$$

§ 3. – Effetto delle perturbazioni.

Se si usano i procedimenti di calcolo indicati nel paragrafo precedente, si trova in parecchi casi che i valori di μ calcolati da diversi termini, per lo stesso nucleo concordano discretamente tra di loro. Altri termini invece dànno valori di μ in completo disaccordo coi precedenti.

Come abbiamo già accennato, discrepanze di questo genere, erano state rilevate varie volte tanto che avevano fatto mettere seriamente in dubbio l'ipotesi che le sif fossero, almeno nelle loro linee generali, spiegabili come effetto di un momento magnetico nucleare. Noi cercheremo di mostrare invece che le divergenze indicate sono attribuibili ad effetti di perturbazioni fra i termini elettronici.

Quando i termini che si perturbano sono energeticamente prossimi, l'effetto è assai appariscente e senz'altro individuabile; tuttavia anche quando i termini perturbatori sono lontani, la perturbazione percentuale può essere grande; ciò avviene in particolare quando il termine perturbato ha una costante di sif piccola rispetto a quella del termine perturbatore.

Si debbono considerare due tipi di perturbazioni: *a*) tra termini della stessa configurazione; *b*) tra termini di configurazioni diverse.

Quelle del primo tipo sono già state considerate da altri Autori [10] ed hanno luogo nei numerosi casi di accoppiamento intermedio fra quello di Russell–Saunders e quello (J , J). Non ci soffermiamo quindi su questo, genere di perturbazioni; e del resto nel caso del Tl II ne troveremo un esempio tipico.

Più interessanti si presentano le perturbazioni fra termini di configurazioni elettroniche differenti, che passiamo adesso a trattare.

L'autofunzione di un atomo con più elettroni si suole scrivere approssimativamente come prodotto antisimmetrizzato delle autofunzioni individuali dei singoli elettroni. In questo senso si suole parlare per esempio di una configurazione $6s^2\ 7p$ intendendosi con ciò che si hanno due elettroni $6s$ ed uno $7p$. Gli anelli chiusi vengono generalmente tralasciati nella scrittura. In realtà le vere autofunzioni sono combinazioni lineari di quelle di approssimazione zero. Sia ψ_1 l'autofunzione imperturbata di un termine di una data configurazione elettronica e sia ψ_i l'autofunzione di un altro termine della stessa o di un'altra configurazione elettronica. Secondo la teoria delle perturbazioni l'autofunzione di prima approssimazione sarà

$$\Psi'_1 = \psi_1 + \sum_i{}' \frac{V_{1i}}{E_1 - E_i} \psi_i \tag{18}$$

dove E_1 ed E_i sono le energie imperturbate e V_{1i} l'elemento della matrice di perturbazione che connette gli stati 1 ed i. Nel caso nostro come perturbazione è da considerarsi essenzialmente l'energia elettrostatica. Per gli elementi di matrice che la rappresentano valgono delle semplici regole di selezione:

a) sono eguali a zero tutti gli elementi di matrice che connettono stati con J differente;

b) sono eguali a zero elementi di matrice che connettono termini con parità diversa rispetto alla regola di Laporte (e cioè per i quali la somma algebrica dei numeri quantici azimutali per i singoli elettroni ha parità diversa).

Quando i due termini 1 e i hanno lo stesso J e la stessa parità, V_{1i} è, almeno in generale, diverso da zero. Siccome esso rappresenta una energia elettrostatica, il suo ordine di grandezza è poi confrontabile con le separazioni che si hanno tra termini di molteplicità differente della stessa configurazione, cioè circa 1 volt per termini profondi di atomi neutri.

Dal punto di vista della teoria delle sif le contaminazioni tra configurazioni diverse hanno importanza per il fatto che esistono configurazioni, anche energeticamente elevate, che sono fortemente accoppiate col nucleo; tali sono particolarmente quelle che contengono un elettrone relativamente profondo isolato in una orbita *s*. Se una autofunzione ψ_2 di questo tipo si mescola secondo la [18] con un'altra ψ_1 che abbia una piccola sif, essa può produrre una perturbazione percentuale assai rilevante della costante di sif anche se $\frac{V_{12}}{E_1 - E_2}$ è piuttosto piccolo.

Supponiamo per esempio che la configurazione perturbatrice abbia una differenza di energia di 5 volt da quella perturbata, e appartenga quindi a una regione spettrale totalmente diversa. Se l'elemento di matrice V_{12} ha valore di 1 volt, l'autofunzione perturbata sarà, secondo la (18)

$$\Psi_1 = \psi_1 + 0{,}2\,\psi_2$$

così che lo stato 1 viene a contenere una contaminazione di $(0{,}2)^2$ pari al 4 % dello stato 2. Se ammettiamo, ciò che può verificarsi in parecchi casi, che la struttura della configurazione 2, contenente un elettrone *s* profondo, sia per esempio 50 volte maggiore di quella dello stato 1, una perturbazione del 4 % potrà triplicare o cambiar di segno la costante di accoppiamento dello stato 1; anche prescindendo dalla azione di termini non diagonali dell'accoppiamento col nucleo, che possono in qualche caso determinare un effetto dell'ordine di grandezza o anche maggiore del precedente.

Esaminando il materiale empirico sulle sif si trovano numerosi esempi di perturbazioni del tipo illustrato; ad esse sono secondo noi da attribuirsi per la massima parte quelle discrepanze, a cui abbiamo accennato, che avevano fatto mettere in dubbio la correttezza delle basi stesse della teoria.

Un esempio assai appariscente dell'effetto di perturbazioni tra configurazioni diverse, si trova nello spettro del Pb II. L'isotopo 207 del Pb ha I = 1/2, ed i termini del suo spettro, secondo Kopfermann e Schüler e Jones, presentano le seguenti separazioni di sif:

Termine			$\Delta\nu$, cm^{-1}
$6s^2$	$7s$	$^2S_{1/2}$	0,352
$6s^2$	$7p$	$^2P_{1/2}$	0,077
$6s^2$	$7p$	$^2P_{3/2}$	0,022
$6s^2$	$6d$	$^2D_{5/2}$	0,713
$6s$	$6p^2$	$^2D_{5/2}$	0,856

Dall'esame di questa tabella si rileva un valore abbastanza grande della separazione del termine $7^2S_{1/2}$ ciò che è naturale per la presenza in questo termine di un elettrone $7s$ non saturo. Più piccole, in accordo qualitativo con la teoria, sono le separazioni dei due termini 7^2P. Il termine

$6s^2\ 6d\ ^2D_{5/2}$ ha invece una separazione assai grande, che a prima vista sorprende, essendo i due elettroni $6s$ compensati tra di loro, e l'altro elettrone, $6d$, in una orbita molto esterna. La ragione del comportamento assai anomalo di questo termine è da ricercarsi nel fatto che esso è fortemente perturbato dall'altro termine $6s\ 6p^2\ ^2D_{5/2}$ che trae la sua grossa separazione dalla presenza dell'elettrone $6s$ non saturo. I due termini distano di circa 0,6 volt, ed avendo lo stesso J e la stessa parità, a norma delle regole di selezione, possono perturbarsi. Data la loro lieve differenza di energia si può anzi valutare che le autofunzioni di prima approssimazione sono combinazioni lineari delle due autofunzioni imperturbate ψ_1 e ψ_2 (corrispondenti alle configurazioni $6s^2\ 6d$ e $6s\ 6p^2$) i cui coefficienti sono, in valore assoluto, approssimativamente eguali; si ha cioè

$$\left\{\begin{aligned} \Psi_1 &= \alpha'\,\psi_1 + \beta'\,\psi \\ \Psi_2 &= \alpha''\,\psi_1 + \beta''\,\psi_2 \end{aligned}\right. \tag{19}$$

dove le α e le β hanno lo stesso ordine di grandezza. Se si ammette, come corrisponde al nostro caso, che l'accoppiamento dello stato ψ_1 col nucleo sia trascurabile, si riconosce facilmente che le sif dei due termini sono rispettivamente date da β'^2 e β''^2 moltiplicate per la sif che avrebbe il termine ψ_2 se esistesse allo stato puro. Deve notarsi tuttavia che la situazione è in realtà un po' più complicata, per l'esistenza, nella stessa configurazione $6s\ 6p^2$ del termine $^4P_{5/2}$ la cui posizione non è conosciuta, ma che verosimilmente si trova non lontano dai due termini considerati, e può quindi anche esso perturbarli notevolmente.

Nello spettro del Tl neutro si trova invece un esempio tipico di una configurazione energeticamente lontana che, pur entrando in piccola misura nelle autofunzioni di prima approssimazione, altera profondamente le sif. È stata rilevata più volte una notevole discrepanza tra la teoria elementare delle sif e i risultati sperimentali per il doppietto fondamentale $6\ ^2P_{1/2,3/2}$ del tallio. I due termini $^2P_{1/2}$ e $^2P_{3/2}$ che lo costituiscono, hanno separazioni iperfine di 0,707 e 0,008 cm rispettivamente. Secondo la teoria elementare la struttura del termine con $J = 3/2$ dovrebbe essere bensì più piccola di quella del termine con $J = 1/2$; il rapporto tuttavia, anche tenuto conto di tutte le possibili correzioni relativistiche, non dovrebbe eccedere 10. Quindi il termine $^2P_{3/2}$, per essere d'accordo col $^2P_{1/2}$ dovrebbe avere una separazione iperfina dell'ordine di grandezza di 0,070, in contrasto col valore sperimentale 8 o 9 volte più piccolo. Per ristabilire l'accordo occorre dimostrare l'esistenza di una perturbazione, capace di alterare di circa 0,060 la sif del termine $^2P_{3/2}$.

Esamineremo dunque se esiste qualche configurazione capace di produrre una perturbazione di questo genere. Essa dovrà avere una sif grossa e quindi contenere un elettrone $6s$ non saturo (se si pensasse di intaccare le configurazioni $5s$ molto stabili, l'energia salirebbe a valori inammissibili), e inoltre dovrà essere dispari come la configurazione perturbata $6s^2\ 6p$. Si riconosce così che le due uniche configurazioni da considerare sono la $6s\ 6p\ 7s$ e la $6s\ 6p\ 6d$. Di queste due energeticamente più bassa è la prima i cui termini possono essere localizzati verso il limite di jonizzazione dell'atomo o poco al di sopra e cioè circa 6 o 7 volt al di sopra del termine fondamentale; l'altra potrà essere ancora un paio di volt più elevata. Entrambe contengono termini con $J = 3/2$ i quali possono produrre perturbazioni. L'effetto maggiore però deriverà dalla configurazione $6s\ 6p\ 7s$ che ha un'azione preponderante sia perché energeticamente più vicina, sia perché gli elementi di matrice V_{ik}, dipendenti dagli integrali elettrostatici, sono per essa più grandi. Una valutazione precisa della perturbazione prodotta da questa configurazione non sembra possibile coi metodi numerici oggi conosciuti, i quali sono in grado di farne prevedere solo l'ordine di grandezza. Rimandando all'appendice II qualche considerazione più precisa su questo argomento, ci limitiamo qui ad osservare che nella configurazione $6s\ 6p\ 7s$ esistono termini la cui azione tende a impicciolire la sif del termine $^2P_{3/2}$ e che calcolando numericamente gli integrali di perturbazione risulta plausibile che quest'azione prevalga su quella degli altri termini in modo da produrre in definitiva una diminuzione delle sif di una novantina di millesimi di cm^{-1} che è appunto l'ordine di grandezza di quanto occorre a ristabilire l'accordo fra teoria ed esperienza.

Nella discussione delle sif elemento per elemento troveremo altri esempi di perturbazioni di tal genere a cui accenneremo via via.

Per il calcolo dei momenti magnetici nucleari è naturalmente essenziale che i termini su cui tale calcolo si basa, siano poco perturbati.

Il riconoscimento di questo fatto presenta spesso delle incertezze. Come regola generale i termini che hanno una perturbazione relativa più piccola, sono quelli che hanno di loro una grossa costante di sif e quindi particolarmente i termini provenienti da una configurazione con un elettrone *s* non saturo. In tutti gli esempi studiati non si è mai presentato il caso che due termini che sembravano a priori poco perturbati, abbiano dato valori del momento magnetico in discordanza considerevole.

Negli spettri in cui vari termini che si presentano a priori come attendibili dànno, con discreta approssimazione, lo stesso valore del momento magnetico nucleare, questo raggiunge un considerevole grado di sicurezza. Con più riserve sono da prendere i risultati per quegli spettri in cui, per mancanza di dati sperimentali, bisogna fondare il calcolo su un solo termine come per esempio nei metalli alcalini in cui il solo termine poco perturbato conosciuto è quello fondamentale dell'atomo neutro e quindi non sono possibili controlli.

§ 4. – Discussione dei dati sperimentali.

Passiamo ora in rivista vari elementi per esaminare le particolarità che presentano le loro sif e per dedurne il valore del momento magnetico nucleare che indicheremo dando il rapporto μ_0/μ tra il magnetone di Bohr e il momento magnetico nucleare.

Li $Z = 3$. – L'isotopo 7 ha $I = 3/2$. Esso è stato trattato esaurientemente da Breit [11], Güttinger e Pauli [12]. Secondo Breit $\mu_0/\mu = 575$.

Na Z = 11. – Esiste un solo isotopo di peso 23. Le sif consistono [13] in una separazione di 0,061 cm^{-1} nel termine fondamentale $3\,^2S_{1/2}$. Il momento meccanico nucleare I non è conosciuto con sicurezza. Schüler dà come probabili i valori 5/2 e 3/2. Applicando la [9] si trova

$$\frac{2\,I}{2\,I+1}\,\frac{\mu_0}{\mu} = 660$$

da cui, se si prende

$$I = 3/2$$

risulta

$$\frac{\mu_0}{\mu} = 880\,,$$

se invece si prende

$$I = 5/2$$

risulta

$$\frac{\mu_0}{\mu} = 790\,.$$

Cu Z = 29. – Le sif di questo spettro provengono in parte da uno spostamento da isotopia e in parte da un momento magnetico nucleare. Ritschl [14] che ha studiato parecchi termini del Cu I ha trovato per entrambi gli isotopi 63, 65 I = 3/2. I termini del Cu I sono di due differenti sistemi: un sistema di doppietti normali $3\,d^{10}\,ms$, $3\,d^{10}\,mp$, $3\,d^{10}\,md$, ecc. e un sistema di termini in cui uno dei 10 elettroni $3\,d$ viene allontanato da questo anello e portato in altri stati quantici; i termini più profondi di questo tipo sono $3\,d^9\,4\,s^2$. Altri termini profondi in cui è intaccato l'anello $3\,d$ appartengono alla configurazione a $(3\,d^9\,4\,s\,4\,p)$.

I termini profondi $3\,d^{10}\,4\,s\,^2S_{1/2}$ e $3\,d^9\,4\,s^2\,^2D$ hanno costanti di sif abbastanza grandi e non sono verosimilmente molto perturbati. Da essi si può calcolare μ_0/μ colle (9) e (10 *bis*), ponendo $Z^* = 19{,}6$.

Si ottiene:

dal termine	$4\,^2S_{1/2}$	$A = 0{,}195\ cm^{-1}$	$\mu_0/\mu = 760$
	$3\,^2D_{3/2}$	$A = 0{,}060$	$\mu_0/\mu = 740$
	$3\,^2D_{5/2}$	$A = 0{,}022$	$\mu_0/\mu = 870$.

Non abbiamo considerato i termini $3\,d^{10}\,4p\,^2P$ perché essi sono senza dubbio perturbati dalla configurazione a. I tre valori precedenti sono in discreto accordo fra di loro, accordo che sarebbe forse migliorato se si tenesse conto delle correzioni relativistiche per i termini D. Come valore probabile di μ_0/μ per i due isotopi del Cu riteniamo 780.

Ga Z = 31. – Isotopi 69 e 71; I = 3/2. Lo spettro del Ga I è stato studiato da Jackson [15]. Le misure non sono molto precise. Per i termini $5\,^2S_{1/2}$ e $4\,^2P_{1/2}$ egli trova separazioni di 0,09 e 0,10 rispettivamente. Jackson non trova differenza tra le strutture dei due isotopi. Dai suoi dati risulta μ_0/μ = circa 800.

Campbell [16] ha studiato le sif del Ga II che sono considerevolmente più grosse di quelle del Ga I; tanto che egli osserva una differenza tra le sif dei due isotopi sfuggita a Jackson. I dati di Campbell si riferiscono al termine 5 $^3S_1$. La costante di sif di questo termine, per l'isotopo 69 è A = 0,259 per l'isotopo 71 si ha invece A = 0,331. La valutazione di μ_0/μ da questi dati si fa con i metodi che saranno illustrati a proposito dello spettro analogo del Tl II. Si trova per l'isotopo 69, $\mu_0/\mu = 860$; per l'isotopo 71, $\mu_0/\mu = 670$. Questi risultati sono alquanto imprecisi per l'incertezza nell'attribuzione del numero quantico efficace dell'elettrone 4 *s*; tuttavia essi coincidono discretamente col risultato precedente.

Rb Z = 37. – Ha due isotopi con massa 85 e I = 5/2 e con massa 87 e I = 3/2 o 5/2. I valori di I sono dedotti dal secondo spettro [17]. Anche per questo elemento il solo termine abbastanza attendibile per la determinazione del momento magnetico nucleare conosciuto è il termine fondamentale 5 $^2S_{1/2}$ del Rb I. La separazione di questo è diversa pei due isotopi. Secondo Kopfermann si ha $\Delta\nu = 0,106$ pel Rb^{85} e 0,242 pel Rb^{87}. Da questi dati si trova $\mu_0/\mu = 1350$ pel Rb^{85} e $\frac{2\,I}{2\,I+1}\mu_0/\mu = 490$ per l'altro isotopo da cui risulta, pel Rb^{87}, $\mu_0/\mu = 590$ o 660 secondo che I = 5/2 o 3/2.

Cd Z = 48. – Gli isotopi pari 110, 112, 114, 116, hanno I = 0; quelli dispari 111 e 113 hanno I = 1/2 e le sif eguali ciò che indica che il loro momento magnetico nucleare è eguale [18]. Questo tipo di spettro è analogo a quello del Hg I e del Tl II e ne parleremo diffusamente a proposito di quest'ultimo pel quale si hanno anche estesi dati sulle sif. Basti qui accennare che le separazioni osservate per termini:

			$\Delta\nu$, cm^{-1}
5*s*	6*s*	$^3S_1$	— 0,395
5*s*	7*s*	$^3S_1$	— 0,375
5*s*	8*s*	$^3S_1$	— 0,365
5*s*	5*p*	$^3P_2$	— 0,300
5*s*	5*p*	$^3P_1$	— 0,223

sono tutte dovute per la maggior parte all'azione dell'elettrone 5 *s* non saturo; si noti che esse sono tutte invertite. Si può valutare che l'elettrone 5*s* deve avere una costante di accoppiamento individuale di circa 0,4 cm^{-1} dal che si ricava $\mu_0/\mu = 3500$. Questo valore è tuttavia assai impreciso per l'incertezza nell'attribuzione di n_{eff} all'orbita 5*s*.

In Z = 49. – Isotopo 115; I = 9/2 secondo Jackson [19]. Sono note le sif dei termini di In I

6 $^2S_{1/2}$	$\Delta\nu = 0,28\ cm^{-1}$
5 $^2P_{1/2}$	0,38

dal primo si ricava $\mu_0/\mu = 370$, dal secondo, tenendo conto di alcune correzioni relativistiche, $\mu_0/\mu = 310$ circa. Come valore più probabile di μ/μ_0 riteniamo 350.

Cs Z = 55. – Isotopo 133. I = 7/2 secondo risultati di Kopfermann [20] dedotti dall'osservazione del Cs II. Nello spettro del Cs I è nota soltanto la sif del termine fondamentale

$$6\,{}^2S_{1/2} = 0{,}315\ \text{cm}^{-1}$$

per essa valgono le stesse considerazioni che per gli altri alcalini e si ha

$$\mu_0/\mu = 700 .$$

Ba Z = 56. – Isotopi 136, 138 I = 0. Isotopo 137 I = 5/2. È conosciuta la sif del termine fondamentale del Ba II [21]. Dalle righe di risonanza risulta pel termine $6\,{}^2S_{1/2}$ la separazione di 0,272. Da questa si calcola $\mu_0/\mu = 1750$.

Le strutture dei termini $6\,{}^2P_{1/2,3/2}$, misurate con poca precisione perché non completamente risolte dànno risultati compatibili col valore precedente.

Au Z = 79. – Isotopo 197 I = 3/2 (?) [22]. Dalle righe di risonanza si trova come separazione del termine fondamentale

$$6\,{}^2S_{1/2} = 0{,}221 \text{ e } 0{,}224 .$$

Se ne deduce

$$\frac{2I}{2I+1}\frac{\mu_0}{\mu} = 760$$

e se

$$I = 3/2 \qquad \mu_0/\mu = 1010 .$$

Hg Z = 80. – Gli isotopi pari 198, 200, 202, 204, hanno I = 0; l'isotopo 199 ha I = 1/2, quello 201 I = 3/2 [23]. Le sif di quest'ultimo sono invertite indicando che per esso μ_0/μ è negativo. Per lo spettro Hg I come per quello del Cd, valgono le considerazioni che faremo a proposito del Tl II. Per Hg I isotopo 199 sono osservate le strutture dei termini:

$6s$	$7s$	$^3S_1$	$\Delta\nu = 1{,}070\ \text{cm}^{-1}$
$6s$	$8s$	$^3S_1$	1,045
$6s$	$9s$	$^3S_1$	1,030
$6s$	$6p$	$^1P_1$	— 0,181
$6s$	$6p$	$^3P_1$	0,727
$6s$	$6p$	$^3P_2$	0,758
$6s$	$6d$	$^1D_2$	0,860
$6s$	$7d$	$^1D_2$	0,496
$6s$	$6d$	$^3D_2$	— 0,470.

Tutte queste strutture si interpretano come dovute essenzialmente all'azione del solo elettrone $6\,s$ per il quale risulterebbe una costante di accoppiamento col nucleo di circa $1{,}15\ \text{cm}^{-1}$. Di qui si ricaverebbe come valore probabile di μ_0/μ per Hg^{199}, 4000, valore alquanto impreciso per le ragioni già addotte a proposito del Cd. Per l'isotopo 201 risulterebbe analogamente $\mu_0/\mu = - 3600$.

Tl Z = 81. - Gli isotopi 203, 205 hanno I = 1/2 e quasi esattamente le stesse sif a meno di uno spostamento di isotopia. Sono conosciuti gli spettri Tl I, Tl II, Tl III [24].

Cominciando da quest'ultimo che ha la struttura dello spettro di un metallo alcalino, ossia un anello 5 *d* completo e un elettrone di valenza, abbiamo calcolato μ_0/μ dai vari termini ottenendo i valori riportati nella seguente tabella:

Termine	$\Delta\nu$ cm^{-1}	μ_0/μ
$7\,s\ ^2S_{1/2}$	1,37	1310
$8\,s\ ^2S_{1/2}$	0,606	1210
$6\,p\ ^2P_{1/2}$	1,21	1060
$7\,p\ ^2P_{1/2}$	0,375	1310

Per i termini P è stata usata la formula (10 *bis*) corretta ancora per tener conto di un effetto relativistico secondo Racah. Il dato che si ricaverebbe dal termini $7\ ^2P_{3/2}$ è discordante coi precedenti, ciò che deve attribuirsi alla perturbazione su di esso esercitata dal termine $5\,d^9\ 6\,s\ 6\,p\ ^2P_{3/2}$ che gli è energeticamente prossimo e che ha una grossa costante di sif. Similmente non abbiano tenuto conto dei dati sui termini F manifestamente perturbati poiché anche le loro strutture grosse sono invertite.

Pel Tl II si hanno configurazioni $6\,s\,ms$, $6\,s\,mp$, $6\,s\,md$, ecc., le cui separazioni provengono per la massima parte dall'elettrone $6\,s$. Ciò è confermato dal paragone dei dati sperimentali con la teoria, svolta da Goudsmit e Bacher [25], delle sif di configurazioni contenenti un solo elettrone *s* non saturo accoppiato con I.

In questo caso vale la formula

$$A = a(s)\,\frac{[3/4 + S(S+1) - S'(S'+1)]\,[S(S+1) + J(J+1) - L(L+1)]}{4S(S+1)\,J(J+1)} \tag{20}$$

che permette di calcolare la sif di qualsiasi termine purché in accoppiamento di Russell-Saunders, per mezzo di un solo parametro, $a(s)$ che rappresenta la costante di accoppiamento che avrebbe l'elettrone *s* se esistesse isolato; in questa formula S' è lo spin risultante di tutti gli elettroni meno quello che dà la sif e quindi $S' = S \pm 1/2$. Le altre lettere hanno il solito significato.

Nello spettro del Tl II si trova che le sif di numerosi termini stanno fra loro approssimativamente nei rapporti indicati dalla (20). Fanno eccezione invece i termini con L = J. Per questi si hanno, in ogni configurazione, un termine di singoletti ed uno di tripletti collo stesso J e che quindi si perturbano tra di loro, dando luogo ad un accoppiamento intermedio tra quello di Russell-Saunders e quello (J , J). Nell'accoppiamento di Russell-Saunders il termine di singoletti dovrebbe sempre avere secondo la (20), una separazione di sif nulla; quella che effettivamente si osserva proviene sostanzialmente dalla perturbazione. Vale tuttavia anche in questo caso la seguente regola: la somma algebrica delle separazioni dei due termini che si perturbano è eguale a quella che risulterebbe dalla (20) per il solo termine di tripletti. Tale regola risulta effettivamente discretamente verificata dall'esperienza.

I termini con L diverso da J, dànno i valori di $a(s)$ riportati nella tabella seguente:

Termine			A	$a(s)$
$6s$	$7s$	$^3S_1$	3,32	6,64
$6s$	$9s$	$^3S_1$	3,02	6,04
$6s$	$6p$	$^3P_2$	1,29	5,16
$6s$	$7p$	$^3P_2$	1,39	5,56
$6s$	$6d$	$^3D_1$	— 1,41	5,64
$6s$	$7d$	$^3D_1$	— 1,47	5,88
$6s$	$8d$	$^3D_1$	— 1,50	6,00
$6s$	$6d$	$^3D_3$	0,95	5,70
$6s$	$7d$	$^3D_3$	0,97	5,82
$6s$	$8d$	$^3D_3$	0,98	5,88
$6s$	$5f$	$^3F_2$	— 0,99	5,94
$6s$	$6f$	$^3F_2$	— 1,03	6,18
$6s$	$5f$	$^3F_4$	0,74	5,92
$6s$	$6f$	$^3F_4$	0,74	5,92

Come si vede questa tabella $a(s)$ pei termini non perturbati è quasi costante il che conferma che le sif provengono per la maggior parte dall'elettrone $6s$. Si vede poi che la costante di accoppiamento aumenta via via che l'elettrone più esterno passa in orbite meno legate. Ciò si spiega col fatto che, allontanandosi l'elettrone luminoso dal nucleo, ne diminuisce l'azione di schermo sull'orbita $6s$ e questa per conseguenza si raccoglie intorno al nucleo, in modo che ne aumenta il valore di $\psi(0)$. I termini $6s\ 7s\ ^3S_1$ e $6s\ 9s\ ^3S_1$ hanno invece comportamento opposto a quello descritto or ora, ma ciò è dovuto al fatto che l'elettrone $7s$ contribuisce esso stesso alla sif, aumentandola di circa il 20 %, mentre l'elettrone $9s$ dà un contributo del 4 % solamente, come può valutarsi mediante la (9). In genere gli $a(s)$ della tabella sono tutti superiori alla vera costante di accoppiamento dell'elettrone $6s$ che si avrebbe se l'elettrone $6s$ si muovesse nel campo corrispondente alla configurazione $6s^2$. Tale costante non è direttamente desumibile dai dati precedenti, dai quali tuttavia si può dedurre, per estrapolazione, che essa abbia il valore di circa 5 cm^{-1}. Per mezzo della (9) si può da qui calcolare μ_0/μ conoscendo n_{eff} per l'orbita in questione. Questo deve ricavarsi per estrapolazione dato che il termine $6s^2$ non esiste nel sistema dei tripletti, si ottiene così $n_{eff} = 1,56$. Risulta allora dalla (9) $\mu_0/\mu = 1400$ in buon accordo coi valori ricavati dal Tl III.

Nel Tl I sono stati analizzati alcuni termini tra cui il fondamento $6s^2\ 6p\ {}^2P_{1/2}$ e i termini $6s^2\ 7p\ {}^2P_{3/2}$ e $6s\ 7s\ {}^2S_{1/2}$. Con le solite formule si può calcolare μ_0/μ dal termine $7\ {}^2S_{1/2}$ e si trova 1380; del termine $6\ {}^2P_{1/2}$ senza tener conto della correzione di Breit, la quale sia come ordine di grandezza che come segno è sufficiente a ristabilire l'accordo, 950. Il termine $6\ {}^2P_{3/2}$ è invece fortemente perturbato come è stato rilevato nel paragrafo precedente e non è utilizzabile per il calcolo di μ_0/μ.

Dal complesso di questi dati sufficientemente concordanti tra di loro, reputiamo come valore probabile pel Tl

$$\mu_0/\mu = 1300. \tag{21}$$

Pb Z = 82. – Gli isotopi pari 204, 206, 208, hanno I = 0 l'isotopo 207 ha I = 1/2 e le sif sono state studiate negli spettri Pb I, Pb, II, Pb III [26]. Di quest'ultimo è nota solo la sif del termine $6s\ 7s\ {}^2S_1$ che ha la separazione di 2,25 cm^{-1}, essa è dovuta, come nel termine analogo del Tl II, in gran parte all'elettrone 6 *s*. L'elettrone 7 *s* vi contribuisce per circa 1/5 della sif stessa. Da questo dato si deduce $\mu_0/\mu = 3500$ circa.

Del Pb II sono stati studiati vari termini delle configurazioni $6s^2\ ms$, $6s^2\ mp$, $6s^2\ md$ e delle configurazioni $6s\ 6p^2$.

Essi sono in gran parte perturbati da termini di una configurazione $6s\ 6p\ 7s$ la quale ha sif assai grandi dovute all'elettrone $6s$ e che quindi fa sentire la sua influenza anche su termini relativamente lontani, analogamente a quanto accade nel Tl I. Un calcolo esatto di tali effetti è praticamente impossibile per modo che le sif di questi termini non si prestano per una valutazione del momento magnetico nucleare.

Similmente inadatto per calcoli quantitativi è lo spettro del Pb I. In conclusione riteniamo 3500 come vale probabile di μ_0/μ pur essendo esso alquanto incerto.

Bi Z = 83. – Isotopo 209 I = 9/2. Di questo elemento sono note le strutture degli spettri Bi I, Bi II, Bi III e Bi V [27]. Quest'ultimo ha struttura analoga a quella degli alcalini ed è nota la separazione di sif del termine fondamentale $6\ {}^2S_{1/2}$ che ha $\Delta\nu = 13,2\ cm^{-1}$. Prendendo secondo McLennan $n_{eff} = 2,35$ si trova da questo dato $\mu_0/\mu = 510$. Benché le sif del Bi siano tra le meglio conosciute, esse non si prestano tuttavia ad una determinazione esatta del momento magnetico nucleare perché le configurazioni elettroniche sono intrecciate tra di loro negli spettri del Bi I e Bi II. Anche lo spettro del Bi III non è adatto a considerazioni quantitative perché la configurazione $6s^2\ 7s$ è certamente perturbata dall'altra $6s\ 6p^2$ che le è energeticamente prossima e nemmeno molto attendibili sono i dati che si ricavano sull'accoppiamento dell'elettrone $6s$ dalla configurazione $6s\ 6p^2$ mancando conoscenze sicure sopra l'accoppiamento intermedio di questa configurazione tra gli schemi di Russel–Saunders e (J , J). Tuttavia anche le costanti di accoppiamento degli elettroni $6s$ e $7s$ sembrano confermare il valore dato testè di μ_0/μ.

Nella Tabella segue raccogliamo i dati sul valore del rapporto μ_o/μ pei vari nuclei. Dove non è conosciuto con sicurezza il valore di I abbiamo indicato anche il valore di $\frac{2\,I}{2\,I+1}\,\frac{\mu_o}{\mu}$ che si ricava indipendentemente dalla conoscenza di I. Nella colonna intestata « Peso » è indicato il numero di termini che dànno, in sufficiente concordanza tra di loro, il valore di μ_o/μ.

Elemento	Z	M	I	μ_o/μ	Peso	Osservazioni
Li	3	7	3/2	575	—	Secondo Breit [28]
Na	11	23	3/2?	880?	1	$\frac{2\,I}{2\,I+1}\,\frac{\mu_o}{\mu} = 660$
Cu	29	63, 65	3/2	780	3	
Ga	31	69	3/2	860	1	
Ga	31	71	3/2	670	1	
Rb	37	85	5/2	1350	1	
Rb	37	87	5/2?	590?	1	$\frac{2\,I}{2\,I+1}\,\frac{\mu_o}{\mu} = 490$
Cd	48	111, 113	1/2	— 3500	5	
In	49	115	9/2	350	2	
Cs	55	133	7/2	700	1	
Ba	56	137	5/2	1750	1	
Au	79	197	3/2?	1010	1	$\frac{2\,I}{2\,I+1}\,\frac{\mu_o}{\mu} = 760$
Hg	80	199	1/2	4000	7	
Hg	80	201	3/2	— 3600	7	
Tl	81	203, 205	1/2	1300	20	
Pb	82	207	1/2	3500	1	
Bi	83	209	9/2	510	1	

Si potrebbe esser tentati di trarre dalla tabella precedente qualche deduzione circa la struttura dei nuclei. Non ci sembra tuttavia che la scarsa precisione dei dati e le numerose lacune possano permettere di arrivare ad alcun risultato conclusivo in questo senso. Una osservazione ovvia è che nella tabella tutti i nuclei di numero atomico pari e peso atomico dispari i quali, secondo la teoria di Heisenberg, contengono un numero pari di protoni e dispari di neutroni hanno un momento magnetico nucleare nettamente più piccolo degli altri con un numero dispari di protoni e pari di neutroni. Nemmeno questa però sembra una regola assoluta perché ad essa quasi certo fa eccezione il K^{39}, non incluso nella tabella; per questo elemento dalle intensità alternate negli spettri di bande [29], risulterebbe $I \neq 0$, mentre le sif che si può supporre debbano esistere in esso, non sono mai state risolute.

APPENDICE I.

Dimostrazione della formula (7).

La determinazione di $\psi_{\text{eff}}(0)$ si riduce sostanzialmente al calcolo di

$$\int_0^\infty \mathrm{FG}\, dr. \tag{22}$$

Integrando le equazioni differenziali di F e G si trova che in prossimità dell'origine F e G si comportano rispettivamente, a meno di un fattore di normalizzazione, come $r^{-\beta}$ e $\frac{Z\alpha}{2-\beta}r^{-\beta}$ dove

$$\beta = 1 - \sqrt{1 - \alpha^2 Z^2} \quad \text{e} \quad \alpha = 2\pi e^2/hc.$$

Raccordandosi nell'origine ai valori sopra indicati si ottengono autofunzioni non normalizzate che chiameremo Φ e Γ. F e G sono legati a Φ e Γ dalle relazioni

$$\mathrm{F} = \rho\Phi \qquad \mathrm{G} = \rho\Gamma \tag{23}$$

in cui ρ è scelto in modo che

$$\rho^2 \int_0^\infty (\Phi^2 + \Gamma^2)\, d\tau = 1. \tag{24}$$

Per calcolare l'integrale (22) occorre quindi conoscere il fattore di normalizzazione ρ. All'uopo osserviamo che a grande distanza dal nucleo G si identifica con l'autofunzione di Schrödinger poiché le correzioni relativistiche diventano trascurabili. Posto $u = \psi r$ in cui ψ è l'autofunzione di Schrödinger si ha per u l'equazione differenziale

$$\frac{d^2 u}{dr^2} + \frac{8\pi^2 m}{h^2}(\mathrm{E} - e\mathrm{V})\, u = 0 \tag{25}$$

che può risolversi col metodo di Wentzel–Brillouin ponendo

$$u = k\mathrm{R} \sin\theta \tag{26}$$

in cui k è un fattore di normalizzazione che viene determinato in modo che la u calcolata col metodo di Wentzel–Brillouin si raccordi colla Γ calcolata a partire dallo sviluppo in serie, e R e $d\theta/dr$ sono funzioni di r che obbediscono alla condizione

$$\mathrm{R}^2 \frac{d\theta}{dr} = 1. \tag{27}$$

Esse si calcolano col metodo di approssimazioni successive di Wentzel–Brillouin il quale non è convergente, ma solo semiconvergente; tuttavia limitandosi ad un numero opportuno di approssimazioni del metodo (4 nel nostro caso) si riesce ad ottenere una precisazione nella definizione di R e $d\theta/dr$ sufficiente per questo calcolo.

Studiamo l'effetto sugli autovalori di (25) di una perturbazione del potenziale V consistente nell'aggiunta di un termine di perturbazione della forma $\lambda\delta(r-\xi)$ in cui δ è la solita funzione impropria che ha valori $\neq 0$ solo per valori nulli dell'argomento e ξ un valore della distanza dal nucleo abbastanza grande perché ivi siano già trascurabili gli effetti delle correzioni relativistiche.

L'energia di perturbazione ΔE può calcolarsi in due modi:

a) col solito metodo delle perturbazioni;

b) osservando che una discontinuità del potenziale produce un salto brusco $\Delta\theta$ nella fase della (26) e quindi sposta la posizione dei nodi di u che sono legati al numero quantico n.

Eguagliando il risultato delle due considerazioni si ottiene senza difficoltà:

$$\Delta E = 4\pi\lambda\rho^2 u^2(\xi) = -\frac{\Delta\theta}{\pi}\frac{dE}{dn}. \tag{28}$$

Perché questa formula abbia un senso occorre naturalmente che E sia una funzione regolare di n in modo che possa definirsi ragionevolmente dE/dn.

Dalla (28) si deduce il valore di ρ non appena sia noto $\Delta\theta$ che passiamo appunto a calcolare. Dalla (25) si ha, indicando con Δu e $\Delta u'$ il salto nella u e nella sua derivata prima indotti dalla perturbazione

$$\Delta u = 0 \tag{29}$$

$$\Delta u' = \frac{8\pi^2 m}{h^2}\lambda u(\xi)$$

e quindi tenendo presenti le [26] e [27]

$$\Delta\theta = -\frac{8\pi^2 m\lambda}{h}\frac{\sin^2\theta}{\frac{d\theta}{dr}} \tag{30}$$

con questo valore del $\Delta\theta$ la [28] dà

$$\rho^2 = \frac{2m}{h^2 k^2}\frac{dE}{dn}. \tag{31}$$

Il coefficiente k che compare in questa formula, è stato ottenuto da estese tabelle numeriche delle autofunzioni ∞s che nella regione interna dell'atomo coincidono sensibilmente con quelle ottiche. Esse dànno direttamente il valore del coefficiente k calcolato raccordando per $x = 0{,}3$ (x è la distanza dal nucleo misurata colla unità μ del metodo statistico) le autofunzioni ottenute con uno sviluppo in serie relativistico con quelle calcolate col metodo di Wentzel–Brillouin. Tale calcolo è stato eseguito per 15 atomi convenientemente distribuiti nel sistema periodico [30].

Conosciuto ρ l'integrale (20) può calcolarsi servendosi degli sviluppi in serie precedentemente citati per Φ e Γ.

Abbiamo eseguito questo calcolo per tre valori di Z ottenendo fattori di correzione

$$\begin{aligned} 3{,}30 &\text{ per } Z = 92 \\ 1{,}93 &\text{ » } Z = 74 \\ 1{,}18 &\text{ » } Z = 42 \end{aligned}$$

rispetto alla formula

$$\psi^2(0) \frac{1}{\pi a^3} \frac{Z}{2\,\mathrm{R}h} \frac{d\mathrm{E}}{dn}$$

che nel caso di termini Rydberghiani si riduce alla formula (6) di Goudsmit-Landé.

I fattori di correzione precedenti si lasciano interpolare con esattezza pari a quella del calcolo numerico dalla formula empirica

$$\frac{1}{\left[1 - \left(\frac{Z}{139}\right)^2\right]^2} . \tag{32}$$

Questa correzione è di origine relativistica e sparirebbe eseguendo i calcoli non relativisticamente. Essa coincide sensibilmente con quella data da Racah e Breit [4]. La nostra formula empirica presenta su quella di questi ultimi Autori il vantaggio di una maggior rapidità di calcolo.

In definitiva si trova la formula (7) del testo (*).

$$\psi^2_{\mathrm{eff}}(0) = \frac{1}{\pi a^3} \frac{1}{\left(1 - \frac{Z^2}{139^2}\right)^2} \frac{Z}{2\,\mathrm{R}h} \frac{d\mathrm{E}}{dn}$$

Ringraziamo il dott. E. Majorana per varie discussioni relative ai calcoli di questa Appendice.

APPENDICE II

Anomalia del termine $6\,^2P_{3/2}$ del Tl I.

Lo scopo dei seguenti calcoli è solo di far vedere che la perturbazione della separazione di sif del termine $6\,^2P_{3/2}$ del Tl I è sufficiente per spiegare le anomalie che si osservano. Essi non hanno una pretesa quantitativa, ma sono dedicati a una sola valutazione dell'ordine di grandezza della perturbazione. In questo ordine di idee ci limitiamo qui a considerare due sole configurazioni $6s^2\, 6p$ a cui appartiene il termine considerato e $6s\, 7s\, 6p$ che produce la perturbazione, benché naturalmente l'intervento delle altre configurazioni possa alterare ulteriormente in maniera considerevole i risultati.

Innanzi tutto osserviamo che, elettrostaticamente, si perturbano tra loro solo stati collo stesso M. Potremo quindi fare il calcolo per un valore fisso opportunamente scelto di M e il risultato è lo stesso che si otterrebbe con qualsiasi altro valore di M.

(*) Il numero 139 deve essere sostituito con 137 (n.' d. Red.).

Indicheremo le autofunzioni individuali dei singoli elettroni con

$$(33) \qquad \begin{cases} sa & \sigma a & p_1 a & p_0 a & p_{-1} a \\ sb & \sigma b & p_1 b & p_0 b & p_{-1} b \end{cases}$$

in cui a e b sono le autofunzioni dello spin corrispondenti ad $m_s = \pm 1/2$ rispettivamente; s e σ autofunzioni delle coordinate per l'elettrone $6s$ e $7s$ e $p_1\, p_0\, p_{-1}$ le autofunzioni delle coordinate per elettroni p con $m_l = +1, 0, -1$ rispettivamente.

Se si sceglie $M = 3/2$ la configurazione $6\, s^2\, 6\, p$ fornisce una sola autofunzione dal simbolo

$$(34) \qquad \{sa\, sb\, p_1\, a\} = \psi_0$$

in cui colle graffe s'intende che il prodotto va antisimmetrizzato. Essa è l'autofunzione imperturbata $6\, {}^2P_{3/2}$.

La configurazione $6s\ 7s\ 6p$ dà 4 autofunzioni con $M = 3/2$ e precisamente:

$$(35) \qquad \begin{aligned} \{sb\, \sigma a\, p_1 a\} &= \psi_1 \\ \{sa\, \sigma b\, p_1 a\} &= \psi_2 \\ \{sa\, \sigma a\, p_1 b\} &= \psi_3 \\ \{sa\, \sigma a\, p_0 a\} &= \psi_4 . \end{aligned}$$

L'autofunzione imperturbata si cambierà in prima approssimazione in

$$(36) \qquad \Psi_0 = \psi_0 + \frac{1}{\Delta E} (V_{01}\, \psi_1 + V_{02}\, \psi_2 + V_{03}\, \psi_3 + V_{04}\, \psi_4)$$

in cui ΔE è la differenza di energia tra le configurazioni $6s^2\ 6p$ e $6s\ 7s\ 6p$ e coincide quindi colla differenza di energia tra l'orbita $6s$ e $7s$ ed è pertanto negativa. La sua grandezza può stimarsi in 7 volt. Le V_{0i} sono gli elementi di matrice dell'energia elettrostatica che connettono lo stato o collo stato i.

Posto:

$$(36) \qquad \begin{cases} G = \left(s\sigma \left| \frac{e^2}{r} \right| ss\right) = \int \frac{e}{r_{12}} \sigma^2 (1)\, \sigma (2)\, s (2)\, d\tau_1\, d\tau_2 \\ H = \left(sp_1 \left| \frac{e^2}{r} \right| \sigma p_1\right) = \int \frac{e}{r_{12}} s (1)\, \sigma (1)\, | p (2) |^2\, d\tau_1\, d\tau_2 \\ K = \left(sp_1 \left| \frac{e^2}{r} \right| p_1 \sigma\right) = \int \frac{e}{r_{12}} s_1 (1)\, p_1 (2)\, p_1 (1)\, \sigma (2)\, d\tau_1\, d\tau \end{cases}$$

si trova che

$$(37) \qquad \begin{cases} V_{01} = -G - H + K \\ V_{02} = G + H \\ V_{03} = -K \\ V_{04} = 0 \end{cases}$$

Una valutazione numerica degli integrali elettrostatici ha dato col metodo statistico

$$G = -0{,}84 \text{ volt} \qquad H = 0{,}43 \qquad K = 0{,}24$$

da cui

$$V_{01} = 1{,}51 \text{ volt} \qquad V_{02} = -1{,}27 \qquad V_{03} = -0{,}24 \qquad V_{04} = 0 .$$

Questi valori sono assai imprecisi perché il metodo statistico non fornisce buoni resultati per il calcolo di integrali elettrostatici, tutta via essi dànno un'indicazione circa gli ordini di grandezza e il comportamento qualitativo della perturbazione.

L'autofunzione normalizzata di prima approssimazione a norma della (36) risulta numericamente:

$$\Psi_0 = 0{,}96\,\psi_0 - 0{,}22\,\psi_1 + 0{,}18\,\psi_2 + 0{,}03\,\psi_3. \tag{38}$$

Partendo ora da questa autofunzione bisogna calcolare il valor medio del campo magnetico nel nucleo o meglio la grandezza $H_z\,\mu/I$ che a norma della (1) determina la sif. All'uopo cominciamo a valutare i suoi elementi di matrice per le autofunzioni individuali. Servendoci della stessa (1) e dei dati sperimentali che dànno una costante di sif di 0,404 cm^{-1} per l'elettrone $7\,s$ si può dedurre che la costante di sif dell'elettrone $6\,s$ sarà circa 3,23 cm^{-1} tenendo presente che la costante di sif è inversamente proporzionale ai cubi dei numeri quantici efficaci. Dal dato sperimentale sull'elettrone $6\,p_{1/2}$ si deduce che l'elettrone $6\,p_{3/2}$ imperturbato avrebbe una costante di sif di 0,035. Con questi dati risulta

$$\begin{aligned} \left(\sigma a \,\middle|\, H_z \frac{\mu}{I} \,\middle|\, \sigma a\right) &= 0{,}20\ \mathrm{cm}^{-1} \\ \left(s a \,\middle|\, H_z \frac{\mu}{I} \,\middle|\, s a\right) &= 1{,}62 \\ \left(p_1 a \,\middle|\, H_z \frac{\mu}{I} \,\middle|\, p_1 a\right) &= 0{,}052. \end{aligned} \tag{39}$$

Per valutare l'elemento non diagonale $\left(sa \,\middle|\, H_z \frac{\mu}{I} \,\middle|\, \sigma a\right)$ si può osservare che in approssimazione non relativistica, gli elementi diagonali sono proporzionali a $s^2(0)$ e $\sigma^2(0)$; l'elemento non diagonale è invece come si verifica facilmente proporzionale a $s(0)\,\sigma(0)$ ed è dato, in questa approssimazione, dalla media geometrica dei due. Siccome poi nel computo degli integrali elettrostatici le autofunzioni s e σ sono state scelte con segni opposti, nella media geometrica si deve prendere il segno meno. Si ha così

$$\left(sa \,\middle|\, H_z \frac{\mu}{I} \,\middle|\, \sigma a\right) = -0{,}57.$$

Da queste formule segue che:

$$\begin{aligned} \left(\frac{\mu}{I} H_z\right) sa &= 1{,}62\, sa - 0{,}57\, \sigma a + \cdots \\ \left(\frac{\mu}{I} H_z\right) sb &= -1{,}62\, sb + 0{,}57\, \sigma b + \cdots \\ \left(\frac{\mu}{I} H_z\right) \sigma a &= 0{,}20\, \sigma a - 0{,}57\, sa + \cdots \\ \left(\frac{\mu}{I} H_z\right) \sigma b &= -0{,}20\, \sigma b + 0{,}57\, sb + \cdots \\ \left(\frac{\mu}{I} H_z\right) p_1 a &= 0{,}052\, p_1 a + \cdots . \end{aligned}$$

I puntini indicano ulteriori termini che non hanno importanza per il nostro calcolo perché non incrociano coi termini a cui abbiamo limitato le nostre considerazioni.

È d'uopo ora passare dalle autofunzioni individuali alle $\psi_0, \psi_1, \psi_2, \psi_3$. Tenendo presenti le loro espressioni (35) e che esse, ripetiamo, rappresentano prodotti antisimmetrizzati si ottiene:

$$\left(\frac{\mu}{I} H_z\right) \psi_0 = 0{,}052\, \psi_0 + 0{,}57\, \psi_1 + 0{,}57\, \psi_2$$

$$\left(\frac{\mu}{I} H_z\right) \psi_1 = 0{,}57\, \psi_0 - 1{,}37\, \psi_1$$

$$\left(\frac{\mu}{I} H_z\right) \psi_2 = 0{,}57\, \psi_0 \qquad\qquad + 1{,}47\, \psi_2$$

$$\left(\frac{\mu}{I} H_z\right) \psi_3 = \qquad\qquad\qquad + 1{,}82\, \psi_3 .$$

L'ultimo coefficiente è inesatto perché è stato trascurato il termine $\left(p_1 b \left| \frac{\mu}{I} H_z \right| p_1 b\right)$ si riconosce subito però che ciò porta uno spostamento insignificante dei risultati dato che ψ_3 interviene nell'autofunzione perturbata (38) nella percentuale dell'1 ‰ circa.

Il calcolo di $\left(\Psi_0 \left| \frac{\mu}{I} H_z \right| \Psi_0\right)$ con questi dati non offre più difficoltà; si trova

$$\left(\frac{\mu}{I} H_z\right) = 0{,}027\, \psi_0 + 0{,}85\, \psi_1 + 0{,}81\, \psi_2 + 0{,}05\, \psi_3$$

e

$$\left(\Psi_0 \left| \frac{\mu}{I} H_z \right| \Psi_0\right) = -0{,}013$$

ciò che darebbe secondo la (1) un $\Delta\nu$ di $-0{,}017\ \mathrm{cm}^{-1}$ cioè una perturbazione di $0{,}087\ \mathrm{cm}^{-1}$ poiché abbiamo assunto $0{,}070\ \mathrm{cm}^{-1}$ come $\Delta\nu$ imperturbato, mentre per ottenere il $\Delta\nu$ sperimentale sarebbe occorsa una correzione di $0{,}062\ \mathrm{cm}^{-1}$ nello stesso senso.

Risulta chiaramente come l'ordine di grandezza delle perturbazioni sia quello delle discrepanze tra teoria elementare ed esperienza. Per le ragioni più volte addotte un calcolo quantitativo trascende le attuali possibilità, tuttavia anche i semplici computi sopra scritti levano ogni solida base alle obiezioni contro l'attribuzione delle sif all'effetto di un momento magnetico nucleare, salvo effetti di un ordine di grandezza più piccolo a cui abbiamo già accennato [31]. Allo stato attuale tutto fa ritenere che la spiegazione delle varie anomalie vada ricercata nell'effetto delle perturbazioni nell'atomo e che bisogna esser guardinghi nell'attribuire ad effetti nucleari le divergenze osservate.

Aggiunta alla correzione: È uscito nel frattempo un lavoro di S. Goudsmit, « Phys. Rev. », *43*, 636 (1933), i cui risultati coincidono con una parte dei nostri. Anche la tabella dei momenti magnetici nucleari data da Goudsmit è sostanzialmente eguale alla nostra.

BIBLIOGRAFIA.

[1] Un ottimo riassunto sullo stato attuale del problema delle strutture iperfini si trova nell'articolo di H. KALLMANN e H. SCHÜLER in «Ergeb. der Exakt. Naturw.», *11*, 134, (1932).
[2] S. GOUDSMIT ed E. BACK, «Zs. f. Phys.», *43*, 321 (1927); P. ZEEMAN, E. BACK, S. GOUDSMIT, ibid., *66*, 1 (1930).
[3] E. FERMI, «Zs. f. Phys.», *60*, 320 (1930); J. HARGREAVES, «Proc. Roy. Soc.», *127*, 141, 407 (1930); H. CASIMIR, non pubblicato.
[4] G. RACAH, «Zs. f. Phys.», *71*, 431 (1931); G. BREIT, «Phys. Rev.», *38*, 463 (1931).
[5] S. GOUDSMIT, «Phys. Rev.», *37*, 663, 1931; S. GOUDSMIT e R. BACHER, «Phys. Rev.», *34*, 1500 (1929).
[6] Un riassunto si trova in S. GOUDSMIT, *Rendiconto del Convegno di fisica nucleare del Convegno Volta*. Roma, 1931.
[7] G. C. DARWIN, «Proc. Roy. Soc.», *118*, 654 (1928).
[8] L. PAULING e S. GOUDSMIT, *Structure of Line Spectra* vedi anche J. C. McLENNAN, A. B. McLAY, M. F. CRAWFORD, «Proc. Roy. Soc.», *133*, 652 (1931).
[9] G. BREIT, «Phys. Rev.», *37*, 51 (1931).
[10] S. GOUDSMIT, «Phys. Rev.», *34*, 1500 (1929); G. RACAH, «Zs. f. Phys.», *71*, 431 (1931).
[11] G. BREIT, «Phys. Rev.», *36*, 1732 (1930).
[12] P. GÜTTINGER, «Zs. f. Phys.», *64*, 749 (1930); P. GÜTTINGER e W. PAULI, ibid., *67*, 743 (1931).
[13] H. SCHÜLER, «Naturw.», *16*, 512 (1928); H. SCHÜLER ed H. BRÜCK, «Zs. f. Phys.», *58*, 735 (1929).
[14] R. RITSCHL, «Zs. f. Phys.», *79*, 1 (1932).
[15] D. A. JACKSON, «Zs. f. Phys.», *75*, 229 (1932) (Ga I).
[16] J. S. CAMPBELL, «Nature», *131*, 204 (1933) (Ga II).
[17] H. KOPFERMANN, «Naturw.», *21*, 24 (1933).
[18] H. SCHÜLER e J. E. KEYSTON, «Zs. f. Phys.», *71*, 413 (1931).
[19] D. A. JACKSON, «Zs. f. Phys.», *80*, 59 (1933).
[20] H. KOPFERMANN, «Zs. f. Phys.», *73*, 437 (1931).
[21] P. GERALD KRUGER, R. C. GIBBS, R. C. WILLIAMS, «Phys. Rev.», *41*, 322 (1932).
[22] R. RITSCHL, «Naturw.», *19*, 690 (1931).
[23] H. SCHÜLER e E. KEYSTON, «Zs. f. Phys.», *72*, 423 (1931).
[24] H. SCHÜLER, J. E. KEYSTON, «Zs. f. Phys.», *70*, 1 (1931) (Tl I e Tl II); J. C. Mc LENNAN, F. CRAWFORD, «Proc. Roy. Soc.», *132*, 10 (1931) (Tl II); J. C. McLENNAN, E. J. ALLIN, «Proc. Roy. Soc.», *129*, 43 (1930) (Tl III).
[25] S. GOUDSMIT e R. BACHER, «Phys. Rev.», *34*, 1500 (1929).
[26] H. KOPFERMANN, «Zs. f. Phys.», *75*, 363 (1932); H. SCHÜLER ed E. G. JONES, «Zs. f. Phys.», *75*, 563 (1932) (Pb I e Pb II); vedi anche J. L. ROSE, L. P. GRANATH, «Phys. Rev.», *40*, 760 (1932); J. C. McLENNAN, M. F. CRAWFORD, L. B. LEPPARD, «Nature», *128*, 301 (1931) (Pb III).
[27] E. BACK, S. GOUDSMIT, «Zs. f. Phys.», *47*, 174 (1928); P. ZEEMAN, E. BACK, S. GOUDSMIT, «Zs. f. Phys.», *66*, 1, (1930) (Bi I); R. FISHER, S. GOUDSMIT, «Phys. Rev.», *37*, 1057, 1674 (1931) (Bi II e III); ARVIDSSON, «Nature», *126*, 565 (1930) (Bi V).
[28] G. BREIT, «Phys. Rev.», *36*, 1732 (1930).
[29] F. W. LOOMIS, R. W. WOOD, «Phys. Rev.», *38*, 854 (1931).
[30] Cfr. un lavoro di prossima pubblicazione di E. AMALDI ed E. FERMI.
[31] Tali potrebbero essere per esempio le anomalie rilevate da H. CASIMIR «Zs. f. Phys.», *77*, 811 (1932) nello spettro del Hg.

N° 76, 80 *a*, 80 *b*, 80 *c*.

The better to appreciate the contribution brought by Fermi with his beta-decay theory, it is necessary to recapitulate the history of ideas on the subject.

Until the neutron was discovered in early 1932, it was assumed that nuclei consisted of protons and electrons, although speculations about the existence of a heavy neutral particle were advanced as early as 1920 (1). Even the name "neutron" had already been used in this sense (2).

The quantum-mechanical difficulties stemming from the confinement of electrons within a volume of nuclear dimensions and the continuous beta-ray spectra were the two great stumbling blocks of nuclear theory. Two possible alternatives were suggested. Bohr (3) assumed that electrons in the nucleus could not be described within the framework of quantum mechanics, that they lost their property of determining the statistics and spin of the nucleus as a whole (it was known that Li^6 and N^{14} nuclei, against expectation, were bosons and had integral spins) and that energy was not conserved when electrons were involved in nuclear processes. On the other hand, Pauli was unwilling to abandon energy conservation and proposed the simultaneous emission of the electron and one (or more) neutral particles in beta-decay. This suggestion was first made informally in a letter addressed by Pauli on December 4, 1930 to the group of nuclear physicists at the University of Tübingen, chiefly H. Geiger and Lise Meitner. It was next mentioned in the discussions at a physics meeting in Pasadena in the spring of 1931, and again at Ann Arbor the following summer, but the author considered the proposal too tentative to justify its appearance in published records (4).

Nevertheless, the idea soon spread among the community of theoretical physicists, and although Pauli had given no name to his hypothetical neutral particle, it became known among them as the "neutron". The use of the same word for a heavy particle was mainly confined to the Rutherford school. In October of the same year (1931), the *Convegno di Fisica nucleare* sponsored by the *Fondazione Volta* of the *Reale Accademia d'Italia* took place in Rome. The meeting was attended by most of the active experimental and theoretical workers in nuclear physics, such as Blackett, Bothe, Bohr, Marie Curie, Ehrenfest, Ellis, Fermi, Geiger, Goudsmit, Heisenberg, Lise Meitner, Pauli, Sommerfeld, and others. Here the beta-decay process was discussed and while Bohr suggested energy non-conservation (3), Fermi was favorably impressed by the Pauli hypothesis. At this time the word "neutron" was being used more and more often for the heavy neutral particle, whose discovery was impending, and it was felt that another term should designate the "Pauli particle" in order to avoid confusion. The name "neutrino" was jokingly suggested by Fermi in a conversation with the other Rome physicists. It must be explained that the endings *-one* and *-ino* can be appended to many Italian nouns and adjectives to denote bigness and smallness, respectively. The Italian word for the neutron, *neutrone*, suggests a compound of *neutro*, neutral, and *-one*, meaning "a large neutral object"; correspondingly *neutrino* would mean "a small neutral object". The word was mentioned at the meeting and soon became the official name for the neutral beta-decay particle.

The next important nuclear conference was the *Septième Conseil de Physique* of the *Institut International de Physique Solvay*, which met in Brussels in October 1933. This was another plenary meeting of the theoretical and experimental authorities on nuclei. Bohr, Blackett, Bothe, Chadwick, Cockcroft, Marie Curie, L. De Broglie, Dirac, Ellis, Fermi, Ga-

(1) E. Rutherford, «Proc. Roy. Soc.», *A-97*, 324 (1920).

(2) J. Chadwick, «Proc. Roy. Soc.», *A-136*, 692 (1932).

(3) N. Bohr, *Convegno di Fisica Nucleare*, «R. Accad. d'Italia», p. 119. Roma 1932.

(4) The writer is indebted for these and other details of the history of beta-decay theories to a correspondence with the late Professor Wolfgang Pauli.

mow, Heisenberg, Irène Joliot-Curie, F. Joliot, Lawrence, Meitner, Pauli, and Rutherford were present among others. By that time, the neutron had been discovered and the model of the nucleus composed of protons and neutrons had been reasonably well established through the work of Heisenberg, Ivanenko and Majorana. In the proceedings of that meeting for the first time the neutrino hypothesis appears in print, in a comment by Pauli (5) to the report of Heisenberg. He leaves the question open whether the neutrino should be attributed a finite rest mass; but definitely suggests a spin 1/2 and Fermi statistics, in order to allow spin and statistics conservation in beta-decay; of course, energy and momentum conservation is also assumed. Thus Fermi was present at these historic discussions on nuclear structure and beta-decay, and although his role at the time was mainly confined to comments on problems of nuclear forces, it is certain that he had given deep thought to beta-decay theory. In fact, a paper (N° 76) announcing his complete results on beta-spectra already appeared in the December 1933 issue of "Ricerca Scientifica" (there is no indication of the date on which the article was submitted to that journal). This was only 2 months after the Solvay conference.

To understand the fundamental contribution brought by Fermi with his beta-decay theory, it must be emphasized that the idea of the neutrino had remained up to that time a rather vague hypothesis, while the construction of a formal theory had never been attempted. When Pauli first made his suggestion in 1930, electrons were believed to exist in the nucleus, and the hypothetical neutral particle was by analogy considered as another nuclear constituent (with small but finite rest mass). Only vague hints at the possible creation of the electron (and the neutrino) in the beta-decay process are found in the proceedings of the 1933 Solvay conference.

While thus well acquainted with the problem, Fermi at the same time acquired the theoretical tools that would lead him to the solution. He had been thoroughly studying the Dirac radiation theory, as evidenced by the celebrated summary of that subject (N° 67) published in 1932. Apparently he had some difficulty with the Dirac-Jordan-Klein method of the second quantization of fields, but eventually also mastered that technique and considered a beta-decay theory as a good exercise on the use of creation and destruction operators. He also made use of the isotopic spin formalism, recently invented by Heisenberg and later to prove of great usefulness in view of the charge independence of the strong interactions.

The theory that he built on these foundations is remarkable for its ability to withstand almost unchanged two and a half decades of revolutionary advances in nuclear physics. One might say that seldom was a physical theory born in such definitive form.

Fermi explicitly pointed out that there were many possible forms of an interaction Hamiltonian that would lead to beta-decay processes. His choice was dictated by simplicity and by analogy with the interaction of charges and currents with the electromagnetic field. Accordingly he chose the (polar) vector interaction. It was later shown by Wigner that the possible relativistically invariant interactions bilinear in the electron and neutrino wave functions (and not containing the gradients of these functions) are of five types (or any linear combination thereof). The particular choice did not affect the main results of Fermi's calculations, the shape of the allowed spectra (entirely determined by the "statistical factor" and Coulomb effects) and the *ft*-values for these same transitions. However, the selection rules obtained were too restrictive. Gamow and Teller later showed that other interaction forms (axial vector and tensor) produce different selection rules for allowed transitions. The experimental evidence has proved that the beta-decay interaction must be a mixture, in about equal parts, of a Fermi (scalar or vector) and a Gamow-Teller term. At present (1959) it seems fairly well established that the Fermi interaction is of vector type (as he originally assumed) and the Gamow-Teller term is axial vector. With this extension, the Fermi theory successfully explains quantitatively a number of phenomena well beyond those calculated in the original papers. One may mention the spectral shapes and lifetimes for transitions of various order of forbiddenness, positron

(5) W. PAULI, *Structure et Propriétés des Noyaux Atomiques*, p. 324, Rapports et discussions du Septième Conseil de Physique, Institut International de Physique Solvay, Gauthier-Villars, Paris 1934.

decay and orbital electron capture, and also processes other than beta-decay of nucleons, but all having the common characteristic of being due to "weak interactions" between four fermions. Among such processes are the decay of the muon and the reaction of a negative muon with a proton. It appears compatible with the present experimental evidence to describe all these processes by the same "universal Fermi interaction". One modification of the beta-decay theory that recent evidence made necessary is the addition of parity-non-conserving terms.

Fermi intended to announce the results of his beta-decay theory in a letter to "Nature", but the manuscript was rejected by the Editor of that journal as containing abstract speculations too remote from physical reality to be of interest to the readers. He then sent a somewhat longer paper (N° 76) to "Ricerca Scientifica" where it was promptly published. This article includes all the essential results, showing that the calculations (including the numerical value of *ft*) had been completed. The longer papers in "Nuovo Cimento" and "Zeitschrift für Physik" were sent to the respective journals very early in 1934.

F. RASETTI.

76.

TENTATIVO DI UNA TEORIA DELL'EMISSIONE DEI RAGGI «BETA»

«Ric. Scientifica», *4* (2), 491–495 (1933).

RIASSUNTO. — Teoria della emissione dei raggi β delle sostanze radioattive, fondata sull'ipotesi che gli elettroni emessi dai nuclei non esistano prima della disintegrazione ma vengano formati, insieme ad un neutrino, in modo analogo alla formazione di un quanto di luce che accompagna un salto quantico di un atomo. Confronto della teoria con l'esperienza.

Mi propongo di esporre qui i fondamenti di una teoria dell'emissione dei raggi β che, benché basata sopra ipotesi delle quali manca al momento presente qualsiasi conferma sperimentale, sembra tuttavia capace di dare una rappresentazione abbastanza accurata dei fatti e permette una trattazione quantitativa del comportamento degli elettroni nucleari che, se pure le ipotesi fondamentali della teoria dovessero risultare false, potrà in ogni caso servire di utile guida per indirizzare le ricerche sperimentali.

È ben noto che nel cercare di costruire una teoria dei raggi β si incontra una prima difficoltà dipendente dal fatto che i raggi β escono dai nuclei radioattivi con una distribuzione continua di velocità che si estende fino a una certa velocità massima: ciò che a prima vista non sembra conciliabile col principio della conservazione dell'energia. Una possibilità qualitativa di spiegare i fatti senza dovere abbandonare il principio della conservazione dell'energia consiste, secondo Pauli, nell'ammettere l'esistenza del così detto «neutrino», cioè di un corpuscolo elettricamente neutro con massa dell'ordine di grandezza di quella dell'elettrone o minore. In ogni disintegrazione β si avrebbe emissione simultanea di un elettrone e di un neutrino; e l'energia liberata nel processo si ripartirebbe comunque tra i due corpuscoli in modo appunto che l'energia dell'elettrone possa prendere tutti i valori da zero fino ad un certo massimo. Il neutrino d'altra parte, a causa della sua neutralità elettrica e della piccolissima massa, avrebbe un potere penetrante così elevato da sfuggire praticamente ad ogni attuale metodo di osservazione. Nella teoria che ci proponiamo di esporre ci metteremo dal punto di vista della ipotesi dell'esistenza del neutrino.

A parte la difficoltà della distribuzione continua delle energie, una teoria dei raggi β incontra un'altra difficoltà essenziale nel fatto che le presenti teorie delle particelle leggere non spiegano in un modo soddisfacente come queste possano venir legate in modo stabile o quasi stabile nell'interno di un nucleo, dato il piccolo volume di questo.

La via più semplice per la costruzione di una teoria che permetta una discussione quantitativa dei fenomeni in cui intervengono gli elettroni nucleari, sembra in conseguenza doversi ricercare nella ipotesi che gli elettroni non esistano come tali nel nucleo prima della emissione β, ma che essi, per così dire, acquistino esistenza nell'istante stesso in cui vengono emessi; allo stesso modo come un quanto di luce emesso da un atomo in un salto quantico non si può in alcun modo considerare preesistente nell'atomo prima del processo di emissione. In questa teoria dunque il numero totale degli elettroni e dei neutrini (al pari del numero totale dei quanti di luce nella teoria della irradiazione) non sarà necessariamente costante, potendosi avere dei processi di creazione o di distruzione delle particelle leggere.

Secondo le idee di Heisenberg, considereremo le particelle pesanti, neutrone e protone, come due stati quantici connessi a due possibili valori di una coordinata interna ρ della particella pesante. Ad essa attribuiremo il valore $+1$ se la particella è un neutrone e -1 se la particella è un protone.

Cercheremo poi una espressione dell'energia di interazione tra le particelle leggere e quelle pesanti che consenta delle transizioni tra i due valori $+1$ e -1 della coordinata ρ, e cioè la trasformazione di un neutrone in protone o viceversa; in modo tale però che alla trasformazione da neutrone a protone sia di necessità connessa la creazione di un elettrone, che si osserva come particella β, e di un neutrino; mentre alla trasformazione inversa da protone a neutrone sia connessa la scomparsa di un elettrone e di un neutrino; così come nella teoria della irradiazione a un certo salto quantico dell'atomo è connessa l'emissione di un quanto di luce, mentre al salto quantico opposto è connesso l'assorbimento di un quanto.

Il formalismo matematico più semplice per costruire una teoria in cui il numero delle particelle leggere (elettroni e neutrini) non sia necessariamente costante, si ha nel metodo di Dirac-Jordan-Klein delle « ampiezze di probabilità quantizzate ». In questo formalismo le ampiezze di probabilità ψ degli elettroni e φ dei neutrini, e le loro complesse coniugate ψ^* e φ^*, vengono considerate come operatori non commutativi che agiscono sopra le funzioni dei numeri di occupazione degli stati quantici degli elettroni e dei neutrini; in modo tale che l'operatore ψ determina transizioni in cui il numero degli elettroni cala di una unità, mentre l'operatore complesso coniugato ψ^* determina transizioni opposte in cui si ha aumento di uno del numero totale degli elettroni. Nelle ordinarie applicazioni del metodo, naturalmente, gli operatori ψ e ψ^* figurano sempre associati uno all'altro, per modo che, nei processi che si considerano, il numero totale dei corpuscoli resti costante. Nella teoria presente invece la possibilità della variazione del numero degli elettroni si ottiene introducendo i due operatori opposti ψ e ψ^* in termini separati della energia di interazione.

Dobbiamo ancora introdurre due altri operatori Q e Q* che operano sulle funzioni della variabile a due valori ρ come le sostituzioni lineari

$$Q = \begin{vmatrix} 0 & 1 \\ 0 & 0 \end{vmatrix} \qquad Q^* = \begin{vmatrix} 0 & 0 \\ 1 & 0 \end{vmatrix}$$

Questi operatori, come facilmente si riconosce, determinano rispettivamente le transizioni da protone a neutrone e da neutrone a protone.

Una espressione dell'energia di interazione che associ necessariamente alle transizioni da neutrone a protone (operatore Q*) la creazione di un elettrone e di un neutrino (operatori ψ^* e φ^*) e alla transizione da protone a neutrone (operatore Q) la scomparsa di un elettrone e di un neutrino (operatori ψ e φ) ha come sua forma più generale la seguente:

$$H = QL\ (\psi \cdot \varphi) + Q^* L^* (\psi^* \cdot \varphi^*)$$

in cui L rappresenta una espressione bilineare in ψ e φ, che può eventualmente contenere anche le coordinate, i momenti e le coordinate di spin della particella pesante. Una limitazione alle possibilità di scelta di L si ottiene cercando quelle espressioni che, al variare del sistema di riferimento, si comportano come la componente temporale di un quadrivettore polare. Si può dimostrare che la più semplice di tali espressioni è la seguente:

$$L\,(\psi \cdot \varphi) = g\,(\psi_2 \varphi_1 - \psi_1 \varphi_2 + \psi_3 \varphi_4 - \psi_4 \varphi_3)$$

dove le ψ e le φ sono le quattro componenti della ampiezza di probabilità relativistica per gli elettroni e i neutrini; e g rappresenta una costante di proporzionalità. Siccome le conseguenze di questa scelta si mostrano in buon accordo con i fatti sperimentali non occorre per il momento ricorrere ad espressioni più complicate.

Prenderemo dunque in definitiva come espressione della interazione la seguente.

$$H = g\,\{Q\,(\psi_2 \varphi_1 - \psi_1 \varphi_2 + \psi_3 \varphi_4 - \psi_4 \varphi_3) + Q^*\,(\psi_2^* \varphi_1^* - \psi_1^* \varphi_2^* + \psi_3^* \varphi_4^* - \psi_4^* \varphi_3^*)\}\,. \tag{1}$$

In questa espressione ψ e le φ (considerate come operatori) debbono prendersi nel punto dello spazio occupato dalla particella pesante. La costante g che figura nella (1) ha le dimensioni $L^5 M T^{-2}$.

Prendendo la (1) come espressione della energia di interazione si può costruire una teoria delle disintegrazioni β, con metodi simili a quelli usati nella teoria dell'irradiazione per calcolare la vita media di uno stato eccitato di un atomo. Senza entrare qui in dettagli matematici a proposito di questa teoria, ci limiteremo solo ad osservare che la lunghezza d'onda di de Broglie, per particelle leggere aventi energie non superiori ad alcuni milioni di volt, è grande a confronto delle dimensioni nucleari. Ne segue che in prima approssimazione potremo trascurare le variazioni di ψ e φ per punti differenti del nucleo; ciò corrisponderebbe nella teoria della radiazione a trascurare l'irradiazione di quadrupolo. Facendo questa approssimazione, si trova che la vita media per un processo di disintegrazione β in cui un neutrone legato in una orbita con autofunzione u_n si trasforma in un protone appartenente a uno stato quantico v_m emettendo un raggio β e un neutrino, è data da:

$$\frac{1}{\tau} = 1{,}65 \cdot 10^{95} g^2 (10^{12} r)^{-0{,}4} \left| \int u_n v_m^* d\tau \right|^2 F(\eta_0) . \tag{2}$$

In questa espressione $mc\eta_0$ rappresenta il massimo momento degli elettroni emessi, che è determinato dalla differenza di energia tra i due stati u_n e v_m della particella pesante, r è una lunghezza dell'ordine di grandezza di 10^{-12} cm il cui valore esatto dipende dalle ipotesi che si fanno sopra l'andamento del potenziale elettrico nell'interno del nucleo. La funzione F ha una espressione analitica alquanto complicata; per piccoli valori dell'argomento si comporta quasi esattamente come $\eta_0^6/24$ mentre per argomenti maggiori prende i valori dati dalla seguente tabella:

η_0	$F(\eta_0)$	η_0	$F(\eta_0)$
1	0,03	5	80
2	1,2	6	185
3	7	7	380
4	29		

Questi valori, come anche la formula (2), sono stati calcolati per il numero atomico Z = 82; essi però non variano considerevolmente entro il piccolo intervallo di numeri atomici rappresentati nelle famiglie radioattive. Nella (2) inoltre la massa del neutrino è stata presa eguale a zero. Poiché da un confronto delle curve teoriche e sperimentali che dànno la distribuzione continua dell'energia delle particelle β, si trova che, affinché esse concordino, è necessario ammettere che la massa del neutrino sia molto minore di quella dell'elettrone; l'ipotesi più semplificativa consiste nel porla addirittura eguale a zero.

L'unico elemento incerto nella (2) è l'elemento di matrice

$$q = \int u_n v_m^* d\tau$$

per calcolare il quale occorrerebbe conoscere le autofunzioni u_n e v_m dei due stati del neutrone e del protone entro il nucleo. Si può tuttavia affermare che q sarà dell'ordine di grandezza dell'unità; solo in casi di particolari simmetrie delle due autofunzioni u_n e v_m, q potrà essere esattamente zero. Questi casi corrispondono alle transizioni proibite dell'ottica, in cui si annulla l'elemento corrispondente alla transizione considerata della matrice che rappresenta il momento elettrico. In questi casi la approssimazione che ha condotto alla (2) non è più sufficiente e si debbono prendere in considerazione anche i termini che dipendono dalle variazioni della ψ e della φ entro l'estensione del nucleo. Un calcolo di questo genere, del tutto analogo alla considerazione dei termini di quadrupolo nella teoria della irradiazione, porta a prevedere in questi casi che le intensità della transizione β siano, come ordine di grandezza, qualche centinaio di volte più piccole che nel caso normale.

Dobbiamo dunque attenderci che se si formano per diversi corpi che si disintegrano emettendo raggi β i prodotti τ F (η_0), questi siano tutti dello stesso ordine di grandezza; salvo i casi in cui la transizione β, nel senso che abbiano spiegato, non possa avvenire in prima approssimazione: in questi casi il prodotto τF (η_0) dovrà prendere valori qualche centinaio di volte più grandi che nei casi normali.

Nella tabella seguente sono riuniti i prodotti τF (η_0) per tutte le sostanze che emettono raggi β per le quali si hanno dati sufficienti; τ è espresso in ore.

Elemento	$\tau F(\eta_0)$	Elemento	$\tau F(\eta_0)$
Ra B	0,62	Ra C	130
Th B	1,7	Ra E	770
Th C″	2,9	Ms Th_2	180
Ac B	0,7	Th C	150
Ac C″	1,4		
UX_1	3,3		
UX_2	3,3		

In questo prospetto sono chiaramente riconoscibili i due gruppi che avevamo attesi. Si noti anche che elementi che occupano posizioni omologhe nelle differenti famiglie radioattive appartengono sempre al medesimo gruppo.

Da questi risultati è anche possibile calcolare almeno l'ordine di grandezza della costante g. Esso risulta $5 \cdot 10^{-50}$.

La teoria permette infine di calcolare la curva di distribuzione delle velocità delle particelle β. Siccome i dati sperimentali, particolarmente per quanto riguarda la parte della curva di piccola energia, sono alquanto contraddittori tra di loro, non è possibile basare su di essi un controllo preciso della teoria; l'accordo qualitativo si mostra tuttavia buono. Dalla teoria risulta in particolare che la curva di distribuzione per piccole velocità dovrebbe tendere a zero più rapidamente per il caso degli elementi in cui la transizione è proibita in prima approssimazione, e cioè per quelli appartenenti al secondo gruppo della tabella precedente che non per gli altri. I dati sperimentali non sembrano essere sufficientemente definiti per un controllo di questo punto.

Una esposizione più estesa di questa teoria e ulteriori suoi risultati verranno pubblicati prossimamente in altro luogo.

77 *b*.

ON THE RECOMBINATION OF ELECTRONS AND POSITRONS

E. FERMI and G. E. UHLENBECK, « Phys. Rev. », *44*, 510–511 (1933)

The discovery of the positron has opened the possibility of a physical interpretation of the negative energy levels in Dirac's relativistic theory of the electron. In particular the formation of a positron-electron pair under a hard γ-radiation, as observed by Anderson, has been interpreted by Oppenheimer and Plesset (1), as a kind of photoeletric effect, whereby the energy of the absorbed γ-quantum raises an electron from a negative energy state to a positive one, forming a hole or positron and an electron. The reverse process, recombination of a positron and an electron under emission of γ-radiation, has been suggested by Blackett and Occhialini (2), as a possible explanation of the radiation observed by Gray and Tarrant (3) in the scattering of hard γ-rays. According to these authors, the scattered radiation contains besides the normal Compton scattering, two fairly monochoromatic components with energies of about 0.5 and 1×10^6 volts, i.e., very nearly mc^2 and $2mc^2$. The intensity of the hard component is smaller than that of the soft component, increases somewhat with the atomic number Z, and is, for lead, 1/3 of the total radiation. The recombination energy of an electron and a positron, both with negligible kinetic energies, is the sum of the rest masses or $2mc^2 \cong 10^6$ volts. Because of the necessary conservation of energy and momentum this energy, in the case that both particles are free, can only be emitted in the form of two quanta of the same energy ($= 0.5 \times 10^5$ volt). This type of radiation would therefore explain the soft component. For the hard component the energy must be emitted in one quantum, which is only possible for a strongly bound electron, where the nucleus can take up the recoil momentum of the quantum.

According to this explanation the mechanism of the scattering would be as follows. First the primary γ-quantum is absorbed forming a pair. The high velocity positron which is formed loses its kinetic energy by collisions, and at the end of its path there is a chance of its being destroyed by either of the two processes mentioned above. The ratio of the intensities of the two components should be equal to the ratio of the probabilities of the two

(1) OPPENHEIMER and PLESSET, « Phys. Rev. », *44*, 53 (1933). See also BECK, « Zeits. f. Physik », *83*, 498 (1933).

(2) BLACKETT and OCCHIALINI, « Proc. Roy. Soc. », *A 139*, 699 (1933).

(3) GRAY and TARRANT, « Proc. Roy. Soc. », *A 136*, 662 (1932).

processes. Now it has been shown by several authors that the rate of destruction of a low velocity positron by the two quanta process is:

$$R_2 = N\pi e^4/m^2 c^3 = 7.5 \times 10^{-15}\, N, \tag{1}$$

here N is the electron density. The nuclear repulsion prevents the positron om reaching the inner parts of the atoms. Therefore not all electrons are ective, so that N will lie between n and nZ, n being the number of atoms unit volume. For lead $n = 3.3 \times 10^{22}$ which gives as lower limit for $= 2.5 \times 10^8$. To explain the intensity of the hard component we must refore expect for the one quantum process a rate of the order 10^8.

We have calculated the cross section for the destruction of the positron y the one quantum process. Since we are dealing with low velocity positrons ne might expect that a nonrelativistic approximation [4] will give at least the right order of magnitude. We find as the contribution to the cross section due to the K-shell of the atom, neglecting the screening by the atomic electrons:

$$\sigma_K = (4/3)\, \pi^2 \alpha^9 a^2 Z^6 (1 + Z^2/W) \{\exp (2\pi Z/W^{1/2}) - 1\}^{-1}, \tag{2}$$

where α is the fine structure constant, a is Bohr's radius and W is the energy of the positron expressed in Rydbergs. The contribution of the L-shell is considerably smaller; we find:

$$\sigma_L = (1/24)\, \pi^2 \alpha^9 a^2 Z^6 (4 + 7 Z^2/W) \{\exp (2\pi Z/W^{1/2}) - 1\}^{-1}. \tag{3}$$

From these cross sections we find the rate of destruction by multiplying with nv. For positrons of low velocity it is not permissible to neglect the screening, because the exponential factor which is the main term in (2) and (3) and is due to the repulsion of the positron by the nuclear charge, is essentially reduced by the screening. Using the statistical model one can show that instead of $\exp(-2\pi Z/W^{1/2})$ one gets:

$$\exp\{-2.66\, Z^{1/3} \varphi (0.442\, W/Z^{4/3})\}, \tag{4}$$

where $\varphi(\xi)$ has the following values: $\varphi(0) = 4.18$; $\varphi(0.0018) = 3.13$; $\varphi(0.0158) = 2.63$; $\varphi(0.122) = 1.96$; $\varphi(0.425) = 1.48$; $\varphi(1.31) = 1.06$ and $\varphi(\xi) = (\pi/2)(\xi + 1)^{-1/2}$ for larger values of ξ. For example in lead ($Z = 82$) we get for respectively: $W = 1$, 100, 10,000, and 75,000 ($= 10^6$ e V.) the following rates for destruction: $R_1 = 2 \times 10^{-5}$, 10, 5×10^7, 10^9. The last value is only given as an indication, since the velocity is so high that (2) is no longer applicable.

The first value shows that the original explanation of the hard component as due to positrons which have completely lost their initial velocity cannot be maintained. We get a rate too small by a factor 10^{13}. One might perhaps still expect a very hard component due to the destruction of positrons at the beginning of their path, since the probability for destruction is then much

(4) Nonrelativistic means here that nonrelativistic radial eigenfunctions are used, and that one keeps only the terms with the smallest power of $1/c$. This corresponds to the first formula of Oppenheimer and Plesset, which we have been able to check except that we got a result larger by a factor 2.

higher. For an estimate, take positrons of 10^6 volts, which have a range in lead of 0.06 cm. The time required is about 3×10^{-12} sec, so that even if we assume during this time the maximum rate of 10^9 we get as total probability of destruction by the one quantum process 0.003. This would give of course radiation of about 2×10^6 volts. The probability is still very small, but one might perhaps hope that the relativistic corrections will increase this result appreciably.

We do not think however that this is very likely, because one needs an increase by a factor 100 to get an observable result. To estimate very roughly what the influence of relativity can be, we have made a numerical calculation of the cross section for destruction of positrons of 10^6 volt in lead. To simplify the calculations we have only considered the contribution due to transitions of s-states of the positron to s-states of the electron. In nonrelativistic approximation the rate is 0.4×10^9. The relativistic calculation gave instead 2.3×10^7. This of course does not necessarily mean that the relativistic result will really be smaller than the one given by the approximate formula (2), because we have many more transitions to take into account. But it makes one doubt if really the exact calculation will considerably increase the nonrelativistic result.

Finally one should of course not forget the probability of destruction by the two quantum process before the positron has lost its kinetic energy. This would give rise to a continuous band above 0.5×10^6 volts. The upper limit for this probability is however for lead and 10^6-volt positron only 5 percent and the actual value is probably less.

In conclusion our results are that it is difficult to reconcile the explanation of the hard component of Gray and Tarrant as due to the destruction of positrons with the Dirac theory.

Department of Physics,
University of Michigan,
August 18, 1933.

N° 78.

Discussion of this paper offers the opportunity of describing Fermi's activities in experimental nuclear physics for the period 1930–34, previous to the discovery of the radioactivity induced by neutrons.

It is not generally known that Fermi not only intended to learn the techniques of nuclear physics, but spent considerable time in the process several years before his name became famous in that field with the discovery of neutron-induced activities.

His intention of entering the nuclear field first became manifest in the winter of 1930–31, although his theoretical work at the time followed entirely different lines. He set himself as a first task the construction and operation of a cloud chamber, with the help of E. Amaldi. The weakest point of the Rome physics department at the time was its poorly equipped and inefficiently staffed machine shop; only the simplest instruments could be built there, and even these at the cost of inordinately long time and much prodding by the physicists. These circumstances induced Fermi to make use of the "do-it-yourself" methods that were characteristic of him both in theoretical and experimental work. In order to minimize shop work and build a cloud chamber with his own hands aided only by the most elementary tools, he first inspected several hardware stores and bought assorted kitchenware and gas plumbing. The cloud chamber was to be of the floating piston type, the seal being formed by water. Accordingly the cylinder and piston were two aluminum pans of slightly different diameters. A gas valve, rotated by a motor and gear, periodically established communication between the space under the piston and a vacuum pump, causing expansions at 15-second intervals. Alpha-particle tracks were soon observed, but their quality was poor owing to the slowness of the expansion and difficulty of clearing the field of old tracks. No attempt to take photographs was made. In the spring of 1931 Fermi gave up the cloud chamber project and went back to theoretical work. It is regrettable that lack of shop facilities prevented him from achieving the success that his experimental skill deserved.

In the fall of 1931 Rasetti went to Berlin-Dahlem for a year to learn nuclear techniques under Professor Lise Meitner. During his residence there, the papers of Bothe and Bekker, Curie-Joliot and Chadwick, leading from the mysterious "Po + Be" radiation of the former authors to Chadwick's identification of the neutron, were published. Rasetti acquired experience in the preparation of neutron sources and in neutron detection both by cloud-chamber and counter methods.

In the fall of 1932, Fermi and Rasetti organized a joint program of research in nuclear physics in Rome. To minimize the drawback of the inadequate machine shop, several instruments were designed and their construction was "farmed out" in a private shop in Rome. A rather large cloud chamber, essentially designed after those in use in Berlin-Dahlem, worked excellently as soon as it was assembled. The same shop built the mechanical parts of the gamma-ray crystal spectrometer described in the present paper.

These developments were made possible by a grant from the Consiglio Nazionale delle Ricerche, which had raised the research budget of the department to an amount of the order of $ 2000 to $ 3000 per year; a fabulous wealth when one considers that the average for physics departments in Italian universities was about one-tenth of that amount.

By the end of 1933 the nuclear instrumentation in Rome was adequate for research along several lines. Rasetti [1] had separated a strong source of RaD from a radium solution, further separated the polonium, and mixing the latter with beryllium powder had prepared a neutron source comparable to the most powerful ones then in use elsewhere. Proton recoils and nitrogen disintegrations were readily observed in the above-mentioned cloud chamber. Various types of counters were also in operation.

(1) F. RASETTI, «Ricerca Scient.», *5–1*, 3 (1934).

In March, 1934, after the discovery of the alpha-induced radioactivity had been announced by Curie and Joliot, Fermi suggested to Rasetti that they try to observe similar effects with neutrons. Near the end of that month several elements were irradiated with the Po + Be source and tested for activity by means of a thin-walled Geiger-Mueller counter, with a totally negative result obviously due to lack of intensity. When the enhancing effect of hydrogenous media was later discovered, activities induced by the Po + Be source could be readily observed.

At the beginning of April, Rasetti left for Morocco while Fermi continued the experiments. The idea, essential to success, then occurred to him that in order to observe a neutron-induced activity it was not necessary to use a Po + Be source; a much stronger Rn + Be source could be employed, since its beta and gamma radiations were no objection to the observation of a delayed effect. Radon sources were already familiar to Fermi since they had been supplied previously by Professor Trabacchi for use with the gamma-ray spectrometer. All one had to do was to prepare a similar source in a glass bulb filled with beryllium powder. Experiments with this type of source led to an almost immediate positive result.

After these introductory remarks there remains little to be mentioned in connection with the paper under discussion. The resolution attained was superior to that previously achieved by using rocksalt crystals. However, it was realized that the method was of little promise as a tool for the investigation of gamma-ray spectra owing to the extremely strong sources required to photograph the lines even with exceedingly long exposures. Attempts to use Geiger-Mueller counters as detectors were not successful. However, the crystal method, as later revived by Du Mond and co-workers using much improved techniques, yielded the most accurate gamma-ray wavelength measurements available at present.

F. Rasetti.

78.

UNO SPETTROGRAFO PER RAGGI «GAMMA» A CRISTALLO DI BISMUTO

E. Fermi e F. Rasetti, «Ric. Scientifica», *4* (2), 299–302 (1933).

Riassunto. — Si dà notizia di una tecnica speciale per la preparazione di grossi cristalli singoli di bismuto, accennando al loro impiego nella spettrografia dei raggi «gamma». Si conclude che l'uso di questo cristallo dà risultati superiori a quelli che si ottengono col salgemma, comunemente adoperato. Sono state fotografate numerose righe γ del Radio B + C, raggiungendo le più alte frequenze già osservate col metodo del cristallo.

Sebbene lo studio dei raggi gamma si faccia più spesso per mezzo dei raggi beta secondari, cioè dei fotoelettroni generati nell'atomo stesso che si disintegra dalla «conversione» del quanto γ, pur tuttavia l'analisi diretta degli spettri γ mediante l'interferenza nei cristalli presenta molto interesse. Infatti, per esempio, mentre dalla misura del numero di elettroni presenti in ciascuna riga dello spettro β secondario non si ha senz'altro la intensità della corrispondente riga γ, poiché vi è di mezzo il cosidetto «coefficiente di conversione» il quale varia da una riga all'altra, dall'altra parte le misure dirette per mezzo di un cristallo permettono di misurare queste intensità. Dal confronto tra i due tipi di misure si può dedurre appunto il detto coefficiente di conversione, dato assai importante per i problemi della struttura nucleare. Inoltre la misura col cristallo si presta meglio alla determinazione dei valori assoluti delle frequenze, ed è scevra da incertezze di interpretazione che talvolta si presentano negli spettri dei raggi β secondari.

Per queste ragioni abbiamo deciso di riprendere lo studio sperimentale degli spettri dei raggi γ, come parte di un programma di ricerche di fisica nucleare che l'Istituto Fisico di Roma sta svolgendo sotto gli auspici del Consiglio Nazionale delle Ricerche. Sebbene questa ricerca non sia ancora compiuta, pure crediamo utile dare qualche notizia sui risultati già raggiunti, i quali possono presentare un certo interesse.

I precedenti ricercatori che si sono occupati degli spettri dei raggi γ, e cioè Rutherford e Andrade (2), Thibaud (3) e Frilley (4) si sono serviti esclusivamente di cristalli di salgemma. Convinti che ormai la tecnica avesse raggiunto il massimo progresso realizzabile con questo cristallo, abbiamo pensato se non fosse possibile ottenere risultati migliori servendosi di cristalli di elementi pesanti, nei quali le radiazioni penetrano soltanto in uno strato sottile della sostanza, con vantaggio della definizione delle righe.

Esaminati i vari cristalli possibili, la scelta è caduta sopra il bismuto metallico, per varie ragioni: alto peso atomico, facilità di ottenere cristalli singoli, facilità di sfaldatura, costante reticolare non troppo grande.

Per ottenere gli spettri dei raggi γ è necessario avere una faccia cristallina di grandi dimensioni, dato l'angolo di incidenza piccolissimo sotto cui si deve lavorare e la necessità di avere la massima intensità possibile. Abbiamo perciò dovuto sviluppare una tecnica per la preparazione di grossi cristalli singoli di Bi e per isolarne una faccia cristallina parallela ai piani reticolari 111 e di grandi dimensioni (2×8 cm.). Dopo molti tentativi abbiamo adottato il procedimento seguente, con ottimi resultati. Se il bismuto fuso in una stufa elettrica viene raffreddato molto lentamente (abbassando la temperatura della stufa pochi gradi sotto il punto di fusione) si ottiene spesso un cristallo unico, purché si abbia cura che l'espansione che ha luogo nella solidificazione non produca sforzi meccanici al contatto con le pareti; ciò si può realizzare usando una forma di steatite o di gesso a sezione triangolare. Occorre ora avere precauzioni speciali perché i piani 111 del cristallo siano orientati parallelamente all'asse longitudinale del blocco di bismuto, in modo da poter poi isolare una faccia delle dimensioni desiderate. Per questo scopo ci siamo serviti del metodo del seme, già studiato accuratamente dal Goetz e dai suoi collaboratori (5). Avendo ottenuto un gran numero di piccoli cristalli singoli di Bi di forma cilindrica, se ne sceglie uno in cui per caso i piani 111 sono paralleli all'asse del cilindro, e questo si usa come germe. Si fonde il bismuto da far cristallizzare nella sua forma in contatto del cristallo–germe, avendo cura che anche una parte di quest'ultimo venga rifusa. Se ora si fa raffreddare il tutto lentamente, disponendo il gradiente termico nella stufa in modo che l'estremità dalla parte del germe sia

(2) E. Rutherford e N. da C. Andrade, «Phil. Mag.», *27*, 854; *28*, 262 (1914).
(3) J. Thibaud, Thèse, Paris 1925.
(4) M. Frilley, Thèse, Paris 1928.
(5) A. Goetz, «Phys Rev.», *35*, 193 (1930).

sempre la più fredda, la cristallizzazione dell'intera massa procede nell'orientazione del germe. Ottenuto il cristallo, il germe vien tagliato e può essere usato per nuovi cristalli indefinitamente.

Ottenuto il cristallo singolo con l'orientazione desiderata, occorre sfaldarlo per liberare una faccia 111. Data la facile deformabilità dei cristalli di Bi, ciò richiede alcune cure particolari. Trovammo che il miglior metodo era di sfaldare successivi strati di circa 1/2 mm di spessore mediante una lama affilata. Queste sottili fette si arricciano nello sfaldarle, senza che si deformi il resto del cristallo. Compiuta la sfaldatura, la faccia presenta però irregolarità dannose per la definizione degli spettri. Esse vennero tolte mediante pulitura della faccia con carta smerigliata fina, operazione che si trovò assai vantaggioso eseguire alla temperatura dell'aria liquida, poiché altrimenti un certo strato superficiale del cristallo resta danneggiato. Ciò non

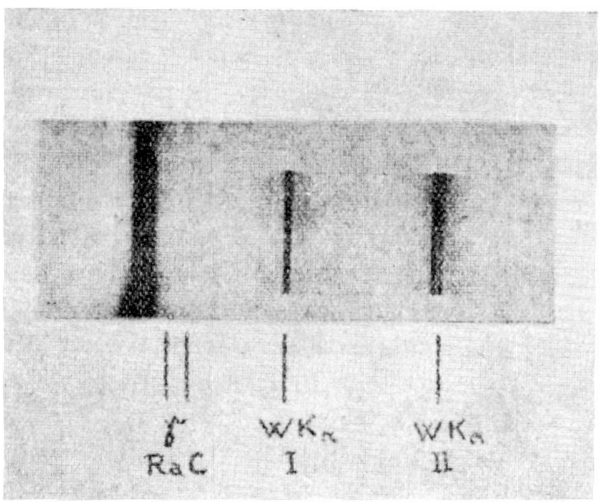

ostante, si ha un aumento di potere riflettente se in ultimo si toglie ancora un sottile strato per mezzo di acido nitrico diluito. Allora la faccia presenta un aspetto sericeo uniforme dovuto alla corrosione dell'acido e, come mostrò il controllo eseguito continuamente con i raggi X in tutte le fasi dell'operazione, dà il massimo di intensità e regolarità nella riflessione.

Il cristallo era fissato sopra un goniometro in modo tale da possedere i necessari gradi di libertà. La fenditura dello spettrografo era costituita da due blocchi di bronzo e aveva la profondità di cm 20; quando si usava un preparato radioattivo tale fenditura serviva esclusivamente da limitazione per proteggere la lastra fotografica dai raggi diretti, il preparato essendo sufficientemente sottile per servire da fenditura esso stesso. Davanti alla fenditura poteva applicarsi un tubo a raggi X del tipo Müller-Metalix (ad anticatodo di tungsteno) il quale serviva per mettere a punto l'apparecchio e per misurare le righe γ servendosi come riferimento delle radiazioni K del tungsteno.

Fu usata una distanza di 65 cm tra preparato e lastra fotografica. L'asse di rotazione del cristallo si trovava a metà di tale distanza conformemente alla condizione di Bragg.

Furono impiegate lastre Lumière «Opta» e uno schermo di rinforzo, a tungstato di calcio, del tipo «Ampli». Le lastre vennero sviluppate con glicina diluita.

Le sorgenti radioattive consistevano in tubetti capillari di vetro, della lunghezza di 15 mm e del diametro interno di 0,2 ÷ 0,3 mm, riempiti di emanazione di radio, appositamente preparati dal prof. G. C. Trabacchi dell'Istituto Fisico della Sanità Pubblica. Senza i mezzi di cui dispone questo Istituto e senza la continua e volenterosa collaborazione del prof. Trabacchi questa ricerca non sarebbe stata possibile, e perciò desideriamo esprimergli anche qui la nostra gratitudine.

L'intensità delle sorgenti disponibili variava tra 100 e 150 millicurie. La lastra era esposta per sei a otto giorni, in modo che praticamente tutta la radiazione era utilizzata. Durante le pose il cristallo non veniva ruotato, perché, date le sue dimensioni, si otteneva in una posizione sola un intervallo spettrale abbastanza grande. Dopo la posa si ruotava il cristallo portandolo nelle posizioni di riflessione delle righe K del W, nel primo e nel secondo ordine, e si registravano queste righe sulla lastra per riferimento (con circa tre minuti di posa, 5 mA, e 110 kV).

Fu dapprima studiata la regione delle righe γ molli (righe K di fluorescenza degli elementi 82 e 83), poi le regioni di più alta frequenza, fino a 20 U. X. Le righe sono molto sottili, in modo che il potere risolutivo appare almeno uguale a quello raggiunto da Frilley, sebbene nel suo caso la distanza sorgente–lastra fosse quasi tripla. Ciò mostra la perfezione notevolmente maggiore del cristallo di bismuto (sebbene non usato in forma di lamina sottile) in confronto a quello di salgemma. Nella figura che precede diamo una riproduzione di uno spettrogramma contenente le righe del RaC di 20 e 35 U. X. Osserviamo che per la particolare disposizione degli atomi nel reticolo cristallino del bismuto, dalla faccia 111 vengono riflessi con grande intensità gli ordini pari, e perciò tale reticolo funziona in pratica come se esso avesse una costante metà di quella vera, che è di 3,94 Å. Tutte le righe γ osservate appartengono infatti al secondo ordine.

Le più alte frequenze osservate, corrispondenti a 16 e 20 U. X., sono anche le più alte raggiunte dai precedenti ricercatori. Tuttavia, a Frilley sono occorse sorgenti assai più forti (700 mc) ed è probabile che se potremo disporre di una sorgente di questa intensità riusciremo a superare il limite attuale. Anche se questo non sarà possibile, resta molto lavoro da compiere. Si tratta di passare ad una fase quantitativa, eseguendo misure precise di lunghezze d'onda, che ancora scarseggiano, e magari misure di intensità, sostituendo alla lastra fotografica una camera di jonizzazione o un contatore di Geiger–Müller.

Ci è grato ringraziare il Consiglio Nazionale delle Ricerche per il largo aiuto concessoci nello svolgimento di questo lavoro.

79.

LE ULTIME PARTICELLE COSTITUTIVE DELLA MATERIA

«Atti Soc. It. Progr. Sci.», 22ª Riunione, Bari 1933-XI vol. 2, 7–14; «Scientia», *55*, 21–28 (1934-XII).

Forse la più essenziale delle differenze tra gli oggetti del mondo macroscopico, cioè gli oggetti comuni e gli oggetti del microscopico mondo degli atomi è la seguente:

Nel mondo macroscopico non si trovano mai due oggetti eguali. Consideriamo per esempio due pezzi di ferro: noi potremo ridurli ad avere lo stesso peso e la stessa forma esteriore; potremo anche cercare di rendere il più possibile eguale la grana della loro struttura microcristallina, lo stato di tempera, il contenuto delle varie impurità e così di seguito. Evidentemente però mai potremo sperare che i due pezzi di ferro siano ridotti ad essere completamente eguali, e la ragione di questa impossibilità è da ricercarsi nella estrema complessità degli oggetti in esame, costituiti da aggregati di miliardi di miliardi di atomi e di molecole: basta che uno di questi atomi in uno dei due pezzi di ferro sia spostato rispetto all'atomo corrispondente dell'altro pezzo, perché i due oggetti non possano più chiamarsi identici. In questo senso dunque la non esistenza di corpi identici nel mondo macroscopico può interpretarsi come un indizio di una struttura molto complessa.

La situazione diventa da questo punto di vista assai diversa se, dalla considerazione degli oggetti comuni, passiamo a quella degli atomi e delle molecole o, più ancora, dei loro costituenti, i nuclei e gli elettroni. Nel mondo atomico infatti si incontrano frequentemente oggetti eguali tra di loro; così per esempio, può affermarsi che due qualsiasi elettroni, o anche due qualsiasi atomi della stessa specie sono eguali tra di loro. Naturalmente, si dirà, l'identità assoluta di due oggetti qualsiasi è impossibile a constatarsi: essa può controllarsi soltanto entro certi limiti di approssimazione, dipendenti dalle nostre possibilità di osservazione. E se pur questi limiti, col progredire dei mezzi tecnici, potranno ridursi sempre più piccoli, non si può certo mai pensare che essi scendano fino a zero. Tutto ciò, naturalmente, è vero, e cosi anche la affermazione che per esempio due elettroni qualsiasi sono eguali tra di loro deve necessariamente intendersi in senso relativo alle nostre attuali possibilità di osservazione. E tuttavia queste, per il caso specifico della constatazione della identità tra due elettroni, sono estremamente accurate. Senza entrare qui in dettagli, che ci porterebbero troppo lontano, mi limiterò ad osservare che la grande precisione con cui può constatarsi l'identità di due elettroni dipende dalla possibilità di osservare gli effetti di una eventuale differenza accumulati per un tempo molto lungo. Così per esempio se tra due elettroni contenuti nello stesso atomo vi fosse una differenza sia pure assai lieve, l'effetto di questa sarebbe pure lieve, se considerato durante un breve periodo di tempo della vita dell'atomo, ma alla lunga finirebbe per alterare in modo essenziale la struttura e le proprietà esterne dell'atomo. Possiamo pertanto affermare l'identità di due elettroni, se non in senso assoluto, per lo meno entro limiti di precisione straordinariamente piccoli.

Il fatto che nel mondo atomico si incontrino frequentemente oggetti eguali ci incoraggia a pensare che la struttura dei corpuscoli atomici non sia estremamente complessa, e che, una volta che si sia riusciti ad analizzare la natura dei corpuscoli atomici, nuclei ed elettroni, non ci si trovi semplicemente ad avere spostato il problema della struttura della materia un gradino più in basso verso elementi più minuti, ma si sia, per così dire, raggiunto un pianerottolo che, se pur non rappresenti la base ultima su cui è costruito l'edificio materiale, e che forse non è raggiungibile dalla intelligenza umana, possa per lo meno considerarsi sufficiente per un tempo molto lungo. Naturalmente tutto questo non è che una semplice supposizione, se pure esistono ragioni serie per farla considerare attendibile.

Oggi che la struttura dell'atomo è conosciuta con notevole sicurezza sappiamo che ogni atomo è fatto da un nucleo e da un numero maggiore o minore di elettroni e possiamo contare quanti sono i tipi di particelle differenti dai cui aggregati sono costituiti tutti i corpi. Si tratta in tutto di poche centinaia di corpuscoli diversi, ciò che, come sopra si diceva, sembrava indicare che non si sia lontani da una analisi completa degli elementi strutturali della materia. Forse i più importanti tra questi diversi tipi di corpuscoli sono gli elettroni, i quali sono presenti in tutti gli atomi; essi sono anche quelli che sono conosciuti da più lungo tempo, essendo stato possibile produrli e studiarli isolatamente nella scarica elettrica nei gas rarefatti; i così detti raggi catodici non sono infatti che una proiezione di elettroni liberi con velocità assai elevata e, come tutti sanno, dallo studio delle proprietà di questi si può determinare la carica e la massa dell'elettrone.

Si ritiene generalmente che l'elettrone sia un vero e proprio corpuscolo elementare, e cioè che esso non sia ulteriormente analizzabile in elementi più semplici; e certamente non si è mai avuta fino ad oggi alcuna indicazione che possa far supporre una struttura complessa di esso. Per contro vi sono delle buone ragioni per ritenere che i nuclei positivi degli atomi non siano corpuscoli semplici ma aggregati, in qualche caso abbastanza numerosi, di elementi più semplici.

Ciò è già indicato in qualche modo dal numero relativamente grande di possibili nuclei, che indica di per sè una struttura di una certa complicazione. Ma un argomento assai più diretto per dimostrare la complessità dei nuclei ci è dato dalle disintegrazioni nucleari, sia quelle spontanee che si osservano nelle sostanze radioattive, sia quelle artificiali prodotte bombardando un nucleo con particelle alfa o con protoni velocissimi. In tutti questi casi assistiamo a uno spezzamento o a un riassettamento della compagine nucleare; un nucleo può emettere dei corpuscoli oppure assorbirne trasformandosi così in un altro nucleo di specie differente. Oggi che la struttura dell'atomo è sostanzialmente conosciuta, il problema di analizzare e comprendere le strutture dei nuclei è venuto ad essere il problema più centrale e profondo degli studi di fisica.

Tra tutti i nuclei certamente il più semplice è quello dell'idrogeno o protone; esso è il nucleo che ha la massa più piccola (circa una unità di peso atomico) e la carica elettrica più piccola (eguale, salvo il segno, alla carica elettronica). Il protone, al pari dell'elettrone, si considera generalmente come una particella elementare, non costituita da elementi ulteriori; semplice o no, esso è certo uno degli elementi fondamentali della struttura nucleare.

Fino a un paio di anni fa, l'elettrone e il protone erano gli unici corpuscoli semplici conosciuti, e si riteneva che tutti i nuclei altro non fossero che aggregati di protoni e di elettroni, in numero tale che il nucleo risultante venisse ad avere il giusto valore per la carica elettrica e il peso atomico. Oggi le possibilità per costruire una teoria dei nuclei si sono assai notevolmente arricchite con la scoperta sperimentale di due nuove particelle elementari o, per lo meno, semplicissime: il neutrone e l'elettrone positivo o positrone.

Di queste la prima ad essere scoperta è stato il neutrone. Per primo Bothe, nel 1931, osservava che il Berillio, bombardato con delle particelle alfa, emetteva una radiazione di potere penetrante notevolmente superiore a quello di tutte le radiazioni gamma allora conosciute; poco dopo F. Joliot e sua moglie Irene Curie, figlia della celebre scopritrice del Radio, osservarono che le radiazioni emesse dal Berillio nelle condizioni accennate avevano la proprietà di proiettare dei protoni fuori dalla paraffina o da altre sostanze contenenti idrogeno. Questa proprietà dimostrava che la radiazione del Berillio certo non era costituita da soli raggi gamma, poiché questi non possono trasmettere che un impulso lievissimo a corpuscoli relativamente pesanti come i protoni. Al principio del 1932 Chadwick, riprendendo le esperienze dei Joliot, potè dimostrare che la radiazione del Berillio era capace anche di trasmettere un impulso a nuclei più pesanti del protone e, interpretando questi risultati, arrivò alla conclusione che la radiazione doveva essere costituita da un nuovo tipo di particelle, la cui esistenza era stata spesse volte sospettata ma mai dimostrata, e che egli chiamò neutroni. Essi sono elettricamente neutri, ed hanno massa assai prossima a quella del protone, cioè di circa una unità di peso atomico. Il forte potere penetrante dei neutroni, che sono capaci di traversare uno spessore di piombo di parecchi centimetri, si spiega facilmente col fatto che essi sono elettricamente neutri; per questa ragione infatti, quando i neutroni traversano la materia, essi non risentono alcun effetto dei campi elettrici dovuti agli elettroni del corpo attraversato, che avrebbero una azione frenante su di essi, ma possono soltanto interagire con un nucleo, quando per caso vengano a passare a una distanza straordinariamente piccola da esso. Dall'essere questi corpuscoli elettricamente neutri, resta giustificata anche la loro efficacia come agenti per produrre disintegrazioni artificiali di nuclei. Se infatti si vuole produrre una disintegrazione nucleare, lanciando contro al nucleo un corpuscolo carico di elettricità positiva, quale un protone o una particella alfa, il campo elettrico del nucleo urtato respinge la particella urtante, ostacolando il suo avvicinarsi; mentre un neutrone, non avendo carica elettrica, non viene respinto e può giungere indisturbato fino al nucleo.

La scoperta dell'altra particella elementare a cui abbiamo accennato, il positrone o elettrone positivo, fu annunciata per la prima volta da Anderson circa un anno fa; dopo pochi mesi la scoperta fu confermata e dimostrata in modo assai più completo da Blackett e Occhialini. Questi fisici, osservando in una camera di Wilson le disintegrazioni della materia prodotte dall'urto dei corpuscoli della radiazione cosmica, osservarono in alcuni casi delle vere e proprie esplosioni in cui un nucleo veniva spezzato in una ventina di corpuscoli proiettati da esso come le pallottole di uno shrapnell. Deflettendo questi proiettili in un campo magnetico, essi poterono constatare che, mentre alcuni di essi venivano incurvati in un verso tale da indicare che si trattava di elettroni negativi, le traiettorie di altri venivano piegate in verso opposto indicando trattarsi di corpuscoli con carica elettrica positiva. Dalla densità della ionizzazione prodotta da questi corpuscoli, e con altre considerazioni che non è qui il caso di esporre, si potè anche riconoscere che la loro massa doveva essere dell'ordine di grandezza della massa dell'elettrone. Si trattava dunque di un vero e proprio elettrone positivo. Dopo la scoperta del positrone la sua presenza fu riscontrata anche in fenomeni indipendenti dalla radiazione penetrante. Importante in particolare fu la scoperta, che irradiando atomi, specialmente di elevato peso atomico, con delle radiazioni gamma assai dure, ha luogo una produzione di elettroni positivi o meglio, almeno nella maggior parte dei casi, di una coppia costituita da un normale elettrone negativo e da uno positivo. Vedremo tra un momento la notevole portata teorica di questo fatto.

La scoperta del neutrone permette, secondo Heisenberg e E. Majorana, di costruire uno schema generale della struttura dei nuclei, esente da molte delle obbiezioni che si potevano elevare contro il modello primitivo secondo cui i nuclei dovevano pensarsi costituiti da soli protoni ed elettroni. Invero, data la piccolissima massa di questi ultimi corpuscoli, un elettrone, costretto entro una orbita con dimensioni dell'ordine di grandezza di un nucleo, dovrebbe acquistare delle energie cinetiche enormi, che dovrebbero essere fortemente sensibili nel difetto di massa dell'atomo; a questa difficoltà di carattere quantitativo se ne aggiungono altre di carattere più qualitativo, ma che non è possibile illustrare completamente date le limitazioni di spazio imposte a questo discorso. Vorrei qui solamente accennare che il comportamento statistico del nucleo dell'azoto, messo per la prima volta in evidenza dalle belle osservazioni di Rasetti sopra l'effetto Raman dell'azoto allo stato gassoso, sembra incompatibile con l'ammettere che esso sia costituito da un numero dispari di particelle elementari (14 protoni e 7 elettroni secondo il vecchio schema); incompatibilità di questo genere, osservate per la prima volta nell'azoto, sono state in seguito incontrate in numerosi altri esempi, tanto da far perdere ogni credito allo schema di struttura per protoni ed elettroni.

Secondo le teorie di Heisenberg e di Majorana invece, come elementi fondamentali della struttura nucleare si debbono considerare protoni e neutroni. Potrebbero poi forse questi ultimi essere pensati a loro volta costituiti dalla unione di un protone con un elettrone, ciò che probabilmente è necessario per spiegare la possibilità di una disintegrazione con emissione di raggi β, si tratterebbe in ogni caso di una aggregazione fatta secondo leggi diverse da quelle della ordinaria meccanica quantistica: come la meccanica ordinaria perde la sua validità per descrivere il comportamento degli elettroni nel cambiamento di scala da quella ordinaria a quella atomica, così il nuovo cambiamento di scala dai fenomeni atomici a quelli nucleari renderebbe necessario un nuovo cambiamento, i cui particolari naturalmente ci sono per ora sconosciuti; sembra invece che la ordinaria meccanica quantistica sia sufficiente per descrivere, anche nell'interno del nucleo, il comportamento di corpuscoli relativamente pesanti, quali i protoni e i neutroni. Questi infatti, per la loro considerevole massa, hanno, anche nelle orbite nucleari, delle velocità abbastanza piccole a confronto di quella della luce, per modo che le correzioni relativistiche al loro movimento restano sempre di importanza secondaria. Se dunque veramente si potrà dimostrare che gli elettroni nucleari, la cui esistenza si deve certo ammettere se si vuol spiegare l'emissione delle particelle β, non esistono tuttavia allo stato libero, ma sono per esempio intimamente associati a dei protoni in modo da formare dei neutroni, resterà aperta una possibilità relativamente agevole di costruire, almeno fino ad un punto abbastanza avanzato, una teoria del nucleo, in quanto sarà possibile, almeno fino a che non si vogliano studiare quei fenomeni nei quali viene intaccata la struttura neutronica, servirsi dei procedimenti e dello schema generale interpretativo della meccanica quantistica. Naturalmente, una volta che si fosse a questo punto, anche a prescindere dalle difficoltà matematiche, resterebbe sempre il problema essenziale di conoscere le leggi delle forze agenti tra i corpuscoli costitutivi del nucleo. Resterebbe poi in un secondo tempo da chiarire la struttura del neutrone, per la quale, come si diceva, non dovebbe verosimilmente essere più applicabile lo schema della meccanica quantistica; si ha innanzi in proposito, nello spettro continuo dei raggi β, qualche indizio, che secondo Bohr indurrebbe a pensare che in queste nuove leggi sconosciute non sia forse più valido nemmeno il principio della conservazione dell'energia; quando almeno non si voglia ammettere con Pauli l'esistenza del così detto « neutrino » e cioè di una ipotetica particella elettricamente neutra ed avente massa dell'ordine di grandezza della massa elettronica. Questa, per il suo enorme potere penetrante, sfuggirebbe praticamente a ogni attuale mezzo di osservazione, e la sua energia cinetica servirebbe a ristabilire la bilancia energetica nelle disintegrazioni β. Certamente su tutte queste questioni sopra la struttura nucleare porterà molta luce l'accumularsi delle nostre conoscenze sopra le varie possibilità di disintegrazione dei nuclei; e dà ragione a bene sperare la rapidità veramente impreveduta con cui si sono succeduti i progressi sperimentali in questo campo negli ultimi due anni.

Ci resta ancora da aggiungere poche parole sopra la importanza teorica della scoperta dell'elettrone positivo. Nei tentativi di costruire una meccanica quantistica per l'elettrone, che fosse compatibile con il principio di relatività, tentativi culminati con i lavori di Dirac, si era sempre incontrata la difficoltà dell'apparire, accanto agli stati normali che descrivono il comportamento dell'elettrone, di altri stati, senza un'apparente corrispondenza fisica, in cui l'elettrone avrebbe avuta una energia cinetica negativa. Dirac propose allora di ammettere che tutti questi stati anomali fossero occupati;

si verrebbe così ad avere un riempimento uniforme di tutto lo spazio che, appunto per la sua uniformità non sarebbe osservabile; potrebbe invece osservarsi solamente una lacuna in questa distribuzione uniforme, e cioè la mancanza di un elettrone da uno degli stati di energia negativa. Dirac dimostrò che una di queste lacune si comporterebbe a tutti gli effetti come un normale corpuscolo, dotato di carica elettrica positiva, e propose originariamente l'ipotesi che i protoni dovessero identificarsi con queste lacune. L'ipotesi originaria di Dirac si mostrò insostenibile, perché incapace di spiegare la grande differenza di massa tra l'elettrone e il protone, e siccome l'esperienza non aveva allora ancora dimostrata l'esistenza di un corpuscolo che avesse le proprietà delle lacune di Dirac, si pensò che esse effettivamente non dovessero avere nessun corrispondente reale. La scoperta dell'elettrone positivo ha sostanzialmente modificata questa situazione: il nuovo corpuscolo ha infatti, almeno per quanto ne sappiamo fino ad oggi, esattamente tutte quelle proprietà che la teoria di Dirac prevede per le lacune; è naturale dunque identificare le lacune con l'elettrone positivo. Se questa veduta è corretta, deve essere possibile, estraendo un elettrone da uno stato di energia negativa e portandolo in uno di energia positiva, dar luogo alla formazione simultanea di una lacuna, e cioè di un elettrone positivo e di un normale elettrone negativo, a sole spese dell'energia necessaria per lo spostamento dell'elettrone. Sembra che un fenomeno di questo genere abbia veramente luogo nella formazione di coppie di un elettrone e di un positrone prodotta dall'azione dei raggi gamma duri. Se sarà dimostrato che così è effettivamente gli elettroni positivi e negativi che così si formano potranno a buon diritto chiamarsi, secondo la proposta di Madame Curie, elettroni di materializzazione, rappresentando la trasformazione completa di energia raggiante in corpuscoli materiali.

80 *a.*

TENTATIVO DI UNA TEORIA DEI RAGGI β [*]

« Nuovo Cimento », *11*, 1–19 (1934).

SUNTO. — Si propone una teoria quantitativa dell'emissione dei raggi β in cui si ammette l'esistenza del « neutrino » e si tratta l'emissione degli elettroni e dei neutrini da un nucleo all'atto della disintegrazione β con un procedimento simile a quello seguito nella teoria della irradiazione per descrivere l'emissione di un quanto di luce da un atomo eccitato. Vengono dedotte delle formule per la vita media e per la forma dello spettro continuo dei raggi β, e le si confrontano coi dati sperimentali.

IPOTESI FONDAMENTALI DELLA TEORIA.

§ 1. Nel tentativo di costruire una teoria degli elettroni nucleari e della emissione dei raggi β, si incontrano, come è noto, due difficoltà principali. La prima dipende dal fatto che i raggi β primari vengono emessi dai nuclei con una distribuzione continua di velocità. Se non si vuole abbandonare il principio della conservazione dell'energia, si deve ammettere perciò che una frazione dell'energia che si libera nel processo di disintegrazione β sfugga alle nostre attuali possibilità di osservazione. Secondo la proposta di Pauli si può per esempio ammettere l'esistenza di una nuova particella, il così detto « neutrino », avente carica elettrica nulla e massa dell'ordine di grandezza di quella dell'elettrone o minore. Si ammette poi che in ogni processo β vengano emessi simultaneamente un elettrone, che si osserva come raggio β, e un neutrino che sfugge all'osservazione portando seco una parte dell'energia. Nella presente teoria ci baseremo sopra l'ipotesi del neutrino.

Una seconda difficoltà per la teoria degli elettroni nucleari, dipende dal fatto che le attuali teorie relativistiche delle particelle leggere (elettroni o neutrini) non dànno una soddisfacente spiegazione della possibilità che tali particelle vengano legate in orbite di dimensioni nucleari.

Sembra per conseguenza più appropriato ammettere con Heisenberg [1] che tutti i nuclei consistano soltanto di particelle pesanti, protoni e neutroni. Per comprendere tuttavia la possibilità dell'emissione dei raggi β, noi tenteremo di costruire una teoria dell'emissione delle particelle leggere da un nucleo in analogia alla teoria dell'emissione di un quanto di luce da un atomo eccitato nell'ordinario processo della irradiazione. Nella teoria dell'irradiazione, il numero totale dei quanti di luce non è costante; i quanti vengono

(*) Cfr. la nota preliminare in « La Ricerca Scientifica », *4* (2), 491 (1933).

(1) W. HEISENBERG, « ZS. für Phys. », *77*, 1 (1932); E. MAJORANA, « ZS. für Phys. », *82*, 137 (1933).

creati all'atto della loro emissione da un atomo eccitato, e spariscono invece quando sono assorbiti. In analogia a ciò cercheremo di fondare la teoria dei raggi β sopra le seguenti ipotesi:

a) Il numero totale degli elettroni e dei neutrini non è necessariamente costante. Elettroni (o neutrini) possono essere creati o distrutti. Questa possibilità non ha per altro alcuna analogia con la possibilità della creazione o della distruzione di una coppia elettrone–positrone; se infatti si interpreta un positrone come un « buco » di Dirac, si può considerare semplicemente quest'ultimo processo come un salto quantico di un elettrone da uno stato di energia negativa a uno di energia positiva, con conservazione del numero totale (infinitamente grande) degli elettroni.

b) Le particelle pesanti, neutrone e protone, possono considerarsi, secondo le vedute di Heisenberg, come due diversi stati quantici interni della particella pesante. Formuleremo questo con la introduzione di una coordinata interna ρ della particella pesante, che sia suscettibile di assumere soltanto due valori: $\rho = +1$, se la particella è un neutrone; $\rho = -1$, se la particella è un protone.

c) La funzione Hamiltoniana del sistema complessivo, costituito da particelle pesanti e leggere, deve scegliersi in modo che ogni transizione da neutrone a protone sia accompagnata dalla creazione di un elettrone e di un neutrino; e che il processo opposto, trasformazione di un protone in un neutrone, sia accompagnato dalla sparizione di un elettrone e di un neutrino. Si noti che con ciò resta assicurata la conservazione della carica elettrica.

Gli operatori della teoria.

§ 2. Un formalismo matematico che permette di costruire una teoria in accordo con i tre punti precedenti può costruirsi nel modo più facile per mezzo del metodo [2] di Dirac–Jordan–Klein detto della « seconda quantizzazione ». Noi considereremo dunque le ampiezze di probabilità ψ e φ degli elettroni e dei neutrini nello spazio ordinario, e le loro complesse coniugate ψ^* e φ^* come degli operatori; mentre per la descrizione delle particelle pesanti useremo la consueta rappresentazione nello spazio delle configurazioni, nella quale naturalmente anche ρ dovrà computarsi tra le coordinate.

Introduciamo dapprima due operatori Q e Q^* che operano sopra le funzioni della variabile a due valori ρ come le sostituzioni lineari

$$Q = \begin{vmatrix} 0 & 1 \\ 0 & 0 \end{vmatrix} \quad ; \quad Q^* = \begin{vmatrix} 0 & 0 \\ 1 & 0 \end{vmatrix} . \tag{1}$$

Si riconosce immediatamente che Q determina le transizioni da protone a neutrone, e Q^* le transizioni opposte da neutrone a protone.

Il significato delle ampiezze di probabilità ψ e φ interpretate come operatori è, come si sa, il seguente. Sia

$$\psi_1 \, \psi_2 \cdots \psi_s \cdots$$

(2) Cfr. per esempio P. JORDAN e O. KLEIN, « ZS. für Phys. », *45*, 751 (1927); W. HEISENBERG, « Ann. d. Phys. », *10*, 888 (1931).

un sistema di stati quantici individuali degli elettroni. Poniamo poi

$$\psi = \sum_s \psi_s a_s \quad ; \quad \psi^* = \sum_s \psi_s^* a_s^* . \tag{2}$$

Le ampiezze a_s e le grandezze complesse coniugate a_s^* sono operatori che agiscono sulle funzioni dei numeri di occupazione $N_1, N_2, \cdots, N_s, \cdots$ degli stati quantici individuali. Se vale il principio di Pauli, ciascuna delle N_s può prendere soltanto uno dei valori, 0, 1; e gli operatori a_s ed a_s^* sono definiti al modo seguente:

$$\left\{\begin{aligned} & a_s^* \Psi (N_1, N_2, \cdots, N_s, \cdots) \\ & \qquad = (-1)^{N_1+N_2+\cdots+N_s-1} (1 - N_s) \Psi (N_1, N_2, \cdots, 1 - N_s, \cdots) \\ & a_s \Psi (N_1, N_2, \cdots, N_s, \cdots) \\ & \qquad = (-1)^{N_1+N_2+\cdots+N_s-1} N_s \Psi (N_1, N_2, \cdots, 1 - N_s, \cdots). \end{aligned}\right. \tag{3}$$

L'operatore a_s^* determina la creazione, mentre l'operatore a_s determina la scomparsa di un elettrone nello stato quantico s.

Corrispondentemente alle (2) porremo per i neutrini:

$$\varphi = \Sigma \varphi_\sigma b_\sigma \quad ; \quad \varphi^* = \Sigma \varphi_\sigma^* b_\sigma^* . \tag{4}$$

Gli operatori complessi coniugati b_σ e b_σ^* operano sulle funzioni dei numeri di occupazione $M_1, M_2, \cdots, M_\sigma, \cdots$ degli stati quantici individuali $\varphi_1, \cdots, \varphi_2, \varphi_\sigma, \cdots$ dei neutrini. Ammettendo che anche per questi corpuscoli valga il principio di Pauli, i numeri M_σ possono prendere soltanto i due valori 0, 1; e si ha

$$\left\{\begin{aligned} & b_\sigma^* \Phi (M_1, M_2, M_\sigma, \cdots) \\ & \qquad = (-1)^{M_1+M_2+\cdots+M_\sigma-1} (1 - M_\sigma) \Phi (M_1, M_2, \cdots, 1 - M_\sigma, \cdots) \\ & b_\sigma \Phi (M_1, M_2, M_\sigma, \cdots) \\ & \qquad = (-1)^{M_1+M_2+\cdots+M_\sigma-1} M_\sigma \Phi (M_1, M_2, \cdots, 1 - M_\sigma, \cdots). \end{aligned}\right. \tag{5}$$

Gli operatori b_σ e b_σ^* determinano rispettivamente la sparizione e la creazione di un neutrino nello stato σ.

La funzione Hamiltoniana.

§ 3. L'energia del sistema complessivo costituito dalle particelle pesanti e leggere, è la somma dell'energia H_{pes} delle particelle pesanti + l'energia H_{leg} delle particelle leggere + l'energia di interazione $\mathcal{H}$ tra le particelle leggere e quelle pesanti.

Limitandoci per semplicità a considerare un solo corpuscolo pesante, scriveremo il primo termine nella forma

$$H_{pes} = \frac{1+\rho}{2} \mathfrak{N} + \frac{1-\rho}{2} \mathfrak{P} \tag{6}$$

in cui $\mathfrak{N}$ e $\mathfrak{P}$ sono gli operatori che rappresentano l'energia del neutrone e del protone. Si osservi infatti che per $\rho = +1$ (neutrone) la (6) si riduce a $\mathfrak{N}$; mentre per $\rho = -1$ (protone) essa si riduce a $\mathfrak{P}$.

Per scrivere l'energia H_{leg} nel modo più semplice, prenderemo per stati quantici ψ_s e φ_σ degli elettroni e dei neutrini degli stati stazionari. Per gli elettroni prenderemo le autofunzioni nel campo coulombiano del nucleo (opportunamente schermato per tener conto dell'azione degli elettroni atomici); per i neutrini potremo prendere semplicemente delle onde piane di De Broglie, poiché eventuali forze agenti sui neutrini sono certamente lievissime. Siano $H_1, H_2, \cdots, H_s, \cdots$ e $K_1, K_2, \cdots, K_\sigma, \cdots$ le energie degli stati stazionari degli elettroni e dei neutrini; avremo allora

$$H_{leg} = \sum_s H_s N_s + \sum_\sigma K_\sigma M_\sigma . \tag{7}$$

Resta ancora da scrivere la energia di interazione. Questa consiste in primo luogo dell'energia coulombiana tra protone ed elettroni; nel caso di nuclei pesanti tuttavia, la attrazione esercitata da un solo protone non ha molta importanza [(3)] e in ogni caso non contribuisce in alcun modo al processo della disintegrazione β. Per non complicare inutilmente le cose non terremo dunque conto di questo termine. Dobbiamo invece aggiungere alla Hamiltoniana un termine tale da soddisfare alla condizione *c*) del § 1.

Un termine che accoppia necessariamente la trasformazione di un neutrone in protone con la creazione di un elettrone e di un neutrino ha, secondo i risultati del § 2, la forma

$$Q^* a_s^* b_\sigma^* \tag{8}$$

mentre l'operatore complesso coniugato

$$Q a_s b_\sigma \tag{8'}$$

accoppia tra di loro i processi inversi (trasformazione di un protone in neutrone e sparizione di un elettrone e di un neutrino).

Un termine di interazione soddisfacente alla condizione *c*) avrà dunque la forma seguente

$$\mathcal{H} = Q \sum_{s\sigma} c_{s\sigma} a_s b_\sigma + Q^* \sum_{s\sigma} c_{s\sigma}^* a_s^* b_\sigma^* \tag{9}$$

dove $c_{s\sigma}$ e $c_{s\sigma}^*$ sono grandezze che possono dipendere dalle coordinate, dai momenti, ecc... della particella pesante.

Una determinazione ulteriore di $\mathcal{H}$ non può farsi altro che col criterio della massima semplicità; naturalmente tuttavia le possibilità di scelta della $\mathcal{H}$ sono limitate dalla necessità che $\mathcal{H}$ sia invariante rispetto a un cambiamento delle coordinate e che inoltre soddisfi la conservazione della quantità di moto.

Se in un primo tempo trascuriamo gli effetti relativistici e lo spin, la più semplice scelta per (9) è la seguente

$$\mathcal{H} = g\,[Q\psi(x)\,\varphi(x) + Q^*\,\psi^*(x)\,\varphi^*(x)] \tag{10}$$

(3) La attrazione coulombiana dovuta ai numerosi altri protoni deve naturalmente tenersi in conto come un campo statico.

in cui g è una costante avente le dimensioni $L^5 MT^{-2}$; x rappresenta le coordinate della particella pesante; $\psi, \varphi, \psi^*, \varphi^*$ sono date da (2) e (4) e debbono prendersi nella posizione x, y, z occupata dalla particella pesante.

La (10) non rappresenta in alcun modo l'unica scelta possibile di $\mathcal{H}$; altrettanto ammissibile sarebbe qualsiasi espressione scalare come

$$L(p)\,\psi(x)\,M(p)\,\varphi(x)\,N(p) + \text{compl. coniug.}$$

in cui $L(p)$, $M(p)$, $N(p)$ rappresentano opportune funzioni del momento della particella pesante. Siccome per altro le conseguenze della (10) si sono fino ad ora mostrate in accordo con l'esperienza, sembra inutile ricorrere per il momento ad espressioni più complicate.

È essenziale invece generalizzare la (10) in modo da poter trattare relativisticamente per lo meno le particelle leggere. Naturalmente, anche in questa generalizzazione, non sembra possibile procedere senza una certa arbitrarietà. La soluzione più naturale del problema sembra tuttavia essere la seguente: Relativisticamente si hanno, in luogo di ψ e di φ, due quaterne $\psi_1\,\psi_2\,\psi_3\,\psi_4$ e $\varphi_1\,\varphi_2\,\varphi_3\,\varphi_4$ di funzioni di Dirac. Consideriamo le 16 combinazioni bilineari indipendenti di $\psi_1\,\psi_2\,\psi_3\,\psi_4$ e $\varphi_1\,\varphi_2\,\varphi_3\,\varphi_4$. Quando il sistema di riferimento subisce una trasformazione di Lorentz, le 16 combinazioni bilineari subiscono una sostituzione lineare, che dà una rappresentazione del gruppo di Lorentz. In particolare le quattro combinazioni bilineari

$$(11)\qquad \begin{cases} A_0 = -\psi_1\varphi_2 + \psi_2\varphi_1 + \psi_3\varphi_4 - \psi_4\varphi_3 \\ A_1 = \psi_1\varphi_3 - \psi_2\varphi_4 + \psi_3\varphi_1 + \psi_4\varphi_2 \\ A_2 = i\psi_1\varphi_3 + i\psi_2\varphi_4 - i\psi_3\varphi_1 - i\psi_4\varphi_2 \\ A_3 = -\psi_1\varphi_4 - \psi_2\varphi_3 + \psi_3\varphi_2 + \psi_4\varphi_1 \end{cases}$$

si trasformano come le componenti di un quadrivettore, e cioè come le componenti del quadripotenziale elettromagnetico. È quindi naturale introdurre nella Hamiltoniana della particella pesante le quattro grandezze

$$g\,(QA_i + Q^*A_i^*)$$

in una situazione corrispondente a quella delle componenti del quadripotenziale. Incontriamo qui una difficoltà dipendente dal fatto che non è conosciuta una equazione d'onda relativistica per le particelle pesanti. Tuttavia, nel caso in cui la velocità della particella pesante sia piccola rispetto a c, si potrà limitarsi al termine che corrisponde ad eV (V potenziale scalare) e scrivere

$$(12)\qquad \begin{aligned} \mathcal{H} = g\,[&Q\,(-\psi_1\varphi_2 + \psi_2\varphi_1 + \psi_3\varphi_4 - \psi_4\varphi_3) + \\ &+ Q^*\,(-\psi_1^*\varphi_2^* + \psi_2^*\varphi_1^* + \psi_3^*\varphi_4^* - \psi_4^*\varphi_3^*)]. \end{aligned}$$

A questo termine debbono aggiungersene altri dell'ordine di grandezza v/c. Per il momento però noi trascureremo questi termini, dato che le velocità dei neutroni e dei protoni nei nuclei sono generalmente piccole in confronto a c (cfr. § 9).

Con la scrittura simbolica delle matrici, la (12) può scriversi

$$\mathcal{H} = g\,[Q\tilde{\psi}^*\,\delta\varphi + Q^*\,\tilde{\psi}\,\delta\varphi^*] \tag{13}$$

dove ψ e φ debbono intendersi scritte come matrici di una sola colonna, e il segno $\sim$ trasforma una matrice nella coniugata trasposta; inoltre è

$$\delta = \begin{vmatrix} 0 & -1 & 0 & 0 \\ 1 & 0 & 0 & 0 \\ 0 & 0 & 0 & 1 \\ 0 & 0 & -1 & 0 \end{vmatrix}. \tag{14}$$

Con queste notazioni si trova, dal confronto di (12) e (9),

$$c_{s\sigma} = g\tilde{\psi}_s^*\,\delta\varphi_\sigma \quad ; \quad c_{s\sigma}^* = g\tilde{\psi}_s\,\delta\varphi_s^* \tag{15}$$

dove ψ_s e φ_σ rappresentano le autofunzioni normalizzate a quattro componenti degli stati s dell'elettrone e σ del neutrino, prese come funzioni della posizione x, y, z, occupata dalla particella pesante.

La matrice di perturbazione.

§ 4. Con la Hamiltoniana che abbiamo stabilita si può svolgere una teoria della disintegrazione β in completa analogia alla teoria dell'irradiazione. In questa, la funzione di Hamilton consiste, come è noto, della somma: Energia dell'atomo + Energia del campo di radiazione + Interazione tra atomo e radiazione; quest'ultimo termine viene considerato come una perturbazione degli altri due. Analogamente prenderemo

$$H_{pes} + H_{leg} \tag{16}$$

come Hamiltoniana imperturbata. La perturbazione è rappresentata dal termine di interazione (13).

Gli stati quantici del sistema imperturbato possono numerarsi al modo seguente:

$$\rho\,,\, n\,,\, N_1\, N_2 \cdots N_s \cdots M_1\, M_2 \cdots M_\sigma \cdots \tag{17}$$

in cui il primo numero ρ prende uno dei valori ± 1, e indica se la particella pesante è un neutrone o un protone. Il secondo numero n indica lo stato quantico del neutrone o del protone. Per $\rho = +1$ (neutrone) la corrispondente autofunzione sia

$$u_n(x) \tag{18}$$

mentre per $\rho = -1$ (protone) la autofunzione sia

$$v_m(x). \tag{19}$$

Gli altri numeri $N_1\, N_2 \cdots N_s \cdots M_1\, M_2 \cdots M_\sigma \cdots$ possono prendere solo i valori 0,1 e indicano quali stati degli elettroni e dei neutrini sono occupati.

Dall'esame della forma generale (9) dell'energia di perturbazione, si vede immediatamente che essa ha elementi di matrice diversi da zero solo per transizioni in cui o la particella pesante passa da neutrone a protone, mentre contemporaneamente si creano un elettrone e un neutrino, o viceversa.

Per mezzo di (1) (3) (5) (9) (18) (19) si trova facilmente che l'elemento di matrice corrispondente è

$$\mathcal{H}^{1\, n\, N_1\, N_2 \cdots 0_s \cdots M_1\, M_2 \cdots 0_\sigma \cdots}_{-1\, m\, N_1\, N_2 \cdots 1_s \cdots M_1\, M_2 \cdots 1_\sigma \cdots} = \pm \int v_m^* \, c_{s\sigma}^* \, u_n d\tau \tag{20}$$

dove l'integrazione deve estendersi a tutto lo spazio delle configurazioni della particella pesante (ad eccezione della coordinata ρ); il segno ± significa più precisamente

$$(-1)^{N_1+N_2+\cdots+N_{s-1}+M_1+M_2+\cdots M_{\sigma-1}}$$

e sparisce del resto dai calcoli che seguiranno. Alla transizione opposta corrisponde un elemento di matrice complesso coniugato di (20).

Tenendo conto delle (15), la (20) diviene:

$$\mathcal{H}^{1\, n\, 0_s\, 0_\sigma}_{-1\, m\, 1_s\, 1_\sigma} = \pm\, g \int v_m^* \, u_n \, \tilde{\psi}_s \, \delta\varphi_\sigma^* d\tau \tag{21}$$

ove per brevità abbiamo omesso di scrivere nel primo membro tutti gli indici che non subiscono una variazione.

Teoria delle disintegrazioni β.

§ 5. Una disintegrazione β consiste in un processo in cui un neutrone nucleare si trasforma in un protone, mentre contemporaneamente, col meccanismo che abbiamo descritto, vengono emessi un elettrone, che si osserva come particella β, e un neutrino. Per calcolare la probabilità di questo processo, ammetteremo che, al tempo $t = 0$, si abbia un neutrone in uno stato nucleare di autofunzione $u_n(x)$, e che inoltre lo stato elettronico s e lo stato di neutrino σ siano liberi, cioè: $N_s = M_\sigma = 0$. Per $t = 0$ dunque si potrà prendere eguale a 1 la ampiezza di probabilità dello stato $(1, n, 0_s, 0_\sigma$. cioè

$$a_{1,n,0_s,0_\sigma} = 1 \tag{22}$$

mentre l'ampiezza di probabilità dello stato $(-1, m, 1_s, 1_\sigma)$, in cui il neutrone si è trasformato in un protone con autofunzione $v_m(x)$ emettendo un elettrone e un neutrino negli stati s e σ è inizialmente nulla.

Applicando le consuete formule della teoria delle perturbazioni, si trova, per un tempo abbastanza corto perché (22) sia ancora approssimativamente valida,

$$\dot{a}_{-1,m,1_s,1_\sigma} = -\frac{2\pi i}{h} \mathcal{H}^{1\, n\, 0_s\, 0_\sigma}_{-1\, m\, 1_s\, 1_\sigma} e^{\frac{2\pi i}{h}(-W+H_s+K_\sigma)t} \tag{23}$$

dove con W si è indicata la differenza di energia tra lo stato del neutrone e quello del protone.

Integrando (23) otteniamo (siccome per $t = 0$, $a_{1m1_s1_\sigma} = 0$)

$$(24) \qquad a_{-1m1_s1_\sigma} = -\mathcal{H}^{1n0_s0_\sigma}_{-1m1_s1_\sigma} \frac{e^{\frac{2\pi i}{h}(-W+H_s+K_\sigma)t} - 1}{-W+H_s+K_\sigma} .$$

La probabilità della transizione considerata è dunque

$$(25) \qquad \left| a_{-1m1_s1_\sigma} \right|^2 = 4 \left| \mathcal{H}^{1n0_s0_\sigma}_{-1m1_s1_\sigma} \right|^2 \frac{\sin^2 \frac{\pi t}{h}(-W+H_s+K_\sigma)}{(-W+H_s+K_\sigma)^2} .$$

Per calcolare la vita media dello stato neutronico u_n occorre sommare (25) rispetto a tutti gli stati non occupati degli elettroni e dei neutrini. Una notevole semplificazione di questa somma si ottiene osservando che la lunghezza d'onda di De Broglie per elettroni o neutrini aventi energie di alcuni milioni di Volt è notevolmente più grande delle dimensioni nucleari. Si può dunque, in prima approssimazione, considerare le autofunzioni ψ_s e φ_σ come costanti entro l'estensione del nucleo (21) diventa così

$$(26) \qquad \mathcal{H}^{1n0_s0_\sigma}_{-1m1_s1_\sigma} = \pm g \tilde{\psi}_s \delta \varphi^*_\sigma \int v^*_m u_n d\tau$$

dove qui e nel seguito ψ_s e φ_σ debbono intendersi prese nel nucleo (cfr. § 8) Da (26) si ricava:

$$(27) \qquad \left| \mathcal{H}^{1n0_s0_\sigma}_{-1m1_s1_\sigma} \right|^2 = g^2 \left| \int v^*_m u_n d\tau \right|^2 \tilde{\psi}_s \delta \varphi^*_\sigma \tilde{\varphi}^*_\sigma \delta \tilde{\psi}_s .$$

Gli stati σ del neutrino sono individuati dal loro impulso p_σ e dalla direzione dello spin. Se, per comodità di normalizzazione, quantizziamo entro un volume Ω, le cui dimensioni si faranno in seguito tendere all'infinito, le autofunzioni normalizzate del neutrino sono onde piane di Dirac con densità $1/\Omega$. Facili considerazioni algebriche permettono allora di effettuare in (27) la media rispetto a tutte le orientazioni di p_σ e dello spin. (In ciò sono da considerarsi solo gli stati di energia positiva; gli stati di energia negativa debbono eliminarsi con un artificio simile alla teoria dei buchi di Dirac). Si trova

$$(28) \qquad \overline{\left| \mathcal{H}^{1n0_s0_\sigma}_{-1m1_s1_\sigma} \right|^2} = \frac{g^2}{4\Omega} \left| \int v^*_m u_n d\tau \right|^2 \left(\tilde{\psi}_s \psi_s - \frac{\mu c^2}{K_\sigma} \tilde{\psi}_s \beta \psi_s \right)$$

in cui μ è la massa di quiete del neutrino, e β è la matrice di Dirac

$$(29) \qquad \left| \begin{array}{cccc} 1 & 0 & 0 & 0 \\ 0 & 1 & 0 & 0 \\ 0 & 0 & -1 & 0 \\ 0 & 0 & 0 & -1 \end{array} \right| .$$

Osservando che il numero degli stati di neutrino con energia positiva e momento compreso tra p_σ e $p_\sigma + dp_\sigma$ è $8\pi\Omega p^2_\sigma dp_\sigma/h^3$; che inoltre $\partial K_\sigma/\partial p_\sigma = v_\sigma$, v_σ essendo la velocità del neutrino nello stato σ; e infine che (25) ha un massimo assai pronunziato per quel valore di p_σ per cui non si ha variazione dell'energia imperturbata, per cui cioè:

$$(30) \qquad -W + H_s + K_\sigma = 0$$

si può effettuare nel modo consueto [4] la somma di (25) rispetto a σ e si trova

$$(31) \qquad t \cdot \frac{8\pi^3 g^2}{h^4} \left| \int v_m^* u_n d\tau \right|^2 \frac{p_\sigma^2}{v_\sigma} \left(\tilde{\psi}_s \psi_s - \frac{\mu c^2}{K_\sigma} \tilde{\psi}_s \beta \psi_s \right)$$

p_σ essendo il valore del momento del neutrino per cui vale la (30).

Elementi determinativi della probabilità di transizione.

§ 6. (31) esprime la probabilità che nel tempo t abbia luogo una disintegrazione β in cui l'elettrone viene emesso nello stato s. Come deve essere, tale probabilità risulta proporzionale al tempo (t è stato supposto piccolo in confronto alla vita media); il coefficiente di t dà la probabilità di transizione per il processo indicato; essa risulta:

$$(32) \qquad P_s = \frac{8\pi^3 g^2}{h^4} \left| \int v_m^* u_n d\tau \right|^2 \frac{p_\sigma^2}{v_\sigma} \left(\tilde{\psi}_s \psi_s - \frac{\mu c^2}{K_\sigma} \tilde{\psi}_s \beta \psi_s \right) .$$

Si osservi che:

a) Per gli stati liberi dei neutrini si ha sempre $K_\sigma \geq \mu c^2$. È dunque necessario, affinché la (30) possa essere soddisfatta, che sia

$$(33) \qquad H_s \leq W - \mu c^2 .$$

Al segno = corrisponde il limite superiore dello spettro dei raggi β.

b) Essendo poi per gli stati elettronici non occupati $H_s \geq mc^2$, si ottiene, perché la disintegrazione β sia possibile, la seguente condizione:

$$(34) \qquad W \geq (m + \mu) c^2 .$$

Nel nucleo si deve dunque avere uno stato neutronico occupato abbastanza elevato al di sopra di uno stato protonico libero perché l'emissione β sia possibile.

c) Secondo (32) P_s dipende dalle autofunzioni u_n e v_m della particella pesante nel nucleo, attraverso all'elemento di matrice

$$(35) \qquad Q_{mn}^* = \int v_m^* u_n d\tau .$$

Questo elemento di matrice ha, nella teoria dei raggi β, una funzione analoga a quella dell'elemento di matrice del momento elettrico nella teoria dell'irradiazione. L'elemento di matrice (35) ha normalmente l'ordine di grandezza 1; accade tuttavia spesso che, per particolari simmetrie delle autofunzioni u_n e v_m, Q_{mn}^* viene ad annullarsi esattamente. In questi casi parleremo di *transizioni β proibite*. Non si deve per altro attendersi che le transizioni proibite siano veramente impossibili, poiché (32) è soltanto una formula approssimata. Ritorneremo sopra questo argomento nel § 9.

(4) Per la descrizione dei metodi che si usano per eseguire tali somme, cfr. qualsiasi articolo riassuntivo sulla teoria dell'irradiazione. Per esempio E. FERMI, « Rev. Mod. Phys. », *4*, 87 (1932).

La massa del neutrino.

§ 7. La probabilità di transizione (32) determina tra l'altro la forma dello spettro continuo dei raggi β. Discuteremo qui come la forma di questo spettro dipende dalla massa di quiete del neutrino, in modo da poter determinare questa massa da un confronto con la forma sperimentale dello spettro stesso. La massa μ interviene in (32) tra l'altro nel fattore p_σ^2/v_σ. La dipendenza della forma della curva di distribuzione dell'energia da μ, è marcata specialmente in vicinanza della energia massima E_0 dei raggi β. Si riconosce facilmente che la curva di distribuzione per energie E prossime al valore massimo E_0, si comporta, a meno di un fattore indipendente da E, come

$$\frac{p_\sigma^2}{v_\sigma} = \frac{1}{c^3}(\mu c^2 + E_0 - E)\sqrt{(E_0 - E)^2 + 2\mu c^2 (E_0 - E)}. \tag{36}$$

Nella fig. 1 la fine della curva di distribuzione è rappresentata per $\mu = 0$, e per un valore piccolo e uno grande di μ. La maggiore somiglianza con le

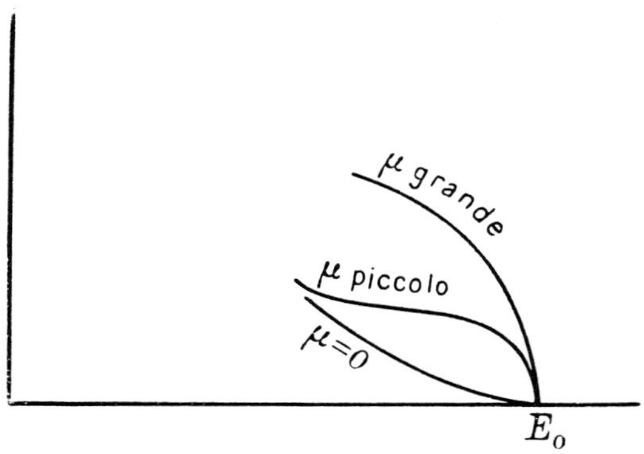

Fig. 1.

curve sperimentali si ha per la curva teorica corrispondente a $\mu = 0$. Arriviamo così a concludere che la massa del neutrino è uguale a zero o, in ogni caso, piccola in confronto della massa dell'elettrone [5]. Nei calcoli che seguono porremo per semplicità $\mu = 0$.

Abbiamo allora, tenendo anche conto della (30),

$$v_\sigma = c \quad ; \quad K_\sigma = c p_\sigma \quad ; \quad p_\sigma = \frac{K_\sigma}{c} = \frac{W - H_s}{c} \tag{37}$$

e le diseguaglianze (33) e (34) diventano:

$$H_s \leq W \quad ; \quad W \geq mc^2. \tag{38}$$

Infine la probabilità di transizione (32) prende la forma

$$P_s = \frac{8\pi^3 g^2}{c^3 h^4} \left| \int v_m^* \, u_n \, d\tau \right|^2 \tilde{\psi}_s \psi_s (W - H_s)^2. \tag{39}$$

(5) In una recente notizia F. Perrin, «C. R.», *197*, 1625 (1933), giunge con argomenti qualitativi a una simile conclusione.

VITA MEDIA E FORMA DELLA CURVA DI DISTRIBUZIONE DELL'ENERGIA PER TRANSIZIONI PERMESSE.

§ 8. Dalla (39) si può ricavare una formula che esprime quante transizioni β, in cui la particella β riceve un impulso compreso tra $mc\eta$ e $mc(\eta + d\eta)$, hanno luogo nell'unità di tempo. Per questo occorre calcolare la somma dei valori di $\tilde{\psi}_s \psi_s$ nel nucleo, estesa a tutti gli stati (del continuo) che appartengono all'intervallo di impulso indicato. Si noti a questo proposito che le autofunzioni relativistiche in campo coulombiano per gli stati con $j = 1/2$ (${}^2s_{1/2}$ e ${}^2p_{1/2}$) diventano infinite nel centro. D'altra parte la legge di Coulomb non è valida fino al centro del nucleo, ma solo fino a una distanza da questo maggiore di R, R essendo il raggio nucleare. Ora un calcolo di orientamento dimostra che, se si fanno ipotesi plausibili sopra l'andamento del potenziale elettrico nell'interno del nucleo, il valore di $\tilde{\psi}_s \psi_s$ nel centro del nucleo risulta assai prossimo a quello che $\tilde{\psi}_s \psi_s$ assumerebbe, nel caso della legge di Coulomb, a distanza R dal centro. Applicando le note formule (6) per le autofunzioni relativistiche dello spettro continuo in campo Coulombiano, si trova con un calcolo un po' lungo ma privo di difficoltà,

$$(40)\qquad \sum_{d\eta} \tilde{\psi}_s \psi_s = d\eta \cdot \frac{32\,\pi m^3 c^3}{h^3 [\Gamma(3+2S)]^2} \left(\frac{4\pi mc R}{h}\right)^{2S} \eta^{2+2S} e^{\pi\gamma \frac{\sqrt{1+\eta^2}}{\eta}} \left| \Gamma\left(1+S+i\gamma \frac{\sqrt{1+\eta^2}}{\eta}\right)\right|^2$$

dove si è posto

$$(41)\qquad \gamma = Z/137 \quad ; \quad S = \sqrt{1-\gamma^2} - 1 .$$

La probabilità di transizione in uno stato elettronico in cui l'impulso abbia valore appartenente all'intervallo $mc\,d\eta$ diventa dunque (39)

$$(42)\qquad P(\eta)\,d\eta = d\eta \cdot g^2 \frac{256\,\pi^4}{[\Gamma(3+2S)]^2} \frac{m^5 c^4}{h^7} \left(\frac{4\pi mc R}{h}\right)^{2S} \left|\int v_m^* u_n \, d\tau\right|^2 \eta^{2+2S}$$
$$\times e^{\pi\gamma \frac{\sqrt{1+\eta^2}}{\eta}} \left|\Gamma\left(1+S+i\gamma\frac{\sqrt{1+\eta^2}}{\eta}\right)\right|^2 \left(\sqrt{1+\eta_0^2} - \sqrt{1+\eta^2}\right)^2$$

dove η_0 rappresenta l'impulso massimo dei raggi β emessi, misurato in unità mc.

Per la valutazione numerica di (42) riferiamoci al valore particolare $\gamma = 0{,}6$, che corrisponde a $Z = 82{,}2$ poiché i numeri atomici delle sostanze radioattive non sono lontani da questo valore. Per $\gamma = 0{,}6$, si ha da (41) $S = -0{,}2$. Si trova inoltre che, per $\eta < 10$, può porsi, con approssimazione sufficiente

$$(43)\qquad \eta^{1,6} e^{0,6\pi \frac{\sqrt{1+\eta^2}}{\eta}} \left|\Gamma\left(0{,}8 + 0{,}6\, i \frac{\sqrt{1+\eta^2}}{\eta}\right)\right|^2 \cong 4{,}5\,\eta + 1{,}6\,\eta^2 .$$

La (42) diviene con ciò, ponendo in essa $R = 9{,}10^{-13}$,

$$(44)\qquad P(\eta)\,d\eta = 1{,}75 \cdot 10^{95} g^2 \left|\int v_m^* u_n\, d\tau\right|^2 (\eta + 0{,}355\,\eta^2)\left(\sqrt{1+\eta_0^2} - \sqrt{1+\eta^2}\right)^2 .$$

(6) R. H. HULME, « Proc. Roy. Soc. », *133*, 381 (1931).

L'inversa della vita media si ottiene integrando (44) da $\eta = 0$ fino a $\eta = \eta_0$; si trova

$$\frac{1}{\tau} = 1{,}75 \cdot 10^{95} g^2 \left| \int v_m^* u_n d\tau \right|^2 F(\eta_0) \tag{45}$$

dove si è posto

$$F(\eta_0) = \frac{2}{3}\sqrt{1+\eta_0^2} - \frac{2}{3} + \frac{\eta_0^4}{12} - \frac{\eta_0^2}{3} \tag{46}$$

$$+ 0{,}355 \left[-\frac{\eta_0}{4} - \frac{\eta_0^3}{12} + \frac{\eta_0^5}{30} + \frac{\sqrt{1+\eta_0^2}}{4} \log\left(\eta_0 + \sqrt{1+\eta_0^2}\right) \right].$$

Per piccoli valori dell'argomento, $F(\eta_0)$ si comporta come $\eta_0^6/24$; per argomenti maggiori i valori di F sono raccolti nella tabella seguente.

TABELLA I.

η_0	$F(\eta_0)$	η_0	$F(\eta_0)$	η_0	$F(\eta_0)$	η_0	$F(\eta_0)$
0	$\eta_0^6/24$	2	1,2	4	29	6	185
1	0,03	3	7,5	5	80	7	380

LE TRANSIZIONI PROIBITE.

§ 9. Prima di passare al confronto della teoria con l'esperienza vogliamo ancora illustrare alcune proprietà delle transizioni β proibite.

Come già abbiamo detto, una transizione è proibita quando è nullo il corrispondente elemento di matrice (35). Se la rappresentazione del nucleo per stati quantici individuali dei protoni e dei neutroni rappresenta una buona approssimazione, l'elemento di matrice Q_{mn}^* si annulla, per ragione di simmetria, quando non sia

$$i = i' \tag{47}$$

i ed i' essendo il momento della quantità di moto, in unità $h/2\pi$, dello stato neutronico u_n e dello stato protonico v_m. Nel caso in cui gli stati quantici individuali non siano una buona approssimazione, alla regola di selezione (47) corrisponde l'altra

$$I = I' \tag{48}$$

dove I e I' rappresentano il momento del nucleo prima e dopo la disintegrazione β.

Le regole di selezione (47) e (48) sono assai meno rigorose delle regole di selezione dell'ottica. Si possono avere delle eccezioni da esse specialmente coi due seguenti processi:

a) La formula (26) è stata ottenuta trascurando le variazioni di ψ_s e φ_s entro l'estensione del nucleo. Se invece si tiene conto di tali variazioni, si ha una possibilità di ottenere transizioni β anche quando Q^*_{mn} si annulla. Si riconosce facilmente che l'intensità di queste transizioni sta, come ordine di grandezza, a quella dei processi permessi, nel rapporto $(R/\lambda)^2$, λ essendo la lunghezza d'onda di De Broglie delle particelle leggere. Si noti che, a parità di energia di un elettrone e di un neutrino, l'energia cinetica del primo, quando si trova in vicinanza del nucleo, è, a causa dell'attrazione elettrostatica, molto maggiore, e quindi l'effetto più importante deriva dalle variazioni di ψ_s. Una valutazione d'ordine di grandezza dell'intensità di questi processi proibiti mostra che essi, a parità dell'energia degli elettroni emessi, debbono essere un centinaio di volte meno intensi dei processi normali.

Oltre alla intensità relativamente piccola, una caratteristica di transizioni proibite di questo tipo è da ricercarsi nella differente forma della curva di distribuzione dell'energia dei raggi β, che, per transizioni proibite, deve dare un numero di particelle con piccola energia minore che nel caso normale.

b) Una seconda possibilità che avvengano transizioni β proibite dalla regola di selezione (48) dipende dal fatto, già notato alla fine del § 3, che, quando la velocità dei neutroni e dei protoni non sia trascurabile di fronte alla velocità della luce, debbono aggiungersi al termine di interazione (12) altri termini dell'ordine v/c. Se per esempio si ammettesse anche per le particelle pesanti una equazione d'onda relativistica del tipo di Dirac, si potrebbero aggiungere a (12) termini come

$$gQ\,(\alpha_x A_1 + \alpha_y A_2 + \alpha_z A_3) + \text{complesso coniugato} \tag{49}$$

$\alpha_x\ \alpha_y\ \alpha_z$ essendo le consuete matrici di Dirac per la particella pesante e $A_1\ A_2\ A_3$ le componenti spaziali del quadrivettore definito dalle (11).

Un termine del tipo (49) permette anche dei passaggi β che non soddisfano la regola di selezione (48), e la loro intensità è, rispetto a quella dei processi normali, dell'ordine di grandezza $(v/c)^2$ e cioè all'incirca 1/100. Troviamo così una seconda possibilità di transizioni proibite circa 100 volte meno intense di quelle normali.

Confronto con l'esperienza.

§ 10. La (45) stabilisce una relazione tra l'impulso massimo η_0 dei raggi β emessi da una sostanza e la sua vita media. In questa relazione interviene invero anche un elemento incognito, l'integrale

$$\int v^*_m\, u_n\, d\tau \tag{50}$$

la cui valutazione richiede la conoscenza delle autofunzioni nucleari u_n e v_m del neutrone e del protone. Nel caso delle transizioni permesse tuttavia (50) è dell'ordine di grandezza dell'unità. Ci attendiamo dunque che il prodotto

$$\tau F(\eta_0) \tag{51}$$

abbia lo stesso ordine di grandezza per tutte le transizioni permesse. Per le transizioni proibite invece la vita media sarà, come ordine di grandezza, cento volte maggiore, e corrispondentemente sarà maggiore anche il prodotto (51). Nella tabella seguente sono raccolti i prodotti $\tau F(\eta_0)$ per tutte le sostanze che si disintegrano emettendo raggi β per le quali si hanno dati sufficientemente esatti.

TABELLA II.

Elemento	τ (in ore)	η_0	$F(\eta_0)$	$\tau F(\eta_0)$
UX_2	0,026	5,4	115	3,0
RaB	0,64	2,04	1,34	0,9
ThB	15,3	1,37	0,176	2,7
ThC″	0,076	4,4	44	3,3
AcC″	0,115	3,6	17,6	2,0
RaC	0,47	7,07	398	190
RaE	173	3,23	10,5	1800
ThC	2,4	5,2	95	230
$MsTh_2$	8,8	6,13	73	640

In questa tabella sono senz'altro riconoscibili i due gruppi che avevamo attesi; del resto una tale divisione degli elementi che emettono raggi β primari in due gruppi, era già stata osservata per via empirica da Sargent [7]. I valori di η_0 sono stati presi dal lavoro citato di Sargent (per confronto si noti che è: $\eta_0 = (H\rho)_{max}/1700$). Oltre ai dati riportati nella tabella, Sargent dà i dati per altri tre elementi, avvertendo che essi non sono altrettanto attendibili come gli altri. Essi sono: UX_1 per cui si ha $\tau = 830$; $\eta_0 = 0,76$; $F(\eta_0) = 0,0065$; $\tau F(\eta_0) = 5,4$; questo elemento sembra dunque doversi attribuire al primo gruppo. Per l'AcB si ha: $\tau = 0,87$; $\eta_0 = 1,24$; $F(\eta_0) = 0,102$; $\tau F(\eta_0) = 0,09$; si trova dunque un valore di $\tau F(\eta_0)$ circa dieci volte più piccolo di quelli del primo gruppo. Infine per il RaD si ha: $\tau = 320\,000$; $\eta_0 = 0,38$ (assai incerto); $F(\eta_0) = 0,00011$; $\tau F(\eta_0) = 35$. Questo elemento viene quindi a essere situato circa a metà tra i due gruppi. Non mi è riuscito trovare dati sopra gli altri elementi che emettono raggi β primari, e cioè $MsTh_1$, UY, Ac, AcC, UZ, RaC″.

Nel complesso si può concludere da questo confronto tra teoria ed esperienza che la concordanza è certo altrettanto buona quanto si sarebbe potuto attendersi. Le divergenze osservate per gli elementi su cui si hanno dati sperimentali incerti, RaD e AcB, possono bene spiegarsi parte con la imprecisione delle misure, parte anche con oscillazioni perfettamente plausibili nel

(7) B. W. SARGENT, «Proc. Roy. Soc.», *139*, 659 (1933).

valore dell'elemento di matrice (50). Deve inoltre osservarsi che il fatto che la maggior parte delle disintegrazioni β sono accompagnate da emissione di raggy γ, sta ad indicare che la maggior parte dei processi β può lasciare il protone in differenti stati di eccitazione, ciò che dà ancora un meccanismo che può determinare delle oscillazioni nel valore di $\tau F(\eta_0)$.

Dai dati della Tabella II si può ricavare una valutazione, se pure assai grossolana della costante g. Se si ammette per esempio che, quando l'elemento di matrice (50) ha valore 1, si abbia $\tau F(\eta_0) = 1$ ora $= 3600$ sec, si trova da (45):

$$g = 4 \cdot 10^{-50}\ \mathrm{cm}^3 \cdot \mathrm{erg}.$$

valore che, naturalmente, non dà nulla più dell'ordine di grandezza.

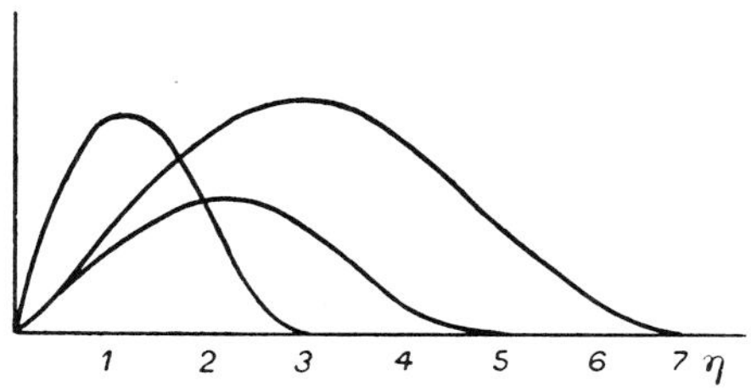

Fig. 2.

Passiamo a discutere la forma della curva di distribuzione delle velocità dei raggi β. Nel caso dei processi permessi, la curva di distribuzione, in funzione di η (cioè, a meno del fattore 1700, di $H\rho$) è rappresentata nella fig. 2, per vari valori dell'impulso massimo η_0. Le curve sono soddisfacentemente simili a quelle sperimentali raccolte da Sargent [(8)]. Solo nel tratto di piccola energia le curve di Sargent sono un po' più basse di quelle teoriche, ciò che è meglio visibile nelle curve della fig. 3 in cui come ascisse sono prese le energie invece degli impulsi. Deve notarsi però che la parte delle curve di piccola energia è assai imperfettamente conosciuta sperimentalmente [(9)]. Inoltre, per le transizioni proibite, anche teoricamente il tratto di piccola energia deve essere più basso che non per le curve relative alle transizioni permesse, rappresentate nelle figg. 2 e 3. Di questo si deve particolarmente tener conto per

Fig. 3.

(8) B. W. SARGENT, « Proc. Camb. Phil. Soc. », *28*, 538 (1932).

(9) Cfr. per esempio RUTHERFORD, ELLIS e CHADWICK, *Radiations from Radioactive Substances*, Cambridge, 1930. Ved. in particolar modo a p. 407.

il caso del RaE, che è quello meglio conosciuto dal lato sperimentale. La emissione dei raggi β di questo elemento, come risulta dal valore anormalmente grande di $\tau F(\eta_0)$ (cfr. Tabella II) è certo proibita, anzi è possibile che essa sia permessa soltanto in seconda approssimazione. Spero di potere, in una prossima comunicazione, precisare meglio l'andamento delle curve di distribuzione per transizioni proibite.

Riassumendo sembra giustificato affermare che la teoria, nella forma in cui la abbiamo qui esposta, si trova in accordo con i dati sperimentali, i quali per altro non sono sempre molto precisi. Se anche del resto, in un ulteriore confronto della teoria con l'esperienza, si dovesse arrivare a delle contraddizioni, sarebbe sempre possibile modificare la teoria senza alterarne essenzialmente i fondamenti concettuali. Si potrebbe precisamente conservare l'equazione (9) scegliendo però in modo diverso le $c_{s\sigma}$. Ciò porterebbe in particolare a una diversa forma della regola di selezione (48), e a una diversa forma della curva di distribuzione dell'energia.

Solo un ulteriore sviluppo della teoria, come anche un aumento nella precisione dei dati sperimentali, potranno indicare se una tale modificazione si renderà necessaria.

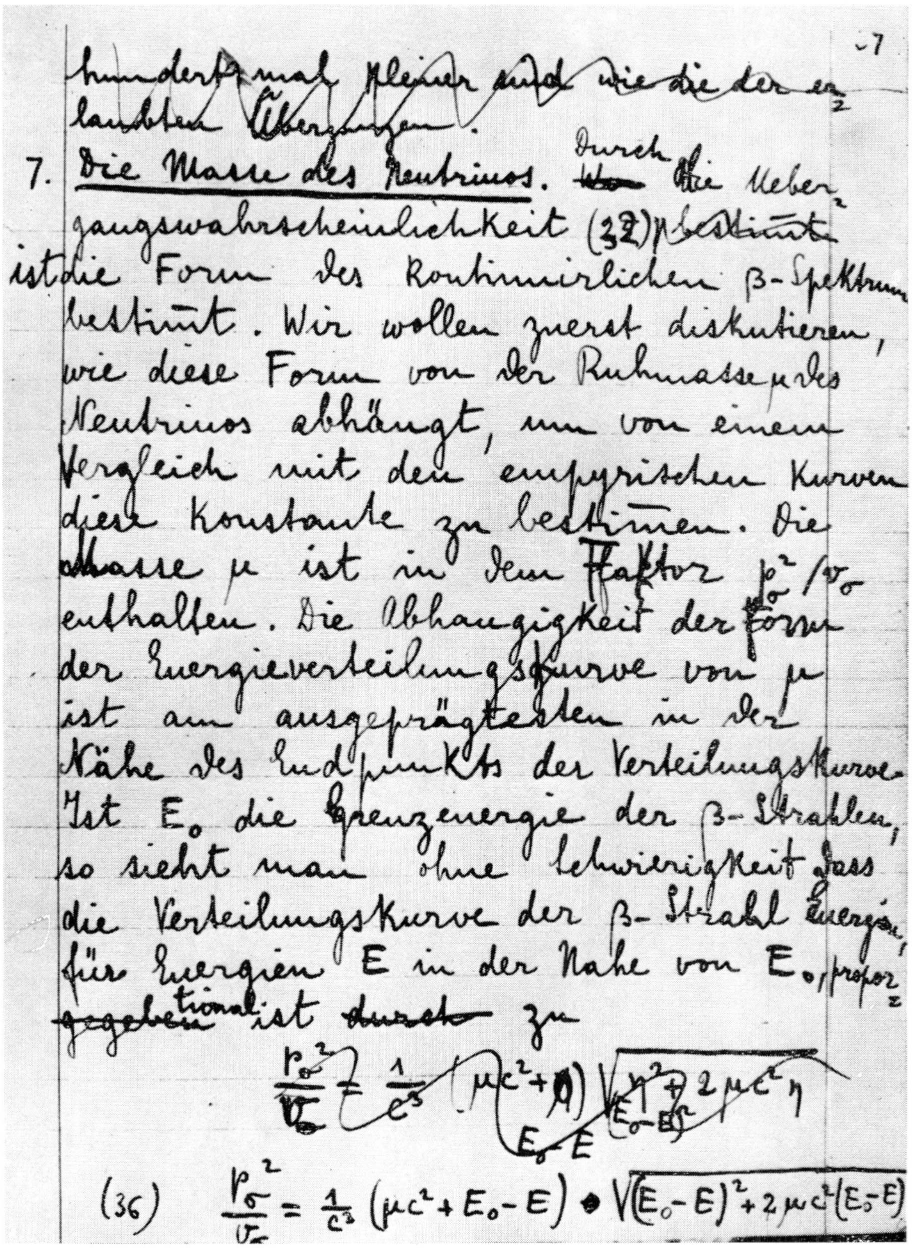

7. Die Masse des Neutrinos. Durch die Uebergangswahrscheinlichkeit (32) ist die Form des kontinuirlichen β-Spektrums bestimmt. Wir wollen zuerst diskutieren, wie diese Form von der Ruhmasse μ des Neutrinos abhängt, um von einem Vergleich mit den empyrischen Kurven diese Konstante zu bestimmen. Die Masse μ ist in dem Faktor p_σ^2/v_σ enthalten. Die Abhangigkeit der Form der Energieverteilungskurve von μ ist am ausgeprägtesten in der Nähe des Endpunkts der Verteilungskurve. Ist E_0 die Grenzenergie der β-Strahlen, so sieht man ohne Schwierigkeit dass die Verteilungskurve der β-Strahl Energien für Energien E in der Nahe von E_0, proportional ist zu

$$\frac{p_\sigma^2}{v_\sigma} = \frac{1}{c^3}\left(\mu c^2 + E_0 - E\right)\sqrt{(E_0-E)^2 + 2\mu c^2 (E_0-E)} \quad (36)$$

Manuscript of Paper N° 80 *b*.

80 *b*.

VERSUCH EINER THEORIE DER β-STRAHLEN. I. (*)

« Z. Physik », *88*, 161–171 (1934).

Eine quantitative Theorie des β–Zerfalls wird vorgeschlagen, in welcher man die Existenz des Neutrinos annimmt, und die Emission der Elektronen und Neutrinos aus einem Kern beim β–Zerfall mit einer ähnlichen Methode behandelt, wie die Emission eines Lichtquants aus einem angeregten Atom in der Strahlungstheorie. Formeln für die Lebensdauer und für die Form des emittierten kontinuierlichen β–Strahlenspektrums werden abgeleitet und mit der Erfahrung verglichen.

1. – Grundannahmen der Theorie.

Bei dem Versuch, eine Theorie der Kernelektronen sowie der β–Emission aufzubauen, begegnet man bekanntlich zwei Schwierigkeiten. Die erste ist durch das kontinuierliche β–Strahlenspektrum bedingt. Falls der Erhaltungssatz der Energie gültig bleiben soll, muss man annehmen, dass ein Bruchteil der beim β–Zerfall frei werdenden Energie unseren bisherigen Beobachtungsmöglichkeiten entgeht. Nach dem Vorschlag von W. Pauli kann man z.B. annehmen, dass beim β–Zerfall nicht nur ein Elektron, sondern auch ein neues Teilchen, das sogenannte « Neutrino » (Masse von der Grössenordnung oder kleiner als die Elektronenmasse; keine elektrische Ladung) emittiert wird. In der vorliegenden Theorie werden wir die Hypothese des Neutrinos zugrunde legen.

Eine weitere Schwierigkeit für die Theorie der Kernelektronen besteht darin, dass die jetzigen relativistischen Theorien der leichten Teilchen (Elektronen oder Neutrinos) nicht imstande sind, in einwandfreier Weise zu erklären, wie solche Teilchen in Bahnen von Kerndimensionen gebunden werden können.

Es scheint deswegen zweckmässiger, mit Heisenberg (1) anzunehmen, dass ein Kern nur aus schweren Teilchen, Protonen und Neutronen, besteht. Um trotzdem die Möglichkeit der β–Emission zu verstehen, wollen wir versuchen, eine Theorie der Emission leichter Teilchen aus einem Kern in Analogie zur Theorie der Emission eines Lichtquants aus einem angeregten Atom beim gewöhnlichen Strahlungsprozess aufzubauen. In der Strahlungstheorie ist die totale Anzahl der Lichtquanten keine Konstante: Lichtquanten entstehen, wenn sie von einem Atom emittiert werden, und verschwinden, wenn sie absorbiert werden. In Analogie hierzu wollen wir der β–Strahlentheorie folgende Annahmen zugrunde legen:

(*) Vgl. die vorläufige Mitteilung: « La Ricerca Scientifica », *2*, Heft 12 (1933).
(1) W. Heisenberg, « ZS. f. Phys. », *77*, 1 (1932).

a) Die totale Anzahl der Elektronen, sowie der Neutrinos, ist nicht notwendigerweise konstant. Elektronen (oder Neutrinos) können entstehen und verschwinden. Diese Möglichkeit hat jedoch keine Analogie zum Entstehen oder Verschwinden eines Paares, aus einem Elektron und einem Positron; falls man das Positron als Diracsches « Loch » interpretiert, kann man in der Tat diesen letzten Prozess einfach als einen Quantensprung eines Elektrons zwischen einem Zustand mit negativer Energie und einem Zustand mit positiver Energie mit Erhaltung der totalen (unendlich grossen) Anzahl der Elektronen auffassen.

b) Die schweren Teilchen, Neutronen und Protonen, können wie bei Heisenberg als zwei innere Quantenzustände des schweren Teilchens betrachtet werden. Wir formulieren dies durch die Einführung einer inneren Koordinate ρ des schweren Teilchens, welche nur zwei Werte annehmen kann: $\rho = 1$, falls das Teilchen ein Neutron ist; $\rho = -1$, falls das Teilchen ein Proton ist.

c) Die Hamilton–Funktion des aus schweren und leichten Teilchen bestehenden Systems muss so gewählt werden, dass jedem Übergang von Neutron zu Proton das Entstehen eines Elektrons und eines Neutrinos zugeordnet ist. Dem umgekehrten Prozess, Verwandlung eines Protons in ein Neutron, soll dagegen das Verschwinden eines Elektrons und eines Neutrinos zugeordnet sein. Man bemerke, dass hierdurch die Erhaltung der Ladung gesichert ist.

2. – Die in der Theorie auftretenden Operatoren.

Ein mathematischer Formalismus der Theorie in Einklang mit diesen drei Forderungen kann am leichtesten mit Hilfe der Dirac–Jordan–Kleinschen Methode [(2)] der « zweiten Quantelung » aufgebaut werden. Wir werden also die Wahrscheinlichkeitsamplituden ψ und φ der Elektronen und der Neutrinos sowie die komplex konjugierten Grössen ψ^* und φ^* als Operatoren auffassen; für die Beschreibung der schweren Teilchen werden wir dagegen die übliche Darstellung im Konfigurationsraum benutzen, wobei natürlich auch ρ als Koordinate mitgezählt werden muss.

Wir führen zuerst zwei Operatoren Q und Q^* ein, welche auf die Funktionen der zweiwertigen Variablen ρ als die linearen Substitutionen

$$Q = \begin{vmatrix} 0 & 1 \\ 0 & 0 \end{vmatrix} \quad ; \quad Q^* = \begin{vmatrix} 0 & 0 \\ 1 & 0 \end{vmatrix} \tag{1}$$

wirken. Man sieht ohne weiteres, dass Q einem Übergang von Proton zu Neutron entspricht und Q^* einem Übergang von Neutron zu Proton.

Die Bedeutung der als Operatoren aufgefassten Wahrscheinlichkeitsamplituden ψ und φ ist bekanntlich die folgende: Sei

$$\psi_1 \, \psi_2 \cdots \psi_s \cdots$$

(2) Vgl. z.B. P. Jordan u. O. Klein, « ZS. f. Phys. », *45*, 751 (1927); W. Heisenberg, « Ann. d. Phys. », *10*, 888 (1931).

ein System individueller Quantenzustände für die Elektronen. Man setze weiter

(2) $$\psi = \sum_s \psi_s a_s \quad ; \quad \psi^* = \sum_s \psi_s^* a_s^* .$$

Die Amplituden a_s und die komplex konjugierten Grössen a_s^* sind Operatoren, welche auf die Funktionen der Besetzungszahlen $N_1, N_2, \cdots, N_s, \cdots$ der individuellen Quantenzustände wirken. Im Falle des Pauli–Prinzips ist jedes der N_s nur der beiden Werte 0 und 1 fähig. Die Operatoren a_s und a_s^* sind dann folgendermassen definiert:

(3) $$\begin{cases} a_s^* \Psi(N_1 N_2 \cdots N_s \cdots) = (-1)^{N_1+N_2+\cdots+N_{s-1}} (1 - N_s) \Psi(N_1 N_2 \cdots 1 - N_s \cdots) \\ a_s \Psi(N_1 N_2 \cdots N_s \cdots) = (-1)^{N_1+N_2+\cdots+N_{s-1}} N_s \Psi(N_1 N_2 \cdots 1 - N_s \cdots). \end{cases}$$

Der Operator a_s^* entspricht der Erzeugung und der Operator a_s dem Verschwinden eines Elektrons im Quantenzustand s.

Entsprechend zu (2) setze man für die Neutrinos

$$\varphi = \sum_\sigma \varphi_\sigma b_\sigma \qquad \varphi^* = \sum_\sigma \varphi_\sigma^* b_\sigma^* .$$

Die komplex–konjugierten Grössen b_σ und b_σ^* sind Operatoren, die auf die Funktionen der Besetzungszahlen $M_1, M_2, \cdots, M_\sigma, \cdots$ der individuellen Quantenzustände $\varphi_1, \varphi_2, \cdots, \varphi_\sigma, \cdots$ der Neutrinos wirken. Nimmt man an, dass auch für die Neutrinos das Pauli–Prinzip gilt, so sind die Zahlen M_σ nur der beiden Werte 0 und 1 fähig. Es ist ferner:

(5) $$\begin{cases} b_\sigma^* \Phi(M_1 M_2 \cdots M_\sigma \cdots) = (-1)^{M_1+M_2+\cdots+M_{\sigma-1}} (1 - M_\sigma) \Phi(M_1 M_2 \cdots 1 - M_\sigma \cdots) \\ b_\sigma \Phi(M_1 M_2 \cdots M_\sigma \cdots) = (-1)^{M_1+M_2+\cdots+M_{\sigma-1}} M_\sigma \Phi(M_1 M_2 \cdots 1 - M_\sigma \cdots). \end{cases}$$

Die Operatoren b_σ und b_σ^* entsprechen dem Verschwinden bzw. dem Entstehen eines Neutrinos im Quantenzustand σ.

3. – Aufstellung der Hamilton–Funktion.

Die Energie des gesamten, aus schweren und leichten Teilchen bestehenden, Systems ist die Summe der Energien H_{schwer} der schweren Teilchen $+ H_{\text{leicht}}$ der leichten Teilchen + der Wechselwirkungsenergie H zwischen schweren und leichten Teilchen.

Das erste Glied schreiben wir, indem wir vorläufig nur ein einziges schweres Teilchen betrachten, in der Form

(6) $$H_{\text{schwer}} = \frac{1+\rho}{2} N + \frac{1-\rho}{2} P ,$$

wo N und P die Energieoperatoren des Neutrons bzw. des Protons darstellen. Für $\rho = 1$ (Neutron) reduziert sich in der Tat (6) auf N; für $\rho = -1$ (Proton) reduziert sich (6) auf P.

Die Energie H_{leicht} der leichten Teilchen nimmt die einfachste Form an, wenn man als Quantenzustände $\psi_1 \psi_2 \cdots \psi_s \cdots$ und $\varphi_1 \varphi_2 \cdots \varphi_s \cdots$ stationäre Zustände für die Elektronen bzw. die Neutrinos nimmt. Für die Elektronen soll man dabei etwa die stationären Zustände im Coulomb-Feld des Kerns, unter Berücksichtigung der Elektronenabschirmung, wählen. Für die Neutrinos kann man einfach ebene de Broglie-Wellen annehmen, da wohl die auf die Neutrinos wirkenden Kräfte keine wesentliche Rolle spielen. Seien $H_1\, H_2 \cdots H_s \cdots$ und $K_1\, K_2 \cdots K_s \cdots$ die Energien der stationären Zustände der Elektronen und der Neutrinos; dann haben wir:

$$H_{\text{leicht}} = \sum_s H_s\, N_s + \sum_\sigma K_\sigma\, M_\sigma . \tag{7}$$

Es bleibt nur noch die Wechselwirkungsenergie zu schreiben. Diese besteht erstens aus der Coulomb-Energie zwischen Proton und Elektronen; bei schweren Kernen spielt jedoch die Anziehung durch ein einziges Proton nur eine untergeordnete Rolle [(3)] und trägt in keinem Falle zum Prozess des β-Zerfalls bei. Wir wollen also dies Glied der Einfachheit halber nicht berücksichtigen. Wir müssen hingegen zur Hamilton-Funktion ein Glied addieren, das die Bedingung *c*) von Ziffer 1 erfüllt.

Ein Glied, das notwendigerweise die Verwandlung eines Protons in ein Neutron mit dem Verschwinden eines Elektrons und eines Neutrinos koppelt, hat nun nach Ziffer 2 die Form

$$Q a_s\, b_\sigma . \tag{8}$$

Der komplex konjugierte Operator

$$Q^*\, a_s^*\, b_\sigma^* \tag{8'}$$

koppelt dagegen die umgekehrten Prozesse (Verwandlung eines Neutrons in ein Proton und Entstehen eines Elektrons und eines Neutrinos).

Ein Wechselwirkungsglied, das die Bedingung *c*) erfüllt, kann also in der folgenden Form geschrieben werden:

$$H = Q \sum_{s\sigma} c_{s\sigma}\, a_s\, b_\sigma + Q^* \sum_{s\sigma} c_{s\sigma}^*\, a_s^*\, b_\sigma^* , \tag{9}$$

wo $c_{s\sigma}$ und $c_{s\sigma}^*$ Grössen darstellen, die von den Koordinaten, Impulsen usw. des schweren Teilchens abhängen können.

Zur näheren Bestimmung von H ist man auf Einfachheitskriterien angewiesen. Eine wesentliche Einschränkung in der Freiheit der Wahl von H ist durch die Erhaltung des Impulses sowie durch die Bedingung gesetzt, dass bei einer Drehung oder einer Translation der Raumkoordinaten (9) invariant bleiben muss.

Sehen wir zunächst von den Relativitätskorrektionen und der Spinwirkung ab, so ist wohl die einfachst mögliche Wahl von (9) die folgende:

$$H = g\, \{ Q \psi(x)\, \varphi(x) + Q^*\, \psi^*(x)\, \varphi^*(x) \} , \tag{10}$$

wo g eine Konstante mit den Dimensionen $L^5 M T^{-2}$ darstellt; x repräsentiert die Koordinaten des schweren Teilchens; ψ, φ, ψ^*, φ^* sind durch (2) und (4) gegeben und sind an dem Orte x, y, z des schweren Teilchens zu nehmen.

(3) Die Coulombsche Wirkung der zahlreichen übrigen Protonen muss natürlich als statisches Feld in Betracht gezogen werden.

(10) stellt keineswegs die einzig mögliche Wahl von H dar. Jeder skalare Ausdruck, wie etwa

$$L(p)\,\psi(x)\,M(p)\,\varphi(x)\,N(p) + \text{kompl. konjug.},$$

wo $L(p)$, $M(p)$, $N(p)$ passende Funktionen des Impulses des schweren Teilchens darstellen, würde ebensogut möglich sein. Da jedoch die Folgerungen aus (10) bisher mit der Erfahrung in Einklang zu sein scheinen, ist es wohl besser, sich vorläufig auf die einfachste Wahl zu beschränken.

Wesentlich ist es jedoch, den Ausdruck (10) derart zu verallgemeinern, dass man mindestens die leichten Teilchen relativistisch behandeln kann. Auch bei dieser Verallgemeinerung ist natürlich eine gewisse Willkür nicht auszuschliessen. Die einfachste Lösung des Problems dürfte die folgende sein:

Relativistisch treten an Stelle von ψ und φ je vier Diracsche Funktionen $\psi_1 \psi_2 \psi_3 \psi_4$ und $\varphi_1 \varphi_2 \varphi_3 \varphi_4$. Wir betrachten nun die 16 unabhängigen bilinearen Kombinationen aus $\psi_1 \psi_2 \psi_3 \psi_4$ und $\varphi_1 \varphi_2 \varphi_3 \varphi_4$. Bei einer Lorentz-Transformation der Koordinaten erfahren diese 16 Grössen eine lineare Transformation, eine Darstellung der Ordnung 16 der Lorentz-Gruppe. Diese Darstellung spaltet sich in verschiedene einfachere Darstellungen; im besonderen transformieren sich die vier bilinearen Kombinationen:

$$(11)\qquad \begin{cases} A_0 = -\psi_1\varphi_2 + \psi_2\varphi_1 + \psi_3\varphi_4 - \psi_4\varphi_3, \\ A_1 = \psi_1\varphi_3 - \psi_2\varphi_4 - \psi_3\varphi_1 + \psi_4\varphi_2, \\ A_2 = i\psi_1\varphi_3 + i\psi_2\varphi_4 - i\psi_3\varphi_1 - i\psi_4\varphi_2, \\ A_3 = -\psi_1\varphi_4 - \psi_2\varphi_3 + \psi_3\varphi_2 + \psi_4\varphi_1, \end{cases}$$

wie die Komponenten eines polaren Vierervektors, also wie die Komponenten des elektromagnetischen Viererpotentials. Es liegt nun nahe, die Grössen

$$g\,(QA_i + Q^* A_i^*)$$

in der Hamilton-Funktion des schweren Teilchens in einer Stellung aufzunehmen, die der Stellung der Komponenten des Viererpotentials entspricht.

Hier begegnen wir einer Schwierigkeit, welche davon herrührt, dass die relativistische Wellengleichung für die schweren Teilchen unbekannt ist. Falls die Geschwindigkeit des schweren Teilchens klein gegenüber c ist, kann man sich jedoch auf den zu eV (V = skalares Potential) analogen Term beschränken und schreiben:

$$(12)\qquad \begin{aligned} H = g\,[&Q\,(-\psi_1\varphi_2 + \psi_2\varphi_1 + \psi_3\varphi_4 - \psi_4\varphi_3) \\ &+ Q^*\,(-\psi_1^*\varphi_2^* + \psi_2^*\varphi_1^* + \psi_3^*\varphi_4^* - \psi_4^*\psi_3^*)]. \end{aligned}$$

Zu diesem Glied sollen noch andere Glieder von der Grössenordnung v/c addiert werden. Da die Geschwindigkeiten der Neutronen und Protonen in den Kernen gewöhnlich klein gegenüber der Lichtgeschwindigkeit sind, wollen wir diese Glieder vorläufig vernachlässigen (vgl. hierzu Ziffer 9).

(12) kann in symbolischer Schreibweise folgendermassen abgekürzt werden:

$$(13)\qquad H = g\,[Q\tilde{\psi}^*\,\delta\varphi + Q^*\,\tilde{\psi}\,\delta\varphi^*],$$

wo ψ und φ als vertikale Matrixspalten zu schreiben sind; das Zeichen $\sim$ verwandelt eine Matrix in die konjugiert transponierte; und es ist

$$\delta = \begin{vmatrix} 0 & -1 & 0 & 0 \\ 1 & 0 & 0 & 0 \\ 0 & 0 & 0 & 1 \\ 0 & 0 & -1 & 0 \end{vmatrix} \tag{14}$$

Mit diesen Bezeichnungen bekommt man durch Vergleich mit (9)

$$c_{s\sigma} = g \tilde{\psi}_s^* \delta \varphi_\sigma \quad ; \quad c_{s\sigma}^* = g \tilde{\psi}_s \delta \varphi_\sigma^* , \tag{15}$$

wo ψ_s und φ_s die normierten vierkomponentigen Eigenfunktionen der Zustände s (des Elektrons) und σ (des Neutrinos) darstellen. ψ und φ sind in (15) an der Stelle des schweren Teilchens, also als Funktionen von x, y, z, zu nehmen.

4. – Die Störungsmatrix.

Die Theorie des β–Zerfalls kann mit Hilfe der aufgestellten Hamilton–Funktion in voller Analogie zur Strahlungstheorie durchgeführt werden. In dieser letzteren besteht die Hamilton–Funktion bekanntlich aus der Summe: Energie des Atoms + Energie des reinen Strahlungsfeldes + Kopplungsenergie. Dieses letzte Glied wird als Störung der beiden anderen aufgefasst. In Analogie hierzu werden wir in unserem Falle die Summe

$$H_{\text{schwer}} + H_{\text{leicht}} \tag{16}$$

als ungestörte Hamilton–Funktion betrachten; hinzu kommt die durch das Kopplungsglied (13) dargestellte Störung.

Die Quantenzustände des ungestörten Systems können folgendermassen numeriert werden:

$$(\rho\,, n\,, N_1 N_2 \cdots N_s \cdots M_1 M_2 \cdots M_\sigma \cdots)\,, \tag{17}$$

wo die erste Zahl ρ einen der beiden Werte ± 1 annimmt und angibt, ob das schwere Teilchen ein Neutron oder ein Proton ist. Die zweite Zahl n numeriert den Quantenzustand des Neutrons oder des Protons. Für $\rho = 1$ (Neutron) sei die entsprechende Eigenfunktion

$$u_n(x)\,, \tag{18}$$

wo x die Koordinaten des schweren Teilchens, bis auf ρ, darstellt. Für $\rho = -1$ (Proton) sei die Eigenfunktion

$$v_m(x)\,. \tag{19}$$

Die übrigen Zahlen $N_1 N_2 \cdots N_s \cdots M_1 M_2 \cdots M_\sigma \cdots$ sind nur der beiden Werte 0 und 1 fähig und geben an, ob der betreffende Zustand des Elektrons oder des Neutrinos besetzt ist.

Fasst man nun die allgemeine Form (9) der Störungsenergie ins Auge, so sieht man, dass sie von Null verschiedene Elemente nur für solche Übergänge hat, bei denen entweder das schwere Teilchen von einem Neutron in einen Protonenzustand übergeht und zugleich ein Elektron und ein Neutrino entstehen, oder umgekehrt.

Mit Hilfe von (1), (3), (5), (9), (18), (19) findet man ohne weiteres das betreffende Matrixelement

$$H^{1\, n\, N_1\, N_2 \cdots 0_s \cdots M_1\, M_2 \cdots 0_\sigma \cdots}_{-1\, m\, N_1\, N_2 \cdots 1_s \cdots M_1\, M_2 \cdots 1_\sigma \cdots} = \pm \int v_m^* \, c_{s\sigma}^* \, u_n \, d\tau \,, \tag{20}$$

wo die Integration über den Konfigurationsraum des schweren Teilchens (bis auf die Koordinate ρ) erstreckt werden muss. Das ± –Zeichen bedeutet genauer

$$(-1)^{N_1+N_2+\cdots+N_{s-1}+M_1+M_2+\cdots+M_{\sigma-1}}$$

und wird übrigens aus den folgenden Rechnungen herausfallen. Dem entgegengesetzten Übergang entspricht ein komplex konjugiertes Matrixelement.

Führt man für $c_{s\sigma}^*$ den Wert (15) ein, so erhält man

$$H^{1\, n\, 0_s\, 0_\sigma}_{-1\, m\, 1_s\, 1_\sigma} = \pm g \int v_m^* u_n \, \tilde{\psi}_s \, \delta \varphi_\sigma^* \, d\tau \,, \tag{21}$$

wo der Kürze wegen im ersten Glied alle gleichbleibenden Indizes fortgelassen worden sind.

5. – Theorie des β–Zerfalls.

Ein β–Zerfall besteht in einem Prozess, bei welchem ein Kernneutron sich in ein Proton verwandelt und gleichzeitig mit dem geschilderten Mechanismus ein Elektron, das als β–Strahl beobachtet wird und ein Neutrino emittiert werden. Um die Wahrscheinlichkeit dieses Prozesses zu berechnen, wollen wir annehmen, dass zur Zeit $t = 0$ ein Neutron in einem Kernzustand mit Eigenfunktion $u_n(x)$ vorhanden ist und $N_s = M_\sigma = 0$, d. h. der Elektronenzustand s und der Neutrinozustand σ leer sind. Dann ist für $t = 0$ die Wahrscheinlichkeitsamplitude des Zustands $(1\,, n\,, 0_s\,, 0_\sigma)$

$$a_{1\, n\, 0_s\, 0_\sigma} = 1 \tag{22}$$

und die des Zustandes $(-1\,, m\,, 1_s\,, 1_\sigma)$, wo das Neutron in ein Proton mit der Eigenfunktion $v_m(x)$ unter Emission eines Elektrons und eines Neutrinos übergegangen ist, gleich Null.

Bei Anwendung der gewöhnlichen Störungsformeln hat man nun für eine Zeit, die kurz genug ist, damit (22) noch angenähert gültig ist:

$$\dot{a}_{-1\, m\, 1_s\, 1_\sigma} = -\frac{2\pi i}{h} H^{1\, n\, 0_s\, 0_\sigma}_{-1\, m\, 1_s\, 1_\sigma} \, e^{\frac{2\pi i}{h}(-W+H_s+K_\sigma)\, t} \,, \tag{23}$$

wo W die Energiedifferenz des Neutronen– und des Protonenzustandes darstellt.

Aus (23) erhält man (da für $t = 0\,, a_{-1\, m\, 1_s\, 1_\sigma} = 0$)

$$a_{-1\, m\, 1_s\, 1_\sigma} = - H^{1\, n\, 0_s\, 0_\sigma}_{-1\, m\, 1_s\, 1_\sigma} \, \frac{e^{\frac{2\pi i}{h}(-W+H_s+K_\sigma)\, t} - 1}{-W + H + K_\sigma} \,. \tag{24}$$

Die Wahrscheinlichkeit des betrachteten Übergangs ist also zur Zeit t

$$(25)\qquad |a_{-1\,m\,1_s\,1_\sigma}|^2 = 4\left|H^{1\,n\,0_s\,0_\sigma}_{-1\,m\,1_s\,1_\sigma}\right|^2 \frac{\sin^2 \frac{\pi t}{h}(-W + H_s + K_\sigma)}{(-W + H_s + K_\sigma)^2}\,.$$

Um die Lebensdauer des Neutronenzustands u_n zu berechnen, hat man den Ausdruck (25) über alle freien Elektronen- und Neutrinozustände zu summieren.

Eine wesentliche Vereinfachung in der Ausführung der Summe erhält man durch die Bemerkung, dass die de Broglie-Wellenlänge für Elektronen und Neutrinos mit Energien von einigen Millionen Volt wesentlich grösser ist als die Kerndimensionen. In erster Näherung kann man also die Eigenfunktionen ψ_s und φ_σ innerhalb des Kerns als Konstante betrachten (21) wird dann:

$$(26)\qquad H^{1\,n\,0_s\,0_\sigma}_{-1\,m\,1_s\,1_\sigma} = \pm\, g\,\tilde{\psi}_s\,\delta\varphi^*_\sigma \int v^*_m u_n\, d\tau\,,$$

wobei hier und im folgenden ψ_s und φ_σ an der Stelle des Kerns zu nehmen sind (vgl. Ziffer 8). Aus (26) hat man

$$(27)\qquad \left|H^{1\,n\,0_s\,0_\sigma}_{-1\,m\,1_s\,1_\sigma}\right|^2 = g^2 \left|\int v^*_m u_n\, d\tau\right|^2 \tilde{\psi}_s\,\delta\varphi^*_\sigma\,\tilde{\varphi}^*_\sigma\,\delta\psi_s\,.$$

Die Zustände σ des Neutrinos sind durch ihren Impuls p_σ und die Spinrichtung bestimmt. Falls wir zu Normierungszwecken in einem Volumen Ω quantisieren, dessen Dimensionen wir nachher ins Unendliche wachsen lassen werden, so sind die normierten Neutrinoeigenfunktionen ebene Dirac-Wellen, mit der Dichte $1/\Omega$. Eine einfache Algebra erlaubt dann in (27) den Mittelwert über alle Richtungen von p_σ und alle Spinrichtungen des Neutrinos zu nehmen. (Zu betrachten sind dabei nur die positiven Eigenwerte; die negativen sind mit einem der Diracschen Löchertheorie analogen Kunstgriff zu beseitigen). Man findet:

$$(28)\qquad \overline{\left|H^{1\,n\,0_s\,0_\sigma}_{-1\,m\,1_s\,1_\sigma}\right|^2} = \frac{g^2}{4\,\Omega}\left|\int v^*_m u_n\, d\tau\right|^2 \left(\tilde{\psi}_s\,\psi_s - \frac{\mu c^2}{K_\sigma}\,\tilde{\psi}_s\,\beta\psi_s\right),$$

wo μ die Ruhemasse des Neutrinos und β die Diracsche Matrix

$$(29)\qquad \beta = \begin{vmatrix} 1 & 0 & 0 & 0 \\ 0 & 1 & 0 & 0 \\ 0 & 0 & -1 & 0 \\ 0 & 0 & 0 & -1 \end{vmatrix}$$

darstellt. Beachtet man nun:

dass die Anzahl der Neutrinozustände positiver Energie mit Impuls zwischen p_σ und $p_\sigma + dp_\sigma$, $\frac{8\,\pi\Omega}{h^3}\,p^2_\sigma\, dp_\sigma$ ist;

dass $\frac{\partial K_\sigma}{\partial p_\sigma} = v_\sigma$, wo v_σ die Geschwindigkeit des Neutrinos im Zustand σ darstellt;

dass (25) ein scharfes Maximum in der Nähe des Wertes von p_σ hat, für den die Variation der ungestörten Energie verschwindet, d.h.

$$-W + H_s + K_\sigma = 0\,, \tag{30}$$

so kann man die Summe von (25) über σ in bekannter Weise [4] ausführen, und man findet:

$$t \cdot \frac{8\pi^3 g^2}{h^4} \left| \int v_m^* u_n \, d\tau \right|^2 \frac{p_\sigma^2}{v_\sigma} \left(\tilde{\psi}_s \psi_s - \frac{\mu c^2}{K_\sigma} \tilde{\psi}_s \beta \psi_s \right), \tag{31}$$

wo p_σ hier den Wert des Neutrinoimpulses bedeutet, für den (30) gültig ist.

6. – Bestimmungsstücke der Übergangswahrscheinlichkeit.

(31) gibt die Wahrscheinlichkeit dafür an, dass während der Zeit t ein β-Zerfall mit Übergang des Elektrons in den Zustand s stattfindet. Wie es sein soll, ist diese Wahrscheinlichkeit proportional der Zeit t (t ist als klein in Bezug auf die Lebensdauer angenommen worden); der Koeffizient von t gibt die Übergangswahrscheinlichkeit für den geschilderten Prozess an. Sie ist:

$$P_s = \frac{8\pi^3 g^2}{h^4} \left| \int v_m^* u_n \, d\tau \right|^2 \frac{p_\sigma^2}{v_\sigma} \left(\tilde{\psi}_s \psi_s - \frac{\mu c^2}{K_\sigma} \tilde{\psi}_s \beta \psi_s \right). \tag{32}$$

Man bemerke:

a) Für die freien Neutrinozustände ist immer $K_\sigma > \mu c^2$. Damit (30) befriedigt werden kann, ist also notwendig, dass

$$H_s \leqq W - \mu c^2. \tag{33}$$

Dem = -Zeichen entspricht die obere Grenze des kontinuierlichen β-Strahlspektrums.

b) Da für die freien Elektronenzustände $H_s > mc^2$ ist, bekommt man die folgende, für die Möglichkeit des β-Zerfalls notwendige Bedingung

$$W \geqq (m + \mu)\, c^2. \tag{34}$$

Ein besetzter Neutronenzustand n im Kerne muss also hoch genug über einem unbesetzten Protonenzustand m liegen, damit der β-Prozess vor sich gehen kann.

c) Nach (32) hängt P_s von den Eigenfunktionen u_n, v_m des schweren Teilchens im Kerne durch das Matrixelement

$$Q_{mn}^* = \int v_m^* u_n \, d\tau \tag{35}$$

ab. Dieses Matrixelement spielt in der β-Strahlungstheorie eine ähnliche Rolle wie das Matrixelement des elektrischen Moments eines Atoms in der Strahlungstheorie. Das Matrixelement (35) hat normalerweise die Grössenordnung 1; durch besondere Symmetrieeigenschaften von u_n und v_m kann es jedoch

(4) Für die genaue Beschreibung der Methode, solche Summen auszuführen, vgl. irgendeinen Aufsatz über Strahlungstheorie; etwa: E. Fermi, « Rev. Mod. Phys. », *4*, 87 (1932).

oft vorkommen, dass Q^*_{mn} verschwindet. In solchen Fällen sprechen wir von *verbotenen β-Übergängen*. Man muss natürlich nicht erwarten, dass die verbotenen Übergänge überhaupt nicht vorkommen, da (32) nur eine Näherungsformel ist. Wir werden in Ziffer 9 etwas über diesen Typ von Übergängen sprechen.

7. – Die Masse des Neutrinos.

Durch die Übergangswahrscheinlichkeit (32) ist die Form des kontinuierlichen β-Spektrums bestimmt. Wir wollen zuerst diskutieren, wie diese Form von der Ruhemasse μ des Neutrinos abhängt, um von einem Vergleich mit den empirischen Kurven diese Konstante zu bestimmen. Die Masse μ ist in dem Faktor p_σ^2/v_σ enthalten. Die Abhängigkeit der Form der Energie-

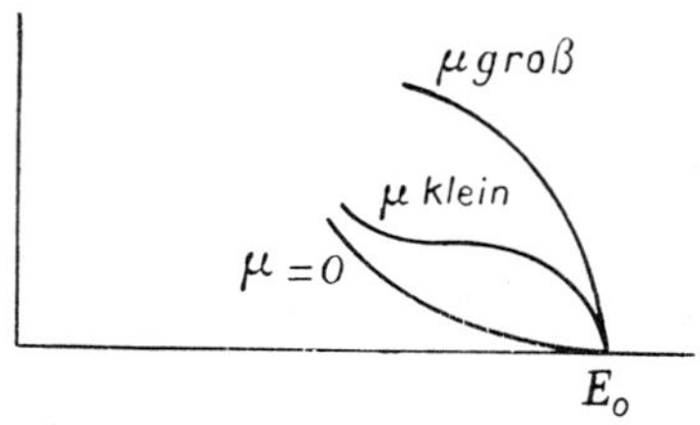

Fig. 1.

verteilungskurve von μ ist am meisten ausgeprägt in der Nähe des Endpunktes der Verteilungskurve. Ist E_0 die Grenzenergie der β-Strahlen, so sieht man ohne Schwierigkeit, dass die Verteilungskurve für Energien E in der Nähe von E_0 bis auf einen von E unabhängigen Faktor sich wie

$$\frac{p_\sigma^2}{v_\sigma} = \frac{1}{c^3} (\mu c^2 + E_0 - E) \sqrt{(E_0 - E)^2 + 2 \mu c^2 (E_0 - E)} \tag{36}$$

verhält.

In der Fig. 1 ist das Ende der Verteilungskurve für $\mu = 0$ und für einen kleinen und einen grossen Wert von μ gezeichnet. Die grösste Ähnlichkeit mit den empirischen Kurven zeigt die theoretische Kurve für $\mu = 0$.

Wir kommen also zu dem Schluss, dass die Ruhemasse des Neutrinos entweder Null oder jedenfalls sehr klein in Bezug auf die Masse des Elektrons ist [5]. In den folgenden Rechnungen werden wir die einfachste Hypothese $\mu = 0$ einführen. Es wird dann (30)

$$v_\sigma = c \quad ; \quad K_\sigma = p_\sigma c \quad ; \quad p_\sigma = \frac{K_\sigma}{c} = \frac{W - H_s}{c}. \tag{37}$$

Die Ungleichungen (33), (34) werden jetzt:

$$H_s \leqq W \quad ; \quad W \geqq mc^2. \tag{38}$$

Und die Übergangswahrscheinlichkeit (32) nimmt die Form an:

$$P_s = \frac{8 \pi^3 g^2}{c^3 h^4} \left| \int v_m^* u_n \, d\tau \right|^2 \tilde{\psi}_s \psi_s (W - H_s)^2. \tag{39}$$

(5) In einer kürzlich erschienenen Notiz kommt F. Perrin, «C. R.», *197*, 1625 (1933), mit qualitativen Überlegungen zu demselben Schluss.

8. – Lebensdauer und Form der Verteilungskurve für « erlaubte » Übergänge.

Aus (39) kann man eine Formel ableiten, welche angibt wieviel β–Übergänge in der Zeiteinheit stattfinden, für welche das β–Teilchen einen Impuls zwischen $mc\eta$ und $mc(\eta + d\eta)$ erhält. Dazu muss man eine Formel für die Summe von $\tilde{\psi}_s \psi_s$ am Orte des Kerns über alle im kontinuierlichen Spektrum liegenden Quantenzustände des betreffenden Intervalls ableiten.

Dabei sei bemerkt, dass die relativistischen Eigenfunktionen im Coulomb–Feld für die Zustände mit $j = 1/2$ (${}^2s_{1/2}$ und ${}^2p_{1/2}$) für $r = 0$ unendlich gross werden. Nun gehorcht aber die Kernanziehung für die Elektronen dem Coulombschen Gesetz nur bis $r > \rho$, wo ρ hier den Kernradius bedeutet. Eine Überschlagsrechnung zeigt, dass, wenn man plausible Annahmen über den Verlauf des elektrischen Feldes innerhalb des Kerns macht, der Wert von $\tilde{\psi}_s \psi_s$ im Mittelpunkt einen Wert hat, der sehr nahe dem Werte liegt, den $\tilde{\psi}_s \psi_s$ im Falle des Coulomb–Gesetzes in der Entfernung ρ vom Mittelpunkt annehmen würde.

Durch Heranziehung der bekannten Formeln (6) für die relativistischen Eigenfunktionen des Kontinuums im wasserstoffähnlichen Falle findet man also nach einer ziemlich langwierigen Rechnung

$$(40)\qquad \sum_{d\eta} \tilde{\psi}_s \psi_s = d\eta \cdot \frac{32\,\pi m^3 c^3}{h^3 [\Gamma(3 + 2S)]^2} \left(\frac{4\,\pi mc\rho}{q}\right)^{2S} \eta^{2+2S} e^{\pi\gamma \frac{\sqrt{1+\eta^2}}{\eta}} \left|\Gamma\left(1 + S + i\gamma \frac{\sqrt{1+\eta^2}}{\eta}\right)\right|^2,$$

wo

$$(41)\qquad \gamma = Z/137 \quad ; \quad S = \sqrt{1 - \gamma^2} - 1 \,.$$

Die Übergangswahrscheinlichkeit in einen Elektronenzustand mit einem Impuls des Intervalls $mc\, d\eta$ wird dann nach (39):

$$(42)\qquad P(\eta)\, d\eta = d\eta \cdot g^2 \frac{256\,\pi^4}{[\Gamma(3 + 2S)]^2} \frac{m^5 c^4}{h^7} \left(\frac{4\,\pi mc\rho}{h}\right)^{2S} \left|\int v_m^* u_n \, d\tau\right|^2 \cdot$$

$$\cdot\, \eta^{2+2S} e^{\pi\gamma \frac{\sqrt{1+\eta^2}}{\eta}} \left|\Gamma\left(1 + S + i\gamma \frac{\sqrt{1+\eta^2}}{\eta}\right)\right|^2 \left(\sqrt{1 + \eta_0^2} - \sqrt{1 + \eta^2}\right)^2$$

wo η_0 den in Einheiten mc gemessenen maximalen Impuls der emittierten β–Strahlen darstellt.

Die numerische Auswertung von (42) kann man etwa für $\gamma = 0{,}6$, d.h. $Z = 81{,}2$ machen, da ja die Atomnummern der radioaktiven Stoffe nicht weit von diesem Wert liegen. Für $\gamma = 0{,}6$ ist nach (41) $S = -0{,}2$. Man findet weiter, dass für $\eta < 10$ die folgende Formel angenähert gilt:

$$(43)\qquad \eta^{1,6} e^{0,6\,\pi \frac{\sqrt{1+\eta^2}}{\eta}} \left|\Gamma\left(0{,}8 + 0{,}6\, i \frac{\sqrt{1+\eta^2}}{\eta}\right)\right|^2 \simeq 4{,}5\,\eta + 1{,}6\,\eta^2.$$

(6) R. H. Hulme, « Proc. Roy. Soc. », London (A), *133*, 381 (1931).

Formel (42) wird damit, wenn man $\rho = 9 \cdot 10^{-13}$ setzt:

$$(44) \quad P(\eta)\,d\eta = 1{,}75 \cdot 10^{95} g^2 \left| \int v_m^* \, u_n \, d\tau \right|^2 (\eta + 0{,}355\,\eta^2) \left(\sqrt{1+\eta_0^2} - \sqrt{1+\eta^2}\right)^2 .$$

Die reziproke Lebensdauer erhält man aus (44) durch Integration von $\eta = 0$ bis $\eta = \eta_0$; man findet:

$$(45) \quad \frac{1}{\tau} = 1{,}75 \cdot 10^{95} g^2 \left| \int v_m^* \, u_n \, d\tau \right| F(\eta_0) ,$$

wo

$$(46) \quad F(\eta_0) = \frac{2}{3}\left(\sqrt{1+\eta_0^2} - 1\right) + $$

$$+ \frac{\eta_0^4}{12} - \frac{\eta_0^2}{3} + 0{,}355 \left[\frac{\eta_0}{4} - \frac{\eta_0^3}{12} + \frac{\eta_0^5}{30} + \frac{\sqrt{1+\eta_0^2}}{4} \log\left(\eta_0 + \sqrt{1+\eta_0^2}\right) \right] .$$

Für kleine Argumente verhält sich $F(\eta_0)$ wie $\eta_0^6/24$; für grössere Argumente sind die Werte von F in der folgenden Tabelle zusammengestellt.

TABELLE I.

η_0	$F(\eta_0)$	η_0	$F(\eta_0)$	η_0	$F(\eta_0)$	η_0	$F(\eta_0)$
0	$\eta_0^6/24$	2	1,2	4	29	6	185
1	0,03	3	7,5	5	80	7	380

9. – Die verbotenen Übergänge.

Bevor wir zum Vergleich mit der Erfahrung übergehen, wollen wir noch einige Eigenschaften der verbotenen β–Übergänge diskutieren.

Wie schon bemerkt, ist ein Übergang verboten, wenn das zugehörige Matrixelement (35) verschwindet. Falls nun die Darstellung des Kerns mit individuellen Quantenzuständen der Neutronen und der Protonen eine gute Näherung ist, verschwindet immer Q_{mn}^* aus Symmetriegründen, wenn nicht

$$(47) \quad i = i' ,$$

wo i und i' die Impulsmomente (in Einheiten $h/2\pi$) des Neutronenzustands u_n und des Protonenzustands v_m darstellen. Der Auswahlregel (47) entspricht, falls die individuellen Zustände keine gute Näherung sind, die allgemeinere

$$(48) \quad I = I' ,$$

wo I und I′ die Impulsmomente des Kerns vor und nach dem β–Zerfall bedeuten.

Die Auswahlregeln (47) und (48) sind bei weitem nicht so scharf wie die Auswahlregeln der Optik. Es gibt hauptsächlich zwei Prozesse, wodurch ein Durchbrechen dieser Auswahlregeln möglich ist:

a) Formel (26) ist durch Vernachlässigung der Variationen von ψ_s und φ_σ innerhalb der Kernausdehnung erhalten worden. Falls man aber ψ_s und φ_σ

im Bereich des Kerns nicht als Konstante betrachtet, so erhält man die Möglichkeit von β-Übergängen auch in Fällen, wo Q^*_{mn} verschwindet.

Es ist leicht einzusehen, dass die Intensität solcher Übergänge zur Intensität der erlaubten Prozesse grössenordnungsmässig im Verhältnis $(\rho/\lambda)^2$ steht, wo λ die de Broglie-Wellenlänge der leichten Teilchen darstellt. Man bemerke hierzu, dass, bei gleicher Energie, die kinetische Energie der Elektronen am Orte des Kerns wegen der elektrostatischen Anziehung erheblich grösser ist als die der Neutrinos; die grösste Wirkung rührt also von der Variation von ψ_s her. Eine Abschätzung der Intensität dieser verbotenen Prozesse zeigt, dass sie rund 100 mal schwächer sein müssen als die nach (48) erlaubten Übergänge, für welche β-Teilchen der gleichen Energie emittiert werden.

Ein Merkmal für verbotene Übergänge dieses Typs könnte man nicht nur in der verhältnismässig längeren Lebensdauer, sondern auch in der verschiedenen Form der Energieverteilungskurve der β-Strahlen erblicken; man findet nämlich, dass für diese Übergänge die Verteilungskurve für kleine Energien tiefer liegen muss als im normalen Falle.

b) Eine zweite Möglichkeit von nach (48) verbotenen Übergängen folgt aus der am Ende von Ziffer 3 bemerkten Tatsache, dass, falls man die Geschwindigkeit der schweren Kernbestandteile nicht gegen die Lichtgeschwindigkeit vernachlässigt, zum Wechselwirkungsglied (12) noch weitere von der Grössenordnung v/c hinzutreten. Falls man etwa auch für die schweren Teilchen eine relativistische Wellengleichung vom Diracschen Typus annimmt, könnte man z. B. zu (12) Terme wie

$$gQ\,(\alpha_x A_1 + \alpha_y A_2 + \alpha_z A_3) + \text{komplex konjugiert} \tag{49}$$

addieren, wo $\alpha_x \alpha_y \alpha_z$ die Diracschen Matrizen für das schwere Teilchen bedeuten und A_1, A_2, A_3 die Raumkomponenten des von (11) definierten Vierervektors sind. Das Glied (49) würde zu (12) in demselben Verhältnis stehen wie die Terme $e\,V$ bzw. $e\,(\alpha, U)$ (V = skalares Potential; U = Vektorpotential) zu der Diracschen Hamilton-Funktion.

Ein Wechselwirkungsglied wie (49) würde natürlich auch verbotene Übergänge ermöglichen, mit einer relativen Intensität von der Grössenordnung $(v/c)^2$ in bezug auf die der erlaubten Übergänge. Dies gibt also eine zweite Möglichkeit für das Vorhandensein von Übergängen, die etwa 100 mal schwächer sind als die normalen.

10. – Vergleich mit der Erfahrung.

Formel (45) gibt eine Beziehung zwischen dem maximalen Impuls der emittierten β-Strahlen und der Lebensdauer der β-strahlenden Substanz. In dieser Beziehung tritt zwar noch ein unbekanntes Element auf, nämlich das Integral

$$\int v^*_m\, u_n\, d\tau, \tag{50}$$

für dessen Auswertung eine Kenntnis der Eigenfunktionen des Protons und des Neutrons im Kern notwendig wäre. Im Falle der erlaubten Übergänge ist jedoch (50) von der Grössenordnung 1. Man kann also erwarten, dass das Produkt

$$\tau F(\eta_0) \tag{51}$$

für alle erlaubten Übergänge dieselbe Grössenordnung hat. Falls aber der betreffende Übergang verboten ist, ist ihre Lebensdauer rund 100 mal grösser als im normalen Falle und auch das Produkt (51) wird entsprechend grösser.

In der Tabelle 2 sind die Produkte (51) für die radioaktiven Elemente zusammengestellt, für welche man genügende Daten über das kontinuierliche β-Spektrum hat.

TABELLE 2.

Element	τ (Stunden)	η_0	$F(\eta_0)$	$\tau F(\eta_0)$
UX_2	0,026	5,4	115	3,0
RaB	0,64	2,04	1,34	0,9
ThB	15,3	1,37	0,176	2,7
ThC″	0,076	4,4	44	3,3
AcC″	0,115	3,6	17,6	2,0
RaC	0,47	7,07	398	190
RaE	173	3,23	10,5	1800
ThC	2,4	5,2	95	230
$MsTh_2$	8,8	6,13	73	640

Aus der Tabelle sind die zwei erwarteten Gruppen ohne weiteres erkennbar; eine solche Einteilung ist übrigens bereits von Sargent [7] auf empirischem Wege festgestellt worden. Die Werte von η_0 sind aus der genannten Arbeit von Sargent genommen (zum Vergleich bemerke man, dass: $\eta_0 = (H\rho)_{max}/1\,700$). Die von Sargent als nicht zuverlässig angegebenen Werte von η_0 passen nicht besonders gut in die Einteilung: für UX_1 hat man $\tau = 830$; $\eta_0 = 0{,}76$; $F(\eta_0) = 0{,}0065$; $\tau F(\eta_0) = 5{,}4$; dieses Element scheint also zur ersten Gruppe zu passen. Für AcB hat man die folgenden Daten: $\tau = 0{,}87$; $\eta_0 = 1{,}24$; $F(\eta_0) = 0{,}102$; $\tau F(\eta_0) = 0{,}09$, also ein τF-Wert etwa zehnmal kleiner als die der ersten Gruppe. Für RaD hat man $\tau = 320\,000$; $\eta_0 = 0{,}38$ (sehr unsicher); $F(\eta_0) = 0{,}00011$; $\tau F(\eta_0) = 35$. RaD liegt also ungefähr in der Mitte zwischen den beiden Gruppen. Ich habe keine Daten über die anderen β-strahlenden Elemente $MsTh_1$, UY, Ac, AcC, UZ, RaC″ gefunden.

(7) B. W. SARGENT, « Proc. Roy. Soc. », London (A), *139*, 659 (1933).

Aus den Daten der Tabelle 2 kann man eine, wenn auch sehr grobe, Abschätzung der Konstante g gewinnen. Nimmt man etwa an, dass in den Fällen wo (50) gleich Eins wird, man $\tau F(\eta_0) = 1$ hat (d.h., in Sekunden, $= 3600$), so bekommt man aus (45):

$$g = 4 \cdot 10^{-50} \text{ cm}^3 \cdot \text{erg}.$$

Dieser Wert gibt natürlich nur die Grössenordnung von g.

Zusammenfassend kann man sagen, dass dieser Vergleich von Theorie und Erfahrung eine so gute Übereinstimmung gibt, wie man nur erwarten konnte. Die bei den als experimentell unsicheren Elementen RaD und AcB festgestellten Abweichungen können wohl teilweise durch Ungenauigkeit der Messungen erklärt werden, teilweise auch durch etwas abnorm grosse aber

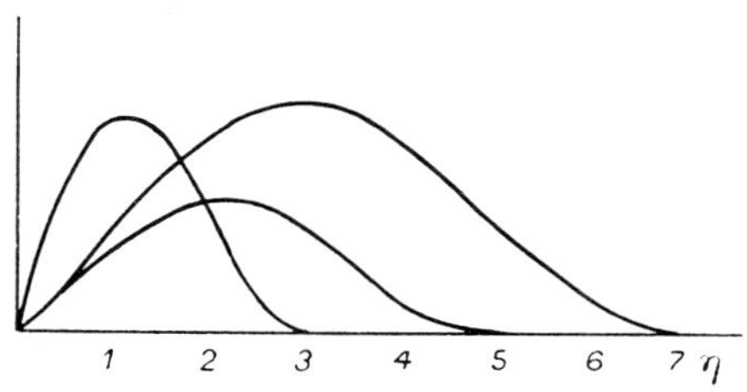

Fig. 2.

gar nicht unplausible Schwankungen des Matrixelements (50). Man hat weiter zu bemerken, dass man aus der den β-Zerfall begleitenden γ-Strahlung schliessen kann, dass die meisten β-Zerfälle zu verschiedenen Endzuständen des Protons führen können, worduch wieder Schwankungen in dem $\tau F(\eta_0)$-Wert erklärt werden können.

Wir wenden uns jetzt zur Frage nach der Form der Geschwindigkeitsverteilungskurve der emittierten β-Strahlen. Für den Fall der erlaubten Übergänge ist die Verteilungskurve als Funktion von η (d.h. bis auf den Faktor 1700, von $H\rho$) durch (44) gegeben. Verteilungskurven für verschiedene Werte von η_0 sind in der Fig. 2 zusammengestellt, wobei für die Bequemlichkeit der Zeichnung die Ordinateneinheit in den verschiedenen Fällen passend gewählt worden ist. Diese Kurven zeigen eine befriedigende Ähnlichkeit etwa zu den von Sargent [(8)] zusammengestellten Verteilungskurven. Nur in dem Teil der Kurve kleiner Energie liegen die Kurven von Sargent

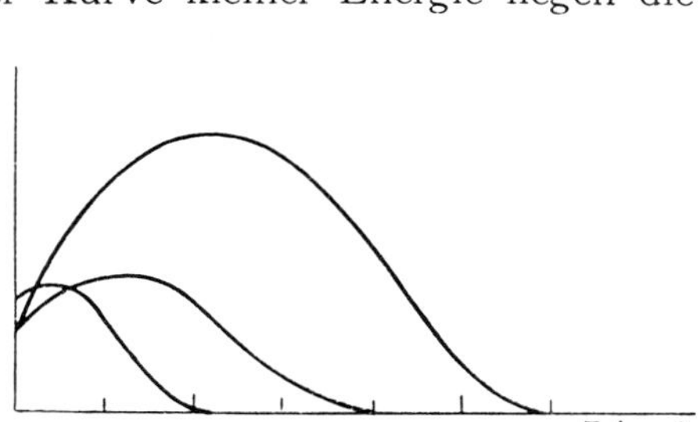

Fig. 3.

(8) B. W. SARGENT, « Proc. Cambridge Phil. Soc. », *28*, 538 (1932).

etwas tiefer als die theoretischen. Dies ist deutlicher in der Fig. 3 zu sehen, wo als Abszisse die Energie an Stelle des Impulses genommen worden ist. (Hierzu muss man jedoch bemerken, dass die experimentelle Kenntnis des Verteilungsgesetzes für kleine Energien besonders unsicher ist). (9) Übrigens hat man für die verbotenen Übergänge auch theoretisch Kurven zu erwarten, die im Gebiet kleiner Energie tiefer liegen als die der Fig. 2 und 3. Dieser letzte Punkt ist besonders für den Fall der experimentell verhältnismässig gut bekannten Kurve des RaE zu beachten. Aus der Tabelle 2 sieht man nämlich, dass RaE einen sehr grossen $\tau F(\eta_0)$–Wert hat; der β–Zerfall des RaE ist also gewiss verboten und wird sogar möglicherweise nur erst in zweiter Näherung erlaubt. Ich hoffe in einer nächsten Mitteilung etwas Genaueres über den Verlauf der Energieverteilungskurven für die verbotenen Übergänge sagen zu können.

Zusammenfassend darf man wohl sagen, dass die Theorie in der hier angegebenen Form in Übereinstimmung mit den allerdings nicht immer besonders genauen experimentellen Daten ist. Sollte man übrigens auch bei einem näheren Vergleich von Theorie und Erfahrung zu Widersprüchen kommen, so wäre es noch möglich, die Theorie abzuändern, ohne ihre begrifflichen Fundamente zu berühren. Man könnte nämlich Gleichung (9) behalten und eine andere Wahl der $c_{s\sigma}$ treffen. Diese könnte insbesondere zu einer Abänderung der Auswahlregel (48) führen und eine andere Form der Energieverteilungskurve sowie der Abhängigkeit der Lebensdauer von der maximalen Energie ergeben. Ob eine solche Änderung notwendig sein wird, kann jedoch erst durch eine weitere Entwicklung der Theorie und möglicherweise auch durch eine Verschärfung der experimentellen Daten gezeigt werden.

(9) Vgl. z.B. E. RUTHERFORD, B. ELLIS u. J. CHADWICK, *Radiations from Radioactive Substances*, Cambridge 1932. Siehe insbesondere S. 407.

81.

ZUR BEMERKUNG VON G. BECK UND K. SITTE (*)

«Z. Physik», *89*, 522 (1934).

Es sei mir gestattet, zur Note von Beck und Sitte einige kurze Bemerkungen hinzuzufügen.

Die Neutrinohypothese ist bekanntlich aus den Schwierigkeiten der Energiebilanz beim β–Zerfall entstanden. Obgleich die Existenz des Neutrinos sehr hypothetisch ist und wohl sicher auch in der nächsten Zukunft einer direkten experimentellen Prüfung unzugänglich sein wird, gibt sie jedoch heute, meines Wissens, die einzige Möglichkeit, der Energieschwierigkeit zu entgehen, ohne die Fundamente der Theorie sehr tief zu berühren. Gerade in diesem Punkte sehe ich die grösste Differenz zwischen der Theorie von Beck und Sitte und der von mir vorgeschlagenen Theorie. Diese letzte setzt uns instand, ein formal vollständiges Bild des β–Zerfalls zu geben, während bei der Theorie von Beck und Sitte der Prozess der Anlagerung des Positrons zum Kerne als ein im Rahmen der heutigen Theorien unbegreifliches Phänomen erscheint, bei dem die Energie nicht erhalten wird.

Was die Frage der Form der Energieverteilungskurve für kleine Energien betrifft, sei bemerkt, dass bei verbotenen Übergängen, welche durch die Variationen der Eigenfunktionen des Elektrons und des Neutrinos am Orte des Kernes ermöglicht werden, praktisch bei weitem die Variation der Eigenfunktion des Elektrons die wichtigste ist, da diese, als Folge der Potentialmulde der Kernanziehung, viel stärker variiert als die Eigenfunktion des Neutrinos. Dieses hat zur Folge, dass die Intensitätsabnahme des β–Spektrums bei kleinen Energien bei verbotenen Übergängen sehr ausgeprägt ist; desto mehr, je grösser die Änderung des Kernmoments ist. Bei der grossen Unschärfe unserer Kenntnisse der Energieverteilungskurve kann man die Theorie an diesem Punkt noch nicht mit der Erfahrung vergleichen.

Was den letzten von Beck und Sitte erwähnten Punkt betrifft, verweise ich auf eine Arbeit von G. C. Wick [1], wo die induzierte Emission von Positronen vom Standpunkt der von mir vorgeschlagenen Theorie diskutiert wird.

(*) G. Beck u. K. Sitte, «Zs. f. Phys.», *89*, 259 (1934).
(1) G. C. Wick, «Rendiconti Lincei», *19*, 319 (1934).

N° 82.

This paper is the last in the series on the application of the statistical method to atomic properties. The possibility of working out a systematic calculation of atomic eigenfunctions had been considered several times at the Institute. Indeed the agreement between calculation and experiment obtained for numerous properties of atoms and ions suggested that such a project might be very useful. There was even talk of preparing a manual containing the eigenfunctions of a large number of orbits of many elements in the form of numerical tables. The manual was jokingly referred to as "Thesaurus Ψ–arum".

Paper N° 82 represents the first step toward the preparation of such a manual. It also represents the last contribution to this field of research since the discovery of artifical radioactivity induced by neutron bombardment, in the winter of 1934, brought about a change in the direction of the Institute's scientific activity.

The paper itself contains the numerical tables of the ∞s–eigenfunctions for 14 values of the atomic number Z ranging between 10 and 92. The arrangement of the numerical results is such as to allow the computation by interpolation of the eigenfunction of the s-terms of any other element with $Z \geq 10$.

This paper differs from all Fermi's previous works, which where based on Schrödinger's equation, because Fermi made use of the relativistic equations in the Darwin approximation.

The actual calculations were carried out by some students and young Ph. D's (A. Biava, F. Colajori, V. Crocco, G. Giovene, E. Medi and R. Nuzzo) who kindly helped with this laborious task.

A notebook entitled "Thesaurus" containing the numerical tables published in paper N° 82 is preserved in the Domus Galilaeana Archives at Pisa.

E. AMALDI.

82.

LE ORBITE ∞s DEGLI ELEMENTI (*)

E. FERMI ed E. AMALDI, «Mem. Accad. d'Italia», *6* (1) (Fis.), 119–149 (1934).

RIASSUNTO. — Lo scopo di questo lavoro è di porre le basi per il calcolo sistematico di tutte le autofunzioni degli elementi chimici col metodo di Thomas–Fermi. Il potenziale viene scelto in modo un po' diverso dai lavori precedenti ed il problema viene trattato colle equazioni relativistiche. Il metodo viene applicato al calcolo delle autofunzioni ∞s.

Secondo la meccanica quantistica, le proprietà di un atomo si possono calcolare facilmente qualora siano note le autofunzioni e gli autovalori degli stati quantici in cui l'atomo si può trovare. Le autofunzioni dipendono, in modo generalmente complicato, dalle coordinate di tutti gli elettroni. Tuttavia una schematizzazione efficace ed in molti casi sufficiente si può ottenere considerando delle autofunzioni individuali per i singoli elettroni; immaginando cioè ciascuno di essi come isolato in un opportuno campo di poten-

(*) Presentata nell'Adunanza del 18 maggio 1934.

ziale che rappresenta l'azione del nucleo e degli altri elettroni. Da queste autofunzioni individuali si risale all'autofunzione dello stato quantico dell atomo formando i prodotti antisimmetrizzati delle autofunzioni degli stati quantici individuali che intervengono nella configurazione considerata.

Il problema si riconduce con questa schematizzazione al calcolo delle autofunzioni e degli autovalori dei singoli elettroni di ciascuna specie atomica.

Per risolvere tale problema bisogna trovare le soluzioni della equazione di Schroedinger nella quale compare esplicitamente il potenziale agente sull'elettrone in questione, dovuto all'azione del nucleo e di tutti gli altri elettroni dell'atomo.

Per la ricerca del potenziale si può procedere con diversi gradi di approssimazione: un primo grado si ottiene col metodo statistico di Thomas–Fermi [1] in cui gli elettroni vengono considerati come un gas degenere in equilibrio per effetto dell'attrazione nucleare. Questo metodo pur non essendo molto preciso presenta tuttavia il vantaggio di una grande semplicità in quanto che, per mezzo di una sola funzione calcolata numericamente una volta per tutte, si può rappresentare il comportamento di tutti gli atomi.

Un metodo numerico più approssimato è quello del « selfconsistent field », di Hartree [2]. In questo metodo tuttavia la complicazione assai maggiore dei calcoli non sembra corrispondere, almeno di solito, alla maggior esattezza dei risultati. Più preciso del metodo di Hartree in quanto tiene conto degli effetti di risonanza, è il metodo proposto da Fock [3] che tuttavia per la sua estrema complicazione non è ancora stato applicato a nessun atomo.

Proponendoci in questo lavoro calcoli numerici sistematici delle autofunzioni di tutti gli elementi, abbiamo per ragioni di praticità, data la preferenza al metodo statistico, anche perché esso in ogni caso fornisce la base per calcoli numerici più approssimati.

Dal lato pratico il metodo presenta anche il vantaggio che le autofunzioni di un elettrone di determinati numeri quantici dipendono con continuità dal numero atomico per modo che non è necessario volendo conoscere per esempio, le autofunzioni ∞s di tutti gli atomi, calcolarle elemento per elemento, ma sarà sufficiente eseguire i calcoli per una quindicina di elementi scelti opportunamente nel sistema periodico, deducendo poi le autofunzioni ∞s di tutti gli altri atomi per interpolazione.

Nel presente lavoro vengono esposti i metodi generali in base ai quali si eseguiscono tali calcoli e la loro applicazione al calcolo delle autofunzioni ∞s. I calcoli sono stati eseguiti per gli elementi:

Ne , Si , K , Fe , Ga , Rb , Mo , Ag , I , Ce , Ho , W , Hg , U

mentre per gli altri si possono ottenere le autofunzioni per interpolazione.

(1) L. H. THOMAS, « Proc. Camb. Phil. Soc. », *23*, 542 (1927); E. FERMI, « Rend. Lincei », *6*, 602 (1927); *7*, 342 (1928); « ZS. f. Phys. », *48*, 73; *49*, 550 (1928).

(2) D. R. HARTREE, « Proc. Camb. Phil. Soc. », *24*, 89, III (1928).

(3) V. FOCK, « ZS. f. Phys. », *61*, 126 (1930).

Rispetto ai calcoli di autofunzioni eseguiti precedentemente pure col metodo statistico, sono stati introdotti due miglioramenti, consistenti l'uno in una differente scelta del potenziale, determinato in modo che anche per un atomo neutro il raggio risulti finito; e l'altro nel fare tutti i calcoli colle equazioni relativistiche anziché coll'equazione di Schroedinger, ciò che ha una importanza non trascurabile specialmente per le autofunzioni *s* degli elementi pesanti. Nel paragrafo 2 verrà esposto il metodo di calcolo del potenziale agente sull'elettrone in questione: nel paragrafo 3 verranno prese in considerazione le equazioni relativistiche e verrà mostrato in qual modo convenga trasformarle per poter eseguire in modo comodo i calcoli numerici.

Vengono quindi date le tabelle generali che sono servite per il calcolo delle autofunzioni ∞ *s* e che servono per il calcolo di molte altre autofunzioni.

Dopo di che vengono date le autofunzioni ∞ *s*.

Vogliamo in fine notare che la conoscenza delle autofunzioni ∞ *s* permette di calcolare assai rapidamente anche le autofunzioni *s* ottiche poiché queste non differiscono sensibilmente dalle ∞ *s* nella parte più interna dell'atomo, ossia nella parte ove è più difficile e gravoso il calcolo numerico.

2. Calcolo del potenziale. – Tutte le considerazioni svolte in questo paragrafo si riferiscono al caso più generale di un atomo di numero atomico Z, jonizzato *z* volte.

Per stabilire l'equazione differenziale che determina il potenziale in funzione della distanza dal nucleo, si procede, come già nei lavori precedenti (4), dalla relazione che collega tra di loro la densità del gas elettronico col potenziale nell'ipotesi che gli elettroni si possano trattare come un gas completamente degenere. Tale relazione è

$$(1) \qquad n = \begin{cases} \dfrac{2^{9/2}\pi m^{3/2} e^{3/2}}{3h^3}(V+\alpha)^{3/2} & \text{per} \quad V+\alpha > 0 \\ 0 & \text{per} \quad V+\alpha < 0 \end{cases}$$

α è una costante additiva da determinarsi colla condizione che il numero totale degli elettroni sia quello dato e cioè

$$\int n d\tau = Z - z$$

dove l'integrale va esteso a tutta la regione dello spazio dove $n \neq 0$.

Occorre qui tener presente tuttavia, che il potenziale su di un elettrone non coincide esattamente col potenziale che si ha in un punto geometrico posto ad egual distanza dal nucleo.

Infatti il primo potenziale rappresenta l'azione del nucleo e di $Z - z - 1$ elettroni, mentre il secondo rappresenta l'azione del nucleo e di tutti gli $Z - z$ elettroni.

Indichiamo con V e V' i due potenziali ora definiti. Avremo

$$V = \frac{Ze}{r} + U \qquad V' = \frac{Ze}{r} + U'$$

(4) E. Fermi, loc. cit.

dove U ed U′ rappresentano l'azione rispettivamente degli $Z - z - 1$ e $Z - z$ elettroni. In prima approssimazione noi ammetteremo dunque U ed U′ proporzionali rispettivamente a $Z - z - 1$ e $Z - z$ e scriveremo

$$U' = \frac{Z - z}{Z - z - 1} U .$$

V′ rappresenta il potenziale dovuto al nucleo ed a tutti gli elettroni: varrà dunque per esso l'equazione di Poisson

$$\Delta V' = \Delta U' = 4 \pi n e$$

da cui

$$\Delta V = \Delta U = \frac{Z - z - 1}{Z - z} \Delta U' = \left(1 - \frac{1}{Z - z}\right) 4 \pi n e .$$

Tenendo conto della (1) e del fatto che α è costante e ponendo

$$v = V + \alpha$$

avremo dunque

$$\Delta v = \begin{cases} \left(1 - \dfrac{1}{Z - z}\right) \dfrac{2^{13/2} \pi^2 m^{3/2} e^{5/2}}{3 h^3} v^{3/2} & \text{per} \quad v > 0 \\ 0 & \text{per} \quad v < 0 . \end{cases}$$

Siccome poi v per ragioni di simmetria dipende dalla sola r, l'equazione precedente diventa

$$\frac{d^2 v}{dr^2} + \frac{2}{r} \frac{dv}{dr} = \begin{cases} \left(1 - \dfrac{1}{Z - z}\right) \dfrac{2^{13/2} \pi^2 m^{3/2} e^{5/2}}{3 h^3} v^{3/2} & \text{per} \quad v > 0 \\ 0 & \text{per} \quad v < 0 . \end{cases}$$

Le costanti di integrazione si determinano colle condizioni

$$\lim_{r=0} v r = Z e$$

$$\int_0^{r_0} 4 \pi r^2 n dr = Z - z$$

dove r_0 rappresenta la distanza per cui v si annulla, alla quale cioè viene a terminare la nuvola elettronica, circostante al nucleo.

Per semplificare le equazioni precedenti conviene porre analogamente a quanto è fatto nelle Note (5)

$$r = \mu x$$

dove

$$\mu = \frac{3^{2/3} h^2}{2^{13/3} \pi^{4/3} m e^2} \frac{(Z - z)^{2/3}}{Z^{1/3} (Z - z - 1)^{2/3}} = \left(\frac{9 \pi^2}{128}\right)^{1/3} a \frac{(Z - z)^{2/3}}{Z^{1/3} (Z - z - 1)^{2/3}} \tag{2}$$

a essendo il raggio della prima orbita dell'idrogeno,

$$v = \frac{eZ}{\mu} \frac{\varphi(x)}{x} . \tag{3}$$

(5) E. FERMI, « Leipziger Vorträge », 95 (1928).

Si trova allora per la funzione $\varphi(x)$ l'equazione differenziale

$$\varphi'' = \frac{\varphi^{3/2}}{\sqrt{x}} \tag{4}$$

con le condizioni ai limiti

$$\varphi(0) = 1 \qquad \int_0^{x_0} \varphi^{3/2} \cdot \sqrt{x}\, dx = 1 - \frac{z+1}{Z} \tag{5}$$

dove $x_0 = \frac{r_0}{\mu}$ e $\varphi(x_0) = 0$.

L'ultimo integrale si può eseguire tenendo conto della (4) e si trova che esso è equivalente a

$$- x_0 \varphi'(x_0) = \frac{z+1}{Z} \cdot \tag{6}$$

L'integrazione della (4) colle condizioni ai limiti indicate, è già stata trattata nella Nota [(6)]: l'unica differenza rispetto al caso allora trattato e che in luogo di z si deve ora scrivere $z + 1$. Si trova in particolare che in approssimazione largamente sufficiente per i casi pratici può porsi

$$\varphi = \begin{cases} \varphi_0 - k\eta & \text{per } x > x_0 \\ -\frac{z+1}{Z}\frac{x - x_0}{x_0} & \text{per } x < x_0 . \end{cases} \tag{7}$$

Le funzioni φ_0 ed η sono già state tabulate nei lavori [(7)] ed i loro valori sono stati riportati nel presente lavoro per comodità: k è una costante il cui valore dipende da $(z + 1)/Z$ in modo da soddisfare la (6): una espressione empirica che la rappresenta con sufficiente approssimazione entro l'intervallo di valori che interessa e che è stata usata nei calcoli è la seguente

$$k = 0.083 \left(\frac{z+1}{Z}\right)^3 . \tag{8}$$

I valori di x_0 sono pure stati calcolati per i vari elementi di cui ci occupiamo nel presente lavoro e i loro valori sono dati nelle singole tabelle.

Il potenziale v dato dalla (3) non si annulla all'infinito: conviene perciò aggiungere ad esso una costante in modo da farlo tendere a zero per $x \to \infty$. Si riconosce immediatamente che tale costante ha il valore

$$\frac{e}{\mu}\frac{z+1}{x_0} .$$

Prenderemo dunque in definitiva come espressione del potenziale la seguente

$$V = \frac{e}{\mu}\left[\frac{Z}{x}\varphi(x) + \frac{z+1}{x_0}\right] . \tag{9}$$

3. RICERCA DELLE SOLUZIONI DELLE EQUAZIONI RELATIVISTICHE. – Stabilito, come si è visto nel paragrafo precedente, il potenziale del campo di forza in cui si muove l'elettrone in questione, prendiamo a considerare le equazioni relativistiche della meccanica quantistica.

(6) E. FERMI, « Acc. d'Ital. », *1*, 149 (1930).

(7) E. FERMI, *Falkenhagen, Quantentheorie und Chemie.* « Leipziger Vorträge », *95*, 111 (1928); « Acc. d'Italia », *1*, 149 (1930).

Limitandoci a considerare termini s ($l = 0$) noi possiamo scrivere tali equazioni sotto la forma

$$(1)\qquad \begin{cases} \dfrac{2\pi}{h}\left(\dfrac{W - eV}{c} + 2\,mc\right)F + \dfrac{dG}{dr} = 0 \\[2ex] -\dfrac{2\pi}{h}\dfrac{W - eV}{c}G + \dfrac{dF}{dr} + \dfrac{2}{r}F = 0 \end{cases}$$

ove F e G sono le autofunzioni, W l'energia e V il potenziale [8].

Le autofunzioni F e G sono legate alle quattro funzioni di Dirac dalle relazioni

$$(1')\qquad \begin{cases} \psi_1 = -iF\cos\theta \\ \psi_2 = -iF\sin\theta e^{i\varphi} \\ \psi_3 = G \\ \psi_4 = 0 \end{cases} \qquad \begin{cases} \psi_1 = -iF\sin\theta e^{-i\varphi} \\ \psi_2 = iF\cos\theta \\ \psi_3 = 0 \\ \psi_4 = G. \end{cases}$$

Le due soluzioni corrispondono alle due orientazioni dello spin. Nel caso non relativistico la unica funzione ψ di Schroedinger si riduce alla G. In particolare a una certa distanza dal nucleo, dove le correzioni relativistiche diventano trascurabili, si può sensibilmente porre $\psi = G$. Ponendo nelle (1) al posto di V la (9) del paragrafo precedente, e

$$W = -\frac{e^2}{2\,a}\varepsilon$$

si ottiene

$$(2)\qquad \begin{cases} \left[\dfrac{2}{\alpha}\dfrac{\mu}{a} - \dfrac{\alpha}{2}\dfrac{\mu}{a}\varepsilon + \alpha Z\left(\dfrac{\varphi}{x} + \dfrac{1}{Zx_0}\right)\right]F + G' = 0 \\[2ex] \left[\dfrac{\alpha}{2}\dfrac{\mu}{a}\varepsilon - \alpha Z\left(\dfrac{\varphi}{x} + \dfrac{1}{Zx_0}\right)\right]G + F' + \dfrac{2}{x}F = 0\,. \end{cases}$$

In queste equazioni è stata introdotta la variabile $x = \dfrac{r}{\mu}$: α è la costante della struttura fina che ha il valore numerico $1/137.3$.

Per trovare le soluzioni delle equazioni (2) conviene trattarle in modo differente in vicinanza dell'origine ed a grande distanza da essa.

Dato che la funzione $\varphi_0(x)$, a mezzo della quale si esprime il potenziale statistico, è rappresentata con approssimazione sufficiente per gli scopi pratici, per $x \leq 0.3$, dalla espressione empirica

$$(3)\qquad \varphi_0(x) = 1 - px + qx^2 = 1 - 1.304\,x + 1.288\,x^2\,,$$

possiamo cercare di soddisfare le equazioni (2) entro l'intervallo $x = 0$ $x = 0{,}3$ a mezzo di sviluppi in serie di potenze di x. Per $x > 0.3$ cercheremo invece le soluzioni delle equazioni (2) col metodo di Wentzel-Brillouin.

Notiamo fin d'ora che tale metodo si presta assai bene all'esecuzione di calcoli in serie poiché esso sostituisce ad una funzione irregolare, due funzioni di andamento regolare: l'ampiezza e la fase; le interpolazioni si devono appunto eseguire su queste ultime.

(8) Vedi per esempio DARWIN, « Proc. Roy. Soc. », A *118*, 654 (1928).

Conviene ora modificare un poco la forma in cui sono state scritte le equazioni (2). Ponendo.

$$F = Z\alpha \mathfrak{R}$$

dove $\mathfrak{R}$ è la nuova funzione incognita, avremo dunque

$$(4)\quad \begin{cases} \left[\dfrac{2\mu Z}{a} - \dfrac{\alpha^2}{2}\dfrac{\mu Z}{a}\varepsilon + \alpha^2 Z^2\left(\dfrac{\varphi}{x} + \dfrac{1}{Zx_0}\right)\right]\mathfrak{R} + G' = 0 \\ \left[\dfrac{\mu}{2aZ}\varepsilon - \left(\dfrac{\varphi}{x} + \dfrac{1}{Zx_0}\right)\right]G + \mathfrak{R}' + \dfrac{2}{x}\mathfrak{R} = 0 \end{cases}$$

dalle quali, ricordando la (3) si ottiene

$$(5)\quad \begin{cases} \left[\dfrac{a}{2\mu Z} - \alpha^2 Z^2\left\{\dfrac{\mu\varepsilon}{2\alpha Z} - \left(\dfrac{1}{x} - p + \dfrac{1}{Zx_0} + qx\right)\right\}\right]\mathfrak{R} + G' = 0 \\ \left[\dfrac{\mu}{2a}\dfrac{\varepsilon}{Z} - \left(\dfrac{1}{x} - p + \dfrac{1}{Zx_0} + qx\right)\right]G + \mathfrak{R}' + \dfrac{2}{x}\mathfrak{R} = 0. \end{cases}$$

Qualora si ponga per semplicità

$$T = p - \frac{1}{Zx_0} + \frac{\mu\varepsilon}{2aZ}$$

avremo

$$(6)\quad \begin{cases} \mathfrak{R}' + \dfrac{2}{x}\mathfrak{R} - \left(\dfrac{1}{x} - T + qx\right)G = 0 \\ G' + \left[\dfrac{2\mu Z}{a} + \alpha^2 Z^2\left(\dfrac{1}{x} - T + qx\right)\right]\mathfrak{R} = 0. \end{cases}$$

Tali equazioni non si prestano però ancora per la ricerca di una soluzione in serie di potenze di x, poiché G ed R divengono infinite nell'origine. Per eliminare tale singolarità poniamo:

$$(7)\qquad G = x^{-\beta}u \quad , \quad \mathfrak{R} = \frac{\beta}{\alpha^2 Z^2}x^{-\beta}v.$$

Introducendo queste espressioni nelle equazioni (6) e ammettendo che u e v siano regolari per $x = 0$, si ottiene per β una equazione di secondo grado:

$$\beta(2 - \beta) = \alpha^2 Z^2.$$

Dei due valori possibili di β si vede facilmente che uno solo soddisfa al nostro scopo perché esso deve essere minore di 3/2, affinché G^2, che nell'origine si comporta come $x^{-2\beta}$, sia integrabile per $x = 0$. Da ciò risulta

$$(8)\qquad \beta = 1 - \sqrt{1 - \alpha^2 Z^2}$$

Otteniamo così in definitiva le equazioni differenziali cercate per le due nuove funzioni incognite u e v

$$(9)\quad \begin{cases} -\dfrac{xv'}{2-\beta} - v + u - Txu + qx^2 u = 0 \\ -\beta u + xu' + \beta v + Lxv + \beta q x^2 v = 0. \end{cases}$$

In queste equazioni si è posto per semplicità

$$L = \frac{2\mu Z}{a(2-\beta)} - T\beta.$$

Poniamo ora

$$(10)\qquad \begin{cases} u = 1 + u_1 x + u_2 x^2 + \cdots \\ v = 1 + v_1 x + v_2 x^2 + \cdots \end{cases}$$

ed introduciamo questi sviluppi in serie nelle equazioni (9) si ottengono in tal modo delle formule ricorrenti che permettono il calcolo numerico dei coefficienti

$$(11)\qquad \begin{cases} u_n = \dfrac{(2-\beta)\beta}{n(2-2\beta+n)}[\mathrm{T}u_{n-1} - qu_{n-2}] \\ \qquad + \dfrac{2-\beta+n}{n(2-2\beta+n)}[-\mathrm{L}v_{n-1} - \beta q v_{n-2}] \\ v_n = \dfrac{(2-\beta)(n-\beta)}{n(2-2\beta+n)}[\mathrm{T}u_{n-1} - qu_{n-2}] \\ \qquad + \dfrac{2-\beta}{n(2-2\beta+n)}[-\mathrm{L}v_{n-1} - \beta q v_{n-2}]\,. \end{cases}$$

Per gli scopi pratici basta calcolare i termini degli sviluppi in serie (10) fino alla quindicesima potenza di x.

Passiamo ora al calcolo delle soluzioni delle equazioni (1) per $x > 0.3$. Qualora si ponga per semplicità

$$(12)\qquad g = -\frac{\mu^2}{a^2}\varepsilon + \frac{2\mu Z}{a}\left(\frac{\varphi}{x} + \frac{1}{Zx_0}\right)$$

le (4) si possono scrivere

$$(13)\qquad \begin{cases} a) \quad \mathfrak{R}' + \dfrac{2}{x}\mathfrak{R} - \dfrac{a}{2\mu Z} g\mathrm{G} = 0 \\ b) \quad \mathrm{G}' + \mathfrak{R}\left[\dfrac{2\mu Z}{a} + \dfrac{aZ}{2\mu}\alpha^2 g\right] = 0\,. \end{cases}$$

Se noi ci limitiamo a considerare termini fino ad α^2 inclusi, noi otterremo dalla (13, *b*))

$$\mathfrak{R} = -\frac{a}{2\mu Z}\mathrm{G}'\left(1 - \frac{a^2\alpha^2}{4\mu^2}g\right)$$

espressione che introdotta nella (13, *a*)) dà

$$g\mathrm{G} + \left(\mathrm{G}'' + \frac{2}{x}\mathrm{G}'\right)\left(1 - \frac{a^2\alpha^2}{4\mu^2}g\right) - \frac{a^2\alpha^2}{4\mu^2}g'\,\mathrm{G}' = 0\,.$$

Se noi poniamo

$$(14)\qquad x\mathrm{G} = \mathrm{U}$$

di modo che

$$\mathrm{G}' = \frac{\mathrm{U}'}{x} - \frac{\mathrm{U}}{x^2} \quad ; \quad \mathrm{U}'' = x\left(\mathrm{G}'' + \frac{2}{x}\mathrm{G}'\right)$$

avremo

$$g\mathrm{U} + \mathrm{U}''\left(1 - \frac{a^2\alpha^2}{4\mu^2}g\right) - \frac{a^2\alpha^2}{4\mu^2}g'\left(\mathrm{U}' - \frac{\mathrm{U}}{x}\right) = 0\,.$$

Limitandoci sempre a considerare termini sino ad α^2 inclusi, noi otteniamo dalla equazione precedente

$$(15)\qquad \mathrm{U}'' + g\mathrm{U} = \gamma^2\left\{g'\,\mathrm{U}' - \frac{g'}{x}\mathrm{U} - g^2\,\mathrm{U}\right\}$$

dove

$$\gamma^2 = \frac{a^2 x^2}{4\mu^2}. \tag{16}$$

Cerchiamo ora la soluzione della equazione (15) col metodo di Wentzel Brillouin. Siccome per le funzioni ∞ s la funzione U ha un andamento oscillante per tutti i valori di x, porremo

$$U = e^{i\int Z dx} \tag{17}$$

dove Z è la nuova funzione incognita. Sia la parte reale che la immaginaria di U sono soluzioni dell'equazione.

Dalla (15) si ottiene a mezzo della (17) l'equazione

$$iZ' - Z^2 + g = \gamma^2 \left\{ igZ - \frac{g'}{x} + g^2 \right\}. \tag{18}$$

In questa equazione il secondo membro rappresenta la correzione relativistica. Essa si può risolvere per approssimazioni successive: eseguendo i calcoli fino alla quarta approssimazione, e prendendo per U l'espressione reale:

$$U = K \cdot R \operatorname{sen} \Theta \tag{19}$$

si ottiene

$$R(x) = \frac{1 + \frac{g''}{16 g^2} - \frac{5}{64}\frac{g'^2}{g^3} + \frac{\gamma^2}{2} g}{\sqrt[4]{g}} \tag{20}$$

$$\Theta(x) = \Theta(0{,}3) \tag{21}$$

$$+ \left[\int \sqrt{g}\, dx - \frac{1}{32}\int \frac{g'^2}{g^{5/2}} dx - \frac{g'}{8 g^{3/2}} + \frac{\gamma^2}{2}\left(\int \frac{g'\, dx}{x\sqrt{g}} + \int g^{3/2} dx \right) \right]_{0,3}^{x}.$$

K e Θ (0,3) sono costanti che vanno determinate in modo che per $x = 0.3$, la funzione U (x) e la sua derivata si raccordino con continuità alla funzione $u(x)\, x^{1-\beta}$ determinata con lo sviluppo in serie valido per $x < 0.3$. Dalla (19) si ricava

$$\begin{cases} K \operatorname{sen} \Theta = \dfrac{U}{R} \\ K \cos \Theta = \dfrac{1}{R\Theta'} \left\{ U' - U \dfrac{d \log R}{dx} \right\}. \end{cases} \tag{22}$$

A mezzo delle (20) (21) si calcolano Θ', R e $\frac{d \log R}{dx}$ per $x = 0.3$ ed i valori così ottenuti si introducono nelle equazioni precedenti.

Nel caso di autofunzioni ∞ s, essendo nullo l'autovalore, potremo scrivere le (20) (21) con sufficiente approssimazione, data la precisione con cui sono stati eseguiti i calcoli nel presente lavoro, nella forma

$$R = A^{-1/4}\left(\frac{x}{\varphi_0}\right)^{1/4} \left\{ 1 + \frac{1}{A}\left[\frac{1}{16}\left(\frac{x}{\varphi_0}\right)^{1/2} + \frac{3}{64}\frac{1}{x\varphi_0} \right.\right. \tag{23}$$

$$\left.\left. - \frac{5}{64}\frac{x\varphi_0'^2}{\varphi_0^3} + \frac{1}{32}\frac{\varphi_0'}{\varphi_0^2} \right] + \frac{\gamma^2 A}{2}\frac{\varphi_0}{x} \right\}$$

$$\Theta(x) = \Theta(0{,}3) + A^{1/2}\left\{\int_{0,3}^{x}\left(\frac{\varphi_0}{x}\right)^{1/2} dx - \frac{k}{2}\int_{0,3}^{x}\frac{\eta}{(x\varphi_0)^{1/2}}\,dx \right. \tag{24}$$

$$\left. + \frac{1}{2\,Zx_0}\int_{0,3}^{x}\left(\frac{x}{\varphi_0}\right)^{1/2} dx\right\} - A^{-1/2}\left\{\frac{1}{32}\left[\int_{0,3}^{x}\frac{x^{1/2}\,\varphi_0'^{2}}{\varphi_0^{5/2}}\,dx + \int_{0,3}^{x}\frac{dx}{\varphi_0^{1/2}\,x^{3/2}}\right.\right.$$

$$\left.\left. - 2\int_{0,3}^{x}\frac{\varphi_0'}{x^{1/2}\,\varphi_0^{3/2}}\,dx\right] + \frac{1}{8}\left(\frac{x^{1/2}\,\varphi_0'}{\varphi_0^{3/2}} - \frac{1}{(x\varphi_0)^{1/2}}\right)_{0,3}^{x}\right.$$

$$\left. + \frac{\gamma^2}{2}\left\{A^{1/2}\left[\int_{0,3}^{x}\frac{\varphi_0'}{x^{3/2}\,\varphi_0^{1/2}}\,dx - \int_{0,3}^{x}\frac{\varphi_0^{1/2}}{x^{5/2}}\,dx\right] + A^{3/2}\int_{0,3}^{x}\left(\frac{\varphi_0}{x}\right)^{3/2} dx\right\}\right\}.$$

In queste espressioni la costante A dipende dal numero atomico ed ha il valore

$$A = 2\,Z\,\frac{\mu}{a}\,.$$

Le (23) (24) sono naturalmente applicabili solo fino ad una certa distanza dal limite dell'atomo, e precisamente finché η è piccolo rispetto a φ_0; in prossimità del limite dell'atomo le autofunzioni ∞ *s* si debbono calcolare elemento per elemento a mezzo di (20) e (21). Sempre nel caso di autofunzioni ∞ *s*, poiché in prossimità dell'origine la funzione η si può senz'altro trascurare, è lecito porre per $x \leq 0.3$

$$g = A\,\frac{\varphi_0}{x}\,.$$

Tenendo conto di questa semplificazione le formule definitive per il calcolo di R, Θ' e $\frac{d \log R}{dx}$ per $x = 0.3$ sono le seguenti:

$$\left\{\begin{aligned} R &= \frac{0.804}{\sqrt[4]{A}}\left\{1 + \frac{0.190}{A} + 1.20\,\gamma^2 A\right\} \\ \Theta' &= 1.55\sqrt{A} - \frac{0.588}{\sqrt{A}} + \frac{\gamma^2}{2}\left(3.73\,A^{3/2} - 2.60\,A^{1/2}\right) \\ \frac{d\log R}{dx} &= 1.063 - \frac{0.519}{A} - 5.11\,\gamma^2 A\,. \end{aligned}\right. \tag{25}$$

I valori di K e di $\Theta(0{,}3)$ calcolati da queste espressioni e dalle (22), vengono dati nelle singole tabelle dei vari elementi.

Nella Tabella I sono riportati i valori delle funzioni φ_0 ed η, e sono altresì date alcune formule empiriche che servono per il calcolo del potenziale per i vari atomi. Nelle Tabelle II, III, riportiamo i valori numerici delle espressioni per il calcolo delle funzioni R e Θ, espressioni che come abbiamo detto son valide solo fino ad una certa distanza dal limite dell'atomo.

Seguono le tabelle relative ai singoli elementi per le autofunzioni ∞ *s*; in ciascuna di esse vengono dati, oltre ai valori numerici di alcune costanti (come per esempio, il raggio atomico x_0 e l'unità di misura μ/a), i coefficenti degli sviluppi in serie di u e v che servono per il calcolo delle autofunzioni fino ad $x = 0.3$; i valori di R, Θ ed U da 0.3 fino ai margini dell'atomo.

Dal margine dell'atomo in poi le autofunzioni ∞s si riducono semplicemente a

$$U = B\sqrt[4]{x}\sin(C\sqrt{x} + D) \tag{26}$$

dove B e D sono due costanti da determinare in modo che la (26) e la sua derivata si raccordino colla U per $x = x_0$ e: $C = 2\sqrt{\frac{2\mu}{a}}$. In ciascuna tabella diamo quindi anche i valori delle costanti B, C, D.

Per l'aiuto prestatoci nella esecuzione dei calcoli numerici, ringraziamo qui i signori A. Biava, F. Colajori, V. Crocco, G. Giovene, E. Medi, R. Nuzzo.

TABELLA I.

$\varphi = \varphi_0 - k\eta$; $k = 0.083\left(\frac{z+1}{Z}\right)^3$; per $x \leq 0.3 : \varphi_0 = 1 - 1.304x + 1.288x^2$

x	φ_0	$\eta \cdot 10^{-4}$	x	φ_0	$\eta \cdot 10^{-4}$
0.3	0.721	—	6	0.059	0.005
0.4	0.660	—	7	0.046	0.0070
0.5	0.607	—	8	0.037	0.0103
0.6	0.562	—	9	0.029	0.0146
0.7	0.521	—	10	0.024	0.0201
0.8	0.485	—	12	0.017	0.0359
0.9	0.453	—	14	0.012	0.0598
1.0	0.425	—	16	0.0093	0.0942
1.2	0.375	—	18	0.0072	0.142
1.4	0.333	—	20	0.0056	0.208
1.6	0.297	—	22	0.0045	0.294
1.8	0.268	—	24	0.0037	0.405
2.0	0.244	—	26	0.0031	0.545
2.5	0.194	0.001	28	0.0026	0.721
3.0	0.157	0.001	30	0.0022	0.941
3.5	0.130	0.001	32	0.0019	1.20
4.0	0.108	0.002	34	0.0017	1.51
4.5	0.093	0.002	36	0.0015	1.88
5	0.079	0.003			

TABELLA II.

x	$\left(\frac{x}{\varphi}\right)^{1/4}$	Q	$\frac{\varphi_0}{x}$
0.3	0.803	0.19	2.40
0.4	0.882	0.15	1.65
0.5	0.953	0.13	1.21
0.6	1.016	0.11	0.937
0.7	1.076	0.10	0.744
0.8	1.133	0.09	0.606
0.9	1.187	0.08	0.503
1.0	1.238	0.08	0.425
1.2	1.338	0.07	0.312
1.4	1.432	0.07	0.238
1.6	1.523	0.06	0.186
1.8	1.609	0.06	0.149
2.0	1.693	0.06	0.122
2.5	1.893	0.06	0.0776
3.0	2.090	0.06	0.0524
3.5	2.276	0.06	0.0371
4.0	2.469	0.06	0.0270
4.5	2.635	0.06	0.0206
5.0	2.817	0.06	0.0158
6	3.175	0.07	0.00980
7	3.510	0.07	0.00658
8	3.846	0.07	0.00458
9	4.174	0.08	0.00329
10	4.509	0.09	0.00242
12	5.178	0.10	0.00139
14	5.872	0.11	0.00840
16	6.441	0.12	0.00580
18	7.071	0.14	0.00040
20	7.738	0.16	0.00028
22	8.362	0.18	0.00020
24	8.980	0.20	0.00015
26	9.578	0.22	0.00012
28	10.15	0.24	0.000093
30	10.85	0.27	0.000073
32	11.40	0.29	0.000059
34	11.86	0.32	0.000050

$$R = A^{-1/4}\left(\frac{x}{\varphi}\right)^{1/4}\left\{1 + \frac{1}{A}\left[\frac{1}{16}\left(\frac{x}{\varphi}\right)^{1/2} + \frac{3}{64}\,\frac{1}{x\varphi} - \frac{5}{64}\,\frac{x\varphi'^2}{\varphi^3} + \frac{1}{32}\,\frac{\varphi'}{\varphi^2}\right] + \frac{\gamma^2}{2}\,A\,\frac{\varphi}{x}\right\}$$

$$= A^{-1/4}\left(\frac{x}{\varphi}\right)^{1/4}\left\{1 + \frac{1}{A}\,Q + \frac{\gamma^2}{x}\,A\,\frac{\varphi}{x}\right\} \quad ; \quad A = 2\,Z\mu/a$$

TABELLA III.

x	$\int_{0,3}^{x}\left(\frac{\varphi_0}{x}\right)^{1/2}dx$	$\int_{0,3}^{x}\frac{\eta\cdot 10^{-4}}{(x\varphi_0)^{1/2}}dx$	$\int_{0,3}^{x}\left(\frac{x}{\varphi_0}\right)^{1/2}dx$	Ω	$\int_{0,3}^{x}\frac{\varphi_0'}{x^{3/2}\varphi_0^{1/2}}dx-\int_{0,3}^{x}\frac{\varphi_0^{1/2}}{x^{5/2}}dx$	$\int_{0,3}^{x}\left(\frac{\varphi_0}{x}\right)^{3/2}dx$
0.3	0.000	0.000	0.0000	0.00	— 0.000	0.000
0.4	0.141	0.000	0.0712	0.05	— 1.552	0.282
0.5	0.260	0.000	0.156	0.08	— 2.377	0.452
0.6	0.363	0.000	0.253	0.11	— 2.863	0.562
0.7	0.454	0.000	0.362	0.12	— 3.195	0.639
0.8	0.563	0.000	0.484	0.14	— 3.412	0.693
0.9	0.611	0.000	0.619	0.15	— 3.584	0.735
1.0	0.678	0.000	0.766	0.17	— 3.701	0.766
1.2	0.799	0.000	1.098	0.18	— 3.872	0.811
1.4	0.904	0.000	1.482	0.20	— 3.981	0.839
1.6	0.995	0.000	1.919	0.21	— 4.056	0.859
1.8	1.077	0.000	2.410	0.22	— 4.109	0.872
2.0	1.150	0.000	2.956	0.23	— 4.152	0.882
2.5	1.306	0.000	4.565	0.25	— 4.217	0.898
3.0	1.432	0.001	6.551	0.27	— 4.251	0.906
3.5	1.538	0.001	8.937	0.28	— 4.276	0.911
4.0	1.627	0.003	11.749	0.29	— 4.289	0.913
4.5	1.703	0.004	15.023	0.30	— 4.299	0.915
5.0	1.771	0.006	18.727	0.31	— 4.306	0.916
6	1.883	0.012	27.728	0.32	— 4.316	0.918
7	1.972	0.022	38.919	0.33	— 4.321	0.919
8	2.047	0.037	52.440	0.34	— 4.325	0.919
9	2.109	0.061	68.555	0.35	— 4.327	0.919
10	2.162	0.095	87.385	0.35	— 4.329	0.919
12	2.248	0.211	134.322	0.35	— 4.331	0.919
14	2.313	0.434	195.416	0.35	— 4.331	0.919
16	2.366	0.819	271.812	0.36	— 4.331	0.919
18	2.410	1.446	362.388	0.36	— 4.331	0.919
20	2.447	2.447	472.728	0.36	— 4.332	0.919
22	2.478	3.989	602.022	0.36	— 4.332	0.919
24	2.504	6.247	752.892	0.36	— 4.332	0.919
26	2.528	9.516	924.858	0.36	— 4.332	0.919
28	2.548	14.057	1120.032	0.36	— 4.332	0.919
30	2.566	20.356	1339.366	0.36	— 4.332	0.919
32	2.582	28.867	1589.066	0.36	— 4.332	0.919
34	2.597	40.136	1858.014	0.36	— 4.332	0.919

$$\Theta=\Theta(0,3)+A^{1/2}\left\{\int_{0,3}^{x}\left(\frac{\varphi_0}{x}\right)^{1/2}dx-\frac{k}{2}\int_{0,3}^{x}\frac{\eta}{(x\varphi_0)^{1/2}}dx+\frac{1}{2Zx_0}\int_{0,3}^{x}\left(\frac{x}{\varphi_0}\right)^{1/2}dx\right\}-A^{-1/2}\,\Omega$$

$$+\frac{\gamma^2}{2}\left\{A^{1/2}\left[\int_{0,3}^{x}\frac{\varphi_0'}{x^{3/2}\varphi_0^{1/2}}dx-\int_{0,3}^{x}\frac{\varphi_0^{1/2}}{x^{5/2}}dx\right]+A^{3/2}\int_{0,3}^{x}\left(\frac{\varphi_0}{x}\right)^{3/2}dx\right\}$$

dove

$$\Omega=\frac{1}{32}\left[\int_{3,0}^{x}\frac{x^{1/2}\varphi_0'^{\,2}}{\varphi_0^{5/2}}dx+\int_{0,3}^{x}\frac{dx}{\varphi_0^{1/2}x^{3/2}}dx-\int_{0,3}^{x}\frac{\varphi_0'}{x^{1/2}\varphi_0^{3/2}}dx\right]+\frac{1}{8}\left(\frac{x^{1/2}\varphi_0'}{\varphi_0^{3/2}}-\frac{1}{(x\varphi_0)^{1/2}}\right)_{0,3}^{x};$$

$$A=2Z\,\mu/a$$

$$Z = 10;\ \mathrm{Ne}$$

$$x_0 = 10.8\ ;\ k = 8.300 \cdot 10^{-5}\ ;\ \frac{\mu}{a} = 0.449\ ;\ \beta = 0.002656\ ;\ K = 0.1902\ ;$$

$$B = 0.197\ ;\ C = 1.88\ ;\ D = 3.71\ ;\ \alpha_s = 1.68.$$

v_0	1.00000	u_0	1.00000
v_1	− 3.80341	u_1	− 4.41286
v_2	7.68997	u_2	8.38845
v_3	− 11.13019	u_3	− 11.30318
v_4	12.5582	u_4	12.2680
v_5	− 11.8498	u_5	− 11.0713
v_6	9.700	u_6	8.705
v_7	− 7.023	u_7	− 6.107
v_8	4.59	u_8	3.87
v_9	− 2.75	u_9	− 2.25
v_{10}	1.5	u_{10}	1.2
v_{11}	− 0.8	u_{11}	− 0.6
v_{12}	0.4	u_{12}	0.3

x	R	Θ	U	x	R	Θ	U
0.3	0.476	2.40	+ 0.0613	2.0	0.971	5.81	− 0.084
0.4	0.521	2.80	+ 0.0331	2.5	1.08	6.30	+ 0.004
0.5	0.561	3.15	− 0.0057	3.0	1.18	6.70	+ 0.092
0.6	0.597	3.45	− 0.0338	3.5	1.26	7.04	+ 0.165
0.7	0.632	3.71	− 0.0643	4.0	1.34	7.34	+ 0.222
0.8	0.661	3.95	− 0.0909	4.5	1.41	7.61	+ 0.260
0.9	0.691	4.17	− 0.115	5	1.47	7.85	+ 0.279
1.0	0.720	4.38	− 0.129	6	1.59	8.28	+ 0.274
1.2	0.779	4.75	− 0.148	7	1.68	8.66	+ 0.220
1.4	0.831	5.06	− 0.148	8	1.75	9.01	+ 0.133
1.6	0.885	5.34	− 0.136	9	1.81	9.34	+ 0.028
1.8	0.931	5.58	− 0.114	10	1.85	9.64	− 0.077

$Z = 14$; Si

$x_0 = 13.0$; $k = 3.025 \cdot 10^{-5}$; $\frac{\mu}{a} = 0{,}3859$; $\beta = 0.005238$; $K = 0.1726$;

$B = 0.185$; $C = 1.76$; $D = 4.84$; $\alpha_s = 2.04$.

v_0	1.00000	u_0	1.00000
v_1	— 4.47174	u_1	— 5.41506
v_2	10.19540	u_2	12.09815
v_3	— 16.39938	u_3	— 18.37971
v_4	20.5019	u_4	22.1655
v_5	— 21.2791	u_5	— 22.1621
v_6	19.085	u_6	19.164
v_7	— 15.116	u_7	— 14.731
v_8	10.78	u_8	10.21
v_9	— 7.02	u_9	— 6.47
v_{10}	4.2	u_{10}	3.8
v_{11}	— 2.3	u_{11}	— 2.1
v_{12}	1.2	u_{12}	1.1
v_{13}	— 1	u_{13}	0

x	R	Θ	U	x	R	Θ	U
0.3	0.451	2.72	+ 0.0317	2.5	1.04	6.99	+ 0.117
0.4	0.494	3.17	— 0.0025	3.0	1.14	7.42	+ 0.179
0.5	0.532	3.55	— 0.0369	3.5	1.23	7.79	+ 0.212
0.6	0.566	3.89	— 0.0660	4.0	1.31	8.10	+ 0.218
0.7	0.599	4.18	— 0.0892	4.5	1.39	8.38	+ 0.206
0.8	0.630	4.45	— 0.105	5.0	1.46	8.63	+ 0.180
0.9	0.660	4.69	— 0.114	6	1.58	9.07	+ 0.093
1.0	0.688	4.91	— 0.116	7	1.69	9.45	— 0.009
1.2	0.743	5.31	— 0.106	8	1.75	9.79	— 0.109
1.4	0.795	5.65	— 0.082	9	1.83	10.11	— 0.201
1.6	0.845	5.95	— 0.048	10	1.91	10.40	— 0.274
1.8	0.892	6.22	— 0.010	12	2.00	10.93	— 0.345
2.0	0.939	6.46	+ 0.029				

$Z = 19$; K

$x_0 = 15.4$; $k = 1.210 \cdot 10^{-5}$; $\frac{\mu}{a} = 0.3440$; $\beta = 0.009620$; $K = 0.1580$;

$B = 0.175$; $C = 1.66$; $D = 6.06$; $\alpha_s = 2.46$.

v_0	1.00000	u_0	1.00000
v_1	— 5.23710	u_1	— 6.56779
v_2	13.47003	u_2	17.17568
v_3	— 24.02038	u_3	— 29.42806
v_4	33.1374	u_4	39.3365
v_5	— 37.7148	u_5	— 43.3937
v_6	36.926	u_6	41.140
v_7	— 31.861	u_7	— 34.515
v_8	24.69	u_8	26.05
v_9	— 17.44	n_9	— 17.94
v_{10}	11.3	u_{10}	11.4
v_{11}	— 6.8	u_{11}	— 6.7
v_{12}	4	u_{12}	4
v_{13}	— 2	u_{13}	— 2
v_{14}	1	u_{14}	1

x	R	Θ	U	x	R	Θ	U
0.3	0.429	3.06	+ 0.0056	2.5	0.985	7.73	+ 0.153
0.4	0.470	3.56	— 0.0299	3.0	1.08	8.21	+ 0.160
0.5	0.507	3.98	— 0.0595	3.5	1.17	8.60	+ 0.136
0.6	0.539	4.35	— 0.0795	4.0	1.26	8.94	+ 0.090
0.7	0.570	4.67	— 0.0901	4.5	1.34	9.23	+ 0.040
0.8	0.600	4.96	— 0.0919	5.0	1.41	9.51	— 0.020
0.9	0.628	5.23	— 0.0861	6	1.55	9.87	— 0.126
1.0	0.655	5.47	— 0.0752	7	1.67	10.35	— 0.212
1.2	0.708	5.91	— 0.0410	8	1.77	10.69	— 0.267
1.4	0.754	6.28	0.0000	9	1.87	10.99	— 0.296
1.6	0.801	6.62	+ 0.0421	10	1.94	11.27	— 0.294
1.8	0.843	6.92	+ 0.0800	12	2.04	11.79	— 0.225
2.0	0.885	7.18	+ 0.110	14	2.14	12.25	— 0.103

$Z = 26$; Fe

$x_0 = 18.2$; $k = 4.722 \cdot 10^{-6}$; $\frac{\mu}{a} = 0.3068$; $b = 0.01809$; $K = 0.1451$;

$B = 0.166$; $C = 1.57$; $D = 7.43$; $\alpha_s = 2.86$.

v_0	1.00000	u_0	1.00000
v_1	— 6.22180	u_1	— 8.05918
v_2	18.31498	u_2	25.01708
v_3	— 36.59013	u_3	— 49.02616
v_4	56.1379	u_4	73.3901
v_5	— 70.5984	u_5	— 90.0127
v_6	75.977	u_6	94.277
v_7	— 71.843	u_7	— 86.914
v_8	60.85	u_8	71.88
v_9	— 46.85	u_9	— 54.09
v_{10}	33.2	u_{10}	37.5
v_{11}	— 21.8	u_{11}	— 24.1
v_{12}	13.3	u_{12}	14.5
v_{13}	— 7.7	u_{13}	— 8.2
v_{14}	4	u_{14}	4
v_{15}	— 2	u_{15}	— 2

x	R	Θ	U	x	R	Θ	U
0.3	0.408	3.46	— 0.0186	3.0	1.04	9.14	+ 0.042
0.4	0.446	4.01	— 0.0496	3.5	1.13	9.58	— 0.026
0.5	0.481	4.48	— 0.0680	4.0	1.21	9.94	— 0.087
0.6	0.512	4.88	— 0.0732	4.5	1.29	10.27	— 0.141
0.7	0.542	5.25	— 0.0677	5	1.32	10.54	— 0.172
0.8	0.570	5.57	— 0.0540	6	1.49	11.02	— 0.216
0.9	0.597	5.87	— 0.0348	7	1.64	11.43	— 0.216
1.0	0.623	6.13	— 0.0137	8	1.74	11.78	— 0.178
1.2	0.673	6.61	+ 0.0318	9	1.86	12.08	— 0.125
1.4	0.720	7.03	+ 0.0710	10	1.95	12.36	— 0.056
1.6	0.765	7.39	+ 0.0994	12	2.10	12.85	+ 0.087
1.8	0.808	7.72	+ 0.116	14	2.20	13.29	+ 0.214
2.0	0.850	8.01	+ 0.122	16	2.26	13.70	+ 0.298
2.5	0.951	8.63	+ 0.098	18	2.36	14.08	+ 0.342

$Z = 31$; Ga

$x_0 = 19.9$; $k = 2.786 \cdot 10^{-6}$; $\frac{\mu}{a} = 0.2881$; $\beta = 0.02582$; $K = 0.1394$;
$B = 0.161$; $C = 1.52$; $D = 8,24$; $\alpha_s = 3.12$.

v_0	1.00000	u_0	1.000000
v_1	— 6.88517	u_1	— 9.07047
v_2	21.97258	u_2	31.13193
v_3	— 46.96643	u_3	— 66.10849
v_4	76.6782	u_4	105.8425
v_5	— 102.2175	u_5	— 138.1137
v_6	— 116.237	u_6	153.302
v_7	— 115.906	u_7	— 149.318
v_8	103.35	u_8	130.21
v_9	— 83.66	u_9	— 103.14
v_{10}	62.2	u_{10}	75.1
v_{11}	— 42.8	u_{11}	— 50.7
v_{12}	27.5	u_{12}	32.0
v_{13}	— 17	u_{13}	— 19
v_{14}	9	u_{14}	11
v_{15}	— 5	u_{15}	— 6

x	R	Θ	U	x	R	Θ	U
0.3	0.396	3.71	— 0.0298	3.5	1.09	10.18	— 0.105
0.4	0.434	4.30	— 0.0555	4.0	1.18	10.51	— 0.146
0.5	0.468	4.79	— 0.0650	4.5	1.26	10.90	— 0.175
0.6	0.498	5.22	— 0.0605	5	1.34	11.15	— 0.184
0.7	0.527	5.60	— 0.0460	6	1.48	11.65	— 0.163
0.8	0.554	5.95	— 0.0252	7	1.62	12.07	— 0.106
0.9	0.580	6.26	— 0.0015	8	1.73	12.42	— 0.034
1.0	0.605	6.54	+ 0.0218	9	1.84	12.74	+ 0.046
1.2	0.654	7.05	+ 0.0634	10	1.95	13.02	+ 0.120
1.4	0.699	7.49	+ 0.0913	12	2.10	13.50	+ 0.236
1.6	0.744	7.88	+ 0.104	14	2.21	13.94	+ 0.303
1.8	0.785	8.22	+ 0.102	16	2.31	14.33	+ 0.315
2.0	0.826	8.53	+ 0.090	18	2.38	14.70	+ 0.279
2.5	0.924	9.19	+ 0.030	20	2.44	15.04	+ 0.210
3.0	1.02	9.72	— 0.042				

$Z = 37$; Rb

$x_0 = 21.8$; $k = 1.639 \cdot 10^{-6}$; $\frac{\mu}{a} = 2.2706$; $\beta = 0.3699$; $K = 0.1345$;

$B = 0.157$; $C = 1.47$; $D = 9.26$; $\alpha_s = 3.45$.

v_0	1.00000	u_0	1.00000
v_1	— 7.65289	u_1	— 10.24861
v_2	26.59775	u_2	39.05686
v_3	— 61.04338	u_3	— 90.25280
v_4	106.3173	u_4	155.0727
v_5	— 150.5607	u_5	— 215.7854
v_6	181.248	u_6	254.380
v_7	— 190.889	u_7	— 262.235
v_8	179.44	u_8	241.45
v_9	— 152.89	u_9	— 201.60
v_{10}	119.5	u_{10}	154.5
v_{11}	— 86.4	u_{11}	— 109.7
v_{12}	58.3	u_{12}	72.7
v_{13}	— 37	u_{13}	— 45
v_{14}	22	u_{14}	27
v_{15}	— 12	u_{15}	— 15

x	R	Θ	U	x	R	Θ	U
0.3	0.385	3.99	— 0.0388	3.5	1.08	10.84	— 0.143
0.4	0.421	4.61	— 0.0564	4.0	1.17	11.24	— 0.153
0.5	0.454	5.13	— 0.0557	4.5	1.24	11.59	— 0.138
0.6	0.484	5.59	— 0.0415	5	1.32	11.90	— 0.109
0.7	0.512	6.00	— 0.0194	6	1.46	12.42	— 0.027
0.8	0.539	6.36	+ 0.0056	7	1.59	12.85	+ 0.062
0.9	0.564	6.69	+ 0.0304	8	1.70	13.22	+ 0.141
1.0	0.588	6.99	+ 0.0514	9	1.83	13.54	+ 0.204
1.2	0.637	7.53	+ 0.0810	10	1.92	13.82	+ 0.246
1.4	0.680	8.00	+ 0.0905	12	2.10	14.31	+ 0.278
1.6	0.722	8.40	+ 0.0828	14	2.23	14.74	+ 0.246
1.8	0.763	8.77	+ 0.0628	16	2.31	15.14	+ 0.165
2.0	0.803	9.09	+ 0.0348	18	2.39	15.49	+ 0.067
2.5	0.898	9.79	— 0.0437	20	2.46	15.83	— 0.043
3.0	0.991	10.36	— 0.107	22	2.52	16.15	— 0.147

Z = 42; Mo

$x_0 = 23.2$; $k = 1.120 \cdot 10^{-6}$; $\frac{\mu}{a} = 0.2588$; $\beta = 0.04793$; K = 0.1322 ;

B = 0.156 ; C = 1.44 ; D = 9.98 ; $\alpha_s = 3.68$.

v_0	1.00000	u_0	1.00000
v_1	— 8.27771	u_1	— 11.21535
v_2	30.66223	u_2	46.17681
v_3	— 74.21031	u_3	— 113.64649
v_4	135.5737	u_4	205.8257
v_5	— 200.7285	u_5	— 300.3176
v_6	251.969	u_6	370.041
v_7	— 276.207	u_7	— 397.697
v_8	269.87	u_8	381.09
v_9	— 238.74	u_9	— 330.67
v_{10}	— 193.5	u_{10}	263.1
v_{11}	— 145.1	u_{11}	— 193.7
v_{12}	101.3	u_{12}	133.0
v_{13}	— 66	u_{13}	— 86
v_{14}	41	u_{14}	52
v_{15}	— 24	u_{15}	— 30

x	R	Θ	U	x	R	Θ	U
0.3	0.377	4.20	— 0.0436	3.5	1.06	11.34	— 0.131
0.4	0.413	4.85	— 0.0540	4.0	1.15	11.76	— 0.109
0.5	0.445	5.40	— 0.0453	4.5	1.22	12.12	— 0.069
0.6	0.474	5.88	— 0.0247	5	1.2[illegible]	12.44	— 0.021
0.7	0.501	6.30	+ 0.0014	6	1.44	12.99	+ 0.079
0.8	0.528	6.68	+ 0.0269	7	1.54	13.43	+ 0.156
0.9	0.552	7.03	+ 0.0495	8	1.69	13.80	+ 0.212
1.0	0.576	7.34	+ 0.0662	9	1.81	14.13	+ 0.239
1.2	0.622	7.90	+ 0.0821	10	1.92	14.41	+ 0.244
1.4	0.666	8.38	+ 0.0758	12	2.09	14.92	+ 0.197
1.6	0.708	8.81	+ 0.0539	14	2.25	15.34	+ 0.104
1.8	0.747	9.19	+ 0.0229	16	2.35	15.71	— 0.031
2.0	0.786	9.53	— 0.0110	18	2.44	16.06	— 0.112
2.5	0.879	10.26	— 0.0859	20	2.50	16.39	— 0.212
3.0	0.971	10.83	— 0.127	22	2.56	16.71	— 0.286

$Z = 47$; Ag

$x_0 = 24.5$; $k = 7.994 \cdot 10^{-7}$; $\frac{\mu}{a} = 0.2489$; $\beta = 0.06041$; $K = 0.1310$;

$B = 0.157$; $C = 1.41$; $D = 10.66$; $\alpha_s = 3.89$.

v_0	1.00000	u_0	1.00000
v_1	— 8.89816	u_1	— 12.18270
v_2	34.96173	u_2	53.85006
v_3	— 88.87276	u_3	— 140.47200
v_4	169.6367	u_4	267.0624
v_5	— 261.5827	u_5	— 406.9550
v_6	341.132	u_6	522.120
v_7	— 387.796	u_7	— 582.809
v_8	392.43	u_8	579.05
v_9	— 359.12	u_9	— 520.29
v_{10}	300.8	u_{10}	428.1
v_{11}	— 232.9	u_{11}	— 325.7
v_{12}	167.9	u_{12}	231.0
v_{13}	— 113	u_{13}	— 154
v_{14}	72	u_{14}	96
v_{15}	— 44	u_{15}	— 57

x	R	Θ	U	x	R	Θ	U
0.3	0.370	4.41	— 0.0463	4.0	1.13	12.25	— 0.045
0.4	0.405	5.08	— 0.0495	4.5	1.20	12.62	+ 0.009
0.5	0.437	5.65	— 0.0336	5	1.27	12.96	+ 0.065
0.6	0.465	6.15	— 0.0082	6	1.42	13.52	+ 0.152
0.7	0.492	6.59	+ 0.0194	7	1.55	13.97	+ 0.200
0.8	0.518	6.98	+ 0.0436	8	1.67	14.36	+ 0.213
0.9	0.542	7.34	+ 0.0619	9	1.80	14.69	+ 0.200
1.0	0.565	7.66	+ 0.0727	10	1.90	14.98	+ 0.163
1.2	0.611	8.25	+ 0.0739	12	2.09	15.49	+ 0.057
1.4	0.653	8.75	+ 0.0535	14	2.24	15.91	— 0 061
1.6	0.695	9.19	+ 0.0211	16	2.36	16.29	— 0.172
1.8	0.734	9.59	— 0.0157	18	2.44	16.63	— 0.256
2.0	0.772	9.94	— 0.0498	20	2.54	16.97	— 0.318
2.5	0.863	10.69	— 0.108	22	2.60	17.26	— 0.340
3.0	0.953	11.30	— 0.119	24	2.66	17.57	— 0.333
3.5	1.04	11.82	— 0.103				

Z = 53; I

$x_0 = 26.1$; $k = 5.575 \; 10^{-7}$; $\frac{\mu}{a} = 0.2387$; $\beta = 0.007750$; $K = 0.1313$;

$B = 0.158$; $C = 1.38$; $D = 11.46$; $\alpha_s = 4.14$.

v_0	1.00000	u_0	1.00000
v_1	— 9.63756	u_1	— 13.34733
v_2	40.42145	u_2	63.78886
v_3	— 108.47731	u_3	— 177.42538
v_4	217.2471	u_4	355.8605
v_5	— 350.1901	u_5	— 568.6572
v_6	476.055	u_6	762.310
v_7	— 562.943	u_7	— 886.774
v_8	591.68	u_8	916.24
v_9	— 561.66	u_9	— 854.82
v_{10}	487.6	u_{10}	729.4
v_{11}	— 390.8	u_{11}	— 575.1
v_{12}	291.5	u_{12}	422.1
v_{13}	— 204	u_{13}	— 290
v_{14}	134	u_{14}	188
v_{15}	— 83	u_{15}	— 115

x	R	Θ	U	x	R	Θ	U
0.3	0.363	4.64	— 0.0476	4.0	1.10	12.80	+ 0.034
0.4	0.397	5.35	— 0.0420	4.5	1.18	13.19	+ 0.090
0.5	0.429	5.94	— 0.0188	5	1.25	13.54	+ 0.135
0.6	0.456	6.45	+ 0.0104	6	1.39	14.11	+ 0.182
0.7	0.483	6.91	+ 0.0373	7	1.52	14.58	+ 0.199
0.8	0.508	7.32	+ 0.0575	8	1.64	14.98	+ 0.142
0.9	0.532	7.70	+ 0.0690	9	1.77	15.33	+ 0.083
1.0	0.554	8.03	+ 0.0716	10	1.89	15.62	+ 0.019
1.2	0.599	8.64	+ 0.0555	12	2.06	16.13	— 0.112
1.4	0.641	9.16	+ 0.0215	14	2.24	16.57	— 0.225
1.6	0.681	9.62	— 0.0175	16	2.36	16.94	— 0.293
1.8	0.719	10.03	— 0.0531	18	2.44	17.29	— 0.320
2.0	0.757	10.40	— 0.0822	20	2.54	17.61	— 0.314
2.5	0.846	11.18	— 0.109	22	2.60	17.91	— 0.273
3.0	0.934	11.82	— 0.084	24	2.68	18.20	— 0.202
3.5	1.02	12.35	— 0.028	26	2.72	18.48	— 0.126

$Z = 58$; Ce

$= 27.2$; $k = 4.254 \cdot 10^{-7}$; $\frac{\mu}{a} = 0.2314$; $\beta = 0.09359$; $K = 0.1324$;

$B = 0.161$; $C = 1.36$; $D = 12.00$; $\alpha_s = 4.32$.

v_0	1.00000	u_0	1.00000
v_1	— 10.26104	u_1	— 14.33995
v_2	45.29885	u_2	72.84333
v_3	— 126.84384	u_3	— 213.04123
v_4	263.7147	u_4	445.5261
v_5	— 439.9598	u_5	— 738.7314
v_6	617.593	u_6	1024.593
v_7	— 752.852	u_7	— 1230.258
v_8	814.63	u_8	1310.01
v_9	— 795.30	u_9	— 1258.10
v_{10}	709.5	u_{10}	1103.9
v_{11}	— 583.9	u_{11}	— 894.1
v_{12}	447.0	u_{12}	673.9
v_{13}	— 320	u_{13}	— 476
v_{14}	216	u_{14}	316
v_{15}	— 138	u_{15}	— 196

x	R	Θ	U	x	R	Θ	U
0.3	0.358	4.84	— 0.0470	4.0	1.09	13.24	+ 0.0900
0.4	0.392	5.56	— 0.0342	4.5	1.16	13.63	+ 0.134
0.5	0.422	6.18	— 0.0060	5	1.24	14.00	+ 0.163
0.6	0.450	6.70	+ 0.0244	6	1.37	14.59	+ 0.162
0.7	0.476	7.17	+ 0.0490	7	1.51	15.07	+ 0.118
0.8	0.500	7.60	+ 0.0649	8	1.64	15.47	+ 0.049
0.9	0.524	7.98	+ 0.0688	9	1.76	15.82	— 0.028
1.0	0.546	8.33	+ 0.0643	10	1.88	16.12	— 0.101
1.2	0.590	8.96	+ 0.0353	12	2.06	16.64	— 0.221
1.4	0.631	9.50	— 0.0060	14	2.23	17.08	— 0.290
1.6	0.671	9.97	— 0.0459	16	2.40	17.45	— 0.313
1.8	0.709	10.39	— 0.0773	18	2.50	17.78	— 0.280
2.0	0.746	10.77	— 0.0962	20	2.59	18.09	— 0.234
2.5	0.834	11.57	— 0.0924	22	2.71	18.38	— 0.157
3.0	0.920	12.23	— 0.0405	24	2.75	18.65	— 0.069
3.5	1.00	12.78	+ 0.0280	26	2.81	18.92	+ 0.300

Z = 67; Ho

$x_0 = 29{,}2$; $k = 2.760 \cdot 10^{-7}$; $\frac{\mu}{a} = 0.2202$; $\beta = 0.1271$; $K = 0.1378$;

$B = 0.170$: $C = 1.33$; $D = 13.05$; $\alpha_s = 4.66$.

v_0	1.00000	u_0	1.00000
v_1	— 11.40902	u_1	— 16.19716
v_2	54.90957	u_2	91.14407
v_3	— 165.13950	u_3	— 289.98981
v_4	365.3936	u_4	650.3858
v_5	— 645.2426	u_5	— 1146.5478
v_6	954.923	u_6	1681.828
v_7	— 1223.606	u_7	— 2127.846
v_8	1388.64	u_8	2380.52
v_9	— 1419.33	u_9	— 2396.89
v_{10}	1323.7	u_{10}	2200.9
v_{11}	— 1137.7	u_{11}	— 1863.2
v_{12}	908.3	u_{12}	1465.7
v_{13}	— 679	u_{13}	— 1079
v_{14}	477	u_{14}	747
v_{15}	— 317	u_{15}	— 489

x	R	Θ	U	x	R	Θ	U
0.3	0.350	5.17	— 0.0433	4.5	1.13	14.40	+ 0.151
0.4	0.383	5.93	— 0.0182	5	1.20	14.77	+ 0.132
0.5	0.413	6.57	+ 0.0163	6	1.34	15.39	+ 0.056
0.6	0.439	7.13	+ 0.0454	7	1.47	15.89	— 0.039
0.7	0.465	7.62	+ 0.0624	8	1.60	16.32	— 0.128
0.8	0.489	8.07	+ 0.0658	9	1.73	16.68	— 0.198
0.9	0.512	8.47	+ 0.0573	10	1.84	16.99	— 0.244
1.0	0.533	8.83	+ 0.0409	12	2.05	17.51	— 0.274
1.2	0.576	9.49	— 0.0054	14	2.24	17.95	— 0.239
1.4	0.616	10.06	— 0.0504	16	2.37	18.32	— 0.161
1.6	0.655	10.55	— 0.0816	18	2.48	18.67	— 0.056
1.8	0.692	11.00	— 0.0954	20	2.59	18.97	+ 0.046
2.0	0.728	11.39	— 0.0925	22	2.67	19.27	+ 0.153
2.5	0.814	12.24	— 0.0360	24	2.73	19.54	+ 0.242
3.0	0.899	12.92	+ 0.0434	26	2.84	19.81	+ 0.323
3.5	0.979	13.50	+ 0.108	28	2.84	20.06	+ 0.368
4.0	1.06	13.99	+ 0.145				

Z = 74; W

$x_0 = 30.6$; $k = 2.048 \cdot 10^{-7}$; $\frac{\mu}{a} = 0.2128$; $\beta = 0.1577$; $K = 0.1444$

$B = 0.179$; $C = 1.30$; $D = 13.89$; $\alpha_s = 4.91$.

v_0	1.00000	u_0	1.00000
v_1	— 12.34398	u_1	— 17.74078
v_2	63.31604	u_2	107.63765
v_3	— 200.65021	u_3	— 364.22553
v_4	464.4998	u_4	859.5821
v_5	— 854.4885	u_5	— 1583.9396
v_6	1313.250	u_6	2418.610
v_7	— 1743.566	u_7	— 3176.116
v_8	2047.03	u_8	3679.78
v_9	— 2161.18	u_9	— 3831.46
v_{10}	2079.5	u_{10}	3633.2
v_{11}	— 1842.5	u_{11}	— 3172.2
v_{12}	1515.3	u_{12}	2571.9
v_{13}	— 1165	u_{13}	— 1949
v_{14}	843	u_{14}	1390
v_{15}	— 575	u_{15}	— 937

x	R	Θ	U	x	R	Θ	U
0.3	0.345	5.42	— 0.0378	4.5	1.11	14.96	+ 0.109
0.4	0.377	6.21	— 0.0040	5	1.18	15.35	+ 0.058
0.5	0.406	6.88	+ 0.0328	6	1.33	15.99	— 0.056
0.6	0.432	7.45	+ 0.0574	7	1.46	16.50	— 0.151
0.7	0.457	7.96	+ 0.0656	8	1.58	16.94	— 0.216
0.8	0.481	8.42	+ 0.0586	9	1.71	17.31	— 0.246
0.9	0.503	8.84	+ 0.0401	10	1.84	17.66	— 0.246
1.0	0.525	9.21	+ 0.0159	12	2.04	18.15	— 0.189
1.2	0.567	9.89	— 0.0369	14	2.23	18.62	— 0.070
1.4	0.606	10.48	— 0.0761	16	2.41	18.97	+ 0.045
1.6	0.645	10.99	— 0.0931	18	2.48	19.34	+ 0.172
1.8	0.681	11.45	— 0.0884	20	2.54	19.62	+ 0.258
2.0	0.716	11.86	— 0.0673	22	2.68	19.95	+ 0.347
2.5	0 801	12.73	+ 0.0191	23	2.73	20.19	+ 0.386
3.0	0 884	13.44	+ 0.0978	26	2.79	20.50	+ 0.402
3.5	0 963	14.03	+ 0.138	28	2.84	20.71	+ 0.392
4.0	1.04	14 54	+ 0.139	30	2.94	21.00	+ 0.353

Z = 80; Hg

$x_0 = 31.7$; $k = 1.621 \cdot 10^{-7}$; $\frac{\mu}{a} = 0.2072$; $\beta = 0.1873$; $K = 0.1529$;

$B = 0.191$; $C = 1.29$; $D = 14.44$; $\alpha_s = 5.11$.

v_0	1.00000	u_0	1.00000
v_1	— 13.18965	u_1	— 19.16232
v_2	71.34381	u_2	123.79120
v_3	— 236.11635	u_3	— 440.82317
v_4	567.3108	u_4	1085.0355
v_5	— 1079.0323	u_5	— 2073.0997
v_6	1710.227	u_6	3270.884
v_7	— 2336.974	u_7	— 4427.212
v_8	2819.23	u_8	5276.88
v_9	— 3055.06	u_9	— 5643.47
v_{10}	3014.5	u_{10}	5490.8
v_{11}	— 2736.5	u_{11}	— 4915.0
v_{12}	2304.6	u_{12}	4082.0
v_{13}	— 1814	u_{13}	— 3167
v_{14}	1341	u_{14}	2311
v_{15}	— 939	u_{15}	— 1592

x	R	Θ	U	x	R	Θ	U
0.3	0.341	5.65	— 0.0307	4.5	1.10	15.45	+ 0.044
0.4	0.372	6.46	+ 0.0101	5	1.18	15.84	— 0.024
0.5	0.401	7.15	+ 0.0466	6	1.31	16.49	— 0.142
0.6	0.427	7.74	+ 0.0648	7	1.44	17.02	— 0.213
0.7	0.451	8.26	+ 0.0632	8	1.57	17.46	— 0.236
0.8	0.475	8.73	+ 0.0462	9	1.69	17.84	— 0.217
0.9	0.497	9.16	+ 0.0197	10	1.81	18.16	— 0.174
1.0	0.518	9.55	— 0.0097	12	2.01	18.71	— 0.040
1.2	0.560	10.24	— 0.0625	14	2.20	19.16	+ 0.106
1.4	0.597	10.85	— 0.0904	16	2.38	19.54	+ 0.235
1.6	0.636	11.37	— 0.0906	18	2.49	19.87	+ 0.326
1.8	0.672	11.84	— 0.0682	20	2.56	20.19	+ 0.381
2.0	0.707	12.26	— 0.0325	22	2.67	20.48	+ 0.407
2.5	0.791	13.16	+ 0.0671	24	2.80	20.76	+ 0.402
3.0	0.873	13.88	+ 0.129	26	2.86	21.01	+ 0.363
3.5	0.950	14.49	+ 0.136	28	2.90	21.26	+ 0.292
4.0	1.03	15.01	+ 0.102	30	2.92	21.49	+ 0.209

Z = 92; U

$x_0 = 34.1$; $k = 1.066 \cdot 10^{-7}$; $\frac{\mu}{a} = 0.1976$; $\beta = 0.2577$; $K = 0.1805$;

$B = 0.228$; $C = 1.26$; $D = 15.60$; $\alpha_0 = 5.46$.

v_0	1.00000	u_0	— 1.00000
v_1	— 15.07648	u_1	— 22.42622
v_2	90.60110	u_2	164.07771
v_3	— 326.47690	u_3	— 645.84395
v_4	843.0255	u_4	1725.1473
v_5	— 1709.4048	u_5	— 3534.0590
v_6	— 2872.335	u_6	5934.641
v_7	— 4143.935	u_7	— 8504.130
v_8	5261.69	u_8	10690.65
v_9	— 5987.33	u_9	— 120274.55
v_{10}	6190.5	u_{10}	1227.9
v_{11}	— 5879.1	u_{11}	— 11506.5
v_{12}	5173.1	u_{12}	9991.9
v_{13}	— 4248	u_{13}	— 8096
v_{14}	3274	u_{14}	6161
v_{15}	— 2381	u_{15}	— 4423

x	R	Θ	U	x	R	Θ	U
0.3	0.334	6.11	— 0.0105	5	1.14	16.79	— 0.182
0.4	0.364	6.69	+ 0.0411	6	1.28	17.47	— 0.227
0.5	0.393	7.68	+ 0.0699	7	1.41	18.02	— 0.189
0.6	0.418	8.30	+ 0.0682	8	1.53	18.48	— 0.099
0.7	0.442	8.85	+ 0.0434	9	1.67	18.87	+ 0.009
0.8	0.464	9.34	+ 0.0070	10	1 78	19.20	+ 0.113
0.9	0.485	9.79	— 0.0314	12	1.99	19.76	+ 0.286
1.0	0.506	10.19	— 0.0636	14	2.22	20.22	+ 0.394
1.2	0.547	10.92	— 0.0985	16	2.36	20.59	+ 0.420
1.4	0.585	11.55	— 0.0894	18	2.53	20.93	+ 0.397
1.6	0.622	12.10	— 0.0501	20	2.66	21.22	+ 0.332
1.8	0.657	12.60	+ 0.0035	22	2.72	21.50	+ 0.227
2.0	0.691	13.04	+ 0.0566	24	2.80	21.76	+ 0.111
2.5	0.773	13.98	+ 0.138	26	2.85	22.02	— 0.020
3.0	0.853	14.73	+ 0.127	28	2.91	22.26	— 0.145
3.5	0.929	15.38	+ 0.055	30	2.96	22.49	— 0.260
4.0	1.01	15.91	— 0.037	32	2.98	22.72	— 0.363
4.5	1.07	16.37	— 0.129	34	3.06	22.94	— 0.453

NOTAZIONI USATE NELLE TABELLE DEI SINGOLI ELEMENTI

a = raggio di Bohr = $5.28 \cdot 10^{-9}$ cm.
μ = unità di misura di lunghezza (V. formula (2)).
x = raggio vettore in unità μ.
x_0 = raggio dell'atomo.
k = costante per il calcolo del potenziale (formule (7) e (8)).

Le due autofunzioni relativistiche F e G (formule (1) e (1')) sono date, per $x < 0.3$ da:

$$G = x^{-\beta} [u_0 + u_1 x + u_2 x^2 + \cdots]$$

$$F = \frac{\beta}{\alpha Z} x^{-\beta} [v_0 + v_1 x + v_2 x^2 + \cdots] .$$

Per $x > 0.3$ si ha:

$$G = U/x \quad ; \quad U = KR \sin \Theta .$$

Per $x > x_0$ si ha:

$$U = B \sqrt[4]{x} \sin (C \sqrt{x} + D) .$$

Nella zona non relativistica G coincide con la autofunzione ψ di Schroedinger.
α_s = correzione di Rydberg.

N° 83.

Fermi, on his arrival in Rome as a professor of theoretical physics at the University, had a very modest salary. In order to improve his earnings he joined the staff of the Enciclopedia Italiana, founded by Giovanni Treccani, and served on it for several years. He contributed a number of articles to the Enciclopedia Italiana, mainly on atomic physics, and had also editorial and organizational duties. He was rather proud of some of his contributed articles. It would be inappropriate to try to collect them all here, but it is found desirable to give a sample of them in order to show also this facet of Fermi's activity. The choice of the following article is mainly dictated by the fact that Fermi occasionally referred to it and suggested it as reading matter to his students.

EMILIO SEGRÈ.

83.
STATISTICA, MECCANICA

«Enciclopedia Italiana di Scienze, Lettere ed Arti», Istituto G. Treccani, Roma, Vol. 32°, 518–523 (1936–XV).

Lo studio della struttura molecolare e atomica della materia ha reso necessario lo sviluppo di metodi particolari, adatti alla discussione delle proprietà di sistemi costituiti da un enorme numero d'individui costituenti. Per la analogia che il problema presenta con lo studio di una popolazione, essa stessa costituita da un grande numero di uomini, si suol dare a questo ramo della fisica il nome di *meccanica statistica*.

È evidente che la complicazione del calcolo matematico del moto di un sistema costituito da miriadi di particelle (e nel più piccolo corpo percettibile ai nostri sensi sono contenute migliaia di miliardi di molecole) è tanto grande da non lasciare alcuna speranza di una trattazione esatta. Conviene per altro osservare che gli elementi del problema che interessa determinare si riferiscono in generale a proprietà di media su un grande numero d'individui. Così, per esempio, si potrà domandare quale sia la velocità media delle molecole; quale la loro energia cinetica media; quale il valore medio della distanza di una molecola da quella che le è più prossima e così di seguito. Ora si trova che l'andamento di queste proprietà di media è estremamente più regolare, e quindi più facilmente accessibile allo studio, che non le proprietà individuali delle singole molecole. Le forti fluttuazioni di queste tendono infatti a livellarsi tra di loro nell'effettuare le medie; e, se il numero degl'individui è abbastanza grande, vengono praticamente a sparire. Così, per esempio, è ben noto che la pressione esercitata da un gas sulle pareti del recipiente che lo contiene è dovuta agli urti delle molecole del gas contro la parete stessa: ciascun urto trasmette un leggiero impulso alla parete e la pressione risulta dall'insieme di questi impulsi elementari. È evidente che gli urti delle varie molecole differiscono di molto in intensità a seconda della velocità della molecola urtante e dell'angolo d'incidenza dell'urto; ma ciò che interessa agli effetti della determinazione della pressione non è l'impulso trasmesso in un singolo urto, bensì il valore medio di questo impulso e il numero (grandissimo) degli urti che hanno luogo nell'unità di tempo. Entrambi questi due fattori sono indipendenti dalle fluttuazioni accidentali nelle proprietà delle singole molecole; e conseguentemente la pressione esercitata da un gas è anch'essa un fenomeno perfettamente regolare.

Il campo della fisica che ha dato origine allo sviluppo della meccanica statistica, e anche quello nel quale la meccanica statistica trova il maggior numero di applicazioni, è lo studio delle proprietà dei gas. Si può anzi dire che meccanica statistica e teoria cinetica dei gas formino un insieme difficilmente divisibile. Alla voce GAS si trova un cenno storico sui fondamenti della teoria cinetica dei gas e un'esposizione dei principali metodi di calcolo; viene anche discussa la connessione tra la probabilità e l'entropia di uno stato di un sistema, che forma il fondamento della termodinamica statistica. Nel seguito di questo articolo supporremo il lettore familiare con queste nozioni e daremo alcune notizie complementari, con particolare riguardo alle connessioni delle teorie statistiche con la teoria dei quanti.

SPAZIO DELLE FASI. - Il problema della statistica di un sistema costituito da un grande numero di costituenti, quale, ad esempio, un gas costituito da un grandissimo numero di molecole si può formulare così. Lo stato di un sistema meccanico si caratterizza, come è ben noto, nella meccanica classica, mediante la conoscenza della posizione e della velocità di tutti i punti materiali che lo costituiscono. Limiteremo per semplicità le nostre considerazioni al caso dei sistemi olonomi a vincoli indipendenti dal tempo (v. CINEMATICA, n. 32), tanto più che a questa categoria appartiene la quasi totalità dei sistemi che occorre considerare nelle applicazioni fisiche. Per questi sistemi, la posizione viene caratterizzata dai valori di certi parametri $q_1, q_2, \cdots, q_f$, detti *coordinate generali*; il numero f di questi parametri è il numero dei gradi di libertà del sistema. La conoscenza delle f coordinate generali determina soltanto le posizioni, e non le velocità dei punti costituenti il sistema; per avere queste ultime occorre, oltre alle $q_1, q_2, \cdots, q_f$, conoscere anche le loro derivate rispetto al tempo $\dot{q}_1, \dot{q}_2, \cdots, \dot{q}_f$. Possiamo dunque affermare che, per definire completamente lo stato di un sistema meccanico occorre dare $2f$ grandezze: $q_1, q_2, \cdots, q_f, \dot{q}_1, \dot{q}_2, \cdots, \dot{q}_f$. È usuale e conveniente, nella meccanica analitica, usare un insieme di coordinate differente da questo: precisamente in luogo delle f grandezze: $\dot{q}_1, \dot{q}_2, \cdots, \dot{q}_f$ se ne introducono altre f: $p_1, p_2, \cdots, p_f$, definite al modo seguente. L'energia cinetica T, come del resto qualsiasi altra funzione dello stato del sistema, è una funzione delle q e delle $\dot{q}$

$$\mathrm{T} = \mathrm{T}(q_1, q_2, \cdots, q_f \quad ; \quad \dot{q}_1, \dot{q}_2, \cdots, \dot{q}_f).$$

La variabile p_r, che prende il nome di *momento coniugato* alla q_r, è data, per definizione, da

$$p_r = \frac{\partial \mathrm{T}}{\partial \dot{q}_r} \qquad (r = 1, 2, \cdots, f) \tag{1}$$

Invece che per mezzo delle coordinate generali e delle loro derivate rispetto al tempo, rappresenteremo lo stato del sistema per mezzo delle $2f$ variabili di stato

$$q_1, q_2, \cdots, q_f, p_1, p_2, \cdots, p_f. \tag{2}$$

Le leggi che determinano come le variabili di stato variano col tempo (equazioni del movimento) si possono scrivere, nella forma canonica di Hamilton,

$$\dot{q}_r = \frac{\partial H}{\partial p_r} \quad , \quad \dot{p}_r = -\frac{\partial H}{\partial q_r} \qquad (r = 1, 2, \cdots, f), \tag{3}$$

dove la funzione H $(q_1, q_2, \cdots, q_f, p_1, p_2, \cdots, p_f)$, detta *funzione di Hamilton*, o semplicemente *Hamiltoniana*, s'identifica, per il caso dei sistemi conservativi, con l'energia del sistema.

È assai opportuno, per la rappresentazione degli stati di un sistema, introdurre la seguente locuzione geometrica. Chiameremo, secondo J. W. Gibbs, *spazio delle fasi* uno spazio di $2f$ dimensioni, avente le $2f$ variabili di stato (2) come coordinate cartesiane ortogonali. Esiste evidentemente una corrispondenza tra i punti dello spazio delle fasi e gli stati del sistema: invero, dato lo stato, sono noti i valori delle q_r e delle p_r, e quindi si può costruire un punto dello spazio delle fasi; viceversa, dato un punto nello spazio delle fasi, se ne conoscono le coordinate q_r, e p_r, e queste definiscono uno stato del sistema. Possiamo dunque affermare che un punto nello spazio delle fasi rappresenta uno stato del sistema e nel seguito parleremo indifferentemente degli stati del sistema oppure dei punti che li rappresentano nello spazio delle fasi.

Come lo stato di un sistema varia col tempo per effetto del movimento del sistema, così il punto che rappresenta lo stato nello spazio delle fasi si muove descrivendo una traiettoria. Il movimento è determinato dalle equazioni di Hamilton (3) che determinano le derivate rispetto al tempo delle coordinate q_r e p_r del punto rappresentativo e cioè le $2f$ componenti della sua velocità nello spazio a $2f$ dimensioni delle fasi.

Dalla forma delle equazioni di Hamilton risulta che, note le q_r e le p_r al tempo $t = 0$, resta determinato lo stato del sistema, e quindi i valori delle q_r e delle p_r, per ogni istante t passato o futuro. Ciò s'interpreta geometricamente dicendo: se è noto il punto P_0 che rappresenta lo stato al tempo zero, si può determinare il punto $P(t)$ che rappresenta lo stato ad un istante qualunque. Per il punto P_0 passa dunque una traiettoria (e una sola), luogo geometrico di tutti i punti $P(t)$; e tutto lo spazio delle fasi è rigato da una famiglia di tali traiettorie. Nel caso comune che il sistema sia conservativo e la Hamiltoniana H indipendente dal tempo, questa famiglia di traiettorie ha le seguenti proprietà:

a) per ogni punto dello spazio delle fasi passa una e una sola traiettoria;

b) ogni traiettoria giace per intero sopra una delle superficie

$$H = \text{costante}$$

(principio della conservazione dell'energia).

La forma di queste traiettorie, che si distendono sulle superficie H = cost., è in generale assai complicata; essa ha notevole importanza nei problemi statistici.

A questo proposito si considerano i sistemi *ergodici*, e cioè quelli per cui le predette traiettorie, avvolgendosi continuamente sopra le superficie di energia costante, finiscono per passare per tutti i punti della detta superficie. Il sistema si dice invece *quasi–ergodico* se, pur non esistendo traiettorie che passino per tutti i punti della superficie H = cost, esistono però delle traiettorie che la riempiono densamente; vale a dire che passano a distanza arbitrariamente piccola da ogni suo punto. Una definizione che sostanzialmente equivale alla precedente è questa: un sistema è quasi–ergodico se, presi sopra ogni superficie H = cost. due elementi di superficie arbitrariamente piccoli esiste sempre almeno una traiettoria che li attraversa entrambi.

È essenziale, per la deduzione delle proprietà statistiche, il sapere se il sistema che si considera possa, oppur no, considerarsi ergodico o quasi–ergodico. Infatti, per arrivare a una deduzione meccanica della legge di Boltzmann, e cioè della legge fondamentale della meccanica statistica, è necessario ammettere che il sistema in considerazione (gas, corpo solido, ecc.) sia per lo meno quasi–ergodico. Vi sono state pertanto lunghe discussioni per decidere fino a che punto fosse legittima l'ipotesi che i sistemi, ai quali ordinariamente si applica la meccanica statistica, si possano considerare ergodici o quasi–ergodici. La conclusione è stata che, mentre non esistono sistemi meccanici ergodici, possono invece esistere sistemi quasi ergodici; sembra anzi che sistemi meccanici molto complicati e senza speciali caratteristiche di simmetria debbano in generale essere quasi–ergodici. E, naturalmente, i sistemi che occorre considerare nelle applicazioni statistiche, sono sempre di eccezionale complicazione, sia per l'enorme numero di costituenti, sia anche perché, a rigor di termini, si deve considerare che faccia parte del sistema anche l'insieme di tutti quei corpi che possono, per esempio attraverso a una trasmissione di calore, influenzare il movimento del corpo in considerazione. È dunque plausibile l'ipotesi che i sistemi ai quali si applicano le considerazioni statistiche siano quasi–ergodici (ipotesi della quasi–ergodicità).

Tra le proprietà dello spazio delle fasi che hanno maggiore importanza per le applicazioni statistiche dobbiamo ancora citare il teorema di Liouville. Consideriamo, nello spazio delle fasi, un volume elementare τ_0; a ogni punto P_0 appartenente a τ_0, facciamo corrispondere un altro punto P dello spazio delle fasi con la regola che P sia il punto rappresentativo dello stato in cui si trova, al tempo t, un sistema che, al tempo zero, era nello stato rappresentato da P_0. Se teniamo fisso il tempo t, l'insieme dei punti P riempirà un altro elemento di volume τ, corrispondente a τ_0; il teorema di Liouville afferma che gl'ipervolumi di τ e τ_0 sono eguali.

LEGGE DI RIPARTIZIONE DI BOLTZMANN. – Il problema fondamentale della meccanica statistica consiste nella ricerca della legge di distribuzione per un sistema che si trovi a una temperatura assegnata. Questo problema ha avuto per la prima volta una soluzione parziale dal Maxwell, per il caso di un gas costituito da molecole puntiformi; la soluzione generale per un sistema qualsiasi è dovuta a L. Boltzmann, e da lui prende il nome di legge di ripartizione di Boltzmann. Il problema della ricerca della legge di distribuzione si può formulare in modi diversi che vogliamo qui considerare.

Riferiamoci dapprima al caso di un gas costituito da molecole che, per semplicità, ammetteremo essere tutte della stessa specie. Ciascuna delle molecole è un sistema meccanico, il cui stato si potrà rappresentare come un punto in uno spazio delle fasi della molecola. Pensiamo di segnare in questo spazio tutti i punti che rappresentano gli stati in cui si trovano le varie molecole a un certo istante. La legge di densità, secondo cui questi punti sono distribuiti nello spazio delle fasi, determina in questo caso direttamente la distribuzione statistica delle molecole.

Accenniamo a un metodo per determinare tale distribuzione che, pur non essendo in alcun modo logicamente rigoroso, presenta il vantaggio di una grande semplicità ed evidenza. Pensiamo di suddividere lo spazio delle fasi in tante cellette, aventi tutte lo stesso ipervolume. Il volume si può scegliere abbastanza grande perché ogni cella contenga un considerevole numero di molecole, ma abbastanza piccolo perché le variazioni della densità di distribuzione dei punti rappresentativi entro una cella siano trascurabili. Indichiamo le varie celle con un numero d'ordine, e siano $N_1, N_2, \cdots N_s, \cdots$ i numeri di molecole i cui punti rappresentativi appartengono alla $1^a, 2^a, \cdots, s^a, \cdots$ cella. Una elementare considerazione di calcolo combinatorio dimostra che il numero di modi di distribuire le N molecole costituenti il gas tra le varie celle, in modo che N_1, appartengano alla prima cella, N_2, alla seconda, ecc. $\cdots$ è dato da

$$\Pi = \frac{N!}{N_1!\,N_2!\cdots N_s!\cdots}. \tag{4}$$

Ammettendo a priori eguale la probabilità di ogni ripartizione delle molecole tra le varie celle (e si noti che questa ipotesi non è in alcun modo giustificata), l'espressione (4) si può ritenere proporzionale alla probabilità della ripartizione $N_1, N_2, \cdots, N_s, \cdots$

Conviene ora ricercare la ripartizione più probabile, quella cioè per cui Π è massimo. Occorre tener conto che le N_s sono vincolate da due relazioni. Una di esse:

$$N_1 + N_2 + \cdots + N_s + \cdots = N \tag{5}$$

esprime che il numero totale delle molecole è N; l'altra esprime il principio della conservazione dell'energia e si può formulare nel modo seguente. La energia di una molecola è una funzione dello stato di essa, e quindi del corrispondente punto nello spazio delle fasi; data però la piccola estensione delle celle, si può ritenere che entro ogni cella l'energia non varii sensibilmente. Chiameremo w_s l'energia delle molecole appartenenti alla s^a cella. Indicando con W l'energia totale del gas avremo allora la relazione

$$N_1 w_1 + N_2 w_2 + \cdots + N_s w_s + \cdots = W. \tag{6}$$

Il problema della ricerca della ripartizione più probabile si riduce così alla determinazione dei numeri $N_1, N_2, \cdots, N_s, \cdots$ in modo che soddisfino le due relazioni (5) e (6) e rendano massimo (4). Si trova così che i numeri N_s sono espressi da

$$N_s = g e^{-\beta w_s}, \tag{7}$$

in cui g e β sono costanti, che si possono determinare in modo da soddisfare (5) e (6) e, al solito, e denota la base dei logaritmi naturali. La costante β si può anche esprimere per mezzo della temperatura assoluta T a cui è legata dalla relazione.

$$\beta = \frac{1}{kT}, \tag{8}$$

dove la costante k, detta costante di Boltzmann, non è altro che il rapporto tra la costante R dei gas perfetti e il numero di Avogadro. Il valore di k è

$$k = 1{,}36 \cdot 10^{-16}\ \text{cm}^2\ \text{gr}\ \text{sec}^{-2}\ \text{gradi}^{-1}\ (*).$$

Siccome tutte le celle in cui abbiamo diviso lo spazio delle fasi hanno egual volume, i numeri N_s sono proporzionali alla densità dei punti rappresentativi nello spazio delle fasi. Possiamo dunque così formulare il risultato precedente:

Il numero delle molecole le cui coordinate e i cui momenti sono rispettivamenti comprese negl'intervalli

$$\begin{cases} q_1 \text{ e } q_1 + dq_1 \ ; \ q_2 \text{ e } q_2 + dq_2 \ ; \cdots ; \ q_f \text{ e } q_f + dq_f \ ; \\ p_1 \text{ e } p_1 + dp_1 \ ; \ p_2 \text{ e } p_2 + dp_2 \ ; \cdots ; \ p_f \text{ e } p_f + dp_f \end{cases} \tag{9}$$

(e cioè i cui punti rappresentativi cadono nell'elemento d'ipervolume dello spazio delle fasi $dq_1 dq_2 \cdots dq_f dp_1 dp_2 \cdots dp_f$) è dato da:

$$dn = A e^{-\frac{w}{kT}} dq_1 dq_2 \cdots dq_f dp_1 dp_2 \cdots dp_f, \tag{10}$$

A essendo una costante (che si può determinare conoscendo il numero totale delle molecole), mentre w rappresenta l'energia di una molecola espressa in funzione delle sue variabili di stato $q_1, \cdots p_1, \cdots$ La formula (10) esprime la legge di ripartizione di Boltzmann, e si può considerare il risultato più importante della meccanica statistica.

Casi particolari e conseguenze importanti della legge di Boltzmann (10) sono:

La *legge di ripartizione delle velocità di Maxwell*: in un gas alla temperatura T il numero di molecole, per le quali le componenti u, v, w della velocità di traslazione sono comprese entro i limiti u ed $u + du$; v e $v + dv$; w e $w + dw$, è dato da

$$dn = \text{cost}\ e^{-\frac{m}{2kT}(u^2+v^2+w^2)} du\, dv\, dw, \tag{11}$$

m essendo la massa di una molecola (v. GAS).

(*) Questo valore di k era quello accettato al tempo della composizione di questo articolo. Un valore più approssimato è $1{,}38 \cdot 10^{-16}$ nelle stesse unità (N. d. R.).

Il *principio della equipartizione dell'energia*, che si può formulare nel modo seguente. Supponiamo che le molecole del gas siano riferite a un sistema di coordinate generali ortogonali, tali cioè che l'energia cinetica, che è in genere una forma quadratica delle $\dot{q}$, sia una forma quadratica ortogonale

$$T = 1/2\, [m_1(q)\, \dot{q}_1^2 + m_2(q)\, \dot{q}_2^2 + \cdots + m_f(q)\, \dot{q}_f^2]. \tag{12}$$

In questo caso l'energia cinetica si spezza nella somma di f addendi, ciascuno dei quali rappresenta l'energia pertinente a uno degli f gradi di libertà della molecola. Il principio della equipartizione dell'energia, che si può dedurre abbastanza semplicemente dalla legge di Boltzmann, afferma che: il valore medio dell'energia cinetica pertinente a ciascuno dei gradi di libertà della molecola è costante ed uguale a $k\mathrm{T}/2$ (per le applicazioni di questo principio v. GAS).

Il problema della ricerca della legge di ripartizione si può formulare anche in una forma differente, che presenta in qualche caso notevoli vantaggi su quella che abbiamo fin qui illustrato. Invece di rappresentare individualmente lo stato di ciascuna molecola come un punto nello spazio delle fasi della singola molecola, riguarderemo il nostro corpo, ed eventualmente anche tutti gli altri corpi con i quali esso può interagire, come un unico sistema meccanico, avente un numero enormemente grande di gradi di libertà, e conseguentemente uno spazio delle fasi di un numero elevatissimo di dimensioni. Lo stato di questo sistema complesso sarà rappresentato da un punto del suo spazio delle fasi; e questo punto, col variare del tempo, si muoverà lungo una traiettoria che, se il sistema è quasi–ergodico, finirà per riempire densamente tutta una ipersuperficie di energia costante.

Accenniamo qui a un procedimento per arrivare con queste considerazioni a stabilire la legge di ripartizione di Boltzmann, in base alla ipotesi della quasi–ergodicità del sistema. Premettiamo perciò le seguenti proposizioni:

a) Si abbia un numero molto grande N di sistemi quasi–ergodici, tutti identici e del tutto indipendenti uno dall'altro. Lo stato di ciascuno di questi sistemi sarà rappresentato da un punto dello spazio delle fasi. Al variare del tempo ogni punto rappresentativo si muove su una superficie di energia costante, descrivendola tutta densamente. Supponiamo ora che le energie degli N sistemi siano ripartite uniformemente in uno strettissimo intervallo di energia da E ad $\mathrm{E} + \Delta\mathrm{E}$. Per conseguenza i punti rappresentativi saranno tutti contenuti nella intercapedine tra le due ipersuperficie

$$\mathrm{H} = \mathrm{E} \quad \text{ed} \quad \mathrm{H} = \mathrm{E} + \Delta\mathrm{E}. \tag{13}$$

Inizialmente i punti siano ripartiti tra queste due ipersuperficie con una legge di densità arbitraria. Col variare del tempo i punti si sposteranno e la densità della distribuzione verrà in generale a variare col tempo. Si può dimostrare, con una facile applicazione del teorema di Liouville, che la sola legge di densità che si mantiene stazionaria col tempo è quella uniforme. Per tale distribuzione di densità il numero di punti rappresentativi appartenenti a un elemento di volume dello spazio delle fasi, contenuto nell'intercapedine tra le due superficie (13), è proporzionale al volume dell'elemento stesso; o, in altre parole, la probabilità che uno dei sistemi si trovi in uno stato appartenente all'elemento di volume è proporzionale al volume dell'elemento.

b) Consideriamo un sistema costituito da due parti, tra le quali si eserciti una interazione debolissima; e supponiamo che ciascuno dei due sistemi parziali 1 e 2 abbia un grandissimo numero di gradi di libertà. (Si pensi, per fissare le idee, a due corpi, contenenti ciascuno miriadi di molecole, e posti in contatto l'uno con l'altro, in modo che possano avvenire scambi di calore). Siano $x_1, x_2, \cdots, x_{N_1}$, le coordinate e i momenti del primo sistema, e siano $y_1, y_2, \cdots, y_{N_2}$ le coordinate e i momenti del secondo sistema parziale. Dovremo allora considerare tre diversi spazi delle fasi: quello del primo sistema parziale, avente coordinate $x_1, x_2, \cdots, x_{N_1}$; quello del secondo sistema parziale con coordinate $y_1, y_2, \cdots, y_{N_2}$; e infine quello del sistema complessivo, costituito dall'insieme dei due sistemi parziali, che ha come coordinate $x_1, x_2, \cdots, x_{N_1}, y_1, y_2, \cdots, y_{N_2}$. Si noti che un elemento di volume dello spazio delle fasi complessivo si può scrivere nella forma $dx_1\, dx_2 \cdots dx_{N_1}\, dy_1\, dy_2 \cdots dy_{N_2}$ e cioè come prodotto di un elemento di volume del primo spazio delle fasi (e cioè $dx_1\, dx_2 \cdots dx_{N_1}$) per un elemento di volume del secondo spazio delle fasi (e cioè $dy_1\, dy_2 \cdots dy_{N_2}$).

Il volume dello spazio delle fasi del primo sistema, corrispondente a stati di energia compresa tra E_1 ed $E_1 + dE_1$, si potrà scrivere

$$d\tau_1 = \omega_1(E_1)\, dE_1\,; \tag{14}$$

e similmente per il secondo sistema avremo

$$d\tau_2 = \omega_2(E_2)\, dE_2\,. \tag{15}$$

È facile convincersi che se, come abbiamo ammesso, entrambi i sistemi hanno un numero grandissimo di gradi di libertà, ω_1 e ω_2 sono funzioni rapidissimamente crescenti dei loro argomenti. Si riconosce, per esempio, senza difficoltà, che se uno dei sistemi è un gas perfetto, contenente N molecole puntiformi, racchiuse in un recipiente di volume costante, si ha

$$\omega(E) = \text{cost}\; E^{\frac{3N}{2}-1}. \tag{16}$$

E siccome N è un numero grandissimo, l'esponente di E è molto grande e quindi la funzione ω rapidissimamente crescente.

Sia:

$$E = E_1 + E_2 \tag{17}$$

l'energia totale del sistema complessivo. La probabilità che l'energia del primo sistema parziale sia compresa entro i limiti E_1 ed $E_1 + dE_1$ (e conseguentemente l'energia del secondo sistema parziale entro i limiti $E - E_1$ ed $E - E_1 - dE_1$) è proporzionale, secondo quanto si è detto in *a*), al volume dello spazio delle fasi complessivo corrispondente agli stati predetti. Un calcolo elementare dimostra che questo è a sua volta proporzionale a:

$$\omega_1(E_1)\, \omega_2(E_2)\, dE_1\,.$$

La ripartizione di energia più probabile si ha cercando il massimo della espressione precedente (o, che è lo stesso, del suo logaritmo) con la condizione di vincolo (17). Si trova così la seguente condizione di massimo:

$$\frac{d \log \omega_1 (E_1)}{dE_1} = \frac{d \log \omega_2 (E_2)}{dE_2} . \tag{18}$$

Siccome d'altra parte le due funzioni ω_1 e ω_2 crescono rapidissimamente al crescere dei loro argomenti, si riconosce che la probabilità di ripartizioni dell'energia totale E sensibilmente diverse da quella definita da (18) è praticamente trascurabile.

Il risultato precedente ci dice dunque che, se si portano in contatto diversi sistemi, aventi ciascuno un numero assai grande di gradi di libertà (in pratica costituito ciascuno da un grandissimo numero di molecole), l'energia si ripartisce tra di loro in modo tale che, per tutti i sistemi, la grandezza

$$\frac{d \log \omega (E)}{dE}$$

abbia lo stesso valore. E ciò significa che la grandezza precedente non dipende dalle particolarità speciali dei singoli sistemi, ma è funzione unicamente della temperatura T comune a tutti i sistemi in contatto. Possiamo dunque scrivere, per ogni sistema contenente un grande numero di molecole,

$$\frac{d \log \omega (E)}{dE} = F(T), \tag{19}$$

dove F è il simbolo di una funzione universale.

La relazione precedente si può usare per una definizione statistica della temperatura assoluta; oppure si può determinare F (T) per mezzo della temperatura assoluta definita da un termometro a gas perfetto nel modo seguente. Applicando la (19) al caso di un gas perfetto, e tenendo conto della (16), si trova (trascurando l'unità in confronto del numero grandissimo N delle molecole)

$$F(T) = \frac{3N}{2E} .$$

D'altra parte l'energia delle N molecole – vedi GAS, formule (8), (8') – è:

$$E = \frac{3}{2} NkT$$

e quindi risulta:

$$F(T) = \frac{1}{kT} .$$

La (19) si può dunque scrivere:

$$\frac{d \log \omega (E)}{dE} = \frac{1}{kT} . \tag{20}$$

Siamo ora in grado di ottenere una semplice dimostrazione della legge di ripartizione di Boltzmann. Dato un sistema che si trovi alla temperatura T, ci proponiamo di trovare la probabilità che il punto rappresentativo del suo stato si trovi entro un elemento di volume $d\tau$ dello spazio delle fasi del sistema dato. Per mantenere il sistema alla temperatura T, possiamo pensare di porlo in contatto con un corpo assai più grande del sistema dato e che si trovi alla temperatura T; questo sistema funziona da « bagno di temperatura » e dovremo ammettere che esso abbia un numero grandissimo di gradi di libertà così che per esso valga la relazione (20). Chiamiamo w l'energia posseduta dal primo sistema quando il suo stato è rappresentato da un punto appartenente a $d\tau$. Se E è l'energia totale, l'energia del secondo sistema « bagno di temperatura » sarà $E - w$; per ipotesi w si può considerare assai piccolo in confronto di E. La probabilità cercata è proporzionale al volume dello spazio delle fasi per il sistema totale (primo + secondo); e questo volume è a sua volta proporzionale a:

$$\omega\,(E - w)\,d\tau.$$

Tenendo conto di (20), e del fatto che w si può considerare infinitesimo rispetto ad E, troviamo:

$$\omega\,(E - w) = \omega\,(E)\,e^{-\frac{w}{kT}}$$

e siccome ω (E) non dipende dallo stato del primo sistema, possiamo concludere che la probabilità che questo si trovi in uno stato appartenente a $d\tau$ è proporzionale a:

$$e^{-\frac{w}{kT}}\,d\tau,$$

che esprime la legge di Boltzmann. La costante di proporzionalità, ove occorra, si può determinare con la condizione che la probabilità totale è eguale ad 1; tale costante è dunque:

$$\frac{1}{\int e^{-\frac{w}{kT}}\,d\tau},$$

dove l'integrale si deve estendere a tutto lo spazio delle fasi.

Un'altra conseguenza notevole della formula (20) è la relazione, stabilita da Boltzmann, tra l'entropia di un sistema e la probabilità del suo stato. Da quanto si è detto risulta infatti che ω (E) si può ritenere proporzionale alla probabilità che il sistema abbia energia E. Integrando la (20) si ricava d'altra parte:

$$k \log \omega\,(E) = \int \frac{dE}{T}\,.$$

Il secondo membro rappresenta l'entropia; la relazione precedente si può dunque leggere:

$$\text{entropia} = k \log \text{probabilità}.$$

La statistica della teoria dei quanti. – Lo sviluppo della teoria dei quanti, iniziatosi per opera di M. Planck negli ultimi anni del secolo XIX, e cioè, poco dopo i classici lavori di Boltzmann, fu profondamente influenzato e, si può anzi dire, promosso dalle considerazioni statistiche; e a sua volta reagì sopra queste, modificandole in molti punti essenziali.

Il principio dell'equipartizione dell'energia, che è una diretta conseguenza della legge di ripartizione di Boltzmann, prende una forma particolarmente espressiva nel caso che il sistema a cui lo si applica sia soggetto a forze di tipo elastico; è noto infatti che in questo caso i valori medî delle energie cinetica e potenziale sono eguali, e quindi l'energia totale è il doppio dell'energia cinetica media; quest'ultima, d'altra parte, è eguale, per il principio della equipartizione, a $(1/2) fk\mathrm{T}$, essendo f il numero dei gradi di libertà del sistema. Il valore medio $\overline{w}$ dell'energia in un sistema elastico risulta dunque:

$$\overline{w} = fk\mathrm{T} \, . \tag{21}$$

Questo risultato, come anche del resto tutte le altre conseguenze del principio dell'equipartizione, si mostra in accordo con i fatti sperimentali solo parzialmente.

Così, per esempio, la (21) si può applicare a un corpo solido contenente N atomi, riguardando in prima approssimazione gli atomi come punti materiali legati tra di loro da forze elastiche; il numero dei gradi è in questo caso 3 N, e quindi la (21) ci dà:

$$w = 3 \, \mathrm{N}k\mathrm{T}$$

come espressione dell'energia termica del corpo solido. Se, per esempio, ci riferiamo ad un grammo-atomo, N è il numero di Avogadro; quindi $\mathrm{N}k = \mathrm{R}$. La relazione diventa:

$$w = 3 \, \mathrm{RT},$$

a cui corrisponde il calore atomico

$$c = \frac{dw}{d\mathrm{T}} = 3 \, \mathrm{R},$$

cioè: il calore atomico di tutti gli elementi solidi è costante e eguale a 3 R (cioè circa 6 calorie-gradi). Questa non è altro che l'espressione della legge trovata empiricamente da Dulong e Petit.

Mentre così da una parte troviamo una conferma sperimentale del principio della equipartizione, lo stesso risultato conduce d'altro lato a serie difficoltà per la teoria. In primo luogo è noto infatti che la legge di Dulong e Petit è valida per la maggioranza dei corpi solidi a temperatura ordinaria; a basse temperature invece il calore atomico dei solidi scende al disotto del valore di Dulong e Petit, e tende ad annullarsi allo zero assoluto. Secondo il principio della equipartizione il calore specifico dei solidi dovrebbe invece essere costante. Altra difficoltà per la teoria si incontra quando si analizza la legittimità della ipotesi che, agli effetti statistici, gli atomi del solido si possono riguardare come punti materiali. È ben noto al contrario che l'atomo è un edificio di notevole complessità, avente un gran numero di gradi di libertà interni, che non sono stati considerati nella computazione del calore specifico del solido. Nella statistica classica non si trova alcuna giustificazione del perché questi gradi di libertà interni si debbano trascurare. Una difficoltà analoga, e cioè che alcuni dei gradi di libertà si debbono computare e altri trascurare agli effetti del calore specifico, si trova anche nella teoria del calore specifico dei gas.

Come ultimo esempio delle difficoltà che s'incontrano per l'applicazione del principio della equipartizione, citiamo il caso dello spettro del corpo nero, cioè della radiazione che si stabilisce entro una cavità, le cui pareti siano mantenute alla temperatura uniforme T. Si può dimostrare che il campo di radiazione entro una cavità di volume V ha proprietà dinamiche simili a quelle di un solido elastico che riempisse la cavità; ed è in particolare suscettibile di vibrare con certe frequenze caratteristiche, ognuna delle quali si può considerare corrispondente a un grado di libertà del sistema. Naturalmente, trattandosi di un sistema continuo, il numero totale dei gradi di libertà è infinito; si trova più precisamente che il numero delle frequenze caratteristiche comprese tra i limiti ν e $\nu + d\nu$ è dato da:

$$dn = \frac{8\pi V}{c^3}\nu^2 d\nu ;$$

(l'espressione precedente, integrata rispetto a ν da o a ∞, diverge; ciò che indica appunto che il numero totale dei gradi di libertà è ∞).

Secondo il principio dell'equipartizione dell'energia si sarebbe indotti ad attribuire a ognuno di questi gradi di libertà, secondo la (21), l'energia media kT. Quindi, nella radiazione del corpo nero, l'energia raggiante di frequenza tra ν e $\nu + d\nu$ dovrebbe essere:

$$dw = \frac{8\pi k}{c^3} VT\nu^2 d\nu. \tag{23}$$

Questa formula, stabilita da Rayleigh e Jeans, si mostra in accordo con l'esperienza soltanto per basse frequenze ed alte temperarure. Che essa non possa esser valida in generale, risulta tra l'altro dal fatto che in questo caso l'energia totale della radiazione nera dovrebbe essere infinita, poiché si trova un risultato infinito integrando la (23) rispetto a ν da o a ∞.

Appunto tentando di risolvere le difficoltà incontrate nella teoria statistica del corpo nero il Planck fu per il primo indotto a introdurre l'ipotesi dei quanti (v. ATOMO).

Il risultato fondamentale della teoria dei quanti, e cioè l'esistenza, per ogni sistema atomico o molecolare, di una successione discreta di valori possibili per l'energia, ha conseguenze statistiche assai notevoli. Siano $w_1, w_2, \cdots, w_r, \cdots$ i livelli energetici di un atomo o molecola. Possiamo domandarci quale è la probabilità che, alla temperatura T, il sistema si trovi in uno stato di energia w_r; questo problema, che è l'analogo quantistico della ricerca della legge di ripartizione di Boltzmann, si può trattare in modo analogo al caso classico; si trova che le probabilità che il sistema sia negli stati $w_1, w_2, \cdots, w_r, \cdots$ sono rispettivamente proporzionali a:

$$e^{-\frac{w_1}{kT}}, e^{-\frac{w_2}{kT}}, \cdots, e^{-\frac{w_r}{kT}}, \cdots$$

Si noti che, affinché il risultato precedente sia valido, i livelli $w_1, w_2, \cdots, w_r \cdots$, debbono essere tutti semplici; e quindi, in caso di degenerazione, ogni livello energetico deve venir ripetuto tante volte quanto è il suo grado di degenerazione.

Questa estensione della legge di Boltzmann ai sistemi quantizzati ha conseguenze molto importanti. Essa permette, per esempio, quando la si applica a un oscillatore armonico, di calcolare il valore medio dell'energia alla temperatura T. Questo risulta:

$$\bar{w} = \frac{h\nu}{e^{\frac{h\nu}{kT}} - 1}, \tag{24}$$

dove ν è la frequenza propria dell'oscillatore e h la costante di Planck. Nel caso limite, in cui $h\nu$ sia trascurabile di fronte a kT, la formula precedente ci dà, come nella statistica classica, $\bar{w} = k\mathrm{T}$; mentre altrimenti $\bar{w}$ risulta minore di quanto corrisponderebbe al principio della equipartizione della energia. Attribuendo a ciascuno degli oscillatori (22) l'energia media (24) si trova, in luogo della formula di Rayleigh e Jeans, la seguente espressione per l'energia della radiazione del corpo nero avente frequenza compresa tra ν e $\nu + d\nu$

$$dw = \frac{8\,\pi \mathrm{V}}{c^3} \frac{h\nu^3}{e^{\frac{h\nu}{kT}} - 1} d\nu. \tag{25}$$

È questa la celebre formula di Planck per lo spettro del corpo nero, che è ottimamente verificata dall'esperienza, e costituì il primo fondamento sperimentale della teoria dei quanti.

La formula (24) fu successivamente applicata da Einstein e da P. Debye al problema dei calori specifici dei corpi solidi, e permise di trovare una spiegazione assai soddisfacente del fatto che la legge di Dulong e Petit cessa di essere valida per temperature molto basse e che il calore specifico dei solidi tende allo zero quando la temperatura scende allo zero assoluto.

Troviamo ancora nell'estensione della legge di Boltzmann ai sistemi quantizzati la giustificazione del fatto che normalmente nel calcolo dei calori specifici gli atomi si possono considerare come punti materiali, trascurando completamente i loro gradi di libertà interni: in genere negli atomi la differenza di energia tra il livello energetico più profondo (livello fondamentale) e gli altri livelli energetici è molto grande a confronto di kT. Perciò risulta dalle estensioni della legge di Boltzmann che la probabilità che l'atomo non si trovi nel livello energetico fondamentale è praticamente nulla. Per conseguenza l'energia dovuta ai gradi di libertà interni dell'atomo non varia al variare della temperatura; e questi dunque non portano alcun contributo al calore specifico della sostanza. Naturalmente risulta dalla spiegazione che abbiamo data, che essa non è valida in senso assoluto, ma solo finché la temperatura non oltrepassa certi limiti. Che del resto il contributo dei varî gradi di libertà del sistema al calore specifico vada decrescendo con la temperatura risulta anche, per il caso dell'oscillatore armonico, della (24); un'elementare discussione di questa formula mostra infatti che, per T tendente a zero, non solo $\bar{w}$, ma anche $d\bar{w}/d\mathrm{T}$ tende a zero.

Relazioni con il principio di Nernst. - Conviene qui notare la connessione dei risultati, a cui abbiamo accennato, col terzo principio della termodinamica o principio di Nernst (v. termodinamica).

Questo, come è noto, fissa la costante additiva che resta indeterminata nell'usuale definizione termodinamica dell'entropia, stabilendo che per tutti i corpi (condensati) l'entropia si annulla allo zero assoluto. L'entropia di un corpo alla temperatura T si ottiene dunque dall'integrale

$$S = \int_0^T \frac{dQ}{T},$$

esteso ad un processo in cui il corpo viene riscaldato in modo reversibile dalla temperatura zero alla temperatura T; dQ è il differenziale della quantità di calore trasmessa al corpo. Se indichiamo con c il calore specifico, potremo scrivere, per l'unità di massa della sostanza, $dQ = c dT$; e l'integrale precedente prende la forma

$$S = \int_0^T \frac{c dT}{T}.$$

Di qui si riconosce la necessità, affinché l'integrale converga nel limite inferiore, che il calore specifico c si annulli per $T = 0$. La statistica dei sistemi quantizzati conduce appunto a questo risultato.

Che, del resto, vi debba essere una correlazione tra l'esistenza di stati quantici discreti e la determinazione della costante dell'entropia, risulta anche dalle considerazioni seguenti.

Riferiamoci a un sistema A, costituito da un grande numero N di sistemi a indipendenti, tutti uguali tra di loro. Classicamente il problema della distribuzione statistica di un tale sistema si tratta suddividendo lo spazio delle fasi di ciascuno dei sistemi a in cellette aventi tutte lo stesso ipervolume. Se $N_1, N_2, \cdots$ sono i numeri dei sistemi a, il cui punto rappresentativo appartiene alla prima, alla seconda,... cella, la probabilità dello stato, definita come numero di modi di realizzare la data distribuzione, è

$$\Pi = \frac{N!}{N_1! N_2! \cdots},$$

e la distribuzione più probabile, che corrisponde alla legge di ripartizione di Boltzmann, è quella che rende massima la probabilità Π.

In questa considerazione, basata sulla meccanica classica, il valore di Π, data la legge di densità della distribuzione nello spazio delle fasi, non è completamente determinato, poiché esso dipende ancora dalla scelta arbitraria del volume delle celle in cui è stato diviso lo spazio delle fasi. Un facile calcolo mostra che, cambiando questo volume per un fattore ϑ, Π si altera approssimativamente di un fattore.

$$\vartheta^{-N}.$$

Abbiamo stabilito d'altra parte, tra entropia S e probabilità Π la relazione di Boltzmann

$$S = k \log \Pi.$$

Al fattore che resta indeterminato in Π, per l'arbitrarietà nella scelta del volume delle celle, corrisponde, secondo questa formula, un'indeterminazione per una costante additiva nell'entropia S. Ciò indica che, servendosi della meccanica statistica classica, si ottiene l'interpretazione del secondo principio della termodinamica, ma non quella del principio di Nernst.

La situazione è del tutto differente, se introduciamo la quantizzazione dei sistemi *a*. In questo caso infatti resta automaticamente introdotta una classificazione discreta degli stati possibili dei sistemi *a*, senza la necessità di un'arbitraria suddivisione in cellette dello spazio delle fasi; e la distribuzione resta definita dalla conoscenza dei numeri N_1, N_2,... dei sistemi, che si trovano rispettivamente nel primo, nel secondo,... stato quantico, Resta così tolto ogni elemento arbitrario dalla definizione della probabilità Π e conseguentemente, attraverso alla relazione di Boltzmann, anche dalla definizione statistica dell'entropia. Si potrebbe facilmente dimostrare che questa definizione dell'entropia corrisponde statisticamente al principio di Nernst.

Secondo quanto abbiamo esposto, alle varie celle, in cui si divide lo spazio delle fasi nella meccanica classica, corrispondono, nel caso dei sistemi quantizzati, gli stati quantici. È interessante notare quali sono le conseguenze di questo fatto nel caso limite di grandi numeri quantici, quando cioè, secondo il principio di corrispondenza, le differenze tra meccanica classica tendono a sparire. Riferiamoci per semplicità ad un sistema a un solo grado di libertà, avente la coordinata generale q e il momento coniugato p. Lo spazio delle fasi è in questo caso a due dimensioni e ha q e p come coordinate.

Nel caso limite dei grandi numeri quantici vengono, come è noto, a coincidere non soltanto i risultati della meccanica quantistica e di quella classica, ma anche quelli ottenuti dalle condizioni di quantizzazione di Bohr–Sommerfeld. La condizione di Sommerfeld

$$\int p dq = nh$$

s'interpreta geometricamente, nel «piano» delle fasi, così: l'area del piano delle fasi racchiusa entro l'orbita corrispondente all'n^{mo} stato quantico è nh. Ne segue che, se tracciamo, nel piano delle fasi, le orbite corrispondenti a tutti gli stati quantici, l'area compresa tra due orbite successive è uguale ad h. E siccome ciascuno stato quantico corrisponde a una cella, concludiamo che: nel caso di sistemi a un grado di libertà lo spazio (piano) delle fasi si deve suddividere in celle di estensione (area) h. Se il sistema avesse invece f gradi di libertà, le celle dovrebbero prendersi d'ipervolume h^f.

Questi risultati si possono altresì mettere in relazione col principio di indeterminazione di Heisenberg

$$\Delta p \cdot \Delta q \sim h.$$

Questo equivale, infatti, a dire che è fisicamente impossibile determinare lo stato di un sistema come un punto nello spazio delle fasi; il margine d'incertezza è, nelle migliori condizioni, entro un'area dell'ordine di grandezza h, pari cioè all'estensione che abbiamo trovato doversi attribuire alle celle dello spazio delle fasi.

LE NUOVE STATISTICHE. – Abbiamo fin qui considerata la legge di ripartizione di Boltzmann da due punti di vista diversi, cioè:

a) esprimere la probabilità che un sistema, mantenuto a una data temperatura T, si trovi in un certo stato quantico (probabilità proporzionale ad $e^{-w/k\mathrm{T}}$);

b) dato un sistema, costituito dall'insieme di molti sistemi elementari eguali tra di loro (atomi, molecole,...), ricercare il numero di sistemi elementari che si trovano in un determinato stato quantico (numero proporzionale ad $e^{-w_r/k\mathrm{T}}$).

Per la validità della legge di ripartizione di Boltzmann in questo secondo caso è necessario che i sistemi elementari che costituiscono il sistema totale siano indipendenti tra di loro (e cioè lo stato in cui si trova uno di essi non sia influenzato dallo stato in cui si trovano gli altri). Così, per esempio, la legge di Boltzmann sarà certamente valida se i sistemi elementari sono completamente isolati uno dall'altro; se invece i sistemi parziali sono distribuiti entro lo stesso volume, come per esempio, gli atomi o le molecole di un gas, esistono in genere delle interdipendenze tra di essi che vengono a infirmare la validità della legge di Boltzmann.

Se anche si esclude il caso che tra le molecole del gas si esercitano delle forze considerevoli, e cioè si considera il caso limite di un gas ideale, le molecole non si possono considerare indipendenti le une dalle altre a causa del fatto che esse sono indistinguibili tra di loro.

È noto che in un sistema contenente un certo numero di corpuscoli identici (quali, per esempio, gli elettroni di un atomo, oppure le molecole di un gas) si osservano dei fenomeni peculiari dovuti all'identità dei corpuscoli. Da questo punto di vista i diversi tipi di corpuscoli si possono classificare in due categorie:

a) corpuscoli, per cui vale il principio di Pauli (per esempio elettroni, protoni): hanno la proprietà che uno stato quantico semplice può essere occupato al più da un solo corpuscolo; uno stato degenere può essere al massimo occupato da tanti corpuscoli, quanto è il grado di degenerazione dello stato (v. ATOMO). Le autofunzioni sono in questo caso caratterizzate dalla proprietà di essere antisimmetriche (e cioè di cambiare di segno e non di valore) rispetto allo scambio delle coordinate di due qualunque delle particelle identiche;

b) corpuscoli, per i quali vale la statistica di Bose–Einstein (la giustificazione di questo nome si vedrà tra breve), quali, ad esempio, il nucleo di elio, l'atomo d'idrogeno o di elio, i quanti di luce, ecc.: sono caratterizzati matematicamente dalla proprietà delle loro autofunzioni di essere simmetriche rispetto allo scambio di due dei corpuscoli. Per quanto interessa le applicazioni statistiche, la proprietà che importa notare è la seguente: siano $\mathrm{N}_1, \mathrm{N}_2, \cdots, \mathrm{N}_r, \cdots$ i numeri di corpuscoli che occupano rispettivamente il 1°, 2°, ..., r^{mo}, ... stato quantico; esiste uno, e un solo, stato del sistema definito dalla conoscenza dei «numeri di occupazione» $\mathrm{N}_1, \mathrm{N}_2, \cdots, \mathrm{N}_r, \cdots$ Per intendere il significato di questo, si tenga presente che, per dedurre la legge di Boltzmann, abbiamo ammesso che lo stato predetto si potesse invece realizzare in

$$\frac{\mathrm{N}!}{\mathrm{N}_1!\,\mathrm{N}_2!\cdots}$$

modi diversi.

È facile convincersi che la probabilità *a priori* dei diversi stati viene considerevolmente alterata da queste proprietà dei sistemi contenenti particelle identiche. Possiamo chiarire la situazione con un esempio particolarmente semplice: si abbia un sistema contenente due particelle identiche, ciascuna suscettibile di occupare due stati quantici 1 , 2. Siano N_1 ed N_2 i numeri di occupazione dei due stati; essi potranno *a priori* avere i valori

$$(2\,,0)\quad(1\,,1)\quad(0\,,2).$$

Il numero di modi per realizzare questi tre stati sono, nel caso di Boltzmann, rispettivamente 1 , 2 , 1 – il fatto che lo stato (1 , 1) si possa realizzare in due modi deriva dalla circostanza che le due particelle possono prendersi in un ordine oppure nell'ordine opposto –. Se invece vale la statistica di Bose–Einstein, i numeri di modi di realizzare i tre stati sono 1 , 1 , 1. Infine, se vale il principio di Pauli, i due stati (2 , 0) e (0 , 2) non si possono realizzare, e quindi i numeri di modi di realizzare i tre stati sono rispettivamente 0 , 1 , 0.

Si comprende agevolmente come queste peculiarità dei sistemi contenenti particelle identiche vengano ad alterare i pesi statistici da attribuirsi ai diversi stati di un gas, e quindi ad alterarne le proprietà.

I principi della statistica di Bose–Einstein furono per la prima volta usati dal fisico indiano Bose nell'applicazione alla teoria della radiazione del corpo nero considerata come un gas di fotoni (1925); successivamente Einstein riconobbe che il principio usato da Bose nella numerazione degli stati era suscettibile di applicazione più generale, e svolse la sua teoria di un gas ideale le cui molecole obbediscano alla statistica di Bose–Einstein.

Le conseguenze statistiche del principio di Pauli furono sviluppate da E. Fermi (1925) (*); la statistica basata su questo principio è perciò chiamata statistica di Fermi.

Successivamente lo sviluppo della meccanica quantistica, particolarmente per opera di Heisenberg e Dirac, permise di riconnettere la validità della statistica di Fermi o di Bose–Einstein per le varie particelle elementari al carattere di antisimmetria oppure di simmetria delle autofunzioni rispetto allo scambio delle coordinate di due delle particelle identiche.

Le proprietà di un gas che obbedisca alla statistica di Fermi o a quella di Bose–Einstein, differiscono assai poco dalle leggi classiche dei gas perfetti (deducibili in base alla statistica di Boltzmann) finché la temperatura è elevata e la densità non troppo grande; le deviazioni diventano invece sensibili per basse temperature e per densità elevate, e sono, per le due nuove statistiche, in verso opposto. Precisamente, nel caso della statistica di Fermi, la pressione risulta più alta di quanto corrisponderebbe alla equazione di stato dei gas perfetti, mentre, se vale la statistica di Bose–Einstein, essa è più bassa. Tuttavia, se si considera un gas ordinario, il campo di temperatura e di pressione in cui queste deviazioni sono sensibili, è quello in cui si hanno già fortissime deviazioni dalle leggi dei gas perfetti, dovute all'azione delle forze intermolecolari e all'estensione finita delle molecole;

(*) Le relative pubblicazioni (N° 30, 31) sono del 1926 (N. d. R.).

queste deviazioni sono tanto grandi da mascherare l'effetto puramente statistico.

Per conseguenza l'importanza delle nuove statistiche nell'applicazione ai gas ordinarî è piuttosto scarsa. La statistica di Fermi ha invece trovato un'applicazione interessante allo studio del comportamento degli elettroni nell'interno di un metallo; è noto che alcuni degli elettroni nell'interno di un metallo possono spostarsi da atomo ad atomo e costituiscono una specie di «gas elettronico». La densità di questo è talmente grande, che le sue leggi statistiche e la sua equazione di stato deviano completamente dalle leggi dei gas perfetti, e obbediscono invece a quelle deducibili dalla statistica di Fermi. Le conseguenze della statistica di Fermi nello studio delle proprietà elettriche dei metalli sono state specialmente sviluppate da Sommerfeld (1927), che ha potuto, in base ad esse, dare un'interpretazione qualitativa della maggior parte delle proprietà dei metalli.

Raccogliamo qui i principali risultati delle nuove statistiche.

Legge di distribuzione: il numero medio di molecole in un stato quantico (completamente definito, sia per quanto riguarda lo stato interno, sia per quanto riguarda il moto del baricentro della molecola) è dato da

$$\frac{1}{Ae^{\frac{w}{kT}}}$$

nel caso di Boltzmann;

$$\frac{1}{Ae^{\frac{w}{kT}} - 1}$$

nel caso di Bose–Einstein;

$$\frac{1}{Ae^{\frac{w}{kT}} + 1}$$

nel caso di Fermi; A è una costante, e w l'energia dello stato considerato. È chiaro che se l'espressione $Ae^{w/kT}$ è molto maggiore di 1, e cioè se la probabilità di occupazione dello stato è molto piccola, i risultati delle tre statistiche vengono praticamente a coincidere; altrimenti si hanno considerevoli deviazioni tra di esse.

La differenza tra la legge di ripartizione di Boltzmann e le leggi di ripartizione delle nuove statistiche ha come conseguenza una differenza nella equazione di stato di un gas ideale (intendendosi con ciò un gas per il quale siano trascurabili le forze che si esercitano tra le varie molecole). Queste deviazioni, e i fenomeni ad esso connessi, si sogliono chiamare *fenomeni di degenerazione*.

L'entità dei fenomeni di degenerazione è determinata dal valore del parametro

$$D = \frac{nh^3}{(2\pi mkT)^{3/2}}, \tag{26}$$

dove n è il numero delle molecole per unità di volume ed m la massa molecolare. I fenomeni di degenerazione sono piccoli o grandi a seconda che D è piccolo o grande in confronto all'unità. Così, ad esempio, per un gas ideale monoatomico, costituito da N molecole alla temperatura T e pressione p, occupanti il volume V, vale, secondo la statistica di Boltzmann, la classica equazione di stato dei gas perfetti

$$pV = NkT .$$

Nel caso delle nuove statistiche si trovano invece equazioni di stato differenti, che nel caso limite di piccola degenerazione, si possono scrivere

$$pV = NkT \left\{ 1 \mp \frac{Nh^3}{2^{5/2} V (2 \pi mkT)^{3/2}} \right\} . \tag{27}$$

Vale il segno superiore nel caso della statistica di Bose–Einstein, e il segno inferiore nel caso della statistica di Fermi.

Dalla (27) risulta che le divergenze dall'equazione classica dei gas perfetti tendono a scomparire, quando il parametro di degenerazione si annulla. Se invece D è comparabile con l'unità, cioè nel caso limite di forte degenerazione, le leggi del gas degenere vengono a differire completamente dalle leggi classiche.

Si trova così, per esempio, che in un gas ideale che obbedisca alla statistica di Fermi, sia la pressione sia l'energia cinetica media delle molecole non si riducono a zero allo zero assoluto; esse tendono invece a valori limiti (detti rispettivamente *pressione* ed *energia allo zero assoluto*) dati da

$$P_0 = \frac{1}{20} \left(\frac{6}{\pi} \right)^{2/3} \frac{h^2 n^{5/3}}{m} \quad ; \quad L_0 = \frac{3}{40} \left(\frac{6}{\pi} \right)^{2/3} \frac{h^2 n^{2/3}}{m} . \tag{28}$$

Parimenti si trova che, per un gas ideale che obbedisca alla statistica di Fermi, il calore specifico C_v a volume costante non è costante, come risulterebbe dalla statistica di Boltzmann, ma decresce con la temperatura, e tende a zero per $T = 0$.

Queste proprietà, come abbiamo già accennato, hanno importanza per la spiegazione delle proprietà elettriche dei metalli. Si trova infatti che il gas di elettroni, a cui sono dovute queste proprietà, si trova in condizioni di degenerazione assai forte (infatti per esso, data la considerevole concentrazione e la piccola massa degli elettroni, il parametro D risulta molto grande anche a temperature elevate) e quindi il suo comportamento è sotto molti punti di vista del tutto differente da quello che avrebbe un gas secondo le leggi classiche.

BIBLIOGRAFIA.

L. BOLTZMANN, *Leçons sur la théorie des gas*. Parigi 1902.

J. H. JEANS, *The Dynamical Theory of Gases*, Cambridge 1925.

R. H. FOWLER, *Statistical Mechanics*, ivi, 1929.

L. BRILLOUIN, *Les statistiques quantiques et leurs applications*. Parigi 1930.

E. FERMI, *Molecole e cristalli*, Bologna 1934.

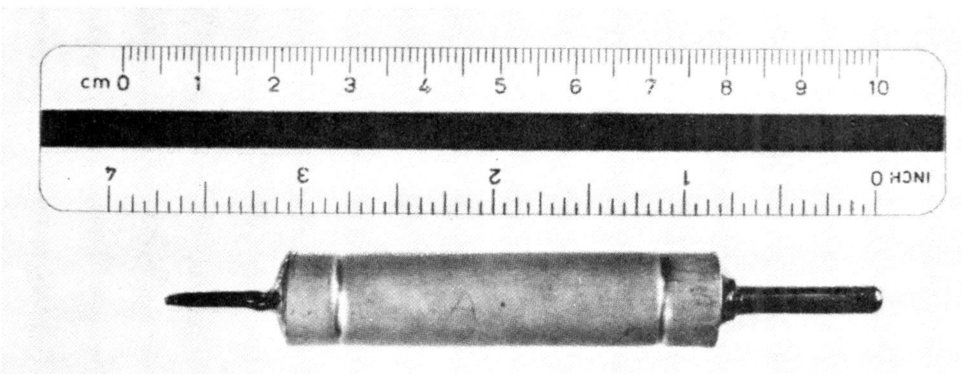

Geiger-Müller counter used by E. Fermi in his early work on artificial radioactivity.

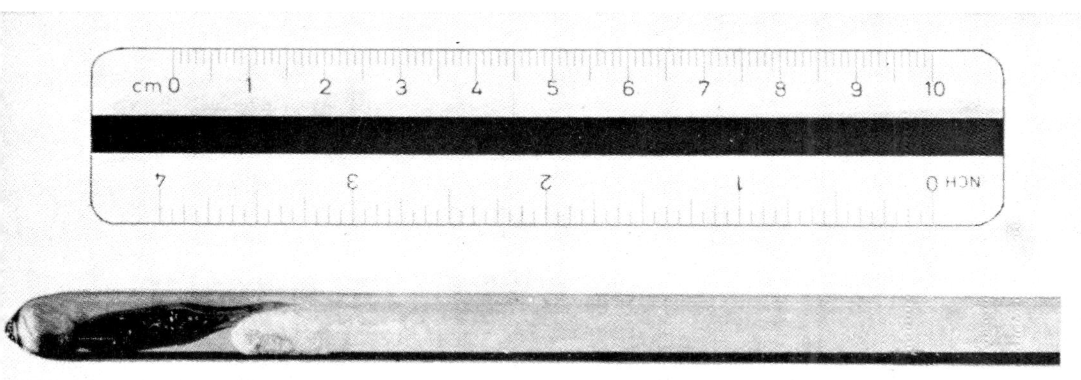

One of the neutron sources used in the work on artificial radioactivity.

N° 84 *a* to 110 (*).

Papers N° 84 to 110 will be considered together. Their results are the culmination of the experimental activity of Fermi in Italy and they embrace one of the most fruitful periods of his career.

Fermi himself in the Nobel lecture of 1938 (N° 128) gave a review of the work on the slow neutrons performed in Italy, and that paper should be kept in mind for a historical appraisal of the work.

Fermi's activity until 1934 had been mainly theoretical, but it is clear that he had already acquired some skill and experience in experimental physics as shown by the papers written up to that date on experimental subjects (e.g. N° 28, 40 *a*, 78).

After Fermi had established himself in Rome and formed a school of young physicists, he had considered many times the possibility of fostering experimental work in Italy. Several of the physicists around him, who later became his close collaborators for years, were interested not only in theory but primarily in experimental physics and often the group had discussed which experimental fields looked most promising. Although the experimental equipment of the Institute of Rome was almost exclusively spectroscopic, and although the spectroscopic tradition was quite strong, there had been the feeling for several years that spectroscopy was approaching its maturity and that one should branch out into nuclear physics, because that was the field which held most promise.

Between 1930 and 1934 several of the group in Rome went to visit laboratories abroad in order to acquire knowledge of experimental techniques which were unknown at the time in Italy. Most of this work was subsidized by grants from the Rockefeller Foundation. Among the laboratories visited at length were those of Millikan in Pasadena (Rasetti), Stern in Hamburg and Zeeman in Amsterdam (Segrè), Debye in Leipzig (Amaldi) and Meitner in Berlin (Rasetti). Several of us spent a year or more in these laboratories learning several experimental techniques and when we returned to Rome we discussed at length what we had seen. After lively debates in which different opinions were discussed with plenty of vigor, it was decided, mainly on the impulse of Fermi, that the laboratory should branch into the nuclear field. It was at this stage that Rasetti, who had learned some of the nuclear techniques at Dahlem, started to prepare several pieces of apparatus and also initiated some experimental work, traces of which are found in these papers (N° 78).

The occasion for a really new departure in the nuclear field occurred in 1934 with the discovery by I. Curie and F. Joliot of artificial radioactivity. Fermi immediately saw that their work could be expanded tremendously by using neutrons as projectiles. Just an occasion of this type was needed to trigger large-scale experimental activity in Rome. The start of the neutron experiments was facilitated by the very lucky circumstance that Professor G. C. Trabacchi in the "Laboratorio della Sanità Pubblica", which was located in the same premises as the Physics Institute in Via Panisperna in Rome, had on hand more than one gram of radium and the plant with which to extract the radium emanation and prepare radon–

(*) In arranging this group of papers it was found advisable to put together those having the same title and subject matter. This has produced some departure from a strictly chronological order. This remark applies especially to papers N° 84 to 92.

Artificial Radioactivity by Neutron Bombardment VI is in "Ricerca Scientifica", 5 (2), 381 (1934) and as it is written by AMALDI, D'AGOSTINO and SEGRÈ it has been omitted from the series.

There is a number of papers: N° 101, 104, 111, 126 which were written mainly in order to satisfy requests for semi-popular expositions or for other reasons important only at the time. They have not been reproduced.

beryllium sources. His wholehearted cooperation was invaluable and all the neutron sources which were used in the years 1934–1937 were prepared at this plant. The neutron work started in 1934 during Easter vacation. Fermi decided to put to the experimental test his idea that neutrons would be powerful projectiles for inducing nuclear transmutation. With his own hands, he built some primitive Geiger-Mueller counters of aluminum which looked very ugly but worked adequately for the purpose, and then started to bombard with radon plus beryllium neutrons all elements in order of increasing atomic number. The first source available was quite weak: 50 millicuries of radon plus beryllium. For a few days trials did not bring success, but Fermi was a systematic man. He stuck to the order of increasing Z. He started with hydrogen, and followed with lithium, beryllium, boron, carbon, nitrogen, oxygen, all with negative results. However, he was finally successful when he tried fluorine and got the expected result.

This was on March 25, 1934 and the letter announcing this result was promptly sent to the "Ricerca Scientifica" (N° 84). Fermi, who wanted to proceed as fast as possible with the work, asked Amaldi and Segrè to help him carry out the experiment. Rasetti was absent in Morocco at the time and was notified to give him the opportunity to come back as soon as possible and participate in the work. D'Agostino was not originally in our group; he was a chemist in the laboratory of the Sanità Pubblica and at the time of this work he had a fellowship in Paris at the laboratory of Madame M. Curie. At the beginning, Fermi, Amaldi and I tried to push the work as fast as we could. We organized our activities in this way: Fermi would do a good part of the experiments and the calculations; Amaldi would take care of what we would now call the electronics, and I would secure the substances to be irradiated, the sources, and other necessary equipment. This division of labor was by no means rigid and each of us participated in all phases of the work. We proceeded at very great speed. It was soon apparent that chemistry would be an essential tool for the work to be performed and we were very glad to be able to get the help of a professional chemist, D'Agostino, when he returned from Paris for the Easter vacation. We barely knew him at the time, but soon became excellent friends. He never used his return ticket to Paris. Soon Rasetti also returned from his trip.

The papers N° 84–88 are a faithful record written at short intervals of time of the work performed during the spring and early summer of 1934. The first thing we decided to do was the obvious: irradiate all the substances we could lay our hands on. Luckily, we obtained a small subsidy from the Consiglio Nazionale delle Ricerche in Rome, a sum of approximately 20,000 lire ($\sim$ \$ 1,000 at the time) which we could spend with complete freedom. The irradiations went on quite regularly and soon we identified the (n, p) and (n, α) reactions. We also found that in many cases the neutron would produce radioactive isotopes of the target and there was great doubt whether this was the result of an $(n, 2n)$ or an (n, γ) reaction. At the time we still had the thought that the more energetic the neutrons, the more effective they would be in producing reactions. We finally reached the irradiation of uranium (N° 86). The activities that we could produce were weak and it was necessary to purify the uranium from all its beta active products which would mask any artificial radioactivity produced by us. This was a very tedious task and allowed us to work only for a short time because the natural growth of beta activity in purified uranium prevented us from following the artificial activity for a long time. We thought that the irradiation of uranium should produce transuranic elements, for which we expected properties similar to those of Re, Os, Ir, Pt. We proceeded to show that uranium irradiated with neutrons did not produce any elements which have an atomic number between that of lead and uranium. The proof of this fact was obtained and was correct, but the possibility of fission somehow escaped us, although it was called specifically to our attention by I. Noddack, who sent us an article in which she clearly predicted the possibility of fission. The reason for our blindness is not quite clear; Fermi said occasionally, years later, that the data on mass defect available at the time were misleading and that on the basis of that information fission did not seem to be possible to him.

The summer vacation followed and Fermi spent it in South America. (See paper N° 101). On his return to Europe he did not go directly to Rome but stopped in England for an International Conference on physics. (See papers N° 102–103) At this conference he reported the results of the work performed in Rome. During the same summer Amaldi and I went to Cambridge, England. Lord Rutherford had followed our work closely and on receiving one of the early letters to the « Ricerca Scientifica », had congratulated Fermi saying:

23rd April, 1934.

" Dear Fermi,

" I have to thank you for your kindness in sending me an account of your recent experiments in causing temporary radioactivity in a number of elements by means of neutrons. Your results are of great interest, and no doubt later we shall be able to obtain more information as to the actual mechanism of such transformations. It is by no means clear that in all cases the process is as simple as appears to be the case in the observations of the Joliots.

" I congratulate you on your successful escape from the sphere of theoretical physics! You seem to have struck a good line to start with. You may be interested to hear that Professor Dirac also is doing some experiments. This seems to be a good augury for the future of theoretical physics!

Congratulations and best wishes,

Yours sincerely,
RUTHERFORD

When we went to Cambridge, we discussed with him, in great detail, our work. The work which had been accomplished up to that date by our group is summarized in a paper which was presented by Lord Rutherford to the Royal Society (paper N° 98). The manuscript of this paper had been prepared in Rome and delivered to him in Cambridge. He read it immediately with great attention, made several corrections to improve our English, and turned it over to the Royal Society. I asked him whether it would be possible to obtain a speedy publication and he immediately answered, " What did you think I was President of the Royal Society for ? " Unfortunately my understanding of Rutherford's English at the time was quite imperfect, and I lost several of his remarks. One of the questions which seemed of great importance to us was whether the reactions in which an isotope of the target was formed were (n, γ) or $(n, 2n)$. We tried to solve this problem by cross bombardments. At Cambridge at the time Bjerge and Westcott were working on similar problems and we joined forces. A reaction was found which we thought was a very clear-cut example of an (n, γ) reaction. The result was considered very important by us and was correct, although part of the argument by which the conclusion was reached was wrong.

In September, Amaldi and I returned to Rome. We immediately tried to reinforce our conclusions on the question of (n, γ) versus $(n, 2n)$ reactions by irradiating other substances besides the one used by Bjerge and Westcott. We thought we had found another clear-cut example of an (n, γ) reaction in aluminum, and we immediately communicated this result to Fermi who was still at the London International Conference. He referred to our experiment at one of the meetings. Shortly afterwards I caught a cold and could not go to the laboratory for several days. Amaldi tried to repeat our experiments and found a different decay period for irradiated aluminum which showed that our so called (n, γ) reaction did not occur. This was hurriedly relayed to Fermi who resented having communicated a result which now looked to be in error. He strongly criticized us and did not conceal his displeasure. The whole business was becoming very troublesome because we could not find any fault with the various experiments which gave inconsistent results. At this time we were joined by B. Pontecorvo who had just finished his Ph. D. thesis during the preceding year and had shown considerable promise. He was asked to help in the work and from there on participated in our research. It was now the beginning of the scholastic season 1934–35, and we started irradiations, again in a systematic fashion. In paper N° 98, published in the " Proceedings of the Royal Society ", the activity induced in various elements had been classified in a qualitative manner as strong (s), medium (m), and weak (w). Such a classification seemed unsatisfactory to us and we decided to establish a quantitative scale of activability, even if it was a purely arbitrary one. Amaldi and Pontecorvo were assigned the study of this problem. They began to search for conditions of irradiation which would give easily reproducible results, using the 2.3 min period induced in silver as a standard.

They immediately encountered some problems because it became apparent that the intensity of the radioactivity that was obtained depended on the conditions of irradiation. In particular, there were certain tables supporting a spectroscope which had miraculous properties, since silver irradiated on those tables became much more active than when it was irradiated on other marble tables in the same room. The explanation was that the spectroscopic tables were of wood, and the source on the wood gave slow neutrons, whereas the source on the marble did not give them.

To clear up the mystery we felt the need for systematic observations. In the observation notebooks (the notebooks are now in the archives of the Domus Galilaeana) the data are recorded beginning with October 18, 1934. The first step is a series of measurements taken by Amaldi in and outside a small lead house (castelletto) whose walls were 5 cm thick. The results of these measurements show clearly that while, outside lead, the activation greatly decreases with the increase in distance, this does not happen in the lead house.

The next day these same measurements and other analogous ones were repeated, always in order to explain the action of the surrounding lead. Particular attention was focused on an absorption measurement in lead. As the next step a lead wedge was prepared to be inserted between the neutron source and the detector with the idea of comparing its absorption with that of a wide brick of lead of the same thickness. The lead was ready the morning of the 20th of October but it was actually used only a few days later. That morning Fermi [1] suddenly decided to try first filters of light elements instead of the prepared lead filter. Paraffin was tried first. Although the experiment was performed during an examination period and several of us were absent, our friends Persico and Rossi were kibitzing on our work. About noon, we were all summoned to watch the miraculous effects of the filtration by paraffin. At first I thought that some counter had gone wrong because such strong activities had seldom appeared before. But it was immediately demonstrated that this strong activation was the result of filtering the radiation producing the radioactivity through paraffin. All this happened around noon. We tried a few more substances as filters and recognized that the powerful effect occurred only with paraffin. We went to lunch extremely puzzled by our observations and came back around three o'clock, after our usual siesta, to find that Fermi in the meantime had found the explanation of the strange behavior of filtered neutrons. He hypothesized that the neutrons could be slowed down by collisions and in this way become more effective, an idea which was contrary to our expectations. If Fermi's explanation was correct, it was clear that several consequences followed: for instance, maybe one could find an explanation of the (n, γ) versus $(n, 2n)$ reaction. During the same afternoon each of us thought of and repeated some past experiment with the slowing down hypothesis in mind. In about half an hour we had the explanation of the disagreement between the results of the different irradiations in aluminum; slow neutrons produced one activity, fast neutrons another. This was a great relief to us because it gave us the explanation to a serious puzzle and showed that we did not need to correct the minutes of the London Conference. We also tested that the radiation which produced the activity (neutrons and slow neutrons) was emitted by radon plus beryllium and not by radon alone, and that the effect of the paraffin was characteristic of hydrogen and not of carbon. Finally, Fermi went to the extreme of thinking that the neutrons could really be thermalized, and devised in the same day an experiment (unsuccessful at the time) to demonstrate this fact by slowing down the neutrons in hot water instead of cold water. All this happened on October 22, 1934, and by the evening of that day a short letter to the " Ricerca

(1) In this connection see the introductory remarks by CHANDRASEKHAR to the paper on *Origin of cosmic rays* to be published in volume 2 of these "Collected papers".

Scientifica" telling of all these surprising effects was on its way (N° 105). The discovery of the hydrogen effect obviously opened a host of problems and we had to re-orient our entire research program. The first step was to measure, for many substances, what we called the "coefficient of aquaticity": namely, how much immersion in water would increase the activity, under specified conditions. This measurement gave us a confirmation that the (n, γ) reactions were the only ones sensitive to hydrogenous substances and by early November we were convinced that the slowing-down was the correct explanation of the phenomena observed (N° 88). Our attention then turned more to the study of the slow neutrons themselves than the substances they produced, and we again tried to see whether slowing down in hot or cold water would change the properties of the neutrons. We had strong suspicions that neutrons were effectively thermalized, but at first we could not show any positive effect. Nevertheless, we kept trying this idea. Ultimately, Moon and Tillman in England succeeded in showing a thermal effect and we eagerly repeated their experiments. We soon found that some substances—for instance cadmium—absorb slow neutrons very strongly and we measured crude cross sections for this effect. We detected the capture gamma rays and we also started crude measurements of the density of slow neutrons in a hydrogenous medium as a function of the distance from the source. Finally, we tried to slow down neutrons by collisions with substances other than hydrogen and found some effects of inelastic collisions which explained our original observations on the effect of surrounding the source and the detector by lead. All this work was initiated and had given significant results by December, 1934; that is, about six weeks after the discovery of the slow neutrons themselves. Papers N° 105 and N° 106 are a chronological record of the experiments performed in those weeks. Approximately at this time, Professor Corbino, director of the laboratory, suggested that slow neutrons might have important practical applications and that it would be advisable to take out a patent on this work. This was done and resulted in Italian patent N° 324458, of October 26, 1935, which was later extended to other countries.

The boron reaction was studied immediately afterwards and interpreted correctly as an (n, α) reaction (N° 90). More systematic work followed in line with previous studies of activations and with considerable emphasis on uranium on which, however, we did not make great progress.

In this connection it may be interesting to recall the following unsuccessful attempt, made in January-February 1935, to explain the great number of new activities induced in thorium and uranium which had been isolated by the group in Rome as well as by other groups. Since, at that time, we tried to interpret all the activities observed as due to transuranic elements and their possible decay products, we thought that besides beta activity there must also be new radioactive alpha emitters. Thus we began to irradiate some foils of uranium or thorium (in the form of oxides), and put them, immediately after the irradiation, in front of a thin-window ionization chamber connected to a linear amplifier. As we failed to observe any alpha activity, we thought that this might be due to the fact that the corresponding half lives were too short, perhaps a fraction of a second. Therefore we set uranium in front of the ionization chamber and irradiated it there with a neutron source surrounded by paraffin. We set a piece of lead between the neutron source and the chamber, to reduce the action of gamma rays on the ionization chamber.

In performing these experiments, we thought that if the alpha emitters existed, and if they had a short half life, they ought to emit (according to Geiger-Nuttall's law) alpha particles of a range considerably greater than that of the particles emitted by uranium or thorium. For this reason the experiments were always carried out with uranium or thorium covered with a layer of aluminium equivalent to 5 or 6 cm of air. The experiments gave a negative result, and today it can be said that the phenomenon of fission of the heavy elements, discovered in 1939 by Hahn and Strassman, was not seen in Rome early in 1935 because of that aluminium layer. The aluminium layer stopped all the fission fragments and thus prevented them from getting inside the ionization chamber. In retrospect, it is difficult to say whether the fission explanation would have been found even if the phenomenon had been seen.

The main results of all the work mentioned above were summarized in a paper for the Royal Society (N° 107) written in February, 1935. In it are the seeds of most of the important ideas and facts of neutron physics. Indeed, by this time Fermi had already made very substantial progress in the study of the theory of the slowing-down process. From the point of view of nuclear physics, there was a serious difficulty caused by the apparent lack of correlation between the scattering and the capture cross sections in cases where the capture cross section was very large, as for instance, cadmium. Fermi made several attempts to explain this phenomenon, but the reason became clear only through Bohr's work on the compound nucleus. On the other hand, the concept of scattering length and the possibility of resonances were very much present in Fermi's mind. Actually, the concept of scattering length with its typical diagram had been developed by Fermi since 1933 in order to explain some spectroscopic phenomena (pressure shift of spectral lines) which had been discovered experimentally by Amaldi and Segrè (N° 95). Fermi had a great insight for discerning analogies in phenomena which were apparently completely unrelated, and much to our surprise he transferred the theory of the shift of the spectral lines to neutron physics without any difficulty. At this time, Fermi also made a number of calculations on the behavior of neutrons in hydrogenous media by what we would call today the Monte Carlo method. At the time he did not speak about it very much and I learned of this only many years later from Fermi himself. This experimenting with the theory of slowing-down neutrons brought its fruit the following year, as we shall see. In the spring of 1935, we also devised a mechanical experiment by which one could compare the velocity obtained by the neutrons with mechanical velocities in a wheel (N° 92 *a*), and soon after the experiments of Moon and Tillman showing the influence of temperature, we could not only confirm their results but also measure by a mechanical device some data relevant to the velocity of the neutrons. A fast rotating wheel had been prepared for this experiment for some time, but it was possible to perform the work only in the very last days before Fermi's departure. The results did not become statistically established until Fermi's luggage was packed and he was about to take a boat from Naples. This is the only case I can remember of a real rush, at the time of the Rome work, and it certainly was very much against our general attitude and method of work which was intensive but never rushed.

With this we had reached the summer of 1935. When we left for our summer vacations our mood was not as happy as in previous years. The political developments of the time, especially the preparations of the Ethiopian war, and the grave deterioration of the situation in Europe, had seriously affected our work. During the summer of 1935 Fermi went abroad and some of us went to the United States to work at Columbia University. When the fall of 1935 came and we all should have returned to Italy, the Ethiopian crisis was at its height. Rasetti remained in the United States; Segrè had been appointed director of the physics laboratory at Palermo, and moved there; Pontecorvo had left Italy for France; and D'Agostino was busy with his work at the Sanità Pubblica. The group was virtually disbanded. Only Amaldi and Fermi were left in Rome with neutron sources. From there on the investigation concentrated mainly on the study of the slowing-down process and on the "groups of neutrons". Although very early it was surmised that the groups might correspond simply to different velocities, as it is in fact, there were many difficulties to be surmounted before this explanation was established. At a certain time it was even thought that there might be several kinds of neutrons, possibly different in spin. This work, however, was carried on under distressing conditions. The Ethiopian war and the sanctions imposed by the League of Nations on Italy for her aggression produced a gloomy atmosphere in Rome. There was a reduction of electric power and fairly severe regulations which required keeping windows sealed after dark. Fermi and Amaldi worked extremely hard, in part to forget the unpleasant situation. I remember visiting Rome from Palermo and finding them in a tired, preoccupied mood, very different from the one which had prevailed only a few months earlier.

EMILIO SEGRÈ.

84 *a*.

RADIOATTIVITÀ INDOTTA DA BOMBARDAMENTO DI NEUTRONI. — I.

« Ric. Scientifica », 5 (1), 283 (1934).

Desidero riferire in questa lettera sopra alcune esperienze destinate ad accertare se un bombardamento di neutroni non determini dei fenomeni di radioattività susseguente analoghi a quelli osservati dai coniugi Joliot con bombardamento di particelle α.

Il dispositivo che ho usato è il seguente: La sorgente di neutroni è costituita da un tubetto di vetro contenente polvere di berillio ed emanazione. Usando circa 50 millicurie di emanazione, che mi sono stati forniti dal professor G. C. Trabacchi che qui desidero ringraziare vivissimamente, si possono così ottenere oltre 100.000 neutroni al secondo, misti naturalmente a una intensissima radiazione γ, che però non dà alcun disturbo per esperienze di questo genere. Dei cilindretti contenenti l'elemento in esame sono sottoposti per un tempo variabile da alcuni minuti ad alcune ore alle radiazioni di questa sorgente.

Essi vengono poi rapidamente disposti attorno ad un contatore a filo, la cui parete esterna è formata da una foglia d'alluminio di spessore di circa 0,2 mm. tale quindi da permettere l'ingresso di eventuali raggi β nel contatore. Fino ad ora l'esperienza ha dato esito positivo per due elementi:

Alluminio. – Un cilindretto di alluminio irradiato dai neutroni per un paio d'ore e posto successivamente attorno al contatore determina nei primi minuti un aumento assai considerevole degli impulsi, che crescono di 30 o 40 al minuto. L'effetto decresce col tempo riducendosi a metà in circa 12 minuti.

Fluoro. – Il fluoruro di calcio, irradiato per pochi minuti e portato poi assai rapidamente accanto al contatore determina nei primi momenti un aumento del numero degli impulsi. L'effetto si smorza rapidamente, riducendosi a metà in circa 10 secondi.

Una possibile interpretazione di questi fenomeni è la seguente. Il fluoro, bombardato coi neutroni, si disintegra emettendo particelle α. La reazione nucleare è probabilmente:

$$F^{19} + n^{1} \rightarrow N^{16} + He^{4}.$$

Si formerebbe così un azoto di peso 16 che, emettendo successivamente una particella β può trasformarsi in O^{16}. Una simile interpretazione potrebbe aversi per l'alluminio, conformemente alla possibile reazione nucleare:

$$Al^{27} + n^{1} \rightarrow Na^{24} + He^{4}.$$

Il Na^{24} così formato sarebbe un nuovo elemento radioattivo e si trasformerebbe in Ca^{24} con emissione di una particella β.

Se queste interpretazioni sono corrette, si avrebbe qui la formazione artificiale di elementi radioattivi che emettono normali particelle β, a differenza di quelli trovati dai Joliot che emettono invece positroni. In particolare nel caso dell'azoto si avrebbero due isotopi radioattivi: N^{13}, trovato dai Joliot, che emettendo un positrone si trasforma in C^{13}; ed N^{16} che, emettendo un elettrone si trasforma in O^{16}.

Sono in corso esperienze per estendere l'esame ad altri elementi e per studiare meglio le particolarità del fenomeno.

Roma, 25 marzo 1934–XII.

85 *a*.

RADIOATTIVITÀ PROVOCATA DA BOMBARDAMENTO DI NEUTRONI. — II.

« Ric. Scientifica », *5* (1), 330–331 (1934).

Proseguendo le esperienze col metodo descritto nella precedente lettera, sono state studiate le radioattività che si producono sotto bombardamento di neutroni ottenuti da un preparato di Emanazione + Berillio in numerosi altri elementi.

Ferro. – Periodo circa 2 ore. Nel caso di questo elemento si è potuta effettuare la separazione chimica. Il prodotto attivo che si forma è probabilmente manganese, perché segue questo metallo nelle sue reazioni caratteristiche.

Silicio. – Effetto molto intenso. Periodo circa 3 minuti. Gli elettroni emessi sono stati fotografati nella camera di Wilson. Si ottiene la stessa vita media sia usando silicio metallico che in forma di quarzo.

Fosforo. – Periodo circa 3 ore. Effetto intenso. Elettroni fotografati nella camera di Wilson. Una separazione chimica dà per il prodotto attivo che si forma i caratteri del silicio.

Cloro. – Dà un effetto con periodo molto più lungo degli altri elementi fino ad ora esaminati.

Vanadio. – Periodo 5 minuti.

Alluminio. – A conferma dei risultati della lettera precedente gli elettroni sono stati fotografati nella camera di Wilson.

Rame. – Effetto non intenso. Periodo circa 6 minuti.

Arsenico. – Periodo circa 2 giorni.

Argento. – Effetto intenso. Periodo circa 2 minuti.

Tellurio. – Periodo circa 1 ora.

Jodio. – Effetto intenso. Periodo circa 30 minuti.

Cromo. – Effetto intenso. Periodo circa 6 minuti. Elettroni fotografati nella camera di Wilson.

Bario. – Piccolo effetto. Periodo circa 2 minuti.

Hanno anche date indicazioni di effetti di varie intensità e periodo Na, Mg, Ti, Zn, Zr, Se, Sb, Br, La.

Alcuni elementi dànno indicazioni di avere due o più periodi che possono in parte essere dovuti a diversi costituenti isotopi, e in parte anche a successive trasformazioni radioattive. Si proseguono le esperienze per precisare meglio questi risultati ed estendere lo studio ad altri elementi.

La reazione nucleare che determina questi fenomeni può essere diversa nei diversi casi. Le separazioni chimiche effettuate nel caso del ferro e del fosforo sembrano indicare che, almeno in questi due casi, venga assorbito il neutrone ed emesso un protone. Il prodotto instabile, rispettivamente manganese e silicio, con emissione di una particella beta ritorna eguale all'elemento di partenza.

Le separazioni chimiche sono state eseguite dal dott. O. D'Agostino; per la parte fisica hanno collaborato i dott. E. Amaldi ed E. Segrè.

Roma, *Istituto Fisico dell'Università.*

86 *a*.

RADIOATTIVITÀ «BETA» PROVOCATA DA BOMBARDAMENTO DI NEUTRONI. — III.

E. AMALDI, O. D'AGOSTINO, E. FERMI, F. RASETTI, E. SEGRÈ

«Ric. Scientifica», 5 (1), 452–453 (1934).

Sono state proseguite ed estese le esperienze di cui alle Note precedenti [1] coi risultati che ricordiamo appresso.

Idrogeno – Carbonio – Azoto – Ossigeno. – Non dànno effetto apprezzabile. Sono stati esaminati paraffina irradiata al solito modo per 15 ore con una sorgente di 220 mC, acqua irradiata per 14 ore con 670 mC e carbonato di guanidina irradiato per 14 ore con 500 mC.

Fluoro. – Il periodo del Fluoro è sensibilmente minore di quanto indicato precedentemente e cioè di pochi secondi.

Magnesio. – Il Magnesio ha due periodi, uno di circa 40 secondi e uno più lungo.

Alluminio. – Oltre al periodo di 12 minuti segnalato precedentemente ve ne è anche un altro dell'ordine di grandezza di un giorno. L'attività corrispondente a questo secondo periodo segue le reazioni chimiche caratteristiche del Sodio. Si tratta probabilmente di un Na^{24}.

Zolfo. – Il periodo dello S è assai lungo, certamente di molti giorni. L'attività si separa con le reazioni caratteristiche del Fosforo.

Cloro. – Si comporta analogamente allo S. Anche qui si può separare un principio attivo; probabilmente si tratta di un P^{32} identico a quello che si ricava dallo S.

Manganese. – Ha un effetto debole con un periodo di circa 15 minuti.

Cobalto. – Ha un effetto di 2 ore. Il principio attivo si comporta come Mn. Data l'identità di periodo e di comportamento chimico si tratta quasi certo di un Mn^{56} identico a quello che si forma irradiando il Fe.

Zinco. – Ha due periodi, uno di 6 minuti e uno assai più lungo.

Gallio. – Periodo 30 minuti.

Bromo. – Ha due periodi, uno di 30 minuti e l'altro di 6 ore. L'attività corrispondente al periodo lungo e probabilmente anche l'altra, seguono chimicamente il Br.

Palladio. – Periodo di alcune ore.

Jodio. – Periodo 30 minuti. L'attività segue chimicamente lo Jodio.

Praseodimio. – Ha due periodi. Uno di 5 minuti e l'altro più lungo.

Neodimio. – Periodo 55 minuti.

Samario. – Ha due periodi uno di 40 minuti e uno più lungo.

Oro. – Periodo dell'ordine di grandezza di 1 o 2 giorni.

(1) E. FERMI, «Ricerca Scientifica», 5 (1) p. 283, p. 330 (1934).

Uranio. – L'U liberato dai prodotti dotati di attività β che lo accompagnano e susseguentemente irradiato dà un effetto intenso con più periodi: uno di circa 1 minuto, uno di 13 minuti oltre a periodi più lunghi non ancora esattamente determinati. I raggi β corrispondenti al periodo di 13 minuti sono notevolmente penetranti.

Si è cercato di riconoscere con operazioni chimiche se l'elemento che si disintegra col periodo di 13 minuti fosse un isotopo di qualcuno degli elementi più pesanti. Possiamo escludere immediatamente che si tratti di un isotopo dell'U (92) o del Th (90) perché il principio attivo si separa facilmente da questi due elementi. Più difficile è la prova che non si tratti di un isotopo del Pa (91) dato che manca una sostanza radioattiva β adatta per queste prove, isotopa del Pa. Ci siamo dovuti servire per questo dell'UX_2 il quale ha un periodo di soli 70 secondi e perciò costringe a eseguire le operazioni chimiche molto rapidamente. Tuttavia operando in circa 2 minuti siamo riusciti a separare insieme a Mn, il principio attivo dall'U senza che venisse trascinata altro che una piccola frazione dell'attività di UX_2 che si sarebbe dovuta avere tenendo conto del decadimento di questa sostanza. Una tale separazione effettuata precipitando con $NaClO_3$ il MnO_2 da una soluzione nitrica fortemente acida è ugualmente completa anche in presenza di Ce e La. Ciò sembrerebbe escludere che il principio attivo abbia numero atomico 89 data l'analogia di questo elemento con le terre rare. La stessa reazione sembra anche escludere che si tratti di un isotopo del Ra (88) e tanto più di un ekacesio (87) o di una emanazione (86).

Questo insieme di conclusioni che stiamo cercando di suffragare con ulteriori esperienze fa sorgere spontanea l'ipotesi che il principio attivo dell'U possa avere numero atomico 93 (omologo del renio); il processo in questa ipotesi potrebbe consistere in una cattura del neutrone da parte dell'U con una formazione di un U^{239} il quale subirebbe successivamente delle disintegrazioni β.

Abbiamo anche esaminato se alcune delle disintegrazioni osservate nei vari elementi fossero accompagnate da emissione di raggi γ. L'osservazione veniva fatta cercando se le sostanze agivano su di un contatore attraverso 2 millimetri di Pb.

Hanno dato un effetto positivo i seguenti elementi:

Al (ambedue i periodi) ; Si ; Fe ; Zn ; As ; U.

Hanno dato effetto negativo:

Mg ; S ; Ag ; Br ; J.

È stato infine costruito un dispositivo magnetico per determinare il segno delle particelle emesse analogo a quello descritto da J. Thibaud [2]. Per ragioni di intensità dell'effetto si sono potuti studiare finora i seguenti elementi:

Al ; Si ; P ; S ; Ga ; Cr ; Fe ; Ag ; Br ; J.

In tutti i casi si sono osservati solo elettroni negativi.

Istituto Fisico della R. Università.
Roma, 10 maggio 1934–XII

(2) J. THIBAUD, «C. R.», *197*, 447 (1933).

87 *a*.

RADIOATTIVITÀ PROVOCATA DA BOMBARDAMENTO DI NEUTRONI. — IV.

E. Amaldi, O. D'Agostino, E. Fermi, F. Rasetti, E. Segrè

« Ric. Scientifica », 5 (1), 652–653 (1934).

A complemento dei risultati comunicati nelle lettere precedenti [1], riferiamo su alcune ulteriori osservazioni.

Fluoro. – Abbiamo rideterminato il periodo del F in 9 secondi. Gli elettroni emessi sono assai penetranti riducendosi a metà in 0,32 gr/cm² di alluminio. Anche nelle seguenti misure gli assorbimenti sono fatti con spessori di alluminio. L'emissione dei raggi β del F è probabilmente accompagnata da raggi γ, poiché 2 mm. di Pb non assorbono completamente la radiazione emessa.

Magnesio. – Oltre al periodo di 40 secondi, abbiamo misurato il periodo lungo che è risultato di circa 15 ore ed è accompagnato da emissione γ. Gli elettroni corrispondenti a questo periodo si riducono a metà (spessore di dimezzamento) con 0,07 gr/cm². Il principio attivo si comporta come il sodio e si identifica probabilmente, poiché ha approssimativamente lo stesso periodo e la stessa penetrazione, col Na^{24} estratto dall'Al. Le reazioni nucleari corrispondenti potrebbero essere le seguenti:

$$Mg_{12}^{24} + n_0^1 = Na_{11}^{24} + H_1^1$$

e

$$Al_{13}^{27} + n_0^1 = Na_{11}^{24} + He_2^4 \cdot$$

Alluminio. – Lo spessore di dimezzamento dell'attività corrispondente, al periodo di 12 minuti è circa 0,09 gr/cm². Il principio attivo del periodo breve non è stato ancora estratto chimicamente, ma è probabilmente Mg_{12}^{27}.

Silicio. – Lo spessore di dimezzamento dei raggi β è 0,2 gr/cm.² L'attività si separa chimicamente coll'Al e ciò rende probabile che la reazione nucleare sia

$$Si_{14}^{28} + n_0^1 = Al_{13}^{28} + H_1^1 .$$

Ciò confermerebbe l'ipotesi dei coniugi Joliot e di Preiswerk che si formi qui l'Al_{13}^{28} ottenuto dai Joliot bombardando con particelle α il Mg.

Fosforo. – Abbiamo confermato il risultato dei Joliot sull'esistenza di un periodo di circa 3 minuti. Lo spessore di dimezzamento dei raggi β corrispondenti al periodo di 3 ore è 0,11 gr/cm².

(1) « Ric. Scientifica », 5 (1), p. 283, 330, 452 (1934).

Zolfo e Cloro. – Lo spessore di dimezzamento risulta 0,12 gr/cm². La coincidenza delle curve di assorbimento della radiazione emessa conferma che si tratta in entrambi i casi della produzione di un P^{32}_{15}.

Cromo. – Spessore di dimezzamento 0,2 gr/cm². Ha raggi γ. La coincidenza dei periodi fa supporre che V e Cr diano luogo a uno stesso isotopo radioattivo V^{52}.

Ferro. – Spessore di dimezzamento 0,2 gr/cm².

Zinco. – È stato separato il principio attivo del periodo di 6 minuti che segue il Cu. L'eguaglianza dei periodi può far pensare che il principio attivo si identifichi con quello del Cu.

Arsenico. – Lo spessore di dimezzamento degli elettroni è di 0,2 gr/cm².

Molibdeno. – Ha almeno due periodi di cui uno dell'ordine di grandezza di un quarto d'ora e l'altro assai più lungo. L'effetto è poco intenso.

Rodio. – È molto attivo. Abbiamo rilevato 2 periodi. Uno di circa 2 minuti e uno molto più lungo. Lo spessore di dimezzamento pel periodo breve è circa 0,12 gr/cm².

Palladio. – Lo spessore di dimezzamento è circa 0,03 gr/cm².

Argento. – Oltre al periodo di due primi abbiamo rilevato un periodo breve, di poche decine di secondi. Lo spessore di dimezzamento è di 0,1 gr/cm².

Jodio. – Lo spessore di dimezzamento è di circa 0,11 gr/cm².

Tungsteno. – Ha un debolissimo effetto con periodo dell'ordine di grandezza di 1 ora che potrebbe forse esser giustificabile con impurità.

Iridio. – Effetto cospicuo. Periodo di circa 20 ore. Lo spessore di dimezzamento è di 0,17 gr/cm².

Oro. – Lo spessore di dimezzamento è 0,03 gr/cm².

Mercurio. – Debolissimo effetto con periodo di poche decine di primi, forse giustificabile con impurità.

I seguenti elementi irradiati per alcune ore con sorgenti dell'ordine di grandezza di mezzo Curie non hanno dato effetto apprezzabile:

Li ; Os ; Pb ; Bi ; Ru ; Tl.

Le sezioni d'urto per gli elementi più attivi sono dell'ordine di grandezza di $10^{-24} \div 10^{-25}$ cm². Naturalmente i neutroni usati da noi hanno una distribuzione continua di energia non ben conosciuta.

Tuttavia, a titolo di esperienza di orientamento per cominciare a farci un'idea dell'effetto che l'energia dei neutroni ha sulla sezione d'urto, abbiamo attivato F, Ag, J, e U con neutroni provenienti da una sorgente di Em+Be nell'intervallo di tempo in cui l'emanazione si stava mettendo in equilibrio coi suoi prodotti che emettono particelle α e in particolare col Ra C.

Dal confronto dell'attività prima e dopo che si fosse raggiunto l'equilibrio non ci sembra di poter dedurre che i neutroni più veloci, almeno nel piccolo intervallo di energia sperimentata, abbiano una sezione d'urto molto più grande di quelli lenti. Un aumento di attività sensibilmente maggiore che negli altri casi si osserva per lo Jodio.

Istituto Fisico della R. Università.

Roma, 23 giugno 1934–XII.

88 *a*.

RADIOATTIVITÀ PROVOCATA DA BOMBARDAMENTO DI NEUTRONI. — V.

E. Amaldi, O. D'Agostino, E. Fermi, F. Rasetti, E. Segrè

« Ric. Scientifica », *5* (2), 21–22 (1934).

Riferiamo su alcune ulteriori osservazioni che completano i risultati già comunicati nelle lettere precedenti.

Sodio. – Il periodo di dimezzamento di questa sostanza è di circa 40 secondi, l'intensità media. Da alcune separazioni chimiche di fluoro dal sodio attivato, sembra che il principio attivo non sia fluoro.

Vanadio – Cromo – Manganese. – Per tutti e tre gli elementi abbiamo riconosciuto un periodo di circa 4 minuti con misure più precise di quelle fatte in precedenza per il vanadio ed il cromo. Separazioni chimiche hanno mostrato che il principio attivo del cromo e del manganese è vanadio. È molto plausibile che si tratti di un V^{52} che si formerebbe anche dal V^{51} bombardato. Per il manganese, oltre il periodo di 4 minuti, se ne è osservato uno di circa 150 minuti il cui principio attivo non si lascia separare dal manganese ed è presumibilmente Mn^{56} che si forma anche dal ferro e dal cobalto.

Arsenico. – Una separazione chimica del principio attivo in presenza di gallio e germanio permette di escludere che si tratti del primo elemento e rende poco verosimile anche il secondo. L'ipotesi più probabile è che l'attività sia dovuta ad As^{76}.

Rubidio. – Ha un effetto molto debole con periodo di circa 20 minuti.

Rodio. – Abbiamo rideterminato i periodi e come valori più probabili diamo 50 secondi e 5 minuti.

Cadmio. – Ha un effetto molto debole con periodo di circa un'ora.

Iridio. – Una separazione chimica del principio attivo in presenza di osmio e renio sembra escludere che l'attività sia dovuta a questi elementi. Probabilmente essa proviene da isotopi dell'iridio.

Oro. – Una separazione chimica del principio attivo in presenza di iridio, platino e mercurio ci induce a ritenere che esso sia isotopo dell'oro.

Torio. – Da saggi preliminari sul torio liberato dai suoi prodotti radioattivi β, risulta una forte attività con almeno due periodi di cui uno minore di 1 minuto e l'altro dell'ordine di grandezza di un quarto d'ora.

Uranio. – Oltre ai periodi già noti di 10 secondi, 40 secondi e 13 minuti, si è determinato uno dei periodi lunghi dell'uranio attivato; esso è di circa un'ora e mezza.

Abbiamo cercato di stabilire con maggior precisione il rendimento dell'estrazione dell'attività di 13 minuti con la reazione del biossido di manganese la quale trascina anche un po' dell'elemento 91 (UX_2). Si è trovato che il rendimento della reazione rispetto ai due elementi è nettamente diverso variando le condizioni di precipitazione. Per l'attività di 13 minuti si riesce a mantenerlo praticamente costante intorno al 15 °/o dell'attività totale, mentre quello per l'UX_2 oscilla tra l'1 ed il 7 °/o.

Si è ripresa anche la reazione di separazione del solfuro di renio in soluzione fortemente acida, con l'impiego di iposolfito sodico. Si è trovato che in questo modo si riesce a separare una quantità assai grande dell'attività di 13 minuti e di 100 minuti. Questa reazione oltre ad avere un rendimento intorno al 50 °/o è tanto selettiva in confronto all'UX_1, e UX_2 che permette di separare le due attività anche da uranio non depurato. Queste reazioni sembrano confermare l'ipotesi che si tratti di elementi con numero atomico maggiore di 92.

Abbiamo anche determinato approssimativamente il numero di atomi dell'elemento attivo con periodo di 13 minuti che si disintegrano ogni minuto, formatisi in un grammo di uranio elemento irradiato in equilibrio alla distanza di 1 cm da una sorgente di un millicurie di radon + berillio. Tale numero è dell'ordine di grandezza di 50.

Istituto Fisico della R. Università.
Roma, 12 luglio 1934–XII.

89 *a*.

RADIOATTIVITÀ PROVOCATA DA BOMBARDAMENTO DI NEUTRONI. — VII.

E. AMALDI, O. D'AGOSTINO, E. FERMI, B. PONTECORVO, F. RASETTI, E. SEGRÈ

« Ric. Scientifica », 5 (2), 467–470 (1934).

1. NUOVE ATTIVITÀ. – Abbiamo determinato i periodi di dimezzamento per l'attività dei seguenti elementi irradiati:

Potassio. – Ha una debole attività con periodi 16 ore (*a*) (1); una separazione chimica di Cl, A e Ca lascia l'attività nel K. Essa è quindi probabilmente dovuta a un nuovo isotopo radioattivo del potassio.

Gallio. – Ha un periodo di un giorno (*a*), oltre quello breve già segnalato.

Cadmio. – Il cadmio irradiato molto a lungo in acqua dà un debolissimo periodo di circa due giorni. Tuttavia, secondo le condizioni di irradiamento, si eccitano attività con periodi diversi.

Indio. – Il periodo breve è di 13 secondi (*a*).

Antimonio. – Oltre a un periodo breve e debole ne ha uno di 2, 5 giorni intenso (*a*).

Praseodimio. – Oltre quello di 5 minuti ha un periodo di un giorno intenso (*a*).

Delle seguenti sostanze sono stati rimisurati con l'elettrometro alcuni dei periodi già indicati, ottenendo risultati piuttosto precisi: Mn–150^m, As–26^h, Rh–44^s.

2. SENSIBILITÀ ALLE SOSTANZE IDROGENATE. – Per cominciare a precisare quantitativamente l'effetto delle sostanze idrogenate sull'attivazione, abbiamo irradiato alcune sostanze sotto forma di cilindri del diametro del contatore, mettendo la sorgente all'interno del cilindro, in modo da ottenere la massima attivazione possibile. L'irradiazione era fatta prima in un blocco di paraffina di dimensioni molto grandi, poi in aria. Il rapporto fra le attività osservate dà grossolanamente un coefficiente di sensibilità α all'azione di sostanze idrogenate che risulta, naturalmente, eguale ad 1 per quelle sostanze che non presentano un aumento di attività, irradiate in paraffina. Si trova per Na (40^s) $\alpha = 1$, Mg (40^s) $\alpha = 1$, Al (12^m) $\alpha = 1$, Al ($2{,}5^m$) α molto grande, Si ($2{,}5^m$) $\alpha = 1$, Cl (45^m) α molto grande, V (4^m) $\alpha = 40$, Cu (6^m) $\alpha = 15$, Rh (44^s) $\alpha = 15$, Ag (2^m) $\alpha = 15$, Ag (20^s) $\alpha = 30$, In (13^s) $\alpha = 12$, U (10^s, 13^m, 90^m) $\alpha = 1{,}6$.

(1) Il simbolo (*a*) significa che il periodo in questione ha un elevato coefficiente di sensibilità alle sostanze idrogenate (vedi N. 2 di questa Nota).

Questi dati sono da intendersi come puramente indicativi, poiché piccole variazioni nelle condizioni geometriche di irradiamento influiscono molto sul risultato.

3. SEPARAZIONE DI ISOTOPI RADIOATTIVI. – Abbiamo separato, con varî metodi, gli isotopi radioattivi degli alogeni, dell'arsenico e del manganese da quelli ordinari. Approfittando del fatto che l'urto neutronico può decomporre molecole o ioni e lasciarli in uno stato diverso da quello iniziale Szilard e Chalmers [2] erano riusciti ad arricchire in isotopo radioattivo lo iodio estratto dallo ioduro di metile e il bromo estratto dal bromoformio.

Abbiamo ripetuto le loro esperienze estendendole anche al cloro (45^m) irradiato sotto forma di varî composti organici. Successivamente abbiamo riconosciuto che lo stesso effetto si può avere in composti inorganici, purché l'elemento sia legato in complessi difficili a formare spontaneamente. Per esempio, abbiamo irradiato iodati, bromati e clorati alcalini sia allo stato solido che in soluzione. Precipitando successivamente *una traccia* di ioduro, bromuro, o cloruro di argento, si trova dal 25 al 50% dell'attività totale nel precipitato.

Analogamente irradiando acido cacodilico, abbiamo potuto separare buona parte dell'attività dell'arsenico, precipitando dalla soluzione tracce di solfuro d'arsenico. Irradiando permanganato potassico, si può raccogliere circa l'80 % dell'attività precipitando una traccia di carbonato manganoso.

Dallo ioduro di metile o di etile gassosi irradiati è stato anche possibile raccogliere lo iodio radioattivo sopra un elettrodo di nichel (inattivabile) posto nel gas che si trovava in un campo elettrico destinato a facilitare la raccolta degli ioni. Il rendimento tuttavia non era molto elevato.

4. VELOCITÀ DEI NEUTRONI NELLA PARAFFINA. – Nell'ipotesi che l'azione di sostanze idrogenate sulla radioattività provocata da bombardamento di neutroni sia dovuta al rallentamento dei neutroni stessi, abbiamo cercato di farci un'idea della velocità di questi neutroni, che chiameremo convenzionalmente « lenti ». Poiché con una ventina di urti successivi con protoni, la velocità dei neutroni si può ridurre all'ordine di grandezza di quella dovuta all'agitazione termica, si potrebbe pensare che la temperatura agisse sulla densità del gas dei neutroni. Per controllare questo punto abbiamo fatto la seguente esperienza: Una lastrina di Rh veniva prima attivata in un miscuglio di idrocarburi (pentano e benzolo) alla temperatura ambiente; quindi a 200° in un altro miscuglio di idrocarburi (paraffina) che aveva, a quella temperatura, la stessa densità e composizione elementare del miscuglio freddo. La lastrina di Rh nelle due esperienze era irradiata nelle identiche condizioni geometriche (1 cm. dal pelo libero del liquido, 2 cm. dalla sorgente).

Non si è constatata alcuna differenza nell'attivazione, dal che sembra si possa concludere che l'energia dei neutroni « lenti » sia maggiore dell'energia di agitazione termica.

(2) L. SZILARD e T. A. CHALMERS, « Nature », *134*, 462 (1934).

5. Assorbimento dei neutroni «lenti». – Continuando le misure di assorbimento dei neutroni lenti abbiamo trovate diverse altre sostanze che presentano sezioni d'urto anormalmente grandi. Tali sezioni d'urto, prendendo per unità 10^{-24} cm² risultano per i seguenti elementi: Cl, 130; Co, 100; Ag, 100; Cd, 10000; In, 450; I, 40; Pr, 100; Au, 100; Hg, 1000.

Queste sezioni d'urto corrispondono ad assorbimenti misurati dal confronto della attività indotta nel rivelatore (Rh), posto in una cavità scavata nella paraffina, con quella indotta nello stesso rivelatore, posto nella medesima cavità, ma protetto da due strati della sostanza assorbente in esame. Occorre osservare che se le esperienze vengono fatte all'esterno della paraffina, anziché in una cavità, si trovano coefficienti di assorbimento minori; questo fatto può interpretarsi, nell'ipotesi che si tratti di neutroni «lenti», osservando che la loro velocità media in una cavità è minore che all'esterno della paraffina.

Abbiamo rilevato la curva di assorbimento, nel cadmio, dei neutroni «rallentati» dalla paraffina: si trova che l'attività trasmessa diminuisce con l'aumentare dello spessore assorbente, seguendo una legge non esponenziale; precisamente si arriva a valori tali dello spessore assorbente che un aumento dello spessore stesso influisce molto poco sull'attività. Questo fatto ha riscontro in misure da noi eseguite attivando del Rh, a distanza fissa dalla sorgente, in un recipiente contenente una soluzione con concentrazione variabile di acido borico (3). Questi risultati possono interpretarsi ricordando che la distribuzione di velocità dei neutroni è continua.

In qualche misura di assorbimento abbiamo usato come rivelatore argento oltre che rodio, ottenendo risultati sostanzialmente eguali.

Per studiare l'assorbimento dei neutroni non «rallentati» da mezzi idrogenati, abbiamo interposto uno spessore di 5 gr/cm² di acido borico tra la sorgente e un cilindretto di Rh: la attivazione non è affatto minore di quella ottenuta in identiche condizioni geometriche, senza assorbitore.

6. Emissione di raggi γ. – Abbiamo istituito esperienze per stabilire se ai forti assorbimenti dei neutroni «lenti», particolari per alcuni elementi, corrispondesse l'emissione di radiazioni secondarie. Per questo scopo ci siamo serviti di una sorgente di Po + Be (circa 60 mC) e di un contatore di Geiger–Müller, immersi entrambi in un grosso blocco di paraffina. Sul percorso diretto si trovava uno spessore di 10 cm. di piombo per assorbire i raggi γ del Po + Be. Il contatore era protetto da una lamina di 2 mm. di piombo. Ponendo all'esterno di questa un sottile strato dei seguenti elementi: Cd, Hg, Cl, Y, Ir, abbiamo notato un aumento del numero degli impulsi, che arrivava fino a un centinaio per minuto. Si tratta probabilmente dell'emissione di raggi γ, che possono interpretarsi come corrispondenti all'energia di legame del neutrone «lento» catturato. Non abbiamo osservato tale radiazione, e neppure una eventuale radiazione secondaria meno penetrante (α, β, H) nel caso del boro, benché anche questo elemento presenti una grande sezione d'urto.

(3) «Ricerca Scientifica», 5 (2), 380 (1934).

Sono in corso esperienze per cercare di chiarire tale punto.

7. Attivazione in funzione della distanza. – Abbiamo studiato la variazione della intensità dell'attivazione del rivelatore di Rh in funzione della distanza dalla sorgente, qualora ambedue siano immersi in una vasca d'acqua di grandi dimensioni. Il risultato è riassunto nella seguente tabella:

Distanza in cm.	1	. . . .	Attività	100
» »	2	. . . .	»	92
» »	3	. . . .	»	86
» »	4	. . . .	»	76
» »	5	. . . .	»	71
» »	6	. . . .	»	54
» »	7	. . . .	»	46
» »	8	. . . .	»	33
» »	9	. . . .	»	26
» »	10	. . . .	»	22

Si vede che l'attivazione decresce piuttosto lentamente con la distanza. Sostituendo l'acqua con una soluzione di acido borico, pur diminuendo, naturalmente, l'attivazione, la forma della curva di dipendenza dalla distanza non cambia sensibilmente.

8. Diffusione dei neutroni «lenti». – Abbiamo fatto inoltre una serie di esperienze per ricercare elementi che diffondessero i neutroni «lenti». Sull'asse di un blocco cilindrico, a 3 cm. dalla base superiore, viene posta la sorgente; sulla base superiore viene collocata una lastrina di Rh; il confronto tra l'attività così ottenuta nel Rh, con quella che si ha ricoprendo la lastrina con le diverse sostanze, dà una idea della diffusione dei neutroni «lenti» da parte delle sostanze stesse.

Si è trovato un aumento ben netto dell'attivazione ricoprendo la lastrina di Rh con carbone; (uno strato di parecchi cm. di spessore arriva a quintuplicare l'attività); un risultato analogo si ottiene con uno strato di basalto. Anche strati sottili di elementi leggeri (Be, C) producono un aumento sensibile della attivazione.

Con lo stesso dispositivo è stata fatta la curva della intensità della attivazione della lastrina di Rh in funzione dei cm di spessore di acqua sovrapposti; uno spessore di 1 cm. di acqua sovrapposto alla lastrina di Rh appoggiata sulla paraffina raddoppia circa la attivazione; per spessori maggiori l'incremento della attività si va facendo più lento.

Analoghi risultati si ottengono sovrapponendo alla lastrina di Rh strati variabili di paraffina.

Abbiamo anche studiato l'assorbimento e la diffusione di alcune sostanze con il seguente dispositivo: nel centro di un blocco cilindrico di paraffina di 11 cm. di altezza e 13 cm. di diametro, si trova la sorgente di neutroni: un cilindretto di argento (rivelatore) è posto a 30 cm. dalla sorgente.

Tra il cilindro di paraffina e il rivelatore, a 6 cm. dal rivelatore, possono interporsi gli strati delle diverse sostanze. Le condizioni geometriche delle esperienze sono tali che l'assorbitore intercetta esattamente i raggi provenienti dal cilindro di paraffina.

La differenza tra le attivazioni dell'argento, con e senza assorbitori, è dovuta a diffusione + assorbimento. Con un diffusore di paraffina si trova che 0,5 gr/cm^2 riducono l'attivazione a metà, ma la curva di assorbimento non è esponenziale. Sembra dunque che il cammino libero medio dei neutroni lenti nella paraffina sia dell'ordine di 7 o 8 mm.

L'esperienza fatta con diversi spessori diffondenti di acqua non ha dato risultati sostanzialmente dissimili.

Nel caso che lo strato assorbente sia di acido borico, l'assorbimento prevale sulla diffusione, e si trovano i risultati già riferiti. Per il carbonio si trova che 5 gr/cm^2 riducono all'incirca a metà; 3,45 gr/cm^2 di piombo riducono nel rapporto 0,83 $\pm$ 0,07; tale valore è da attribuirsi alla diffusione, perché sappiamo che uno strato di 3,45 gr/cm^2 di Pb non assorbe praticamente i neutroni « lenti ».

9. Effetto di sostanze non idrogenate sull'attivazione. – Abbiamo anche cercato se sostanze non idrogenate producano incrementi nella attivazione analoghi a quelli dovuti, per esempio all'acqua e alla paraffina. Nelle poche sostanze fin qui esaminate (piombo, grafite, basalto), abbiamo trovato sempre un effetto, notevolmente minore tuttavia di quello dovuto, per esempio, all'acqua:

Piombo. – Un cilindretto di Rh è posto a 10 cm. dalla sorgente, in un interstizio praticato in mezzo al piombo; la quantità usata di piombo è tale che, grossolanamente, il punto medio del segmento sorgente–Rh può ritenersi al centro di una sfera di piombo di 50 cm. di diametro.

L'attività indotta nel Rh è circa 3 volte maggiore di quella che si ottiene attivando il Rh nelle identiche condizioni geometriche, ma senza piombo. Il fattore per cui si moltiplica l'attività diminuisce, naturalmente, quando diminuisce la distanza della sorgente dal rivelatore. Un effetto dovuto al piombo si trova anche nella attivazione dell'argento. Usando il silicio come rivelatore, per contro, non si osserva alcun effetto, analogamente a quanto avviene nei mezzi idrogenati. Quest'ultimo fatto suggerisce l'ipotesi che il fenomeno abbia la stessa natura di quello che si osserva nelle sostanze idrogenate: abbiamo quindi cercato se la radiazione che attiva il Rh in ambiente di Pb sia considerevolmente assorbita dal Cd, proteggendo il cilindretto di Rh con spessori variabili di Cd: si trova che il Cd assorbe notevolmente (riduzione a metà in circa 1 gr/cm^2), in misura minore tuttavia di quanto assorbirebbe in ambiente idrogenato. Si noti che nel caso di un elemento pesante quale il Pb, non sembra possibile spiegare l'effetto come un rallentamento di neutroni dovuto a urti elastici. Se veramente si tratta di neutroni rallentati, potrebbe pensarsi all'influenza di urti anelastici.

L'effetto della *grafite* sull'attivazione del Pb è più intenso dell'effetto dovuto al Pb; l'intensità dell'attivazione, nel silicio, non aumenta in ambiente di grafite. La radiazione attivante il Rh è assorbita dal Cd, in misura almeno altrettanto grande che nel caso del Pb. La massa di grafite a disposizione era assai piccola; un suo aumento potrebbe forse produrre un aumento considerevole degli effetti osservati. Con *basalto* si ottengono effetti analoghi a quelli sopra descritti.

È in corso una ricerca sistematica sull'effetto delle diverse sostanze sulla radioattività provocata da bombardamento di neutroni, dato che queste azioni sembrano essere generali e che spesso si producono involontari aumenti della attivazione per effetto degli oggetti circostanti: appunto per questa ragione le misure descritte in questo paragrafo non sono molto precise.

Istituto Fisico della R. Università.
Roma, 6 dicembre 1934–XIII.

90 *a*.

RADIOATTIVITÀ PROVOCATA DA BOMBARDAMENTO DI NEUTRONI. — VIII.

E. Amaldi, O. D'Agostino, E. Fermi, B. Pontecorvo, F. Rasetti, E. Segrè

« Ric. Scientifica », *6* (1), 123–125 (1935).

1. Nuove attività. – Abbiamo determinato i periodi di dimezzamento per i seguenti elementi irradiati:

Magnesio. – Irradiando MgO purissimo di Kahlbaum esente da alluminio, oltre alle attività già note ne abbiamo rilevata una debole con periodo di circa 10 minuti (*a*). Essa è probabilmente dovuta all'isotopo radioattivo ^{27}Mg (periodo 12 minuti) che si forma anche dall'alluminio irradiato con neutroni. In questo caso si tratterebbe di un nuovo esempio di cattura del neutrone.

Silicio. – Quarzo fuso irradiato ha mostrato una debole attività con periodo di qualche ora. Riteniamo si tratti anche qui di cattura del neutrone con formazione del ^{31}Si (periodo 145^m) che si ottiene anche dal P irradiato.

Bario. – Oltre al periodo breve abbiamo rilevato un periodo di circa 80^m (*a*).

Gadolinio. – Ossido di Gd purissimo, gentilmente donatoci dal professor L. Rolla mostra un'attività con periodo di circa 8 ore (*a*).

Tungsteno. – Tungsteno mostra irradiato un'attività con periodo di circa 1 giorno (*a*).

2. Determinazione precisa di periodi. – Servendoci di una camera di jonizzazione (vedi 4), abbiamo determinato con maggior precisione i periodi di dimezzamento seguenti: Si, 137^s; P, 145^m; V, 225^s; Ga, 23^h; Br, 18^m e 255^m; Ag, 22^s e 140^s; J, 25^m; Pr, 19^h; Re, 24^h (oltre a un periodo più breve); Ir, 19^h; Au, 65^h.

3. Sensibilità alle sostanze idrogenate. – Col metodo descritto nella lettera precedente abbiamo determinato il coefficiente di sensibilità α alle sostanze idrogenate per i seguenti elementi: P (137^s), Cr (225^s), Mn (225^s), Fe (150^m), Zn (6^m), Ba (3^m), Pr (5^m), Th (1^m e 15^m) hanno $\alpha = 1$. Se (35^m), Ba (80^m) hanno α molto grande.

4. – Assorbimento degli elettroni emessi – Raggi γ. – Per le misure di assorbimento ci siamo serviti di una camera di jonizzazione riempita di CO_2 alla pressione di 3 Atm. e chiusa con una foglia di alluminio di 0,1 mm. di spessore. Le sostanze irradiate erano disposte in strato piano e

l'assorbimento era misurato interponendo foglie di alluminio. Per alcune sostanze che presentano un'intensa radiazione γ, (rivelata dalla jonizzazione residua che si ha schermando la camera con 2 mm. di Pb) abbiamo tenuto conto di questa attività nel valutare il coefficiente di assorbimento dei raggi β. Una misura di controllo eseguita su UX_2 e RaE ha dato per i valori dei coefficienti di assorbimento quelli ordinariamente ammessi.

La seguente tabella dà gli spessori di dimezzamento in gr/cm² di alluminio per le stanze indicate.

P	145^m	0,15	In	54^m	0,045
V	225^s	0,17	Sb	$2,5^d$	0,09
Mn	150^m	0,14	J	25^m	0,10
Ga	30^m	0,17	Pr	19^h	0,13
As	26^h	0,16	Re	24^h	0,12
Br	18^m	0,12	Ir	19^h	0,12
Br	255^m	0,12	Au	65^h	0,04
Rh	44^s	0,15	U	13^m	0,14

Data la maggiore intensità di molti prodotti attivi attualmente disponibili in seguito alla esaltazione del potere attivante dei neutroni in sostanze idrogenate, abbiamo potuto rilevare una emissione di raggi γ per alcuni elementi, (Rh 44^s, Br 18^m e 255^m), nei quali ci era sfuggita in ricerche precedenti. Si è inoltre osservata emissione di raggi γ in V (225^s), Mn (150^m), Cu (6^m), Ga (23^h), In (54^m), Sb ($2,5^d$), J (25^m), W (1^d).

5. Separazioni chimiche. – *Antimonio* metallico irradiato è stato sciolto in acqua regia. Aggiunto Sn e separatolo dall'antimonio col metodo di Clarke, l'attività seguiva l'antimonio. Allo stesso Sb abbiamo aggiunto, e quindi separato da esso, indio, approfitttando della diversa solubilità dei solfuri negli acidi diluiti. L'In è stato trovato inattivo. Anche Te e J, aggiunti e separati, rispettivamente l'uno per riduzione, l'altro come AgJ, sono stati trovati inattivi. L'attività dell'antimonio di periodo $2,5^d$ è quindi da attribuirsi a un suo isotopo.

Anidride tungstica irradiata è stata disciolta in soda caustica e dalla soluzione è stato separato il tantalio come pentossido, inattivo. Alla soluzione di tungstato è stato poi aggiunto renio in soluzione nitrica e separato il W come anidride tungstica. Il Re rimasto in soluzione era inattivo. Non disponendo di hafnio abbiamo separato da una soluzione ammoniacale di anidride tungstica l'idrato di zirconio inattivo. È perciò molto probabile che l'attività sia dovuta a un isotopo del W.

Dal *renio* irradiato sciolto in HNO_3 si sono separati successivamente Ta come pentossido e W come anidride tungstica e trovati inattivi. Le attività sono dovute probabilmente ad isotopi del Re.

6. Azione dei neutroni lenti sul boro. – Abbiamo accennato in precedenti lettere che:

a) i neutroni rallentati dalla paraffina o dall'acqua sono anomalmente assorbiti dal B (riduzione a metà da 0,004 gr/cm²);

b) che a differenza di quanto accade per parecchi altri elementi che hanno un assorbimento intenso per neutroni lenti, il B (come anche il Li) non emette raggi γ all'atto dell'assorbimento del neutrone. Mentre preventivamente non eravamo riusciti ad osservare in questo caso l'emissione di particelle, abbiamo potuto ora rilevare, colle esperienze che passiamo a descrivere, una radiazione α.

1. – Una camera di jonizzazione è stata riempita di BF_3 gassoso. Ponendola in presenza di una sorgente di 50 mC, di Po + Be e circondandola di paraffina per rallentare i neutroni, si osserva una jonizzazione molto intensa. Il valore di tale jonizzazione si riduce circa a metà schermando la camera con Cd in strato di 0,02 gr/cm²; ciò indica che l'effetto è specifico dei neutroni lenti. Un'altra conferma si ha dal fatto che il Po + Be senza paraffina non dà un effetto paragonabile col precedente.

2. – Lo stesso effetto in misura naturalmente minore si può osservare con una camera di jonizzazione piena d'aria disponendo del boro elemento sul fondo della camera. Variando la pressione nello strumento e determinando la differenza tra la jonizzazione senza e con uno schermo di Cd, (cioè la frazione della jonizzazione dovuta ai neutroni lenti), si può determinare per quale pressione il «range» delle particelle jonizzanti emesse dal boro in queste condizioni diventa paragonabile colle dimensioni della camera di jonizzazione. Si trova così un percorso di 0,7 o 0,8 cm.

3. – Abbiamo osservato lo stesso fenomeno per mezzo di un amplificatore proporzionale: probabilmente nelle esperienze preliminari questa radiazione ci era sfuggita a causa della sua piccola penetranza. Anche le misure coll'amplificatore proporzionale sono in accordo con le precedenti e dànno un percorso di 7 od 8 mm. Queste osservazioni hanno permesso anche di misurare il valore massimo della jonizzazione totale prodotta in un singolo processo. Questa (40.000 coppie di joni) corrisponde alla jonizzazione prodotta da una particella α del percorso di 0,8 cm.; poiché tale valore coincide con quello osservato, concludiamo che si tratta molto probabilmente di particelle α di circa 14×10^5 e V.

Da ciò e dal fatto che nel B non si osserva radioattività indotta, si è portati a ritenere come la reazione nucleare più plausibile:

$$^{10}_{5}\mathrm{B} + ^{1}_{0}n = ^{7}_{3}\mathrm{Li} + ^{4}_{2}\mathrm{He}$$

Computando anche l'energia del nucleo di Li di rinculo, l'energia che si libera nella reazione è di 22×10^5 eV.

Calcolata in base ai difetti di massa, l'energia di reazione, trascurando la piccola energia del neutrone primario, risulterebbe di 2×10^6 eV. più grande.

Anche il Li irradiato con neutroni lenti emette particelle pesanti di percorso di alcuni centimetri. Sono in corso ulteriori esperienze in proposito.

7. Effetto di sostanze non idrogenate sull'attivazione. – Abbiamo studiato l'effetto del silicio e del ferro sulla attivazione provocata da bombardamento di neutroni. A 10 cm. di distanza dalla sorgente veniva

attivato il Rh, prima in aria, poi in un recipiente contenente silice. L'attività del rodio viene moltiplicata circa per un fattore 2,5 in un ambiente di SiO_2. Se si attiva del Si per contro, non si ottiene nessun aumento di attività in ambiente di SiO_2.

Abbiamo invece constatato che grosse masse di ferro circostanti alla sorgente e il rivelatore non producono un aumento notevole della attivazione, sia nel Rh che nel Si.

8. SUI NEUTRONI PRODOTTI DA RAGGI γ + Be. – L. Szilard e T. A. Chalmers (1), L. Meitner (2) hanno osservato che, usando come sorgente di neutroni del Be sottoposto all'azione dei raggi γ, si attivano soltanto gli elementi che rimangono isotopi di sé stessi. Poiché i neutroni rallentati nella paraffina godono, per quanto finora noto, della stessa proprietà, abbiamo ricercato se i neutroni di Szilard e Chalmers fossero assorbiti notevolmente da un elemento, come il Cd, che presenta per i neutroni lenti una sezione d'urto elevatissima. Per questo abbiamo usato come sorgente di neutroni un preparato di 100 mgr di Ra circondato da Be; l'attivazione in un cilindretto di Rh non diminuiva sensibilmente quando la sorgente era protetta da 0,5 gr/cm^2. di Cd. Ciò può interpretarsi ammettendo che i neutroni di Szilard e Chalmers siano notevolmente più veloci di quelli rallentati in paraffina.

Istituto Fisico della R. Università.
Roma 11 gennaio 1935–XIII.

(1) L. SZILARD e T. A. CHALMERS, «Nature», *134*, 494 (1934).
(2) L. MEITNER, «Naturw.», *22*, 759 (1934).

91 *a*

RADIOATTIVITÀ INDOTTA DA BOMBARDAMENTO DI NEUTRONI (*). — IX.

E. AMALDI, O. D'AGOSTINO, E. FERMI, B. PONTECORVO, E. SEGRÈ

« Ric. Scientifica », *6* (1), 435–437 (1935).

1. DETERMINAZIONE PRECISA DI PERIODI. – Abbiamo determinato con precisione per mezzo di una camera di ionizzazione collegata con un elettrometro i periodi di dimezzamento di alcune sostanze bombardate con neutroni:

Al 137 secondi e 10 minuti; S 14 giorni; Cl 37 minuti; Cu 5 minuti; Zn 10 ore; Ga 20 minuti; Rh 3,9 minuti; Sb 2,5 giorni; Th 24 minuti.

2. SENSIBILITÀ ALLE SOSTANZE IDROGENATE. – Col metodo descritto precedentemente abbiamo determinato il coefficiente di sensibilità alle sostanze idrogenate nei seguenti elementi:

F (9 secondi e 45 secondi) $\alpha = 1$, Mn (150 minuti) $\alpha = 23$,
Zn (10 ore) $\alpha = 1$, Ga (20 minuti) $\alpha = 2{,}5$, As (26 ore) $\alpha = 6$,
Se (35 minuti) $\alpha = 5$, Br (18 minuti) $\alpha = 8$, Rh (3,9 minuti) $\alpha = 25$,
Sb (2,5 giorni) $\alpha = 4$, J (25 minuti) $\alpha = 5$, Ba (80 minuti) $\alpha = 9$.

3. ASSORBIMENTO DEGLI ELETTRONI EMESSI. – Col metodo descritto precedentemente abbiamo misurato lo spessore di dimezzamento in gr/cm^2 di Al per gli elettroni emessi nelle seguenti attività:

Si (137 secondi) 0,19; P (145 minuti) 0,15 (nella nota precedente per errore di stampa era stato dato 0,07); S (14 giorni) 0,10; Ag (140 secondi) 0,08; J (25 minuti) 0,11.

4. SEPARAZIONI CHIMICHE. – *Fosforo*. – L'attività con periodo 137 secondi è stata separata insieme ad alluminio. Una soluzione di acido fosforico irradiato è stata neutralizzata con carbonato sodico; quindi si è aggiunta una traccia di cloruro di alluminio. Il fosfato di alluminio separato ha trascinato tutta l'attività.

Cloro. – Ad acido cloridrico irradiato e neutralizzato con soda abbiamo aggiunto acido nitrico facendo poi bollire. Aggiungendo alla soluzione acido solforico e fosforico abbiamo precipitato solfato di bario e fosfato ammonico magnesiaco ambedue inattivi. Il cloruro sodico residuo conteneva l'attività (37 minuti) che deve quindi attribuirsi a un isotopo del cloro.

(*) Cfr. « Ricerca Scientifica », *5* (1), pp. 283, 330, 452, 652 (1934); *5* (2), pp. 21, 381 467 (1934); *6* (1), p. 123 (1935).

Rame. – Rame metallico irradiato è stato sciolto in acido cloridrico e vi si sono aggiunte tracce di cobalto nichel e zinco. Dalla soluzione si è separato il rame come solfuro, che è stato trovato attivo. Lo zinco, il cobalto e il nichel separati anch'essi come solfuri per neutralizzazione del filtrato sono stati trovati inattivi. Il tempo impiegato per questa prova permette di concludere solo per la vita di 10 ore che essa si deve attribuire a un isotopo del rame.

Zinco. – Da zinco metallico irradiato e sciolto in acido cloridrico abbiamo separato come sopra il rame in presenza di nichel e cobalto. Il rame trascinava l'attività, mentre gli altri metalli rimasti in soluzione, separati, erano inattivi. Quindi anche la vita di 10 ore dello zinco è da attribuirsi a rame.

Gallio. – Con lo stesso procedimento indicato altra volta per l'attività di 20 minuti si è esaminata l'attività di 23 ore: anche questa si deve attribuire ad un isotopo del gallio.

Selenio. – Anidride selenica irradiata è stata sciolta in acido cloridrico al 30 °/₀ e alla soluzione è stata aggiunta anidride arseniosa. Dalla soluzione abbiamo separato selenio metallico con anidride solforosa. Il selenio separato era attivo, mentre l'arsenico precipitato dal filtrato come solfuro era inattivo. Siccome anche il germanio non è precipitato dall'anidride solforosa, esso precipita come solfuro; questa prova indica quindi che l'attività del selenio è dovuta a un isotopo del selenio.

Torio. – Le prove chimiche si sono limitate a considerare l'attività con periodo di 24 minuti. Il torio è stato depurato dai suoi prodotti radioattivi precipitandolo più volte, con acqua ossigenata, in presenza di bario, lantanio, piombo e bismuto.

Dalla soluzione molto diluita di nitrato di torio irradiato è stato separato il torio come fluoruro in presenza di uranio e rispettivamente tantalio. Il torio separato conservava l'attività mentre l'uranio o il tantalio precipitati come idrati dal filtrato, erano inattivi.

In altre prove è stato separato il torio dall'uranio come carbonato con carbonato di ammonio. L'uranio separato dalla soluzione ammoniacale era quasi inattivo. La piccola attività trovata è probabilmente giustificabile con le piccole quantità di torio precipitate insieme all'uranio. Da una soluzione acida di nitrato di torio irradiato è stato separato zirconio come fosfato. Il precipitato era quasi inattivo, mentre il torio conservava l'attività. Da queste prove sembra doversi escludere che l'attività sia dovuta a isotopi dell'uranio e del protoattinio.

Dalla soluzione diluita di nitrato di torio irradiato abbiamo precipitato il bario come solfato che abbiamo trovato inattivo. In altre prove abbiamo separato dalla soluzione diluita di nitrato di torio, il torio con acqua ossigenata in presenza di bario e lantanio. Dal filtrato è stato separato il bario come solfato e il lantanio come idrato, che abbiamo trovato inattivi, mentre l'attività rimaneva nel torio. Sembra quindi che l'attività non sia dovuta nemmeno ad isotopi del Pb, Bi, Ra, Ac e Po. È quindi probabile che l'attività con periodo di 24 minuti del torio sia dovuta a un suo isotopo.

In tutte le operazioni chimiche su sostanze radioattive naturali, accade talvolta che i residui di un'imperfetta depurazione si concentrino in qualcuna delle frazioni separate, il che richiede particolari cure per l'identificazione dell'attività dal periodo di decadimento ed è una delle cause che rendono tali operazioni chimiche notevolmente delicate ed impongono grandi cautele nella interpretazione dei risultati.

Uranio. – Abbiamo già detto in note precedenti che i coefficienti di sensibilità alle sostanze idrogenate dei periodi di 15 secondi, 13 minuti e 100 minuti sono leggermente maggiori di 1 e sensibilmente eguali. Il coefficiente di sensibilità per la attività con periodo di 40 secondi è invece inferiore a quello delle attività precedenti. È evidente che tutti i prodotti che derivano da uno stesso processo primario debbano avere lo stesso coefficiente di sensibilità: ne concludiamo quindi che l'attività con periodo di 40 secondi è dovuta a un processo primario a sè mentre le altre tre attività son dovute probabilmente allo stesso processo primario. In questa ipotesi le tre attività potrebbero essere o prodotte in catena o avere relazioni più complicate attraverso ramificazioni. Un indizio a favore della prima ipotesi, almeno per quanto riguarda i periodi di 13 minuti e di 100 minuti è dato dal seguente esperimento: abbiamo misurato con l'elettrometro la curva di decadimento di uno strato spesso di uranio irradiato; questa curva analizzata in esponenziali con i periodi di 13 minuti e 100 minuti mostra che le attività iniziali sono nel rapporto di 100 a 45; d'altra parte lo spessore di dimezzamento per i raggi β dell'attività di 13 minuti è di 0,14 gr/cm^2 di Al; quello del periodo più lungo non si è potuto misurare con precisione ma è certamente più piccolo del precedente e probabilmente circa la metà. Questo risultato è in accordo con l'ipotesi che il numero iniziale delle disintegrazioni per i due periodi, con la sostanza irradiata a saturazione, sia lo stesso, il che è necessario per ammettere che le due attività siano in catena.

In una lettera precedente abbiamo messo in evidenza con prove chimiche che l'attività con periodi di 13 minuti e 100 minuti non possono essere attribuite ad isotopi di elementi con $Z < 92$. Von Grosse e Agruss [1] hanno criticato queste nostre prove deducendo da esperimenti eseguiti solamente su sostanze non irradiate con neutroni, che l'attività con periodo di 13 minuti era dovuta ad un isotopo del protoattinio. Noi abbiamo perciò ripetuto i nostri vecchi esperimenti confermandoli ed abbiamo eseguito anche alcune nuove prove.

La precipitazione dell'attività con un solfuro è stata ripetuta usando diversi metalli (Ag, Cu , Pb , Hg): l'acidità della soluzione cloridrica era di circa il 20 %. Il rendimento della separazione era generalmente buono (circa il 50 %) e variava secondo le condizioni di precipitazione. La presenza di acido nitrico abbassa molto il rendimento della separazione. In presenza di acido fluoridrico e tantalio invece si è ottenuto un elevato rendimento; il protoattinio come è noto segue, il tantalio nelle sue reazioni e quest'ultimo precipita solo con grande difficoltà da soluzioni fluoridriche.

(1) VON GROSSE e AGRUSS, « Phys. Rev. », vol. 46, p. 241 (1934).

Abbiamo anche fatta la prova esplicitamente consigliata da v. Grosse come la più caratteristica del protoattinio. Abbiamo sciolto ossido di uranio purificato e irradiato in HCl al 25 °/₀: abbiamo aggiunto nitrato di zirconio e acido fosforico; il precipitato di fosfato di zirconio era inattivo mentre nel filtrato si poteva raccogliere l'attività con la reazione dei solfuri col rendimento solito.

Secondo von Grosse e Agruss questa reazione deve essere considerata una prova che le attività di 13 minuti e 100 minuti non sono dovute a isotopi del protoattinio. Con differenti prove chimiche, da noi stessi successivamente ripetute e confermate, O. Hahn e L. Meitner [2] concludono anche essi che le attività di 13 minuti e 100 minuti sono dovute a elementi transuranici.

Le due attività hanno certo caratteri chimici molto simili: un indizio di separazione fu ottenuto soltanto col seguente esperimento; ossido di uranio depurato fu irradiato e sciolto in HCl. La soluzione fu versata in carbonato di ammonio fino a ridisciogliere completamente il precipitato di carbonato di uranile. Aggiungendo alla soluzione nitrato di manganese o piombo e raccogliendo il carbonato corrispondente si trovava in questo una parte delle attività di 13 minuti e 100 minuti. Nel filtrato dopo averlo acidificato abbiamo precipitato solfuro di rame; anche questo era attivo. Sembra però che il rapporto delle due attività sia diverso nei due precipitati e precisamente che l'attività con periodo di 13 minuti sia più abbondante nel precipitato dei solfuri.

Da tutte le prove eseguite, la nostra ipotesi che le attività di 13 minuti e 100 minuti si debbano attribuire ad elementi transuranici riceve una nuova conferma. La più semplice interpretazione, in accordo con i fatti conosciuti, è che le attività di 10 secondi, 13 minuti, 100 minuti, siano in catena e che si debbano attribuire ad elementi di numero atomico 92, 93, 94 e peso atomico 239.

ASSORBIMENTO DEI NEUTRONI LENTI ED EMISSIONE DI RAGGI GAMMA SOTTO L'AZIONE DEI NEUTRONI LENTI. – Abbiamo studiato l'assorbimento per cattura del cloro, usando come rivelatore sia il rodio che il vanadio; nei due casi abbiamo trovato sostanzialmente la stessa sezione d'urto.

Abbiamo anche determinato il libero cammino medio dei neutroni lenti nella paraffina, usando come rivelatore il rodio, e abbiamo rilevato una curva analoga a quella che avevamo trovato precedentemente usando per rivelatore l'argento.

In una lettera precedente abbiamo descritto un dispositivo sperimentale atto a rilevare l'emissione di raggi γ da parte di alcune sostanze, sotto l'azione dei neutroni lenti. Con lo stesso dispositivo abbiamo compiuto una ricerca sistematica, che ha confermato le previsioni secondo le quali agli elementi con grande sezione d'urto corrisponde l'emissione di raggi γ (fanno eccezione il boro e il litio, di cui è già stato detto): oltre agli elementi già elencati, si riscontra una notevole emissione di raggi γ in cobalto, rame, oro.

Istituto Fisico della R. Università
Roma, 3 aprile 1935–XIII.

(2) HAHN e MEITNER, «Naturw.», vol. 23, p. 37 (1935).

92 *a*.

RADIOATTIVITÀ PROVOCATA DA BOMBARDAMENTO DI NEUTRONI (*). — X.

E. AMALDI, O. D'AGOSTINO, E. FERMI, B. PONTECORVO, E. SEGRÈ

« Ric. Scientifica », *6* (1), 581–584 (1935).

1. RADIOATTIVITÀ INDOTTA NEL BROMO. – Abbiamo constatato anche noi l'esistenza di un terzo isotopo radioattivo del Br con periodo di circa 36 ore, già rilevato da Kourtchatow, etc. (1), ed abbiamo confermato le loro misure sui raggi β. Abbiamo constatato che questa attività è sensibile, come le altre due del Br, all'azione delle sostanze idrogenate, il che rende molto inverosimile la spiegazione data dai citati Autori che il radioelemento relativo a questa attività sia ottenuto da un isotopo stabile con l'espulsione di un neutrone, poiché un neutrone lento non ha l'energia necessaria per provocare questo processo.

Poiché sembra esclusa, secondo dati gentilmente fornitici da Aston, l'esistenza di un terzo isotopo stabile del Br, la natura di questa terza attività presenta un interessante problema, probabilmente simile a quello dell'indio (2,3).

2. INFLUENZA DELLA TEMPERATURA. – Moon e Tillman (4) hanno recentemente osservato una differenza di attivazione nell'argento, rodio e iodio rallentando i neutroni con paraffina a temperatura ambiente o alla temperatura dell'aria liquida. Alcuni mesi or sono avevamo anche noi tentato, senza successo (5), di mettere in evidenza un simile effetto di temperatura. Ripetendo ora l'esperienza nelle condizioni gentilmente comunicateci da Moon e Tillman e analoghe a quelle da loro descritte, abbiamo potuto constatare anche noi un aumento di intensità nell'attivazione di vari elementi col calare della temperatura della paraffina. Abbiamo trovato aumenti tra il 10 e il 20 % nell'attivazione di Ag ($2{,}3^{m}$), V, Al ($2{,}3^{m}$): nel rodio invece non abbiamo trovato alcun aumento apprezzabile di attività.

(*) Cfr. « Ricerca Scientifica », *5* (1), pp. 283, 330, 452, 652 (1934); *5* (2), pp. 21, 381, 467 (1934); *6* (1), pp. 123, 435 (1935).

(1) KOURTCHATOW, MYSSOWSKY, ROUSSINOW, « C. R. », *200*, 1201 (1935).

(2) SZILARD e CHALMERS, « Nature », *135*, 98 (1935).

(3) « Proc. Roy. Soc. », *149*, 592 (1935).

(4) MOON e TILLMAN, « Nature », *135*, 904 (1935).

(5) « Proc. Roy. Soc. », *149*, 592 (1935); « Ricerca Scientifica », *5* (2), 467 (1934).

Se queste differenze di comportamento di vari elementi, osservate anche da Moon e Tillman, saranno confermate, dovrà ritenersi che il processo di cattura dei neutroni lenti sia più complicato di quanto finora si potrebbe ritenere basandosi sulla legge di forza comunemente ammessa tra neutroni e nuclei (buca di potenziale delle dimensioni del nucleo).

Le differenze si potrebbero, forse, interpretare ammettendo che esista anche tra neutrone e nucleo una forza debolissima con raggio di azione relativamente grande.

3. Trascinamento meccanico dei neutroni lenti. – Per confermare con un metodo indipendente i risultati delle esperienze di Moon e Tillman abbiamo cercato di determinare con una esperienza meccanica per quanto tempo i neutroni si trattengono nella paraffina.

Una ruota d'acciaio di 80 cm di diametro porta sul suo contorno una sorgente di Em + Be e accanto a questa a circa 5 cm di distanza, rispettivamente a destra e a sinistra, due rivelatori eguali dei neutroni lenti. I rivelatori erano in alcune esperienze di manganese in altre di vanadio.

La ruota veniva fatta girare tra due grossi anelli di paraffina di sezione rettangolare (10×20 cm circa) disposti con gli assi coincidenti con l'asse della ruota uno posteriormente, l'altro anteriormente al piano della ruota, e ad una distanza di circa 6 cm dal piano stesso: il peso degli anelli era complessivamente di circa un quintale.

Si faceva girare la ruota con una velocità angolare di circa 1600 giri al minuto per un tempo conveniente a seconda del rivelatore usato e si misurava quindi il rapporto tra le attività acquistate dai due rivelatori. Per la misura della attività i rivelatori venivano staccati dalla ruota: dopo questa prima misura si rimettevano i rivelatori nelle stesse condizioni, si faceva girare la ruota in senso inverso, e si rimisurava il rapporto tra le attività dei due rivelatori.

Il risultato è che il rivelatore che segue, nel senso del moto, la sorgente, acquista una attività circa del 10 % superiore a quella presentata dal rivelatore che precede la sorgente.

Questo risultato mostra anzitutto che i neutroni raggiungono veramente velocità dell'ordine di quelle dell'agitazione termica. Si può precisare inoltre il significato di questa esperienza colla seguente considerazione: la stessa differenza del 10 % tra le attività dei due rivelatori si ottiene spostando, con la ruota ferma, la sorgente dal punto di mezzo tra i due rivelatori, di circa 0,5 cm (questa taratura è stata fatta usando come rivelatore il rodio). Questo spazio è percorso dalla periferia della ruota in $7 \cdot 10^{-5}$ secondi e quindi questo deve essere, approssimativamente, il tempo di indugio dei neutroni lenti nella paraffina. Assumendo che i neutroni lenti abbiano la velocità di agitazione termica (2500 m/sec) e un cammino libero medio [(6)] di circa 0,5 cm, quale risulta da esperienze di diffusione, se ne deduce che un neutrone

(6) « Proc. Roy. Soc. », loc. cit. e Westcott e Bjerge, « Proc. Camb. Phil. Soc. », *31*, 145 (1935).

subisce in media circa 30 urti dopo esser pervenuto alla velocità dell'agitazione termica e prima di essere diffuso fuori dalla paraffina o catturato dai nuclei presenti.

4. Emissione di neutroni da parte di diverse sostanze. – Poiché il rodio in ambiente idrogenato costituisce un rivelatore particolarmente sensibile per i neutroni, abbiamo ricercato in alcuni elementi una emissione di neutroni sotto l'azione delle particelle α.

Sia il silicio che il fosforo, mescolati a 200 mC di emanazione, hanno mostrato di attivare un cilindretto di rodio con una intensità che con i nostri contatori ammontava a circa 100 impulsi iniziali al minuto: l'emissione dei neutroni da parte di questi due elementi è molto debole rispetto alla emissione di neutroni in tutti i casi fino a oggi osservati, come abbiamo constatato con esperimenti eseguiti sul boro, litio, alluminio, ecc. Nel caso del fosforo l'emissione osservata è probabilmente legata alla formazione di cloro radioattivo (7) per positroni. Con molte sostanze si ha una piccola emissione di neutroni, probabilmente dovuta a impurità superficiali; tra le sostanze da noi sperimentate il nichel è quello che dà l'effetto minore.

Abbiamo anche cercato di determinare alcune proprietà dei neutroni emessi dal litio, boro, alluminio, fluoro e magnesio. Per questo abbiamo costruito sorgenti simili a quelle solite, sostituendo il berillio con l'elemento da studiare.

Boro. – Abbiamo misurato, con una sorgente di Em + B, il coefficiente di sensibilità α alle sostanze idrogenate delle attività delle sostanze sottosegnate. Per confronto diamo anche gli α che corrispondono ad una sorgente di Em + Be.

	α_B	α_{Be}
Si (135^s)	1	1
V (225^s)	120	40
Rh (44^s)	50	15
Ag (140^s)	20	15
J (25^m)	10	5

Si vede da questi dati, che però hanno un valore più che altro di indicazione, che i coefficienti α_B risultano maggiori dei coefficienti α_{Be} per un fattore che varia grosso modo da 3 a 1.

È inoltre interessante confrontare gli effetti di radioattività provocata, senza rallentare i neutroni da due sorgenti di Em + B e di Em + Be di eguale forma geometrica e contenenti la stessa quantità di emanazione; irradiando il V in aria si trova un rapporto tra l'attività con la sorgente con Be e l'attività ottenuta con la sorgente con B di circa 12. Tale rapporto, irradiando il Si (135^s) è maggiore di 15. Questo insieme di fatti potrebbe spiegarsi con la seguente ipotesi: nello spettro dei neutroni emessi da una sor-

(7) Frisch, « Nature », *133*, 721 (1934).

gente Em + Be vi è una banda di neutroni molto lenti, all'esistenza della quale sono dovuti i valori relativamente bassi dei coefficienti α e una banda di neutroni molto veloci, che sono quelli già noti per altre esperienze, responsabile della relativamente intensa attivazione del Si; nello spettro di una sorgente Em + B per contro mancherebbero i neutroni molto lenti e quelli molto veloci. Considerazioni di questo genere possono essere d'altra parte suggerite anche dai risultati della seguente esperienza: abbiamo costruito la curva di attivazione in funzione della distanza dalla sorgente con una sorgente di Em + B; il rivelatore era di Rh, la irradiazione avveniva in acqua in una vasca di 40 × 30 × 80 cm e le misure erano fatte con una camera di jonizzazione: si trova una curva che ha una forma leggermente diversa da quella dell'analoga curva ottenuta con una sorgente Em + Be [(8)]; precisamente per piccole distanze la attivazione con Em + B decresce più lentamente di quella con Em + Be; per distanze maggiori di 10 cm. le due curve hanno la stessa forma.

Abbiamo infine misurato, col dispositivo descritto in una precedente lettera, il coefficiente di assorbimento dei neutroni lenti nel Cl, usando una sorgente di Em + B : si trova lo stesso valore già ottenuto con una sorgente di Em + Be (prima riduzione a metà in 0,3 gr/cm^2).

Fluoro. – Una sorgente di circa 300 mC di CaF_2 + Em non attiva sensibilmente il Si, mentre si riusciva ad osservare una notevole attivazione del V e del Rh soltanto se l'irradiazione avveniva in acqua.

Magnesio. – Neutroni di una sorgente di Mg + Em di circa 600 mC attivano il P ($2{,}4^h$), il Si (135^s), l'Al (10^m), il Rh (44^s) e il V (225^s). I coefficienti di sensibilità α_{Mg} sono notevolmente più grandi dei coefficienti α_{Be} per Rh (44^s) e V (225^s).

Alluminio. – Neutroni di una sorgente Al + Em di circa 600 mC non attivano il Si in modo apprezzabile ma attivano il Rh (44^s). Il coefficiente α_{Al} per il Rh (44^s) è maggiore del corrispondente α_{Be}.

Una conoscenza più precisa ed estesa di dati simili a quelli raccolti in questo paragrafo fornirà utili indicazioni, da un lato sullo spettro di velocità dei neutroni emessi dalle varie sostanze bombardate con particelle α, dall'altro, sulla energia necessaria per l'attivazione dei vari elementi a mezzo di neutroni.

5. Esperimenti sulla produzione di neutroni a mezzo di raggi γ. – Abbiamo confrontato le attività provocate nel rodio in paraffina usando come sorgenti MsTh + Be e MsTh + D_2O (questa sorgente contiene circa 35 mC di RdTh); gli esperimenti sono stati fatti usando 25 grammi di berillio e 25 grammi di D_2O (98 °/₀) disposti nelle stesse condizioni geometriche: in tali condizioni abbiamo trovato che il numero dei neutroni emessi dal D_2O sotto l'azione dei raggi γ del MsTh è circa 3,5 volte maggiore di quelli emessi dal berillio. Abbiamo eseguito le stesse misure sostituendo il preparato di MsTh con uno di 100 mgr di Ra: nel caso del berillio, nelle nostre condizioni

(8) « Proc. Roy. Soc. », l. c.

geometriche, si ottenevano usando come rivelatore il rodio, circa 200 impulsi al minuto iniziali, mentre con il D_2O non si riusciva ad avere una attività apprezzabile. Questo insieme di fatti si interpreta facilmente qualora si tenga presente che il neutrone è legato più debolmente nel berillio che nel deuterio. Notiamo infine che il numero di neutroni emessi dal berillio è circa lo stesso usando sorgenti equivalenti di mesotorio e radio.

Sempre con una sorgente MsTh + D_2O e MsTh + Be abbiamo rilevato le curve di attivazione in funzione della distanza dalla sorgente. L'irradiazione era fatta in una vasca di acqua di dimensioni $40 \times 30 \times 80$ cm, il rivelatore era un cilindretto di rodio (misure fatte con il contatore).

Le due curve trovate nel caso del Be e del D_2O hanno sostanzialmente la stessa forma. Nella seguente tabella è l'attività (numero di impulsi) corrispondente alle diverse distanze nel caso per esempio che la sorgente sia MsTh + D_2O.

Distanza cm	Attivazione
4	1100 ± 50
6	962 ± 40
9	388 ± 20
12	188 ± 14
14	107 ± 7

I primi numeri indicano in cm le distanze misurate dal rivelatore all'asse del tubetto contenente il MsTh.

Istituto Fisico della R. Università.
Roma, 14 giugno 1935–XIII.

84 *b*.

RADIOACTIVITY INDUCED BY NEUTRON BOMBARDMENT.—I.

Translated from « Ric. Scientifica », 5 (1), 283 (1934) (*).

In this letter I want to report on several experiments undertaken to determine whether a bombardment with neutrons will produce phenomena of induced radioactivity similar to those observed by M. and Mme. Joliot when the bombardment was done with α–particles.

I used the following apparatus: The source of neutrons was a small glass tube containing beryllium powder and emanation. Using about 50 millicurie of emanation (which was given to me by Professor G. C. Trabacchi, to whom I extend here my cordial thanks), I could obtain more than 100,000 neutrons per second, mixed, of course, with a very intense γ–radiation; however, the latter does not influence experiments of this kind. Small cylindrical containers filled with the substances tested were subjected to the action of the radiation from this source during intervals of time varying from several minutes to several hours.

Immediately after being irradiated, the targets were placed in the vicinity of a Geiger-Müller counter, whose wall was formed of aluminum sheet about 0.2 mm thick, allowing β–rays to enter the counter. Positive results have been obtained, so far, with the two following elements:

Aluminum. – A small aluminum cylinder, irradiated by neutrons for about two hours, gives rise, in the first few minutes after the end of the irradiations, to a considerable increase in the rate of pulses from the counter, the rate increasing by about 30–40 pulses per minute. A decrease follows, the rate reducing to half of its initial value in about 12 minutes.

Fluorine. – Calcium fluoride, irradiated for a few minutes and rapidly brought into the vicinity of the counter, causes in the first few moments an increase of pulses; the effect decreases rapidly, reaching the half-value in about 10 seconds.

These phenomena can possibly be explained in the following way. Fluorine under neutron bombardment disintegrates with the emission of an α–particle, the probable nuclear reaction being:

$$F^{19} + n^1 \rightarrow N^{16} + He^4 .$$

The isotope N^{16} may then, by emitting a β–ray, transmute into O^{16}. A similar interpretation can be given to the case of aluminum, the possible nuclear reaction being:

$$Al^{27} + n^1 \rightarrow Na^{24} + He^4 .$$

(*) The translations of papers N° 84 *b* to 92 *b* were issued by John Crerar Library. The Editors have made some changes in the original translations.

The atom Na^{24} must be a new radioactive isotope, which, through the emission of a β–particle, transforms into Ca^{24}.

If these interpretations are correct, we have here an artificial formation of radioactive elements emitting ordinary β–particles, in contradistinction to the substances discovered by Joliot, which emit positrons. In the case of nitrogen, we would have two radioactive isotopes: N^{13}, found by Joliot, which transforms into C^{13} by positron emission, and N^{16}, which, emitting an electron, transmutes into O^{16}.

Experiments are in progress, extending the investigation to other elements, and studying the details of the phenomenon.

Rome, March 25, 1934–XII.

85 *b*.

RADIOACTIVITY PRODUCED BY NEUTRON BOMBARDMENT.—II.

Translated from « Ric. Scientifica », 5 (1), 330–331 (1934).

Continuing the experiments with the method described in the preceding letter, radioactivities were studied which are produced by neutron bombardment from a preparation of emanation + beryllium, on numerous other elements.

Iron. – Period: about 2 hours. With this element a chemical separation could be obtained. The active product formed is probably manganese, because it gave all the characteristic reactions of this metal.

Silicon. – A very strong effect was observed. Period: about 3 minutes. The electrons produced were photographed in a Wilson chamber. By using the pure metallic element or quartz, the same mean life is obtained.

Phosphorus. – Period: about 3 hours. Strong effect. The electrons have been photographed in a Wilson chamber. The chemical separation of the active product formed exhibits the characteristics of silicon.

Chlorine. – Gives the effect with the longest period of all elements investigated so far.

Vanadium. – Period: 5 minutes.

Aluminum. – In order to confirm the results of the preceding letter, the electrons have been photographed in a Wilson chamber.

Copper. – No strong effect. Period: about 6 minutes.

Arsenic. – Period: about 2 days.

Silver. – Strong effect. Period: about 2 minutes.

Tellurium. – Period: about 1 hour.

Iodine. – Strong effect. Period: about 30 minutes.

Chromium. – Strong effect. Period: about 6 minutes. The electrons have been photographed in a Wilson chamber.

Barium. – Small effect. Period: about 2 minutes.

Indications of effects of different intensities and periods were given by Na, Mg, Ti, Zn, Zr, Se, Sb, Br, La.

Some elements seem to possess two or more periods which may be due to different isotopes, or to consecutive radioactive transformations. The experiments are being continued in order to determine better these results and to extend the investigation to other elements.

The nuclear reaction which determines these phenomena can be different in different cases. The chemical separations in the case of iron and phosphorus appear to indicate, at least in these two cases, that the neutron is absorbed and one proton is emitted. The unstable product, manganese or silicon, emitting a β–particle, transforms into the original element.

The chemical separations were carried out by Dr. O. D'Agostino; for the physical part, Dr. E. Amaldi and Dr. E. Segrè have collaborated in this work.

Rome, *Istituto Fisico dell'Università*.

86 *b*.

BETA-RADIOACTIVITY PRODUCED BY NEUTRON BOMBARDMENT.—III.

Translated from E. AMALDI, O. D'AGOSTINO, E. FERMI, F. RASETTI, E. SEGRÈ
« Ric. Scientifica », *5* (1), 452–453 (1934).

The experiments reported in the preceding letters [1] have been continued and extended. We communicate the results obtained:

Hydrogen-Carbon-Nitrogen-Oxygen do not give appreciable effects. Paraffin irradiated in the usual manner for 15 hours with a source of 220 mC, water irradiated for 14 hours with 670 mC and guanidine carbonate irradiated for 14 hours with 500 mC have been examined.

Fluorine. – The period of fluorine is remarkably smaller than previously indicated, that is, a few seconds.

Magnesium. – Magnesium has two periods: one of about 40 seconds, the other longer.

Aluminum. – Besides the period of 12 minutes previously announced there is also another period about one day long. The activity corresponding to the second period has the characteristic chemical reactions of sodium. Probably it is due to Na^{24}.

Sulphur. – The period of sulphur is very long, certainly many days. The activity is distinguished by the characteristic reaction of phosphorus.

Chlorine. – Cl behaves like S. Here also an active substance can be separated; probably it is P^{32}, identical to that obtained from S.

Manganese. –Mn has a small effect with a period of approximately 15 minutes.

Cobalt. – Co has an effect of 2 hours. The active substance behaves like Mn. The period and chemical behavior indicate that we have a Mn^{56}, identical with that which is obtained by irradiating Fe.

Zinc. – Zn has two periods: one 6 minutes, the other much longer.

Gallium. – Ga has a period of 30 minutes.

Bromine. – Br has two periods: one 30 minutes, the other 6 hours. The activity corresponding to the longer period and probably the other activity also, show the chemical reactions of Br.

Palladiun. – Period of several hours.

Iodine. – Period of 30 minutes. The activity shows the chemical reactions of iodine.

Praseodymium. – Has two periods: one 5 minutes, the other longer.

Neodymium. – Period 55 minutes.

Samarium. – Has two periods: one 40 minutes, the other longer.

(1) FERMI, « Ric. Scientifica », *5* (1), pp. 283, 330 (1934).

Gold. – Period of 1 or 2 days.

Uranium. – Uranium was freed from β–active products and subsequently irradiated. It gives a very intense effect with several periods: one period of about 1 minute, another of 13 minutes besides longer periods not yet exactly determined. The β–rays corresponding to the 13 minute period are considerably penetrating.

Using chemical operations, an attempt was made to determine whether the element which disintegrates with the period of 13 minutes is an isotope of one of the heavier elements. We can immediatly conclude that it is not an isotope of U (92) or Th (90) because the active substance is easily separated from these two elements. Much more difficult is the proof that it is not an isotope of Pa (91), since there is not a β–radioactive substance suitable for this proof, i.e., an isotope of Pa of convenient period. For this reason we used UX_2, which has a period of only 70 seconds and which therefore forces to speed up the chemical operations. However, in about 2 minutes we succeeded in separating (together with Mn) the active substance from U by carrying along only a small fraction of the UX_2, which should have been present taking into account the decay of this substance. This separation was accomplished by precipitating MnO_2 with $NaClO_3$ from a strong nitric acid solution, and is equally complete even in the presence of Ce and La. This fact seems to exclude the possibility that the active substance has atomic number 89, since this element is analogous to the rare earth elements. The same reaction also appears to exclude the possibility that we deal with an isotope of Ra (88) and all the more of eka-cesium (87) or emanation (86).

This complex of conclusions which we are trying to corroborate by means of further experiments gives rise to the spontaneous hypothesis that the active substance of U might have atomic number 93 (homologous with rhenium); the process in this hypothesis could consist in the capturing of the neutron by U with the formation of U^{239}, which subsequently would undergo β–disintegration.

We also tried to find out whether some disintegration, which was observed in different elements, was accompanied by γ–ray production. The experiment was carried out by trying to determine whether the substances acted upon a counter through 2 mm of Pb.

A positive effect was given by: Al (both periods), Si, Fe, Zn, As, U.

A negative effect was observed with: Mg, S, Ag, Br, I.

A magnetic device similar to that described by Thibaud [(2)] was constructed in order to determine the charge of the particles produced. For intensity reasons, only the following elements have been studied up to now: Al, Si, P, S, Ga, Cr, Fe, Ag, Br, I. In all cases only negative electrons could be observed.

Istituto Fisico della R. Università.
Rome, May 10, 1934–XII.

(2) J. THIBAUD, « C. R. », *197*, 447 (1933).

87 *b.*

RADIOACTIVITY PRODUCED BY NEUTRON BOMBARDMENT.—IV.

Translated form E. AMALDI, O. D'AGOSTINO, E. FERMI, F. RASETTI, E. SEGRÈ
« Ric. Scientifica », 5 (1), 652–653 (1934).

In order to supplement the results previously communicated [1], additional observations have been made and we report them here:

Fluorine. – We redetermined the period of F as 9 seconds. The electrons emitted are very penetrating and are reduced to one-half in 0.32 gr/cm^2 of aluminum. The absorption measurements reported below were also performed with aluminum absorbers. The emission of β–rays of F is probably accompanied by γ–rays, because 2 mm of Pb do not completely absorb the radiation emitted.

Magnesium. – In addition to the 40-second period, we also have measured a long period, which is about 15 hours and is accompanied by γ-emission. The electrons corresponding to this period are reduced to one-half by 0.07 gr/cm^2 of absorber (half-value thickness). The active substance behaves like sodium and can probably be identified with the Na^{24} extracted from Al, because it has approximately the same period and the same absorption curve for the electrons. The corresponding nuclear reactions could be the following:

$$_{12}Mg^{24} + {}_0n^1 = {}_{11}Na^{24} + {}_1H^1$$

$$_{13}Al^{27} + {}_0n^1 = {}_{11}Na^{24} + {}_2He^4 .$$

Aluminum. – The half-value thickness of the activity corresponding to the 12-minute period is about 0.09 gr/cm^2. The active substance of the short period has not yet been extracted chemically, but probably it is $_{12}Mg^{27}$.

Silicon. – The half-value thickness for the β–rays is 0.2 gr/cm^2. The activity is chemically separated together with Al and it seems probable that the nuclear reaction is:

$$_{14}Si^{28} + {}_0n^1 = {}_{13}Al^{28} + {}_1H^1 .$$

This would confirm Joliots' and Preiswerk's hypothesis that the same $_{13}Al^{28}$ is formed, which was obtained by the Joliots bombarding Mg with α–particles.

Phosphorus. – We confirmed the Joliots' result concerning the existence of a 3-minute period. The half-value thickness of the β–rays corresponding to the 3–hour period is 0.11 gr/cm^2.

(1) « Ric. Scientifica », 5 (1), pp. 283, 330, 452 (1934).

Sulphur and *Chlorine.* – The half-value thickness is 0.12 gr/cm². The identity of the absorption curves for the radiations emitted confirms that in both cases the production of $_{15}P^{32}$ is involved.

Chromium. – The half-value thickness is 0.2 gr/cm² and also γ–radiations are emitted. The identity in the periods observed suggests that V and Cr produce the same radioactive isotope, V^{52}.

Iron. – The half-value thickness is 0.2 gr/cm².

Zinc. – The active substance of a 6-minute period has been separated, and chemically it follows copper. The equality of the periods suggests that the active substance is identical with that of Cu.

Arsenic. – The half-value thickness of the electrons is 0.2 gr/cm².

Molybdenum. – Has at least two periods: one of about a quarter of an hour, the other much longer. The effect is of low intensity.

Rhodium. – It is very active. We found two periods: one of about 2 minutes, the other much longer. The half-value thickness for the short period is about 0.12 gr/cm².

Palladium. – The half-value thickness is about 0.03 gr/cm²

Silver. – Besides showing a period of 2 minutes, we found a second period of the order of twenty seconds. The half-value thickness is 0.1 gr/cm².

Iodine. – The half-value thickness is about 0.11 gr/cm².

Tungsten. – Has a very small effect with a period of about one hour. This effect could be ascribed to the presence of impurities.

Iridium. – Has a large effect. The period is about 20 hours. The half-value thickness is 0.17 gr/cm².

Gold. – The half-value thickness is 0.03 gr/cm².

Mercury. – Very small effect with a period of a few dozen minutes, probably due to impurities.

The following elements were irradiated for several hours with one-half Curie sources, and they did not show any appreciable effect:

Li, Os, Pb, Bi, Ru, Tl.

The cross-sections for the most active elements are of the order of magnitude 10^{-24}–10^{-25} cm². Of course the neutrons we used have a continuous energy distribution which is poorly known.

However, in order to orient ourselves on the effect, which the neutron energy has on the cross-sections, we performed a preliminary experiment, activating F, Ag, I, and U with neutrons coming from a source of Em+Be in the time interval in which the emanation was not in equilibrium with its α–emitting products, especially with RaC.

By comparing the activity before and after equilibrium has been reached it seems that we may conclude that the faster neutrons, at least in this small energy interval, do not have a much larger cross-section than the slower ones. An increase in activity appreciably larger than for other elements was observed for iodine.

Istituto Fisico della R. Università.
Rome, June 23, 1934-XII.

88 *b*.

RADIOACTIVITY PRODUCED BY NEUTRON BOMBARDMENT.—V.

Translated from E. AMALDI, O. D'AGOSTINO, E. FERMI, F. RASETTI, E. SEGRÈ
« Ric. Scientifica », 5 (2), 21-22 (1934).

We report some further observations which supplement the results communicated in preceding letters:

Sodium. – The half-life of this substance is about 40 seconds, and the intensity is average. From chemical separations of fluorine from activated sodium it appears that the active substance is not fluorine.

Vanadium. – *Chromium*. – *Manganese*. – For all three elements we have determined a period of 4 minutes with more accurate measurements than those carried out previously for vanadium and chromium. Chemical separations have shown that the active substance of chromium and manganese is vanadium. It is very likely that V^{52} is involved, which also could be obtained by bombarding V^{51}. For manganese, besides the period of 4 minutes, another of about 150 minutes was observed, the active substance of which cannot be separated from manganese and is probably Mn^{56}, which also is obtained from iron and cobalt.

Arsenic. – A chemical separation of the active substance in presence of gallium and germanium enables us to exclude the possibility that it is gallium and makes it very unlikely that it is germanium. The most probable hypothesis is that the activity is due to As^{76}.

Rubidium. – Has a very small effect with a period of about 20 minutes.

Rhodium. – We redetermined the periods of rhodium and found 50 seconds and 5 minutes to be the most probable values.

Cadmium. – Has a very small effect with a period of one hour.

Iridium. – A chemical separation of the active substance in the presence of osmium and rhenium seems to exclude the possibility that the activity is due to these elements. Probably it is due to iridium isotopes.

Gold. – A chemical separation of the active substance in the presence of iridium, platinum, and mercury leads us to believe that an isotope of gold is involved.

Thorium. – From preliminary experiments on thorium, separated from its β–radioactive products, a strong activity results with at least two periods: one period of less than one minute and another of about 15 minutes.

Uranium. – Besides the known periods of 10 seconds, 40 seconds, and 13 minutes, one of the longer periods of uranium has been determined: it is about one hour and a half.

We tried to determine with higher precision the extraction yield of the 13-minute activity using the manganese dioxide reaction, which, however, carries along some of the element 91 (UX_2). On varying the precipitation conditions it appears that the yield for the two elements is completely different. For the 13-minute activity it is possible to maintain the yield practically constant: about 15 per cent of the total activity, while that for UX_2 varies between 1 per cent and 7 per cent.

The reaction for the separation of rhenium sulphide in a strong acid solution, by using sodium thiosulphate, was repeated. It has been found that in this way it is possible to separate a very high amount of the 13-minute and 100-minute activity. These reactions, besides giving a yield of about 50 per cent, are so specific with respect to UX_1 and UX_2 that they also allow the separation of the two activities from unpurified uranium, in equilibrium with its products. These two reactions appear to confirm the hypothesis that we have elements of atomic number higher than 92.

We also determined the approximate number of atoms of the active element with the 13-minute period which disintegrate each minute and which were formed irradiating to equilibrium one gram of uranium at a distance of 1 cm from a source of 1 mC of radon + beryllium. The number was found to be of the order of magnitude of 50.

Istituto Fisico della R. Università.
Rome, July 12, 1934–XII.

89 *b*.

RADIOACTIVITY PRODUCED BY NEUTRON BOMBARDMENT.—VII.

Translated from E. AMALDI, O. D'AGOSTINO, E. FERMI, B. PONTECORVO, F. RASETTI, E. SEGRÈ
« Ric. Scientifica », *5* (2), 467–470 (1934).

1. NEW ACTIVITIES. – We have determined the half-lives of the substances produced by the irradiation of the following elements:

Potassium. – It shows a weak activity with periods of 16 hours (*a*) [1]. The chemical separation of Cl, A and Ca does not remove the activity from K. The activity is probably due to a new radioactive isotope of potassium.

Gallium. – Besides the short period previously reported it also has another one of one day (*a*).

Cadmium. – Cadmium which has been irradiated for a long time in water has a very weak activity corresponding to a period of about two days. However, depending on the conditions of irradiation, activities with different periods are produced.

Indium. – The short period is 13 seconds (*a*).

Antimony. – Besides a short and weak period it also has an intensive one of 2.5 days (*a*).

Praseodymium. – Besides the period of 5 minutes it also has an intensive one of one day (*a*).

Some of the previously quoted periods of the following substances were remeasured with the electrometer, and the results obtained are rather precise: Mn–150 min; As–26 hrs; Rh–44 sec.

2. SENSITIVITY TO HYDROGENOUS SUBSTANCES. – In order to begin to determine quantitatively the effect of hydrogenous substances on the activation, we irradiated some substances by putting the sources inside a cylinder of the same diameter as the counter so as to obtain the maximum possible activation. First the irradiation was carried out in a paraffin block of very large dimensions and then in air. The ratio of the activities observed gives, roughly, a coefficient of sensitivity α, to the action resulting from the hydrogenous substances. This coefficient is one for those substances which do not show any increase in activity when irradiated in paraffin. We find for: Na (40 sec) $\alpha = 1$; Mg (40 sec) $\alpha = 1$; Al (12 min) $\alpha = 1$; Al (2.5 min) $\alpha =$ very high; Si (2.5 min) $\alpha = 1$; Cl (45 min) $\alpha =$ very high; V (4 min)

(1) The letter (*a*) means that the period in question has a high sensitivity coefficient towards hydrogenous substances (see N. 2 of this paper).

$\alpha = 40$; Cu (6 min) $\alpha = 15$; Rh (44 sec) $\alpha = 15$; Ag (2 min) $\alpha = 15$; Ag (20 sec) $\alpha = 30$; In (13 sec) $\alpha = 12$; U (10 sec, 13 min, 90 min) $\alpha = 1.6$.

These figures should be considered only as an indication, because small variations in the geometrical conditions of irradiation have an important influence on the result.

3. SEPARATION OF RADIOACTIVE ISOTOPES. – Using a variety of methods, we separated the radioactive isotopes of the halogens, arsenic, and manganese from the ordinary isotopes. Szilard and Chalmers have succeeded in enriching iodine and bromine in radioactive isotopes, extracting them from irradiated methyl iodide and from bromoform, respectively. They took advantage of the fact that neutron bombardment can decompose molecules and ions and leave them in conditions differing from those which prevailed before the bombardment [(2)].

We repeated their experiments extending them also to chlorine (45 min) which was irradiated in different organic compounds. Subsequently it was recognized that the same effect can be obtained with inorganic compounds, provided the element is present as a complex which cannot form spontaneously. For instance, we irradiated alkaline iodates, bromates, and chlorates both in solutions and in the solid state. In subsequent precipitations of *only traces* of silver chloride, bromide, or iodide, the precipitate exhibited about 25–50 per cent of the total activity.

In an analogous manner, irradiating cacodylic acid we succeeded in separating a good deal of arsenic activity, by precipitating traces of arsenic sulphide from the solution. Irradiating potassium permanganate, about 80 per cent of the activity can be obtained by precipitating a trace of manganous carbonate.

Radioactive iodine could be obtained from gaseous methyl– or ethyl–iodide by collecting ions from the gas with the help of an electric field. Nickel electrodes, which are not activated, were placed in the gas. However, the yield was not very high.

4. VELOCITY OF THE NEUTRONS IN PARAFFIN. – Assuming that the action of the hydrogenous substances on radioactivity induced by neutron bombardment is due to the slowing-down of the neutrons, we tried to get an idea about the velocity of the neutrons, which by convention we shall call "slow." Since the velocity of the neutrons which have suffered about 20 successive proton collisions can be reduced to a value of the order of that of thermal agitation, it could be expected that the temperature acts on the density of the neutron gas. In order to check this point, we carried out the following experiment: A rhodium plate was first activated in a mixture of hydrocarbons (pentane and benzene) at room temperature; then it was activated at 200° in another hydrocarbon mixture (paraffin) which, at this temperature, had the same density and composition as the cold mixture.

(2) «Nature», *134*, 462 (1934).

In both experiments the Rh-plate was irradiated using the same geometrical arrangement (one cm from the liquid surface and two cm from the source).

No difference in the activation was observed, and hence it seems we may conclude that the energy of the " slow " neutrons is larger than the energy of thermal agitation.

5. ABSORPTION OF THE " SLOW " NEUTRONS. – Continuing the measurements of absorption of " slow " neutrons, we found other substances which have abnormally large collision cross-sections. If 10^{-24} cm² is taken as unity, the effective cross-sections for the following elements are: Cl, 130; Co, 100; Ag, 100; Cd, 10,000; In, 450; I, 40; Pr, 100; Au, 100; Hg, 1,000.

These effective cross-sections correspond to absorptions measured by comparing the activity produced in a Rh detector, located in a cavity in paraffin, with the activity induced in the same detector located in the same cavity but protected by two layers of the absorbing substance being examined. We remark that if the experiments are carried out outside the paraffin, instead of in a cavity, smaller absorption coefficients are obtained. This fact can be explained, assuming that the activation is due to " slow " neutrons, by observing that their average velocity in a cavity is smaller than outside of the paraffin.

We obtained the absorption curve, for cadmium, of the neutrons " slowed down " by paraffin. It has been found that the transmitted activity on increasing the absorbing thickness does not decrease according to an exponential law. Precisely we can reach absorber thicknesses such that an increase in thickness has little influence on the activity. This fact is paralleled by the measurements we made activating Rh at a fixed distance from the source in a vessel containing a boric acid solution of variable concentration (3). These results can be explained by taking into account the fact that the velocity distribution of the neutrons is continuous.

In some absorption measurements, silver instead of rhodium has been used as a detector, but the results obtained were essentially the same.

In order to study the absorption of neutrons which are not " slowed down " by hydrogenous media, we inserted boric acid with a thickness of 5 gr/cm² between the source and a small rhodium cylinder; the activation is not smaller than the one obtained under identical geometrical conditions without the absorber.

6. EMISSION OF γ–RAYS. – We began experiments to determine whether strong absorption of " slow " neutrons, which is characteristic for some elements, corresponds to the emission of secondary radiations. For this purpose we used Po+Be (about 60 mC) as a source and a Geiger-Müller counter. Both were put in a big paraffin block. In the direct path there was a lead filter 10 cm thick to absorb the γ–rays of Po+Be. The counter was protected by a lead plate two millimeters thick. When using a thin layer of the elements Cd, Hg, Cl, Y, and Ir outside the counter, we noticed an increase

(3) « Ric. Scientifica », *5* (2), 380 (1934).

in the number of impulses up to 100 per minute. Probably this has to do with the emission of γ–rays, which can be interpreted as having an energy corresponding to the capture of " slow " neutrons. We did not observe such radiation nor even a secondary, less penetrating, radiation (α, β, H) in boron, although this element also has a large absorption cross-section.

Experiments are now in progress to clear up this point.

7. ACTIVATION AS A FUNCTION OF DISTANCE. – We studied the variation in the activation of an Rh detector as a function of the distance from the source, when both are immersed in a large water container. The following table gives the results:

Distance in cm.		Activity	
Distance in cm.	1	Activity	100
» » »	2	»	92
» » »	3	»	86
» » »	4	»	76
» » »	5	»	71
» » »	6	»	54
» » »	7	»	46
» » »	8	»	33
» » »	9	»	26
» » »	10	»	22

It can be seen that the activation decreases rather slowly with the distance. By substituting a boric acid solution for water, although we diminish the activation, the curve showing the distance dependence does not show a marked change of shape.

8. SCATTERING OF " SLOW " NEUTRONS. – We carried out a series of experiments to search for elements which scatter " slow " neutrons. The source was put on the axis of a cylindrical block at a distance of 3 cm from the upper base. An Rh–plate was placed on the upper base. The comparison between the activity thus obtained and the one obtained by covering the Rh–plate with different substances gives an indication of the scattering of the " slow " neutrons due to the covering substances.

By covering the Rh–plate with carbon a remarkable increase in activation was observed (a layer of several cm increases the activity by a factor five); a similar result was obtained with a basalt layer. Also, thin layers of light elements (Be, C) produce a remarkable increase in activity.

With the same arrangements, the curve of activation intensity of the Rh–plate was determined, as a function of thickness of an overlying water layer increasing in 1 cm steps. The thickness of one cm of water over the Rh–plate, lying on the paraffin, approximately doubles the activation; for larger thicknesses the increase in activity is smaller.

Analogous results are obtained by covering the Rh–plate with paraffin layers of variable thickness.

We also studied absorption and scattering of some substances with the following device: a neutron source was placed at the center of a cylindrical paraffin block 11 cm high and 13 cm in diameter; a small silver cylinder (detector) was placed a distance of 30 cm from the source. Between the paraffin cylinder and the detector, 6 cm from the detector, layers of different substances were inserted. The geometrical conditions of the experiment were such that the absorber intercepted exactly the rays coming from the paraffin cylinder.

The difference between the activation of silver, with and without the absorber, is due to scattering plus absorption. With a paraffin scatterer it is found that 0.5 gr/cm^2 reduces the activation to one-half, but the absorption curve is not exponential. It appears, therefore, that the mean free path of slow neutrons in paraffin is 7–8 mm.

The experiment made with different thicknesses of water gives essentially similar results.

When using boric acid as the absorbing layer, absorption prevailed over scattering and the results previously indicated were found. Using carbon 5 gr/cm^2 caused a reduction of about one-half; 3.45 gr/cm^2 of lead caused a reduction in the ratio of 0.83 ± 0.07; this value is due to scattering, because it is known that a layer of 3.45 gr/cm^2 of lead absorbs hardly any " slow " neutrons.

9. EFFECT OF NON-HYDROGENOUS SUBSTANCES ON ACTIVATION. - We also tried to determine whether non-hydrogenous substances produce increases in activation as do water and paraffin. With the few substances examined so far (lead, graphite, basalt) we always obtained much smaller effects than, for example, with water.

Lead. - A small Rh cylinder is put into a slot in a block of lead at a distance of 10 cm from the source; the quantity of lead used is such that roughly the center of the line source -Rh detector could be considered at the center of a lead sphere with a diameter of 50 cm.

The activity induced in Rh is about three times higher than that obtained activating Rh under identical geometrical conditions, but in the absence of the lead. The factor, of course, by which the activity is multiplied decreases if the distance of the source from the detector is smaller. An effect due to lead is also found in the activation of silver. However, when silicon is used as a detector, no effect is observed, just as with hydrogenous media. This fact suggests the hypothesis that the phenomenon is of the same nature as that observed in hydrogenous substances; therefore, we tried to determine whether the radiation which activates Rh in a lead medium is considerably absorbed by Cd, by protecting the small Rh cylinder with variable thicknesses of Cd; it was found that the absorption due to the Cd was considerable (reduction of one-half by about 1 gr/cm^2), although less than in a hydrogenous medium. It is of interest to note that in the case of a heavy metal like Pb, it is not possible to explain the effect as a slowing down of the neutrons due to elastic collisions. If the effect is actually due to slow neutrons, we should rather ascribe it to inelastic collisions. The effect of graphite on the activation of Rh is stronger than that of Pb; the activation intensity in silicon does not increase in a graphite medium. The radiation which activates Rh is absorbed by Cd at least to the same extent as in the case of Pb. We had at our disposal only a small amount of graphite; an increase in the amount of graphite probably could considerably increase the observed effects. With basalt the same effects are obtained as those described above.

Systematic research into the effect of different substances on radioactivity induced by neutron bombardment is in progress, since it appears that these results are general, and often incidental increases in activation are produced by surrounding objects. Just for these reasons the measurements described in this paragraph are not very precise.

Istituto Fisico della R. Università.
Rome, December 6, 1934–XIII.

90 *b*.

RADIOACTIVITY PRODUCED BY NEUTRON BOMBARDMENT.—VIII.

Translated from E. AMALDI, O. D'AGOSTINO, E. FERMI, B. PONTECORVO, F. RASETTI, E. SEGRÈ
« Ric. Scientifica », *6* (1), 123–125 (1935).

1. NEW ACTIVITIES. – We have determined the half-lives of the following irradiated elements:

Magnesium. – By irradiating very pure Kahlbaum's MgO, aluminum free, we discovered, besides the activities already known, a weak one with a period of about 10 minutes (*a*). It is probably due to the radioactive isotope Mg^{27} (period: 12 minutes) which is also produced by irradiating Al with neutrons. This case is a new example of neutron-capture.

Silicon. – Irradiated fused quartz has shown a weak activity with a period of several hours. We believe that this is also a case of neutron-capture with the formation of Si^{31} (period: 145 minutes), which is also obtainable from irradiated phosphorus.

Barium. – Besides the short period we found a period of about 80 minutes (a).

Gadolinium. – Gd oxide, kindly given to us by Professor L. Rolla, shows an activity with a period of about 8 hours (*a*).

Tungsten. – Irradiated tungsten shows an activity with a period of about one day (*a*).

2. PRECISE DETERMINATION OF PERIODS. – By using an ionization chamber (see 4, below) we determined with greater precision the half-lives of the following elements: Si, 137 seconds; P, 145 minutes; V, 225 seconds; Ga, 23 hours; Br, 18 minutes and 255 minutes; Ag, 22 seconds and 140 seconds; I, 25 minutes; Pr, 19 hours; Re, 24 hours (besides a shorter period); Ir, 19 hours; Au, 65 hours.

3. SENSITIVITY TO HYDROGENOUS SUBSTANCES. – With the method described in the preceding letter we determined the sensitivity coefficient α towards hydrogenous substances for the following elements: P, 137 seconds; Cr, 225 seconds; Mn, 225 seconds; Fe, 150 minutes; Zn, 6 minutes; Ba, 3 minutes; Pr, 5 minutes; Th, 1 minute and 15 minutes; they all have $\alpha = 1$. Se, 35 minutes, and Ba, 80 minutes, have a very high α.

4. ABSORPTION OF THE ELECTRONS. GAMMA RAYS. – For absorption measurements we used an ionization chamber filled with CO_2 at 3 atm pres-

sure and closed with an Al foil having a thickness of 0.1 mm. The irradiated substances were arranged in flat layers and the absorption was measured by interposing Al foils. For some substances which give intensive γ–radiations (detected by the residual ionization obtained by screening the chamber with 2 mm Pb) we took into account this residual activity in evaluating the absorption coefficient of the β–rays. A control measurement carried out on UX_2 and RaE gave values of the absorption coefficient equal to those ordinarily accepted.

The following table gives the half-value thickness in gr/cm^2 of Al for the substances indicated:

P	145	min	0.15	In	54	min	0.045
V	225	sec	0.17	Sb	2.5	days	0.09
Mn	150	min	0.14	I	25	min	0.10
Ga	30	min	0.17	Pr	19	hrs	0.13
As	26	hr	0.16	Re	24	hrs	0.12
Br	18	min	0.12	Ir	19	hrs	0.12
Br	255	min	0.12	Au	65	hrs	0.04
Rh	44	sec	0.15	U	13	min	0.14

Since many active products are presently at our disposal with higher intensity on account of the increase of the activating power of neutrons in hydrogenous substances, we could detect an emission of γ–rays for some elements (Rh 44 sec, Br 18 min and 255 min) where it escaped us in previous experiments. In addition, the emission of γ–rays was observed in V (225 sec), Mn (150 min), Cu (6 min), Ga (23 hrs), In (54 min), Sb (2.5 days), I (25 min), W (1 day).

5. Chemical separations. – Irradiated metallic antimony has been dissolved in aqua regia. By adding Sn and separating it by Clarke's method, the activity followed antimony. To the same Sb we added and subsequently separated indium, by taking advantage of the different solubilities of the sulphides in diluted acids. We found In inactive. We also added to active Sb, Te and I and separated Te by reduction and I as AgI. They were found to be inactive. The activity of Sb with a period of 2.5 days is due therefore to one of its isotopes.

Irradiated tungstic anhydride was dissolved in sodium hydroxide and inactive tantalum was separated as pentoxide. To this tungstic solution rhenium in nitric acid solution was added, and the tungsten was separated as tungstic anhydride. Rhenium which remained in solution was inactive. Not having Hf at our disposal, we separated inactive Zr hydroxide from an ammonium solution containing active tungstic anhydride. Therefore, it is very likely that the activity is due to an isotope of tungsten.

From irradiated Re dissolved in HNO_3, Ta and W were separated: the former as pentoxide, the latter as tungstic anhydride. Both were found to be inactive. The activities therefore are probably due to isotopes of Re.

6. Action of slow neutrons on boron. – In previous letters we mentioned that:

a) The neutrons slowed down by paraffin or water are abnormally absorbed by B (reduction to one-half by 0.004 gr/cm^2);

b) Boron (as well as Li) differs from several other elements, which have a strong slow-neutron absorption, in that it does not emit γ–rays during neutron absorption. While in previous experiments we could not observe any emission of particles in this specific case, now with the experiments we are going to describe, we could detect α–radiation.

1. – An ionization chamber was filled with gaseous BF_3. By placing it near a source of 50 mC of Po+Be and by surrounding it with paraffin to slow down the neutrons, a very intense ionization could be observed. The value of such ionization is reduced to about one-half if the chamber is screened with Cd in a layer of 0.02 gr/cm^2 thickness; this indicates that the effect is specifically due to slow neutrons.

Another confirmation of this interpretation results from the fact that Po+Be without paraffin does not give any effect comparable to the preceding one.

2. – The same effect, but in smaller proportions, can be observed with an ionization chamber filled with air and containing boron at the bottom of the chamber. By varying the pressure in the chamber and determining the difference between the ionization with and without a Cd absorber (that is, the fraction of the ionization due to slow neutrons), one can determine at which pressure the range of the ionizing particles, which are emitted by boron under these conditions, becomes comparable to the dimensions of the ionization chamber. The range was found to be 0.7 or 0.8 cm.

3. – The same phenomenon was observed using a proportional amplifier; this radiation probably escaped us in our preliminary experiments because of its small range. The measurements with the proportional amplifier agree with the preceding measurements and give a range of about 7 or 8 mm. These observations allowed also the measurement of the maximum value of the total ionization produced in a single process. This ionization (40,000 ion pairs) corresponds to the ionization produced by a single α–particle having a range of 0.8 cm; since this value agrees with that observed, we conclude that the radiation is due to α–particles of about $14 \cdot 10^5$ eV.

From this fact and from the fact that no induced radioactivity is observed in boron we believe that the most plausible nuclear reaction is the following:

$$_5B^{10} + {}_0n^1 = {}_3Li^7 + {}_2He^4 \,.$$

Taking into account also the recoil energy of Li, the reaction energy is 22×10^5 eV.

If we neglect the small amount of energy of the primary neutron, on the basis of the mass defects, the calculated reaction energy should be 2×10^6 eV higher.

Li irradiated with slow neutrons emits also heavy particles of several cm range. Further experiments are in progress with regard to this reaction.

7. Effect of non-hydrogenous substances on activation. – We have been studying the effect of silicon and iron on activation induced by

neutron bombardment. At a distance of 10 cm from the source, Rh was activated, first in air and then in a box containing silicon dioxide. The activity of Rh is multiplied by a factor of about 2.5 in a SiO_2 medium. If silicon is activated, no increase in activity due to the SiO_2 medium is obtained.

However, we noticed that surrounding the source and detector with large masses of iron produces no noticeable increase in the activation in either Rh or Si.

8. NEUTRONS PRODUCED BY γ–RAYS + Be. – L. Szilard and T. A. Chalmers [1] and L. Meitner [2] have observed that by using Be irradiated with γ–rays as a neutron source the only elements activated are those whose activity is due to isotopes of the target. Since neutrons slowed down in paraffin, as far as it is known at the present time, have the same property, we tried to find out whether Szilard and Chalmers' neutrons are noticeably absorbed by an element, like Cd, which shows a very high absorption cross-section for slow neutrons.

For this experiment we used as a neutron source 100 mgr of Ra surrounded by Be; the activation of a Rh cylinder did not show any appreciable decrease, when the source was screened by 0.5 gr/cm^2 of Cd. This can be explained assuming that Szilard and Chalmers' neutrons are much faster than those slowed down in paraffin.

Istituto Fisico della R. Università.
Rome, January 11, 1935–XIII.

(1) « Nature », *134*, 494 (1934).
(2) « Naturwissenschaften », *22*, 759 (1934).

91 *b*.

RADIOACTIVITY INDUCED BY NEUTRON BOMBARDMENT(*).—IX.

Translated from E. AMALDI, O. D'AGOSTINO, E. FERMI, B. PONTECORVO, E. SEGRÈ
« Ric. Scientifica », *6* (1), 435–437 (1935).

1. PRECISE DETERMINATION OF PERIODS. – We determined the half-life of some substances activated by neutron bombardment using an ionisation chamber connected with an electrometer; the measurements are rather precise:

Al, 137 seconds and 10 minutes; S, 14 days; Cl, 37 minutes; Cu, 5 minutes; Zn, 10 hours; Ga, 20 minutes; Rh, 3.9 minutes; Sb, 2.5 days; Th, 24 minutes.

2. SENSITIVITY TO HYDROGENOUS SUBSTANCES. – With the method previously described we determined the sensitivity coefficient to hydrogenous substances for the following elements:

F (9 seconds and 45 seconds) $\alpha = 1$, Mn (150 minutes) $\alpha = 23$, Zn (10 hours) $\alpha = 1$, Ga (20 minutes) $\alpha = 2.5$, As (26 hours) $\alpha = 6$, Se (35 minutes) $\alpha = 5$, Br (18 minutes) $\alpha = 8$, Rh (3.9 minutes) $\alpha = 25$, Sb (2.5 days) $\alpha = 4$, I (25 minutes) $\alpha = 5$, Ba (80 minutes) $\alpha = 9$.

3. ABSORPTION OF EJECTED ELECTRONS. – With the method previously described we measured the half–value thickness in gr/cm^2 of Al for the electrons emitted by the following substances:

Si (137 seconds), 0.19; S (14 days), 0.10; Ag (140 second), 0.08; I (25 minutes), 0.11.

4. CHEMICAL SEPARATIONS. – *Phosphorus*. – The activity with a period of 137 seconds was separated together with Al. An irradiated phosphoric acid solution was neutralized with sodium carbonate; then a trace of aluminum chloride was added. Aluminum phosphate was separated and it carried along all the activity.

Chlorine. – Irradiated hydrochloric acid was neutralized with sodium carbonate. Nitric acid was added and the solution was boiled. Sulphuric and phosphoric acids were added and we precipitated barium sulphate and ammonium magnesium phosphate. Both were inactive. The remaining

(*) « Ric. Scientifica », *5* (1), pp. 283, 330, 452, 652 (1934); *5* (2), pp. 21, 381, 467 (1934); *6* (1), p. 123 (1935).

sodium chloride showed an activity (37 minutes) which must be due to an isotope of chlorine.

Copper. – Irradiated metallic copper was dissolved in hydrochloric acid and traces of cobalt, nickel, and zinc were added. Copper was separated from the acid solution as a sulphide, and was found to be active. Zn, Co, and Ni were also separated as sulphides by neutralizing the filtrate, and were found to be inactive. The time required for this test allows us to conclude, for the 10 hour activity only, that it is due to an isotope of copper.

Zinc. – From irradiated zinc dissolved in hydrochloric acid we separated, in the manner described above, copper in the presence of nickel and cobalt. Copper carried along the activity, while the other elements which remained in solution, after further precipitation, did not show any activity. Therefore, the 10-hour life in zinc is to be attributed to copper.

Gallium. – Chemical properties of the 23-hour activity were determined with the same procedure as that used for 20-minute activity; here also the activity must be assigned to an isotope of gallium.

Selenium. – Irradiated selenic anhydride was dissolved in 30 per cent hydrochloric acid, and arsenious anhydride was added to it. From the solution metallic selenium was separated with sulphur dioxide. The separated selenium was active, while arsenic precipitated from the solution as sulphide was inactive. Since germanium is also not precipitated by sulphur dioxide, it would be precipitated as sulphide; this test indicates that the activity of selenium is due to an isotope of selenium.

Thorium. – The chemical tests were limited to investigate the 24-minute activity. Thorium was purified from its radioactive products by repeated precipitations with hydrogen peroxide in the presence of barium, lanthanum, lead, and bismuth.

From a very dilute solution of irradiated thorium nitrate, thorium was separated as a fluoride in the presence of uranium and tantalum. The separated thorium conserved the activity, while uranium and tantalum, precipitated as hydroxides, were inactive.

In other tests thorium was separated from uranium as a carbonate using ammonium carbonate. Uranium separated from the ammonium solution was almost inactive. The small activity found in uranium is probably due to small quantities of thorium precipitated together with uranium. From an irradiated thorium nitrate solution, zirconium was separated as phosphate. The precipitate was almost inactive, while thorium retained its activity. From these experiments it seems that the activity cannot be attributed to isotopes of uranium and protactinium.

From the irradiated, dilute, thorium nitrate solution we precipitated barium as a sulphate; it was found to be inactive. In other tests we precipitated thorium from the thorium nitrate solution with hydrogen peroxide in presence of barium and lanthanum. From the filtrate barium and lanthanum were separated as sulphate and hydroxides, respectively. Both were found to be inactive, while the activity remained in thorium. It appears, therefore, that the activity is not due to isotopes of Pb, Bi, Ra, Ac, and Po either.

Therefore, it is probable that the activity of thorium with a period of 24 minutes is due to one of its isotopes.

In all chemical operations with naturally radioactive substances it sometimes happens that active residues due to incomplete purification concentrate themselves in some fractions. This fact requires special attention for the identification of the activity from its disintegration period and it is one of the reasons which make chemical operations quite delicate. Great caution is therefore indicated in the interpretation of the results.

Uranium. – We have already said that the sensitivity coefficients to hydrogenous substances of the 15-second, 13-minute, and 100-minute periods are slightly higher than 1 and practically equal. The sensitivity coefficient for the 40-second activity, however, is smaller than that of the activities mentioned above. It is obvious that all products resulting from the same primary process must have the same sensitivity coefficient; we conclude, therefore, that the activity with the period of 40 seconds is due to one primary process, while the other activities probably are due to a different primary process, the same for all three. In this hypothesis the three activities could be either genetically related or have more complicated relations through branchings. The following experiment favors the first hypothesis, at least for the 13-minute and 100-minute periods. We measured the decay curve of an irradiated thick uranium layer with an electrometer; by analysing this curve in exponentials of 13-minute and 100-minute periods, it appears that the initial activities are in the ratio of 100 : 45; on the other hand, the half-thickness value for the β–rays of the 13-minute activity is 0.14 gr/cm^2 of Al; the half-value thickness connected with the longer period could not be measured exactly, but certainly it is smaller than the other and probably about one-half of it. This result agrees with the hypothesis that the initial number of disintegrations for the two periods, when the substance is irradiated to saturation, is the same. This fact would be necessary in order to admit that the two activities are genetically related.

In a previous communication we showed by chemical tests that the activity with periods of 13 and 100 minutes cannot be attributed to isotopes of elements with $Z < 92$ [1]. Von Grosse and Agruss have criticized our tests, concluding, from experiments carried out by them only on substances which were not irradiated by neutrons, that the 13-minute activity was due to an isotope of protactinium. Therefore, we repeated our old experiments and confirmed our results. At the same time some new tests were carried out.

The precipitation of the activity with a sulphide has been repeated by using different metals (Ag, Cu, Pb, Hg); the concentration of the hydrochloric acid solution was about 20 per cent. The yield of the separation generally was good (about 50 per cent) and depended on the conditions of precipitation. The presence of nitric acid substantially lowers the yield of separation. In the presence of hydrofluoric acid and tantalum a high yield was obtained; protactinium, as is well known, follows tantalum in its reactions, and the latter precipitates only with great difficulty from hydrofluoric acid solutions.

(1) VON GROSSE and AGRUSS, « Phys. Rev. », *46*, 241 (1934).

We also carried out the test suggested by von Grosse as the most characteristic reaction for protactinium. We dissolved irradiated and purified uranium oxide in 25 per cent HCl; we added zirconium nitrate and phosphoric acid; the precipitated zirconium phosphate was inactive, while the activity could be collected from the filtrate as sulphide with the usual yield.

According to von Grosse and Agruss this reaction should be considered a test to determine that the activities of 13 and 180 minutes are not due to isotopes of protactinium. By different chemical tests, which were repeated and confirmed by us, O. Hahn and L. Meitner (2) conclude, too, that the activities of 13 and 100 minutes are due to transuranic elements.

Certainly the two activities have very similar chemical characteristics. An indication of separation was obtained only with the following experiment: uranium oxide was purified, irradiated, and dissolved in HCl. To the solution, ammonium carbonate was added until complete solution of the uranyl carbonate, which precipitates at first, was obtained. By adding manganese nitrate or lead nitrate to the solution and separating the corresponding carbonates, a part of the 13-minute and 100-minute activities was found in them; the filtrate was acidified and copper was precipitated as copper sulphide; the copper sulphide was active; it seems, however, that the ratio of the two acitivities is different in the two precipitates, i.e., the activity with the period of 13 minutes is more abundant in the sulphide precipitate.

From all our experiments the hypothesis that the 13-minute and 100-minute activities are to be attributed to transuranic elements is again confirmed. According to the known facts the simplest interpretation is that the activities of 10 seconds, 13 minutes, and 100 minutes are genetically related and must be attributed to elements having the atomic numbers 92, 93, 94, and the atomic weight 239.

5. ABSORPTION OF SLOW NEUTRONS AND EMISSION OF GAMMA RAYS BY THE ACTION OF SLOW NEUTRONS. - We studied the neutron absorption by chlorine, using either rhodium or vanadium as detectors; in both cases the same cross-section was found.

We also determined the mean free path of slow neutrons in paraffin by using rhodium as a detector. The curve obtained was analogous to that found previously using silver as a detector.

In a previous letter we described an experimental arrangement suited to detecting γ-ray emission produced in some substances by the action of slow neutrons. With the same apparatus we carried out a systematic experiment, which confirmed the expectation that elements with large cross-sections emit γ-rays (we mentioned that boron and lithium are exceptions). Besides the elements already listed, cobalt, copper, and gold emit considerable amounts of γ-rays.

Istituto Fisico della R. Università.
Rome, April 3, 1935-XIII.

(2) HAHN and MEITNER, « Naturwissenschaften », *23*, 37 (1935).

92 *b*.

RADIOACTIVITY INDUCED BY NEUTRON BOMBARDMENT (*).—X.

Translated from E. AMALDI, O. D'AGOSTINO, E. FERMI, B. PONTECORVO, E. SEGRÈ
« Ric. Scientifica », *6* (1), 581–584 (1935).

1. RADIOACTIVITY INDUCED IN BROMINE. – We also noticed the existence of a third radioactive isotope of Br with a period of 36 hours, mentioned by Kurtchatow and others, [1] and we confirmed their measurements on the β–rays. We noticed that this activity is sensitive to the action of hydrogenous substances as are the other two Br activities. This fact renders very unlikely the explanation proposed by the above-mentioned authors postulating that this activity is obtained from a stable bromine isotope by the emission of one neutron, because a slow neutron has insufficient energy to bring this process about.

Since, according to information kindly given to us by Aston, the existence of a third stable isotope of bromine can be excluded, the nature of this third activity presents an interesting problem, probably similar to that of indium [2].

2. INFLUENCE OF TEMPERATURE. – Moon and Tillman [3] recently observed a difference in the activation of silver, rhodium and iodine by slowing down neutrons with paraffin either at room temperature or at the temperature of liquid air. Several months ago [4] we tried to detect a similar temperature effect without success. By repeating the experiment under the conditions kindly communicated to us by Moon and Tillman, we too noticed an increase in the activation intensity of different elements by lowering the temperature of the paraffin. We noticed an increase from 10 to 30 per cent in the activation of Ag (2.3 minutes), V, Al (2.3 minutes); in rhodium, however, we did not find any appreciable increase in activity.

If the different behavior of various elements, also observed by Moon and Tillman, is confirmed, the process of slow neutron-capture must be considered more complicated than is now assumed, on the basis of the commonly accepted force law between neutron and nuclei (potential hole of nuclear dimensions).

(*) Cfr. « Ricerca Scientifica », *5* (1), pp. 283, 330, 452, 652 (1934); *5* (2), pp. 21, 381, 467 (1934); *6* (1), pp. 123, 432 (1935).

(1) KURTCHATOW, MYSSOWSKY, ROUSSINOW, « C. R. », *200*, 1201 (1935).

(2) SZILARD and CHALMERS, « Nature », *135*, 98 (1935); « Proc. Roy Soc. », *149*, 592 (1935).

(3) MOON and TILLMAN, « Nature », *135*, 904 (1935).

(4) « Proc. Roy. Soc. », *149*, 592 (1935); « Ricerca Scientifica », *5* (2), p. 467 (1934).

The differences in behavior probably could be interpreted by assuming that there exists a very weak force between neutron and nucleus having a relatively large radius of action.

3. MECHANICAL TRANSPORT OF SLOW NEUTRONS. – In order to confirm by an independent method the results of Moon and Tillman's experiments, we tried to determine with a mechanical experiment how long the neutrons remain in paraffin.

A steel wheel of 80 cm diameter has on its edge a source of Em+Be and near to it, 5 cm to the right and to the left respectively, two equal slow neutron detectors. In some of the experiments the detectors were of manganese, in some of vanadium.

The wheel was rotated between two large paraffin rings with rectangular cross section (10×20 cm), the axes of which coincided with the axis of the wheel. One of the rings was in the front, the other in the rear of the plane of the wheel at distance of 6 cm from said plane; the weight of the rings was about 100 kg.

The wheel was rotated with an angular velocity of about 1,600 revolutions per minute for a certain time, according to the detector used, and subsequently the ratio of the activities acquired by the two detectors was measured. For the measurement of the activity the detectors were removed from the wheel; after the first measurement the detectors were placed again in the same conditions as before, the wheel was now rotated in the opposite direction and the ratio between the activities of both detectors was measured again.

The result is that the detector which follows the source in the direction of motion acquires about 10 per cent more activity than the detector which precedes the source.

This result shows first of all that the neutrons indeed acquire a velocity corresponding to that of thermal agitation. The meaning of this experiment can be made more precise with the following explanation.

The same difference of 10 per cent between the activities of the two detectors is obtained by displacing the source from the center of the two detectors by about 0.5 cm and by keeping the wheel fixed (this check was made by using rhodium as a detector). This distance is passed by the periphery of the wheel in 7×10^{-5} seconds and therefore it must approximately coincide with the time slow neutrons survive in paraffin. Assuming that slow neutrons have the velocity of thermal agitation (2500 m/sec) and an average mean free path of 0.5 cm [(5)], as results from scattering experiments, we can infer that a neutron undergoes on the average about 30 collisions after it has reached thermal velocity and before it diffuses out of the paraffin or before it is captured by the surrounding nuclei.

4. NEUTRON EMISSION BY DIFFERENT SUBSTANCES. – Since rhodium in a hydrogenous medium is a particularly sensitive neutron detector, we in-

(5) « Proc. Roy. Soc. », *149*, 592 (1935), and WESTCOTT and BJERGE, « Proc. Cambridge Phil. Soc. », *31*, p. 145, (1935).

vestigated with it some elements for neutron emission under α–particle bombardment.

Both silicon and phosphorus mixed with 200 mC of emanation activated a rhodium cylinder with an initial activity of about 100 counts per minute on our counters. In all experimental cases observed up to now the emission of neutrons from silicon and prosphorus is very low compared with the emission of neutrons from boron, lithium, aluminum, etc. In the case of phosphorus the emission observed is probably associated with the formation of radioactive chlorine [(6)]. With many substances a small emission of neutrons is obtained, probably due to surface impurities; among the substances examined, nickel shows the smallest effect.

We also tried to determine some properties of neutrons emitted by lithium, boron, aluminum, fluorine, magnesium. For this purpose we constructed sources similar to the usual ones but replacing beryllium with the element to be studied.

Boron. – We measured with a source Em + B the sensitivity coefficient to hydrogenous substances, α, of the elements listed below. For purposes of comparison we also list the value of α corresponding to a source Em + Be.

		α_B	α_{Be}
Si	(135 sec)	1	1
V	(225 sec)	120	40
Rh	(44 sec)	50	15
Ag	(140 sec)	20	15
I	(25 min)	10	5

From these data, which are however only indicative, it can be seen that the coefficient α_B is higher than α_{Be} by a factor which roughly varies from 3 to 1.

Furthermore it is interesting to compare the activities produced without slowing down the neutrons by two sources of Em + Be and Em + B of the same geometrical form and containing the same amount of emanation. By irradiating V in air the ratio between the activity obtained with the Be source and the activity obtained with the B source is found to be about 12. The similar ratio obtained by irradiating Si (135 sec) is larger than 15. These facts could be explained by the following hypothesis: in the spectrum of the neutrons emitted by an Em + Be source, there is a band of very slow neutrons to which the relatively low value of the coefficient α is attributed, and a band of very fast neutrons, already known from other experiments, responsible for the relatively strong activation of Si; on the other hand, in the spectrum of an Em + B source very slow and very fast neutrons are absent. Such considerations are suggested also by the results obtained from the following experiments: we measured the activation as a function of the distance from the source with a source Em + B; the detector used was Rh, the irradiation was carried out in a water tank of dimensions 40 × 30 × 80

(6) FRISCH, «Nature», *133*, 721 (1936).

cm, and the measurements were made with an ionization chamber: a curve is obtained which differs slightly from that obtained by using as source Em + Be [7]; for small distances the activation by Em + B decreases slower than that by Em + Be; for distances larger than 10 cm both curves have the same form.

Finally we measured with the apparatus described in a preceding communication the absorption coefficient of slow neutrons in chlorine using an Em+B source; we obtained the same value as with an Em + Be source (first reduction to halfvalue in 0.3 g/cm^3).

Fluorine. – A source of about 300 mC of CaF_2+Em does not appreciably activate silicon, but a noticeable activation of V and Rh was observed when the irradiation was carried out in water.

Magnesium. – Neutrons from a 600 mC source of Mg + Em activate P (2.4 hrs), Si (135 sec), Al (10 min), Rh (44 sec) and V (225 sec). The sensitivity coefficients α_{Mg} are remarkably larger than the coefficients α_{Be} for Rh (44 sec) and V (225 sec).

Aluminum. – Neutrons of a source Al + Em of about 600 mC do not activate Si appreciably, but activate Rh (44 sec). The coefficient α_{Al} for Rh (44 sec) is larger than the corresponding α_{Be}.

A more exact and more detailed knowledge of data similar to those collected above should give valuable information about the velocity spectrum of the neutrons emitted by various substances under α–particle bombardment and, on the other hand, about the energy necessary for the activation of various elements by neutrons.

5. EXPERIMENTS CONCERNING THE PRODUCTION OF NEUTRONS BY γ–RAYS. – We compared the activities produced in rhodium surrounded by paraffin by the two sources MsTh + Be and MsTh + D_2O (this source contains about 35 mC RdTh); the experiments were carried out by using 25 g beryllium and 25 g D_2O (98 %), in the same geometry. In such conditions we found that the number of neutrons emitted by D_2O under the action of γ–rays coming from MsTh is about 3.5 times as great as that emitted by Be. We carried out the same measurements by substituting the MsTh source with one containing 100 mg Ra. In the case of Be, under our geometrical conditions, using rhodium as detector, we obtained about 200 impulses per minute, while with D_2O we did not succeed in obtaining any appreciable activity. These facts can be easily interpreted if we consider that the neutron is bound more loosely in beryllium than in deuterium. Finally, we noticed that the number of neutrons emitted by Be was about the same as when we used the equivalent sources of mesothorium and radium.

With a source MsTh + D_2O and MsTh + Be we measured the activation as a function of the distance from the source. The irradiation was carried out in a water tank of the dimensions 40 × 30 × 80 cm; the detector was a small rhodium cylinder (measurements have been made with counters).

(7) « Proc. Roy. Soc. », loc. cit.

The two curves obtained with Be and D_2O have substantially the same form.

The following table shows the activity (number of impulses) corresponding to different distances with the source MsTh + D_2O.

Distance in cm	Activation
4	1100 ± 50
6	962 ± 40
9	388 ± 20
12	188 ± 14
14	107 ± 7

The numbers to the left indicate in cm the distances from the detector to the axis of the tube containing MsTh.

Istituto Fisico della R. Università.
Rome, June 14, 1935–XIII.

N° 93.

This paper is a summary and in part a translation of papers N° 84 and 85.

93.

RADIOACTIVITY INDUCED BY NEUTRON BOMBARDMENT

« Nature », (London) *133*, 757 (Letter) (1934).

Experiments have been carried out to ascertain whether neutron bombard ment can produce an induced radioactivity giving rise to unstable products which disintegrate with emission of β–particles. Preliminary results have been communicated in a letter to « La Ricerca Scientifica », *5* (1), 283 (1934).

The source of neutrons is a sealed glass tube containing radium emanation and beryllium powder. The amount of radium emanation available varied in the different experiments from 30 to 630 millicuries. We are much indebted to Prof. G. C. Trabacchi, Laboratorio Fisico della Sanità Pubblica, for putting at our disposal such strong sources.

The elements, or in some cases compounds containing them, were used in the form of small cylinders. After irradiation with the source for a period which varied from a few minutes to several hours, they were put around a Geiger counter with walls of thin aluminium foil (about 0.2 mm thickness) and the number of impulses per minute was registered.

So far, we have obtained an effect with the following elements:

Phosphorus. – Strong effect. Half-period about 3 hours. The disintegration electrons could be photographed in the Wilson chamber. Chemical separation of the active product showed that the unstable element formed under the bombardment is probably silicon.

Iron. – Period about 2 hours. As the result of chemical separation of the active product, this is probably manganese.

Silicon. – Very strong effect. Period about 3 minutes. Electrons photographed in the Wilson chamber.

Aluminium. – Strong effect. Period about 12 minutes. Electrons photographed in the Wilson chamber.

Chlorine. – Gives an effect with a period much longer than that of any element investigated at present.

Vanadium. – Period about 5 minutes.

Copper. - Effect rather small. Period about 6 minutes.

Arsenic. – Period about two days.

Silver. – Strong effect. Period about 2 minutes.

Tellurium. – Period about 1 hour.

Iodine. – Intense effect. Period about 30 minutes.

Chromium. – Intense effect. Period about 6 minutes. Electrons photographed in the Wilson chamber.

Barium. – Small effect. Period about 2 minutes.

Fluorine. – Period about 10 seconds.

The following elements have also given indication of an effect: sodium, magnesium, titanium, zirconium, zinc, strontium, antimony, selenium and bromine. Some elements give indication of having two or more periods, which may be partly due to several isotopic constituents and partly to successive radioactive transformations. The experiments are being continued in order to verify these results and to extend the research to other elements.

The nuclear reaction which causes these phenomena may be different in different cases. The chemical separation effected in the cases of iron and phosphorus seems to indicate that, at least in these two cases, the neutron is absorbed and a proton emitted. The unstable product, by the emission of a β-particle, returns to the original element.

The chemical separations have been carried out by Dr. O. D'Agostino. Dr. E. Amaldi and Dr. E. Segrè have collaborated in the physical research.

Physical Institute, Royal University.
Rome, April, 10.

94.

SULLA POSSIBILITÀ DI PRODURRE ELEMENTI DI NUMERO ATOMICO MAGGIORE DI 92

E. FERMI, F. RASETTI, O. D'AGOSTINO

« Ric. Scientifica », *5* (1), 536-537 (1934).

In una precedente lettera è stata data notizia di esperienze preliminari per studiare la attivazione dell'Uranio per effetto del bombardamento con neutroni [1]. Desideriamo qui esporre con maggiore dettaglio le osservazioni fatte e altre ulteriori.

Le curve di decadimento della attività β, che si ottengono dall'Uranio, preventivamente liberato dai suoi ordinari prodotti di decadimento, i quali renderebbero impossibile ogni misura, e successivamente bombardato con neutroni per un tempo variabile da pochi secondi a una dozzina di ore, possono analizzarsi in esponenziali con i seguenti periodi: 10^s, 40^s, 13^m, oltre ad almeno altri due più lunghi. Questi fatti indicano che il processo è certamente complicato e che hanno luogo disintegrazioni successive; tuttavia, per la necessaria imprecisione delle misure, dovuta alle fluttuazioni statistiche, non è ancora possibile decidere quali siano processi alternativi e quali processi in catena. Come già abbiamo detto nella precedente lettera, si è cercato di riconoscere la natura chimica dell'elemento con periodo di 13^m, che per ragioni pratiche si presenta il più conveniente. Lo schema generale della ricerca è stato il seguente. Alla soluzione di Uranio in forma di nitrato, prima purificato radioattivamente e quindi irradiato con neutroni, si aggiungeva una certa quantità di un ordinario elemento dotato di attività β tale da dare un migliaio di impulsi per minuto nel contatore. Se si riesce a dimostrare che l'attività indotta, riconoscibile dal suo periodo caratteristico, può essere separata chimicamente dalla attività aggiunta, è ragionevole ammettere che le due attività non siano dovute a isotopi. Si è già detto che l'attività di 13^m è trascinata da un precipitato di biossido di Manganese in soluzione acida. Questa reazione ci ha permesso di separare questa attività da molti elementi pesanti. Sono state ripetute e confermate le separazioni già precedentemente annunziate da U, Pa, Th. Inoltre si è effettuata anche la separazione dal Ra e dall'Ac, che sono stati aggiunti nella forma di Ms Th 1 e di Ms Th 2; conviene anche, affinché la separazione avvenga in modo completo, aggiungere alla soluzione piccole quantità di Bario e di Lantanio a sostegno degli elementi precedenti. Si è trovato infine che le condizioni di precipitazione del biossido di Manganese si possono regolare in modo che il precipitato si formi in presenza

(1) « La Ricerca Scientifica », *5* (1), p. 452, (1934).

di Piombo e Bismuto inattivi senza trascinarli e portando seco invece la attività di 13^m.

In questo modo sembra doversi escludere che la attività di 13^m sia dovuta a un isotopo di U (92), (Pa) (91), Th (90), Ac (89), Ra (88), Bi (83), Pb (82). Il suo comportamento esclude anche Ekacesio (87) ed Emanazione (86).

Da questo complesso di prove negative sembra plausibile considerare la possibilità già enunciata che il numero atomico dell'elemento in questione sia maggiore di 92. Se fosse un elemento 93, esso sarebbe omologo del Manganese e del Renio. Questa ipotesi è confermata in qualche misura dal fatto osservato che la attività di 13^m è trascinata da un precipitato di solfuro di Renio insolubile in HCl. Tuttavia, poiché molti elementi pesanti precipiterebbero in questa forma, questa prova non può considerarsi molto dimostrativa.

La possibilità di un numero atomico 94 o 95, non sarebbe facile da distinguere dalla precedente, poiché le proprietà chimiche di questi elementi sono probabilmente abbastanza simili. Utili informazioni sui processi che si svolgono si potrebbero probabilmente ricavare da un esame di una eventuale emissione di particelle pesanti. Naturalmente non è però possibile osservare eventuali disintegrazioni di lunga vita; e nemmeno disintegrazioni rapide, poiché l'osservazione di particelle pesanti richiede di necessità manipolazioni chimiche per portare la sostanza attiva in uno strato sottile. In queste condizioni appare pertanto prematuro formare ipotesi troppo definite sulla serie di disintegrazioni che hanno luogo, prima che altre esperienze abbiano ulteriormente chiarito i fenomeni.

Istituto Fisico della R. Università.
Roma, 6 giugno 1934-XII.

N° 95.

This paper originated in connection with some experimental work by Amaldi and Segrè on the absorption spectrum of the alkali metals near the limit of the *s–p* series. It was found experimentally that the presence of a foreign gas under pressure shifted the absorption lines without broadening them in an excessive way, with the result that even the terms between 20 *p* and 30 *p* were observable. The absence of the expected broadening and the shift of the energy levels were rather puzzling and were discussed extensively with Fermi. It occurred to him that the outer electron might feel the influence of the dielectric constant of the gas (a similar idea recurs later in studies on the stopping power) and he made a calculation on this hypothesis. The order of magnitude of the effect agreed with observation, but when several gases were tried, it was found that nitrogen gave an effect of the wrong sign. The dielectric effect always shifts the lines towards the long wave lengths and here the shift was towards the short wave lengths. Further meditation by Fermi led him to discover another cause for the shift of the levels, which could have both signs and was comparable in magnitude with the dielectric effect.

This is described in the second part of the paper, in which we meet for the first time the concept of scattering length and of "pseudo-potential". The theoretical development is remarkable because it shows for the first time the application of Born's method to a deep and narrow potential and gives a technique that Fermi used many times in connection with his neutron work. Indeed Fig. 1 was a sort of trademark in many of his later theoretical studies.

E. SEGRÈ.

95.

SOPRA LO SPOSTAMENTO PER PRESSIONE DELLE RIGHE ELEVATE DELLE SERIE SPETTRALI

«Nuovo Cimento», *11*, 157–166 (1934).

SUNTO. – Lo spostamento dei termini elevati delle serie di assorbimento dei vapori alcalini, osservato da Amaldi e Segrè quando il vapore si trova in una atmosfera di gas estraneo, viene studiato teoricamente. Si stabilisce una relazione tra lo spostamento delle righe e la sezione d'urto limite per elettroni di velocità piccolissima contro le molecole del gas perturbatore.

§ 1. E. Amaldi ed E. Segrè hanno descritto nella Nota precedente [1] il considerevole spostamento che subiscono le righe di ordine elevato delle serie di assorbimento dei vapori alcalini, quando questi si trovano immersi in un gas estraneo. Le caratteristiche più notevoli del fenomeno osservato da questi Autori possono così riassumersi:

(1) E. AMALDI ed E. SEGRÈ, «Nuovo Cimento», *11*, 145 (1934).

Lo spostamento dei termini elevati della serie di assorbimento è, a parità di pressione del gas estraneo, assai maggiore di quello che subiscono i primi termini della serie.

Lo spostamento, al crescere del numero d'ordine della riga, sembra convergere verso un limite approssimativamente proporzionale alla pressione del gas estraneo.

Nei due vapori esaminati, sodio e potassio, lo spostamento è eguale, entro i limiti della precisione delle misure.

La grandezza dello spostamento, e anche il suo verso, dipendono dalla natura del gas estraneo.

Mi propongo in questa Nota di discutere teoricamente il comportamento dei termini elevati di un atomo in atmosfera di un gas estraneo nel caso, che corrisponde alle condizioni realizzate sperimentalmente da Amaldi e Segrè, in cui le orbite siano tanto estese da contenere, entro il raggio dell'orbita, alcune migliaia di atomi perturbatori.

Dal punto di vista qualitativo due cause principali intervengono a determinare uno spostamento per pressione delle righe di ordine elevato, e cioè:

a) Quando eccitiamo un atomo, trasportando un suo elettrone in una orbita molto lontana, tutti gli atomi del gas perturbatore che si trovano a una distanza dal nucleo minore del raggio dell'orbita dell'elettrone allontanato, vengono a trovarsi in un campo elettrico dovuto al fatto che, alla azione di schermo della carica nucleare da parte degli elettroni, manca il contributo dell'elettrone che è stato allontanato. Questo campo determina una polarizzazione degli atomi estranei, e l'energia necessaria a produrla dà luogo a uno spostamento dei termini. Faremo vedere nel § 2 che questa causa agisce sempre nel senso di produrre uno spostamento delle righe verso il rosso.

b) Gli atomi del gas perturbatore agiscono sopra l'elettrone luminoso come buche di potenziale disseminate sul suo cammino; si potrebbe pensare che l'effetto di queste buche potesse valutarsi per mezzo della prima approssimazione della teoria delle perturbazioni, considerando che esse vengono ad abbassare il livello medio del potenziale in cui si muove l'elettrone esterno. Se così fosse, ci si dovrebbe attendere in ogni caso un abbassamento dei livelli energetici elevati rispetto al termine fondamentale, e quindi uno spostamento delle righe verso il rosso. Una considerazione più accurata dimostra invece che la prima approssimazione non solo non è sufficiente dal lato quantitativo, ma può perfino dare errato il verso dello spostamento. Faremo vedere infatti che l'effetto delle buche di potenziale può, a seconda dei casi, consistere in uno spostamento verso il rosso o verso il violetto, e che l'ordine di grandezza dello spostamento è, almeno di solito, più grande di quello dello spostamento dovuto alla causa *a*); per modo che, in accordo con quanto si osserva dovremo attenderci che lo spostamento totale delle righe avvenga in alcuni casi verso il rosso e in altri verso il violetto. Faremo anche vedere che, almeno nel caso che il gas estraneo sia monoatomico, esiste una semplice relazione tra lo spostamento delle righe spettrali e la sezione d'urto di elettroni lentissimi contro gli atomi del gas estraneo; relazione che, nel caso dei gas monoatomici studiati da Amaldi e Segrè, sembra effettivamente essere verificata.

Nei §§ 2 e 3 studieremo gli spostamenti dovuti alle due cause *a*) e *b*). Nel § 4 discuteremo poi come si possa giustificare il fatto che, nonostante la presenza di un grande numero di atomi perturbatori nell'interno dell'orbita, i termini restino tuttavia abbastanza definiti perché si possano osservare le righe spettrali fino a un ordine assai elevato.

§ 2. Lo spostamento delle righe dovuto alla polarizzazione degli atomi del gas estraneo si calcola nel modo seguente. Quando l'elettrone luminoso viene eccitato, esso si porta sopra una orbita di raggio tanto grande, che nel suo interno vengono ad essere contenuti molti atomi del gas estraneo; questi si trovano ora ad essere soggetti all'azione della carica elettrica e del resto atomico. Sia r_i la distanza di uno degli atomi estranei dal centro dell'atomo alcalino; esso verrà a trovarsi in un campo elettrico di intensità e/r_i^2. Indichiamo con α il coefficiente di polarizzabilità di un atomo del gas estraneo, legato alla costante dielettrica ε del gas dalla relazione:

$$\varepsilon = 1 + 4\pi n\alpha \tag{1}$$

dove n è il numero di atomi per unità di volume. Si calcola elementarmente che l'energia dell'atomo, per effetto della sua polarizzazione, quando lo si pone in un campo elettrico E è data da $-\alpha E^2/2$. Ponendo $E = e/r_i^2$ si trova dunque, per effetto della polarizzazione degli atomi del gas estraneo, una variazione dell'energia

$$\Delta W_\varepsilon = -\frac{\alpha e^2}{2} \Sigma \frac{1}{r_i^4} . \tag{2}$$

La somma si deve estendere a tutti gli atomi del gas estraneo che si trovano entro una distanza dal nucleo dell'atomo considerato minore della distanza media dell'elettrone eccitato; o anche praticamente, data la grande estensione dell'orbita, la somma in (2) potrà estendersi a tutti gli atomi del gas estraneo.

Lo spostamento delle righe dovuto a questa causa dipende dunque essenzialmente dalla somma degli inversi delle quarte potenze della distanza degli atomi perturbatori dal nucleo dell'atomo perturbato. Si riconosce facilmente che il valor medio di una tale somma quando almeno si trascuri l'estensione finita degli atomi, risulta infinito; tuttavia il valore più probabile della somma stessa è finito. Ponendo

$$z_i = \frac{4\pi}{3} r_i^3 n$$

si ha

$$\Sigma \frac{1}{r_i^4} = \left(\frac{4\pi n}{3}\right)^{4/3} \Sigma \frac{1}{z_i^{4/3}} .$$

Inoltre, se gli atomi sono distribuiti con la concentrazione n, le variabili z_i possono prendere tutti i valori positivi e la probabilità che uno dei valori z_i cada nell'intervallo tra z e $z + dz$ è eguale a dz. In queste condizioni la legge di distribuzione statistica dell'espressione $\Sigma z_i^{-4/3}$ è stata calcolata recentemente da G. C. Wick [2], che ha trovato che il valore più probabile di essa è approssimativamente 2,6. Troviamo dunque, come valore più probabile di $\Sigma \frac{1}{r_i^4}$ il seguente

$$\left(\Sigma \frac{1}{r_i^4}\right)_0 = \left(\frac{4\pi n}{3}\right)^{4/3} 2{,}6 \cong 17{,}5\, n^{4/3}\,. \tag{3}$$

Per la causa che stiamo esaminando dunque la riga si allarga e si sposta; e lo spostamento del punto in cui la intensità è massima è dato, secondo la (3), da

$$\Delta W_\varepsilon = -\, 17{,}5 \frac{\alpha e^2}{2} n^{4/3}\,.$$

Siccome dall'altra parte la legge di distribuzione statistica di $\Sigma \frac{1}{r_i^4}$ non è in alcun modo simmetrica rispetto al suo massimo, ma decresce assai più rapidamente per valori piccoli che per valori grandi, riteniamo che, in pratica, il punto che si osserva effettivamente non coincida col punto della riga in cui l'intensità è massima ma sia più spostato verso il rosso. Probabilmente un dato un po' più corrispondente alla pratica delle osservazioni si ottiene sostituendo a 17,5 il valore 20. La differenza del resto non produce altro che una alterazione quasi insensibile dei risultati. Tenendo conto anche di (1) scriveremo:

$$\Delta W_\varepsilon = -\, 20 \frac{(\varepsilon - 1)\, e^2}{8\pi} n^{1/3}\,. \tag{4}$$

La causa che stiamo esaminando produce dunque un abbassamento dei termini di ordine elevato rispetto al termine fondamentale. Essa dà pertanto origine a uno spostamento delle righe spettrali verso il rosso che, espresso in numeri di onde, è dato da:

$$\Delta_\varepsilon = +\, 20 \frac{(\varepsilon - 1)\, e^2}{8\pi hc} n^{1/3} = +\, 0{,}00093\, (\varepsilon - 1)\, n^{1/3} \tag{5}$$

(il segno + significa spostamento verso il rosso). Lo spostamento dovuto a questa causa, per gas estranei a pressione dell'ordine di grandezza di una atmosfera, risulta dell'ordine di un cm^{-1}. Deve notarsi infine che tutti i risultati precedenti sono validi solo nel caso che il gas estraneo non abbia un dipolo permanente. Le osservazioni di Amaldi e Segrè si limitano appunto a casi di questo tipo.

§ 3. Passiamo ad esaminare l'effetto della causa *b*) del § 1 che, come si è già detto, è, almeno nella maggior parte dei casi, più importante della precedente. Tratteremo il caso che il gas perturbatore sia monoatomico.

Il potenziale che agisce sopra l'elettrone luminoso quando questo si trova in uno stato di numero quantico assai elevato è costituito dalla somma del potenziale U del resto atomico, e cioè da una funzione che, ove si eccettui la immediata vicinanza del resto atomico, varia assai regolarmente e lentamente, più le buche di potenziale V_i che rappresentano l'azione esercitata

(2) G. C. WICK, « Rend. Lincei », *19*, 27 (1934).

sull'elettrone dai singoli atomi del gas perturbatore. Queste buche di potenziale sono notevolmente profonde ma hanno dimensioni piccole a confronto della lunghezza d'onda di De Broglie dell'elettrone poiché gli atomi perturbatori sono neutri. La autofunzione ψ dell'elettrone, sotto l'azione di questo potenziale, avrà evidentemente un andamento medio con variazione lenta e regolare, al quale si sovrappongono delle irregolarità localizzate in una piccola zona attorno a ciascuna delle buche di potenziale. Converrà pertanto considerare, accanto alla ψ, un'altra funzione, che indicheremo con $\overline{\psi}$ (x, y, z), e che rappresenta il valor medio di ψ in un intorno del punto x, y, z piccolo a confronto della lunghezza d'onda di De Broglie, ma abbastanza grande per contenere un numero considerevole di atomi del gas estraneo. La funzione $\overline{\psi}$ riproduce l'andamento generale di ψ, senza avere le irregolarità di questa ultima funzione in corrispondenza delle buche di potenziale.

Dalla definizione di $\overline{\psi}$ risulta, con calcoli elementari, che

$$\Delta\overline{\psi} = \overline{\Delta\psi} \tag{6}$$

dove Δ è l'operatore di Laplace.

L'equazione di Schrödinger per il nostro elettrone ha la forma

$$\Delta\psi + \frac{8\pi^2 m}{h^2}(\mathrm{W} - \mathrm{U} - \Sigma \mathrm{V}_i)\,\psi = 0\,. \tag{7}$$

Prendiamo la media della (7) in un intorno come quello definito precedentemente del punto x, y, z, che sia abbastanza piccolo perché U non subisca in esso considerevoli variazioni. Otteniamo, tenendo conto anche di (6)

$$\Delta\overline{\psi} + \frac{8\pi^2 m}{h^2}(\mathrm{W} - \mathrm{U})\,\overline{\psi} - \frac{8\pi^2 m}{h^2}\overline{\Sigma \mathrm{V}_i\,\psi} = 0\,. \tag{8}$$

Per calcolare l'ultimo valor medio si osservi che, entro una delle buche di potenziale, W — U è trascurabile di fronte al potenziale V della buca. Inoltre la ψ in una regione circostante alla buca di potenziale, a distanza da questa grande in confronto al raggio ρ della buca, ma piccola in confronto alla lunghezza d'onda di De Broglie, ha valore praticamente costante ed eguale in prima approssimazione a $\overline{\psi}$; si riconosce dunque che la ψ nell'interno della buca e nelle sue vicinanze dipende praticamente solo dalla distanza r dal centro della buca. Ponendo perciò in questa regione

$$\psi = \frac{u(r)}{r} \tag{9}$$

la funzione $u\,(r)$ soddisferà in prima approssimazione all'equazione differenziale:

$$u''(r) = \frac{8\pi^2 m}{h^2}\,\mathrm{V}(r)\,u(r)\,. \tag{10}$$

All'esterno della buca, dove V (r) si annulla, la u è dunque una funzione lineare di r. E siccome il valore di ψ lontano dalla buca

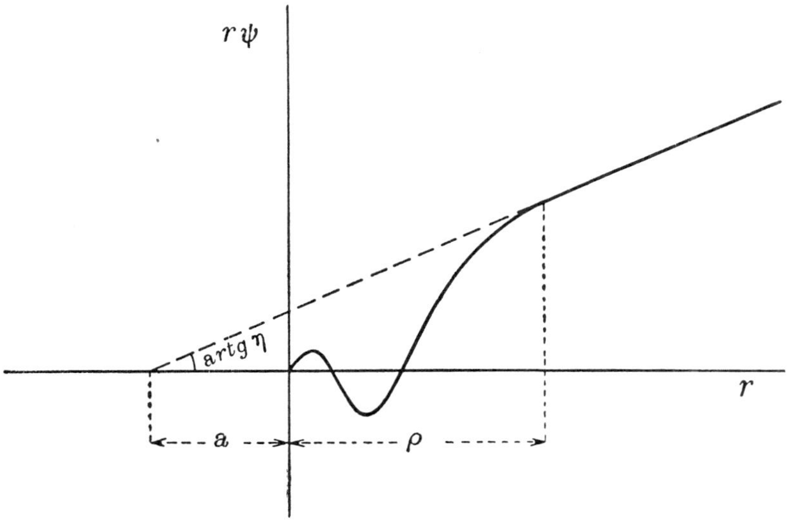

Fig. 1.

deve tendere approssimativamente al valore $\overline{\psi}$, si potrà porre, all'esterno della buca:

$$u(r) = (a + r)\,\overline{\psi} \tag{11}$$

dove a è una lunghezza, il cui significato è chiarito nella Fig. 1. In essa sono riportati in ascisse i valori di r e in ordinate quelli di u. La u, come risulta dalla (9), è nulla per $r = 0$, mentre, per r maggiore di ρ ha per grafico una retta. Prolunghiamo questa retta fino ad incontrare l'asse delle ascisse; a è la distanza del punto di intersezione dall'origine delle coordinate.

Tenendo conto di (9), (10) e (11) troviamo

$$\begin{aligned} \frac{8\pi^2 m}{h^2}\int \mathrm{V}\,\psi d\tau &= 4\pi\,\frac{8\pi^2 m}{h^2}\int \mathrm{V}\,urdr = 4\pi\int u''\,rdr \\ &= 4\pi \,\big|\, u'\,r - u\,\big|_0^r = -\,4\pi a\overline{\psi} \end{aligned} \tag{12}$$

e siccome nell'unità di volume sono contenute n buche di potenziale, ricaviamo infine

$$\frac{8\pi^2 m}{h^2}\,\overline{\Sigma \mathrm{V}_i\psi} = -\,4\pi a n\overline{\psi}\,. \tag{13}$$

Con ciò la (8) diventa

$$\Delta\overline{\psi} + \frac{8\pi^2 m}{h^2}(\mathrm{W}_0 - \mathrm{U})\,\overline{\psi} = 0 \tag{14}$$

dove si è posto

$$\mathrm{W}_0 = \mathrm{W} + \frac{h^2 a n}{2\pi m}\,. \tag{15}$$

L'equazione (14) per ψ ha la forma dell'equazione di Schrödinger per il caso in cui non esista il gas estraneo. Segue dunque da essa che W_0 prenderà valori eguali ai livelli energetici dell'atomo imperturbato. Secondo la (15) l'energia W dell'elettrone in presenza del gas estraneo differisce da W_0 per la costante

$$-\frac{h^2 an}{2\pi m}. \tag{16}$$

Per a positivo i termini elevati si abbassano rispetto al termine fondamentale, mentre per a negativo essi si alzano. Nel primo caso le righe si spostano verso il rosso e nel secondo si ha invece spostamento verso il violetto. Lo spostamento, in cm^{-1} è dato da

$$\Delta_\sigma = \frac{han}{2\pi mc}. \tag{17}$$

La costante a che figura in questa formula e che ha l'ordine di grandezza delle dimensioni atomiche, potrebbe calcolarsi, integrando l'equazione differenziale (10) se fosse conosciuta la forma del potenziale V dovuto a un atomo del gas estraneo. In realtà la conoscenza di questo potenziale è troppo imperfetta per permettere anche una valutazione grossolana della costante.

È interessante però osservare che la grandezza di a, non però il suo segno, si possono ricavare quando si conosca la sezione d'urto di elettroni lentissimi contro gli atomi del gas estraneo. Per elettroni lentissimi infatti la sezione d'urto è dovuta, come è noto, solamente alla diffusione delle onde s. In questo caso essa è data da (3)

$$\sigma = \frac{h^2}{\pi p^2} \sin^2 \delta_0 \tag{18}$$

dove p è la quantità di moto (per ipotesi assai piccola) dell'elettrone e δ_0 misura la differenza di fase tra onde s corrispondenti ad elettroni liberi di impulso p e onde corrispondenti ad elettroni su cui agisce il potenziale V dell'atomo. Dalla (11) si vede facilmente che, al limite per p molto piccolo, si ha

$$\delta_0 = \frac{2\pi pa}{h}. \tag{19}$$

Con ciò la (18), sostituendo l'arco al seno, diventa

$$\sigma = 4\pi a^2. \tag{20}$$

Questa formula permette di ricavare il valore assoluto di a, non però il suo segno. Sostituendo nella (17) si ottiene

$$\Delta_\sigma = \pm n \frac{h\sqrt{\sigma}}{4\pi^{3/2} mc}. \tag{21}$$

Lo spostamento complessivo delle righe è dato dalla somma di Δ_ε e di Δ_σ; esso risulta dunque, secondo la (5) e la (17),

$$\Delta = 20 \frac{(\varepsilon - 1) e^2}{8\pi hc} n^{1/3} + \frac{han}{2\pi mc}. \tag{22}$$

(3) Cfr. per esempio G. WENTZEL, «Handbuch der Physik», XXIV/1, p. 711, 2ª edizione, Berlino 1933.

Nella maggior parte dei casi, il secondo termine è prevalente rispetto al primo, e siccome esso può essere per alcuni gas positivo e per altri negativo, ne risulta che anche lo spostamento totale delle righe può essere sia verso il rosso che verso il violetto, come effettivamente si osserva.

Per il confronto delle formule qui ottenute con l'esperienza rimandiamo al citato lavoro di Amaldi e Segrè.

Terminiamo questo paragrafo ricapitolando le condizioni necessarie per la validità dei risultati qui raggiunti. Esse sono:

a) che il gas estraneo sia monoatomico, in modo che il suo potenziale possa considerarsi avere simmetria sferica;

b) che la densità del gas estraneo sia abbastanza grande perché entro un cubo di lato eguale alla lunghezza d'onda di De Broglie dell'elettrone luminoso siano contenuti molti atomi del gas; questa condizione è effettivamente verificata per gas a pressione dell'ordine di una atmosfera e per termini aventi numero quantico totale di alcune decine;

c) che il gas estraneo sia abbastanza rarefatto perché la distanza media tra i suoi atomi risulti grande sia rispetto al raggio degli atomi stessi, che rispetto al raggio d'urto per elettroni di piccola velocità.

§ 4. Vogliamo da ultimo precisare le condizioni necessarie perché la perturbazione esercitata dagli atomi del gas estraneo non sia tale da determinare una confusione delle righe tra di loro, che renderebbe impossibile ogni osservazione. Trascuriamo in ciò le comuni cause di allargamento delle righe spettrali (effetto Doppler, urti di tutto l'atomo contro gli atomi del gas estraneo, ecc.) limitandoci solo a tener conto di quelle che sono peculiari del fenomeno che stiamo esaminando. Esse sono sostanzialmente due, una delle quali è da ricondursi all'effetto della polarizzabilità degli atomi del gas estraneo, che è stato esaminato nel § 2; mentre l'altra dipende dall'azione delle buche di potenziale esaminata nel § 3.

La prima causa di allargamento dipende dal fatto che lo spostamento ΔW_ε del termine, come risulta dalla (2) è proporzionale alla grandezza $\Sigma \frac{1}{r_i^4}$. Il valore di questa dipende dalla distribuzione degli atomi del gas estraneo attorno al resto atomico dell'atomo considerato, ed i suoi valori possono, a seconda delle possibili distribuzioni, oscillare entro limiti piuttosto larghi. Dai calcoli di Wick sopra la legge di distribuzione della grandezza $\Sigma \frac{1}{z_i^{4/3}}$, risulta che la funzione che dà la distribuzione statistica di questa grandezza ha valori eguali alla metà del massimo della funzione stessa in corrispondenza degli argomenti 2 e 5 circa. Da ciò si deduce, al modo seguito nel § 2 che la riga ha, per questa ragione, una larghezza di circa

$$20 \frac{(\varepsilon - 1) e^2}{8\pi hc} n^{1/3} \text{ cm}^{-1} \tag{23}$$

che in condizioni pratiche può spesso superare un cm^{-1}.

Una seconda causa di allargamento delle righe è la seguente. Consideriamo classicamente il moto dell'elettrone luminoso lungo la sua grande orbita di numero quantico assai elevato. Esso traversa nel suo percorso gli atomi del gas estraneo, muovendosi abbastanza lentamente attraverso ad essi. Ora è evidentemente necessario, affinché le righe possano apparire distinte, che l'elettrone possa percorrere almeno alcuni giri lungo la sua orbita prima che un urto contro una delle molecole del gas estraneo ne disturbi il movimento. Se σ è la sezione d'urto di un elettrone lentissimo contro gli atomi del gas estraneo, il suo cammino libero medio è

$$l = 1/\sigma n \,. \tag{23}$$

Perché le righe possano apparire distinte è dunque necessario che questo cammino libero medio sia più grande del perimetro dell'orbita. Tale perimetro, per orbite p di numero quantico totale n^* elevato, è dato approssimativamente da

$$4\, a n^{*2} \tag{24}$$

dove a rappresenta qui il raggio della prima orbita di Bohr. La condizione che il perimetro dell'orbita sia piccolo a confronto del cammino libero medio dell'elettrone ci dà dunque

$$4\, a n^{*2} < \frac{1}{\sigma n} \,. \tag{25}$$

Dal che si ricava che, per questa ragione, non è praticamente possibile osservare righe di numero d'ordine superiore a

$$n_0^* = \frac{1}{2\sqrt{a\sigma n}} \,. \tag{26}$$

Per esempio nel caso che il gas estraneo sia Argon alla concentrazione normale di $2{,}7 \cdot 10^{19}$ atomi per cm^3 prendendo per la sezione d'urto il valore trovato da Amaldi e Segrè $\sigma = 11{,}6 \cdot 10^{-16}\ cm^2$ si trova da (26) $n_0^* = 39$. In pratica naturalmente, per effetto delle altre cause di allargamento, il numero d'ordine delle ultime righe fino alle quali si riesce a seguire la serie di assorbimento in presenza della concentrazione indicata di atomi di Argon sarà alquanto minore di questo limite.

96.

RADIOATTIVITÀ PRODOTTA DA BOMBARDAMENTO DI NEUTRONI (*)

«Nuovo Cimento», *11*, 429–441 (1934).

SUNTO. – Vengono descritte esperienze eseguite per ricercare se si possano produrre radioattività artificiali per mezzo del bombardamento con neutroni. Segue un elenco di alcuni elementi per i quali si è potuto riscontrare l'effetto cercato e alcune considerazioni teoriche sulle particolarità del fenomeno.

INTRODUZIONE.

§ 1. I fenomeni di disintegrazione artificiale dei nuclei atomici che si producono per effetto del bombardamento con particelle pesanti (particelle α, protoni, deutoni, neutroni, ecc...) conducono per solito alla formazione di nuove specie atomiche differenti da quella bombardata. Si riteneva generalmente, fino agli ultimi tempi, che gli atomi così formati corrispondessero sempre a comuni isotopi stabili. Si deve a F. Joliot e I. Curie (1) l'aver dimostrato per primi che, almeno in alcuni casi, il prodotto della disintegrazione può essere radioattivo e subire una disintegrazione successiva con vita media relativamente lunga. In questi casi dunque il processo si può considerare costituito da due fasi:

a) in una prima fase l'atomo A colpito dalla particella bombardante subisce in un tempo estremamente breve una trasmutazione in una nuova specie atomica B;

b) se l'atomo B così formato non corrisponde ad alcun isotopo stabile, esso può disintegrarsi ulteriormente con una vita media relativamente lunga, così che questa seconda disintegrazione può seguirsi nel suo svolgimento anche dopo allontanata la sorgente delle particelle bombardanti.

Nelle loro esperienze i coniugi Joliot si sono serviti, per bombardare i diversi elementi, delle particelle α del polonio. Queste particelle, a causa della repulsione coulombiana, possono, come è ben noto, esercitare un effetto soltanto sopra elementi leggeri. Infatti Joliot e altri (2) poterono constatare che alcuni elementi leggeri (B, N, Mg, Na, Al, P) divengono radioattivi per azione del bombardamento con particelle α. In tutti questi casi, ad eccezione del

(*) Confronta le note preliminari in «Ricerca Scientifica», *5* (1), pp. 283, 330, 452, 533, 650; (2) p. 21 (1934); «Nature», *133*, 757, 898 (1934).

(1) I. CURIE e F. JOLIOT, «C. R.», *198*, 254, 561 (1934).

(2) I. CURIE e F. JOLIOT, loc. cit., L. WERTENSTEIN, «Nature», *133*, 564 (1934), O. FRISCH, «Nature», *133*, 721 (1934).

magnesio, di cui si parlerà in seguito, il corpuscolo emesso è un positrone. Cockcroft, Gilbert, Walton e altri (3) ottennero fenomeni analoghi a quelli osservati dai Joliot bombardando il carbonio con protoni accelerati artificialmente; infine Crane, Lauritsen e Harper (4) si servirono come proiettili di deutoni.

Mi propongo di riferire nel presente lavoro sopra alcune esperienze da me fatte per ricercare radioattività artificiali prodotte da bombardamento con neutroni. L'uso dei neutroni come corpuscolo bombardante presenta lo svantaggio che il numero dei neutroni di cui si può praticamente disporre è enormemente inferiore al numero delle particelle α che si possono ottenere da sorgenti radioattive, o al numero di protoni o di deutoni che si possono ricavare nei tubi ad alta tensione; questo inconveniente è per altro compensato in parte dalla maggiore efficacia dei neutroni come agenti per produrre disintegrazioni artificiali. I neutroni presentano anche il vantaggio di poter produrre effetti sopra gli elementi pesanti, per modo che il numero di elementi che è stato possibile attivare per mezzo di essi è molto maggiore del numero di elementi attivi che si sono ottenuti usando altre specie di proiettili.

Nel § 2 è contenuta una descrizione del dispositivo usato in queste ricerche; nel § 3 sono riassunti alcuni dei risultati; nel § 4 i risultati sono discussi dal punto di vista teorico.

Dispositivo sperimentale.

§ 2. La sorgente dei neutroni era costituita da un tubetto di vetro saldato, avente il diametro esterno di circa 6 mm e lunghezza di circa 15 mm contenente polvere di berillio ed emanazione di radio. Si può ritenere che una tale sorgente emetta all'incirca 1000 neutroni al secondo per ogni millicurie di emanazione; ho usato quantità di emanazione variabili da 50 a 800 millicurie. La distribuzione energetica dei neutroni non è ben conosciuta; si tratta di una distribuzione continua di energie estendentesi da zero fino a 7 o 8 milioni di volt; è inoltre probabile la presenza di una piccola percentuale di neutroni aventi energie fino a 12 o 15 milioni di volt.

Oltre ai neutroni, una sorgente del tipo che abbiamo descritto emette naturalmente anche una grande quantità di raggi γ. Essi tuttavia non producono alcun disturbo in esperienze di questo genere, essendosi dimostrato che la emanazione da sola (senza il berillio) non produce attivazioni; ciò, ben inteso, quando le pareti del recipiente che la racchiude assorbono completamente le particelle α. Deve infine tenersi presente che i neutroni del berillio sono anche accompagnati da una radiazione γ assai dura (5 o 6 milioni di volt; circa un quanto γ per ogni neutrone); sembra tuttavia improbabilissimo che i fenomeni che descriveremo siano connessi con la presenza di questa radiazione γ, poiché radiazioni γ enormemente più intense e di durezza non molto inferiore non producono alcun effetto osservabile.

(3) J. D. Cockcroft, G. W. Gilbert, E. T. S. Walton, «Nature», *133*, 328 (1934); Neddermeyer e C. D. Anderson, «Phys. Rev.», *45*, 498 (1934).

(4) Lauritsen, Crane e Harper, «Science», *79*, 234 (1934).

Il dispositivo usato per rivelare eventuali attivazioni delle sostanze bombardate consisteva in un contatore a filo di Geiger–Müller. Il tubo di questo era costituito da una foglia di alluminio dello spessore di 0,1 o 0,2 mm tale da permettere l'ingresso nel contatore anche ad elettroni di energia non molto grande. Le dimensioni dei contatori erano generalmente circa 5 cm di lunghezza e 1,4 cm di diametro. Il filo di alluminio del diametro di 0,1 o 0,2 mm era connesso nel modo consueto a un sistema amplificatore degli impulsi che faceva agire un numeratore su cui si leggeva di tempo in tempo il numero degli impulsi. L'aria nell'interno dei contatori era alla pressione di 5 a 10 cm di mercurio, in modo da avere tensioni di funzionamento da 1000 a 1500 volt.

Le sostanze da esaminare erano molto spesso disposte in forma di cilindretti di diametro tale da poter venire infilate sul contatore, in modo che le perdite per fattori geometrici fossero ridotte al minimo possibile. Dapprima i cilindretti venivano posti attorno alla sorgente dei neutroni sicché la sostanza rimaneva sottoposta per un certo tempo a un intenso bombardamento dei neutroni. Successivamente il cilindretto era rapidamente portato sul contatore, che si trovava in un'altra stanza, e si registravano gli impulsi. Le sostanze pulverulente erano spesso irradiate in recipienti di carta di forma opportuna e dei quali si era preventivamente constatata la inattivabilità. Talvolta sostanze di questo tipo erano irradiate introducendole in una provetta nella quale era posta la sorgente, in modo da aumentare l'intensità della irradiazione; questo metodo naturalmente è praticabile solo per sostanze di vita non troppo breve, poiché altrimenti manca il tempo necessario a travasare la sostanza in un recipiente adatto a portarla sul contatore. Infine alcune volte, e particolarmente quando la sostanza doveva essere sottoposta ad operazioni chimiche dopo l'irradiazione, essa veniva irradiata in soluzione concentrata.

Con questi metodi ho potuto constatare che numerosi elementi in seguito al bombardamento acquistano la proprietà di emettere elettroni. Nei casi più semplici il numero degli elettroni emessi decresce esponenzialmente col tempo ed ha una vita media caratteristica dell'elemento irradiato; talvolta invece la curva di decremento della attività può analizzarsi in due o più esponenziali con vite medie diverse, ciò che è probabilmente dovuto in molti casi a diversi costituenti isotopi dell'elemento in esame, che dànno ciascuno indipendentemente un proprio effetto. In altri casi si possono avere per lo stesso isotopo diversi processi di eccitazione con formazione di altrettanti elementi attivi, avente ciascuno la propria vita media, o infine si possono avere più disintegrazioni in catena.

L'intensità della attivazione è molto diversa nei diversi elementi; talora l'effetto è appena misurabile, e si hanno nel contatore pochi impulsi al minuto. In altri casi si hanno invece delle attivazioni molto intense corrispondenti ad alcune migliaia di impulsi per minuto, tanto che ponendo le sostanze troppo vicine ai contatori non si riesce più, coi nostri apparecchi, a risolvere gli impulsi uno dall'altro. Una misura precisa della intensità della attivazione nei diversi elementi richiederebbe un lavoro abbastanza lungo, poiché sarebbe necessario precisare in modo rigoroso le particolarità geometriche sia durante la irradiazione della sostanza che durante la misura al contatore; sarebbe inoltre necessaria una conoscenza precisa della sensibilità del contatore per elettroni delle diverse durezze e infine del coefficiente di assorbimento degli elettroni emessi dalla sostanza in esame. In considerazione di ciò mi sono limitato nella maggior parte dei casi a indicazioni di carattere piuttosto qualitativo; in altri casi ho cercato di esprimere quantitativamente l'intensità ma anche per questi le misure dànno poco più che l'ordine di grandezza. Nei casi in cui l'intensità è espressa con un numero (indicato sempre con la lettera i) essa è definita nel modo seguente:

i = numero di disintegrazioni che si producono ogni secondo in un grammo dell'elemento posto a un cm di distanza da una sorgente contenente un millicurie di emanazione in equilibrio coi suoi prodotti e polvere di berillio La sostanza deve essere inoltre irradiata a saturazione, per il che è necessario esporla al bombardamento per un tempo abbastanza lungo in confronto della vita media.

Probabilmente l'elemento più incerto nella misura di i è la sensibilità del contatore. Con i contatori e la disposizione geometrica da me usati, si può calcolare che in media venissero registrate una su venti disintegrazioni (a prescindere dall'assorbimento degli elettroni entro la sostanza attivata, di cui si deve tener conto a parte). Questa sensibilità non molto elevata è giustificabile in parte con fattori geometrici, in parte con l'assorbimento degli elettroni entro le pareti del contatore e in parte con la sensibilità non totale del contatore anche rispetto agli elettroni che arrivano a penetrare nel suo interno. La sensibilità è stata determinata misurando il numero di impulsi che si ottengono avvicinando al contatore quantità misurate di uranio o di potassio.

Conoscendo l'intensità i della attivazione definita nel modo precedente, si può immediatamente risalire alla sezione d'urto, σ, per il processo di attivazione quando si conosca ancora il numero di neutroni emessi da una sorgente di un millicurie. Indicando questo numero con n, e chiamando M il peso atomico dell'elemento si ha infatti, come immediatamente si verifica,

$$\sigma = 2{,}1 \cdot 10^{-23} \frac{\mathrm{M}i}{n} .$$

Se si prende n approssimativamente eguale a 1000 si ha dunque

$$\sigma = 2 \cdot 10^{-26} \mathrm{M}i . \tag{1}$$

Per la discussione della reazione nucleare che dà origine all'elemento attivo è essenziale poter individuare la natura chimica di questo. Lo schema generale secondo cui può condursi questa ricerca è il seguente. È naturale ammettere che, bombardando un elemento chimico, questo possa trasformarsi in un elemento di numero atomico prossimo. Il numero di atomi attivi che si formano è estremamente piccolo (nelle nostre esperienze è probabile che non se ne siano mai formati più di un miliardo) e sarebbe quindi impossibile isolarli con metodi ordinari della chimica analitica. Pertanto la sostanza irradiata viene prima disciolta e alla soluzione si aggiungono piccole quantità degli elementi che si sospetta possano essere isotopi dell'elemento attivo. Con i procedimenti della analisi chimica si separano poi di nuovo questi elementi e quello originario; le varie frazioni vengono separatamente provate sui contatori per cercare con quale di esse si è accompagnata la attività. Le operazioni chimiche sono state eseguite dal Dr. O. D'Agostino che riferirà dettagliatamente su di esse in altro luogo. Ci limiteremo qui pertanto a brevi cenni in proposito.

RISULTATI DELLE ESPERIENZE.

§ 3. Raccolgo qui i risultati ottenuti per alcuni elementi da me esaminati. Un secondo elenco di elementi sarà pubblicato in un lavoro dello scrivente con Amaldi, Rasetti e Segrè.

9. *Fluoro.* – Il fluoro è, fra tutte le sostanze esaminate fino ad ora, quella che ha la vita media più breve. Il periodo di dimezzamento è di circa 9 secondi. Il fluoro è stato irradiato come fluoruro di calcio; siccome il calcio non viene attivato dai neutroni, l'effetto è senz'altro da attribuirsi al fluoro. Data la brevissima vita media di questo elemento non si può naturalmente pensare a una separazione chimica del prodotto attivo che si forma per effetto del bombardamento. Siccome d'altra parte è noto che il fluoro bombardato con neutroni si disintegra emettendo particelle α, può ritenersi probabile che avvenga la reazione nucleare seguente

$$F_9^{19} + n_0^1 = N_7^{16} + He_2^4 .$$

Si formerebbe così un N^{16} instabile che, disintegrandosi ulteriormente con emissione di una particella β si trasformerebbe in O^{16}. Il fatto che il prodotto finale che si forma sia l'ossigeno, che, come è ben noto, corrisponde a una configurazione nucleare assai stabile, trova riscontro nel fatto che i raggi β emessi sono assai duri, come risulta dalle misure del loro coefficiente di assorbimento.

La attività del fluoro è molto intensa e corrisponde approssimativamente a $i \sim 0{,}7$; l'effetto è reso ancora più appariscente dal fatto che i raggi β emessi sono molto duri, e permettono in conseguenza di utilizzare un forte spessore della sostanza.

13. *Alluminio.* – L'alluminio bombardato coi neutroni si attiva intensamente emettendo elettroni. La analisi del decadimento della attività che si ottiene sia irradiando alluminio metallico che allumina, dimostra che vi sono due periodi di decremento. Un primo con riduzione a metà in circa 12 minuti ed un secondo in circa 15 ore (questo secondo dato è poco preciso). Al primo periodo corrisponde una intensità $i \sim 0{,}8$; il secondo è un po' meno intenso; e si ha per esso approssimativamente $i = 0{,}5$.

È stata possibile una separazione chimica del prodotto che dà origine alla attività di periodo lungo. All'alluminio irradiato e portato in soluzione sono state aggiunte piccole quantità di sodio e di magnesio; sono stati poi precipitati gli idrati e i fosfati di alluminio e di magnesio e si è constatato che essi erano inattivi. Nella soluzione resta il sodio; portando a secco ed avvicinando il residuo al contatore si trova che esso contiene la attività che decade col periodo di 15 ore. Ciò rende probabile che la disintegrazione che dà origine all'elemento attivo con vita di 15 ore avvenga secondo lo schema:

$$\mathrm{Al}_{13}^{27} + n_0^1 = \mathrm{Na}_{11}^{24} + \mathrm{He}_2^4 .$$

Il Na^{24} si disintegrerebbe successivamente con emissione di un elettrone trasformandosi nell'isotopo stabile Mg^{24}.

Non è stata fatta una separazione dell'elemento attivo con la vita di 12 minuti. Si può tuttavia ritenere plausibile che esso sia un Mg^{27}; infatti le altre possibilità, Al^{28} e Al^{26} sono probabilmente da escludersi, la prima perché Al^{28}, come vedremo tra un momento, è un isotopo radioattivo avente periodo di circa 3 minuti, la seconda perché Al^{26}, se esiste, dovrebbe verosimilmente disintegrarsi emettendo dei positroni.

14. *Silicio.* - Anche il silicio irradiato coi neutroni si attiva con notevole intensità ($i \sim 0,7$). L'effetto è reso particolarmente appariscente dal fatto che gli elettroni emessi sono molto penetranti. Il periodo di dimezzamento della attività, dedotto dalle curve di decadimento sia per il silicio metallico che per il quarzo, risulta di circa 3 minuti; forse leggermente inferiore.

La separazione chimica dell'elemento attivo è stata fatta evaporando la silice irradiata con acidi fluoridrico e solforico, dopo aver aggiunti solfati di alluminio e di magnesio. Il silicio si elimina come fluoruro e nel residuo si precipita l'alluminio che trascina con sè l'attività. Resta con ciò assai probabile che la reazione nucleare avvenga secondo lo schema

$$\mathrm{Si}_{14}^{28} + n_0^1 = \mathrm{Al}_{13}^{28} + \mathrm{H}_1^1 .$$

Si formerebbe così un isotopo radioattivo dell'alluminio, Al^{28}, che disintegrandosi con emissione di elettroni, si trasforma di nuovo in Si^{28}. Resta con ciò confermata l'ipotesi dei Joliot e di Preiswerk (5) che questo sia lo stesso Al^{26} che si forma bombardando il Mg^{25} con particelle α a che, entro i limiti dell'errore, ha la stessa vita media.

15. *Fosforo.* - Anche il fosforo è uno degli elementi che si attivano con maggiore intensità. Esso presenta due vite medie; una di circa 3 ore (intensità $i \sim 0,6$) e una breve di circa 3 minuti, segnalata dai coniugi Joliot; l'intensità di questa è circa 10 volte minore di quella della precedente.

Il prodotto attivo corrispondente alla vita di 3 ore è stato separato chimicamente e si comporta come silicio. Lo si può separare facilmente irradiando il fosforo in forma di pentossido in soluzione acquosa concentratissima. Dopo l'irradiazione si riprende con acido solforico diluito, aggiungendo una piccola quantità di silicato di sodio; si tira a secco in modo da rendere la silice insolubile e si filtra dopo aver ripreso con acqua. La silice trascina con sè l'attività.

La reazione nucleare che dà origine al prodotto con la vita di 3 ore è dunque probabilmente

$$\mathrm{P}_{15}^{31} + n_0^1 = \mathrm{Si}_{14}^{31} + \mathrm{H}_1^1 .$$

Il Si^{31} così formato, con emissione di un elettrone, si trasforma di nuovo nell'elemento di partenza P^{31}.

(5) I. Curie, F. Joliot, P. Preiswerk, «C. R.», *198*, 2089 (1934).

Non è stata fatta, fino ora, una separazione chimica del prodotto con il periodo di 3 minuti. Sembra tuttavia probabile, data la eguaglianza delle vite medie, che si formi lo stesso Al^{28} che si ottiene dal silicio.

17. *Cloro*. – L'attività presentata da questo elemento per bombardamento con neutroni è notevole per la sua lunga durata. Il periodo di dimezzamento è infatti di circa 13 giorni. Intensità dell'ordine di 0,1.

Il prodotto attivo che si forma segue le reazioni del fosforo. Esso può separarsi irradiando il cloro in forma di cloruro sodico; dopo la irradiazione il cloruro viene disciolto in acido nitrico diluito; si aggiunge una piccola quantità di fosfato sodico e si precipita quindi il fosfomolibdato ammonico aggiungendo molibdato di ammonio. Il precipitato trascina con sè l'attività.

La reazione nucleare che dà origine alla attivazione del cloro sembra in conseguenza essere

$$Cl_{17}^{35} + n_0^1 = P_{15}^{32} + He_2^4 .$$

Il P^{32} formato, con emissione di elettroni, si trasforma nell'isotopo stabile S^{32}.

26. *Ferro*. – Il ferro presenta un effetto non molto intenso ($i \sim 0{,}05$) con periodo di dimezzamento di circa 2,5 ore.

Il prodotto attivo che si forma ha le reazioni del manganese; lo si può separare per esempio sciogliendo il ferro irradiato in acido nitrico; aggiungendo piccole quantità di un sale di manganese e precipitando poi questo elemento come biossido per aggiunta di clorato sodico. Il biossido di manganese trascina la attività. La reazione corrispondente a questa attivazione sembra di conseguenza essere

$$Fe_{26}^{56} + n_0^1 = Mn_{25}^{56} + H_1^1 .$$

L'isotopo instabile Mn^{56} con emissione di un elettrone ritorna Fe^{56}.

27. *Cobalto*. – Questo elemento si attiva debolmente con la stessa vita media del ferro. Anche in questo caso il prodotto attivo si comporta come manganese e può separarsi con lo stesso procedimento descritto per il caso del ferro. L'isotopo instabile che si forma è probabilmente lo stesso Mn^{56} che si ricava dal ferro. La reazione nucleare sarebbe:

$$Co_{27}^{59} + n_0^1 = Mn_{25}^{56} + He_2^4 .$$

47. *Argento*. – L'attività provocata nell'argento ha due periodi; uno più breve di circa 20 secondi, ed un altro di circa 2 minuti; l'intensità corrispondente a quest'ultimo periodo è $i \sim 0{,}5$.

53. *Jodio*. – Il periodo di dimezzamento dello jodio è di 30 minuti; si ottiene lo stesso periodo irradiando jodio elemento oppure joduro di ammonio. L'intensità è notevole ($i \sim 0{,}4$).

Tentativi di identificare la natura chimica del prodotto attivo con uno dei due elementi precedenti, antimonio e tellurio, hanno dato risultato negativo. La attività segue lo jodio, sia precipitandolo come joduro di argento, sia ricavando lo jodio elemento da joduro di ammonio irradiato, per azione dell'acido nitrico, in presenza di antimonio e tellurio. Sembra con ciò assai probabile che lo I^{127} si trasformi in un suo isotopo radioattivo I^{128} oppure I^{126} a seconda che il neutrone viene assorbito oppure produce l'emissione di un neutrone nucleare.

CONSIDERAZIONI TEORICHE.

§ 4. Vogliamo qui esporre alcuni punti di vista teorici sopra i vari processi che possono aver luogo per effetto del bombardamento con neutroni. Allo stato attuale della teoria del nucleo tali considerazioni sono necessariamente assai vaghe e malsicure; può tuttavia essere opportuno tenerle presenti nella valutazione dei risultati. Terremo conto in queste considerazioni, oltre che dei risultati esposti nel presente lavoro, anche di altri ottenuti in questo laboratorio e attualmente in corso di pubblicazione [6].

I fatti empirici che sono emersi nello studio delle radioattività artificiali prodotte dal bombardamento con neutroni possono riassumersi nei seguenti punti:

a) una larga percentuale degli elementi, sia leggeri che pesanti, sono suscettibili di attivarsi per effetto del bombardamento; non si nota da questo punto di vista una particolare differenza tra elementi leggeri e pesanti;

b) le sezioni d'urto per gli elementi che presentano il fenomeno con notevole intensità arrivano ad essere dell'ordine di grandezza della sezione geometrica del nucleo, e cioè una notevole percentuale dei neutroni che colpiscono il nucleo produce l'effetto;

c) il prodotto attivo che si forma sotto l'azione del bombardamento in qualche caso è isotopo dell'elemento bombardato; in qualche altro caso esso ha numero atomico inferiore di una o due unità. A questo riguardo sembra vi sia una differenza di comportamento fra elementi leggeri e pesanti. Mentre infatti per i primi della maggior parte dei casi l'elemento attivo ha un numero atomico inferiore a quello dell'elemento originario, ciò non è più vero per elementi pesanti; nei cinque casi esaminati chimicamente di elementi aventi numero atomico maggiore di 32 e non spontaneamente radioattivi si è sempre trovato che il prodotto attivo è isotopo dell'elemento bombardato;

d) gli elettroni emessi dagli elementi attivati sono sempre negativi; o, per lo meno, non si sono mai trovati casi di emissioni di elettroni positivi con intensità sufficiente per poter essere rivelati.

Non sembra che vi siano speciali difficoltà per rendersi conto del meccanismo del fenomeno per il caso degli elementi leggeri; il processo più frequente sembra in questi casi consistere nell'assorbimento del neutrone urtante, immediatamente seguito dalla espulsione di una particella α o di un protone. Se l'energia con la quale viene espulsa la particella α o il protone è di alcuni milioni di volt, risulta dalla teoria di Gamow che, per elementi leggeri, il tempo necessario per espellere il corpuscolo è estremamente breve, così che la probabilità che la espulsione possa avvenire prima che il nucleo abbia il tempo di liberarsi dal neutrone temporaneamente assorbito è effettivamente considerevole. Dopo finito questo processo, che può durare un tempo dell'ordine di

(6) E. AMALDI, E. FERMI, F. RASETTI, E. SEGRÈ, «Nuovo Cimento», *11*, 442 (1934); E. AMALDI, E. SEGRÈ, «Nuovo Cimento», *11*, 452 (1934); O. D'AGOSTINO, «Gazz. Chimica Italiana», in stampa, 1934.

grandezza di 10^{-20} secondi, il nucleo si trova trasformato in un nuovo isotopo avente, almeno in media, peso atomico maggiore di quello che competerebbe al suo numero atomico (si noti che tanto l'assorbimento di un neutrone, quanto la espulsione di un protone o di una particella α sono processi che tendono ad elevare il numero dei neutroni nucleari in paragone del numero dei protoni). Questa è verosimilmente la ragione per cui si osserva in genere emissione di elettroni negativi; infatti l'emissione di un elettrone nucleare, che può interpretarsi connessa alla trasformazione di un neutrone in un protone, viene a ristabilire il rapporto normale tra i numeri dei neutroni e dei protoni entro il nucleo, e determina il passaggio ad un isotopo stabile.

Via via che cresce il peso atomico degli elementi bombardati, l'ostacolo opposto dalla barriera di potenziale che circonda il nucleo alla espulsione di particelle α o di protoni va rapidamente crescendo; è dunque comprensibile che processi nei quali si ha espulsione di una di queste particelle, con conseguente diminuzione del numero atomico dell'elemento, vadano diventando sempre meno probabili col crescere del peso atomico.

Maggiori difficoltà per la interpretazione teorica presentano i casi in cui l'elemento attivo resta isotopo dell'elemento bombardato; questi casi, come si è osservato, sembrano ricorrere normalmente fra gli elementi pesanti. L'ipotesi più ovvia consisterebbe nell'ammettere che il neutrone urtante venisse catturato dal nucleo urtato, formando un prodotto instabile avente peso atomico di una unità maggiore di quello dell'elemento primitivo, e isotopo di questo. Questa interpretazione, che giustificherebbe anche il fatto che normalmente vengano emessi degli elettroni negativi, incontra tuttavia serie difficoltà teoriche quando si cerchi di spiegare con quale meccanismo il neutrone possa essere stabilmente o quasi stabilmente catturato dal nucleo. Se infatti ammettiamo la opinione corrente che i neutroni siano attratti dai nuclei con forze agenti solo quando la loro distanza dal centro del nucleo è dell'ordine di grandezza di 10^{-12} cm, dobbiamo concludere che un neutrone avente un eccesso di energia di alcuni milioni di volt può trattenersi nell'interno di un nucleo (e cioè trovarsi in intensa interazione con i corpuscoli che costituiscono il nucleo) soltanto per un tempo dell'ordine di grandezza di 10^{-21} secondi, pari all'incirca al tempo che il neutrone impiegherebbe classicamente a traversare il nucleo. Perché il neutrone possa venir catturato sembrerebbe dunque necessario che durante questo tempo brevissimo il neutrone riesca a perdere la sua energia (emettendo per esempio un quanto γ) in modo da restare legato stabilmente. Ora la probabilità della emissione di un quanto γ in un tempo così breve, almeno se la si calcola con i consueti procedimenti, è molto piccola e sembra assolutamente insufficiente a spiegare la intensità delle radioattività artificiali di questo tipo; se si vuol conservare l'ipotesi che il neutrone urtante resti aderente al nucleo, bisognerebbe dunque ammettere o che la emissione di un quanto γ o altro processo equivalente (quale per esempio la formazione di una coppia elettrone–positrone) possa avvenire con una rapidità maggiore di quanto non sia generalmente ammesso; oppure che, per ragioni che sfuggono alla attuale teoria dei nuclei, un nucleo possa mantenersi per un tempo di almeno 10^{-16} secondi in uno stato di energia abbastanza alta da poter espellere un neutrone di alcuni milioni di volt.

Un'altra ipotesi che potrebbe sostituirsi a quella che abbiamo ora discussa consiste nell'ammettere che il neutrone urtante, senza venire catturato, possa produrre l'espulsione di uno dei neutroni del nucleo urtato. Il processo potrebbe descriversi nel modo seguente. Il neutrone primitivo, passando vicino al nucleo, cede ad esso una parte della sua energia eccitandolo e subendo una specie di urto anelastico. Non vi sono difficoltà teoriche per ammettere che processi di questo genere possano avvenire assai frequentemente e che gli urti anelastici possano rappresentare una elevata percentuale degli urti tra neutrone e nucleo. Se la energia ceduta al nucleo è abbastanza grande perché il processo di emissione di uno dei suoi neutroni diventi esoenergetico, è assai probabile che, prima che il nucleo ritorni allo stato normale emettendo un quanto γ, venga espulso un neutrone, e si formi così un nuovo nucleo isotopo dell'originario ma con peso minore di una unità.

Si può obbiettare contro questa ipotesi che se in questi casi il numero dei neutroni decresce invece di crescere, dovrebbe a priori essere più probabile una susseguente emissione di positroni che non una di elettroni negativi quale invece si osserva. A questa obbiezione si può tuttavia rispondere che, nei pochi casi studiati di elementi pesanti i quali attivandosi, per bombardamento di neutroni, rimangono isotopi di sè stessi e per i quali è conosciuta la costituzione isotopica degli elementi contigui, esiste sempre un isotopo stabile dell'elemento $Z + 1$ avente peso atomico di una unità inferiore a quello dell'elemento bombardato. L'elemento perdendo un neutrone e successivamente emettendo un elettrone potrebbe andare a finire in questo isotopo. Fa eccezione l'uranio in cui probabilmente si hanno disintegrazioni successive. Deve infine tenersi presente che se per un nucleo instabile esiste energicamente la doppia possibilità di disintegrarsi sia emettendo un elettrone negativo che uno positivo, la teoria dei raggi β [7] dà che, a parità di altre condizioni, è più probabile la emissione di un elettrone negativo.

In conclusione, allo stato attuale di incertezza delle teorie sul nucleo, sembra prematuro il volersi formare una opinione definitiva sul meccanismo della disintegrazione per gli elementi pesanti; e converrà attendere che nuove esperienze portino maggiori dati per questa discussione.

Le sorgenti di emanazione per tutte le ricerche compiute in questo Laboratorio sono state preparate dal prof. G. C. Trabacchi, direttore del Laboratorio fisico della Sanità Pubblica. Senza questo suo aiuto tutte le ricerche descritte sarebbero state impossibili. Tengo pertanto a esprimergli anche a nome dei miei collaboratori la più viva gratitudine. Larga parte dei mezzi impiegati nella ricerca ci sono stati messi a disposizione dal Consiglio Nazionale delle Ricerche.

Ringrazio infine i proff. N. Parravano e L. Rolla per averci procurato alcune sostanze rare.

(7) E. FERMI, «Nuovo Cimento», *II*, 1 (1934).

97.

NUOVI RADIOELEMENTI PRODOTTI CON BOMBARDAMENTO DI NEUTRONI (*)

E. AMALDI, E. FERMI, F. RASETTI ed E. SEGRÈ

«Nuovo Cimento», *11*, 442–447 (1934).

SUNTO. – In questo lavoro viene dato un secondo elenco di elementi che diventano radioattivi per bombardamento con neutroni. Alla fine si trova una tabella che riassume i risultati di tutte le ricerche compiute in questo campo.

Le prime ricerche sulla radioattività provocata dal bombardamento di neutroni hanno mostrato l'interesse di esaminare sistematicamente il comportamento del maggior numero possibile di elementi.

Il metodo generale seguito in questa ricerca è stato già descritto nella Nota precedente (1). Le operazioni chimiche, che sono state eseguite dal dott. O. D'Agostino, sono qui semplicemente accennate perché egli ne riferirà in un prossimo lavoro.

Abbiamo esaminato i seguenti elementi:

1. *Idrogeno*. – Acqua irradiata 14 ore con una sorgente di emanazione più berillio di 670 mC non ha dato effetto misurabile.

3. *Litio*. – Litina irradiata 14 ore con 750 mC non ha dato effetto misurabile.

4. *Berillio*. – Abbiamo usato il berillio allo stato di elemento, fornitoci da Fraenkel & Landau, puro al 98 %; si è notata una leggera attività con periodo di alcuni minuti che però potrebbe essere dovuta ad impurità di sostanze fortemente attive (Si, Al).

5. *Boro*. – Abbiamo trovato una traccia di attività che però non riteniamo molto significativa data la poca purezza della sostanza a disposizione.

6. *Carbonio*. – Paraffina irradiata 15 ore con 220 mC non ha dato effetto misurabile.

7. *Azoto*. – Carbonato di guanidina, che contiene circa il 35 % di azoto, irradiato 15 ore con 500 mC non ha dato effetto misurabile.

8. *Ossigeno*. – Inattivo, vedi idrogeno.

11. *Sodio*. – Questo elemento si attiva con un periodo di circa 40 secondi. Intensità dell'ordine di grandezza 0,1. Abbiamo tentato di separare chimicamente il principio attivo nell'ipotesi che fosse fluoro aggiungendo ad una soluzione di nitrato sodico irradiato acido fluoridrico e precipitando fluoruro cerico. Il precipitato esaminato 2 minuti dopo a fine dell'irradiamento

(*) Vedi le comunicazioni preliminari sulla «Ricerca Scientifica», *5* (1), pp. 283, 330, 452, 533, 650 (1934).

(1) E. FERMI, «Nuovo Cimento», *11*, 429 (1934).

del nitrato sodico si è dimostrato inattivo, il che ci fa ritenere poco probabile che il principio attivo del sodio sia fluoro.

12. *Magnesio.* - Il magnesio ha due periodi; uno di circa 40 secondi ed uno di circa 15 ore. Intensità media. Dalla soluzione cloridrica del magnesio irradiato, previa aggiunta di alluminio e sodio, sono stati separati il magnesio e l'alluminio come fosfati. Essi sono inattivi. Nel filtrato contenente il sodio aggiunto si ritrova il principio attivo con periodo di 15 ore. La relazione nucleare è probabilmente la seguente:

$$\mathrm{Mg}_{12}^{24} + n_0^1 = \mathrm{Na}_{11}^{24} + \mathrm{H}_1^1$$

e quindi dà luogo allo stesso Na^{24} radioattivo che si ottiene irradiando l'alluminio. Questa identificazione è confermata anche dalle misure di assorbimento dei raggi β [(2)].

16. *Zolfo.* - Ha un periodo di circa 13 giorni. Il principio attivo si può separare dall'acido solforico irradiato aggiungendo una traccia di fosfato sodico e separando il fosforo come fosfomolibdato ammonico. Riteniamo che si tratti di un P^{32} ottenuto dallo zolfo secondo la reazione nucleare seguente.

$$\mathrm{S}_{16}^{32} + n_0^1 = \mathrm{P}_{15}^{32} + \mathrm{H}_1^1 .$$

Esso si identifica col fosforo estratto dal cloro; questa identificazione, che non potrebbe farsi con grande sicurezza attraverso il confronto dei periodi di dimezzamento i quali sono noti con poca precisione, è invece confermata da misure di assorbimento dei raggi β discretamente esatte.

20. *Calcio.* - Non ha dato effetto sensibile.

22. *Titanio.* - Ha un debolissimo effetto con periodo di circa 3 minuti. Non riteniamo del tutto escluso che possa essere dovuto a impurità.

23. *Vanadio.* - Ha un periodo di 4 minuti, intensità media. La coincidenza del periodo, entro gli errori di osservazione, fa pensare che il principio attivo sia lo stesso V^{52} che sembra formarsi dal cromo e dal manganese.

24. *Cromo.* - Ha un effetto intenso con periodo di circa 4 minuti. Per identificarne il principio attivo abbiamo irradiato del cromato ammonico, aggiunto vanadato sodico e separato il vanadio per aggiunta di cloruro ammonico. L'attività è stata ritrovata nel vanadio. Abbiamo anche aggiunto titanio, in presenza di vanadio, e separato il titanio per idrolisi; l'attività sembra non segua il titanio. Pertanto riteniamo probabile che il principio attivo sia V^{52} formatosi secondo la reazione

$$\mathrm{Cr}_{24}^{52} + n_0^1 = \mathrm{V}_{23}^{52} + \mathrm{H}_1^1 .$$

25. *Manganese.* - Questo elemento irradiato sotto forma di biossido si attiva con media intensità e due periodi: uno di circa 4 minuti l'altro di circa 150 minuti. L'attività con periodo lungo non si separa dal manganese aggiungendo cromo e vanadio e precipitandoli come cromato e vanadato di piombo. È da ritenersi perciò probabile che il principio attivo sia Mn^{56}. Ciò è convalidato dall'identità del periodo lungo del manganese con quelli del

(2) E. AMALDI ed E. SEGRÈ, «Nuovo Cimento», *11*, 452 (1934).

ferro e cobalto irradiati, da cui si estrae un manganese attivo, e da misure di assorbimento.

La reazione nucleare sembra dunque essere:

$$Mn^{55}_{25} + n^{1}_{0} = Mn^{56}_{25}.$$

Per studiare l'attività con periodo di 4 minuti abbiamo irradiato nitrato di manganese, aggiunto un sale di vanadio e separato il vanadio come vanadato di piombo. L'attività è stata ritrovata nel precipitato di vanadato. Una reazione analoga eseguita col cromo ha pure trascinato un po' dell'attività ma con rendimento molto minore. Riteniamo perciò probabile che il principio attivo del periodo breve sia V^{52} formatosi secondo le reazioni

$$Mn^{55}_{25} + n^{1}_{0} = V^{52}_{23} + H^{4}_{2}$$

e quindi identico al vanadio radioattivo che si ottiene per irradiazione del cromo e del vanadio. I rapporti genetici tra ferro, cobalto, manganese e vanadio sono schematicamente illustrati nella Fig. 1.

28. *Nichel.* – Irradiato 13 ore con sorgente di 250 mC non ha dato effetto misurabile.

29. *Rame.* – Ha un periodo di circa 6 minuti e intensità media. La coincidenza del periodo con quello dello zinco fa ritenere possibile che il principio attivo sia un isotopo del rame.

30. *Zinco.* – Ha un periodo di 6 minuti con piccola intensità e uno assai più lungo debolissimo. L'attività del periodo breve è stata separata col rame in presenza di nichel sia facendo precipitare solfuro rameoso che facendo depositare elettrochimicamente il rame su zinco inattivo. Il nichel è stato precipitato con dimetilgliossima e trovato inattivo. Anche lo zinco, dopo estratto il rame, è inattivo. Riteniamo dunque probabile che l'attività con periodo di 6 minuti provenga da un isotopo del rame.

31. *Gallio.* – Ha un effetto abbastanza intenso con un periodo di mezz'ora.

33. *Arsenico.* – Ha un effetto intenso con un periodo di circa un giorno. Si è cercato di separare il principio attivo insieme a gallio e germanio, precipitando il primo come ferrocianuro il secondo come solfuro. Per il germanio si è anche fatta una distillazione di cloruro germanico. L'attività è però sempre rimasta nell'arsenico. Riteniamo escluso che si tratti di gallio, mentre la prova per il germanio non può considerarsi definitiva. La reazione nucleare più plausibile ci sembra comunque:

$$As^{75}_{33} + n^{1}_{0} = As^{76}_{33}.$$

34. *Selenio.* – Ha un debole effetto con periodo di circa 35 minuti.

35. *Bromo.* – Ha un effetto intenso con due periodi uno di circa 30 minuti e l'altro di circa sei ore. Il principio attivo di ambedue i periodi non si lascia separare né coll'arsenico né col selenio. Il primo è stato separato come solfuro ed il secondo elettrochimicamente. Precipitando bromuro di argento da una soluzione di bromuro ammonico irradiato si trova l'attività nel precipitato. Riteniamo che l'attività derivi da Br^{80} e Br^{82}.

37. *Rubidio.* – Ha un effetto molto debole con periodo di circa 20 minuti.

38. *Stronzio*. – Irradiato allo stato di idrato per 13 ore con sorgente di 400 mC non ha dato effetto apprezzabile.

39. *Ittrio*. – Irradiato allo stato di ossido per 30 minuti con sorgente di 750 mC è risultato inattivo.

40. *Zirconio*. – Ha un debolissimo effetto con periodo di qualche minuto. Non escludiamo che possa trattarsi di impurità.

42. *Molibdeno*. – Ha due periodi con debole intensità. Uno di circa 15 minuti e l'altro di giorni.

4. *Rutenio*. – Irradiato 40 minuti con 750 mC non ha dato effetto.

45. *Rodio*. – Il rodio è stato irradiato in lamina metallica. Ha due periodi il primo di circa 50 secondi, il secondo di 5 minuti. L'attività del periodo breve è intensa; l'altra media.

46. *Palladio*. – Si attiva con media intensità. Sembra che abbia un periodo intorno a 6 ore e forse anche uno più breve.

48. *Cadmio*. – Abbiamo irradiato del cadmio metallico; esso si attiva debolmente con periodo di circa 70 minuti.

50. *Stagno*. – Ha mostrato una traccia di attività di cui non abbiamo potuto misurare il periodo. Forse è dovuta ad impurità.

51. *Antimonio*. – Come l'elemento precedente.

52. *Tellurio*. – Ha una debole attività che decade con un periodo dell'ordine di grandezza di mezz'ora.

55. *Cesio*. – Ha un debolissimo effetto con periodo non ancora determinato.

56. *Bario*. – Abbiamo irradiato idrato di bario. Esso ha una debole attività che decade con un periodo di 3 minuti.

57. *Lantanio*. – Non abbiamo potuto rilevarne un'attività.

58. *Cerio*. – Nitrato di cerio irradiato per 30 minuti con 400 mC non ha dato effetto misurabile.

59. *Praseodimio*. – Abbiamo irradiato ossido purissimo generosamente donatoci dal prof. Rolla insieme a ossidi di Nd, Sm, La, Gd, per il che teniamo ringraziarlo anche qui. Il praseodimio ha una debole attività con periodo di di circa 5 minuti e forse anche un periodo più lungo.

60. *Neodimio*. – Abbiamo irradiato l'ossido; ha un debole effetto con un periodo di circa un'ora.

62. *Samario*. – Abbiamo irradiato l'ossido. Ha una debole attività con un periodo di circa 40 minuti.

73. *Tantalio*. – Irradiato due ore e mezzo con 750 mC non ha dato effetto misurabile.

74. *Tungsteno*, – Irradiato 12 ore con sorgente di 650 mC ha mostrato una debolissima attività, di periodo probabilmente piuttosto lungo.

75. *Renio*. – Irradiato 10 minuti con sorgente di 750 mC non ha mostrato effetto.

76. *Osmio*. – Irradiato 15 ore con 450 mC non dà effetto.

77. *Iridio*. – Irradiato in lamina ha un effetto intenso con periodo di 20 ore. Osmio e renio aggiunti ad una soluzione di cloruro di iridio irradiato e poi separati, il primo per distillazione del tetrossido, il secondo come

solfuro, sono stati trovati inattivi. L'iridio invece rimane attivo facendo supporre che l'attività sia dovuta a un isotopo di questo elemento.

78. *Platino.* – Platino metallico fisicamente puro di Heraeus ha mostrato una traccia di attività.

79. *Oro.* – Abbiamo irradiato oro in lamina; esso ha un periodo di circa 2 giorni e una discreta attività. Sciolto l'oro irradiato in acqua regia, con platino e iridio, abbiamo precipitato questi ultimi due elementi come cloroplatinati e cloroiridiati. Essi erano inattivi. L'oro invece era rimasto attivo e conservava questa attività anche dopo sciolto in mercurio ed evaporato quest'ultimo. Reputiamo che l'attività sia dovuta ad un isotopo dell'oro.

80. *Mercurio.* – Ha un effetto debolissimo forse dovuto ad impurità.

81. *Tallio.* – Come l'elemento precedente.

82. *Piombo.* – Irradiato per una diecina di ore con sorgente su i 500 mC non mostra attività.

83. *Bismuto.* – Come l'elemento precedente.

90. *Torio.* – Da saggi preliminari sul torio liberato dai suoi prodotti radioattivi β risulta una forte attività con almeno due periodi di cui uno più breve di un minuto e l'altro dell'ordine di grandezza di un quarto d'ora.

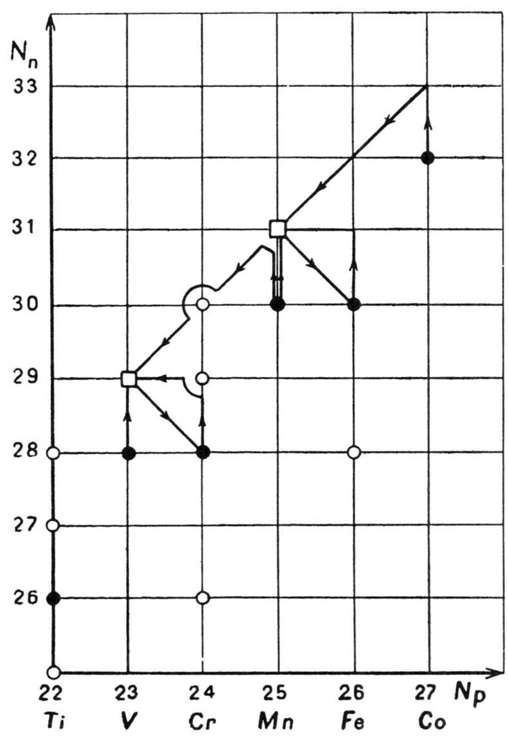

Fig. 1. – In questa figura sono illustrate le trasformazioni degli elementi con Z compreso tra 23 e 27. Le ascisse dànno il numero di protoni e le ordinate quello dei neutroni dei nuclei. Gli isotopi stabili sono rappresentati da circoletti pieni se l'isotopo costituisce più del 20 % dell'elemento, vuoti altrimenti. I quadratini denotano gli isotopi radioattivi e le linee con le frecce indicano le successive trasformazioni.

92. *Uranio.* – Questo elemento dà luogo a fenomeni più complessi dei precedenti e ha formato oggetto di una Nota a parte (3).

RIASSUNTO DEI RISULTATI ESPOSTI IN QUESTO E IN ALTRI LAVORI (4).

I risultati sono ricapitolati nella seguente tabella.

La prima colonna dà il simbolo degli elementi studiati e il loro numero atomico. La seconda i pesi degli isotopi; sono scritti in grassetto gli isotopi che rappresentano più del 20 % dell'elemento. La terza colonna dà i periodi di dimezzamento, un tratto indica che l'elemento è stato sperimentato e trovato inattivo. La quarta dà l'intensità dell'attivazione. Essa è indicata con tre lettere: *f* (forte), *m* (media), *d* (debole). Per una definizione più precisa della intensità vedi (5). I dati di questa colonna e delle successive si riferiscono ordinatamente ai diversi periodi indicati nella terza colonna. La quinta colonna dà l'energia media degli elettroni in milioni di volt. Questi dati generalmente alquanto incerti sono dedotti da misure di assorbimento, qualche volta con forti estrapolazioni (6). La sesta colonna indica se sono stati trovati raggi γ. Un tratto significa che sono stati cercati e non trovati. La settima colonna dà l'isotopo a cui è dovuta l'attività. Nei casi dubbi in cui non si sa se il neutrone sia stato assorbito od emesso si è sempre supposto che esso sia stato assorbito.

Numero Atomico ed Elemento	Isotopi	Periodi di dimezzamento	Intensità	Energia media elettroni in 10^6 e. volt	Raggi γ	Isotopo attivo
1 H	**1**, 2	—	—			
3 Li	6, **7**	—	—			
4 Be	**9**	?	?			
5 B	**10**, **11**	?	?			
6 C	**12**, 13	—	—			
7 N	**14**, 15	—	?			
8 O	**16**, 17, 18	—	—			
9 F	**19**	9^s	*f*	2	si	N^{16} (?)
11 Na	**23**	40^s	*m*			
12 Mg	**24**, 25, 26	40^s; 15^h	*m*; *m*	—; 0,5	(?); si	—; Na^{24}
13 Al	**27**	12^m; 15^h	*f*; *f*	0,6; 0,5	si; si	—; Na^{24}
14 Si	**28**, 29, 30	3^m	*f*	1,3	si	Al^{28}
15 P	**31**	3^m; 3^h	*m*; *f*	—; 0,7	?	Si^{31}
16 S	**32**, 33, 34	13^d	*m*	0,8	—	P^{32}
17 Cl	**35**, 37	13^d	*m*	0,8		P^{32}
20 Ca	**40**, 42, 43, 44	—	—			
22 Ti	46, 47, **48**, 49, 50	3^m	*d*			
23 V	**51**	4^m	*m*	1,3		V^{52} (?)
24 Cr	50, **52**, 53, 54	4^m	*m*	1,3	si	V^{52}
25 Mn	**55**	4^m; 150^m	*m*; *m*	—; 1,3		V^{52}; Mn^{56}
26 Fe	54, **56**	150^m	*m*	1,3	si	Mn^{56}
27 Co	**59**	150^m	*d*			Mn^{56}
28 Ni	**58**, **60**, 61, 62	—	—			
29 Cu	**63**, **65**	6^m	*m*			
30 Zn	**64**, **66**, 67, 68, 70	6^m ; ?	*d*; *d*			Cu; —
31 Ga	**69**, **71**	30^m	*m*			
33 As	**75**	1^d	*f*	1,3	si	As^{76}
34 Se	74, 76, 77, **78**, **80**, 82	35^m	*d*			
35 Br	**79**, **81**	30^m; 6^h	*f*; *f*	—; 0,7 (?)	—	Br^{80}; Br^{82}
37 Rb	**85**, **87**	20^m	*d*			
38 Sr	86, 87, **88**	—	—			
39 Y	**89**	—	—			

(3) E. FERMI, F. RASETTI, O. D'AGOSTINO, «Ricerca Scientifica», 5, (1), p. 533 (1934). Vedi anche «Nature», *133*, 898 (1934).

(4) E. FERMI, loc. cit.; E. AMALDI ed E. SEGRÈ, loc. cit.

(5) E. FERMI, «Nuovo Cimento», loc. cit.

Numero Atomico ed Elemento	Isotopi	Periodi di dimezzamento	Intensità	Energia media elettroni in 10^6 e. volt	Raggi γ	Isotopo attivo
40 Zr	**90**, 91, **92**, 94, 96	?	*d*			
42 Mo	92, 94, 95, 96, 97, **98**, 100	15^m ; (?)	*d*; *d*			
44 Ru	96, 98, 99, 100, **101**, **102**, 104	—	—			
45 Rh	—	50^s; 5^m	*f*; *m*	—; 0,8	—	
46 Pd	—	6^h (?)	*d*	0,3	—	
47 Ag	**107**, **109**	20^s ; 2^m	*f*; *f*	—; 0,7	—	
48 Cd	110, 111, **112**, 113, **114**, 116	70^m	*d*			
50 Sn	112, 114, 115, 116, 117, **118**, 119, **120**, 121, 122, 124	—	—			
51 Sb	**121**, **123**	?	?			
52 Te	122, 123, 124, 125 **126**, (127), **128**, **130**	30^m (?)	*d*			
53 I	**127**	30^m	*f*	0,7	—	I^{128}
55 Cs	**133**	(?)	*d* (?)			
56 Ba	135, 136, 137, **138**	3^m	*d*			
57 La	**139**	—	—			
58 Ce	**140**, 142	—	—			
59 Pr	**141**	5^m	*d*			
60 Nd	**142**, 143, **144**, 145, 146	1^h	*d*			
62 Sm	144, 147, 148, 149, 150, **152**, 154	40^m	*d*			
73 Ta	**181**	—	—			
74 W	**182**, 183, **184**, **186**,	(?)	*d* (?)			
75 Re	**185**, **187**	—	—			
76 Os	186, 187, 188, 189, **190**, **192**	—	—			
77 Ir	—	20^h	*f*	1,1	si	Ir
78 Pt	—	(?)	*d* (?)			
79 Au	—	2^d	*f*	0,3	—	Au
80 Hg	196, 197, 198, 199, **200**, 201, **202**, 203, 204	(?)	(?)			
81 Tl	**203**, **205**	(?)	(?)			
82 Pb	203, 204, 205, **206**, **207**, **208**, 209	—	—			
83 Bi	**209**	—	—			
90 Th	**232**	1^m (?); 15^m?	*f*; *f*			
92 U	**238**	15^s; 40^s 13^m 100^m	*f*; *f* *f*; *f*; *f*; *f*		si	(7)

(6) E. AMALDI ed E. SEGRÈ, «Nuovo Cimento», loc. cit.

(7) Vedi E. FERMI, F. RASETTI, O. D'AGOSTINO, «Ricerca Scientifica», loc. cit. «Nature», loc. cit.

98.

ARTIFICIAL RADIOACTIVITY PRODUCED BY NEUTRON BOMBARDMENT

E. FERMI, E. AMALDI, O. D'AGOSTINO, F. RASETTI, and E. SEGRÈ

(*Communicated by Lord Rutherford, O. M., F.R.S.—Received July 25, 1934*)

« Proc. Roy. Soc. » (London) Series A, *146*, 483–500 (1934).

1. – INTRODUCTION.

This paper aims at giving a fuller account of experiments made in the Physical Laboratory of the University of Rome, on new radio-elements produced by neutron bombardment. Preliminary results have already been announced in short communications [1].

Curie and Joliot [2] first discovered that the product atom of an artificial disintegration need not always correspond to a stable isotope, but could also disintegrate with a relatively long mean life with emission of light particles. As bombarding particles they used α–particles from polonium, and found that the light particle emitted was generally a positron. Similar results were obtained on several elements by other experimenters using α–particles, and artificially accelerated protons and deutons [3].

The use of charged particles for the bombardment, limits the possibility of an activation only to light elements. Indeed, only about ten elements up to the atomic number 15 could be activated by these methods.

It seemed therefore convenient to try the effect of a neutron bombardment, as these particles can reach the nucleus even of the heaviest elements. Available neutron sources are, of course, much less intense than α–particles or proton or deuton sources. But it was reasonable to assume that this factor would be partly compensated by the higher efficiency of neutrons in producing disintegrations. Indeed, experiment showed that more than forty elements out of about sixty investigated could be activated by this method.

(1) FERMI, « Ric. Scient. », vol. 1, pp. 283, 330 (1934); AMALDI, D'AGOSTINO, FERMI, RASETTI and SEGRÈ, « Ric. Scient. », vol. 1, pp. 452, 652, 21 (1934); FERMI, RASETTI and D'AGOSTINO « Ric. Scient. », vol. 1, p. 533 (1934); FERMI, « Nature », vol. 133, pp. 757, 898 (1934). See also FERMI, « Nuovo Cim. », vol. 11, p. 429 (1934); AMALDI, FERMI, RASETTI and SEGRÈ, « Nuovo Cim. », vol. 11, p. 442 (1934); AMALDI and SEGRÈ, « Nuovo Cim. », vol. 11, p. 452 (1934); D'AGOSTINO « Gazz. Chim. Ital. », in press (1934).

(2) « C. R. Acad. Sci. Paris », vol. 198, pp. 254, 561 (1934).

(3) FRISCH, « Nature », vol. 133, p. 721 (1934); WERTENSTEIN, « Nature », vol. 133, p. 564 (1934), COCKCROFT, GILBERT and WALTON, « Nature », vol. 133, p. 328 (1934), NEDDERMEYER and ANDERSON, « Phys. Rev. ». vol. 45, p. 498 (1934); LAURITSEN, CRANE and HARPER, « Science », vol. 79, p. 234 (1934).

2. – The Experimental Method.

The neutron source consisted of a sealed glass tube about 6 mm in diameter and 15 mm in length, containing beryllium powder and radon in amounts up to 800 millicuries. According to the ordinarily assumed yield of neutrons from beryllium, the number of neutrons emitted by this source ought to be of the order of 1000 neutrons per second per millicurie. These neutrons are distributed over a very wide range of energies from zero up to 7 or 8 million volts, besides a very small percentage having energies about twice as high as this limit.

The neutrons are mixed with a very intensive γ–radiation. This does not, however, produce any inconvenience, as the induced activity is tested after irradiation, and it was shown that radon without beryllium produced no effect. The neutrons from beryllium are accompanied by a γ–radiation harder than any emitted by the radon products (5 to 6 million volts—about one γ–quantum per neutron). It seems, however, most unlikely that the observed effects are in any way connected with this γ–radiation, as a γ–radiation of enormously greater intensity and only slightly lower energy produces no effect.

The emission of electrons from the activated substances was tested with Geiger-Müller counters about 5 cm in length and 1.4 cm diameter. The walls of the counter were of thin aluminium foil, 0.1 to 0.2 mm in thickness. The applied voltage ranged between 1000 and 1500 volts. The amplified impulses were counted on a mechanical meter worked by a thyratron.

The substances to be investigated were generally put into form of cylinders, which could be fitted round the counter in order to minimize the loss in intensity through geometrical factors. During irradiation the material was located as close as possible round the source. Substances which had to be treated chemically after irradiation were often irradiated as concentrated water solutions in a test tube.

The decay curves of the induced activity were for many elements simple exponentials. Sometimes they could be analysed into two or more exponentials; it was then convenient to irradiate the substance for different lengths of time in order to activate the various components with different intensity. The existence of several mean lives is, sometimes, certainly due to different isotopic constituents of the element; when a single isotope is present it may be attributed to alternative processes of disintegration, and sometimes (uranium) to a chain of disintegrations. The intensity of activation varies within a wide range among the different elements. In some the effect is hardly measurable, the number of impulses produced by the irradiated substance being of the order of magnitude of the number of spontaneous impulses in the counter. In others the activation is so strong that when the substance is placed too near the counter the number of impulses is of the order of some thousands per minute, so that they cannot be counted because of lack of resolving power.

No accurate measurement of the intensity of activation of the different elements was carried out, as it would require experimenting in well-defined geometrical conditions, and a knowledge of the efficiency of our counters in counting electrons, and of the absorption in the substance and in the aluminium foil. However, a very rough evaluation of these factors was made, and for some elements a number expressing the intensity of activation (i) is given. This intensity is defined as the number of disintegrations per second which take place in 1 gm of the element, placed at the distance of 1 cm from a neutron source consisting of one millicurie of radon (in equilibrium with its decay products) and beryllium powder. The substance was always irradiated until saturation of the active product was reached. The efficiency of our counters (including absorption in the aluminium foil and geometrical factors) was about 1/20, as determined by the measurement of the impulses from known quantities of potassium and uranium.

From this number expressing the intensity it is easy to obtain the cross-section for the activating neutron impact, if the number of neutron emitted per second by a one millicurie source is known. Assuming this number to be 1000, one finds immediately the cross-section

$$\sigma = 2 \cdot 10^{-26} \, i \cdot A \, ,$$

A being the atomic weight of the element.

In order to be able to discuss the nuclear reaction giving rise to the active element, it is essential to identify it chemically. It is reasonable to assume that the atomic number of the active element should be close to the atomic number Z of the bombarded element. As the amount of the active substance is exceedingly small (in the most favourable cases about 10^9 atoms), there is no hope of separating it by ordinary methods. The irradiated substance was therefore dissolved, and small amounts of the inactive elements, which are suspected of being isotopic with the active product, were added. These added elements and the irradiated element were then chemically separated from each other, and separately tested for activity. It is generally found that the activity follows definitely one element. The active product can then be considered as identified with this element.

A preliminary investigation of the penetrating power of the β-rays of the new radio-elements has been carried out. For this purpose counters of the standard type were used, and the substance, instead of being put quite close to the counter, was shaped in the form of a cylinder of inner diameter somewhat larger than the diameter of the counter in order to allow cylindrical aluminium screens of different thicknesses to be interposed. In this way absorption curves of a more or less exponential type were obtained. As the geometrical conditions of this absorption measure are different from the standard ones, and moreover, the number of impulses instead of the total ionization is computed, we checked the method by measuring the absorption coefficients for known radioactive substances; as expected, we found a difference (about (20 %). The data are corrected for this factor.

In several cases the absorption by 2 mm of lead was not complete; this was assumed as a proof of the existence of a γ-radiation.

It was very important to determine whether the emitted particles were positive or negative electrons. Owing to the weakness of the radiation it seemed convenient to use for this purpose Thibaud's [4] method of the inhomogeneous magnetic field. Even by this intensive arrangement this investigation had to be limited to elements which could be strongly activated (Al, Si, P, S, Cr, As, Br, Rh, Ag, I, Ir, U). In every case only negative electrons were observed. This, however, does not exclude that a small percentage (up to about 15 %) of the emitted particles might be positrons.

For a few very strongly activated elements the emitted electrons could be also photographed in a Wilson chamber.

3. – Experimental Results.

The elements investigated are here arranged in order of atomic number; a summary of the results to be found in a table at the end of the paper.

1. *Hydrogen.* – Shows no effect when water is irradiated 14 hours with a 670-millicuries source.

3. *Lithium.* – The hydroxide irradiated 14 hours with 750 millicuries is inactive.

4. *Beryllium.* – Shows an extremely weak activity which might well be due to impurities.

5. *Boron.* – Same as beryllium.

6. *Carbon.* – Paraffin irradiated 15 hours with 220 millicuries is inactive.

7. *Nitrogen.* – Guanidine carbonate (about 35 % N) irradiated 14 hours with 500 millicuries is inactive.

8. *Oxygen.* – No activity: see hydrogen.

9. *Fluorine.* – This element irradiated as calcium fluoride can be strongly activated ($i = 0.7$). As calcium proves to be inactive, the effect is due to fluorine. The activity decreases with a very short half period, about 9 seconds. No chemical separation was possible in this case. However, as it is known that fluorine disintegrates under neutron bombardment with emission of an α–particle, the active nucleus is probably N^{16}. This unstable isotope goes over to stable O^{16} with emission of an electron. The remarkable stability of the latter nucleus agrees with the observed very high energy of the β–rays; the intensity reduces to half value in 0.24 gm/cm² Al. This and all the following absorption data are given for aluminium.

11. *Sodium.* – This element has been irradiated as carbonate. Sodium shows fairly strong activation, decreasing with a period of about 40 seconds.

12. *Magnesium.* – This element can be fairly strongly activated, and the decay curves show the existence of two periods, of about 40 seconds and 15 hours. Half-value thickness for the long period 0.06 gm/cm².

The active element decaying with the 15 hours' period could be chemically separated. The irradiated magnesium was dissolved, and a sodium salt was added. The magnesium was then precipitated as phosphate and found

(4) «C. R. Acad. Sci. Paris», vol. 197, p. 447 (1933).

to be inactive, while the sodium which remains in the solution carries the activity. The active atom is thus proved not to be an isotope of magnesium, and as neon also can be excluded, we assume it to be an isotope of sodium, formed according to the reaction:

$$Mg_{12}^{24} + n_0^1 = Na_{11}^{24} + H_1^1 .$$

13. *Aluminium.* – This element acquires a strong activity under neutron bombardment. The decay curves indicate two periods of about 12 minutes ($i = 0.8$) and 15 hours ($i = 0.5$). Half-value thickness respectively, 0.07 and 0.06 gm/cm².

The long-period activity could be chemically separated by dissolving the irradiated aluminium and adding to it small quantities of sodium and magnesium. Aluminium and magnesium are then precipitated as hydroxides and phosphates, and are found to be inactive. The solution containing sodium is then dried up, and shows an activity decaying with the 15 hours' period.

The active element is probably the same as in the former case of magnesium, as the identity of the periods and of the half-value thickness suggests. In the present case the nuclear reaction is

$$Al_{13}^{27} + n_0^1 = Na_{11}^{24} + He_2^4 .$$

This active isotope Na^{24} then goes over to the stable isotope Mg^{24} with emission of an electron.

The active product with the 12-minute period has not been separated. However, we consider it likely to be Mg^{27}, as the other two possible cases, Al^{28} and Al^{26}, are probably to be excluded, the first because Al^{28}, as we shall next see, is a radioactive isotope with a period of 3 minutes, and the latter because Al^{26} should probably disintegrate with emission of positrons.

14. *Silicon.* – Silicon is also strongly active ($i = 0.7$), and has a period probably somewhat shorter than 3 minutes. Half-value absorption thickness of the β–rays, 0.16 gm/cm².

The chemical separation of the active element has been performed by evaporating irradiated silica with hydrofluoric and sulphuric acids, after addition of aluminium and magnesium. Silicon is eliminated as fluoride, and aluminium precipitated from the residue is found to contain the activity. The active product is therefore probably an isotope of aluminium, and the nuclear reaction is

$$Si_{14}^{28} + n_0^1 = Al_{13}^{28} + H_1^1 .$$

This is in accordance with the hypothesis of Curie, Joliot and Preiswerk (5) about the identity of this active isotope with the one which is formed by the impact of α–particles on Mg^{25}, and which has actually the same period.

15. *Phosphorus.* – This element shows a strong activity ($i = 0.6$) decaying with a period of about 3 hours, and also an activity ten times less intense with a period of 3 minutes, first noticed by Curie, Joliot and Preiswerk. The half-value thickness for the β–rays of the 3 hours' product is 0.09 gm/cm².

(5) «C. R. Acad. Sci. Paris», vol. 198, p. 2089 (1934).

The 3 hours' active product could be chemically separated. For this purpose phosphorus was irradiated as a concentrated solution of phosphoric acid. This solution was afterwards diluted with water, adding sulphuric acid and a small amount of sodium silicate. The substance is dried up to render silica insoluble, and then dissolved in water and filtered. The activity is found with the silica.

The nuclear reaction is then probably

$$P^{31}_{15} + n^{1}_{0} = Si^{31}_{14} + H^{1}_{1}.$$

The 3 minutes' active product has not been chemically separated. The identity of period suggests that it might be the same Al^{28} obtained from silicon.

16. *Sulphur.* - Sulphur shows a fairly strong activity, decaying with a period of about 13 days (rather inaccurately measured). Half-value absorption thickness of the β-rays 0.10 gm/cm².

A chemical separation of the active product was carried out as follows: irradiated sulphuric acid was diluted, a trace of sodium phosphate added, and phosphorus precipitated as phosphomolybdate by addition of ammonium molybdate. The activity was found in the precipitate.

We think, in consequence, that the nuclear reaction is

$$S^{32}_{16} + n^{1}_{0} = P^{32}_{15} + H^{1}_{1}.$$

This active P^{32} is transformed by emission of an electron into the stable isotope S^{32}.

17. *Chlorine.* - Half-period and penetration of the β-rays of this element are about the same as for sulphur. Intensity, $i = 0.1$. The active substance was separated with a method quite similar to that used for sulphur. Irradiated ammonium chloride was dissolved in diluted nitric acid, and then phosphorus added and separated as in the former case. This element carried the activity as before.

The nuclear reaction which gives rise to the same active phosphorus as obtained from sulphur is

$$Cl^{35}_{17} + n^{1}_{0} = P^{32}_{15} + He^{4}_{2}.$$

20. *Calcium.* -- No activity could be detected.

22. *Titanium.* - A very weak effect, with a period of a few minutes, could be observed. However, it cannot be excluded that it might be due to impurities.

23. *Vanadium.* - This element shows a medium activity. The half-period is about 4 minutes, and coincides within experimental error with those observed in chromium and manganese, which are due to isotopes of vanadium. This suggests the hypothesis that the active element might be V^{52}. The half-value thickness is also the same as for chromium, 0.16 gm/cm².

24. *Chromium.* - Metallic chromium becomes fairly strongly active under neutron irradiation. The half-period, is, as in the former case, about 4 minutes. Half-value absorption thickness, 0.16 gm/cm².

In order to identify the active element, we proceeded as follows: to irradiated ammonium chromate some sodium vanadate was added, and vanadium precipitated by addition of ammonium chloride. The activity being found in the precipitate, it is certainly not due to an isotope of chromium. To see whether it was an isotope of titanium, a titanium salt was added besides vanadium to the irradiated chromium compound, and titanium precipitated by hydrolysis. The precipitate showed no activity. In consequence, we consider the active substance to be probably the same isotope V^{52} as before, formed according to the nuclear reaction

$$Cr_{24}^{52} + n_0^1 = V_{23}^{52} + H_1^1 .$$

25. *Manganese.* – We irradiated manganese dioxide and found a fairly intense activity decaying with two periods: about four minutes and two and a half hours. Half-value absorption thickness for the electrons of the 2.5 hours' product, 0.16 gm/cm².

The activity with the long period cannot be separated from manganese by adding chromium and vanadium and precipitating them respectively as chromate and vanadate of lead. It is therefore probably due to an isotope of manganese, Mn^{56}, the same which is extracted from irradiated iron and cobalt, as the identity of the half-life periods suggests.

In order to identify the 4-minute active element, we irradiated manganese nitrate, added a vanadium compound and precipitated vanadium as vanadate of lead. The activity was carried down in the precipitate. A similar reaction was carried out with chromium and also in this case the precipitate was found to be active, but apparently with a definitely lower yield. The active principle is probably the isotope V^{52}, and the nuclear reaction is

$$Mn_{25}^{55} + n_0^1 = V_{23}^{52} + He_2^4 .$$

The same active vanadium is thus probably obtained from chromium, vanadium and manganese, fig. 1.

26. *Iron.* – This element shows a fairly intense activity ($i = 0.05$), decaying with a period of 2.5 hours (6). Half-value absorption thickness, 0.16 gm/cm². The active product can be separated as follows: the irradiated iron is dissolved in nitric acid, a small amount of a soluble manganese salt is added, and then the Mn is precipitated as MnO_2 by addition of sodium chlorate. The activity is found in the manganese precipitate. The active element is probably formed according to the reaction

$$Fe_{26}^{56} + n_0^1 = Mn_{25}^{56} + H_1^1 .$$

and is the same as obtained from manganese.

27. *Cobalt.* – Cobalt can be activated, and the decay curves show the same half-period as iron. The active product can be chemically separated with manganese by the same method as that described for iron. This sug-

(6) See also FLEISCHMANN, «Naturw.», vol. 22, p. 434 (1934).

gests that it may be again the isotope Mn^{56}. In the present case the nuclear reaction would be

$$Co_{27}^{59} + n_0^1 = Mn_{25}^{56} + He_2^4.$$

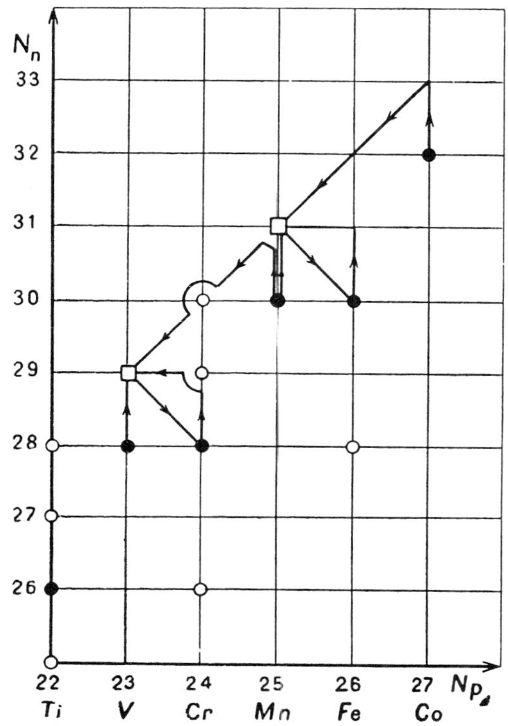

Fig. 1. – The figure shows a possible scheme of the transformations which take place in the elements of atomic number 23 to 27, plotted in a proton-neutron diagram. Stable isotopes are indicated by dots (●) when representing more than 20 % of the element, otherwise by circles (○). Active isotopes are indicated by small squares (□). The arrows indicate the transformations.

The active isotope Mn^{56} appears to originate from cobalt, iron and manganese (see fig. 1).

28. *Nickel.* – The metal, irradiated 13 hours with 250 millicuries, was found to be inactive.

29. *Copper.* – This element shows a medium activity, decaying with a period of about 6 minutes. We suggest as a possibility that the active nucleus might be the same as that which is produced from zinc, that is, an isotope of copper.

30. *Zinc.* – Shows a weak activity, whose decay curve is composed of two periods, one of about 6 minutes and the other much longer and not yet measured.

The active element with the 6-minute period has been chemically separated, dissolving the irradiated metal, adding copper and nickel, and then precipitating copper as sulphide, or depositing it electrochemically on a zinc plate. In both cases copper carries the activity. The added nickel, precipitated with dimethylglyoxime, was found to be inactive. The active element is thus probably an isotope of copper.

31. *Gallium.* – This element shows a moderately intense activity, decaying with a period of about 30 minutes.

33. *Arsenic.* – Has a strong effect, the activity decaying with a period of about one day. Half-value absorption thickness, 0.16 gm/cm². We have tried to separate the active substance by adding gallium and germanium and precipitating the former as ferrocyanide, the latter as sulphide after elimination of the arsenic. While the precipitation of the gallium appeared to be complete, in the case of germanium we are not sure that this element had been completely separated. In both cases the substance separated was found to be inactive, suggesting as probable that the activity belongs to an isotope of arsenic. However, as the reaction with germanium gave a somewhat uncertain result, we cannot give this conclusion as proved beyond doubt.

34. *Selenium.* – This element could be weakly activated. The half-period is about 35 minutes.

35. *Bromine.* – Ammonium bromide was found to be strongly activated by neutron irradiation. The decay curves show two periods, of about 30 minutes and 6 hours. In order to identify the active substance we added arsenic and selenium, and separated the former as sulphide, the latter electrochemically. Both were found to be inactive. The active elements seem thus to be isotopes of bromine. As a control we added arsenic and selenium and precipitated bromine as silver bromide, and found that the precipitate carried down the activity. We suggest that the two periods may be due to two active isotopes, Br^{80} and Br^{82}, formed from the two ordinary isotopes, Br^{79} and Br^{81}.

37. *Rubidium.* – Rubidium nitrate showed a very weak activity, decaying with a period of about 20 minutes.

38. *Strontium.* – This element irradiated 13 hours with 400 mC was found inactive.

39. *Yttrium.* – This element irradiated 30 minutes with 750 mC was found inactive.

40. *Zirconium.* – Zirconium oxide showed an extremely weak activity with a period of a few minutes. It is possibly due to impurities.

42. *Molybdenum.* – This element shows a very weak activity. The decay curves indicate at least two periods, one of about 15 minutes and one longer than a day.

44. *Ruthenium.* – This element irradiated 40 minutes with 750 mC was found inactive.

45. *Rhodium.* – Metallic rhodium could be strongly activated, and the decay curves found are consistent with the existence of two periods, one of about 50 seconds and the other of 5 minutes. Most of the activity belongs to the shorter period. Half-value thickness for the electrons of the 5 minutes' active product, 0.10 gm/cm².

46. *Palladium.* – This element becomes moderately active under neutron bombardment. The decay curves are not very accurate, but indicate a period of the order of six hours. Half-value thickness, 0.03 gm/cm².

47. *Silver.* - A strongly active element. The decay curves show a period of about 20 seconds and one of about 2 minutes (intensity of the latter, $i = 0.5$; half-value thickness, 0.08 gm./cm[2]).

48. *Cadmium.* - Metallic cadmium could be only weakly activated. The decay curves indicate a period slightly longer than one hour.

50. *Tin.* - This metal showed an extremely weak activity, perhaps due to impurities.

51. *Antimony.* - The same as tin.

52. *Tellurium.* - Shows a weak activity decaying with a period of about half an hour.

53. *Iodine.* - We irradiated both the element and ammonium iodide. Both showed a strong activity, decaying with a period of 30 minutes. Intensity, $i = 0.4$. Half-value thickness, 0.09 gm/cm[2]. The active element follows the iodine when this substance is precipitated by adding nitric acid to the solution of the irradiated ammonium iodide, to which tellurium and antimonium had been added. It also follows the same element when precipitated as silver iodide in presence of tellurium and antimonium. It is very likely that we have here the formation of an active isotope of iodine.

55. *Cæsium.* - Cæsium (tartrate) showed a very weak activity, whose period has not yet been measured with any accuracy.

56. *Barium.* - We irradiated the hydroxide and found it weakly active. The curves indicate a period of about 3 minutes.

57. *Lanthanum.* - This element irradiated 40 minutes with 400 mC was found inactive.

58. *Cerium.* - This element irradiated 30 minutes with 400 mC was found inactive.

59. *Praseodymium.* - This element, as well as the other rare earths—Nd, Sm, and Pr—were kindly given to us by Professor Rolla, of the University of Florence. They are very pure, and in the form of oxides.

Praseodymium shows a rather weak activity. The decay curves indicate a period of about 5 minutes, and perhaps a longer one.

60. *Neodymium.* - Shows a rather weak activity. Period about one hour.

62. *Samarium.* - Also has a rather weak activity. Period about 40 minutes.

73. *Tantalum.* - This element irradiated 2.5 hours with 750 mC was found inactive.

74. *Tungsten.* - This element shows an extremely weak activity of probably long period. This might possibly be due to impurities.

75. *Rhenium.* - This element irradiated 10 minutes with 750 mC was found inactive.

76 *Osmium.* - This element irradiated 15 hours with 450 mC was found inactive.

77. *Iridium.* - This element has a strong activity, decaying with a period of about 20 hours. Half-value thickness, 0.13 gm/cm[2].

In order to establish the chemical nature of the activity, we irradiated the tetrachloride, added osmium and rhenium, and separated the former by distillation of the tetroxide, and the latter as sulphide. Both were found to be inactive. Iridium still contained the activity, which appears to be due to an isotope of this element.

78. *Platinum.* – Pure platinum from Heræus showed an extremely weak activity of short period.

79. *Gold.* – We irradiated the metal, which showed a fairly intense activity. The period is about two days. Half-value thickness 0.33 gm/cm^2.

We tried to separate the active substance by dissolving the irradiated gold foil in aqua regia, adding platinum and iridium, and precipitating these elements as respectively chloroplatinates and chloroiridiates. Both were found to be inactive, while the activity was still found in gold, thus indicating an isotope of this element as the active nucleus. The activity remains with gold also when mercury is added and evaporated.

80. *Mercury.* – This element, strongly irradiated, showed an extremely weak activity which might be due to impurities.

81. *Thallium.* – The same as mercury.

82. *Lead.* – This element irradiated 10 hours with 500 mC was found inactive.

83. *Bismuth.* – The same as lead.

90. *Thorium.* – This element has not been investigated sufficiently at present. Thorium nitrate, freed of its ordinary β–active decay products and irradiated, becomes strongly active. The decay curves indicate at least two periods, of less than 1 minute and about 15 minutes.

92. *Uranium.* – We give here only the main results on this element, as its behaviour has been discussed recently elsewhere [(7)].

Besides the half-periods of 10 seconds, 40 seconds and 13 minutes, we have identified later one more of about 1.5 hours. The intensity of the activation is of the order of 0.5 for each of these lives. We have already discussed experiments which appear to prove that the 13 minutes' active product is not isotopic with any of the elements with atomic number from 86 to 92 (emanation, ekacæsium, radium, actinium, thorium, protactinium, uranium). These experiments have been repeated under different conditions, chiefly in order to obtain a negative proof of the identity of the 13-minute product with protactinium, this proof being the most difficult to establish on account of the short period of the available Pa isotope, UX_2. The manganese reaction which has already been described gives a yield of about 15 °/₀ for the 13-minute product. Its yield for UX_2 depends widely upon the conditions of the reaction, and may be varied between 2 °/₀ and 10 °/₀, account being taken of the natural decay of this substance. A more effective reaction for obtaining the 13-minute active product is the following: irradiated uranium nitrate is dissolved in diluted hydrochloric acid; some rhenium nitrate is added, and then rhenium is precipitated as sulphide by addition of sodium thiosulphate. This precipitate carries about 50 °/₀ of the activity, and sometimes more. The percentage of UX_1 and of UX_2 found in the rhenium precipitate varies also with the conditions of the reaction (particularly with

(7) FERMI, « Nature », vol. 133, p. 898 (1934).

the acidity), but can be made very low, probably less than 1 %. It was actually possible to separate the 13-minute active product and to measure its period using uranium which had not been purified at all from UX. The 90-minute active product has apparently chemical properties very analogous to those of the 13-minute active product, as in every type of reaction they are always obtained in about the same percentage. These activities seem, therefore, both to be due to products with atomic number higher than 92, and possibly to isotopes of a same element.

4. – Theoretical Discussion.

We want here to discuss, from the theoretical point of view, the processes that may take place under neutron bombardment. At the present state of the nuclear theory these considerations can have only a provisional character. We can resume the empirical results of the preceding sections in the following points.

a) A large percentage of elements of any atomic weight can be activated. From this point of view no special difference can be noticed between light and heavy elements.

b) The cross-sections for neutron impact for the elements which can be most intensely activated are of the order of the geometrical cross-section of the nucleus. This means that a large percentage of the neutrons which hit the nucleus produce an active atom.

c) The active product is sometimes an isotope of the original atom (atomic number Z); in other cases its atomic number is lower by one or two units. In this respect there appears to be a difference between light and heavy elements. For light elements the atomic number of the active products is usually lower than Z, while in the five cases investigated for heavy, non-spontaneously radioactive elements, the active product is always an isotope of the bombarded element.

d) The emitted electrons always have a negative charge, or at least no positrons could ever be detected.

There seems to be no special difficulty in explaining the general mechanism of the activation for light elements. This seems to consist usually in the capture of the impinging neutron, followed immediately by the expulsion of an α–particle or a proton. If the energy of the emitted α–particle or proton is of some million volts, it results from Gamow's theory that the time which is necessary to emit the particle is extremely short, and there is therefore a fairly high probability for the process to happen before the neutron has left the nucleus. After this process, which may last a time of the order of 10^{-20} seconds, the nucleus has been transformed into a new one having, on the average, an atomic weight higher than would belong to its nuclear charge, as the processes of absorbing a neutron and emitting an α–particle or a proton increase the neutron/proton ratio in the nucleus. This is probably the reason why an emission of negative electrons is always observed. The process of the electron emission re–establishes the correct value for neutron/proton ratio, and corresponds to the formation of a stable isotope.

As the atomic weight of the bombarded element increases, the potential barrier around the nucleus becomes an increasingly strong obstacle to the emission of heavy, positively–charged particles; it is therefore easy to understand why processes with emission of protons and α–particles become very improbable.

The reactions whose theoretical interpretation seems to meet with difficulties are those, normally occurring among heavy elements, in which the activated atom is isotopic with the original atom. The simplest hypothesis would be to assume a capture of the impinging neutron, giving rise to an unstable isotope of the bombarded element with an atomic weight higher by one unit than before the process. This hypothesis, which would be in agreement with the observed fact of the emission of negative electrons, gives rise, however, to serious theoretical difficulties when one tries to explain how a neutron can be captured by the nucleus in a stable or quasi-stable state. It is generally admitted that a neutron is attracted by a nucleus only when its distance from the centre of the nucleus is of the order of 10^{-12} cm. It follows that a neutron of a few million volts' energy can remain in the nucleus (*i.e.*, have a strong interaction with the constituent particles of the nucleus) only for a time of the order of 10^{-21} seconds; that is, of the classical time needed to cross the nucleus. The neutron is captured if, during this time, it is able to lose its excess energy (*e.g.*, by emission of a γ–quantum). If one evaluates the probability of this emission process by the ordinary methods one finds a value much too small to account for the observed cross-sections. In order to maintain the capture hypothesis, one must then either admit that the probability of emission of a γ–quantum (or of an equivalent process, as, for example, the formation of an electron-positron pair) should be much larger than is generally assumed; or that, for reasons that cannot be understood in the present theory, a nucleus could remain for at least 10^{-16} seconds in an energy state high enough to permit the emission of a neutron.

An alternative hypothesis is to admit that the impinging neutron is not captured, but causes the expulsion of another neutron from the nucleus. This process could be described as follows: the primary neutron loses part of its energy, bringing the nucleus into an excited state by a sort of inelastic impact. One can easily understand, theoretically, that these processes may take place in a large percentage of the collisions between nuclei and neutrons. If the excitation energy is large enough, a neutron can be emitted before the nucleus loses its energy by emission of a γ–quantum. The atom formed by such a process is an isotope of the original one, with atomic weight lower by one unit.

An objection which may be raised against this hypothesis is that if the number of neutrons decreases instead of increasing, it is, *a priori*, more likely that the atom, in the following disintegration, should emit a positron than a negative electron as observed. However, in the few cases investigated of heavy elements which are activated by neutron bombardment and are transformed into their isotopes, when the isotopic constitution of the neighbouring elements is known, there always exists a stable isotope of the element Z — 1, having atomic weight one unit less than the original element, as a possible end-product of the transformation.

One has, moreover, to bear in mind that if an unstable nucleus has energetically the possibility of emitting both an electron and a positron the theory of the β–rays [8] gives *ceteris paribus* the emission of an electron as the most probable.

In conclusion, the choice between these two alternatives seems at present rather uncertain, and further experiments must be performed to test this point.

The radon sources were supplied by Professor G. C. Trabacchi, Laboratorio Fisico della Sanità Pubblica, Rome, without whose kind help these researches would have been impossible. To him, therefore, our warmest thanks are due. A fund for this work was granted to us by the Consiglio Nazionale delle Ricerche, Rome. Our thanks are due also to Professor N. Parravano and Professor L. Rolla, who have supplied us some rare chemicals.

(8) FERMI, « Z. Physik », vol. 88, p. 161 (1934).

5. – Tabular Summary.

The main result of this investigation are summarized in the table. Column 1 contains the atomic number and symbol of the elements investigated. Column 2 gives the isotopic constitution, numbers in bold type refer to isotopes which represent more than 20 % of the element. Column 3 gives the observed half-periods, a heavy—means that the activity was sought for and not found. Column 4 gives a rough evaluation of the intensity *s* (strong), *m* (medium), *w* (weak). Column 5 gives the average energy of the electrons in million volts. This was obtained from the absorption measurement by a rather rough extrapolation based on the absorption coefficients of ordinary β-active substances without strong γ-rays (Ra E and UX_2). Column 6 shows whether γ-rays were observed or not; a line means that γ-rays have been sought for and not observed. Column 7 gives a probable active product. For simplicity we have always assumed that the neutron is captured; if, instead, a neutron was emitted, the corresponding atomic weights should be decreased by two units. When two or more periods are present, the data of columns 4, 5, 6 and 7 refer to the different periods in their order.

Atomic number	Isotopes	Half-period	Intensity	Mean energy of β-rays in 10^6 volts	γ-rays	Active isotope
1 H	**1**, 2	—	—			
3 Li	6, **7**	—	—			
4 Be	**9**	?	?			
5 B	**10**, **11**	?	?			
6 C	**12**, 13	—	—			
7 N	**14**, 15	—	—			
8 O	**16**, 17, 18	—	—			
9 F	**19**	9 s.	*s*	2	yes	N^{16} (?)
11 Na	**23**	40 s.	*m*			
12 Mg	**24**, 25, 26	40 s.; 15 h.	*m*; *m*	—;0.5	?; yes	—; Na^{24}
13 Al	**27**	12 m.; 15 h.	*s*; *s*	0.6;0.5	yes; yes	—; Na^{24}
14 Si	**28**, 29, 30	3 m.	*s*	1.3	yes	Al^{28}
15 P	**31**	3 m. ; 3 h.	*m*; *s*	—;0.7	?	Si^{31}
16 S	**32**, 33, 34	13 d.	*m*	0.8	—	P^{32}
17 Cl	**35**, 37	13 d.	*m*	0.8		P^{32}
20 Ca	**40**, 42, 43, 44	—	—			
22 Ti	46, 47, **48**, 49, 50	3 m.	*w*			
23 V	**51**	4 m.	*m*	1.3		V^{52} (?)
24 Cr	50, **52**, 53, 54	4 m.	*m*	1.3	yes	V^{52}
25 Mn	**55**	4 m.; 150 m.	*m*; *m*	—;1.3		V^{52}; Mn^{56}
26 Fe	54, **56**	150 m.	*m*	1.3	yes	Mn^{56}
27 Co	**59**	150 m.	*w*			Mn^{56}
28 Ni	**58**, **60**, 61, 62	—	—			
29 Cu	**63**, **65**	6 m.	*m*			
30 Zn	**64**, **66**, 67, 68, 70	6 m. ; ?	*w*; *w*			Cu; —
31 Ga	**69**, **71**	30 m.	*m*			
33 As	**75**	1 d.	*s*	1.3	yes	As^{76}
34 Se	74, 76, 77, **78**, **80**, 82	35 m.	*w*			
35 Br	**79**, **81**	30 m. ; 6 h.	*s*; *s*	—; 0.7 (?)	—	Br^{80} ; Br^{82}
37 Rb	**85**, **87**	20 m.	*w*			
38 Sr	86, 87, **88**	—	—			
39 Y	**89**	—	—			
40 Zr	**90**, 91, **92**, 94, 96	?	*w*			
42 Mo	92, 94, 95, 96, 97, **98**, 100	15 m. ; (?)	*w*; *w*			
44 Ru	96, 98, 99, 100, **101**, **102**, 104	—	—			
45 Rh	—	50 s. ; 5 m.	*s*; *m*	0.8	—	
46 Pd	—	6 h. (?)	*w*	0.3	—	
47 Ag	**107**, **109**	20 s.; 2 m.	*s*; *s*	—; 0.7	—	

Tabular Summary–(continued).

Atomic number	Isotopes	Half-period	Intensity	Mean energy of β-rays in 10^6 volts	γ-rays	Active isotope
48 Cd	110, 111, **112**, 113, **114**, 116	70 m.	*w*			
50 Sn	112, 114, 115, 116, 117, **118**, 119, **120**, 121, 122, 124	—	—			
51 Sb	**121**, **123**	?	?			
52 Te	122, 123, 124, 125, **126**, (127), **128**, **130**	30 m. (?)	*w*			
53 I	**127**	30 m.	*s*	0.7	—	I^{128}
55 Cs	**133**	(?)	*w* (?)			
56 Ba	135, 136, 137, **138**	3 m.	*w*			
57 La	**139**	—	—			
58 Ce	**140**, 142	—	—			
59 Pr	**141**	5 m.	*w*			
60 Nd	**142**, 143, **144**, 145, 146	1 h.	*w*			
62 Sm	144, 147, 148, 149, 150, **152**, 154	40 m.	*w*			
73 Ta	**181**	—	—			
74 W	**182**, 183, **184**, **186**	(?)	*w* (?)			
75 Re	**185**, **187**	—	—			
76 Os	186, 187, 188, 189, **190**, **192**	—	—			
77 Ir	—	20 h.	*s*	1.1	yes	Ir
78 Pt	—	(?)	*w* (?)			
79 Au	—	2 d.	*s*	0.3	—	Au
80 Hg	196, 197, 198, 199, **200**, 201, **202**, 203, 204	(?)	(?)			
81 Tl	**203**, **205**	(?)	(?)			
82 Pb	203, 204, 205, **206**, **207**, **208**, 209	—	(?) —			
83 Bi	**209**	—	—			
90 Th	**232**	1 m. (?); 15m.	*s*; *s*			
92 U	**238**	15 s.; 40 s.; 13 m. 100 m.	*s*; *s*; *s*; *s*		yes	(*)

(*) See under "uranium" in section 3. of this paper.

99.

POSSIBLE PRODUCTION OF ELEMENTS OF ATOMIC NUMBER HIGHER THAN 92

« Nature » (London), *133*, 898-899 (1934).

Until recently it was generally admitted that an atom resulting from artificial disintegration should normally correspond to a stable isotope. M. and Mme. Joliot first found evidence that it is not necessarily so; in some cases the product atom may be radioactive with a measurable mean life, and go over to a stable form only after emission of a positron.

The number of elements which can be activated either by the impact of an α–particle (Joliot) or a proton (Cockcroft, Gilbert, Walton) or a deuton (Crane, Lauritsen, Henderson, Livingston, Lawrence) is necessarily limited by the fact that only light elements can be disintegrated, owing to the Coulomb repulsion.

This limitation is not effective in the case of neutron bombardment. The high efficiency of these particles in producing disintegrations compensates fairly for the weakness of available neutron sources as compared with α–particle or proton sources. As a matter of fact, it has been shown (1) that a large number of elements (47 out of 68 examined until now) of any atomic weight could be activated, using neutron sources consisting of a small glass tube filled with beryllium powder and radon up to 800 millicuries. This source gives a yield of about one million neutrons per second.

All the elements activated by this method with intensity large enough for a magnetic analysis of the sign of the charge of the emitted particles were found to give out only negative electrons. This is theoretically understandable, as the absorption of the bombarding neutron produces an excess in the number of neutrons present inside the nucleus; a stable state is therefore reached generally through transformation of a neutron into a proton, which is connected to the emission of a β–particle.

In several cases it was possible to carry out a chemical separation of the β–active element, following the usual technique of adding to the irradiated substance small amounts of the neighbouring elements. These elements are then separated by chemical analysis and separately checked for the β–activity with a Geiger-Müller counter. The activity always followed completely a certain element, with which the active element could thus be identified.

(1) E. FERMI, « Ricerca Scientifica », *5* (1), 283, 330 (1934), « Nature » *133*, 757, May 19 (1934); E. AMALDI, O. D'AGOSTINO, E. FERMI, F. RASETTI, E. SEGRÈ, « Ricerca Scientifica », *5* (1), 452 (1934).

In three cases (aluminium, chlorine, cobalt) the active element formed by bombarding the element of atomic number Z has atomic number Z — 2. In four cases (phosphorus, sulphur, iron, zinc) the atomic number of the active product is Z — 1. In two cases (bromine, iodine) the active element is an isotope of the bombarded element.

This evidence seems to show that three main processes are possible: (*a*) capture of a neutron with instantaneous emission of an α–particle; (*b*) capture of the neutron with emission of a proton; (*c*) capture of the neutron with emission of a γ–quantum, to get rid of the surplus energy. From a theoretical point of view, the probability of processes (*a*) and (*b*) depends very largely on the energy of the emitted α– or H–particle; the more so the higher the atomic weight of the element. The probability of process (*c*) can be evaluated only very roughly in the present state of nuclear theory; nevertheless, it would appear to be smaller than the observed value by a factor 100 or 1,000.

It seemed worth while to direct particular attention to the heavy radioactive elements thorium and uranium, as the general instability of nuclei in this range of atomic weight might give rise to successive transformations. For this reason an investigation of these elements was undertaken by the writer in collaboration with F. Rasetti and O. D'Agostino.

Experiment showed that both elements, previously freed of ordinary active impurities, can be strongly activated by neutron bombardment. The initial induced activity corresponded in our experiments to about 1,000 impulses per minute in a Geiger counter made of aluminium foil of 0.2 mm thickness. The curves of decay of these activities show that the phenomenon is rather complex. A rough survey of thorium activity showed in this element at least two periods.

Better investigated is the case of uranium; the existence of periods of about 10 sec, 40 sec, 13 min, plus at least two more periods from 40 minutes to one day is well established. The large uncertainty in the decay curves due to the statistical fluctuations makes it very difficult to establish whether these periods represent successive or alternative processes of disintegration.

Attempts have been made to identify chemically the β–active element with the period of 13 min. The general scheme of this research consisted in adding to the irradiated substance (uranium nitrate in concentrated solution, purified of its decay products) such an amount of an ordinary β–active element as to give some hundred impulses per minute on the counter. Should it be possible to prove that the induced activity, recognizable by its characteristic period, can be chemically separated from the added activity, it is reasonable to assume that the two activities are not due to isotopes.

The following reaction enables one to separate the 13 min-product from most of the heaviest elements. The irradiated uranium solution is diluted in 50 per cent nitric acid; a small amount of a manganese salt is added and then the manganese is precipitated as dioxide (MnO_2) from the boiling solution by addition of sodium chlorate. The manganese dioxide precipitate carries a large percentage of the activity.

This reaction proves at once that the 13 min-activity is not isotopic with uranium. For testing the possibility that it might be due to an element 90 (thorium) or 91 (protactinium), we repeated the reaction at least ten times, adding an amount of uranium $X_1 + X_2$ corresponding to about 2,000 impulses per minute; also some cerium and lanthanum were added in order to sustain uranium X. In these conditions the manganese reaction carried only the 13 min-activity; no trace of the 2,000 impulses of uranium X_1, (period 24 days) was found in the precipitate; and none of uranium X_2, although the operation had been performed in less than two minutes from the precipitation of the manganese dioxide, so that several hundreds of impulses of uranium X_2 (period 75 sec) would have been easily recognizable.

Similar evidence was obtained for excluding atomic numbers 88 (radium) and 89 (actinium). For this, mesothorium–1 and–2 were used, adding barium and lanthanum; the evidence was completely negative, as in the former case. The eventual precipitation of uranium–X_1 and mesothorium–1, which do not emit β–rays penetrating enough to be detectable in our counters, would have been revealed by the subsequent formation respectively of uranium–X_2, and mesothorium–2.

Lastly, we added to the irradiated uranium solution some inactive lead and bismuth, and proved that the conditions of the manganese dioxide reaction could be regulated in such a way as to obtain the precipitation of manganese dioxide with the 13 min-activity, without carrying down lead and bismuth.

In this way it appears that we have excluded the possibility that the 13 min-activity is due to isotopes of uranium (92), protactinium (91), thorium (90), actinium (89), radium (88), bismuth (83), lead (82). Its behaviour excludes also ekacæsium (87) and emanation (86).

This negative evidence about the identity of the 13 min-activity from a large number of heavy elements suggests the possibility that the atomic number of the element may be greater than 92. If it were an element 93, it would be chemically homologous with manganese and rhenium. This hypothesis is supported to some extent also by the observed fact that the 13 min-activity is carried down by a precipitate of rhenium sulphide insoluble in hydrochloric acid. However, as several elements are easily precipitated in this form, this evidence cannot be considered as very strong.

The possibility of an atomic number 94 or 95 is not easy to distinguish from the former, as the chemical properties are probably rather similar. Valuable information on the processes involved could be gathered by an investigation of the possible emission of heavy particles. A careful search for such heavy particles has not yet been carried out, as they require for their observation that the active product should be in the form of a very thin layer. It seems therefore at present premature to form any definite hypothesis on the chain of disintegrations involved.

100.

ARTIFICIAL RADIOACTIVITY PRODUCED BY NEUTRON BOMBARDMENT

« Nature » (London), *134*, 668 (1934).

E. Fermi, E. Amaldi, O. D'Agostino, F. Rasetti and E. Segrè have recently published an account of their experiments on the artificial production of radioactive elements by bombarding various elements with neutrons (« Proc. Roy. Soc. », A, Sept.). Preliminary announcements have appeared in « Nature » and in Italian papers. The neutrons can enter the nuclei of even the heaviest elements, and more than forty elements have been activated. The activity decays with halving periods varying from a few minutes to some days. In all the cases where the activity was sufficiently strong to be tested by Thibaud's inhomogeneous magnetic deflection method, the particles emitted were found to be β–particles. In many cases the artificial radio-elements could be separated out by chemical precipitations after addition of an isotopic element, and the chemical nature of the radio-elements could therefore be securely placed. The nuclear reactions studied are of the following types:

$$ {}_{13}\mathrm{Al}^{27} - {}_{0}n^{1} \rightarrow {}_{11}\mathrm{Na}^{24} + {}_{2}\mathrm{He}^{4} , $$

$$ {}_{14}\mathrm{Si}^{28} + {}_{0}n^{1} \rightarrow {}_{13}\mathrm{Al}^{28} + {}_{1}\mathrm{H}^{1} , $$

$$ {}_{25}\mathrm{Mn}^{55} + {}_{0}n^{1} \rightarrow {}_{25}\mathrm{Mn}^{56} . $$

102.

NATURAL BETA DECAY

« International Conference on Physics », London 1934, Vol. I, Nuclear Physics, 66–71 Physical Society (London).

DISCUSSION

Prof. E. FERMI. – In connexion with the theory of β–rays I should like to point out that there is apparently some disagreement between theory and experiment as to the drop in intensity of the continuous β spectrum for low energies.

I do not think that experimental results are completely definite on this point, as the small penetrating power of low energy electrons greatly increases the difficulties of investigating them; if however a disagreement should be established beyond doubt, it would still be possible to alter the particular form of interaction between protons and neutrons on one hand and electrons and neutrinos on the other, in order to reestablish the agreement. The form of this interaction that has been given in my theory of the β–rays was chosen only on account of its simplicity, but there is a large variety of other special forms for the interaction which might be tried if needed. This refers not only to the energy distribution in the continuous β-spectrum, but also to the selection rule of forbidden transitions, although the evidence which has been presented in Dr. Ellis' report seems to be definitely in favour of the selection rule in its present form.

I heartily agree with Dr. Gamow, that both his empirical evidence and the more general argument of symmetry strongly support the possibility of the existence of a negative proton. I should like nevertheless to point out some difficulties for his view-point, which are chiefly connected with the theoretical interpretation of the fact that the atomic weights are approximately equal to twice the atomic numbers. In Heisenberg's theory the interpretation of this fact is based on the simple assumption that the proton-neutron interaction is the main force in determining the nuclear binding, and is quite independent of more special assumptions on this interaction. If we now assume the existence inside the nucleus of a new kind of particle into which neutrons can transform with emission of a positive electron, it seems difficult to understand why the number of these negative protons always remains very small and is not noticeable in the relation of atomic weight to atomic number.

I agree with Dr. Gamow that this difficulty is in no way a fundamental one, as one might assume that there are strong repulsive forces between protons of different signs, acting only at close distances; but if we assume these forces, it seems to me that they must be very delicately adjusted in order to prevent the formation of negative protons in considerable number, and yet to allow for their presence in nuclei of very different atomic weights, as beryllium and lead.

103.

ARTIFICIAL RADIOACTIVITY PRODUCED BY NEUTRON BOMBARDMENT

« International Conference on Physics », London 1934, Vol. I, Nuclear Physics, 75–77
Physical Society (London).

ABSTRACT. – The advantages of neutrons as bombarding particles are pointed out, and experiments are described in which radioactive elements are produced by this process. About 40 cases were found, out of 60 examined.

In general, the atomic number of the radioelement formed is either that of the parent element, or is less than this by either 1 or 2 units, the first case being commonest among the heavier elements.

It is assumed that the first stage is the capture of the neutron, and that this is followed by the ejection of a quantum of γ-radiation, a proton or an α-particle. The first stage presents no theoretical difficulty, but the subsequent ejection is not easy to understand. The theoretical difficulties presented here are briefly discussed.

After the discovery by Mons. and Mme Joliot that radioactive substances could be obtained by bombardment with α–particles, I undertook experiments in the Physical Institute of the University of Rome to investigate whether phenomena analogous to those observed by the Joliots could also be obtained as a result of neutron bombardment. Although the neutron sources which it is possible to employ are inferior to the sources of α–particles or of swift protons, yet right from the beginning we could expect a certain compensation owing to the fact that neutrons, being uncharged, are more efficient in producing artificial disintegrations than charged particles; in particular we could hope for results even in the case of heavy atoms which up to the present have been unaffected by bombardment with charged particles.

Experience has fully confirmed these expectations, neutron bombardment having given rise to the formation of radioactive products in about 40 cases (out of 60 elements studied up to the present). This high fraction of elements which can be activated is very roughly the same for light elements as for heavy ones.

Without describing in detail the results obtained to date, which have been published [1], I must give a little space to certain problems relating to the interpretation of the results themselves.

One of the points of greatest interest is that of ascertaining what is the nuclear reaction which gives rise to the formation of the radio-element ob-

(1) For a résumé of this work see FERMI, AMALDI, D'AGOSTINO, RASETTI and SEGRÈ, « Proc. Roy. Soc. », A, *146*, 483 (1934).

tained by bombarding a given element of atomic number Z with neutrons. An important factor in discussing this question is the chemical examination of the radioactive element, an examination which it has been possible to carry out in a fairly large number of cases. The result has been to show that the atomic number of the element formed may be (except for the case of uranium, where probably several successive transformations are involved) one of the following: Z — 2 , Z — 1 , Z.

The first two cases are common among the light elements, whilst for heavy elements only the last case has been encountered up to the present.

In interpreting these facts we may assume that the process of formation of radioelements consists in the absorption of neutrons by the bombarded nucleus, with the emission either of an α–particle (radio-element with atomic number Z — 2) or a proton (radio-element with atomic number Z — 1) or a quantum of γ–radiation, necessary because the neutron loses its initial energy (radio-element of atomic number Z). In the cases examined up to now, this hypothesis is compatible with the facts; in particular, if the atomic weight of the radioactive isotope which would be formed by the bombardment is calculated on this hypothesis, it never corresponds to that of a known stable isotope; further, whenever on the basis of these three possible processes, it is found that isotopic isobars should be obtained by bombardment of different elements, then they are actually found to have the same mean life, which confirms that the same radioactive element is in this case obtained in different ways.

Turning now to the consideration of these processes from the point of view of present theories of nuclear structure, it appears that there is no theoretical difficulty in understanding how it may happen that the neutron is captured in a large percentage of cases with the expulsion of an α–particle or a proton; remembering that these particles, in order to escape from the atom, must penetrate Gamow's potential barrier, it is found, in agreement with experiment, that the probability of processes of this kind decreases fairly rapidly with increasing atomic number of the bombarded element; in such a way that unless we accept rather unlikely values for the energy of the emitted α–particle or proton, the process would become practically impossible for the heavy elements. On the other hand, there are considerable theoretical difficulties in understanding the simple capture of the neutron, which would lead to the formation of a radio-element isotopic with the original element but with an atomic weight greater by unity. As a matter of fact, in this case the neutron at the moment of capture has an energy surplus equal to its kinetic energy, i.e. several millions of electron volts. In this condition, the neutron would escape from the nucleus in an extremely short time which, if we accept the usual views on nuclear structure, would be of the order of 10^{-20} 10^{-21} sec; to ensure the capture of the neutron a quantum of γ–radiation of sufficient frequency to carry off the excess energy of the neutron must be emitted in this short interval of time, so that the probability of simple capture of the neutron would equal the probability of emission of the quantum of γ-radiation in a time interval of the order mentioned. If we accept the usual data for the probability of emission of γ–radiation by an excited nucleus, the calculated probability comes out much less than that observed. If we wish to retain the hypothesis of the possibility of this process of neutron capture, we must admit that for some reason which we do not know the neutron can remain in the excited nucleus for a much longer time than that quoted; or else the process of expelling the excess energy, either by the emission of a quantum of γ–radiation or by some other unknown process, can occur much more rapidly. In view of this position of the theory, the observations of Bjerge and Westcott on the activation of sodium, and those of Amaldi and Segrè on aluminium, are to be considered particularly important; they seem to prove that, at least in one of the processes of activation, a radioactive isotope of the same element is formed, having an atomic weight increased by one unit. Finally we must note that completely analogous difficulties of interpretation to those just discussed present themselves in the explanation of the synthesis of heavy hydrogen observed by Lea when hydrogen is bombarded with neutrons.

105 *a*.

AZIONE DI SOSTANZE IDROGENATE SULLA RADIOATTIVITÀ PROVOCATA DA NEUTRONI. — I.

E. Fermi, E. Amaldi, B. Pontecorvo, F. Rasetti, E. Segrè

« Ric. Scientifica », *5* (2), 282–283 (1934).

Nel corso di esperienze sulla radioattività provocata nell'argento da bombardamento di neutroni si sono notate anomalie nella intensità della attivazione; uno spessore di alcuni centimetri di paraffina interposto fra la sorgente e l'argento invece di diminuire l'attivazione la aumenta. In seguito abbiamo potuto constatare che la presenza di grossi blocchi di paraffina circondanti la sorgente e l'oggetto irradiato esalta l'intensità della attivazione per un fattore che, a seconda delle condizioni geometriche, varia da alcune decine ad alcune centinaia.

In seguito a questa constatazione abbiamo cercato di riconoscere, in modo per ora sommario, le circostanze in cui si presenta questo fenomeno. I fatti che sono emersi fino ad ora sono i seguenti:

a) un preparato di radio senza berillio non produce effetto, ciò che induce ad attribuire i fenomeni ai neutroni e non ai raggi γ;

b) un effetto approssimativamente della stessa intensità di quello ottenuto colla paraffina si ha coll'acqua. Riteniamo molto probabile che esso dipenda dalla presenza dell'idrogeno perché sostanze ossigenate prive di idrogeno ($NaNO_3$) non producono un aumento di attività, almeno nello stesso ordine di grandezza;

c) il fenomeno osservato nel caso dell'argento non si presenta in tutti gli elementi che si attivano con i neutroni. Abbiamo finora constatato che per il silicio, zinco e fosforo, non si ha un aumento apprezzabile di intensità, mentre il rame, l'argento e lo iodio dànno luogo ad effetti enormemente maggiori di quelli che si avrebbero senza la presenza dell'acqua.

Da questi pochi casi sembra valga la regola che siano sensibili solo quegli elementi che per bombardamento dánno luogo a sostanze radioattive isotope con l'elemento di partenza.

Notevole è il caso dell'alluminio, il quale si attiva nell'acqua con un periodo di poco inferiore a tre minuti che corrisponde a quello del Al^{28} estratto dal silicio irradiato. Questa attività, prodotta in condizioni normali, è così debole che quasi sparisce di fronte alle altre dello stesso elemento.

Parimenti lo zinco ed il rame, che dànno origine agli stessi prodotti attivi [1] isotopi del rame, in condizioni normali hanno attività dello stesso ordine di grandezza mentre nell'acqua il rame lascia a grande distanza lo zinco.

(1) T. Bjerge e C. H. Westcott, « Nature », *134*, 286 (1934).

Una possibile spiegazione di questi fatti sembra essere la seguente: i neutroni per urti multipli contro nuclei di idrogeno perdono rapidamente la propria energia. È plausibile che la sezione di urto neutrone–protone cresca al calare della energia e può quindi pensarsi che dopo alcuni urti i neutroni vengano a muoversi in modo analogo alle molecole diffondentesi in un gas, eventualmente riducendosi fino ad avere solo l'energia cinetica competente alla agitazione termica. Si formerebbe così intorno alla sorgente qualcosa di simile ad una soluzione di neutroni nell'acqua o nella paraffina. La concentrazione di questa soluzione in ogni punto dipenderebbe dalla intensità della sorgente, dalle condizioni geometriche della diffusione e da eventuali processi di cattura del neutrone da parte dell'idrogeno o di altri nuclei presenti.

Non è escluso che un simile punto di vista possa avere importanza nella spiegazione degli effetti osservati da Lea [2].

Sono in corso indagini su tutto questo complesso di fenomeni.

Istituto Fisico della R. Università.
Roma, 22 ottobre 1934-XII.

(2) D. E. LEA, «Nature», *133*, 24 (1934).

106 *a.*

EFFETTO DI SOSTANZE IDROGENATE SULLA RADIOATTIVITÀ PROVOCATA DA NEUTRONI. — II.

E. FERMI, B. PONTECORVO, F. RASETTI

« Ric. Scientifica », 5 (2), 380–381 (1934).

In una precedente comunicazione abbiamo riferito sulla azione di sostanze idrogenate sulla radioattività provocata da bombardamento di neutroni, accennando anche alla ipotesi che l'esaltazione dell'effetto fosse dovuta al rallentamento dei neutroni emessi dalla sorgente.

Abbiamo ora eseguito le seguenti esperienze:

1. – Osservato che l'attivazione diretta da parte di una sorgente di 350 mC di Em + Be posta a 20 cm di distanza da un cilindretto di argento è trascurabile, abbiamo immerso la sorgente al centro di un vaso cilindrico di acqua del diametro di 14 cm. Si nota allora che l'argento, nelle stesse condizioni geometriche si attiva fortemente dando, nei nostri contatori, un centinaio di impulsi al minuto, cioè circa quanti se ne otterrebbero ponendo il cilindretto a contatto della sorgente, senza acqua.

Nella ipotesi che l'azione delle sostanze idrogenate sia effettivamente dovuta alla diffusione e conseguente rallentamento dei neutroni, questa esperienza può interpretarsi al modo seguente: i neutroni, o almeno molti di essi, prima di uscire dal vaso di acqua, sono rallentati così che pur rimanendo costante, o se mai diminuito, il flusso dei neutroni uscenti, la loro velocità media è considerevolmente diminuita. La esperienza permette allora di confrontare l'azione dello stesso numero di neutroni con velocità differente e rivela la enorme efficacia di quelli lenti.

2. – Il fatto che i neutroni lenti siano così efficaci nell'attivare le sostanze, fa pensare che essi debbano anche essere molto assorbiti. Per verificare questo punto abbiamo fatto una seconda serie di esperienze: una lastrina di rodio in una cavità praticata in un blocco di paraffina viene attivata una volta senza schermi e una volta circondando il rodio con assorbitori di spessore fino a 0,5 cm; la ricerca è stata finora limitata solo a forti assorbimenti. È stato scelto come rivelatore dei neutroni lenti il rodio perché è attivissimo, così che si possono comodamente seguire i fenomeni all'elettroscopio anche usando una lamina di pochi cm². In ambiente idrogenato si attiva praticamente solo il periodo breve del rodio, che è di circa 44 secondi, ed ha dunque un valore comodo per le misure. Da tali esperienze è risultato che esistono enormi

disparità nel potere assorbente dei diversi elementi per i neutroni lenti; per citare alcuni estremi noteremo che l'attivazione vien ridotta a metà da uno strato di circa 4 milligrammi per centimetro quadrato di boro o da 15 milligrammi di ittrio: si hanno dunque poteri assorbenti paragonabili addirittura con quelli per le particelle α. Invece, per esempio, per produrre lo stesso assorbimento col piombo, occorrono spessori di parecchi centimetri. Questi risultati si possono indicare esprimendo il potere assorbente per mezzo di una sezione di urto per atomo. I valori che si trovano per tali sezioni di urto nel caso degli elementi che assorbono più intensamente sono dell'ordine di 10^{-21} cm^2, ossia un migliaio di volte maggiori delle sezioni geometriche dei nuclei. Le sezioni di urto più grandi finora trovate sono $3 \cdot 10^{-21}$ pel B, $7 \cdot 10^{-21}$ per l'Y, $0{,}8 \cdot 10^{-21}$ per l'Ir, $0{,}4 \cdot 10^{-21}$ pel Rh, $0{,}16 \cdot 10^{-21}$ pel Li. Sezioni di urto considerevoli si hanno anche pel Cl, Co, Ag. È in corso una ricerca sistematica su questo argomento, dato che simili sezioni di urto sono inusuali per assorbimenti nucleari.

Alcune delle sostanze con grandi sezioni di urto sono tra quelle che si attivano più energicamente in ambiente idrogenato. Per queste, da un computo sommario del numero totale di attivazioni ottenibili all'esterno di un recipiente di acqua contenente la sorgente, risulta che, almeno come grossolana approssimazione, il numero di atomi attivati coincide con quello dei neutroni emessi. Invece per altre sostanze, per esempio il Li ed il B, al forte coefficiente di assorbimento non corrisponde una apprezzabile attivabilità. Ciò può interpretarsi o ammettendo che si formino in questi casi dei prodotti attivi che ci siano sfuggiti a causa del loro periodo (molto lungo o molto corto), o della scarsa penetranza dei loro raggi β; ovvero, colla formazione per cattura del neutrone di isotopi stabili, per esempio trasformazione del Li^6 in Li^7 o del B^{10} in B^{11}.

Abbiamo anche ricercato con esito negativo se degli assorbimenti anomali si avessero nella attivazione diretta, senza l'uso di sostanze idrogenate. Ciò fa pensare che l'azione sia specifica per neutroni lenti.

Si è fatta infine una esperienza sulla attivazione del rodio tenuto ad una distanza fissa dalla sorgente, in un vaso contenente soluzioni di acido borico di concentrazione variabile. Come era da prevedere bastano concentrazioni assai piccole per ridurre grandemente l'attivazione. La presenza di boro elemento al 2‰ riduce l'attivazione a metà, nelle nostre condizioni geometriche (Rh a circa 3 cm dalla sorgente). Concentrazioni maggiori non producono una ulteriore riduzione nello stesso rapporto. Per esempio per concentrazione tripla l'assorbimento è circa del 70 %.

Istituto Fisico della R. Università.

Roma, 7 novembre 1934–XII.

105 *b*.

INFLUENCE OF HYDROGENOUS SUBSTANCES ON THE RADIOACTIVITY PRODUCED BY NEUTRONS.—I.

Translated from E. Fermi, E. Amaldi, B. Pontecorvo, F. Rasetti and E. Segrè
« Ric. Scientifica », 5, (2) 282–283 (1934) (*)

In performing some experiments on the neutron-induced radioactivity of silver we noticed the following anomaly in the intensity of the activation: a layer of paraffin a few centimeters thick inserted between the neutron source and the silver increases the activity rather than diminishing it. Following this experiment we have ascertained that the presence of large paraffin blocks sorrounding the neutron source and the target increases the activation intensity by a factor which according to the geometry used ranges from a few tens to a few hundreds.

Following these observations we tried to ascertain, for the time being only summarily, the circumstances under which this phenomenon occurs. The facts which have been ascertained up to now are the following:

a) a radium source, without beryllium, does not produce the effect: a circumstance which makes us attribute the effect to neutrons and not to gamma rays;

b) an effect of approximately the same intensity as with paraffin, is obtained with water. We deem it very probable that the effect depends on the presence of hydrogen because substances containing oxygen, but not hydrogen ($NaNO_3$) do not produce an increase of activity, at least of the same order of magnitude;

c) the effect observed in silver is not shown by all elements which can be activated by neutrons. We have observed, up to now, that silicon, zinc and phosphorus do not show an appreciable increase in activity, whereas copper, silver and iodine give rise to activities much larger than one would obtain without the presence of water.

From these few cases it seems that there is the rule that only those elements are sensitive to hydrogenous substances which under bombardment give rise to activities due to isotopes of the starting element.

The case of aluminum is noteworthy. In water it acquires an activity showing a period slightly shorter then 3 minutes, corresponding to that of Al^{28} which can be extracted from irradiated silicon. This activity under normal conditions is so weak that it almost disappears compared to other activities generated in the same element.

(*) Translation by E. Segrè.

Similarly zinc and copper, which give rise to the same active products (1), isotopic with copper, under normal conditions acquire activities of the same order of magnitude, while in water copper shows a much larger effect than zinc.

A possible explanation of these facts seems to be the following: neutrons rapidly lose their energy by repeated collisions with hydrogen nuclei. It is plausible that the neutron-proton collision cross section increases for decreasing energy and one may expect that after some collisions the neutrons move in a manner similar to that of the molecules of a diffusing gas, eventually reaching the energy corresponding to thermal agitation. One would form in this way something similar to a solution of neutrons in water or paraffin, surrounding the neutron source. The concentration of this solution at each point depends on the intensity of the source, on the geometrical conditions of the diffusion process and on possible neutron-capture processes due to hydrogen or to other nuclei present.

It is not ruled out that this point of view may be important in explaining the effects observed by Lea (2).

The investigation of all these phenomena is being continued.

Istituto Fisico della R. Università.
Rome 22 October 1934–XII.

(1) T. BJERGE and C. H. WESTCOTT, « Nature », *134*, 286 (1934).
(2) D. E. LEA « Nature », *133*, 24 (1934).

106 *b*.

INFLUENCE OF HYDROGENOUS SUBSTANCES ON THE RADIOACTIVITY PRODUCED BY NEUTRONS.—II.

Translated from E. Fermi, B. Pontecorvo and F. Rasetti
« Ric. Scientifica », *5* (2), 380–381 (1934) (*)

In a preceding Letter we reported on the action of hydrogenous substances on the radioactivity produced by neutron bombardment, also mentioning the hypothesis that the increased effect might be due to slowing down of the neutrons emitted by the source.

We have now carried out the following experiments:

1.—Having observed that the direct activation of a small silver cylinder by a 350–mC Em + Be source placed at 20 cm distance is negligible, we immersed the source at the center of a water-filled cylindrical container of 14 cm diameter. One then observes that the silver, under the same geometrical conditions, becomes strongly active, yielding in our counters about one hundred counts per minute, roughly as many as would be recorded by placing the small cylinder in contact with the source in the absence of water.

On the hypothesis that the action of hydrogenous substances is actually due to diffusion and consequent slowing down of the neutrons, this experiment may be interpreted as follows: the neutrons, or at least a large fraction of them, before leaving the water container, are slowed down in such a way that, although the outcoming flux remains constant or possibly decreases, their average velocity is considerably lowered. The experiment then enables one to compare the effects of the same number of neutrons with different velocities and reveals the enormous effectiveness of the slower ones.

2.—The fact that slow neutrons appear so effective in activating various substances induces one to believe that they must also be strongly absorbed. To verify this point we performed a second series of experiments. A small rhodium plate, located in a cavity inside a paraffin block, is activated once without filters, and once surrounded by absorbers up to 0.5 cm in thickness. Up to now we have investigated only strong absorbers. Rhodium was chosen as a slow-neutron detector because of its very strong activity, such that the effects may be conveniently observed on an electroscope even using a rhodium plate of a few square centimeters surface. In a hydrogenous

(*) Translation by F. Rasetti.

medium practically only the short-lived activity of rhodium is excited. It has a 44 seconds half-life, a value convenient for the measurements. The result of these experiments was that there are enormous differences in the absorbing power of different elements for slow neutrons. To mention a few extreme cases, we note that the activity is halved by a boron layer of about 4 mg/cm^2 or an yttrium layer of 15 mg/cm^2. Hence one has absorbing powers even comparable with those for alpha-particles. Instead, in order to produce the same absorption in lead, thicknesses of the order of several centimeters are required. These results may be described by expressing the absorbing power through an atomic cross section. The values of the cross sections found for the most strongly absorbing elements are of the order of magnitude of 10^{-21} cm^2, hence about one thousand times larger than the geometric nuclear cross sections. The largest cross sections yet found are 3×10^{-21} for boron, 7×10^{-21} for yttrium, 0.8×10^{-21} for iridium, 0.4×10^{-21} for rhodium, and 0.16×10^{-21} for lithium. Large cross sections are also shown by chlorine, cobalt, and silver. A systematic investigation of this topic is under way, as such cross sections are unusual for nuclear absorption processes.

Some of the elements with the largest cross sections are among those that show the strongest activities in hydrogenous media. For these elements a crude estimate of the total number of active nuclei obtainable outside of a water-filled vessel containing the source, yields the result that, at least approximately, the number of activated atoms equals the number of emitted neutrons. For other elements, instead, such as lithium and boron, no appreciable activity corresponds to the large absorption coefficient. This fact may be interpreted either by assuming that the radioactive nuclei formed escaped us because of their half-lives (either very long or very short) or of the low penetration of their beta-rays; or that neutron capture leads to the formation of stable isotopes, such as the transformation of Li^6 into Li^7 or B^{10} into B^{11}.

We also searched, with negative results, for anomalous absorptions in direct activation, without the use of hydrogenous media. This fact leads one to believe that the effect is specific for slow neutrons.

Finally, an experiment was performed on the activation of rhodium kept at a fixed distance from the source, in a vessel containing boric acid solutions of different concentrations. As it might be foreseen, very low concentrations are sufficient greatly to reduce the activation. The presence of a concentration of 0.2 per cent of boron element reduces the activity to one-half under our geometric conditions (rhodium at about 3 cm from the source). Higher concentrations do not produce a further reduction in the same ratio. For example, for a concentration three times higher the absorption is about 70 per cent.

Physics Laboratory of the University.
Rome, 7 November 1934-XII.

107.

ARTIFICIAL RADIOACTIVITY PRODUCED BY NEUTRON BOMBARDMENT.—II.

E. Amaldi, O. D'Agostino, E. Fermi, B. Pontecorvo, F. Rasetti, and E. Segrè

(*Communicated by Lord Rutherford, O. M., F.R.S.—Received February 15, 1935*)
« Proc. Roy. Soc. » (London), Series A, *149*, 522–558 (1935)

Introduction.

We describe in this paper some further results on artificial radioactivity induced by neutron bombardment, which have been obtained in the Physical Laboratory of the University of Rome, after the publication of our first paper on the same subject. (1) Preliminary reports containing the main results have been already published by several of us. (2)

By far the most interesting new result concerns the effect of hydrogenated substances on the activation of several elements by neutron bombardment (§ 1). We give an interpretation of this phenomenon by assuming that neutrons are slowed down by impacts against hydrogen nuclei (§ 2). Some anomalously large absorption cross-sections for slow neutrons are discussed in § 3. In many cases the anomalous absorption is connected with the emission of γ–rays (§ 4). § 5 describes some attempts to get information about the energy of the slow neutrons. Scattering experiments on slow neutrons are described in § 6, while § 7 deals with the production of slow neutrons through the action of non-hydrogenated substances. In § 8 the above results are discussed from a theoretical point of view. §§ 9 and 10 are dedicated to chemical methods for the separation of radioactive isotopes, and to improvements in the technique of the measurements. Results of a systematic investigation of the different elements are given in § 11. These results are collected in a table at the end of the paper.

(1) Fermi, Amaldi, D'Agostino, Rasetti, Segrè, « Proc. Roy. Soc. », A, vol. 146, p. 483 (1934).

(2) Fermi, Amaldi, Pontecorvo, Rasetti, Segrè, « Ric. Scient. », vol. 2, p. 280 (1934); Fermi, Pontecorvo, Rasetti, « Ric. Scient. », vol. 2, p. 380 (1934); Amaldi, D'Agostino, Segrè, « Ric. Scient. », vol. 2, p. 381 (1934); Amaldi, D'Agostino, Fermi, Pontecorvo, Rasetti, Segrè, « Ric. Scient. », vol. 2, p. 467 (1934); vol. 1, p. 123 (1935). Some of our experiments have been repeated with analogous results by Bjerge and Westcott, « Proc. Camb. Phil. Soc. », vol. 31, p. 145 (1935); we thank them for having submitted their manuscript to us.

§ 1. – Effect of Hydrogenated Substances on the Activation.

In our previous work we had noticed some irregularities in the intensity of the activation of silver by neutrons from a radon + beryllium source, which apparently depended upon some not very clear geometrical factors. Further investigation showed that the activation was strongly influenced by objects surrounding the neutron source, and in particular that the activation could be enormously increased by surrounding the source and the activated substance with a large amount of water or paraffin wax. This effect appeared at once to be due to the presence of hydrogen, as other substances not containing hydrogen failed to give comparable effects (see § 7).

To ascertain whether these large activations were due to the neutrons or to the γ-rays emitted very strongly from our source, we repeated the experiment using as a source 100 mg radium, without beryllium, and found no induced radioactivity. It follows that the effect is actually connected with the neutrons. As a check on this point, we observed the same hydrogen effect with a Po + Be neutron source with an intensity in accordance with the number of neutrons emitted.

Not every substance which is activated by neutrons shows an increase in activity when irradiated under water. Among the strongly influenced activities are: Na (15 h); Al (2.3 m); V (3.75 m); Ag (22 s, 2.3 m); Cu (5 m); Rh (44 s, 3.9 m). The activation of other elements, or possibly of single decay periods, is not influenced by water; among these are: Si (2.3 m); Al (10 m); Mg (40 s); Mn (3.75 m); Zn (5 m). We have observed that in every case where the active element is known to be an isotope of the bombarded one (about 20 cases), the activation is increased by the presence of water.

In order to measure approximately the sensitivity of the different activations to the action of water or paraffin, we express it in a conventional scale. A cylinder (about 2 cm in diameter and 5 cm in height) of the substance under investigation is irradiated by putting the source in its centre. Source and cylinder are sustained by thin metallic supports at some distance from other objects. The activity reached after a convenient time of irradiation is measured. Afterwards the same cylinder is irradiated with the same source in the same relative position for an equal length of time surrounding both with a large cylindrical block of paraffin 27 cm in diameter and 20 cm in height. The ratio of the activities with and without paraffin is taken as a measure of the sensitivity to hydrogenated substances and denoted by α. In this scale $\alpha = 1$ means that the substance shows no increase in activity when irradiated in paraffin. Of course this definition of the sensitivity coefficient is only an empirical one, as it depends somewhat on the geometrical conditions.

Sensitivity coefficients different from one have been observed ranging from $\alpha = 1.6$ for U (13 m, 100 m) to $\alpha = 40$ for V (3.75 m). However, it must be pointed out that the last figure may not be the largest, because many substances fail to show some activities when irradiated in air, and these activities appear only by irradiation under water, and then rather weakly.

The ratio of increase in activation due to the presence of paraffin or water is much larger if source and irradiated substance are kept some centimeters apart (see § 5).

§ 2. – Interpretation in Terms of Slow Neutrons.

The experiments described in the preceding section can be explained on the hypothesis that the effect of water, or better of hydrogen, surrounding the source is due to scattering and slowing down of the primary neutrons by elastic collisions with hydrogen nuclei.

It is easily shown that an impact of a neutron against a proton reduces, on the average, the neutron energy by a factor $1/e$. From this it follows that 10 impacts reduce the energy to about 1/20,000 of its original value. Assuming the initial energy to be $4 \cdot 10^6$ electron volts, the energy after 10 impacts would be about 200 electron volts: and less than 20 impacts would be necessary to reduce the energy to thermal equilibrium values. The phenomena that we have described can now be explained on the assumption that slow neutrons are more easily captured by some nuclei than fast ones. In this and in the following sections we shall discuss our experiments in terms of this hypothesis.

The increase in activity through the action of hydrogen may be ascribed both to the scattering which causes an increased neutron flow through the substance to be activated and to the higher efficiency of the impact of a slow neutron as compared with that of a fast one. In order to show that this second factor is largely important, we performed the following experiment.

A silver cylinder was irradiated with a Rn + Be source containing 350 mC; the distance between source and cylinder was 20 cm. If nothing but air surrounded the source and the cylinder, no activity could be detected in this last after irradiation. Keeping the distance between the source and the irradiated cylinder constant, we then put round the source a cylinder (14 cm diameter and 14 cm depth) filled with water. A strong activity is shown by the silver irradiated in these conditions (about 100 impulses per minute in our counters). This experiment shows that the slow neutrons are much more effective than the fast ones. Indeed, the numbers of neutrons impinging on the silver per second is not increased (possibly slightly lowered through absorption) by the water surrounding the source. As the activation is much increased by the water in these conditions, we must conclude that the yield of activation per impinging neutron is very much enhanced by the slowing down of the neutrons.

§ 3. – Absorption of Slow Neutrons.

From the result of the preceding section, that the collision cross-section of slow neutrons for the activation of many elements is much larger than that of fast neutrons, one is led to investigate whether slow neutrons are strongly absorbed by the elements which are most strongly activated by them. We therefore made a systematic investigation of the absorption of slow neutrons in the different elements.

The main purpose of this investigation was to find elements with anomalously large absorption coefficients for the slow neutrons, and hence we generally used fairly thin layers of absorbing material. The arrangement for these absorption measures is shown in fig. 1. The source of neutrons, a Rn + Be tube (S) is put inside a paraffin cylinder P (24 cm diameter, 14 cm height) about 2 cm under the upper surface. A second paraffin cylinder P′ was put on P; in its lower surface a hole a few centimetres diameter and 2 or 3 cm in depth was excavated. The detector of the slow neutrons, a rhodium plate (sometimes also a silver plate), was put inside this hole and its activation measured, after irradiation during a standard length of time, once without absorbing layers and once interposed between two layers A of the absorbing material, as shown in the figure. The ratio of the activities without or with absorbing layers gives the absorption. In these as well as in many other experiments, we generally used rhodium as detector of the slow neutrons, because the activity induced in this element is very large and can be measured very exactly with an ionization chamber; moreover, of the two periods of Rh, 44 s and 3.9 m, practically the first only is of importance, which makes the reduction of the measurements very easy.

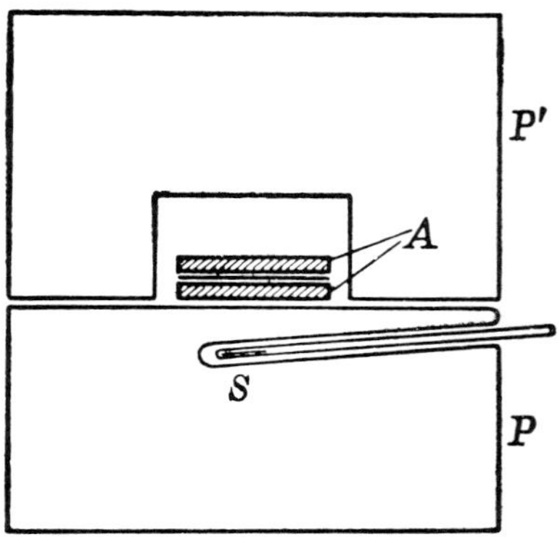

Fig. 1.

As a result of these absorption measures, we found that the half-value thickness δ for the absorption of slow neutrons for the different elements varies over a very wide range. For several elements it is exceedingly small. As an example we found for boron $\delta = 0.004$ gm/cm^2; for yttrium $\delta = 0.015$ gm/cm^2; for cadmium $\delta = 0.014$ gm/cm^2; for some other elements instead, δ is several thousands times larger; for instance, several centimetres of lead absorb less than a few milligrams per cm^2 of boron.

Expressing the absorption coefficients in terms of a nuclear cross-section σ for the activating impact of a slow neutron, we find in some cases surprisingly large values, *e.g.*, $\sigma = 3000 \cdot 10^{-24}$ for B; $\sigma = 7000 \cdot 10^{-24}$ for Y; $10{,}000 \cdot 10^{-24}$ for Cd. This last is the largest cross-section as yet found. It is remarkable how much larger these cross-sections are than the geometrical cross-sections of nuclei, while, as is well known, fast neutrons have cross-sections comparable to the geometrical cross-sections. [(3)] Indeed, we found directly that the absorption of boron for ordinary fast neutrons is at least 1000 times less than the absorption found for slow neutrons in the same element.

These absorption measurements do not refer to homogeneous neutrons. Indeed, the absorption curves are by no means exponentials; the absorption

(3) DUNNING, « Phys. Rev. », vol. 45, p. 586 (1934).

coefficient decreases with increasing thickness, as shown, for instance, by the absorption curve of cadmium, fig. 2. It must also be noticed that the half-

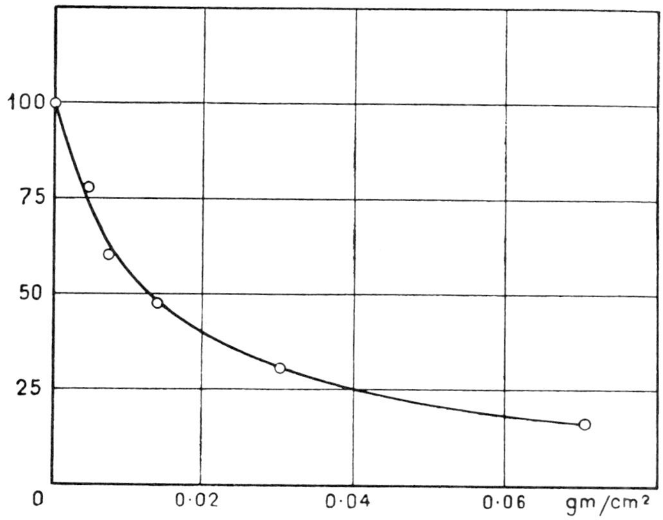

Thickness of absorber

Fig. 2.

value thickness depends to some extent upon the arrangement of the paraffin, as this affects the average velocity of the neutrons. For instance absorption of the neutrons inside a hole in paraffin is larger than outside a paraffin block.

§ 4. – Emission of γ-rays Through the Capture of the Neutron.

In connection with the anomalous absorption of slow neutrons, one is led to investigate more closely the absorption process. The simplest assumption is to admit that the neutron is captured by the nucleus with formation of an isotope heavier by one mass unit. If this heavier isotope is unstable, we might expect a strong induced radioactivity for the strongly absorbing elements. This occurs, for example, for indium and iridium, which are known to go over into radioactive isotopes. A rough evaluation of the intensity of the activation of the absorbing substance and of the number of the absorbed neutrons, shows that approximately an activated atom is formed in each absorption process. In other cases it is found that no activation, or at least no strong activation, corresponds to an anomalously large absorption (B, Y, Cd). In these cases we might expect that the capture of the neutron leads to the formation of a stable nucleus. It is obvious that this should happen more easily for elements with many stable isotopes differing in atomic weight by one unit (Cd, Hg).

In both cases, if the slow neutron is simply captured, we might expect that the absorption process should be accompanied by the emission of a

γ–radiation, with energy corresponding to the binding energy of the neutron. For fast neutrons a γ–radiation of this type has already been announced by Lea [4].

In the case of the absorption of slow neutrons we were able to show for some elements the emission of a relatively strong γ–radiation of this type by the following experiment.

A Po + Be neutron source S, fig. 3, of about 60 millicuries was immersed in a paraffin block together with one of our standard Geiger Müller counters C.

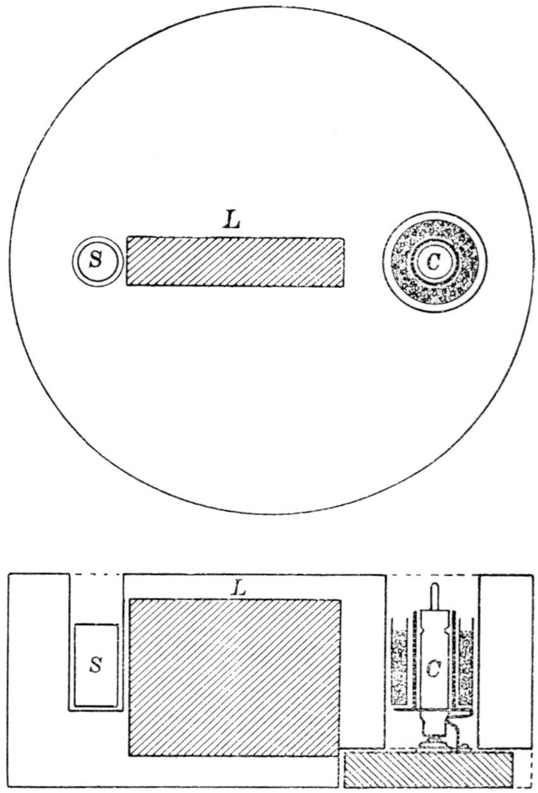

Fig. 3.

A lead absorber L, 10 cm thick, protected the counter from the direct γ–rays of the source. The counter was surrounded by a lead foil 2 mm thick. The counter registered in these conditions about 30 counts per minute. Small cylindrical layers of the various substances were placed around the counter, outside the lead foil. We generally observed a very marked increase in the number of counts for the strongly absorbing substances. This was observed, for instance, for Co, Cd, Y, Cl, Ir, Au, Hg, when the number of counts was sometimes more than doubled. An exception to this behaviour is shown by

(4) « Nature », vol. 133, p. 24 (1934).

boron and lithium, which, although showing a strong absorption for the slow neutrons, do not emit γ–rays. In these elements the absorption of the slow neutrons has been found to be connected instead to the emission of heavy particles (see § 11). This alternative possibility to the emission of γ–rays appears theoretically likely only for very light elements owing to the low potential barrier.

That the emission of γ–rays, as described before for some substances, was actually due to slow neutrons was shown by the fact that no emission was observed by removing the paraffin. Furthermore, the effect was very much reduced by surrounding substances and counters with a boron screen.

§ 5. – Energy of the Slow Neutrons.

It would be very important to evaluate somehow the mean energy of the slow activating neutrons, and we describe in this section some attempts to do this.

We have already stated that the mean energy of the neutrons is reduced by a factor $1/e$ for each elastic collision against a proton, provided the values of the energy are large compared with the thermal energy. If the energy of the slow neutrons were really reduced down to this limit, one would expect that the diffusion process might be influenced by temperature. We made an attempt to find such an effect by the following experiment.

A rhodium or a silver detector was irradiated with the Rn + Be neutron source under the same geometrical conditions, once in hot paraffin at 200° C, and once at 20° C in a mixture of benzene and penthane having at this temperature the same density and elementary composition as paraffin at 200° C. No difference in the activation was found in the two tests, within the accuracy of 2 %. The mixtures filled a cylinder 26 cm in diameter and 15 cm in height. The detector was on the axis of the cylinder 1 cm under the liquid surface; the source, also on the axis, 2 cm lower. It must be concluded that temperature, at least in these conditions, does not affect the activation, which might be interpreted to mean that the energy of the activating neutrons is higher than the thermal energy. This is, however, not quite conclusive, as the dependency of both the cross-section for activation and the mean free path upon the velocity for very slow neutrons is unknown.

A direct method of measuring or at least of establishing an upper limit to the energy of the slow neutrons would be to measure the average ionization produced in each collision of a slow neutron against a hydrogen nucleus. This might be done either by measuring the total ionization in a hydrogen-filled chamber or by directly measuring the ionization in a single process by means of a linear amplifier. In both methods the action of the slow neutrons can be separated from that of the fast ones, by screening the chamber with a thin boron layer. Experiments of this kind are being attempted but have not yet led to definite results.

We have investigated whether the increase in activation produced by paraffin is also to be found with different sources of neutrons. We have done this with the neutrons emitted by beryllium irradiated with the γ–rays of radium, [5] and found a large effect with this source. This shows that the slow neutrons of paraffin are slower than the neutrons emitted in this nuclear reaction.

§ 6. – Scattering and Diffusion of the Slow Neutrons.

In connection with the absorption measurements described in § 4, we have made a rough survey of the scattering properties of the different substances. The experimental arrangement is shown in fig. 4. The source S is enclosed in the cylindrical paraffin block P as described in § 4. The rhodium plate used as detector was activated by placing it on the top of this paraffin block. The activation was measured once with the substance to be investigated placed in a layer above the rhodium plate, and once without it. In some cases an increased activation was found when a substance was superimposed, indicating an action from slow neutrons scattered back from the substance. For instance, a layer of carbon a few centimetres thick increases the activation about five times. Similar results are found with several light elements (Be, C, Si). Heavy elements generally produce a much smaller effect. Boron, which has a large absorption coefficient, does not produce any appreciable increase of activity. This shows that the anomalously large absorption of this element is mainly due to real absorption and not to scattering.

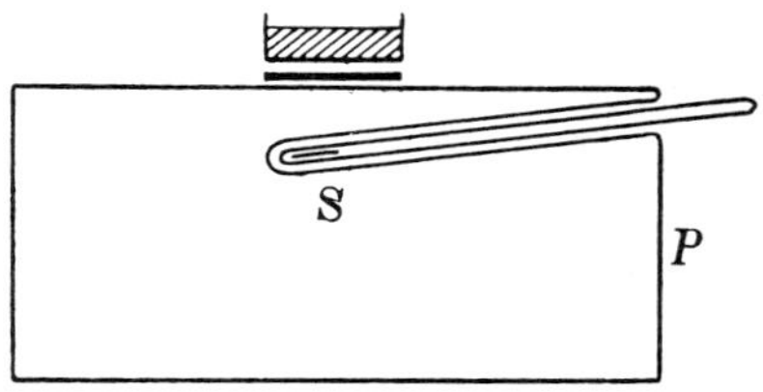

Fig. 4.

We have repeated the same experiment with water, obtaining the curve reproduced in fig. 5 for the intensity of activation as a function of the thickness of the scattering layer. A similar result is obtained employing paraffin instead of water.

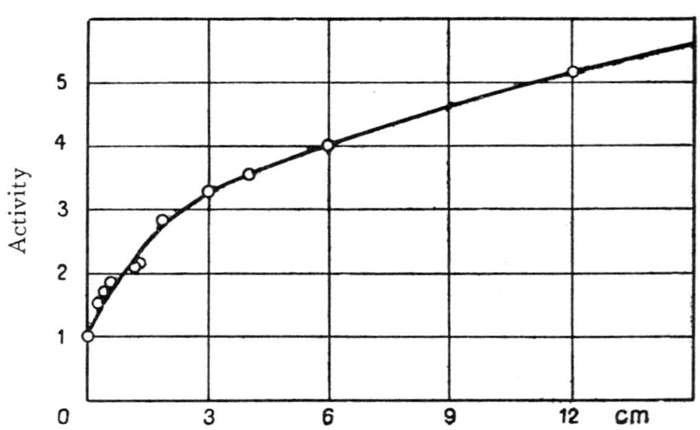

Fig. 5.

(5) Szilard and Chalmers, «Nature», vol. 134, p. 494 (1934).

Other experiments on scattering have been made by the arrangement shown in fig. 6. A paraffin cylinder, 13 cm diameter and 11 cm high, contains in its centre the neutron source S. The detector, a silver or rhodium cylinder R, is kept at a distance of 30 cm from the source. A screen D of the scattering material is interposed between the source and the detector; its dimensions are such as to intercept exactly all the neutrons which might reach the detector from the paraffin block. The difference in the activity of the detector measured when irradiated with or without the screen D gives a measure of the effect of scattering + absorption. We found by this apparatus that a paraffin layer 0.5 gm/cm² thick reduces the intensity of the silver activation to about one-half; here also we found that the absorption is not exponential. A similar result is obtained, using water as a scatterer. From this we might deduce the order of magnitude of the mean free path of slow neutrons in water or paraffin, this being comparable with the observed half-value thickness.

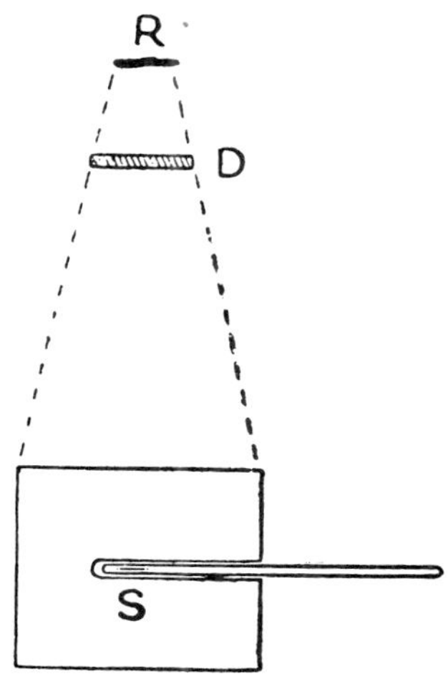

Fig. 6.

For carbon the half-value thickness is about 5 gm/cm². Lead of a thickness of about 4 gm/cm² gave a slight decrease in intensity, which is mainly due to scattering.

Another set of experiments was made in order to determine how the intensity of the activation of rhodium in water decreases with the distance from the source. The source was immersed in a water basin of the dimensions

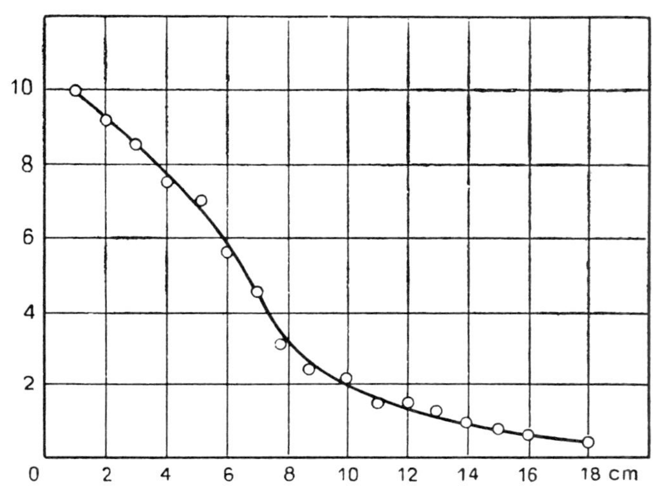

Fig. 7.

40 × 40 × 100 cm and the activity induced in rhodium at different distances measured with an ionization chamber. The result is shown in fig. 7, where the activity is plotted as a function of the distance. A similar experiment

was made with 2 % boric acid dissolved in the water, obtaining a curve similar to the former, only very much reduced in intensity, owing to the absorption of boron.

The general picture of the slowing down process, which one gets from these experiments is the following. The largest part of the diffusion process takes place when the primary neutron still retains much of its original energy, having collided only a few times with hydrogen nuclei; as the velocity is reduced down to the value for which the absorption by boron becomes appreciable, the mean free path in hydrogen appears to be fairly small, so that the diffusion process after this stage is contained in a small volume. This explains also the fact that the absorption coefficient of boron for the neutrons in water is fairly independent of the distance from the source.

§ 7. – Effect of Non-Hydrogenated Substances on the Activation.

We have tried if effects similar to that of water or paraffin could be obtained also with non-hydrogenated substances. Owing to the large amount of materials necessary for these experiments, only a few substances have been tested: Pb, SiO_2, C, Fe. In all these, with the exception of Fe, a definite increase of the activation of rhodium was observed. Under the geometrical conditions to be described later, the effect was increased by a factor two to five, while in water the factor would have been of some hundreds. That this increase in activity is not simply due to scattering of the neutrons but mainly to their loss in velocity, is proved by the fact that the activation of silicon under the same circumstances is not increased; moreover, very thin cadmium absorbers considerably reduced the activation of rhodium. These facts seem rather surprising chiefly for lead, as it would be difficult to explain the slowing down of neutrons as due to elastic collisions.

The details of these experiments were as follows. We built with lead blocks a cube of side of about 50 cm. The neutron source was placed in the centre and the rhodium detector 10 cm apart. The activity is about three times larger than in the same geometrical conditions without lead. A similar result is obtained with silver as detector instead of rhodium, while a silicon detector shows no increase in the activity. This fact is to be taken as an indication that the velocity of the neutrons is reduced by the impact against lead nuclei. This view is confirmed by the observation that the activation of rhodium in lead is reduced to half value by an absorber of cadmium of 1 gm/cm^2. Comparing this absorption of cadmium with that of the same element for the slow neutrons in water (half-value thickness 0.014 gm/cm^2), one is led to assume that the average velocity of the neutrons in lead is not so low as in water.

A similar experiment was performed using a graphite cylinder of 3 litres volume. The source was placed in the centre and the rhodium detector at 5 cm distance. The increase in activation was about the same as that produced by lead in the former experiment and the absorption of the neutrons in cadmium was certainly not smaller. Similar results were obtained with silica in the form of " Kieselguhr ". The failure to observe a similar effect with iron is probably due to its relatively large absorption.

We conclude from these experiments that the property of intensifying the activation is shown by most materials. The possible action of surrounding objects, possibly even of air, must always be considered, when experimenting quantitatively on the activation of some substances. For these reasons, the figures given above for the effect of some materials are to be taken only as an indication.

§ 8. - THEORETICAL CONSIDERATIONS ON THE PROPERTIES OF SLOW NEUTRONS.

In our first paper we left undecided the question whether whenever the neutron bombardment gives rise to a radio element isotopic with the original one, the neutron is captured or causes the expulsion of a neutron from the struck nucleus, *i.e.*, whether the atomic weight of the radioactive product is A + 1 or A — 1. We believe that the evidence accumulated since then speaks all in favour of the first hypothesis. The main points of this evidence are the following.

(*a*) Two new weak activities have been recorded: one, a 15-hour period in sodium, the other a 2·3-minute period in aluminium. Both their identity with known periods, in the first case of aluminium and magnesium, in the second case of silicon and phosphorus, and the chemical evidence in the case of sodium (see § 11), show that the radio elements are respectively isotopes of Na and Al. As both these elements have only one stable isotope, *i.e.*, ^{23}Na and ^{27}Al, the choice remains for radio sodium between ^{24}Na and ^{22}Na, for radio aluminium between ^{28}Al and ^{26}Al. Now both the lighter isotopes ^{22}Na and ^{26}Al are known from other nuclear reactions to have different periods and to emit positrons instead of electrons, so that it seems well founded to admit that, at least in these cases, the neutron is captured.

(*b*) While there is no theoretical difficulty in understanding how a neutron with a negligible kinetic energy can be captured by a nucleus, it seems unlikely that it could knock out a stably bound neutron from the nucleus. Still more difficult it would be to understand that in the latter case energy could be left over to account for the emission of γ-rays.

We shall, then, discuss the experimental facts on the standpoint that the neutrons, and particularly slow ones, are easily captured by many nuclei.

There are, however, some theoretical difficulties in understanding this capture process, or at least to account for the large cross-sections that have been experimentally observed in some cases. Nevertheless it may be worth while to state some general consequences of the theory, which must be always kept in mind in the discussion of this problem.

Let us admit, as it has been generally assumed, that the forces acting between the neutron and a nucleus extend about as far as the nuclear radius itself. If this is so, the de Broglie wave-length is, for fast neutrons, of the order of the radius of action, and consequently for slow neutrons is much larger. The well-known theory of the impact, in which the nucleus is treated as a potential hole, takes in this case an extremely simple form. Let ψ be the s-eigenfunction corresponding to zero energy. In fig. 8 the product $r\psi$ is plotted as a function of the radius vector r, and

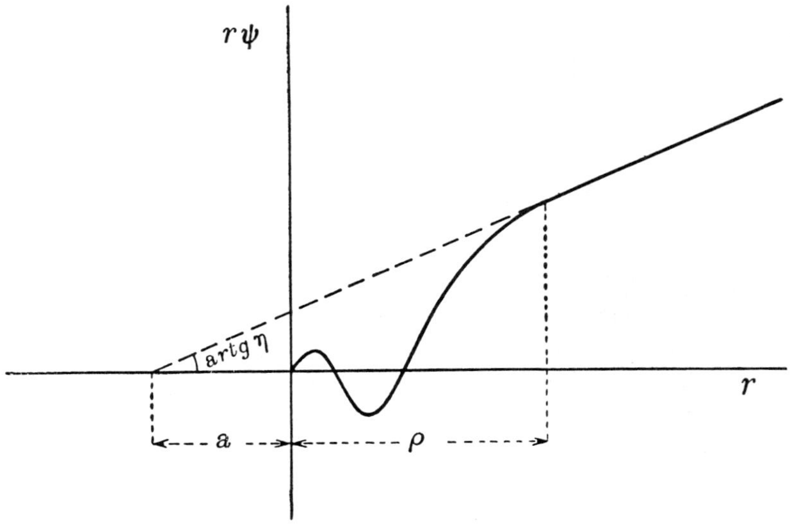

Fig. 8.

ρ represents the radius of action. The curve has an irregular shape for $r < \rho$, while for $r > \rho$ it becomes a straight line. We normalize ψ taking $\psi(0) = 1$. Let then the equation of the straight line into which $r\psi$ goes over for large values of r be

$$r\psi \rightarrow \eta\,(a + r),$$

where the geometrical meaning of η and a is clear from the figure. The values of these two quantities could be easily calculated if one knew the form of the potential hole representing the nucleus. The cross-section for elastic collisions in the limit for low velocities is found to be

$$\sigma_{el} = 4\,\pi a^2, \tag{1}$$

while the density of probability for the neutron to be found in the centre of the nucleus is

$$\mathrm{P} = n/\eta^2, \tag{2}$$

where n is the density of the neutrons outside of the nucleus.

Whatever the mechanism of capture may be, it is natural to assume that, at least to a certain approximation, the probability of capture of the neutron by a nucleus per unit time will be proportional to P, *i.e.*, that this probability will be given by kn/η^2, where k is a constant for each nucleus.

This probability can be expressed in terms of a cross-section σ_{ca} for the capture process through the relation

$$\sigma_{ca} = k/\eta^2\,v, \tag{3}$$

where v is the velocity of the neutron. A limit to the validity of (3) is set by the fact that σ_{ca} can evidently be at most of the order of magnitude of the square of de Broglie wave-length. This allows one, considering the largest cross-sections found experimentally (Cd, $\sigma_{ca}=10^{-20}$), to set an upper limit for the energy of the slow neutrons. This limit turns out to be a few hundred volts. However, it must be borne in mind that this deduction is valid only under the explicitly stated assumptions.

Formula (3), valid only for low velocities, gives a cross-section for capture for a given nucleus inversely proportional to the velocity of the neutron. This explains why the cross-section for capture is generally larger for slow neutrons than for fast ones. This result may also be expressed by saying that the mean life of a slow neutron in a substance is independent of its velocity. While the capture cross-section is inversely proportional to the velocity, the cross-section for elastic impact expressed by (1) is independent of the velocity, which means that the mean free path for this type of collisions is also independent of the velocity.

One more feature of (3) is that σ_{ca} is inversely proportional to η^2. The straight line of fig. 8 may occasionally be nearly parallel to the abscissae axis. In this case η is very small and consequently the capture cross-section becomes very large. This behaviour of the eigenfunction is probably responsible for the anomalously large cross-sections observed for a few nuclei.

In order to deduce from (3) the absolute values of the cross-sections, it would be necessary to know also the values of k, which quantity depends on the physical mechanism of the capture. We know from the experiments that there are two different processes. In some light elements (Li, B) the capture of the neutron is followed by the emission of heavy particles; while in the heavier elements the normal process seems to be capture with emission of a γ-ray. The binding of an additional neutron to a nucleus sets free an amount of energy which, on the average, has a value about 7 million electron volts. This surplus energy might in some cases produce the emission of an α-particle, provided the potential barrier surrounding the nucleus is low enough to allow for a quick escape of the particle. Therefore this process is expected to occur only for the lightest elements, whereas in the activation by fast neutrons, owing to the kinetic energy of the impinging neutron which is to be added to the binding energy, the emission of charged particles is possible also in elements of higher atomic weight. (6)

In the case of capture with emission of a γ-ray, which is observed for elements of any atomic weight up to the highest, the ordinarily accepted mechanism of irradiation gives apparently rather too small a value for k. k depends upon two factors which are rather difficult to evaluate: matrix elements and energy of the emitted γ-quanta. As the probability of emission of a γ-quantum is *ceteris paribus* proportional to ν^3, it is to be assumed that processes by which the binding energy of the neutron is very large ought to be preferred; this binding energy might in several

(6) MEITNER, «Naturwiss.», vol. 45, p. 789 (1934).

cases considerably exceed 10 million volts. Nevertheless one ought probably to assume rather too high a value for the matrix elements in order to get a plausible interpretation of the relative frequency of the anomalously large cross-sections. From the theory one would also expect that generally anomalous cross-sections for the capture process were connected to anomalous elastic cross-sections. No experimental evidence has yet been found for this fact.

In this section we shall also discuss briefly the velocity distribution law for the slow neutrons in hydrogen. From the above theoretical considerations (formulae (1) and (3)), it seems plausible to assume that neutrons having velocities smaller than a certain limit have a constant mean free path λ for elastic collisions, and a constant mean life-time τ before they are captured by the nuclei present in the medium. One can then easily show that the number of neutrons in a hydrogenated substance, having velocities between v and $v + dv$, is proportional, for v less than the above-stated limit to

$$\frac{v\,dv}{(v+\lambda/\tau)^3}\,. \tag{4}$$

This distribution law of the slow neutrons might be used for interpreting the fact that the absorption curves of the slow neutrons are not exponentials.

§ 9. – Separation of Radioactive Isotopes.

Szilard and Chalmers (7) were the first to separate radioactive iodine from usual iodine by chemical methods. We have extended their procedure to some other cases, using also inorganic instead of organic compounds. The principle of the method is the following: let us suppose that the element before irradiation is bound in a molecule or radical which, once decomposed, has a practically negligible probability of being rebuilt. As the neutron strikes the atom, this is generally knocked out from the molecule, and has a tendency to remain in an atomic or ionic state. It follows that, after irradiation, the modified atoms are in a different chemical state from the main bulk of the unmodified ones, and can thus be separated by convenient reactions. The energy of the chemical bond is a few volts, and, even admitting that the impinging neutron might have a negligible energy, the recoil of the γ–quantum emitted in the capture of the neutron would be sufficient to overcome the binding chemical forces.

Szilard and Chalmers separated radioactive iodine from irradiated, ethyl iodide adding a trace of free iodine, reducing and precipitating I with $AgNO_3$. The same method was applied by us to bromoform, chloroform, carbon tetrachloride and some other halogenic organic compounds.

(7) « Nature », vol. 134, p. 462 (1934).

We were always able to separate almost completely the radioactive halogen from the bulk of the inactive substance.

We have also separated radioactive chlorine (35 m) starting from sodium chlorate. The chlorine atom is knocked out of the ClO_3^- ion by the neutron impact; adding a small trace of Cl^- to carry the radioactive isotope and precipitating carefully with $AgNO_3$ with addition of HNO_3 to prevent precipitation of $AgClO_3$, one finds the activity concentrated in the AgCl precipitate. Similar results are obtained also with bromates and iodates. In these last compounds the nitric acid was substituted with ammonia. From 70 to 90 % of the total activity is concentrated in the precipitate.

Cacodylic acid $(CH_3)_2AsOOH$ is a good starting substance for concentrating radioactive arsenic. By precipitating arsenic sulphide from irradiated cacodylic acid, it is possible to concentrate the activity.

Potassium permanganate irradiated and then filtered through an ordinary paper filter left much of the activity on the filter with manganese dioxide formed in the oxidation of the paper. About 80 % of the total activity was separated, adding a trace of a manganous salt and precipating the manganous carbonate. The MnO_4^- ion is dissociated by the neutron bombardment and the manganese atom is left in lower states of oxidation; these are collected with manganese dioxide on the filter, or precipitated by carbonates. We tried this separation starting either from solid salts or from solutions; there is no very remarkable difference between the two cases. Also no marked influence of the acidity or alkalinity of the irradiated solution was observed.

A physical method for separating radioactive isotopes is analogous to the well-known method of collecting the active deposit from the emanation. We tried this method with a gaseous iodine compound. A glass cylinder of about 1 litre volume with a large cylindrical aluminium electrode close to the wall and a nickel electrode along the axis was filled with methyl or ethyl iodide saturated vapour. The temperature of the cylinder was regulated so that the vapour pressure was about an atmosphere. The whole apparatus was immersed in hot water and a potential difference of 3000 volts applied to the electrodes. A Rn + Be source was placed outside the wall. After irradiation the nickel electrode was removed and showed an activity decaying with the iodine period. Nickel was used as a support because it does not become radioactive under neutron bombardment. The yield was rather a poor one. Inversion of the polarity of the electrodes gave no consistent results.

§ 10. - Methods of Measurements.

The observed fact of the intense activating effect of the slow neutrons opened the possibility of obtaining stronger sources of the artificial radio elements than were avaibable last year. This allows more accurate measurements of the constants of these substances by using an ionization chamber instead of Geiger Müller counters.

The ionization chamber was built of steel; the inner electrodes were a wire net and a brass rod. The β–rays entered the chamber through a circular window of 6 cm diameter on the top, closed by an aluminium foil 0.01 cm thick. The chamber was filled with CO_2 at 3 atmospheres, pressure. The chamber was connected to a Perucca electrometer, the total capacity of the system being about 20 cm, the electrometer sensitivity 0.02 volts per scale division. The system was perfectly constant both in its zero effect and in its sensitivity (as controlled by a uranium standard preparation). We reproduce in figs. 9 and 10 some decay curves obtained with this apparatus.

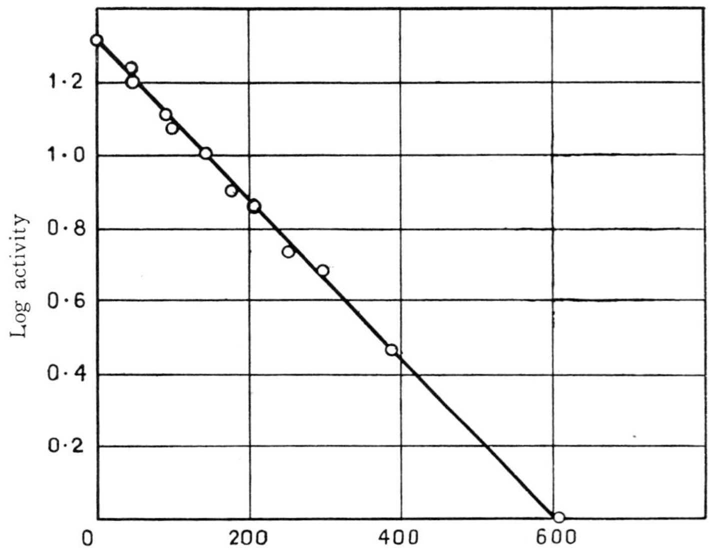

Fig. 9.

The same apparatus was used for measurements of the absorption of the β–rays. The window was protected with aluminium foils and the activity plotted as a function of the aluminium thickness. The absorption curve was roughly exponential. Sometimes substances show besides the β–activity a strong γ–radiation. We conventionally attributed to β–rays the residual ionization when the chamber was screened with 2 mm of lead. In the evaluation of the absorption coefficients for β–rays allowance has been made for the γ–radiation. To check the measurements, the absorption coefficients of the β–rays of RaE and UX_2 have been measured and results obtained in accordance with the values commonly admitted.

§. 11 – Systematic Investigation of Elements.

In this section we shall report all the new data that we have found about each element, both as regards the induced activities and the properties with respect to slow neutrons. Some data differ slightly from our previous ones, owing to the increased precision of our measurements.

1 – *Hydrogen* – No activity could be detected either in water or in paraffin irradiated in a large can of water with 500 millicuries Rn + Be for several days.

3 – *Lithium* – Lithium hydroxide was found to be inactive after irradiation with slow neutrons (14 hours, 400 millicuries). Although lithium remains inactive, it strongly absorbs the slow neutrons; half-value thickness $\delta = 0.05$ gm/cm². This absorption is not accompanied by a γ–radiation. It was shown independently by Chadwick and Goldhaber [8] and by us that when the slow neutrons are absorbed, heavy charged particles

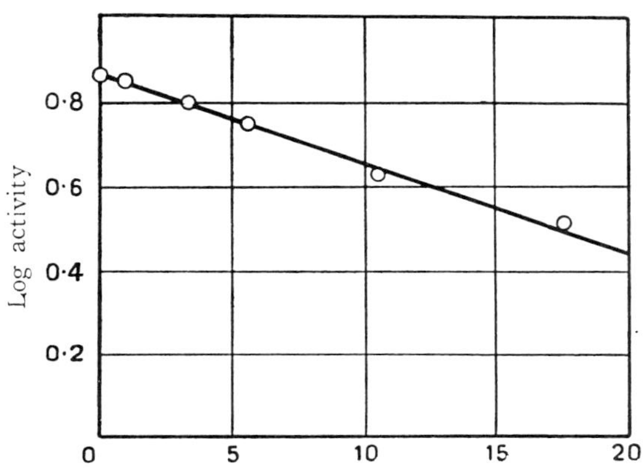

Fig. 10.

are emitted. According to Chadwick and Goldhaber, the nuclear process is represented by the following reaction,

$$^{6}_{3}\text{Li} + ^{1}_{0}n = ^{4}_{2}\text{He} + ^{3}_{1}\text{H} .$$

4 – *Beryllium* – Metallic beryllium (purity 99 %), strongly irradiated with slow neutrons, showed only an extremely weak activity possibly due to impurities. Owing to the very strong activation of several elements when irradiated under water, impurities might easily be misleading.

5 – *Boron* – Metallic boron irradiated 14 hours under water with 500 millicuries was found inactive. Boron has the highest absorption coefficient as yet found for slow neutrons, $\delta = 0.004$ gm/cm², corresponding to a cross section of about $3 \cdot 10^{-20}$ cm². No γ–rays have been found to accompany this absorption: instead of a γ–radiation in this case as well as for lithium, α–particles are emitted, as was shown by Chadwick and Goldhaber [8] and by us. This effect can be easily detected by the strong discharge in an ionization chamber filled with boron trifluoride surrounded by paraffin and irradiated with a Po+Be neutron source. Screening the ionization chamber with a thin cadmium foil in order to absorb the slow neutrons, reduces con-

(8) «Nature», vol. 135, p. 65 (1935).

siderably the ionization current. The same effect was observed with the ionization chamber filled with air, some boron being spread on its floor. The emission of α-particles was also detected with a small ionization chamber connected to a linear amplifier, either spreading some boron on its walls or filling it with boron trifluoride. In order to explain these phenomena we have proposed the nuclear reaction,

$$^{10}_{5}\mathrm{B} + ^{1}_{0}n = ^{7}_{3}\mathrm{Li} + ^{4}_{2}\mathrm{He}.$$

Chadwick and Goldhaber have proposed instead the reaction,

$$^{10}_{5}\mathrm{B} + ^{1}_{0}n = 2\,^{4}_{2}\mathrm{He} + ^{3}_{1}\mathrm{H}.$$

We do not think that there is at present sufficient evidence to decide between these two possibilities, and we are now experimenting to try to get a more exact measurement of the number of ions formed in each process in an ionization chamber containing boron either in a gaseous form (total process) or spread on its walls (effect of only one or two particles). We are also trying to observe the disintegration in a Wilson chamber containing a gaseous compound of boron [(9)].

6 – *Carbon* – No activity; see hydrogen. For the scattering properties see § 6.

7 – *Nitrogen* – Ammonium nitrate irradiated 12 hours with 600 millicuries under water showed no activity.

8 – *Oxygen* – No activity, see hydrogen.

9 – *Fluorine* – Both activities of this element (periods 9 seconds and 40 seconds) are not sensitive to hydrogenated substances.

11 – *Sodium* – This element has two activities: one of these (period 40 seconds) is not sensitive to hydrogenated substances. A very weak, activity with a long period was reported by Bjerge and Westcott [(10)]. As this activity is strongly enhanced by water, we were able to measure its period with reasonable accuracy and found it to be 15 hours. Owing to the theoretical importance of this activity (see § 8), we compared very carefully its decay curve with that of the long period of aluminium in order to check their identity. For a chemical investigation of the active substance we irradiated pure sodium carbonate (Kahlbaum). We dissolved the irradiated substance in hydrochloric acid and added aluminium and magnesium chlorides. A precipitate of the hydroxides of the latter elements obtained by adding ammonia, was found inactive. Afterwards we added some sodium fluoride to the solution, and precipitated the fluorine as barium fluoride; this precipitate was also found inactive. The solution containing the original sodium was then evaporated and ignited gently, in order to eliminate neon, where an active isotope of this element would have been formed. The activity was found in the dried sodium salt. We conclude that the active prod-

(9) [*Note added in proof*–TAYLOR and GOLDHABER, « Nature », vol. 135, p. 341 (1935), have shown that the reaction takes place according to the first scheme].

(10) BJERGE and WESTCOTT, « Nature », vol. 134, p. 286 (1934).

uct is an isotope of sodium, ^{24}Na. The same isotope was produced by us last year by bombarding magnesium or aluminium with neutrons. ^{24}Na has also been produced recently in considerable amounts and studied very completely by Lawrence [11] bombarding several elements with artificially accelerated particles.

12 – *Magnesium* – Pure magnesium oxide (Kahlbaum), especially tested by us in order to ensure that it was aluminium free, was irradiated under water. The substance was kept at some distance from the source in order to prevent the activation of the periods not sensitive to water. A new very weak activity with a period of about 10 minutes was found. As this period coincides with the 10-minute period of aluminium, which is known to be due to ^{27}Mg (see aluminium), it is very likely that it is due to the same isotope formed by the capture of a neutron by ^{26}Mg present in an amount of 11 % in ordinary magnesium.

The 40-second period is insensitive to water.

13 – *Aluminium* – Aluminium irradiated in water shows a fairly strong new activity decaying with a period of 2.3 minutes (measured with an ionization chamber). Irradiated outside of water, this activity is extremely weak. As the period of the new activity coincides with the 2-3-minute period of silicon, which is due to ^{28}Al, we assume that this activity is also due to the same isotope formed by capture of a neutron from ^{27}Al.

The second period of aluminium has been measured with the ionization chamber and found to be 10 minutes instead of 12. This activity is insensitive to water. A chemical separation of the carrier of this activity has been performed. Irradiated metallic aluminium was dissolved in a caustic soda solution and magnesium chloride was added. The precipitate of magnesium hydroxide carried the 10-minute activity. We assume that the active isotope is ^{27}Mg formed according to the reaction,

$$^{27}_{13}\text{Al} + {}^{1}_{0}n = {}^{27}_{12}\text{Mg} + {}^{1}_{1}\text{H}\,.$$

14 – *Silicon* – We have determined with the ionization chamber the short period of this element, finding it to be 2.3 minutes. This activity is insensitive to water.

Besides this activity, we found a new longer period of some hours in fused silica irradiated in water. This activity is very weak and very sensitive to water. We think probably that its carrier is ^{31}Si which is obtained by irradiated phosphorus and has a period of 2.4 hours. ^{31}Si could be formed by capture of a neutron from ^{30}Si present in an amount of 3 %.

15 – *Phosphorus* – The short-period activity of this element (2.3 minutes) is not enhanced by water. Curie, Joliot and Preiswerk [12] ascribe this period to ^{28}Al. A chemical test in favour of this hypothesis is the following: we irradiated phosphoric acid, neutralized the solution with sodium carbonate, and added aluminium chloride; the activity was found to be concentrated in the precipitated aluminium.

(11) « Phys. Rev. », vol. 47, p. 17 (1935).

(12) « C.R. Acad. Sci. Paris », vol. 198, p. 2089 (1934).

We have observed, with the aid of the ionization chamber, the decay-curve of the longer period of phosphorus. Its period is 2.4 hours instead of 3 as given before. We have also measured, with the ionization chamber, the half-value thickness of the corresponding β–rays and found it to be 0.15 gm/cm^2 Al.

16 – *Sulphur* – We have determined in the ionization chamber the period of phosphorus extracted from irradiated sulphur. We found: period 14 days, half-value thickness of the β–rays 0.10 gm/cm^2 Al.

17 – *Chlorine* – Chlorine irradiated under water showed a new period of 35 minutes measured electrometrically. For the chemical tests on the carrier of this activity see § 9.

Chlorine absorbs fairly strongly the slow neutrons (half-value thickness $\delta = 0.3$ gm/cm^2). The process of absorption is accompanied by emission of γ–rays.

19 – *Potassium* – We have found in irradiated potassium an induced activity strongly sensitive to water, decaying with a period of 16 hours. A chemical investigation of the carrier of the activity, performed by the same method described for sodium, excluded the elements Cl, A, Ca. We conclude therefrom that the activity is probably carried by an isotope of potassium. According to von Hevesy [13], this isotope is to be identified with a ^{42}K, that was obtained by him by neutron bombardment of scandium, and has the same decay period.

20 – *Calcium* – No activity was found in calcium fluoride irradiated 14 hours in water with a 600 millicuries source.

23 – *Vanadium* – The decay of the activity induced in vanadium has been measured in the ionization chamber with the following results: half-value period 3.75 minutes; half-value thickness of the β–rays 0.17 gm/cm^2 Al. The β–rays are accompanied by a γ–radiation. The activation of vanadium is strongly sensitive to hydrogenated substances; with the definition of § 1, $\alpha = 40$.

24 – *Chromium* – The activity of chromium is insensitive to water.

25 – *Manganese* – The activity with short period (3.75 minutes) is insensitive to water ($\alpha = 1$). On the other hand, the activity with longer period (2.5 hours measured in the ionization chamber) is strongly enhanced by water ($\alpha = 23$). Half-value thickness of β–rays measured electrometrically is 0.14 gm/cm^2 Al; the disintegration is accompanied by γ–rays. The 2.5-hour product is known to be an isotope of manganese; in § 9 a method for concentrating the activity has been described. In order to get new evidence in favour of the fact that the active product is really an isotope of manganese, we first concentrated the activity obtained in irradiated manganese permanganate by a precipitation of manganese carbonate. The carbonate containing the activity was then dissolved in hydrochloric acid, and large amounts of chromium, vanadium and iron salts were added to the solution. Afterwards the manganese was separated once again as diox-

(13) «Nature», vol. 135, p. 96 (1935).

ide, with nitric acid and sodium chlorate. The manganese precipitate carried the activity, while the fractions containing chromium, vanadium and iron were found to be inactive.

26 – *Iron* – The activity of this element (period 2.5 hours) is insensitive to water. Half-value thickness for the absorption of slow neutrons 8 gm/cm^2.

27 – *Cobalt* – This element absorbs fairly strongly the slow neutrons; half-value thickness 0.7 gm/cm^2. The absorption is accompanied by the emission of a γ–radiation.

28 – *Nickel* – Strongly irradiated nickel showed only a dubious trace of activity.

29 – *Copper* – Both induced activities of this element (periods 5 minutes, measured electrometrically, and 10 hours) are strongly enhanced by water. For the first $\alpha = 15$. Copper absorbs the slow neutrons with a half-value thickness of about 3 gm/cm^2; this absorption is accompanied by a weak γ–radiation.

Irradiated metallic copper was dissolved in hydrochloric acid, and small quantities of cobalt, nickel and zinc salts were added. Copper sulphide was precipitated from the acid solution and found to be active. The precipitates of the zinc, cobalt and nickel sulphides, obtained by neutralizing the solution and adding ammonium sulphide, were inactive. As the time employed for this test was rather long, the test refers only to the longer period. The carrier of this activity can then be assumed to be an isotope of copper, as suggested by Bjerge and Westcott (*loc. cit.*).

30 – *Zinc* – The activity of the short period of zinc is not enhanced by water. The longer period was measured electrometrically and found to be 10 hours. The carrier of this activity has been investigated by means of the following test: irradiated metallic zinc was dissolved in hydrochloric acid, and a small quantity of copper, nickel and cobalt salts added. Copper was precipitated partially by reduction on small traces of undissolved metallic zinc and partially as sulphide in the acid solution. The collected copper was strongly active. Neutralizing the solution and adding ammonium sulphide, the other elements were precipitated and found to be inactive. This confirms the results of Bjerge and Westcott that the long period of zinc is due to copper, probably to the same isotope of copper which is responsible for the longer period of this element. There is only a certain difference between these authors and us with respect to the period (6 hours according to Bjerge and Westcott).

31 – *Gallium* – The 20-minute period (measured electrometrically) is not very sensitive to water ($\alpha = 3$). Half-value thickness of the corresponding β–rays in 0.17 gm/cm^2 Al. The carrier of this activity is probably an isotope of gallium. In order to test this point, we irradiated gallium nitrate and afterwards added to the solution traces of copper and zinc. Copper was separated as a metallic deposit on zinc powder and zinc as zinc mercury sulphocyanate after adding mercury sulphocyanate. Both elements were found to be inactive.

Besides this 20-minute activity, we have also found, irradiating under water, a new activity which is accompanied by a rather strong γ–radiation; it decays with a period of 23 hours (measured electrometrically).

33 – *Arsenic* – The activity of this element is strongly sensitive to water ($\alpha = 6$). We have measured electrometrically its period (26 hours) and its half-value thickness of the β–rays (0.16 gm/cm² Al). For a concentration of the activity, see § 9.

34 – *Selenium* – The activity of this element (period 35 minutes) is sensitive to water ($\alpha = 4$). Irradiated selenious anhydride was dissolved in 30 % hydrochloric acid and some arsenious anhydride added to the solution. We precipitated metallic selenium by reduction with gaseous sulphurous anhydride, and found it strongly active. We precipitated from the solution arsenic sulphide and found it inactive. This test seems to rule out also germanium, and we conclude that the activity is due to an isotope of selenium.

35 – *Bromine* – Both activities of this element are sensitive to water. The short-period activity has $\alpha = 10$. The periods have been measured electrometrically; they are 18 minutes and 4.2 hours. The half-value thickness of the γ–rays is for both activities 0.12 gm/cm² Al, and both are accompanied by γ–rays. For the concentration of the activity see § 9.

38 – *Strontium* – No activity was found after a long and strong irradiation under water.

39 – *Yttrium* – Strongly irradiated yttrium oxide showed only a very weak activity possibly due to impurities. Yttrium absorbs very intensively the slow neutrons (half-value thickness $\delta = 0.015$ gm/cm²). This absorption is accompanied by γ–rays.

40 – *Zirconium* – Strongly irradiated zirconium nitrate showed only a very weak activity probably due to impurities.

41 – *Niobium* – The same as zirconium.

43 – *Rhodium* – The short-period activity is sensitive to water ($\alpha = 15$). Period and half-value thickness of the β–rays have been determined electrometrically (44 seconds; 0.15 gm/cm² Al). We also made a more accurate measurement in the ionization chamber of the longer period and found it to be 3.9 minutes. The activity is accompanied by a weak γ-radiation. Rhodium absorbs fairly strongly the slow neutrons (half-value thickness 0.3 gm/cm²): the absorption probably corresponds to the formation of the active isotopes.

46 – *Palladium* – Also the activities of this element are sensitive to water. We find at least two periods: a short one of about a quarter of an hour and one of about 12 hours. McLennan, Grimmett, and Read record a period of 14 hours, which is consistent with our precision.

47 – *Silver* – The two periods have been redetermined with the ionization chamber. They are 22 seconds and 2.3 minutes. They are both very sensitive to water, having $\alpha = 30$ and 15 respectively. To the strong activation of this element corresponds a considerable absorption for slow neutrons (half-value thickness $\delta = 1.2$ gm/cm²).

We added palladium nitrate and rhodium chloride to a solution of irradiated silver nitrate. Adding hydrochloric acid, we precipitated silver which was found active. From the filtered solution we precipitated palladium with dimethylglioxime and rhodium by reduction. Both were inactive. This test is valid only for the longer period, owing to the time employed, and shows that the carrier of the activity is probably an isotope of silver.

48 – *Cadmium* – Cadmium irradiation under different conditions showed several weak activities with various periods not yet identified. Cadmium absorbs with great intensity the slow neutrons. (Half-value thickness 0.013 gm/cm^2). The corresponding cross-section is the largest as yet found for slow neutrons ($\sigma = 10^{-20}$ cm^2). The absorption is accompanied by an intensive γ–radiation and probably corresponds to the transformation of a stable isotope of cadmium into another stable isotope of the same element.

49 – *Indium* – The activity induced in indium shows three periods: the shortest period (13 seconds) corresponds to an activity sensitive to water ($\alpha = 12$). Also the second period (54 minutes, measured electrometrically) is very sensitive to water. Magnetic deflection experiments show that the corresponding electrons are negative; their half-value thickness is 0.045 gm/cm^2 Al. A still longer period of some hours is recorded by Szilard and Chalmers; [14] this last activity is either insensitive to water or is only moderately sensitive.

Chemical tests have been made in order to identify the carriers of the last two activities. To a solution of irradiated indium nitrate, silver was added and precipitated as silver chloride; the precipitate was inactive. Afterwards we added to the solution tin, antimony and cadmium and precipitated them as sulphides with sulphuretted hydrogen. The acidity of the solution must be adjusted in such a way as to leave the indium in solution while precipitating the other metals. This precipitate also was inactive; neutralizing the solution, we precipitated the indium sulphide which carried the activity.

Corresponding to the strong activation of indium, it is found that this element has a considerable absorption power for the slow neutrons: half-value thickness $\delta = 0.3$ gm/cm^2.

50 – *Tin* – Tin strongly irradiated under water showed no activity.

51 – *Antimony* – We have found an induced activity in this element, decaying with a period of 2.5 days; the activation is sensitive to hydrogenated substances. The half-value thickness for the emitted β–rays is 0.09 gm/cm^2 Al. The following chemical test indicates that the carrier of this activity is probably an isotope of antimony. We dissolved metallic irradiated antimony in aqua regia and added some tin to the solution; after separation of tin as a sulphide according to Clarke, we found the activity in a precipitate of sulphide of antimony. The antimony sulphide was then dissolved again; indium was added to the solution and antimony separated as a sulphide in a moderately acid solution; the solution was neutralized and indium precipi-

(14) «Nature», vol. 135, p. 493 (1935).

tated and found to be inactive. To a new solution of the antimony we added tellurium and iodine and separated the first by reduction and the second as a silver iodide. Both were inactive.

52 – *Tellurium* – Shows a weak activity sensitive to water; the period resulted 45 minutes instead of 30 as given in our former paper.

53 – *Iodine* – Period and half-value thickness of the β–rays were determined electrometrically: period 25 minutes; half-value thickness 0.11 gm/cm² Al. The activation is moderately sensitive water ($\alpha = 5$). For the concentration of the activity see § 9.

56 – *Barium* – A new activity sensitive to water ($\alpha = 8$) with a period of 80 minutes has been found. The following chemical test is in favour of the assumption that the carrier of this activity is an isotope of barium. We dissolved irradiated barium hydroxide in hydrochloric acid, and added a small quantity of sodium chloride and precipitated barium sulphate. The activity was carried by the precipitate; we evaporated the solution and found the residual sodium to be inactive.

57 – *Lanthanum* – No activity was found after strong irradiation under water.

58 – *Cerium* – Same as lanthanum.

59 – *Praseodymium* – The short-period activity (5 minutes) is insensitive to water. Irradiating under water we have found a new water-sensitive activity decaying with a period of 19 hours; half-value thickness of the corresponding β–rays 0.12 gm/cm² Al (both measured electrometrically).

64 – *Gadolinium* – We irradiated under water a very pure sample of gadolinium oxide kindly given to us, together with the other rare earths, by Professor L. Rolla. We found an activity, decaying with a period of 8 hours.

73 – *Tantalum* – Only a dubious activity was found after 12 hours' irradiation under water with 500 millicuries.

74 – *Tungsten* – Metallic tungsten was irradiated under water and showed an activity ($\alpha = 15$) decaying with a period of about 1 day. [15]

We irradiated tungstic anhydride, dissolved it in caustic soda and then added and separated tantalum pentoxide which was found to be inactive. To the tungstic solution we added a nitric rhenium solution and precipitated the tungstic anhydride adding hydrochloric acid. The precipitate carried the activity, while the rhenium, precipitated from the filtrate as sulphide, was inactive. As we have no hafnium, we have made the following experiment in order to exclude an isotope of this element as carrier of the activity. From a solution of irradiated tungstic anhydride in ammonia, we precipitated zirconium hydroxide. The precipitate was inactive. We conclude that the activity of tungsten is probably due to an isotope of this element.

75 – *Rhenium* – We irradiated pure metallic rhenium under water; its activity is enhanced by water and decays with a period of about 20 hours. The half-value thickness of the electrons is 0.12 gm/cm² Al. The activity

(15) Cf. McLENNAN, GRIMMETT, and READ, «Nature», vol. 135, p. 147 (1935).

is probably carried by an isotope of rhenium. Irradiated rhenium was dissolved in nitric acid; we added tantalum and tungsten and separated them as tantalum pentoxide and tungstic anhydride. Both were inactive, while rhenium conserved the activity.

77 – *Iridium* – The activity induced in this element is strongly sensitive to water. Period and half-value thickness of the β–rays have beenmeasur ed in the ionization chamber; period 19 hours, half-value thickness 0.12 gm/cm² Al. To the strong activation of iridium corresponds a strong absorption of the slow neutrons; half-value thickness 0.3 gm/cm²; the absorption is accompanied by the emission of γ–rays.

78 – *Platinum* – Very pure metallic platinum (Heraeus 4th purity standard) irradiated under water showed an activity decaying with a period of about 50 minutes. McLennan, Grimmett, and Read (*loc. cit.*) record a period of 36 minutes.

79 – *Gold* – The activity of this element is sensitive to water; its period has been measured electrometrically and is 2.7 days. The β–rays were magnetically deflected and found to be negative. They have a very small penetrating power: half-value thickness 0.04 gm/cm² Al [16].

80 – *Mercury* – No activity was found after strong irradiation. This element absorbs intensely the slow neutrons, half-value thickness 0.2 gm/cm². γ–rays are emitted during the absorption.

81 – *Thallium* – No activity was found after strong irradiation.

82 – *Lead* – The same as thallium.

83 – *Bismuth* – The same as thallium.

90 – *Thorium* – The 1-minute and 24-minute (measured electrometrically) periods are scarcely sensitive to water.

92 – *Uranium* – We have also studied the influence of hydrogenated substances on the induced activities of this element. (Periods 15 seconds, 40 seconds, 13 minutes, 100 minutes). The result was that while the activities corresponding to the first, third and fourth period are slightly increased by water, no increase was found for the activity corresponding to the 40-second period. We have measured the increase in activity for the 15-second, 13-minute, and 100-minute activities. We have found for all these $\alpha =$ about 1.6. For the 15-second activity the measurement was made with the counters and, owing to the shortness of the period, is not very exact. The α values for the other two activities have been measured in the ionization chamber, special care being taken in order to verify that the α values for these two periods are equal. For this we compared three decay curves of the activity obtained by irradiating the same amount of uranium oxide for 14 hours, once in air placing the uranium around the source inside a test tube; once in the same geometrical disposition surrounding the test tube with paraffin, and once inside a large mass of paraffin placing the source at about 5 cm from the test tube containing the uranium. In all these tests we found that the decay curves were proportional, *i.e.*, showed the same ratio of the two activities. We think, therefore, that the test on the

(16) Gold bombarded with slow neutrons emits a strong γ–radiation.

identity of the sensitivities to water for these two periods is rather more accurate than the absolute value of the sensitivity coefficient and also than the identity in sensitivity coefficient with the 15-second activity. It is now evident that all the active products arising from the same primary process must have the same sensitivity coefficient. We conclude, therefore, that the 40-second activity is due to an independent primary process while the other three activities are probably due to the same primary process. This conclusion is limited by the possibility of a chance coincidence of the sensitivity coefficients within the rather wide limits of our precision. In this assumption the three activities could either be chain products (the short-period activities being parents of the long-period activities) or their relations could be complicated by branching phenomena. A certain amount of evidence in favour of the former assumption, at least as regards the 13-minute and the 100-minute periods, is given by the following test. We have measured on the electrometer the decay curves of a thick layer of irradiated uranium; these curves analysed in exponentials with the periods 13 minutes and 100 minutes show that the initial activities are in the ratio of about 100:45. The half-value thickness for the β–rays of the 13-minute activity is 0.14 gm/cm^2 Al; for the longer period it could not be measured with any accuracy, but is definitely less than the former value and probably about a half. These results are consistent with the assumption that the number of disintegrations for the 13-minute and the 100-minute periods is equal.

In our former paper we gave some chemical evidence which seemed to indicate that the carriers of the 13-and the 100-minute activities were not isotopes of any of the known heaviest elements, and that they were probably due to transuranic elements. Our point of view has in the meantime been criticized by von Grosse and Agruss [17] who, although never having experimented on activated uranium, deduced from our chemical tests the opposite conclusion, that these activities were due to isotopes of protactinium. We have therefore performed some new chemical experiments on the behaviour of these activities.

The precipitation of the activity with a sulphide was repeated, precipitating several metals (silver, copper, lead, mercury); the acidity of the solution (hydrochloric acid) was about 20 %; sometimes slightly varied in order to facilitate the precipitation of the sulphide of the metal used. The yield in activity of the precipitate was generally good—about 50 %—and varied according to the conditions of the precipitation. Nitric acid lowers the yield of the reaction very much. The usual high yield of the sulphide reaction is also obtained in presence of a hydrofluoric solution of tantalum. We also made a test in order to see whether the induced activities presented a reaction which is given by von Grosse as the most characteristic of protactinium.

We dissolved in a 25 % hydrochloric acid solution uranium oxide which had been purified and irradiated; we added to the solution zirconium nitrate

(17) « Phys. Rev. », vol. 46, p. 241 (1934).

and phosphoric acid; the precipitate of zirconium phosphate was inactive. After the separation of zirconium we precipitated a sulphide from the filtered solution, and collected the activity in the sulphide with the usual yield. According to von Grosse and Agruss, this reaction must be considered a proof of the non-identity of the carrier of the activity with a protactinium isotope.

By different chemical experiments, Hahn and Meitner [18] conclude also that the 13-and 100-minute activities are very probably due to transuranic elements; we have repeated some of their experiments finding the same results.

The two activities have beyond doubt a similar chemical behaviour. Some slight evidence of a possible separation was obtained only in the following experiment. Carefully purified uranium oxide was irradiated and dissolved in hydrochloric acid. The solution was poured into an ammonium carbonate solution until the uranium precipitate was completely dissolved again. Adding lead or manganese nitrate, we collected the precipitate of carbonates and found it to carry a fraction of the 13-and 100-minute activities. In the filtrate we precipitated copper sulphide, and this also carried a fraction of the activities. It seems that the ratio of the two activities was somewhat different in the two precipitates, the 13-minute activity being more abundant in the sulphide precipitate.

Through these experiments our hypothesis that the 13-minute and 100-minute induced activities of uranium are due to transuranic elements seems to receive further support. The simplest interpretation consistent with the known facts is to assume that the 15-second, 13-minute and 100-minute activities are chain products, probably with atomic number 92, 93 and 94 respectively and atomic weight 239.

We express our warmest thanks to Professor G. C. Trabacchi who supplied us with the radon sources. Our thanks are due also to Professor L. Rolla who placed at our disposal some rare earths, to Dr. G. Fea for help in the experiments and to the Consiglio Nazionale delle Ricerche for making a grant.

(18) «Naturwis.», vol. 23, p. 37 (1935).

Element	Isotopes	Half-periods	Half-value thickness gm/cm² Al	γ-rays	Sensitivity to hydrogen	Carrier of the activity	δ gm/cm²
1 H	**1**, 2, 3	—					
2 He	3, **4**						
3 Li	6, **7**	—					0.05
4 Be	**9**	—					>3
5 B	**10**, **11**	—					0.004
6 C	**12**, 13	—					>3
7 N	**14**, 15	—					>0.5
8 O	**16**, 17, 18	—					
9 F	**19**	9 s; 40 s	0.24; —	yes	1 ; 1	^{16}N (?)	>3
10 Ne	**20**, 21, 22						
11 Na	**23**	40 s; 15 h	—; 0.12	yes	1 ; *a*	^{23}Ne (?) ^{24}Na	>4
12 Mag	**24**, 25, 26	40 s; 10 m; 15 h	—; 0.07; 0.12	yes	1 ; *a*	^{23}Ne (?)^{27} Mg; ^{24}Na	>0.5
13 Al	**27**	2.3 m; 10 m; 15 h	0.16; 0.07; 0.12	yes	*a* ; 1	^{28}Al; ^{27}Mg; ^{24}Na	>7
14 Si	**28**, 29, 30	2.3 m; 2.4 h	0.16; —	yes	1 ; *a*	^{28}Al; ^{31}Si	>5
15 P	**31**	2.3 m; 2.4 h	0.16; 0.15		1 ; —	^{28}Al; ^{31}Si	>3
16 S	**32**, 33, 34	14 d	0.10			^{32}P	>2
17 Cl	**35**, **37**	35 m; 14 d	—; 0.10		*a* ; —	Cl; ^{32}P	0.3
18 A	36, 38, **40**						
19 K	**39**, 41	16 h			*a*	^{42}K	>1
20 Ca	**40**, 42, 43, 44	—					>3
21 Sc	**45**	16 h				^{42}K	
22 Ti	46, 47, **48**, 49, 50.	3 m					>2
23 V	**51**	3.75 m	0.17	yes	40	^{52}V	>1
24 Cr	50, **52**, 53, 54.	3.75 m	0.17	yes	1	^{52}V	>2
25 Mn	**55**	3.75 m; 2.5 h	0.17; 0.14	yes	1 ; 23	^{52}V; ^{56}Mn	>3
26 Fe	54, **56**	2.5 h	0.14	yes	1	^{56}Mn	8
27 Co	**59**	2.5 h	0.14	yes	—	^{56}Mn	0.7
28 Ni	56, **58**, **60**, 61, 62, 64	—					>3
29 Cu	**63**, **65**	5 m; 10 h		yes	15 ; *a*	Cu; Cu	3
30 Zn	**64**, **66**, 67, 68, 70	5 m, 10 h		yes	1 ; —	Cu; Cu	>10
31 Ga	**69**, **71**	20 m; 23 h	0.17; —	yes	3 ; *a*	Ga; —	>5
32 Ge	**70**, **72**, 73, **74**, 76	30 m (?)					
33 As	**75**	26 h	0.16	yes	6	^{76}As	>3
34 Se	74, 76, 77, **78**, **80**, 82	35 m			4	Se	4
35 Br	**79**, **81**	18 m, 4.2 h	0.12; 0.12		10 ; *a*	Br; Br	3
36 Kr	78, 80, 82, 83, **84**, 86						
37 Rb	**85**, **87**	? ?					>2
38 Sr	86, 87, **88**	—					>2
39 Y	**89**	—					0.015

44 Ru	96, 98, 99, 100, **101**, **102**, 104	—					>3
45 Rh		44 s; 3.9 m	0.15; —		15 ; *a*		0.3
46 Pd		15 m; 12 h			— ; *a*		<2
47 Ag	**107**, **109**	22 s; 2.3 m	—; 0.08	yes	30 ; 15	—; Ag	1.2
48 Cd	106, 108, 110, 111, **112**, 113, **114**, 115, 116	? ?					0.013
49 In	113, **115**	13 s; 54 m; 3 h (?)	—; 0.045; —	yes	12 ; *a* ; —	— ; In; In	0.3
50 Sn	112, 114, 115, 116, 117, **118**, 119, **120**, 121, 122, 124	—					>10
51 Sb	**121**, **123**	2.5 d	0.09	yes	*a*	Sb	>10
52 Te	122, 123, 124, 125, **126**, (127), 128, 130.	45 m			*a*		
53 I	**127**	25 m	0.11	yes	5	^{128}I	4
54 Xe	124, 126, 128, **129**, 130, **131**, **132**, 134, 136						
55 Cs	**133**	1.5 h (?) (?)					
56 Ba	135, 136, 137, **138**	3 m; 80 m			1 ; 8	— ; Ba	>3
57 La	**139**	—					
58 Ce	**140**, 142	—					
59 Pr	**141**	5 m; 19 h	—; 0.12		1 ; *a*		1.5
60 Nd	**142**, 143, **144**, 145, 146	1 h					
61							
62 Sm	144, 147, 148, 149, 150, **152**, **154**	40 m					
63 Eu	**151**, **153**						
64 Gd	**155**, **156**, 157, **158** 160	8 h			*a*		
65 Tb	**159**						
66 Dy	**161**, **162**, **163**, **164**						
67 Ho	**165**						
68 Er	**166**, **167**, **168**, 170						
69 Tu	**169**						
70 Yb	171, **172**, 173, **174**, 176						
71 Lu	**175**						
72 Hf	176, 177, 178, 179, 180						
73 Ta	**181**	?					
74 W	**182**, 183, **184**, **186**	1 d			15	W	
75 Re	**185**, **187**	20 h	0.12		*a*	Re	
76 Os	**186**, 187, 188, 189, **190**, **192**	—					
77 Ir		19 h	0.12		*a*	Ir	0.3
78 Pt		50 m			*a*		
79 Au		2.7 d	0.04	yes	*a*	Au	2
80 Hg	196, 197, 198, 199, **200**, 201, **202**, 203, 204	—					0.2
81 Tl	**203**, **205**	—					
82 Pb	203, 204, 205, **206**, 207, **208**, 209, 210.	—					>6
83 Bi	**209**	—					>10
90 Th	**232**	1 m; 24 m			~1 ; ~1		
92 U	**238**	15 s; 40 s; 13 m 100 m	—; — ; 0.14 0.070	yes	1.6;~1 ;1.6 1.6	See § 11	

§ 12. – Tabular Summary.

The main results of radioactivity induced by neutron bombardment are summarized in the table. Column 1 contains the atomic numbers and symbols of the elements investigated. Column 2 gives the isotopic constitution; numbers in bold type refer to isotopes which represent more than 20 % of the element. Column 3 gives the observed half-value periods in order of increasing magnitude. Column 4 gives the half-value thickness of the β–rays in gm/cm^2 of aluminium; the mean energy of the β–rays in millions volts might be very roughly obtained by multiplying the figure of column 4 by 8. Column 5 indicates whether γ–rays have been observed to accompany the β–disintegration. Column 6 indicates the carrier of the activity. Column 7 gives the sensitivity of the activation to hydrogenated substances; this is given either by the numerical value of the sensitivity coefficient (for definition see § 1; sensitivity coefficient 1 means that the activation is not enhanced by hydrogenated substances) or by the letter *a*, which means that the activation is increased by hydrogenated substances, but the sensitivity coefficient has not been measured. Column 8 gives the half-value thickness for the absorption of the slow neutrons.

In a few cases the periods of products, known from chemical evidence to be equal, but obtained by bombardment of different atoms are given as equal, even when measured very accurately only in one case. The same holds for some half-value thicknesses of β–rays.

108.

RICERCHE SUI NEUTRONI LENTI

E. Fermi e F. Rasetti

« Nuovo Cimento », *12*, 201–210 (1935).

Sunto. — È stato osservato che la presenza di sostanze idrogenate intensifica fortemente le radioattività artificiali provocate dai neutroni in alcuni elementi. Questa azione è attribuita al rallentamento subito dai neutroni per urti multipli contro i nuclei di idrogeno. Vengono descritte alcune proprietà dei neutroni lenti, e in particolare gli assorbimenti anomali che alcune sostanze presentano per essi.

Nel corso di esperienze eseguite nell'Istituto Fisico della R. Università di Roma dagli scriventi in collaborazione con vari assistenti dell'Istituto stesso si sono notate delle anomalie nella intensità della attivazione di un cilindretto di argento da parte di una sorgente di neutroni [1]. Una ricerca sistematica delle circostanze di queste anomalie ha condotto al risultato che, circondando la sorgente e la sostanza da attivare con grossi pezzi di paraffina, l'intensità dell'attivazione resta moltiplicata per un fattore assai grande; lo stesso effetto è presentato dall'acqua e, in genere, dalle sostanze che contengono molto idrogeno. Questa osservazione come pure le ricerche sistematiche che ad essa hanno fatto seguito sono state eseguite con la più larga collaborazione, per modo che non sarebbe possibile attribuire ai singoli ricercatori i vari risultati ottenuti; si può anzi dire che a ciascuno di essi tutti hanno contribuito in misura maggiore o minore. Cio non ostante in questa nota e nelle seguenti relative allo stesso argomento verrà reso conto dettagliato delle diverse esperienze eseguite, da coloro che maggiormente hanno contribuito alla esecuzione. Notizie preliminari su queste esperienze sono già state pubblicate altrove.

Effetto di sostanze idrogenate sulle attivazioni.

1. Descriviamo anzi tutto una esperienza tipica che permette di osservare la intensificazione della attivazione prodotta dalle sostanze idrogenate. Un cilindretto (circa 2 cm di diametro × 5 cm di altezza) della sostanza da attivare col bombardamento dei neutroni, viene attivato collocando nel suo centro per un tempo conveniente la sorgente di emanazione e berillio. Portando poi la sostanza ad un contatore si misura la intensità della attivazione prodotta. Si ripete poi l'esperienza ponendo durante l'irradiazione

(1) E. Fermi, E. Amaldi, B. Pontecorvo, F. Rasetti ed E. Segrè, « Ric. Scient. », 2, 280 (1934) e Note successive sullo stesso giornale.

la sorgente dei neutroni e la sostanza da attivare in una piccola cavità scavata nel centro di un blocco di paraffina (cilindro di 27 cm di diametro × 20 cm di altezza). Confrontando l'intensità della attivazione che si ottiene in questo secondo caso con quella avuta prima, si trova per molte sostanze un fortissimo aumento della attivazione.

Per accertare se queste intense attivazioni erano dovute ai neutroni o ai raggi γ emessi dalla sorgente, abbiamo ripetuta l'esperienza usando come sorgente 100 mg di radio senza berillio, senza trovare nell'argento alcuna attività indotta. L'effetto sembra dunque doversi attribuire ai neutroni; ciò è confermato dal fatto che una sorgente di polonio più berillio, che non emette altro che una debolissima radiazione γ, si comporta per questi fenomeni come una sorgente di emanazione più berillio.

Non tutte le sostanze che sono attivate da neutroni mostrano un aumento di attività quando vengano irradiate in paraffina o in acqua. Anche per lo stesso elemento, quando questo dà attività indotte con più periodi diversi, accade spesso che solo alcune delle attività siano sensibili alla azione delle sostanze idrogenate. Così per esempio sono fortemente influenzate le seguenti attività: Na (15^h); Al (2^m,3); V (3^m,7); Ag (22^s; 2^m,3); Cu (5^m); Rh (44^s; 3^m,9); I (25^m). Invece per esempio le seguenti attivazioni non subiscono aumento: Si (2^m,3); Al (10^m); Mg (40^s); Mn (3^m,7); Zn (5^m). Da tutti i casi studiati appare la regola che le attivazioni sensibili alla azione delle sostanze idrogenate sono quelle per le quali l'elemento attivo formatosi è isotopo dell'elemento bombardato.

L'esperienza che abbiamo precedentemente descritta si presta a dare una misura approssimativa della sensibilità della attivazione alla azione delle sostanze idrogenate; chiameremo coefficiente di sensibilità il rapporto α tra le intensità delle attivazioni che si ottengono effettuando il bombardamento nelle stesse condizioni entro il blocco di paraffina oppure nell'aria. Con questa definizione $\alpha = 1$ corrisponde al caso di attivazioni che non vengono aumentate. Per le attivazioni sensibili abbiamo osservato coefficienti di sensibilità variabili entro limiti abbastanza larghi (minimo $\alpha = 1,6$ per l'uranio (15^s, 13^m, 100^m); massimo $\alpha = 40$ per il vanadio (3^m,7)). Si noti che i valori dei coefficiente di sensibilità sono relativi alle condizioni geometriche che abbiamo specificate. Cambiando le condizioni il fattore di incremento della attivazione può essere anche notevolmente più grande; ciò ha luogo in particolare quando la sostanza da attivare si pone alla distanza di 5 o 10 centimetri dalla sorgente dei neutroni.

L'ipotesi dei neutroni lenti.

2. Gli esperimenti descritti si possono spiegare con l'ipotesi che l'effetto delle sostanze idrogenate sia dovuto alla diffusione e al rallentamento dei neutroni primari per urti elastici contro i nuclei di idrogeno.

Si trova facilmente che, in urto contro un protone, l'energia di un neutrone si riduce in media per un fattore $1/e$. Ne segue che 10 urti riducono l'energia a circa 1/10000 del valore iniziale. Se questo è per esempio $4 \cdot 10^6$

e–volt, l'energia dopo 10 urti sarà ridotta a circa 200 e–volt. 20 urti sarebbero già sufficienti a ridurre l'energia a valori corrispondenti alla agitazione termica. È abbastanza naturale pensare che, quando il processo della attivazione non richiede somministrazione di energia da parte del neutrone, un neutrone lento sia catturato più facilmente dal nucleo; ciò spiegherebbe in modo semplice l'effetto delle sostanze idrogenate. Nella interpretazione delle esperienze che descriveremo in seguito ci metteremo dal punto di vista di questa ipotesi dei neutroni lenti.

Nella esperienza tipica che abbiamo descritta l'aumento della attivazione può essere in parte attribuito alle diffusioni multiple dei neutroni, i quali possono in conseguenza attraversare più volte la sostanza da attivare. Che questa però non sia la causa principale è dimostrato dalla esperienza seguente: Un cilindretto di argento era irradiato con una sorgente di emanazione + berillio di 350 mC; la distanza tra la sorgente e il cilindro era di 20 cm. Quando non c'era nessun corpo salvo l'aria nell'intorno della sorgente e del cilindro l'attività indotta nell'argento era troppo debole per poter essere rivelata. Mantenendo costante la distanza tra la sorgente e il cilindretto abbiamo poi immerso la sorgente in un recipiente contenente acqua (14 cm di diametro e 14 cm di profondità). L'argento irradiato in queste condizioni mostra una attività indotta abbastanza intensa (circa 100 impulsi per minuto). Nelle condizioni descritte il numero totale dei neutroni che colpiscono l'argento non può essere aumentato dalla presenza dell'acqua; esso sarà anzi semmai diminuito dall'assorbimento. Si deve dunque concludere che l'efficacia di un neutrone nel produrre la attivazione dell'argento è aumentata dal passaggio attraverso alle sostanze idrogenate.

Assorbimento dei neutroni lenti.

3. Dalle esperienze precedenti risulta che la attivazione di molti elementi è più grande per neutroni lenti che per neutroni veloci. È perciò naturale ricercare se a queste forti attivazioni faccia riscontro un forte assorbimento dei neutroni lenti nella sostanza attivata. Abbiamo perciò fatta una ricerca sistematica sopra l'assorbimento dei neutroni lenti nei diversi elementi.

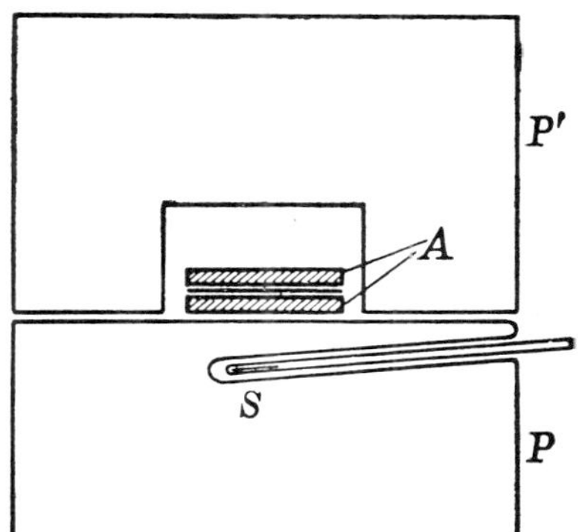

Fig. 1.

Lo studio è stato rivolto specialmente alla ricerca di sostanze con coefficiente di assorbimento anomalmente grande. Si sono perciò usati di solito spessori assorbenti piuttosto sottili. Il dispositivo per le misure è mostrato nella figura 1. La sorgente dei neutroni era introdotta in un foro praticato in un cilindro di paraffina (24 cm di diametro e 14 di altezza), circa due centimetri al di sotto della faccia superiore. Un secondo cilindro di paraffina si sovrapponeva al primo. In esso era scavata una cavità in cui veniva posto un rivelatore dei neutroni (costituito per solito da una lastra di rodio di cui si misurava la attività indotta). La lastra veniva interposta tra due strati eguali del materiale assorbente; e le misure consistevano in un confronto tra le attività con e senza assorbitori. In questi, come pure in molti altri esperimenti, abbiamo assai spesso usato il rodio come rivelatore dei neutroni lenti, poiché la attività indotta in questo elemento è molto intensa e può venir misurata in modo assai preciso con una camera di ionizzazione. Inoltre dei due periodi del rodio, 44^s e $3{,}9^m$, ha praticamente importanza solo il primo, ciò che rende particolarmente spedita la riduzione delle misure.

Dalle esperienze di assorbimento che abbiamo descritte è risultato che lo spessore di riduzione a metà δ per l'assorbimento dei neutroni lenti varia entro limiti assai larghi ed irregolarmente per i diversi elementi. In qualche

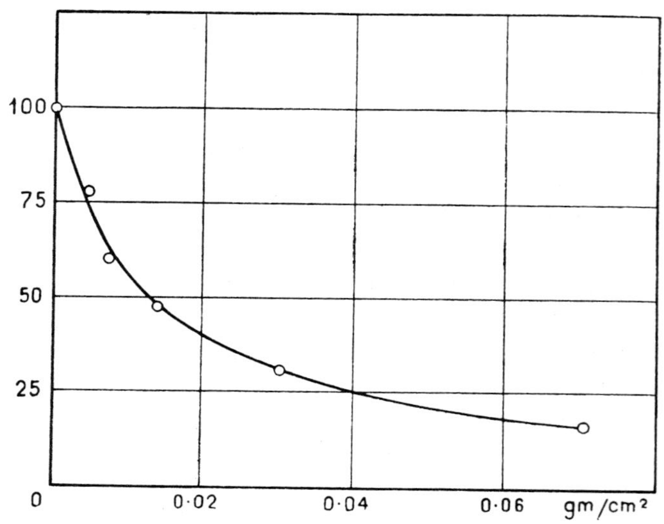

Fig. 2.

caso questo spessore è estremamente piccolo. Come casi estremi citiamo il boro ($\delta = 0{,}004$ gr/cm²); l'ittrio ($\delta = 0{,}015$ gr/cm²); il cadmio ($\delta = 0{,}014$ gr/cm²). Questi elementi assorbono dunque i neutroni lenti con intensità comparabile all'assorbimento per le particelle α. Invece molti altri elementi, per esempio zinco e piombo, presentano per i neutroni lenti assorbimenti di qualche migliaio di volte inferiori, aventi lo stesso ordine di grandezza di quelli noti per i neutroni veloci.

I coefficienti di assorbimento si possono dare espressivamente per mezzo di una sezione d'urto σ_{ca} per la cattura del neutrone. Si trovano così, nei casi sopra citati i valori di σ_{ca} $3\,000 \cdot 10^{-24}$ per il boro; $7\,000 \cdot 10^{-24}$ per l'ittrio; $10\,000 \cdot 10^{-24}$ per il cadmio. È notevole che queste sezioni d'urto risultino qualche migliaio volte maggiori della sezione geometrica del nucleo.

Le curve di assorbimento dei neutroni lenti non sono esponenziali ma si trova che il coefficiente di assorbimento decresce col crescere dello spessore attraversato. Diamo come esempio nella figura 2 una curva di assorbimento nel cadmio. Questo fatto deve ricondursi alla notevole inomogeneità dei neutroni rallentati dagli urti nella paraffina.

IL PROCESSO DELLA CATTURA DEL NEUTRONE.

4. I risultati descritti sopra l'assorbimento dei neutroni lenti rendono spontaneo ricercare il meccanismo dell'assorbimento. L'ipotesi più semplice consiste nell'ammettere che il neutrone sia catturato dal nucleo formando un isotopo più pesante di una unità. Se questo isotopo è instabile dobbiamo aspettarci che gli elementi fortemente assorbenti presentino anche una intensa attivabilità. Queste circostanze si verificano per esempio nell'indio e nell'iridio, per i quali è noto da prove chimiche che dànno luogo a loro isotopi radioattivi. Una valutazione grossolana della attivazione della sostanza assorbente e del numero di neutroni assorbiti mostra che si ha verosimilmente la formazione di un atomo radioattivo per ciascun neutrone assorbito. In altri casi si trova invece che ad un assorbimento anomalmente intenso non corrisponde alcuna attivazione di notevole intensità (boro, ittrio, cadmio, mercurio). In questi casi possiamo supporre che la cattura del neutrone dia luogo alla formazione di un nucleo stabile. Evidentemente queste circostanze potranno verificarsi più facilmente negli elementi che posseggono numerosi isotopi stabili che differiscono tra di loro di una unità di peso atomico (cadmio, mercurio).

In qualsiasi caso, se il neutrone lento è semplicemente catturato, dobbiamo aspettarci che il processo di assorbimento sia accompagnato dalla emissione di una radiazione γ, con una energia corrispondente alla energia di legame del neutrone. Per neutroni veloci una radiazione γ di questo tipo è stata descritta da Lea [(2)].

Nel caso dell'assorbimento dei neutroni lenti questa radiazione è stata da noi osservata mediante l'esperimento seguente. Una sorgente di neutroni di Po + Be, della intensità di circa 60 mC, era immersa in un blocco di paraffina insieme ad un contatore di Geiger–Mueller. Uno schermo di piombo dello spessore di 10 cm proteggeva il contatore dai raggi γ diretti della sorgente. Il contatore era circondato da uno strato di piombo di 2 mm; in queste circostanze esso registrava circa 30 impulsi per minuto. Sottili strati cilindrici delle sostanze in esame erano posti intorno al contatore al di fuori della foglia di piombo. Abbiamo trovato un aumento nel numero degli impulsi per tutte le sostanze che assorbono fortemente i neutroni lenti, ad eccezione del boro e del litio. L'anomalia del comportamento di questi due elementi è dovuta al fatto che l'assorbimento del neutrone è in essi accompagnato dalla emissione di particelle pesanti invece che di raggi γ [(3)]. Che la emissione di questi raggi γ sia dovuta alla azione dei neutroni lenti è provato dal fatto

(2) LEA, « Nature », *133*, 24 (1934).

(3) J. CHADWICK e M. GOLDHABER, « Nature », *135*, 65 (1935); E. AMALDI, O. D'AGOSTINO, E. FERMI, B. PONTECORVO, F. RASETTI ed E. SEGRÈ, « Ricerca Scient. », *1*, 123 (1935).

che togliendo la paraffina cessa il fenomeno; inoltre esso è ridotto di intensità proteggendo i vari elementi con un sottile strato di boro.

Considerazioni teoriche sulle proprietà dei neutroni lenti.

5. In un lavoro precedente (4) è stata discussa la questione se, nel caso che il bombardamento di neutroni dia luogo a un radioelemento isotopo dell'elemento di partenza, il neutrone sia catturato oppure produca la espulsione di un altro neutrone dal nucleo; cioè se il peso atomico del prodotto attivo sia A + 1 oppure A — 1. Riteniamo che i fatti osservati in seguito tendano tutti a confermare la prima ipotesi. Infatti:

a) si sono trovate due nuove deboli attività corrispondenti ad un periodo di 15 ore nel sodio e ad uno di 2,3 minuti nell'alluminio. Sia l'identità con periodi conosciuti, nel primo caso dell'alluminio e del magnesio, nel secondo del silicio e del fosforo, come pure la prova chimica nel caso del sodio, mostrano che i radioelementi sono rispettivamente isotopi del sodio e dell'alluminio. Siccome entrambi questi elementi hanno un solo isotopo stabile, cioè ^{23}Na e ^{27}Al, il radiosodio formato può a priori essere ^{24}Na o ^{22}Na, e il radioalluminio ^{28}Al o ^{26}Al. Ora entrambi gli isotopi più leggeri ^{22}Na e ^{26}Al sono conosciuti da altre reazioni nucleari e si sa che hanno periodi diversi ed emettono positroni invece che elettroni. In questi casi non resta pertanto altra ipotesi che ammettere la cattura del neutrone;

b) mentre non vi sono difficoltà teoriche per comprendere come un neutrone di energia trascurabile possa essere catturato da un nucleo, sembra invece difficile che esso possa espellere dal nucleo un neutrone legato stabilmente, e tanto meno che oltre a ciò possa restare una energia da emettere come quanto γ.

Discuteremo pertanto i fatti sperimentali ammettendo che i neutroni e più facilmente quelli lenti sono catturati da numerosi nuclei. Benché ci siano alcune difficoltà teoriche per la comprensione di questo processo di cattura, o almeno per interpretare i valori molto grandi delle sezioni d'urto osservate in certi casi, riteniamo utile sviluppare alcune conseguenze generali della teoria da tener presenti nella discussione di questo problema.

Ammettiamo, come si assume generalmente, che le forze agenti tra un neutrone e un nucleo si estendano circa fino alla distanza del raggio nucleare. Se così stanno le cose, la lunghezza d'onda di De Broglie è, per neutroni veloci, dell'ordine del raggio d'azione e quindi per neutroni lenti molto maggiore. La nota teoria dell'urto, nella quale il nucleo è trattato come una buca di potenziale, prende in questo caso una forma particolarmente semplice. Sia ψ l'autofunzione *s* corrispondente alla energia zero. Nella fig. 3 il prodotto $r\psi$ è rappresentato come funzione del raggio vettore r, e ρ rappresenta il raggio d'azione. La curva ha una forma irregolare per $r < \rho$, mentre che per $r > \rho$ diviene una linea retta. Supponiamo di normalizzare ψ in modo

(4) E. Fermi, «N. Cimento», *9*, 429 (1934).

che sia $\psi(0) = 1$. Sia allora l'equazione della linea retta che rappresenta $r\psi$ per grandi valori di r:

$$r\psi = \eta\,(a + r)$$

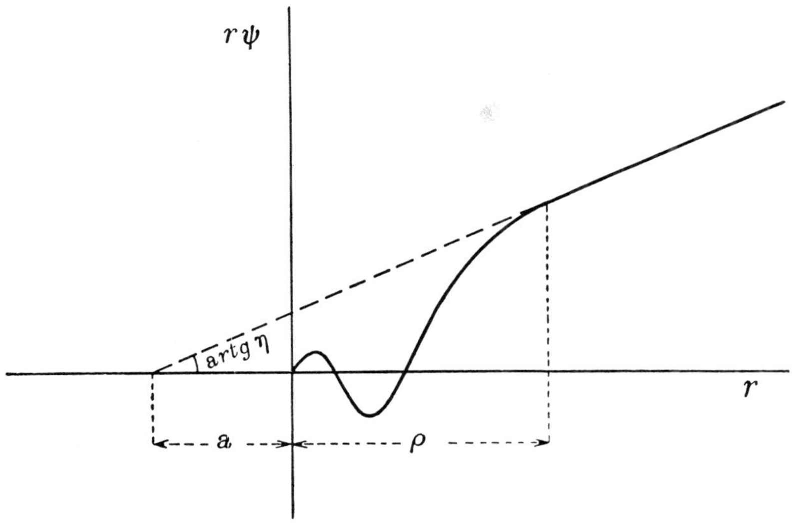

Fig. 3.

ove il significato geometrico di η e di a è chiaro dalla figura. Queste due grandezze potrebbero calcolarsi facilmente conoscendo la forma della buca di potenziale che rappresenta il nucleo. Si trova allora che la sezione d'urto elastico, al limite per basse velocità, è

$$\sigma_{el} = 4\,\pi a^2 \tag{1}$$

mentre che la densità di probabilità di trovare il neutrone nel centro del nucleo è

$$\mathrm{P} = n/\eta^2 \tag{2}$$

essendo n la densità dei neutroni all'esterno del nucleo.

Qualunque sia il meccanismo della cattura è naturale ammettere che, almeno in una certa approssimazione, la probabilità di cattura del neutrone da parte del nucleo per unità di tempo sia proporzionale a P, cioè che questa probabilità sia data da kn/η^2, dove k è costante per ciascun nucleo.

Questa probabilità può esprimersi per mezzo di una sezione d'urto per cattura, σ_{ca} mediante la relazione.

$$\sigma_{ca} = k/\eta^2\,v \tag{3}$$

dove v è la velocità del neutrone. Un limite alla validità di questa formula è posto dal fatto che σ_{ca} non può evidentemente superare l'ordine di grandezza del quadrato della lunghezza d'onda di de Broglie. Considerando le massime sezioni d'urto trovate sperimentalmente ($\sigma_{ca} = 10^{-20}$ cm^2) questa considerazione permette di stabilire un limite superiore per l'energia dei neutroni lenti, limite che risulta di alcune centinaia di e–volt. Tuttavia è bene ricordare che questa conclusione è valida soltanto nelle ipotesi che abbiamo fatte.

La formula (3), valida soltanto per basse velocità, dà una sezione d'urto per cattura da parte di un dato nucleo inversamente proporzionale alla velocità del neutrone. Questo spiega perché la sezione d'urto per cattura è generalmente più grande per neutroni lenti che per neutroni veloci.

Dalla (3) risulta ancora che σ_{ca} è inversamente proporzionale a η^2. Siccome in qualche caso la linea retta della fig. 3 può casualmente risultare quasi parallela all'asse delle ascisse, η può essere molto piccolo, e conseguentemente la sezione di cattura molto grande. In questo comportamento delle autofunzioni si deve probabilmente ricercare la causa delle sezioni di cattura anomalmente grandi che si osservano per alcuni elementi.

Per dedurre dalla (3) i valori assoluti delle sezioni d'urto, occorre poi in ogni caso una ipotesi sul meccanismo fisico della cattura dal quale dipende il valore del coefficiente k. Si sa dall'esperienza che i processi di cattura possono essere due. Per alcuni elementi molto leggeri (Li, B) la cattura del neutrone è seguita dalla emissione di particelle pesanti; mentre negli elementi più pesanti il processo normale sembra essere la cattura con emissione di raggi γ. Il legame di un ulteriore neutrone a un nucleo mette in libertà una energia avente in media un valore di 7 milioni di e–volt. Questo eccesso di energia può dunque in alcuni casi determinare la espulsione di una particella α, purché la barriera di potenziale che circonda il nucleo sia abbastanza bassa per consentire alla particella di sfuggire rapidamente. Si può dunque attendere che questo processo si abbia solo per gli elementi più leggeri; mentre nella attivazione con neutroni veloci, l'aggiunta alla energia di legame del neutrone, della sua energia cinetica, rende il processo di emissione di una particella α o di un protone possibile anche per elementi di peso un po' maggiore.

Nel caso della cattura del neutrone con emissione di raggi γ, che si osserva per elementi di tutti i pesi atomici, fino ai più elevati sembra che il meccanismo ordinario della irradiazione dia dei valori un po' troppo bassi per k. Questo coefficiente dipende da due fattori entrambi di difficile valutazione: la energia del quanto γ emesso e l'elemento di matrice. Siccome, a parità di altre condizioni, la probabilità di emissione di un quanto γ è proporzionale a ν^3, si deve ritenere che siano preferiti i processi con forte energia di legame del neutrone; questa energia di legame può probabilmente in parecchi casi essere considerevolmente maggiore di 10 milioni di volt.

Anche a nome dei nostri collaboratori esprimiamo i più vivi ringraziamenti al prof. G. C. Trabacchi che ci ha fornite le sorgenti di emanazione. Ringraziamo anche il prof. L. Rolla per il dono di alcune terre rare; il dottor G. Fea per aiuto nelle esperienze e il Consiglio Nazionale delle Ricerche per aiuti finanziari. Ringraziamo anche i tecnici dell'Istituto, Cav. A. Diotallevi e fratelli R. e G. Berardo per il loro solerte ed abile aiuto.

109.

ON THE VELOCITY DISTRIBUTION LAW FOR THE SLOW NEUTRONS

Zeeman, Verhandelingen p. 128–130, Martinus Nijhoff, the Hague, 1935

We shall deal in this paper with the problem of finding the velocity distribution law for an ensemble of neutrons projected through an hydrogenated substance. The problem has become of a certain interest, as it was shown (1) that several elements, which can be transformed by a neutron impact into unstable isotopes, showing the phenomenon of the artificial radioactivity, are by far more strongly activated, when irradiation with the neutrons takes place inside an hydrogenated medium, such as water or paraffine wax. An interpretation of this phemonenon, which seems to fit fairly well with the known facts, is afforded by the hypothesis, that neutrons inside an hydrogenated medium are rapidly slowed down by the impacts against the hydrogen nuclei. There are some theoretical reasons for admitting that, at least in some cases, slow neutrons are more easily captured by nuclei than fast ones; the high intensity of the activation obtained by neutron bombardment inside a mass of water or paraffine could be thus understood as due to the action of the slow neutrons.

In order to calculate the velocity distribution for the neutrons, we will admit the following points:

a) The slowing down of the neutrons is only due to the elastic collisions against the hydrogen nuclei. Let $\lambda(v)$ be the mean free path for elastic collision against hydrogen for a neutron with velocity v.

b) In an impact between a neutron and a proton, the probability of scattering is uniformly distributed in all directions, when referred to a system in which the centre of gravity of the neutron and the proton is at rest.

c) The velocity of the protons due to the thermal agitations is negligible.

d) A source of neutrons projects in the medium Q neutrons per second, all having the same velocity v_0.

e) Neutrons can be captured by nuclei; let $\tau(v)$ be the mean life of a neutron with velocity v before it is captured by a nucleus.

Let n_0 be the number of neutrons having velocity v_0. This number increases per unit time by Q through the effect of the source; and decreases by $n_0(v_0/\lambda_0 + 1/\tau_0)$ due to collisions against hydrogen and to capture processes. For the statistical equilibrium:

$$Q = n_0\left(\frac{v_0}{\lambda_0} + \frac{1}{\tau_0}\right). \tag{1}$$

(1) E. FERMI, E. AMALDI, B. PONTECORVO, F. RASETTI ed E. SEGRÈ, «La Ricerca Scientifica», 5 (2), 280 (1934).

Let $n(v)\,dv$ be the number of neutrons with veloctity between v and $v + dv$. Collisions against the protons and capture processes produce a decrease of $n(v)$ by

$$n(v)\left[\frac{v}{\lambda(v)} + \frac{1}{\tau(v)}\right]$$

per unit time.

This decrease is compensated by the effect of the collisions of the neutrons which have before the impact a velocity $v' > v$, and whose velocity is reduced by the collision to a value between the limits v and $v + dv$. It is easily shown, that the probability for an impact of a neutron with velocity v' to reduce the velocity to a value $< v$ is given (see point *b*) by v^2/v'^2; hence follows, that the probability for the velocity to be reduced to a value between v and $v + dv$ is $2\,v dv/v'^2$. We deduce herefrom, that the increase of $n(v)$ per unit time, due to collisions of the $n(v')\,dv'$ neutrons with velocity between v' and $v' + dv'$ is:

$$n(v')\,dv' \cdot \frac{v'}{\lambda(v')} \cdot \frac{2\,v}{v'^2}\,.$$

This must be integrated with respect to v' between the limits v and v_0. To this increase we must add the contribution of the impacts of the neutrons with velocity v_0, which is similarly given by:

$$n_0 \cdot \frac{v_0}{\lambda_0} \cdot \frac{2\,v}{v_0^2}\,.$$

As a condition for the statistical equilibrium we find at last

$$n(v)\left[\frac{v}{\lambda(v)} + \frac{1}{\tau(v)}\right] = 2\,v\int_v^{v_0}\frac{n(v')\,dv'}{v'\,\lambda(v')} + \frac{2\,v n_0}{v_0\,\lambda_0} \tag{2}$$

In order to determine $n(v)$, we divide by $2\,v$ and derivate with respect to v; we find thus the following differential equation:

$$\frac{d}{dv}\left\{\frac{n(v)}{2}\left[\frac{1}{\lambda(v)} + \frac{1}{v\tau(v)}\right]\right\} + \frac{n(v)}{v\lambda(v)} = 0$$

whose integral is

$$n(v) = c\,\frac{v\lambda(v)}{v + \lambda(v)/\tau(v)}\,e^{2\int_v^{v_0}\frac{dv}{v+\lambda(v)/\tau(v)}} \tag{3}$$

where c is the integration constant. Its value, when needed, can be determined from (1) and (2) and results:

$$c = \frac{2\,n_0}{\lambda_0\,v_0} = \frac{2\,Q}{v_0\,(v_0 + \lambda_0/\tau_0)} \tag{4}$$

(3) and (4) completely solve our problem.

The result takes a very simple form, if we assume that, at least under a certain speed limit, λ and τ are independent of the velocity. These assumptions are supported by a quantum mechanical consideration of the collision process, and are probably valid when the de Broglie wave length is large compared to the nuclear dimensions. With these assumptions (3) takes the simple form:

$$n(v) = \text{const.}\,\frac{v}{(v + \lambda/\tau)^3}\,. \tag{5}$$

With this distribution we find as the most probable velocity of the slow neutrons

$$v_p = \frac{\lambda}{2\tau}\,. \tag{6}$$

For this velocity the mean free path for elastic collisions is of the same order of magnitude (more exactly the double) of the mean free path $v_p\tau$ for the capture process.

February 15, 1935.

110.

ON THE RECOMBINATION OF NEUTRON AND PROTONS

« Phys. Rev. », *48*, 570 (1935).

The process of the protoelectric disintegration of the deuteron, and the reverse process (recombination of a neutron and a proton with formation of a deuteron and emission of a γ quantum) have been treated theoretically by Bethe and Peierls (1). These authors consider the emission or absorption of γ-rays by the ordinary mechanism of the electric dipole radiation. For the recombination cross section they obtain an expression which vanishes in the limit of low relative velocities of the neutron and the proton. This last result seems to be in contradiction with general experimental evidence, that slow neutrons are captured in hydrogenated materials such as paraffin or water. Recently the lifetime of a slow neutron in paraffin has been evaluated by a direct experiment and found to be of the order of magnitude of 10^{-4} second (2).

I wish to point out that if we take into account also the radiation processes due to oscillations of the magnetic dipole of the neutron-proton system, one finds a probability of capture which is in satisfactory agreement with the above value. The theory is based on the customary assumption that the fundamental state of the deuteron is a ^{3}S (neutron and proton with parallel spins).

The ordinary theory of impact shows that in the case of slow neutrons only the capture from S states in the continuum is of importance. The usual radiative processes of electric dipole and quadripole are forbidden for these S—S transitions. Instead transitions from a ^{1}S state in the continuum to the fundamental ^{3}S state can occur by the mechanism of magnetic dipole radiation.

The ordinary electromagnetic theory gives for the inverse mean life $1/\tau$ of a neutron with velocity v moving in a medium containing n protons per unit volume the following formula:

$$\frac{1}{\tau} = n \cdot \frac{64\,\pi^4\,h\nu^3}{c^3\,m^2\,v^2} \cdot \mu_0\,(g_p - g_n)^2 \left| \int f\varphi\, r^2\, dr \right|^2 .$$

(1) H. Bethe and R. Peierls, « Proc. Roy. Soc. », *A 148*, 146 (1935).

(2) E. Amaldi, O. D'Agostino, E. Fermi, B. Pontecorvo and E. Segrè, « Ric. Scientifica », *6* (1), 581 (1935).

m is the neutron mass: ν is the emitted frequency; μ_0 is the nuclear magneton; g_p and g_n are the nuclear g–factors for proton and neutron. The last integral contains the normalized eigenfunction $f(r)$ of the fundamental state and the continuum eigenfunction $\varphi(r)$. This last is normalized in such a way that for large r it goes over into $(1/r)\sin(2\pi pr/h + \text{const.})$.

This last integral can be evaluated numerically by methods similar to those used by Bethe and Peierls. The parameter needed for the description of the ¹S eigenfunction can be obtained on the assumption that the high elastic scattering cross section of slow neutrons in hydrogen is mainly due to the ¹S eigenfunction (real or virtual ¹S level close to energy zero). We find for slow neutrons

$$\frac{1}{\tau} = \frac{16\pi^3 W^{5/2}}{h^3 c^3 m^{1/2}} \frac{\mu_0^2 (g_p - g_n)^2}{l},$$

where l is the mean free path for the elastic scattering of slow neutrons (for paraffin about 0.5 cm) (3), and $W = 2.1 \times 10^6$ e. v. is the binding energy of the deuteron. Assuming 3 and 1, nuclear magnetons, as magnetic moments of the proton and the deuteron (4), we get $g_p = 6$ and $g_n = -4$. This gives $1/\tau = 5.2 \times 10^3$ in agreement with the experimental result.

The magnetic dipole radiation affects also the process of the photoelectric disintegration of the deuteron. The cross section for the photoelectric absorption of a γ–ray by a deuteron results on the same assumptions

$$\sigma_\gamma = \frac{4\pi^2 a^2 \alpha^2 \mu_0^2 (g_p - g_n)^2}{3hc} \frac{(\gamma - 1)^{1/2}}{\gamma\{1 + a^2\alpha^2(\gamma - 1)\}},$$

where

$$a^2 = \frac{1}{\pi nl}, \quad \alpha^2 = \frac{4\pi^2 mW}{h^2}, \quad \gamma = \frac{h\nu}{W}.$$

The cross section must be added to that already calculated by Bethe and Peierls in order to give the total cross section.

Details of this theory will be published elsewhere.

University of Michigan.
August 27, 1935.

(3) E. AMALDI, O. D'AGOSTINO, E. FERMI, B. PONTECORVO, F. RASETTI and E. SEGRÈ, « Proc. Roy. Soc. », *A 149*, 522 (1935).

(4) O. FRISCH and O. STERN, « Z. Physik », *85*, 4 (1933); I. ESTERMANN and O. STERN, « Z. Physik », *85*, 17 (1933); « Phys. Rev. », *45*, 761 (1934); I. I. RABI, J. M. B. KELLOGG, J. R. ZACHARIAS, « Phys. Rev. », *45*, 761, 769 (1934).

N° 112–119.

These papers are a collection of Fermi's work from October 1935, to May 1936.

Paper N° 119 is a theoretical one. In addition to many new results it contains some results obtained earlier but previously unpublished or published only in preliminary form. All of the purely experimental work was done in collaboration with me. The original notebooks which record the data are preserved in the Archives of the Domus Galilaeana at Pisa.

After the summer vacation of 1935, Fermi and I found ourselves alone in Rome. Most of the group had dispersed by now. The general atmosphere in Italy was chiefly to blame for this as the country prepared for the Ethiopian war. Rasetti had gone to the U.S.A. and planned to stay at Columbia University for at least a year. Segrè, too, had left for a summer in the United States, and in the meantime had been appointed professor at the University of Palermo. Upon his return to Italy, he also left Rome to go to Sicily. D'Agostino no longer worked with us and had taken a position at the Istituto di Chimica del Consiglio Nazionale delle Ricerche. Pontecorvo had returned to Rome shortly after us and for a few months worked with Wick; later he won a Ministero dell'Educazione Nazionale scholarship for study abroad and left Rome in the spring of 1936 to work with the Joliots at the Curie Laboratory in Paris. Thereafter his visits to Rome were very brief and infrequent.

Upon resuming work, Fermi and I turned our attention to some results of Bjerge and Westcott, and of Moon and Tillman, who had observed that the absorption of slow neutrons by various elements differed slightly depending on the element used as detector. This fact was not explained by the current theory of the absorption of neutrons by nuclei. This theory predicted a capture cross section inversely proportional to the velocity of the neutrons for all nuclei. This energy dependence was supposed to be valid for such large energy intervals as to certainly cover the energy range of slow neutrons.

We went to work with even greater energy than in the past, as if by our own more intensive efforts we wanted to compensate for the loss of manpower in our group.

We had prepared a systematic plan of attack which we jokingly summarized by saying that we would measure the absorption coefficient of all 92 elements combined in all possible ways with the 92 elements used as detectors. In jest we added that after combining all the elements two by two, we would also combine them three by three. By this we meant that we would also study the absorption properties of the neutron radiation filtered in several ways.

Actually, after having measured the absorption coefficient of eleven different elements in all possible combinations with seven detectors, we were convinced that the observations of the two English research groups were correct, and in general that the rule that the absorption coefficient of a certain element was greater when the element itself was used as detector was valid. We began to study the particular cases of silver, rhodium and cadmium in great detail. The absorption properties of cadmium were investigated more thoroughly. We performed on the neutrons filtered by a cadmium layer, absorption measurements of different elements with various detectors, as we had already done in studying the unfiltered neutron radiation. Thus, early in November 1935, we established that, if the neutrons were previously filtered with cadmium, the self-absorption effect mentioned above was considerably greater (N° 112). Indeed, this idea was already contained in the papers of the two English groups, but our measurements, which were more extensive and complete, made this interpretation compelling.

I think it is interesting to note that on this occasion I, as many others, tended to make a simple picture of the phenomenon; I tried to interpret the different groups of neutrons as different bands of energy. Fermi, however, did not want to accept this description. He, too, was convinced that this was obviously the simplest hypothesis, but maintained that it was not strictly necessary, at least for the moment, and was therefore harmful if introduced into our mental picture. He insisted that one must proceed by reasoning with the observed experimental facts only. The correct interpretation of the nature of the neutron groups would finally emerge as a necessary consequence of the data. He was afraid that a preconceived interpretation, however plausible it sounded, would sidetrack us from an objective appraisal of the phenomenon that confronted us.

Therefore we began a systematic study of the absorption and diffusion properties of the various neutron groups, labelling them with letters, both for brevity and in order to avoid any trace of interpretation. The expression «group *C*» was used for neutrons strongly absorbed by cadmium; «group *D*» for neutrons strongly absorbed by rhodium, but not by cadmium; «group *A* and *B*» for the two components which we believed we had characterized in the radiation strongly absorbed by silver, but not absorbed by cadmium.

In a second letter to the «Ricerca Scientifica», dated December 12, 1935, (N° 113), the groups of slow neutrons are clearly defined and their absorption and other properties are studied. In this paper there are reported experiments which were performed to establish the number of neutrons belonging to each group (numerosity) among those that emerge from the surface of the moderator (paraffin or water), the reflection coefficient of the various groups (albedo) (1), and their diffusion distance.

Thus we showed evidence that the group *C*—the neutrons strongly absorbed by cadmium—had very different properties from the neutrons that passed through cadmium. The albedo of the neutrons of group *C* was very high (0.83), while that of groups *D*, *A*, etc. was negligible. The diffusion length in paraffin was about 3 cm for group *C*, while it was about six times less for the other groups.

These latter results were further clarified by a more accurate experiment performed in January 1936, and published in paper N° 114, in which the diffusion length was determined from the escape probability of a neutron originally found at a depth x within a medium filling a halfspace. The expression of such a probability is derived from the diffusion equation which had been adopted for the description of the properties of group *C*. In this same paper, the interpretation of the neutron groups, as due to energy differences, is discussed as the most likely interpretation, without, however, precluding other interpretations. The work concludes with reference to an experiment well under way, but not yet completed, whose purpose is to clarify this point: if various groups differ only in energy, the neutrons which at a certain moment belong to a group, as a result of further slowing down, must transform, themselves into neutrons of another group.

At that time we had already learned from several experiments by other workers quoted in paper N° 114 that group *C* included thermal neutrons. Therefore, if the interpretation of the groups in terms of different bands of energy was correct, all the other groups ought to be transformable by slowing down into group *C*. The definitive results of these experiments are recorded in paper N° 117. An experiment of the same type had already been published by Halban and Preiswerk. Plate VIII reproduces p. 114, notebook B3, dated February 20 1936, in which the results of an experiment of this type are recorded.

In order to establish precisely the diffusion properties of group *C* neutrons, it was necessary to supplement the measurements of the diffusion length mentioned above with a measurement of the mean free path. The result of a first measure of this quantity is reported in the letter to the «Ricerca Scientifica», dated March 20, 1936 (N° 115), in which a clear difference is shown between the values of the group *C* neutrons' mean free path and the mean free path of all other neutron groups.

The explanation of neutron groups in terms of differences in energies had in the meantime been imposed both by the experiments of the type referred to above—the transformability of various groups into group *C*—and as a result of a work by Bohr (2) presented at the Danish Academy, January 27, 1936. In this paper a mechanism of the nuclear processes was suggested, explaining the existence of many thin levels of nuclei of intermediate and high atomic number.

(1) The expression albedo is used by astronomers for the fraction of incident light diffusely reflected from the surface of a planet or satellite. The most common use refers to the moon. It is, however, also used for other surfaces: for instance, for snow.

(2) N. BOHR, «Nature» (London), *137*, 344 (1936).

However, the difference in mean free path reported in paper N° 115, was interpreted by Fermi as due to the chemical bond, since he could not imagine any nuclear phenomenon that could possibly give rise to a similar effect. The theory of this phenomenon is given in paper N° 119.

At this point, perhaps it is useful to remember that while this work progressed, measurement techniques were becoming considerably more refined. During the first period of investigation by the group at the University of Rome, the activity measurements were taken exclusively by means of Geiger counters with thin aluminum walls (one of these counters is preserved at the Domus Galilaeana). However, after the discovery of the effect of hydrogenous substances, the activity had become so high that it was frequently possible to use an ionization chamber connected to an electrometer. This technique was then developed and perfected by experimenting with new types of ionization chambers and new ways of using the electrometers. These had been calibrated with great care in order to know their characteristics well and utilize them to their maximum potentialities. The preparation of many nomograms and graphs allowed a rapid computation, from the readings made on the electrometer scale, of the activity of the radioactive body being measured.

Once the interpretation of the phenomena observed on the basis of the compound nucleus' resonance levels, according to Bohr's hypothesis, was accepted, the problem of determining the width and energy of these resonance lines naturally arose. It is treated in paper N° 116 (dated April 13, 1936) in which use was made of various formulas whose derivation is reported in paper N° 119. This work demonstrates how the mean value of the square of the distance travelled by the neutrons, before they reach the resonance energy of the detector, increases as the resonance energy of the detector decreases. In this way a quantitative relation between spatial distribution of resonance neutrons and their energy was established.

Paper N° 117 reports a measure of group C (thermal neutrons) mean free path, based on the extrapolation distance, and re-examines all the diffusion properties of slow neutrons in paraffin.

All of these and various other measurements, including the determination of the total number of neutrons emitted by a source based on the space integration of slow neutrons in an hydrogenous medium of large dimensions, are summarized and discussed in paper N° 118, dated May 29, 1936.

The academic year, which had slipped by in an atmosphere of frenzied work and isolation, was by now drawing to a close. Rasetti wrote to us every now and then about what was happening at Columbia University; reprints by Halban and Preiswerk kept us informed on the work in Paris; and a correspondence with Placzek kept us in contact with Copenhagen. Through this latter correspondence we learned of Bohr's work, as well as that of Frisch and Placzek, concerning the $1/v$ absorption law in boron. Through this correspondence the joke that, as the captain's age can be determined from the length of the ship's mast, so the energy of a neutron group can be determined by the distance it travels as it slows down, spread from Rome. The expression " age ", used later by Fermi to represent the magnitude $\overline{r^2}/6$, might date back to this period. At first the expression " the age of the captain " was used to refer to experiments concerning the transformation of one group into another of lower energy (fig. 1).

Paper N° 119 is theoretical; it includes, besides the calculation for $\overline{r^2}$ as a function of the energy, the calculations for the albedo, for the extrapolation length, for the angular distribution of neutrons leaving the surface of the moderator, and numerous other problems treated with the « one dimensional medium model », which, according to Fermi, even many years later, was sufficient for dealing with the majority of diffusion problems. The most refined mathematical methods were pure delusion, according to Fermi, since the uncertainties inherent in physical hypotheses, which were at the base of any treatment, were much more than the mathematical crudities inherent in this model.

The second part of paper N° 119 includes the theory of the effect of chemical bond on the neutron-proton collision and the theory of the radiative capture of protons by hydrogen on which he had already published a brief note (N° 110).

These theoretical activities took place in the early hours of the morning before Fermi arrived at the Institute. He was always there before nine and, at that time, often at eight.

This paper contains the seeds of nearly all of the important ideas on neutrons that Fermi developed in succeding years.

Besides Fermi and me, Wick and Pontecorvo were also working at the Institute. For some months they measured the backwards scattering of the neutrons of different groups on different substances. Then Pontecorvo left for Paris, and Wick went back to theoretical physics. There was Ettore Majorana, and occasionally Segrè came to visit us from Palermo.

We worked with incredible stubborness. We would begin at eight o'clock in the morning and take measurements almost without a break until six or seven in the evening, and often later. The measurements were taken with a chronometric schedule as we had studied the minimum time necessary for all the operations. They were repeated every three or four minutes, according to need, for hours and hours and for as many successive days as were necessary to reach a conclusion on a particular point.

Having solved one problem we immediately attacked another without a break or feelings of uncertainty: " Physics as soma " (*) was the phrase we used to refer to our work performed while the general situation in Italy grew more and more bleak, first as a result of the Ethiopian campaign, and then as Italy took part in the Spanish Civil War.

In June, Fermi left for the United States where he was to give a thermodynamics course at Columbia University. I followed him there shortly afterwards, having received a grant from the Volta Foundation of the Accademia d'Italia with which I intended to study the construction of a 1 Mev proton accelerator.

On my way to Washington D.C., I stayed in New York for about a month where, at Columbia University, I translated paper N° 118 into English for publication in the " Physical Review " (N° 118 *bis*). I was helped in the work of translation by H. H. Goldsmith, who was taking Fermi's course at Columbia. I also tried to convince Fermi to translate his theoretical paper N° 119 into English, but he said that he didn't want to waste the time, and if someone was interested in studying slow neutrons, he would have to read it, even if it was published only in Italian.

E. AMALDI.

112.

SULL'ASSORBIMENTO DEI NEUTRONI LENTI. — I.

E. AMALDI ed E. FERMI

« Ric. Scientifica », *6* (2), 344–347 (1935).

1. *Assorbimenti selettivi.* – Bjerge e Westcott, Moon e Tillman e altri [1] hanno osservato che l'assorbimento delle diverse sostanze per i neutroni lenti risulta differente a seconda della sostanza usata come rivelatrice. Ci siamo proposti uno studio sistematico su questo argomento e ne esponiamo qui i primi risultati.

(*) This expression comes from the Aldous Huxley novel " Brave New World " and refers to a pill with a sexual hormone base used by men in the year 2000 to combat spleen.

(1) T. BIERGE e C. H. WESTCOTT, « Proc. Roy. Soc. », *A 150*, 709 (1935); J. R. TILLMAN e P. B. MOON, « Nature », *136*, 66 (1935); B. PONTECORVO, « Ric. Scientifica », *6* (2), 145 (1935); MITCHELL, DUNNING, SEGRÈ, PEGRAM, « Phys. Rev. », *48*, 175 (1935).

Nella Tabella I, che è analoga a quella riportata già da altri Autori, le colonne corrispondono agli elementi usati come assorbitori e le righe ai rivelatori. I numeri rappresentano l'assorbimento in percentuale. Le caratteristiche dei rivelatori sono le seguenti: Manganese (periodo 150 min; 30 g/25 cm^2 di MnO_2); Bromo (periodo 18 min; 26 g/25 cm^2 di NaBr); Rodio (periodo 44 s; 10 g/28 cm^2 di elemento); Argento (periodo 22 s; 8 g/27 cm^2 di elemento); Indio (periodo 54 min; 3,8 g/17 cm^2 di elemento); Iodio (periodo 25 min; 25 g/25 cm^2 di elemento); Iridio (periodo 19 ore; 14,5 g/25 cm^2 di elemento).

Le caratteristiche degli assorbitori sono: Boro (0,17 g/cm^2 di elemento); Cloro (1 g/cm^2 di NaCl); Manganese (1 g/cm^2 di MnO_2); Bromo (1 g/cm^2 di NaBr); Rodio (0,36 g/cm^2 di elemento); Argento (0,96 g/cm^2 di elemento); Cadmio (0,27 g/cm^2 di elemento); Iodio (1 g/cm^2 di elemento); Iridio (0,58 g/cm^2 di elemento); Mercurio (8,2 g/cm^2 di elemento).

In queste misure come nelle altre che saranno descritte in seguito, la sorgente dei neutroni di Emanazione e Berillio (fino a circa 700 millicurie) veniva posta entro un blocco cilindrico di paraffina (diametro cm 25,5, altezza cm 16) sull'asse tre centimetri sotto la base superiore. Le sostanze, in forma di strati il più possibile sottili, venivano appoggiate sul centro della base superiore ponendo gli strati assorbenti a diretto contatto della paraffina e sopra di essi i rivelatori.

Quando gli assorbitori non erano sufficientemente sottili, i risultati venivano corretti per tener conto della altezza del rivelatore sopra la paraffina.

TABELLA I.

Rilevatori	Assorbitori: B	Cl	Mn	Br	Rh	Ag	Cd	I	Ir	Hg
Mn	13	59	73	92	59	59	15	98	35	9
Br	50	68	81	61	79	79	69	86	84	43
Rh	17	67	88	96	54	67	27	97	38	11
Ag	30	60	86	91	68	45	40	89	48	27
In	23	69	91	97	53	67	40	97	38	10
I	54	73	79	84	91	70	75	55	68	48
Ir	29	62	83	87	49	71	49	—	34	—

Dall'esame di questa tabella si conferma il risultato, già osservato da altri Autori, che di regola l'assorbimento di un elemento è più forte quando l'elemento stesso è usato come rivelatore, ciò che fa pensare a delle selettività nell'assorbimento dei neutroni. Si nota inoltre che alcuni elementi hanno un comportamento analogo come per esempio le coppie Br e I, In e Rh.

2. *Curve di assorbimento.* – Ulteriori informazioni su questi fenomeni si hanno dallo studio delle curve di assorbimento. Fino ad ora abbiamo studiato dettagliatamente gli assorbimenti del Cd con rivelatori Rh ed Ag e

dell'Ag con rivelari Rh, In ed Ag di due differenti spessori. Nelle nostre condizioni sperimentali la curva di assorbimento che si avrebbe nel caso di una radiazione omogenea non è esponenziale poiché molti neutroni attraversano obliquamente lo strato assorbente; di questo fatto è stato tenuto conto nell'interpretazione dei risultati.

I risultati non corretti per questa causa sono i seguenti:

TABELLA II.

Assorbitore Cd g/cm²	0	0,00076	0,003	0.0071	0,0193	0,042
Rh 0,36 g/cm²	100	97,3	93,9	86,3	68,7	51,2
Ag 0,057 g/cm²	100	—	95,4	89	75,3	64,4

Assorbitore Cd g/cm²	0,070	0,112	0,139	0,27	0,54	0,79	1,05	4
Rh 0,36 g/cm²	41,4	33,5	31,9	27,0	25,4	24	24	17,5
Ag 0,057 g/cm²	57	53	53	50	50	49	48	—

TABELLA III.

Assorbitore Ag g/cm²	0	0,0018	0,0037	0,0096	0,021	0,057
Rh 036 g/cm²	100	—	—	—	—	98
Ag 0,057 g/cm²	100	96,7	93,5	89,4	83,9	75
Ag 0,30 g/cm²	100	—	—	—	—	85
In 0,22 g/cm²	100	—	—	—	—	—

Assorbitore Ag g/cm²	0,306	0,612	0,925	1,78	2,63	3,48
Rh 036 g/cm²	85,5	77	68,3	53,8	43,8	36
Ag 0,057 g/cm²	56	45	36	—	—	—
Ag 030 g/cm²	65,3	55,3	46,5	32	25,2	19,4
In 0,22 g/cm²	86	—	—	—	43	36

Gli spessori assorbenti più sottili sono stati fatti laminando leghe di Sn contenenti opportune percentuali di Ag e Cd, avendo verificato preventivamente che strati sottili di Sn non assorbono apprezzabilmente.

Di tutti questi assorbimenti i più semplici sono quelli dell'Ag rivelatore il Rh e l'In; tali curve, corrette per l'obliquità dei neutroni corrispondono ad assorbimenti esponenziali con coefficiente di assorbimento massico eguale in entrambe i casi a 0,13 cm²/g.

La curva di assorbimento del Cd sul rivelatore Rh, anch'essa corretta per l'obliquità, si può analizzare in un esponenziale con grande coefficiente di assorbimento (coefficiente di assorbimento massico 10) oltre ad una componente pochissimo assorbibile (circa 300 volte più penetrante dell'altra) che rappresenta circa il 25 % della intensità totale. Anche la curva di assorbimento del Cd con rivelatore Ag (0,057 g/cm²) ha caratteristiche simili; in questo caso la componente molle ha coefficiente di assorbimento massico 13 cm^2/g e rappresenta circa il 50 % dell'attività totale.

La curva di assorbimento dell'Ag con rivelatore Ag è stata fatta usando due rivelatori con differente spessore poiché essendosi subito notato che la curva di assorbimento, particolarmente per spessori di assorbimento sottili, deviava fortemente da un esponenziale, si è ritenuto conveniente usare anche un rivelatore molto sottile in modo da accentuare l'intensità relativa della porzione più molle della radiazione. La curva di assorbimento che si ottiene con un rivelatore sottile, corretta per l'obbliquità, può analizzarsi entro i limiti degli errori sperimentali in due esponenziali: il primo rappresenta il 25 % dell'intensità totale ed ha un coefficiente di assorbimento massico di 19 cm^2/g; il secondo, che rappresenta il 75 % dell'intensità totale, ha coefficiente di assorbimento massico di 0,3 cm^2/g; naturalmente questa analisi non è in alcun modo univoca ma indica chiaramente che una parte della radiazione viene fortemente assorbita mentre l'altra è notevolmente più penetrante.

Le cose si presentano come se la banda d'assorbimento corrispondente alla attivazione dell'Ag avesse un massimo assai accentuato che non si manifesta nell'assorbimento dell'elemento stesso usando come rivelatori il Rh e l'In.

3. *Assorbimenti multipli*. - Abbiamo descritto nel paragrafo precedente l'assorbimento del Cd con rivelatori Rh ed Ag ed abbiamo osservato che da spessori di pochi decigrammi di Cd filtra una radiazione che non è ulteriormente assorbita dal Cd che in misura minima. Fatti analoghi si osservano anche con altri rivelatori. Ci siamo proposti uno studio delle proprietà di questa radiazione che filtra attraverso ad alcuni decigrammi di Cd (più precisamente nelle nostre esperienze 0,27 g/cm^2 di Cd).

La Tabella IV è analoga alla I solo che si riferisce alla radiazione filtrata da 0,27 g/cm^2 di Cd.

TABELLA IV.

Rivelatori	Assorbitori: Rh	Ag	Ir
Mn	87	80	—
Br	84	78	74
Rh	31	81	37
Ag	88	29	53
In	43	84	—
I	82	76	—

Le caratteristiche degli assorbitori e dei rivelatori usati in questa tabella sono le stesse di quelli usati nella Tabella I.

La Tabella IV si presta al alcune interessanti osservazioni: così per esempio gli assorbimenti del Rodio e dell'Argento usati nei quattro modi possibili come assorbitori e come rivelatori, mostrano una selettività assai maggiore che non usando radiazione non filtrata. Anzi gli assorbimenti dell'Ag con rivelatore Ag e del Rh con rivelatore Rh sono maggiori sulla radiazione filtrata che su quella naturale come risulta dal confronto delle Tabelle I e IV. Questo complesso di fatti indica che la radiazione filtrata dal Cd ha le proprietà generali dei neutroni lenti; nell'ipotesi che le selettività corrispondano a bande di assorbimento comprendenti un certo intervallo di velocità dei neutroni, questi fatti suggerirebbero che la banda del Cd sia interposta fra quelle del Rh e dell'Ag e che si sovrapponga alla parte comune di queste due. Allo stato presente della ricerca tali considerazioni non possono avere che un carattere estremamente provvisorio. Analoghi fenomeni sono stati osservati sulla radiazione filtrata da B oppure dal Hg.

Per studiare più dettagliatamente le proprietà della radiazione filtrata dal Cd abbiamo rilevato la curva di assorbimento dell'Ag con rivelatore Ag (0,057 g/cm²) filtrando prima la radiazione con 0,27 g/cm² di Cd. La Tabella V dà i risultati:

TABELLA V.

Assorbitore Ag g/cm²	0	0,0018	0,0037	0,0096	0,021	0,057
Ag 0,057 g/cm²	100	96,8	92,5	81	70,7	54,4

Assorbitore Ag g/cm²	0,306	0,925	3,48
Ag. 0,057 g/cm²	33	22	12

Come si vede da questa tabella la curva di assorbimento sulla radiazione filtrata dal Cd dell'Ag con rivelatore Ag, pur avendo le stesse caratteristiche generali dell'analoga curva fatta con la radiazione non filtrata dal Cd, ne differisce in quanto i rapporti fra le intensità della parte molle e della parte dura sono alterati a vantaggio della prima; ciò fa pensare che il Cd assorba intensamente la radiazione poco assorbita dall'Ag e viceversa. Questa circostanza trova del resto una conferma nel fatto che le curve di assorbimento del Cd con rivelatori di Ag sottile o spesso differiscono nel senso che con rivelatore sottile l'intensità della componente poco assorbita dal Cd è maggiore che con rivelatore spesso.

Roma, *Istituto Fisico della R. Università.*

113

SULL'ASSORBIMENTO DEI NEUTRONI LENTI. — II.

E. FERMI ed E. AMALDI

« Ric. Scientifica », *6* (2), 443-447 (1935).

1. *Gruppi di neutroni lenti*. - In una precedente lettera [1] abbiamo esposto i primi risultati dello studio delle curve di assorbimento dei neutroni lenti, usando varii rivelatori ed assorbitori.

Cercheremo di riassumere i dati già esposti ed altri, di cui parleremo in seguito, servendoci del classico metodo consistente nel fare un'analisi delle curve di assorbimento in curve esponenziali, in modo da identificare nel complesso delle radiazioni neutroniche, che escono da un blocco di paraffina (contenente la sorgente di Em + Be), delle componenti omogenee nel senso che per ognuna di esse ed ogni assorbitore si possa assegnare un coefficiente di assorbimento indipendente dal rivelatore usato.

È ben noto che l'analisi in componenti esponenziali di una curva di assorbimento può essere ritenuta attendibile solo quando la precisione delle misure singole è assai elevata e inoltre la disposizione geometrica dell'assorbitore e rivelatore rispetto alla sorgente, corrisponde a condizioni esattamente calcolabili. Nel nostro caso queste condizioni non sono ben verificate, e quindi ci manca la possibilità di garantirci in modo preciso della omogeneità di una componente della radiazione. Ci siamo quindi accontentati di caratterizzare dei gruppi di componenti approssimativamente omogenee. Si capisce quindi che se si esaminano i coefficienti di assorbimento di un gruppo di neutroni lenti nella stessa sostanza assorbente usandone vari spessori e servendosi di differenti rivelatori, si trovino delle piccole differenze, attribuibili del resto in parte ad errori di misura. Peraltro le differenze di comportamento tra gruppo e gruppo sono così marcate da potere assegnare a ciascuno di essi una individualità nonostante le piccole inomogeneità interne.

L'analisi in gruppi è resa praticamente eseguibile dalle seguenti circostanze:

a) per alcuni assorbitori esistono delle fortissime differenze nel coefficiente di assorbimento fra gruppo e gruppo:

b) è possibile usare molti assorbitori che si comportano in modo molto differente;

c) è possibile filtrare i neutroni lenti con assorbitori scelti in modo da ridurne la inomogeneità.

(1) « Ric. Scientifica », *6* (2), 344 (1935).

2. *Sistematica dei gruppi di neutroni lenti.* – Abbiamo già osservato nella precedente lettera che, nelle condizioni sperimentali là descritte, la curva di assorbimento del Cd usando come rivelatori Rh od Ag si scinde in una componente approssimativamente esponenziale con forte coefficiente di assorbimento (assorbimento massico 13 cm^2/g rivelatore Ag; 10 cm^2/g rivelatore Rh) ed in una componente assai poco assorbita. La parte molle è una radiazione che, come vedremo, è approssimativamente omogenea, e costituisce quindi un gruppo che chiameremo gruppo *C* (iniziale di Cadmio). Questo gruppo può studiarsi in modo particolarmente facile poiché un opportuno spessore di Cd (in pratica nelle nostre misure 0,27 g/cm^2) lo assorbe totalmente mentre non riduce sensibilmente le intensità delle altre radiazioni.

Per determinare quindi quanta parte dell'attività di un rivelatore è dovuta a neutroni del gruppo *C* basta fare la differenza fra le attivazioni che si osservano senza e con interposizione di uno strato di Cd di 0,27 g/cm^2. I risultati finora ottenuti sui coefficienti di assorbimento massico del gruppo *C* sono riassunti nella Tabella I. La disposizione sperimentale è la stessa descritta nella lettera precedente; di regola ci siamo limitati a rivelatori abbastanza intensi in modo da poter eseguire le misure di attività con una camera di ionizzazione. Alcuni dei dati sono ottenuti da un solo spessore assorbente, mentre altri sono dedotti da più punti della curva di assorbimento.

TABELLA I.

Gruppo *C*

Rivelatori / Assorbitori:	Rh	Ag	Cd	In	Ir	Au	Hg
Mn	0,68	0,24	—	—	—	—	—
Rh	0,45	0,2	10	0,5	0,8	0,15	0,6
Ag	0,55	0,2	13	1	0,6	0,2	0,8
In	0,5	0,2	12	0,7	0,8	0,15	0,5
Ir	0,86	0,25	12	—	—	—	0,5

Si noti che i coefficienti di assorbimento massico scritti in verticale concordano abbastanza bene tra loro.

Altri dati sui coefficienti di assorbimento del gruppo *C* con rivelatore una camera di ionizzazione a Li collegata ad un amplificatore proporzionale si possono ricavare dal lavoro di Dunning, Pegram, Fink e Mitchell [(2)] nonché per rivelatori B e Li dal lavoro di Mitchell, Dunning, Segrè e Pegram [(3)]; i dati calcolati da questo lavoro concordano con i nostri risultati per quanto ci si può attendere data la differenza del dispositivo.

Passiamo ora a studiare la radiazione che filtra attraverso al Cd. L'esame della Tabella IV della lettera precedente, tabella che si riferisce appunto alla

(2) « Phys. Rev. », *48*, 265 (1935).
(3) « Phys. Rev. », *48*, 774 (1935).

radiazione filtrata da 0,27 g/cm² di Cd, mostra senz'altro che questa radiazione non è omogenea. Nello studio preliminare che ci è stato possibile fare fino ad ora abbiamo scisso questa radiazione in tre gruppi, che abbiamo denominato *A* (iniziale di Argento), *B* e *D*.

Il gruppo *A* è quello meglio caratterizzabile in quanto viene fortemente assorbito dall'Argento (coefficiente di assorbimento massico 17 cm²/g). La esistenza di questa radiazione è stata già dimostrata nella lettera precedente dove si è fatto vedere che circa il 25 °/₀ della attività di una sottile lastra di argento (0,057 g/cm²) è dovuto a questa radiazione fortemente assorbita dall'Ag e pochissimo assorbita dal Cd. Sono rari i rivelatori che vengono eccitati dal gruppo *A*; è noto fino ad ora oltre all'Ag soltanto l'Ir. Il gruppo si osserva bene come differenza fra l'attività senza e con un sottile strato di Ag (nelle nostre esperienze 0,057 g/cm² di Ag); le misure dei coefficienti di assorbimento di questo gruppo sono in corso di esecuzione.

Dell'attività totale di un rivelatore di Ag di 0,057 g/cm² il 50 °/₀ è dovuto al gruppo *C*, il 25 °/₀ al gruppo *A*; il resto sembra costituire in parte un terzo gruppo, che abbiamo provvisoriamente chiamato B, il quale ha coefficiente di assorbimento piccolissimo nel Cd e oltre 0,2 nell'Ag. Poco è noto sulle proprietà di questo gruppo data la difficoltà di isolarlo. Le percentuali di attivazione dell'Ag date sopra si riferiscono alla vita di 22 s. La vita più lunga (2,3 min) è attivata con percentuali differenti; in particolare essa non è attivata o ben poco dal gruppo *A*.

La vita di 44 s del Rh (0,36 g/cm²) è attivata per il 72 °/₀ dal gruppo *C*; il 28 °/₀ residuo è attivato da una radiazione che sembra costituire un gruppo diverso dai precedenti, che abbiamo chiamato *D*. Lo stesso gruppo attiva abbastanza fortemente la vita di 54 min dell'In, la cui attività per un rivelatore di 0,084 g/cm² è dovuta al gruppo *C* per il 60 °/₀ ed al gruppo *D* per il rimanente 40 °/₀. Abbiamo studiato questo gruppo usando appunto rivelatori In e Rh e filtrando la radiazione col Cd in modo da eliminare il gruppo *C*. Segue una tabella contenente alcuni coefficienti di assorbimento massico del gruppo *D*:

TABELLA II.

Gruppo D

Rivelatori	Assorbitori: Rh	Ag	Cd	In	Ir	Au	Hg
Rh	—	0,04	0,04	2,3	—	0,03	0,05
In	1,2	0,06	—	3	0,9	0,03	0,03

Naturalmente non si deve ritenere che i gruppi ora descritti li esauriscano tutti; si ha per esempio indicazione di un nuovo gruppo che attiva in parte lo iodio. Parimenti potrebbe darsi che alcuni dei gruppi qui elencati avessero in realtà una struttura complessa.

3. *Numerosità dei gruppi*. - Oltre ai coefficienti di assorbimento dei gruppi ha anche importanza determinare il numero di neutroni lenti che

escono dalla paraffina appartenenti ad ognuno di essi. In misura relativa una valutazione di questa grandezza, che chiameremo numerosità del gruppo, può ottenersi col metodo seguente. Supponiamo di appoggiare nel centro del solito cilindro di paraffina (vedi lettera precedente) un rivelatore assai sottile; ammettiamo che in esso tutti i neutroni assorbiti diano luogo alla formazione di atomi radioattivi. Il numero di disintegrazioni al secondo che si ottengono al tempo zero dopo irraggiamento infinito è dato da 2 N K P, dove N è il numero di neutroni che escono dalla paraffina per secondo e per centimetro quadrato di superfice, K il coefficiente di assorbimento massico dei neutroni e P la massa del rivelatore. Il fattore 2 deriva dal fatto che i neutroni escono in tutte le direzioni. Se la attivazione è dovuta a varii gruppi di neutroni, si dovrà porre per N e per K i dati relativi ai singoli gruppi. Si intende facilmente che la numerosità di un gruppo possa ricavarsi dalla misura dell'attività di un rivelatore irradiato nelle condizioni precedenti; naturalmente bisognerà in pratica applicare parecchie correzioni per tener conto dello spessore finito dei rivelatori e del conseguente assorbimento dei neutroni e dei raggi β. Conviene anche dividere le numerosità ottenute per l'intensità della sorgente in modo da ottenere dei dati indipendenti da quest'ultima.

In tal modo dallo studio dell'attività dell'Ag (22 s) si ottengono come numerosità dei gruppi *A*, *B*, *C* i seguenti valori (in misura relativa) 1, 20, 80; analogamente dallo studio del Rh risultano per i gruppi *C* e *D* le numerosità 50 e 10. Per i due stessi gruppi si ottiene dall'In, sommando le attivazioni della vite di 16 s e 54 min, 60 e 10. Il Manganese dà per il gruppo *C* numerosità 60. L'Iridio invece dà per il gruppo *C* una numerosità molto più bassa ciò che verosimilmente indica l'esistenza di un processo di assorbimento dei neutroni del gruppo *C* in questo elemento senza formazione di un radioelemento conosciuto.

Si vede dunque che il gruppo *C* è molto più numeroso degli altri e a ciò si deve probabilmente attribuire il fatto che esso determina in quasi tutti i rivelatori grosse percentuali dell'attivazione totale; il gruppo *A* è invece assai poco numeroso e il fatto che ciò non ostante lo si osservi abbastanza bene nell'Ag deriva dall'enorme coefficiente di assorbimento di questo elemento nel gruppo *A*.

Dato che attualmente è sconosciuta l'origine dei gruppi e non si sa se la distinzione in gruppi sia determinata da differenze nella velocità oppure da differenze nello stato interno dei neutroni, abbiamo fatto alcune prove per cercar di stabilire se usando diverse sorgenti di neutroni le percentuali dei varii gruppi venissero alterate. Queste esperienze, che hanno dato fino ad ora risultato negativo, sono state fatte con una sorgente di Em + Be appena preparata, seguendo le attivazioni dell'Ag e del Rh con e senza opportuni filtri di Cd e di Ag nel periodo di formazione del RaC. E anche con una sorgente di RaTh + D_2O e Rh come rivelatore; quest'ultima esperienza non è molto precisa a causa della debole intensità.

4. *Coefficiente di riflessione dei neutroni dei varii gruppi.* – Le misure esposte fino ad ora si riferiscono ad esperienze eseguite all'esterno di un

blocco di paraffina contenente la sorgente. Ha però interesse anche studiare ciò che accade nell'interno della paraffina nei riguardi dei varii gruppi.

Abbiamo eseguito alcune misure servendoci del seguente dispositivo: la sorgente di Em + Be si trovava sull'asse di un cilindro di paraffina (diametro cm 25,5, altezza cm 16) 3 cm sotto la base superiore; il rivelatore (Rh 10 g/28 cm^2 oppure Ag 2 g/39 cm^2) e gli eventuali assorbitori (Cd 0,27 g/cm^2; Ag 0,057 g/cm^2) erano appoggiati nel centro della base superiore: un secondo cilindro di paraffina poteva esser sovrapposto al tutto. Le attività ottenute nelle varie esperienze sono riassunte nella Tabella III; la prima colonna indica schematicamente il dispositivo usato, con la convenzione di indicare con S la sorgente, con P la paraffina, con R il rivelatore e coi simboli Cd e Ag gli assorbitori, così che per esempio S P Cd R Cd P significa una esperienza nella quale al disopra della sorgente (S) si trovano prima 3 cm di paraffina (P) per giungere alla base superiore del cilindro di paraffina, poi uno strato di Cadmio (Cd), il rivelatore (R), un secondo strato di Cadmio (Cd) ed infine il secondo cilindro di paraffina (P).

TABELLA III.

Dispositivo	Argento	Rodio
S P R	71	290
S P Cd R	35	82
S P Cd Ag R	18	80
S P R P	438	1055
S P Cd R Cd P	66	144
S P Cd Cd R Cd Cd P	66	130
S P Cd R P	90	251
S P R Cd P	111	382
S P Cd Ag R Ag Cd P	34	148
S P Cd Ag R Cd P	44	—
S P Cd R Ag Cd P	57	—

Confrontando fra di loro le esperienze S P R ed S P R P si nota il forte aumento dell'attività dovuto al sovrapporre il secondo blocco di paraffina. Nell'esperienza S P Cd R Cd P si vede che il porre il rivelatore fra due strati di Cd riduce l'attività dell'Ag al 15 % e quella del Rh al 14 %; si ha cioè nell'interno della paraffina una riduzione della attività dell'Ag e del Rh molto maggiore che all'esterno (nelle esperienze S P R ed S P Cd R il Cd riduce sull'Ag al 50 % e sul Rh al 28 %. Come all'esterno della paraffina anche all'interno (esperienze S P Cd R Cd P ed S P Cd Cd R Cd Cd P) il raddoppiare lo strato di Cd non riduce considerevolmente l'attività.

Ciò può interpretarsi ammettendo che anche nell'interno della paraffina il Cd porti via quasi completamente i neutroni del gruppo *C*, senza assorbire o quasi quelli degli altri gruppi, i quali, come vedremo fra poco, nell'interno della paraffina hanno una numerosità relativa minore.

Si osservi il risultato dell'esperienza S P Cd Ag R Ag Cd P che mostra come l'aggiunta di due sottili assorbitori di Ag a quelli di Cd determini una considerevole riduzione (da 66 a 34) quando si usi l'Ag come rivelatore. Ciò deve interpretarsi attribuendo tale riduzione all'assorbimento dei neutroni del gruppo *A*. Invece l'aggiunta dei sottili assorbitori di Argento non riduce ulteriormente l'attività del Rodio, poiché questo elemento non è eccitato dai neutroni del gruppo *A*.

La differenza fra le esperienze S P Cd R P ed S P R Cd P si interpreta facilmente tenendo conto del fatto che i neutroni lenti nell'interno della paraffina hanno una distribuzione di densità che decresce allontanandosi dalla sorgente in modo che uno strato di Cd che protegge il rivelatore dalla parte verso la sorgente arresta più neutroni di uno strato posto dalla parte opposta. Per riportarci al caso semplice in cui i neutroni fossero distribuiti uniformemente nella paraffina, si può prendere la media dei risultati delle esperienze predette come intensità di un rivelatore da un lato del quale sia posto uno strato di Cd. Si trova così per l'Ag senza Cd 438, con uno strato di Cd in media 100 e fra due strati di Cd 66. Sottraendo 66 da 438 e da 100 si trovano le attività dovute al solo gruppo *C* e cioè: senza Cd 372, con un solo strato di Cd 34. Se i neutroni potessero attraversare una volta sola il rivelatore, il rapporto di questi due numeri dovrebbe essere 2; il fatto che si trovi un rapporto molto più grande significa che un neutrone che ha attraversato il rivelatore senza venire assorbito ha una elevata probabilità di ricaderci sopra. Questa probabilità esprime il coefficiente di riflessione per i neutroni del gruppo *C* che cadano su di una parete piana che limita un blocco di paraffina. Un esame più dettagliato di questi esperimenti dimostra che i risultati possono interpretarsi soddisfacentemente ammettendo un coefficiente di riflessione per i neutroni del gruppo *C* pari a circa 0,83; deve anche tenersi conto dell'assorbimento dei neutroni nel rivelatore, assorbimento che è diverso nel caso del Rh e dell'Ag in modo tale da giustificare il differente comportamento di questi rivelatori.

Esperienze analoghe sul coefficiente di riflessione sono state fatte anche per i neutroni del gruppo *A*, usando sia il rivelatore di Ag di 0,057 g/cm^2 che un rivelatore più sottile (0,010 g/cm^2) in modo da ridurre il più possibile l'assorbimento dei neutroni lenti nel rivelatore. Tali esperienze hanno messo in evidenza che il coefficiente di riflessione dei neutroni del gruppo *A* è nullo o molto piccolo.

5. *Distanza di diffusione dei neutroni dei vari gruppi.* – Questo diverso comportamento dei gruppi *A* e *C* è confermato dalla seguente esperienza. Un rivelatore di Ag (0,057 g/cm^2) viene posto nell'interno di una massa di paraffina fra due strati di Cd (0,27 g/cm^2) i quali possono o essere a diretto contatto dell'Ag oppure distare simmetricamente da esso di un tratto x, essendo in questo caso lo spazio fra i Cd ed il rivelatore pieno di paraffina. Nella Tabella IV sono date le attività iniziali dell'Ag in funzione di x.

TABELLA IV.

x cm	0	0,5	1,0	1,5	2	3	∞
Ag	46	70	109	154	183	259	358

Si vede dunque che i Cd a 3 cm di distanza dal rivelatore determinano ancora un notevole abbassamento dell'intensità, ciò che dimostra che i neutroni del gruppo *C* diffondono per un tratto di oltre 3 cm.

Un analogo esperimento sul gruppo *A* si può fare sostituendo ai due assorbitori di Cd due lastre di Ag di 0,057 g/cm^2 e ponendo sempre il rivelatore di Ag fra due lastre di Cd (0,27 g/cm^2) a contatto in modo da assorbire completamente il gruppo *C*. Nella Tabella V sono date le attività iniziali dell'Ag in funzione della distanza *x* delle lastre di Ag.

TABELLA V.

x cm	0	0,5	1,0	1,5	2	∞
Ag	25	40,5	45	46	46	46

Come si vede da questa tabella gli strati di Ag ad 1 cm di distanza non producono alcun abbassamento sensibile dell'attività del rivelatore. La saturazione viene quasi raggiunta già per $x = 0,5$ cm. Ciò indica che il tratto di diffusione dei neutroni del gruppo *A* è molto più piccolo di quello del gruppo *C* e probabilmente si riduce ad un solo cammino libero medio, ciò che è in accordo col basso potere riflettente della paraffina per neutroni di questo gruppo.

Anche la dipendenza dalla distanza dalla sorgente è diversa per i varii gruppi nel senso che il decremento con la distanza dell'intensità del gruppo *C* è minore che per gli altri gruppi.

Le esperienze che abbiamo esposte ci sembrano indicare che i gruppi di neutroni si differenziano uno dall'altro non solo per i diversi assorbimenti, ma anche per le proprietà di diffusione. Esse non permettono invece di decidere quale sia la causa fisica di questo diverso comportamento. Forse l'ipotesi più ovvia è di attribuirlo a considerevoli differenze di velocità; potrebbe anche pensarci che le differenze fossero dovute invece a diversi stati interni del neutrone.

Roma, *Istituto fisico della R. Università*
14 dicembre 1935-XIV.

114.

SULL'ASSORBIMENTO DEI NEUTRONI LENTI. — III.

E. AMALDI ed E. FERMI

«Ric. Scientifica», 7 (1), 56–59 (1936).

In una lettera precedente [1] abbiamo discusso l'esistenza di gruppi di neutroni lenti che si differenziano gli uni dagli altri, oltre che per le proprietà di assorbimento, anche per quelle di diffusione. Nel § 1 di questa lettera preciseremo ulteriormente le proprietà di diffusione dei gruppi; nel § 2 raccoglieremo alcune proprietà speciali di vari elementi in relazione ai gruppi. Nell'ultimo paragrafo discuteremo le varie ipotesi sull'origine dei gruppi di neutroni.

1. *Diffusione dei singoli gruppi di neutroni lenti.* – Già nella lettera precedente abbiamo accennato alle differenze tra le proprietà di diffusione dei neutroni dei gruppi *A* e *C*. Abbiamo cercato di precisare quantitativamente queste differenze, estendendo inoltre le ricerche al gruppo *D*. Discutiamo in un primo tempo le esperienze sul gruppo *C*.

Consideriamo un neutrone di questo gruppo che si trovi nell'interno di un blocco di paraffina limitato da una superficie piana, a distanza x da questa; sia $p(x)$ la probabilità che il neutrone esca dalla superficie prima di esser stato distrutto con un meccanismo qualsiasi (per esempio cattura da parte dell'idrogeno). Nell'ipotesi che i neutroni del gruppo *C* nella paraffina obbediscano alle solite leggi della diffusione, ed abbiano in più la possibilità di esser distrutti con vita media τ, la loro densità n soddisfa all'equazione differenziale

$$\frac{dn}{dt} = D\Delta n - \frac{n}{\tau} \tag{1}$$

dove D è il coefficiente di diffusione dato da $1/3\,\lambda\bar{v}$, essendo λ il cammino libero medio e v la velocità media; questa espressione di D vale nella ipotesi che il cammino libero medio sia indipendente dalla velocità e gli atomi di idrogeno possano considerarsi come centri fissi di diffusione isotropa, ipotesi verosimili per i neutroni del gruppo *C*, poiché i quanti elastici del moto di vibrazione dell'idrogeno sono grandi rispetto alla energia cinetica media dei neutroni C (agitazione termica; vedi paragrafo 3).

(1) «Ric. Scientifica», *6* (2), 443 (1935); vedi anche «Ric. Scientifica», *6* (2), 344 (1935).

È facile dimostrare dall'equazione (1) che la probabilità $p(x)$ sopra definita è data da

$$p(x) = e^{-\frac{x}{\sqrt{D\tau}}}$$

Si noti che $D\tau = \frac{1}{3}\lambda^2 N$ dove $N = \frac{\bar{v}\tau}{\lambda}$ è il numero dei cammini liberi fatti in media da un neutrone del gruppo *C* prima di esser distrutto. Determinando l'andamento di $p(x)$ verremo dunque a conoscere la costante $D\tau$ ovvero $\lambda^2 N$.

La determinazione di $p(x)$ può farsi nel modo seguente: Immaginiamo un rivelatore R del gruppo *C* che ricopra tutta la superficie piana di un blocco di paraffina contenente la sorgente. Se ad una profondità x distruggiamo q neutroni C, l'attività del rivelatore si abbasserà proporzionalmente a $qp(x)$; la distruzione dei q neutroni si può ottenere ponendo nella posizione x un assorbitore del gruppo *C*; misurando la attività indotta in questo si determina q. Dunque $p(x)$ è proporzionale al rapporto della diminuzione della attività del rivelatore R alla attività dell'assorbitore. Misurando tale rapporto per diversi valori di x si può dunque verificare l'andamento esponenziale di $p(x)$ e dedurre il valore di $D\tau = 1/3\,(\lambda^2 N)$.

In pratica abbiamo usato come rivelatore una lastra di rodio di 10 g/28 cm². Siccome questa non copriva che una piccola parte della superficie della paraffina le misure sono state fatte ponendo il rivelatore successivamente in 9 posizioni diverse e sommando opportunamente i risultati. Come assorbitore del gruppo *C* abbiamo usata una lastra di lega di stagno e cadmio contenente 0,019 g/cm² di cadmio; siccome il cadmio, in piccoli spessori, assorbe praticamente solo i neutroni del gruppo *C*, la riduzione della attività del rivelatore ha luogo solo in questo gruppo, come può del resto anche verificarsi direttamente.

Siccome l'assorbimento dei neutroni da parte del cadmio non dà luogo ad attivazione, non era possibile misurare direttamente il numero q di neutroni assorbiti; si è perciò misurata l'attività indotta dai neutroni del gruppo *C* in una lastra di rodio equivalente a quella di cadmio in questo gruppo.

Le misure sono state eseguite per $x = 0, 1, 2, 3$ cm ed hanno permesso di verificare che l'andamento di $p(x)$ è effettivamente esponenziale entro i limiti del nostro errore; è risultato inoltre $\sqrt{D\tau} = 2{,}0$ cm, dal che si ricava $\lambda^2 N = 12$.

Analoghi esperimenti sono stati fatti per i gruppi *D* ed *A*, usando nel primo caso un rivelatore di indio schermato da cadmio (in modo da eliminare il gruppo *C*) ed un assorbitore di rodio; nel secondo caso rivelatore ed assorbitore di argento determinando la sola frazione della attività dovuta al gruppo *A*.

Naturalmente in questi casi la teoria della diffusione usata per il gruppo C non è più applicabile, poiché (vedi § 3) i neutroni di questi gruppi percorrono un solo cammino libero o poco più. Tuttavia, riducendo anche in questi casi i risultati con le stesse formule usate per il gruppo *C*, si ottiene per il gruppo *D*: $\sqrt{D\tau} = 0{,}44$ cm; e per il gruppo *A*: $\sqrt{D\tau} = 0{,}41$ cm. Si noti la marcata differenza di comportamento di questi due gruppi rispetto al gruppo *C*; il comportamento dei gruppi *D* ed *A* in queste esperienze è invece molto simile; la piccola differenza è compresa entro i limiti dell'errore sperimentale.

La teoria della diffusione è applicabile, come abbiamo già notato, nell'ipotesi che i neutroni percorrano molti cammini liberi; nell'ipotesi opposta, ammettendo cioè che i neutroni dei gruppi *D* ed *A* percorrano un solo cammino libero, si può, con esperienze molto simili alle precedenti che qui non descriveremo in dettaglio, ricavare il valore del cammino libero medio per i neutroni di questi gruppi. Esso risulta per entrambi compreso tra 0,6 e 0,7 cm. Questi valori sono più grandi di quelli generalmente ammessi per i neutroni del gruppo *C*. Ciò può esser dovuto all'essere errata l'ipotesi che i neutroni di questi gruppi percorrano un solo cammino libero, oppure ad una effettiva dipendenza di λ dal gruppo, imputabile eventualmente all'effetto del legame chimico dell'idrogeno nella paraffina.

2. *Dati particolari su alcuni gruppi ed elementi.* – Già nella lettera precedente abbiamo osservato che buona parte della attività dello iodio è probabilmente dovuta a un nuovo gruppo di neutroni. Abbiamo precisate ulteriormente queste osservazioni in modo da stabilire l'esistenza di un nuovo gruppo, che indicheremo con la lettera *I*, al quale è dovuta l'attività indotta nello iodio per circa il 70 %.

Questo gruppo è pochissimo assorbito da quasi tutti gli elementi: di quelli studiati fino ad ora soltanto lo iodio lo assorbe in misura notevole (coefficiente massico di assorbimento 1,1 cm^2/g). La numerosità di questo gruppo (vedi lettera precedente, § 3) è di circa 4.

Abbiamo già osservato che la numerosità del gruppo *C* dedotta dalla attività di un rivelatore di iridio (periodo 19 ore) risulta anomalmente bassa (circa 20, invece di una media di circa 70), ed espressa l'ipotesi che ciò fosse da attribuirsi alla esistenza di un secondo processo di assorbimento dei neutroni C in questo elemento. Abbiamo pertanto irradiato per alcune notti di seguito una lastra di iridio e, esaminando il decadimento della attività in essa indotta, abbiamo potuto riconoscere, oltre al periodo già noto, anche, un nuovo periodo di circa 2 mesi; gli elettroni corrispondenti a questo nuovo periodo sono molto molli. Benché non possiamo ancora dare risultati quantitativi, possiamo affermare che la intensità di questa nuova attività è tale da giustificare la bassa numerosità calcolata dal solo periodo di 19 ore.

Dobbiamo a questo proposito rettificare le numerosità date nella lettera precedente, poiché erano falsate da un errore di calcolo, che fortunatamente non aveva prodotto altro che lievi variazioni. I valori corretti sono i seguenti.

Gruppo *C*: Rivelatore argento, 60; rodio, 80, indio, 70; manganese, 70; oro, 60.

Gruppo *A*: Rivelatore argento, 1.

Gruppo *D*: Rivelatore rodio, 12; indio, 14.

3. *Ipotesi sull'origine dei gruppi di neutroni.* – Abbiamo già osservato nella lettera precedente che l'ipotesi più naturale sull'origine dei gruppi di

neutroni lenti è che essi siano dovuti a differenze nella velocità. Contro questa ipotesi si può obbiettare che secondo le ordinarie teorie del nucleo la sezione d'urto per la cattura di un neutrone lento dovrebbe essere inversamente proporzionale alla velocità, almeno per velocità non troppo grandi. Se così fosse non potrebbero evidentemente esistere assorbimenti selettivi determinati da differenti velocità dei neutroni. Per altro, data la incertezza delle teorie nucleari, ed in particolare di quelle relative ai processi di cattura dei neutroni, non bisogna dare eccessivo peso a questo argomento, e ciò particolarmente in vista delle esperienze sulla misura diretta della dipendenza della sezione di cattura dalla velocità [2], dalle quali sembra risultare che la dipendenza prevista dalla teoria non sia valida.

Un'altra ipotesi a cui già avevamo accennato, è che i neutroni dei vari gruppi differiscano uno dall'altro per qualche caratteristica interna, quale potrebbe essere per esempio un diverso stato di eccitazione. Questa ipotesi, per quanto possa sembrare meno spontanea, presenterebbe il vantaggio di non essere in diretta contraddizione coi risultati della teoria. Naturalmente anche in questo caso il diverso stato interno del neutrone sarebbe probabilmente accompagnato da differenze nella velocità. Varie esperienze concordano nel far ritenere che i neutroni del gruppo C abbiano la velocità di agitazione termica; di queste la più evidente è la misura diretta della velocità fatta con un selettore [3] in cadmio, e tale quindi da agire soltanto sul gruppo C. Per quanto riguarda gli altri gruppi vi sono invece indizi per ritenere che essi corrispondano a velocità differenti. Così per esempio Preiswerk e von Halban [4] non hanno trovato un effetto termico nella attività indotta nell'argento dalla radiazione filtrata dal cadmio. Alla stessa conclusione conducono anche i nostri risultati sui coefficienti di riflessione (albedo) della paraffina per i vari gruppi. L'elevata albedo per il gruppo C indica infatti che i neutroni di questa specie possono percorrere successivamente molti cammini liberi; mentre la bassa albedo del gruppo A, confermata successivamente anche per il gruppo D, indica che i neutroni di questi gruppi possono percorrere solo un cammino libero o poco più (cfr. anche § 1). È improbabile che neutroni che appartengono a un gruppo per un solo cammino libero, abbiano velocità di agitazione termica.

Qualunque sia l'origine dei gruppi, ha interesse determinare se, per urti successivi nella paraffina, i neutroni di un gruppo si trasformano in neutroni di un gruppo diverso. Una tale trasformazione dovrebbe evidentemente accadere nel caso che i gruppi differissero per la velocità, nel qual caso tutti i gruppi dovrebbero andare a finire nel gruppo C. L'accertare un simile processo di trasformazione di un gruppo nel gruppo C, porterebbe dunque un argomento, se pure non decisivo, a favore dell'ipotesi che i vari gruppi differiscano solamente per la velocità. Naturalmente l'argomento non sarebbe

(2) RASETTI, SEGRÈ, FINK, DUNNING and PEGRAM, « Phys. Rev. », *49*, 104 (1936).

(3) DUNNING, PEGRAM, FINK, MITCHELL and SEGRÈ, « Phys. Rev. », *48*, 704 (1935).

(4) PREISWERK and VON HALBAN, « Nature » (London), *136*, 1027 (1935). Vedi anche « C. R. », *202*, 133 (1936).

del tutto conclusivo, poiché, anche con altre ipotesi, è possibile giustificare simili trasformazioni.

Abbiamo pertanto iniziata una esperienza allo scopo di decidere se i neutroni del gruppo *D* si trasformano dopo alcuni urti in neutroni C. Il principio su cui è basata questa esperienza può facilmente comprendersi se ammettiamo per un momento di disporre di un rivelatore R che sia attivato esclusivamente dai neutroni del gruppo *C* e di un assorbitore A che assorba esclusivamente i neutroni del gruppo *D*. Si pensi il rivelatore R appoggiato nel centro della base superiore di un cilindro di paraffina contenente la sorgente S; se l'assorbitore A viene inserito tra la faccia della paraffina e il rivelatore, non si osserverà alcuna riduzione nella attività di quest'ultimo, poiché i neutroni C che escono dalla paraffina non vengono assorbiti dall'assorbitore A. Se ora pensiamo di sprofondare A entro la paraffina di 1 o 2 cm, potrà accadere che esso assorba alcuni dei neutroni D che in seguito ad ulteriori urti contro la paraffina avrebbero potuto trasformarsi in neutroni C prima di raggiungere R. Nel caso dunque che i neutroni D possano trasformarsi in C si deve attendere che A non produca alcun effetto messo a contatto di R, mentre ne riduca l'attività quando sia incluso nella paraffina 1 e 2 cm al di sotto della superficie.

L'esperienza, i cui dettagli verranno dati in una prossima comunicazione, richiede una elevatissima precisione delle misure. Pertanto non possiamo ancora ritenere definitivi i risultati attuali, che pur sembrano dare qualche indizio che la trasformazione D–C avvenga.

Roma, *Istituto Fisico della R. Università.*

115.

SUL CAMMINO LIBERO MEDIO DEI NEUTRONI NELLA PARAFFINA

E. AMALDI ed E. FERMI

« Ric. Scientifica », 7 (1), 223–225 (1936).

In alcune lettere precedenti [1] abbiamo riconosciuto che i neutroni che escono da un blocco di paraffina contenente la sorgente di Em + Be possono classificarsi in gruppi caratterizzati dalle loro proprietà di assorbimento nei diversi elementi. Tra questi gruppi il più notevole è quello che abbiamo chiamato gruppo *C* il quale da diverse esperienze, risulta molto verosimilmente costituito da neutroni aventi velocità di agitazione termica. Questi neutroni si separano facilmente dagli altri poiché essi sono assorbiti selettivamente dal cadmio. Dalle proprietà di diffusione dei neutroni C nella paraffina risulta che essi possono subire un numero considerevole di urti elastici contro gli atomi di idrogeno prima di essere catturati da un protone con formazione di idrogeno pesante. In una lettera precedente abbiamo anzi determinato per i neutroni C la costante λ^2 N, dove λ è il cammino libero medio ed N il numero medio di urti che un neutrone C subisce prima di essere catturato secondo il meccanismo precedentemente indicato. Il risultato è stato

$$\lambda^2 \mathrm{N} = 12 \, \mathrm{cm}^2.$$

Descriveremo ora delle esperienze intese a misurare direttamente λ; dal valore di λ si potrà quindi ricavare anche N grazie al risultato precedente.

Per misurare λ abbiamo seguìto questo metodo: un cilindro di paraffina di 12 cm di diametro e di 13 cm di altezza conteneva la sorgente di Em + Be 2 cm sotto il centro della base superiore. Tale cilindro era completamente rivestito di uno strato di cadmio di 0,5 g/cm^2; una finestra di $5 \times 5{,}5$ cm nel centro della base superiore del rivestimento di cadmio si poteva aprire o chiudere. Misurando l'attività di un rivelatore dei neutroni lenti, posto all'esterno di un simile blocco di paraffina, una volta con la finestra aperta ed una volta con la finestra chiusa e facendo la differenza delle due attività ottenute, si ricava evidentemente la parte di attività dovuta ai soli neutroni C che escono dall'apertura. Tale differenza equivale dunque alla attività che sarebbe dovuta ad una sorgente di soli neutroni C avente le

(1) E. AMALDI, E. FERMI, « Ric. Scientifica », *6* (2), 344 (1935); E. FERMI, E. AMALDI, « Ric. Scientifica », *6* (2), 443 (1935); E. AMALDI, E. FERMI, « Ric. Scientifica », *7* (1), 56 (1936).

dimensioni dell'apertura. In questo senso parleremo nel seguito di una « sorgente di neutroni C », o semplicemente « sorgente C ».

Come rivelatore dei neutroni C ci siamo serviti di una lastra di rodio di 0,35 g/cm², avente le stesse dimensioni della finestra sopra nominata. Questa lastra di rodio era posta 10 cm al disopra della base superiore della paraffina e l'attività indotta in essa era sufficientemente intensa per poter venire misurata con una camera di ionizzazione. A metà distanza tra il rivelatore e la sorgente C si potevano interporre degli strati di paraffina di vari spessori aventi le stesse dimensioni del rivelatore e della sorgente C stessa; tali diffusori erano disposti in modo da occultare esattamente il rivelatore dalla sorgente C. L'attività del rivelatore (differenza fra le attività senza e con finestra di cadmio) è stata misurata interponendo i seguenti spessori di paraffina: 0,036; 0,099; 0,203; 0,38; 0,68; 1,41 cm.

Se ragioni di intensità non ci avessero costretto a porre il rivelatore ad una distanza dalla sorgente eguale solo al doppio delle sue dimensioni e si fosse potuto aumentare molto questa distanza, l'attività del rivelatore dovuta ai neutroni C avrebbe dovuto decrescere in funzione dello spessore x del diffusore di paraffina, con legge esponenziale $e^{-x/\lambda}$. Invece nelle nostre condizioni geometriche la legge da attendersi è più complicata, poiché un neutrone diffuso dallo strato di paraffina ha una probabilità tutt'altro che trascurabile di raggiungere il rivelatore. Abbiamo valutato quale correzione debba applicarsi alla curva esponenziale per tenere conto della nostra disposizione geometrica; il calcolo è stato eseguito nell'ipotesi che la diffusione dei neutroni C sia isotropa rispetto ai centri diffondenti; tale ipotesi sarebbe certamente inesatta per neutroni veloci, ma corrisponde probabilmente al caso dei neutroni C per le ragioni esposte alla fine della presente lettera. Oltre a ciò sono state introdotte alcune semplificazioni per rendere materialmente possibile il calcolo.

È stata così ottenuta la legge di decremento dell'intensità in funzione di x/λ essendo x lo spessore diffondente di paraffina. Abbiamo quindi determinato λ in modo da ottenere l'accordo migliore possibile tra la curva teorica ed i punti misurati sperimentalmente.

Nella figura 1 i punti rappresentano i valori sperimentali dell'attività C per i vari spessori del diffusore con i relativi errori quadratici medi. La curva è quella teorica tracciata per $\lambda = 0{,}15$ cm.

Riteniamo dunque che il cammino libero medio dei neutroni C nella paraffina sia prossimo a 0,15 cm. Da questo valore e dalla (1) ricaviamo che il numero N di cammini liberi percorsi in media dai neutroni C è circa 500, ciò che corrisponde ad un percorso totale di 75 cm.

Nell'ipotesi assai plausibile che sia la diffusione che la cattura dei neutroni C nella paraffina siano dovute quasi esclusivamente all'idrogeno, ed ammettendo per esssi la velocità di agitazione termica risultano i seguenti dati per i neutroni C:

Sezione d'urto elastica . . . . . $\sigma_e = 86 \times 10^{-24}$ cm²

Sezione d'urto con cattura . . . . $\sigma_c = 0{,}17 \times 10^{-24}$ cm²

Vita media . . . . $\tau = 3{,}8 \times 10^{-4}$ s.

Il valore del cammino libero medio dei neutroni C ottenuto con le presenti misure è notevolmente inferiore al valore comunemente ammesso per i neutroni lenti [2], ciò che è in parte giustificabile col fatto che le nostre misure si riferiscono al solo gruppo *C* anziché ad un miscuglio di vari gruppi.

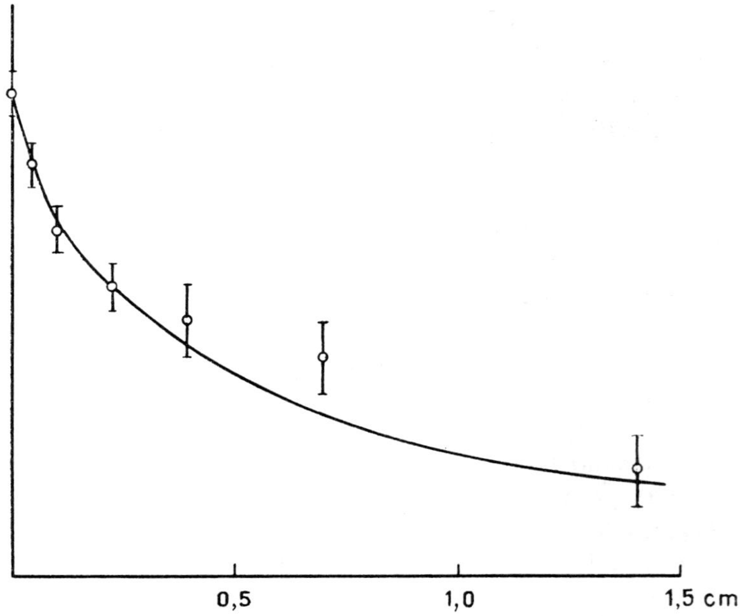

Fig. 1.

Dalla teoria quantistica dell'urto elastico dei neutroni lenti contro atomi di idrogeno, risulta, come è noto, che la sezione d'urto è indipendente dalla velocità. Tuttavia il valore della sezione d'urto come anche la distribuzione angolare dei neutroni diffusi, sono differenti a seconda che gli idrogeni diffondenti possano oppure no considerarsi come liberi; nel primo caso la diffusione è isotropa nel sistema di riferimento in cui è fermo il centro di gravità del neutrone e dell'atomo di idrogeno, mentre nel sistema di riferimento in cui l'idrogeno è inizialmente fermo, la diffusione avviene in avanti. Quando invece gli idrogeni diffondenti possono considerarsi come fissi, la diffusione è isotropa e la sezione d'urto elastica è quattro volte maggiore che nel caso precedente.

È ovvio che gli idrogeni della paraffina si potranno oppure no considerare come liberi a seconda che l'energia W del neutrone urtante è grande o piccola rispetto al quanto $h\nu$ delle frequenze di legame elastico dell'idrogeno nella paraffina. Questo quanto è abbastanza grande a confronto dell'energia

(2) E. AMALDI, O. D'AGOSTINO, E. FERMI, B. PONTECORVO, F. RASETTI and E. SEGRÈ, « Proc. Roy. Soc. » A *149*, 522 (1935); T. BJERGE and C. H. WESTCOTT, « Proc. Roy. Soc. » A *150*, 709 (1935).

di agitazione termica per modo che per i neutroni C gli atomi di idrogeno debbono considerarsi come centri fissi. Resta con ciò giustificata l'ipotesi che la diffusione dei neutroni C sia isotropa.

Ci si deve inoltre attendere che neutroni lenti aventi energia dell'ordine o superiore ad un volt abbiano nella paraffina un cammino libero medio quattro volte maggiore di quello trovato per i neutroni C e cioè $4 \times 0{,}15 = 0{,}6$ cm.

In una lettera precedente abbiamo osservato che nell'ipotesi che il numero di cammini liberi percorsi dai neutroni dei gruppi *A* e *D* sia eguale ad 1, risultano per questi dei cammini liberi medi compresi tra 0,6 e 0,7 cm. Questo fatto potrebbe dunque interpretarsi attribuendo ai neutroni dei gruppi *A* e *D* un'energia dell'ordine di grandezza o maggiore di un volt.

Istituto fisico della R. Università
Roma, 20 marzo 1936-XIV.

116.

SUI GRUPPI DI NEUTRONI LENTI

E. AMALDI ed E. FERMI

« Ric. Scientifica », 7 (1), 310–315 (1936).

1. In alcune lettere precedenti [1] abbiamo studiato le proprietà dei gruppi di neutroni lenti che si formano nella paraffina contenente una sorgente di Em–Be. In un recente articolo Bohr [2] ha mostrato la possibilità di interpretare l'esistenza di tali gruppi attribuendoli a neutroni di differente velocità.

Tutte le esperienze concordano nel far ritenere che i neutroni del gruppo *C* siano quelli dotati di velocità di agitazione termica, mentre gli altri gruppi sarebbero costituiti da neutroni che non hanno ancora subìto un rallentamento fino a questo stadio. Poco è noto sia sulla posizione che sulla larghezza delle bande di energia corrispondenti a questi gruppi di neutroni.

La posizione è stata valutata da vari Autori [3] ammettendo l'ipotesi che il coefficiente di assorbimento dei neutroni lenti nel boro sia inversamente proporzionale alla velocità e determinando la costante di proporzionalità per mezzo dell'assorbimento dei neutroni del gruppo *C* che sono dotati di velocità di agitazione termica. In questa ipotesi si trova che i gruppi *D*, *A*, *B*, *I*, sono in ordine di energia crescente e le relative energie risultano approssimativamente eguali a 3, 8, 30, 70 volt.

Date le ipotesi su cui è basata questa valutazione è evidentemente desiderabile un controllo indipendente. In questo lavoro ci proponiamo di esporre un metodo che consente di determinare i rapporti delle velocità dei gruppi *D*, *A*, *B*, *I* e inoltre anche le loro larghezze; tale metodo è basato esclusivamente sulle proprietà di diffusione dei neutroni lenti nelle sostanze idrogenate. Mentre il metodo permette di determinare con una discreta precisione le larghezze dei gruppi, la determinazione dei loro rapporti di velocità è alquanto più incerta; tale determinazione potrà tuttavia venire migliorata in seguito.

(1) E. AMALDI, E. FERMI, « Ric. Scientifica », *6* (2), 344 (1935); E. FERMI, E. AMALDI, « Ric. Scientifica », *6* (2), 443 (1935); E. AMALDI, E. FERMI, « Ric. Scientifica », *7* (1), 56 (1936); *7* (1), 223 (1936).

(2) N. BOHR, « Nature » (London), *137*, 344 (1936); vedi anche G. BREIT e E. WIGNER, Phys. Rev. *49*, 519 (1936).

(3) O. R. FRISCH, G. PLACZEK, « Nature » (London), *137*, 357 (1936), D. F. WEEKES, M. S. LIVINGSTON, H. A. BETHE, « Phys. Rev. », *49*, 471 (1936).

2. DETERMINAZIONE DELLA LARGHEZZA DEI GRUPPI. – Sia $\mathfrak{N}(v)\,dv$ il numero totale di neutroni di velocità tra v e $v + dv$ presenti in tutto un blocco di paraffina di dimensioni infinite contenente una sorgente di neutroni veloci che emette Q neutroni al secondo. La funzione $\mathfrak{N}(v)$ corrisponde alla distribuzione maxwelliana per i neutroni del gruppo *C*: essa può anche valutarsi per neutroni aventi energia $W \gg h\nu$ dove ν è la frequenza del legame elastico dell'idrogeno. La determinazione di $\mathfrak{N}(v)$ in questo caso è stata fatta da uno di noi [4]; l'espressione trovata (formule 3 e 4 del lavoro citato) prende una forma semplicissima nell'ipotesi assai plausibile che il processo di distruzione dei neutroni per assorbimento da parte dell'idrogeno, non abbia importanza per neutroni di energia superiore ad 1 volt. Si trova in questo caso

$$\mathfrak{N}(v) = \frac{2\,Q\lambda(v)}{v^2} \tag{1}$$

dove $\lambda(v)$ è il cammino libero medio.

Partendo dalla conoscenza di queste due distribuzioni si intende come sia possibile stabilire una relazione tra l'attività di un rivelatore nel gruppo *C* e l'attività del rivelatore stesso dovuta a neutroni appartenenti ad una banda di energia compresa tra le energie W_{min} e W_{max}, purché queste siano entrambe superiori ad 1 volt. Eseguendo i calcoli [5] nell'ipotesi di un rivelatore molto sottile, si trova la relazione

$$\log \frac{W_{max}}{W_{min}} = 2\,\frac{\sqrt{\dfrac{\lambda_c^2\,N}{3}}}{\lambda}\,\frac{K_c}{K}\,\frac{\int A\,d\tau}{\int B_c\,d\tau} \tag{2}$$

dove λ_c è il cammino libero medio dei neutroni del gruppo *C* ed *N* il numero medio di cammini liberi percorsi da un neutrone C (indipendentemente dalla misura di λ_c ed N separatamente, l'espressione $\sqrt{\dfrac{\lambda_c^2\,N}{3}}$ che figura nella formula è stata da noi misurata direttamente [6] ed è risultata eguale a 2,0 cm); λ è il cammino libero medio dei neutroni del gruppo in questione nella paraffina; K_c e K sono i coefficienti di assorbimento dei neutroni del gruppo *C* e del gruppo in esame nel materiale usato come rivelatore; A è l'attività indotta dai neutroni del gruppo in esame nel rivelatore; tale attività va misurata in molti punti entro la sostanza idrogenata in modo da potere eseguire su di essa un'integrazione estesa a tutto il volume; B_c è l'attività del rivelatore dovuta ai soli neutroni del gruppo *C* che si ottiene quando il rivelatore è protetto da una parte da una lastra di cadmio abbastanza spessa da assorbire totalmente i neutroni del gruppo *C*. Anche B_c si deve integrare su tutto il volume prendendo in ogni punto la media tra le attivazioni che si ottengono col cadmio da una parte oppure dall'altra.

(4) E. FERMI, « Zeeman Verhandelingen », p. 128, Martinus Nijhoff, s'Gravenhage (1935).

(5) Dettagli su queste come su altre questioni matematiche connesse alla teoria dei neutroni lenti verranno pubblicate in dettaglio da E. Fermi altrove.

(6) « Ric. Scientifica », 7 (1), 56 (1936).

In pratica non conviene usare un rivelatore molto sottile per non ridurre troppo la intensità; con un rivelatore spesso è necessario apportare alcune correzioni alla formula precedente.

La (2) permette di calcolare il $\log \frac{W_{max}}{W_{min}}$ che chiameremo larghezza logaritmica; tale espressione rappresenta il numero medio di urti che un neutrone subisce entro il gruppo dato che in un urto l'energia si riduce in media di un fattore $1/e$.

Abbiamo eseguito le misure della larghezza logaritmica per i gruppi *D*, *A*, *I*. Per tutti i gruppi abbiamo preso $\lambda = 0{,}65$ cm; abbiamo poi rilevato le curve di attivabilità in funzione della distanza per i vari gruppi come verrà descritto più dettagliatamente nel paragrafo seguente. Per il gruppo *D* abbiamo usato un rivelatore di rodio di 0,36 g/cm² ed abbiamo preso $K_c = 0{,}46$, $K_D = 1{,}7$ cm²/g valori risultati da misure recenti più accurate delle precedenti. Per il gruppo *A* abbiamo usato un rivelatore di argento di 0,057 g/cm² ed abbiamo preso $K_c = 0{,}2$, $K_A = 17$ cm²/g. Notiamo che in questo caso per il calcolo di $\int A d\tau$ abbiamo misurato l'attività dell'argento dovuta ai neutroni A ad una distanza di 3 cm dalla sorgente ed abbiamo ammesso che la dipendenza dell'attività dalla distanza per il gruppo *A* fosse la stessa che si trova per i gruppi $A + B$, ciò che porta un errore trascurabile. Infine per il gruppo *I* abbiamo usato un rivelatore di PbI_2 di 0,68 g/cm². Abbiamo assunto $K_I = 0{,}63$ cm²/g di PbI_2 ed abbiamo usato per il gruppo *C* i dati relativi al rodio.

In tal modo abbiamo trovato per i gruppi *D*, *A*, *I*, le seguenti larghezze logaritmiche: 0,8; 0,06; 0,2.

Come si vede il gruppo *D* è quello che ha larghezza maggiore, ciò che corrisponde alla relativamente elevata numerosità di questo gruppo. Si deve anzi notare che ad una larghezza logaritmica di 0,8 corrisponde una probabilità non trascurabile che un neutrone D rimanga appartenente allo stesso gruppo anche dopo un urto. Questo fatto non appare in accordo completo con i risultati delle esperienze di diffusione [7], secondo i quali sembra che i neutroni del gruppo *D* percorrano in media un solo cammino. La divergenza non è peraltro considerevole, e potrebbe facilmente ricondursi o ad imprecisione di misura o anche all'essere il gruppo *D* costituito da due bande.

Il gruppo *A* è invece particolarmente stretto, poiché corrisponde ad una banda di energia per cui $\frac{\Delta W}{W}$ è del 6 %.

3. Rapporti di energia tra i vari gruppi. – Abbiamo misurato in una vasca cilindrica di acqua profonda 95 cm e del diametro di 90 cm la dipendenza dell'attività prodotta dai neutroni dei vari gruppi, dalla distanza del rivelatore dalla sorgente. Come rivelatore per i gruppi *C* e *D* abbiamo usato una lastra di rodio di 0,36 g/cm² schermata o no da cadmio. Per il gruppo *I* un rivelatore PbI_2 di 0,68 g/cm² schermato da cadmio e per i gruppi

(7) « Ric. Scientifica », 7 (1), 56 (1936).

$A + B$, che non sono stati misurati separatamente, un rivelatore di argento di 0,057 g/cm^2 schermato da cadmio. Le misure sono state fatte con una camera di ionizzazione, per il gruppo C fino a una distanza di 40 cm (in queste condizioni l'attività era ridotta a circa 1/1000 dell'attività a piccola distanza), per il gruppo D fino a 20 cm, per i gruppi $A + B$ fino a 15 e per il gruppo I fino a 10 cm. Per piccole distanze i valori sperimentali sono stati corretti per tener conto delle dimensioni finite del rivelatore e della sorgente. Per ridurre al minimo anche a grande distanza l'errore dovuto alle dimensioni finite del recipiente d'acqua, le misure venivano fatte mantenendo nel centro del cilindro d'acqua il punto di mezzo fra la sorgente e il rivelatore. Queste curve sono state utilizzate nel calcolo degli integrali del paragrafo precedente.

Un'altra applicazione è la seguente: da ciascuna curva è possibile calcolare la media quadratica $\overline{r^2}$ della distanza dalla sorgente a cui giungono i neutroni dei varii gruppi.

I valori ottenuti sono i seguenti:

Gruppo	$\overline{r^2}$ (in cm^2)
C	327
D	277
$A+B$	271
I	262

Le curve in funzione della distanza sono state rilevate con una notevole precisione; tuttavia per ragioni di intensità non ci è stato possibile spingere le misure sui gruppi D, $A + B$ ed I a grande distanza dalla sorgente in modo che si è dovuto estrapolare l'andamento di queste curve; ciò si può fare abbastanza bene poiché, a grande distanza, la legge di decremento diventa assai simile per tutti i gruppi, ma naturalmente l'estrapolazione può introdurre un certo errore nei valori degli $\overline{r^2}$.

È chiaro che, col procedere del rallentamento dei neutroni in seguito ad urti contro gli atomi di idrogeno, cresce il corrispondente valore di $\overline{r^2}$; l'ordine dei gruppi corrispondente a valori decrescenti di $\overline{r^2}$ deve dunque interpretarsi come ordine di energia crescente. Si può quindi dire, in accordo coi risultati ottenuti dall'assorbimento del boro, che i gruppi $C, D, A + B, I$ sono in ordine di energia crescente.

Tale considerazione si può precisare in modo da ricavare dalle differenze degli $\overline{r^2}$ dei vari gruppi, i rapporti delle loro energie, purché queste siano superiori ad 1 volt.

Si può dimostrare infatti, tenendo presente che l'angolo di deviazione di un neutrone in seguito ad un urto è sempre acuto, che

$$\log \frac{W'}{W} = \frac{\overline{r^2} - \overline{r'^2}}{6\lambda^2} \tag{3}$$

dove W e W' sono le energie dei due gruppi in esame ai quali corrispondono le distanze quadratiche medie $\overline{r^2}$ ed $\overline{r'^2}$.

Prendendo $\lambda = 0{,}65$ cm una differenza degli $\overline{r^2}$ di 2,5 cm² corrisponde ad un rapporto e nelle energie.

Dai dati precedenti risulterebbe dunque che le energie dei gruppi D, $A + B$, I sono nei rapporti 1 : 11 : 400.

Questi rapporti sono sensibilmente maggiori di quelli dedotti dall'assorbimento del boro. La differenza non è però forse tanto grande da non potersi imputare all'imprecisione della nostra valutazione basata sulla misura del logaritmo del rapporto delle energie.

La differenza $\overline{r_C^2} - \overline{r_D^2}$ è di 50 cm²; di questi 24 $(= 2\lambda_c^2 N)$ sono dovuti alla diffusione dei neutroni del gruppo C, gli altri 26 sono probabilmente in gran parte attribuibili al processo di rallentamento relativamente lungo che ha luogo nell'intervallo di energia tra 1 volt e l'energia di agitazione termica; tale differenza non è quindi incompatibile con l'energia di 3 volt trovata per il gruppo D dall'assorbimento del boro.

Istituto fisico della R. Università
Roma, 13 aprile 1936-XIV.

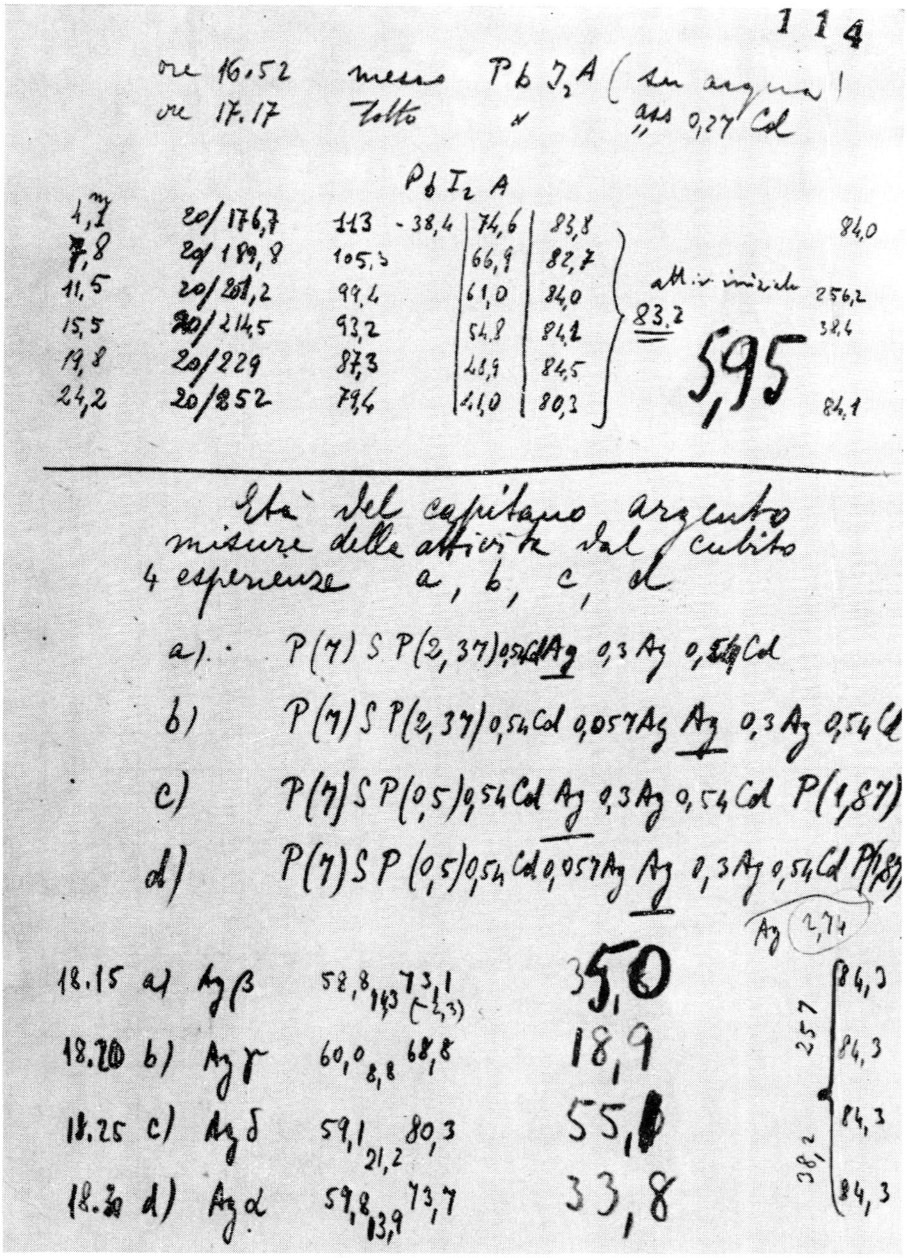

Laboratory notes referring to an experiment reported in Paper N° 117. The expressions " età del capitano " and " cubito " belong to the jargon and jokes current at that time at the Institute. The names "neutron age" or " Fermi age" were derived later from the first of these expressions.

117.

SULLE PROPRIETÀ DI DIFFUSIONE DEI NEUTRONI LENTI

E. AMALDI ed E. FERMI

« Ric. Scientifica », 7 (1), 393–395 (1936).

In questa lettera ci proponiamo di raccogliere e confrontare tra di loro i diversi dati sulle proprietà di diffusione dei neutroni lenti che abbiamo riportato in alcune lettere precedenti [1]; vengono inoltre rettificati i dati sui cammini liberi medii che sono stati nel frattempo da noi determinati con maggiore accuratezza.

Una esposizione dettagliata di tutte le nostre esperienze sui gruppi di neutroni lenti sarà pubblicata prossimamente.

I dati di questa lettera differiscono lievemente da quelli riportati in precedenza poiché questi ultimi erano sempre stati ridotti nell'ipotesi che la distribuzione angolare dei neutroni che escono dalla paraffina obbedisce alla legge del coseno. Ragioni teoriche [2] fanno invece ritenere che il numero di neutroni che escono dalla superficie della paraffina entro un angolo solido $d\omega$ inclinato dell'angolo θ rispetto alla normale, anziché proporzionale a $\cos\theta\, d\omega$ (legge del coseno) sia proporzionale a

$$(\cos\theta + \sqrt{3}\cos^2\theta)\, d\omega\,. \tag{1}$$

Si noti che questa legge di distribuzione angolare dà, rispetto alla legge del coseno, una maggiore concentrazione in avanti dei neutroni uscenti.

Oltre a ciò nell'eseguire una revisione di tutte le nostre misure, ci siamo accorti di un errore materiale incorso nella deduzione del cammino libero medio dei gruppi D ed A dalle proprietà di diffusione dei neutroni di questi gruppi entro la paraffina (vedi « Ric. Scientifica », 7 (1), 56 (1936)).

Da quelle misure, trattando in modo corretto i dati sperimentali, si deduce infatti un cammino libero medio di circa 1,1 cm anziché di 0,6 ÷ 0,7 cm. Ammettendo questo valore si dovrebbe trovare per i neutroni del gruppo C un cammino libero medio λ_c, 4 volte [3] più piccolo e cioè 0,275 cm invece del valore 0,15 cm dato di una lettera precedente [4].

(1) E. AMALDI, E. FERMI, « Ric. Scientifica », 6 (2), 344, 443 (1935), 7 (1), 56, 223, 310 (1936).

(2) E. FERMI, in corso di pubblicazione.

(3) E. FERMI, loc. cit.

(4) E. AMALDI, E. FERMI, « Ric. Scientifica », 7 (1), 223 (1936).

Un altro argomento che induce a ritenere troppo basso il valore di 0,15 cm per λ_c si deduce dalla discussione dei valori della albedo del gruppo *C* come sarà specificato in seguito.

Abbiamo per ciò ritenuto opportuno riprendere le misure del cammino libero medio λ_c con un metodo che consentisse di raggiungere una maggior precisione.

La maggiore difficoltà in esperienze di diffusione coi neutroni lenti è che per avere delle condizioni geometriche abbastanza ben definite, è necessario porre il rivelatore ad una distanza dalla paraffina in cui si producono i neutroni lenti, grande rispetto alle dimensioni della sorgente e del rivelatore. Se si soddisfa questa condizione si è costretti a lavorare con intensità assai piccole in modo che le misure risultano imprecise. Altrimenti è necessario apportare forti correzioni ai dati sperimentali per tener conto della non esponenzialità dell'attività del rivelatore in funzione dello spessore del diffusore.

Si comprende quindi la necessità di usare un rivelatore molto sensibile.

A tale scopo abbiamo costruito delle piccole camere di jonizzazione cilindriche del diametro interno di 3,5 cm e lunghe 10 cm piene di ossigeno alla pressione di 75 atmosfere. L'elemento da irradiare (Rh 0,125 g/cm^2) si trovava in forma di lamina cilindrica di 5 cm di altezza e 3,5 cm di diametro nell'interno della camera e costituiva esso stesso l'elettrodo a potenziale elevato; l'altro elettrodo, connesso all'elettrometro, era una bacchetta metallica sull'asse della camera. La parete era di acciaio dello spessore di 4 mm e non assorbiva sensibilmente i neutroni lenti i quali dall'esterno penetravano nella camera ed attivavano il rivelatore.

Questo dispositivo presenta il vantaggio rispetto ai metodi da noi precedentemente usati, che a causa della elevata pressione, quasi tutta l'energia dei raggi β viene utilizzata entro la camera; inoltre l'elettrodo ha una notevole estensione e le condizioni geometriche sono assai buone. Si può infatti con questo dispositivo raggiungere quasi la sensibilità di osservazioni fatte coi contatori e si ha il vantaggio della maggiore stabilità caratteristica delle camere di jonizzazione.

La camera connessa all'elettrometro veniva irradiata per 2 minuti e quindi, dopo aver allontanata la sorgente, si effettuava la lettura del numero di divisioni percorse in 2 minuti a partire da 15 secondi dopo la fine dell'irraggiamento.

Usando lo stesso dispositivo già descritto (5) abbiamo fatto due serie di misure; nella prima la distanza fra la superfice della paraffina e l'asse della camera di jonizzazione era di 10 cm e nella seconda di 20 cm; in entrambe i diffusori di paraffina (densità 0,9) venivano posti nel punto di mezzo fra la sorgente ed il rivelatore.

Nella prima serie di misure, pur essendo assai grande la precisione delle singole letture (2 %) vi era un notevole fattore di incertezza dovuto alle forti correzioni da apportarsi ai risultati sperimentali a causa della imperfezione della geometria.

(5) E. Amaldi, E. Fermi, « Ric. Scientifica », 7 (1), 223 (1936).

La seconda serie di misure era fatta in condizioni geometriche assai migliori, ma naturalmente la precisione delle letture era inferiore.

Correggendo i dati di entrambe le misure per tener conto della geometria, abbiamo ottenuto dalla prima serie $\lambda_c = 0{,}34$ cm e dalla seconda serie $\lambda_c = 0{,}27$ cm. Da queste misure risulterebbe quindi come valore più probabile $\lambda_c = 0{,}3$ cm.

Più che il risultato di questa misura specifica, ha interesse considerare nel loro insieme tutte le grandezze da noi misurate, nelle quali interviene il cammino libero medio λ_c dei neutroni termici.

Si può dimostrare infatti che quattro diverse grandezze da noi misurate si possono esprimere, tenendo conto dei risultati della teoria della diffusione dei neutroni lenti [(6)], in funzione delle due sole grandezze λ_c ed N (numero di cammini liberi percorsi in media da un neutrone termico prima di venir catturato nella paraffina). Esse sono:

1° λ_c che come abbiamo visto sopra si può misurare direttamente;

2° λ_D, cammino libero medio dei neutroni D ed A che come si è accennato, è legato teoricamente a λ_c dalla relazione

$$\lambda_D = 4\lambda_c \, ;$$

3° l'albedo β [(7)] che si dimostra essere legata al numero N dalla relazione [(8)]

$$\beta = 1 - \frac{2}{\sqrt{N}} \, ;$$

4° $\sqrt{D\tau}$ (D coefficente di diffusione dei neutroni C e τ loro vita media nella paraffina). $\sqrt{D\tau}$ è legato a λ_c ed N dalla relazione [(9)]

$$\sqrt{D\tau} = \lambda_c \sqrt{\frac{N}{3}} \, .$$

Nella tabella i valori di queste quattro grandezze ottenuti dalle misure dirette (seconda colonna) vengono confrontati coi valori calcolati con le formule indicate assumendo $\lambda_c = 0{,}27$ cm ed $N = 150$.

Come si vede l'accordo fra i valori calcolati e quelli misurati è assai buono.

Assumendo $\lambda_D = 1{,}1$ cm viene anche notevolmente migliorato l'accordo fra i rapporti di energia dei varii gruppi calcolati col metodo esposto nella nostra lettera « Ric. Scientifica », 7 (1), 310 (1936), oppure col metodo dedotto dall'assorbimento del boro [(10)].

(6) E. FERMI, loc. cit.

(7) E. FERMI, E. AMALDI, « Ric. Scientifica », *6* (2), 443 (1935).

(8) E. FERMI, loc. cit.

(9) E. AMALDI, E. FERMI, « Ric. Scientifica », *7* (1), 56 (1936).

(10) O. R. FRISCH, G. PLACZEK, « Nature » (London), *137*, 357 (1936); D. F. WEEKES, M. S. LIVINGSTON, H. A. BETHE, « Phys. Rev. », *49*, 471 (1936).

	Misurato	Calcolato
λ_c	0,3	0,27
λ_D	1,1	1,08
β	0,83	0,837
$\sqrt{D\tau}$	2,1	1,91

Le larghezze dei gruppi nella stessa lettera risultano col nuovo valore del cammino libero medio un poco minori.

Infine notiamo che assumendo $\lambda_c = 0{,}27$ ed $N = 150$ risultano i seguenti dati per i neutroni termici nella paraffina:

sezione d'urto elastica coll'idrogeno	$\sigma_e = 48 \cdot 10^{-24}\ \text{cm}^2$
sezione d'urto con cattura	$\sigma_c = 0{,}32 \cdot 10^{-24}\ \text{cm}^2$
vita media	$\tau = 1{,}6 \cdot 10^{-4}\ \text{s}.$

Istituto fisico della R. Università
Roma, 14 maggio 1936-XIV.

118 *a*.

SOPRA L'ASSORBIMENTO E LA DIFFUSIONE DEI NEUTRONI LENTI

E. AMALDI ed E. FERMI

« Ric. Scientifica », 7 (1), 454-503 (1936).

RIASSUNTO. — Lo studio delle proprietà di assorbimento dei neutroni lenti ha permesso di stabilire l'esistenza di bande di assorbimento corrispondenti ad intervalli di energia abbastanza delimitati, che permettono di isolare dei gruppi di neutroni relativamente omogenei. In questo lavoro vengono esposte alcune esperienze e considerazioni intese a caratterizzare le proprietà di assorbimento e diffusione dei varii gruppi. — *Sommario*: 1. Introduzione. - 2. Misure e loro riduzione. - 3. Assorbimenti selettivi. - 4. Gruppi di neutroni lenti. - 5. Misure sistematiche di assorbimento. - 6. Albedo. - 7. Diffusione dei neutroni di un gruppo. - 8. Cammino libero medio dei neutroni termici. - 9. Gruppi in funzione della distanza dalla sorgente, rapporti di energia. - 10. Passaggio dei neutroni da un gruppo ad un altro. - 11. Comportamento dei gruppi in vicinanza del bordo di paraffina. - 12. Estensione delle bande di energia corrispondenti ai gruppi. - 13. Riassunto e discussione.

1. INTRODUZIONE.

Lo scopo di questo lavoro è di esporre sistematicamente le ricerche da noi eseguite sulle proprietà di assorbimento e di diffusione dei neutroni lenti [1].

È noto che le teorie più semplici [2] sulla probabilità di cattura di un neutrone da parte di un nucleo conducevano a prevedere che la sezione di cattura fosse, per piccole velocità del neutrone, inversamente proporzionale alla velocità; la costante di proporzionalità avrebbe potuto invece variare entro larghissimi limiti da elemento a elemento. Comunque, se così fosse, spessori di due diversi elementi inversamente proporzionali alle dette costanti, dovrebbero essere perfettamente equivalenti come assorbitori dei neutroni lenti, indipendentemente dalla velocità di questi ultimi, e quindi anche dalla sostanza usata come rivelatore dei neutroni lenti.

Che le cose non andassero in modo così semplice è stato osservato ben presto da varii Autori [3] i quali hanno dimostrato che di regola l'assorbimento di un elemento è più intenso qualora i neutroni lenti si rivelino a mezzo dell'attività indotta nell'elemento stesso.

A simili conclusioni conducono anche le esperienze termiche di Moon e Tillman ed altri [4], dalle quali risulta che è diversa la sensibilità di diversi rivelatori a variazioni di temperatura della paraffina. Infine, più recentemente, Rasetti, Segrè, Fink, Dunning, Pegram [5] hanno misurato con un dispositivo meccanico la dipendenza del coefficiente di assorbimento dei neutroni lenti dalla velocità trovando che non vale, almeno in alcuni casi, la legge di proporzionalità inversa alla velocità.

In uno studio sistematico di questi fenomeni abbiamo riscontrato che le selettività di assorbimento dei neutroni lenti vengono accentuate qualora essi vengano filtrati con uno strato di cadmio abbastanza spesso [6]; questo fatto, osservato indipendentemente da Szilard [7], fa sospettare l'esistenza di bande di assorbimento relativamente ristrette caratteristiche dei varii elementi.

Queste circostanze sono state da noi precisate mediante una analisi del complesso dei neutroni lenti in gruppi approssimativamente omogenei rispetto alle loro proprietà di assorbimento [8]; tale analisi è stata eseguita decomponendo le curve di assorbimento in curve esponenziali.

Seguendo questa via abbiamo cercato di caratterizzare il più possibile questi gruppi di neutroni lenti particolarmente rispetto alle loro proprietà di diffusione nelle sostanze idrogenate.

Nel paragrafo 2 vengono descritti i criteri seguiti nella esecuzione delle esperienze e nella riduzione delle misure. I paragrafi 3, 4, 5 contengono una raccolta di dati sistematici sulle proprietà di assorbimento e sulla intensità con cui i gruppi attivano i diversi rivelatori. Nel paragrafo 6 viene determinato il coefficiente di riflessione della paraffina per neutroni lenti di varii gruppi (albedo); i paragrafi 7 ed 8 sono dedicati alla determinazione dei cammini liberi medii e del numero di urti che un neutrone subisce entro un gruppo. Nel paragrafo 9 si studia come l'intensità dei varii gruppi dipende dalla distanza dalla sorgente e se ne ricavano la successione dei gruppi in ordine di energia, nonché una valutazione dei loro rapporti di energia. Nei paragrafi 10 e 11 sono descritte altre esperienze di diffusione; nel paragrafo 12 viene descritto un metodo per la determinazione della larghezza delle bande di energia corrispondenti ai varii gruppi: infine nel paragrafo 13 i dati precedenti vengono raccolti e discussi insieme.

Nel corso del presente lavoro verranno spesso usate relazioni dedotte teoricamente da uno di noi e che fanno oggetto di un altro lavoro in corso di pubblicazioni [9] che verrà nel seguito indicato con F.

2. Misure e loro riduzione.

Come sorgente di neutroni ci siamo sempre serviti di ampolline di vetro di circa 4 mm di diametro per 15 di lunghezza contenenti emanazione (fino a 800 mC) e polvere di berillio. Dato che le nostre ricerche si sono estese per parecchi mesi è stato evidentemente necessario fissare delle norme in modo da rendere confrontabili fra di loro risultati ottenuti con diverse sorgenti a notevole distanza di tempo.

Passiamo qui a descrivere i criteri seguiti a questo scopo.

Le misure di attività sono state eseguite con una camera di jonizzazione rappresentata schematicamente nella fig. 1, dove è indicata anche la scala. La camera era riempita di anidride carbonica a tre atmosfere, ed era chiusa superiormente con una finestra di alluminio di 0,1 mm di spessore e del diametro di 7 cm. Siccome la finestra di alluminio è fortemente incurvata per la pressione, al di sopra di questa vi è un foglio di cellofan che ha il duplice scopo di proteggere la camera contro eventuali contaminazioni e di fornire una superficie piana su cui appoggiare i preparati.

La jonizzazione veniva misurata con un elettrometro Edelmann, il cui filo era proiettato su di una scala; la sensibilità e l'ingrandimento variavano da 5 a 250 divisioni per volt.

Per tarare le letture abbiamo usato un preparato di ossido di uranio ripartito su di una superficie di 5×5 cm^2 racchiuso in un pacchetto di alluminio. Indicheremo d'ora in poi questo preparato e la sua attività con U.

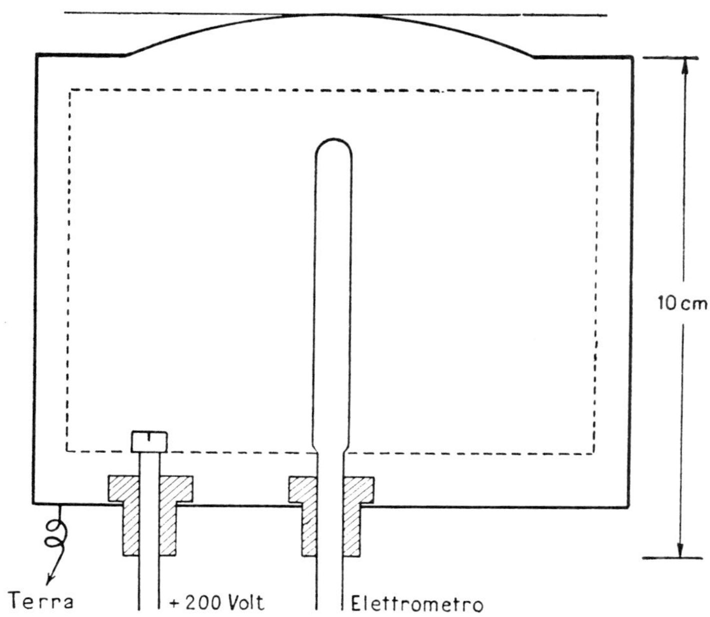

Fig. 1.

La jonizzazione prodotta nella nostra camera dall'U equivale a quella che si ottiene appoggiando sulla camera una vaschetta di alluminio di 0,1 mm di spessore, di base 5×5 cm^2, contenente 0,96 g di nitrato di uranile ($UO_2(NO_3)_2 + 3\,H_2O$) sciolto in 25 cm^3 di acqua. Correggendo l'attività di questa soluzione per l'assorbimento nella soluzione stessa e nella vaschetta si trova che essa equivale a 0,073 g di uranio elemento, cioè a 920 disintegrazioni al secondo.

Per rendere più rapide le misure e poter usare diverse sensibilità dell'elettrometro e diversi intervalli della scala, ci siamo costruiti un grafico a doppia entrata che ci permetteva di ricavare direttamente, data la sensibilità dell'elettrometro ed il numero di divisioni su cui era stata eseguita la lettura, il tempo che avrebbe impiegato l'U a dare la stessa jonizzazione. In tal modo veniva eseguita automaticamente la correzione per la non linearità dello strumento e per la dipendenza della capacità dell'elettrometro dalla sensibilità.

Naturalmente le letture venivano corrette per l'effetto di zero che rappresentava il 15 °/₀ dell'U.

L'attività di un dato rivelatore posto in una determinata posizione rispetto alla sorgente ed agli altri oggetti che la circondano, è evidentemente proporzionale al numero di neutroni emessi dalla sorgente. Per avere dati confrontabili le misure di attività debbono dunque venir sempre divise per la intensità della sorgente misurata in neutroni (*intensità neutronica*). Per determinarla non è sufficiente ammetterla proporzionale alla quantità di emanazione contenuta nella sorgente, quale può misurarsi, per esempio dalla attività γ; infatti lievi differenze nella grana del berillio o nella confezione della sorgente, possono produrre delle variazioni abbastanza forti nel numero dei neutroni emessi. Conviene perciò misurare la intensità neutronica riferendosi direttamente alla radioattività indotta in un opportuno rivelatore; le norme che abbiamo seguito sono le seguenti:

una lastra di rodio delle dimensioni di $5 \times 5{,}6 \text{ cm}^2$ e del peso di 10,15 g, veniva attivata appoggiandola nel centro della base superiore di un cilindro di paraffina di

(1) 25 cm di diametro per 15 cm di altezza;

al di sopra della lastra di rodio veniva appoggiato un secondo cilindro di paraffina di 13 cm di diametro e di 10 cm di altezza; la sorgente era posta sull'asse del cilindro inferiore 3 cm al di sotto della base superiore.

Indichiamo in S l'attività iniziale della vita di 44 s del rodio che, si ottiene in queste condizioni dopo irraggiamento infinito. Chiameremo intensità neutronica I il rapporto fra questa attività iniziale S e l'U:

$$\text{I} = \frac{\text{S}}{\text{U}} \cdot \tag{2}$$

Si noti che questa definizione dipende, se pur lievemente, dalla camera di jonizzazione usata. La sorgente più intensa di cui abbiamo potuto disporre ha avuto una intensità neutronica di 144,9. L'intensità neutronica 1 corrisponde ad una sorgente contenente da 5 a 6 mC di emanazione.

Siamo ora in grado di definire che cosa si intende per *attivabilità* A di un dato rivelatore (riferendoci, quando occorra, ad una sola delle sue vite medie) posto in una determinata posizione rispetto alla sorgente ed all'ambiente circostante.

Sia a l'attività iniziale di un rivelatore irraggiato infinito tempo; definiamo come attivabilità A del detto rivelatore in una certa posizione la espressione

$$\text{A} = 1000 \frac{a}{\text{S}} = 1000 \frac{a}{\text{I} \times \text{U}} \cdot \tag{3}$$

In pratica naturalmente si determinano le attivabilità con l'ultima formula; l'intensità neutronica viene determinata per ogni sorgente una volta per tutte con una notevole accuratezza e quindi calcolata istante per istante a mezzo del periodo di decadimento dell'emanazione.

L'attività iniziale a viene misurata irradiando per un tempo finito e riducendosi a tempo infinito con ovvie formule.

Nel caso di elementi a vita breve conviene fissare delle norme per la misura dell'attività iniziale in modo da rendersi il più possibile indipendenti, almeno per le misure relative, da eventuali errori nel valore della vita media.

Così per il rodio (44 s) che per l'argento (22 s), che sono stati usati come rivelatori in moltissime esperienze, abbiamo effettuato irradiazioni di 1 min; nel caso dell'argento le letture venivano fatte da 20 s a 80 s dopo la fine dell'irraggiamento e l'attività così misurata moltiplicata per un opportuno fattore, dava l'attività iniziale. In realtà una misura di questo genere non dà soltanto la vita di 22 s, poiché vi è una perturbazione dovuta alla vita di 2,3 min; questa perturbazione non è tuttavia grande e l'abbiamo di solito trascurata (vedi anche § 5).

Per il rodio abbiamo sempre irradiato 1 min ed abbiamo eseguito la lettura fra circa 30 s e 60 s dopo la fine dell'irraggiamento; l'attività iniziale veniva calcolata rapidamente a mezzo di un nomogramma. Anche in questo caso abbiamo trascurato la piccola perturbazione dovuta al periodo di 4,2 min (vedi § 5).

Nel corso di questo lavoro abbiamo eseguito un gran numero di misure di assorbimento dei neutroni lenti in varie sostanze. I criteri seguiti per queste misure sono stati i seguenti.

Le misure sono state fatte all'esterno della paraffina, poiché altrimenti una elevata frazione dei neutroni passa più volte attraverso all'assorbitore (vedi § 6). La sorgente di Em + Be era normalmente inserita nel cilindro di paraffina (1), 3 cm sotto il centro della base superiore sulla quale venivano appoggiati i varii rivelatori. Nella maggior parte dei casi essi erano quadrati di circa 5 cm di lato.

Chiamiamo *attivabilità normale* A_n di un rivelatore la sua attivabilità in queste condizioni. Essa fornisce una indicazione sulla efficienza dei varii rivelatori.

A titolo di esempio nella Tabella I sono riportate le attivabilità normali di alcuni dei rivelatori da noi usati.

Per le misure di assorbimento gli assorbitori venivano inseriti in strati il più possibile sottili fra il rivelatore e la paraffina.

Per correggere l'errore dovuto all'altezza h del rivelatore sopra la superficie della paraffina, è stato rilevato come varia l'attività di un rivelatore di rodio in funzione di h; la regola pratica così dedotta per ridurre le attivabilità ad $h = 0$, consiste nel dividere i risultati della misura per $1 - (h/7)$, dove h è l'altezza in cm; questa regola è sufficientemente corretta per h fino a poco oltre 1 cm.

A rigor di termini la correzione per l'altezza dovrebbe prendersi differente a seconda dell'assorbitore interposto ed anche a seconda del gruppo di neutroni con cui si lavora. Per il momento abbiamo trascurato questi raffinamenti.

Anche se i neutroni che escono dalla paraffina fossero omogenei le curve di assorbimento ottenute nelle condizioni che abbiamo descritto non sarebbero esponenziali, poiché i neutroni che escono con diverse inclinazioni θ attraversano differenti spessori dell'assorbitore. Nel calcolo dei coefficienti di assorbimento dai dati sperimentali si deve naturalmente tener conto di questa circostanza; per molto tempo abbiamo calcolato questa correzione ammettendo che i neutroni escano dalla paraffina con la legge del coseno.

TABELLA I.

Sostanza	Periodo	Peso (in g)	Area (in cm²)	A_n
MnO_2	2,5 h	30	26	30
Cu	5 min	13	38	2
Ga	20 min	2,6	7	1,3
As	26 h	7	20	24
NaBr	18 min	26	25	9,2
Rh	44 s	10,15	28	282
Ag	22 s	8	27	135
In	54 min	0,76	9	21,2
In	16 s	0,76	9	14,2
I	25 min	25	25	16,9
Ir	19 h	14,5	25	78
Au	2,7 d	6,6	25	26

In queste condizioni la curva di assorbimento con rilevatore sottile, in funzione di $K\delta$ (K = coefficiente di assorbimento dei neutroni e δ = spessore dell'assorbitore) anziché essere l'esponenziale $e^{-K\delta}$ (curva *a* della figura 2), è la funzione

$$b(K\delta) = \int_0^1 e^{-\frac{K\delta}{x}} dx \tag{4}$$

rappresentata dalla curva *b* della fig. 2.

Riteniamo oggi che per i neutroni del gruppo *C* (cfr. *F.*, § 6) sia più plausibile ammettere che il numero di neutroni uscenti per unità di angolo solido anziché essere proporzionale a cos θ, sia approssimativamente proporzionale a

$$\cos\vartheta + \sqrt{3}\cos^2\vartheta\,. \tag{5}$$

Se si prende questa distribuzione angolare dei neutroni, la curva di assorbimento in funzione di $K\delta$ è data dalla funzione

$$c(K\delta) = \frac{2}{2+\sqrt{3}} \int_0^1 e^{-\frac{K\delta}{x}} (1 + \sqrt{3}\,x)\, dx \tag{6}$$

rappresentata dalla curva *c* della fig. 2.

Questa rappresenta probabilmente abbastanza bene l'andamento dell'assorbimento dei neutroni C; per i neutroni degli altri gruppi è probabile che la curva di assorbimento sia intermedia fra *b* e *c*; comunque le riduzioni sono state fatte servendoci nella curva *c*.

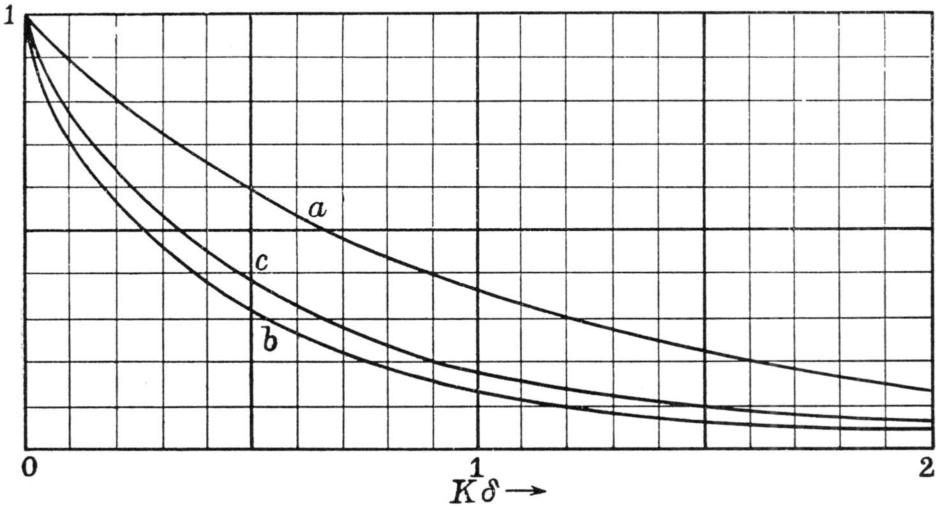

Fig. 2.

Nel caso che il rivelatore sia spesso, e cioè che l'assorbimento dei neutroni entro lo spessore utile del rivelatore sia considerevole, i dati ricavati dalla curva *c* vanno ulteriormente corretti.

Le differenze tra i coefficienti di assorbimento dati in questo lavoro e nelle comunicazioni preliminari sono dovute in gran parte all'aver effettuato le riduzioni con la curva *c* anziché con la curva *b* della fig. 2.

3. Assorbimenti selettivi.

Abbiamo già ricordato che, come è stato osservato da vari Autori, di regola l'assorbimento dei neutroni lenti da parte di un elemento è maggiore quando si usa un rilevatore fatto dell'elemento stesso anziché di un altro elemento.

Così, per esempio, la Tabella II indica le percentuali di attività trasmesse da assorbitori di 0,36 g/cm^2 di rodio e 0,96 g/cm^2 di argento, usando come rivelatori una lastra di rodio 0,36 g/cm^2 oppure una lastra di argento di 0,30 g/cm^2. Dati di questo genere su un maggiore numero di elementi sono stati pubblicati, oltre che da altri Autori, anche da noi [10].

Tabella II.

Assorbitore	Rh	Ag
Rivelatore		
Rh	54	67
Ag	68	45

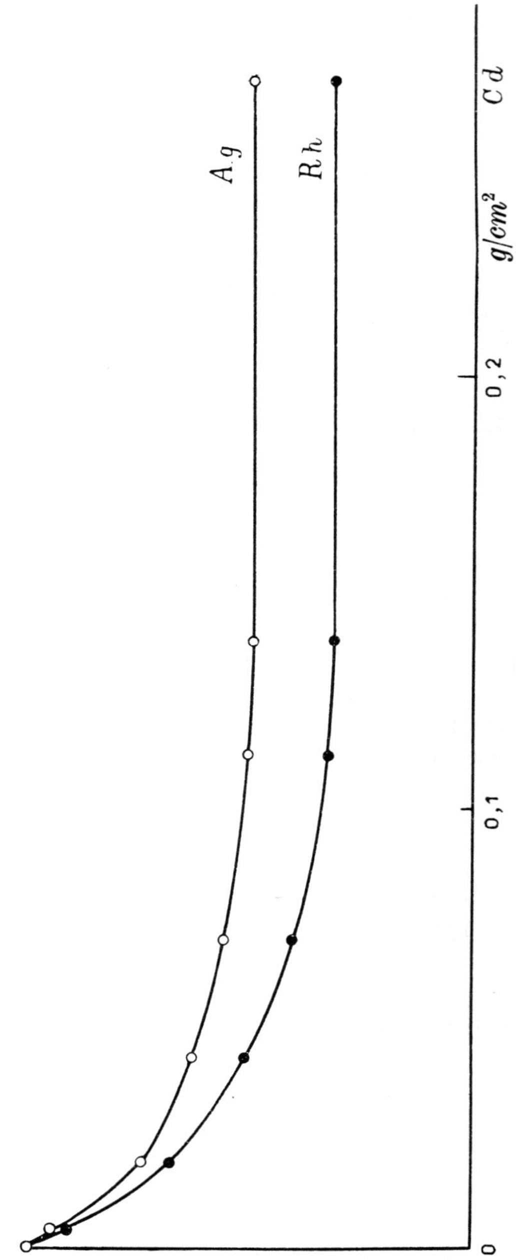

Fig. 3.

Questi fatti inducono a pensare che vi siano delle selettività negli assorbimenti dei neutroni.

Il modo migliore per studiarle consiste nel rilevare ed analizzare delle curve di assorbimento. Nella fig. 3 sono riportate le curve di assorbimento del cadmio con rivelatori rodio (0,36 g/cm²) e argento (0,057 g/cm²).

È evidente da queste curve che le radiazioni che attivano entrambi questi rivelatori non sono omogenee, ma consistono di una frazione assai assorbita dal cadmio ed una frazione pochissimo assorbita. Tenendo conto del fatto che col nostro dispositivo l'assorbitore viene attraversato obliquamente dai neutroni, come è precisato al paragrafo 2, si può calcolare il coefficiente di assorbimento della frazione più assorbibile nel cadmio da ciascuna delle due curve.

Si ottiene così, dalla curva con rivelatore argento K = 16 cm²/g, e dalla curva con rivelatore rodio K = 13,5 cm²/g. Il coefficiente di assorbimento della frazione dura è certamente qualche centinaio di volte minore; così, per esempio, nel caso del rivelatore rodio si trova per essa K = 0,05 cm²/g.

Dall'esame delle curve della fig. 3 si vede che nel nostro rivelatore di argento l'attività è dovuta per metà alla componente fortemente assorbita dal cadmio e metà alla componente poco assorbita. Per il nostro rivelatore di rodio invece il 72 % dell'attività è dovuta alla componente fortemente assorbita dal cadmio ed il resto alla componente poco assorbita.

L'assorbimento del cadmio ci permette dunque di dividere i neutroni lenti che escono da un blocco di paraffina, contenente una sorgente di neutroni, in due frazioni: di queste quella fortemente assorbita dal cadmio sarà indicata d'ora in poi col nome di gruppo *C*. Nel paragrafo 5 ne discuteremo dettagliatamente le proprietà di assorbimento.

Passiamo ora a studiare le proprietà della radiazione che filtra attraverso ad uno spessore di cadmio sufficiente per assorbire totalmente il gruppo *C*.

A tale scopo ripetiamo sulla radiazione filtrata da 0,27 g/cm² di cadmio alcune esperienze di assorbimento usando il rodio e l'argento sia come assorbitori che come rivelatori [11]. La Tabella III, analoga alla II, viene data come esempio dei risultati di una simile ricerca: gli assorbitori ed i rivelatori usati sono gli stessi della Tabella II.

TABELLA III.

Assorbitore	Rh	Ag
Rivelatore		
Rh	33	84
Ag	88	29

Il confronto delle Tabelle II e III mostra che le selettività sono assai più accentuate per la radiazione filtrata dal cadmio, ciò che indica che la componente poco assorbita da questo elemento non è omogenea.

Allo scopo di investigare ulteriormente le proprietà della radiazione filtrata dal cadmio, abbiamo rilevato le curve di assorbimento dell'argento con rivelatore argento (2,2 g/38 cm^2) sia sulla radiazione non filtrata che su quella filtrata da 0,27 g/cm^2 di cadmio. Nella fig. 4 sono date queste curve, riportando in ascissa gli spessori dell'assorbitore di argento ed in ordinate l'attivabilità del rivelatore. La curva superiore è relativa alla radiazione non filtrata e quella inferiore alla radiazione filtrata da 0,27 g/cm^2 di cadmio.

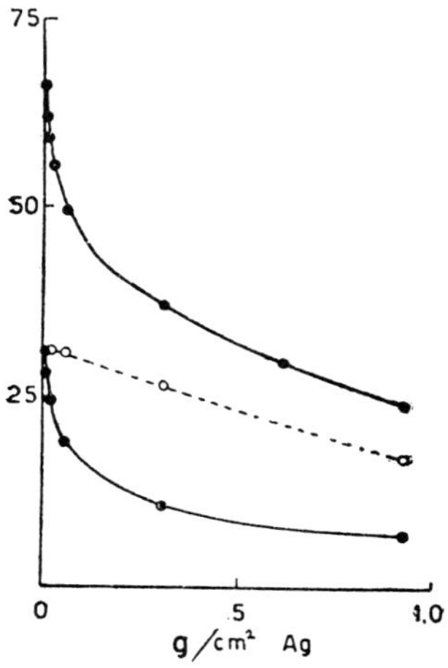

Fig. 4.

In entrambe le curve si riconosce che la radiazione che attiva l'argento contiene una componente fortemente assorbita nell'elemento stesso, la quale si presenta con uguale intensità nella radiazione filtrata e in quella non filtrata. Ciò significa che questa radiazione è assai poco assorbita dal cadmio. Ad essa abbiamo dato il nome di gruppo *A*. Il coefficiente di assorbimento dei neutroni del gruppo *A* nell'argento è K = 20 cm^2/g.

Curve di aspetto simile a quelle date nella fig. 4 si trovano anche usando un rivelatore di argento più spesso; in questo caso però la percentuale di attività dovuta alla radiazione molto assorbita nell'argento è naturalmente minore, poiché essa si satura in meno di 0,1 mm di spessore del rivelatore.

La curva tratteggiata della fig. 4 è la differenza delle altre due curve; essa rappresenta dunque il comportamento di quella parte dell'attività dell'argento che è assorbita da 0,27 g/cm^2 di cadmio. Essa è dunque la curva di assorbimento dell'argento con rivelatore argento dovuta ai soli neutroni del gruppo *C*. Da essa si ricava come valore del coefficiente di assorbimento dell'argento per i neutroni C , K = 0,3 cm^2/g.

4. Gruppi di neutroni lenti.

Abbiamo visto, sugli esempi del paragrafo precedente, come l'analisi delle curve di assorbimento permetta di classificare i neutroni lenti in componenti diversamente assorbite da vari elementi.

È ben noto che l'analisi in componenti eponenziali delle curve di assorbimento può essere ritenuta attendibile solo quando la precisione delle misure singole è assai elevata, ed inoltre la disposizione geometrica dell'assorbitore e del rivelatore, rispetto alla sorgente, corrisponde a condizioni esattamente calcolabili.

Nel nostro caso queste condizioni non sono esattamente verificate, e ci manca quindi la possibilità di garantirci il modo preciso della omogeneità di una componente della radiazione.

In quel che segue ci siamo quindi limitati a caratterizzare dei gruppi di componenti approssimativamente omogenei. Le differenze di comportamento fra gruppo e gruppo sono così marcate da poter assegnare a ciascuno di essi una individualità, nonostante le piccole inomogeneità interne.

L'analisi in gruppi è resa praticamente eseguibile dalle seguenti circostanze:

a) per alcuni assorbitori esistono delle fortissime differenze nel coefficiente di assorbimento tra gruppo e gruppo;

b) è possibile usare svariati assorbitori e rivelatori che si comportano in modo molto differente;

c) è possibile filtrare i neutroni lenti con assorbitori che ne riducono la inomogeneità.

L'ipotesi più spontanea sulla natura fisica delle differenze fra gruppo e gruppo consiste nell'attribuirle a differenze di velocità. In un blocco di paraffina i neutroni veloci che escono dalla sorgente subiscono un progressivo rallentamento, in seguito ad urti successivi contro gli atomi di idrogeno, fino a raggiungere l'energia di agitazione termica; nella paraffina sono quindi presenti in ogni istante neutroni di tutte le velocità, comprese tra l'energia di agitazione termica e l'energia con cui essi vengono emessi dalla sorgente. Parimenti neutroni di tutte queste velocità diffondono all'esterno di un blocco di paraffina. Si può quindi pensare che i vari gruppi corrispondano a diversi intervalli di energia dei neutroni. Le selittività di assorbimento debbono allora interpretarsi come dovute ad irregolari variazioni del coefficiente di assorbimento in funzione della velocità dei neutroni.

La plausibilità di una dipendenza irregolare del coefficiente di assorbimento dalla velocità dei neutroni è stata mostrata da recenti considerazioni di Bohr e di Breit e Wigner [12]. Torneremo in seguito a discutere l'origine fisica dei gruppi dopo avere esposto, oltre alle esperienze di assorbimento, anche il complesso delle proprietà di diffusione dei vari gruppi.

Già nel paragrafo precedente abbiamo indicato col nome di gruppo *C* le radiazioni che vengono fortemente assorbite dal cadmio (coefficiente di assorbimento K $= 13{,}5 \div 16$ cm^2/g) e col nome di gruppo *A* quelle fortemente assorbite dall'argento (K $= 20$ cm^2/g). Abbiamo inoltre veduto che solo il 50 % dell'attività dei nostri rivelatori di argento (0,057 g/cm^2) è dovuto al gruppo *C*; del rimanente 50 % circa la metà è dovuta al gruppo *A*, mentre il resto è dovuto ad una radiazione che è poco assorbita sia dal cadmio che dall'argento (vedi fig. 4); questa radiazione è stata da noi chiamata gruppo *B*; essa è però quasi certamente complessa; fino ad ora non abbiamo potuto analizzarla in componenti pressoché omogenee, sopratutto a causa del fatto che non abbiamo trovato fra gli elementi da noi provati nessun assorbitore con elevato coefficiente di assorbimento per il gruppo *B*.

Nel paragrafo precedente abbiamo mostrato che il 72 % dell'attività del nostro rivelatore di rodio (0,36 g/cm^2) è dovuto al gruppo *C*; il rimanente 28 % costituisce un nuovo gruppo, che chiameremo D, il quale è assai poco assorbito dal cadmio.

Dallo studio di un rivelatore di PbI_2 (19 g/25 cm^2, attivabilità normale 11,4) abbiamo dedotto che solo il 25 % della sua attività è dovuto al gruppo *C* e meno del 10 % è dovuto al gruppo *A*; il rimanente è dovuto ad una componente assai poco assorbita in tutti gli elementi fino ad ora provati, salvo che nello jodio; abbiamo quindi indicato col nome di gruppo *I* questa radiazione che produce circa il 70 % dell'attività del nostro rivelatore.

Frisch, Hevesy e McKay [13] hanno studiato l'assorbimento dell'oro con rivelatore oro sulla radiazione filtrata da cadmio ed hanno trovato una curva assai simile a quelle dell'argento con rivelatore argento da noi date nella fig. 4. Noi abbiamo cercato di analizzare l'attività di un rivelatore di oro di 1,75 g/25 cm^2 nei gruppi di neutroni fino ad ora caratterizzati; si è così riconosciuto che l'attività dell'oro è dovuta in parte ai gruppi *C*, *A* e *B*; l'elevato coefficiente di assorbimento dell'oro con rivelatore oro è però dovuto ad una radiazione distinta da tutte le precedenti.

5. Misure sistematiche di assorbimento.

Gruppo C. – Tra tutti i gruppi quello meglio conosciuto è il gruppo *C*, poiché esso determina in quasi tutti i rivelatori elevate percentuali dell'attivazione; si hanno molte ragioni per ritenere (cfr. § 13) che questo gruppo sia costituito, in gran parte almeno, da neutroni aventi l'energia di agitazione termica. Per questa ragione lo indicheremo spesso anche come gruppo termico.

Lo studio del gruppo *C* è reso particolarmente agevole dal fatto che esiste un assorbitore praticamente esclusivo di questo gruppo quale è il cadmio; uno spessore di $0{,}3 \div 0{,}4$ g/cm^2 di questo elemento assorbe in modo quasi completo il gruppo *C* senza assorbire sensibilmente gli altri gruppi. Per determinare quindi la porzione di attività di un rivelatore dovuta al gruppo *C* basta fare la differenza fra le attività che si osservano senza e con un filtro di cadmio dello spessore indicato.

Nella Tabella IV sono raccolti i coefficienti di assorbimento in cm^2/g per i neutroni del gruppo *C*, osservati con vari rivelatori (facendo sempre la differenza fra le attività senza e con un filtro di cadmio) e vari assorbitori. Si noti che i coefficienti di assorbimento dello stesso elemento, misurati con diversi rivelatori, sono approssimativamente costanti. Le differenze sono certamente in parte dovute ad errori di misura; riteniamo tuttavia che essi non siano totalmente imputabili a questa causa. Comunque, se si tengono presenti le grandi differenze dei coefficienti di assorbimento di un elemento per i neutroni di diversi gruppi, si riconosce che il gruppo *C* ha effettivamente una notevole omogeneità interna.

Tabella IV.

Gruppo C.

Assorbitore	Rh	Ag	Cd	In	Ir	Au	Hg
Mn 2,5 h . .	0,9	0,3	—	—	—	—	—
Rh 44 s . .	0,6	0,2	13,5	0,6	1,0	0,2	0,7
Ag 22 s . .	0,7	0,3	16	—	0,9	0,25	0,9
In 54 min . .	0,6	0,2	14	0,9	1,0	0,2	0,6
Ir 19 h . . .	1,0	0,3	—	—	—	—	0,6

Gruppo D. – Il gruppo *D* si osserva comodamente usando come rivelatori il rodio oppure l'indio; entrambi questi elementi si prestano ad eseguire misure precise.

Il fatto che il gruppo *D* si possa caratterizzare usando sia l'uno che l'altro di questi due rivelatori, porta a concludere che le corrispondenti bande di assorbimento sono quasi coincidenti; naturalmente non vi sono ragioni per credere che la coincidenza debba essere completa. Vedremo nel paragrafo 9 che fra tutti i gruppi da noi studiati, il gruppo *D* è quello di minima energia dopo il gruppo *C*.

Abbiamo già detto che il 28 °/o dell'attività dei nostri rivelatori di rodio di 0,36 g/cm² è dovuto al gruppo *D*. In un rivelatore di indio di 0,76 g/9 cm² il 42 °/o dell'attività della vita di 54 min è dovuto al gruppo *D*; il rimanente 58 °/o è dovuto al gruppo *C*.

Per studiare il gruppo *D* basta quindi usare uno di quei rivelatori filtrando la radiazione con 0,3 ÷ 0,4 g/cm² di cadmio in modo da eliminare il gruppo *C*.

Nella Tabella V sono riportati alcuni coefficienti di assorbimento in cm²/g dei neutroni del gruppo *D*.

TABELLA V.

Gruppo D.

Assorbitore	B	Rh	Ag	Cd	In	Ir	Au	Hg
Rivelatore								
Rh 44 s	4,7	2,0	0,06	0,05	3,0	1,0	0,03	0,07
In 54 min. . . .	—	1,6	0,09	—	3,8	1,0	0,04	0,04

Gruppo A. – Questo gruppo caratterizzato dal forte coefficiente di assorbimento nell'argento (K = 20 cm²/g), si osserva di solito usando come rivelatore un sottile strato di argento. Come si è già detto il 25 °/o dell'attivabilità normale di un rivelatore di 0,057 g/cm² è dovuto al gruppo *A*. Per isolare questo gruppo basta fare la differenza delle attivabilità di un rivelatore di argento senza e con un filtro di argento sufficiente ad assorbire il gruppo *A* e tale da non attenuare notevolmente gli altri gruppi. In pratica conviene eseguire le misure eliminando il gruppo *C* con un filtro di cadmio. Il gruppo *A* si presenta anche nell'oro; un rivelatore di oro di 1,535 g/25 cm² ha una attivabilità normale di 21 di cui il 13 °/o è dovuto al gruppo *A*; il coefficiente di assorbimento dell'oro per questo gruppo è 4 cm²/g. Il coefficiente di assorbimento del gruppo *A* nel boro è 3 cm²/g.

Poco è noto sulle proprietà degli altri gruppi dei quali è stato detto qualche cosa nel paragrafo precedente. Abbiamo effettuate alcune misure dei coefficienti di assorbimento che qui riportiamo:

gruppo *B*: boro K = 2,3 cm²/g ; oro K = 1,7 cm²/g ;

gruppo *I*: boro K = 1 cm²/g ; iodio K = 0,7 cm²/g .

Abbiamo visto che mentre il gruppo *C* è presente con percentuali quasi sempre superiori al 50 % nei diversi rivelatori, posti all'esterno della paraffina, gli altri gruppi si incontrano in un minor numero di rivelatori e determinano percentuali minori dell'attività.

Questo fatto non è dovuto all'essere i coefficienti di assorbimento dei vari elementi per i neutroni del gruppo *C* maggiori che per gli altri gruppi; abbiamo anzi già visto numerosi esempi del contrario. Esso è dovuto principalmente al fatto che il numero dei neutroni *C* che escono dalla paraffina è assai maggiore del numero di neutroni degli altri gruppi.

Un dato importante per caratterizzare un gruppo è dunque la sua *numerosità*. Per dare un significato ben determinato a questa grandezza la definiamo al modo seguente: numerosità N_0 di un gruppo è il numero di neutroni di esso che escono per secondo, centimetro quadrato e unità di intensità neutronica della sorgente, in media dai 25 cm^2 centrali della superficie superiore del blocco di paraffina (1), contenente la sorgente (intensità neutronica I) 3 cm sotto il centro della base superiore.

Il principio su cui è basata la misura della numerosità di un gruppo è il seguente: appoggiamo sul centro del blocco (1) un rivelatore sottile di superficie s; il numero di neutroni che colpiscono questo rivelatore sarà

$$N_0 \, I s .$$

Se con K indichiamo il coefficiente di assorbimento del rivelatore, supposto assai sottile, e con δ il suo spessore, si può vedere che il numero di neutroni catturati dal rivelatore è

$$\sqrt{3} \, N_0 \, I s \, K \delta$$

dove il fattore $\sqrt{3}$ è dovuto all'obliquità dei neutroni uscenti, di cui è stato tenuto conto con la legge (5).

Se indichiamo con η l'efficienza della camera di jonizzazione, e cioè la corrente di jonizzazione dovuta ad una disintegrazione al secondo in superficie, l'attività normale (cfr. formula 3) del nostro rivelatore è

$$A = 1000 \sqrt{3} \, s N_0 \, K \delta \frac{K_1}{K} \frac{\eta}{920 \, \eta_U} \tag{7}$$

dove η_U è l'efficienza della camera per i raggi β del preparato di uranio da noi usato, che come si è già detto corrisponde a 920 disintegrazioni al secondo in superficie.

Il fattore K_1/K è stato introdotto per tener conto del fatto che quando un rivelatore assorbe i neutroni di un dato gruppo (con coefficiente di assorbimento K) è possibile che una frazione di essi dia luogo o a formazione di isotopi non radioattivi oppure ad isotopi radioattivi con periodo diverso da quello in esame; con K_1 abbiamo indicato il coefficiente di assorbimento dei neutroni, dovuto al solo processo responsabile dell'attività in esame; è evidente che mentre K è il coefficiente che determina l'assorbimento dei neutroni nel rivelatore, K_1 determina l'attività.

Dalla (7) si ricava la numerosità del gruppo in esame

$$N_0 = 0{,}53 \frac{\eta_U}{\eta} \frac{K}{K_1} \frac{A}{s K \delta} . \tag{8}$$

In questa espressione di N_0 le grandezze A, s, δ, sono direttamente misurabili; K_1 può porsi nella maggior parte dei casi eguale a K; in altri, casi, quando in un elemento si hanno attivazioni con più vite medie, il rapporto delle loro attivabilità, corrette per l'assorbimento dei raggi β e per diversa efficienza η, permette di valutare K/K_1.

Una incertezza non trascurabile nella misura di N_0 dipende dalla scarsa conoscenza del rapporto delle efficienze della camera per raggi β di diversa durezza. Dalle variazioni relativamente piccole che si trovano nelle numerosità calcolate usando differenti rivelatori, nell'ipotesi di efficienza η costante, sembra doversi concludere che tale grandezza non varia entro limiti molto vasti. Nei calcoli che seguono abbiamo sempre assunto $\eta = \eta_U$.

Nella Tabella VI sono riportati i valori delle numerosità dei vari gruppi calcolati usando diversi rivelatori; come coefficienti di assorbimento abbiamo introdotto nei calcoli le medie di quelli dati nelle tabelle precedenti. I valori dati nella Tabella VI sono stati calcolati con una formula analoga alla (8) in cui però si è tenuto conto dell'assorbimento dei neutroni e degli elettroni entro lo spessore del rivelatore; si è inoltre apportata una piccola correzione per tener conto della riflessione degli elettroni sugli strati superiori del rivelatore.

TABELLA VI.

Numerosità N_0.

Gruppo	*C*	*D*	*A*	*I*
Rivelatore				
Rh	47	10	—	—
Ag	57	—	0,6	—
In	34	7	—	—
I	—	—	—	5
Au	37	—	0,5	—

I dati relativi all'indio sono stati calcolati sommando i contributi delle due vite di 16 s e 54 min e trascurando la debole vita di 4 h. Per il rodio e per l'argento abbiamo eseguito i calcoli tenendo conto oltre che delle vite di 44 s e 22 s anche delle vite di 4,2 min. e 2,3 min.

Se si calcola la numerosità del gruppo *C* dall'attività dell'iridio (19 h) si trova un valore circa 10; questo valore ci è sembrato troppo piccolo per poter imputare la differenza a errori sperimentali. Abbiamo quindi cercato se non ci fosse un'altra attività abbastanza intensa per poter giustificare la piccolezza del valore precedente della numerosità. Effettivamente il sospetto è risultato fondato poiché esiste una seconda attività dell'iridio con periodo di 68 giorni, la cui attivabilità, per quanto non ancora misurata, è dell'ordine di grandezza dell'attivabilità del periodo di 19 h. I raggi β corrispondenti a questo nuovo periodo hanno un potere penetrante assai piccolo.

Ha interesse determinare se negli elementi che si attivano con due o più periodi, entrambi dovuti a neutroni lenti, l'attività delle diverse vite sia oppure no determinata dagli stessi gruppi.

Abbiamo fatto quindi uno studio dei rilevatori: rodio (0,36 g/cm^2; 44 s e 4,2 min); argento (0,057 g/cm^2; 22 s e 2,3 min); indio 0,065 g/cm^2; 16 s e 54 min); bromuro di sodio (1 g/cm^2; 18 min e 4,2 h) confrontando per ciascuno di essi le attivabilità delle due vite senza e con opportuni filtri.

Nel caso del rodio il rapporto delle attivabilità del periodo di 44 s e di quello di 4,2 min resta lo stesso (circa 10) con e senza un filtro di cadmio sufficiente per assorbire il gruppo *C* [14].

Nell'argento invece si notano delle differenze marcate di comportamento delle due vite; nel gruppo *C* l'attivabilità della vita lunga è risultata il 24 °/₀ dell'attivabilità totale; nel gruppo *B* il 19 °/₀; infine nel gruppo *A* è risultato dalle misure il 3 °/₀ che entro i limiti dell'errore sperimentale potrebbe essere zero; si può quindi dire che il gruppo *A* è caratteristico della vita dell'argento di 22 s.

Si noti fin d'ora che la banda di energia corrispondente al gruppo *A* (cfr. § 12) è assai ristretta; così che per interpretare questo comportamento basterebbe ammettere una lieve differenza fra le energie delle bande di assorbimento dei due isotopi.

Nel caso dell'indio le percentuali di attivazione delle due vite dovute ai gruppi *C* e *D* sono approssimativamente uguali.

Similmente per il bromo un filtro di cadmio non altera considerevolmente il rapporto tra le attivabilità delle due vite.

Abbiamo anche provato a sostituire il cilindro di paraffina su cui vengono fatte le misure normali, con un eguale cilindro di acqua, e usando rivelatori rodio, argento e ioduro di piombo ed opportuni filtri abbiamo determinato le percentuali di attivazione dovute ai vari gruppi. Abbiamo potuto in tal modo riconoscere che i rapporti fra le numerosità dei vari gruppi sono gli stessi con l'acqua e con la paraffina. Anche i valori assoluti non variano considerevolmente. Si noti tuttavia che variando la distanza dalla sorgente i rapporti fra le intensità dei vari gruppi subiscono delle variazioni, come sarà precisato al paragrafo 9.

Infine abbiamo cercato di vedere se i rapporti fra le intensità dei vari gruppi si spostano usando diverse sorgenti di neutroni veloci [15]. Abbiamo perciò determinato le attivabilità di rivelatori di rodio e argento con filtri di cadmio e argento, usando una sorgente appena preparata, in cui il radio C non era ancora in equilibrio, e seguendo la crescita delle attivabilità durante la formazione del radio C. Sia per il rodio che per l'argento non si sono tro-

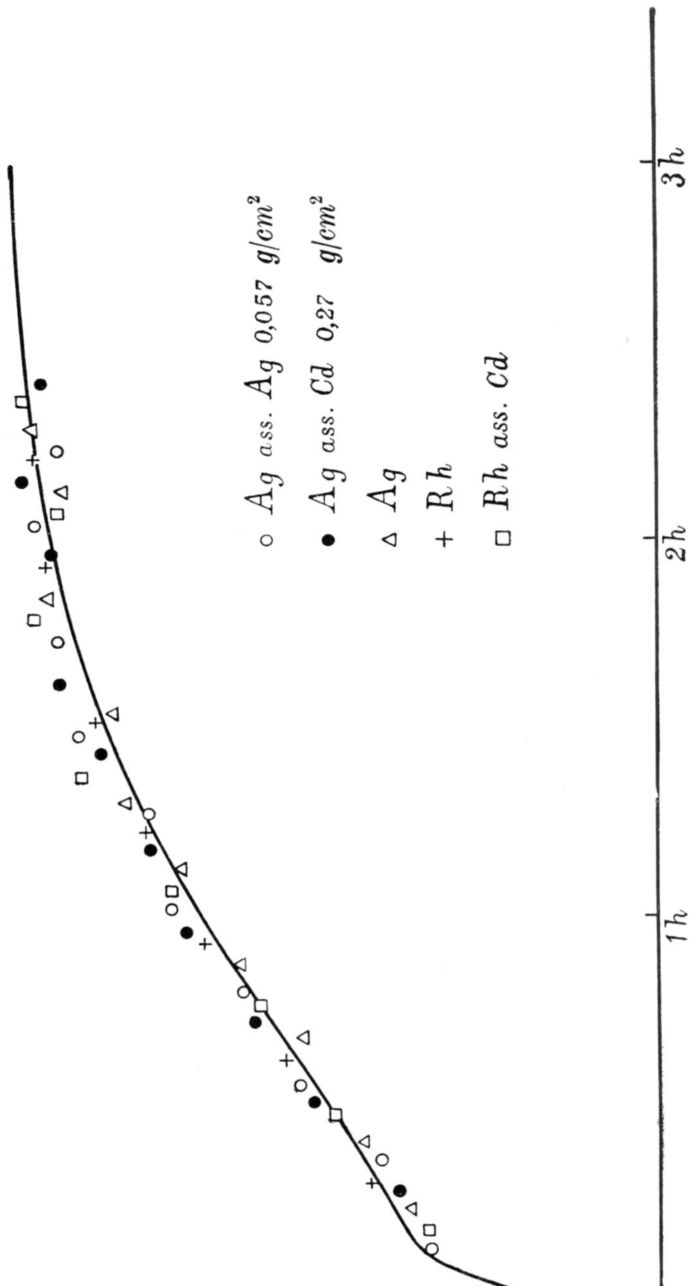

Fig. 5.

vate variazioni nei rapporti dei gruppi durante la formazione del radio C, benché le attivabilità crescessero da 40 (effetto dovuto alle particelle α dell'Em + RaA) a 100 (effetto delle particelle α dell'Em + RaA + RaC).

Nella fig. 5 sono riportate le curve dell'accrescimento dell'attività in funzione del tempo rilevate in questa esperienza. Si noti la elevata efficienza delle particelle α del RaC nel produrre neutroni dal berillio: la loro efficienza è tre volte maggiore di quella delle particelle α dell'emanazione e del radio A.

6. Albedo.

Le misure di attivabilità normale come anche le misure di assorbimento di cui abbiamo parlato fin'ora sono eseguite all'esterno di un blocco di paraffina. In questo paragrafo studieremo invece le proprietà dei neutroni lenti nell'interno della paraffina.

Il rivelatore (rodio 10 g/28 cm^2, oppure argento 2,2 g/38,5 cm^2) veniva posto nel centro della base superiore del cilindro di paraffina normale (1) ed al di sopra veniva appoggiato un secondo cilindro di paraffina eguale. Il rivelatore poteva inoltre venir posto fra opportuni strati assorbenti.

Per indicare rapidamente le posizioni relative della sorgente, della paraffina, del rivelatore e degli assorbitori, è conveniente fissare la seguente convenzione: rappresentiamo con le lettere S la sorgente, P la paraffina, R il rivelatore e con i simboli Cd ed Ag gli assorbitori di cadmio e di argento aventi rispettivamente gli spessori 0,27 g/cm^2 e 0,057 g/cm^2. Un simbolo come S P Cd R Cd P significa quindi una esperienza nella quale al di sopra della sorgente (S) si trovano prima 3 cm di paraffina (P) per giungere alla base superiore del cilindro di paraffina, poi uno strato di cadmio (Cd), il rivelatore (R), un secondo strato di cadmio (Cd) ed infine il secondo cilindro di paraffina (P).

Tabella VII.

Dispositivo	Argento	Rodio
S P R	66,1	282
S P Cd R	32,2	80
S P Cd Ag R	17,5	78
S P R P	417	1025
S P Cd R Cd P	66,5	140
S P Cd R P	92,2	244
S P R Cd P	113,1	371
S P Cd Ag R Ag Cd P	64,5	144
S P Cd Ag R Cd P	45,6	—
S P Cd R Ag Cd P	54,8	—

Nella Tabella VII sono riportate le attivabilità dei due rivelatori di rodio e argento in diverse esperienze.

Confrontando fra di loro le esperienze S P R ed S P R P si nota il forte aumento dell'attività dovuto al sovrapporre il secondo blocco di paraffina. Questo forte aumento è dovuto al fatto che i neutroni sia veloci che lenti, che in assenza del secondo blocco uscirebbero dalla paraffina, possono invece venire riflessi all'indietro da questo.

Ponendo il rivelatore tra opportuni assorbitori si può fare un'analisi in gruppi della radiazione che lo attiva. Così per esempio il confronto delle esperienze S P R P ed S P Cd R Cd P ci mostra che nel caso del rivelatore argento la frazione $\frac{417 - 66{,}5}{417} = 0{,}84$ dell'attività è dovuta al gruppo C e nel caso del rivelatore rodio la frazione $\frac{1025 - 140}{1025} = 0{,}86$ è dovuta al gruppo C. Si noti che il rapporto tra il gruppo C e gli altri gruppi è molto maggiore entro la paraffina che non all'esterno; vedremo tra poco la ragione di questa differenza.

Un controllo della correttezza dell'analisi in gruppi dell'attività di un rivelatore anche nell'interno della paraffina, si ha dal fatto che raddoppiando lo spessore degli assorbitori di cadmio non si ha altro che una piccolissima riduzione ulteriore dell'attività. Nel caso del rivelatore argento invece l'aggiunta di assorbitori di argento a quelli di cadmio, determina una riduzione della intensità a circa $^1/_2$ (esperienze S P Cd R Cd P ed S P Cd Ag R Ag Cd P); questo fatto mostra che anche nell'interno della paraffina si può parlare di gruppo A intensamente assorbito dall'argento. In conformità a quanto si osserva fuori dalla paraffina gli assorbitori sottili di argento, aggiunti a quelli di cadmio non provocano una ulteriore riduzione nell'attività del rodio.

Studiamo ora il comportamento del gruppo C. Osserviamo i risultati delle misure S P R P , S P Cd R P , S P R Cd P , S P Cd R Cd P : questa ultima misura ci indica quale parte dell'attività non è dovuta al gruppo C, per modo che detraendo questo dato dai risultati delle altre tre misure si ottiene la parte dell'attività dovuta al solo gruppo C.

La differenza fra le misure S P Cd R P ed S P R Cd P si interpreta facilmente tenendo conto del fatto che i neutroni lenti nell'interno della paraffina hanno una densità che decresce allontanandosi dalla sorgente in modo che uno strato di cadmio che protegge il rivelatore dalla parte della sorgente arresta più neutroni di uno strato messo dalla parte opposta. Per riportarci al caso semplice in cui i neutroni fossero distribuiti uniformemente nella paraffina, basterà dunque prendere la media dei risultati delle due esperienze precedenti.

Troviamo dunque per il rivelatore di argento, come attivabilità dovuta al solo gruppo C: $A_c = 417 - 66{,}5 = 350{,}5$. L'attivabilità dovuta al solo gruppo C quando l'argento è protetto solo da una parte da uno strato di cadmio è invece: $B_c = \frac{92{,}2 + 113{,}1}{2} - 66{,}5 = 36{,}1$.

I corrispondenti valori per il rodio sono $A_c = 885$, $B_c = 167{,}5$. Il rapporto A_c/B_c è per l'argento 9,7 e per il rodio 5,3.

Una considerazione elementare permette di rendersi conto dei valori di questo rapporto.

Indichiamo con β la probabilità che un neutrone che incide su di una superficie piana che limita un blocco molto esteso di paraffina ne riesca dopo avere eventualmente subìto alcuni urti nella paraffina.

Chiameremo *albedo* questa grandezza β che corrisponde al coefficiente di riflessione diffusa dei neutroni lenti da parte di una superficie di paraffina. In realtà l'albedo dipende dall'angolo di incidenza dei neutroni lenti ed è evidentemente tanto minore quanto più la incidenza è normale (cfr. *F.*, § 5). In questa considerazione elementare trascureremo però questa dipendenza.

Indichiamo con ζ la probabilità che un neutrone che attraversa il rivelatore venga da esso catturato. Anche per ζ trascureremo la dipendenza dall'angolo di incidenza.

Il metodo per la determinazione della albedo è basato sul confronto fra le attivabilità A_c e B_c sopra definite.

Sia $\mathfrak{N}$ il numero di neutroni C che incidono su di un rivelatore esclusivo del gruppo *C* quando esso è protetto da un lato da un assorbitore totale del gruppo *C* stesso. Il numero dei neutroni catturati dal rivelatore sarà $\mathfrak{N}\zeta$ e ad esso sarà proporzionale l'attivabilità B_c.

Supponiamo ora di togliere l'assorbitore e cerchiamo di determinare di quanto crescerà l'attivabilità del rivelatore.

In primo luogo dovremo tener conto del fatto che i neutroni possono colpire il rivelatore da ambo le parti ciò che determina un fattore 2.

Ma un aumento ancora più importante è dovuto al fatto che quando accanto al rivelatore è posto uno strato totalmente assorbente, un neutrone C che colpisca il rivelatore lo attraversa una volta sola ed ha quindi una probabilità ζ di venir catturato in esso; quando invece non vi è l'assorbitore il neutrone può attraversare più volte il rivelatore.

Infatti al primo passaggio si ha la probabilità ζ che il neutrone venga catturato nel rivelatore ed $1 - \zeta$ che lo attraversi; la probabilità che il neutrone torni sul rivelatore è $\beta(1 - \zeta)$ e la probabilità che sia catturato al secondo passaggio è $\beta(1 - \zeta)\zeta$; così proseguendo si trova che la probabilità di cattura è

$$\zeta + \zeta\beta(1 - \zeta) + \zeta\beta^2(1 - \zeta)^2 + \cdots = \frac{\zeta}{1 - \beta(1 - \zeta)} = \frac{\zeta}{1 - \beta + \beta\zeta}.$$

Il numero dei neutroni catturati è quindi

$$\frac{2\mathfrak{N}\zeta}{1 - \beta + \beta\zeta},$$

e quindi il rapporto A_c/B_c risulta

$$\frac{A_c}{B_c} = \frac{2}{1 - \beta + \beta\zeta} \tag{9}$$

che per ζ piccolissimo, cioè per rivelare molto sottile, si riduce a

$$\frac{A_c}{B_c} = \frac{2}{1 - \beta}. \tag{10}$$

La misura del rapporto A_c/B_c permette dunque di determinare l'albedo β.

Nel caso del nostro rivelatore di argento ζ è assai piccolo ma pur tuttavia non trascurabile. Il suo valore si può valutare dallo spessore del rivelatore, dal coefficiente di assorbimento dell'argento tenendo conto naturalmente dell'obliquità dei neutroni. Si trova così $\zeta = 0{,}03$ e di conseguenza

$$\beta = 0{,}82.$$

Il rivelatore di rodio è molto più spesso; per esso ζ è prossimo a 0,35; dato questo elevato valore di ζ un piccolo errore nella sua valutazione porterebbe ad un forte errore nella determinazione di β. Per conseguenza non si possono usare le misure col rodio per determinare β, ma piuttosto calcolare per questo rivelatore il rapporto A_c/B_c usando il valore dell'albedo ottenuto dall'argento; si trova così

$$\frac{A_c}{B_c} = 4{,}3$$

in sufficiente accordo col valore sperimentale 5,3.

Abbiamo già osservato che il concetto di albedo come noi l'abbiamo introdotto, non è interamente ben definito, poiché la percentuale di neutroni riflessi da una superficie di paraffina dipende dalla legge di distribuzione angolare dei neutroni entranti. Per precisarne il concetto si può definire l'albedo β per mezzo della (10), scritta per un rivelatore infinitamente sottile (albedo sperimentale cfr. *F.*, § 7). Le nostre misure si riferiscono appunto all'albedo così definita.

Essa è legata al numero N di cammini liberi che un neutrone C percorre in media prima di venire catturato da un protone della paraffina.

Si può dimostrare (*F.*, formula 63) che la relazione che lega queste due grandezze è:

$$\beta = 1 - \frac{2}{\sqrt{N}} \cdot \tag{11}$$

Questa espressione è esatta nell'ipotesi che l'energia di agitazione termica sia trascurabile rispetto al quanto $h\nu$ del legame elastico dell'idrogeno nella paraffina; altrimenti è necessario apportare una piccola correzione (cfr. § 13).

Prendendo $\beta = 0{,}82$ si ricaverebbe (cfr. § 13)

$$N = 124.$$

Abbiamo eseguito delle esperienze analoghe alle precedenti per determinare l'albedo dei neutroni del gruppo *A* usando un rivelatore di argento di 0,010 g/cm², in modo che per esso ζ anche nel gruppo *A* non fosse troppo grande. A differenza del gruppo *C*, che ha albedo prossima ad 1, abbiamo trovato per il gruppo *A* un'albedo sensibilmente nulla. Ciò non significa che i neutroni del gruppo *A* non vengano affatto riflessi, ma piuttosto che quando vengono riflessi la loro velocità varia per effetto degli urti subìti nella paraffina in modo che all'uscita essi non appartengono più alla banda di energia corrispondente al gruppo *A*.

La bassa albedo del gruppo *A* trova riscontro nel fatto che l'attività di un rivelatore provocata dai neutroni di questo gruppo aumenta soltanto di un fattore prossimo a 2 quando si pone al di sopra del rivelatore un secondo cilindro di paraffina. Tale aumento non è dovuto ad un effetto di albedo ma sostanzialmente al fatto che i neutroni lenti penetrano nel rivelatore da entrambe le facce.

Un comportamento analogo si osserva anche per i neutroni dei gruppi *D*, *B* ed *I*. Può dunque concludersi che ad eccezione del gruppo *C* i neutroni di tutti i gruppi hanno una piccolissima albedo, ciò che significa che dopo un solo urto o poco più essi cambiano di velocità tanto da cessare di appartenere al gruppo (cfr. § 12).

Questa notevole differenza di comportamento fra il gruppo *C* e gli altri gruppi porta come conseguenza che nell'interno della paraffina le percentuali di attivazione dovute al gruppo *C* sono sempre molto maggiori che all'esterno.

Nello stesso senso vanno interpretate le esperienze di Tillman [16] sul diverso comportamento di vari rivelatori appoggiati su di un blocco di paraffina, in funzione dello spessore di un secondo strato di paraffina sovrapposto. Un forte aumento percentuale dell'attività di un rivelatore dovrà attendersi per rivelatori molto sensibili al gruppo *C*; dovrà naturalmente tenersi conto nell'interpretazione di queste esperienze anche del coefficiente ζ del rivelatore secondo la formula (9).

7. Diffusione dei neutroni di un gruppo.

Nel paragrafo precedente abbiamo concluso dalle misure di albedo che i neutroni del gruppo *C* percorrono successivamente un considerevole numero di cammini liberi. Il movimento di questi neutroni nella paraffina si potrà quindi assimilare con buona approssimazione ad un moto di diffusione in cui naturalmente si tenga conto della distruzione dei neutroni per cattura da parte dei protoni.

Se indichiamo con N il numero di cammini liberi fatti in media da un neutrone C e con λ il suo cammino libero medio, è evidente che la distanza media a cui diffonde questo neutrone è dell'ordine di grandezza di $\lambda\sqrt{\mathrm{N}}$. Quindi delle esperienze di diffusione permettono di determinare $\lambda\sqrt{\mathrm{N}}$ [17]. Chiameremo tratto di diffusione la grandezza

$$l = \lambda\sqrt{\frac{\mathrm{N}}{3}} \tag{12}$$

che, come vedremo, può ottenersi direttamente dalle misure.

La misura del tratto di diffusione può rendersi quantitaviva usando un dispositivo basato su questo principio: consideriamo un neutrone C che si trova entro un blocco di paraffina, limitato da una superficie piana, a distanza x da questa. In base alla teoria della diffusione (cfr. *F.*, § 3) si calcola che la probabilità $p(x)$ che il neutrone arrivi ad uscire dalla paraffina, prima di venir catturato da un protone, è

$$p(x) = e^{-\frac{x}{\sqrt{D\tau}}} \tag{13}$$

dove D è il coefficiente di diffusione dei neutroni termici e τ la loro vita media nella paraffina.

Per i neutroni C la cui energia cinetica è piccola rispetto al quanto delle frequenze elastiche degli idrogeni nella paraffina, ammetteremo che gli idrogeni possano considerarsi come centri fissi di diffusione isotropa e che il cammino libero medio λ sia indipendente dalla velocità (cfr. *F.*, § 10 e 11). In realtà questa ipotesi non è soddisfatta esattamente come sarà discusso nel § 13.

In queste ipotesi il coefficiente di diffusione è legato a λ dalla relazione

$$D = \frac{1}{3} \lambda \bar{v} \tag{14}$$

($\bar{v}$ = velocità media dei neutroni); si ha d'altra parte

$$N = \frac{v\tau}{\lambda} \tag{15}$$

e quindi

$$D\tau = \frac{1}{3} \lambda^2 N . \tag{16}$$

Si ha dunque (cfr. *F.*, formula 34)

$$p(x) = e^{-x \Big/ \sqrt{\frac{\lambda^2 N}{3}}} = e^{-\frac{x}{l}} \tag{17}$$

Una determinazione della probabilità $p(x)$ permette dunque la misura del tratto di diffusione l e quindi di $\lambda^2 N$.

La determinazione di $p(x)$ si può fare, in linea di principio, nel modo seguente: immaginiamo un rivelatore R del gruppo *C* che ricopra tutta la superficie piana di un blocco di paraffina da cui escono i neutroni C; tale rivelatore assorba tutti i neutroni che escono dalla paraffina in uno spessore piccolo rispetto all'assorbimento degli elettroni. L'attività di un simile rivelatore sarebbe proporzionale al numero di neutroni che lo investono indipendentemente dalla loro distribuzione angolare. Se ad una profondità x distruggiamo Q neutroni C, l'attività di questo rivelatore calerà proporzionalmente a $Q \cdot p(x)$; la distruzione dei Q neutroni si può realizzare ponendo nella posizione x un assorbitore del gruppo *C*, e Q si può dedurre direttamente dall'attività indotta nell'assorbitore.

L'esperienza è stata eseguita in pratica ponendo il rivelatore su di un blocco cilindrico di paraffina di 25 cm di diametro e 15 di altezza, contenente la sorgente sull'asse 3,5 cm al disotto della base superiore; questo, cilindro di paraffina era tagliato a fette in modo da poter inserire l'assorbitore a varie profondità.

Abbiamo usato come rivelatore una lastra di rodio di 10 g/28 cm²; e siccome questa non copriva che una piccola parte della superficie della paraffina, le misure sono state fatte ponendo il rivelatore successivamente in nove posizioni diverse e sommando i risultati.

Come assorbitore del gruppo *C* abbiamo usata una lastra di lega di stagno e cadmio contenente 0,019 g/cm² di cadmio; siccome il cadmio in piccoli spessori assorbe praticamente solo i neutroni del gruppo *C*, la riduzione dell'attività del rivelatore ha luogo solo in questo gruppo, come si può del resto verificare direttamente.

Poiché l'assorbimento dei neutroni da parte del cadmio non dà luogo ad attivazione, non era possibile misurare direttamente il numero Q di neutroni assorbiti; si è perciò misurata l'attività indotta dai neutroni del gruppo C in una lastra di rodio equivalente a quella di cadmio agli effetti dell'assorbimento del gruppo C.

L'attività di questa lastra di rodio è stata sempre misurata su tutte e due le faccie ed i risultati sono stati sommati in modo da tenere conto egualmente dei neutroni entrati dalle due faccie (vedi § 11).

L'assorbitore è stato posto alle distanze $x = 0, 1, 2, 3$ cm. Nella Tabella VIII sono riassunti i risultati della misura; la prima colonna indica le profondità a cui si trovava l'assorbitore; la seconda la diminuzione della attivabilità del rivelatore dovuta alla presenza dell'assorbitore; la terza colonna dà la somma delle attivabilità della lastra equivalente all'assorbitore, determinate nel solo gruppo C su entrambe le faccie. La quarta colonna dà il rapporto della seconda alla terza. Nella quinta colonna i valori di questo rapporto sono corretti per tener conto del fatto che il nostro rivelatore non soddisfa alla condizione di assorbire totalmente i neutroni in uno spessore piccolo rispetto all'assorbimento degli elettroni (coefficiente di assorbimento degli elettroni 7,3 cm^2/g di rodio; coefficiente di assorbimento dei neutroni 0,7 cm^2/g; spessore 0,36 g/cm^2). Di conseguenza il nostro rivelatore è tanto più sensibile ai neutroni che escono dalla paraffina quanto più grande è la loro obliquità. Siccome l'assorbitore posto in posizione $x = 0$ assorbe prevalentemente i neutroni che escono con forte obliquità, la diminuzione dell'attività del rivelatore è relativamente maggiore di quello che si ha quando l'assorbitore è ad una certa profondità nella paraffina, poiché in questo caso, per effetto degli urti, si perde la coerenza fra la direzione dei neutroni che attraversano l'assorbitore, e la direzione che essi avrebbero avuto all'uscita dalla paraffina.

Si comprende dunque come, per tener conto di questo fatto, il valore del rapporto per $x = 0$ vada diminuito in modo da renderlo confrontabile con gli altri rapporti. Calcolando quantitativamente questa correzione si trova che esso va moltiplicato per 0,832.

TABELLA VIII.

x (in cm)	Diminuzione di attivabilità	Q	Rapporto	Rapporto corretto
0	89	358	0,249	0,207
1	139	1135	0,123	0,123
2	128	1580	0,081	0,081
3	98	1950	0,050	0,050

I valori del rapporto corretto, dati nell'ultima colonna, sono dunque proporzionali a $p(x)$; effettivamente il loro andamento in funzione di x è esponenziale con buona approssimazione. Il tratto di diffusione l è l'inverso del coefficiente di questo esponenziale, e risulta

$$l = 2,1 \text{ cm}$$

ossia, ricordando la (12),

$$\lambda^2 \text{N} = 13 \text{ cm}^2.$$

Abbiamo eseguito anche sui gruppi *D* ed *A* delle esperienze del tipo di quelle ora descritte.

Nel caso del gruppo *D* la sorgente si trovava 3,4 cm al disotto del centro della base superiore di un cilindro di paraffina simile a quello usato nella esperienza precedente; per il gruppo *A* la distanza era 2,4 cm.

Come rivelatori sono stati usati rispettivamente nei gruppi *D* ed *A*, indio (0,065 g/cm^2) e argento (0,057 g/cm^2), entrambi schermati con filtri di cadmio di circa mezzo grammo per cm^2. Come assorbitori abbiamo usato rispettivamente rodio (0,36 g/cm^2) e argento (0,057 g/cm^2).

Nella Tabella IX sono raccolti i risultati di queste misure; nella prima colonna sono date le profondità x dell'assorbitore nella paraffina; la seconda e la quinta colonna contengono rispettivamente per il gruppo *D* e per il gruppo *A* le diminuzioni di attivabilità del rivelatore, dovuta alla presenza dell'assorbitore. Nella terza colonna è riportata la somma delle attivabilità misurate alle diverse profondità sulle due faccie dell'assorbitore di rodio quando questo è schermato con cadmio, in modo da misurare la sola attività del gruppo *D*; la sesta colonna contiene i dati analoghi per l'assorbitore di argento del gruppo *A*. Le colonne quarta e settima sono le attivabilità degli assorbitori misurate sulla faccia inferiore e dovute ai soli neutroni del gruppo *D*, oppure *A*, che entrano nell'assorbitore dalla faccia inferiore. Sui procedimenti seguiti per misurare queste grandezze confronta il paragrafo 11. I numeri tra parentesi dell'ultima colonna sono stati interpolati in base a misure del tipo di quelle esposte al paragrafo 11.

TABELLA IX.

	Gruppo *D*			Gruppo *A*		
x	Diminuzione di attivabilità	Attivabilità assorbitore	Attivabilità inferiore	Diminuzione di attivabilità	Attivabilità assorbitore	Attivabilità inferiore
0	12,0	102	68	15,3	27,3	17,4
0,16	9,6	105	71	11,6	31,2	(18)
0,34	6,4	124	75	8,3	34,6	(19)
0,87	2,7	173	93	3,8	51,5	(22)
1,87	0,5	244	101	0,5	61,5	23,1

La piccola albedo dei gruppi D ed A indica, come abbiamo già osservato nel paragrafo precedente, che i neutroni di questi percorrono in media un solo cammino libero o poco più; in conseguenza non si può evidentemente pensare di applicare a questi gruppi la teoria della diffusione. Pur tuttavia da un punto di vista formale è interessante ridurre i dati delle misure sui gruppi D ed A allo stesso modo in cui sono stati trattati i dati sul gruppo C. Per fare questo basta esaminare il comportamento in funzione di x dei rapporti della seconda alla terza colonna oppure della quinta alla sesta. Si trova in tal modo che questi rapporti hanno un andamento abbastanza approssimativamente esponenziale; la riduzione ad $1/e$ avviene per entrambi i gruppi in uno spessore di 0,42 cm.

Si vede dunque il comportamento analogo di questi due gruppi in contrasto col gruppo C in cui lo spessore corrispondente è di 2,1 cm.

I dati della Tabella IX si prestano tuttavia ad una riduzione più corretta nell'ipotesi estrema che i neutroni dei gruppi D ed A percorrano un solo cammino libero.

In questa ipotesi si può determinare, come ora vedremo, il cammino libero medio di questi gruppi; se l'ipotesi fatta non fosse vera i valori che daremo per i cammini liberi medi rappresenterebbero dei limiti superiori.

Confrontiamo le attività del rivelatore del gruppo D quando a profondità x nella paraffina vi è oppure no l'assorbitore. Entrambe queste attività si possono scindere nella somma di due termini:

a) contributo dovuto ai neutroni D che non hanno attraversato il piano a profondità x;

b) contributo dovuto ai neutroni D che come tali hanno attraversato il piano a profondità x.

Dall'ipotesi fatta che i neutroni D percorrano un solo cammino libero, segue che a questo secondo termine contribuiscono solo i neutroni che attraversano il piano a profondità x dal basso verso l'alto.

Mentre il termine *a*) non viene evidentemente perturbato dalla presenza dell'assorbitore nel piano a profondità x, e quindi scompare nella differenza delle attività senza e con assorbitore, il termine *b*) diminuisce per la presenza dell'assorbitore.

Per brevità indicheremo nel corso di questa discussione come neutroni di classe b i neutroni del gruppo D che incidono dal disotto sul piano a profondità x.

La diminuzione dell'attività del rivelatore data nella colonna seconda della Tabella IX, si può quindi calcolare tenendo conto solo dei neutroni di classe b.

Nell'ipotesi fatta che un neutrone D cessi di essere tale dopo un urto, lo strato di paraffina che i neutroni di classe b debbono attraversare per raggiungere il rivelatore, deve per essi considerarsi come uno strato assorbente (e non diffondente) di coefficiente di assorbimento $1/\lambda_D$.

Quando non vi sia assorbitore l'attività del rivelatore dovuta ai soli neutroni di classe b (il cui numero indichiamo con $\mathfrak{N}_b$) sarà dunque proporzionale a

$$\mathfrak{N}_b \, c\left(\frac{x}{\lambda_D}\right)$$

dove c è la funzione data dalla (6) (curva di assorbimento tenendo conto dell'obliquità dei neutroni).

Vi è in ciò l'ipotesi che la distribuzione angolare dei neutroni di classe b sia indipendente dalla profondità x, ciò che in realtà non è completamente esatto, ma che riteniamo non introduca un errore importante.

Quando invece vi sia l'assorbitore, l'attività del rivelatore dovuta ai soli neutroni di classe b sarà similmente proporzionale a

$$\mathfrak{N}_b \, c\left(\frac{x}{\lambda_D} + K_D \, \delta\right)$$

dove δ è lo spessore dell'assorbitore e K_D il suo coefficiente di assorbimento per i neutroni D.

La differenza delle due attività sarà dunque proporzionale a

$$\mathfrak{N}_b \left\{ c\left(\frac{x}{\lambda_D}\right) - c\left(\frac{x}{\lambda_D} + K_D \, \delta\right) \right\}.$$

Questa espressione corrisponde ai dati della colonna seconda della Tabella IX.

Nella colonna quarta è data invece l'attivabilità dell'assorbitore dovuta ai soli neutroni di classe b; essa è evidentemente proporzionale ad $\mathfrak{N}_b$ (cfr. § 11).

Quindi i rapporti dei dati corrispondenti delle colonne seconda e quarta saranno proporzionali a

$$c\left(\frac{x}{\lambda_D}\right) - c\left(\frac{x}{\lambda_D} + K_D \, \delta\right). \tag{18}$$

Per il nostro assorbitore di rodio si ha $K_D = 1{,}8$ cm²/g; $\delta = 0{,}36$ g/cm², quindi $K_D \, \delta = 0{,}65$.

Una analoga discussione si può fare per il gruppo A ($K_A = 20$ cm²/g, $\delta = 0{,}057$ g/cm²).

Nella fig. 6 sono riportate le curve (18) per i gruppi D ed A ridotte al valore 100 per $x = 0$; i punti indicano i valori sperimentali del rapporto calcolati assumendo $\lambda_D = \lambda_A = 1{,}1$ cm.

Sembra dunque che questi due gruppi abbiano lo stesso cammino libero medio che risulta di poco superiore ad 1 cm.

Abbiamo già osservato che il metodo è basato sull'ipotesi che i neutroni di questi gruppi percorrano un solo cammino libero. Questa ipotesi, molto plausibile per il gruppo A, è un po' più discutibile per il gruppo D (cfr. § 12). Tuttavia il fatto che si trovi approssimativamente lo stesso valore per λ_D e λ_A può considerarsi un argomento in favore della legittimità della ipotesi.

8. Cammino libero medio dei neutroni termici.

Nel paragrafo precedente abbiamo dato il valore del cammino libero medio dei gruppi D ed A.

Il valore trovato dipende dall'ipotesi che i neutroni di questi gruppi percorrano come tali un solo cammino libero (cfr. § 12). Sarebbe desiderabile controllare l'esattezza di questa ipotesi con una misura diretta dei cammini liberi medi. Fino ad ora non ci è stato possibile eseguire una tale misura per ragioni di intensità. Abbiamo potuto invece misurare direttamente il cammino libero medio λ dei neutroni termici nella paraffina.

Questo risultato, preso in connessione con la misura di λ^2 N del paragrafo precedente, permette di dedurre il valore del numero N di cammini liberi percorsi in media da un neutrone termico.

La misura del cammino libero medio λ si può fare usando una sorgente ed un rivelatore dei neutroni termici posti ad una distanza il piu possibile grande in confronto alle loro dimensioni ed interponendo fra di essi vari spessori di paraffina di dimensioni tali da occultare esattamente il rivelatore dalla sorgente.

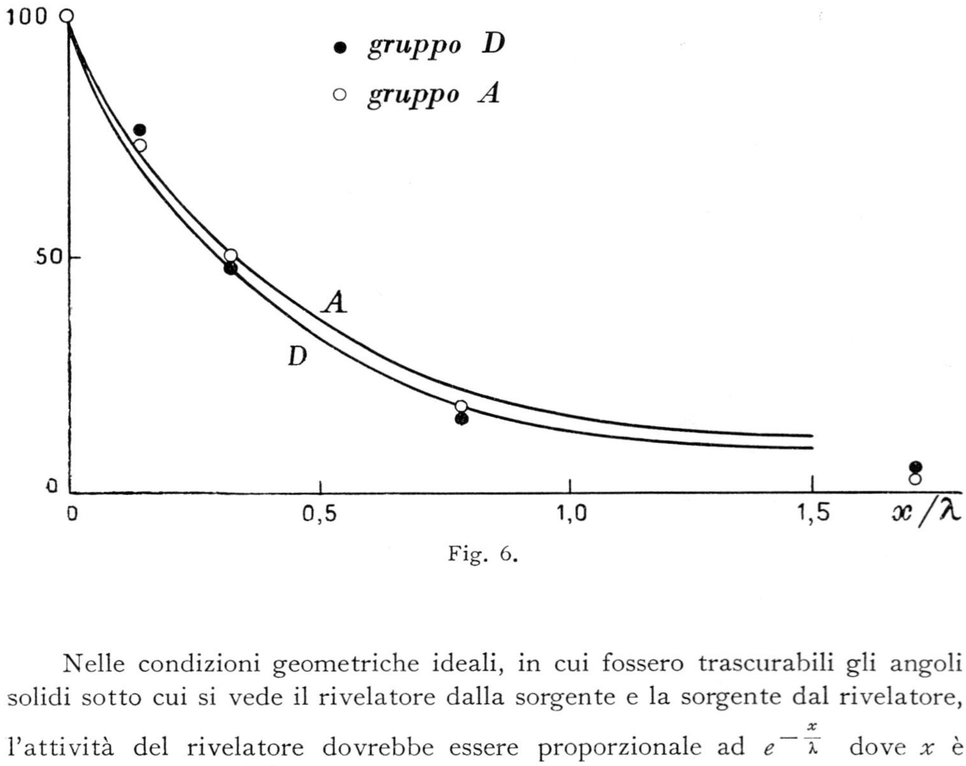

Fig. 6.

Nelle condizioni geometriche ideali, in cui fossero trascurabili gli angoli solidi sotto cui si vede il rivelatore dalla sorgente e la sorgente dal rivelatore, l'attività del rivelatore dovrebbe essere proporzionale ad $e^{-\frac{x}{\lambda}}$ dove x è lo spessore della paraffina diffondente.

In pratica, per ragioni di intensità, si è abbastanza lontani da queste condizioni ideali ed occorre quindi correggere i risultati per tener conto della imperfetta geometria.

Il dispositivo usato nelle misure è stato il seguente:

Sorgente. - Un cilindro di paraffina di 12 cm di diametro e 13 di altezza conteneva la sorgente di Em + Be, 2 cm sotto il centro della base superiore. Tale cilindro era completamente rivestito di cadmio di 0,5 g/cm^2; una finestra di $5 \times 5,5$ cm^2 nel centro della base superiore del rivestimento di cadmio si poteva aprire o chiudere. Misurando l'attività di un rivelatore dei neutroni lenti, posto all'esterno di un simile blocco di paraffina, una volta con la finestra aperta ed una volta con la finestra chiusa, e facendo la differenza delle due attività ottenute, si ricava evidentemente la parte di attività dovuta ai soli neutroni C che escono dalla apertura. Tale differenza equivale dunque all'attività che sarebbe dovuta ad una sorgente di soli neutroni C avente le dimensioni dell'apertura. In questo senso parleremo nel seguito di una sorgente di neutroni C o semplicemente sorgente C.

Rivelatore. – Per poterci mettere in condizioni geometriche non troppo sfavorevoli ed avere tuttavia una intensità sufficiente, abbiamo costruito delle piccole camere di jonizzazione cilindriche del diametro interno di 3,5 cm e lunghe 10 cm, piene di ossigeno alla pressione di 75 atmosfere. L'elemento da irradiare (rodio 0,125 g/cm²) si trovava, in forma di lamina cilindrica di 5 cm di altezza e 3,5 cm di diametro, nell'interno della camera e costituiva esso stesso l'elettrodo o potenziale elevato; l'altro elettrodo, connesso all'elettrometro, era una bacchetta metallica sull'asse della camera. La parete era di acciaio dello spessore di 4 mm e non assorbiva sensibilmente i neutroni lenti, i quali penetrando nella camera attivavano il rivelatore.

Questo dispositivo presenta il vantaggio, rispetto ai metodi da noi precedentemente usati, che, a causa della elevata pressione, quasi tutta l'energia dei raggi β viene utilizzata entro la camera; inoltre l'elettrodo ha una notevole estensione e le condizioni geometriche sono assai buone. Si può infatti con questo dispositivo raggiungere quasi la sensibilità di osservazioni fatte coi contatori e si ha il vantaggio della maggiore stabilità caratteristica delle camere di jonizzazione.

La camera, connessa all'elettrometro, veniva irradiata per 2 minuti e quindi, dopo aver allontanata la sorgente, si effettuava la lettura del numero di divisioni percorse in 2 minuti a partire da 15 secondi dopo la fine dell'irraggiamento.

Abbiamo eseguito due serie di misure; nella prima la distanza dalla sorgente C all'asse della camera di jonizzazione era di 20 cm, nella seconda di 10 cm; in entrambe le serie i diffusori di paraffina erano posti a metà distanza fra la sorgente C ed il rivelatore. Per evitare diffusioni secondarie, la camera di jonizzazione e tutto il percorso da questa alla sorgente C erano protetti da uno strato di cadmio di 0,5 g/cm².

Nella Tabella X sono riportati, in unità arbitrarie, i risultati delle due serie di misure (ciascun valore media di tre osservazioni); l'errore è di circa una unità. Nella prima colonna sono dati gli spessori dei diffusori di paraffina (densità 0,9) in cm. Nella seconda e terza colonna sono date le misure di attività a 20 cm di distanza con la finestra di cadmio della sorgente rispettivamente aperta e chiusa. L'attività dovuta al solo gruppo *C* è, come si è già detto, la differenza fra queste due colonne. La quarta e la quinta colonna sono analoghe alle due precedenti per la distanza di 10 cm.

Per dedurre da queste misure il valore del cammino libero medio dei neutroni termici è indispensabile tener conto delle correzioni dovute alla imperfezione della geometria, che sono particolarmente notevoli nelle misure a 10 cm di distanza.

Il calcolo delle correzioni per spessori notevoli del diffusore presenta alquanta incertezza, mentre e possibile una valutazione abbastanza attendibile delle correzioni per piccoli spessori.

TABELLA X.

x	20 cm		10 cm	
	senza Cd	con Cd	senza Cd	con Cd
0	45,1	23,6	175	92,5
0,047	40,5	22,1	162	89,5
0,099	36,8	21,7	158	90
0,203	29,9	19,3	140,5	87
0,38	24,8	17,8	127	86
0,68	21,2	16,7	116,5	87

Nel caso in cui la distanza è di 20 cm la valutazione della correzione è stata fatta, per piccoli spessori, nel modo seguente: la sezione utile del rivelatore investita dai neutroni è di 17,5 cm² e l'area del diffusore è di 22,3 cm²; i neutroni che giungono sul rivelatore sono costituiti da quelli che vi arrivano direttamente senza avere subito urti nel diffusore e da quelli che vi arrivano dopo aver subìto un urto nel diffusore (il caso di urti multipli si può trascurare se, come supponiamo, lo spessore x del diffusore è piccolo).

Il primo numero è evidentemente proporzionale all'area (17,5 cm²) del rivelatore ed alla probabilità $e^{-\frac{x}{\lambda}}$ che un neutrone non subisca urti nel diffusore. Per x piccolo il numero di questi neutroni è proporzionale a

$$17{,}5\left(1 - \frac{x}{\lambda}\right).$$

Si è trascurato l'effetto molto piccolo della obliquità delle traiettorie dei neutroni rispetto al diffusore. D'altronde il numero di neutroni che colpiscono il diffusore è proporzionale al quadruplo della sua area poiché la sua distanza dalla sorgente è la metà di quella del rivelatore; tale numero è dunque proporzionale a

$$4 \cdot 22{,}3 = 89{,}2.$$

Di questi la frazione x/λ subisce un urto entro il diffusore; nella ipotesi che per i neutroni C la diffusione da parte dell'idrogeno della paraffina possa considerarsi sferica (vedi *F.*, § 10 e 11, cfr. § 13), la frazione dei neutroni diffusi che colpisce il rivelatore è uguale al rapporto dell'area di quest'ultimo all'area di una sfera di raggio eguale alla distanza diffusore–rivelatore (10 cm).

In conclusione il numero totale di neutroni diffusi che colpiscono il rivelatore è proporzionale a

$$89{,}2 \cdot \frac{x}{\lambda} \frac{17{,}5}{4\pi\, 10^2}\,.$$

Aggiungendo a questo numero quello dei neutroni che colpiscono il rivelatore senza avere subìto urti, si ottiene in funzione di x, a meno del solito fattore di proporzionalità, il numero totale dei neutroni che colpiscono il rivelatore

$$17,5\left(1 - 0,929\frac{x}{\lambda} + \cdots\right).$$

Assumendo come 1 l'attività per $x = 0$ si trova dunque che l'attività con diffusore sottile x è data da

$$1 - 0,929\frac{x}{\lambda} + \cdots$$

Di qui risulta che la tangente alla curva di diffusione nel punto $x = 0$ interseca l'asse delle ascisse nel punto

$$x_1 = \frac{\lambda}{0,929}.$$

Da un grafico dei dati sperimentali abbiamo ricavato, col metodo delle secanti, la tangente del punto iniziale; questa interseca l'asse nelle ascisse per

$$x_1 = 0,29 \text{ cm}.$$

Si ricava dunque

$$\lambda = 0,29 \cdot 0,929 = 0,27 \text{ cm}.$$

Abbiamo inoltre verificato che tutto l'andamento della curva sperimentale di diffusione si potesse portare a coincidere, anche per forti spessori, con una curva calcolata. (Il calcolo di cui non diamo i dettagli, è stato fatto per mezzo dei procedimenti illustrati in *F.*, § 5).

Un po' più complicata è la valutazione della correzione per la esperienza a 10 cm, poiché in questo caso non è più lecito trascurare l'obliquità dei neutroni. Il calcolo è stato eseguito con un procedimento numerico da cui è risultato, come espressione del comportamento iniziale della curva di diffusione, l'espressione

$$1 - 0,84\frac{x}{\lambda} + \cdots$$

Dalla curva sperimentale si è ricavato come sopra

$$x_1 = 0,4 \text{ cm},$$

per modo che da questa misura risulta

$$\lambda = 0,84 \cdot 0,4 = 0,34 \text{ cm}.$$

Dal complesso di queste due misure si può ricavare, come valore piu probabile del cammino libero medio dei neutroni termici (cfr. § 13).

$$\lambda = 0,3 \text{ cm}.$$

9. Gruppi in funzione della distanza dalla sorgente. Rapporti di energie.

In un lavoro precedente [18] è stata misurata l'attività di un rivelatore di rodio in una vasca d'acqua in funzione della distanza dalla sorgente. I dati allora ottenuti si riferivano ad un miscuglio dei gruppi *C* e *D* entrambi capaci di attivare il rivelatore di rodio.

Ha evidentemente interesse studiare, in funzione della distanza dalla sorgente, il comportamento dei vari gruppi separatamente.

Le misure sono state eseguite in una vasca cilindrica di acqua profonda 95 cm e del diametro di 90 cm. Come rivelatori sono stati usati: per i gruppi *C* e *D* una lastra di rodio di 0,36 g/cm² e delle dimensioni di 5 × 5,5 cm². Questo rivelatore, schermato da 0,5 g/cm² di cadmio, è sensibile al solo gruppo *D*; mentre le differenze tra le attività che si misurano senza e con lo schermo di cadmio dànno l'attività dovuta al solo gruppo *C*.

Un rivelatore di argento di 0,057 g/cm² di 5 × 5 cm² di area schermato da cadmio, è stato usato come rivelatore dei gruppi $A + B$, che non sono stati studiati separatamente. Infine per il gruppo *I* abbiamo usato un rivelatore di 0,68 g/cm² di PbI_2, schermato da cadmio avente una area di 5 × 5 cm².

Le attività dei rivelatori sono state misurate a varie distanze dalla sorgente su entrambe le facce; la somma delle due attività è proporzionale al numero di neutroni dei vari gruppi che entrano nel rivelatore (cfr. § 11).

Per ridurre al minimo, anche per grandi distanze, l'errore dovuto alle dimensioni finite del recipiente di acqua, le misure sono state eseguite mantenendo nel centro del cilindro di acqua il punto di mezzo fra la sorgente e il rivelatore.

Date le dimensioni dei nostri rivelatori e della ampollina contenente l'Em + Be, non si può, particolarmente per piccole distanze, identificare la distanza sorgente–rivelatore, con la distanza tra i centri di questi due oggetti. Ci siamo perciò riferiti a delle distanze medie. I risultati sperimentali per i gruppi *C*, *D*, $A + B$ ed *I* sono riportati nelle Tabelle XI, XII, XIII, XIV. I numeri riportati sono le somme delle attività misurate sulle due faccie dei rivelatori. I dati sui gruppi *C*, *D* ed $A + B$ sono ottenuti come media di tre serie di misure su ciascuna faccia; per il gruppo *I*, data la vita media relativamente lunga dello jodio, abbiamo fatto una sola serie di misure su ogni faccia.

Tabella XI.

Gruppo C (rivelatore rodio; differenza senza e con uno schermo di cadmio).

r (cm) . . .	2,5	3,6	5,4	7,3	10,2	15,1	20,1	25	30	35	40
Attività . .	2121	1898	1441	1006	523	167	58	23	9	3,3	1,7

TABELLA XII.

Gruppo D (rivelatore rodio schermato da cadmio).

r (cm) . . .	2,5	3,6	5,4	7,3	10,2	15,1	20,1
Attività . .	306	252	166	96	40	10,6	3,8

TABELLA XIII.

Gruppi A + B (rivelatore argento schermato da cadmio).

r (cm). . .	2,5	3,6	5,4	7,3	10,2	15,1
Attività . .	144	119	73	39,6	15,7	4,3

TABELLA XIV.

Gruppo I (rivelatore ioduro di piombo schermato da cadmio).

r (cm) . . .	2,62	3,36	5,1	7,0	9,9
Attività . .	24,9	19,7	12,0	6,28	2,46

Nella fig. 7 sono riportate graficamente le attività in funzione della distanza, date nelle Tabelle XI, XII, XIII e XIV, moltiplicate per r^2; l'unità sulle ordinate è stata alterata in modo che tutte e quattro le curve racchiudano la stessa area.

Benché le differenze tra le curve D , A + B, I siano piccole, la precisione delle misure è sufficiente per distinguerle con sicurezza.

Dall'ispezione delle curve si riconosce subito che la legge di decremento in funzione della distanza dell'attività dovuta ai vari gruppi, è differente e precisamente il decremento in funzione della distanza è via via più rapido per i gruppi nell'ordine *C*, *D*, *A* + *B*, ed *I*.

Questo fatto è suscettibile di una semplice interpretazione che permette di fissare l'ordine di energia dei gruppi. Si osservi infatti che in media i neutroni che hanno subìto più urti, dopo essere stati emessi dalla sorgente, hanno perduto più energia; d'altra parte i neutroni che hanno subìto più urti saranno diffusi in media ad una distanza maggiore dalla sorgente. Ne segue che i gruppi per i quali il decremento dell'intensità in funzione della distanza è più rapido, sono quelli di energia maggiore.

L'ordine di energia crescente dei gruppi studiati è quindi

$$C\,,D\,,A+B\,,I.$$

Questa considerazione può precisarsi quantitativamente nel seguente modo: un parametro conveniente per caratterizzare la rapidità del decremento dell'attività dei vari gruppi in funzione della distanza è la media dei quadrati $\overline{r^2}$ delle distanze dalla sorgente a cui giungono i neutroni dei vari gruppi. Questo parametro è stato da noi calcolato in base ai dati precedenti. Nella Tabella XV sono riportati i valori di $\overline{r^2}$ per i diversi gruppi.

TABELLA XV.

Gruppo	$\overline{r^2}$ in (cm^2)
C	326,7
D	276,6
$A+B$	270,6
I	262,2

Benché le curve in funzione della distanza siano state rilevate con una notevole precisione, ragioni di intensità ci hanno impedito di spingere le misure sui gruppi D , $A+B$ ed I a notevole distanza, in modo che è stato necessario estrapolare l'andamento di queste curve. Ciò si può fare abbastanza bene poiché a grandi distanze la legge di decremento diventa assai simile per tutti i gruppi e può rappresentarsi con molta accuratezza assimilando l'estremo delle curve della fig. 7 ad esponenziali con coefficiente 0,106 cm^{-1}.

Ciò non ostante questa estrapolazione può introdurre un sensibile errore nei valori degli $\overline{r^2}$.

Si può dimostrare (cfr. *F.*, § 2) che, almeno per gruppi di energia superiore a circa 1 volt, e cioè maggiore sia dell'energia di agitazione termica (circa 1/30 di volt) che del quanto $h\nu$ del legame elastico dell'idrogeno nella paraffina (circa $^1/_2$ volt), la differenza $\overline{r'^2}-\overline{r''^2}$ delle medie dei quadrati delle distanze sopra considerate è legata al rapporto W''/W' delle energie dei due gruppi dalla relazione

$$\log\frac{W''}{W'}=\frac{\overline{r'^2}-\overline{r''^2}}{6\lambda_g^2} \tag{19}$$

dove λg è il cammino libero medio nella paraffina (supposto eguale, cfr. § 7) dei neutroni dei due gruppi.

Si può facilmente giustificare qualitativamente la relazione precedente, salvo il fattore 6 nel denominatore, osservando che il primo membro rappresenta il numero medio di urti necessari perché la energia dei neutroni si riduca da W″ a W′; d'altronde il quadrato medio dello spostamento del neutrone durante un cammino libero, è $2\lambda_g^2$ (il fattore 2 proviene dall'avere eseguito la media dei quadrati e non il quadrato della media); se si ammettesse l'incoerenza delle orientazioni dei successivi cammini liberi risulterebbe che per un urto $\overline{r^2}$ cresce di $2\lambda_g^2$ e quindi per effetto di log W″/W′ urti esso dovrebbe crescere di $2\lambda_g^2 \log \frac{W''}{W'}$. In questa ipotesi si troverebbe dunque una formula analoga alla (19) col fattore 2 invece del fattore 6 del denomi-

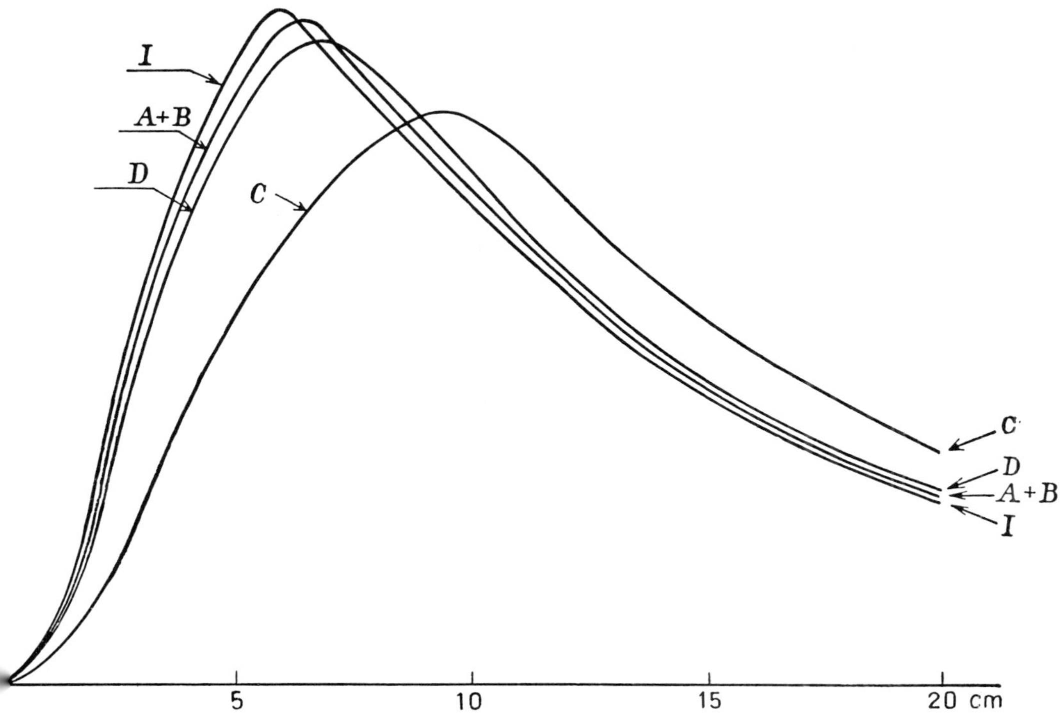

Fig. 7.

natore. Il fattore 6 si trova tenendo conto in modo esatto della coerenza dei cammini liberi successivi, nonché delle fluttuazioni del numero di cammini liberi necessari per ridurre l'energia da W″ a W′.

La relazione (19), può essere usata per determinare i rapporti di energia fra due gruppi (ad eccezione del gruppo *C* per cui evidentemente la formula non è valida). Si deve tuttavia notare che il metodo è assai poco preciso, poiché, come si riconosce subito dalla struttura della formula, piccoli errori sui valori di λ_g e di $\overline{r^2}$, alterano in modo assai notevole il rapporto W″/W′.

Dato che i valori degli $\overline{r^2}$ sono stati misurati nell'acqua e non nella paraffina, si deve porre per λ_g il cammino libero medio nell'acqua; possiamo dedurlo da quello nella paraffina (1,1 cm per i gruppi non termici; cfr. § 7) supponendolo inversamente proporzionale alla concentrazione dell'idrogeno. Si trova così $\lambda_g = 1{,}27$ cm. Per modo che ad una differenza degli $\overline{r^2}$ di $6\lambda_g^2 = 9{,}7$ cm² corrisponde il rapporto e delle energie.

Dalla Tabella XV si possono quindi ricavare i seguenti rapporti di energia per i gruppi $I, A + B, D$

$$W_I : W_{A+B} : W_D = 4{,}4 : 1{,}85 : 1 .$$

Come si è avvertito non bisogna dare troppo peso a questi rapporti data la grande imprecisione del metodo.

È forse più significativo ricavare dalla (19) i valori del cammino libero medio nell'acqua assumendo che le energie dei gruppi siano inversamente proporzionali ai loro coefficienti di assorbimento nel boro [19] (cfr. anche § 13).

Prendendo per i coefficienti di assorbimento del boro i valori seguenti

$$K_I = 1 \quad ; \quad K_B = 2{,}3 \quad ; \quad K_A = 3 \quad ; \quad K_D = 4{,}7 \quad ; \quad K_C = 38 \text{ cm}^2/\text{g}$$

si hanno i seguenti rapporti tra le energie dei vari gruppi

$$W_I \,;\, W_B : W_A : W_D : W_C = 1440 : 270 : 160 : 65 : 1$$

dai quali si ottiene ponendo

$$W_C = 3/2\, kT = 0{,}037 \text{ volt},$$

$$W_I = 55 \quad ; \quad W_B = 10 \quad ; \quad W_A = 6 \quad ; \quad W_D = 2{,}4 \text{ volt}.$$

Da questi valori si ricava a mezzo della (19) il cammino libero medio nell'acqua

$$\lambda = 0{,}87 \text{ cm}$$

invece di 1,27 cm. Questa differenza può essere ampiamente giustificata cogli errori sperimentali.

Dalla Tabella XV si riconosce che il valore di $\overline{r^2}$ del gruppo C è fortemente distanziato da quelli degli altri gruppi:

$$\overline{r_C^2} - \overline{r_D^2} = 50 \text{ cm}^2.$$

Questa differenza deve in parte attribuirsi all'elevato numero di cammini liberi percorsi dai neutroni già rallentati alla velocità di agitazione termica; il contributo di questa diffusione è (cfr. § 7).

$$2\lambda^2 N = 26 \text{ cm}^2.$$

Gli altri 24 cm² corrispondono al processo di rallentamento dal gruppo D alla velocità di agitazione termica. Questo processo non è calcolabile in modo semplice perché complicato dall'effetto del legame chimico dell'idrogeno (cfr. *F.*, § 11).

10. Trasformazione di un neutrone da un gruppo ad un altro.

Nel paragrafo precedente abbiamo stabilito l'ordine di energia decrescente dei gruppi. È evidente che un neutrone di un gruppo di energia elevata, dopo un conveniente numero di urti, avrà una energia inferiore e potrà quindi appartenere successivamente ad altri gruppi di energia sempre più piccola.

Dobbiamo anzi attenderci che i neutroni di tutti i gruppi finiscano per trasformarsi in neutroni termici. Abbiamo cercato quindi di verificare direttamente la trasformazione dei neutroni D in neutroni C in seguito ad alcuni urti nella paraffina [20].

Il principio su cui è basata questa esperienza può facilmente comprendersi se ammettiamo per un momento di disporre di un rivelatore (R) che sia attivato esclusivamente dal gruppo *C e di un assorbitore* (A) che assorba esclusivamente i neutroni del gruppo *D*. Si pensi il rivelatore R appoggiato nel centro della base superiore di un cilindro di paraffina contenente la sorgente S; se l'assorbitore A viene inserito fra la faccia superiore della paraffina ed il rivelatore, non si osserverà alcuna riduzione della attività del rivelatore poiché i neutroni C che escono dalla paraffina non vengono assorbiti dall'assorbitore A. Se ora pensiamo di sprofondare A entro la paraffina uno o due cm, potrà accadere che esso assorba alcuni dei neutroni D che in seguito ad ulteriori urti contro la paraffina avrebbero potuto trasformarsi in neutroni C prima di raggiungere il rivelatore. Se dunque i neutroni D si trasformano in C, si deve attendere che l'assorbitore A non produca alcun effetto messo a contatto di R mentre ne riduca l'attività quando sia incluso nella paraffina uno o due cm al di sotto di R.

In pratica l'esperienza non si può eseguire in queste condizioni ideali poiché manca sia un rivelatore del gruppo *C* che un assorbitore esclusivo del gruppo *D*.

Abbiamo perciò proceduto nel modo seguente: come rivelatore abbiamo usato una lastra di rodio di 0,36 g/cm² della cui attività il 70 % circa è, dovuto al gruppo *C* ed il rimanente al gruppo *D*. Per studiare l'attività dovuta al solo gruppo *C* tutte le misure venivano eseguite senza e con uno strato di cadmio di 0,54 g/cm² immediatamente sottostante alla lastra di rodio; le differenze fra le attività senza e con cadmio dànno l'attività dovuta al solo gruppo *C*.

Come assorbitore del gruppo *D* abbiamo usato una lastra di indio di 1,83 g/28,5 cm². Il coefficiente di assorbimento dell'indio nel gruppo *D* è 3,4 cm²/g e nel gruppo *C* 0,7 cm²/g. Ne segue che l'assorbimento della nostra lastra di indio nel gruppo *C* non è affatto trascurabile specialmente se si tien conto della elevata numerosità e grande albedo di questo gruppo.

Per correggere da questo inconveniente, abbiamo preparato una lastra di lega di stagno e cadmio equivalente alla lastra di indio per l'assorbimento nel gruppo *C*, la quale non assorbe praticamente i neutroni del gruppo *D*. Facendo quindi le differenze fra gli effetti prodotti nel gruppo *C* dalla lastra di indio e da quella di lega di stagno e cadmio, si ottiene quanto è dovuto all'assorbimento del solo gruppo *D* dell'indio.

Nella Tabella XVI sono riportate le attività indotte con vari dispositivi nella lastra di rodio; per indicare le varie esperienze usiamo qui una notazione simile a quella usata nel paragrafo 6, con l'aggiunta di un numero tra parentesi dopo il simbolo P della paraffina che indica lo spessore in cm degli strati di paraffina; siccome abbiamo usato due diversi strati di cadmio aventi uno lo spessore di 0,54 g/cm^2 allo scopo di assorbire completamente il gruppo *C*, e l'altro di circa 0,0036 g/cm^2, equivalente, come assorbitore del gruppo *C*, allo strato di indio, abbiamo indicato questi due assorbitori rispettivamente con Cd e cd.

I numeri dati nella tabella sono le medie di 10 letture.

TABELLA XVI.

S P (3,5) R	280,4 ± 0,7
S P (3,5) In R	230,2 ± 0,7
S P (3,5) In Cd R	45,6 ± 0,4
S P (3,5) Cd R	68,7 ± 0,5
S P (3,5) cd I	254,8 ± 0,7
S P (2,5) cd P (1) R	253,0 ± 0,7
S P (2,5) In P (1) R	241,5 ± 0,7
S P (2,5) cd P (1) Cd R	71,1 ± 0,5
S P (2,5) In P (1) Cd R	64,9 ± 0,5
S P (1,5) cd P (2) R	254,0 ± 0,7
S P (1,5) In P (2) R	248,9 ± 0,7
S P (1,5) cd P (2) Cd R	69,2 ± 0,5
S P (1,5) In P (2) Cd R	68,1 ± 0,5

Le prime cinque misure hanno lo scopo di controllare l'identità di assorbimento nel gruppo *C* del cadmio sottile (cd) e dell'indio (In). Da queste è risultato che l'assorbimento dell'indio è forse un pochino maggiore di quello del cadmio. La differenza $280,4 - 68,7 - (230,2 - 45,6) = 27,1, \pm 1,2$ rappresenta l'attività del rodio nel gruppo *C*, assorbita dall'indio, mentre la differenza $280,4 - 254,8 = 25,6 \pm 1$ rappresenta l'analoga riduzione prodotta invece dal cadmio sottile (cd).

Le quattro misure successive sono eseguite ponendo l'assorbitore (cd oppure In) 1 cm dentro la paraffina. La differenza $253,0 - 71,1 = 181,9 \pm 0,9$ rappresenta l'attività del rodio nel gruppo *C* coll'assobitore sottile di cadmio 1 cm dentro la paraffina. La differenza $241,5 - 64,9 = 176,6 \pm 0,9$ è l'analoga attività coll'assorbitore di indio invece di quello di cadmio.

Fra queste due attività si nota una differenza di $5,3 \pm 1,2$, notevolmente maggiore di quanto sembra doversi attribuire alla piccola differenza che abbiamo notato fra i due assorbitori quando sono posti a contatto del rivelatore.

Le quattro misure successive sono analoghe a queste salvo che gli assorbitori sono posti 2 cm sotto la superficie della paraffina. In questa ultima serie di esperienze di nota ancora una differenza analoga a quella ora discussa che è in questo caso pari a $4{,}0 \pm 1{,}2$.

Elaborando questi dati si giunge alla conclusione che allo stato attuale delle misure, l'effetto osservato è nel senso che corrisponderebbe alla trasformazione del gruppo D in C per urti, e risulta 2,3 volte maggiore dell'errore quadratico medio.

Benché dunque ci sia una seria presunzione di un effetto reale in questo senso, non si possono trarre da questa esperienza delle conclusioni definitive.

Una valutazione dell'effetto da attendersi in questa esperienza dimostra per altro che esso è effettivamente dell'ordine di grandezza di quello osservato; essa è stata fatta al modo seguente.

Abbiamo misurato le attività indotte nella lastra di indio della esperienza precedente, posta nella paraffina 2 cm sotto la superficie, separatamente nei gruppi C e D ed abbiamo trovato rispettivamente 257 e 47,5. Questi numeri si possono ritenere proporzionali ai numeri di neutroni dei due gruppi assorbiti dall'indio.

Indichiamo con p_1 la probabilità che un neutrone C che si trova nella paraffina nella posizione dell'assorbitore di indio giunga alla superficie attivando ivi il rilevatore di rodio; analogamente sia p_2 la probabilità che un neutrone D che si trova nella posizione dell'indio giunga alla superficie già trasformato in neutrone C ed attivi il rodio. Le riduzioni dell'attività del rodio nel gruppo C dovute rispettivamente all'assorbimento da parte dell'indio dei neutroni D e C staranno nel rapporto $47{,}5\ p_2 : 257\ p_1$. Ammettendo che p_1 e p_2 siano eguali od almeno prossimi, risulta dunque che tali riduzioni stanno all'incirca nel rapporto $47{,}5/257$ e siccome la riduzione, dovuta all'assorbimento dei neutroni C risulta pari a 26,9 (vedi Tabella XVI), l'effetto da attendersi nella esperienza precedente è pari a 5 e quindi dell'ordine di grandezza che effettivamente si osserva.

Analoghe esperienze sono state eseguite da Preiswerk e von Halban [21] che hanno trovato un rapporto genetico fra i gruppi I ed A.

11. Comportamento dei gruppi vicino al bordo della paraffina.

Nel paragrafo 9 abbiamo studiato l'attività dovuta ai vari gruppi nell'interno di una massa di acqua talmente grande da poter trascurare le perturbazioni ai bordi.

Studieremo ora invece come si comporta l'attività dovuta ai diversi gruppi in vicinanza della superficie che limita la paraffina.

Occupiamoci prima di tutto del comportamento del gruppo C. Risulta dalla teoria della diffusione dei neutroni termici che, indipendentemente dalla posizione della sorgente, la densità dei neutroni termici, in vicinanza della superficie che limita la paraffina, decresce verso l'esterno (cfr. *F.*, § 6, formula 57); si può anzi dimostrare che se si indica con x la profondità al di sotto della superficie della paraffina, l'andamento della densità $n(x)$ è per x piccolo, approssimativamente proporzionale ad

$$x + \frac{\lambda}{\sqrt{3}}$$

di modo che estrapolando l'andamento di $n(x)$ verso l'esterno della paraffina $(x < 0)$, $n(x)$ si annulla ad una distanza dal bordo

$$x_0 = -\frac{\lambda}{\sqrt{3}} .$$

È appunto questo andamento della densità dei neutroni termini in vicinanza del bordo che determina la legge di distribuzione angolare (5) dei neutroni C uscenti. Se la densità in vicinanza del bordo fosse invece costante, si otterrebbe la legge del coseno.

Per rilevare sperimentalmente $n(x)$ si deve usare un rivelatore del gruppo *C* abbastanza sottile in modo da non perturbare la distribuzione dei neutroni. La sua attività sarà allora proporzionale alla densità $n(x)$ dei neutroni C.

Abbiamo usato due rivelatori ottenuti per deposito elettrolitico di rodio sopra una lastra di nichel (che serve solo come supporto inattivabile) di 29 cm^2 di area; i pesi di rodio depositati sui due rivelatori erano rispettivamente 0,087 e 0,169 g. Tenendo conto del coefficiente di assorbimento del rodio per il gruppo *C* (0,7 cm^2/g) $K\delta$ per questi due assorbitori risulta rispettivamente eguale a 0,002 e 0,004; a questi valori corrisponde all'incirca una probabilità media ζ di cattura per un neutrone termico che attraversi il rivelatore (vedi § 6) rispettivamente di 0,004 e 0,008. Tali probabilità, come si vede, sono assai piccole, e quindi i rivelatori usati si possono considerare molto sottili.

Le misure sono state eseguite su di un cilindro di paraffina del diametro di 24 cm e dell'altezza di 10 cm, contenente la sorgente 3,2 cm sotto il centro della base superiore. Lo strato di paraffina di 3,2 cm compreso tra la sorgente e la base superiore era tagliato a fette in modo da potere inserire il rivelatore a diverse profondità. Per fare le misure sul gruppo *C* si facevano al solito le differenze tra le attività del rivelatore senza e con due schermi di cadmio di 0,44 g/cm^2, fra i quali poteva venire posto il rivelatore.

Nella fig. 8 sono riportate in funzione della profondità x, le attività dovute al solo gruppo *C* dei due rivelatori sopra nominati. È accennata anche l'estrapolazione dell'andamento dell'attività verso valori negativi di x. Da essa risulta la intersezione con l'asse delle ascisse in vicinanza di

$$x_0 = -0,18 \text{ cm}.$$

Da ciò si ricava

$$\lambda = 0,31 \text{ cm}$$

in buon accordo col valore ottenuto dalla misura diretta (cfr. § 8).

Abbiamo già notato che questa esperienza consente una interpretazione così semplice purché il rivelatore sia sottile. Con un rivelatore spesso invece, la stessa presenza di questo, perturba fortemente la densità dei neutroni termici in modo che l'andamento dell'attività in funzione della profondità x è alquanto diverso. Così usando un rivelatore di rodio di 0,36 g/cm^2 e sommando le attività nel gruppo *C* misurate su entrambe le faccie a varie profondità in modo da ottenere un dato paragonabile ai dati dell'esperienza precedente, si trova che l'andamento dell'attività, estrapolato verso valori negativi di x, taglia l'asse delle ascisse per

$$x_0 = -0{,}4\,\mathrm{cm}.$$

L'annullamento si ha quindi ad una distanza dalla superficie più che doppia che non nel caso di rivelatore sottile.

In questo esperimento abbiamo sommato le attività misurate sulle due faccie del rivelatore poiché questa somma è proporzionale al numero totale di neutroni termici penetrati nel rivelatore. Se indichiamo infatti con N_1

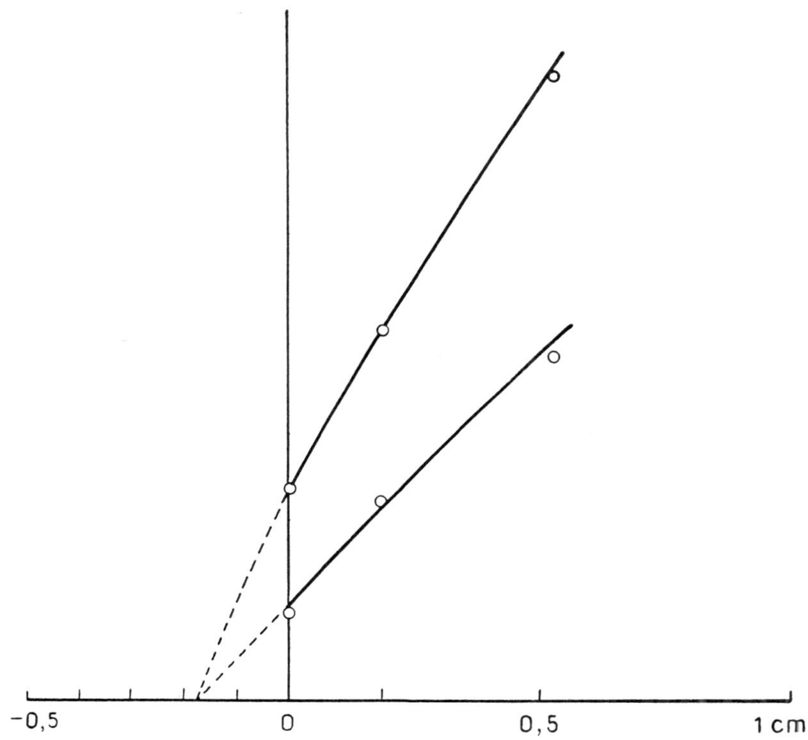

Fig. 8.

ed N_2 i numeri di neutroni che entrano dalle faccie 1 e 2 del rivelatore, le attività misurate sulle faccie 1 e 2 sono evidentemente date da

$$(20)\qquad \begin{cases} A_1 = N_1 a + N_2 b \\ A_2 = N_1 b + N_2 a \end{cases}$$

dove a e b sono costanti caratteristiche del rivelatore dipendenti dai suoi coefficienti di assorbimento per i neutroni e gli elettroni e dal suo spessore. Il rapporto a/b si ottiene misurando il rapporto A_1/A_2 quando il rivelatore è appoggiato all'esterno della paraffina in modo che $N_2 = 0$.

Nel rivelatore di rodio di 0,36 g/cm², $a/b = 4/3$ per il gruppo C; l'analogo rapporto per il gruppo D è 2. La differenza dipende dal maggior coefficiente di assorbimento del rodio per il gruppo D. Per un rivelatore sottile invece si avrebbe $a = b$.

Sommando membro a membro le (20) si ottiene

$$A_1 + A_2 = (a + b)(N_1 + N_2)$$

e quindi appunto la somma delle attività sulle due faccie è proporzionale al numero totale $N_1 + N_2$ dei neutroni che attraversano il rivelatore. Dalle (20) si ricava

$$N_1 = \frac{a A_1 - b A_2}{a^2 - b^2} \quad : \quad N_2 = \frac{a A_2 - b A_1}{a^2 - b^2} . \tag{21}$$

Misurando dunque A_1 ed A_2 e conoscendo il rapporto a/b si ricavano dei valori proporzionali ad N_1 ed N_2 (questo artificio per misurare separatamente N_1 ed N_2 è stato usato nel § 7).

Nella Tabella XVII sono riportati gli andamenti di N_1 ed N_2 così misurati con un rivelatore di rodio di 0,36 g/cm², per i gruppi C e D separatamente. Per rendere confrontabili i due dati si è assunto eguale all'unità in entrambi i gruppi il valore di N_1 per $x = 0$. La sorgente in queste esperienze si trovava a $x = 2,4$ cm.

TABELLA XVII.

x	Gruppo C		Gruppo D	
	N_1	N_2	N_1	N_2
0	1	0	1	0
0,16	1,10	0,28	1,07	0,07
0,34	1,12	0,63	1,17	0,14
0,53	1,28	0,91	1,25	0,28
0,87	1,60	1,19	1,31	0,45
1,87	2,19	2,00	1,35	1,00

I dati sul gruppo C non consentono una interpretazione semplice poiché il rivelatore non si può considerare come sottile. Nel gruppo D invece, pur essendo sempre il rivelatore spesso, il fatto che i neutroni di questo gruppo percorrano in media come tali un solo cammino libero (cfr. §§ 6, 7, 12), porta come conseguenza che ognuno di essi attraversa una sola volta il rivelatore, e quindi non vi è perturbazione dell'attività del rivelatore del gruppo D dovuta alla presenza del rivelatore stesso. Estrapolando per il gruppo D l'andamento di $N_1 + N_2$ per valori negativi di x si trova l'intersezione con l'asse delle x per

$$x_0 = -0,9 \text{ cm}.$$

Il fatto che questa intersezione avvenga notevolmente più lontano dal bordo della paraffina che per il gruppo C ($x_0 = -0{,}18$ cm con rivelatore sottile), è dovuto da un lato al maggior cammino libero medio (cfr. § 7) dall'altro alla coerenza tra le orientazioni di successivi cammini liberi, che ha luogo con notevole entità finché l'energia è superiore al quanto del legame elastico dell'idrogeno della paraffina.

Dal confronto delle colonne seconda e quarta della Tabella XVII si nota che l'andamento di N_1 in vicinanza della superficie, per il nostro rivelatore spesso, non è molto dissimile nei gruppi C e D, mentre a profondità maggiori di gruppo C cresce assai più rapidamente del gruppo D. Ciò si può, giustificare tenendo presente la crescita dell'albedo dello strato di paraffina soprapposto al rivelatore in funzione dello spessore dello strato.

12. Estensione delle bande di energia corrispondenti ai gruppi. Numero totale dei neutroni.

Discuteremo in questo paragrafo un metodo per determinare la estensione delle bande di energia che costituiscono i vari gruppi, e più precisamente il rapporto W_{max}/W_{min} tra la massima e la minima energia che limitano la banda [22].

L'ultima parte del paragrafo sarà invece dedicata alla valutazione del numero totale di neutroni emessi da una sorgente di Em + Be.

Il metodo per determinare la larghezza della banda che costituisce un gruppo g di energia superiore ad 1 volt, è basato sul confronto delle due seguenti grandezze: la attivabilità A_g di un rivelatore dovuta ai neutroni del gruppo g in esame, e la attivabilità B_c dello stesso rivelatore dovuta ai soli neutroni C ed ottenuta proteggendo il rivelatore da una parte con uno strato di cadmio di spessore sufficiente per assorbire totalmente i neutroni termici. B_c si ottiene naturalmente come differenza fra le attivabilità del rivelatore con uno strato di cadmio da una sola parte oppure incluso tra due strati di cadmio.

Se la distribuzione dei neutroni veloci entro una grande massa di paraffina fosse uniforme, è ovvio che A_g e che B_c avrebbero valore indipendente dalla posizione e dall'orientazione del rivelatore. Siccome invece la sorgente dei neutroni veloci è concentrata, si debbono confrontare i valori medi di A_g e B_c su tutte le posizioni ed orientazioni entro la paraffina, o, ciò che è lo stesso, confrontare gli integrali di volume

$$\int \bar{A}_g \, d\tau \quad \text{e} \quad \int \overline{B}_c \, d\tau$$

i cui integrandi $\bar{A}_g$ e $\overline{B}_c$ sono ottenuti facendo la media delle attivabilità del rivelatore misurate con due orientazioni opposte; tale media è praticamente identica alla media eseguita sull'intero angolo solido. Riferendosi ai detti integrali verremo ad eliminare le complicazioni dovute al fatto che i neutroni non sono distribuiti uniformemente.

Supponiamo che nella paraffina, in cui si trova il nostro rivelatore si producano q neutroni veloci per cm^3 e per s. Questi si andranno rallentando per modo che ad ogni istante vi saranno nell'ambiente neutroni di tutte le velocità.

Si può dimostrare (cfr. *F.*, § 1) che, per energie superiori ad 1 volt, il numero di neutroni di velocità fra v e $v + dv$ è

$$\frac{2\,q\lambda(v)}{v^2}\,dv. \tag{22}$$

Da questo, è possibile calcolare l'attività A_g in funzione delle energie W_{min} e W_{max} che limitano la banda di energia g, del cammino libero medio λ_g, del coefficiente di assorbimento K_g del rilevatore per il gruppo g, della sua superficie s e del suo spessore δ. Eseguendo i calcoli si trova (cfr. *F.*, § 8)

$$A_g = \eta\, s\, q\, \lambda_g K_g \log\frac{W_{max}}{W_{min}} \times \frac{\int_0^\delta b(K_g x)\, e^{-\mu x}\, dx + \int_0^\delta b(K_g x)\, e^{-\mu(\delta - x)}\, dx}{2} \tag{23}$$

dove μ è il coefficiente di assorbimento degli elettroni del rivelatore nel rivelatore stesso, e $b(K_g x)$ è la funzione (4); η è l'efficienza della camera di jonizzazione per i raggi β del rivelatore (cfr. § 2).

L'ultimo fattore rappresenta l'effetto dell'assorbimento dei neutroni e dei raggi β nel rivelatore e si ridurrebbe a δ per rivelatore sottile.

Nel calcolo di B_c (neutroni termici) intervengono, oltre agli elementi geometrici del rivelatore, il tratto di diffusione $\sqrt{D\tau} = \sqrt{\lambda^2 \frac{N}{3}}$ (cfr. § 7) e il coefficiente di assorbimento K_c del rivelatore per i neutroni termici. Eseguendo i calcoli e tenendo conto che in questo caso la legge di distribuzione angolare dei neutroni incidenti sul rivelatore è la (5), si trova

$$B_c = \eta\, s\, q\, \lambda \sqrt{N}\, K_c \int_0^\delta c(K_c x)\, e^{-\mu x}\, dx \tag{24}$$

dove $c(K_c x)$ è la funzione (6). L'integrale, che per rivelatore sottile si riduce a δ, rappresenta l'effetto dell'assorbimento dei neutroni e degli elettroni nel rivelatore.

Integrando, come si è detto, le (23) (24) si trovano le espressioni

$$\int A_g\, d\tau = \eta\, s\, Q\, K_g \lambda_g \log\frac{W_{max}}{W_{min}} \times \frac{\int_0^\delta b(K_g x)\, e^{-\mu x}\, dx + \int_0^\delta b(K_g x)\, e^{-\mu(\delta - x)}\, dx}{2} \tag{25}$$

$$\int B_c\, d\tau = \eta\, s\, Q\, K_c \lambda \sqrt{N} \int_0^\delta c(K_c x)\, e^{-\mu x}\, dx$$

dove $Q = \int q d\tau$ è il numero totale di neutroni emessi dalla sorgente per secondo. Da queste si trova

$$\log \frac{W_{max}}{W_{min}} = \frac{K_c}{K_g} \frac{\lambda \sqrt{N}}{\lambda_g} \frac{\int A_g d\tau}{\int B_c d\tau} \tag{26}$$

$$\times \frac{2 \int_0^\delta c(K_c x) e^{-\mu x} dx}{\int_0^\delta b(K_g x) e^{-\mu x} dx + \int_0^\delta b(K_g x) e^{-\mu(\delta - x)} dx}$$

Abbiamo eseguito le misure per il gruppo *D* usando un rivelatore di rodio (0,36 g/cm²), per il gruppo *A* un rivelatore di argento (0,057 g/cm²) e per il gruppo *I* un rivelatore di joduro di piombo (0,76 g/cm²); siccome non è ben conosciuto il piccolo coefficiente di assorbimento dello jodio nel gruppo *C*, abbiamo utilizzato i dati relativi al rodio per calcolare B_c anche per il gruppo *I*.

Per valutare gli integrali

$$\int \bar{A}_g d\tau \quad \text{e} \quad \int \bar{B}_c d\tau$$

abbiamo determinato $\bar{A}_g$ e $\bar{B}_c$ ad una distanza fissa deducendo i valori a distanza qualsiasi a mezzo delle curve della fig. 7.

Nelle espressioni seguenti compaiono i valori numerici usati nello stesso ordine in cui sono scritte le varie grandezze nella (26):

gruppo *D*

$$\log \frac{W_{max}}{W_{min}} = \frac{0{,}7}{1{,}8} \frac{\sqrt{13}}{1{,}1} \frac{4{,}58 \cdot 10^5}{1{,}5 \cdot 10^6} \frac{2 \cdot 0{,}108}{0{,}087 + 0{,}052} = 0{,}60$$

gruppo *A*

$$\log \frac{W_{max}}{W_{min}} = \frac{0{,}25}{20} \frac{\sqrt{13}}{1{,}1} \frac{7{,}8 \cdot 10^4}{2{,}04 \cdot 10^5} \frac{2 \cdot 0{,}048}{0{,}021 + 0{,}019} = 0{,}038$$

gruppo *I*

$$\log \frac{W_{max}}{W_{min}} = \frac{0{,}7}{0{,}38} \frac{\sqrt{13}}{1{,}1} \frac{4{,}16 \cdot 10^4}{1{,}5 \cdot 10^6} \frac{2 \cdot 0{,}108}{0{,}085 + 0{,}049} = 0{,}27.$$

Da questi valori si possono ricavare i rapporti W_{max}/W_{min} della energia massima alla energia minima di un gruppo. La grandezza $\log W_{max}/W_{min}$ che è data dalle formule precedenti e che chiameremo larghezza logaritmica del gruppo, ha significato fisico immediato in quanto rappresenta il numero medio di urti subiti da un neutrone entro il gruppo in questione.

Così, per esempio, il gruppo *A*, che ha larghezza logaritmica 0,04, è una banda di energia così ristretta che solo il 4 % dei neutroni, durante il rallentamento nella paraffina, passano da essa.

Le larghezze logaritmiche del gruppo I e sopratutto del gruppo D, sono notevolmente maggiori, ciò che trova riscontro nella maggiore numerosità (cfr. § 5) di questi gruppi. Tuttavia anche nel gruppo D è relativamente raro il caso che un neutrone percorra entro il gruppo più di un cammino libero; dalla larghezza logaritmica 0,6 si deduce che tale probabilità è del 27 per cento.

Ciò giustifica il fatto che per tutti questi gruppi è praticamente nulla l'albedo.

Ci si dovrebbe invece attendere nelle esperienze sulla diffusione (cfr. § 7) dei gruppi A e D, che il tratto di diffusione dei neutroni del gruppo D fosse leggermente maggiore di quello dei neutroni del gruppo A. Nel paragrafo 7 non abbiamo trovato una differenza di questo genere, ciò che del resto si può attribuire alla imprecisione delle misure.

Ammettendo che le energie dei vari gruppi siano quelle dedotte nel paragrafo 9, col metodo del boro si troverebbero per i gruppi D, A, e I le seguenti larghezze espresse in volt:

$$1{,}5 \quad ; \quad 0{,}23 \quad ; \quad 15 .$$

Si può infine utilizzare la seconda delle (25) per valutare il numero totale Q dei neutroni emessi dalla sorgente. B_c nella (25), è l'attività iniziale del rivelatore usato. Per passare all'attivabilità (3) bisogna moltiplicare per il fattore

$$\frac{1000}{\mathrm{IU}} .$$

Il preparato U di uranio, d'altra parte, equivale a 920 disintegrazioni al secondo, di modo che chiamando al solito η_U l'efficienza della camera di jonizzazione per i raggi dell'uranio, si può porre $U = 920\,\eta_U$. Dalla (25) si ricava dunque, esprimendo B_c in attivabilità,

$$\frac{Q}{I} = 0{,}92\,\frac{\eta_U}{\eta}\,\frac{\int B_c\,d\tau}{s\lambda\sqrt{N}\,K_c\int_0^\delta c\,(K_c x)\,e^{-\mu x}\,dx} . \tag{27}$$

In questa formula figurano gli stessi elementi che si hanno nel calcolo delle larghezze logaritmiche. Utilizzando, per esempio, i dati della attivabilità del rodio, si ottiene

$$\frac{Q}{I} = 180.009\,\frac{\eta_U}{\eta} .$$

Ammettendo inoltre le efficienze η ed η_U eguali si ricaverebbe che una delle nostre unità neutroniche equivale a 180.000 neutroni al secondo, e ricordando che una unità neutronica equivale a circa 6 mC di Em (+ Be) si trova una emissione di

30.000 neutroni al secondo per millicurie.

Questo valore è notevolmente maggiore dei valori trovati con altri metodi [23], ciò che ci sembra giustificabile solo in parte con una differenza fra η ed η_U.

13. Riassunto e discussione.

Dai risultati che abbiamo esposto sembra che si debba concludere che negli elementi sensibili ai neutroni lenti, la sezione d'urto per cattura sia spesso una funzione irregolare dell'energia dei neutroni lenti che presenta dei massimi accentuati, tali da costituire delle vere e proprie bande di assorbimento.

L'analisi delle curve di assorbimento dei vari elementi con diversi rivelatori, permette di riconoscere e differenziare fra di loro le varie bande di assorbimento (gruppi *A*, *B*, *C*, *D*, *I*).

Vari argomenti concordano nel far ritenere che il gruppo *C* (radiazione intensamente assorbita dal cadmio) corrisponda, almeno prevalentemente, a neutroni di velocità di agitazione termica, mentre gli altri gruppi a velocità superiori.

L'argomento più diretto a favore di questa interpretazione, è costituito dall'esperienza col selettore di velocità di cadmio [24] (e quindi tale da agire solo sul gruppo *C*), che ha permesso di controllare direttamente che la velocità dei neutroni C è quella di agitazione termica. Alla stessa conclusione portano le osservazioni di Preiswerk e von Halban ed altri [25], secondo le quali la radiazione filtrata da cadmio non presenta effetto termico, ciò che indica che i neutroni non appartenenti al gruppo *C* non hanno velocità di agitazione termica.

Infine le nostre misure di albedo (§ 6) dimostrano che i soli neutroni del gruppo *C* percorrono molti cammini liberi entro questo gruppo a differenza di ciò che avviene negli altri gruppi. Questo fatto si spiega facilmente ammettendo che il gruppo *C* sia costituito da neutroni in equilibrio termico con l'ambiente, poiché allora i successivi urti non ne alterano in media la energia; mentre un neutrone appartenente ad un gruppo costituito da una banda di energia $> k\mathrm{T}$, avrà una elevata probabilità di uscirne in seguito ad un urto.

In tutti i casi studiati una frazione notevole dell'attività (quasi sempre maggiore del 50 %) è dovuta ai neutroni termici; da ciò non si deve concludere che il coefficiente di assorbimento per neutroni termici sia di regola maggiore di quello per neutroni di energia di alcuni volt; il fatto dipende in buona parte anche dall'elevato numero dei neutroni di energia termica che escono da un blocco di paraffina contenente la sorgente (cfr. Tabella VI).

Il problema di localizzare le bande di energia corrispondenti agli altri gruppi (non termici) può trattarsi al modo seguente.

Frisch e Placzek e Weekes, Livingston e Bethe si sono basati sull'ipotesi che la sezione d'urto per cattura di un neutrone lento da parte del boro sia inversamente proporzionale alla velocità v del neutrone. In questa ipotesi le energie dei vari gruppi risultano dalle nostre misure:

gruppo *C*	:	0,037 volt ($= 3/2\ k\mathrm{T}$)
gruppo D	:	2,24 volt
gruppo *A*	:	6 volt
gruppo *B*	:	10 volt
gruppo *I*	:	55 volt.

Si intende che l'errore in misure di questo genere può essere considerevole.

Un controllo dell'ipotesi che sta a base di questa valutazione si avrebbe dal confronto dei coefficienti di assorbimento del litio per i vari gruppi con quelli del boro, poiché i rapporti dovrebbero essere eguali dato che il meccanismo di assorbimento dei neutroni lenti in questi due elementi è il medesimo. Dalle misure di von Halban e Preiswerk [26] sembrerebbe che un tale accordo non vi sia, benché le differenze possano essere forse dovute a cause secondarie.

Una determinazione indipendente della successione energetica dei gruppi ed anche una valutazione dei loro rapporti di energia (quest'ultima solo per i gruppi non termici) è stata da noi data dal paragrafo 9 in base alla determinazione del comportamento dei vari gruppi in funzione della distanza dalla sorgente. L'ordine di energia da noi trovato coincide con quello dedotto dagli assorbimenti del boro ed i rapporti di energia, per quanto un po' più piccoli non sono incompatibili coi valori dati precedentemente. Il nostro metodo, pur non essendo molto preciso, presenta il vantaggio di essere più diretto.

In base a considerazioni analoghe alle precedenti, abbiamo anche valutate le estensioni delle bande (propriamente $\Delta W/W$) corrispondenti ai vari gruppi non termici. Abbiamo trovato che la banda più ristretta è quella detta gruppo *A* (radiazione fortemente assorbita dall'argento) per cui $\Delta W/W = 0{,}04$. Gli altri gruppi (cfr. § 12) sono alquanto più larghi, sempre però abbastanza ristretti in modo che è piuttosto raro il caso che un neutrone percorra più di un cammino libero restando sempre appartenente alla stessa banda di energia.

Collegata alla larghezza della banda è la numerosità dei vari gruppi, che rappresenta essenzialmente il numero di neutroni che escono per secondo da una superficie di paraffina (per la definizione precisa cfr. § 5). Il più numeroso di tutti è il gruppo termico (numerosità 45), mentre il meno numeroso è il gruppo *A* (numerosità 0,5) che è anche, come si è visto, il più ristretto.

Questi fatti sono qualitativamente in accordo con le considerazioni di Bohr e di Breit e Wigner.

Qualche ulteriore informazione sulle proprietà di queste bande di assorbimento si può avere dal confronto del comportamento di due isotopi, entrambi sensibili ai neutroni lenti. Il risultato di queste ricerche, che non abbiamo potuto estendere a molti casi, può riassumersi dicendo che sembra esistere una certa tendenza per diversi isotopi di uno stesso elemento ad avere bande di assorbimento prossime (§ 5); non mancano tuttavia le eccezioni.

I paragrafi dal 6 all'11 sono dedicati allo studio delle proprietà di diffusione dei neutroni lenti nelle sostanze idrogenate. Il processo di diffusione si può analizzare in due fasi successive. In un primo tempo (fase di rallentamento) il neutrone perde energia in seguito a successivi urti fino a raggiungere l'energia di agitazione termica; dopo di ciò (fase di diffusione termica) l'energia in media non decresce ulteriormente ed il neutrone seguita a diffondere fino a che non viene catturato dai protoni o da altri nuclei presenti.

Durante la fase di rallentamento il cammino libero medio scende ben presto ad un valore dell'ordine di 1 cm, dopo di che si mantiene approssimativamente costante fino a che l'energia del neutrone non diventa comparabile con il quanto $h\nu$ delle frequenze delle oscillazioni elastiche dell'idrogeno nella paraffina.

Si può ritenere che tutti i gruppi, ad eccezione di quello termico, appartengano all'intervallo di energia in cui il cammino libero medio λ_g nella paraffina è circa 1 cm. Risulta dalla teoria dell'urto tra neutroni lenti ed atomi di idrogeno, tenendo conto del legame chimico di questi ultimi (cfr. *F.*, §§ 10, 11) che quando l'energia dei neutroni lenti scende da valori superiori ad $h\nu$ a valori piccoli, il cammino libero medio decresce e tende ad un limite eguale a $1/4\,\lambda_g$. A questa variazione del cammino libero medio è connessa anche una differente distribuzione angolare dei neutroni dopo un urto; quando $W \gg h\nu$ l'angolo fra le direzioni del moto del neutrone prima e dopo l'urto è sempre acuto, mentre, per $W \ll h\nu$ la distribuzione angolare dopo un urto tende a diventare isotropa.

In accordo con questi risultati teorici si osserva effettivamente una notevole riduzione del cammino libero medio nel passaggio dai gruppi non termici al gruppo *C*. Per quest'ultimo si può ritenere che $W/h\nu$ sia all'incirca $1/10$; non ostante che questo valore sia abbastanza piccolo, non si possono confondere i neutroni termici con neutroni di energia praticamente nulla. Così risulta (cfr. *F.*, formula 102) che per $W/h\nu = 1/10$ il cammino libero medio è

$$\lambda = \lambda_g/3{,}3$$

mentre al limite per $W = 0$ si avrebbe $\lambda = \lambda_g/4$.

Si può anche valutare fino a che punto è mantenuta la coerenza tra le direzioni del moto di un neutrone termico prima e dopo un urto.

Come espressione quantitativa della coerenza si può prendere $\overline{\cos\Theta}$, e cioè il valor medio del coseno dell'angolo di diffusione dei neutroni dopo l'urto. $\overline{\cos\Theta} = 1$ significa coerenza completa, mentre per diffusione isotropa $\overline{\cos\Theta} = 0$. Per neutroni di energia superiore ad un volt si trova $\overline{\cos\Theta} = 2/3 = 0{,}67$; per neutroni termici invece, ammettendo $W/h\nu = 1/10$, si deduce (cfr *F.*, formula 103) $\overline{\cos\Theta} = 0{,}067$, e cioè un decimo del valore precedente. Tale risultato giustifica l'approssimazione fatta più volte nel corso di questi lavori, di considerare la diffusione dei neutroni termici come isotropa. Si noti tuttavia che in tal modo si viene ad introdurre un piccolo errore che è nel senso che il tratto di diffusione, dato dalla (12), è troppo piccolo, mentre l'albedo, data dalla (11), è troppo grande. Sarebbe anche facile valutare queste correzioni, una volta noto con sicurezza $\overline{\cos\Theta}$; questo valore è però subordinato alla conoscenza di $W/h\nu$ del quale si può dare solo l'ordine di grandezza.

Nel corso dei paragrafi dal 6 all'11 sono state misurate 5 grandezze diverse le quali si possono esprimere in funzione di due sole, grazie alle relazioni ottenute dalla teoria della diffusione dei neutroni (cfr. *F.*). Queste 5 grandezze sono i cammini liberi medi λ e λ_g dei neutroni termici (cfr. § 8) e dei neutroni dei gruppi *D* ed *A* (cfr. § 7); l'abedo β dei neutroni termici (cfr. § 6), il tratto di diffusione l dei neutroni termici (cfr. § 7) e la lunghezza $|x_0|$ relativa ai neutroni termici, considerata nel paragrafo 11, che è legata a λ dalla relazione

$$|x_0| = \frac{\lambda}{\sqrt{3}}\,. \tag{28}$$

Per mezzo delle (28), (11), (12), (29) le 5 grandezze ora citate si possono esprimere in funzione di λ_g ed N (numero di cammini liberi percorsi in media da un neutrone termico).

Nella Tabella XVIII viene dato il confronto dei valori misurati delle 5 grandezze con quelli calcolati per mezzo delle formule sopra citate assumendo

$$\lambda_g = 1{,}0 \text{ cm} \quad ; \quad N = 140.$$

Come si vede l'accordo è molto soddisfacente.

TABELLA XVIII.

	Misurato	Calcolato
λ	0,3	0,30
λ_g	1,1	1,0
β	0,82	0,83
l	2,1	2,05
$\lvert x_0 \rvert$	0,18	0,174

Da questi valori di λ_g e di N risultano i seguenti dati per i neutroni termici nella paraffina:

sezione d'urto elastica coll'idrogeno . $\sigma_e = 43 \cdot 10^{-24}$ cm^2

sezione d'urto con cattura . . . . . . $\sigma_c = 0{,}31 \cdot 10^{-24}$ cm^2

vita media . . . . $\tau = 1{,}7 \cdot 10^{-4}$ sec.

Questo valore è in ottimo accordo con la misura di τ eseguita con un dispositivo meccanico (27) e col valore teorico dedotto dalla teoria svolta in *F.*, § 12.

Esprimiamo i nostri più vivi ringraziamenti all'Istituto di Sanità Pubblica e in particolare al prof. Trabacchi, per averci fornito le sorgenti di emanazione con le quali è stato eseguito questo lavoro. Larga parte dei mezzi impiegati nelle ricerche ci sono stati messi a disposizioni dal Consiglio Nazionale delle Ricerche.

Istituto fisico della R. Università
Roma, 29 maggio 1936-XIV.

BIBLIOGRAFIA.

[1) E. AMALDI, E. FERMI, « Ric. Scientifica », *6* (2), 334, 443 (1935); *7* (1), 56, 223, 310, 410 (1936), ved. anche E. FERMI, E. AMALDI, O. D'AGOSTINO, F. RASETTI, E. SEGRÈ, « Proc. Roy. Soc. » A, *146*, 483 (1934); E. AMALDI, O. D'AGOSTINO, E. FERMI, B. PONTECORVO, F. RASETTI, E. SEGRÈ, « Proc. Roy. Soc. » A, *149*, 522 (1935).

[2] E. AMALDI, O. D'AGOSTINO, E. FERMI, B. PONTECORVO, F. RASETTI, E. SEGRÈ, « Proc. Roy. Soc. » A, *149*, 522 (1935); H. A. BETHE, « Phys. Rev. », *47*, 747 (1935); F. PERRIN, W. M. ELSASSER, « C. R. », *200*, 450 (1935).

[3] P. B. MOON, J. R. TILLMAN, « Nature » (London), *135*, 904; *136*, 136 (1935); T. BJERGE, C. H. WESTCOTT, « Proc. Roy. Soc. » A, *150*, 709 (1935); L. ARSIMOVITCH, I. KOURTSCHATOW, L. MISSOWSKI, P. PALIBIN, « C. R. », *200*, 2159 (1935); B. PONTECORVO, « Ric. Scientifica », *6* (2), 145 (1935); L. N. RIDENOUR, DON M. YOST, « Phys. Rev. », *48*, 383 (1935).

[4] P. B. MOON, J. R. TILLMAN, « Nature » (London), *135*, 904 (1935); DUNNING, PEGRAM, FINK, MITCHELL, « Phys. Rev. », *48*, 265 (1935); P. B. MOON, J. R. TILLMAN, « Proc. Roy. Soc. » A, *153*, 476 (1936).

[5] F. RASETTI, E. SEGRÈ, G. A. FINK, J. R. DUNNING, G. B. PEGRAM, « Phys. Rev. », *49*, 104 (1936), « Rend. Lincei. », *23*, 343 (1936).

[6] E. AMALDI, E. FERMI, « Ric. Scientifica », *6* (2), 344 (1935).

[7] L. SZILARD, « Nature » (London), *136*, 950 (1935).

[8] E. FERMI, E. AMALDI, « Ric. Scientifica », *6* (2), 443 (1935).

[9] E. FERMI, *Sul moto dei neutroni nelle sostanze idrogenate*, « Ric. Scientifica », 7 (2), 13 (1936).

[10] P. B. MOON, J. R. TILLMAN, « Nature », *136*, 66 (1935); B. PONTECORVO, « Ric. Scientifica », *6* (2), 145 (1935); E. AMALDI, E. FERMI, « Ric. Scientifica », *6* (2), 344 (1935).

[11] E. AMALDI, E. FERMI, « Ric. Scientifica », *6* (2), 344 (1935), L. SZILARD, « Nature » (London), *136*, 950 (1935).

[12] N. BOHR, « Nature », *137*, 344 (1936); G. BREIT, E. WIGNER, « Phys. Rev. », *49*, 519 (1936).

[13] O. R. FRISCH, G. HEVESY, H. A. C. MCKAY, « Nature » (London), *137*, 149 (1936).

[14] E. FERMI, E. AMALDI, « Ric. Scientifica », *6* (2), 443 (1935); E. SEGRÈ, « Ric. Scientifica », 7 (1), 389 (1936).

[15] E. FERMI, E. AMALDI, « Ric. Scientifica », *6* (2), 443 (1935).

[16] J. R. TILLMAN, « Nature » (London), *137*, 107 (1936).

[17] T. BJERGE, C. H. WESTCOTT, « Proc. Roy. Soc. » A, *150*, 709 (1935); E. FERMI, E. AMALDI, « Ric. Scientifica », *6* (2), 443 (1935), 7 (1), 56 (1936).

[18] E. AMALDI, O. D'AGOSTINO, E. FERMI, B. PONTECORVO, F. RASETTI, E. SEGRÈ, « Proc. Proy. Soc. » A, *149*, 522 (1935).

[19] O. R. FRISCH, G. PLACZEK, « Nature » (London), *137*, 357 (1936); D. F. WEEKES, M. S. LIVINGSTONE, H. A. BETHE, « Phys. Rev. », *49*, 471 (1936).

[20] E. AMALDI, E. FERMI, « Ric. Scientifica », 7 (1), 56 (1936).

[21] P. PREISWERK, H. VON HALBAN, « C. R. », *202*, 840 (1936).

[22] E. AMALDI, E. FERMI, « Ric. Scientifica », 7 (1), 310 (1936).

[23] Cfr. p. es. R. JAECKEL, « Z. Physik », *91*, 493 (1934); F. A. PANETH, H. LOLEIT « Nature » (London), *136*, 950 (1935).

[24] J. R. DUNNING, G. B. PEGRAM, G. A. FINK, D. P. MITCHELL, E. SEGRÈ, « Phys. Rev. », *48*, 704 (1935); « Rend. Lincei », *23*, 340 (1936).

[25] P. PREISWERK, H. VON HALBAN, « Nature » (London), *136*, 1027 (1935), F. RASETTI, G. FINK, « Phys. Rev. », *49*, 642 (1936).

[26] H. VON HALBAN, P. PREISWERK, « Nature » (London), *137*, 905 (1936).

[27] E. AMALDI, O. D'AGOSTINO, E. FERMI, B. PONTECORVO, E. SEGRÈ, « Ric. Scientifica », *6* (1), 581 (1935).

118 *b*.

ON THE ABSORPTION AND THE DIFFUSION OF SLOW NEUTRONS

E. AMALDI and E. FERMI
« Phys. Rev. », *50*, 899–928 (1936).

The study of the absorption properties of slow neutrons establishes the existence of absorption bands which correspond to energy intervals sufficiently narrow to allow one to isolate relatively homogeneous neutron groups. In this work, we describe several experiments and considerations designed to disclose the characteristics of the absorption and scattering properties of the various groups. Summary: 1. Introduction. – 2. Measurements and their reduction. – 3. Selective absorptions. – 4. Slow neutron groups. – 5. Systematic absorption measurements. – 6. Albedo. – 7. Scattering of the neutrons of a single group. – 8. Mean free path of thermal neutrons. – 9. Groups as a function of the distance from the source; energy ratios. – 10. Passage of the neutrons from one group to another. – 11. Behavior of the groups in the neighborhood of the surface of the paraffin. – 12. Width of the energy bands corresponding to the groups. Total number of neutrons. – 13. Summary and discussion.

1. INTRODUCTION.

The purpose of this work is to describe systematically the researches we have carried out on the absorption and scattering properties of slow neutrons [1]. It is known that the simplest theory [2] on the probability of capture of a neutron by a nucleus leads to the prediction that the capture cross section is, for small neutron velocities, inversely proportional to the velocities. However, the constant of proportionality can vary between large limits from element to element.

If this were true, thicknesses of two different elements inversely proportional to their constants would be exactly equivalent as abosrbers of slow neutrons independently of neutron velocity, and of the substance used as detector. That the phenomenon was not so simple was early observed by various authors [3], who showed that generally the absorption by an element

(1) E. AMALDI and E. FERMI, « Ric. Scientifica », *6* (2), 334, 443 (1935); *7* (1), 56, 223, 310, 393, 454 (1936); see also: E. FERMI, E. AMALDI, O. D'AGOSTINO, F. RASETTI and E. SEGRÈ, « Proc. Roy. Soc. », *A 146*, 483 (1934); E. AMALDI, O. D'AGOSTINO, E. FERMI, B. PONTECORVO, F. RASETTI and E. SEGRÈ, « Proc. Roy Soc. », *A 149*, 522 (1935).

(2) E. AMALDI, O. D'AGOSTINO, E. FERMI, B. PONTECORVO, F. RASETTI and E. SEGRÈ, « Proc. Roy. Soc. », *A 149*, 522 (1935); H. A. BETHE, « Phys. Rev. », *47*, 747 (1935); F. PERRIN and W. M. ELSASSER, « C. R. », *200*, 450 (1935).

(3) P. B. MOON and J. R. TILLMAN, « Nature » (London), *135*, 904; *136*, 136 (1935); T. BJERGE and C. H. WESTCOTT, « Proc. Roy. Soc. », *A 150*, 709 (1935); L. ARSIMOVITCH, I. KOURTSCHATOW, L. MISSOWSKI and P. PALIBIN, « C. R. », *200*, 2159 (1935); B. PONTECORVO, « Ric. Scientifica », *6* (2), 145 (1935); L. N. RIDENOUR and DON M. YOST, « Phys. Rev. », *48* 383 (1935).

was greater when the slow neutrons were detected by means of the activity induced in the same element.

The thermal experiments of Moon and Tillman and others (4), which showed that different detectors exhibit different sensitivities with a variation of the temperature of the paraffin, lead to a similar conclusion. Finally, more recently, Rasetti, Segrè, Fink, Dunning and Pegram (5) have measured the dependence of the absorption coefficient of thermal neutrons on their velocity by means of a mechanical arrangement, and have found that, at least for Cd, the $1/v$ law is not fulfilled. In a systematic study of these phenomena, we have found that the selective absorption of slow neutrons is accentuated by filtering with a sheet of cadmium of sufficient thickness (6). This fact, observed independently by Szilard (7), pointed to the possible existence of relatively narrow absorption bands characteristic of various elements.

A systematic description of these phenomena can be obtained by means of an analysis of the slow neutron radiation into groups which are approximately homogeneous with respect to their absorption properties (8). Such analysis could be performed by decomposing the absorption curves into exponentials. We also studied different groups of neutrons by means of their scattering properties in hydrogenated substances.

In § 2, we describe the methods used in performing the experiments and the reduction of the results. In §§ 3, 4, 5, we have collected some systematic data on the absorption properties and on the intensity of the activation produced by the various groups in several detectors. In § 6, we describe the measurement of the reflection coefficient (albedo) for the neutrons of different groups; in §§ 7 and 8, we study the mean free paths and the average number of impacts of a neutron of a given group in paraffin. In § 9, we show how the intensity of the activation due to several groups varies as a function of the distance from the source inside a large mass of water; from this relation we can deduce the energy sequence of the various groups and we are also able to evaluate the ratio of the energies of different groups. In §§ 10 and 11, we describe some diffusion experiments. In § 13, all the results are collected and discussed.

In this work, we will often make use of some theoretical relationships that have been derived by one of us (9). We shall refer to this paper in what follows by F.

(4) P. B. MOON and J. R. TILLMAN, «Nature» (London), *135*, 904 (1935); J. R. DUNNING, G. B. PEGRAM, G. A. FINK and D. P. MITCHELL, «Phys. Rev.», *48*, 265 (1935); P. B. MOON and J. R. TILLMAN, «Proc. Roy. Soc.», *A 153*, 476 (1936).

(5) F. RASETTI, E. SEGRÈ, G. A. FINK, J. R. DUNNING and G. B. PEGRAM, «Phys. Rev.», *49*, 104 (1936); «Rend. Lincei», *23*, 343 (1936).

(6) E. AMALDI, E. FERMI, «Ric. Scientifica», *6* (2), 344 (1935).

(7) L. SZILARD, «Nature» (London), *136*, 950 (1935).

(8) E. FERMI and E. AMALDI, «Ric. Scientifica», *6* (2), 443 (1935).

(9) E. FERMI, *Sul moto dei neutroni nelle sostanze idrogenate*, «Ric. Scientifica», *7* (2), 13 (1936).

2. Measurements and Their Reduction.

As neutron source, we always used small glass bulbs about 4 mm in diameter and 15 mm in length which contained Rn (up to 800 mc) and powdered beryllium. Since our researches were extended over a period of several months, it was clearly necessary to establish some standard method of comparing results obtained with different sources and at different times.

We now describe the criteria employed for this purpose: the measurements of activity were made with an ionization chamber represented schematically in Fig. 1, which also contains the scale. The chamber was filled with carbon dioxide at three atmospheres and was closed by a window of aluminum 0.1 mm thick and 7 cm in diameter. Since the aluminum window is greatly curved by the pressure, it was covered by a cellophane foil which served the double purpose of protecting the chamber from possible contamination and providing a flat surface upon which to place the detectors. The ionization was measured with an Edelmann electrometer, the wire of which was projected on a scale; it was possible to vary the sensitivity of the electrometer and the magnification of the image of the wire so that we obtained from 5 to 250 divisions per volt. To check the readings we used a uranium oxide preparation contained in a packet of aluminum 5 × 5 cm². We will designate this preparation and its activity by U.

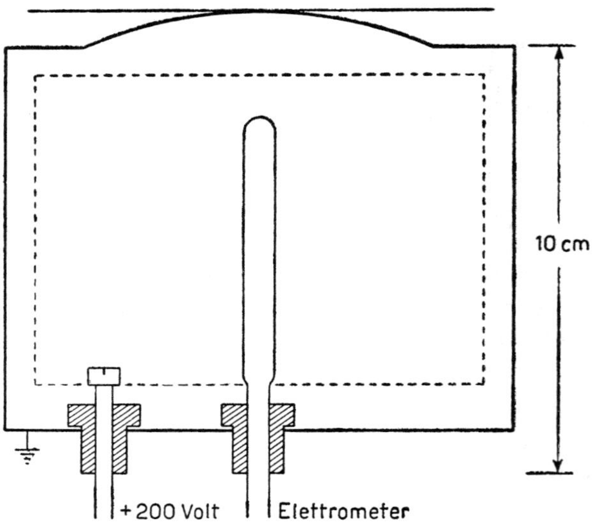

Fig. 1. – Ionization chamber.

The ionization produced in our chamber by the U is equivalent to that obtained from a solution of 0.96 g of uranyl nitrate ($UO_2(NO_3)_2$ + 6 H_2O) in 25 cm³ of water contained in a dish of aluminum 0.1 mm thick and 5 × 5 cm² base. Correcting the activity of this solution for the absorption in the solution and in aluminum, we find that it is equal to 0.066 g of uranium element, that is 840 disintegrations per second.

In order to make the readings more rapidly and to be able to use different sensitivities of the electrometer and different scale intervals, we constructed a nomogram which permitted us to find directly, given the sensitivity of the electrometer and the number of divisions used in the reading, the time it would take the U to produce the same ionization. In this way one obtains automatically the correction for the nonlinearity of the instrument and for the dependence of the capacity on the sensitivity. Naturally the readings were corrected for the zero effect which was about 15 percent of the U. The activity of a given detector placed in a definite position with respect to the source and the surrounding objects is evidently proportional to the number of neutrons emitted by the source. In order to have comparable data, it is always necessary to divide the activity by the intensity of the source measured in neutrons (neutron intensity). In order to measure the neutron intensity, it is not sufficient to assume it to be proportional to the quantity of radon contained in the source, which can be measured for example by the γ activity; in fact small differences in the size of the beryllium granules or in the preparation of the source, produce considerable variations in the number of emitted neutrons. Because of this it is convenient to measure the neutron intensity directly by means of the radioactivity induced in a suitable detector. Our usual procedure was as follows: a sheet of rhodium of $5 \times 5.6\ \text{cm}^2$ and 10.15 g was activated by placing it in the center of the top surface of a paraffin cylinder

(1) 25 cm diameter, 15 cm height;

on top of the sheet of rhodium was placed a second cylinder of paraffin of 13 cm diameter and 10 cm height; the source was placed on the axis of the lower cylinder 3 cm below its upper surface.

We will designate by S the initial activity of the 44 s period of Rh obtained under these conditions after irradiation for an infinite time. We will call neutron intensity I the ratio of this initial activity S to the U:

$$\text{I} = \text{S}/\text{U}. \tag{2}$$

It should be noted that this definition depends, though not very markedly, on the ionization chamber used. The strongest source we have used had a neutron intensity of 144.9. A neutron intensity 1 corresponds to a source which contains from 5 to 6 mc radon.

We are now able to define what we shall call *activability* A of a given detector (referring, when necessary, to one of its periods) placed in a definite position with respect to the source and the surrounding objects.

Let a be the initial activity of a detector irradiated for an infinite time; we define the activability A of this detector in a given position by the expression

$$\text{A} = 1000\, a/\text{S} = 1000\, a/\text{I} \times \text{U}. \tag{3}$$

In practice, naturally, we have calculated the activability by means of this last formula; the neutron intensity was measured once for each source with very high accuracy, and from this value the intensity at any later time was calculated by means of the period of radon.

The initial intensity a was measured by irradiating for a finite time, and was reduced to infinite time by well-known formulae.

In the case of elements of short period, it is convenient to establish a standard for the measurement of the initial activity in order to be as independent as possible, at least for relative measurements, from the possible errors in the value of the period. Therefore, in the case of rhodium (44 s) and of silver (22 s), which we have used as detectors in a very large number of experiments, we have performed the irradiation for 1 minute; in the case of silver, the readings were made from 20 s to 80 s after the end of the irradiation, and the activity so measured multiplied by an appropriate factor, gave the initial activity. In reality, a measurement of this kind gives not only the 22 s period, because there is a disturbance due to the 2.3 min period; this disturbance was not very large and ordinarily was not corrected for (see also § 5).

In the case of rhodium we have always irradiated 1 min and we have taken the readings from about 30 s to 60 s after the end of irradiation; the initial activity was rapidly calculated by means of a nomogram. Also in this case we have neglected the small disturbance due to the 4.2 min period.

In the course of this work we have performed a large number of measurements of absorption of slow neutrons in different substances. The criteria adopted in these measurements have been the following: The measurements were performed outside of the paraffin because otherwise a large fraction of the neutrons pass through the absorber many times (see § 6). The source of Rn + Be was usually inserted in the paraffin cylinder (1), 3 cm below the center of the upper face on which were placed the various detectors. Most of the detectors were squares of 5 cm side.

We will call normal activability A_n of a detector its activability under these conditions. This furnishes an indication of the efficiency of the different detectors. As examples, in Table I we give the normal activabilities of some of the detectors which we have used.

For absorption measurements, the absorbers were inserted in layers as thin as possible between the paraffin and the detectors.

In order to correct the error due to the height h of the detector from the top of the paraffin, we have measured the variation of the activity of a rhodium detector as a function of h; the practical rule found for reducing the activability to $h = 0$, consists in dividing the results of the measurements by $1 - (h/7)$, where h is the height in cm; this rule is sufficiently correct for h up to a little more than 1 cm.

Rigorously, the correction for height would depend on the absorber used and on the group of neutrons with which we work. For the present, we have neglected these refinements. Even if the neutrons leaving the paraffin were homogeneous, the absorption curves obtained under the conditions which we have described, would not be exponential, because the neutrons which come out with different inclination θ traverse different thicknesses of the absorber. In the calculation of the absorption coefficients from the experimental data, it is necessary to take these facts into account; for a long time we have calculated this correction on the assumption that the neutrons leave the paraffin according to the cosine law.

TABLE I.

Activability A_n *of various detectors.*

Substance	Period	Weight (in g)	Area (cm^2)	A_n
MnO_2	2,5 hr	30	26	30
Cu	5 min	13	38	2
Ga	20 min	2.6	7	1.3
As	26 hr	7	20	24
NaBr	18 min	26	25	9.2
Rh	44 s	10.15	28	282
Ag	22 s	8	27	135
In	54 min	0.76	9	21.2
In	16 s	0.76	9	14.2
I	25 min	25	25	16.9
Ir	19 hr	14.5	25	78
Au	27 d	6.6	25	26

On this assumption, the absorption curve with a thin detector, as a function of $K\delta$ (K = absorption coefficient of the neutrons, δ = thickness of the absorber) instead of being the exponential $e^{-K\delta}$ (curve *a* of Fig. 2) is the function

$$b\,(K\delta) = \int_0^1 e^{-K\delta/x}\,dx \tag{4}$$

represented by the curve *b* of Fig. 2. We now believe that for the neutrons of group *C* (see *F*., § 6) it is better to assume that the number of neutrons emerging per unit solid angle instead of being proportional to cos θ, is approximately proportional to

$$\cos\theta + \sqrt{3}\cos^2\theta\,. \tag{5}$$

If we take this angular distribution of the neutrons, the absorption curve as a function of $K\delta$ is given by the expression

$$c(K\delta) = \frac{2}{2+\sqrt{(3)}} \int_0^1 e^{-K\delta/x} (1 + \sqrt{3}\,x)\, dx \tag{6}$$

represented by the curve *c* of Fig. 2. This curve probably represents the behavior of the absorption of the C neutrons fairly well; for the neutrons of other groups, it is probable that the absorption curve will be between *b* and *c*; however all reductions have been made with curve *c*. In the case of thick detectors, in other words when the absorption of the neutrons in the

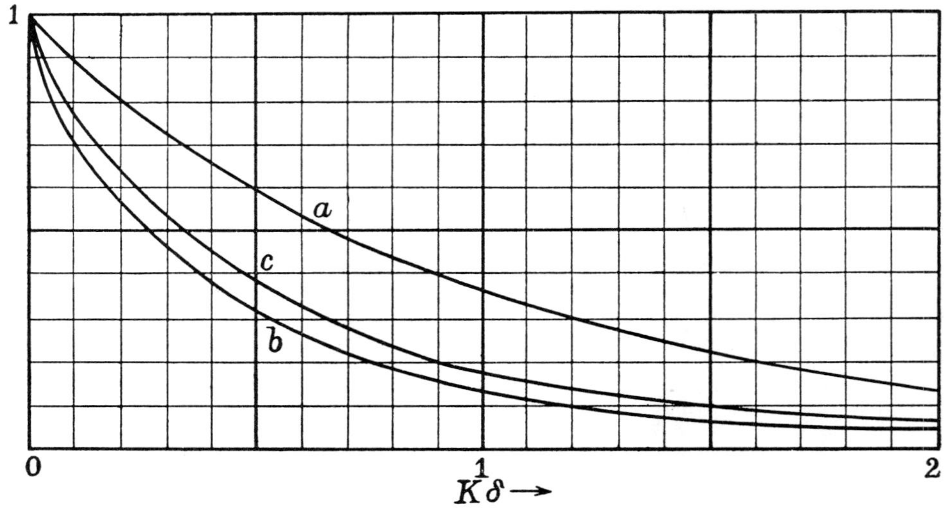

Fig. 2. – Absorption curves following different laws. Curve (a): $a(K\delta) = e^{-K\delta}$; curve (b): $b(K\delta) = \int_0^1 e^{-K\delta/x} dx$; curve (c): $c(K\delta) = [2/(2+\sqrt{3})] \int_0^1 e^{-K\delta/x} (1+\sqrt{3}\,x)\, dx$.

useful thickness was considerable, the data obtained from curve *c* were further corrected.

The differences between the absorption coefficients given in this work and those reported in preliminary communications, are due, in large part, to the fact that the reductions have been made with curve *c* instead of curve *b* of Fig. 2.

3. Selective Absorptions.

We have already mentioned that, as was observed by different authors, the absorption of slow neutrons in an element is generally stronger when one uses a detector of the same element. For instance, Table II indicates the percentages of the activity transmitted by absorbers of 0.36 g/cm² of rhodium and 0.96 g/cm² of silver when the detector is a rhodium target of

0.36 g/cm² or a silver target of 0.80 g/cm². Data of this kind concerning a larger number of elements have been published, by various authors [10].

TABLE II.

Absorption of slow neutrons as measured by different detectors.

Absorber / Detector	Rh	Ag
Rh	54	67
Ag	68	45

These facts suggest the existence of selective absorptions of the slow neutrons. The most obvious method of studying these selective absorptions consists in measuring and analyzing the absorption curves. In Fig. 3 are given the absorption curves of cadmium with rhodium detector (0.36 g/cm²) and silver detector (0.057 g/cm²). It is clear from these curves that the radiations which are responsible for the activity of both these detectors are not homogeneous, but consist of a fraction which is absorbed very strongly in cadmium, and a fraction which is very weakly absorbed. Taking into account the fact that with our geometrical disposition the absorber is traversed obliquely by the neutrons (see § 2), it is possible to calculate the absorption coefficient of the strongly absorbed fraction from each of these curves.

In this way we obtain, from the curve with silver detector K = 16 cm²/g, and from the curve with rhodium detector K = 13.5 cm²/g. The absorption coefficient of the hard fraction is certainly less than 1/100 of these values: so, for instance, in the case of the rhodium detector we find K = 0.05 cm²/g.

From the examination of the curves of Fig. 3, it can be seen that in our silver detector half of the activity is due to the component which is strongly absorbed in cadmium and half to the hard component. Instead, in our rhodium detector 72 percent of the total activity is due to the soft component and the remainder to the hard component. Therefore the absorption by cadmium makes it possible to divide the slow neutrons which come out of a paraffin block containing a source of neutrons into two fractions: that fraction which is strongly absorbed by cadmium will be hereafter designated as group *C*. We will discuss its absorption properties in more detail in § 5.

We will now study the properties of the radiations which traverse a thickness of cadmium sufficient to totally absorb the *C* group. For this purpose we have repeated some absorption experiments on the radiations

(10) P. B. MOON and J. R. TILLMAN, « Nature » (London), *136*, 66 (1935); B. PONTECORVO, « Ric. Scientifica », *6* (2), 145 (1935); E. AMALDI and E. FERMI, « Ric. Scientifica », *6* (2), 344 (1935).

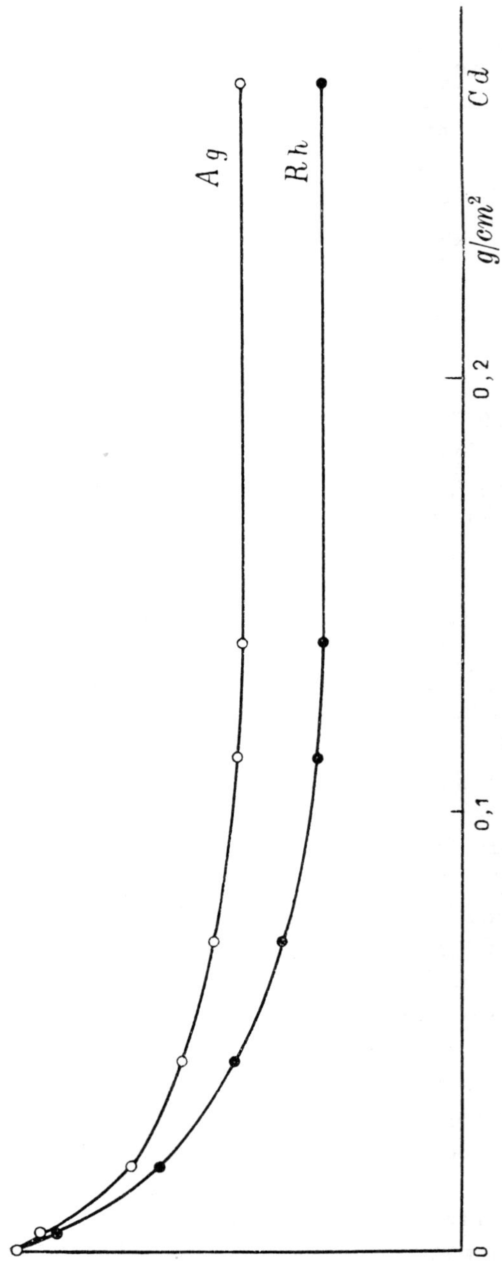

Fig. 3. – Absorption curves for Cd as determined by Ag and Rh detectors.

filtered by 0.27 g/cm² cadmium using rhodium and silver as absorbers as well as detectors (11). Table III, similar to Table II, is given as an example of the results of such experiments; the results summarized in Tables II and III were obtained with the same absorbers and detectors.

TABLE III.

Absorption of radiations filtered by 0.27 g/cm² Cd.

Absorber	Rh	Ag
Detector		
Rh	33	84
Ag	88	29

Comparison of Tables II and III shows that the selectivities are accentuated for the radiation filtered by cadmium, which indicates that the component weakly absorbed by cadmium, is not homogeneous. In order to further investigate the properties of the radiation filtered by cadmium, we have measured the absorption curve of silver with silver detector (2.2 g/28 cm²) both for the unfiltered radiation and for the radiation filtered by 0.27 g/cm² of cadmium. These curves are given in Fig. 4 where the abscissa represents the thickness of the silver absorber and the ordinate represents the activability of the detector. The upper curve refers to the unfiltered radiation and the lower curve to the radiation filtered by 0.27 g/cm² of cadmium.

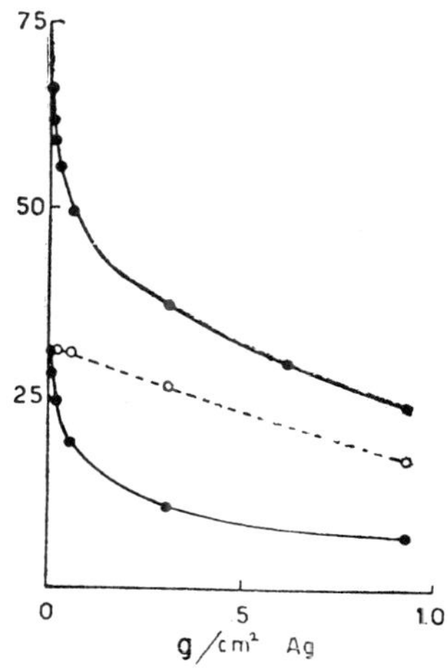

Fig. 4. – Upper curve: absorption curve for Ag, unfiltered radiation. Lower curve: absorption curve for Ag, radiation filtered by 0.27 g/cm² of Cd. Ordinates are the activability of Ag detector. Dotted curve is difference of upper and lower curves.

In both curves, we see that the radiation which is responsible for the activity of silver contains a component strongly absorbed in the same element, which component is present with the same intensity both in the filtered and in the unfiltered radiations. This means that this component is not strongly absorbed by cadmium. We will call this component group *A*. The absorption coefficient of the neutrons of group *A* in silver is K = 20 cm²/g.

Curves of similar character have also been found with a thicker silver detector; in this case however the percentage of activity due to the radiation

(11) E. AMALDI and E. FERMI, « Ric. Scientifica », *6* (2), 344 (1935); L. SZILARD, « Nature », *136*, 950 (1935).

strongly absorbed in silver is naturally smaller, since it is saturated in less than 0.1 mm of thickness of the detector.

The dotted curve of Fig. 4 is the difference of the two other curves; it therefore represents the behavior of that fraction of the silver activity that is absorbed by 0.27 g/cm^2 of Cd. It is, therefore, the absorption curve of silver with silver detector due only to the neutrons of group *C*. From this we obtain as value of the absorption coefficient of silver for the C neutrons, K = 0.3 cm^2/g.

4. Slow Neutron Groups.

We have seen from the examples of the preceding paragraph, that the analysis of the absorption curves permits us to classify slow neutrons into components which are differently absorbed in different elements. It is well known that the analysis of absorption curves into exponentials can be considered reliable only when the accuracy of the individual measurements is very high, and also the geometrical disposition of the absorber and detector with respect to the source, corresponds to accurately calculable conditions. In our case, neither of these conditions were exactly satisfied, and therefore we seek the possibility of checking accurately the homogeneity of each component. For this reason, we have limited ourselves to the description of the properties of groups of approximately homogeneous neutrons.

The differences of behavior from group to group are so marked that it is possible to assign to each of them an individuality, despite the small internal inhomogeneity. The analysis into groups is made possible by the following circumstances:

(*a*) For some absorbers there exist very large differences in the absorption coefficients from group to group.

(*b*) It is possible to use different absorbers and detectors which behave quite differently.

(*c*) It is possible to filter the slow neutrons with absorbers so as to reduce the inhomogeneity.

The most obvious assumption concerning the physical nature of the difference between the various groups is that it is due to a difference in velocity. In a paraffin block the fast neutrons which escape from the source suffer a progressive slowing down due to the successive collisions with the hydrogen atoms until they arrive to the energy of thermal agitation; therefore, in paraffin there are present at each instant neutrons of all velocities, from the energy of thermal agitation to the energy with which they leave the source. Likewise, neutrons of all these velocities diffuse out of the paraffin block. Hence we may think that the different groups correspond to neutrons of different energy intervals. Selective absorptions must be interpreted as due to irregular variations of the absorption coefficient as a function of the neutron velocity.

The possibility of an irregular dependence of the absorption coefficient on the neutron velocity has been demonstrated by recent considerations of

Bohr, Breit and Wigner (12). We will return to the dicussion of the physical origin of the groups after we have described the absorption experiments and also the scattering properties of the various groups.

Already in the preceding paragraph, we have called the radiation strongly absorbed in cadmium (K = 13.5 ÷ 16 cm²/g) group *C*; and the radiation strongly absorbed in silver (K = 20 cm²/g) group *A*.

We have also seen that only 50 percent of the activity of our silver detector (0.057 g/cm²) is due to group *C*; of the remaining 50 percent about one-half is due to group *A*, while the rest is due to a radiation which is only slightly absorbed in cadmium and in silver (see Fig. 4); this radiation we have called group *B*, which is very probably complex. Up to now we have not been able to analyze it into approximately homogeneous components, since we have not found any elements which absorb group *B* with a high absorption coefficient.

In the preceding section, we have shown that 72 percent of the activity of our rhodium detector (0.36 g/cm²) is due to group *C*; the remaining 28 percent constitutes a new group, which we have called group *D*, which is very slightly absorbed by cadmium. In studying a PbI_2 detector (19 g/25 cm², normal activability 11.4), we have found that only about 25 percent of its activity is due to group *C* and less than 10 percent is due to group *A*; the remainder is due to a component which is very weakly absorbed in all elements that we have studied up to date, except iodine; this radiation which is responsible for about 70 percent of the activity of our detector has been called the I group.

Frisch, Hevesy and McKay (13) have studied the absorption of gold with gold detector using the cadmium filtered neutrons and have obtained results similar to those we obtained for silver (see Fig. 4). We have tried to analyze the activity of a gold detector (1.75 g/25 cm²) in terms of the groups already mentioned; we have found that the activity of gold is due in part to group *C*, *A* and *B*; the high absorption coefficient of gold with gold detector is probably due to a radiation distinct from all the preceding groups.

5. Systematic Absorption Measurements.

Group C. – Of all groups the group *C* is the best known, because it is responsible in almost all detectors for a large percentage of the activity; there are many reasons for believing (see § 13) that this group is composed, in large measure, of neutrons having energy of thermal agitation; for this reason we will often refer to it as the thermal group. The study of the *C* group is particularly easy because of the fact that a thickness of 0.3 ÷ 0.4 g/cm² of cadmium absorbs it practically completely without appreciably

(12) N. Bohr, «Nature» (London), *137*, 344 (1936); G. Breit and E. Wigner, «Phys. Rev.», *49*, 519 (1936).

(13) O. R. Frisch, G. Hevesy and H. A. C. McKay, «Nature» (London), *137*, 149 (1936).

absorbing any of the other groups. In order to obtain the fraction of the activity due to the *C* group, it is sufficient to obtain the difference in the activity observed without and with a filter of cadmium of the above thickness.

In Table IV we give the absorption coefficients in cm^2/g for the C neutrons, observed with various detectors (always forming the difference between the activity without and with a cadmium filter) and various absorbers. It is to be noted that the absorption coefficients of the same element measured with different detectors are approximately the same. The differences are due in part to errors in measurements; however we believe that they are not completely accounted for by this reason.

Despite this, if we remember the large differences of the absorption coefficients of an element for the neutrons of different groups, we see that the C group actually has a notable internal homogeneity.

TABLE IV.

Absorption coefficients in cm^2/g *for group C neutrons (those strongly absorbed by* Cd).

Absorber	Rh	Ag	Cd	In	Ir	Au	Hg
Mn 2.5 ha . .	0.9	0.3	—	—	—	—	—
Rh 44 s . . .	0.6	0.2	13.5	0.6	1.0	0.2	0.7
Ag 22 s . . .	0.7	0.3	16	—	0.9	0.25	0.9
In 54 min . .	0.6	0.2	14	0.9	1.0	0.2	0.6
Ir 19 hr . .	1.0	0.3	—	—	—	—	0.6

Group D. – The *D* group is easily observed by using rhodium or indium as detector; with both these elements precise measurements are possible.

The fact that it is possible to characterize the *D* group by using one or the other of these detectors indicates that the corresponding absorption bands are nearly coincident; naturally there is no reason to believe that this coincidence is complete; we will see in § 9 that of all the groups we have studied, the *D* group has smaller energy than any, except the C group.

We have already mentioned that 28 percent of the activity of our detectors of rhodium is due to the *D* group. In an indium detector of 0.76 g/9 cm^2, 42 percent of the activity of the 54 min period is due to the *D* group; the remaining 58 percent is due to the *C* group. In order to study group *D*, it is therefore sufficient to use one of these detectors, filtering the radiation by 0.3÷0.4 g/cm^2 of cadmium in order to eliminate the *C* group. In Table V some absorption coefficients in cm^2/g of the D neutrons are given.

TABLE V.

Absorption coefficients in cm²/g for group D neutrons (those strongly absorbed in Rh).

Absorber	B	Rh	Ag	Cd	In	Ir	Au	Hg
Detector								
Rh 44 s	4.7	2.0	0.06	0.05	3.0	1.0	0.03	0.07
In 54 min	—	1.6	0.09	—	3.8	1.0	0.04	0.04

Group A. – This group, characterized by the large absorption coefficient in silver (K = 20 cm²/g), is ordinarily studied by using a thin sheet of silver as detector.

As was mentioned above, 25 percent of the normal activability of a silver detector of 0.057 g/cm² is due to the A neutrons. In order to isolate this group it is sufficient to form the difference of the activabilities of a silver detector without and with a silver filter sufficiently thick to absorb completely the *A* group and not thick enough to appreciably absorb the other groups. In practice it is convenient to eliminate the *C* group by means of a cadmium filter.

The *A* group is also detected by gold; a gold detector of 1.535 g/25 cm² has a normal activability of 21, of which 13 percent is due to group *A*; the absorption coefficient of gold for this group is 4 cm²/g. The absorption coefficient of the *A* group in boron is 3 cm²/g.

Not much is known concerning the properties of the other groups mentioned in the preceding section. We have performed a few measurements of absorption coefficients which we give here:

Group *B* : boron K = 2.3 cm²/g ; gold K = 1.7 cm²/g.

Group *I* : boron K = 1 cm²/g ; iodine K = 0.7 cm²/g.

We have seen that despite the fact that the *C* group almost always constitutes more than 50 percent of the activity of the different detectors, placed outside of the paraffin, the other groups are found in a smaller number of detectors and are responsible for a smaller percentage of the activities.

This is not due to the fact that the coefficient of absorption of the different elements are greater for the neutrons of group *C* than for the neutrons of the other groups; we have already encountered numerous examples to the contrary. It is rather due to the fact that the number of C neutrons which come out of paraffin is much larger than the number of neutrons of the other groups.

Therefore, a very important datum for characterizing a group is its *numerosity*. We will define this quantity for practical purposes as follows:

the numerosity N_0 of a group is the number of neutrons which are emitted per second per square centimeter and per unit of neutron intensity of the source, averaged over 25 cm² at the center of the top surface of the paraffin (1) containing the source (neutron intensity I) 3 cm below the center of the top.

The principle on which the measurement of the numerosity of a group is based is as follows: We place on the center of block (1) a thin detector of area s; the number of neutrons which strike this detector is

$$N_0 I s .$$

If we designate the absorption coefficient of the detector by K, assuming that its thickness δ is very small, it is possible to see that the number of neutrons captured by the detector is

$$\sqrt{3}\, N_0 I s K \delta ,$$

where the factor $\sqrt{3}$ is due to the obliquity of the neutrons, which is taken into account according to Eq. (5).

If we designate by η the efficiency of the ionization chamber, that is the ionization current due to one disintegration per second on the surface, the normal activability (see Eq. (3)) of our detector is

$$A = 1000 \sqrt{3}\, s N_0 K \delta \,(K_1/K)\,(\eta/840\, \eta_U) , \tag{7}$$

where η_U is the efficiency of the chamber for the β–rays emitted by the uranium preparation we used, which, as already has been mentioned, corresponds to 840 disintegrations per second at the surface.

The factor K_1/K is introduced in order to take into account that when a detector absorbs neutrons of a given group (with an absorption coefficient K) it is possible that a fraction of these gives rise either to nonradioactive isotopes or to radioactive isotopes with periods different from that studied; we indicate by K_1 the absorption coefficient due only to the proces which is responsible for the studied activity; it is evident that although K is the coefficient which determines the absorption of the neutron in the detector, K_1 determines its activity.

From (7) we find the numerosity of the group under examination

$$N_0 = 0.485 \frac{\eta_U}{\eta} \frac{K}{K_1} \frac{A}{s K \delta} . \tag{8}$$

In this expression for N_0, the quantities A, s, K, δ are directly measurable; in most cases it is possible to put K_1 equal to K; in the other cases where an element has more than one mean life, the ratio of their activabilities corrected for the absorption of the β–rays and for the different efficiencies η, allows one to evaluate K/K_1.

A considerable error in the measurement of N_0 is due to the insufficient knowledge of the ratio of the efficiency of the chamber for β–rays of different hardness. From the relatively small variations which are found in the calculated numerosities for different detectors on the assumption of a constant efficiency η, it can be concluded that this quantity does not vary between very large limits. In the following calculations we have always put $\eta = \eta_U$.

In Table VI we give the values of the numerosities of the various groups calculated using different detectors; as absorption coefficients we have used the averages of the data of the preceding tables. The values given in Table VI have been calculated with a formula similar to (8) in which however the absorption of the neutrons and the electrons in the detector have been taken into account. We have also introduced a small correction in order to take into account the reflection of the electrons from the upper layers of the detector.

TABLE VI.

Numerosity N_0 *for various neutron groups.*

Detector	Group			
	C	*D*	*A*	*I*
Rh	43	9	—	—
Ag	51	—	0.6	—
In	31	6	—	—
I	—	—	—	5
Au	33	—	0.5	—

The data for indium have been calculated by adding the contributions of the two periods of 16 s and 54 min and by neglecting the weak period of 4 hours. For rhodium and for silver we have made the calculations by taking into account not only the 44 s and the 22 s periods, but also the 4.2 min and 2.3 min periods.

If we try to calculate the numerosity of group *C* from the activity of iridium (19 hr), we find a value of about 10; this value seemed to us much too small to be attributed to errors in measurements. Therefore we have tried to find another activity strong enough to justify the previous small value of the numerosity. In reality this suspicion was well grounded since we have found a second iridium activity with a period of 68 days whose activability although not yet measured, is of the order of magnitude of the activability of the 19 hr period. The β–particles of this new activity have very small penetrating power.

It is very interesting to determine for elements which have two or more periods, both due to slow neutrons, whether or not the activities of the different periods are due to the same groups.

Therefore we have studied the following detectors: rhodium (0.36 g/cm^2; 44 s, 4.2 min), silver (0.057 g/cm^2; 22 s, 2.3 min), indium (0.065 g/cm^2; 16 s, 54 min), sodium bromide (1 g/cm^2; 18 min, 4.2 hr) and we have compared for each of them the activability of the two periods with and without appropriate filters.

In the case of rhodium, the ratio of the activabilities of the 44 s and of the 4.2 min periods remains the same (about 10) with and without a filter of cadmium sufficient to absorb the *C* group [14]. In the case of silver, however, we have found a marked difference in the behavior of the two periods; in group *C* the activability of the long period is 24 percent of the total activity; in group *B*, 19 percent; finally in group *A* we have found 3 percent which within the limits of experimental error could be zero; we can therefore say that the group *A* is characteristic of the 22 s period of silver.

In the case of indium the percentage of activation of the two periods due to group *C* and *D* are approximately equal. Similarly for bromine, a cadmium filter does not considerably change the ratio of the activability of the two periods.

We have also replaced the paraffin cylinder on which the measurements were usually made, by an equal cylinder of water and using rhodium, silver and lead iodide detectors and appropriate filters, we have determined the percentages of activation due to the different groups. In this way we have been able to show that the ratios of the numerosities of the various groups are the same in water as in paraffin. Also the absolute values do not vary considerably. However it should be noted that the ratios of the intensities of the various groups undergo a variation at different distances from the source' as will be discussed in more detail in § 9.

Finally we have tried to see whether the ratio of intensity of the various groups is affected by using different source of fast neutrons. [15]. For this purpose we have determined the activability of rhodium and silver detectors with and without cadmium and silver filters, using a source newly prepared, in which the radium C was not yet in equilibrium, and we have followed the increase of activability during the formation of radium C. We have not found any difference in the ratio of the groups for silver and rhodium during the formation of radium C, although the activability increased from 40 (effect due to the α–particles of Rn + RaA) to 100 (effect due to the α–particles of Rn + RaA + RaC). In Fig. 5 we give the curve of the increase of the activity as a function of the time. One should note the high efficiency of the radium C α–particles for producing neutrons from beryllium; their efficiency is three times greater than that of the α–particles of Rn and RaA.

6. Albedo.

The measurements of normal activability as also the absorption measurements of which we have spoken up to now are made outside of the paraffin block. In this section we will study the properties of the slow neutrons inside the paraffin.

(14) E. Fermi and E. Amaldi, « Ric. Scientifica », *6* (2), 443 (1935); E. Segrè, « Ric. Scientifica », *7* (1), 389 (1936).

(15) E. Fermi and E. Amaldi, « Ric. Scientifica », *6* (2), 443 (1935).

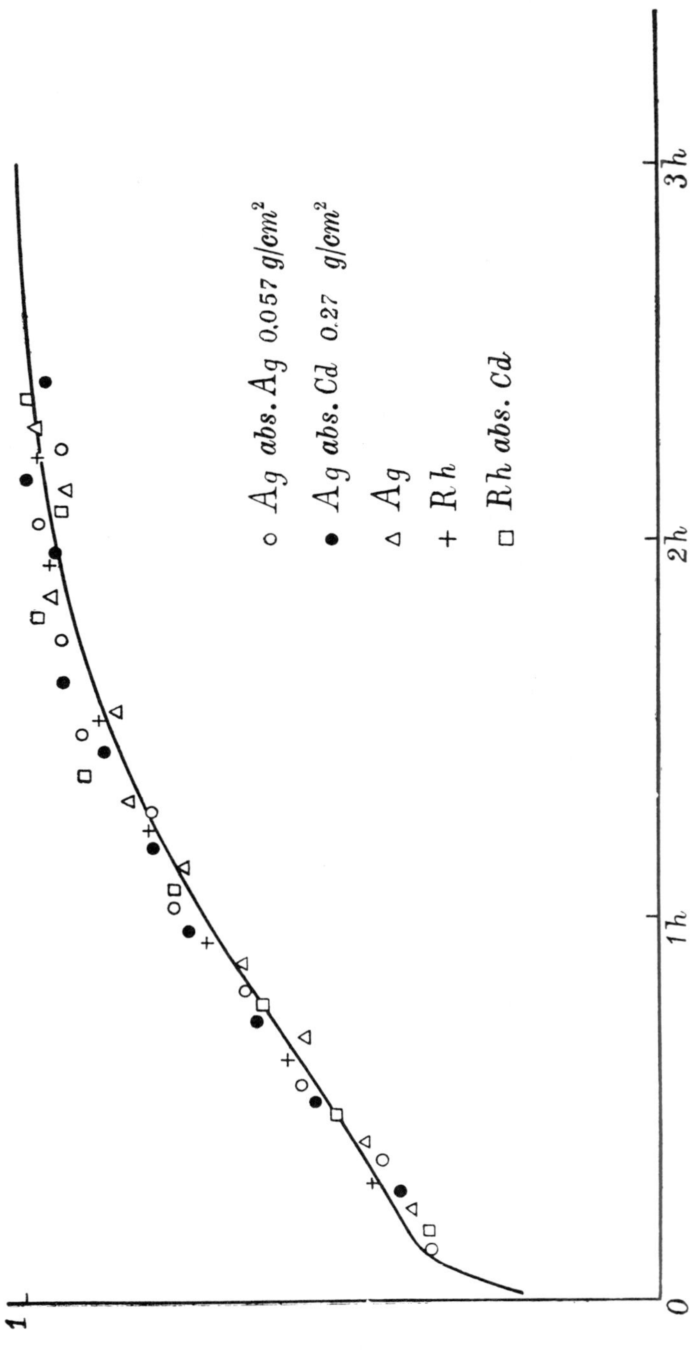

Fig. 5. – Increase of activity as a function of time.

The detector (rhodium 10 g/28 cm², or silver 2.2 g/38.5 cm²) was placed in the center of the top of the normal paraffin block (1) and upon it was placed a second equal paraffin cylinder. The detector could be placed between appropriate filters.

In order to indicate rapidly the relative position of source, paraffin, detector and absorbers, it is convenient to set up the following convention: we will indicate with the letter S the source, P the paraffin, R the detector, and with the symbols Cd and Ag the cadmium (0.27 g/cm²) and silver (0.057 g/cm²) absorbers. Thus S P Cd R Cd P means an experiment in which above the source (S) there are 3 cm of paraffin (P), followed by a layer of cadmium (Cd), the detector (R), a second cadmium layer (Cd), and finally the second paraffin cylinder (P).

In Table VII we give the activabilities of the two detectors of silver and rhodium in various experiments. If we compare the experiment S P R and S P R P we note the strong increase of the activity due to the superposition of the second block of paraffin. This large increase is due to the fact that neutrons, both fast and slow, which in the absence of the second block would escape, can now be reflected back from this block.

TABLE VII.

Activabilities for various arrangements.

S = source, P = paraffin, Cd = cadmium, Ag = silver, R = detector.

Arrangement	Silver	Rhodium
S P R	66.1	282
S P Cd R	32.2	80
S P Cd Ag R	17.5	78
S P R P	417	1025
S P Cd R Cd P	66.5	140
S P Cd R P	92.2	244
S P R Cd P	113.1	371
S P Cd Ag R Ag Cd P	34.5	144
S P Cd Ag R Cd P	45.6	—
S P Cd R Ag Cd P	54.8	—

If we put the detector between appropriate absorbers, it is possible to analyze the radiation responsible for the activity into groups. Thus the comparison of experiments S P R P and S P Cd R Cd P shows that in the case of silver detectors the fraction $(417 - 66.5)/417 = 0.84$ of the activity is due to group C and in the case of rhodium detector, the fraction $(1025 - 140)/1025 = 0.86$ is due to group C. It is to be noted that the ratio of group C to the other groups is much larger inside of the paraffin than outside; we will soon see the reason for this behavior. As a check of the correctness of the analysis into groups of the activity of a detector, also inside of the paraffin, we have doubled the thickness of the cadmium absorbers in the experiment S P Cd R Cd P, and we have found only a very small further decrease in the activity. In the case of the silver detector, the addition of silver absorbers to those of cadmium causes a reduction of the activity to one-half (experiments S P Cd R Cd P and S P Cd Ag R Ag Cd P); this fact shows that also inside the paraffin it is possible to speak of a group A intensely absorbed by silver. In agreement with what is obtained outside the paraffin, the thin absorbers of silver added to those of cadmium do not produce a further reduction of the rhodium activity.

Let us now study the behavior of group C. Let us consider the results of the measurements S P R P, S P Cd R P, S P R Cd P, S P Cd R Cd P; this last measurement indicates what part of the activity is not due to the group C, so that if we subtract this value from the results of the other three measurements, we will obtain the fraction of the activity due only to group C.

The difference between S P Cd R P and S P R Cd P can easily be interpreted by taking into account the fact that the density of slow neutrons inside the paraffin decreases with the distance from the source in such a way that a sheet of cadmium which screens the detector on the side of the source stops more neutrons than an equal sheet on the other side. The mean value of these two experiments may be considered as the value which would be obtained in the case of a uniform density distribution of slow neutrons inside the paraffin, when the detector is screened on one side by cadmium.

Therefore we find for the silver detector, as activability due only to group C, $A_c = 417 - 66.5 = 350.5$. The activability due only to group C when the silver is screened only on one side by a layer of cadmium is $B_c = (92.2 + 113.1)/2 - 66.5 = 36.1$.

The corresponding values for rhodium are $A_c = 885$, $B_c = 167.5$. The ratio A_c/B_c is for silver 9.7 and for rhodium 5.3.

An elementary consideration allows us to understand the values of this ratio. We will call β the probability that a neutron that is incident on a flat surface which limits a very large paraffin block, comes out after having suffered several collisions in the paraffin.

We call this value β *albedo*, which corresponds to the coefficient of diffuse reflection of slow neutrons from a paraffin surface. Actually it depends upon the angle of incidence of the slow neutrons, and it is evidently smaller, the smaller the angle of incidence (see *F.*, § 5).

In this elementary consideration we will neglect this dependence.

We call ζ the probability that a neutron which traverses the detector is captured by it. Also for ζ we will neglect the dependence on the angle of incidence.

The method for the measurement of the albedo is based on the comparison of the activabilities A_c and B_c defined above.

Let $\mathfrak{N}$ be the number of C neutrons which are incident on a detector of group *C* when it is screened on one side by a total absorber of C neutrons. The number of neutrons captured by the detector is $\mathfrak{N}\zeta$ and its activability B_c will be proportional to $\mathfrak{N}\zeta$.

Let us now remove the absorber and let us calculate the increase of the activity of the detector.

First, we must take into account the fact that the neutrons can impinge on the detector on both sides; this gives rise to a factor 2.

But a much more important increase is due to the fact that when one places immediately behind the detector a sheet which absorbs all neutrons, a C neutron which impinges on the detector traverses it only once and therefore has the probability ζ of being captured; when the absorber is not present the neutron can traverse the detector many times.

Actually, the first time the neutron has the probability ζ to be captured and the probability $1 - \zeta$ to go through; the probability that the neutron goes back to the detector is therefore $\beta(1 - \zeta)$ and the probability that it is captured in the second passage is $\beta(1 - \zeta)\,\zeta$ and so on. Thus we find that the probability of capture is

$$\zeta + \zeta\beta(1 - \zeta) + \zeta\beta^2(1 - \zeta)^2 + \cdots = \frac{\zeta}{1 - \beta(1 - \zeta)} = \frac{\zeta}{1 - \beta + \beta\zeta}\,.$$

The number of captured neutrons is

$$2\,\mathfrak{N}\zeta/(1 - \beta + \beta\zeta)$$

and hence the ratio A_c/B_c is

$$A_c/B_c = 2/(1 - \beta + \beta\zeta)\,. \tag{9}$$

If ζ is very small, i.e., in the case of very thin detector, this reduces to

$$A_c/B_c = 2/(1 - \beta)\,. \tag{10}$$

Therefore the measurement of the ratio A_c/B_c allows us to determine the albedo β.

In the case of our silver detector, ζ is very small but not negligible. We can evaluate it from its thickness and absorption coefficient, also taking into account the obliquity of the neutrons. Thus we find $\zeta = 0.03$ and therefore $\beta = 0.82$.

The rhodium detector is much thicker; for it ζ is about 0.35; it is clear that a small error in this large value of ζ gives rise to a great error in the value of β. Therefore we cannot use the measurements with rhodium in order to determine β, but we can rather calculate for this detector the ratio A_c/B_c, using the value of the albedo found by silver measurements. We have $A_c/B_c = 4.3$ in sufficient agreement with the experimental value 5.3.

We have observed that the concept of albedo, as we have introduced it, is not completely defined, since the percentage of neutrons reflected by a paraffin surface depends on the angular distribution of the incident neutrons. In order to give a more definite meaning to the concept of albedo, we can define β by means of Eq. (10), which holds in the case of an infinitely thin detector (experimental albedo; see *F.*, § 7).

The experimental albedo depends on the number N of free paths which a neutron C can traverse on the average before it is captured by protons in paraffin.

It is possible to show (*F.*, formula 63) that the relation connecting these two values is

$$\beta = 1 - 2/\sqrt{N}. \tag{11}$$

This equation holds on the assumption that the energy of thermal agitation is negligible with respect to the quantum $h\nu$ of the elastic bond of hydrogen atom in paraffin; otherwise it is necessary to introduce a small correction (see § 13).

From the value $\beta = 0.82$, we find from Eq. (11) $N = 124$.

We have performed some experiments similar to those just described, in order to determine the albedo of the neutrons of group *A*, using a detector of silver of 0.010 g/cm², so that ζ for the *A* group is not too large.

The behavior of the group *A* is, in this respect, quite different from that of group *C*, as we found that the albedo of group *A* is practically zero. This fact must not be interpreted in the sense that the A neutrons are not at all reflected from paraffin, but it shows that when A neutrons are reflected they undergo such a change in velocity that they no longer belong to the *A* group.

Because of the low albedo of group *A*, the increase in activity produced by A neutrons when a second block of paraffin is placed on the detector is only of the order of 2. This increase is not due to the albedo but to the fact that the neutrons can enter the detector on both sides.

Groups *B*, *D* and *I* also have a negligible albedo. Hence it follows that for all the groups, except group *C*, the change in the velocity due to one impact is normally sufficient to remove the neutron from the energy band corresponding to the group (see § 12).

The high value of the albedo of the group *C* compared to the albedo of the other groups is responsible for the fact that inside the paraffin the percentages of the activations due to the *C* group are always much higher than outside the paraffin.

Similarly, one can interpret the experiments of Tillmann [16] on the different behavior of various detectors lying on a block of paraffin, as a function of the thickness of a second superimposed layer of paraffin. A strong relative increase of the activity of a detector is to be expected for detectors which are very sensitive to group *C*; naturally it is necessary to take into account in the interpretation of these experiments also the coefficient ζ of the detector according to relation (9).

(16) J. R. TILLMAN, «Nature» (London), *137*, 107 (1936).

7. Scattering of the Neutrons of a Single Group.

From the results of the preceding paragraph we deduced that the C neutrons can traverse a large number of free paths. Their motion in paraffin is therefore analogous to a diffusion, provided that the destruction of the neutrons by proton capture is taken into account. Let N be the average number of free paths of a C neutron and λ its mean free path; evidently the neutron will diffuse through a length of the order of magnitude of $\lambda\sqrt{N}$; therefore the diffusion experiments will enable us to measure this magnitude [17]. We shall call the expression

$$l = \lambda\,(N/3)^{1/2} \tag{12}$$

diffusion length; this magnitude can be obtained directly from the measurements we are going to describe.

Let us consider a C neutron inside a paraffin block limited by a plain surface and let x be the distance of the neutron from this surface.

It can be proved by the diffusion theory (see *F.*, § 3) that the probability $p(x)$ for the neutron to get out of the paraffin before it is captured is

$$p(x) = e^{-x/(D\tau)^{1/2}}\,, \tag{13}$$

where D is the diffusion coefficient for thermal neutrons in paraffin, and τ is their mean life for the capture process. In the case of thermal neutrons the kinetic energy is small compared to the quantum of the frequencies due to the elastic bond of the hydrogen in paraffin. We shall therefore make the assumption that the hydrogen atoms can be considered as fixed centers of isotropic diffusion, and that the mean free path λ is independent of the velocity (see *F.*, §§ 10, 11). These assumptions are not quite correct, as shall be discussed in detail in § 13.

In the case of isotropic diffusion, the diffusion coefficient D is given by

$$D = \frac{1}{3}\lambda v\,, \tag{14}$$

v being the average velocity of the neutrons; we have furthermore

$$N = v\tau/\lambda\,; \tag{15}$$

hence

$$D\tau = \frac{1}{3}\lambda^2 N\,. \tag{16}$$

We find finally (see *F.*, formula 34)

$$p(x) = e^{-x/(\lambda^2 N/3)^{1/2}} = e^{-x/l}\,. \tag{17}$$

By determining the probability $p(x)$ we shall be able to measure l and therefore also $\lambda^2 N$.

(17) T. Bjerge and C. H. Westcott, « Proc. Roy. Soc. », *A 150*, 709 (1935); E. Fermi, and E. Amaldi, « Ric. Scientifica », 6 (2), 443 (1935); 7 (1), 56 (1936).

The principle on which the determination of $p(x)$ is based is the following: let R be a detector of the *C* group, large enough to cover all the plane surface of a large paraffin block from which C neutrons can emerge. We will assume that this detector absorbs all the outcoming thermal neutrons in a thickness small compared to the absorption of the β-rays. The activity of this detector is proportional to the number of neutrons emerging from the paraffin, and is independent of their angular distribution. If we now destroy Q thermal neutrons at a depth x below the paraffin surface, the activity of the detector shall decrease by an amount proportional to $Q \cdot p(x)$; in order to destroy these neutrons we can put an absorber of thermal neutrons at the depth x; we shall then be able to deduce Q from the activity induced in the absorber.

The experiment was performed by putting the detector on a cylinder of paraffin 24 cm in diameter and 15 cm high. The source was 3.5 cm below the center of the top face. The upper part of the cylinder was sectioned into slices so as to enable us to insert the absorber at the various depths. As detector we used a rhodium plate (10 g/28 cm^2); as this plate covered only a small part of the upper paraffin surface, we took measurements placing the detector in nine different positions and adding the results.

As absorber of thermal neutrons, we used a plate of Cd–Sn alloy containing 0.019 g/cm^2 cadmium. As a small thickness of cadmium practically absorbs only the C neutrons, the decrease in activity of the detector is due to C neutrons only, as can be easily verified. Since the absorption of neutrons in cadmium does not activate this element we could not measure directly the number Q of the absorbed neutrons; therefore we measured the activity induced by C neutrons in a rhodium plate having for the *C* group the same absorption as our Cd absorber. We measured the activity on the two faces of this rhodium plate and added the results in order to take into account that the neutrons entered from both sides as explained in § 11.

The absorber was put at the distances $x = 0, 1, 2, 3$ cm. In Table VIII are collected the results of the measurements. Column one gives the depth of the absorber; columm two gives the decrease in the activability of the detector due to the presence of the absorber; column three gives the sum of the activabilities of the rhodium plate equivalent to the absorber, measured on both sides for the group *C* only; column four gives the ratio of column two to column three. In column five the values of these ratios are corrected in order to take into account the fact that our detector does not satisfy the condition of totally absorbing the C neutrons in a thickness small compared to the absorption of electrons (absorption coefficient for electrons 7.3 cm^2/g in rhodium; absorption coefficient for neutrons 0.7 cm^2/g; thickness 0.36 g/cm^2). From this it follows that our detector is more sensitive to the neutrons of larger obliquity. The absorber, when in the position $x = 0$, absorbs more intensely those neutrons that get out very obliquely; the decrease in activity of the detector is therefore in this case relatively larger than in the case when the absorber is placed deep inside the paraffin, since in this latter case there is no coherence between the direction of the emergent neutrons and the direction that they had when traversing the absorber. In order to take into account this factor we must reduce the value of the ratio for $x = 0$ in order that it may be compared to the other ratios.

TABLE VIII.

Data for the determination of the probability $p(x)$.

x (cm)	Decrease of Activability	Q	Ratio	Corrected Ratio
0	89	358	0.249	0.207
1	139	1135	0.123	0.123
2	128	1580	0.081	0.081
3	98	1950	0.050	0.050

A quantitative calculation shows that the first ratio must be corrected by a factor 0.832.

The values of the corrected ratios given in the last column are proportional to $p(x)$; it can be seen that they depend with good approximation exponentially on x. The diffusion length is the inverse of the coefficient of this exponential; thus we find

$$l = 2.1 \text{ cm}.$$

From this we obtain according to (12),

$$\lambda^2 N = 13 \text{ cm}^2.$$

We performed similar experiments also for groups D and A. In the case of the D group, the source was 3.4 cm below the center of the upper face of a cylinder of paraffin similar to that used in the preceding experiments. In the case of group A the distance was 2.4 cm.

As detectors, we used, respectively, for the group D and A indium (0.065 g/cm^2) and silver (0.057 g/cm^2), both screened with cadmium filters of about 0.5 g/cm^2. As absorbers we used for the D group rhodium, (0.36 g/cm^2) and for the A group silver (0.057 g/cm^2).

In Table IX we have collected the results of these measurements; in column one we give the depth x of the absorber; columns two and five give for the groups D and A the decreases in activability of the detectors; column three is the sum of the activability measured at various depths on both faces of the rhodium absorber screened by cadmium so as to measure the activity due to the D group only; column six contains similar data for the silver absorbers of group A. Columns four and seven give the activabilities of the absorber measured on the lower face and due only to the neutrons of group D or A that enter the absorber from the lower face (for the method of performing this measurement see § 11). The numbers in brackets of the last column have been interpolated using some results similar to those mentioned in § 11.

TABLE IX.

Data for D *and* A *neutrons.*

x = depth of absorber.

x	Group D			Group A		
	Decrease of Activability	Activability of Absorber	Lower Activability	Decrease of Activability	Activability of Absorber	Lower Activability
0	12.0	102	68	15.3	27.3	17.4
0.16	9.6	105	71	11.6	31.2	(18)
0.34	6.4	124	75	8.3	34.6	(19)
0.87	2.7	173	93	3.8	51.5	(22)
1.87	0.5	244	101	0.5	61.5	23.1

From the low albedo of the groups D and A, we have already deduced in the preceding paragraph that the neutrons of these groups generally traverse as such only one free path. We cannot therefore apply to these groups the theory of diffusion. It is nevertheless illuminating to reduce the data of the measurements on the groups D and A by the same method that we used for the C group. For this, we examine, as a function of x, the ratios of the second to the third column, and the ratios of the fifth to the sixth column. One finds that both ratios decrease exponentially with a good approximation; in both cases the inverse coefficient of the exponential is 0.42 cm. We notice the analogous behavior of these two groups in contrast with the markedly different behavior of the C group.

It is more correct, however, to reduce the data of Table IX on the extreme assumption that the neutrons of groups D and A can traverse only one free path. Assuming this, we can deduce the mean free path of the neutrons of these two groups. If this assumption is not correct, the values that we shall obtain would represent only upper limits for the mean free path.

Let us compare the activities of the detector of group D with or without the absorber at a depth x. Both activities are the sum of two terms: (*a*) activity due to D neutrons that have not crossed the plane at depth x; (*b*) activity due to D neutrons that have crossed this plane.

From our assumption that the D neutrons can traverse only one free path, it follows that all the neutrons that contribute to the part (*b*) of the activity have traversed the plane x from below. The term (*a*) is not changed by the presence of the absorber in the position x, while the term (b) is decreased by the absorber. The total decrease in activity is hence equal to the decrease of the term (b) only. For simplicity we shall in this discussion denote as neutrons of class b the D neutrons that arrive on the plane x from below.

The decrease of the activity of the detector given in column two, Table IX, can thus be calculated by taking into account only the neutrons of class b. From our assumption, it follows that the layer of paraffin of thickness x that the neutrons of class b must traverse in order to reach the detector, must be considered as an absorbing (and not as a scattering) layer, having absorption coefficient $1/\lambda_D$.

When there is no absorber, the activity of the detector due to the neutrons of class b only is thus proportional to

$$N_b\, c\,(x/\lambda_D)\,,$$

where N_b is the number of the neutrons of class b, and c is the function defined by (6) (absorption curve taking into account the obliquity of the neutrons).

We have implicitly made the assumption that the angular distribution of the neutrons of class b is independent of the depth x; in reality this assumption is not quite correct, but we believe that it will not introduce an appreciable error.

Instead, when the absorber is in the plane x, the activity of the detector due only to the neutrons of class b is proportional to

$$N_b\, c\left(\frac{x}{\lambda_D} + K_D\,\delta\right),$$

where δ is the thickness of the absorber and K_D its absorption coefficient for the D neutrons.

Therefore the difference of the two activities is proportional to

$$N_b\left\{c\left(\frac{x}{\lambda_D}\right) - c\left(\frac{x}{\lambda_D} + K_D\,\delta\right)\right\}.$$

This expression corresponds to the data of column two Table IX.

In column four is given the activability of the absorber due only to the neutrons of class b; this is evidently proportional to N_b (see § 11). Therefore the ratios of the corresponding data of the second and fourth columns are proportional to

$$c\left(\frac{x}{\lambda_D}\right) - c\left(\frac{x}{\lambda_D} + K_D\,\delta\right). \tag{18}$$

For our absorber of rhodium $K_D = 1.8$ cm^2/g; $\delta = 0.36$ g/cm^2 so that $K_D\,\delta = 0.65$. Similar argument holds for group A ($K_A = 20$ cm^2/g; $\delta = 0.057$ g/cm^2).

In Fig. 6 are plotted the curves (18) for groups A and D, assuming their values equal to 100 for $x = 0$; the points represent the experimental values of the ratio of columns two and four, Table IX. The abscissae were calculated assuming $\lambda_D = \lambda_A = 1.1$ cm.

Therefore it seems that the neutrons of these two groups have the same mean free path which is a little larger than 1 cm. We have already noted

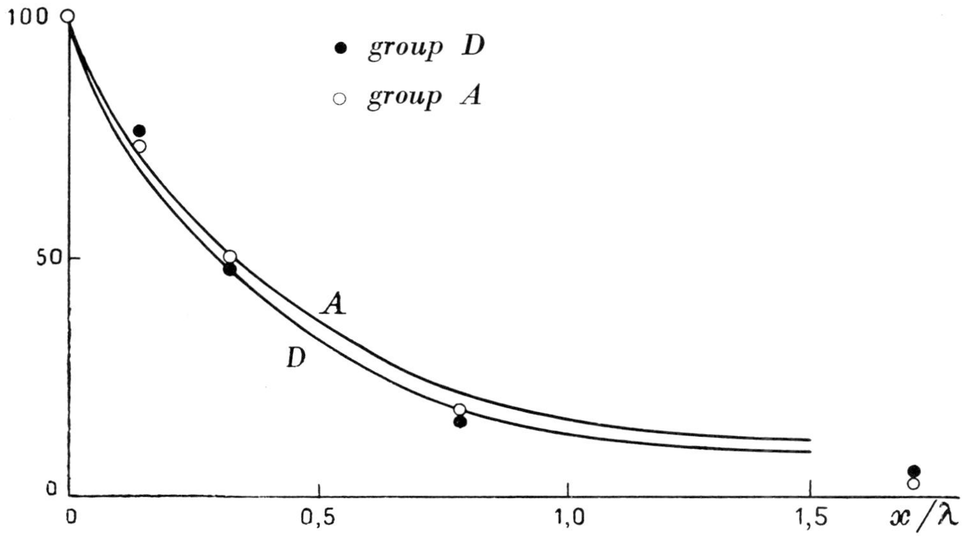

Fig. 6. – Ratio of activities (Eq. 18). Dots are the experimental values. $\lambda_A = \lambda_D = 1.1$ cm.

that this method is based on the assumption that the neutrons of these groups traverse only one mean free path. Though this assumption is quite reliable for group *A*, it is more doubtful for group *D* (see § 12). However, the fact that we have found approximately the same value for λ_D and λ_A can be considered an argument for the validity of this assumption.

8. – Mean Free Path of Thermal Neutrons.

In the preceding paragraphs, we have found the value of the mean free path of the groups *D* and *A*. The value found depends on the assumption that the neutrons of these groups traverse only one mean free path (see § 12).

In order to check this assumption, one could measure directly the mean free path of these groups. Up to now, we have not been able to perform such a measurement because of feeble intensity. However, we were able to measure directly the mean free path λ of thermal neutrons in paraffin.

This result, plus the measured value of λ^2 N of the preceding paragraph, allows one to calculate the average number N of mean free paths of a thermal neutron. We must remember that the albedo measurements also permit one to calculate the value of N (see § 6).

The measurement of the mean free path λ can be made using a source and a detector of thermal neutrons placed at a distance as large as possible with respect to their dimensions, and interposing between them various thicknesses of paraffin, the dimensions of which are just sufficient to screen the detector from the source. In ideal geometric conditions, when the solid angles under which the detector is seen from the source, and the source from the detector are negligibly small, the activity of the detector would be proportional to $e^{-x/\lambda}$ where x is the thickness of the scatterer of paraffin.

In practice, in order not to get too small an intensity, we were very far from these ideal conditions, and therefore it is necessary to correct the results in order to take into account the imperfection of the geometry.

The disposition used was as follows:

Source. – A paraffin cylinder of 12 cm diameter and 13 cm in height contained the source of Rn + Be, 2 cm below the center of the upper face. This cylinder was completely covered with a cadmium sheet of 0.5 g/cm^2; in the center of the upper face there was a window of 5×5.5 cm^2 in the cadmium, which could be opened or closed. If we measure the activity of a detector of slow neutrons, placed outside such a paraffin block, once with open window and once with closed window, and we take the difference of the two activities so measured, we will obtain the part of the activity due only to neutrons *C* coming out through the window. This difference is therefore equivalent to the activity that would be due to a source of only neutrons C having the same position and dimensions as the window. In this sense we will speak in the following paragraphs of source of C neutrons, or simply source C.

Detector. – In order to be able to have reasonably good geometrical conditions, and sufficient intensity, we have constructed [18] a few small cylindrical ionization chambers of 3.5 cm inside diameter and 10 cm in length, full of oxygen at a pressure of 75 atmospheres. The element which must be irradiated (Rh 0.125 g/cm^2) was placed, as a cylindrical sheet of 5 cm length and 3.5 cm diameter, inside the chamber, and was itself the electrode at high potential. The other electrode, connected to the electrometer, was a metal rod placed on the axis of the chamber. The wall of the chamber was of steel 4 mm in thickness and it practically absorbed none of the slow neutrons which, going through the wall of the chamber, could impinge on the detector.

This arrangement has the advantage, with respect to the methods we used up to now, that due to the high pressure, almost all the energy of the β–rays is spent inside the chamber; besides the electrode has a large useful surface and the geometrical conditions are very good. In fact, with this disposition it is possible to obtain almost the same sensitivity of observation as made with counters, and it has the advantage of the greater stability of the ionization chambers.

The chamber connected to the electrometer was irradiated for 2 minutes, and 15 seconds after having taken away the source, we started the readings; i.e., we measured the number of scale divisions covered in 2 minutes.

We have performed two series of measurements: In the first one the distance of the source C from the axis of the ionization chamber was 20 cm;

(18) E. AMALDI and E. FERMI, « Ric. Scientifica », 7 (1), 393 (1936).

in the second one, 10 cm; in both series the paraffin scatterers were put at half distance between the source C and the detector. In order to avoid random diffusions, the ionization chamber and all the path between the source C and the detector was screened by a cadmium layer of 0.5 g/cm².

In Table X are given in arbitrary units the results of the two series of measures (each value is the average of three readings); the error is about one unit.

TABLE X.

Measurements of activity for different thicknesses (x) of paraffin with detector at 10 cm and 20 cm from the source.

x	20 cm		10 cm	
	without Cd	with Cd	without Cd	with Cd
0	45.1	23.6	175	92.5
0.047	40.5	22.1	162	89.5
0.099	36.8	21.7	158	90
0.203	29.9	19.3	140.5	87
0.38	24.8	17.8	127	86
0.68	21.2	16.7	116.5	87

In the first column are given the thicknesses of the scatterers of paraffin (density 0.9) in centimeters. In columns two and three are given the measurements of activity at the distance of 20 cm, with the cadmium window of the source opened and closed. The activity due only to the neutrons *C* coming out from the window is, as we have said above, the difference of the values of these columns. Columns four and five are similar to columns two and three, for the case that the distance between the source and the detector is 10 cm.

In order to deduce from these measurements the value of the mean free path of the thermal neutrons, it is necessary to take into account the corrections due to the imperfections of the geometry, which are especially large in the measurements at 10 cm distance. Calculation of the corrections for thick scatterers are rather unreliable; instead, it is possible to obtain a fairly accurate evaluation of the correction for thin scatterers.

For the distance of 20 cm and a thin scatterer, the correction was calculated as follows: the cross section of the detector was 17.5 cm², and the area of the scatterer was 22.3 cm²; the neutrons that reach the detector are in part those that did not undergo collision in the scatterer, and in part those that had collided once in the scatterer (the case of multiple scattering can be neglected when the thickness x of the scatterer is small).

The first number of neutrons is evidently proportional to the area (17.5 cm²) of the detector, and to the probability $e^{-x/\lambda}$ that a neutron did not collide in the scatterer. For small x, the number of these neutrons is proportional to

$$17.5\,(1 - x/\lambda).$$

The small effect due to the obliquity of the neutrons' path in the scatterer has been neglected.

On the other hand, the number of neutrons that strike the scatterer is proportional to 4 times its area, because its distance from the source is one-half the distance of the detector from the source. This number is hence proportional to

$$4 \cdot 22.3 = 89.2\,.$$

The fraction x/λ of these neutrons has a collision in the scatterer; assuming that the C neutrons are scattered isotropically by the paraffin hydrogens (see § 13 and *F.*, §§ 10 and 11), we find that the fraction of the scattered neutrons that impinge on the detector is equal to the ratio of the area of the detector to the area of a sphere with radius equal to the distance between the scatterer and the detector (10 cm).

The total number of scattered neutrons which hit the detector is hence proportional to

$$89.2 \cdot (x/\lambda) \cdot (17.5/4\,\pi\,10^2)\,.$$

Adding this number to the number of neutrons that hit the detector without being scattered by the scatterer, we obtain the total number of neutrons that impinge on the detector as a function of x. This number is proportional to

$$17.5\,(1 - 0.929\,(x/\lambda) + \cdots).$$

Therefore, assuming as 1 the activity for $x = 0$, we find that the activity of the detector as a function of the thickness x of the scatterer is given by

$$1 - 0.929\,(x/\lambda) + \cdots.$$

From this relation we obtain that the tangent of the diffusion curve for $x = 0$ cuts the abscissae axis at

$$x_1 = (\lambda/0.929)\ \text{cm}\,.$$

From a graph of the experimental data we have found, by means of the method of the secants, the tangent for $x = 0$; this cuts the axis of the abscissae at the point

$$x_1 = 0.29\ \text{cm}\,.$$

Thus we find

$$\lambda = 0.29 \cdot 0.929 = 0.27\ \text{cm}\,.$$

We have also verified that all the entire experimental diffusion curve could be brought to coincide, also for thick scatterer, with the calculated curve (calculation, of which we do not give details, is performed by means of the methods explained in *F.*, § 5).

A little more complicated is the calculation of the corrections for the experiments at 10 cm, because in this case one cannot neglect the obliquity of the neutrons.

The calculation was performed with a numerical procedure from which we have found as expression of the initial shape of the diffusion curve

$$1 - 0.84\,(x/\lambda) + \cdots .$$

From the experimental curve we have found, as above,

$$x_1 = 0.84 \text{ cm}$$

so that from this measurement we have

$$\lambda = 0.84 \cdot 0.4 = 0.34 \text{ cm}.$$

From these two combined measurements we can obtain, as the more probable value of the mean free path of thermal neutrons (see § 13),

$$\lambda = 0.3 \text{ cm}.$$

9. Groups as a Function of the Distance from the Source; Energy Ratios.

In a preceding work [19] we have measured the activity of a rhodium detector in a water tank as a function of the distance from the source. The data given there correspond to a mixture of groups *C* and *D*, which are both efficient in activation of rhodium. It is evidently interesting to study, as a function of the distance from the source, the behavior of each single group.

The measurements were performed in a cylindrical water tank 95 cm deep and 90 cm in diameter. As detectors, we used:

For groups *C* and *D*: rhodium (0.36 g/cm², area 5×5.5 cm²): this detector, screened by 0.5 g/cm² of cadmium, is sensitive only to the D neutrons; instead the difference between the activities measured without and with cadmium gives the activity due only to group *C*. A detector of silver of 0.057 g/cm² and 5×5 cm² area, screened by cadmium, was used as a detector of groups $A + B$, which were not studied separately. Finally, for the group *I* we have used a detector of 0.068 g/cm² of PbI_2 screened by cadmium, the area of which was 5×5 cm².

The activities of the detectors were measured on both sides at various distances from the source; the sum of the two activities is proportional to the number of neutrons of the various groups which enter the detector (see § 11).

In order to keep as small as possible the errors due to the finite dimensions of the tank, the measurements were performed, maintaining in the center of the water cylinder the mean point between the source and the detector.

(19) E. Amaldi, O. D'Agostino, E. Fermi, B. Pontecorvo, F. Rasetti and E. Segrè, « Proc. Roy. Soc. », *A 149*, 522 (1935).

The dimensions of our detectors, and also the bulb containing the Rn + Be, are not small enough that one can consider the distance between the centers of the source and the detectors equal to the distance between these two objects, especially for small distances. Therefore we have used average distances. The experimental results for groups *C*, *D*, *A* + *B*, and *I* are given in Tables XI, XII, XIII, XIV.

TABLE XI.

Group C (difference of the activity of Rh *detector without and with cadmium).*

r (cm)	2.5	3.6	5.4	7.3	10.2	15.1	20.1	25	30	35	40
Activity	2121	1898	1441	1006	523	167	58	23	9	3 3	1 7

TABLE XII.

Group D (Rh *detector screened by cadmium*).

r (cm)	2.5	3.6	5.4	7.3	10.2	15.1	20.1
Activity	306	252	166	96	40	10.6	3.8

TABLE XIII.

Groups A + B (Ag *detector screened by cadmium*).

r (cm)	2.5	3.6	5.4	7.3	10.2	15.1
Activity	144	119	73	39.6	15.7	4.3

TABLE XIV.

Group I (PbI_2 *detector screened by cadmium*).

r (cm)	2.62	3.36	5.1	7.0	9.9
Activity	24.9	19.7	12.0	6.28	2.46

The numbers are the sum of the activities measured on both sides of the detectors. The data on groups *C*, *D* and *A* + *B* are obtained as average values of three series of measurements on each side; for group *I*, we have made only one series of measurements on each side, since the period of iodine is long enough to get sufficiently accurate results.

In Fig. 7 the activations multiplied by r^2 (r = distance from the source) are plotted against r. The data are taken from Tables XI, XII, XIII, XIV; the units of the ordinates have been chosen so that each curve incloses the same area.

Although the differences among the curves D, A + B, and I are small, the accuracy of the measurements is sufficiently high to distinguish among them. By inspecting these curves, one can see that the activity due to the various groups decreases as a function of r according to different laws; the activity decreases more rapidly for the groups in the following order: C, D, A + B, I. This fact enables us to dispose the groups in order of the increasing energy. We notice that neutrons of lower energy have, as an average, collided more times after they have been emitted, than neutrons of high energy; on the other hand, neutrons that have collided more times will as an average have diffused to a larger distance from the source. It follows that groups for which the intensity decreases more rapidly as a function of r, are those of greatest energy. The order of increasing energy is therefore, for the groups that we have investigated, the following:

$$C , D , A + B , I .$$

These considerations can be made quantitative, as follows: a convenient parameter, in order to characterize the rapidity of the decrement of the various groups as a function of the distance, is the mean value $(r^2)_{av}$ of the distances from the source for the neutrons of the various groups. $(r^2)_{av}$ was calculated for each group by

$$\int_0^\infty f(r)\, r^2\, dr \Big/ \int_0^\infty f(r)\, dr$$

where $f(r)$ is the curve of Fig. 7 referring to the group under consideration. The values of $(r^2)_{av}$ are given in Table XV.

TABLE XV.

Mean values of r^2.

Group	$(r^2)_{av}$ (in cm^2)
C	326.7
D	276.6
A + B	270.6
I	262.2

Although the curves of the activities as a function of the distance have been measured very accurately, we were not able to measure the intensity of groups D, $A + B$, and I up to a long distance. In the calculation for $(r^2)_{av}$ we had therefore to extrapolate the intensities of these groups for great distances. These extrapolations are relatively reliable as for great distances the law of decrease becomes quite similar for all the groups so that the end part of the curves of Fig. 7 can be represented satisfactorily by an exponential with coefficient 0.106 cm^{-1}.

Nevertheless, a considerable error might be due to this extrapolation.

It can be proved (see *F.*, § 2) that for groups of energy larger than the

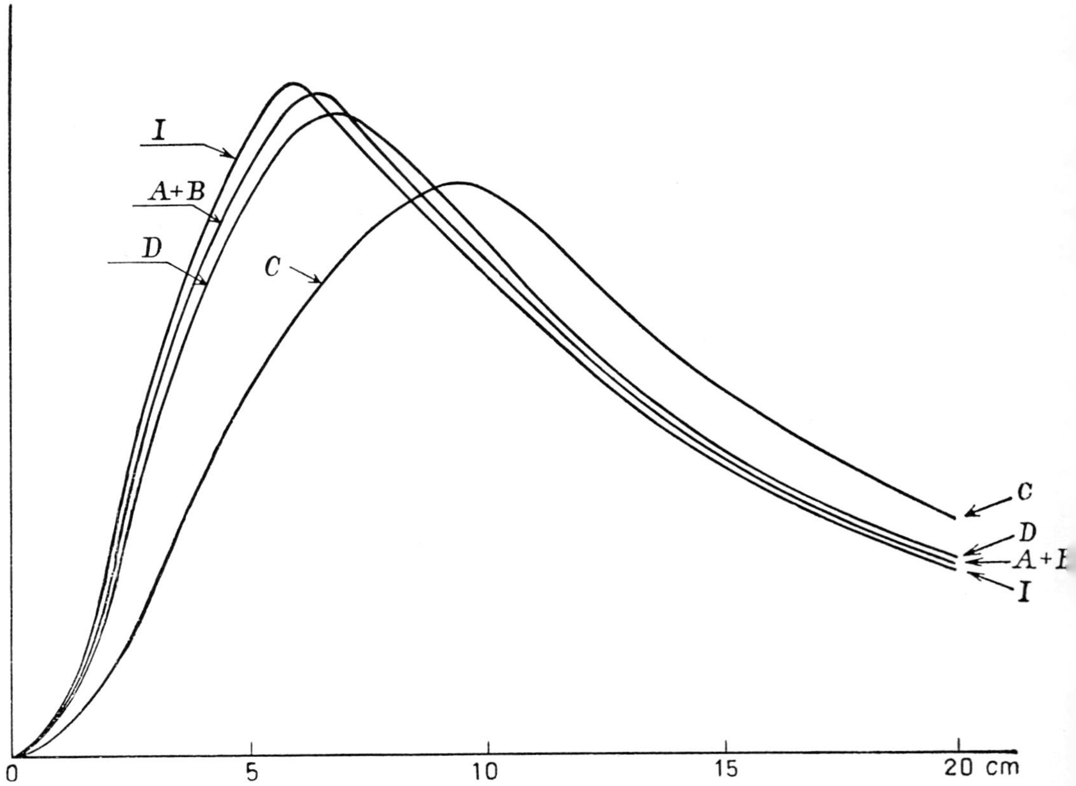

Fig. 7. – Ordinates = activation $\times r^2$. Abscissae = r. The units of the ordinates are chosen so that each curve includes the same area.

quantum $h\nu$ of the frequence of the elastic bond of hydrogen in paraffin (in practice, for groups of energy larger than 1 volt) the difference $(r'^2)_{av} - (r''^2)_{av}$ in the average square distances of two groups is connected to the ratio W″/W′ of the corresponding energies by the equation

$$\log \frac{W''}{W'} = \frac{(r'^2)_{av} - (r''^2)_{av}}{6\lambda_g^2}, \tag{19}$$

λ_g being the mean free path of the neutrons of the two groups in paraffin. The assumption has been made that λ_g is equal for the two groups (see § 7).

The preceding equation can easily be understood qualitatively, but for the numerical factor in the denominator; the left-hand side represents the average number of impacts necessary for the reduction of the energy from W″ to W′. On the other hand, the mean square of the displacement of a neutron in a free path is $2\lambda_g^2$ (the factor 2 is due to having performed mean square and not the square of the mean). If the orientations of successive free paths were incoherent, $(r^2)_{av}$ would then increase by $2\lambda_g^2$ for each impact, and then log W″/W′ impacts would produce an increase by $2\lambda_g^2 \log W''/W'$; this assumption would then give a formula similar to (19) with a factor 2 instead of 6 in the denominator. The factor 6 is obtained by taking into account the coherence of successive free paths, and also the fluctuations in the number of impacts necessary for reducing the energy from W″ to W′.

We can use (19) in order to determine the energy ratio of two groups (except group *C*, for which this formula does not hold). This method is, however, very inaccurate, since small errors in λ_g and $(r^2)_{av}$ affect the ratio W″/W′ considerably.

As we measured $(r^2)_{av}$ in water, and not in paraffin, we shall have to put for λ_g the mean free path in water; we can deduce this from the value in paraffin (1.1 cm, for nonthermal neutrons, see § 7) by assuming the mean free path to be inversely proportional to the concentration of hydrogen; we find then $\lambda_g = 1.27$ cm.

Assuming this value, we find from (19) that the difference in $(r^2)_{av}$ corresponding to a ratio *e* in the energies is $6\lambda_g^2 = 9.7\ \text{cm}^2$.

From Table XV we obtain the following ratios of the energies for the groups *I*, *A* + *B* and *D*:

$$W_I : W_{A+B} : W_D = 4.4 : 1.85 : 1 .$$

It has been already mentioned that the error in these ratios can be very large.

It is perhaps more appropriate to deduce from (19) the values of the mean free path in water by assuming that the energies of the groups are inversely proportional to the square of their absorption coefficients in boron [(20)] (see also § 13).

Taking the following absorption coefficients in boron

$$K_I = 1 \quad ; \quad K_B = 2.3 \quad ; \quad K_A = 3 \quad ; \quad K_D = 4.7 \quad ; \quad K_C = 38\ \text{cm}^2/\text{g}$$

we obtain the following energy ratios:

$$W_I : W_B : W_A : W_D : W_C = 1440 : 270 : 160 : 65 : 1 .$$

Assuming [(21)] $W_C = kT = 0.025$ volts, we obtain

$$W_I = 36 \quad ; \quad W_B = 7 \quad ; \quad W_A = 4 \quad ; \quad W_D = 1.6 \text{ volts.}$$

From these values and (19), we find the mean free path in water

$$\lambda_g = 0.87\ \text{cm}$$

(20) O. R. FRISCH and G. PLACZEK, « Nature » (London), *137*, 357 (1936); D. F. WEEKES, M. S. LIVINGSTON and H. A. BETHE, « Phys. Rev. », *49*, 471 (1936).

(21) H. H. GOLDSMITH and F. RASETTI, « Phys. Rev. », *50*, 328 (1936) show that it is more correct to use kT instead of $3kT/2$ as thermal energy in this calculation.

instead of 1.27 cm. The difference is within the limits of the experimental error.

We notice further that $(r^2)_{av}$ for the C group given in Table XV is considerably different from the values of $(r^2)_{av}$ for the other groups. For instance,

$$(r_C^2)_{av} - (r_D)_{av} = 50 \text{ cm}^2.$$

This difference is partly due to the large number of free paths of the neutrons when their energy is already reduced to the thermal energy; the amount due to this process of diffusion is (see § 7)

$$2 \lambda^2 N = 26 \text{ cm}^2.$$

The remaining 24 cm² correspond to the slowing down of the neutrons from the energy corresponding to the D group to the energy of thermal agitation. This process cannot be easily calculated, because it is complicated by the effect due to the chemical bond of hydrogen (see *F.*, § 19).

10. Passage of the Neutrons from One Group to Another.

In the preceding section, we arranged the groups in order of decreasing energy. It is now obvious that a neutron of a group of high energy, after some impacts, will have a smaller energy, and may thus go over into a group of less energy. We must, to be more precise, expect that the neutrons of all the groups, after a certain number of impacts, become thermal neutrons.

We tried, therefore, to observe the transformation of D neutrons into C neutrons [22].

The principle on which these experiments are based can be clearly understood if we assume that we have a detector R that can be activated only by C neutrons and an absorber A for the D neutrons only. We put the detector R on the center of the upper face of a paraffin cylinder, containing the source S; if the absorber is inserted between the upper face of the paraffin and the detector, no reduction of the activity of the detector will occur, since the C neutrons that come out of the paraffin are not absorbed by A. If we put the absorber A inside the paraffin, 1 or 2 cm below the upper face, it can happen that the absorber captures some of the D neutrons which could be transformed into C neutrons by impacts in the paraffin before reaching the detector.

If D neutrons can be really transformed into C neutrons, we must expect that the absorber A produces no effect when put between the paraffin and detector, but that it reduces the activity of A when put 1 or 2 cm inside the paraffin.

The experiment cannot be performed in this ideal way, because we have no detector of group C only, and no absorber of group D only. We have, therefore, used as detector a rhodium plate, 0.36 g/cm² thick; about 70 percent of its activity is due to the C group and the rest to the D group.

(22) E. Amaldi, E. Fermi, « Ric. Scientifica », 7 (1), 56 (1936).

In order to measure only the activity due to the *C* group, all measurements were performed without and with a cadmium absorber 0.54 g/cm^2 thick, that was put immediately below the rhodium plate. The difference in activities without and with cadmium gives the activity due to the *C* group only.

As absorber for the *D* group, we used an indium plate 1.83 g/28.5 cm^2; the absorption coefficient of indium for the *D* group is 3.4 cm^2/g and for the *C* group 0.7 cm^2/g. It follows that the absorption of C neutrons in our indium plate is far from negligible, the more so if we keep in mind the high numerosity and albedo of this group. In order to correct for this, we prepared a plate of Sn—Cd alloy that was equivalent to the indium plate as absorber for the *C* group, and practically did not absorb the D neutrons. Taking the difference of the activities produced by group *C* in the detector with the indium plate or the cadmium plate, we obtain the effect due only to the absorption of the *D* group by indium.

In Table XVI are collected the activities induced in the rhodium plate with different set-ups; the different arrangements are denoted by a notation similar to that used in § 6; the number added in parentheses after the symbol P of paraffin, gives in centimeters the thickness of paraffin layers; since we used two different cadmium sheets, one of 0.54 g/cm^2 thickness for complete absorption of group *C*, and the other 0.0036 g/cm^2 thickness equivalent to the indium sheet as absorber in the *C* group, we denoted these two absorbers, respectively, with Cd and cd. The figures given in the table are averages of ten readings.

TABLE XVI.

Activities induced in Rh *plate under different arrangements.*
S = source, P = paraffin (cm thick), R = detector.

Arrangement	Activity
S P(3.5) R	280.4 ± 0.7
S P(3.5) In R	230.2 ± 0.7
S P(3.5) In Cd R	45.6 ± 0.4
S P(3.5) Cd R	68.7 ± 0.5
S P(3.5) cd R	254.8 ± 0.7
S P(2.5) cd P(1) R	253.0 ± 0.7
S P(2.5) In P(1) R	241.5 ± 0.7
S P(2.5) cd P(1) Cd R	71.1 ± 0.5
S P(2.5) In P(1) Cd R	64.9 ± 0.5
S P(1.5) cd P(2) R	254.0 ± 0.7
S P(1.5) In P(2) R	248.9 ± 0.7
S P(1.5) cd P(2) Cd R	69.2 ± 0.5
S P(1.5) In P(2) Cd R	68.1 ± 0.5

The first five measurements are performed in order to control the identity of the thin cadmium (cd) and the indium (In) as absorbers of the C neutrons. Their result is that the absorption of the indium plate is probably slightly larger than the absorption of the cadmium sheet; the difference 280.4 — 68.7 — (230.2 — 54.6) = 27.1 ± 1.2 represents the activity of the rhodium detector due to group *C* that is absorbed by the indium plate; the difference 280.4 — 254.8 = 25.6 ± 1 is the analogous absorption of the thin cadmium sheet.

The next four measurements are performed with the absorber (Cd or In) 1 cm inside the paraffin. The difference 253.0 — 71.1 = 181.9 ± 0.9 is the activity of the detector due to the group *C* with the absorber Cd 1 cm inside the paraffin. The difference 241.5 — 64.9 = 176.6 ± 0.9 is the similar activity with the indium absorber. The difference between these two activities is 5.3 ± 1.2 and is considerably larger than the difference that we might expect from the small difference between the two absorbers when placed close to the detector.

The next four measurements are analogous to these last, with the only difference that the absorbers are 2 cm below the paraffin surface. Also in this set of measurements there is a difference analogous to that observed in the preceding set; this difference is now 4.0 ± 1.2.

By discussing together all these data, one might conclude that the observed effect has a sign corresponding to a transformation of D neutrons into C neutrons, and is 2.3 times larger than the quadratic error.

Although we have a serious presumption of a real effect in this sense, we cannot deduce from this experiment a final conclusion.

An evaluation of the effect that one must expect in this experiment shows that it is really of the order of magnitude of the observed effect; the evaluation was made as follows:

We have measured the induced activities in the indium sheet of the above experiment, placed inside the paraffin 2 cm below the upper face, separately in group *C* and *D*, and we have found, respectively, 257 and 47.5. These numbers are proportional to the numbers of neutrons of the two groups absorbed by our indium sheet.

Let p_1 be the probability that a neutron C which is in paraffin in the place of the indium absorber comes out from the upper face and is captured by the rhodium detector; similarly, let p_2 be the probability that a neutron D which is in the place of the indium sheet, comes out from the upper face transformed into a C neutron, and is captured by the rhodium.

The decrements of the activities of the rhodium sheet in group *C*, due, respectively, to the absorption by indium sheet of neutrons D and C are in the ratio $47.5\ p_2/257\ p_1$. Assuming p_1 and p_2 equal or at least of the same order of magnitude, we find that these decrements are about in the ratio 47.5/257, and, since the decrement due to the absorption of C neutrons is 26.9 (see Table XVI), the effect which we must expect in the preceding experiment is equal to 5, and therefore of the same order of magnitude as the really observed effect.

Similar experiments were performed by Preiswerk and von Halban (23), who found a genetic relation between groups I and A.

11. Behavior of the Groups in the Neighborhood of the Surface of the Paraffin.

In § 9, we have studied the activity due to the various groups inside a water tank so large that we could neglect the disturbances due to the limitations of size. We will now study the behavior of the activity due to different groups close to the surface of the paraffin.

First, we shall consider the behavior of group C. From the theory of diffusion of thermal neutrons, it follows that, independently of the position of the source, the density of thermal neutrons close to the surface of the paraffin decreases toward the outside (see *F.*, § 6, formula 57); one can furthermore show that if we indicate with x the depth below the surface of the paraffin, the density $n(x)$ is, for small x, approximately proportional to

$$x + \lambda/\sqrt{3},$$

so that if we extrapolate the curve of $n(x)$ towards the outside of the paraffin $(x < 0)$, $n(x)$ is zero at a distance from the surface of the paraffin equal to

$$x_0 = - \lambda/\sqrt{3}.$$

This is just the behavior of the density of thermal neutrons close to the surface which determines the law of angular distribution (5) of the outcoming C neutrons. If the density would be constant close to the surface of the paraffin, one would find the cosine law.

In order to measure $n(x)$, it is necessary to use a detector of group C thin enough not to disturb the neutrons' distribution. Its activity will then be proportional to the density $n(x)$ of thermal neutrons.

We have used two detectors which were obtained by electrolytic deposit of rhodium on a nickel plate (which served only as an inactive carrier) of 29 cm² area; the weights of rhodium deposited on the two detectors were respectively 0.087 and 0.169 g. Taking into account the absorption coefficient of rhodium for the C group (0.7 cm²/g) $K\delta$ for these two detectors is equal, respectively, to 0.002 and 0.004; to these values there corresponds a mean probability ζ for the capture of a thermal neutron which traverses the detector, of about 0.004 and 0.008, respectively (see § 6). These probabilities are very small, and therefore the detectors used can be considered as thin.

The measurements have been performed using a paraffin cylinder 24 cm in diameter and 10 cm in height, containing the source 3.2 cm below the center of the upper face. The layer of paraffin of 3.2 cm thickness be-

(23) PREISWERK and VON HALBAN, «C. R.», *202*, 840 (1936).

tween the source and the upper surface was cut into sheets in order to be able to put the detector at various depths.

In order to find the activity due to the group C only, we have always made the difference between the activities of the detectors without and with two cadmium screens 0.44 g/cm² thick, between which could be placed the detectors.

In Fig. 8 the activity due only to the C group of our two detectors is plotted against the depth x. We give also the curve extrapolated towards negative values of x. The intersection, with the axis of the abscissa is close to the point

$$x_0 = -0.18 \text{ cm}.$$

From this value we find

$$\lambda = 0.31 \text{ cm}$$

in good agreement with the values obtained from the direct measurements (see § 8).

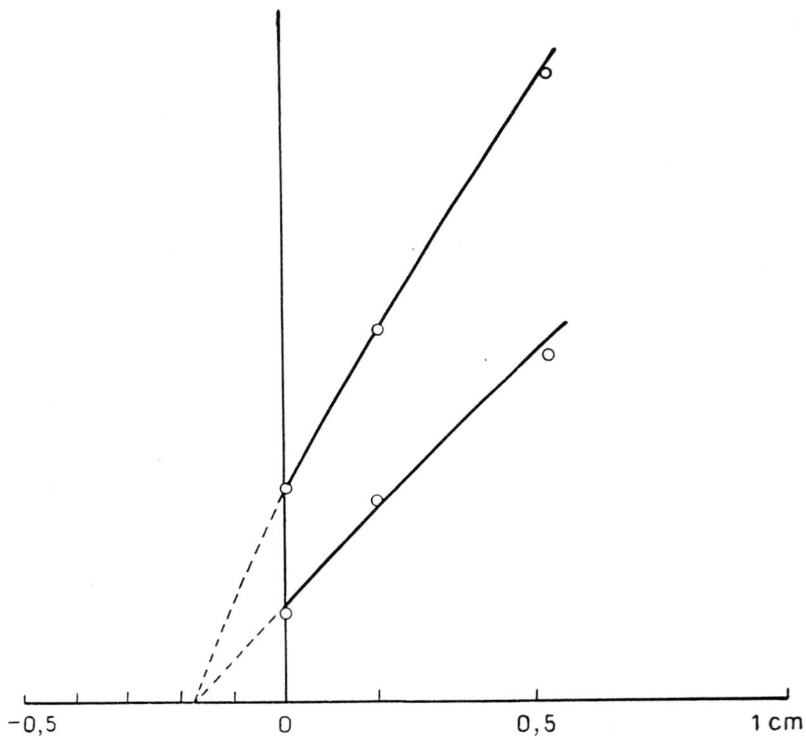

Fig. 8. – Activity of C group plotted against the depth (x) in paraffin.

We have already noticed that this experiment has a very simple interpretation, provided the detector is very thin. Instead, a thick detector disturbs strongly the density of thermal neutrons, so that the curve of the activity as a function of x is very different.

Thus, using a rhodium detector 0.36 g/cm² thick, and adding the activities due to *C* group measured on both sides, in order to get data comparable to that of the preceding experiment, we have found that the curve of the activity, extrapolated toward negative values of x, cuts the x axis at

$$x_0 = -0.4 \text{ cm},$$

i.e., at a distance more than double than in the case of thin detectors.

In this experiment, we have added the activities measured on both sides of the detector, since this sum is proportional to the total number of thermal neutrons which entered the detector. Let N_1 and N_2 be the numbers of neutrons which enter the detector, respectively, through the face 1 and face 2; the activities measured on the faces 1 and 2 are, respectively, given by

$$A_1 = N_1 a + N_2 b, \quad A_2 = N_1 b + N_2 a, \tag{20}$$

where a and b are constants characteristic of the detector, which depend on its absorption coefficients for neutrons and electrons, and on its thickness. The ratio a/b can be obtained by measuring the ratio A_1/A_2 when the detector is placed outside the paraffin so that $N_2 = 0$.

For the rhodium detector, 0.36 g/cm² thick, $a/b = 4/3$ in the *C* group, and 2/1 in the *D* group. This difference depends on the larger absorption coefficient for D neutrons. For a thin detector, one would have $a = b$.

Summing up the two equations in (20), we find

$$A_1 + A_2 = (a + b)(N_1 + N_2),$$

and therefore the sum of the activities measured on both sides of a detector is always proportional to the total number $N_1 + N_2$ of neutrons which traverse the detector.

By (20) we find

$$N_1 = \frac{a A_1 - b A_2}{a^2 - b^2} \quad ; \quad N_2 = \frac{a A_2 - b A_1}{a^2 - b^2}. \tag{21}$$

Therefore, measuring A_1 and A_2, and the ratio a/b, we can find values proportional to N_1 and N_2 (this method was used in § 7 in order to measure separately N_1 and N_2). In Table XVII are given the values of N_1 and N_2 measured with our rhodium detector, 0.36 g/cm² thick for groups *C* and *D* separately. In order to be able to compare these two sets of data, we have taken as 1 in both groups the value of N_1 for $x = 0$. The source in these experiments was at $x = 2.4$ cm. The data on group *C* do not permit simple interpretation, since the detector cannot be considered as thin.

Instead, in group *D* the detector, although thick, does not disturb the distribution of the D neutrons, since the neutrons of this group traverse on the average only one mean free path (see §§ 6, 7, 12), and therefore each of them hits only once against the detector.

By extrapolation of the shape of the curve $N_1 + N_2$ for *D* group towards negative values of x, we find the intersection with the x axis at

$$x_0 = -0.9.$$

The fact that this intersection occurs much farther from the surface of the paraffin than for group *C* ($x_0 = -0.18$ cm with thin detector) is due on one hand to the greater mean free path (see § 7), and on the other hand to the coherence of the orientation of successive free paths, which is very large for neutrons of energy larger than the quantum of the elastic bond of hydrogen in paraffin.

TABLE XVII.

Number of neutrons that enter faces (1) *and* (2) *of detectors.*
x = thickness of paraffin.

x	Group *C*		Group *D*	
	N_1	N_2	N_1	N_2
0	1	0	1	0
0.16	1.10	0.28	1.07	0.07
0.34	1.15	0.63	1.17	0.14
0.53	1.28	0.91	1.25	0.28
0.87	1.60	1.19	1.31	0.45
1.87	2.19	2.00	1.35	1.00

By comparing columns two and four, Table XVII, we notice that the shape of the curve N_1 close to the surface, for our detector, is not very different in groups *C* and *D*; instead, at greater depth the *C* group increases much more rapidly than group *D*. One can justify this behavior by taking into account the increase of the albedo of the layer of paraffin, superimposed on the detector, as a function of its thickness.

12. WIDTH OF THE ENERGY BANDS CORRESPONDING TO THE GROUPS. TOTAL NUMBER OF NEUTRONS.

We shall discuss in this section a method for determining the width of the bands [24] of energy which correspond to the various groups; more precisely, the ratio W_{max}/W_{min} of the maximum to the minimum energy which limits a band. In the calculation the bands were assumed to be sharply defined. It would be possible to perform a similar calculation, assuming for the band a resonance form; however, this seems scarcely worth while, owing to the small accuracy of the experimental data.

(24) E. AMALDI and E. FERMI, « Ric. Scientifica », 7 (1), 310 (1936).

In the last part of this section we will give an evaluation of the total number of neutrons emitted by the source of Rn + Be.

The method for determining the width of the band which corresponds to a group g of energy larger than 1 volt, is based on the comparison of the two following values: the activability A_g of a detector due to the neutrons of the considered group g, and the activability B_c of the same detector due to the thermal neutrons and obtained by screening the detector on one side with a cadmium sheet thick enough to absorb completely the thermal neutrons hitting on it. B_c can naturally be obtained as the difference of the activabilities of our detector with a cadmium sheet on one face only or on both sides.

If the distribution of the fast neutrons inside a large block of paraffin were uniform, it is clear that A_g and B_c would have values independent of the position and orientation of the detector. However, the source of fast neutrons is small and therefore we must compare the mean values of A_g and B_c averaged over all positions and orientations inside the paraffin; i.e. we must compare

$$\int \bar{A}_g \, d\tau \quad ; \quad \int \bar{B}_c \, d\tau ,$$

where the integrands $\bar{A}_g$ and $\bar{B}_c$ are the mean values of the activabilities of the detector measured with the two opposite orientations; these mean values are practically identical with the average taken over all directions. By using the said integrals we eliminate all the complication arising from the nonuniform distribution of the neutrons.

We suppose now that in the paraffin around our detector, q fast neutrons are produced per cubic centimeter and per second. These neutrons are slowed down in such a way that we shall have in the neighborhood of our detector neutrons of all velocities.

It can be shown (see *F.*, § 1) that for energies larger than 1 volt, the number of neutrons having velocity between v and $v + dv$ is

$$(2\,q\lambda\,(v)/v^2)\,dv. \tag{22}$$

From this we can easily calculate the activity A_g as a function of the energy W_{max} and W_{min} limiting the energy band g, of the mean free path λ_g and the absorption coefficient K_g of the detector for the group g and of the surface s and thickness δ of the detector. We find (see *F.*, § 8)

$$A_g = \eta\, s q\, \lambda_g\, K_g \log \frac{W_{max}}{W_{min}} \times \frac{1}{2} \left\{ \int_0^\delta b\,(K_g\, x)\, e^{-\mu x}\, dx + \int_0^\delta b\,(K_g\, x)\, e^{-\mu(\delta - x)}\, dx \right\}, \tag{23}$$

where μ is the absorption coefficient of the β-rays of the detector in the detector itself and $b\,(K_g\, x)$ is the function (4); η is the efficiency of the ionization chamber for the β rays of the detector. The last factor in (23) represents the effect of the absorption of the neutrons and of the β-rays in the detector. This factor would be equal to δ for a very thin detector.

Also the activity B_c due to the thermal neutrons can be calculated in terms of the diffusion length $(D\tau)^{1/2} = (\lambda^2\, N/3)^{1/2}$ (see § 7), of the absorption coefficient K_c of the detector; taking into account the angular distribution (5) of the neutrons impinging on the detector, we find

$$B_c = \eta s\, q\lambda\, (N)^{1/2}\, K_C \int_0^\delta c\,(K_C\, x)\, e^{-\mu x}\, dx \tag{24}$$

where $c\,(K_C\, x)$ is the function (6). We have disregarded the variation of the absorption coefficient of our detector for thermal neutrons of different velocities. For a very thin detector, the integral is equal to δ.

We must now integrate (23) and (24), as already mentioned, and we obtain

$$\int A_g\, d\tau = \eta\, s Q\, K_g\, \lambda_g \log \frac{W_{max}}{W_{min}} \times \frac{1}{2} \left\{ \int_0^\delta b\,(K_g\, x)\, e^{-\mu x}\, dx + \int_0^\delta b\,(K_g\, x)\, e^{-\mu x\,(\delta - x)}\, dx \right\}, \tag{25}$$

$$\int B_c\, d\tau = \eta\, s Q\, K_C \lambda\, (N)^{1/2} \int_0^\delta c\,(K_C\, x)\, e^{-\mu x}\, dx\,,$$

where $Q = \int q d\tau$ is the total number of neutrons emitted by the source per second. From (25) we obtain

$$\log \frac{W_{max}}{W_{min}} = \frac{K_C}{K_g}\, \frac{\lambda\, (N)^{1/2}}{\lambda_g}\, \frac{\int A_g\, d\tau}{\int B_c\, d\tau} \times \frac{2 \int_0^\delta c\,(K_C\, x)\, e^{-\mu x}\, dx}{\int_0^\delta b\,(K_g\, x)\, e^{-\mu x}\, dx + \int_0^\delta b\,(K_g\, x)\, e^{-\mu\,(\delta - x)}\, dx}\,. \tag{26}$$

We performed the experiments on the groups *D*, *A*, and *I*. For D group we used a rhodium detector (0.36 g/cm²); for the group *A* a silver detector (0.057 g/cm²); and for group *I* a lead iodide detector (0.76 g/cm²); since the small absorption coefficient of iodine in the *C* group is not too well known, we used the rhodium data also in the calculation of B_c for the *I* group, assuming equal efficiency of the chamber for β–rays of rhodium and iodine. In order to calculate the integrals

$$\int \bar{A}_g\, d\tau \quad , \quad \int \bar{B}_c\, d\tau$$

we determined $\bar{A}_g$ and $\bar{B}_c$ for a fixed distance, and deduced the values for all the distances by using the curves of Fig. 7.

In the following equations, the numerical values are written in the same order in which the various magnitudes appear in formula (26)

Group *D*

$$\log \frac{W_{max}}{W_{min}} = \frac{0.7}{1.8}\, \frac{\sqrt{13}}{1.1}\, \frac{4.58 \cdot 10^5}{1.5 \cdot 10^6}\, \frac{2 \cdot 0.108}{0.087 + 0.052} = 0.60\,,$$

Group A

$$\log \frac{W_{max}}{W_{min}} = \frac{0.25}{20}\, \frac{\sqrt{13}}{1.1}\, \frac{7.8 \cdot 10^4}{2.04 \cdot 10^5}\, \frac{2 \cdot 0.048}{0.021 + 0.019} = 0.038\,,$$

Group *I*

$$\log \frac{W_{max}}{W_{min}} = \frac{0.7}{0.38}\, \frac{\sqrt{13}}{1.1}\, \frac{4.16 \cdot 10^4}{1.5 \cdot 10^6}\, \frac{2 \cdot 0.108}{0.085 + 0.049} = 0.27\,.$$

From these values we may derive the ratios W_{max}/W_{min} for each group; we shall call the magnitude log W_{max}/W_{min} logarithmic width; this magnitude has a simple physical meaning because it represents the average number of impacts of a neutron while it belongs to a given group. For example, group *A* that has a logarithmic width 0.04 is such a narrow energy band that only 4 percent of the neutrons pass through it during the process of slowing down. In the case of group *A*, the detector used cannot be considered as thin with respect to the absorption of the neutrons; if one would take into account the resonance form of the band, one would find, by a simple numerical evaluation, that the logarithmic width of group *A* as calculated from (26) is too large by a factor of about 2. For the other detectors, this correction is much smaller.

The logarithmic width of group *I*, and even more of group *D*, is considerably larger, as shown also by the larger numerosity (see § 5) of these groups. Even for group D, the probability that a neutron belongs to the group for more than one free path is relatively small and amounts to 0.27. These facts explain our results that the albedo of all nonthermal groups is practically zero.

Instead, one might expect a small difference in the behavior of groups *D* and *A* with respect to the diffusion experiments descriped in § 7. The fact that we failed to find any such difference might, however, be explained by inaccuracy of our measurements, as the expected difference is fairly small.

If we assume that the energy of the groups are those deduced in § 9 by the boron method, we obtain for the width of group *D*, *A* and *I*:

$$1 \quad ; \quad 0.15 \quad ; \quad 10 \text{ volts.}$$

Finally, we can use the second of (25) in order to calculate the total number *Q* of the neutrons emitted from the source. In (25), B_c is the initial activity of the detector used. In order to get the activability (3), it is necessary to multiply B_c given in (25) by the factor

$$1000/I \cdot U = 1000/I \cdot 840\, \eta_U .$$

So that we find

$$\frac{Q}{I} = 0.84 \frac{\eta_U}{\eta} \frac{\int B_c\, d\tau}{s\lambda\, (N)^{1/2}\, K_C \int_0^\delta c\,(K_C\, x)\, e^{-\mu x}\, dx} . \tag{27}$$

In this formula, the same elements enter which we have used in the calculation of the logarithmic width. Using, for instance, the data on the activation of the rhodium, we find

$$Q/I = 160{,}000\, (\eta_U/\eta) .$$

Assuming $\eta = \eta_U$ we have that one of our neutronic units corresponds to 160,000 neutrons per second, and taking into account that 1 neutronic unit corresponds to about 6 mc of Rh + Be, we have finally

27,000 neutrons per s and per mc.

This value is considerably larger than the values found with different methods [25]; this fact can be only partially accounted for by a difference between η and η_U.

13. Summary and Discussion.

From the results we have explained, it seems that one must conclude that in elements which are sensitive to slow neutrons, the capture cross section is often an irregular function of the energy of slow neutrons, with sharp maxima representing some sort of absorption bands.

The analysis of the absorption curve of the various elements with several detectors enables us to identify several absorption bands (groups *A*, *B*, *C*, *D*, *I*).

There are several reasons for believing that group *C* (radiation strongly absorbed by cadmium) corresponds, at least for its largest part, to neutrons having a velocity of thermal agitation, while the other groups correspond to larger velocities.

The most direct proof of this statement is the experiment with a velocity selector of cadmium [26] (so as to observe just group *C*) that enabled a direct measurement of the velocity of the C neutrons that was found equal to the thermal velocity.

The same result can be derived from the experiments by Preiswerk and von Halban and others [27], who found that the radiation filtered by cadmium is not sensitive to temperature variations; this shows that the neutrons not belonging to group *C* have energy larger than thermal agitation energy.

Finally, our albedo measurements (see § 6) show that only the C neutrons can traverse many free paths while belonging to this group. This fact can be easily understood on the assumption that the *C* group consists of neutrons in thermal equilibrium, since then successive impacts do not alter the average energy; instead, a neutron belonging to an energy band $\gg kT$, has a high probability of getting out of the band after one impact.

In all the cases investigated by us, a large percentage of the activity (in most cases larger than 50 percent) is due to thermal neutrons; this does not mean, however, that the absorption coefficient for thermal neutrons is, as a rule, larger than the absorption coefficient for neutrons having an energy of some volts. This fact can be explained to some extent by the large number of thermal neutrons that get out of a paraffin block containing the source (see Table VI).

The problem of determining the energy bands corresponding to the known neutrons groups can be attacked by the following method, first used by Frisch

(25) Cf., e. g., R. Jaeckel, « Z. Physik », *91*, 493 (1934); F. A. Paneth and H. Loleit, « Nature » (London), *136*, 950 (1935).

(26) J. R. Dunning, G. B. Pegram, G. A. Fink, D. P. Mitchell and E. Segrè, « Phys. Rev. », *48*, 704 (1935); « Rend. Lincei », *23*, 340 (1936).

(27) P. Preiswerk and H. von Halban, « Nature » (London), *136*, 1027 (1935); F. Rasetti and G. Fink, « Phys. Rev. », *49*, 642 (1936).

and Placzek and by Weekes, Livingston and Bethe. They assumed that the capture cross section of a slow neutron in boron is inversely proportional to the velocity v of the neutron. Assuming this, we obtained from our measurements the following enegries:

Group	*C*	*D*	*A*	*B*	*I*
Volts	0.037	1.6	4	7	36.

The error in this type of measurements can be, however, fairly large.

A quite independent determination of the energy order of the groups, and also an estimate of the energy ratio (this last only for nonthermal groups) was given by us in § 9 by considerations based on the behavior of the intensity of the various groups as a function of the distance from the source. The energy order found by us coincides with the one deduced from the boron absorption, and the values of the energy ratios found by us, although smaller than the ratios obtained by the boron method, are not inconsistent because of the fairly large error involved in these determinations.

Our method, although very inaccurate, has the advantage of being very direct.

We were also able to derive, by analogous methods, the width of the energy bands (more exactly, the relative width $\Delta W/W$) for the nonthermal groups; we found that the narrowest band is group *A* (radiation strongly absorbed by silver), for which $\Delta W/W = 0.04$. The other groups (see § 12) are somewhat wider; nevertheless, the probability that a neutron remains in the same group after one free path is fairly low.

Connected to the width of the absorption bands, is the numerosity of the various groups, which corresponds essentially to the number of neutrons coming out per second from the surface of a paraffin block containing the source (see § 5). The most numerous group of all is the thermal one (numerosity 40), while the least numerous is group *A* (numerosity 0.5), which, as we have seen, is the narrowest.

These facts are qualitatively in agreement with the consideration of Bohr and Breit and Wigner.

Some further information on the properties of these absorption bands can be obtained by comparing the behavior of two periods of the same element, both sensitive to slow neutrons. We were able to investigate in this respect only silver, rhodium, indium and bromine. Only in the case of silver we found a different behavior for the two periods of this element. In all the other cases, no difference was found within the limits of our accuracy. It may be noticed that rhodium, indium and bromine are the three cases for which the number of observed periods produced by slow neutrons is larger than the number of known isotopes.

Sections 6 to 11 are devoted to a study of the diffusion properties of slow neutrons in hydrogenated substances. The diffusion process can be analyzed in two successive phases. First (slowing down phase), the neutron loses energy by means of successive impacts until it reaches the energy of thermal agitation; afterwards (diffusion phase), the energy does not decrease further on the average and the neutron in scattered until it is captured by the protons or other nuclei.

During the phase of slowing down, the mean free path decreases very soon to a value of the order 1 cm, afterwards it remains approximately constant until the energy of the neutron is comparable with the quantum $h\nu$ of the frequencies of the elastic bond of the hydrogen atom in paraffin.

We can say that all the studied groups except the thermal one, belong to the interval of energy where the mean free path λ_g in paraffin is about 1 cm.

From the theory of the impact of slow neutrons and hydrogen atoms, taking into account the chemical bond of these last (see *F.*, §§ 10 and 11), one finds that when the energy of the slow neutrons passes from values larger than $h\nu$ to values smaller, the mean free path decreases and tends to a limit equal to $\lambda_g/4$. This variation of the mean free path is connected also to a different angular distribution of the neutrons after an impact; when $W \gg h\nu$ the angle between the directions of the motion of the neutrons before and after the impact is always sharp, while for $W \gg h\nu$ the angular distribution after an impact tends to become isotropic.

In agreement with this theoretical result, one observes a really remarkable decrease of the mean free path in going from the nonthermal to the thermal group.

For this last, one can take $W/h\nu$ equal to about 1/10; although this value is pretty small, one cannot consider the thermal neutrons as neutrons of energy practically zero. Thus one finds (see *F.*, formula 102) that for $W/h\nu = 1/10$ the mean free path is

$$\lambda = \lambda_g/3.3 \tag{28}$$

while at the limit for $W = 0$, one would have $\lambda = \lambda_g/4$.

We can also evaluate how much the coherence is conserved between the directions of the motion of a thermal neutron before and after an impact. As a quantitative expression of the coherence, we can take $(\cos\theta)_{av}$, i.e., the mean value of the cosine of the diffusion angle of the neutrons after an impact; $(\cos\theta)_{av} = 1$ means complete coherence, while for an isotropic diffusion, $(\cos\theta)_{av} = 0$. For the neutrons of energy larger than 1 volt, we find $(\cos\theta)_{av} = 2/3 = 0.67$; for the thermal neutrons, assuming $W/h\nu = 1/10$, we deduce (see *F.*, formula 103) $(\cos\theta)_{av} = 0.067$, i.e., one-tenth of the preceding value.

These results justify the approximations that were made sometimes in this work, in considering the diffusion of thermal neutrons as isotropic. We must, however, notice that in such a way we have introduced a small error which is in the sense that the diffusion length, given by (12), is too small, while the albedo, given by (11), is too large. It would also be possible to calculate these corrections, if one knew the value of $(\cos\theta)_{av}$; this value is, however, subordinate to the knowledge of $W/h\nu$ of which we can give only the order of magnitude.

In §§ 6–11 five different magnitudes have been measured, which can be expressed by means of only two, using the relations obtained from the theory of the diffusion of the neutrons (see *F.*).

These five magnitudes are: the mean free paths λ and λ_g of thermal neutrons (see § 8) and of neutrons of groups *D* and *A* (see § 7); the albedo β of thermal neutrons (see § 6); the diffusion length l of thermal neutrons (see § 7); and the length $|x_0|$ relative to the thermal neutrons, considered in Section 11, which is bound to λ by the relation

$$|x_0| = \lambda/\sqrt{3}.$$

By means of (28), (11), (12), and (29), the five above-mentioned magnitudes can be expressed as functions of λ and N (number of mean free paths traversed on the average by thermal neutrons).

TABLE XVIII.

Comparison of measured and calculated values.

$\lambda =$ and $\lambda_g =$ mean free paths of thermal neutrons and of neutrons of group *D* and *A*, respectively. $\beta =$ albedo for thermal neutrons, $l =$ diffusion length, $|x_0| = \lambda/\sqrt{3}$.

λ	0.3	0.30
λ_g	1.1	1.0
β	0.82	0.83
l	2.1	2.05
$\|x_0\|$	0.18	0.174

In Table XVIII we compare the measured values of these five magnitudes with the values calculated by means of the above-mentioned formulae, assuming

$$\lambda = 1.0 \text{ cm} \quad ; \quad N = 140.$$

As we see, the agreement is good.

From these values of λ and N, we find the following values for thermal neutrons in paraffin:

elastic cross section with hydrogen	$\sigma_e = 43 \cdot 10^{-24}$ cm^2
capture cross section	$\sigma_c = 0.31 \cdot 10^{-24}$ cm^2
mean life	$\tau = 1.7 \cdot 10^{-4}$ sec.

This last value is in very good agreement with the measure of τ performed with a mechanical set-up (28) and with the theoretical value deduced from the theory given in *F.*, § 12, based on the assumption that the capture of slow neutrons by protons is due to the emission of a γ-quantum by magnetic

(28) E. AMALDI, O. D'AGOSTINO, E. FERMI, B. PONTECORVO and E. SEGRÈ, « Ric. Scientifica », *6* (1), 581 (1935).

dipole radiation. (29) It is shown there that according to whether the ^1S state of the deuteron ($\pm$ 120,000 volts) is assumed to be real or virtual, the theoretical value of τ is 6.5×10^{-4} or 2.6×10^{-4}. The satisfactory agreement of this last value with the experimental result, seems to indicate that the ^1S state is virtual.

We express our warmest gratitude to the Istituto di Sanità Pubblica of Rome and in particular to Professor G. C. Trabacchi for having supplied us with the radon sources used in this research. Our thanks are also due to the Consiglio Nazionale delle Ricerche of Italy for a grant.

(29) A preliminary account of this theory has been given by E. FERMI, « Phys. Rev. », *48*, 570 (1935).

119 *a*.

SUL MOTO DEI NEUTRONI NELLE SOSTANZE IDROGENATE

« Ric. Scientifica », 7 (2), 13–52 (1936).

In questo lavoro verranno discusse, dal punto di vista teorico, alcune proprietà dei neutroni lenti. Nella prima parte il rallentamento e la diffusione dei neutroni nelle sostanze idrogenate verranno studiate prescindendo da una analisi meccanico–quantistica del processo di urto, elastico o no, tra i neutroni e i protoni della sostanza idrogenata. Allo studio del meccanismo dell'urto sarà dedicata invece la seconda parte. In questo lavoro verranno anche date le giustificazioni matematiche di diverse formule che sono state usate da Amaldi e Fermi [1] nella interpretazione di alcune misure sopra i neutroni lenti. Questi lavori, ai quali dovremo spesso far riferimento nel seguito per illustrare le connessioni tra le teorie svolte e l'esperienza, verranno indicati con *AF*.

PARTE I

Diffusione e rallentamento dei neutroni.

Quando un neutrone viene proiettato da una sorgente in un ambiente ripieno di una sostanza idrogenata (in pratica quasi sempre paraffina o acqua; per brevità nel seguito parleremo normalmente di paraffina) esso subisce un rapido processo di rallentamento dovuto agli urti contro gli atomi di idrogeno [2]. Si dimostra elementarmente che, ad ogni urto, il logaritmo dell'energia del neutrone decresce in media di una unità (riduzione della energia a $1/e$) per modo che, dopo una ventina di urti, l'energia del neutrone, dal valore iniziale di alcuni milioni di volt, si riduce alla energia di agitazione termica. Nel processo di diffusione dei neutroni nella sostanza idrogenata dovremo dunque distinguere:

a) una fase di rallentamento, durante la quale l'energia del neutrone si riduce dal valore iniziale alla energia di agitazione termica;

b) una fase di diffusione termica, durante la quale il neutrone, avendo ormai raggiunta l'energia di agitazione termica, in media non perde più energia negli urti successivi contro gli atomi di idrogeno, e prosegue il suo moto di diffusione fino a quando non esce dalla sostanza idrogenata oppure vien catturato dall'idrogeno o da qualche altro degli atomi presenti nell'ambiente.

1. *La fase di rallentamento*. – Nello studio del processo di rallentamento ammetteremo in un primo tempo che gli atomi di idrogeno si possano considerare come fermi, e che si possa anche prescindere dal fatto che gli idrogeni non sono liberi, ma sono vincolati da forze di legame chimico. Queste ipotesi saranno evidentemente legittime fino a che l'energia w del neutrone è ancora notevolmente maggiore dell'energia dei legami chimici; più precisamente la condizione di validità è che sia $w \gg h\nu$, ν essendo la frequenza degli atomi di idrogeno dovuta al legame chimico. L'ordine di grandezza di queste frequenze, nelle comuni sostanze idrogenate, è da 1000 a 4000 cm^{-1}, per modo che potremo senz'altro ritenere la nostra ipotesi ammissibile finché l'energia del neutrone non discende al di sotto di un volt. Il processo ulteriore di rallentamento da questo limite fino alla energia di agitazione termica ($\sim 1/30$ di volt) è complicato da effetti dovuti al legame chimico dell'idrogeno e sarà discusso nel seguito.

Fino a che l'energia del neutrone è superiore a un volt, ammetteremo che la direzione del movimento del neutrone dopo un urto sia distribuita con probabilità uniforme in tutte le direzioni, purché il moto venga riferito a un sistema in cui è fermo il baricentro del neutrone e del protone che si urtano. Riferendosi invece allo spazio consueto, in cui il protone è inizialmente fermo, e tenendo conto del fatto che neutrone e protone hanno eguale massa, si riconosce subito, con una immediata applicazione dei principi della conservazione della energia e della quantità di moto, che l'angolo ϑ formato tra le traiettorie del neutrone prima e dopo l'urto è sempre acuto; più precisamente si trova, chiamando w e w' le energie del neutrone prima e dopo l'urto, la relazione

$$\frac{w'}{w} = \cos^2 \vartheta \,. \tag{1}$$

Dalla ipotesi che abbiamo ammessa sulla distribuzione della direzione del movimento dopo l'urto segue poi subito che la probabilità che per effetto dell'urto, l'energia si riduca ad un valore compreso tra w' e $w' + dw'$ è

$$dp = \frac{dw'}{w} \cdot \tag{2}$$

Vogliamo prima di tutto determinare il numero di neutroni dotati di velocità compresa entro i limiti v *e* $v + dv$ che sono compresi entro un blocco di paraffina di dimensioni praticamente infinite in cui sia contenuta una sorgente di neutroni veloci che produca Q neutroni al secondo animati da una velocità iniziale v_0.

La distribuzione di velocità in queste circostanze potrebbe ottenersi come caso particolare da quella che è stata da noi dedotta in un'altra pubblicazione [3]. Nel caso presente la si può tuttavia ricavare anche direttamente come segue:

Sia $\mathfrak{N}(v)\,dv$ il numero cercato dei neutroni aventi velocità tra v e $v + dv$, presenti in tutta la massa della paraffina. Questo numero tende a diminuire in quanto che alcuni degli $\mathfrak{N}(v)\,dv$ neutroni subiscono degli urti e conseguentemente la loro velocità diminuisce; tende a crescere perché alcuni dei neutroni di velocità maggiore di v, in seguito a un urto, possono assumere una velocità tra v e $v + dv$. In condizioni stazionarie la diminuzione dovrà essere compensata dalla crescita. Tenendo presente la (2), che introducendo la velocità in luogo dell'energia, si scrive:

$$dp = \frac{2\,v'\,dv'}{v^2} \tag{3}$$

ed osservando che la probabilità che un neutrone subisca un urto nell'unità di tempo è

$$\frac{v}{\lambda(v)} \tag{4}$$

$\lambda(v)$ essendo il cammino libero medio, si trova dunque, come condizione di stazionarietà:

$$\frac{v}{\lambda(v)}\,\mathfrak{N}(v)\,dv = \frac{2\,v\,dv}{v_0^2}\,Q + 2\,v\,dv\int_v^{v_0}\frac{1}{u^2}\,\mathfrak{N}(u)\,\frac{u}{\lambda(u)}\,du$$

cioè

$$\frac{\mathfrak{N}(v)}{\lambda(v)} = \frac{2\,Q}{v_0^2} + 2\int_v^{v_0}\frac{1}{u}\,\frac{\mathfrak{N}(u)}{\lambda(u)}\,du\,. \tag{5}$$

Derivando questa equazione rispetto a v si ricava:

$$\frac{d}{dv}\,\frac{\mathfrak{N}(v)}{\lambda(v)} = -\,\frac{2}{v}\,\frac{\mathfrak{N}(v)}{\lambda(v)}$$

che, integrata, ci dà:

$$v^2\,\frac{\mathfrak{N}(v)}{\lambda(v)} = \text{cost.}$$

La costante, che si determina risostituendo in (5), risulta eguale a $2\,Q$. Otteniamo dunque:

$$\mathfrak{N}(v) = \frac{2\,Q\lambda(v)}{v^2}\,. \tag{6}$$

Questa legge di distribuzione è valida naturalmente solo per energie maggiori di un volt. Nel § 8 ne vedremo l'applicazione ai metodi per determinare la larghezza delle bande di energia che costituiscono i varii gruppi di neutroni lenti.

2. *Ampiezza della diffusione dei neutroni durante il rallentamento.* – La (6) ci dà il numero totale dei neutroni appartenenti a un dato intervallo di energia; essa non ci dice però nulla circa la distribuzione spaziale dei neutroni stessi. Si abbia una sorgente S di neutroni veloci (energia iniziale w_0) nel centro di un blocco di paraffina di dimensioni grandissime. Via via che i neutroni subiscono urti contro gli atomi di idrogeno, essi perdono energia; così che, in media, i neutroni che hanno subìto più urti avranno una energia minore. D'altra parte i neutroni che hanno subìto più urti sono quelli la cui distanza media dalla sorgente è più grande; segue dunque che la distanza media dei neutroni di una data energia dalla sorgente sarà tanto maggiore quanto più piccola è la loro energia.

Ci proponiamo di calcolare la media $\overline{r^2}$ dei quadrati della distanza dalla sorgente in funzione della energia (limitandoci sempre al caso di energie maggiori di un volt). Nel calcolo terremo conto solamente degli urti contro gli atomi di idrogeno; ciò è legittimo, quando la sostanza idrogenata sia acqua o paraffina, poiché può valutarsi che in questo caso, per neutroni veloci, la percentuale di urti contro l'idrogeno sia già del 90 °/₀; mentre per neutroni lenti essa sale notevolmente al di sopra di questo valore. Del resto, alla fine del paragrafo, daremo anche, senza dimostrazione, le formule per il caso che si tenga conto anche degli urti contro atomi pesanti presenti nella sostanza idrogenata in cui diffondono i neutroni.

Il calcolo di $\overline{r^2}$ è complicato dalla circostanza che la direzione del moto di un neutrone dopo un urto non è indipendente da quella che esso aveva prima, essendo sempre acuto l'angolo formato tra le due direzioni. Siano $\boldsymbol{\lambda}_0, \boldsymbol{\lambda}_1, \cdots, \boldsymbol{\lambda}_n$ i percorsi del neutrone durante i primi $n+1$ cammini liberi; lo spostamento totale è la somma vettoriale

$$\boldsymbol{r} = \boldsymbol{\lambda}_0 + \boldsymbol{\lambda}_1 + \cdots + \boldsymbol{\lambda}_n . \tag{7}$$

Supponiamo che, con l'n–esimo urto, l'energia prenda il valore assegnato w_a, e proponiamoci di calcolare $\overline{r^2}$ in funzione di w_a. Nell'eseguire i calcoli conviene introdurre una scala logaritmica per le energie; in luogo dell'energia w useremo la variabile

$$x = \log \frac{w_0}{w} \tag{8}$$

che spesso rappresenteremo graficamente come ascissa di un punto su una retta. Il cammino libero medio, che dipende dalla energia, potrà anche scriversi come funzione di x, $\lambda = \lambda(x)$. Durante il movimento del neutrone la sua energia, dal valore iniziale w_0, prende, per effetto dei successivi urti, energie sempre più piccole

$$w_0 > w_1 > w_2 > \cdots > w_{n-1} > w_a$$

e in corrispondenza la x prenderà i valori

$$0 < x_1 < x_2 < \cdots < x_{n-1} < x_n = \log \frac{w_0}{w_a} .$$

I punti di ascisse $x_1, x_2, \ldots$ sono distribuiti con una legge assai semplice. Infatti la (2) equivale a dire che la probabilità che

$$\xi_r = x_r - x_{r-1} \tag{9}$$

abbia valore compreso tra ξ_r e $\xi_r + d\xi_r$ è

$$e^{-\xi_r} d\xi_r .$$

Questa legge di probabilità per l'intervallo tra due punti successivi in una distribuzione di punti sopra una retta, è quella che si ottiene quando i punti sono distribuiti indipendentemente uno dall'altro e in modo tale che la probabilità che uno di essi cada in un assegnato intervallo dx della retta sia eguale a dx. Tale sarà dunque la legge di distribuzione dei punti x.

Sia ϑ_r l'angolo formato tra i cammini liberi successivi λ_{r-1} e λ_r; si ha dalle (1) (8) (9)

(10)
$$\cos \vartheta_r = e^{-\frac{1}{2}\xi_r} = e^{-\frac{1}{2}(x_r - x_{r-1})}.$$

Poniamo inoltre

(11)
$$a = \log \frac{w_0}{w_a};$$

a rappresenta il numero medio degli urti necessari per ridurre l'energia da w_0 a w_a. Faremo la media di r^2 in due stadi successivi.

I) Supporremo dati i punti

(12)
$$x_1, x_2, \cdots, x_{n-1};$$

e medieremo rispetto a tutte le possibili grandezze dei cammini liberi e a tutte le direzioni compatibili con (10) e (12).

II) Medieremo poi rispetto a tutti i possibili valori delle (12) e di n.

I) Elevando al quadrato la (7) si trova

(13)
$$r^2 = \sum_0^n \lambda_r^2 + 2 \sum_{r=0}^{n-1} \sum_{s=r+1}^{n} \lambda_r \lambda_s \cos \widehat{\lambda_r \lambda_s}.$$

Di questa espressione dobbiamo fare la media ammettendo date le (12). Si trova senza difficoltà, tenendo presente (10),

$$\overline{\lambda_r^2} = 2\lambda^2(x_r) \quad ; \quad \overline{\lambda_r} = \lambda(x_r)$$

(14)
$$\overline{\cos \widehat{\lambda_r \lambda_s}} = \cos \vartheta_{r+1} \cos \vartheta_{r+2} \cdots \cos \vartheta_s = e^{-\frac{1}{2}(x_s - x_r)}$$

e quindi

(15)
$$\overline{r^2} = 2 \sum_0^n \lambda^2(x_r) + 2 \sum_{0 \leq r < s \leq n} \lambda(x_r) \lambda(x_s) e^{-\frac{1}{2}(x_s - x_r)}.$$

In questa formula la media si intende fatta per dati valori delle (12).

II) Dobbiamo ora mediare rispetto ai valori delle x e di n. Tenendo presente che la probabilità che una delle x cada nell'intervallo dx (tra 0 ed a) è eguale a dx, si trova subito

$$\overline{\sum_0^n \lambda^2(x_r)} = \lambda^2(0) + \int_0^a \lambda^2(x)\, dx + \lambda^2(a).$$

Similmente si ha:

$$\overline{\sum_1^n \lambda(0) \lambda(x_s) e^{-\frac{x_s}{2}}} = \lambda(0) \int_0^a \lambda(x) e^{-\frac{1}{2}x} dx + \lambda(0) \lambda(a) e^{-\frac{1}{2}a};$$

$$\overline{\sum_1^{n-1} \lambda(x_r) \lambda(a) e^{-\frac{1}{2}(a - x_r)}} = \lambda(a) \int_0^a \lambda(x) e^{-\frac{1}{2}(a-x)} dx.$$

Resta da calcolare la media di

$$\sum_{1 \leq r < s \leq n-1} \lambda(x_r) \lambda(x_s) e^{-\frac{1}{2}(x_s - x_r)}.$$

Per questo effettuiamo dapprima la media rispetto ai valori possibili di x_s compreso tra i limiti x_r ed a; essa è:

$$\sum_r \lambda(x_r) \int_{x_r}^{a} \lambda(x) e^{-\frac{1}{2}(x-x_r)} dx = \sum_r \lambda(x_r) \int_0^{a-x_r} \lambda(x_r+x) e^{-\frac{1}{2}x} dx .$$

Mediando da ultimo rispetto ai valori di x troviamo infine:

$$\int_0^a \lambda(\xi) d\xi \int_0^{a-\xi} \lambda(\xi+x) e^{-\frac{1}{2}x} dx .$$

Combinando queste varie medie otteniamo il risultato cercato:

$$(16) \qquad \overline{r^2} = 2\lambda^2(0) + 2\lambda^2(a) + 2\int_0^a \lambda^2(x) dx + 2\lambda(0) \int_0^a \lambda(x) e^{-\frac{1}{2}x} dx$$

$$+ 2\lambda(0)\lambda(a) e^{-\frac{a}{2}} + 2\lambda(a) \int_0^a \lambda(x) e^{-\frac{1}{2}(a-x)} dx$$

$$+ 2\int_0^a \lambda(\xi) d\xi \int_0^{a-\xi} \lambda(\xi+x) e^{-\frac{1}{2}x} dx .$$

Questa formula ci descrive come la distanza quadratica media dei neutroni dalla sorgente che li ha emessi va crescendo man mano che la energia decresce (e cioè a cresce). Da essa si ricava:

$$(17) \qquad \frac{d\overline{r^2}}{da} = 4\lambda^2(a) + 4\lambda(a)\lambda'(a) + \lambda(0)[\lambda(a) + 2\lambda'(a)] e^{-\frac{a}{2}}$$

$$+ [\lambda(a) + 2\lambda'(a)] \int_0^a \lambda(x) e^{-\frac{1}{2}(a-x)} dx .$$

Il cammino libero medio $\lambda(0)$ dei neutroni veloci emessi da una sorgente di emanazione + berillio nella paraffina è eguale a 5 o 6 cm. Diminuendo l'energia esso decresce fino a ridursi probabilmente a circa 1 cm quando l'energia, pur essendo maggiore di 1 volt, ha già un valore tale (probabilmente non oltre 1000 volt) che il neutrone si possa chiamare lento. In questo ultimo intervallo di energia (da 1 a 1000 volt) λ è probabilmente costante (cfr. anche la Parte II). Per esso la (17) si semplifica notevolmente. Possiamo infatti trascurare il terzo termine del secondo membro che contiene il fattore piccolissimo $e^{-\frac{a}{2}}$ e sostituire l'integrale, i cui valori sono concentrati vicino al limite superiore, con

$$\int_{-\infty}^a \lambda(x) e^{-(a-x)/2} dx = 2\lambda(a);$$

otteniamo con ciò, ponendo anche, per quanto si è detto, $\lambda'(a) = 0$.

$$(18) \qquad \frac{d\overline{r^2}}{da} = 6\lambda^2(a).$$

Questa formula è stata applicata (cfr. *AF*) alla determinazione del rapporto di energia tra due gruppi di neutroni lenti. Siano w' e w'' le loro energie; ammettiamo poi che nell'intervallo di energia da w' a w'', λ sia costante. Siano infine $\overline{r'^2}$ ed $\overline{r''^2}$ le medie dei quadrati della distanza dei neutroni dei due gruppi dalla sorgente. Tenendo presente che, secondo la (11), la variazione della variabile a nel passaggio tra i due gruppi è $\log w''/w'$, si trova dalla (18)

$$\log \frac{w''}{w'} = \frac{\overline{r'^2} - \overline{r''^2}}{6\lambda^2} \, . \tag{19}$$

che è la formula usata per determinare w''/w'.

Se ammettiamo che il cammino libero medio, nell'intervallo di energia dei gruppi in esame, sia di 1 cm, risulta dalla formula precedente che ad una differenza di 6 cm^2 nelle medie dei quadrati delle distanze dalla sorgente, corrisponde un rapporto e tra le energie di due gruppi. Per una discussione dettagliata di questo argomento, fatta in base ai risultati sperimentali, vedi *AF*.

Con un procedimento simile a quello seguito per dedurre la (16), si può facilmente ottenere una formula analoga valida nel caso che i neutroni diffondano e vengano rallentati in un mezzo contenente, oltre che atomi di idrogeno, anche degli atomi pesanti. Conservando le notazioni (8) e (11), indicheremo con $\rho(x)$ il rapporto che si ha, per neutroni di energia corrispondente a x, tra le probabilità di urto contro uno degli atomi pesanti o contro uno degli atomi di idrogeno. Ammettiamo inoltre che gli urti contro gli atomi pesanti si possano considerare come perfettamente elastici; che data la massa degli atomi urtati la energia trasferita a questi nell'urto possa trascurarsi, e che infine la direzione della velocità dopo un urto contro un atomo pesante sia distribuita con legge di probabilità uniforme in tutte le direzioni, perdendo ogni coerenza con la direzione del movimento prima dell'urto. Si trova allora che la media dei quadrati della distanza dalla sorgente per neutroni di energia w_a è [1]:

(1) N.d.R. Questa equazione è errata, probabilmente per ragioni tipografiche. L'espressione corretta è stata dedotta da vari Autori ed è riportata nell'articolo di R. E. Marshak («Rev. Mod. Phys.», *19*, 185 (1947)). Invece del parametro $\rho = \dfrac{\mathfrak{N}_{ep}\,\sigma_{ep}}{\mathfrak{N}_H\,\sigma_H}$, Marshak usa il parametro $c = \dfrac{\mathfrak{N}_H\,\sigma_H}{\mathfrak{N}_H\,\sigma_H + \mathfrak{N}_{ep}\,\sigma_{ep}}$. Con $c = \dfrac{1}{1+\rho}$ e $\overline{r^2} = 6\,L_s^2$ l'equazione (108) a p. 211 dell'articolo di Marshak diventa

$$\overline{r^2} = 2\lambda^2(0)[1+\rho(0)] + 2\lambda^2(a)[1+\rho(a)] + 2\lambda(0)\lambda(a)\, e^{-\int_0^a \left[\frac{3}{2} - \frac{1}{1+\rho(x')}\right] dx'}$$

$$+ 2\int_0^a \lambda^2(x)[1+\rho(x)]\,dx + 2\lambda(a)\int_0^a \lambda(x)\,dx\, e^{-\int_x^a \left[\frac{3}{2} - \frac{1}{1+\rho(x')}\right] dx'}$$

$$+ 2\lambda(0)\int_0^a \lambda(x)\,dx\, e^{-\int_0^x \left[\frac{3}{2} - \frac{1}{1+\rho(x')}\right] dx'} + 2\int_0^a \lambda(x)\,dx \int_0^x \lambda(x')\,dx'\, e^{-\int_{x'}^x \left[\frac{3}{2} - \frac{1}{1+\rho(x'')}\right] dx''} \, .$$

$$\overline{r^2} = 2\lambda^2(0)[1+\rho(0)] + 2\lambda^2(a)[1+\rho(a)] + 2\int_0^a \lambda^2(x)[1+\rho(x)]\,dx \tag{20}$$

$$+ \frac{2\lambda(0)}{1+\rho(0)}\int_0^a \lambda(x)\,e^{-\frac{x}{2}}\,dx\,e^{-\int_0^x \varrho(\xi)\,d\xi} + 2\lambda(a)\int_0^a \frac{\lambda(x)}{1+\rho(x)}\,e^{-\frac{x}{2}-\int_x^a \varrho(\xi)\,d\xi}\,dx$$

$$+ 2\int_0^a \frac{\lambda(x)\,dx}{1+\rho(x)}\int_0^{a-x}\lambda(x+\xi)\,e^{-\frac{\xi}{2}-\int_0^\xi \varrho(x+\xi)\,d\xi}\,dx.$$

Questa formula, naturalmente, si riduce alla (16) quando si trascuri l'effetto degli urti contro gli atomi pesanti, ponendo in essa $\rho(x) = 0$.

3. *I neutroni termici.* – Ci siamo fino ad ora occupati di neutroni appartenenti all'intervallo di energia dalla massima fino a un volt, e cioè sempre superiore non solo alla energia del moto di agitazione termica ($\sim 1/30$ di volt), ma anche al quanto $h\nu$ delle forze chimiche che legano gli atomi di idrogeno alla sostanza (da $1/2$ a $1/10$ di volt). In questo intervallo di energia abbiamo potuto prescindere dai moti di agitazione termica e dal legame chimico dell'idrogeno. Quando però, con urti successivi, l'energia del neutrone scende al di sotto di un volt, la legge con cui avviene l'ulteriore rallentamento diventa alquanto più complicata (cfr. Parte II); si deve tuttavia ritenere che, dopo una decina di urti, essa sia ridotta alla energia della agitazione termica. Esisterà dunque un notevole numero di neutroni aventi la energia di agitazione termica e la cui distribuzione di velocità sarà più o meno approssimativamente quella Marxwelliana. Il numero di questi neutroni termici è limitato dal fatto che, mentre se ne producono continuamente per rallentamento dei neutroni più veloci, se ne vengono anche continuamente a distruggere, in massima parte, probabilmente, per cattura da parte dei nuclei di idrogeno e formazione di deuterio (cfr. anche Parte II). Rappresenteremo l'effetto di questo processo di cattura introducendo una vita media τ dei neutroni termici; per modo che, se la sorgente produce Q neutroni termici al secondo entro un blocco praticamente illimitato di paraffina, il numero totale dei neutroni termici presenti a un dato istante sarà

$$\mathfrak{N}_T = Q\tau.$$

Anche rispetto agli urti elastici il comportamento dei neutroni termici è verosimilmente diverso da quello dei neutroni di energia superiore a un volt. Ragioni teoriche (cfr. Parte II) inducono infatti a ritenere che, quando l'energia w del neutrone passa da valori grandi a valori piccoli rispetto ad $h\nu$ (ν = frequenza del legame chimico dell'idrogeno nella paraffina) il cammino libero medio si riduca nel rapporto $4:1$, mentre la legge di distribuzione angolare dei neutroni dopo un urto tenda, al limite per w molto piccolo, a diventare isotropa. Benché per neutroni termici $w/h\nu$ sia abbastanza piccolo (circa $1/10$) non sono ancora interamente raggiunte le condizioni di applicabilità delle leggi limite indicate (cfr. § 11). Ciò nonostante nei paragrafi 3, 5, 6, 7 sulla diffusione dei neutroni termici semplificheremo le considerazioni ammettendo che la diffusione sia isotropa; il piccolo errore che così si introduce è evidentemente nel senso di dare un tratto di diffusione troppo corto ed un albedo troppo grande.

Se λ è il cammino libero medio dei neutroni termici nella paraffina (0,3 cm, cfr. *AF*), v la loro velocità media (a temperatura ordinaria $v = 2{,}5 \cdot 10^5$ cm/s) ed N il numero medio dei cammini liberi percorsi da un neutrone termico prima di venir catturato (N = 140; cfr. *AF*, § 13), valgono evidentemente le relazioni:

$$\tau = \frac{\lambda N}{v} \quad ; \quad \mathfrak{N}_T = Q\tau . \tag{22}$$

Da esse si ricava $\tau = 1{,}7 \cdot 10^{-4}$ s (cfr. anche § 12).

Dato il considerevole numero di urti che un neutrone termico subisce in media prima di venir catturato, è chiaro che il suo moto potrà assimilarsi con notevole accuratezza a un moto di diffusione. Si scrive facilmente una equazione differenziale per la densità n dei neutroni termici, in funzione di x, y, z, t. Indicando con $q(x, y, z, t)$ il numero di neutroni termici che si producono per unità di volume e di tempo dal rallentamento di neutroni più veloci, si scrive subito

$$\frac{\partial n}{\partial t} = \frac{\lambda v}{3} \Delta n - \frac{1}{\tau} n + q \tag{23}$$

dove, nel secondo membro, il primo termine rappresenta l'effetto della diffusione (si sa infatti dalla teoria cinetica classica che $\lambda v/3$ è il coefficiente di diffusione); il secondo termine esprime l'effetto della distruzione dei neutroni per cattura; il terzo quello della produzione di neutroni termici per rallentamento. Nel caso stazionario, che è evidentemente quello più importante, la (23) si riduce, tenendo presente anche (22), a:

$$n = \frac{\lambda^2 N}{3} \Delta n + \frac{q \lambda N}{v} . \tag{24}$$

A titolo di esempio mostriamo l'applicazione della (24) alla risoluzione del seguente problema:

Si abbia un blocco di paraffina, limitato a sinistra dal piano $x = 0$, il quale occupi tutto lo spazio a destra di questo piano. Trovare la probabilità che un neutrone termico che si trova nella posizione di ascissa positiva a arrivi ad uscire dalla paraffina attraverso al piano $x = 0$, prima di venir catturato. Il problema equivale evidentemente a quest'altro: supponiamo che, in un punto di ascissa a si producano Q neutroni termici al secondo; trovare la frazione di questi che esce dalla paraffina attraverso al piano $x = 0$.

Per risolvere questo problema riduciamoci dapprima dalla equazione (24) con le tre variabili indipendenti x, y, z, ad una contenente la sola variabile indipendente x; basta per questo moltiplicare i due membri di (24) per $dy\, dz$ ed integrare rispetto a y e a z tra $-\infty$ e $+\infty$. Ponendo

$$\rho(x) = \iint n\, dy\, dz \quad ; \quad f(x) = \iint q\, dy\, dz \tag{25}$$

e tenendo presente le relazioni immediate

$$\iint \frac{\partial^2 n}{\partial y^2} dy\, dz = \iint \frac{\partial^2 n}{\partial z^2} dy\, dz = 0,$$

si trova allora

$$\rho(x) - \frac{\lambda^2 N}{3} \frac{d^2 \rho}{dx^2} = \frac{\lambda N}{v} f(x). \tag{26}$$

Nel caso nostro la funzione $f(x)$ è evidentemente nulla per $x \neq a$; mentre per $x = a$ essa è infinita in modo tale che

$$\int f(x)\, dx = Q \tag{27}$$

quando l'integrale sia esteso a un intervallo contenente a. Segue da queste proprietà di $f(x)$ che, ad eccezione del punto a, $\rho(x)$ soddisfa all'equazione differenziale

$$\rho - \frac{\lambda^2 N}{3} \frac{d^2 \rho}{dx^2} = 0. \tag{28}$$

Per $x = 0$ e per $x = \infty$, la $\rho(x)$ deve evidentemente soddisfare le condizioni

$$\rho(0) = 0 \quad ; \quad \rho(\infty) = 0. \tag{29}$$

Da (26) e (27) risulta poi che nel punto singolare a si ha una discontinuità del valore della derivata di ρ, mentre la funzione stessa resta continua; si ha precisamente:

$$\delta\rho(a) = 0 \qquad \delta\left(\frac{d\rho}{dx}\right)_{x=a} = -\frac{3}{v\lambda} Q. \tag{30}$$

L'integrale generale di (28) è dato da una combinazione lineare dei due integrali fondamentali

$$e^{\sqrt{\frac{3}{\lambda^2 N}}\, x} \quad ; \quad e^{-\sqrt{\frac{3}{\lambda^2 N}}\, x}.$$

Tenendo presenti le (29) si trova immediatamente che le espressioni di $\rho(x)$ a sinistra e a destra del punto a sono della forma

$$\begin{cases} \rho(x) = A \left\{ e^{\sqrt{\frac{3}{\lambda^2 N}}\, x} - e^{-\sqrt{\frac{3}{\lambda^2 N}}\, x} \right\} & \text{per} \quad 0 < x < a \\ \rho(x) = B e^{-\sqrt{\frac{3}{\lambda^2 N}}\, x} & \text{per} \quad x > a \end{cases} \tag{31}$$

in cui A e B sono delle costanti che si determinano subito per mezzo delle (30). Si trova precisamente

$$A = \frac{\sqrt{3N}}{2v} Q e^{-\sqrt{\frac{3}{\lambda^2 N}}\, a} \quad ; \quad B = A\left(e^{2\sqrt{\frac{3}{\lambda^2 N}}\, a} - 1\right). \tag{32}$$

Il numero di neutroni che escono dalla paraffina nell'unità di tempo è dato evidentemente dal prodotto del coefficiente di diffusione $\lambda v/3$ per la derivata di ρ rispetto ad x presa per $x = 0$. Tale numero risulta dunque

$$\frac{\lambda v}{3} \cdot 2A \sqrt{\frac{3}{\lambda^2 N}} = Q e^{-\sqrt{\frac{3}{\lambda^2 N}}\, a}. \tag{33}$$

La probabilità cercata $p(a)$ che un neutrone partente da un punto di ascissa a riesca ad uscire dalla paraffina è evidentemente eguale al rapporto tra l'espressione (33) e Q; e cioè

$$p(a) = e^{-\sqrt{\frac{3}{\lambda^2 N}}\,a} . \tag{34}$$

Questa formula è stata usata (cfr. *AF*) per la misura di $\lambda^2 N$.

Si noti che questo risultato, come tutti gli altri che si possono dedurre delle equazioni differenziali (23) e (24), non è valido esattamente altro che nel caso limite che N sia molto grande e λ molto piccolo, e ciò in modo tale $\lambda^2 N$ si mantenga finito. Si vedrà infatti nei paragrafi seguenti che, in vicinanza della superficie che limita la paraffina, e precisamente entro una distanza da questa dell'ordine di grandezza di λ, la densità dei neutroni presenta delle anomalie.

Per studiarle converrà dapprima semplificare il problema come è indicato nel paragrafo seguente.

4. *Il caso unidimensionale.* – In questo paragrafo semplificheremo lo studio della diffusione riferendoci al caso fittizio che i neutroni siano mobili in una dimensione lungo una retta, invece di essere liberi nello spazio. Il procedimento presenta il vantaggio che, in questo caso semplice, si possono poi senza difficoltà ricavare soluzioni assai approssimate, valide per il caso reale di tre dimensioni.

Nel caso unidimensionale ammetteremo che ad ogni urto (cammino libero medio λ) il neutrone abbia probabilità $1/N$ di essere catturato; se invece l'urto non dà luogo a cattura, il neutrone avrà probabilità eguale di ripartire verso destra o verso sinistra.

Vogliamo prima di tutto cercare per il caso unidimensionale la soluzione esatta di un problema analogo a quello trattato alla fine del paragrafo precedente; precisamente: Si abbia un pezzo di paraffina compreso tra le ascisse 0 ed a; un neutrone termico parte da un punto di ascissa x entro la paraffina; trovare la probabilità $p(x)$ che il neutrone esca dalla paraffina dalla parte di sinistra. Questo problema, nel caso limite $a = \infty$ è l'analogo unidimensionale di quello studiato nel paragrafo precedente.

Per risolverlo stabiliremo una equazione per $p(x)$ al modo seguente: La probabilità $p(x)$ può analizzarsi nella seguente somma:

I) Probabilità che il neutrone, partendo da x, inizi il suo moto verso sinistra ed esca direttamente dalla paraffina senza aver subìto alcun urto. Questa probabilità è

$$\frac{1}{2} e^{-x/\lambda} .$$

II) Probabilità che il neutrone subisca un primo urto tra ξ e $\xi + d\xi$ $\left[\text{e cioè } \frac{1}{2} e^{-|x-\xi|/\lambda} d\xi/\lambda\right]$; che questo urto sia elastico [e cioè $(N - 1)/N$]; e infine che il neutrone, dopo di esso, finisca per uscire dalla paraffina dalla parte sinistra [e cioè $p(\xi)$]. La probabilità di questo evento è il prodotto delle probabilità componenti, e cioè

$$\frac{1}{2} e^{-|x-\xi|/\lambda} \frac{d\xi}{\lambda} \cdot \frac{N-1}{N} \cdot p(\xi)$$

ed essa deve integrarsi rispetto a ξ da 0 ad a. Sommando I e II troviamo per $p(x)$ la seguente equizione integrale:

$$p(x) = \frac{1}{2} e^{-x/\lambda} + \frac{N-1}{2\lambda N} \int_0^a p(\xi) e^{-|x-\xi|/\lambda} d\xi . \tag{35}$$

Si verifica facilmente che la soluzione di (35) può mettersi nella forma:

$$p(x) = Ae^{\alpha x} + Be^{-\alpha x} \tag{36}$$

in cui A, B, ed α sono costanti; basta perciò sostituire (36) in (35); si trova allora che è possibile soddisfare la (35), purché si ponga

$$\left\{ \begin{aligned} A &= -\frac{(N-\sqrt{N})\, e^{-a/\lambda\sqrt{N}}}{(\sqrt{N}+1)^2 e^{a/\lambda\sqrt{N}} - (\sqrt{N}-1)^2 e^{-a/\lambda\sqrt{N}}} ; \\ B &= \frac{(N+\sqrt{N})\, e^{a/\lambda\sqrt{N}}}{(\sqrt{N}+1)^2 e^{a/\lambda\sqrt{N}} - (\sqrt{N}-1)^2 e^{-a/\lambda\sqrt{N}}} \quad ; \quad \alpha = \frac{1}{\lambda\sqrt{N}} . \end{aligned} \right. \tag{37}$$

Queste formule si semplificano notevolmente nel caso che la paraffina sia illimitata dalla parte di destra ($a = \infty$); esse diventano allora:

$$\left. \begin{aligned} A = 0 \quad ; \quad B &= \frac{\sqrt{N}}{\sqrt{N}+1} \quad ; \quad \alpha = \frac{1}{\lambda\sqrt{N}} \\ p(x) &= \frac{\sqrt{N}}{\sqrt{N}+1} e^{-x/\lambda\sqrt{N}} \end{aligned} \right\} \text{per} \quad a = \infty . \tag{38}$$

Il problema ora risolto ci permette di determinare facilmente, sempre per il caso unidimensionale, il coefficiente di riflessione (albedo) dei neutroni termici che incidono sopra uno strato di paraffina. Supponiamo che sul nostro strato di paraffina, esteso come prima da $x = 0$ ad $x = a$, cada un neutrone termico proveniente da sinistra; chiameremo albedo $\beta(a)$ dello strato di spessore a la probabilità che il neutrone, dopo aver subìto uno o più urti, riesca dalla stessa parte da cui è entrato. Per determinare la albedo osserviamo che la probabilità che il neutrone, entrato nella paraffina, subìsca un primo urto tra x ed $x + dx$, e non venga ivi catturato, è data, come sopra, da:

$$\frac{N-1}{N} e^{-x/\lambda} \frac{dx}{\lambda} .$$

Abbiamo visto prima che un neutrone che parte da x ha la probabilità $p(x)$ di uscire dalla paraffina dalla parte sinistra; e quindi la probabilità totale di una uscita a sinistra, e cioè l'albedo, risulta nel nostro caso

$$\beta(a) = \frac{N-1}{N\lambda} \int_0^a p(x)\, e^{-x/\lambda} dx ;$$

da (36) e (37) si ricava, effettuando l'integrazione,

$$(39) \qquad \beta(a) = \frac{(N-1)\,[e^{a/\lambda\sqrt{N}} - e^{-a/\lambda\sqrt{N}}]}{(\sqrt{N}+1)^2\, e^{a/\lambda\sqrt{N}} - (\sqrt{N}-1)^2\, e^{-a/\lambda\sqrt{N}}} = \frac{N-1}{N+1+2\sqrt{N}\,\mathrm{cotgh}\,\frac{a}{\lambda\sqrt{N}}}.$$

Ha particolare importanza il caso che lo spessore a della paraffina sia infinito. La albedo risulta in questo caso da (39), o anche direttamente da (38),

$$(40) \qquad \beta(\infty) = \frac{\sqrt{N}-1}{\sqrt{N}+1} = 1 - \frac{2}{\sqrt{N}} + \cdots.$$

L'ultima espressione, approssimata per N molto grande, è quella che si usa più frequentemente.

Scriviamo ancora l'espressione del coefficiente di trasmissione di uno strato di spessore a, e cioè della probabilità che un neutrone che incide su di esso da sinistra ne riesca dalla destra. Tale coefficiente, con considerazioni analoghe a quelle fatte per ottenere l'albedo, risulta:

$$(41) \qquad \gamma(a) = e^{-a/\lambda} + \frac{N-1}{N\lambda}\int_0^a p(a-x)\, e^{-x/\lambda}\, dx = \frac{4\sqrt{N}}{(\sqrt{N}+1)^2\, e^{a/\lambda\sqrt{N}} - (\sqrt{N}-1)^2\, e^{-a/\lambda\sqrt{N}}}.$$

Per $N = \infty$, $\gamma(a)$ si riduce a:

$$(42) \qquad \gamma_\infty(a) = \frac{2\lambda}{a+2\lambda}.$$

In vista di future applicazioni dobbiamo ancora trattare, sempre per il caso unidimensionale, la seguente questione. Trascuriamo il processo di distruzione dei neutroni ($N = \infty$) e supponiamo che Q neutroni termici al secondo si producano in un pezzo di paraffina esteso da $x = 0$ ad $x = \infty$, a distanza dal bordo $x = 0$ grande (in confronto a λ). I neutroni diffonderanno attraverso al bordo e sfuggiranno dalla paraffina; essendo, per ipotesi, trascurabile il processo di distruzione dei neutroni, è chiaro che, nello stato di regime, il flusso di neutroni dall'interno verso l'esterno della paraffina deve essere di Q al secondo. Per mantenere questo flusso costante la densità $n(x)$ dei neutroni deve crescere lineamente in funzione di x; avremo dunque

$$(43) \qquad n(x) = A + Bx.$$

Le costanti A e B si determinano facilmente scrivendo che il flusso dei neutroni attraverso a qualsiasi traguardo x verso sinistra è eguale a Q. Q deve essere dunque eguale alla differenza tra i numeri di neutroni che attraversano in un secondo il punto x verso sinistra e verso destra; si trova cioè

$$Q = \frac{1}{2}\frac{v}{\lambda}\int_0^\infty n(x+\xi)\, e^{-\xi/\lambda}\, d\xi - \frac{1}{2}\frac{v}{\lambda}\int_0^x n(x-\xi)\, e^{-\xi/\lambda}\, d\xi.$$

Introducendo in questa equazione l'espressione (43) di $n(x)$, si verifica subito che essa può soddisfarsi, purché si abbia:

$$Q = \lambda v \mathrm{B} \tag{44}$$

$$\mathrm{A} = \mathrm{B}\lambda . \tag{45}$$

La (43) si scrive dunque

$$n(x) = (\lambda + x)\,\mathrm{B} \tag{46}$$

mentre da (34) ricaviamo il coefficiente di diffusione (unidimensionale)

$$\mathrm{D} = \lambda v . \tag{47}$$

Osserviamo in particolare che da (46) risulta che la densità dei neutroni non si annulla sul bordo della paraffina; se si estrapolasse l'andamento di $n(x)$ verso l'esterno della paraffina, si troverebbe invece che la densità $n(x)$ verrebbe ad annullarsi a distanza dal bordo eguale al cammino libero medio λ.

5. *Albedo e probabilità di trasmissione in tre dimensioni.* – Dai risultati esatti ottenuti nel paragrafo precedente per il caso unidimensionale, possono facilmente dedursene altri assai approssimati, validi per il caso reale di una diffusione di neutroni in tre dimensioni.

Consideriamo perciò la proiezione sull'asse delle ascisse del moto di un neutrone termico che subisca successivi urti nella paraffina. Il moto proiezione sarà un moto di avanti e indietro, abbastanza analogo a quello che si avrebbe nel caso unidimensionale. La differenza sta nel fatto che mentre, nel caso unidimensionale, la probabilità di subire un urto dopo un dato percorso segue la legge esponenziale, nel caso della proiezione di un moto tridimensionale invece, la legge corrispondente, pur essendo qualitativamente non molto dissimile da una esponenziale, ne differisce tuttavia. Inoltre poi, naturalmente, il cammino libero medio λ' del moto proiezione è più corto di quello λ del moto vero.

Per scegliere in modo conveniente λ', possiamo per esempio scrivere per il caso unidimensionale l'equazione della diffusione analoga alla (24); tenendo presente la espressione (47) del coefficiente di diffusione unidimensionale (introducendo λ' e v' in luogo di λ e v esso risulta $\mathrm{D}' = \lambda' v'$) essa risulta

$$n - \lambda'^2 \mathrm{N} \frac{d^2 n}{dx^2} = \frac{q\lambda' \mathrm{N}}{v'} .$$

Il confronto di questa equazione con la (24) ci suggerisce di porre

$$\lambda' = \frac{\lambda}{\sqrt{3}} . \tag{48}$$

Un altro modo per arrivare a questa stessa scelta di λ' consiste nell'osservare che, nel caso tridimensionale, la media dei quadrati $\overline{r^2}$ dello spostamento di un punto che subisca m urti con cammino libero medio λ è

$$\overline{r^2} = 2\lambda^2 m$$

(il fattore 2 deriva dal fatto che la media dei quadrati del cammino libero è il doppio del quadrato della media del cammino libero stesso). La media dei quadrati della proiezione δx dello spostamento r, è evidentemente 1/3 di $\overline{r^2}$ e cioè

$$\overline{\delta x^2} = \frac{2}{3} \lambda^2 m .$$

D'altra parte in un moto unidimensionale si avrebbe

$$\overline{\delta x^2} = 2 \lambda'^2 m .$$

Le due espressioni si eguagliano appunto con la posizione (48).

Concludiamo dunque che la proiezione di un moto di diffusione in tre dimensioni è assai simile a un moto di diffusione unidimensionale con cammino libero medio λ' dato dalla (48); in questo modo, come ora vedremo, gran parte dei risultati del paragrafo precedente possono essere utilizzati con buona approssimazione anche per il caso di tre dimensioni.

Nel paragrafo precedente abbiamo stabilita la espressione della probabilità $p(x)$ che un neutrone termico che parta dalla posizione x esca da uno strato di paraffina di spessore a limitato da due piani di ascissa 0 ed a. Il problema analogo nel caso tridimensionale è il seguente: Si abbia uno strato di paraffina di spessore a limitato da due faccie piane di ascissa 0 ed a: trovare la probabilità $p(x)$ che un neutrone termico, muovendosi inizialmente da una posizione di ascissa x, esca dallo strato di paraffina attraverso la faccia sinistra.

Come espressione approssimata di $p(x)$ potremo prendere la [35], dove A, B, α sono date da formule analoghe alle (37); nelle quali tuttavia, in luogo di λ, deve scriversi λ', dato da (48). Si ha dunque:

$$(49) \quad \begin{cases} p(x) = \mathrm{A} e^{\alpha x} + \mathrm{B} e^{-\alpha x} \\ \mathrm{A} = -\dfrac{(\mathrm{N} - \sqrt{\mathrm{N}})\, e^{-\alpha a}}{(\sqrt{\mathrm{N}} + 1)^2 e^{\alpha a} - (\sqrt{\mathrm{N}} - 1)^2 e^{-\alpha a}} \\ \mathrm{B} = \dfrac{(\mathrm{N} + \sqrt{\mathrm{N}})\, e^{\alpha a}}{(\sqrt{\mathrm{N}} + 1)^2 e^{\alpha a} - (\sqrt{\mathrm{N}} - 1)^2 e^{-\alpha a}} \quad ; \quad \alpha = \dfrac{1}{\lambda}\sqrt{\dfrac{3}{\mathrm{N}}} . \end{cases}$$

Nel caso che lo spessore a della paraffina sia infinito, queste formule si semplificano in modo analogo alle (38) e si trova

$$(50) \qquad p(x) = \frac{\sqrt{\mathrm{N}}}{\sqrt{\mathrm{N}} + 1} e^{-\sqrt{\frac{3}{\mathrm{N}}}\frac{x}{\lambda}} \qquad (\text{per} \quad a = \infty).$$

Questa formula deve confrontarsi con la (34) con la quale viene a coincidere (quando si ponga $a = x$) nel caso limite indicato alla fine del § 3. La differenza consiste nel fattore $\sqrt{\mathrm{N}}/(\sqrt{\mathrm{N}} + 1)$ che rappresenta l'effetto delle perturbazioni in superficie dell'equazione differenziale (24).

La conoscenza di $p(x)$ permette di risolvere facilmente i problemi relativi alla albedo della paraffina nel caso di tre dimensioni. Nel caso particolare che la paraffina sia infinitamente spessa e per il caso limite di N molto grande, il problema della albedo è stato trattato anche da Wick [4], con un procedimento, che in qualche caso consente di trovare il risultato esatto, in altri casi fornisce un metodo di approssimazione la cui precisione può, in principio, essere arbitrariamente grande. I risultati che qui daremo sono invece soltanto approssimati, pur corrispondendo a casi più generali di quelli trattati da Wick; come si vedrà del resto la approssimazione è in genere assai elevata.

A differenza del caso di una sola dimensione, per il quale la albedo è definita in modo univoco, nel caso di tre dimensioni essa dipende ovviamente dall'angolo di incidenza dei neutroni, essendo tanto più piccola quanto più l'incidenza è normale. Affinché si possa dare alla albedo un significato univoco, sarà dunque necessario indicare l'angolo di incidenza ovvero la legge di distribuzione angolare dei neutroni incidenti, oppure specificare altrimenti come si intenda eseguire la misura del coefficiente di riflessione. Preciseremo così nel § 7 che cosa intendiamo per « albedo sperimentale » che è il valore del coefficiente di riflessione che si ottiene quando ci si metta in condizioni particolarmente convenienti per effettuarne la misura e che verranno allora indicate esattamente; l'importanza della albedo sperimentale deriva dal fatto che dalle misure di albedo si ottiene direttamente questa grandezza e quindi tutti i confronti fra teoria ed esperienza sono fatti in base ad essa. La teoria della albedo sperimentale sarà data nel § 7.

Diamo qui invece le formule per il calcolo della albedo per neutroni in funzione dell'angolo di incidenza ϑ. La albedo di un neutrone che incida sopra uno strato di paraffina di spessore a con un angolo di incidenza ϑ, può evidentemente scriversi come integrale dei prodotti della probabilità

$$\frac{N-1}{N} e^{-\frac{x}{\lambda \cos \vartheta}} \frac{dx}{\lambda \cos \vartheta}$$

che il neutrone, entrato nella paraffina subisca un primo urto alla profondità tra x ed $x + dx$, e non venga ivi catturato, moltiplicata per la probabilità che il neutrone, dopo aver subìto questo primo urto, esca dallo strato di paraffina per la faccia da cui è entrato; quest'ultima probabilità è data da $p(x)$. La albedo cercata è dunque:

$$\beta(a, \vartheta) = \frac{N-1}{N} \int_0^a e^{-\frac{x}{\lambda \cos \vartheta}} \frac{p(x)}{\lambda \cos \vartheta} dx. \tag{51}$$

Ponendo per $p(x)$ l'espressione (49) l'integrazione si effettua senza difficoltà; per brevità scriviamo il risultato esplicitamente solo nel caso che lo spessore a dello strato di paraffina sia infinitamente grande, nel qual caso $p(x)$ è dato dalla (50). In questo caso si trova da (51), effettuando l'integrazione,

$$\beta(\infty, \vartheta) = \frac{\sqrt{N} - 1}{\sqrt{N} + \sqrt{3} \cos \vartheta} = 1 - \frac{1 + \sqrt{3} \cos \vartheta}{\sqrt{N}} + \cdots \tag{52}$$

dove l'ultima espressione è valida per N assai grande.

Dalla (52) si può ricavare la espressione della albedo per ogni assegnata legge di distribuzione angolare dei neutroni incidenti; così per esempio se i neutroni incidenti soddisfano alla legge del coseno, la probabilità che un neutrone incidente formi con la normale un angolo compreso tra ϑ e $\vartheta + d\vartheta$

è $2 \cos \vartheta \sin \vartheta \, d\vartheta$; e quindi la albedo, quando valga la legge del coseno, risulta:

$$\beta(\infty) = \int_0^{\pi/2} \frac{\sqrt{N} - 1}{\sqrt{N} + \sqrt{3} \cos \vartheta} \, 2 \cos \vartheta \sin \vartheta \, d\vartheta \tag{53}$$

$$= \frac{2}{\sqrt{3}} (\sqrt{N} - 1) \left[1 - \sqrt{\frac{N}{3}} \log \left(1 + \sqrt{\frac{3}{N}} \right) \right] = 1 - \frac{1 + \frac{2}{\sqrt{3}}}{\sqrt{N}} + \cdots .$$

Per giudicare il grado di approssimazione del metodo, si può osservare che, nel caso che valga la legge del coseno, Wick (loc. cit.) ha data la espressione asintotica esatta della albedo, ed ha trovato

$$\beta(\infty) = 1 - \frac{4}{\sqrt{3}} \frac{1}{\sqrt{N}} + \cdots . \tag{54}$$

Il valore esatto del coefficiente il $1/\sqrt{N}$ è dunque $4/\sqrt{3} = 2{,}31$; mentre dalla (53) esso risulta $1 + 2/\sqrt{3} = 2{,}15$. La differenza è soltanto del 7 %.

In modo simile si può calcolare il coefficiente di trasmissione attraverso a una lastra di paraffina di spessore a per neutroni termici incidenti con un angolo ϑ. In modo analogo a quello usato per dedurre (51) si trova che esso è dato da

$$\gamma(a, \vartheta) = e^{-\frac{a}{\lambda \cos \vartheta}} + \frac{N - 1}{N \lambda \cos \vartheta} \int_0^a e^{-\frac{x}{\lambda \cos \vartheta}} p(a - x) \, dx . \tag{55}$$

Sostituendo per $p(x)$ la (49) l'integrazione si effettua senza difficoltà; diamo il risultato esplicito soltanto per il caso limite $N = \infty$:

$$\gamma_\infty(a, \vartheta) = \frac{\lambda}{a + \frac{2\lambda}{\sqrt{3}}} \left[\cos \vartheta + \frac{1}{\sqrt{3}} - \left(\cos \vartheta - \frac{1}{\sqrt{3}} \right) e^{-\frac{a}{\lambda \cos \vartheta}} \right] . \tag{56}$$

6. *Legge di distribuzione angolare dei neutroni termici emergenti dalla paraffina.* – Dalla superficie di un blocco di paraffina contenente una sorgente di neutroni, escono continuamente verso l'esterno dei neutroni termici. Ci proponiamo di determinarne approssimativamente la legge di distribuzione angolare, la cui conoscenza è necessaria per la riduzione delle misure dei coefficienti di assorbimento e per numerosi altri problemi (cfr. *AF*). Un semplice metodo per dedurre questa legge è il seguente:

Abbiamo dimostrato nel § 4, che, nel caso unidimensionale, la densità dei neutroni in prossimità del bordo $x = 0$ della paraffina si comporta come (46). Anche questo risultato può estendersi approssimativamente al caso di tre dimensioni, purché λ si sostituisca con λ', dato dalla (48). Concludiamo dunque che la densità $n(x)$ dei neutroni termici in un blocco di paraffina limitato a sinistra dal piano $x = 0$ in vicinanza di questo piano si comporta come

$$n(x) = \text{cost} \left(\frac{\lambda}{\sqrt{3}} + x \right) . \tag{57}$$

Questa legge può esprimersi anche dicendo che se si estrapola verso l'esterno della paraffina (e cioè verso i valori negativi della x) l'andamento che la densità dei neutroni ha nell'interno della paraffina, si trova che essa si annullerebbe a distanza $\lambda/\sqrt{3}$ dal bordo della paraffina. Per una verifica sperimentale di questa proprietà confronta *AF*, § 11.

Siccome la frequenza degli urti subiti dai neutroni termici in ogni elemento di volume è evidentemente proporzionale a $n(x)$, troviamo che il numero degli urti elastici dei neutroni termici per unità di tempo entro il volume $d\tau$ è:

$$\mu\left(\frac{\lambda}{\sqrt{3}}+x\right)d\tau$$

μ essendo una costante. La probabilità che un neutrone che ha subìto uno di questi urti esca dalla paraffina attraverso al piano $x=0$, con una inclinazione sulla normale compresa tra ϑ e $\vartheta+d\vartheta$, e senza subire alcun altro urto, è evidentemente:

$$e^{-\frac{x}{\lambda\cos\vartheta}}\frac{\sin\vartheta\, d\vartheta}{2}.$$

Si trova dunque che il numero dei neutroni che escono da una porzione di area S della superficie che limita la paraffina, con inclinazione tra ϑ e $\vartheta+d\vartheta$ ed avendo subìto l'ultimo urto in una posizione di ascissa tra x ed $x+dx$ è:

$$\mu\left(\frac{\lambda}{\sqrt{3}}+x\right)\mathrm{S}\, dx\, e^{-\frac{x}{\lambda\cos\vartheta}}\frac{\sin\vartheta\, d\vartheta}{2}.$$

Questa espressione, integrata rispetto ad x da 0 a ∞ ci dà il numero totale dei neutroni termici uscenti dalla superficie S con inclinazione tra ϑ e $\vartheta+d\vartheta$; esso risulta:

$$\frac{\lambda^2\mathrm{S}\mu}{2\sqrt{3}}(\cos\vartheta+\sqrt{3}\cos^2\vartheta)\sin\vartheta\, d\vartheta;$$

concludiamo dunque che il numero dei neutroni termici che escono dalla paraffina con direzione entro l'elemento di angolo solido $d\omega$ è proporzionale a:

$$(\cos\vartheta+\sqrt{3}\cos^2\vartheta)\,d\omega. \tag{58}$$

La legge di distribuzione angolare dei neutroni termici uscenti dalla paraffina è stata calcolata numericamente da Wick; i suoi risultati concordano in modo assai soddisfacente con la (58).

7. *Albedo sperimentale.* – Il metodo seguito sperimentalmente per la determinazione della albedo (cfr. *AF*) è essenzialmente il seguente. Si abbia un rivelatore R dei neutroni lenti, costituito da una lastra piana assai sottile, capace di attivarsi per assorbimento dei neutroni termici e soltanto di questi. Ammetteremo che il rivelatore sia molto sottile, oltre che geometricamente, anche nel senso che la probabilità che un neutrone termico che lo attraversa venga catturato (probabilità determinata dal prodotto dello spessore per il coefficiente di assorbimento per neutroni termici) sia

assai piccola. Si abbia inoltre una lastra di materiale capace di assorbire totalmente i neutroni termici che la attraversano, e soltanto questi.

Effettuiamo due misure dell'attività che si induce nel rivelatore ponendolo una volta (caso *a*) in una data posizione entro un blocco di paraffina assai grande contenente la sorgente; ed una volta (caso *b*) ponendolo di nuovo nella stessa posizione, ma mettendogli accanto da una parte la lastra assorbente. Siano A e B le attività misurate nelle posizioni *a* e *b*; ci proponiamo di far vedere che il rapporto A/B determina la albedo.

Sia infatti $\mathfrak{N}$ il numero dei neutroni termici che attraversano il rivelatore nel caso *b*; nel caso *a* il numero dei neutroni termici che attraversano il rivelatore almeno una volta (contando come uno solo anche un neutrone che attraversi il rivelatore più volte) è eguale a $2\,\mathfrak{N}$, poiché i neutroni che colpiscono il rivelatore possono, nel caso *a* essersi originati da ambo le parti di questo. Inoltre mostriamo che un neutrone che traversa una volta il rivelatore, lo attraversa in media $1/(1-\beta)$ volte, β essendo la albedo. Infatti la probabilità che un neutrone, dopo attraversato una volta il rivelatore, lo attraversi una seconda volta, è evidentemente eguale alla albedo β; e, quindi il numero medio degli attraversamenti è appunto

$$1+\beta+\beta^2+\cdots=\frac{1}{1-\beta}\,.$$

In conclusione dunque il numero degli attraversamenti nel caso *a* risulta:

$$\frac{2\,\mathfrak{N}}{1-\beta}\,.$$

Ammettendo che le attività A e B siano proporzionali ai numeri di attraversamenti che si hanno nei due casi *a* e *b*, si trova dunque

$$\frac{A}{B}=\frac{2}{1-\beta} \tag{59}$$

e quindi, dalla misura del rapporto A/B si può ricavare la albedo β. Abbiamo già più volte osservato che il concetto di albedo non è univocamente definito, poiché il valore del coefficiente di riflessione dipende dalla distribuzione angolare dei neutroni incidenti. Così per esempio la albedo per neutroni incidenti con la legge del coseno risulterebbe diversa dalla albedo per neutroni incidenti normalmente, oppure incidenti con la legge (58). Possiamo dunque intendere che la relazione (59) precisi una possibile definizione della albedo. Chiameremo la albedo definita dalla relazione (59) « albedo sperimentale » poiché il suo valore è quello che si ricava direttamente dalle misure. Dalla (59) si ricava come espressione della albedo sperimentale

$$\beta_{sp}=1-\frac{2B}{A}\,. \tag{60}$$

Per i dettagli delle misure effettive della albedo sperimentale rimandiamo ad *AF*. Notiamo qui che, dato che la albedo che risulta dalle misure è proprio la albedo sperimentale, è importante stabilire la relazione che esprime questa grandezza in funzione del numero medio N degli urti che un neutrone subisce prima di venir catturato dai protoni della paraffina.

Stabiliremo questa relazione nel caso limite, praticamente interessante. di N molto grande.

Sia q il numero di neutroni termici che si producono per secondo e per centimetro cubo nella regione in cui si trova il rivelatore R. Nel caso a, essendo per ipotesi R assai sottile, la distribuzione dei neutroni termici circostanti non sarà perturbata dalla sua presenza. Si osservi ora che il cammino medio totale di ciascun neutrone termico nella paraffina è evidentemente $N\lambda$; e quindi il percorso totale dei neutroni termici prodotti per unità di tempo e di volume è $q\,N\lambda$. Concludiamo dunque che la somma delle lunghezze dei tratti percorsi da neutroni termici nell'unità di tempo attraverso al nostro rivelatore (avente area S, spessore δ, e quindi volume $S\delta$) sarà $S\delta q\,N\lambda$. Indicando con K il coefficiente di assorbimento dei neutroni termici nel materiale che costituisce il rivelatore e tenendo presente che, per ipotesi, il rivelatore è molto sottile (cioè $K\delta \ll 1$), concludiamo che il numero di neutroni termici catturati per unità di tempo nel rivelatore è: $K\delta\, Sq\, N\lambda$. L'attività A sarà evidentemente proporzionale a questo numero; essa potrà dunque scriversi

$$A = \eta K\delta\, Sq\, N\lambda\,, \tag{61}$$

η essendo il coefficiente di proporzionalità.

Per calcolare anche B, osserviamo che, quando da una parte del rivelatore è appoggiata una lastra che assorba completamente i neutroni termici, questi possono raggiungere il rivelatore soltanto se si sono originati dalla parte scoperta di R; inoltre ogni neutrone potrà attraversare R non più di una volta poiché appena traversato R, esso sarà catturato dall'assorbitore. Il numero di neutroni che colpiscono il rivelatore si calcola osservando che di fronte al rivelatore (di area S) e a distanza da esso tra x ed $x + dx$ si producono $Sq\,dx$ neutroni termici al secondo. La probabilità che uno di questi raggiunga il rivelatore è data nel nostro caso (N molto grande) dalla (34) in cui si deve scrivere x in luogo di a, e cioè da

$$p(x) = e^{-\sqrt{\frac{3}{\lambda^2 N}}\,x}\,.$$

Il numero di neutroni che colpiscono il rivelatore è dunque

$$Sq\int_0^\infty e^{-\sqrt{\frac{3}{N}}\,\frac{x}{\lambda}}\,dx = Sq\lambda\sqrt{\frac{N}{3}}\,.$$

Per calcolare quanti di questi neutroni restano catturati in R, occorre conoscerne la legge di distribuzione angolare; ammetteremo che questa sia data dalla (58). È infatti evidente che, nel caso b, la presenza di uno strato che assorbe totalmente i neutroni termici, modifica la distribuzione di questi nelle sue vicinanze in modo tale da renderla paragonabile alla distribuzione in vicinanza di una superficie che limita il blocco di paraffina. In base alla (58) si calcola elementarmente che il valor medio del tratto percorso da un neutrone termico che colpisce il rivelatore (di spessore δ) nell'attraversarlo è

$$\sqrt{3}\,\delta$$

e quindi il numero totale dei neutroni assorbiti entro il rivelatore sarà $K\delta\, Sq\, \lambda \sqrt{N}$. L'attività B sarà proporzionale a questo numero, con lo stesso coefficiente η della (61); avremo dunque:

$$B = \eta K \delta\, Sq\, \lambda \sqrt{N}\,. \tag{62}$$

Dalla (60), tenendo presenti (61) e (62) ricaviamo ora il valore della albedo sperimentale

$$\beta_{sp} = 1 - \frac{2}{\sqrt{N}}\,. \tag{63}$$

Questa formula coincide con il risultato trovato da Wick (loc. cit.) con un procedimento di approssimazione numerica.

Le misure effettuate nella paraffina hanno dato come valore della albedo sperimentale (cfr. *AF*) $\beta = 0{,}82$. Con questo dato si può ricavare dalla (63) una misura di N; si ottiene precisamente $N = 123$. Se si tenesse conto del fatto che, anche per neutroni termici, esiste una piccola coerenza tra le direzioni di cammini liberi successivi, il valore di N risulterebbe un po' più grande.

8. *Larghezza logaritmica dei gruppi*. – I gruppi di neutroni lenti, ad eccezione del gruppo C, costituito essenzialmente dai neutroni termici, corrispondono, secondo quanto si ritiene, a bande di energia relativamente ristrette. In questa ipotesi essi sono caratterizzati dai valori $w_{\min}$ e $w_{\max}$ delle energie che delimitano la banda. Ci proponiamo qui di esporre i principi su cui è basato il metodo per la determinazione della grandezza

$$\log \frac{w_{\max}}{w_{\min}}$$

che chiameremo, per intenderci, larghezza logaritmica del gruppo. La descrizione della esecuzione effettiva della misura e delle varie correzioni che devono apportarsi si trova in *AF*.

Si noti che la larghezza logaritmica ha un significato fisico immediato; in base a quanto abbiamo detto nella prima parte del § 2, essa rappresenta infatti il numero medio degli urti che un neutrone subisce mentre appartiene al gruppo in questione; ciò è vero naturalmente solo per gruppi aventi energia maggiore di un volt, in modo che siano verificate le premesse fatte nei paragrafi 1 e 2. In tutta la considerazione che segue ci limiteremo a questi casi.

Il procedimento per determinare la larghezza logaritmica è basato sul confronto tra l'attività indotta in un rivelatore G del gruppo g in esame, e quella indotta in un rivelatore R del gruppo termico, avente area eguale a G, e disposto, come nel caso *b* del paragrafo precedente, con un assorbitore totale del gruppo termico da un lato. Supporremo per semplicità che entrambi i rivelatori siano assai sottili nel senso stesso precisato al paragrafo precedente; altrimenti bisogna introdurre delle correzioni per tener conto dell'assorbimento degli elettroni e dei neutroni nel rivelatore stesso.

Per comprendere facilmente il principio del metodo, supporremo dapprima che in un blocco illimitato di paraffina si producano dappertutto q

neutroni veloci per unità di volume e di tempo. Questi, in seguito agli urti, verranno successivamente a rallentarsi fino alla velocità di agitazione termica e saranno infine catturati dai protoni presenti nella paraffina. L'attività B indotta nel rivelatore R dei neutroni termici, quando esso è protetto da un lato da un assorbitore totale dei neutroni termici, sarà data, come nel caso *b* del paragrafo precedente, dalla (62). Per determinare anche l'attività del rivelatore G del gruppo g, osserviamo che dalla (6) risulta in questo caso che la densità $n(v)\,dv$ dei neutroni aventi velocità v e $v + dv$ ed appartenenti al gruppo g è

$$n(v)\,dv = \frac{2q\lambda_g}{v^2}\,dv\,,$$

λ_g essendo il cammino libero medio dei neutroni del gruppo in esame. Da ciò si calcola subito il numero di neutroni di velocità tra v e $v + dv$ che attraversano nell'unità di tempo il rivelatore G formando con la normale ad esso un angolo tra ϑ e $\vartheta + d\vartheta$; tale numero è:

$$2q\lambda_g\,S\,\frac{dv}{v}\sin\vartheta\cos\vartheta\,d\vartheta\,.$$

Se δ è lo spessore del rivelatore e K_g il suo coefficiente di assorbimento per i neutroni del gruppo g. La frazione dei neutroni assorbita (tenendo conto che per ipotesi $K_g\,\delta \ll 1$) sarà $K_g\,\delta/\cos\vartheta$; e quindi il numero totale dei neutroni g assorbiti risulta:

$$2qS\,\lambda_g\,K_g\,\delta\int_{v_{\min}}^{v_{\max}}\frac{dv}{v}\int_0^{\pi/2}\sin\vartheta\,d\vartheta = 2qS\,\lambda_g\,K_g\,\delta\log\frac{v_{\max}}{v_{\min}}\,,$$

dove $v_{\min}$ e $v_{\max}$ sono le velocità corrispondenti alle energie $w_{\min}$ e $w_{\max}$, per modo che $\log v_{\max}/v_{\min} = 1/2 \log w_{\max}/w_{\min}$. L'attività indotta nel rivelatore G è proporzionale al numero dei neutroni assorbiti per unità di tempo; se η_G è il coefficiente, essa risulta dunque

$$A_g = \eta_G\,q\,S\,\lambda_g\,K_g\,\delta\,\log\frac{w_{\max}}{w_{\min}}\,. \tag{64}$$

Nella pratica delle esperienze si usa di solito lo stesso corpo, con e senza opportuni filtri assorbenti, sia come rivelatore R dei neutroni termici che come rivelatore G del gruppo g; in queste condizioni si ha evidentemente $\eta_G = \eta$. Se quindi si potessero realizzare delle condizioni in cui il numero q dei neutroni veloci prodotti per unità di volume e di tempo fosse costante in tutta una grande massa di paraffina, si ricaverebbe, da (62) e (64)

$$\log\frac{w''}{w'} = \frac{K\lambda\sqrt{N}\,A_g}{\lambda_g\,K_g\,B}\,. \tag{65}$$

In pratica la determinazione di $\log w_{\max}/w_{\min}$ non può farsi direttamente per mezzo di questa formula, poiché non è facile produrre una ripartizione uniforme dei neutroni veloci, i quali di solito provengono tutti da una sorgente praticamente puntiforme. E siccome la distanza a cui diffondono i neutroni termici è maggiore di quella a cui diffondono i neutroni degli altri gruppi, si troverebbe che il rapporto A_g/B non è costante, ma decresce con la distanza dalla sorgente. Un modo semplice per evitare questo inconveniente consiste nell'integrare le equazioni (62) e (64) su tutto il volume (e, occorrendo, anche su tutte le orientazioni del rivelatore). Si trova così un risultato che evidentemente non dipende più dalla uniformità o meno di q, ma soltanto da $Q = \int q \, d\tau$, e cioè dal numero totale dei neutroni veloci prodotti nella unità di tempo. Si trova precisamente, ponendo anche, come si è detto, $\eta_G = \eta$:

$$(66) \qquad \begin{cases} \int B \, d\tau = \eta K \delta \, SQ\lambda \sqrt{N} \\ \int A_g \, d\tau = \eta \, K_g \, \delta \, SQ \, \lambda_g \log \dfrac{w_{max}}{w_{min}} \end{cases}$$

dalle quali si ricava

$$(67) \qquad \log \frac{w_{max}}{w_{min}} = \frac{K\lambda \sqrt{N}}{\lambda_g \, K_g} \frac{\int A_g \, d\tau}{\int B \, d\tau} .$$

Questa formula corretta ancora per tener conto dell'assorbimento nel rivelatore, è stata usata in *AF* per il calcolo delle larghezze logaritmiche.

PARTE II

Meccanismo dell'urto tra neutroni e atomi di idrogeno.

Questa parte sarà dedicata allo studio meccanico-quantistico dell'urto tra i neutroni lenti e i protoni della paraffina. Verrà in particolare esaminato il meccanismo degli urti, tenendo conto del fatto che gli atomi di idrogeno non sono liberi, ma possono considerarsi legati elasticamente a certe posizioni di equilibrio; e verranno discussi i processi che portano alla cattura del neutrone da parte del protone, con formazione di un nucleo di deuterio.

9. *Urto tra neutroni lenti e protoni trascurando l'effetto del legame chimico.* – È evidente che, fino a quando l'energia del neutrone è grande rispetto alla energia del legame chimico dell'idrogeno nella paraffina, si può prescindere, nello studio dell'urto, dall'effetto di questo legame, considerando i protoni come liberi; più precisamente, come abbiamo già osservato, questo procedimento sarà lecito fino a che l'energia w del neutrone sia $\gg h\nu$, e cioè del quanto della frequenza ν del legame elastico degli atomi di idrogeno. Siccome $h\nu$ è al massimo dell'ordine di 0,5 volt, potremo praticamente ritenere i protoni come liberi quando i neutroni hanno energie maggiori di 1 volt. In questo caso il problema degli urti tra neutroni e protoni è abbastanza semplice; poiché, riferendosi alle coordinate relative dei due corpuscoli, lo si riconduce immediatamente al problema dell'urto di un corpuscolo contro un centro di diffusione fisso, problema che si tratta coi classici metodi della

teoria quantistica dell'urto. Come massa, nel calcolo del moto relativo, deve naturalmente prendersi la massa ridotta M/2, M essendo il valore comune della massa del protone e del neutrone.

Il dettaglio della teoria dell'urto, anche in questo caso semplice, dipende, come è ovvio, dalle ipotesi che si fanno sulla legge di forza tra neutrone e potrone. Si dovrà anche tener conto delle due possibili orientazioni relative (parallela o antiparallela) degli spin del neutrone e del protone.

Ammetteremo, come generalmente si ritiene [5], che le forze neutrone-protone abbiano un raggio d'azione ρ alquanto inferiore a 10^{-12} cm. Esse potrebbero poi derivare da un ordinario tipo di potenziale, oppure rappresentare una interazione di scambio secondo le ipotesi di Heisenberg e di Majorana.

È noto che il nucleo del deuterio ha spin 1; sembra dunque che il suo termine fondamentale debba classificarsi (con le notazioni della spettroscopia degli atomi) come un termine ^3S, il cui spin sia la somma degli spin del protone e del neutrone; la energia di legame di questo termine è $2{,}2 \cdot 10^6$ volt. Oltre al termine fondamentale ^3S si può considerare anche un termine di singoletti ^1S nel quale gli spin del protone e del neutrone siano antiparalleli. Se le forze neutrone–protone non dipendessero dalla orientazione degli spin, questi due termini dovrebbero evidentemente coincidere; se vi è invece una differenza tra le forze nei due casi, il termine ^1S deve essere situato al di sopra del termine fondamentale ^3S; esso sarà reale o virtuale a seconda che la sua energia di legame è positiva o negativa.

Nell'ipotesi che il raggio d'azione delle forze neutrone–protone sia molto piccolo a confronto della lunghezza d'onda di de Broglie, il problema dell'urto può trattarsi col semplice procedimento usato da Bethe e Peierls (loc. cit.); si arriva così alla seguente formula per la sezione d'urto [6]

$$\sigma = \frac{h^2}{\pi M} \left\{ \frac{3}{4} \frac{1}{|w_0| + \frac{1}{2} w} + \frac{1}{4} \frac{1}{|w_1| + \frac{1}{2} w} \right\} \tag{68}$$

dove w_0 e w_1 rappresentano le energie dei due termini ^3S ed ^1S e w è l'energia del neutrone. Quando w è molto piccolo (si intenda: piccolo rispetto a w_0 e w_1, pur essendo sempre maggiore di un volt) risulta dalla (68) una sezione d'urto indipendente da w

$$\sigma_1 = \frac{h^2}{\pi M} \left\{ \frac{3}{4} \frac{1}{|w_0|} + \frac{1}{4} \frac{1}{|w_1|} \right\} . \tag{69}$$

Risulta effettivamente dall'esperienza (cfr. *AF*) che, almeno per due gruppi aventi energia maggiore di un volt, il cammino libero medio nella paraffina è approssimativamente lo stesso, e precisamente circa

$$\lambda_g = 1 \text{ cm}. \tag{70}$$

A questo valore del cammino libero medio corrisponde una sezione di urto $\sigma_1 = 12{,}8 \cdot 10^{-24}$ cm². Assumendo questo valore, e prendendo $w_0 = -2{,}2 \cdot 10^6$ volt, si ricava dalla (69)

$$|w_1| = 1{,}86 \cdot 10^{-7} \text{ erg} = 1{,}16 \cdot 10^5 \text{ volt}.$$

La sezione d'urto relativamente grande dei neutroni lenti per l'idrogeno conduce dunque, secondo questa teoria, a concludere che il termine ^1S, reale o virtuale, abbia una energia relativamente assai piccola in valore assoluto.

Ponendo nella (68) questo valore per w_1, e calcolando la sezione d'urto per neutroni veloci, si trovano dei valori troppo piccoli; ciò è probabilmente dovuto alla inesattezza del procedimento di calcolo di Bethe e Peierls per neutroni aventi energie di alcuni milioni di volt. Il valore della sezione di urto dedotto dalla misura di cammino libero medio effettuata da Goldhaber (loc. cit.) per neutroni di circa 200.000 volt ottenuti per effetto fotoelettrico nucleare da deuterio irradiato con i raggi γ del Th C è invece più piccolo di quello che si deduce dalla (68). La precisione delle misure non è però forse sufficiente per trarre delle deduzioni sicure.

10. *Urto di neutroni contro atomi di idrogeno legati.* – Abbiamo più volte osservato che, se l'energia dei neutroni è inferiore a un volt, non è più lecito trascurare il fatto che gli atomi di idrogeno sono legati chimicamente nella paraffina. Ci proponiamo di studiare ora l'urto tra un neutrone e un nucleo di idrogeno, tenendo conto delle forze chimiche che tendono a mantenere l'idrogeno nell'intorno di una posizione di equilibrio.

Il problema si può discutere coi seguenti criteri di approssimazione. Sia ρ il raggio d'azione delle forze neutrone–protone; a l'ordine di grandezza del raggio d'urto tra neutrone e protone; λ infine l'ordine di grandezza della lunghezza d'onda di de Broglie corrispondente ai momenti del neutrone e del protone. Le approssimazioni che ci proponiamo di usare sono legittime se si può scegliere una lunghezza R che soddisfi simultaneamente le diseguaglianze

$$R \ll \lambda \quad ; \quad R \gg \rho \quad ; \quad R \gg a. \tag{71}$$

Queste condizioni si possono soddisfare assai bene per il caso dell'urto tra i neutroni lenti e i protoni nella paraffina. Indicando con x , y , z le coordinate del neutrone e con X , Y , Z quelle del protone, e trascurando l'effetto dello spin dei due corpuscoli, che non ha importanza per i fenomeni che ci interessano presentemente, la Hamiltoniana del sistema potrà scriversi come somma delle energie cinetiche dei due corpuscoli + l'energia potenziale U (X , Y , Z) delle forze chimiche che si esercitano sull'idrogeno + la energia di interazione tra neutrone e protone. Rappresenteremo quest'ultima come una energia potenziale $g(r)$ funzione della distanza r tra neutrone e protone; se anche del resto l'interazione fosse del tipo di Heisenberg o di Majorana, il calcolo si potrebbe svolgere sostanzialmente con lo stesso procedimento.

L'equazione di Schroedinger del sistema può scriversi:

$$-\frac{h}{2\pi i}\dot{\psi} = -\frac{h^2}{8\pi^2 M}\left(\frac{\partial^2\psi}{\partial x^2}+\frac{\partial^2\psi}{\partial y^2}+\frac{\partial^2\psi}{\partial z^2}+\frac{\partial^2\psi}{\partial X^2}+\frac{\partial^2\psi}{\partial Y^2}+\frac{\partial^2\psi}{\partial Z^2}\right) + U\psi + g(r)\,\psi. \tag{72}$$

Lo scalare di campo $\psi(x , y , z , X , Y , Z)$ sarà evidentemente una funzione con comportamento regolare (nel senso che le sue variazioni saranno piccole per variazioni degli argomenti piccoli in rapporto a λ) dappertutto ad eccezione delle zone in cui r è piccolo. È facile convincersi, tenendo presenti le (71), che le irregolarità avranno luogo per $r < \mathrm{R}$ (cfr. anche la discussione sulla fig. 1). Introduciamo una nuova funzione $\overline{\psi}$ che si deduca dalla ψ prendendo la media seguente

$$(73) \qquad \overline{\psi}(x, y, z, \mathrm{X}, \mathrm{Y}, \mathrm{Z}) = \frac{1}{\frac{4}{3}\pi \mathrm{R}^3} \iiint \psi(\xi, \eta, \zeta, \mathrm{X}, \mathrm{Y}, \mathrm{Z})\, d\xi\, d\eta\, d\zeta$$

dove l'integrale è esteso a una sfera di raggio R con centro nel punto x, y, z. La funzione $\overline{\psi}$ avrà evidentemente andamento regolare anche per piccoli valori di r.

Ci proponiamo di stabilire per $\overline{\psi}$ una equazione analoga a quella di Schroedinger. Per questo prenderemo la media dei due membri di (72) sopra i punti di una sfera di raggio R con centro in x, y, z; indicando questa operazione, analoga alla (73), con una sopralineatura, otteniamo, con ovvie considerazioni,

$$(74) \qquad -\frac{h}{2\pi i}\dot{\overline{\psi}} = -\frac{h^2}{8\pi^2 \mathrm{M}}\left(\frac{\partial^2 \overline{\psi}}{\partial x^2} + \cdots + \frac{\partial^2 \overline{\psi}}{\partial \mathrm{X}^2} + \cdots\right) + \mathrm{U}\overline{\psi} + \overline{g(r)\psi}.$$

Per calcolare l'ultimo termine si osservi che, nella zona singolare $r < \mathrm{R}$, le variazioni della ψ sono essenzialmente dovute all'effetto dell'energia potenziale $g(r)$ la quale, pur essendo diversa da zero solo in una regione assai piccola, prende in essa dei valori grandissimi. Per $r < \mathrm{R}$ si potrà anzi prescindere dalle variazioni della ψ dovute agli altri termini della Hamiltoniana, scrivendo la ψ nella forma

$$(75) \qquad \psi = c\chi(r)$$

dove la funzione $\chi(r)$ è soluzione dell'equazione

$$(76) \qquad \chi''(r) + \frac{2}{r}\chi'(r) = \frac{4\pi^2 \mathrm{M}}{h^2} g(r)\chi(r)$$

e cioè dell'equazione di Schroedinger (per le sole onde s, poiché a piccola distanza queste sole hanno importanza) per il moto relativo delle due particelle (coordinata r; massa ridotta M/2) e considerando la sola energia potenziale $g(r)$; c rappresenta una funzione variabile regolarmente.

Se normalizziamo la $\chi(r)$ in modo che per r grande essa prenda il valore 1, il raccordo tra l'espressione (75) e i valori di ψ per $r \geq \mathrm{R}$ si ottiene evidentemente identificando c con la funzione lentamente variabile $\overline{\psi}$. Potremo dunque porre in generale

$$(77) \qquad \psi(x, \mathrm{X}) = \overline{\psi}(x, \mathrm{X})\chi(r).$$

Con la posizione $\chi(r) = \frac{1}{r}v(r)$, la (76) diventa

$$(78) \qquad v''(r) = \frac{4\pi^2 \mathrm{M}}{h^2} g(r)\, v(r).$$

L'andamento di $v(r)$ è rappresentato nella fig. 1. Per $r > \rho$ il grafico di $v(r)$ è rettilineo; per la condizione di normalizzazione di poi, l'inclinazione della retta asintoto della curva $v(r)$ deve rendersi eguale a 45°; per modo che si avrà

$$v(r) = a + r \qquad \text{per } r > \rho . \tag{79}$$

Il significato della costante a è chiaro dalla figura.

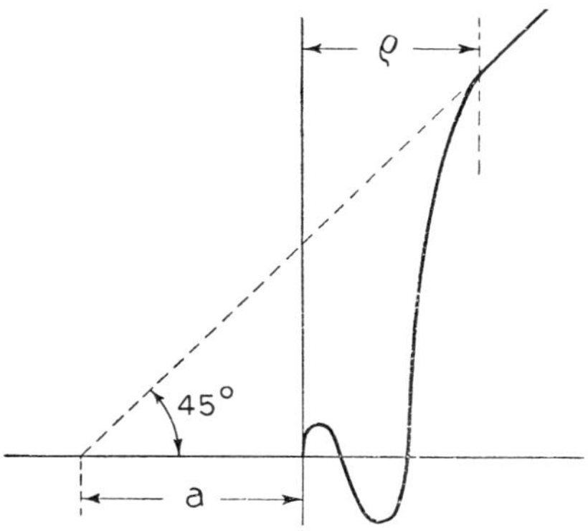

Fig. 1.

Tenendo presenti le (71) e trascurando ρ in confronto a R si riconosce che $\overline{g(r)\psi}$ sarà diverso da zero se i punti x, y, z e X, Y, Z distano meno di R. Se ciò è, si trova, (tenendo presente che $\rho \ll$ R, e che $\bar{\psi}$ non varia sensibilmente entro regioni delle dimensioni R)

$$\overline{g(r)\psi} = \frac{1}{\frac{4}{3}\pi R^3} \int g(r)\,\bar{\psi}(x, X)\,\chi(r)\,d\tau$$

$$= \frac{4\pi\bar{\psi}(x, X)}{\frac{4}{3}\pi R^3} \int_0^\infty g(r)\,v(r)\,r\,dr .$$

L'integrale si calcola facilmente tenendo presenti (78), (79) ed il fatto che $v(0) = 0$ (cfr. fig. 1). Si trova:

$$\int_0^\infty g(r)\,v(r)\,r\,dr = \frac{h^2}{4\pi^2 M}\int_0^\infty v''(r)\,r\,dr = \frac{h^2}{4\pi^2 M}\Big| v' r - v \Big|_0^\infty = -\frac{h^2 a}{4\pi^2 M} .$$

In conclusione otteniamo infine

$$\overline{g(r)\psi} = -\frac{h^2 a}{\pi M}\,\bar{\psi}(x, X)\,\delta_R(r)$$

dove la funzione $\delta_R(r)$ è eguale a zero se $r > R$; ed altrimenti è eguale $3/4\pi R^3$ (per modo che il suo integrale di volume esteso a tutto lo spazio è eguale a 1). Sostituendo nella (74) si ottiene l'equazione cercata per $\bar{\psi}$:

$$-\frac{h}{2\pi i}\dot{\bar{\psi}} = -\frac{h^2}{8\pi^2 M}\left(\frac{\partial^2 \bar{\psi}}{\partial x^2} + \cdots + \frac{\partial^2 \bar{\psi}}{\partial X^2} + \cdots\right) \tag{80}$$

$$+ U(X, Y, Z)\,\bar{\psi} - \frac{h^2 a}{\pi M}\,\delta_R(r)\,\bar{\psi} .$$

L'urto tra neutrone e protone è determinato dalla presenza dell'ultimo termine che rappresenta la interazione tra i due corpuscoli. È facile convincersi che, con l'equazione (80), l'urto si può calcolare mediante la prima approssimazione del metodo di Born (infatti il termine di interazione determina solo una piccola irregolarità in ψ); inoltre, nel calcolare gli elementi di matrice del termine di interazione, la funzione δ_R potrà identificarsi con la funzione δ di Dirac (tridimensionale), poiché le grandezze contenute in (80) variano lentamente entro il campo in cui δ_R è diverso da zero.

Applicando il classico procedimento di Born si trova così senza difficoltà la seguente espressione per la sezione d'urto differenziale $\sigma_{m,n,d\omega}$ relativa a un processo d'urto in cui l'atomo di idrogeno passa dallo stato iniziale m allo stato finale n, mentre il neutrone viene diffuso entro una direzione appartenente all'elemento di angolo solido $d\omega$:

$$\sigma_{m,n,d\omega} = 4a^2 \frac{p}{p_0} d\omega \left| \int u_m^* u_n e^{\frac{2\pi i}{h}((\mathbf{p}-\mathbf{p}_0)\cdot\mathbf{X})} d\tau \right|^2 \tag{81}$$

dopo $\mathbf{p}_0$ rappresenta il momento del neutrone prima dell'urto; $u_m(X, Y, Z)$ ed $u_n(X, Y, Z)$ sono le autofunzioni che descrivono lo stato del protone prima e dopo l'urto ed alle quali corrispondono le energie w_m e w_n; $\mathbf{p}$ è la quantità di moto del neutrone dopo l'urto, la cui grandezza, per il principio della conservazione dell'energia, è data da

$$\frac{p^2}{2M} = \frac{p_0^2}{2M} + w_m - w_n \tag{82}$$

mentre la direzione è quella dell'elemento $d\omega$.

Prima di passare ad ulteriori applicazioni discutiamo la (81) in due casi estremi:

1° *Caso che gli idrogeni siano fortemente legati*. (Più precisamente che la regione, in cui la autofunzione u_m dello stato iniziale del protone è diversa da zero, sia piccola in confronto alla lunghezza d'onda di de Broglie del neutrone). – In questo caso l'esponenziale nell'integrale che figura in (81) può ritenersi costante entro la regione in cui il fattore u_m non è nullo. Segue allora, dalla ortogonalità delle autofunzioni del protone, che l'integrale differisce da zero solo per $n = m$, nel qual caso il quadrato del suo modulo è $= 1$. Concludiamo dunque che, nel caso presente, si hanno solo urti nei quali non varia l'energia dell'idrogeno e quindi, secondo (82), anche il neutrone conserva intera la sua energia cinetica ($p = p_0$); la sezione d'urto per processi in cui il neutrone è diffuso in $d\omega$ risulta in questo caso da (81)

$$\sigma_{d\omega}\, d\omega = 4\, a^2\, d\omega \tag{83}$$

che ci mostra che la diffusione è isotropa. La sezione totale si trova integrando la precedente rispetto all'angolo solido totale e risulta

$$\sigma = 16\,\pi\, a^2. \tag{84}$$

2° *Caso che gli idrogeni siano liberi*. – In questo caso le autofunzioni u_m ed u_n del protone si possono caratterizzare mediante i rispettivi momenti

$\mathbf{p}_m$ e $\mathbf{p}_n$; esse si scrivono nella consueta forma esponenziale. Si deduce allora senza difficoltà che l'integrale in (81) è diverso da zero solo se

$$\mathbf{p}_n + \mathbf{p} = \mathbf{p}_m + \mathbf{p}_0 \tag{85}$$

che esprime la legge di conservazione della quantità di moto; se (85) è soddisfatta, il quadrato del modulo del predetto integrale risulta $= 1$. La (82) prende inoltre la forma

$$p_n^2 + p^2 = p_m^2 + p_0^2. \tag{86}$$

Riferendoci ad un sistema in cui l'idrogeno sia inizialmente fermo, in cui cioè si abbia $p_m = 0$, (85) e (86) prendono le forme

$$\mathbf{p}_n + \mathbf{p} = \mathbf{p}_0 \quad ; \quad p_n^2 + p^2 = p_0^2 . \tag{87}$$

Da queste relazioni si ricava con considerazioni elementari

$$p = p_0 \cos \vartheta \tag{88}$$

ϑ essendo l'angolo tra $\mathbf{p}$ e $\mathbf{p}_0$. Siccome p e p_0 sono essenzialmente positive risulta intanto che la diffusione avviene solo sotto angoli acuti. Da (81) ricaviamo poi la sezione d'urto differenziale

$$\sigma_{d\omega} d\omega = 4\, a^2 \cos \vartheta \, d\omega . \tag{89}$$

La sezione d'urto totale risulta, integrando sulla semisfera $\cos \vartheta > 0$

$$\sigma = \int_0^{\pi/2} 4\, a^2 \cos \vartheta \cdot 2\, \pi \sin \vartheta \, d\vartheta = 4\, \pi\, a^2 . \tag{90}$$

Il confronto di questa equazione con la (84) mostra che la sezione d'urto per idrogeni fortemente legati è quadrupla che per idrogeni liberi. Ciò significa che quando l'energia cinetica del neutrone, da valori superiori a un volt (per i quali gli idrogeni si possono praticamente considerare come liberi), scende a valori molto piccoli (per i quali invece gli idrogeni si debbono considerare fortemente legati) la sezione d'urto diventa quadrupla e quindi il camino libero medio si riduce a 1/4. Nel caso dei neutroni termici, come sarà meglio precisato nel paragrafo seguente, si è abbastanza prossimi a questo valore limite, il quale tuttavia non è ancora completamente raggiunto. Si deve pertanto attendere che il cammino libero medio, pur essendo, per neutroni termici, notevolmente minore del cammino libero medio per neutroni lenti di energia maggiore di un volt, sia tuttavia un po' più grande di un quarto di quest'ultimo. Risulta inoltre da (83) che, nel caso limite in cui l'energia del neutrone è piccolissima, la diffusione avviene isotropicamente; anche questo risultato è vero solo approssimativamente per neutroni termici, poiché per essi, come sarà precisato nel prossimo paragrafo, esiste una leggera prevalenza delle diffusioni sotto angoli piccoli.

11. *Idrogeno legato elasticamente.* – Per studiare i casi intermedi tra i due casi limite ultimamente discussi, specializzeremo le nostre ipotesi ammettendo che l'atomo di idrogeno sia legato ad una posizione di equilibrio

(supposta nell'origine delle coordinate) da una forza elastica isotropa. In questo caso le espressioni delle varie sezioni d'urto possono, come vedremo, ottenersi in forma chiusa. Ammetteremo anche che lo stato di oscillazione in cui si trova inizialmente l'atomo di idrogeno sia quello di energia minima.

Se ν è la frequenza delle oscillazioni dell'atomo di idrogeno, l'energia dei suoi stati di oscillazione (misurandola per comodità a partire dall'energia dello stato fondamentale presa eguale a zero) sarà

$$w_n = nh\nu \tag{91}$$

con n intero. Decomponendo il moto dell'oscillatore nelle sue tre componenti X, Y, Z, il numero quantico totale n risulta dato dalla somma dei tre numeri quantici parziali

$$n = n_1 + n_2 + n_3 \tag{92}$$

che caratterizzano completamente uno stato di oscillazione. Lo stato iniziale m ha per ipotesi i tre numeri quantici eguali allo zero.

La funzione corrispondente ai numeri quantici n_1, n_2, n_3 è il prodotto di tre autofunzioni dell'oscillatore unidimensionale scritte per le tre coordinate X, Y, Z; e cioè:

$$u_{n_1 n_2 n_3} = \left(\frac{4\pi m\nu}{h}\right)^{3/4} \frac{H_{n_1}(\xi)\, H_{n_2}(\eta)\, H_{n_3}(\zeta)}{\sqrt{2^n\, n_1!\, n_2!\, n_3!}}\, e^{-\frac{1}{2}(\xi^2+\eta^2+\zeta^2)}$$

dove si è posto

$$\xi = X\sqrt{\frac{4\pi^2 m\nu}{h}} \quad ; \quad \eta = Y\sqrt{\frac{4\pi^2 m\nu}{h}} \quad ; \quad \zeta = Z\sqrt{\frac{4\pi^2 m\nu}{h}} \tag{94}$$

e le H rappresentano i polinomi di Hermite. La autofunzione dello stato iniziale è semplicemente

$$u_{000} = \left(\frac{4\pi m\nu}{h}\right)^{3/4} e^{-\frac{1}{2}(\xi^2+\eta^2+\zeta^2)} .$$

L'integrale di volume che figura nel secondo membro di (81) si riconduce al prodotto di tre integrali semplici del tipo seguente:

$$\int_{-\infty}^{+\infty} H_s(x)\, e^{-x^2}\, e^{iQ_s}\, dx = i\sqrt{\pi}\, Q_s\, e^{-Q_s^2/4} . \tag{95}$$

Per mezzo di questa formula otteniamo da (81) con facili calcoli

$$\sigma_{n_1 n_2 n_3\, d\omega} = 4\, a^2 \frac{P}{P_0}\, d\omega\, \frac{(P_x - P_{0x})^{2n_1}}{n_1!}\, \frac{(P_y - P_{0y})^{2n_2}}{n_2!} \frac{(P_z - P_{0z})^{2n_3}}{n_3!}\, e^{-(\mathbf{P}-\mathbf{P}_0)^2} . \tag{96}$$

Si è posto

$$\mathbf{P}_0 = \frac{\mathbf{p}_0}{\sqrt{2\, M h\nu}} \quad ; \quad \mathbf{P} = \frac{\mathbf{p}}{\sqrt{2\, M h\nu}} \tag{97}$$

per modo che P_0^2 e P^2 rappresentano le energie del neutrone prima e dopo l'urto divise per $h\nu$. La (82) diventa

$$P^2 = P_0^2 - n . \tag{98}$$

Sommando la (96) rispetto a tutte le soluzioni intere e non negative della (92) corrispondenti a un dato valore di n, otteniamo la sezione d'urto parziale per processi in cui il neutrone viene diffuso in $d\omega$ e l'idrogeno viene eccitato in uno degli stati di oscillazione di energia (91). La somma si effettua facilmente in forma chiusa (ricordando le formule per lo sviluppo della potenza n-esima del trinomio) e si trova:

$$\sigma_{n,d\omega} = 4\,a^2 \frac{P}{P_0} d\omega \frac{(\mathbf{P} - \mathbf{P}_0)^{2n}}{n!} e^{-(\mathbf{P} - \mathbf{P}_0)^2} . \tag{99}$$

Questa, per mezzo della (98) può anche scriversi

$$\sigma_{n,d\omega} = 4\,a^2 \frac{\sqrt{P_0^2 - n}}{P_0} \frac{(2\,P_0^2 - n - 2\,P_0 \sqrt{P_0^2 - n}\cos\vartheta)^n}{n!} e^{-(2\,P_0^2 - n - 2\,P_0\sqrt{P_0^2 - n}\cos\vartheta)} \tag{100}$$

ϑ essendo l'angolo tra $d\omega$ e la direzione di $\mathbf{P}_0$.

La (100) permette di risolvere i diversi problemi sull'urto. Integrandola su tutto l'angolo solido si ricavano i valori della sezione d'urto par-

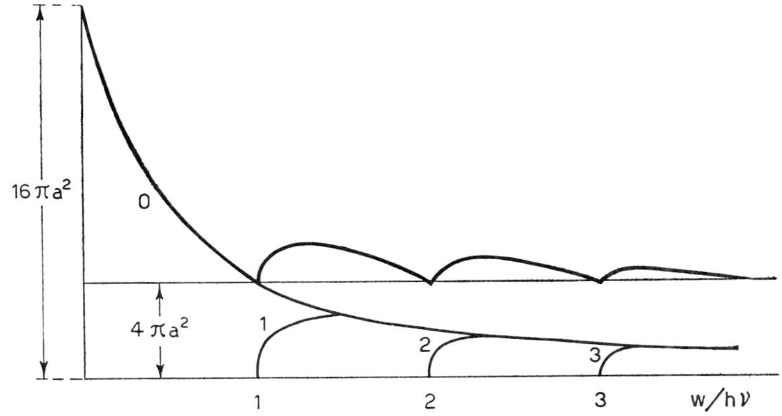

Fig. 2.

ziale σ_n relativa a urti in cui l'atomo di idrogeno viene eccitato nell'n-esimo stato di oscillazione. Essa risulta

$$\sigma_n = \frac{4\,\pi\,a^2}{n!\,P_0^2} \int_{2\,P_0^2 - n - 2\,P_0\sqrt{P_0^2 - n}}^{2\,P_0^2 - n + 2\,P_0\sqrt{P_0^2 - n}} \xi^n e^{-\xi} d\xi \quad ; \tag{101}$$

naturalmente poi σ_n deve porsi eguale allo zero se $P_0^2 < n$, e cioè se il neutrone incidente non ha energia sufficiente per eccitare lo stato n.

L'andamento di σ_n in funzione di $P_0^2 = w/h\nu$ è rappresentato nella fig. 2 (curve 0, 1, 2, 3, $\cdots$). La curva a tratto grosso è la sezione d'urto totale, somma delle sezioni d'urto parziali. La curva asintotica comune delle curve che rappresentano $\sigma_0, \sigma_1, \sigma_2, \cdots$ è, come si riconosce dalla (101), l'iperbole equilatera

$$\frac{4\,\pi a^2}{w/h\nu} .$$

Si noti l'andamento della sezione d'urto totale: per energia nulla il suo valore è $16\pi a^2$. Per $w = h\nu$ essa si è già ridotta ad un valore di pochissimo inferiore a un quarto del precedente; col crescere ulteriore della energia essa si mantiene con oscillazioni sempre meno ampie nell'intorno del valore limite $4\pi a^2$. Questi risultati concordano con le conclusioni generali del paragrafo precedente.

Quando l'energia w è minore di $h\nu$ la sezione d'urto totale si riduce alla sola σ_0; da (101), ricordando che $P_0^2 = w/h\nu$, questa risulta

$$\sigma_0 = 4\pi a^2 \frac{1 - e^{-4w/h\nu}}{w/h\nu} . \tag{102}$$

Mentre la dipendenza angolare è determinata dalla (100), che si riduce a

$$\sigma_{0,d\omega} = 4a^2 e^{-\frac{2w}{h\nu}(1-\cos\vartheta)} . \tag{103}$$

Da questa equazione si riconosce che, conformemente ai risultati del paragrafo precedente, la diffusione tende a diventare isotropa per $w \ll h\nu$; mentre se w è di poco inferiore ad $h\nu$ la anisotropia è considerevole. Per neutroni termici ($w \sim 0{,}037$ volt) $w/h\nu$ è dell'ordine di $1/10$; ciò non ostante la anisotropia non è ancora del tutto trascurabile; parimenti la sezione di urto è in realtà ancora alquanto più piccola del quadruplo della sezione di urto limite per $w \gg h\nu$, essendo il rapporto più prossimo a 3 che a 4.

Notiamo infine che, se gli atomi di idrogeno fossero legati elasticamente a delle posizioni fisse di equilibrio, come abbiamo supposto nel corso di questa trattazione, appena l'energia w del neutrone, con successivi urti, fosse scesa al di sotto di $h\nu$, essa non potrebbe ulteriormente decrescere fino al valore dell'agitazione termica. La possibilità di un ulteriore rallentamento del neutrone, anche dopo che la sua energia è discesa al di sotto del quanto delle vibrazioni elastiche, dipende dalla circostanza che gli atomi di idrogeno nel reticolo cristallino di un corpo solido non sono legati a posizioni di equilibrio fisse, ma agli atomi circostanti i quali sono essi stessi mobili. Per conseguenza, come è ben noto, le frequenze di oscillazione formano uno spettro continuo che si estende fino alla frequenza zero, ciò che permette la riduzione dell'energia fino al valore dell'agitazione termica. Naturalmente però, quando l'energia è già ridotta ad un valore dell'ordine di grandezza o inferiore al valor medio del quanto delle oscillazioni elastiche, il processo di rallentamento è notevolmente meno rapido che per energie maggiori, di modo che per ridurre l'energia a $1/e$ del suo valore occorre in media più di un urto.

Una valutazione del numero di urti necessario per ridurre l'energia da $h\nu$ fino alla energia di agitazione termica (riduzione all'incirca di un fattore 10) conduce a ritenere che per questo processo possano occorrere approssimativamente 10 urti; mentre se il rallentamento avvenisse con la stessa legge che si ha per neutroni di energia maggiore di un volt ne basterebbero 2 o 3.

12. *Il processo di cattura dei neutroni*. – Abbiamo più volte detto che i neutroni termici nella paraffina hanno una vita media finita τ, poiché essi

possono venir catturati dai protoni presenti nell'ambiente riunendosi ad essi per formare dei nuclei di idrogeno pesante. L'ordine di grandezza di τ è stato valutato con una esperienza diretta [7] dalla quale è risultato $\tau \sim 10^{-4}$ s. Altrimenti τ può dedursi dalla conoscenza del cammino libero medio λ dei neutroni termici, e del numero medio N degli urti che un neutrone termico subisce prima di venir catturato. Da questo metodo (cfr. *AF*, § 13) risulta, in buon accordo con la valutazione diretta $\tau = 1{,}7 \cdot 10^{-4}$ s.

In questo paragrafo discuteremo il meccanismo dei processi di cattura dei neutroni da parte dei protoni [8], arrivando così teoricamente a una determinazione di τ che è in ottimo accordo con le determinazioni sperimentali.

La teoria che segue è basata sull'ipotesi che le transizioni in cui un neutrone lento si combina con un protone per formare un deutone (emettendo l'energia che si libera in forma di un quanto gamma) siano determinate dalle oscillazioni del momento magnetico del sistema neutrone–protone. Si trova infatti che le transizioni ordinarie, dovute alle oscillazioni del dipolo elettrico, sono proibite nel caso dei neutroni lenti: ammettiamo di transizioni da uno stato ^{1}S del continuo allo stato fondamentale ^{3}S perché, come risulta dalla ordinaria teoria dell'urto, nel caso dei neutroni lenti, la cattura può avvenire soltanto a partire da stati S del continuo (poiché negli stati P , D ,··· i due corpuscoli restano troppo distanziati per poter interagire efficacemente). Ora è noto che gli ordinari processi di irradiazione per dipolo o per quadrupolo elettrico sono proibiti per le transizioni S→S. Vi è invece, come preciseremo meglio tra poco, la possibilità di transizione da uno stato ^{1}S del continuo allo stato fondamentale ^{3}S del deutone col meccanismo della radiazione per dipolo magnetico.

Le formule per il calcolo della irradiazione dovuta alle vibrazioni di un dipolo magnetico sono sostanzialmente analoghe (data la simmetria tra campo elettrico e magnetico nelle equazioni di Maxwell) a quelle valide per il caso del dipolo elettrico; la probabilità di transizione tra due stati i e j è

$$\frac{64\,\pi^4\,\nu^3}{3\,hc^3}\,|\,\mu_{ij}\,|^2 \tag{104}$$

dove μ_{ij} è l'elemento di matrice i, j del momento magnetico μ del sistema e ν la frequenza emessa nella transizione.

Indichiamo con gli indici 0,1 i due termini ^{3}S (termine fondamentale del deutone, avente $w_0 = -2{,}2 \cdot 10^6$ volt) e ^{1}S (termine del continuo rappresentante uno stato del neutrone lento avente energia w piccolissima in confronto a w_0). Per il calcolo dell'elemento di matrice μ_{01}, si procede come segue:

La autofunzione ψ del sistema neutrone–protone dipende dalle coordinate x , y , z , s di posizione e di spin del neutrone e delle coordinate X , Y , Z , S di posizione e di spin del protone. Riferendoci ad un sistema in cui è fisso il centro di gravità neutrone–protone, le autofunzioni dei termini S, ^{1}S e ^{3}S, dipenderanno, oltre che dalle coordinate di spin s ed S, soltanto dalla distanza r dal protone al neutrone. Siccome poi ciascuna delle due coordinate di spin è suscettibile di prendere solo due valori (+ 1 e — 1) segue che ogni autofunzione sarà l'insieme di quattro funzioni delle sole coordinate di posizione. Le disponiamo nel quadro

$$\psi = \begin{vmatrix} \psi_{11} & \psi_{1-1} \\ \psi_{-11} & \psi_{-1-1} \end{vmatrix}$$

intendendo che il primo ed il secondo indice si riferiscano rispettivamente ai valori di s ed S. Si riconosce facilmente che, con questa notazione, la unica autofunzione di un termine ^{1}S si scrive nella forma

$$\psi\,(^1\mathrm{S}) = \begin{vmatrix} 0 & -v\,(r)/\sqrt{2}\,r \\ v\,(r)/\sqrt{2}\,r & 0 \end{vmatrix} \tag{105}$$

in cui la funzione $v\,(r)$ soddisfa all'equazione

$$\frac{d^2 v}{dr^2} + \frac{4\,\pi^2\,\mathrm{M}}{h^2}\,[w - g_1\,(r)]\,v\,(r) = 0\,. \tag{106}$$

$g_1\,(r)$ rappresenta l'energia potenziale della interazione neutrone–protone quando i due corpuscoli hanno spin antiparalleli; M/2 è la massa ridotta. Nel caso nostro, in cui lo stato ^{1}S rappresenta il neutrone lento, la funzione $v\,(r)$, per valori di r piccoli in confronto alla lunghezza d'onda di de Broglie è del tipo rappresentato nella fig. 1.

Le tre autofunzioni del termine fondamentale ^{3}S saranno date similmente da

$$\psi\,(^3\mathrm{S}_1) = \begin{vmatrix} u\,(r)/r & 0 \\ 0 & 0 \end{vmatrix} \;;\; \psi\,(^3\mathrm{S}_0) = \begin{vmatrix} 0 & u\,(r)/\sqrt{2}\,r \\ u\,(r)/\sqrt{2}\,r & 0 \end{vmatrix}\;; \tag{106}$$

$$(\psi\,^3\mathrm{S}_{-1}) = \begin{vmatrix} 0 & 0 \\ 0 & u\,(r)/r \end{vmatrix}$$

dove la $u\,(r)$ suddisfa all'equazione

$$\frac{d^2 u}{dr^2} + \frac{4\,\pi^2\,\mathrm{M}}{h^2}\,[w_0 - g_3\,(r)]\,u\,(r) = 0$$

$g_3\,(r)$ rappresenta l'energia potenziale dell'interazione quando gli spin sono paralleli; $g_3\,(r)$ sarà in genere diverso da $g_1\,(r)$ (cfr. § 9).

Per calcolare gli elementi di matrice del momento magnetico μ osserviamo che questo è la somma vettoriale dei momenti magnetici intrinseci dei due corpuscoli. Siano μ_n e μ_p i valori dei momenti magnetici del neutrone e del protone (da prendersi con segno + o — a seconda che il momento magnetico è parallelo o antiparallelo rispetto al momento meccanico intrinseco del corpuscolo; in realtà il momento magnetico del neutrone è negativo e quello del protone positivo). Le componenti dei momenti magnetici vettoriali $\boldsymbol{\mu}_n$ e $\boldsymbol{\mu}_p$ si ottengono moltiplicando μ_n e μ_p per gli operatori di Pauli, operanti rispettivamente sulle coordinate s ed S. Si scrivono così senza difficoltà gli operatori che rappresentano le componenti μ_x, μ_y, μ_z del momento magnetico totale $\boldsymbol{\mu} = \boldsymbol{\mu}_p + \boldsymbol{\mu}_n$. Per esempio μ_z opera al modo seguente:

$$\mu_z \begin{vmatrix} \psi_{11} & \psi_{1-1} \\ \psi_{-11} & \psi_{-1-1} \end{vmatrix} = \mu_n \begin{vmatrix} \psi_{11} & \psi_{1-1} \\ -\psi_{-11} & -\psi_{-1-1} \end{vmatrix} + \mu_p \begin{vmatrix} \psi_{11} & -\psi_{1-1} \\ \psi_{-11} & -\psi_{-1-1} \end{vmatrix}.$$

Da (105) e (107) gli elementi di matrice di μ_z risultano allora

$$\mu_z({}^1S\,,{}^3S_1) = \mu_z({}^1S\,,{}^3S_{-1}) = 0 \quad ; \quad \mu_z({}^1S\,,{}^3S_0) = 4\pi(\mu_p - \mu_n)\int_0^\infty uv\,dr\,.$$

In modo analogo si trova

$$-\mu_x({}^1S\,,{}^3S_1) = \mu_x({}^1S\,,{}^3S_{-1}) = 4\pi\frac{\mu_p - \mu_n}{\sqrt{2}}\int_0^\infty uv\,dr \quad ; \quad \mu_x({}^1S\,,{}^3S_0) = 0$$

$$\mu_y({}^1S\,,{}^3S_1) = \mu_y({}^1S\,,{}^3S_{-1}) = 4\pi i\frac{\mu_p - \mu_n}{\sqrt{2}}\int_0^\infty uv\,dr \quad ; \quad \mu_y({}^1S\,,{}^3S_0) = 0\,.$$

La probabilità di transizione si trova ponendo in (104) al posto di $|\mu_{ij}|^2$ la somma dei quadrati dei moduli di tutti questi elementi di matrice; essa risulta dunque

$$\frac{1024\,\pi^6\,\nu^3}{hc^3}(\mu_p - \mu_n)^2\left(\int_0^\infty uv\,dr\right)^2. \tag{109}$$

Per applicare questa formula al problema della cattura dei neutroni lenti la $u(r)$ (u/r = autofunzione radiale dello stato fondamentale del deutone) deve normalizzarsi in modo che

$$4\pi\int_0^\infty u^2\,dr = 1\,. \tag{110}$$

Per quanto riguarda invece la $v(r)$ (v/r = autofunzione radiale dello stato ${}^1S$ del continuo che rappresenta il neutrone lento), la sua normalizzazione, scelta nel § 10 (cfr. formula (79) e fig. 1) corrisponde al caso che la densità dei neutroni lenti che si trovano in uno stato di singoletti rispetto al protone sia = 1; siccome i pesi statistici degli stati di singoletti e di tripletti stanno come 1 : 3, la densità totale dei neutroni lenti che corrisponde alla detta normalizzazione è = 4. Dato questo, ed indicando con V la velocità dei neutroni lenti e con σ_{ca} la sezione d'urto per i processi di cattura, la probabilità di transizione per il processo di cattura risulta evidentemente $= 4\,V\sigma_{ca}$; dal confronto con (109) ricaviamo dunque

$$\sigma_{ca} = \frac{1}{V}\,\frac{256\,\pi^6\,\nu^3}{hc^3}(\mu_p - \mu_n)^2\left(\int_0^\infty uv\,dr\right)^2. \tag{111}$$

Si noti in particolare in questa formula che la sezione d'urto di cattura risulta inversamente proporzionale alla velocità. Indicando con τ la vita media di un neutrone lento in un ambiente che contenga idrogeno alla concentrazione di n atomi per cm³ si ha $\tau V\,n\sigma_{ca} = 1$; si trova dunque

$$\frac{1}{\tau} = n\,\frac{256\,\pi^6\,\nu^3}{hc^3}(\mu_p - \mu_n)^2\left(\int_0^\infty uv\,dv\right)^2. \tag{112}$$

Per il calcolo effettivo dell'integrale contenuto in queste formule useremo il procedimento di Bethe e Peierls (loc. cit.) consistente nel trascurare le irregolarità delle due funzioni u e v per $r < \rho$ (ρ = raggio d'azione) dato che, essendo ρ assai piccolo, il contributo ai vari integrali della regione $r < \rho$ è poco importante. Per $r > \rho$ risulta integrando la (108) e tenendo conto della condizione di normalizzazione (110)

$$(113) \qquad u(r) = \frac{1}{\sqrt{2\pi b}} e^{-\frac{r}{b}} \quad , \quad b = \sqrt{\frac{h^2}{4\pi^2 M |w_0|}} = 0{,}43 \cdot 10^{-12} .$$

(La normalizzazione è stata calcolata ammettendo, per le ragioni indicate, che la (113) valga fino ad $r = 0$; il valore numerico della lunghezza b corrisponde all'energia di legame $-w_0 = 2{,}2 \cdot 10^6$ volt).

Secondo la (79) la $v(r)$ per $r > \rho$ è data da

$$(114) \qquad v(r) = a + r \quad ; \quad a = \pm \sqrt{\frac{h^2}{4\pi^2 M |w_1|}} = 1{,}88 \cdot 10^{-12}$$

dove w_1 rappresenta l'energia dello stato reale o virtuale ^1S di cui si è parlato nel § 9 (da non confondersi con l'energia w dell'altro stato ^1S che rappresenta invece il neutrone lento). Il valore numerico di $|w_1|$ è stato preso equivalente a $1{,}16 \cdot 10^5$ volt (cfr. § 9). Calcolando l'integrale in (112), con l'ammettere (113) e (114) valide fino ad $r = 0$, si trova subito:

$$(115) \qquad \frac{1}{\tau} = n \frac{128\pi^5 \nu^2}{hc^3} (\mu_p - \mu_n)^2 (a + b)^2 b .$$

Introducendo in questa equazione i valori numerici $n = 7{,}8 \cdot 10^{22}$; $\nu = -w_0/h = 5{,}38 \cdot 10^{29}$ otteniamo, a seconda che si prende il segno + o — nella seconda (114)

$$(116) \qquad \frac{1}{\tau} = 61 \left(\frac{\mu_p - \mu_n}{\mu_0}\right)^2 \text{oppure} \ \frac{1}{\tau} = 154 \left(\frac{\mu_p - \mu_n}{\mu_0}\right)^2$$

dove μ_0 è il magnetone nucleare. La prima corrisponde ad a negativo (livello ^1S del § 9 reale); la seconda ad a positivo (livello ^1S virtuale).

Per il momento magnetico del protone ammetteremo un valore di $\mu_p = 2{,}88\ \mu_0$ (media dei risultati delle misure di Stern, Estermann e Frisch e di Rabi, Kellogg e Zacharias [9]). Il momento magnetico del neutrone si ottiene come differenza tra quello del deutone e quello del protone; prendendo come momento del deutone $0{,}75\ \mu_0$ si trova $\mu_n = -2{,}13\ \mu_0$ (si noti il segno —). Dalle (116) si ottiene così

$$\tau = 6{,}5 \cdot 10^{-4} \quad \text{oppure} \quad \tau = 2{,}6 \cdot 10^{-4}\ s .$$

Il secondo di questi due valori si accorda in modo ottimo con il risultato delle esperienze sui neutroni lenti ($\tau = 1{,}7 \cdot 10^{-4}$). Ciò sembrerebbe costituire una indicazione che valga la seconda delle (116) e cioè che il livello ^1S del deutone sia virtuale.

BIBLIOGRAFIA.

[1] E. AMALDI, E. FERMI, *Sopra l'assorbimento e la diffusione dei neutroni lenti*, « Ric. Scientifica », *7* (1), 454 (1936). In questo lavoro sono citate le pubblicazioni preliminari su questo argomento.

[2] E. AMALDI, O. D'AGOSTINO, E. FERMI, B. PONTECORVO, F. RASETTI, E. SEGRÈ, « Proc. Roy Soc. », *A 149*, 522 (1935).

[3] E. FERMI, *Zeeman Festschrift*, p. 128, L'Aia, Martinus Nijhoff, 1935.

[4] G. C. WICK, « Rend. Lincei », *23*, 774 (1936).

[5] H. A. BETHE, R. PEIERLS, « Proc. Roy. Soc. », *A 148*, 146 (1934); *149*, 176 (1934).

[6] Deve tenersi conto del fatto che il rapporto dei pesi statistici degli strati di singoletti e tripletti è 1 : 3 ; cfr. anche M. GOLDHABER, « Nature » (London), *137*, 824 (1936).

[7] E. AMALDI, O. D'AGOSTINO, E. FERMI, B. PONTECORVO, E. SEGRÈ, « Ric. Scientifica », *6* (1), 581 (1935).

[8] E. FERMI, « Phys. Rev. », *48*, 570 (1935).

[9] R. FRISCH, O. STERN, « Z. Physik », *85*, 4 (1933); I. ESTERMANN, O. STERN, « Z. Physik », *85*, 17; *86*, 132 (1933); I. I. RABI, J. M. KELLOGG, J. R. ZACHARIAS, « Phys. Rev. », *46*, 157, 163 (1934); per i segni dei momenti magnetici del protone e del deutone vedi gli stessi autori, « Nature », (London), *137*, 658 (1936).

119 *b*.

ON THE MOTION OF NEUTRONS IN HYDROGENOUS SUBSTANCES

« Ric. Scientifica », 7 (2), 13 (1936).
(Translated from the Italian by G. M. TEMMER).

In this paper we shall discuss some of the properties of slow neutrons from the theoretical viewpoint. In the first part, the slowing down and diffusion of neutrons in hydrogenous media will be studied omitting the quantum-mechanical analysis of the collision process (elastic or inelastic) between the neutrons and the protons of the hydrogenous substance. The second part will be devoted to the study of the collision mechanism instead. In this work will also be given the mathematical proofs of several formulae which were used by Amaldi and Fermi [1] in their interpretation of certain slow neutron measurements. These investigations, to which we shall often refer in the following pages to illustrate the connections between existing theories and experiment, will be denoted by *AF*.

PART I.

DIFFUSION AND SLOWING DOWN OF NEUTRONS.

When a neutron is emitted from a source in an hydrogenous medium (which in practice will always be paraffin or water; for convenience in what follows we shall normally imply paraffin) it will rapidly undergo a slowing down process due to the collisions with the hydrogen atoms [2]. It can easily be shown that with each impact the logarithm of the neutron energy decreases on the average by one unit (reduction of the energy by $1/e$) in such a way that after about twenty collisions the neutron energy is reduced, from an initial value of several million volts, to the energy of thermal agitation. In the diffusion process of the neutrons in the hydrogenous medium, we must therefore distinguish:

(*a*) a slowing down phase during which the neutron energy is reduced from its initial value to the energy of thermal agitation;

(*b*) a thermal diffusion phase during which the neutron, having now attained thermal energies, no longer loses any energy on the average in the successive collisions with the hydrogen atoms, and continues its motion of diffusion until it escapes from the hydrogenous medium or is captured by hydrogen or by some other atom present in the medium.

1. *The slowing down phase.* – In the study of the slowing down process we assume at first that the hydrogen atoms may be considered to be at rest, and that one can also overlook the fact that they are not free but are bound by chemical bonds. These hypotheses will evidently be justified as long as the energy w of the neutron is still appreciably greater than the energy of the chemical bonds; more precisely the condition of validity is that $w \gg h\nu$, ν being the frequency of the hydrogen atoms due to the chemical bond. The order of magnitude of this frequency in common hydrogenous substances is between 1000 and 4000 cm^{-1}, so that we can maintain our hypothesis as long as the neutron energy does not decrease to a value below 1 volt. The further slowing down process, from this limit down to thermal energies ($\sim$ 1/30 volt) is complicated by effects due to the chemical binding of hydrogen and will be discussed in the following.

As long as the neutron energy is greater than 1 volt, we shall assume that the direction of motion of the neutron after a collision will be distributed with equal probability in all directions, provided that the motion be referred to a system in which the center of mass of the colliding neutron and proton is at rest. If instead we refer to the laboratory system in which the proton is initially at rest, and taking into account the fact that the neutron and proton have equal masses, it is easily recognized by an immediate application of the principles of conservation of energy and momentum, that the angle ϑ included between the trajectories of the neutron before and after the impact is always acute; more precisely one finds, calling w and w' the energies of the neutron before and after the collision, the relation

$$\frac{w'}{w} = \cos^2 \vartheta. \tag{1}$$

According to our assumption concerning the distribution of the direction of motion after a collision it follows immediately that the probability that the energy will be reduced by the collision to a value lying between w' and $w' + dw'$ is

$$dp = \frac{dw'}{w} \cdot \tag{2}$$

First of all we wish to determine the number of neutrons having a velocity lying between the limits v and $v + dv$ which are contained in a block of paraffin of practically infinite dimensions in which there is located a source of fast neutrons producing Q neutrons per second at an initial velocity v_0.

The velocity distribution under these circumstances may be obtained as a particular case of the one we have derived in another paper [3]. In the present case this distribution may however also be obtained directly as follows:

Let $\mathfrak{N}(v)\,dv$ be the number of neutrons having velocity between v and $v + dv$, which are present in the entire volume of paraffin. This number tends to decrease inasmuch as some of the $\mathfrak{N}(v)\,dv$ neutrons undergo collisions and consequently their velocity decreases; it tends to increase because some of the neutrons with velocities greater than v can, as a result of a collision, take on a velocity between v and $v + dv$. Under stationary conditions the decrease must be compensated for by the increase. Expressing the energy in equation (2) in terms of the velocity, we can write:

$$dp = \frac{2\,v'\,dv'}{v^2} \tag{3}$$

and observing that the probability that a neutron undergo a collision in unit time is

$$\frac{v}{\lambda(v)}, \tag{4}$$

$\lambda(v)$ being the mean free path, one then finds, as a condition for a stationary state:

$$\frac{v}{\lambda(v)}\,\mathfrak{N}(v)\,dv = \frac{2\,v\,dv}{v_0^2}\,Q + 2\,v\,dv\int_v^{v_0}\frac{1}{u^2}\,\mathfrak{N}(u)\,\frac{u}{\lambda(u)}\,du$$

that is,

$$\frac{\mathfrak{N}(v)}{\lambda(v)} = \frac{2\,Q}{v_0^2} + 2\int_v^{v_0}\frac{1}{u}\,\frac{\mathfrak{N}(u)}{\lambda(u)}\,du\,. \tag{5}$$

Differentiating this expression with respect to v we find

$$\frac{d}{dv}\,\frac{\mathfrak{N}(v)}{\lambda(v)} = -\,\frac{2}{v}\,\frac{\mathfrak{N}(v)}{\lambda(v)}$$

which, upon integration, yields:

$$v^2\,\frac{\mathfrak{N}(v)}{\lambda(v)} = \text{const.}$$

The constant, which is determined by substituting again into (5), has the value $2\,Q$. We therefore obtain

$$\mathfrak{N}(v) = \frac{2\,Q\lambda(v)}{v^2}\,. \tag{6}$$

This distribution law of course only holds for energies greater than one volt. In Section 8 we shall encounter an application of this law to methods for determining the width of the energy bands which constitute the various groups of slow neutrons.

2. *Distance of diffusion of neutrons during the slowing down process.* Equation (6) gives us the total number of neutrons belonging to a given interval of energy; however, it does not tell us anything about the spatial distribution of these neutrons. Let us assume that we have a source S of fast neutrons (initial energy w_0) at the center of a block of paraffin of very large dimensions. As the neutrons undergo collisions with the hydrogen atoms they lose energy so that, on the average, those neutrons which have undergone more collisions will have a lower energy. On the other hand, neutrons which have undergone more collisions are those whose average distance from the source is larger; therefore it follows that the mean distance from the source of neutrons of a given energy will be the larger the lower their energy.

We now propose to calculate the mean square $\overline{r^2}$ of the distance from the source as a function of the energy (still limiting ourselves to the case of energies greater than about one volt). In the calculation we shall take into account collisions with hydrogen atoms only. This is justified when the hydrogenous medium is water or paraffin because in that case, for fast neutrons, one can estimate that 90 % of the collisions occur with hydrogen atoms, whereas for slow neutrons this percentage rises even further. At the end of this section, we shall also give without proof the formulae for the case which takes into account the collisions with the heavy atoms present in the hydrogenous medium into which the neutrons diffuse.

The calculation of $\overline{r^2}$ is complicated by the fact that the direction of motion of a neutron after collision is not independent of the direction before collision, since there is always an acute angle between them. Let $\boldsymbol{\lambda}_0, \boldsymbol{\lambda}_1, \cdots \boldsymbol{\lambda}_n$ be the distances travelled by the neutron during its first $n+1$ free paths; the total displacement is the vector sum

$$\boldsymbol{r} = \boldsymbol{\lambda}_0 + \boldsymbol{\lambda}_1 + \cdots + \boldsymbol{\lambda}_n . \tag{7}$$

Let us suppose that after the nth collision the energy takes on the particular value w_a, and let us attempt to calculate $\overline{r^2}$ as a function of w_a. In order to carry out this calculation it is convenient to introduce a logarithmic scale for energies. Instead of the energy w we shall use the variable

$$x = \log \frac{w_0}{w} \tag{8}$$

which we shall always represent graphically as the abscissa of a point along a straight line. The mean free path, which depends on the energy, may also be written as a function of x, $\lambda = \lambda(x)$. During the motion of the neutron, its energy, with initial value w_0, takes on smaller and smaller values as a result of successive collisions

$$w_0 > w_1 > w_2 > \cdots > w_{n-1} > w_a$$

and correspondingly, x assumes the values

$$0 < x_1 < x_2 < \cdots < x_{n-1} < x_a = \log \frac{w_0}{w_a} .$$

The abscissa points $x_1, x_2, \cdots$ are distributed according to a rather simple law. As a matter of fact, equation (2) is equivalent to stating that the probability that

$$\xi_r = x_r - x_{r-1} \tag{9}$$

has a value lying between ξ_r and $\xi_r + d\xi_r$ is

$$e^{-\xi_r} d\xi_r .$$

This probability law for the interval between two successive points in a distribution of points along a line is the one which results when these points are distributed independently of each other and in such a way that the probability of one of them falling in a given interval dx of the line is just equal to dx. Thus we have the distribution law of the points x.

Let ϑ be the angle formed by the successive free paths λ_{r-1} and λ_r; we have from relations (1), (8), (9)

$$\cos\vartheta_r = e^{-\frac{1}{2}\xi_r} = e^{-\frac{1}{2}(x_r - x_{r-1})}. \tag{10}$$

Furthermore, let us put

$$a = \log\frac{w_0}{w_a}, \tag{11}$$

where a represents the average number of collisions necessary to reduce the energy from w_0 to w_a. We shall compute the average of r^2 in two successive steps.

(I) Suppose that the points

$$x_1, x_2, \cdots, x_{n-1} \tag{12}$$

are given and let us take the average over all possible magnitudes of the free paths, and all directions compatible with (10) and (12).

(II) We shall then average over all possible values of (12) and of n.

(I) Squaring expression (7) one finds

$$r^2 = \sum_0^n \lambda_r^2 + 2\sum_{r=0}^{n-1}\sum_{s=r+1}^{n} \lambda_r\lambda_s\cos\widehat{\lambda_r\lambda_s}. \tag{13}$$

We must average this expression, assuming that the values (12) are given. Keeping relation (10) in mind one easily finds

$$\overline{\lambda_r^2} = 2\lambda^2(x_r) \quad ; \quad \overline{\lambda_r} = \lambda(x_r)$$

$$\overline{\cos\widehat{\lambda_r\lambda_s}} = \cos\vartheta_{r+1}\cos\vartheta_{r+2}\cdots\cos\vartheta_s = e^{-\frac{1}{2}(x_s - x_r)} \tag{14}$$

and hence

$$\overline{r^2} = 2\sum_0^n \lambda^2(x_r) + 2\sum_{0\le r<s\le n}\lambda(x_r)\lambda(x_s)e^{-\frac{1}{2}(x_s-x_r)}. \tag{15}$$

In this equation the average is understood to be taken over given values of (12).

(II) We must now average with respect to the values of x and n. Remembering that the probability of one of the values of x falling into the interval dx (between 0 and a) is equal to dx, one immediately finds

$$\overline{\sum_0^n \lambda^2(x_r)} = \lambda^2(0) + \int_0^a \lambda^2(x)\,dx + \lambda^2(a).$$

Similarly one obtains:

$$\overline{\sum_1^n \lambda(0)\lambda(x_s)e^{-\frac{x_s}{2}}} = \lambda(0)\int_0^a \lambda(x)e^{-\frac{1}{2}x}dx + \lambda(0)\lambda(a)e^{-\frac{1}{2}a};$$

$$\overline{\sum_1^{n-1}\lambda(x_r)\lambda(a)e^{-\frac{1}{2}(a-x_r)}} = \lambda(a)\int_0^a \lambda(x)e^{-\frac{1}{2}(a-x)}dx.$$

There remains to calculate the average value of

$$\sum_{1 \leq r < s \leq n-1} \lambda(x_r)\,\lambda(x_s)\,e^{-\frac{1}{2}(x_s - x_r)}.$$

For this purpose we shall first perform the average over the possible values of x_s contained between the limits x_r and a, which is

$$\sum_r \lambda(x_r) \int_{x_r}^{a} \lambda(x)\,e^{-\frac{1}{2}(x - x_r)}\,dx = \sum_r \lambda(x_r) \int_0^{a - x_r} \lambda(x_r + x)\,e^{-\frac{1}{2}x}\,dx\,.$$

Finally, averaging over the values of x we find

$$\int_0^a \lambda(\xi)\,d\xi \int_0^{a-\xi} \lambda(\xi + x)\,e^{-\frac{1}{2}x}\,dx\,.$$

Combining these various averages we find the desired result:

$$\begin{aligned}(16)\qquad \overline{r^2} = {} & 2\,\lambda^2(0) + 2\,\lambda^2(a) + 2\int_0^a \lambda^2(x)\,dx + 2\,\lambda(0)\int_0^a \lambda(x)\,e^{-\frac{1}{2}x}\,dx \\ & + 2\,\lambda(0)\,\lambda(a)\,e^{-\frac{a}{2}} + 2\,\lambda(a)\int_0^a \lambda(x)\,e^{-\frac{1}{2}(a-x)}\,dx \\ & + 2\int_0^a \lambda(\xi)\,d\xi \int_0^{a-\xi} \lambda(\xi + x)\,e^{-\frac{1}{2}x}\,dx\,.\end{aligned}$$

This formula shows how the mean square distance of the neutrons from the source increases gradually as the energy decreases (that is as a increases). From this expression we find:

$$\begin{aligned}(17)\qquad \frac{d\overline{r^2}}{da} = {} & 4\,\lambda^2(a) + 4\,\lambda(a)\,\lambda'(a) + \lambda(0)\,[\lambda(a) + 2\,\lambda'(a)]\,e^{-\frac{a}{2}} \\ & + [\lambda(a) + 2\,\lambda'(a)]\int_0^a \lambda(x)\,e^{-\frac{1}{2}(a-x)}\,dx\,.\end{aligned}$$

The mean free path $\lambda(0)$ of the fast neutrons emitted from a source of Ra + Be in paraffin is about 5 to 6 cm. Upon decreasing the energy the mean free path decreases until it is reduced probably to around 1 cm, when the energy, while still lying above 1 volt, already has a value such that the neutron may be termed slow (probably not over 1000 volts). In this last energy interval (between 1 and 1000 volts) λ is probably constant (see also Part II). Therefore equation (17) may be considerably simplified. In fact, we can neglect the third term of the right hand member which contains the very small factor $e^{-a/2}$, and can replace the integral, whose values are concentrated near its upper limit, by

$$\int_{-\infty}^{a} \lambda(a)\,e^{-(a-x)/2}\,dx = 2\,\lambda(a)\,.$$

We thus obtain, setting also $\lambda'(a) = 0$ because of what we have stated,

$$\frac{d\overline{r^2}}{da} = 6\lambda^2(a)\,. \tag{18}$$

This formula has been applied (see *AF*) to the determination of the ratio of the energies of two slow neutron groups. Let w' and w'' be their energies; we shall also assume that over the energy interval between w' and w'' λ is constant. Let $\overline{r'^2}$ and $\overline{r''^2}$ be the mean square of the neutrons of the two groups from the source. Keeping in mind that according to (11) the variation of the variable a in passing from one group to the other is log w''/w', one finds from (18) that

$$\log \frac{w''}{w'} = \frac{\overline{r'^2} - \overline{r''^2}}{6\lambda^2} \tag{19}$$

which is the formula used to determine w''/w'.

If we assume that the mean free path, over the energy interval of the groups under study, is 1 cm, it follows that to a difference of 6 cm² in the mean square distances from the source there corresponds a ratio of e between the energies of the two groups. For a detailed discussion of this question, based on experimental results, see *AF*.

By a procedure similar to the one used to derive equation (16) one can easily obtain an analogous formula valid for the case where the neutrons diffuse and are slowed down in a medium containing heavy atoms in addition to hydrogen atoms. Keeping the notation of equations (8) and (11), we shall indicate by $\rho(x)$ the ratio, for neutrons of energy corresponding to x, of the probabilities of collision with a heavy atom to that with a hydrogen atom. Let us further assume that the collisions with heavy atoms are perfectly elastic, and that in view of the mass of these atoms the energy transferred to them in the collision may be neglected, and finally that the direction of the velocity after a collision with a heavy atom is distributed with uniform probability in all directions, losing all coherence with the direction of motion before collision. One then finds that the mean square distance from the source of neutrons with energy w_a is: [1]

$$\begin{aligned}\overline{r^2} = {} & 2\lambda^2(0)\,[1+\rho(0)] + 2\lambda^2(a)\,[1+\rho(a)] + 2\int_0^a \lambda^2(x)\,[1+\rho(x)]\,dx \\ & + \frac{2\lambda(0)}{1+\rho(0)}\int_0^a \lambda(x)\, e^{-\frac{x}{2}}\,dx\, e^{-\int_0^x \rho(\xi)\,d\xi} + 2\lambda(a)\int_0^a \frac{\lambda(x)}{1+\rho(x)}\, e^{-\frac{x}{2}-\int_x^a \rho(\xi)\,d\xi}\,dx \\ & + 2\int_0^a \frac{\lambda(x)\,dx}{1+\rho(x)}\int_0^{a-x} \lambda(x+\xi)\, e^{-\frac{\xi}{2}-\int_0^\xi \rho(x+\xi)\,d\xi}\,d\xi\,.\end{aligned} \tag{20}$$

(1) Note of the Editor: this equation is not correct, probably because of typographical mistakes. The correct expression has been derived by various authors and is given in the review article by MARSHAK (« Rev. Mod. Phys. », *19*, 185 (1947)). Instead of

$$\rho = \frac{\mathcal{N}_{\text{heavy element}}\,\sigma_{\text{heavy element}}}{\mathcal{N}_{\text{H}}\,\sigma_{\text{H}}}$$

Of course, this formula reduces to (16) when we neglect the collisions with heavy atoms by setting $\rho(x) = 0$.

3. *Thermal neutrons*. – Until now we have considered neutrons belonging to the energy interval from the maximum down to 1 volt, i.e. always above not only to the energy of thermal agitation, ($\sim 1/30$ volt) but also above the quantum of energy $h\nu$ of the chemical forces which bind the hydrogen atoms to the substance (between 1/2 and 1/10 volt). In this interval of energy we were able to neglect thermal motion and the chemical bonds of the hydrogen. However when upon successive collisions, the neutron energy falls below 1 volt, the law according to which the further slowing down takes place becomes somewhat more complicated (cf. Part II). However, it must be remembered that after about 10 collisions this energy will be reduced to thermal motion. There will thus exist a considerable number of neutrons possessing thermal energies, whose velocity distribution will be more or less Maxwellian. The number of these thermal neutrons is limited by the fact that, while they are continuously produced by the slowing down of faster neutrons, they are also continuously disappearing mainly because of their capture by hydrogen nuclei and formation of deuterium (cf. Part II). We shall express the effect of this capture process by introducing a mean life τ for thermal neutrons, so that, if the source produces Q thermal neutrons per second in a practically infinite block of paraffin, the total number of thermal neutrons present at a given instant will be

$$\mathfrak{N}_T = Q\tau\,.$$

Also, as far as elastic collisions are concerned, the behavior of thermal neutrons is indeed different from that of neutrons of energies above 1 volt. Theoretical reasons (cf. Part II) cause us to expect that when the energy w of the neutron passes from values which are large, to values which are small compared to $h\nu$ (ν = frequency of the chemical bond of hydrogen in paraffin) the mean free path is reduced in the ratio 4 : 1, while the angular

Marshak uses the parameter

$$c = \frac{\mathfrak{N}_H\,\sigma_H}{\mathfrak{N}_H\,\sigma_H + \mathfrak{N}_{\text{heavy element}}\,\sigma_{\text{heavy element}}}$$

With $c = \dfrac{1}{1+\rho}$ and $\overline{r^2} = 6\,L_S^2$, equation (108) p. 211 of the Marshak article becomes:

$$\overline{r^2} = 2\lambda^2(0)[1+\rho(0)] + 2\lambda^2(a)[1+\rho(a)] + 2\lambda(0)\lambda(a)\,e^{-\int_0^a \left[\frac{3}{2} - \frac{1}{1+\varrho(x')}\right]dx'}$$
$$+ 2\int_0^a \lambda^2(x)[1+\rho(x)]\,dx + 2\lambda(a)\int_0^a \lambda(x)\,dx\,e^{-\int_x^a \left[\frac{3}{2} - \frac{1}{1+\varrho(x')}\right]dx'}$$
$$+ 2\lambda(0)\int_0^a \lambda(x)\,dx\,e^{-\int_0^x \left[\frac{3}{2} - \frac{1}{1+\varrho(x')}\right]dx'} + 2\int_0^a \lambda(x)\,dx\int_0^x \lambda(x')\,dx'\,e^{-\int_{x'}^x \left[\frac{3}{2} - \frac{1}{1+\varrho(x'')}\right]dx''}$$

distribution law of the neutrons after collision tends to become isotropic in the limit of very small w. Although for thermal neutrons $w/h\nu$ is sufficiently small (about 1/10) the conditions under which the above-mentioned limiting laws are applicable do not quite exist (cf. Section 11). Nevertheless, in Sections 3, 5, 6, and 7 on thermal neutron diffusion we shall simplify our considerations assuming that the diffusion is isotropic. The small error thus introduced will evidently tend to make the *diffusion length* too small and the *albedo* too large.

Let λ be the mean free path of thermal neutrons in paraffin (0.3 cm, cf. *AF*), v their average velocity (at ordinary temperatures $v = 2.5 \times 10^5$ cm/s) and N the average number of free paths traversed by a thermal neutron before capture (N = 140; cf. *AF* Sec. 13); the following relations will then evidently hold:

$$\tau = \frac{\lambda N}{v} \quad ; \quad \mathfrak{N}_T = Q\tau . \tag{22}$$

From these one finds that $\tau = 1.7 \times 10^{-4}$ seconds (cf. also Sec. 12).

In view of the considerable number of collisions that a thermal neutron suffers in a medium before capture, it is clear that its motion may be likened with success to a motion of diffusion. One can easily write down a differential equation for the density n of thermal neutrons as a function of x, y, z, and t. Indicating by $q\,(x\,,y\,,z\,,t)$ the number of thermal neutrons produced in unit time by the slowing down of faster neutrons, one can write immediately

$$\frac{\partial n}{\partial t} = \frac{\lambda v}{3}\,\Delta n - \frac{1}{\tau}\,n + q \tag{23}$$

where the first term on the right hand side represents the effect of diffusion (in fact, one knows from classical kinetic theory that $\lambda v/3$ is the diffusion coefficient); the second term expresses the effect of destruction of neutrons by capture, and the third term that of the production of thermal neutrons by slowing down. In the stationary case, which is evidently the most important one, equation (23) reduces to:

$$n = \frac{\lambda^2 N}{3}\,\Delta n + \frac{q\lambda N}{v} \tag{24}$$

when making use of (22).

As an example we shall show the application of (24) to the solution of the following problem:

Suppose that we have a block of paraffin, bounded on the left by the plane $x = 0$ and occupying the entire space to the right of that plane. One has to find the probability that a thermal neutron, located at the point of positive abscissa a, succeeds in escaping from the paraffin across the plane $x = 0$ before being captured. This problem is obviously equivalent to another one: suppose that at a point of abscissa a Q thermal neutrons are produced per second; find the fraction of these neutrons which escape from the paraffin across the plane $x = 0$.

To solve this problem let us first reduce equation (24) involving three independent variables x, y, z, to one containing only the independent variable x. To do this it suffices to multiply both sides of (24) by $dy\,dz$ and to integrate with respect to y and z from $-\infty$ to $+\infty$. Setting

$$\rho(x) = \iint n\,dy\,dz \quad ; \quad f(x) = \iint q\,dy\,dz, \tag{25}$$

and making use of the obvious relations

$$\iint \frac{\partial^2 n}{\partial y^2}\,dy\,dz = \iint \frac{\partial^2 n}{\partial z^2}\,dy\,dz = 0,$$

one finds

$$\rho(x) - \frac{\lambda^2 N}{3}\frac{d^2\rho}{dx^2} = \frac{\lambda N}{v} f(x). \tag{26}$$

In our case the function $f(x)$ is evidently zero for $x \neq a$, while for $x = a$ it is infinite in such a way that

$$\int f(x)\,dx = Q \tag{27}$$

when the integral extends over an interval containing a. It follows from this property of $f(x)$ that with the exception of point a, $\rho(x)$ satisfies the differential equation

$$\rho - \frac{\lambda^2 N}{3}\frac{d^2\rho}{dx^2} = 0. \tag{28}$$

For $x = 0$ and $x = \infty$, $\rho(x)$ must obviously satisfy the conditions

$$\rho(0) = 0 \quad ; \quad \rho(\infty) = 0. \tag{29}$$

From (26) and (27) it follows that at the singular point a there is a discontinuity of the derivative of ρ, while the function itself remains continuous. One has precisely

$$\delta\rho(a) = 0 \quad ; \quad \delta\left(\frac{d\rho}{dx}\right)_{x=a} = -\frac{3}{v\lambda}Q. \tag{30}$$

The general solution of (28) is given by a linear combination of the two fundamental solutions

$$e^{\sqrt{\frac{3}{\lambda^2 N}}\,x} \quad ; \quad e^{-\sqrt{\frac{3}{\lambda^2 N}}\,x}.$$

Keeping the relations (29) in mind one immediately finds that the expressions for $\rho(x)$ to the left and to the right of point a are of the form

$$\begin{cases} \rho(x) = A\left\{e^{\sqrt{\frac{3}{\lambda^2 N}}\,x} - e^{-\sqrt{\frac{3}{\lambda^2 N}}\,x}\right\} & \text{for } 0 < x < a \\ \rho(x) = B\,e^{-\sqrt{\frac{3}{\lambda^2 N}}\,x} & \text{for } x > a, \end{cases} \tag{31}$$

where A and B are constants which may easily be determined by means of relations (30). One finds

$$A = \frac{\sqrt{3N}}{2v}\,Q\,e^{-\sqrt{\frac{3}{\lambda^2 N}}\,a} \quad ; \quad B = A\left(e^{2\sqrt{\frac{3}{\lambda^2 N}}\,a} - 1\right). \tag{32}$$

The number of neutrons escaping from the paraffin in unit time is evidently given by the diffusion coefficient $\lambda v/3$ times the derivative of ρ with respect to x taken at the point $x = 0$. This product is

$$\frac{\lambda v}{3}\cdot 2\,A\sqrt{\frac{3}{\lambda^2 N}} = Q\,e^{-\sqrt{\frac{3}{\lambda^2 N}}\,a}. \tag{33}$$

The desired probability $p(a)$ that a neutron, starting out from a point of abscissa a, succeeds in escaping from the paraffin is evidently equal to the ratio of expression (33) to Q, namely

$$p(a) = e^{-\sqrt{\frac{3}{\lambda^2 N}}\,a}. \tag{34}$$

This formula has been used (cf. *AF*) to measure $\lambda^2 N$.

One notes that this result, like all others which can be deduced from the differential equations (23) and (24) does not hold exactly except in the limiting case of N very large and λ very small, such that $\lambda^2 N$ remains finite. In fact, we shall see in the following sections that in the neighborhood of the boundary surface of the paraffin, and more precisely within a distance of the order of λ from that surface, the neutron density exhibits anomalies.

In order to investigate these it will first be convenient to simplify the problem as shown in the next section.

4. *The one-dimensional case.* – In this section we shall simplify the study of diffusion by considering the fictitious case in which the neutrons move in one dimension along a line, instead of being free in space. This procedure has the advantage that we can easily obtain approximate expressions valid for the actual three-dimensional case.

In the one-dimensional situation we shall assume that in each collision (mean free path λ) the neutron has a probability $1/N$ of being captured; if the collision does not lead to capture, on the other hand, the neutron will have an equal probability of departing toward the right or toward the left.

First of all, we want to find for the one-dimensional case the exact solution of the problem analogous to the one discussed toward the end of the last section, namely: we have a slab of paraffin contained between the abscissae 0 and a. A thermal neutron starts out from a point x inside the paraffin. We are to find the probability $p(x)$ that the neutron escapes from the paraffin to the left. This problem in the limiting case $a = \infty$ is the one–dimensional analogue of the problem discussed before.

In order to solve it we shall establish an equation for $p(x)$ in the following manner. The probability $p(x)$ may be decomposed into the following components:

(I) The probability that the neutron, starting out from x, begins its motion toward the left and directly escapes from the paraffin without having suffered any collisions is

$$\frac{1}{2}\,e^{-x/\lambda}.$$

(II) The probability that the neutron suffers a first collision between ξ and $\xi + d\xi$ (which is $\frac{1}{2} e^{-|x-\xi|/\lambda} d\xi/\lambda$), that this collision be elastic (which is $(N-1)/N$), and finally, that the neutron, after all this, succeeds in escaping from the paraffin to the left (which probability is $p(\xi)$) is the product of the component probabilities, namely

$$\frac{1}{2} e^{-\frac{|x-\xi|}{\lambda}} \cdot \frac{d\xi}{\lambda} \cdot \frac{N-1}{N} \cdot p(\xi).$$

This expression must be integrated with respect to ξ from 0 to a. Adding (I) and (II) we find the following integral equation for $p(x)$:

$$p(x) = \frac{1}{2} e^{-x/\lambda} + \frac{N-1}{2\lambda N} \int_0^a p(\xi)\, e^{-\frac{|x-\xi|}{\lambda}} d\xi. \tag{35}$$

It may easily be verified that the solution of (35) can be put into the form:

$$p(x) = A\, e^{\alpha x} + B\, e^{-\alpha x} \tag{36}$$

where A and B, and α are constants. It suffices to substitute (36) into (35) to find that it is possible to satisfy (35) provided we let

$$\left\{ \begin{aligned} A &= -\frac{(N-\sqrt{N})\, e^{-a/\lambda\sqrt{N}}}{(\sqrt{N}+1)^2 e^{a/\lambda\sqrt{N}} - (\sqrt{N}-1)^2 e^{-a/\lambda\sqrt{N}}} \\ B &= \frac{(N+\sqrt{N})\, e^{a/\lambda\sqrt{N}}}{(\sqrt{N}+1)^2 e^{a/\lambda\sqrt{N}} - (\sqrt{N}-1)^2 e^{-a/\lambda\sqrt{N}}} \quad ; \quad \alpha = \frac{1}{\lambda\sqrt{N}}. \end{aligned} \right. \tag{37}$$

These formulae may be considerably simplified in the case where the paraffin is unlimited to the right ($a = \infty$). They then become

$$\left. \begin{aligned} &A = 0 \quad ; \quad B = \frac{\sqrt{N}}{\sqrt{N}+1} \quad ; \quad \alpha = \frac{1}{\lambda\sqrt{N}} \\ &p(x) = \frac{\sqrt{N}}{\sqrt{N}+1} e^{-\frac{x}{\lambda\sqrt{N}}} \end{aligned} \right\} \text{for } a = \infty. \tag{38}$$

The problem solved here permits us to determine with ease (still for the one–dimensional case) the reflection coefficient (*albedo*) for thermal neutrons incident upon a slab of paraffin. Suppose that upon this layer of paraffin, extending as before from $x = 0$, to $x = a$, a thermal neutron impinges from the left. We shall call the albedo $\beta(a)$ of the slab of thickness a the probability that the neutron after having suffered one or more collisions, escapes through the same face through which it entered. In order to determine the albedo we note that the probability that the neutron, upon entering the paraffin, suffers a first collision between x and $x + dx$ and is not captured there, is given as above by

$$\frac{N-1}{N} e^{-x/\lambda} \frac{dx}{\lambda}.$$

We have seen above that a neutron starting from x has a probability $p(x)$ of escaping from the paraffin toward the left; hence the total probability of an escape to the left, i.e. the albedo, turns out to be in this case

$$\beta(a) = \frac{N-1}{N\lambda}\int_0^a p(x)\, e^{-x/\lambda}\, dx;$$

making use of (36) and (37) and performing the integration, one obtains

$$\beta(a) = \frac{(N-1)\left[e^{a/\lambda\sqrt{N}} - e^{-a/\lambda\sqrt{N}}\right]}{(\sqrt{N}+1)^2 e^{a/\lambda\sqrt{N}} - (\sqrt{N}-1)^2 e^{-a/\lambda\sqrt{N}}} = \frac{N-1}{N+1+2\sqrt{N}\coth\dfrac{a}{\lambda\sqrt{N}}}. \tag{39}$$

Of particular importance is the case where the thickness a of the paraffin is infinite. In this case the albedo is found from (39), or directly from (38), to be

$$\beta(\infty) = \frac{\sqrt{N}-1}{\sqrt{N}+1} = 1 - \frac{2}{\sqrt{N}} + \cdots. \tag{40}$$

The last expression, which is an approximation for N large, is most frequently used.

Let us also write the expression for the *transmission coefficient* for a layer of thickness a, i.e. for the probability that a neutron incident upon this layer from the left will escape from it to the right. By considerations analogous to those necessary to obtain the albedo, this coefficient turns out to be

$$\gamma(a) = e^{-a/\lambda} + \frac{N-1}{\lambda N}\int_0^a p(a-x)\, e^{-x/\lambda}\, dx \tag{41}$$

$$= \frac{4\sqrt{N}}{(\sqrt{N}+1)^2 e^{a/\lambda\sqrt{N}} - (\sqrt{N}-1)^2 e^{-a/\lambda\sqrt{N}}}.$$

For $N = \infty$, $\gamma(a)$ reduces to

$$\gamma_\infty(a) = \frac{2\lambda}{a+2\lambda}. \tag{42}$$

In view of future applications we must also investigate the following question, still for the one-dimensional case. Let us neglect the destruction process of neutrons ($N = \infty$) and suppose that Q thermal neutrons per second are produced in a piece of paraffin extending from $x = 0$ to $x = \infty$, at a distance from the boundary $x = 0$ which is large compared to λ. The neutrons will diffuse across this boundary and escape from the paraffin. Since by hypothesis the destruction process of neutrons is negligible, it is clear that under these circumstances the flux of neutrons from the interior toward the exterior of the paraffin must be Q per second. In order to maintain this flux constant the density of neutrons $n(x)$ must increase linearly as a function of x. We will therefore have

$$n(x) = A + Bx. \tag{43}$$

The constants A and B are easily determined by writing that the neutron flux across any depth x toward the left is equal to Q. Q must therefore be equal to the difference between the number of neutrons traversing the point x per second toward the left and toward the right. We will therefore have

$$Q = \frac{1}{2}\frac{v}{\lambda}\int_0^\infty n(x+\xi)\, e^{-\xi/\lambda}\, d\xi - \frac{1}{2}\frac{v}{\lambda}\int_0^x n(x-\xi)\, e^{-\xi/\lambda}\, d\xi .$$

Introducing expression (43) into this equation, one immediately verifies that the latter is satisfied provided that we have

$$Q = \lambda v B \tag{44}$$

$$A = B\lambda . \tag{45}$$

Eq. (43) may therefore be written

$$n(x) = (\lambda + x)\, B \tag{46}$$

while (44) furnishes the (one-dimensional) diffusion coefficient

$$D = \lambda v . \tag{47}$$

In particular, we observe that as a consequence of (46) the neutron density does not vanish on the boundary of the paraffin; if we were to extrapolate the curve of $n(x)$ toward the exterior of the paraffin, we would find that the density $n(x)$ would go to zero at a distance from the boundary equal to the mean free path λ.

5. *Albedo and transmission probability in three dimensions*. – From the exact results obtained in the previous section for the one-dimensional case we can easily derive other very approximate results valid for the actual case of neutron diffusion in three dimensions.

For that purpose, we shall consider the projection upon the x–axis of the motion of a thermal neutron which suffers successive collisions in the paraffin. This projected motion will be a back and forth motion rather analogous to the motion obtained in the one-dimensional case. The difference lies in the fact that while in the one-dimensional case the probability of suffering a collision after travelling a given distance follows an exponential law, the corresponding law for the case of the projection of a three-dimensional motion is not strictly exponential. Furthermore, the mean free path λ' of the projected motion is of course shorter than the λ of the true motion.

In order to choose λ' conveniently, we can for instance write for the one-dimensional case the diffusion equation analogous to (24); keeping in mind the expression (47) for the diffusion coefficient in one dimension (introducing λ' and v' instead of λ and v the latter is $D' = \lambda' v'$) we have

$$n - \lambda'^2 N \frac{d^2 n}{dx^2} = \frac{q\lambda' N}{v'} .$$

Comparison of this equation with (24) suggests the definition

$$\lambda' = \frac{\lambda}{\sqrt{3}} . \tag{48}$$

Another way for arriving at the same choice for λ' consists in noting that in the three-dimensional case the mean square displacement $\overline{r^2}$ of a point suffering m collisions with mean free path λ is

$$\overline{r^2} = 2\,\lambda^2\,m$$

(the factor 2 comes from the fact that the mean square of the free path is twice the square of the mean free path). The mean square of the projection δx of the displacement r is evidently $1/3$ of $\overline{r^2}$, i.e.

$$\overline{\delta x^2} = \frac{2}{3}\,\lambda^2\,m\,.$$

On the other hand, for one-dimensional motion one would have

$$\overline{\delta x^2} = 2\,\lambda'^2\,m\,.$$

The two expressions are precisely equal with the assumption (48).

We conclude therefore that the projection of three-dimensional diffusion is quite similar to one-dimensional diffusion of mean free path λ' as given by (48). In this way we shall see that a large part of the results of the preceding section may be utilized to a good degree of approximation for the three-dimensional case as well.

In the previous section we have established the expression for the probability $p(x)$ that a thermal neutron, starting from position x, escapes from a slab of paraffin of thickness a limited by two plane faces $x = 0$ and $x = a$. The analogous problem for the three dimensional case is the following: given a slab of paraffin of thickness a bounded by the planes $x = 0$ and $x = a$, find the probability $p(x)$ that a thermal neutron, moving initially from a position of abscissa x, escapes from the slab of paraffin across the left face.

As an approximate expression of $p(x)$ we can take (36), where A, B, and α are given by formulae analogous to (37). However, instead of λ we must write λ', given by (48). We therefore have

$$(49)\qquad \left\{\begin{aligned} p(x) &= \mathrm{A}\,e^{\alpha x} + \mathrm{B}\,e^{-\alpha x} \\ \mathrm{A} &= -\frac{(\mathrm{N}-\sqrt{\mathrm{N}})\,e^{-\alpha a}}{(\sqrt{\mathrm{N}}+1)^2\,e^{\alpha a} - (\sqrt{\mathrm{N}}-1)^2\,e^{-\alpha a}} \\ \mathrm{B} &= \frac{(\mathrm{N}+\sqrt{\mathrm{N}})\,e^{\alpha a}}{(\sqrt{\mathrm{N}}+1)^2\,e^{\alpha a} - (\sqrt{\mathrm{N}}-1)^2\,e^{-\alpha a}}\quad;\quad \alpha = \frac{1}{\lambda}\sqrt{\frac{3}{\mathrm{N}}}\,. \end{aligned}\right.$$

In the case where the thickness a of the paraffin is infinite, these formulae simplify in a manner analogous to equations (38) and we find

$$(50)\qquad p(x) = \frac{\sqrt{\mathrm{N}}}{\sqrt{\mathrm{N}}+1}\,e^{-\sqrt{\frac{3}{\mathrm{N}}}\frac{x}{\lambda}} \qquad (\text{for } a = \infty)\,.$$

This formula should be compared to (34) with which it becomes identical (if we put $a = x$) in the limiting case indicated at the end of Section 3. The difference consists in the factor $\sqrt{\mathrm{N}}/(\sqrt{\mathrm{N}}+1)$ which represents the effect of the perturbations due to the boundary in the differential equation (24).

The knowledge of $p(x)$ permits us to solve the problems related to the albedo of the paraffin in the three-dimensional case with ease. For the particular case of paraffin with infinite thickness and for the limiting case of very large N, the problem of the albedo has also been treated by Wick [4] by a procedure which in some cases enables us to find the exact result, in other cases furnishes a method of approximation whose precision can in principle be made arbitrarily great. The results which we shall give here are only approximate, but correspond to more general cases than those treated by Wick; as we shall see, the approximation is in general quite good.

In contrast to the one-dimensional case where the albedo is uniquely defined, the latter obviously depends upon the angle of incidence of the neutrons in the three-dimensional situation, being the smaller the more normal the incidence. In order that the albedo may be assigned a unique meaning, it will therefore be necessary to specify the angle of incidence, or else to give the angular distribution law of incident neutrons, or finally to specify the way in which one intends to perform a measurement of the reflection coefficient. We shall define in Section 7 what we mean by "experimental albedo", which is the value of the reflection coefficient obtained when one measures it under particularly convenient conditions (which will also be stated later). The importance of the experimental albedo derives from the fact that from measurements of the albedo one obtains that quantity directly, and hence all comparisons between theory and experiment are based upon these measurements. The theory of the experimental albedo will be given in Section 7.

We shall now give the formulae for calculating the albedo for neutrons as a function of the angle of incidence ϑ. The albedo of a neutron incident upon a slab of paraffin of thickness a with angle of incidence ϑ may evidently be written as integral of the products of the probability

$$\frac{N-1}{N} e^{-\frac{x}{\lambda\cos\vartheta}} \frac{dx}{\lambda\cos\vartheta}$$

that the neutron, upon having entered the paraffin, suffers a first collision at a depth between x and $x+dx$ and does not get captured there, multiplied by the probability that the neutron, having suffered this collision, leaves the paraffin across the face through which it entered. This latter probability is given by $p(x)$. The desired albedo is therefore:

$$\beta(a, \vartheta) = \frac{N-1}{N} \int_0^a e^{-\frac{x}{\lambda\cos\vartheta}} \frac{p(x)}{\lambda\cos\vartheta} dx. \tag{51}$$

Substituting for $p(x)$ the expression (49), the integration may be performed without difficulty. For the sake of brevity we shall only write the explicit result for the case where the thickness a of the paraffin is infinitely large, so that $p(x)$ is given by (50). In that case, we find upon integrating (51)

$$\beta(\infty, \vartheta) = \frac{\sqrt{N}-1}{\sqrt{N}+\sqrt{3}\cos\vartheta} = 1 - \frac{1+\sqrt{3}\cos\vartheta}{\sqrt{N}} \tag{52}$$

where the last expression holds for N sufficiently large.

From (52) one can obtain the expression for the albedo for any given angular distribution law of the incident neutrons. Thus, for instance, if the incident neutrons obey the cosine law, the probability that an incident neutron makes an angle with the normal between ϑ and $\vartheta + d\vartheta$ is $2\cos\vartheta \sin\vartheta d\vartheta$. Hence when the cosine law holds, the albedo turns out to be

$$(53)\qquad \begin{cases} \beta(\infty) = \displaystyle\int_0^{\pi/2} \frac{\sqrt{N}-1}{\sqrt{N}+\sqrt{3}\cos\vartheta}\, 2\cos\vartheta \sin\vartheta d\vartheta \\[2ex] \quad = \dfrac{2}{\sqrt{3}}(\sqrt{N}-1)\left[1-\sqrt{\dfrac{N}{3}}\ln\left(1+\sqrt{\dfrac{3}{N}}\right)\right] \\[2ex] \quad = 1 - \dfrac{1+\dfrac{2}{\sqrt{3}}}{\sqrt{N}} + \cdots. \end{cases}$$

To examine the degree of approximation of the method one may note that for the case of the cosine law, Wick (loc. cit.) gives the exact asymptotic expression for the albedo and finds

$$(54)\qquad \beta(\infty) = 1 - \frac{4}{\sqrt{3}}\frac{1}{\sqrt{N}} + \cdots.$$

The exact value of the coefficient of $1/\sqrt{N}$ is therefore $4/\sqrt{3} = 2.31$, while we get from (53) that this coefficient is $1 + 2/\sqrt{3} = 2.15$. The difference is only seven percent.

In a similar manner one can calculate the transmission coefficient across a layer of paraffin of thickness a for thermal neutrons incident with angle ϑ. In a manner analogous to that used to derive (51) one finds that the coefficient is given by

$$(55)\qquad \gamma(a, \vartheta) = e^{-\frac{a}{\lambda\cos\vartheta}} + \frac{N-1}{N\lambda\cos\vartheta}\int_0^a e^{-\frac{x}{\lambda\cos\vartheta}}\, p(a-x)\, dx.$$

Substituting (49) for $p(x)$ the integration may be carried out without difficulty. We shall give the explicit result only for the limiting case $N = \infty$:

$$(56)\qquad \gamma_\infty(a, \vartheta) = \frac{\lambda}{a + \dfrac{2\lambda}{\sqrt{3}}}\left[\cos\vartheta + \frac{1}{\sqrt{3}} - \left(\cos\vartheta - \frac{1}{\sqrt{3}}\right) e^{-\frac{a}{\lambda\cos\vartheta}}\right].$$

6. *Angular distribution of thermal neutrons emerging from paraffin.* – From the surface of a block of paraffin containing a neutron source, there continuously emerge thermal neutrons toward the outside. We propose to determine approximately the angular distribution law, the knowledge of which is necessary for a reduction of the observations of absorption coefficients, and for numerous other problems. (cf. *AF*). A simple method to find this law is as follows.

We have proved in Section 4 that in the one-dimensional case, the neutron density in the vicinity of the boundary $x = 0$ of the paraffin behaves as given by expression (46). This result can also be approximately extended to the three-dimensional case, provided that λ is replaced by λ' as given by (48). We thus conclude that the thermal neutron density $n(x)$ in the vicinity of the plane $x = 0$ bounding a block of paraffin to the left behaves as

$$n(x) = \text{const.} \left(\frac{\lambda}{\sqrt{3}} + x\right). \tag{57}$$

This law can also be stated by saying that if one extrapolates toward the outside of the paraffin (i.e. toward negative values of x) the neutron density which exists inside the paraffin, one finds that it vanishes at a distance from the edge of the paraffin of $\lambda/\sqrt{3}$. For an experimental verification of this property see *AF*, Section 11.

Since the frequency of collisions suffered by the thermal neutrons in each volume element is evidently proportional to $n(x)$, we find that the number of elastic collisions of thermal neutrons in unit time within the volume $d\tau$ is

$$\mu\left(\frac{\lambda}{\sqrt{3}} + x\right) d\tau$$

μ being a constant. The probability that a neutron having suffered one of these collisions escapes from the paraffin across the plane $x = 0$, making an angle with the normal between ϑ and $\vartheta + d\vartheta$ and without making any further collisions, is evidently

$$e^{-\frac{x}{\lambda \cos\vartheta}} \frac{\sin\vartheta \, d\vartheta}{2}.$$

Hence one finds that the number of neutrons which escape from an element of area S of the surface bounding the paraffin with a direction making an angle between ϑ and $\vartheta + d\vartheta$ and having suffered its last collision at a position of abscissa between x and $x + dx$ is

$$\mu\left(\frac{\lambda}{\sqrt{3}} + x\right) \mathrm{S}\, dx\, e^{-\frac{x}{\lambda \cos\vartheta}} \frac{\sin\vartheta \, d\vartheta}{2}.$$

This expression, when integrated with respect to x from 0 to ∞ yields the total number of thermal neutrons issuing from the surface S at an angle between ϑ and $\vartheta + d\vartheta$; it is

$$\frac{\lambda^2 \mathrm{S}\mu}{2\sqrt{3}} (\cos\vartheta + \sqrt{3}\cos^2\vartheta) \sin\vartheta \, d\vartheta;$$

we conclude therefore that the number of thermal neutrons emerging from the paraffin at an angle contained in the element of solid angle $d\omega$ is proportional to:

$$(\cos\vartheta + \sqrt{3}\cos^2\vartheta)\, d\omega. \tag{58}$$

The angular distribution law for thermal neutrons escaping from paraffin has been numerically calculated by Wick; his results agree in a rather satisfactory manner with (58).

7. *Experimental albedo.* – The method used experimentally for the determination of the albedo (cf. *AF*) is essentially the following. One has a detector R of slow neutrons, consisting of a thin plane foil capable of being activated by the absorption of thermal neutrons only. We shall assume that the detector is very thin, not only in the geometrical sense, but also in the sense that the probability of a thermal neutron being captured upon traversing it (a probability determined by the product of the thickness and the absorption coefficient for thermal neutrons) is sufficiently small. In addition, one has a foil capable of totally absorbing all thermal neutrons which traverse it.

We shall carry out two measurements of the activity induced in the detector by placing it once (case (*a*)) at a given position inside a block of paraffin of large size containing the source, and once (case (*b*)) by placing it again in the same position but covering one side of it by an absorbing foil. Let A and B stand for the activities measured in cases (*a*) and (*b*); we propose to show that the ratio A/B determines the albedo.

Let $\mathfrak{N}$ be the number of thermal neutrons which cross the detector in case (*b*); in case (*a*) the number of thermal neutrons going through the detector at least once (counting a neutron which has crossed the detector several times only once) equals $2\,\mathfrak{N}$, since the neutrons which strike the detector in case (*a*) can originate on either side of it. Moreover we can show that a neutron which traverses the detector once will traverse it on the average $1/1-\beta$ times, β being the albedo. In fact the probability that a neutron after having gone through the detector once, will go through it a second time is evidently equal to the albedo; hence the average number of crossings is exactly

$$1 + \beta + \beta^2 + \cdots = \frac{1}{1-\beta} \cdot$$

Then finally, the number of traversal in case (*a*) turns out to be

$$\frac{2\,\mathfrak{N}}{1-\beta} \cdot$$

Assuming that the activities A and B are proportional to the number of traversals which occur in the two cases (*a*) and (*b*), one thus finds

$$\frac{A}{B} = \frac{2}{1-\beta} \tag{59}$$

and hence from a measurement of the ratio A/B one can obtain the albedo β. We have already observed several times that the concept of the albedo is not uniquely defined, since the value of the reflection coefficient depends upon the angular distribution of the incident neutrons. Thus, for instance, the albedo for neutrons incident distributed according to a cosine law would be different from the albedo for normally incident neutrons, or incident with a distribution law like (58). We can therefore appreciate that relation (59) specifies but one possible definition of the abedo. We shall call the albedo defined by relation (59) the " experimental albedo " since its value is the one found directly from measurements. From (59) one finds for the expression of the experimental albedo

$$\beta_{\exp} = 1 - \frac{2\,B}{A}. \tag{60}$$

For a discussion of the details of the actual measurement of the experimental albedo we refer to *AF*. We find there that, since the albedo resulting from the measurements is the experimental albedo, it is important to establish the relation which expresses this quantity in terms of the mean number of collisions N which a neutron suffers before being captured by the protons of the paraffin.

We shall derive this relation for the limiting case of practical interest, namely for N very large.

Let q be the number of thermal neutrons which are produced per second and per cubic centimeter in the region of the detector R. In the case (*a*), since R is by hypothesis very thin, the thermal neutron distribution around it will not be disturbed by its presence. One notes then that the average total path length of each thermal neutron in the paraffin is evidently $N\lambda$. Hence the total distance covered by the thermal neutrons produced per unit time and volume is $qN\lambda$. We conclude therefore that the sum of the path lengths travelled by the thermal neutrons in unit time across our detector (having an area S, thickness δ and hence volume $S\delta$) will be $S\delta qN\lambda$. Denoting by K the absorption coefficient for thermal neutrons in the material of which the detector is made, and keeping in mind that the detector is assumed to be very thin (i.e. $K\delta \ll 1$), we conclude that the number of thermal neutrons captured per unit time in the detector is $K\delta SqN\lambda$. The activity A will evidently be proportional to this number; it may therefore be written

$$A = \eta K\delta\, SqN\lambda, \tag{61}$$

η being the constant of proportionality.

To calculate B, we note that when one side of the detector is covered by a foil which completely absorbs the thermal neutrons, the latter can reach the detector only if they originate from the uncovered side of R. Furthermore, any neutron cannot traverse R more than once, since after one crossing of R it will be captured by the absorber. The number of neutrons striking the absorber is found by noting that in front of the detector (of area S) and at a distance from it between x and $x + dx$ there are produced S $q\,dx$ thermal neutrons per second. The probability that one of these reach the detector is given in our case (N very large) by (34) in which one must write x instead of a, i.e.

$$p(x) = e^{-\sqrt{\frac{3}{\lambda^2 N}}\,x}.$$

The number of neutrons striking the detector is therefore

$$Sq\int_0^\infty e^{-\sqrt{\frac{3}{N}}\frac{x}{\lambda}}\,dx = Sq\lambda\sqrt{\frac{N}{3}}.$$

To calculate how many of these neutrons will be captured in R, one must know the angular distribution law. We shall assume that this is given by (58). In fact, it is evident that in case (*b*) the presence of a layer which totally absorbs thermal neutrons, will modify the neutron distribution in its neighborhood in such a way as to make it similar to the distribution found near the boundary of a paraffin block. On the basis of (58) an elementary calculation shows that the mean value of the distance travelled by a thermal neutron which strikes the detector (of thickness δ) through this detector is

$$\sqrt{3}\,\delta$$

and hence the total number of neutrons absorbed in the detector will be $K\delta\, Sq\, \lambda \sqrt{N}$. The activity B will be proportional to this number, with the same constant η as in (61). We thus have

$$B = \eta K \delta S q \lambda \sqrt{N}\,. \tag{62}$$

From (60), using (61) and (62), we then obtain the value of the experimental albedo:

$$\beta_{exp} = 1 - \frac{2}{\sqrt{N}}\,. \tag{63}$$

This formula coincides with the result found by Wick (*loc. cit.*) by a method of numerical approximation.

Measurements carried out in paraffin have yielded $\beta = 0.82$ for the value of the experimental albedo (cf. *AF*). From this datum one can find a value of N from (63); one finds $N = 123$. If we had taken into account the fact that even for thermal neutrons there exists a slight correlation between the directions of successive free paths, the value of N would have turned out to be a little larger.

8. *Logarithmic width of groups*. – The groups of slow neutrons, with the exception of group *C*, which is essentially made up of thermal neutrons, correspond–according to our present knowledge–to relatively narrow energy bands. They may be characterized by the values w_{min} and w_{max} of the energies which bound the energy band. We shall here discuss the principles upon which is based the method for the determination of the quantity

$$\ln \frac{w_{max}}{w_{min}}$$

which we shall call the *logarithmic width* of the group. The description of the effective performance of the measurement and of the various corrections which must be applied are to be found in *AF*.

One notes that the logarithmic width is of immediate physical significance because of what we have said in the first part of Section 2; this width represents in fact the average number of collisions suffered by a neutron while belonging to the group in question. This is of course only true for groups having energies above one volt, so that the premises stated in Sections

1 and 2 hold true. In all the considerations which follow we shall limit ourselves to these cases.

The procedure for the determination of the logarithmic width is based on the comparison between the activity induced in a detector G of the group g under study, and the activity induced in a detector R of the thermal group, having an area equal to G and arranged as in case (*b*) of the preceding section, with a total absorber of the thermal group on one side. We shall assume for simplicity that both detectors are sufficiently thin in the same sense as specified in the previous section. Otherwise it would be necessary to introduce corrections to take into account the absorption of electrons and neutrons in the detector itself.

In order to understand the principle of the method, we shall suppose at first that in an unlimited block of paraffin there are produced every where q fast neutrons per unit volume and unit time. These neutrons, as a result of collisions, will be successively slowed down until they reach thermal velocities and will finally be captured by the protons present in the paraffin. The activity B induced in the detector R by thermal neutrons, when the latter is protected by a total absorber for thermal neutrons on one side, will be given by (62) as in case (*b*) of the preceding section. In order to determine also the activity in the detector G of group g, we note that according to (6) in this case there is a neutron density $n(v)\,dv$ of neutrons having velocities between v and $v + dv$ and belonging to group g, given by

$$n(v)\,dv = \frac{2\,q\lambda_g}{v^2}\,dv\,,$$

λ_g being the mean free path of the neutrons in that group. From this one immediately calculates the number of neutrons with velocities between v and $v + dv$ which traverse the detector G per unit time, making an angle with the normal lying between ϑ and $\vartheta + d\vartheta$. This number is
i

$$2\,q\lambda_g\,\mathrm{S}\,\frac{dv}{v}\sin\vartheta\cos\vartheta\,d\vartheta\,;$$

f δ is the thickness of the detector and K_g its absorption coefficient for neutrons of group g, the fraction of neutrons absorbed (remembering that by hypothesis $\mathrm{K}_g\delta \ll 1$) will be $\mathrm{K}_g\,\delta/\cos\vartheta$; hence the total number of g–neutrons absorbed is

$$2\,q\mathrm{S}\,\lambda_g\,\mathrm{K}_g\,\delta\int_{v_{\min}}^{v_{\max}}\frac{dv}{v}\int_0^{\pi/2}\sin\vartheta\,d\vartheta = 2\,q\mathrm{S}\,\lambda_g\,\mathrm{K}_g\,\delta\ln\frac{v_{\max}}{v_{\min}}$$

where $v_{\min}$ and $v_{\max}$ are the velocities corresponding to the energies $w_{\min}$ and $w_{\max}$, so that $\ln v_{\max}/v_{\min} = 1/2\,\ln w_{\max}/w_{\min}$. The activity induced in detector G is proportional to the number of neutrons absorbed per unit time. If η_{G} is the constant of proportionality, this activity is

$$\mathrm{A}_g = \eta_{\mathrm{G}}\,q\mathrm{S}\,\lambda_g\,\mathrm{K}_g\,\delta\ln\frac{w_{\max}}{w_{\min}}\,. \tag{64}$$

In experimental practice one usually uses the same object, with and without appropriate absorbing filters, for detector R of thermal neutrons as well as for detector G of the group g. Under these conditions we evidently have $\eta_G = \eta$. Hence if it were possible to set up conditions where the number q of thermal neutrons produced in unit volume and unit time is constant over a large mass of paraffin, we would obtain from (62) and (64).

$$\ln \frac{w''}{w'} = \frac{K\lambda \sqrt{N}\, A_g}{\lambda_g\, K_g\, B}\,. \tag{65}$$

In practice the determination of $\ln (w_{max}/w_{min})$ cannot be achieved directly by means of this formula, since it is not easy to produce a uniform distribution of fast neutrons since they usually originate from practically a point source. Since the distance to which thermal neutrons diffuse is larger than that through which neutrons from the other groups travel, one would find that the ratio A_g/B is not a constant, but decreases with increasing distance from the source. A simple way to avoid this inconvenience is to integrate equations (62) and (64) over the entire volume (and, if necessary, over all orientations of the detector). In this way one finds a result which evidently no longer depends on the uniformity of q, but only upon $Q = \int q\, d\tau$, i.e. upon the total number of fast neutrons produced per second. One finds precisely, again setting $\eta_G = \eta$:

$$\left\{ \begin{aligned} &\int B\, d\tau = \eta K\, \delta\, SQ\, \lambda \sqrt{N} \\ &\int A_g\, d\tau = \eta K_g\, \delta\, SQ\, \lambda_g \ln \frac{w_{max}}{w_{min}} \end{aligned} \right. \tag{66}$$

from which one obtains

$$\ln \frac{w_{max}}{w_{min}} = \frac{K\lambda \sqrt{N}}{\lambda_g\, K_g} \frac{\int A_g\, d\tau}{\int B\, d\tau}\,. \tag{67}$$

This formula, corrected to take into account the absorption in the detector, has been used in AF for the calculation of the logarithmic widhts.

PART II

The collision mechanism between neutrons and hydrogen atoms.

This part will be devoted to the quantum-mechanical study of the collision between slow neutrons and the protons in paraffin. In particular, we shall examine the mechanism of these collisions, taking into account the fact that the hydrogen atoms are not free, but may be considered to be bound elastically in certain equilibrium positions; we shall also discuss the capture process of a neutron by a proton which leads to the formation of a deuterium nucleus.

9. *Collisions between slow neutrons and protons, neglecting the effect of the chemical bond.* – It is evident that as long as the energy of the neutron is large compared to the energy of the chemical bond of hydrogen in paraffin, one can neglect the effect of this bond in the study of the collisions and consider the proton to be free; more precisely, we have already remarked that this procedure will be justified as long as the energy w of the neutrons is $\gg h\nu$, i.e. much larger than the quantum of frequency ν of the elastic bond of the hydrogen atoms. Since $h\nu$ is at most of the order of 0.5 volts, we can call the protons practically free when the neutrons have energies above 1 volt. In this case the problem of collision between neutron and proton is quite simple, since by referring to the relative coordinates of the two particles one is immediately led back to the problem of the collision of one particle with a fixed scattering center, a problem which is treated by standard methods of the quantum theory of collisions. In the calculation of the relative motion one must of course take M/2 as the mass, M being the mass value common to neutron and proton.

The details of the collision theory obviously depend, even in this simple case, upon the assumptions made concerning the force law between neutron and proton. One must further take into account the two possible relative spin orientations (parallel or antiparallel) of neutron and proton.

We shall suppose that the neutron-proton force, as is generally done [5], has a range ρ somewhat below 10^{-12} cm. This force might be derived from an ordinary type of potential, or might represent an exchange interaction according to the hypotheses of Heisenberg and Majorana.

It is well known that the deuteron has a spin of unity; it seems therefore that its fundamental configuration must be classified (in the notation of atomic spectroscopy) as a ^3S, whose spin is the sum of the neutron and proton spins. The binding energy of this term is 2.2×10^6 volts. In addition to the fundamental term ^3S one may also consider a singlet term ^1S in which the spins of proton and neutron are antiparallel. If the neutron-proton force did not depend upon the spin orientation, these two terms should coincide. If however there exists a difference between the forces in the two cases, the ^1S term should be situated above the ground term ^3S; it will be real or virtual according to whether its binding energy is positive or negative.

Under the assumption that the range of the neutron-proton force is very small compared with the de Broglie wavelength, the problem of the collision may be treated by the simple procedure used by Bethe and Peierls (*loc. cit.*). One then arrives at the following expression for the collision cross section [6]:

$$\sigma = \frac{h^2}{\pi M}\left[\frac{3}{4}\frac{1}{|w_0| + \frac{1}{2}w} + \frac{1}{4}\frac{1}{|w_1| + \frac{1}{2}w}\right] \tag{68}$$

where w_0 and w_1 stand for the energies of the two terms ^3S and ^1S and w is the neutron energy. When w is very small (small meaning as compared to

w_0 and w_1, while still being greater than 1 volt) there results from (68) a collision cross section independent of w

$$\sigma_1 = \frac{h^2}{\pi M}\left[\frac{3}{4}\frac{1}{|w_0|} + \frac{1}{4}\frac{1}{|w_1|}\right]. \tag{69}$$

It is found experimentally (cf. *AF*) that at least for two neutron groups having energies above 1 volt, the mean free path in paraffin is approximately the same, namely around

$$\lambda_g = 1 \text{ cm}. \tag{70}$$

This value of the mean free path corresponds to a collision cross section $\sigma_1 = 12.8 \times 10^{-24}$ cm². Assuming this value and taking $w_0 = -2{,}2 \times 10^6$ volts, one finds from (69)

$$|w_1| = 1.86 \; 10^{-7} \text{ ergs} = 1.16 \; 10^5 \text{ volts}.$$

The relatively large collision cross section of slow neutrons for hydrogen therefore leads, according to this theory, to the conclusion that the ^1S term, real or virtual, has a comparatively small binding energy in absolute value.

Inserting into (68) this value for w_1 and evaluating the collision cross section for fast neutrons, one finds values which are too small; this is probably due to the inaccuracy of the method of calculation of Bethe and Peierls for neutrons having energies of several million volts. The value of the collision cross section derived from measurements on the mean free path performed by Goldhaber (*loc. cit.*) for neutrons of about 200,000 volts, produced by the nuclear photoelectric effect in deuterium irradiated by gamma-rays from ThC, is on the other hand smaller than what is found from (68). The precision of the measurements is however not sufficient to draw definite conclusions.

10. *Scattering of neutrons with bound hydrogen atoms.* – We have noted several times that if the energy of the neutrons is below 1 volt, it is no longer legitimate to neglect the fact that the hydrogen atoms are bound chemically in the paraffin. We now propose to study the collision between a neutron and a hydrogen nucleus, taking into account the chemical forces which tend to maintain the proton in the vicinity of an equilibrium position.

The problem can be discussed with the following criteria of approximation. Let ρ be the range of the neutron-proton forces, a the order of magnitude of the scattering length between neutron and proton, λ the order of magnitude of the de Broglie wavelength corresponding to the momenta of neutron and proton. The approximations which we intend to use presently are legitimate if we can choose a length R which simultaneously satisfies the inequalities

$$R \ll \lambda \quad ; \quad R \gg \rho \quad ; \quad R \gg a. \tag{71}$$

These conditions are fairly well satisfied in the case of collision between slow neutrons and protons in paraffin. Let us indicate by x, y, z the coordinates of the neutron and by X, Y, Z the coordinates of the proton, and let us neglect the effect of the spins of the two particles, which is of no importance for the effects we are here interested in. The Hamiltonian of the system can be written as the sum of the kinetic energies of the two particles, the potential energy U (X, Y, Z) of the chemical forces exerted on the hydrogen atom, and the interaction energy between neutron and proton. We shall represent the latter as a potential energy $g(r)$ which is a function of the separation r between neutron and proton. Even if the interaction were of the Heisenberg or Majorana type, the calculation could be developed by essentially the same procedure.

The Schroedinger equation of the system may be written:

$$(72) \qquad -\frac{h}{2\pi i}\dot{\psi} = -\frac{h^2}{8\pi^2 M}\left(\frac{\partial^2\psi}{\partial x^2}+\frac{\partial^2\psi}{\partial y^2}+\frac{\partial^2\psi}{\partial z^2}+\frac{\partial^2\psi}{\partial X^2}+\frac{\partial^2\psi}{\partial Y^2}+\frac{\partial^2\psi}{\partial Z^2}\right) + U\psi + g(r)\,\psi.$$

The field scalar $\psi(x, y, z, Y, X, Z)$ will evidently be a function with regular behavior (in the sense that its variations will be small for changes of the arguments small compared to λ) everywhere except in regions where r is small. It may easily be verified, keeping in mind the inequalities (71), that the irregularities will occur for $r < R$ (see also the discussion on fig. 1). Let us introduce a new function $\bar{\psi}$ which is obtained from ψ by taking the following average:

$$(73) \qquad \bar{\psi}(x, y, z, X, Y, Z) = \frac{1}{\frac{4}{3}\pi R^3}\iiint \psi(\xi, \eta, \rho, X, Y, Z)\, d\xi\, d\eta\, d\rho$$

where the integral is extended over a sphere of radius R with its center at the point x, y, z. The function $\bar{\psi}$ will evidently have regular behavior even for small values of r.

We propose to establish an equation for $\bar{\psi}$ analogous to the Schroedinger equation. For this purpose we shall take the average of both sides of (72) over the volume of a sphere of radius R with center at x, y, z. Denoting this operation by a bar as in (73), we obtain after obvious transformations,

$$(74) \qquad -\frac{h}{2\pi i}\dot{\bar{\psi}} = -\frac{h^2}{8\pi^2 M}\left(\frac{\partial^2\bar{\psi}}{\partial x^2}+\cdots+\frac{\partial^2\bar{\psi}}{\partial X^2}+\cdots\right) + U\bar{\psi} + \overline{g(r)\,\psi}.$$

In order to calculate the last term we note that in the singular region $r < R$, the variations of ψ are essentially due to the effect of the potential energy $g(r)$ which, while being different from zero only in a rather small region, takes on very large values in that region. For $r < R$ one may also neglect the variations in ψ due to the other terms of the Hamiltonian, writing ψ in the form

$$(75) \qquad \psi = c\chi(r)$$

where the function $\chi(r)$ is the solution of the equation

$$(76) \qquad \chi''(r) + \frac{2}{r}\chi'(r) = \frac{4\pi^2 M}{h^2}\, g(r)\,\chi(r),$$

i.e. a solution of the Schroedinger equation (only for S–waves, since these are the only important ones at short distances) for the relative motion of two particles (coordinate r, reduced mass M/2) and considering only the potential energy $g(r)$; c represents a smoothly varying function.

If we normalize $\chi(r)$ so that for large r it approaches the value unity, the connection between (75) and the values of ψ for $r \geq R$ is evidently obtained by identifying the function c with the slowly varying function $\overline{\psi}$. We may therefore write in general:

$$\psi(x, X) = \overline{\psi}(x, X)\,\chi(r). \tag{77}$$

By setting $\chi(r) = \frac{1}{r} v(r)$, Eq. (76) becomes

$$v''(r) = \frac{4\pi^2 M}{h^2} g(r)\, v(r). \tag{78}$$

The behavior of $v(r)$ is represented in Figure 1. For $r > \rho$ the graph

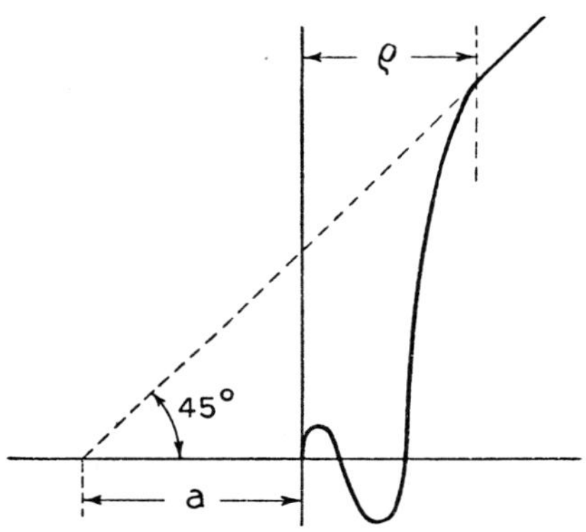

Fig. 1.

of $v(r)$ is a straight line. Because of the normalization condition the slope of the asymptote to the curve $v(r)$ must become 45 degrees, so that we obtain

$$v(r) = a + r \qquad \text{for } r > \rho. \tag{79}$$

The significance of the constant a is clear from the figure.

Keeping (71) in mind and neglecting ρ in comparison to R, we recognize that $\overline{g(r)\psi}$ will be different from zero if the distance between the points x, y, z and X, Y, Z is less than R. If this condition holds, one finds (remembering that $\rho \ll R$, and that $\overline{\psi}$ does not vary appreciably over regions of the dimensions of R)

$$\overline{g(r)\psi} = \frac{1}{\frac{4}{3}\pi R^3}\int g(r)\,\overline{\psi}(x, X)\,\chi(r)\,d\tau$$

$$= \frac{4\pi\overline{\psi}(x, X)}{\frac{4}{3}\pi R^3}\int_0^\infty g(r)\,v(r)\,r dr.$$

The integral is easily calculated with the aid of (78) and (79) and the fact that $v(0) = 0$ (see fig. 1). We find

$$\int_0^\infty g(r)\, v(r)\, r dr = \frac{h^2}{2\pi^2 M} \int_0^\infty v''(r)\, r dr = \frac{h^2}{4\pi^2 M} \Big| v' r - v \Big|_0^\infty = -\frac{h^2 a}{4\pi M}.$$

Finally we have

$$\overline{g(r)\psi} = -\frac{h^2 a}{\pi M} \overline{\psi}(x, X)\, \delta_R(r)$$

where the function $\delta_R(r)$ is equal to zero if $r > R$, and is equal to $3/(4\pi R^3)$ everywhere else (so that its volume integral extended over all space is equal to 1). Substituting into (74) we obtain the desired equation for $\overline{\psi}$:

$$(80) \qquad -\frac{h}{2\pi i} \dot{\overline{\psi}} = -\frac{h^2}{8\pi^2 M} \left(\frac{\partial^2 \overline{\psi}}{\partial x^2} + \cdots + \frac{\partial^2 \overline{\psi}}{\partial X^2} + \cdots \right) + U(X, Y, Z)\, \overline{\psi} - \frac{h^2 a}{\pi M} \delta_R(v)\, \overline{\psi}.$$

The collision between neutron and proton is determined by the presence of the last term which represents the interaction between the two particles. It is easy to convince ourselves that with the aid of equation (80) the collision may be calculated by means of the first Born approximation (indeed, the interaction term gives rise to only a small irregularity in $\overline{\psi}$). Furthermore, when calculating the matrix elements of the interaction term, the function δ_R may be identified with the Dirac δ–function in three dimensions, since the quantities contained in (80) vary slowly over the region in which δ_R is different from zero.

Applying the classical procedure of Born one thus finds without difficulty the following expression for the differential collision cross section $\sigma_{m,n,d\omega}$ for a process of collision in which a hydrogen atom passes from the initial state m to the final state n, the neutron being scattered in a direction lying within the element of solid angle $d\omega$:

$$(81) \qquad \sigma_{m,n,d\omega} = 4a^2 \frac{p}{p_0} d\omega \left| \int u_m^* u_n e^{\frac{2\pi i}{h} (\mathbf{p} - \mathbf{p}_0) \cdot \mathbf{X}} d\tau \right|^2$$

where $\mathbf{p}_0$ stands for the neutron momentum before the collision, $u_m(X, Y, Z)$ and $u_n(X, Y, Z)$ are the eigenfunctions describing the state of the proton before and after the collision with corresponding eigenvalues of the energy w_m and w_n; $\mathbf{p}$ is the neutron momentum after the collision, whose magnitude according to the principle of the conservation of energy, is given by

$$(82) \qquad \frac{p^2}{2M} = \frac{p_0^2}{2M} + w_m - w_n$$

while its direction is that of the element $d\omega$.

Before proceeding to further applications, we shall discuss the expression (81) for two extreme cases:

(1) *Case in which the hydrogen atoms are strongly bound*. (More precisely, the case in which the region over which the eigenfunction u_m of the

initial proton state is different from zero, is small compared to the de Broglie wavelength of the neutron). In this case the exponential term in the integral occurring in (81) can be considered constant over the region where the factor u_m is different from zero. It then follows from the orthogonality of the proton eigenfunctions that the integral differs from zero only for $n = m$, in which case the square of its absolute value is equal to 1. We thus conclude that in the present case, the only collisions which occur are those in which the energy of the hydrogen atom does not change and hence, according to (82), the neutron retains its entire kinetic energy ($p = p_0$). The collision cross section for processes wherein the neutron is scattered into an element of solid angle $d\omega$ follows in this case from (81):

$$\sigma_{d\omega}\, d\omega = 4\, a^2\, d\omega \tag{83}$$

which shows that the scattering is isotropic. The total cross section is found by integrating the preceding expression over all the solid angle; we get

$$\sigma = 16\, \pi\, a^2 \tag{84}$$

(2) *Case in which the hydrogen atoms are free.* – In this case the eigenfunctions u_n and u_m of the proton can be characterized by their respective momenta $\mathbf{p}_m$ and $\mathbf{p}_n$; they are written in the usual exponential form. One can then easily derive that the integral in (81) is different from zero only if

$$\mathbf{p}_n + \mathbf{p} = \mathbf{p}_m + \mathbf{p}_0 \tag{85}$$

which expresses the law of conservation of momentum. If (85) is satisfied, the square of the modulus of the aforementioned integral is equal to 1. Furthermore (82) takes the form

$$p_n^2 + p^2 = p_m^2 + p_0^2. \tag{86}$$

Referring to the system in which the hydrogen atom is initially at rest, i.e. where one has $p_m = 0$, (85) and (86) take the form

$$\mathbf{p}_n + \mathbf{p} = \mathbf{p}_0 \quad ; \quad p_n^2 + p^2 = p_0^2. \tag{87}$$

From these relations one obtains by elementary considerations

$$p = p_0 \cos \vartheta \tag{88}$$

ϑ being the angle between p and p_0. Since p and p_0 are positive definite, it turns out that scattering occurs only through acute angles. From (81) we then obtain the differential scattering cross section

$$\sigma_{d\omega}\, d\omega = 4\, a^2 \cos \vartheta\, d\omega. \tag{89}$$

The total scattering cross section is obtained by integrating over the hemisphere $\cos \vartheta > 0$;

$$\sigma = \int_0^{\pi/2} 4\, a^2 \cos \vartheta\; 2\, \pi \sin \vartheta\, d\vartheta = 4\, \pi a^2. \tag{90}$$

A comparison of this equation with (84) shows that the scattering cross section for strongly bound hydrogen atoms is four times that for free hydrogen atoms. This implies that when the kinetic energy of the neutron decreases from values above 1 volt (for which the hydrogen atoms may practically be considered free) to much smaller values (for which on the other hand the hydrogen atoms must be considered strongly bound) the scattering cross section quadruples and hence the mean free path is reduced to one fourth. In the case of thermal neutrons, as will be better specified in the following section, we are sufficiently close to this limiting value, although it is not entirely reached as yet. One must in fact expect that the mean free path, although considerably smaller for thermal neutrons than for slow neutrons with energies above 1 volt, is nevertheless a little larger then one fourth of the latter value. Furthermore it follows from (83) that in the limit where the energy of the neutron is extremely small the scattering becomes isotropic; this result as well is only approximately true for thermal neutrons, since for these there exists a slight preponderance of small angle scattering, as will be shown in the following section.

11. *Elastically bound hydrogen atoms.* – In order to study the situation intermediate between the two limiting cases discussed above, we shall restrict our hypotheses by supposing that the hydrogen atom is bound to an equilibrium position (the origin of the coordinates) by an isotropic elastic force. In this case the expressions for the different scattering cross sections may be obtained in closed form, as we shall see. We shall further assume that the state of oscillation in which the hydrogen atom is found initially is the one of minimum energy.

If ν is the frequency of oscillation of the hydrogen atom, the energy of its oscillation states (measuring for convenience from the energy of the ground state, which is taken to be zero) will be given by

$$w_n = nh\nu \,, \tag{91}$$

n being an integer. Upon decomposing the motion of the oscillator into its three components X, Y, Z, the total quantum number n will be given by the three partial quantum numbers:

$$n = n_1 + n_2 + n_3 \tag{92}$$

which completely define a state of oscillation. The initial state m has its quantum numbers equal to zero by definition.

The function corresponding to the quantum numbers n_1, n_2, n_3 is the product of the three oscillator eigenfunctions in one dimension written for the three coordinates X, Y, Z, namely

$$u_{n_1 n_2 n_3} = \left(\frac{4\pi m\nu}{h}\right)^{3/4} \frac{H_{n_1}(\xi)\, H_{n_2}(\eta)\, H_{n_3}(\rho)}{\sqrt{2^n\, n_1!\, n_2!\, n_3!}}\, e^{-\frac{1}{2}(\xi^2+\eta^2+\zeta^2)} \tag{93}$$

where we have set

$$\xi = X\sqrt{\frac{4\pi^2 m\nu}{h}} \quad ; \quad \eta = Y\sqrt{\frac{4\pi^2 m\nu}{h}} \quad ; \quad \zeta = Z\sqrt{\frac{4\pi^2 m\nu}{h}} \tag{94}$$

and the H stands for the Hermite polynomials. The eigenfunction of the initial state is simply

$$u_{000} = \left(\frac{4\pi m\nu}{h}\right)^{3/4} e^{-\frac{1}{2}(\xi^2+\eta^2+\zeta^2)}.$$

The volume integral occurring on the right hand side of (81) may be reduced to the product of three simple integrals of the type:

$$\int_{-\infty}^{+\infty} H_s(x)\, e^{-x^2} e^{i Q_s n}\, dx = i\sqrt{\pi}\, Q_s\, e^{-Q_s^2/4}. \tag{95}$$

By means of this formula we obtain from (81) by simple calculations

$$\sigma_{n_1 n_2 n_3} d\omega = 4a^2 \frac{P}{P_0} d\omega \frac{(P_x - P_{0x})^{2n_1}}{n_1!} \frac{(P_y - P_{0y})^{2n_2}}{n_2!} \frac{(P_z - P_{0z})^{2n_3}}{n_3!} e^{-(\mathbf{P}-\mathbf{P}_0)^2}. \tag{96}$$

We have put

$$\mathbf{P}_0 = \frac{p_0}{\sqrt{2M h\nu}} \quad ; \quad \mathbf{P} = \frac{p}{\sqrt{2M h\nu}} \tag{97}$$

so that P_0^2 and P^2 represent the energies of the neutron before and after the collision, divided by $h\nu$. Equation (82) becomes

$$P^2 = P_0^2 - n. \tag{98}$$

Upon summing (96) over all integral and non-negative solutions of (92) corresponding to a given value of n, we obtain the partial collision cross section for processes in which the neutron is scattered into $d\omega$ and the hydrogen atom is excited to one of the oscillation states of energy (91). The summation is easily carried out in closed form (by use of the formulae for the development of the nth power of a trinomial) and one finds

$$\sigma_{n,d\omega} = 4a^2 \frac{P}{P_0} d\omega \frac{(\mathbf{P}-\mathbf{P}_0)^{2n}}{n!} e^{-(\mathbf{P}-\mathbf{P}_0)^2}. \tag{99}$$

The latter may also be written, because of (98),

$$\sigma_{n,d\omega} = 4a^2 \frac{\sqrt{P_0^2 - n}}{P_0} \frac{\left(2P_0^2 - n - 2P_0\sqrt{P_0^2-n}\cos\vartheta\right)^n}{n!} e^{-\left(2P_0^2 - n - 2P_0\sqrt{P_0^2-n}\cos\vartheta\right)} d\omega, \tag{100}$$

ϑ being the angle between $d\omega$ and the direction of $\mathbf{P}_0$.

Equation (100) enables us to solve the various collision problems. Upon integrating it over all solid angles we obtain the value of the partial collision cross section σ_n for collisions in which the hydrogen atom is excited to the nth oscillation state. The latter turns out to be

$$\sigma_n = \frac{4\pi a^2}{n!\, P_0^2} \int_{2P_0^2 - n - 2P_0\sqrt{P_0^2-n}}^{2P_0^2 - n + 2P_0\sqrt{P_0^2-n}} \xi^n e^{-\xi}\, d\xi . \tag{101}$$

Of course σ_n must be set equal to zero if $P_0^2 < n$, i.e. if the incident neutron does not have sufficient energy to excite state n.

The shape of σ_n as a function of $P_0^2 = w/h\nu$ is shown in Figure 2 (curves 0, 1, 2, 3, ···). The curve traced in solid line is the total collision cross section, sum of the partial cross sections. The asymptotic curve common to the curves representing $\sigma_0, \sigma_1, \sigma_2, \cdots$ is, as may be seen from (101), the equilateral hyperbola

$$\frac{4\pi a^2}{w/h\nu} .$$

Note the behavior of the total collision cross section; for zero energy its value is $16\pi a^2$. For $w = h\nu$ it is already reduced to a value very slightly less than one fourth the latter; with a further increase of the energy the value remains around the limiting value $4\pi a^2$, with oscillations of ever

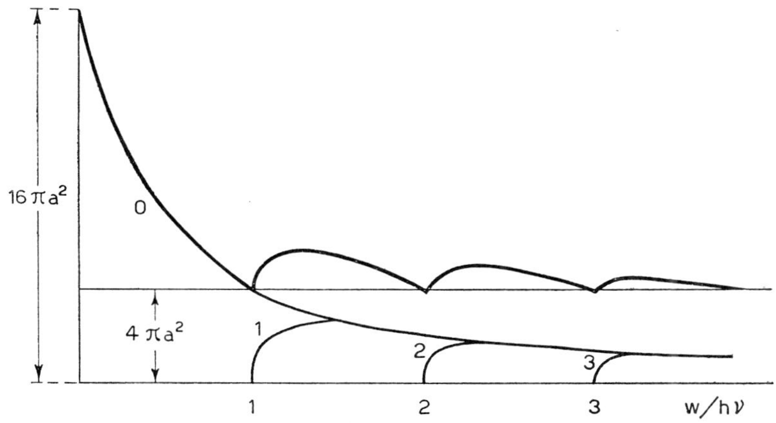

Fig. 2.

decreasing amplitude. These results agree with the general conclusions reached in the preceding section.

When the energy w is smaller than $h\nu$, the total collision cross section is reduced to just σ_0; from (101), recalling that $P_0^2 = w/h\nu$, this turns out to be

$$\sigma_0 = 4\pi a^2 \frac{1 - e^{-4w/h\nu}}{w/h\nu} . \tag{102}$$

The angular dependence is determined by (100), which reduces to

$$\sigma_0 \, d\omega = 4a^2 \, e^{-\frac{2w}{h\nu}(1-\cos\vartheta)} \, d\omega . \tag{103}$$

From this equation one recognizes that in conformity with the results of the last section, the scattering tends to approach isotropy for $w \ll h\nu$; however, for w just slightly less than $h\nu$ the anisotropy is considerable. For thermal neutrons ($w \sim 0.037$ volts) $w/h\nu$ is of the order of $1/10$. Nevertheless the anisotropy is not at all negligible as yet; likewise the cross section is really still somewhat smaller than four times the limiting cross section for $w \gg h\nu$, the ratio being closer to three than to four.

Finally we note that if the hydrogen atoms were elastically bound in stable equilibrium positions as we have assumed in the course of this discussion, as soon as the energy w of the neutron falls below $h\nu$ by successive collisions, it could not decrease further down to values of thermal agitation. The possibility of a further slowing down of the neutron, after its energy has fallen below the quantum of energy of elastic vibrations, depends upon the fact that the hydrogen atoms in the crystal lattice of a solid are not bound to fixed equilibrium positions, but to the surrounding atoms which are themselves in motion. As a consequence we know that the frequencies of oscillation form a continuous spectrum extending down to frequency zero, which fact makes the reduction of the energy down to thermal values possible. However, when the energy is already reduced to a value of the order of the mean quantum of elastic oscillation or below, the slowing down process is considerably less rapid than for higher energies, so that more than one collision takes place on the average to reduce the energy to $1/e$ of its value.

An evaluation of the number of collisions necessary to reduce the energy from $h\nu$ down to thermal energy (a reduction of roughly a factor of ten) shows that for this process approximately ten collisions may be needed, while if the slowing down obeyed the same law as the neutrons with energies above 1 volt, two or three collisions would suffice.

12. *The neutron capture process*. – We have mentioned several times that thermal neutrons in paraffin have a finite mean life τ since they may be captured by protons present in the vicinity, by uniting with them to form nuclei of heavy hydrogen. The order of magnitude of τ has been evaluated by direct experiment [7] to be $\tau \sim 10^{-4}$ seconds. τ can also be derived from a knowledge of the mean free path λ of thermal neutrons, and from the average number of collisions N which a thermal neutron suffers before capture. From this method (cf. *AF*, Sec. 13) one finds $\tau = 1.7 \times 10^{-4}$ s, in good agreement with the direct evaluation.

In this section we shall discuss the mechanism of the capture process of neutrons by protons [8], thus arriving theoretically at a determination of τ which is in excellent accord with the experimental determinations.

The theory which follows is based on the assumption that the transitions in which a slow neutron combines with a proton to form a deuteron (emitting the liberated energy in the form of a gamma ray quantum) are determined by the oscillations of the magnetic moment of the neutron-proton system. In fact, one finds that the ordinary transitions which are due to the oscillations of the electric dipole moment, are forbidden in the case of slow neutrons. We assume transitions from a ^1S state of the continuum to the ^3S ground state, because according to ordinary collision theory, slow neutrons can only be captured from S–states of the continuum, since in P, D, . . . states the two particles remain too far apart to interact effectively. Now it is known that the ordinary processes of radiation by electric dipole or quadrupole are forbidden for $S \to S$ transitions. There is instead, as we shall explain better below, the possibility of transition from a ^1S state of the continuum to the ^3S ground state of the deuteron through magnetic dipole radiation.

The formulae for the calculation of the radiation due to the oscillations of a magnetic dipole are essentially analogous (in view of the symmetry of the electric and magnetic fields in Maxwell's equations) to those which hold for electric dipole radiation. The probability for a transition between two states i and j is

$$\frac{64\,\pi^4\,\nu^3}{3\,h\,c^3}\left|\mu_{ij}\right|^2 \tag{104}$$

where μ_{ij} is the matrix element $(i\,,j)$ of the magnetic moment μ of the system and ν the frequency emitted in the transition.

Let us denote by the indices 0 , 1 the two terms ^3S (ground state of the deuteron, having an energy $w_0 = -2.2 \times 10^6$ volts) and ^1S (state of the continuum representing a slow neutron having an energy w which is very small compared to w_0). In order to calculate the matrix element μ_{01} one proceeds as follows.

The eigenfunction ψ of the neutron-proton system depends on the coordinates $x\,,y\,,z\,,s$ of position and spin of the neutron, and on the coordinates X , Y , Z , S, of position and spin of the proton. Referring to a system in which the center of mass of neutron and proton is at rest, the eigenfunctions of the S-states ^1S and ^3S will depend, in addition to the spin coordinates s and S, only on the distance r between neutron and proton. Since each of the two spin coordinates is capable of taking on only two values (+ 1 and —1), it follows that every eigenfunction will be an ensemble of four functions of the position coordinates only. We shall write them in the square array

$$\psi = \begin{vmatrix} \psi_{1,1} & \psi_{1,-1} \\ \psi_{-1,1} & \psi_{-1,-1} \end{vmatrix}$$

with the understanding that the first and second indices refer to the values of s and S, respectively. It is easily seen that in this notation the unique eigenfunction of a ^1S term is written in the form

$$\psi\,(^1S) = \begin{vmatrix} 0 & -\dfrac{v\,(r)}{\sqrt{2}\,r} \\ \dfrac{v\,(r)}{\sqrt{2}\,r} & 0 \end{vmatrix} \tag{105}$$

in which the function $v\,(r)$ satisfies the equation

$$\frac{d^2 v}{dv^2} + \frac{4\,\pi^2\,M}{h^2}\,[w - g_1\,(r)]\,v\,(r) = 0. \tag{106}$$

$g_1\,(r)$ represents the potential energy of the neutron-proton interaction when the two particles have antiparallel spins and M/2 is the reduced mass. In our case, where the ^1S state represents the slow neutron, the function $v\,(r)$, for values of r small compared to the de Broglie wavelength, is of the type shown in Fig. 1.

The three eigenfunctions of the ground state term ^{3}S will similarly be given by

$$(107)\qquad \psi\,(^3S_1) = \begin{vmatrix} \frac{u\,(r)}{r} & 0 \\ 0 & 0 \end{vmatrix} \; ; \; \psi\,(^3S_0) = \begin{vmatrix} 0 & \frac{u\,(r)}{\sqrt{2}\,r} \\ \frac{u\,(r)}{\sqrt{2}\,r} & 0 \end{vmatrix} \; ; \; \psi\,(^3S_{-1}) = \begin{vmatrix} 0 & 0 \\ 0 & \frac{u\,(r)}{r} \end{vmatrix}$$

where $u\,(r)$ satisfies the equation

$$(108)\qquad \frac{d^2 u}{dr^2} + \frac{4\pi^2 M}{h^2}\,[w_0 - g_3\,(r)]\,u\,(r) = 0.$$

$g_3\,(r)$ represents the potential energy of the interaction when the spins are parallel; $g_3\,(r)$ will in general be different from $g_1\,(r)$ (see Sec. 9).

To calculate the matrix elements of the magnetic moment μ we observe that the latter is the vector sum of the intrinsic magnetic moments of the two particles. Let μ_n and μ_p be the values of the magnetic moments of neutron and proton (to be taken with the positive or negative sign according to whether the magnetic moment is parallel or antiparellel to the intrinsic angular momentum of the particle; actually the magnetic moment of the neutron is negative and that of the proton positive). The components of the magnetic moment vectors $\boldsymbol{\mu}_n$ and $\boldsymbol{\mu}_p$ are obtained by multiplying μ_n and μ_p by the Pauli operators, operating on the coordinates s and S, respectively. One may therefore write without difficulty the operators representing the components μ_x, μ_y, μ_z, of the total magnetic moment $\boldsymbol{\mu} = \boldsymbol{\mu}_p + \boldsymbol{\mu}_n$. For instance, μ_z operates as follows:

$$\mu_z \begin{vmatrix} \psi_{1,1} & \psi_{1,-1} \\ \psi_{-1,1} & \psi_{-1,-1} \end{vmatrix} = \mu_n \begin{vmatrix} \psi_{1,1} & \psi_{1,-1} \\ -\psi_{-1,1} & -\psi_{-1,-1} \end{vmatrix} + \mu_p \begin{vmatrix} \psi_{1,1} & -\psi_{1,1} \\ \psi_{-1,1} & -\psi_{-1,1} \end{vmatrix}.$$

From (105) and (107) we then find the matrix elements of μ_z to be

$$\mu_z\,(^1S\,,\,^3S_1) = \mu_z\,(^1S\,,\,^3S_{-1}) = 0 \quad ; \quad \mu_z\,(^1S\,,\,^3S_0) = 4\pi\,(\mu_p - \mu_n)\int_0^\infty uv\,dr.$$

In a similar manner we find

$$-\mu_x\,(^1S\,,\,^3S_1) = \mu_x\,(^1S\,,\,^3S_{-1}) = 4\pi\,\frac{\mu_p - \mu_n}{\sqrt{2}}\int_0^\infty uv\,dr \quad ; \quad \mu_x\,(^1S\,,\,^3S_0) = 0\,,$$

$$\mu_y\,(^1S\,,\,^3S_1) = \mu_y\,(^1S\,,\,^3S_{-1}) = 4\pi i\,\frac{\mu_p - \mu_n}{\sqrt{2}}\int_0^\infty uv\,dr \quad ; \quad \mu_y\,(^1S\,,\,^3S_0) = 0\,.$$

The transition probability is found by replacing μ_{ij} in (104) by the sum of the squares of the moduli of all these matrix elements. This probability is therefore

$$(109)\qquad \frac{1024\,\pi^6\,\nu^3}{hc^3}\,(\mu_p - \mu_n)^2\left(\int_0^\infty uv\,dr\right)^2.$$

In order to apply this formula to the problem of slow neutron capture, $u(r)$ (u/r = radial eigenfunction of the ground state of the deuteron) must be normalized in such a way that

(110) $$4\pi \int_0^\infty u^2\, dr = 1 .$$

However, as far as $v(r)$ (v/r = radial eigenfunction of the ^1S state of the continuum representing the slow neutron) is concerned, its normalization as chosen in Sec. 10 (cf. formula (79) and Fig. 1) corresponds to the case where the density of neutrons in a singlet state with respect to the proton equals unity. Since the statistical weights of singlet and triplet states are in the ratio of 1 : 3, the total slow neutron density corresponding to the stated normalization equals 4. Taking this into account, and denoting by V the velocity of slow neutrons and by σ_{ca} the capture cross section, the transition probability for the capture process will evidently be $= 4 V \sigma_{ca}$. By comparison with (109) we thus obtain

(111) $$\sigma_{ca} = \frac{1}{V} \frac{256\,\pi^6\,\nu^3}{hc^3} (\mu_p - \mu_n)^2 \left(\int_0^\infty uv\, dr \right)^2 .$$

We note in particular in this formula that the capture cross section is inversely proportional to the velocity. Denoting by τ the mean life of a slow neutron in a medium containing hydrogen at a concentration of n atoms per cubic centimeter, one has $\tau V n \sigma_{ca} = 1$. We thus find

(112) $$\frac{1}{\tau} = n \frac{256\,\pi^6\,\nu^3}{hc^3} (\mu_p - \mu_n)^2 \left(\int_0^\infty uv\, dr \right)^2 .$$

For an effective calculation of the integral contained in these formulae we shall employ the procedure of Bethe and Peierls (*loc. cit.*) which consists in neglecting the irregularities of the two functions u and v for $r < \rho$ (ρ = range) in view of the fact that the contributions to the various integrals from the region $r < \rho$ is of little importance. For $r > \rho$ we obtain from the integration of (108) and taking into account the normalization condition

(113) $$u(r) = \frac{1}{\sqrt{2\pi b}} e^{-\frac{r}{b}} \quad , \quad b = \sqrt{\frac{h^2}{4\pi^2 M |w_0|}} = 0.43 \times 10^{-12} .$$

(The normalization has been calculated assuming, for the reasons indicated, that (113) holds up to $r = 0$; the numerical value of the length b corresponds to the binding energy $-w_0 = 2.2 \times 10^6$ volts).

According to (79) $v(r)$ for $r > \rho$ is given by

(114) $$v(r) = a + r \quad ; \quad a = \pm \sqrt{\frac{h^2}{4\pi^2 M |w_1|}} = 1.88 \times 10^{-12}$$

where w_1 represents the energy of the real or virtual ^1S state which we have mentioned in Sec. 9 (not to be confused with the energy w of the other ^1S

state which represents the slow neutron). The numerical value of $|w_1|$ has been taken as 1.16×10^5 volts (cf. Sec. 9). Calculating the integral in (112) with the assumption that (113) and (114) hold down to $r = 0$, one immediately finds

$$\frac{1}{\tau} = n \frac{128 \pi^6 \nu^2}{hc^3} (\mu_p - \mu_n)^2 (a + b)^2 b. \tag{115}$$

Introducing into this equation the numerical values $n = 7.8 \times 10^{22}$, $\nu = w_0/h = 5.38 \times 10^{29}$, we obtain, according to whether we take the positive or negative sign in the second equation (114)

$$\frac{1}{\tau} = 61 \left(\frac{\mu_p - \mu_n}{\mu_0}\right)^2 \quad \text{or} \quad \frac{1}{\tau} = 154 \left(\frac{\mu_p - \mu_n}{\mu_0}\right)^2 \tag{116}$$

where μ_0 is the nuclear Bohr magneton. The first corresponds to negative a (^{1}S level of Sec. 9 real); the second corresponds to positive a (^{1}S level virtual).

For the magnetic moment of the proton we shall take the value $\mu_p = 2.88\, \mu_0$ (average of the results of measurements by Stern, Estermann and Frisch, and by Rabi, Kellogg and Zacharias [9]). The magnetic moment of the neutron is obtained as the difference between that of the deuteron and the proton. Taking for the magnetic moment of the deuteron $0.75\, \mu_0$, one finds $\mu_n = -2.13\, \mu_0$ (note the minus sign.) Thus we obtain from (116)

$$\tau = 6.5 \times 10^{-4}\ \text{s} \quad \text{or} \quad \tau = 2.6 \times 10^{-4}\ \text{s}.$$

The second one of these two values agrees very well with the results of slow neutron experiments ($\tau = 1.7 \times 10^{-4}$). This seems to indicate that the second relation (116) is valid, i.e. that the ^{1}S level of the deuteron is virtual.

BIBLIOGRAPHY.

[1] E. AMALDI, E. FERMI, *On the absorption and diffusion of slow neutrons*, « Ric. Scientifica », *7* (1), 154 (1936). In this paper the preliminary publications on this question are quoted.

[2] E. AMALDI, O. D'AGOSTINO, E. FERMI, B. PONTECORVO, F. RASETTI, E. SEGRÈ, « Proc. Roy. Soc. », *A 149*, 522 (1935).

[3] E. FERMI, *Zeeman Festschrift*, p. 128; Martinus Nijhoff, The Hague, 1935.

[4] G. C. WICK, « Rend. Lincei », *23*, 774 (1936).

[5] H. A. BETHE, R. PEIERLS, « Proc. Roy. Soc. », *A 148*, 146 (1934); *A 149*, 176 (1934).

[6] One must take into account the fact that the ratio of the statistical weights of singlet and triplet states is 1 : 3; see also M. GOLDHABER, « Nature » (London), *137*, 824 (1936).

[7] E. AMALDI, O. D'AGOSTINO, E. FERMI, B. PONTECORVO, E. SEGRÈ, « Ric. Scientifica », *6* (1), 581 (1935).

[8] E. FERMI, « Phys. Rev. », *48*, 570 (1935).

[9] R. FRISCH, O. STERN, « Z. Physik », *85*, 4 (1933); I. ESTERMANN, O. STERN, « Z. Physik », *85*, 17 (1933); *86*, 132 (1933); I. I. RABI, J. M. KELLOGG, J. R. ZACHARIAS, « Phys. Rev. », *46*, 157, 163 (1934); for the sign of the magnetic moments of proton and deuteron, see the same authors, « Nature » (London), *137*, 658 (1936).

N° 120.

On January 23rd 1937 Professor Corbino died of pneumonia after a very short illness. He was only sixty-one years old and the loss was keenly felt by all members of the Institute in Rome. There was the feeling that an era was coming to an end and indeed with the change of direction of the Institute and the deteriorating political situation, catastrophy was impending. Fermi wrote this brief eulogy in which characteristically he puts the main emphasis on the scientific work of Corbino. However, his warm personality and human traits emerge very clearly from this short memorial.

E. SEGRÈ.

120.

UN MAESTRO: ORSO MARIO CORBINO

« Nuova Antologia », *72*, 313–316 (1937).

Conobbi il senatore Orso Mario Corbino quando tornai a Roma appena laureato, nel 1922. Io avevo allora venti anni e il Corbino quarantasei; era senatore, era stato già ministro della Pubblica Istruzione ed inoltre era universalmente noto come una delle personalità più eminenti nel campo degli studi. Fu perciò con una spiegabile titubanza che mi presentai a Lui; ma la titubanza sparì subito di fronte al modo insieme cordiale e interessante con cui Egli prese a discutere con me dell'argomento dei miei studi. Avemmo in quel periodo conversazioni e discussioni quasi quotidiane, per effetto delle quali non solo mi si chiarirono molte idee che avevo confuse, ma nacque in me la profonda e sentita venerazione del discepolo verso il maestro, venerazione che andò sempre crescendo negli anni che ebbi la fortuna di passare nel suo laboratorio.

Credo di poter affermare che questi sentimenti siano comuni a tutti quanti lo hanno avvicinato: la sua affabilità, il modo intelligente ed arguto con cui riusciva talvolta a dire anche verità spiacevoli senza menomamente offendere, la sua assoluta sincerità, il reale interesse che Egli provava per le questioni sia scientifiche che umane gli conquistavano subito simpatia e ammirazione.

Il Corbino ricordava spesso con un certo compiacimento di dovere sopra tutto a sè stesso la posizione che aveva saputo conquistarsi. Egli era nato in Augusta (il 30 aprile 1876) da una famiglia di modesti artigiani. Il suo amore e la sua eccezionale disposizione per gli studi, riconosciuta e apprezzata dalla famiglia, fecero sì che il padre, a costo di gravi sacrifici, gli facesse seguire gli studi. Fin da ragazzo Egli ebbe gran passione per la fisica, e mentre frequentava le classi ginnasiali soleva unirsi ai giovani del liceo per ascoltare le lezioni di fisica del professor Stracciati.

Laureatosi in fisica a Palermo fu per alcuni anni contemporaneamente professore di liceo e assistente del professor Macaluso nell'istituto fisico dell'università di Palermo.

In questo periodo il Corbino, ancor giovanissimo, eseguì in collaborazione col suo maestro Macaluso una ricerca assai significativa nel campo della magnetoottica, scoprendo il fenomeno che fu poi detto di Macaluso–Corbino. Macaluso e Corbino osservarono che il piano di polarizzazione della luce subisce una intensa rotazione attraversando un vapore posto in un campo magnetico, se la frequenza della luce è prossima a quella di una riga di assorbimento del vapore. Dopo lo sviluppo della teoria elettromagnetica della luce, la fisica andava in quel tempo orientandosi verso la interpretazione dei fenomeni ottici in base alle concezioni elettroniche. L'« effetto » Macaluso–Corbino trovò il suo posto naturale nel quadro delle nuove teorie e contribuì in tal modo a chiarirne i fondamenti.

Sempre in questo periodo il Corbino ebbe discussioni e polemiche varie su argomenti di magnetoottica e dimostrò in esse la acutezza e la profondità delle sue vedute, in modo da farsi notare fin da allora anche fuori d'Italia, come uno dei ricercatori più promettenti. Particolarmente esse gli valsero la stima e l'amicizia di Augusto Righi, allora nel fiore della sua produttività.

Accanto agli studi di fisica pura il Corbino coltivò fin dagli inizi quelli di elettrotecnica. Particolarmente ingegnose in questo campo le sue osservazioni sulle proprietà di un circuito contenente una dinamo con eccitazione in serie. Egli stabilì infatti che la dinamo in queste condizioni si comporta come una resistenza ohmica negativa, per modo che nel circuito vengono a prodursi, in opportune circostanze, spontanee oscillazioni elettriche permanenti. Egli studiò altresì il comportamento del condensatore elettrolitico, del rocchetto di induzione, dell'arco di Duddel, portando chiarezza su questioni controverse relative al funzionamento di questi apparecchi.

Nel 1904 il Corbino vinse contemporaneamente due concorsi universitari, uno per l'elettrotecnica ed uno per la fisica, e scelse il posto di professore di fisica sperimentale nell'università di Messina. Miracolosamente illeso, insieme alla sua famiglia, nel terremoto del 1908 (la sua casa fu una delle poche restate in piedi in quella tragica circostanza), venne a Roma, chiamato da Blaserna a succedere al professor Sella nella cattedra di fisica complementare. Alla morte del Blaserna, nel 1918, gli successe nella direzione dell'Istituto fisico dell'università di Roma, che tenne fino alla morte.

A Roma l'attività del Corbino non si è limitata al solo campo universitario, ma si è svolta anche in quello industriale e nella vita pubblica. Il Corbino fu presidente del Consiglio Superiore delle Acque e di quello dei Lavori Pubblici. Fu nominato senatore nel 1920 e fu ministro della Pubblica Istruzione nel 1921 e dell'Economia Nazionale nel 1923–24. Ebbe numerosi riconoscimenti accademici e civili: Socio Nazionale dei Lincei, e di altre accademie; cavaliere dell'Ordine Civile di Savoia; ebbe il Premio Reale dei Lincei nel 1914 e il premio Mussolini nel 1933.

Nell'Istituto fisico di Roma il Corbino indirizzò le sue ricerche a problemi assai vari. Proseguendo l'indirizzo elettrotecnico egli ideò e realizzò col professor Trabacchi, che fu suo collaboratore in moltissimi lavori, un dispositivo per la produzione di alte tensioni raddrizzate destinate alla alimentazione dei tubi a raggi X. Il dispositivo Corbino–Trabacchi rappresentò un notevole progresso sugli altri fino allora in uso. Importante è anche la realizzazione di un convertitore di correnti alternate in correnti unidirezionali, che trovò utili applicazioni per la carica degli accumulatori.

Una delle più pregevoli caratteristiche dell'opera scientifica del Corbino consiste nell'avere Egli saputo raggiungere risultati importanti con l'impiego di mezzi semplicissimi. Ne è un bell'esempio la sua elegante ricerca sul calore specifico dei metalli (tungsteno e platino) ad altissime temperature, basato sulla osservazione delle fluttuazioni di resistenza elettrica del filamento di una comune lampadina ad incandescenza alimentata da una corrente alternata. Questi studi, i cui risultati furono in seguito confermati da molti altri ricercatori, condussero il Corbino a constatare che, ad altissime temperature, il calore specifico prende un valore considerevolmente superiore a quello previsto con la legge di Dulong e Petit. Ciò ha notevole importanza teorica e trova la sua interpretazione nel fatto che le forze che si esercitano tra gli atomi della sostanza non sono esattamente forze elastiche e ne deviano sensibilmente quando, per l'elevata temperatura, le oscillazioni degli atomi diventano relativamente assai ampie.

In questa come in svariate altre ricerche di fisica pura ed applicata il Corbino si servì del tubo di Braun per rendere visibile la forma delle oscillazioni elettriche. Il tubo di Braun può considerarsi il suo strumento favorito, ed Egli mise in evidenza, con numerose ed elegantissime esperienze, le svariate possibilità di applicazione di questo apparecchio, assai lontano a quel tempo dalla perfezione oggi raggiunta.

Un gruppo di lavori molto numeroso fu dedicato allo studio della corrente elettrica in lamine metalliche immerse in un campo magnetico ad esse perpendicolare. Tra questi deve notarsi lo studio delle linee di corrente in un disco di bismuto percorso da una corrente radiale. Corbino osservò che le linee di corrente, rettilinee in assenza di campo magnetico, assumono la forma di spirale sotto l'azione del campo. Il fenomeno è generalmente noto col nome di «effetto Corbino», quantunque Egli modestamente non gradisse questa denominazione, preferendo considerarlo come una variante dell'effetto Hall. Vari altri aspetti dell'interdipendenza fra correnti elettriche, termiche e campo magnetico, furono oggetto di un gruppo di numerose ricerche, anche di carattere teorico.

Per qualche tempo il Corbino si è dedicato, non senza un certo rischio personale, a studi sulle proprietà fisiche degli esplosivi e in particolare alla determinazione della tensione di vapore della nitroglicerina, fatta in collaborazione con l'ingegnere Chiaraviglio.

Fin dai primi tempi dello sviluppo della radiotelefonia il Corbino si interessò vivamente ai suoi problemi, e fu in particolare uno dei primi ad intuire le grandi possibilità di applicazione della valvola termoionica. Egli seguì con amore i primi passi delle radiodiffusioni, alle quali si interessava non solo dal punto di vista fisico, ma anche quale appassionato intenditore di musica. Ben degnamente perciò Egli ha presieduto nei suoi ultimi anni la Commissione di Vigilanza Artistica e Tecnica sulle Radio–audizioni.

Al congresso internazionale dei fisici tenuto a Como nel 1927 Egli presentò un complesso studio sulla teoria della pila elettrica, nel quale analizzava quali fossero le origini della forza elettromotrice. L'opera del Corbino in questo campo può dirsi aver dato una definitiva sistemazione alle discussioni in proposito; Egli arrivò alla conclusione che la differenza di potenziale risiede, per la massima parte, al contatto tra i due metalli, ravvicinandosi così, sia pure in una nuova forma, a quanto aveva già intuito Alessandro Volta, oltre un secolo fa.

Negli ultimi anni il Corbino si è dedicato con passione a studi di elettroacustica ed ha fondato e organizzato l'Istituto di Elettroacustica del Consiglio Nazionale delle Ricerche. Esso ha sede nei locali di via Panisperna, resi vacanti col trasferimento dell'Istituto Fisico dell'Università alla Città Universitaria, e nei quali per tanti anni il Corbino ha svolto la sua attività di ricercatore e di maestro.

L'ultimo suo lavoro scientifico data da pochi giorni prima della sua prematura scomparsa. È una elegantissima ricerca sopra un nuovo metodo per la taratura del microfono condensatore, basato sulla analisi delle armoniche prodotte dalle variazioni di capacità che hanno luogo quando il microfono oscilla.

Parallelamente alla attività scientifica il Corbino ha svolta la sua attività di maestro. Egli ha sempre amato essere circondato da giovani, ai quali era largo di consigli dettati dalla sua esperienza e dalla sua dottrina. Sapeva infondere in essi non solo il suo acuto senso critico e la sua profonda onestà scientifica, ma anche il sincero entusiasmo che Egli metteva in ogni sua ricerca. Amava discutere con loro le esperienze sia progettate che eseguite e ne coglieva sempre i punti essenziali, sfrondandole dai particolari superflui. Amava ricercare al di sotto delle complesse trattazioni analitiche quale fosse l'essenza fisica dei fatti in esame che riusciva sempre a chiarire con semplici ed espressivi modelli.

Era prontissimo nell'assimilare e approfondire qualsiasi argomento, così che spesso chi gli esponeva una teoria o una ricerca si accorgeva alla fine di aver ricavato dal colloquio una sorprendente chiarificazione delle proprie idee. Espositore eccezionalmente brillante ed arguto, animava le sue lezioni e le sue conferenze rendendo piacevoli e facilmente intelligibili gli argomenti più astrusi.

Al rimpianto per lo scienziato che onorò il suo Paese si accompagna pertanto in tutti noi che abbiamo avuto la ventura di vivergli vicino, il dolore per la scomparsa di un amico e di un consigliere sicuro e il ricordo del debito incancellabile di riconoscenza per quanto per merito di Lui abbiamo appreso della scienza e della vita.

N° 121.

The considerable effort expended in the preceding years to study artificial radioctivity induced by neutrons and the absorption and diffusion properties of slow neutrons, had the effect of leaving the Institute's experimental equipment deficient, at least with regard to artificial neutron sources. By now other institutes had linear accelerators and cyclotrons that furnished much higher intensities than those attainable with a maximum of 800 millicuries of radon plus berillium.

Therefore, as a first attempt, we thought of constructing an accelerator which, utilizing the D + D reaction, would enable us to attain large neutron intensities, even with a very modest accelerating voltage (200 KV). Once completed, however, this apparatus was used very little because soon afterwards the construction of a 1.1 million volt accelerator began at the Istituto Superiore di Sanità. The latter was never used by Fermi since he left Italy shortly after construction began.

E. AMALDI.

121.

UN GENERATORE ARTIFICIALE DI NEUTRONI

E. AMALDI, E. FERMI, F. RASETTI

« Ric. Scientifica », *8* (2), 40–43 (1937).

RIASSUNTO. — Si descrive un impianto per la produzione di neutroni a mezzo di joni di deuterio accelerati a 200 KV, che colpiscono un composto ricco in deuterio. Il rendimento fino ad ora ottenuto è pari a quello di una sorgente di emanazione+berillio di 2,5 Curie.

Lo sviluppo acquistato negli ultimi anni dalle ricerche sulle reazioni nucleari prodotte da neutroni ha reso assai importante il problema di ottenere forti sorgenti di queste particelle.

In precedenti ricerche abbiamo usato, per produrre neutroni, tubetti contenenti emanazione di radio e polvere di berillio. Essi emettono circa 20 mila neutroni al secondo per millicurie. Disponendo di emanazione fino ad un massimo di 800 mC, si possono ricavare $1{,}6 \cdot 10^7$ neutroni al secondo. Riuscendo ad utilizzare il 50 °/₀ dei neutroni per la produzione di una sostanza radioattiva artificiale, ciò che può corrispondere a condizioni eccezionalmente favorevoli, si può così arrivare a preparare al massimo una attività di 1/6 di millicurie.

Attività assai più considerevoli si possono ottenere quando si usino sorgenti artificiali di neutroni. Vari elementi leggeri, bombardati con deutoni accelerati artificialmente, emettono neutroni. La convenienza di bombardare l'uno o l'altro elemento dipende dal potenziale con cui vengono accelerati i deutoni. Per bassi potenziali (fino a poco oltre 600 KV) conviene usare una sostanza contenente una elevata percentuale di deuterio. Per potenziali maggiori si ottiene un rendimento più elevato usando il Li o il Be (in pratica conviene usare quest'ultima sostanza per la sua elevata resistenza meccanica e termica).

Ci proponiamo di riferire qui i risultati preliminari da noi ottenuti con una sorgente artificiale di neutroni realizzata in questo laboratorio. Dispo-

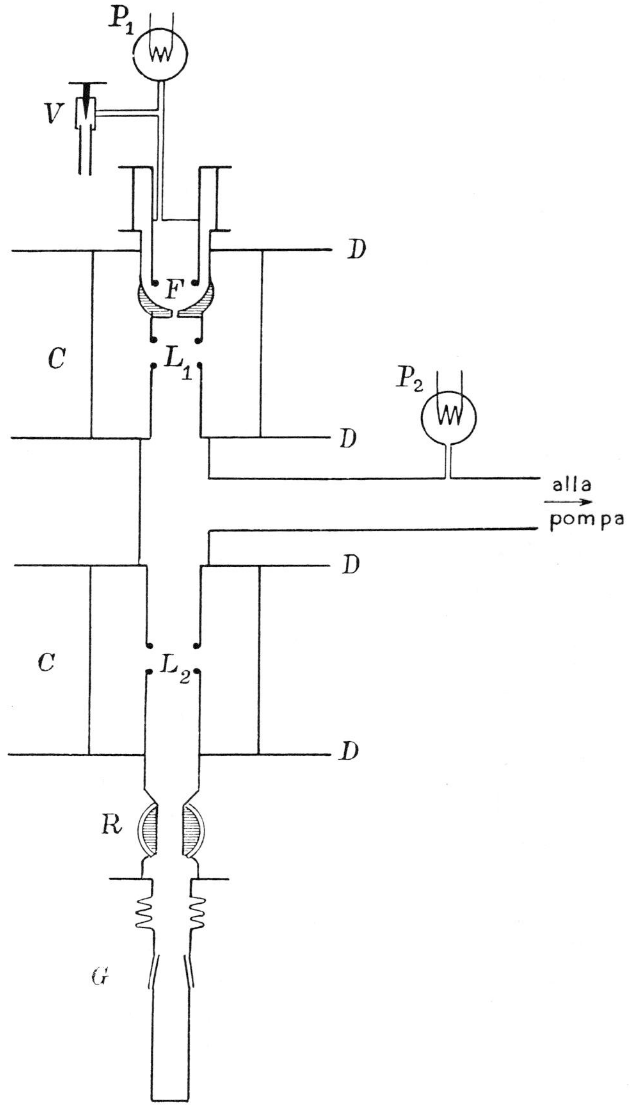

Fig. 1.

nendo di una tensione di soli 200 KV, abbiamo usato la reazione nucleare [1]

$$ {}^{2}_{1}\mathrm{D} + {}^{2}_{1}\mathrm{D} = {}^{2}_{3}\mathrm{H} + {}^{1}_{0}n \, . $$

Il generatore di tensione consiste di un dispositivo trifase in cui vengono utilizzate entrambe le semionde che caricano due capacità rispettivamente a + 100 e — 100 KV, rispetto alla terra. Abbiamo pertanto ritenuto conve-

(1) M. L. E. Oliphant, P. Harteck, Lord Rutherford, «Proc. Soc. Roy.», *A 144*, 692 (134).

niente costruire il nostro tubo in due sezioni attraverso le quali gli ioni subiscono successivamente due accelerazioni di 100 KV ciascuna. Lo schema del tubo è rappresentato nella fig. 1; la fig. 2 mostra una veduta generale dell'impianto. Non avendo ottenuto buoni risultati in numerosi tentativi con sorgenti di ioni ad arco a bassa pressione con catodo incandescente, siamo ricorsi ad una sorgente a raggi canale del tipo sviluppato da Oliphant. Il foro F nel catodo della sorgente dal quale esce il fascio di ioni positivi ha 3 mm di diametro e 7 di lunghezza. Gli ioni subiscono le due accelerazioni successive nelle due lenti elettriche L_1 ed L_2 e sono raccolti in un tubo all'estremità inferiore dell'apparecchio. Questa estremità inferiore, essendo al potenziale di — 100 KV, è sostenuta da una incastellatura isolata da terra mediante isolatori di porcellana. L'estremità inferiore del tubo, nella quale si raccolgono gli

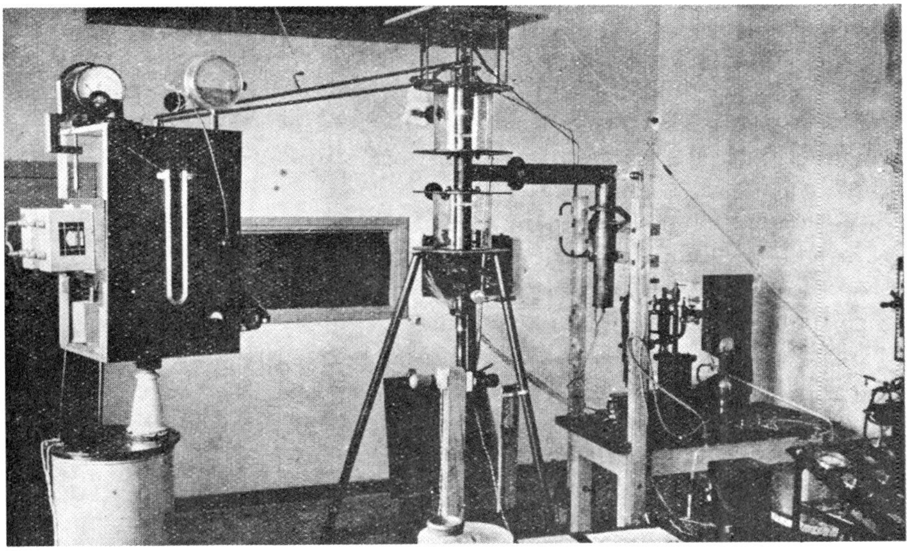

Fig. 2.

ioni, può venir staccata dal tubo (per esempio, per cambiare la sostanza da bombardare) senza far rientrare in esso l'aria, grazie a un grosso rubinetto R e ad un giunto conico G. Le varie parti del tubo sono riunite mediante flangie metalliche con guarnizioni di gomma, che pur assicurando un'ottima tenuta permettono un rapido smontaggio del tubo. I due cilindri di vetro C sono masticiati sui rispettivi dischi di ferro D mediante piceina. Il tubo è evacuato da una pompa a diffusione ad olio del diametro di 100 mm costruita in questo laboratorio su disegni di Sloan. Il tiraggio della pompa, misurato all'imboccatura, è di 50 litri/s mentre il tiraggio effettivo sul tubo è di 28 litri/s a causa della resistenza della condotta. Questo tiraggio è largamente sufficiente per mantenere un buon vuoto nel tubo malgrado la continua ammissione di deuterio attraverso il foro della sorgente (circa 1/20 di cm^3/s). Due vacuometri di Pirani P_1 e P_2 permettono di conoscere la pressione sia nell'interno della sorgente che nel tubo. Per alimentare la sorgente abbiamo costruito un trasformatore in olio il cui secondario è isolato dal primario per 100 KV, e fornisce una tensione di 25 KV. Questa tensione, raddrizzata a mezzo di un diodo e livellata con capacità, viene applicata, attraverso a resistenze di protezione, tra i due elettrodi della sorgente, e vi mantiene una corrente di circa 5 mA. Per mantenere nella sorgente la pressione più conveniente per la scarica, il deuterio vi viene introdotto attraverso ad una valvola a spillo V finemente regolabile e comandata a distanza, in modo da stabilire il giusto equilibrio con l'efflusso attraverso il foro dei raggi canali. La sorgente è raffreddata con una circolazione di olio.

Dato il notevole flusso di deuterio attraverso all'apparecchio, si è reso necessario un sistema di ricupero. Per questo il gas, dopo aver attraversato le pompe, anziché essere scaricato nell'atmosfera è raccolto in un gasometro, e da questo filtrato attraverso a una parete di palladio incandescente per liberarlo dalle impurità a cui si possa essere mescolato.

Il gas così purificato viene condotto da una tubazione alla valvola a spillo e di nuovo immesso nella sorgente.

La sostanza bombardata deve contenere un'alta percentuale di deuterio ed essere stabile anche sotto il bombardamento degli ioni. Abbiamo trovato conveniente usare il ghiaccio di acqua pesante mantenuto a bassissima temperatura con un raffreddamento ad aria liquida.

Riportiamo brevemente i risultati ottenuti fino ad ora, che hanno tuttavia un carattere assolutamente preliminare.

Nelle condizioni di funzionamento della sorgente già descritta, la massima corrente di ioni che siamo riusciti a raccogliere è di 40 microampere. Anche con una corrente così debole, e con la tensione relativamente bassa da noi usata, si ottiene una sorgente di neutroni equivalente a circa 2,5 Curie di emanazione + berillio; pari cioè a circa il triplo delle più forti sorgenti da noi usate fino ad ora. Riteniamo tuttavia che sarà possibile, con modificazioni di non grande entità, aumentare di un fattore 10 la corrente di ioni e quindi anche il numero dei neutroni. Ciò potrà ottenersi aumentando fino a 40 o 50 KV la tensione di alimentazione della sorgente (2). Questo perfezionamento renderà necessario di ricostruire la sorgente di ioni ed il suo sistema di alimentazione.

Se queste modificazioni daranno i risultati che è lecito attenderne, il nostro tubo potrà produrre $5 \cdot 10^8$ neutroni per secondo, ciò che potrà permettere, nei casi più favorevoli, la produzione di radioelementi artificiali con attività di alcuni millicurie. Sarà allora possibile eseguire molte ricerche, per le quali le sorgenti fin qui avute a disposizione sarebbero del tutto insufficienti.

Naturalmente attività enormemente più grandi si otterrebbero se potessimo disporre di tensioni dell'ordine di 1000 KV. Un vantaggio sensibile sarebbe anche in tal caso il poter usare il berillio come sostanza bombardata, invece del ghiaccio pesante. Si renderebbe così inutile il raffreddamento con aria liquida e si ridurrebbe in conseguenza l'ingombro intorno alla sorgente dei neutroni, permettendone una migliore utilizzazione.

Roma, *Istituto Fisico dell'Università.*

(2) A. BOUWERS, F. A. HEYN, A. KUNTKE, « Physica », *4*, 153 (1937).

N° 122.

Paper presented by Fermi at the 29th Reunion of the Italian Physical Society and the International Congress of Physics at Bologna in October 1937, in which he summarized some of the results of the slow neutron experiments performed in Rome and related them to the statistical model of the nucleus proposed by Bohr.

It was during this congress that word came of the sudden death of Lord Rutherford.

E. AMALDI.

122.

NEUTRONI LENTI E LIVELLI ENERGETICI NUCLEARI

« Nuovo Cimento », *15*, 41–32 (1938).

Sunto.

L'A. desidera discutere qui le conclusioni sopra la struttura dei livelli energetici nucleari che possono ricavarsi dallo studio delle proprietà di assorbimento dei neutroni lenti. L'analisi delle curve di assorbimento dei neutroni lenti in curve approssimativamente esponenziali, nonché lo studio delle proprietà di diffusione nelle sostanze idrogenate dei gruppi di neutroni approssimativamente omogenei che è stato così possibile di isolare, hanno permesso di riconoscere che i vari elementi capaci di assorbire i neutroni lenti presentano in genere una, o qualche volta più, bande di assorbimento caratteristiche, corrispondenti ad intervalli di energia dei neutroni relativamente assai ristretti.

Lo studio dei coefficienti di assorbimento per ciascuno di questi gruppi di neutroni nel boro, e l'esame dettagliato dei vari stadi del processo di rallentamento che i neutroni subiscono nella paraffina, hanno permesso di valutare non solo le energie, ma anche le larghezze delle varie bande di assorbimento. Le energie sono distribuite da zero fino a un centinaio di volt. Le larghezze sono in alcuni casi dell'ordine di 0,1 volt ed arrivano fino ad alcuni volt.

Bohr, Breit e Wigner hanno interpretato tali bande caratteristiche come dovute alla risonanza tra l'energia del neutrone lento e un livello energetico virtuale del nucleo. Bohr ha dato in più una suggestiva interpretazione della elevata probabilità con cui si presentano le bande nei diversi elementi. Egli la riconduce infatti alla esistenza nel nucleo di un numero assai elevato di gradi di libertà ed alla conseguente rapidissima crescita della densità dei livelli energetici al crescere della energia di eccitazione del nucleo.

Secondo queste idee di Bohr, in un nucleo di peso medio, i livelli energetici di bassa eccitazione sarebbero in genere distanziati tra di loro di alcune decine o centinaia di migliaia di volt. Mentre per energie di eccitazione di 8 o 10 milioni di volt la distanza tra livelli energetici consecutivi diventerebbe enormemente più piccola, riducendosi probabilmente a qualche decina di volt o anche meno. Per energie di eccitazione anche più grandi, i livelli energetici si farebbero sempre più fitti, fino a che la loro larghezza finirebbe per diventare maggiore della distanza media tra livelli consecutivi.

Lo studio delle bande di assorbimento dei neutroni lenti permette di determinare le larghezze dei livelli che corrispondono ai processi di cattura dei neutroni. Tale larghezza è dovuta al fatto che questi livelli energetici hanno una vita media limitata, potendosi distruggere sia per emissione di raggi gamma, sia per un processo analogo all'effetto Auger, in cui il neutrone, temporaneamente assorbito dal nucleo, ne viene riemesso. È possibile determinare sperimentalmente le probabilità di transizione per questi due processi; esse risultano, come ordine di grandezza, rispettivamente di 10^{14}, e di 10^{11} sec^{-1}, ciò che indica che, in questo caso il processo di gran lunga più probabile è la emissione di un raggio gamma.

Deve invece ritenersi che, quando l'energia totale di eccitazione del nucleo è molto maggiore, la probabilità di riemissione del neutrone debba diventare molto maggiore della probabilità di emissione di un quanto gamma, benché generalmente il neutrone abbandoni il nucleo con un'energia di molto inferiore alla massima che sarebbe energeticamente possibile. Ciò permette di interpretare il fatto che nell'urto tra neutroni veloci e nuclei di medio peso atomico, sembrano prevalere di gran lunga gli urti di seconda specie sugli urti elastici, come è indicato dalla osservazione che quasi tutte le sostanze producono un considerevole rallentamento dei neutroni veloci per urti successivi, benché, naturalmente, in misura assai minore delle sostanze idrogenate.

123.

TRIBUTE TO LORD RUTHERFORD

« Nature » (London), *140*, 1052 (1937).

The unexpected news of Lord Rutherford's death reached me at Bologna, when I was taking part in a meeting for the bicentennial celebration of Galvani's birth. A large group of physicists from all nations were assembled there, and it was quite apparent how deeply everybody felt the loss that science had suffered, by the passing away of a man whose efforts had opened up to physics one of the widest and as yet unfathomable fields of investigation.

Lord Rutherford certainly belonged to that highest class of experimenters —very few in the history of human thought—who appear to their admirers to be led by some sort of instinct always towards the successful attack of fundamental problems. If we consider most of his experiments, we are impressed by the fact that they are conceived so simply as to be easily understood and appreciated by a layman: their performance does not require a complicated piece of machinery, nor even often exceptional experimental skill. But it is not exaggeration to state that such simple experiments, as for example the discovery of the positive nucleus inside its cloud of electrons, or the method for producing artificial disintegrations by α-particle bombardment, are milestones in our knowledge of Nature.

Lord Rutherford will be remembered in the history of science not only on account of his personal contributions but also as a teacher, in the highest meaning of this word. One of the largest and most successful groups of investigators developed around him and learned from him not only the principles and the methods of research, but also the necessity of endurance and steadiness as essential requirements of the man of science.

124.

AZIONE DEL BORO SUI NEUTRONI CARATTERISTICI DELLO IODIO

E. FERMI e F. RASETTI

« Ric. Scientifica », *9* (2), 472-473 (1938).

Recentemente J. L. Michiels [1] ha descritto alcune esperienze sulla attivazione dello iodio con neutroni lenti filtrati da uno schermo di cadmio, in modo da eliminare i neutroni termici. In questo caso, l'attivazione è dovuta a neutroni di energie caratteristiche, valutate (per lo iodio) intorno a un centinaio di volt-elettroni [2].

Michiels misura la diminuzione dell'attività del rivelatore di iodio, interponendo sul cammino dei neutroni schermi di iodio e di boro, disposti: (*a*) prima il filtro di iodio e poi quello di boro; e (*b*) prima quello di boro e poi quello di iodio. Michiels trova che l'attività nel caso (*b*) è minore (circa 75 %) che nel caso (*a*).

In quello che segue descriviamo alcune esperienze che ci hanno permesso di verificare questo risultato e ne proponiamo una spiegazione.

Abbiamo, prima di tutto, ripetuto le esperienze in condizioni molto simili a quelle descritte da Michiels. La sorgente di neutroni, costituita da un tubetto di emanazione con berillio, era posta nell'interno di un cilindro di paraffina di circa 25 cm. di diametro e 15 cm. di altezza, 3 cm. sotto alla base superiore. Il cilindro era interamente avvolto da una foglia di cadmio dello spessore di 0,3 mm., salvo la faccia superiore in cui lo spessore era doppio. Gli assorbitori di boro e iodio e il rivelatore venivano appoggiati uno sull'altro sulla faccia superiore. Ci siamo accertati, con esperienze apposite, che gli assorbitori erano sufficientemente più larghi del rivelatore perché l'attivazione dovuta a neutroni che non avessero attraversato gli assorbitori fosse trascurabile.

L'assorbitore di boro era costituito da una lastra di anidride borica dello spessore di 13 mm, contenente 0,76 g/cm^2 di boro. L'assorbitore di iodio era costituito da uno strato di PbI_2 contenente 0,72 g/cm^2 di iodio. Il rivelatore era un pacchetto di iodio elemento contenuto in carta paraffinata.

Le misure sono state effettuate mediante una camera di ionizzazione a pressione ed elettrometro tipo Edelmann, con dispositivo descritto in precedenti lavori.

Le misure della attività nel caso (*a*) hanno dato, in unità arbitrarie, una attività di 24,3 mentre nel caso (*b*) si trova, conformemente ai risultati di Michiels, una attività minore e precisamente 19,1. Il rapporto è 0,79.

(1) « Nature », *142*, p. 431 (1938).

(2) Cfr. per esempio E. AMALDI ed E. FERMI, « Ricerca Scientifica », VII, 1, p. 454 (1936).

Ottenuta così una conferma del risultato, abbiamo pensato che ciò potesse essere determinato da una variazione di energia che i neutroni subirebbero passando attraverso al boro, per effetto di urti elastici. Normalmente non si osserva la diffusione dei neutroni lenti da parte del boro, perché questa viene mascherata dal fortissimo assorbimento, che (per neutroni termici) è circa cento volte più importante. Per neutroni di energia maggiore, il rapporto si sposta a favore della diffusione, poiché il coefficiente di diffusione resta approssimativamente costante, mentre quello di assorbimento è inversamente proporzionale alla velocità dei neutroni. Tra gli elementi che mostrano assorbimenti selettivi, lo iodio ha certamente un'energia di risonanza tra le più elevate (dell'ordine di 100 volt) ed è plausibile perciò ammettere che per queste energie le sezioni d'urto elastico e di cattura siano dello stesso ordine di grandezza. Ciò sarà tanto più vero per l'anidride borica, in cui all'effetto diffondente del boro deve aggiungersi quello dell'ossigeno. Ci si possono quindi spiegare i fatti osservati al modo seguente.

Il primo filtro (di iodio) elimina dallo spettro continuo dei neutroni una (o più) sottili bande corrispondenti allo spettro caratteristico di questo elemento. Nel successivo filtro (di boro) ha luogo un assorbimento vero dei neutroni, che tende a diminuire l'intensità. Hanno luogo però anche urti elastici dei neutroni contro gli atomi di boro. Se questi ultimi avessero massa praticamente infinita, l'energia dei neutroni non verrebbe alterata da queste diffusioni, e quindi la distribuzione spettrale resterebbe invariata. Invece, la massa dell'atomo di boro è solo dieci volte maggiore di quella del neutrone, così che anche in un urto elastico si ha una sensibile diminuzione dell'energia del neutrone. Questo effetto fa sì che alcuni neutroni di energia di poco superiore a quella della banda dello iodio e che avevano pertanto attraversato senza assorbimento il primo filtro, vengono (da un urto elastico) portati entro tale banda e contribuiscono quindi alla attivazione del rivelatore. Questo effetto non ha invece evidentemente luogo quando gli assorbitori di boro e di iodio sono posti in ordine inverso (caso *b*) ciò che giustifica qualitativamente come l'intensità osservata sia maggiore nel caso *a* che nel caso *b*.

Per controllare la plausibilità di questa ipotesi, abbiamo fatto una esperienza allo scopo di verificare che effettivamente il boro presenti una apprezzabile diffusione per neutroni dell'energia caratteristica dello iodio. A questo scopo abbiamo confrontato le attività che si ottengono usando il solo assorbitore di iodio con o senza la lastra di B_2O_3 appoggiata al disopra del rivelatore. Le attività misurate sono state rispettivamente 44,9 e 37,7. La differenza è dovuta alla diffusione all'indietro da parte della lastra di anidride borica. Ciò sembra rendere assai verosimile la spiegazione proposta.

Ringraziamo vivamente il prof. G. C. Trabacchi, Capo del Laboratorio Fisico dell'Istituto di Sanità Pubblica per averci fornita la sorgente di neutroni. Siamo grati anche al C. N. R. per averci messi a disposizione i mezzi per questa ricerca.

Istituto di Fisica della R. Università
Roma, 17 ottobre 1938.

N° 125.

In this letter to the Editor of the « Physical Review », the point is made that the word albedo meant different things to different authors. The albedo as defined by Halpern, Lueneburg and Clark, refers to the relation between reflected current and current incident on the surface of a medium (paraffin). This definition is the one used to impose—in diffusion theory—the boundary conditions on the separation surface of two media. Another definition is that adopted by Fermi in paper N° 119, and by Fermi and Amaldi in paper N° 118 and preceding ones. This definition corresponds to the so-called experimental albedo and refers to the practical case of measurements taken with a thin detector having a collision cross section inversely proportional to the velocity of the neutrons. Hence, in this case, the $1/v$ factor due to the sensitivity of the detector, compensates the factor v which multiplies the neutron density in the expression of the flux.

E. AMALDI.

125.

ON THE ALBEDO OF SLOW NEUTRONS

E. FERMI and E. AMALDI (*), G. C. WICK (**)
« Phys. Rev. », *53*, 493 (1938).

Halpern, Lueneburg and Clark [1] have recently given a rigorous solution of the integrodifferential equation describing the diffusion of thermal neutrons in paraffin. Although the method used by these authors is very elegant and involves the overcoming of serious mathematical diffculties, the criticism given by them of our interpretation [2] of the experimental results does not seem to be correct.

The definition of albedo as " reflection coefficient for neutrons impinging on a plane surface limiting a large paraffin block " is not quite unique. We introduced, therefore, what we called " experimental albedo " by formula (10) of our paper in the « Physical Review » [3]. All measurements of the albedo are to be compared with this definition, and to the same definition the relationship

$$\beta = 1 - 2/(N)^{1/2}$$

refers, as proved in Fermi's paper.

(*) R. Università - Istituto Fisico, Roma, Italy.

(**) R. Università - Istituto Fisico, Palermo, Italy.

(1) O. HALPERN, R. LUENEBURG and O. CLARK, « Phys. Rev. », *53*, 173 (1938).

(2) E. AMALDI, E. FERMI, « Phys. Rev. », *50*, 899 (1936); E. FERMI, « Ric. Scientifica », *7* 13 (1936).

(3) See e.g. reference 2, p. 910, column 2, line 11. [In the present volume: paper N° 118 *b*, p. 915, line 5].

Halpern, Lueneburg and Clark have instead calculated a different quantity viz., the reflection coefficient defined as the ratio of the number of reflected neutrons to the number of incident neutrons. Our definition rests on the measurement of the activity of a thin detector; now this activity is not simply a measure of the number of incident neutrons, but it also depends on their angular distribution, neutrons falling normally on the detector being less effective than neutrons falling at a large incidence angle (the probability of capture is proportional to $1/\cos\vartheta$). Since the angular distributions of the incident and reflected neutrons are different, the difference between Halpern's result and ours is easily explained (4).

We must therefore maintain, that the average number of impacts of a thermal neutron in paraffin as given in our papers is correct, apart from the uncertainty of the experimental data.

January 14, 1938.

(4) It may be pointed out that in a paper by WICK, « Rend. Lincei », *23*, 775 (1936), the whole problem was based on the same integrodifferential equation as in Halpern's paper and a result in accordance with Fermi was obtained by a numerical method. The reflection coefficient, as defined by Halpern a.o., was also considered under the assumption of a cosine distribution of the incident neutrons, and a result in full accord with Halpern's result for this case was obtained.

127.

GUGLIELMO MARCONI E LA PROPAGAZIONE DELLE ONDE ELETTROMAGNETICHE NELL'ALTA ATMOSFERA

«Soc. It. Progr. Sci.», Collectanea Marconiana, Roma 1–5 (1938–XVI).

Quando nel 1901 Guglielmo Marconi riusciva per la prima volta a trasmettere i tre punti della lettera S, da Poldhu in Cornovaglia a San Giovanni di Terranova, attraverso a 3500 km di oceano, dava non soltanto la definitiva dimostrazione delle grandiose possibilità che la sua scoperta offriva per risolvere i problemi delle comunicazioni transoceaniche, ma veniva anche ad aprire un nuovo ed affascinante campo di indagini sulle proprietà elettriche dell'alta atmosfera. Questi studi, che sono tuttora in corso, e che anzi hanno preso nuovo impulso particolarmente negli ultimi anni, hanno permesso di formarsi un quadro generale dei complessi fenomeni di ionizzazione atmosferica, se pure vi è tuttora discussione su punti fondamentali relativi alle cause e alle modalità con le quali il fenomeno si presenta.

La trasmissione delle radioonde fra due stazioni lontane.

È noto a tutti che le scoperte di Marconi furono in un primo tempo accolte con un certo scetticismo negli ambienti scientifici. Lo scetticismo era basato sulla convinzione che non fosse possibile la trasmissione delle radioonde tra stazioni situate una oltre l'orizzonte dell'altra. Si ragionava, infatti, press'a poco nel modo seguente: le onde elettromagnetiche usate nelle trasmissioni radio sono sostanzialmente analoghe alle onde luminose, dalle quali si differenziano solo per la grande lunghezza d'onda; e la terra, grazie alla sua conducibilità elettrica, si comporta per esse come un corpo opaco. Le radiazioni emesse da una stazione, propagandosi in linea retta, debbono lasciare in ombra tutte le stazioni situate al di sotto dell'orizzonte della stazione trasmittente; e ciò salvo una non grande correzione dovuta a fenomeni di diffrazione. Fu una fortuna per l'umanità che queste argomentazioni, che a priori potevano sembrare ragionevoli e ben fondate, non abbiano distolto Marconi dagli esperimenti sulle trasmissioni a grande distanza. La storia di questi primi successi delle radiotrasmissioni costituisce una riconferma del fatto che nello studio dei fenomeni naturali teoria ed esperimento debbono andare di pari passo. Raramente può l'esperienza, non guidata da un concetto teorico, raggiungere risultati di larga portata; ed è certo uno dei più significativi successi per la teoria che l'esistenza stessa e le proprietà essenziali delle onde elettromagnetiche fossero state previste matematicamente da Maxwell prima della verificazione sperimentale della loro esistenza e prima che esse, attraverso alla geniale intuizione di Marconi, trovassero il loro terreno di pratica applicazione; d'altra parte una fiducia eccessivamente spinta nelle previsioni teoriche avrebbe sconsigliato di insistere in esperimenti che erano destinati a rivoluzionare la tecnica delle comunicazioni.

Nelle discussioni teoriche sulle modalità della trasmissione delle radioonde tra due stazioni lontane era stato, infatti, omesso un fattore che è risultato in seguito di importanza fondamentale; e cioè l'influenza che gli altissimi strati dell'atmosfera, grazie al loro stato di ionizzazione, esercitano sulla propagazione delle onde elettromagnetiche. Che un possibile stato di ionizzazione dell'alta atmosfera potesse avere una influenza sopra certi fenomeni geomagnetici era stato supposto molti anni prima delle scoperte di Marconi, nel 1878, da B. Stewart. Ma solo dopo che Marconi ebbe provato sperimentalmente che le onde possono propagarsi fino a limiti bene al di là dell'orizzonte, Kennelly ed Heaviside pensarono indipendentemente alla possibilità che le trasmissioni a grande distanza fossero rese possibili da riflessioni determinate dalla ionizzazione atmosferica. Il loro ragionamento è schematicamente il seguente: ammettiamo che, ad una altezza dell'ordine di grandezza di un centinaio di chilometri al di sopra del suolo, l'aria sia intensamente ionizzata ed acquisti così le proprietà di un conduttore elettrico. Questo alto strato ionizzato avrebbe allora la proprietà di riflettere le radioonde le quali, invece di disperdersi nello spazio, verrebbero ad essere come convogliate tra i due strati riflettenti costituiti dalla terra e dall'alta atmosfera. Con successive riflessioni su questi due strati resta allora evidentemente possibile alle radiazioni raggiungere un punto qualsiasi della superficie terrestre.

La jonosfera e la propagazione delle radioonde.

Non è questo il luogo per ricordare le innumerevoli ricerche e discussioni di carattere sia teorico che pratico, che si sono sviluppate in seguito alle prime esperienze fondamentali sulle radiotrasmissioni; mi limiterò, invece, a riassumere nelle linee generali quelli che oggi possono considerarsi risultati acquisiti sulle proprietà di ionizzazione dell'alta atmosfera (la così detta jonosfera) e sulla sua influenza sopra la propagazione delle radioonde.

Una considerazione elementare dimostra che vi è una semplice relazione tra l'indice di rifrazione di un gas ionizzato per le onde elettromagnetiche e la densità degli ioni in esso contenuti. La relazione è la seguente:

$$n = \sqrt{1 - \frac{e^2 N}{\pi m \nu^2}}, \tag{1}$$

dove n rappresenta l'indice di rifrazione; e ed m rispettivamente la carica elettrica e la massa degli ioni presenti nel gas; ν la frequenza delle onde in esame; N il numero di ioni per unità di volume.

Dalla ispezione della formula si riconosce:

a) che l'indice di rifrazione è minore di 1, ossia che lo stato di ionizzazione determina un abbassamento della rifrangibilità. Come conseguenza la radiazione elettromagnetica che, provenendo dalla bassa atmosfera, entra nella ionosfera può subire una riflessione totale venendo così respinta verso la terra. È appunto questa la proprietà fondamentale che si deve attribuire alla ionosfera per spiegare la sua influenza sulla propagazione delle onde a grandi distanze;

b) che, a parità di altre condizioni, l'effetto di una determinata quantità di ioni è tanto maggiore quanto più piccola è la loro massa; esso sarà quindi assai più notevole se il gas ionizzato contiene elettroni liberi, che non se essi sono legati sotto forma di ioni atomici o molecolari, di massa molto maggiore. La estrema rarefazione della altissima atmosfera, oltre cento chilometri di altezza, rende d'altra parte probabile che elettroni liberi, una volta formatisi, vi possano rimanere per un tempo relativamente lungo. Per questo, e più ancora per la verificazione diretta fatta in base alla azione che il campo magnetico terrestre esercita sulla rifrazione delle radioonde, deve ritenersi che, almeno nei più alti strati della atmosfera, le onde si rifrangano per effetto prevalente degli elettroni liberi;

c) la radiazione elettromagnetica, che si propaga verticalmente verso l'alto, subirà una riflessione totale raggiungendo uno strato della atmosfera in cui la densità della ionizzazione sia tanto elevata da rendere eguale a 0 l'indice di rifrazione. Si riconosce, dalla formula (1), che ciò si verificherà quando si abbia:

$$N = \frac{\pi m \nu^2}{e^2}, \qquad (2)$$

e cioè per esempio una onda di 1000 m. di lunghezza si rifletterà totalmente incontrando uno strato in cui la concentrazione degli elettroni sia di 1100 per centimetro cubo. Per riflettere una onda di 100 m di lunghezza occorreranno invece 110.000 elettroni per cm^3, mentre per una onda di 10 m. di lunghezza ne occorrerebbero 11.000.000.

Così per esempio se in determinate condizioni la massima concentrazione degli elettroni nella verticale al di sopra di una stazione fosse di 110.000 per cm^3, le onde elettromagnetiche aventi lunghezza d'onda eguale o maggiore di 100 m verrebbero riflesse totalmente verso il basso; mentre onde di lunghezza d'onda minore non subirebbero riflessione. Si comprende da ciò come l'esame delle riflessioni di onde inviate verso l'alto, in dipendenza dalla loro frequenza, possa permettere di determinare la concentrazione degli elettroni negli strati che determinano la riflessione; si può anche del resto determinare l'altezza di questi strati misurando il tempo che intercorre tra la emissione delle onde e l'arrivo delle onde riflesse e che è dell'ordine di grandezza di un millesimo di secondo.

Lo studio sperimentale della jonosfera.

Da vari anni vengono eseguite sistematicamente osservazioni di questo genere in molti osservatori; esse furono iniziate in America da Breit e Tuve; e tra i primi a dedicarsi a questo genere di ricerche fu in Italia il Ranzi.

L'osservazione sistematica di tali riflessioni ha permesso di concludere che esistono nella atmosfera diversi strati ionizzati che vengono usualmente indicati con le lettere dell'alfabeto. I più notevoli sono lo strato E ad una altezza di circa 100 km. al di sopra del suolo e lo strato F, spesso diviso a sua volta in due strati F_1 e F_2 ad altezze tra i 180 e i 300 km.

La densità della ionizzazione in questi strati arriva, come ordine di grandezza, fino a circa 1.000.000 di elettroni per cm^3; essa varia moltissimo dal giorno alla notte ed è tanto più intensa di regola quanto più alto è il Sole sull'orizzonte. Ciò indica l'origine solare dell'agente ionizzante, che si ritiene essere in principale misura la porzione ultravioletta della radiazione solare che non raggiunge la superficie della Terra appunto perché assorbita dall'alta atmosfera. Vi sono però varie esperienze che fanno sospettare che i raggi ultravioletti solari non siano gli unici agenti ionizzanti e che possano forse avere importanza anche radiazioni corpuscolari provenienti dal Sole. Deve anche notarsi la dipendenza della ionizzazione atmosferica dai cicli di attività solare: la ionizzazione è di regola massima nei periodi di massima attività solare, pur non essendo stata riconosciuta alcuna diretta dipendenza tra la ionizzazione e il passaggio di determinati gruppi di macchie solari al meridiano solare rivolto verso la Terra.

Sono infine estremamente appariscenti le relazioni tra la ionizzazione atmosferica e quegli stessi fenomeni solari che dànno origine alle tempeste magnetiche e alle aurore boreali. In casi particolari le perturbazioni della ionizzazione atmosferica che accompagnano le tempeste magnetiche raggiungono valori così grandi da interrompere praticamente le comunicazioni radiotelegrafiche su regioni assai vaste.

La presenza di elettroni o ioni nella atmosfera non produce soltanto una variazione dell'indice di rifrazione per le onde elettromagnetiche, ma determina anche in genere un assorbimento di esse. Tale assorbimento è prodotto dal fatto che gli elettroni, messi in moto oscillatorio dal campo elettrico dell'onda elettromagnetica, subiscono ogni tanto degli urti contro le molecole dell'aria, trasferendo ad esse una parte della loro energia, che viene trasmutata in calore. Si comprende da ciò come, a parità di altre condizioni, l'assorbimento sia tanto più intenso quanto più frequenti sono gli urti che subiscono gli elettroni e cioè quanto più elevata è la densità della atmosfera attraverso alla quale l'onda si propaga. Per conseguenza l'assorbimento ha luogo in misura tanto maggiore, quanto più bassi sono gli strati ionizzati mentre è praticamente trascurabile negli altissimi strati della atmosfera.

Accenneremo infine al notevole effetto esercitato dal campo magnetico terrestre sulla rifrazione delle onde elettromagnetiche. Gli elettroni liberi degli alti strati ionizzati, per effetto del campo magnetico terrestre, invece di muoversi in linea retta tra un urto e un altro, tendono a descrivere delle spirali attorno alle linee di forza magnetiche, girando attorno ad esse con una frequenza che risulta proporzionale alla intensità del campo magnetico. A causa di questa frequenza propria degli elettroni, hanno luogo dei fenomeni di risonanza per cui la rifrazione delle onde elettromagnetiche presenta delle anomalie, quando la frequenza dell'onda è vicina alla frequenza propria degli elettroni. Le lunghezze d'onda per le quali hanno luogo queste specie di risonanze si aggirano attorno ai 200 m e dipendono naturalmente dalla intensità del campo magnetico terrestre nella località. Il fenomeno viene di solito osservato come una specie di birifrangenza e si possono distinguere un raggio ordinario e un raggio straordinario aventi diversa rifrangibilità, in modo analogo a quanto avviene per la propagazione della luce in un mezzo birifrangente.

CONCLUSIONE.

Io spero di avere mostrato con questi brevi cenni quanto sia vasto ed interessante il campo che lo studio della propagazione delle onde elettromagnetiche nell'alta atmosfera, originato dalle prove di Marconi sulle radiocomunicazioni a grandi distanze, ha aperto alla investigazione. L'interesse è accentuato dal fatto che queste proprietà dell'alta atmosfera si ricollegano a molti altri fenomeni, per ora soltanto parzialmente compresi, di fisica terrestre e di fisica solare; ed è da augurarsi che uno studio metodico di questi fenomeni e delle loro interdipendenze possa portare ad un notevole approfondimento delle nostre conoscenze sullo stato della atmosfera da un lato, e che possa dall'altro essere fecondo di risultati pratici per garantire la stabilità e la efficienza delle radiocomunicazioni in condizioni atmosferiche avverse.

N° 128

The 1938 Nobel Prize for Physics was attributed to Fermi. The official proclamation reads: " för hans påvisande av nya radioaktiva grundämnen, framställda genom neutronbestrålning, och hans i anslutning därtill gjorda upptäckt av kärnreaktioner, åstadkomna genom långsamma neutroner ", " for his demonstrations of the existence of new radioactive elements produced by neutron irradiation, and for his related discovery of nuclear reactions brought about by slow neutrons. "

He received the prize in Stockholm on December 10, 1938 from the King of Sweden in a solemn ceremony according to tradition.

The official speech conferring the Prize was given by Professor H. Pleijel of the Swedish Academy who concluded with the following words in Italian:

" Professor Fermi,

La Reale Accademia Svedese delle Scienze ha assegnato a Voi il premio Nobel per la Fisica Millenovecentotrentotto, per la Vostra scoperta di nuove sostanze radioattive appartenenti all'intero campo degli elementi e per la scoperta, da Voi fatta nel corso di tali studi, del potere selettivo dei neutroni lenti.

Vi presentiamo le nostre congratulazioni e Vi esprimiamo la più viva ammirazione per le Vostre geniali ricerche, che gettano nuova luce sulla costituzione dei nuclei atomici ed aprono nuovi orizzonti all'ulteriore sviluppo delle indagini atomiche.

Vi prego ora di ricevere il Premio Nobel dalle mani di Sua Maestà il Re. "

Fermi went to Stockholm on this occasion, having decided that he would not return to Italy. The political situation had forced this decision which he communicated only to a few intimate friends. From Stockholm he sailed for New York where he arrived on January 2nd, 1939.

It will be noted that the Nobel speech composed before the discovery of fission contains the names ausonium and hesperium for elements 93 and 94. The discovery of fission occurred between December 10, 1938 and the printing of the speech. It is mentioned in a footnote and, as is well known, required a reappraisal of the work on the activities induced in uranium.

E. Segrè

128.

ARTIFICIAL RADIOACTIVITY PRODUCED BY NEUTRON BOMBARDMENT

«Les Prix Nobel en 1938». Les Conférences Nobel Stockholm p. 1–8 (1939).

Although the problem of transmuting chemical elements into each other is much older than a satisfactory definition of the very concept of chemical element, it is well known that the first and most important step towards its solution was made only nineteen years ago by the late Lord Rutherford who started the method of the nuclear bombardments. He showed on a few examples that, when the nucleus of a light element is struck by a fast α–particle, some disintegration process of the struck nucleus occurs, as a consequence of which the α–particle remains captured inside the nucleus and a different particle, in many cases a proton, is emitted in its place. What remains at the end of the process is a nucleus different from the original one; different in general both in electric charge and in atomic weight.

The nucleus that remains as disintegration product coincides sometimes with one of the stable nuclei, known from the isotopic analysis; very often, however, this is not the case. The product nucleus is then different from all " natural " nuclei; the reason being that the product nucleus is not stable. It disintegrates further, with a mean life characteristic of the nucleus, by emission of an electron (positive or negative), until it finally reaches a stable form. The emission of electrons that follows with a lag in time the first practically instantaneous disintegration, is the so-called artificial radioactivity, and was discovered by Joliot and Irène Curie at the end of the year 1933.

These authors obtained the first cases of artificial radioactivity by bombarding boron, magnesium and aluminum with α-particles from a polonium source. They produced thus three radioactive isotopes of nitrogen, silicon and phosphorus, and succeeded also in separating chemically the activity from the bulk of the unmodified atoms of the bombarded substance.

The Neutron Bombardment.

Immediately after these discoveries, it appeared evident that α-particles very likely did not represent the only type of bombarding projectiles for producing artificial radioactivity. I decided therefore to investigate from this point of view the effects of the bombardment with neutrons.

Compared with α-particles, the neutrons have the obvious drawback that the available neutron sources emit only a comparatively small number of neutrons. Indeed neutrons are emitted as products of nuclear reactions, whose yield is only seldom larger than 10^{-4}. This drawback is, however, compensated by the fact that neutrons, having no electric charge, can reach the nuclei of all atoms, without having to overcome the potential barrier, due to the Coulomb field that surrounds the nucleus. Furthermore, since neutrons practically do not interact with electrons, their range is very long, and the probability of a nuclear collision is correspondingly larger than in the case of the α-particle or the proton bombardment. As a matter of fact, neutrons were already known to be an efficient agent for producing some nuclear disintegrations.

As source of neutrons in these researches I used a small glass bulb containing beryllium powder and radon. With amounts of radon up to 800 millicuries such a source emits about 2.10^7 neutrons per second. This number is of course very small compared to the yield of neutrons that can be obtained from cyclotrons or from high voltage tubes. The small dimensions, the perfect steadiness and the utmost simplicity are, however, sometimes very useful features of the radon + beryllium sources.

Nuclear Reactions produced by Neutrons.

Since the first experiments I could prove that the majority of the elements tested became active under the effect of the neutron bombardment. In some cases the decay of the activity with time corresponded to a single mean life; in others to the superposition of more than one exponential decay curve.

A systematic investigation of the behaviour of the elements throughout the periodic table was carried out by myself, with the help of several collaborators, namely Amaldi, D'Agostino, Pontecorvo, Rasetti and Segrè. In most cases we performed also a chemical analysis, in order to identify the chemical element that was the carrier of the activity. For short living substances, such an analysis must be performed very quickly, in a time of the order of one minute.

The results of this first survey of the radioactivities produced by neutrons can be summarized as follows: Out of 63 elements investigated, 37 showed an easily detectable activity; the percentage of the activable elements did not show any marked dependence on the atomic weight of the element. Chemical analysis and other considerations, mainly based on the distribution of the isotopes, permitted further to identify the following three types of nuclear reactions giving rise to artificial radioactivity:

$$ {}^{M}_{Z}A + {}^{1}_{0}n = {}^{M-3}_{Z-2}A + {}^{4}_{2}He \tag{1} $$

$$ {}^{M}_{Z}A + {}^{1}_{0}n = {}_{Z-1}{}^{M}A + {}^{1}_{1}H \tag{2} $$

$$ {}^{M}_{Z}A + {}^{1}_{0}n + {}^{M+1}_{Z}A \tag{3} $$

where ${}^{M}_{Z}A$ is the symbol for an element with atomic number Z and mass number M and n is the symbol of the neutron.

The reactions of the types (1) and (2) occur chiefly among the light elements, while those of the type (3) are found very often also for heavy elements. In many cases the three processes are found at the same time in a single element. For instance neutron bombardment of aluminum that has a single isotope ^{27}Al, gives rise to three radioactive product: ^{24}Na, with a period of 15 hours by process (1); ^{27}Mg, with a period of 10 minutes by process (2); and ^{28}Al with a period of 2.3 minutes by process (3).

As mentioned before, the heavy elements usually react only according to the process (3) and therefore, but for certain complications to be discussed later, and for the case in which the original element has more than one stable isotope, they give rise to an activity decaying exponentially. A very striking exception to this behaviour is found for the activities induced by neutrons in the naturally active elements thorium and uranium. For the investigation of these elements it is necessary to purify first the element as thoroughly as possible from the daughter substances that emit β–particles. When thus purified, both thorium and uranium emit spontaneously only α–particles, that can be immediately distinguished by absorption from the β–activity induced by the neutrons.

Both elements show a rather strong induced activity when bombarded with neutrons; and in both cases the decay curve of the induced activity shows that several active bodies with different mean lives are produced. We attempted since the spring of 1934 to isolate chemically the carriers of these activities, with the result that the carriers of some of the activities of uranium are neither isotopes of uranium itself, nor of the elements lighter than uranium down to the atomic number 86. We concluded that the carrier was one or more elements of atomic number larger than 92; we use to call the elements 93 and 94 in Rome with the names of Ausonium and Hesperium respectively. It is known that O. Hahn and L. Meitner have investigated very carefully and extensively the decay products of irradiated uranium, and were able to trace among them elements up to the atomic number 96 [(1)].

It should be noticed here, that besides the processes (1), (2) and (3) for the production of artificial radioactivity with neutrons, neutrons of sufficiently high energy can react also as follows, as was first shown by Heyn: The primary neutron does not remain bound in the nucleus, but knocks off instead one of the nuclear neutrons out of the nucleus; the result is a new nucleus, that is isotopic with the original one and has atomic weight less by one unit. The final result is therefore identical with the products obtained by means of the nuclear photoeffect (Bothe), or by bombardment with fast deuterons. One of the most important results of the comparison of the active products obtained by these processes, is the proof, first given by Bothe, of the existence of isomeric nuclei, analogous to the isomers UX_2 and UZ recognized long since by O. Hahn in his researches on the uranium family. The number of well established cases of isomerism appears to increase rather rapidly, as investigation goes on and represents an attractive field of research.

THE SLOW NEUTRONS.

The intensity of the activation as a function of the distance from the neutron source shows in some cases anomalies apparently dependent on the objects that surround the source. A careful investigation of these effects led to the unexpected result that surrounding both source and body to be activated with masses of paraffin, increases in some cases the intensity of activation by a very large factor (up to 100). A similar effect is produced by water, and in general by substances containing a large concentration of hydrogen. Substances not containing hydrogen show sometimes similar features, though extremely less pronounced.

The interpretation of these results was the following. The neutron and the proton having approximately the same mass, any elastic impact of a fast neutron against a proton initially at rest, gives rise to a partition of the available kinetic energy between neutron and proton; it can be shown that a neutron having an initial energy of 10^6 volts, after about 20 impacts against hydrogen atoms has its energy already reduced to a value close to that corresponding to thermal agitation. It follows that, when neutrons of high

(1) The discovery by Hahn and Strassmann of barium among the disintegration products of bombarded uranium, as a consequence of a process in which uranium splits into two approximately equal parts, makes it necessary to reexamine all the problems of the transuranic elements, as many of them might be found to be products of a splitting of uranium.

energy are shot by a source inside a large mass of paraffin or water, they very rapidly lose most of their energy and are transformed into "slow neutrons". Both theory and experiment show that certain types of neutron reactions, and especially those of type (3), occur with a much larger cross-section for slow neutrons than for fast neutrons, thus accounting for the larger intensities of activation, observed when irradiation is performed inside a large mass of paraffin or water.

It should be remarked furthermore that the mean free path for the elastic collisions of neutrons against hydrogen atoms in paraffin, decreases rather pronouncedly with the energy. When therefore, after three or four impacts, the energy of the neutron is already considerably reduced, its probability of diffusing outside of the paraffin, before the process of slowing down is completed, becomes very small.

To the large cross section for the capture of slow neutrons by several atoms, there must obviously correspond a very strong absorption of these atoms for the slow neutrons. We investigated systematically such absorptions, and found that the behaviour of different elements in this respect is widely different; the cross section for the capture of slow neutrons varies, with no apparent regularity for different elements from about 10^{-24} cm^2 or less, to about a thousand times as much. Before discussing this point, as well as the dependence of the capture cross section on the energy of the neutrons weshall first consider how far down the energy of the primary neutrons can be reduced by the collisions against the protons.

The Thermal Neutrons.

If the neutrons could go on indefinitely diffusing inside the paraffin, their energy would evidently reach finally a mean value equal to that of thermal agitation. It is possible, however, that, before the neutrons have reached this lowest limit of energy, either they escape by diffusion out of the paraffin, or are captured by some nucleus. If the neutron energy reaches the thermal value, one should expect the intensity of the activation by slow neutrons to depend upon the temperature of the paraffin.

Soon after the discovery of the slow neutrons, we attempted to find a temperature dependence of the activation, but, owing to insufficient accuracy, did not succeed. That the activation intensities depend upon the temperature was proved some months later by Moon and Tillman in London; as they showed, there is a considerable increase in the activation of several detectors, when the paraffin, in which the neutrons are slowed down, is cooled from room temperature to liquid air temperature. This experiment definitely proves that a considerable percentage of the neutrons actually reaches the energy of thermal agitation. Another consequence is that the diffusion process must go on inside the paraffin for a relatively long time.

In order to measure directly at least the order of magnitude of this time, an experiment was attempted by myself and my collaborators. The source of neutrons was fastened at the edge of a rotating wheel, and two identical detectors were placed on the same edge, at equal distances from the source, one in front and one behind with respect to the sense of rotation. The wheel was then spun at a very high speed inside a fissure in a large paraffin block. We found that, while, with the wheel at rest, the two detectors became equally active, when the wheel was in motion during the activation, the detector that was behind the source became considerably more active than the one in front. From a discussion of this experiment was deduced, that the neutrons remain inside the paraffin for a time of the order of 10^{-4} seconds.

Other mechanical experiments with different arrangements were performed in several laboratories. For instance Dunning, Fink, Mitchell, Pegram and Segrè in New York, built a mechanical velocity selector, and proved by direct measurement, that a large amount of the neutrons diffusing outside of a block of paraffin, have actually a velocity corresponding to thermal agitation.

After their energy is reduced to a value corresponding to thermal agitation, the neutrons go on diffusing without further change of their average energy. The investigation of this diffusion process, by Amaldi and myself, showed that thermal neutrons in paraffin or water can diffuse for a number of paths of the order of 100 before being captured. Since, however, the mean free path of the thermal neutrons in paraffin is very short (about 0.3 cm) the total displacement of the thermal neutrons during this diffusion process is rather small (of the order of 2 or 3 cm). The diffusion ends when the thermal neutron is captured, generally by one of the protons, with production of a deuteron. The order of magnitude for this capture probability can be calculated, in good agreement with the experimental value, on the assumption that the transition from a free neutron state to the state in which the neutron is bound in the deuteron is due to the magnetic dipole moments of the proton and the neutron. The binding energy set free in this process, is emitted in the form of γ–rays first observed by Lea.

All the processes of capture of slow neutrons by any nucleus are generally accompanied by the emission of γ–rays. Immediately after the capture of the neutron, the nucleus remains in a state of high excitation and emits one or more γ–quanta, before reaching the fundamental state. γ–rays emitted by this process were investigated by Rasetti and by Fleischmann.

Absorption Anomalies.

A theoretical discussion of the probability of capture of a neutron by a nucleus, under the assumption, that the energy of the neutron is small compared with the differences between neighbouring energy levels in the nucleus, leads to the result that the cross-section for the capture process should be inversely proportional to the velocity of the neutron. While this result is in qualitative agreement with the high efficiency of the slow neutron bombardment, observed experimentally, it fails on the other hand to account for several features of the absorption process, that we are now going to discuss.

If the capture probability of a neutron were inversely proportional to its velocity, one would expect two different elements to behave in exactly the same way as absorbers of the slow neutrons, provided the thicknesses of the two absorbers were conveniently chosen, so as to have equal absorption for neutrons of a given energy. That the absorption obeys instead more complicated laws, was soon observed by Moon and Tillman and other authors who showed that the absorption by a given element appears, as a rule, to be larger, when the slow neutrons are detected by means of the activity induced in the same element. That the simple law of inverse proportionality does not hold, was also proved by a direct mechanical experiment by Dunning, Pegram, Rasetti and others in New York.

In the winter of 1935–36 a systematic investigation of these phenomena was carried out by Amaldi and myself. The result was, that each absorber of the slow neutrons has one or more characteristic absorption bands, usually for energies below 100 volts. Besides this or these absorption bands, the absorption coefficient is always large also for neutrons of thermal energy. Some elements, especially cadmium, have their characteristic absorption band overlapping with the absorption in the thermal region. This element absorbs therefore very strongly the thermal neutrons, while it is almost transparent to neutrons of higher energies. A thin cadmium sheet is therefore used for filtering the thermal neutrons out of the complex radiation that comes out of a paraffin block containing inside a neutron source.

Bohr and Breit and Wigner proposed independently to explain the above anomalies, as due to resonance with a virtual energy level of the compound nucleus (i.e. the nucleus composed by the bombarded nucleus and the neutron). Bohr went much farther in giving also a qualitative explanation of the large probability for the existence of at least one such level, within an energy interval of the order of magnitude of 100 volts corresponding to the energy band of the slow neutrons. This band corresponds, however, to an excitation energy of the compound nucleus of many million volts, representing the binding energy of the neutron. Bohr could show that, since nuclei, and especially heavy nuclei, are systems with a very large number of degrees of freedom, the spacing between neighbouring energy levels decreases very rapidly with increasing excitation energy. An evaluation of this spacing shows that, whereas, for low excitation energies, the spacing is of the order of magnitude of 10^5 volts, for high excitation energies, of the order of ten million volts, it is reduced, for elements of mean atomic weight, to less than one volt. It is therefore a very plausible assumption that one (or more) such level lies within the slow neutron band thus explaining the large frequency of the cases in which absorption anomalies are observed.

Before concluding this review of the work on artificial radioactivity produced by neutrons, I feel it as a duty to thank all those who have contributed to the success of these researches. I must thank in particular all my collaborators that have already been mentioned; the Istituto di Sanità Pubblica in Rome and especially prof. G.C. Trabacchi, for the supply of all the many radon sources that have been used; the Consiglio Nazionale delle Ricerche for several grants.